Sedimentary Environments and Facies

Sedimentary Environments and Facies

EDITED BY

H. G. READING

Department of Geology and Mineralogy
University of Oxford

ELSEVIER · NEW YORK

First published 1978 by
Blackwell Scientific Publications
Osney Mead, Oxford, OX2 0EL, England

Sole distributors in U.S.A. and Canada:
Elsevier North-Holland, Inc.
52 Vanderbilt Avenue, New York, New York 10017

Library of Congress Cataloguing in Publication Data

Main entry under title:
Sedimentary environments and facies.
Bibliography: p.
Includes index.
1. Rocks, Sedimentary. 2. Facies (Geology)
3. Sedimentation and deposition. I. Reading, H.G.
QE471.S378 551.7 78–17666

ISBN 0–444–00276–6
ISBN 0–444–00293–6 pbk

Printed in Great Britain

Contents

Authors

HAROLD G. READING, *Department of Geology and Mineralogy, Parks Road, Oxford, U.K.*

JOHN D. COLLINSON, *Department of Geology, Keele University, Keele, Staffs., U.K.*

MARC B. EDWARDS, *Continental Shelf Institute, Trondheim, Norway. (Now with Bureau of Economic Geology, Austin, Texas, U.S.A.).*

TREVOR ELLIOTT, *Department of Geology, Singleton Park, Swansea, U.K.*

HUGH C. JENKYNS, *Department of Geology and Mineralogy, Parks Road, Oxford, U.K.*

HOWARD D. JOHNSON, *Koninklikje/Shell Exploratie en Produktie Laboratorium, Volmerlaan 6, Rijswijk, Netherlands.*

ANDREW H.G. MITCHELL, *United Nations Development Project, P.O. Box 650, Rangoon, Burma.*

NICHOLAS A. RUPKE, *Department of Geology and Mineralogy, Parks Road, Oxford, U.K.*

BRUCE W. SELLWOOD, *Department of Geology, Whiteknights, Reading, U.K.*

ROGER TILL, *Department of Geology, Whiteknights, Reading, U.K. (now with British Petroleum, London, U.K.).*

Preface

This book was conceived in late-1974 to provide a single comprehensive text, covering modern and ancient environments, suitable for advanced university students, research workers and professional geologists. To cover all environments and facies with the authority of an active researcher, we formed a group of authors who knew each other well and shared a common philosophical view. We could thus criticize, amend and integrate each others' contributions, while retaining our individual styles and responsibility for each chapter.

The names of the authors are listed in the Contents and we expect that reference made to a chapter will be according to the name of the individual author.

As with all textbooks, our main problems have been the selection of material and the need to strike a balance between comprehensiveness and cost. Many chapters had to be reduced to half their original length and references pruned. The inevitable loss of data is balanced by a more selective and readable text.

No textbook is solely the product of its authors. In this book we have incorporated the facts, ideas, philosophies and prejudices of many others; some are quoted and acknowledged; others have been absorbed by us over many years from teachers, colleagues, friends and students. The person who has been particularly influential on many of us has been Maurits de Raaf, formerly Professor of Sedimentology at the University of Utrecht. He has taught us to combine careful facies analysis and an examination of every detail in a rock with an unceasing search for sedimentary models, to doubt constantly any hypothesis we may be defending, and to beware of becoming dogmatic. This philosophy, we hope, will be followed by all who read this book and look at rocks.

Many persons have read part of the book and it has benefited from their criticisms and suggestions. In particular thanks go to Ed S. Belt, Mike R. Leeder and Roger G. Walker for their comments on the early synopses and specimen chapters. For reading one or more individual chapters we extend thanks to D. Graham Bell, Ed S. Belt, C.G. Bennet, Bernard M. Besley, Geoffrey S. Boulton, Paul H. Bridges, John C. Crowell, Robert W. Dalrymple, Graham Evans, Alfred G. Fischer, Robert E. Garrison, Joseph H. Hartshorn, Alan P. Heward, Franklyn B. van Houten, Colin M. Jones, Mike R. Leeder, Alan Lees, Bruce K. Levell, I. Nick McCave, Alayne Street, David B. Thompson, Roger Vernon, Roger G. Walker, N. Lewis Watts, E.L. Winterer and Andrew Wood. Carol J. Pudsey assisted with the compilation of the index and the final editing of the book. Responsibility, however, for omissions, lack of balance or for errors must remain with us.

In addition we wish to thank those who typed drafts of individual chapters and in particular Valerie Miles and Kathy Webb who between them typed so many early drafts and the final manuscript.

Special acknowledgement is made to the many authors, societies and publishers who permitted illustrations from their articles, journals and books to be used as the basis for our figures; in particular to the Geological Society of America, the Geological Society of London, the Society of Economic Paleontologists and Mineralogists, the American Association of Petroleum Geologists, the American Geophysical Union and Elsevier Publishing Company.

Finally no book can be completed without the help and encouragement of wives and families or of girl friends who patiently put up with so many evenings 'on the book'.

Oxford, June 1978 Harold G. Reading

CHAPTER 1 Introduction

1.1 DEVELOPMENT OF SEDIMENTOLOGY

Sedimentology is concerned with the composition and genesis of sediments and sedimentary rocks. It includes sedimentary petrology, which is concerned with the nature and relationships of the constituent particles, and it differs from stratigraphy in that time is not of prime importance. It overlaps with other geological disciplines such as geochemistry, mineralogy, palaeontology and tectonics. In addition sedimentology takes from and contributes to chemistry, biology, physics, geomorphology, oceanography, soil science, civil engineering, climatology, glaciology and fluid dynamics.

Before 1950 the sciences of stratigraphy, concerned primarily with correlation and broad palaeogeographic reconstructions, and sedimentary petrology, had evolved more or less independently with the exception of a few notable contributions such as those of Sorby (1859, 1879). Modern sedimentology can be said to have started with the publication of Kuenen and Migliorini's (1950b) paper on turbidity currents as a cause of graded bedding (see Sect. 12.1).

The turbidite concept developed from Daly's (1936) hypothesis that turbidity currents might be the agent of erosion of submarine canyons and from the physical model flume experiments of Kuenen (1937, 1950b). Under the impact of this concept, geologists, who for years had been working on 'flysch' began to realize that an actual mechanism of flow could be envisaged as the agent of transport and deposition of graded sand beds. Geologists could now look at sedimentary rocks as sediments that had modern analogues, some aspects of which could be simulated by experiment. Familiar rocks could be examined with new insight and such features as sole marks, previously largely undetected because they were not understood, could be documented and perhaps explained.

Over the last 25 years, data on the composition, texture and structures of sedimentary rocks have grown enormously. Analysis of these data and understanding of the processes of sediment deposition have been greatly aided by observation and experiments, often conducted by scientists such as hydraulic engineers operating outside geology. In addition a conscious effort has been made by many workers to compare their observations with those predicted by models of particular processes and environments. Many of these comparative models are founded on observations that can be made in now active environments. Others are the result of a creative blend of experience and imagination. Matching process with the corresponding sedimentary product is often difficult. In present-day environments, processes are readily studied and measured, but data on their products are difficult to collect. In the Ancient, composition, texture and sedimentary structures are normally easily observed and in outcrop a bed can be followed laterally, but the processes which produced the various features cannot be directly measured. A prime aim of sedimentology is to narrow the gap between modern process and past product, aided also by an understanding of diagenesis.

The many books on sedimentology which have appeared in the past decade reflect the surge of new concepts. Revised editions of earlier textbooks such as the third edition (1975) of Pettijohn's now classic *Sedimentary Rocks* and Greensmith's *Petrology of the Sedimentary Rocks* (1965) concentrate on the analysis and interpretation of sediments as rocks, rather than on their sedimentation. Sedimentary structures and their use in basinal reconstruction were emphasized by Potter and Pettijohn (1963). Physical processes of sedimentation and their importance in understanding sedimentary structures were first brought to the attention of geologists in Middleton (Ed.) (1965) *Primary Sedimentary Structures and their Hydrodynamic Interpretation* and a deeper understanding of some of the processes was developed by Allen in *Current Ripples* (1968) and in *Physical Processes of Sedimentation* (1970c).

In the carbonate field the first book to reflect the progress in matching process with product in the 1950s was that edited by Ham (1962) *Classification of Carbonate Rocks – a Symposium*. However it scarcely mentioned diagenesis, understanding of which advanced rapidly in the 1960s to culminate in the most important book in the limestone field, that by Bathurst (1971) *Carbonate Sediments and their Diagenesis*. Chemical processes were treated by Degens (1965) in *Geochemistry of Sediments*, by Berner in *Principles of Chemical Sedimentology* (1971) and by Garrels and Mackenzie in *Evolution of Sedimentary Rocks* (1971).

The importance of biological processes has been

underestimated by most sedimentologists and few textbooks mention organisms except as sources or disturbers of sediment. German authors, however, such as Seilacher and the Tübingen school, and Schäfer in *Ecology and Palaeoecology of Marine Environments* (1972) have cultivated the science of analysing faunas and the effect of their life and death patterns upon sediments.

The genesis of sediments was stressed particularly by Blatt, Middleton and Murray in *Origin of Sedimentary Rocks* (1972) which emphasizes the mechanisms and processes of physical and chemical sedimentation.

Environmental analysis is not discussed at length in these textbooks. However, there are several volumes, especially those developed out of symposia, that deal with sedimentary facies and environments, e.g. Stanley (Ed.) (1969) *New Concepts of Continental Margin Sedimentation*, Morgan (Ed.) (1970) *Deltaic Sedimentation Modern and Ancient*, Rigby and Hamblin (Eds.) (1972) *Recognition of Ancient Sedimentary Environments*, Dott and Shaver (Eds.) (1974) *Modern and Ancient Geosynclinal Sedimentation*, Hsü and Jenkyns (Eds.) (1974) *Pelagic Sediments: on Land and Under the Sea*, Jopling and McDonald (Eds.) (1975) *Glaciofluvial and Glaciolacustrine Sedimentation*, Broussard (Ed.) (1975) *Deltas, Models for Exploration*, Davis and Ethington (Eds.) (1976) *Beach and Nearshore Sedimentation*, Stanley and Swift (Eds.) (1976) *Marine Sediment Transport and Environmental Management*.

Reineck and Singh's (1973) *Depositional Sedimentary Environments* covered both physical and biological sedimentary processes and structures and also modern clastic sedimentary environments, with particular emphasis on the shallow-marine and Wilson (1975) does the same for carbonate facies, emphasizing the impact of organic evolution on carbonate build-ups. Selley (1970) in *Ancient Sedimentary Environments* has written a lively book for undergraduates showing how an analysis of sedimentary facies can be used to interpret the ancient environment whilst Harms, Southard, Spearing and Walker (1975) have brought the interpretation and use of sedimentary structures and sequences up to date by emphasizing how they may be used to interpret the facies of certain clastic environments.

1.2 SCOPE AND PHILOSOPHY OF THIS BOOK

The purpose of this book is to show how ancient environments may be reconstructed by interpreting first the process or processes which gave rise to facies and then the environment in which the processes operated.

The recognition of environments requires:

(1) A thorough field description of the rocks with additional laboratory data based on samples collected to answer specific problems. Since time is always limited, rock description is inevitably selective, emphasizing some features, underplaying others and rejecting yet others as quite unimportant. The selection depends on the judgment, experience and purpose of the investigator.

Judgment and experience take time to acquire and can be gained only by seeing lots of rocks. The absence of certain features is often as important as their presence. For example the consistent absence of shallow-water features, rather than any positive evidence for great depth, leads sedimentologists to infer that most turbidites were deposited in deep water. The utilization of negative evidence requires a familiarity with a wide range of sedimentary rocks and environments.

(2) An awareness of processes so that, simultaneously with rock description, the strength or direction of the current or the type of flow which carried and deposited each grain is being considered. Such questions as 'What was the oxidation state, salinity or pH of the water?' or 'What forms of life were extant?' can also be asked. We also have to consider the later alteration or diagenetic processes which may have changed not only the colour of rocks but also their grain-size and composition. Particular processes are seldom confined to one environment, though they may be excluded from some, and therefore similar rocks may form in different environments.

(3) An understanding of the relationship of rocks to their vertical and lateral neighbours, the shape of facies bodies and their contacts with neighbouring facies (Chapter 2). These relationships are used as constraints to eliminate certain environments and thereby reduce the number of options. Facies sequences introduce an element of time and environmental change.

(4) A knowledge of present-day environments and the processes which operate within them. We need to know how environments evolve as sea-level, climate, tectonic activity or sediment supply change. Our understanding of environments is bounded not only by the limits on knowledge of the present day, whole regions still being virtually unexplored, but also by the uniqueness of the present. For example the recent rise of sea level allows us readily to develop models of transgressive sedimentation in shallow seas, but makes it difficult to develop models for periods of relatively stable or falling sea-level. It is also salutary to consider how difficult it would be to conceive a model of glacial sedimentation had the human race developed in an entirely non-glacial period. It would take a courageous scientist to postulate, from a limited knowledge of sea ice and snow falls, the hypothesis of large ice caps and glaciers which could erode and deposit large quantities of sediment.

Thus the emphasis in this book will be on: (1) *environments*, reviewing modern environments, with their associated processes and products, (2) *processes* concentrating on those that occur in each environment and showing how they relate to the resultant sediment. They will not be discussed on their own as there are already several good textbooks on processes and the genesis of sedimentary rocks and structures, (3) *facies*, stressing field data, facies relationships, sequences and associations, (4) *geological applications*, illustrating how sedimentary rocks are related to their geological background and how the recognition of sedimentary processes and environments illuminates our understanding of past climates, the chemistry of the oceans and the land, the development of life and world tectonics.

Lithofacies and environment can seldom be linked directly without passing through the process stage.

1.3 ORGANIZATION OF THE BOOK

There is no unique division of environments; in addition there is no simple match between environment and facies. An environment is a geographical unit with a particular set of physical, chemical and biological variables; a facies is a body of rock with specified characteristics (see Chapter 2).

Matching environment with facies is seldom easy and frequently decisions have had to be made between subdividing the book on a basis of environment or of facies. In most cases it was decided to subdivide on environment. Consequently many facies cut across several parts of a chapter or across several chapters. Evaporites, for example, are now known to have occurred in almost as many environments as have sandstones, ranging from deserts to lakes, coastal flats and deep seas. However, environmental interpretation of the evaporite facies is in its infancy, and it is very difficult to place many ancient evaporites within a particular environment.

Inevitably environments have had to be arbitrarily divided, with artificial boundaries to individual chapters. Divisions are based mainly on the geographical environment but in some cases emphasis has been given to the facies as in the separation of siliciclastic shallow seas from shallow-water carbonate environments where, to some extent, the processes producing the facies are different. In the case of 'Deserts' and 'Glacial Environments', climate, with consequent distinctive processes, is the prime factor. Environments do, however, overlap each other and the reader needs to make the links by reference to the Contents page and the Index. Though cross referencing between chapters helps, it has been kept to the minimum, to avoid breaking of the text.

Since no two environments are the same in that our knowledge of them is uneven and dependent on various forms of data, the treatment given in the book varies from chapter to chapter within a common philosophy. For example facies integration is easier in clastic sediments than in organic ones where time and evolution make a stratigraphic context more important. Consequently ancient examples of the latter are considered in the context of case histories. In deep-sea clastics our knowledge comes from the ancient rather than from the present and there is a long history of study. Shallow-marine siliciclastics have been less studied and our knowledge of ancient examples is in its infancy.

1.4 APPEAL OF THE BOOK

It is hoped that this book will be of value to undergraduates who have completed an elementary course of geology and to post-graduates specializing both in sedimentology and other geological sciences. It should also be of use to professional geologists who need a working reference in the field or who require an up-to-date knowledge of environments. They may be working for an oil company, on coal, as survey geologists or be employed by a mining company. The physical geographer, geomorphologist or oceanographer may also find it useful because a knowledge of the past so often aids our understanding of the present.

CHAPTER 2 Facies

2.1 FACIES CONSTRUCTION

2.1.1 Facies definition

The concept of facies has been used ever since geologists, engineers and miners recognized that features found in particular rock units were useful in correlation and in predicting the occurrence of coal, oil or mineral ores. The term was introduced by Gressly (1838) and has been the subject of considerable debate, well summarized by Teichert (1958b) and Krumbein and Sloss (1963, pp. 316–331).

A facies is a body of rock with specified characteristics. In the case of sedimentary rocks, it is defined on the basis of colour, bedding, composition, texture, fossils and sedimentary structures. 'Lithofacies' should thus refer to an objectively described rock unit. However, the term 'facies' is used in many different senses: (1) in the strictly observational sense of a rock product, e.g. 'sandstone facies'; (2) in a genetic sense for the products of a *process* by which a rock is thought to have formed, e.g. 'turbidite facies' for the products of turbidity currents; (3) in an environmental sense for the *environment* in which a rock or suite of mixed rocks is thought to have formed, e.g. 'fluvial facies' or 'shallow marine facies'; and (4) as a tecto-facies, e.g. 'post-orogenic facies' or 'molasse facies'.

These different uses of the term 'facies' are justified so long as we are aware of the sense in which the word is being used. For example, we can attempt to define objectively a sedimentary product, e.g. 'red, rippled sandstone facies'; or we can subjectively interpret a process, e.g. 'turbidite facies', meaning that we *believe* it to have been deposited by turbidity currents, not that it *was* deposited by turbidity currents. A term like 'fluvial facies' is best avoided when a 'fluvial environment' is intended and should be used only for the products of that environment.

The selection of features to define facies and the weight attached to each of them are dependent on a subjective personal evaluation, based on the material to be sampled, the type of outcrop, time available and research objective. Nevertheless, each facies must be defined objectively on observable, and possibly measurable features. It is very difficult to lay down exact rules for facies selection as each set of rocks is different and the facies boundaries chosen will vary accordingly. Nevertheless, in group studies and in industry, uniformity must be attempted to obtain consistent results.

A facies should ideally be a distinctive rock that forms under certain conditions of sedimentation, reflecting a particular process or environment. Facies may be subdivided into sub-facies or grouped into facies associations or assemblages.

2.1.2 Facies relationships

Individual facies vary in interpretative value. A rootlet bed and coal seam, for example, indicates that the depositional surface was very close to or above water level. A current-rippled sandstone implies that deposition took place in the lower part of the lower flow regime from a current that flowed in a particular direction. However, it indicates little about depth, salinity or environment.

Even a rootlet bed cannot be said to have formed in any one environment. It may have formed in a backswamp, on an alluvial fan, on a river levee or at a shoreline. We therefore have to recognize at the outset the limitations of individual facies, taken in isolation. A knowledge of the context of a facies is essential before proposing an environmental interpretation.

The importance of facies relationships has long been recognized, at least since Walther's *Law of Facies* (1894) which states that 'The various deposits of the same facies area and, similarly, the sum of the rocks of different facies areas were formed beside each other in space, but in a crustal profile we see them lying on top of each other . . . it is a basic statement of far-reaching significance that only those facies and facies areas can be superimposed, without a break, that can be observed beside each other at the present time' (translation from Blatt, Middleton and Murray, 1972, pp. 187–188). It has been taken to indicate that facies occurring in a conformable vertical sequence were formed in laterally adjacent environments and that facies in vertical contact must be the product of geographically neighbouring environments. This principle has long been used to explain how a prograding delta yields a coarsening-upward sequence (Figs. 6.25A, 6.30A).

It follows that the vertical succession of facies, laid on its side, reflects the lateral juxtaposition of environments.

Conversely, a borehole sequence through a modern prograding delta or sabkha may be predicted if the geographical distribution of its constituent depositional environments is known.

However, as Middleton (1973) has pointed out, Walther stressed that the law applies only to successions without major breaks. A break in the succession, perhaps marked by an erosive contact, may represent the passage of any number of environments whose products were subsequently removed. This may occur, for example, in a prograding delta (Figs. 6.25B, 6.28, 6.30B); in transgressive sequences the chances of complete preservation are even smaller (Fig. 7.39).

CONTACTS

Walther's warning has often been ignored by geologists who have failed to describe the type of contact between facies. Non-erosional contacts indicate that the facies immediately followed each other in time, probably by the migration of depositional environments. If contacts are sharp, even when erosion cannot be demonstrated, the facies may have been formed in depositional environments which were widely separated in space.

The three main types of contact are gradational, sharp and erosive, though sometimes one needs to differentiate those that are abruptly gradational, where a transition occurs over a few centimetres. Some contacts show extensive boring, burrowing, penecontemporaneous deformation or diagenesis of the underlying sediments so that the adjacent facies have become mixed or even inverted.

CYCLES

The idea that patterns of facies repeated each other, or the concept of cyclic sedimentation, has been one of the most fruitful in sedimentary geology. It enabled geologists to bring order out of apparent chaos, and to describe concisely a thick pile of complexly interbedded sedimentary rocks. They could compare their cycles, cyclothems or rhythms (here used synonymously) with those found elsewhere. This in turn leads to discussions of the causes of cyclicity. Was it due to repeated subsidence of the basin, uplift of the source area, changes in climate or of sea level or oscillating sedimentary supply?

The concept of cyclic sedimentation has, however, been criticized on two grounds. *Firstly,* the establishment of cycles is too often subjective and pleas have been made for a more rigorous analysis of the sequences (Duff and Walton, 1962). *Secondly,* regardless of the sophistication of the techniques used to establish a cycle, significant facts essential to sedimentological interpretation are omitted, usually due to concentration on selected features or the desire to establish an 'ideal' cycle. Thus the cycle becomes more important than the rocks of which it is composed.

Since the use of cycles is based on the idea that there is a regularity to sedimentary sequences and that sedimentation is a normal steady process apparently random events are commonly neglected, although, in some environments, they may dominate sedimentation.

ASSOCIATIONS AND SEQUENCES

A valuable result of the cyclic concept was to focus attention on the relationship of facies to each other, that is on facies which tend to occur together (associations) and on sedimentary sequences. It demonstrated the advantage of interpreting a facies by reference to its neighbours. This concept of facies associations is fundamental to all environmental interpretation.

Facies associations are groups of facies that occur together and are considered to be genetically or environmentally related. For example, thick-bedded turbidites may be interbedded with grain-flows, slumps and mudstone while thin-bedded turbidites are interbedded with mudstone alone. Each grouping would then be identified as a distinct association. The association provides additional evidence which makes environmental interpretation easier than treating each facies in isolation.

In some successions the facies within an association are interbedded, so far as we can tell, randomly. In others, the facies may lie in a preferred order with vertical transitions from one facies into another occurring regularly and more often than we would expect in a random succession. Where only random interbedding is apparent, a more sophisticated facies analysis may later reveal a preferred sequence. However, random interbedding is often genuine, being the result of randomly occurring events.

A facies sequence is a series of facies which pass gradually from one into the other. The sequence may be bounded at top and bottom by a sharp or erosive junction, or by a hiatus in deposition indicated by a rootlet bed, reworking or early diagenesis. A sequence may occur only once, or it may be repeated (i.e. cyclic). While the concept of cycles tends to emphasize similarities within successions, the sequence approach stresses small differences within broadly similar successions (Sect. 6.2).

In clastic environments, two important kinds of sequence are those in which the grain-size coarsens upward from a sharp or erosive base (coarsening-upward sequence) and those in which the grain-size becomes finer upward to a sharp or erosive top (fining-upward sequence). Both are easily recognizable in field sections or boreholes, including electric logs, and can be objectively defined. They also have a simple process interpretation. Grain-size is normally a measure, though a simple one, of the flow power at the time of deposition and consequently a coarsening-upward sequence indicates an increase in flow power. Conversely, a fining-upward sequence indicates a decrease in flow power. The former may be due to shallowing as a delta, shoreline or river crevasse builds out into deeper water, or it may be due to the progradation of an alluvial fan or submarine fan. The fining-upward sequence may be on the scale of a bed, such as a graded bed in a turbidite, or it may be the thicker fining-upward sequence such as is formed by a migrating point bar in a river or by filling of an abandoned channel. Within a succession, sequences may occur on several scales and there are many kinds of fining-upward and coarsening-upward sequences, whose interpretation is controversial.

Some carbonate rocks lend themselves to the same treatment as clastic sediments because sedimentary structures and textures are clearly visible. However, facies sequences are modelled less on grain-size changes related to physical processes and more on a whole spectrum of biological, chemical and physical factors related to specific environments (see Sect. 10.2.2 and 10.2.5). In addition many carbonate rocks are so modified by diagenesis that facies can only be distinguished by petrographic study and by the definition of micro-facies.

Fig. 2.1. Facies relationship diagram for the deltaic Abbotsham Formation showing type of boundary between facies and the number of times the facies are in vertical contact with each other (after de Raaf, Reading and Walker, 1965).

ESTABLISHING FACIES RELATIONSHIPS

Visual appraisal of successions in the field, in measured sections or in borehole logs, is the simplest way to determine facies relationships. However, where one worker may recognize sequences, another may not. It may therefore be necessary to determine whether or not a particular facies tends to pass into another more often than one would expect in a naturally random arrangement. Not only may this help to detect and define a cyclic arrangement of facies but it may also bring out genetic relationships between facies which might otherwise have been missed.

Statistical techniques are valuable with large amounts of

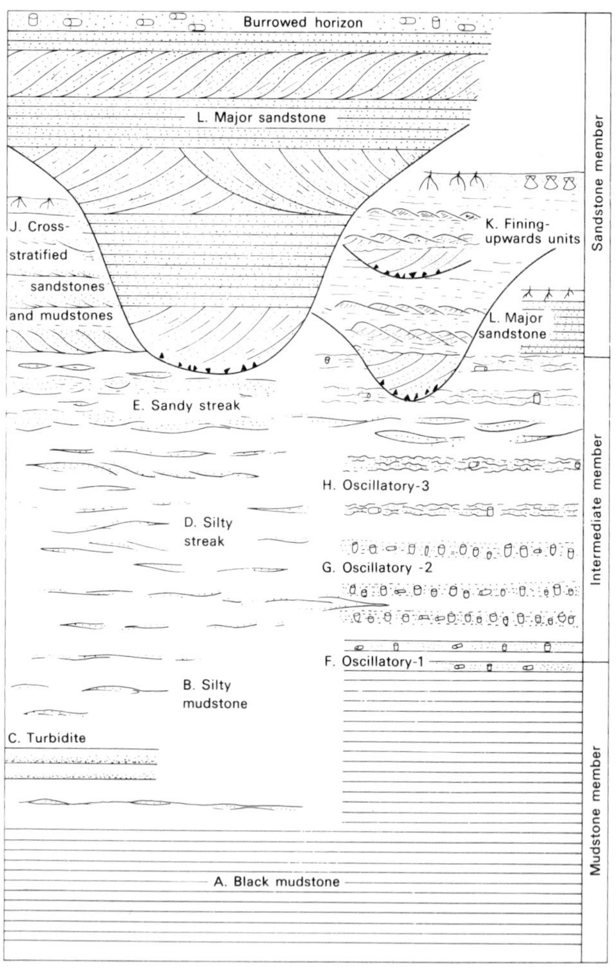

Fig. 2.2. Pictorial representation of the Abbotsham Formation facies and their relationship to each other as determined from Fig. 2.1 (from de Raaf, Reading and Walker, 1965).

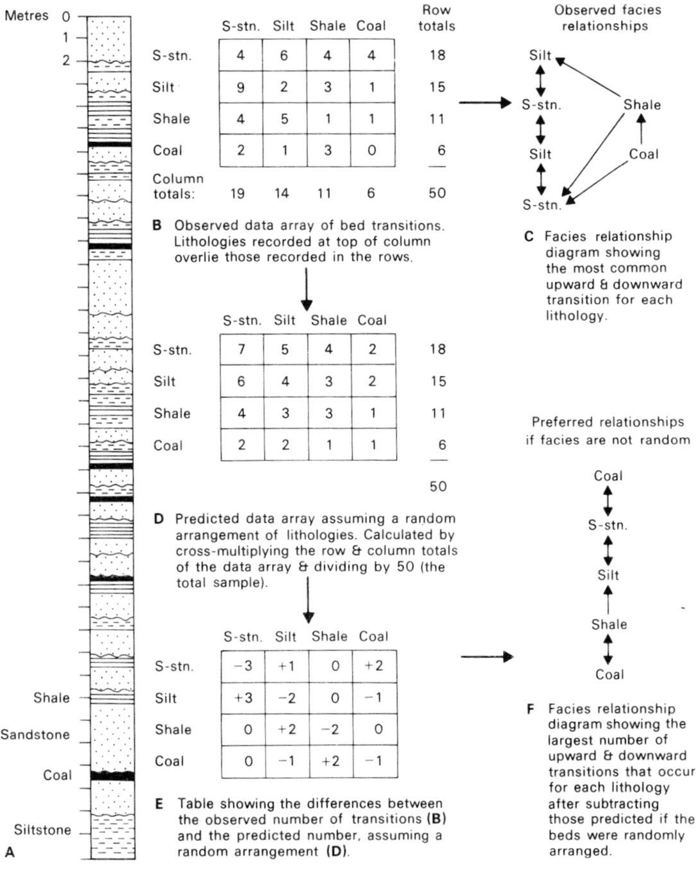

Fig. 2.3. Facies relationship diagrams based on a data array for a hypothetical Coal Measure sequence (after Selley, 1969).

data, when visual inspection of logs is impossible and as a test for sequential or non-sequential facies relationships. A simple visual method is the facies relationship diagram (Fig. 2.1). It was used by de Raaf, Reading and Walker (1965) to show the number of times the facies in a vertical succession are in vertical contact with each other and it formed the basis for the assemblage of sequences shown in Fig. 2.2. The facies can be arranged in such a way that a pictorial representation of the sequence or sequences can be given and the type of contact between facies shown. This example demonstrates clearly the sharp contacts below the black mudstone, fining-upward units, major sandstones and cross-stratified sandstones and mudstones in contrast to the gradational contacts between the remaining facies. The contrasts between the sequences were later used by Elliott (1976a) to construct a detailed model for an ancient fluvial-dominated delta (Fig. 6.48).

Selley (1969) has described a relatively simple method of recording the data in a table, termed the data array, where the observed bed transitions are compared with a predicted data array, assuming a random arrangement of lithologies. The differences between the observed and predicted numbers may then be shown (Fig. 2.3). A facies relationship diagram can also be constructed to show the transitions which occur more and less frequently than would be expected in a random process.

Markov chain analysis is a comparatively simple statistical technique for the detection of repetitive processes in space and time. It has been used to analyse Coal Measure cyclothems (Doveton, 1971), alluvial sediments (Gingerich, 1969; Miall, 1973; Cant and Walker, 1976) and deltaic sediments (Read, 1969). A first-order Markov process is one 'in which the probability of the process being in a given state at a particular time may be deduced from knowledge of the immediately preceding state' (Harbaugh and Bonham-Carter, 1970, p. 98). Miall (1973) gives a clear exposition of Markov analysis of a

A Observed transition probabilities

	SS	A	B	C	D	E	F	G
SS		0·800	0·133	0·067				
A	0·154		0·462	0·231		0·077		0·077
B	0·308	0·077		0·154	0·154	0·077	0·077	0·154
C		0·286	0·571		0·143			
D	0·333						0·667	
E				0·500			0·500	
F	0·667							0·333
G	1·000							

B Transition probabilities for random sequence

	SS	A	B	C	D	E	F	G
SS		0·320	0·245	0·151	0·075	0·038	0·075	0·094
A	0·280		0·260	0·160	0·080	0·040	0·080	0·100
B	0·259	0·315		0·148	0·074	0·037	0·074	0·093
C	0·237	0·288	0·220		0·068	0·034	0·068	0·085
D	0·222	0·270	0·206	0·127		0·032	0·063	0·079
E	0·215	0·262	0·200	0·123	0·062		0·062	0·077
F	0·222	0·270	0·206	0·127	0·063	0·032		0·079
G	0·226	0·274	0·210	0·129	0·065	0·032	0·065	

C Observed minus random transition probabilities

	SS	A	B	C	D	E	F	G
SS		+0·480	−0·112	−0·084	−0·075	−0·038	−0·075	−0·094
A	−0·126		+0·202	+0·071	−0·080	+0·037	−0·080	−0·023
B	+0·049	−0·238		+0·006	+0·080	+0·040	+0·003	+0·061
C	−0·237	−0·002	+0·351		+0·075	−0·034	−0·068	−0·085
D	+0·111	−0·270	−0·206	−0·127		−0·032	+0·604	−0·079
E	−0·215	−0·262	−0·200	+0·377	−0·062		+0·438	−0·077
F	+0·445	−0·270	−0·206	−0·127	−0·063	−0·032		+0·254
G	+0·774	−0·274	−0·210	−0·129	−0·063	−0·032	−0·065	

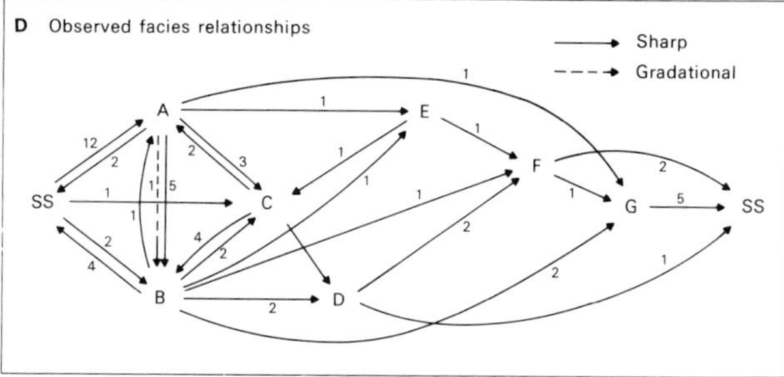

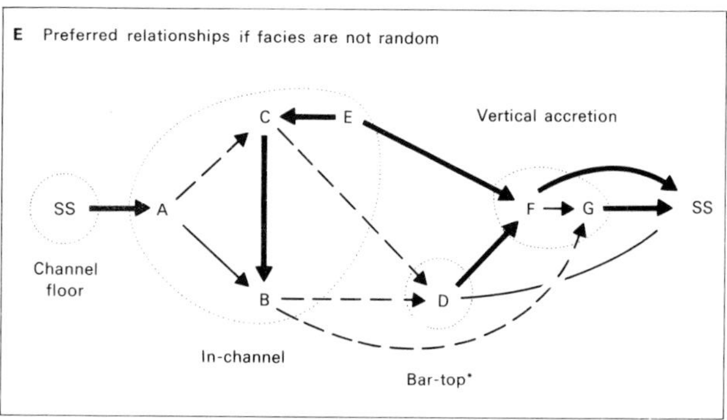

Fig. 2.4. Facies relationship diagrams based on transition probability matrices for the braided river Battery Point Sandstone (after Cant and Walker, 1976).

variety of sedimentary facies deposited in environments ranging from alluvial fans to marginal marine. He shows how, knowing the environmental and process framework of a facies association, the analysis can aid in determining the causes of sedimentation and selecting from possible alternative hypotheses.

Three particular criticisms of statistical methods of analysis are that: (1) The type of contact between facies, which is of prime importance, is frequently omitted, though some methods include this in the analysis (e.g. Selley, 1969). Thus non-sequences and gaps in sedimentation are not considered. (2) The need to reduce the number of facies for statistical analysis may simplify the original data to such an extent that sedimentological sophistication is sacrificed to statistical convenience. (3) The more common sequences are stressed while those that are statistically insignificant, even though they may be geologically significant, are subordinated or eliminated.

These criticisms can be largely circumvented by returning to the raw or semi-processed data after completing the analyses. Statistical anomalies and data sacrificed in the analysis may then be apparent and may illuminate the geological processes.

In addition, the problem of failing to show the nature of the transitions between facies has been partly overcome by Cant and Walker (1976), who treated a scoured surface together with the overlying intraclasts as if it were a facies (Figs. 2.4, 3.51). Since they observed only one gradational transition, all others being sharp or erosive, this method worked in their case, but the inability to show the type of transition is a serious objection to the use of transition probability matrices. Cant and Walker extracted four environmentally significant associations and, although they had rather few transitions, they have shown how such a method might be used in the future to extract facies associations and sequences from an extremely complex and apparently random succession.

2.2 INTERPRETATION OF FACIES

2.2.1 Hypotheses, models and theories

An *hypothesis* is an untested explanation of observed phenomena tentatively adopted so that we may deduce what further critical facts are needed to test its validity. It should have considerable predictive power. A scientific *theory* is a co-ordinated set of hypotheses which are found to be consistent with one another, have been specially tested and to which no exceptions are known. It is therefore a so far unrefuted explanation that encompasses or supersedes a number of hypotheses. *Models* are idealized simplifications set up to aid our understanding of complex natural phenomena and processes. *Scaled experimental models* help detect the factors responsible for particular features, for example when sediment transport and deposition are modelled in flumes under controlled conditions. *Mathematical models* simulate complex geological processes so that, for example, the results of the interaction of a rise of sea-level or an increase in rainfall and sediment supply can be predicted. *Visual models* of environments help us to see the relationships of environments to each other and to picture the processes and resulting products that we should expect to find in these environments. They are simply pictorial representations of working hypotheses. We may draw up either an *actual model* from a modern environment – a North Sea model – or an *inductive* one based on the North Sea such as a 'tidally dominated shallow-sea model'. We may base our model on a facies association or facies sequence of ancient sedimentary rocks. These models may emphasize the environments, the processes or the products (the facies).

The use of sedimentary models in geology goes back to the earliest appreciation that the present can be a key to the past. Models are used both to interpret facies distributions and also to predict where, as yet, undiscovered facies may be found. *Static models* attempt to portray lateral facies patterns at a particular time as in palaeogeographical reconstructions where the distribution of environments is shown. They are predictive in that the position of source areas or other depositional environments for which there is as yet no data can be foretold from the model. The model can then be tested by searching for data that will support or disprove it or lead to modification and refinement.

Dynamic models show a changing pattern of environments or processes. They may be either on a local scale where the vertical sequence of a prograding barrier island (Fig. 7.30) or migrating point bar (Fig. 3.24) is taken to indicate an evolving sequence of environments or on a regional scale, for example where the changes from 'detrital cycles' to 'chemical cycles' in the Lockatong Formation of New Jersey are inferred to be the result of increasing rainfall (Van Houten, 1964) (see Sect. 4.3.3).

The establishing of simplified models has contributed greatly to sedimentology. A limited number of facies models has been developed, each representing a particular environment. For example, Selley (1976a, pp. 253–313), in an excellent review of sedimentary facies and models, correlates each major depositional environment with a single sedimentary model. Walker (1976) has developed facies models from which local details are 'distilled' until the 'pure essence' of the environment remains. The summary model acts as a norm for purposes of comparison and as a framework and guide for future observation. It also acts as a predictor in new geological situations and is the basis for hydrodynamic interpretation of the environment. Divergence from the norm can then be used to enlighten our understanding of particular formations.

At some stage in the sedimentological investigation of a complex succession, simplified facies models are essential to understanding what previously appeared to be randomly interbedded and incomprehensible. The definition of the Bouma sequence (1962) was the basis for the interpretation of turbidites in terms of flow regime (Walker, 1965; Harms and Fahnestock, 1965) and for R.G. Walker's (1967) concept of proximality (Sect. 12.6.3). Deltaic successions were made intelligible by the concept of cyclotherms which assisted in both stratigraphical correlation and environmental interpretation (Sect. 6.2). The formulation of the fining-upward point bar sequence for meandering rivers gave an impetus to studies of fluvial sediments by suggesting that lateral accretion could deposit apparently continuous sheets of sediment within a river channel whose margins are no longer visible (Allen, 1964; Visher, 1965) (Sect. 3.5.2). The discovery of the Sabkha sequence in the Persian Gulf led to a complete re-appraisal of the genesis of evaporitic facies by showing that some facies might be formed by very early diagenetic processes rather than by direct precipitation from saline waters (Shearman, 1966) (Sect. 8.4).

In furthering our understanding and in the early teaching of students these relatively easily grasped models are invaluable. However, a too rigid attachment to a particular model or the dogmatic assertion of its applicability may subordinate facts that do not fit the model. A field or economic geologist may become disenchanted with a model if it cannot be applied to his particular rocks. It can also delay the modification of the model or the creation of new models.

While appreciating that many environments can be described by one general model, the emphasis of this book is on the variability of facies models and on the complexity of environments. It is only by concentrating on differences rather than on similarities that the importance of various processes which combine to make the facies in each environment can be evaluated.

In the interpretation of ancient sequences, there are three

stages of interpretation. At first, unless there is a good regional framework, a specific palaeogeographical reconstruction may be impossible. Perhaps only one borehole or outcrop section is available; the structure is complex; palaeocurrent measurements are lacking; stratigraphical control between localities is too poor for facies variations to be determined. In these cases a limited objective is legitimate and an *initial working hypothesis* developed. It is not related to any specific locality or time and if palaeocurrent measurements are not available it may even lack an orientation.

The next stage is the development of a *palaeogeographical interpretation* which shows the orientation and approximate location of environmental belts. For example, the general area of reefs or channels can be identified, but not the precise location of individual reefs or channels. When these have been located a *realistic interpretation* has been developed.

Working hypotheses help a geologist to decide what to look for. They serve as a constant guide in making observations which may lead to the rejection or modification of the initial hypothesis. However, if care is not taken a geologist may allow subjectivity to influence observation and no longer record the facts objectively. This is a real danger, for, when exposure is limited, it is not impossible to believe that a contact is sharp or that a burrowed bed has rootlets, if this is what one wants to see.

Geologists have to use the method of multiple working hypotheses because they work with incomplete, often uncertain data and because several different processes may have contributed to the final product. Neglecting alternative hypotheses is most likely when a new and easily grasped model has been formulated or when we have hit upon one ourselves and inevitably become over-enthusiastic about it. At times emphasis on a new hypothesis may be necessary because others are unaware of it, but overstatement should be avoided.

While it is true that, other things being equal, the best sedimentologist is the one who has seen most sediments, it is also true that the best sedimentologist is the one who knows the most models and keeps an open mind when confronted with new data. The conviction that one's own hypothesis is right is frequently the mark of one who is poorly informed about alternatives.

2.2.2 Normal v catastrophic sedimentation; abundant and rare sediments; exceptional events

Before the advent of modern sedimentology, geologists tended to ascribe most sedimentary facies and sequences to abnormal catastrophic processes such as floods, earthquakes and tectonic movements. Carboniferous cyclothems, at least in Europe, were usually considered to be the response to some form of intermittent tectonic movement (e.g. Trueman, 1946). The interbedding of sandstone and shale in flysch successions was generally interpreted as the result of alternations of depositional environment between deep water shales and shallow water sandstones, due to major tectonic movements (e.g. Vassoevich, 1951).

However, in the last 20 years, the increasing emphasis on studies of modern processes has led sedimentologists to emphasize the normality of sedimentary processes and to ascribe most sedimentary facies and sequences to the relatively slowly-acting phenomena which can be observed at the present day. For example, there has been the interpretation of deltaic cycles as due to delta switching (Moore, 1959) and fining-upward fluvial cycles as due to point bar migration in meandering rivers (Allen, 1965a).

There is now an increasing appreciation of the importance of both normal and catastrophic sedimentation and of the necessity to distinguish between them. The distinction may not be easy, however, because it is the product which is noted, not the process. The volumetric proportion of normal to catastrophic deposits is observed rather than the duration of processes or the frequency of events. These proportions depend on rates of subsidence, the erosive power of both normal and catastrophic processes and their relative rates of deposition. For example, in the proximal zones of alluvial or submarine fans, each, possibly infrequent, catastrophic event results in a thick deposit, while normal sedimentary processes are very slow and their deposits are usually removed. In the resulting accumulation catastrophic products make up the bulk of the sequence and catastrophic processes appear to have been prevalent.

To resolve these difficulties the types of process should be considered separately from the types of resulting product. *Normal* sedimentary processes persist for the greater proportion of time. Net sedimentation is usually slow. It may be nil or even negative if erosion dominates. Normal processes include pelagic settling, organic growth, diagenesis, tidal and fluvial currents. Some of these processes deposit very slowly; others deposit very fast but, because they erode almost as much as they deposit, have a low net sedimentation rate. Normal processes may or may not produce a large proportion of the total sediment.

Catastrophic sedimentary processes occur almost instantaneously. They frequently involve 'energy' levels several orders of magnitude greater than normal sedimentation. They may deposit a small proportion of the total rock and give rise to only an occasional bed, or they may deposit a large proportion of the total rock and so become the dominant process of deposition.

Sedimentary facies may be divided into *abundant* and *rare*. *Abundant* sedimentary facies are those which make up the major proportion of a sequence. They may result from the *normal* sedimentation of pelagic muds, as in abyssal plains, or be due to catastrophic processes such as turbidity currents on

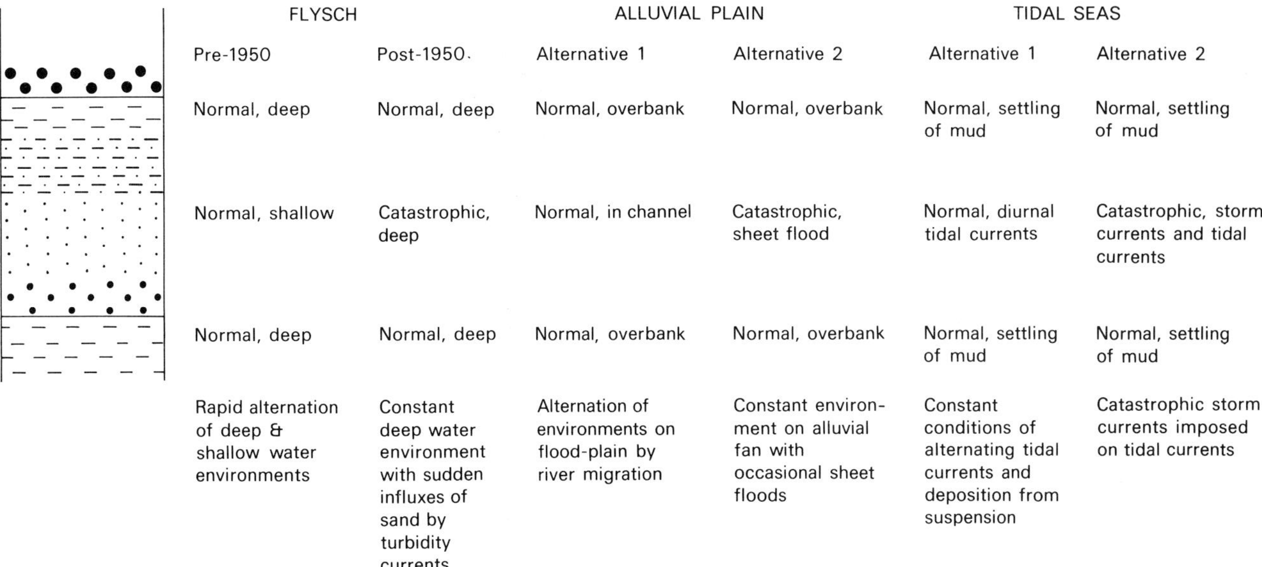

Fig. 2.5. Differing interpretations of graded beds and fining-upward sequences in flysch, alluvial plains and tidal seas depend on distinction between normal and catastrophic sedimentation.

submarine fans. *Rare* sediments may be the result of *catastrophic* processes, for example turbidites in an abyssal plain, or of *normal* processes, for example pelagic muds on submarine fans.

A final consideration is the *exceptional* event or process which produces a single deposit of unique character. While the process is usually catastrophic, the bed produced is so different from abundant catastrophic deposits that it stands out and is frequently a good stratigraphic marker. For example, big beds or megabeds (Sect. 12.7.2) may appear haphazardly within a sequence of normal pelagic mudstones and catastrophic turbidites. Bentonite horizons in a generally non-volcanic sequence may indicate exceptional ashfalls.

To sum up, the terms *normal* and *catastrophic* qualify processes and the sediments formed by these processes. *Abundant* and *rare* refer to the proportion of facies in a sequence. *Exceptional* may be used for an event, a process or a unique deposit (Table 2.1).

Table 2.1. Relationships of terms used for the process of sedimentation and the resulting product

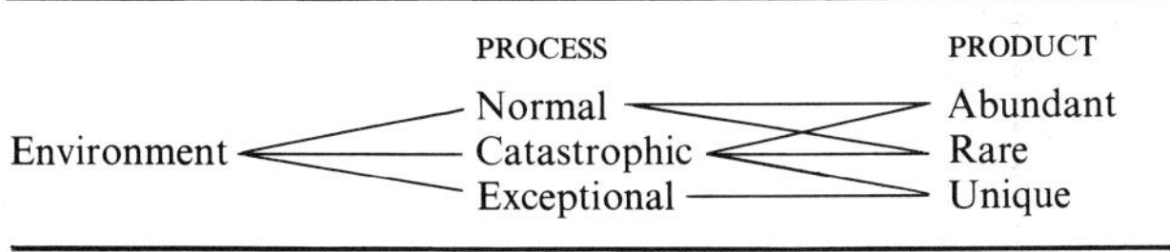

The reason for discussing these questions is that normal, catastrophic and exceptional processes together make up the environment and practical problems devolve from trying to distinguish the three types of process. While it is easy to separate abundant from rare sediments by measuring proportions, it is more difficult to determine the process because particular deposits have first to be interpreted.

In examining a sequence, whenever a change of lithology is reached, the question arises 'Does the facies change indicate a catastrophic event that happened geologically instantaneously with no alteration in normal depositional processes or does it signify a change in physiographic environment?' If the contact between each facies is gradational, the probability is that the depositional processes were gradually changing along with the environment. If, however, a facies unit has a sharp lower and upper contact, is unique and is overlain by a facies identical to the one below, then it was probably deposited by an exceptional event which did not alter the environment.

In many deposits, however, for example where fining-upward sequences dominate an alluvial succession, interpretation is extremely difficult because both environmental switching and catastrophic flows can occur, each producing similar looking sedimentary sequences (Fig. 2.5). One is then faced with the problem of whether normal processes are operating in a changing environment or catastrophic processes are operating in a constant environment. Though an answer may not always be ascertained, posing the question focuses one's mind on the problems.

2.2.3 **Preservation potential**

Few deposits survive. Most sediments are removed by erosion soon after deposition and in many environments most deposits have little chance of preservation.

Thus preservation or fossilization potential (Goldring, 1965) is an important factor in any interpretation, especially in shallow-water and sub-aerial environments. The chances of preserving individual facies vary considerably and an assessment of preservation potential is necessary for comparing modern sediments and ancient rocks. Allowance should be made for their preservation potentials if numerical calculations of the relative abundance of facies are to be used to deduce the importance of various processes.

In addition, the preservation potential of certain tectonic environments should be considered before judging their usefulness as analogues of the past. If the chances of loss down a subduction zone or by uplift and erosion are high then a modern model may have little significance in understanding the rocks in a now uplifted mountain belt or continental region.

The main factors governing preservation potential are the magnitude and frequency of the 'energy' levels of the environment or, on a longer time scale, the rates of subsidence and sedimentation.

The rate of subsidence directly governs preservation potential in those environments which, unlike deep-water basins, have a depositional base level. In most environments, deposits cannot build up above a certain level and thus increased sediment input is relieved by the lateral extension of the area of deposition.

In these environments, although preservation potential is generally low, it is greatly increased by rapid subsidence enabling sediments to be preserved which would otherwise be reworked or removed in a slowly subsiding area. The preservation of thick piedmont fan deposits along fault lines contrasts with the thin veneer of gravel which is all that is preserved during pedimentation in a more stable area.

2.3 FACTORS CONTROLLING THE NATURE AND DISTRIBUTION OF FACIES

Although for many purposes a facies analysis that results in an environmental and palaeogeographical reconstruction may suffice, those geologists who are concerned with the dynamics of the earth, past climates and the evolution of life need also to consider the underlying controls which govern the formation of facies and their vertical and lateral distribution.

This has been frequently attempted in the past. Geologists, trying to explain the causes of cyclic sedimentation, usually argued for one particular theory which may have been intermittent tectonic subsidence, climatic control, eustatic changes of sea level or control by the growth or death of vegetation (Duff, Hallam and Walton, 1967). Facies within geosynclinal belts were related to tectonics; flysch being considered syn-orogenic, and molasse post-orogenic.

Sedimentologists today are critical of these attempts to explain facies in terms of general factors because the actual processes of deposition and the detailed environments were seldom taken into consideration. Concentration on these fields in the past 20 years has been an essential stage in the development of sedimentology since it has enabled processes to be understood and environments identified with reasonable certainty. It is now possible to use sedimentological analysis to widen our understanding of broader geological processes.

Facies distribution and changes in distribution are dependent on a number of interrelated controls:

1 Sedimentary processes
2 Sediment supply
3 Climate
4 Tectonics
5 Sea level changes
6 Biological activity
7 Water chemistry
8 Volcanism

The relative importance of these factors varies between different environments. Probably the two universal factors are climate and tectonics. Climate is critical in continental and shallow marine environments. Its influence is generally less direct in deeper marine basins. Tectonics are very important in continental and in deep marine environments. Sedimentary factors are best documented in deltaic and fluvial environments. Sea-level changes inevitably affect shallow seas and shorelines more than the continental and deep-marine environments, though, even there, their effects are not negligible.

2.3.1 **Sedimentary processes**

Processes intrinsic to sedimentation in a given environment may themselves be responsible for facies distribution and facies change. The progradation of distributaries of an elongate delta so lowers the gradient that the river eventually finds a steeper, shorter route to the sea and a new 'cycle' of deposition is initiated (Sect. 6.6). High sinuosity rivers aggrade above the surrounding flood plain because deposition is largely confined to the channel and its levees. Thus, sooner or later they break through their banks to find a new course. On a delta slope or submarine fan, sediment build-up may cause either diversion or slumping to take place when the sediment load exceeds the strength of the sediment.

By the very nature of the sedimentary environment these changes are inevitable, though their exact timing is usually governed by an unusual event such as an exceptionally violent flood, storm or seismic shock. This 'triggering' mechanism must be distinguished from the fundamental cause which may be delta progradation, river aggradation or sediment instability.

Differential compaction of varied underlying sediments and subsurface sediment movement such as that associated with salt domes and growth faults leads to differential subsidence (Sect. 6.8). This affects the overlying sediment in the same way as vertical tectonic movements. The effects associated with salt domes and growth faults may persist for such long periods of time that they pass gradually into and may be impossible to separate from those due to tectonism.

2.3.2 Sediment supply

The type of sediment available is fundamental to the production of facies. If particular grain-sizes are absent, certain sedimentary structures cannot form since they are dependent not only on flow regime, but also on grain-size. Whether sand or mud is deposited in a particular environment may depend as much on which is available as on the processes which can operate in that environment. Whilst this is obvious in modern environments, particularly deep-sea basins, it is often forgotten when reconstructing ancient environments. The composition is also important. Similar environments may have quite different compositional petrographies in a stable, cratonic region and in an active, orogenic region with abundant volcanic activity.

The availability of sediment is one control on the thickness of depositional facies; it may also govern water depth and environment. Sediment is supplied from two sources: (1) Extrabasinal, which is mainly terrigenous. The type is governed by geology, topography, climate and tectonics. (2) Intrabasinal, which is mainly biochemical and derived from chemical precipitation, plant or animal growth, the erosion of materal previously deposited within the basin, or from sediment extruded upwards from below as sand or mud volcanoes. The type is governed by climate, water composition, tectonism and eustasy.

Within any depositional environment the effect of sediment supply depends on its availability, on subsidence and sea-level change (see also Chapters 7 and 10). Two situations may be envisaged:

1 *Transgressive:* when subsidence and rise of sea-level are more important than the supply of terrigenous sediment, the environment is starved of sediment. This results in reworking, erosion and diagenesis of deposits, transgression and deepening of the environment, together with an increase in chemically and biologically formed sediments.

2 *Regressive:* when subsidence and rise of sea level are less important than terrigenous sediment supply, progradation and an increase in the proportion of continental facies result.

In the construction of models different constraints apply according to whether the supply of sediment is abundant and available to fill any depositional void or is intermittent so that certain areas are temporarily starved.

On flood plains, sediment is generally available. The relatively starved interfluvial areas are local and short-lived and sedimentation quickly switches back. Nevertheless, local starvation may result in the formation of lakes and coal swamps.

In deltaic areas sediment supply is very abundant but its effects are localized since subsidence is also considerable. Consequently the availability of sediment is critical to the development of sedimentary models (Chapter 6).

Present day shallow seas generally lack active sediment supply because of the recent rise of sea-level. In the past, however, shallow seas were apparently stable for long periods during which sediment was abundant (Chapter 9).

2.3.3 Climate

Facies are affected mainly by temperature and rainfall, though wind levels may be locally important. Not only are average temperature and rainfall important but also their seasonal extremes and sporadic fluctuations. Temperature indicators include evaporites, palaeosols, vegetation, some oolites and tillites. Rainfall indicators include vegetation, palaeosols, evaporites and dune-bedded sandstones, fluvial and lake morphology. In many regions the climate of the source area may differ considerably from that of the depositional basin.

A warm climate has a strong effect on the formation of limestones, evaporites and coals. These sediments can aggrade within a depositional basin and maintain a surface close to lake or sea level in spite of no terrigenous input. In climates unfavourable for the formation of limestone or coal, deepening occurs more easily.

2.3.4 Tectonics

The broad geographical framework of sediment supply, climate and environment is dependent on tectonics. The major effects are discussed under other headings and a global tectonic framework is given in Chapter 14. However, tectonics also cause local facies changes through the relative vertical movement and tilting of fault blocks (e.g. Fig. 3.34). The swell and basin facies of geosynclines and of epeiric seas are controlled by relative crustal movements (see Chapters 10 and 11).

2.3.5 Sea-level changes

Relative sea-level changes can be brought about by sediment progradation, by vertical movement or tectonic tilting of crustal blocks or lithospheric plates, by isostatic depression and rebound, by changes in the volume of oceanic waters or by global tectonic changes, such as alteration in the volume of oceanic ridges. Sediment progradation and vertical tectonism are discussed elsewhere. Global tectonic changes may give rise to major transgressions or regressions affecting whole systems such as the extensive Lower Carboniferous or Lower Cretaceous transgressions. They are unlikely to produce frequent, rapid sea-level changes.

Absolute sea-level changes in the form of rapid eustatic changes have resulted from the locking up and freeing of water in polar ice caps. The rapidity of these effects masks those due to tectonic movement. The average rise of sea-level from about 15,000 B.P. to 5,000 B.P. was almost 1 m per century with a maximum of 2.4 m per century. Isostatic rebound resulting from the removal of ice sheets has the reverse effect. Because, in some areas, uplift of the land is faster than the eustatic sea-level rise a local relative drop of sea-level results. Rates of uplift are commonly 1 m per century and may be as much as 4 m per century on the west coast of Baffin Island (Pitty, 1971).

2.3.6 Biological activity

The building of coral, bryozoan, algal and other reefs (Chapter 10.4) and the development of thick plant accumulations are the principal constructive elements in organic sedimentation. Animals and tree roots, by inhibiting current flow and erosion, may trap sediment. On land, plant cover assists in soil development and moderates the erosive effects of rainfall, run-off and wind. Micro-organisms such as foraminifera, radiolaria, algae and diatoms which commonly live in near-surface waters may provide a constant rain of pelagic sediment in oceans and lakes. Bacteria are particularly important in soil formation, as weathering agents in the oxidation and reduction of iron and as reducers of sulphate.

Organisms are closely associated with chemical precipitates. They have a strong effect on the pH and Eh of sediment pore waters. Plant roots disturb the soil and concentrate solutions around them to form concretions. Similarly, burrowing organisms not only destroy sedimentary structures and homogenize sediment, but also act as sediment and chemical sorters. Many concretions are the result of sediment sorting by the lining of burrows.

Since organisms have evolved through geological time, the type, amount and sites of biological activity have continually changed. An understanding of the contemporary biosphere is necessary if ancient and modern facies or ancient facies from different systems are to be compared. Although all environments may be affected to some degree, this factor is of prime importance in both pelagic environments and in the carbonate build-ups of shallow seas (see Sect. 10.4, Chapter 11).

2.3.7 Water chemistry

The salinity and composition of sea and lake waters varies from place to place and over geological time. Water chemistry governs the formation of carbonates and other chemical and biochemical sediments. Variations in temperature and salinity are largely the result of climatic zonation and fluctuation. Oceanic circulation, resulting in upwelling of nutrient-rich waters, is responsible for local accumulation of some oozes, phosphates and diatomites (Chapter 11). The level of saturation with $CaCO_3$ governs whether calcareous skeleta will be corroded, dissolved or preserved, possibly with additional precipitation of carbonate deposits. In lakes, water chemistry is a prime control of depositional facies (Chapter 4).

2.3.8 Volcanism

Volcanic activity provides a local, intrabasinal source of sediment and of ions in solution. Leaching of hot pillow lavas by sea water, formation of clay minerals by chemical exchange with sea water and associated hydrothermal discharge of metal-rich fluids has an important effect upon sedimentation in pelagic seas (Chapter 11). In lakes there may be a close connection between the composition of the volcanic source and lake precipitation (Chapter 4). In addition the creation and foundering of volcanic hills and islands cause rapid changes of environment, particularly water depth.

CHAPTER 3 Alluvial Sediments

3.1 INTRODUCTION

Although rivers have long been recognized as major transporters of sediment, recognition of their importance in the deposition of sedimentary rocks is a comparatively recent development. Prior to the early 1960s, while many channellized sediments were interpreted as fluvial in origin, many sequences now also recognized as fluvial were interpreted as lacustrine deposits.

Geomorphological work on channel types and processes in the 1950s (e.g. Sundborg, 1956; Leopold and Wolman, 1957) was not immediately applied to ancient sediments. Although Bersier (1959), Bernard and Major (1963) and Allen (1964) recognized the presence of fining-upward units and equated these with the lateral migration of point bars in meandering streams, the main stimulus to the interpretation of alluvial sediments was the SEPM Special Publication No. 12, 'Primary Sedimentary Structures and their Hydrodynamic Interpretation' (Middleton, 1965) which provided a basis for the physical interpretation of the sedimentary structures of many sandstones. The emphasis on channel migration and lateral accretion, in particular in relation to meandering streams, throughout the late 1960s, meant that other types of river channel pattern and behaviour were largely ignored. Fining-upward units with sharp bases were almost uncritically interpreted as laterally migrating point bar deposits and, with one or two exceptions, such as Ore (1963), little attempt was made to apply the work of Leopold and Wolman (1957) on channel patterns to ancient sequences.

In the last few years alternative channel models have been developed as interest in low sinuosity streams has increased (see Miall, 1977, for review of the braided river depositional environment) and detailed investigations of pedogenic processes and products have led to a more critical appraisal of inter-channel deposits. Thus a wider view of alluvial sediments is now being taken and interpretations are based on the whole spectrum of systems which exist at the present day rather than on a few restricted models. Such an approach does not make for easy answers, but if answers can be obtained they will lead to the solution of the more fundamental problems of palaeohydrology and palaeoclimates.

In this chapter we deal first with the processes and products of present-day alluvial systems, which are subdivided for convenience into four major groups. We then examine the facies models developed from ancient alluvial sediments and discuss the extent to which these can be interpreted in terms of present-day systems. The underlying controls on the development of particular morphological systems which might enable us to attempt palaeohydrological and palaeoclimatic interpretations form a final section.

3.2 PRESENT-DAY ALLUVIAL FANS

3.2.1 Setting

Alluvial fans are localized deposits whose shape approximates a segment of a cone (Fig. 3.1). They develop in areas of high relief where there is an abundant supply of sediment. Alluvial fans are best known in poorly vegetated arid and semi-arid regions where rainfall is infrequent but violent and erosion is rapid. However, fans also develop in humid climates where relief and sediment supply are suitable, for example the Kosi River Fan of the Himalayan foothills (Gole and Chitale, 1966) and in the Arctic (e.g. Leggett *et al.*, 1966). On many large humid fans, it is difficult to separate the processes from those of a braided river. Fans commonly coalesce to form a bajada and are commonly developed along active fault scarps, so that they frequently give thick sequences of syntectonic sediments on the downthrown side of major faults.

Alluvial fans pass downslope into a variety of other sedimentary environments. In semi-arid settings the environments may be playa lakes, where the drainage system is internal, or an alluvial plain where the drainage system is more extensive. Elsewhere, fans build out into lakes, alluvial or deltaic plains, tidal flats or beaches or even directly into deep lakes or the sea as submarine fans. For example, the fans of the Northwest Territories are truncated at their distal margins by a meandering distributary of the Mackenzie Delta (Leggett *et al.*, 1966) and in East Greenland they build directly into a fjord (Collinson, 1972).

Fig. 3.1. A semi-arid alluvial fan showing radial pattern of surface channels; Canyon Range, Nevada (photo: B.J. Bluck).

3.2.2 Gross Morphology

The size of an alluvial fan is controlled largely by the size of its catchment basin, though the nature of the source area lithologies and the climate also play a part. For semi-arid fans the relationship between the fan area (A_f) and the area of the drainage basin (A_d) is of the general form:

$$A_f = cA_d^n$$

where c and n are empirically derived exponents (Bull, 1964, 1968; Denny, 1965, 1967; Hooke, 1965; Beaumont, 1972). General experience suggests that n is between 0.8 and 1.0 while c varies between 0.15 and 2.1, in response to the source area lithologies, tectonic activity and climate. In the San Joaquin Valley, fans supplied by drainage basins consisting predominantly of mudstones had values of c about twice as large as those for similar basins with a predominance of sandstone (Bull, 1964).

The relief of a fan surface from apex to distal fringe varies with the size of the fan. Large active fans may have a relief of many hundreds of metres implying large thicknesses of sediment in proximal areas. This will be increased by the active subsidence at a marginal fault line.

In radial section, most fans have a concave upwards profile. Larger fans tend to have lower slopes, but mudstone dominated drainage basins produce fans 35–75% steeper than those of similar size with a sandstone dominated source (Bull, 1964).

In detail, the upward concavity need not be a simple curve, but may be a series of evenly sloping segments which decrease in inclination distally (Bull, 1964). Three or four segments were recognized in San Joaquin Valley fans and attributed to intermittent source area uplift.

3.2.3 **Channel patterns**

Channels radiate from the fan apex and dissect the fan surface. The main channel is commonly incised into the fan and the channel base does not meet the fan surface until a lower level on the fan surface, the 'intersection point' of Hooke (1967) (Fig. 3.2). Incision may result from climatic changes, decrease in sediment supply, base level changes or variability of storm events. Other channels originate on the fan surface where infiltrated drainage re-emerges. In humid fans a braided channel system runs from apex to toe and sweeps systematically over the fan surface (Fig. 3.3) (Gole and Chitale, 1966).

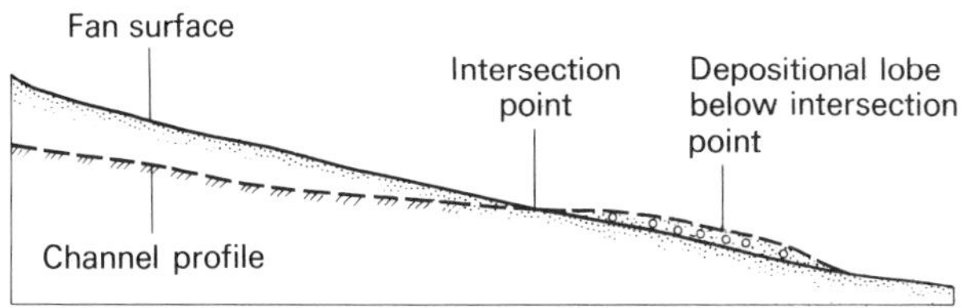

Fig. 3.2. Idealized radial profile through the intersection point of an alluvial fan channel (after Hooke, 1967).

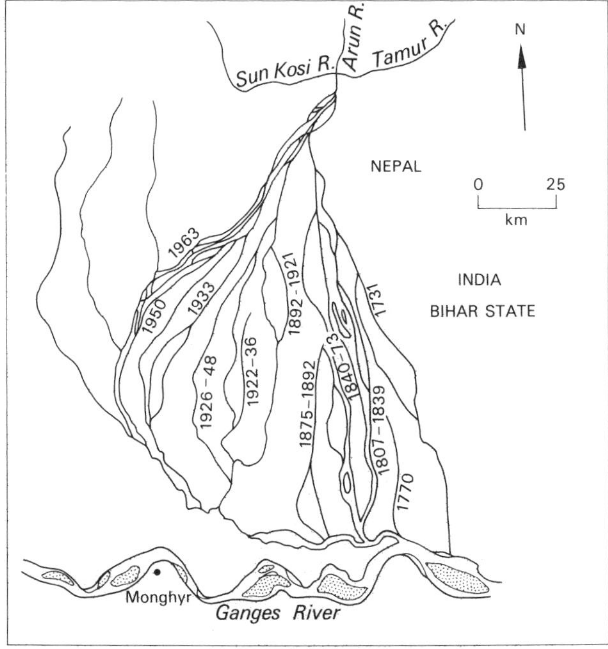

Fig. 3.3. The alluvial fan of the Kosi River on the southern flanks of the Himalayas. This large humid fan is dominated by channels which have migrated from east to west over a period of 230 years (after Gole and Chitale, 1966).

3.2.4 **Depositional processes and products**

At the fan apex, the slope of the canyon or valley floor is generally continuous with the slope of the channel floor on the upper fan. This suggests that deposition is not initiated by a break of slope, but by lateral expansion at the fan apex where shallowing leads to a fall in shear stress and a loss of competence and capacity.

Since most alluvial fans occur in regions of intense ephemeral flood discharge, flood events have seldom been directly observed. Some experimental work on fans has been carried out (Hooke, 1967) but processes mostly have been inferred from the sedimentary products.

In a summary of many observations of fan deposits, Bull (1972) recognized four main types of product which were thought to be related to distinct but intergradational processes based on Blissenbach (1954), Bull (1964) and Hooke (1967). These were:

Debris flow deposits (high viscosity)
Sheetflood deposits }
Stream channel deposits } Fluid flow (low viscosity)
Sieve deposits }

DEBRIS FLOW DEPOSITS

The role of high density, high viscosity debris flows or mud flows on alluvial fans has been known for a long time (Blackwelder, 1928). More recent work (e.g. Blissenbach, 1954; Croft, 1962; Bluck, 1964; Beatty, 1974) has mentioned the process and its distribution on fans but only Sharp and Nobles (1953), Johnson (1970), Hooke (1967) and Carter (1975) have described the processes in detail. 'Debris flow' and 'mud flow' are more or less synonymous but some authors suggest that debris flows should contain a higher proportion of coarser clasts.

Prerequisites of debris flows are source rocks which weather to produce a substantial proportion of fine debris including some clay, and steep slopes which promote rapid erosion and run-off. Debris flows move as dense, viscous masses in which matrix strength is large enough to support and transport clasts up to boulder size. The flows may become rigid or 'frozen' on their edges and in a central plug, where shear is insufficient to overcome the strength of the deforming mass (Johnson, 1970; Carter, 1975; Middleton and Hampton, 1976). The lateral rigid zones may be preserved as levees when the flow overtops a channel or as terraces on the channel sides when flow is contained by the channel. The whole flow will stop when the central plug expands to the full thickness of the flow. The small-scale flows reported by Carter (1975b) involved well sorted mud-free sand, but the basic mechanism seems comparable. Sharpe and Nobles (1953) provided a graphic description of flow surges caused by snow melt in the mountains of Southern California. Here, flows lasted for ten days with the fluidity changing throughout the day as melt rate changed. The surge frequently increased during the hottest part of the day and surges moved at up to 4.5 m sec^{-1} (average 3 m sec^{-1}) on slopes as low as 0.014. Material dumped by one surge was sometimes remobilized by the next. In coarser surges a boulder-rich front was pushed along by the following flow and, as this material moved aside, levees formed and confined later stages of the flow. Surge fronts were up to 1.5 m high while the body of the flow was thinner. As flows decelerated, fronts built to greater heights and the flows became more viscous. Deposits up to 2 m thick were laid down by individual surges.

Fig. 3.4. Small experimental mudflow showing pressure ridges on the flow surface and the sharp steep boundary to the flow. A muddy lobe with less coarse debris is advancing over the principal coarser flow. Flow is from right to left. Length of photograph 1.2 m (photo: B.J. Bluck).

Experimental debris flows have sharply defined steep rounded fronts and produce elongate lobes (Fig. 3.4). Flows sometimes overtop fan-head channels to give minor lateral lobes and levees, but the main depositional site is below the intersection point. Field observations on both the small and large scale suggest comparable results (Carter, 1975; Beatty, 1974).

Because of the high viscosity, a decelerating flow cannot deposit its load selectively, and all sizes will stop together giving a very poorly-sorted deposit with larger clasts floating in a finer matrix. Recent flows are flat-topped apart from protruding boulders and produce rather parallel-sided unstratified beds. These are laterally extensive parallel to the slope but have rather restricted widths, particularly where confined in channels. The upper surfaces of debris flows may be reworked by stream processes.

SHEETFLOOD DEPOSITS

Sheetfloods are flood flows of relatively low viscosity which expand at the downstream ends of channels, usually below the intersection point (Bull, 1972). Blissenbach (1954) originally used 'sheetflood' as a synonym for 'debris flow' as used here. Sheetfloods develop into shallow sheet flows which generally develop upper flow regime conditions and which do not persist very far (Rahn, 1967). They deteriorate into patterns of braided channels and bars which dissect the upper surface of the sediment sheet. The result is a layer of fairly well-sorted sand and gravel with small-scale internal lenticularity and scouring. Cross-bedding and cross-lamination may occur, but are not ubiquitous.

STREAM CHANNEL DEPOSITS

Stream channels are probably most important in the upper part of the fan (Bull, 1972) since below the intersection point floods become unconfined and may develop into sheet floods. However, channels which have their heads on the fan surface and which are fed by infiltrated drainage may persist to lower areas of the fan. The deposits are generally lenticular bedded poorly-sorted gravels and sands resulting from low viscosity flows. The coarser layers may be imbricated while sandier beds may be cross-bedded. These features and their channelled contact with surrounding sediments are probably the most diagnostic criteria for recognition of these deposits.

SIEVE DEPOSITS

Sieve deposition occurs just below the intersection point when the sediment load of the flood is deficient in fine-grained sediment (Fig. 3.5). A highly permeable older deposit causes the flow to diminish rapidly as infiltration of water occurs. As a result a clast-supported gravel lobe is deposited. Sieve deposits were first described by Hooke (1967). Wasson (1974) has described how the lobes may have clearly defined downstream margins, and has suggested that sieve deposits develop from the earliest, clear water stages of a flood and form a mid-channel obstruction at the intersection point around which later sheetfloods bifurcate.

Sieve deposit gravels are probably rather well-sorted and poorly imbricated and may be composed of angular fragments. With burial, the interstices are slowly filled with finer, infiltrating, sediment giving the final sediment a markedly bimodal grain size distribution.

3.2.5 Post-depositional processes

At any particular time, only a relatively small area of an alluvial fan surface receives sediment. Elsewhere sediment-starved areas undergo a variety of post-depositional processes

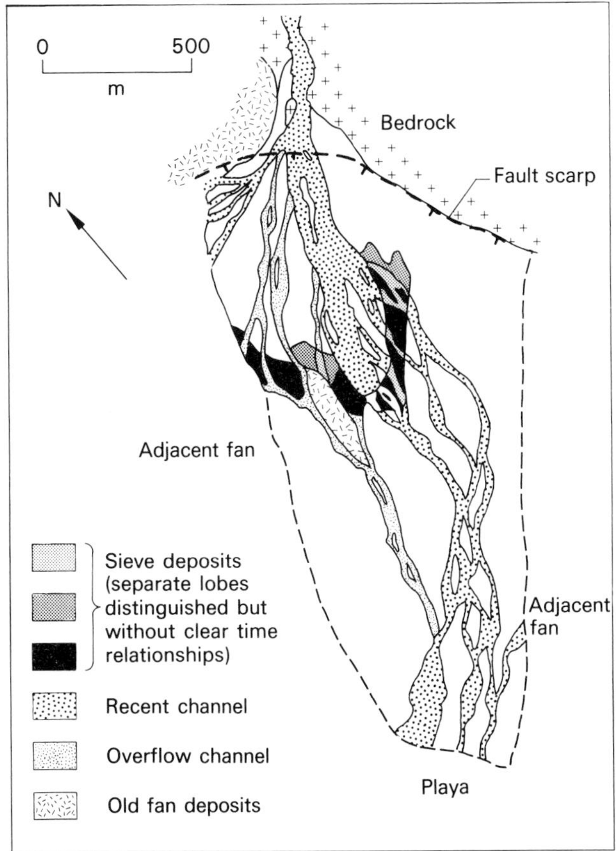

Fig. 3.5. The Gorak Shep Fan, Death Valley, showing the concentration of sieve deposits at the intersection points of the fan channels in the mid-fan area. There is a relief of about 250 m from fan apex to playa surface (after Hooke, 1967).

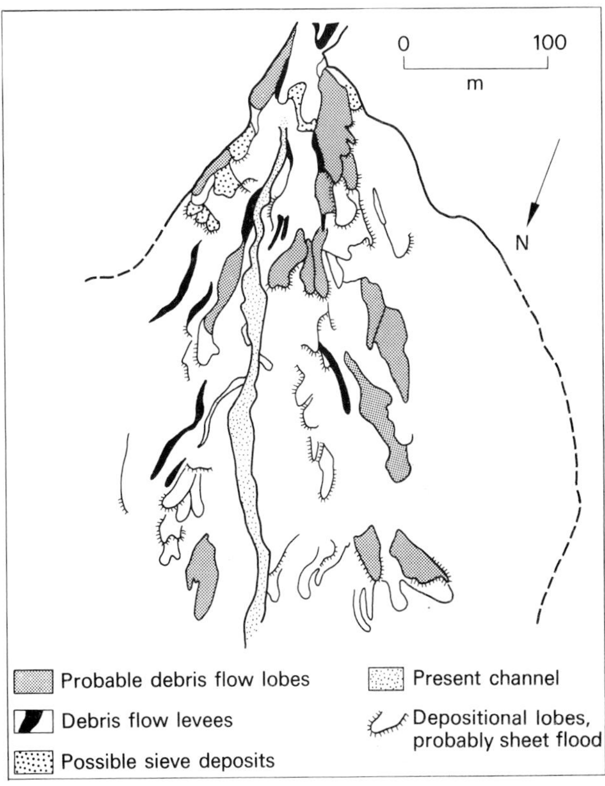

Fig. 3.6. The upper part of the Trollheim Fan, Death Valley, showing the distribution of recent debris flow lobes and levees. Other lobes are of sieve deposit or sheetflood origin. The undecorated areas are fragments of older fan surfaces and of deposits of indeterminate origin. There is a relief of about 100 m over the area shown (after Hooke, 1967).

which may continue for hundreds of years and substantially modify original depositional features (Denny, 1967). Weathering and run-off are the main post-depositional agents. Unstable clasts continue to break down both at and below the sediment surface (T.R. Walker, 1967), and the products are washed either into the underlying sediment or over the fan surface into more distal environments. Some material is also removed by wind deflation. Gullying may develop on the fan surface while the inter-gully areas develop a desert pavement in semi-arid environments. Desert pavements have closely packed angular fragments, commonly coated with desert varnish. They protect the fan surface from erosion and may overlie a silty layer. The smooth pavement surface may sometimes develop small terraces, a few centimetres high, running parallel to contours, owing to down-slope movement.

In semi-arid environments, fan sediments are often coloured red as weathering breaks down ferro-magnesian minerals and biotite into clays and hematite. This process takes thousands of years and is facilitated by periods of wetting and drying (T.R. Walker, 1967). A fuller account of red-bed development is given in the section on inter-channel areas (Sect. 3.6).

3.2.6 Distribution of fan processes and products

Debris flow deposits tend to occur near the fan apex, while sheetflood deposits occur in more distal areas (Fig. 3.6). Sieve deposits are concentrated just below the intersection point (Fig. 3.5) and channels occur mainly above that point, though post-depositional gullying and emergent groundwater may also give channel deposits at lower levels.

The size of the larger clasts decreases down-fan, the rate of change of maximum clast size with down-fan distance being greater for mudflows than for stream deposits (Bluck, 1964). Rate of change of maximum clast size may also vary with lithology (Tanner, 1976). Debris flows dominate where the source area provides abundant fine material and where erosion is rapid. They are therefore most important in semi-arid fans. Stream channel deposits are favoured where sediments are not so fine-grained and run-off is less ephemeral. On more humid fans, the channels comb the fan surface and lower gradients are developed. These generalizations hide the fact that adjacent fans in a semi-arid setting may show widely differing balances between stream and debris flow processes because of source rock differences. Croft (1962) showed that quite minor damage to the soil and vegetation cover of a source area can trigger large increases in deposition rate, and similar changes could also affect the balance between processes.

3.2.7 The semi-arid fan model

The fan sequence as a whole becomes thinner and finer distally. Channelled debris flows should pass distally into sheetflood deposits which, in turn, may pass into playa lake or floodplain deposits. Lateral changes within the fan are gradational, but the transition from the fan to the more distal environments may be so sharp that the deposits interfinger, reflecting extension and retreat of the fan due to tectonic or climatic changes.

At any one site in a fan the products of the different processes are likely to be randomly interbedded because of the spasmodic lateral shifting of the locus of deposition on the fan surface. However, the repeated incision and segmentation often associated with tectonic activity may result in a vertical sequence which coarsens or fines upwards or which shows an overall change from the products of a distal to proximal environment or vice versa.

From distribution maps of the products of present-day fans it is difficult to suggest any firm criteria whereby axial deposits might be distinguished from lateral ones. The radial nature of the distribution pattern means that lateral deposits appear similar to distal axial ones, though the nature of any interbedded sediments may give an indication. Lateral deposits, laid down close to a fault scarp may include scree breccias while axial distal deposits may be interbedded with lake or floodplain sediments. Most deductions of 'proximality' refer to the fan apex rather than the controlling topographic break or fault line. Divergence of palaeocurrents from the mean direction is probably the most useful clue towards resolving this problem.

Palaeocurrent directions, which might prove difficult to collect, particularly in debris flow deposits, should in any one place show a fairly close grouping, though over the whole fan a spread of up to 180° might be expected. Isolated fans might produce a radial pattern, but in a bajada the overlapping of fans will make this less likely.

3.2.8 Humid fans

In humid fans, dominated as they are by channel processes, the deposits will be more comparable to those of braided streams than to those of semi-arid fans and they will therefore be dealt with in Sect. 3.3. Humid fans can be regarded as deposits of braided streams which develop without any lateral confinement.

3.3 PRESENT-DAY PEBBLY BRAIDED RIVERS AND HUMID FANS

There is some ambiguity in the use of the words 'braided' and 'anastomosing' as applied to river channel patterns. For many authors, these terms are synonymous, but Schumm (1971a) suggested that 'braided' should refer to rivers in which flow diverges and rejoins around bars on a scale of the order of channel width, whereas 'anastomosing' is best used where the river divides into more permanent and larger sub-channels separated from each other by floodplain. Each anastomosing sub-channel then shows its own pattern of braiding and sinuosity. Schumm's usage will be followed here.

Settings which commonly produce pebbly braided channels are glacial outwash areas, humid fans and the wadis of semi-arid regions. In most cases, the large-scale morphology is either a shallow fan or a valley train where valley walls confine the flow. Some outwash plains, however, have multiple or shifting sources and therefore have no well-defined apex. Though climates vary and the processes and deposits of interfluvial areas may differ, those of the channels are similar.

Most of our knowledge of pebbly braided streams comes from studies of proglacial outwash areas. This is mainly because the predictable seasonal discharge pattern makes their study easier than that of semi-arid rivers where flash floods are quite unpredictable.

3.3.1 Bedforms and processes

The hydrology of proglacial areas is complex with discharge fluctuations on a number of time scales (Church, 1972; Fahnstock, 1963). The seasonal fluctuation giving high spring and summer and low winter discharges is the dominant medium-term feature. Upon this may be superimposed diurnal variations of weather-dependent amplitude and the whole pattern may be superimposed upon longer-term climatically controlled fluctuations. Small-scale morphology responds rapidly to changing discharge while the larger-scale disposition of channels remains fairly constant over a period of years and only changes with major floods.

Fig. 3.7. Braided pattern of gravel bars, Alakratiak Fjord Valley, Washington Land, Greenland. The main flow is concentrated in a zone within flanking areas of abandoned channels and bars. Emergent bars have superimposed patterns of smaller bars and channels.

There have been many descriptions of glacial outwash areas, mainly from the 'Sandar' (sing. sandur) of Iceland (Bluck, 1974; Hjulström, 1952; Klimek, 1972; Krigström, 1962) and from north-western Canada and Alaska (Boothroyd, 1972; McDonald and Banerjee, 1971; Rust, 1972a; N.D. Smith, 1974; Williams and Rust, 1969). Outwash fans and plains show complex channel patterns with the main flow generally concentrated in a fairly well-defined zone. The rest of the area is made up of abandoned channels and bars, with no separate floodplain (Fig. 3.7).

The braided pattern itself is caused by ephemeral bars which develop on several scales. They show a variety of forms, owing to complex histories of deposition and erosion. While Leopold and Wolman's (1957) experimental results suggest that braid bars may form and split the flow under steady discharge, it is probable that most of the emergence of bars in natural streams is related to discharge fluctuations. The bars of braided streams may be subdivided in various ways (e.g. Church, 1972; Krigström, 1962; N.D. Smith, 1974). Here we will confine ourselves to three major types: (1) longitudinal bars, (2) bars in curved channel reaches and (3) transverse bars.

LONGITUDINAL BARS

These are the diamond or lozenge-shaped bars which are the most obvious form in pebbly braided streams (Fig. 3.7). They correspond to the 'spool bars' of Krigström (1962). They form initially by the segregation of coarser clasts as thin rhomboidal gravel sheets. Such early sheets ('sheet bars') occur in the coarse upstream parts of outwash fans and probably grow into higher relief longitudinal bars by the development of downstream slipfaces, by vertical accretion to the bar tops and by the erosion of other margins (Boothroyd, 1972).

The upper surfaces often carry shallow 'transverse ribs' (McDonald and Banerjee, 1971) (equivalent of 'clast stripes' of Boothroyd, 1972) of coarser clasts with a well-developed imbrication and a transverse clast long axis orientation. They have a regular spacing related to grain size, and Boothroyd considers them to be the product of upper flow regime antidune transport which is common on the bars. Bar surfaces become finer-grained downstream and a superimposed system of smaller bars and channels develops as the bar emerges (Krigström, 1962). Rippled sand mantles much of the bar-top gravel when the bars emerge, owing to the continued mobility of sand at reduced shear stress.

Downstream depositional margins are either slip faces, if the grain-size is finer than cobble grade, or riffles if the grains are coarser. At low water stage, bar margins may be draped with a sand wedge reflecting the continued bar growth after the shear stress had fallen below that needed to transport pebbles (Fig. 3.8) (Rust, 1972a). Other margins are erosional.

Internally, longitudinal bars are composed of massive or crudely horizontally-bedded gravel frequently fining upwards and forming a framework which may be matrix-filled. Smith (1974) reported gravels with alternating matrix-filled and openwork layers which he thought reflected normal (diurnal?) discharge fluctuations (Fig. 3.9). Rust (1972a) observed bedding dipping at less than 3° either upstream or downstream and inferred that the bar surface is the preferred depositional site and that the bars consequently migrated upstream. No direct observations of upstream bar growth are known however. Cross-bedded gravels should result from avalanche faces, while riffle margins will give structureless gravels. Within cross-bedding, reactivation surfaces (Collinson, 1970b) may occur sometimes with silty drapes as a product of falling stage modification of the bar front. Some horizontally-bedded gravels should pass downstream into cross-bedded gravels and sands if sand wedges had developed.

BARS IN CURVED CHANNEL REACHES

Bars may be attached to either bank of a curve (Krigström, 1962). They are commonly extensions and modifications to the flanks of larger longitudinal bars, and have downstream

Fig. 3.8. A sand wedge at the lateral margin of a gravel bar at falling stage has been later dissected by water in minor channels draining off the bar. The water in the main channel is now stagnant and is depositing a thin film of mud on underlying ripples (after Rust, 1972a).

margins which are strongly oblique to the channel trend (Fig. 3.10). Bluck (1974) refers to such bars as 'transverse' or 'lateral', but both terms are ambiguous and 'diagonal bar' is preferred.

When the upstream end of a bar is attached to the bank, flow is concentrated in a channel on the outside, the 'riffle reach', which ends at a steeper 'riffle face' (Fig. 3.11), extending across the channel at an acute angle (Bluck, 1974, 1976). The bar platform is a gravel pavement which grows as the channel shifts laterally. Its downstream end is a steeper extension of the riffle face. Sand lobes and small deltas extend this face into the 'slough channel' reflecting waning flow due either to channel migration or to falling discharge.

Slough channels accumulate fine sediment which may overlie coarse sediment deposited when the channels were active. Diagonal bars with their downstream ends attached to the

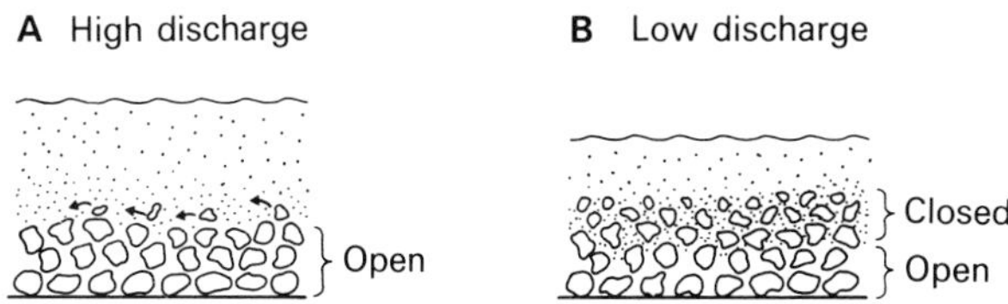

Fig. 3.9. Origin of alternating openwork and matrix-filled gravel layers; A, transport and deposition of coarse gravel during high discharge; fines remain in suspension; B, deposition of finer gravel through waning flow; suspended fines settle into voids and are trapped by surface layer. Process is repeated with next discharge cycle (after N.D. Smith, 1974).

bank lack slough channels (Fig. 3.10B).

Lateral migration of the riffle reach leads to deposition of a parallel-sided gravel sheet, probably coarsest at its base and with an upwards increase in sand. The thickness corresponds to the scour depth of the riffle reach and may increase downstream as the slough channel and the pool in front of the riffle face deepen. The gravel may be imbricated, and massive or horizontal bedding passes downstream into cross-bedded sands. The upper parts of the sheet may include ripple cross-laminated sand bodies elongated parallel to the current.

The internal structure is therefore similar to that of longitudinal bars and it is difficult to distinguish them by their deposits.

TRANSVERSE BARS

Transverse bars, as originally recognized by Ore (1963) and Smith (1970), are most common in sandy low-sinuosity

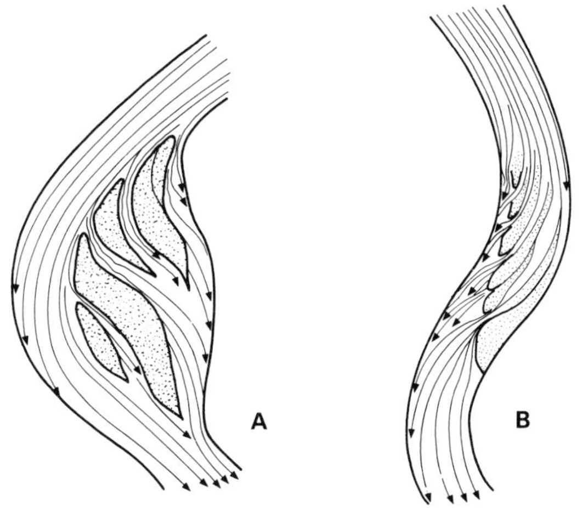

Fig. 3.10. Diagonal bars in curved channel reaches; A, attached to the inside bank; B, attached to the outside bank (after Krigström, 1962).

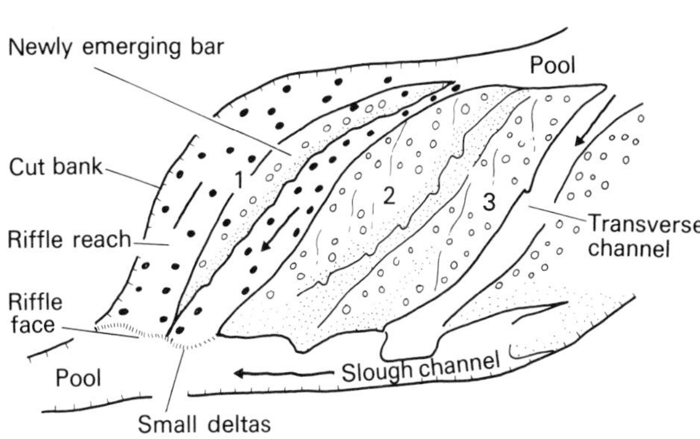

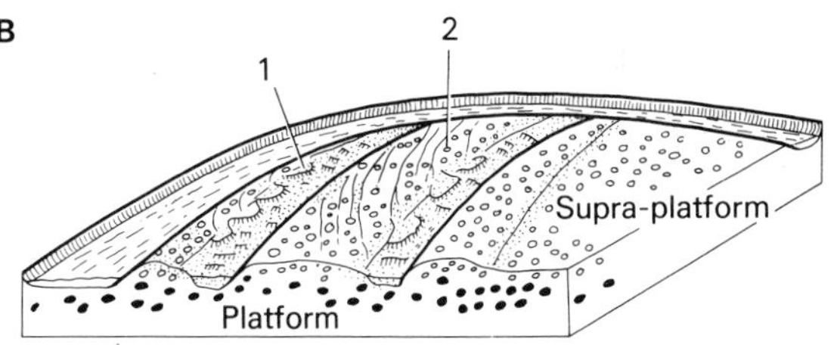

Fig. 3.11. Morphology, terminology and structure of a typical diagonal bar. (1), (2) and (3) are individual units of the bar where (1) is the newly emerging bar which will eventually become attached to (2) (after Bluck, 1974).

streams, but also occur in pebbly ones. In a pebbly, proglacial stream, N.D. Smith (1974) recognized a subsidiary population of transverse bars which included lobate, straight and sinuous fronted forms. The downstream ends were usually slip faces, suggesting an analogy with transverse bars of sandy streams.

Transverse bars with slip faces are likely to produce laterally extensive cross-bedding if there is substantial sand in the sediment population though gravels may be structureless (Bluck, 1974).

CHANNELS

The channels between the bars are active for much longer periods than the bars and develop deposits of their own as well as modifying bar margins. In the upper parts of an outwash fan, channels are straighter and deeper, but become shallower and more braided as the gradient decreases (Boothroyd, 1972). Active channels around bars are floored by the coarsest gravel. The channel floors grade into the lower parts of bar surfaces, particularly at their upstream ends, or where riffles run into channel pools. Where channels are abandoned, either completely or partially as slough channels, the gravel floor may be buried in finer sediment which migrates as dunes or ripples under the reduced flow. Silty sands and silts sometimes mantle not only the floors of abandoned channels but also the tops of some bars (Williams and Rust, 1969; Doeglas, 1962). Williams and Rust also showed how a channel produced a fining-upward fill of decreasing areal extent so that muddy sediment occurs only in a zone some distance from the still active channel.

3.3.2 Directional properties

Structures of directional significance occur on a variety of scales. The main structures with preservation potential are channels, pebble imbrication, cross-bedding, ripples and ripple cross-lamination, and sand lineations. Channels are the largest structures and have the lowest variability (Allen, 1966; Bluck, 1974). Of the smaller structures, imbrication, associated with the early development of bars, shows a unimodal pattern of low variability (Bluck, 1974; Rust, 1972a (Figs. 3.12, 3.13)). Elongated pebbles larger than 2–3 cm have a strongly preferred orientation of their long axes perpendicular to the channel trend (McDonald and Banerjee, 1971; Rust, 1972a and b). Cross-bedding, developed at the downstream ends of bars, reflects the later stages of bar development and is the most variable with a bimodal distribution more or less symmetrical about the channel trend, reflecting the skewed disposition of the bar fronts. Transverse bars produce downstream-dipping foresets. At lower water stage highly dispersed cross-bedding continues to develop and ripples and lineations also form on the bar surface, reflecting the current directions at lower shear stress (Collinson, 1971b). Their directional dispersion is greater than that of imbrication, but it is less than that of cross-bedding.

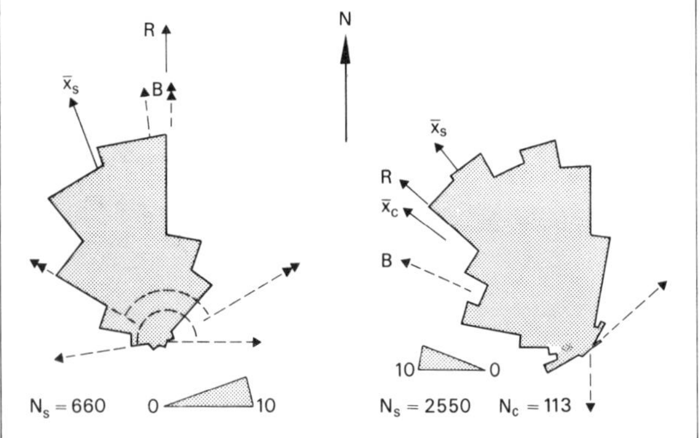

Fig. 3.12. Directional data from two areas of the braided pebbly Donjek River, Canada. The histograms are for small scale structures. R = river trend; $\bar{x}_s$ = vector mean of small scale structures; $\bar{x}_c$ = vector mean of channel orientation; B = channel arc bisector. Double arrows refer to channels, single arrows to small structures (after Rust, 1972a).

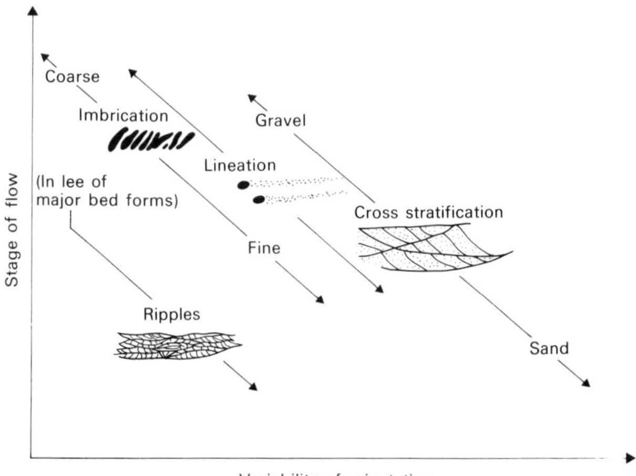

Fig. 3.13. Possible relationships between the flow stage and the type and variability in orientation of structures. Note that ripples lie off the main trend. They tend to be caught in major channel troughs and orientated along them (after Bluck, 1974).

3.3.3 Sedimentation model for pebbly braided streams

The sediments of pebbly braided streams do not fall into any well-ordered pattern on a local scale. Because channels and bars shift rapidly the deposits will show a largely random vertical and lateral distribution of fragmented bars and channel fills. Any single bar will have had a complex history of deposition and erosion. The sequence is likely to be a series of erosionally based lenses of sand and gravel of variable lateral extent, which will probably extend further parallel to the current direction than perpendicular to it. The gravels will commonly be imbricated; possibly fining upwards, near horizontally bedded, and sometimes cross-bedded. Sands will be cross-bedded in tabular sets with ripple cross-lamination sometimes preserved above the sets. The cross-bedding may show reactivation surfaces. Sections parallel to the current may show a downcurrent change over short distances from horizontally bedded imbricated gravels into cross-bedded gravels and sands. Channel fill remnants will show silts and clays resting sharply upon gravels which themselves sit on an erosion surface.

On a larger scale, braided outwash fans and channel systems show downstream changes. Boothroyd (1972) showed that both the surface slope and the size of the largest clast decreased down the fan, along with a change of channel and bar morphology (Fig. 3.14). The increased braiding gives higher small-scale channel sinuosity and diagonal bars may be more

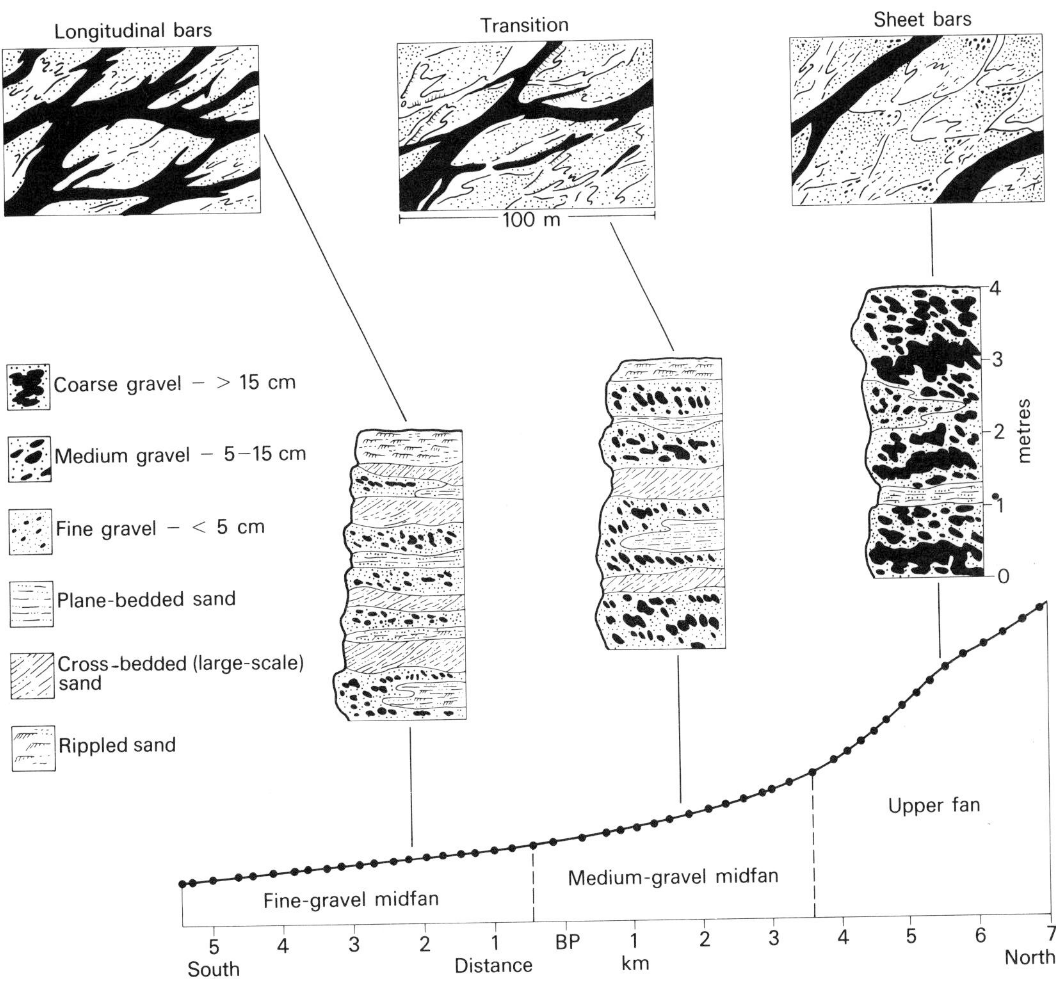

Fig. 3.14. Changes in bar type, slope and structure along a braided outwash fan. The changes are gradational and transverse bars become important in the lowest reaches of the system. The transition from sheet to longitudinal bars is associated with the dying out of boulders (after Boothroyd, 1972).

important (cf. Bluck, 1976).

These result in lateral facies variations (Fig. 3.14). Proximal fan deposits are dominated by crudely horizontally-bedded imbricated gravels while cross-bedding increases distally as grain size reduces.

3.4 PRESENT-DAY SANDY LOW-SINUOSITY RIVERS

Such rivers constitute a spectrum of types from those which are gradational with pebbly braided rivers to those which are gently meandering. They commonly lie downstream of pebbly braided reaches and the change of grain size is often accompanied by a change in the dominant bedform from longitudinal bar to transverse bar. Many sandy low-sinuosity rivers are braided in a random fashion owing to the development of mid-channel bars and other large bedforms; others show meandering of the major thalweg without splitting of the flow by braiding but with the development of alternating, bank-attached bars. In most cases, the channel pattern becomes increasingly braided at low discharge when flood-formed bed forms are stranded and dissected.

The low sinuosity of these rivers, probably caused by the

relatively low levels of suspended sediment (see Sect. 3.10), means than channels are rarely cut off. Thus there are few resultant clay plugs to limit channel migration (cf. Sect. 3.5) and channels may sweep across their floodplains more or less continuously. The extreme example of this is the Kosi River which sweeps the area of its large humid fan in a matter of two or three hundred years (Fig. 3.3) (Gole and Chitale, 1966). However, many of the rivers from which more detailed information is derived occur in areas of net erosion and their channels are confined in valleys. It seems likely that the channels of these rivers would behave in the same way, but we are not sure.

Major rivers of this type include the Niger–Benue (NEDECO, 1959), Brahmaputra (Coleman, 1969), Yellow River (Chien, 1961), rivers of the South American plains (e.g. Rio Segundo (Rich, 1942)) and the major rivers draining the Siberian plains. Rather smaller rivers of the mid-western United States (e.g. Loup (Brice, 1964); Platte (N.D. Smith, 1970, 1971)) and of sub-Arctic areas (Lower Red Deer (Neill, 1969); Tana (Collinson, 1970b)) are also included and have provided much of the detailed information on processes and products. In addition there are the small sandy ephemeral streams of semi-arid regions.

3.4.1 Bed features of low-sinuosity rivers

The great variability of type, scale and distribution of bed features in low-sinuosity rivers has led to various schemes of classification, with considerable confusion over the nomenclature and the hydrodynamic status of the bedforms. Whilst it is possible to erect a hierarchy of bed features for individual rivers, it is very difficult to assemble them into a universally applicable scheme. However, the following scheme covers most features.

SIDE BARS

As the thalweg meanders within the confines of a straight channel, bars develop on the insides of the bends. These 'side bars' (Allen, 1968; Collinson, 1970) are attached to the banks and usually emerge at low stage. They have no slip faces and descend gradually into the channels, usually carrying superimposed dunes or sandwaves. The bars are up to 1.5 km long and 1 km wide in the Niger (NEDECO, 1959) and about 500 m long and 300 m wide in the Tana (Collinson, 1970b). Their internal organization is largely unknown but it is presumably a complex assemblage of the products of smaller bedforms.

MID-CHANNEL BARS AND ISLANDS

In some larger rivers, mid-channel bars and islands of considerable durability are formed and are often stabilized by vegetation. The Brahmaputra, for example, has a channel complex 15 km wide with sub-channels up to 2 or 3 km wide divided by mid-channel bars, locally called 'chars' (Coleman, 1969). These are up to several kilometres long and wide but may shift their position rapidly during floods. They are presumably composite products of smaller bedforms though their upper surfaces may be accreting vertically, mainly by overbank processes. In the Niger, islands with a life of the order of a hundred years divide the flow in places (NEDECO, 1959), while in the Volga, large islands show evidence in morphology of having grown by lateral accretion (Shantzer, 1951).

ALTERNATE BARS

These are similar to side bars in being triangular in plan and attached with regular spacing to alternate banks. They differ in having their own slip faces which may be convex downstream (Harms and Fahnstock, 1965; Maddock, 1969). They are best described and illustrated from straight reaches of relatively small channels where slip faces are up to only 1 m high. The bars probably produce tabular cross-bedded sets.

DUNES, SANDWAVES AND TRANSVERSE BARS

Two major groups of large-scale, asymmetrical bedforms, namely sandwaves and dunes, are now widely recognized (Harms, Southard, Spearing and Walker, 1975). Attempts at understanding the sediment dynamics of sandy river beds should be made with reference to these groups. However, most descriptions were made prior to the full realization that two groups existed and there is therefore a profusion of overlapping and ambiguous names in the literature.

The two major groups are distinguished by both their shape and their relationship to the prevailing hydrodynamic conditions.

(1) *Sandwaves* have high length to height ratios, continuous crest lines and a general absence of localized scour in their leeside areas. They normally have rather sharp-based slip faces and their upper surfaces carry superimposed ripples. These forms have been called 'linguoid bars' (Allen, 1968; Collinson, 1970) and 'transverse bars' (N.D. Smith, 1970, 1971). They are commonly lobate in plan and vary in size even within a particular river. In the Tana River they are mostly 200–300 m long, 200 m wide and up to 2 m high (Fig. 3.15). In the Brahmaputra, the largest bedforms described by Coleman (1969) and termed 'sandwaves' are up to 16 m high and 1 km long, though it is impossible to be sure if these are true sandwaves.

Sandwaves are developed when the depth and velocity of the flow are above those appropriate for ripples but below those for dunes. At low water stage many sandwaves emerge to split

Fig. 3.15. Sandwaves (linguoid or transverse bars) in the Tana River, Finnmark, Norway, emerging at intermediate discharge during falling river stage. Contrast the smooth crest-line of the submerged bar in the foreground with the dissected emergent bars.

the flow into a braided pattern and they become locally modified by the falling water stage (see later).

Internally, they give tabular sets of cross-bedding which are often inter-bedded with units of ripple cross-lamination. Ripple morphologies may sometimes be preserved on the interface between ripple cross-lamination and an overlying cross-bedded set.

(2) *Dunes* have low length to height ratios and form at depths and velocities above those of sandwaves. They have sinuous, discontinuous crests and deep scour troughs on the leeside, usually localized in front of a crestline saddle, suggesting a much stronger coupling between the shape of bedform and the turbulence structure of the whole flow. Ripples may be superimposed on their stoss sides. Sizes are highly variable, depending mainly on the depth of the water under which the dunes develop. In medium-sized rivers, such as the Tana, dunes are commonly of the order of 5–10 m long and 30–50 cm high at the scour trough. The 'dunes' of the Brahmaputra, however, are 1.5–8 m high and 40–500 m long (Coleman, 1969) though it is unclear whether the forms should be considered dunes or sandwaves as the terms are used here.

Because of their association with strongly developed lee-side scour troughs, dunes can normally be expected to produce trough cross-bedding on a variety of scales.

LARGE-SCALE LINEATIONS

These have only been described in rivers from the Brahmaputra (Coleman, 1969) though they are common in tidal settings (e.g. Imbrie and Buchanan, 1965). In the Brahmaputra, they are ridges and troughs elongated parallel to the current with transverse spacings of 8 to 30 m and heights of up to 1.5 m. They are hundreds of metres long, occur near channel edges, and seem to be the result of upper flow regime conditions involving turbulence cells elongated parallel to the flow. They show little internal structure though horizontal lamination was seen in sections cut parallel to the current direction (Coleman, 1969).

3.4.2 **Effects of water stage fluctuations**

Sandy low-sinuosity streams are often subjected to substantial discharge fluctuations either of a seasonal or less predictable nature. The shape of the hydrograph is often closely related to the river's climatic thermal regime. The fluctuations have a wide variety of important effects on channels, bedforms and their internal structures which can be summarized under three headings:

LARGE-SCALE CHANNEL CHANGES

The rise and fall of water over a flood cycle normally involves considerable scour and fill and this is particularly important where the bed and bank material is sandy. At the largest scale, the comparative echograms of the Brahmaputra (Fig. 3.16) show that huge volumes of sediment are eroded and deposited over a flood peak as channels, bars and islands shift. Similar changes take place in smaller rivers though only an exceptional flood can fundamentally remould the bed.

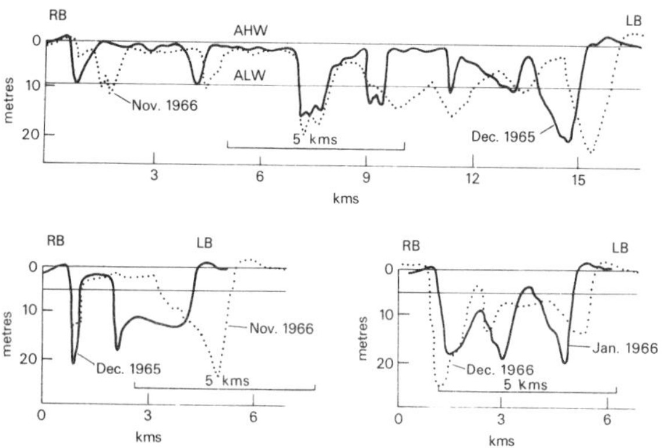

Fig. 3.16. Transverse profiles of the bed of the Brahmaputra made before and after a flood. Over this short time interval huge volumes of sediment are eroded and deposited and the channel pattern radically altered (after Coleman, 1969).

BEDFORM SUPERIMPOSITION

At low water stage, when the bed is exposed, it is often possible to identify several sizes of dunes and/or sandwaves in superimposed relationships, adding to the difficulties of classification. Echograms from flood cycles of several rivers show that there is an important lag between the changes of flow and the bedform response (Fig. 3.17) (Pretious and Blench, 1951; Carey and Keller, 1957; Coleman, 1969; Neill, 1969). Allen and Collinson (1974) argued that superimposition of different scales of bedform is a function of changing flow within the dune field but the recognition of sandwaves as a separate class complicates the issue further. The important point is that rapidly fluctuating discharge is more likely to produce superimposed relationships than are slow changes and this may sometimes be recognizable as a mixture of scales of preserved cross-bedding.

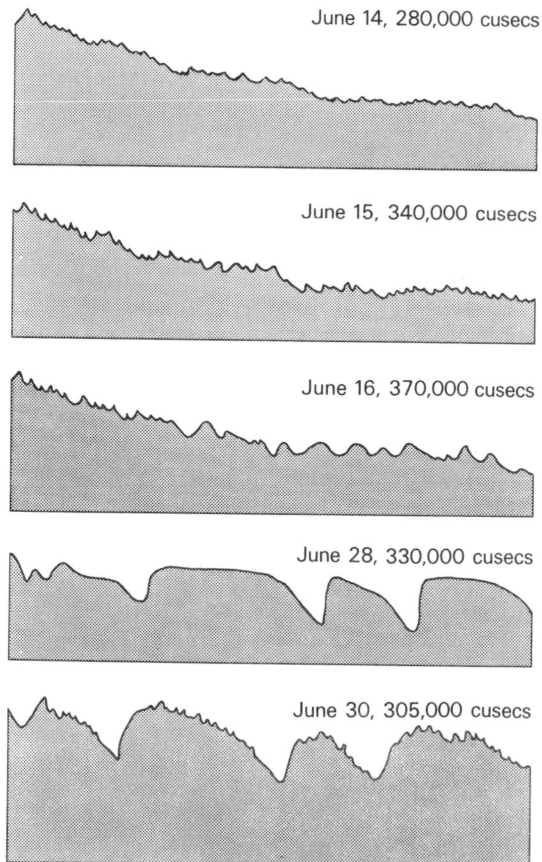

Fig. 3.17. Changing character during a flood of the bedforms in a reach of the River Fraser, British Columbia. Length of the reach is 670 m (after Pretious and Blench, 1951).

FALLING STAGE MODIFICATION

As water level falls and the bed is partially exposed, dunes and, more importantly, sandwaves, are modified. They split the flow and control its pattern because instead of flowing over the top of the bars in a broadly radial pattern, flow is concentrated in the inter-bar depressions which then function as channels (Fig. 3.18). Ripples both on the bar top and in the lee area are reorientated (Collinson, 1970) and flow over the top of the bar may split so that minor channels dissect the bar (N.D. Smith, 1971). This can lead to the progradation of delta lobes into inter-bar channels whilst other bar margins may be eroded.

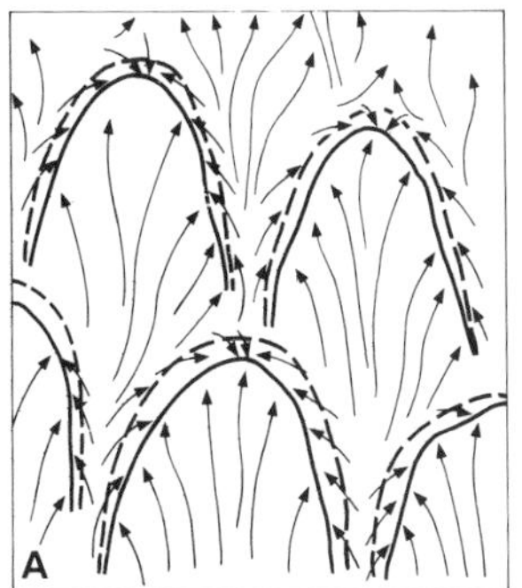

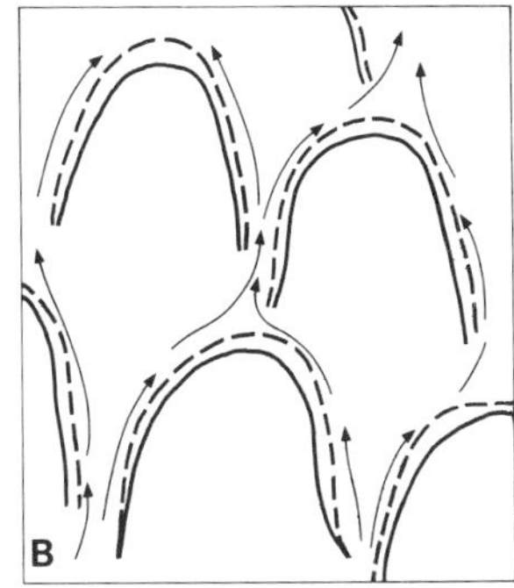

Fig. 3.18. Flow patterns associated with (A) high and (B) low discharge over a river bed dominated by linguoid bars. At high stage, flow is predominantly over the bars. At low stages, flow is concentrated between the bars (after Collinson, 1970b).

As bars emerge, waves rework slip faces and build fringing ridges on the bar crestline. Wave modification may be erosive or it may simply rework the slipface to a lower angle, sometimes accompanied by heavy mineral concentration. In sheltered areas in front of bars, silts and clays are deposited from suspension on the slip face itself and on the rippled area at its foot, a situation similar to slough channels in pebbly braided rivers.

Falling stage modifications to sandwave margins are reflected in their internal structures Collinson, 1970b; N.D. Smith, 1971; Boothroyd, 1972; Boothroyd and Ashley, 1975). The usual tabular cross-bedding is complicated by low-angle erosion surfaces (reactivation surfaces) which mainly reflect slipface reworking (Fig. 3.19). Similar erosion, affecting mainly the upper parts of the set, may sometimes result from the arrival at the crestline of superimposed bedforms (McCabe and Jones, 1977). Normal cross-bedding above reactivation surfaces is the product of subsequent high-stage flow. Flow parallel to the slip face during falling stage may lead to lateral accretion of ripple cross-laminated sand. Silty drapes on the foresets are generally well spaced and extend over the full height of the foresets and sometimes out on to the toeset ripples.

The details and extent of bedform modification depend on both the regime of the river and upon the topographic level of any particular bedform on the channel floor. High rates of falling stage, as in the Tana River, favour abandonment of bedforms and the development of simple reactivation surfaces. Slower rates, as in the Platte River, favour bar dissection and the growth of delta lobes, producing more complex cross-bedding with a wider spread of directions. It is possible therefore that details of internal structure might be a guide to a river's climatic thermal regime (see Sect. 3.10) (Jones, 1977). However, bedforms on topographically high areas of the channel floor will emerge and be abandoned before those on lower areas with similar effects on the internal structures. This equation between internal structure and regime should only be applied to the interpretation of ancient sediments when the position of the structures in a channel sequence is understood.

3.4.3 **Overall organization and current vectors**

The organization of the deposits of rivers of this type is not easily predicted. The repeated scour and fill associated with flood events will lead to random preservation of fragments of channel deposits. Systematic upwards change of grain size or size of cross-bedding set through the channel deposits is unlikely except during channel abandonment when both would diminish. The overall pattern is probably a random interbedding of trough and tabular sets with occasional interbedded units of ripple cross-lamination.

Smith (1972) argued for high dispersion of cross-bedding azimuths because of a high lateral component in the migration of transverse bars in the Platte River. Banks and Collinson (1974), however, thought that low dispersion was likely, based on the predominantly downstream movement of linguoid bars in the Tana River and because larger volumes of downstream-directed cross-bedding were produced during the flood stage. This difference may be related to the shape of the falling stage hydrograph and therefore to river regime as extensive lobe development, found in the Platte, will tend to produce a greater spread of directions (Jones, 1977). In the Brahmaputra, cross-bedding measured in excavations had closely grouped down-channel azimuths with more divergent examples attributed to lower water stages (Coleman, 1969).

3.4.4 **Semi-arid ephemeral streams**

These show many of the features of more continuously flowing streams with bars, dunes, ripples and plane beds producing the expected internal structures (e.g. Williams, 1971; Karcz, 1972; Picard and High, 1973). However, they also show some differences due in part to vertical percolation of water into the

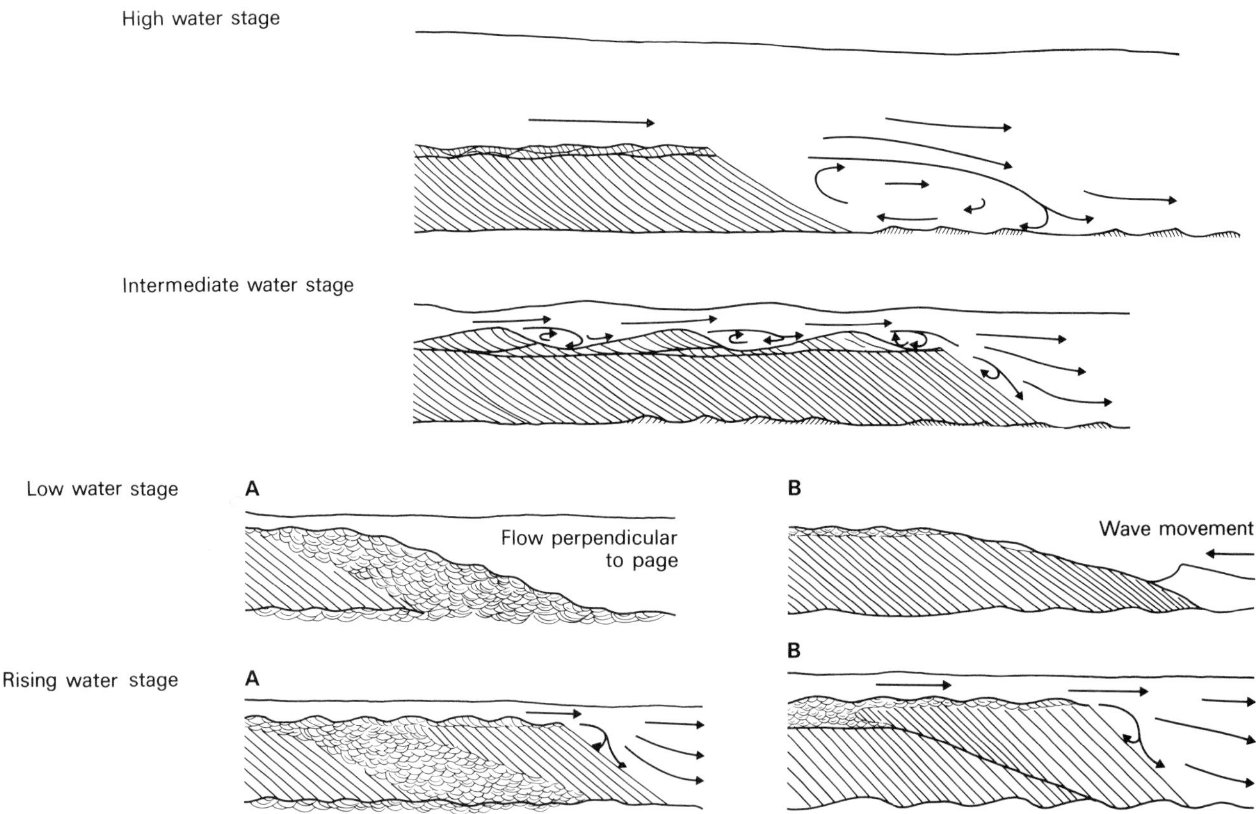

Fig. 3.19. Development of reactivation with changing water stage over a linguoid bar. At high stage, a strong separation eddy gives asymptotic foresets and counter-current ripples. Lowering of the water stage reduces the strength of the eddy but does not immediately stop the advance of the slip face. The foresets now have angular bases without counter-current ripples. Further lowering may give currents parallel to the side of the bar, capable of depositing a laterally accreted increment (A), while wave activity during emergence can reduce the angle of the foreset slope by erosion (B). Rising water stage may reactivate the cross-bedded set and cause it to bury the falling stage features, with or without an associated vertical accretion of the topset (after Collinson, 1970b).

bed, analogous to the 'sieve' effect of alluvial fans. Where the channel is unable to cope with abnormal discharge, sheet flows, associated with major floods, may cover virtually the whole valley floor with a sheet deposit between 1 and 4 m thick as happened in Bijou Creek, Colorado (McKee, Crosby and Berryhill, 1967). Here the importance of upper flow regime conditions is suggested by the dominance of parallel lamination which forms 90–95% of all deposits. In many floodplain sections, the whole of the sheet exhibited parallel lamination, though some sections showed planar cross-bedding in the upper parts, suggestive of late stage waning of the flood. Other sections had irregular interbedding of horizontal lamination, ripple-drift cross-lamination and medium-scale planar cross-bedding, sometimes with convolute lamination at the top. The last stages of the flood were confined to the main channel which cut into the floodplain sands. The channel sediments are dominated by trough cross-bedding, with some parallel-lamination and planar cross-bedding.

As discharge falls, silt and clay are often deposited as a surface coating which, on drying, cracks in patterns. The cracks may be filled by wind-blown sand which may be either preserved or eroded by later floods and the mudflakes reworked into intraformational conglomerates (Glennie, 1970; Karcz, 1972). Wind-blown sand may block channels and divert subsequent floods, and wind erosion may lead to pebble-strewn deflation pavements on stream beds (Glennie, 1970). Thus one feature which may distinguish the deposits of ephemeral streams from those of more permanent streams is aeolian reworking: interbedding of fluvial and aeolian sediments may be characteristic of ephemeral streams (see also Chapter 5).

3.5 PRESENT-DAY MEANDERING RIVERS

Meandering rivers (cf. Shelton and Noble, 1974) show a more organized distribution of channel processes and a clearer separation of channel and overbank environments than the low sinuosity and braided rivers into which they grade. A meandering channel occupies only a small part of its alluvial plain at any one time. It lies within a meander belt which is a complex of active channel, abandoned channels and near channel subenvironments. The meander belt shifts its position on the alluvial plain through time and its mode of shifting is largely a function of channel sinuosity (Allen, 1965c). With high sinuosity, probably related to high suspended load (see Sect. 3.10.1), the position of the meander belt becomes stabilized by clay plugs generated by frequent channel cut-offs (see Sect. 3.5.3). Sedimentation is then concentrated close to the meander belt and an 'alluvial ridge' is built above the level of the floodplain (Fisk, 1952b) (Fig. 3.20). This increasingly unstable situation is periodically relieved by the breaching of a channel bank during flood and the sudden shifting of the meander belt to a new position on the alluvial plain, a process known as 'avulsion'. The new course captures an increasing proportion of the flow and the old meander belt is abandoned.

Meandering rivers of lower sinuosity have fewer cut-offs and clay plugs and consequently the positions of their meander belts are less stable. This allows them to sweep across their alluvial plains in a more continuous manner, a process which achieves its extreme development in very low-sinuosity braided streams like the Kosi River (Gole and Chitale, 1966) (see Sect. 3.2.3).

Channel processes of fairly high-sinuosity rivers will be dealt with here though the gradation into lower sinuosity types should be borne in mind. Floodplain sediments which are common to other river types are considered in Sect. 3.6.

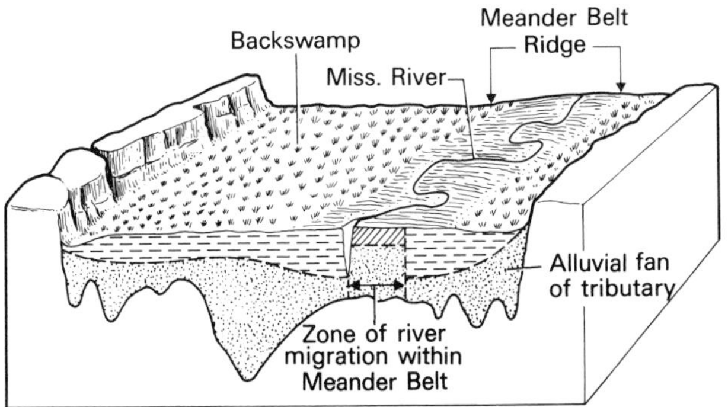

Fig. 3.20. The alluvial plain of the Mississippi, showing the main channel occupying a meander belt on the top of an alluvial ridge, flanked by floodplains. The plain is about 150 km wide and the alluvial ridge about 15 km wide. Large vertical exaggeration (after Schumm, 1971b).

3.5.1 Meander belts

Meandering tends to be favoured by relatively low slopes, by a high suspended load/bed load ratio, and by cohesive bank materials (Leopold and Wolman, 1957; Schumm and Kahn, 1972). Relatively steady discharge may also help. The relationships between meander geometry and other channel and discharge parameters have been studied theoretically and empirically.

Meander curves approximate to a sine function of channel distance and this represents an adjustment of depth, velocity and slope to minimize the variance of shear and frictional resistance (Langbein and Leopold, 1966). Meander wavelength (λ) has a nearly linear relationship with channel width (w), for example:

$$\lambda = 10.9w^{1.01}$$

(Leopold and Wolman, 1960) and with the radius of curvature of the meander (r):

$$\lambda = 4.7r^{0.98}$$

An analysis of published data by Carlston (1965) shows that the relationship between wavelength and discharge is complicated by the need to decide the most appropriate discharge parameter. Mean annual discharge ($\bar{Q}$) and the mean of month of maximum discharge ($\bar{Q}_{mm}$) give the closest correlations:

$$\lambda = 106.1\bar{Q}^{0.46} \quad \text{(Standard error} = 11.8\%)$$
$$\lambda = 80.0\bar{Q}_{mm}^{0.46} \quad \text{(Standard error} = 15\%)$$

This suggests that the effective control is a range of discharges between these two means values, possibly associated with a falling discharge pattern, and that meander width (W_m) and channel width (w) are related to the mean annual discharge by the following equations:

$$W_m = 65.8\bar{Q}^{0.47}$$
$$w = 7\bar{Q}^{0.46}$$

It is worth noting here that several models for meander sedimentation are based on an assumption of bankfull discharge.

Leeder (1973) showed that when sinuosity is greater than 1.7, bankfull channel depth (h) and bankfull width (w) are related by:

$$w = 6.8h^{1.54}$$

thereby giving a means of estimating width from depth, a parameter more commonly measurable in the rock record. Equations of the type given here should form the basis for attempted palaeohydrological reconstruction provided the type of river channel has first been established.

3.5.2 **Channel processes**

Deposition on point bars of meandering streams has been the major theme of sedimentation in fluvial channels until quite recently and an elegant process/response model has been developed for it. In the last few years, however, several studies have cast doubt on the generality of this model and pointed to a rather more complex picture. In this account, the classical model will first be described before the implications of complicating factors are discussed.

THE CLASSICAL POINT-BAR MODEL

It has long been recognized that flow in meander bends is helicoidal, with a surface component towards the outer bank and a bottom component towards the inner bank, and that maximum velocities occur near the outer bank (Fisk, 1947; Sundborg, 1956; Leopold and Wolman, 1960). The thalweg of the channel follows the line of maximum velocity, with scour pools near the outside bank. The velocity asymmetry and the position of the thalweg cross over in the inflection point reach as the helicoidal flow changes its sense of rotation. The flow pattern means that the inner, convex bank is the site of deposition while the outer, concave bank is the site of erosion. The channel as a whole migrates transversely to the flow depositing a unit of sediment by lateral accretion.

(a) *Erosion.* Erosion of the concave bank is influenced by the nature of the bank material. Floodplain silts and clays of low porosity and high cohesive strength tend to resist erosion unless earlier channel sands underlie them. Thick cohesive sediments are eroded as blocks which founder into the channel by undercutting and by the development of shear planes subparallel to the bank (Fig. 3.21) (Sundborg, 1956; Laury, 1971; Klimek, 1974a). Such planes are curved in plan and give scalloped erosion scarps. If they penetrate deeply and involve sufficiently large volumes of sediment, they may emplace material below the level of the channel base by rotational slumping (Turnbull, Krinitzsky and Weaver, 1966; Laury, 1971). Sands are helped to slough into the channel by changes in hydrostatic pressure owing to water stage fluctuations and the loading of overlying sediments. This causes lateral flowage towards the channel and hastens the undercutting.

Lateral changes in lithology influence the shape of the developing meander as well as that of the slump scar (Fisk, 1952b; Laury, 1971). Meander cut-off loops are often filled with cohesive clays and such clay plugs, when exposed in an erosional bank, influence the shape of the bank.

The material eroded from the concave bank usually falls or slides into the deepest part of the channel where it is winnowed to give a lag conglomerate of cohesive blocks and concretions of intrabasinal origin and perhaps coarser extra-basinal clasts. Blocks, emplaced below the level of the thalweg by large-scale basal sliding, avoid this reworking.

(b) *Point-bar deposition.* The sediment body enclosed by the meander loop is termed the *point-bar*. It has an essentially horizontal upper surface at about the level of the surrounding floodplain and it slopes gradually on its channel margin, towards the thalweg. This *point-bar surface* is the site of channel deposition and the classical model develops a pattern for the distribution of grain size and bedforms over this surface and consequently in the vertical sequence produced by the lateral migration.

Both the upper surface and the active point-bar surface are complicated by superimposed *scroll bars* as well as by smaller

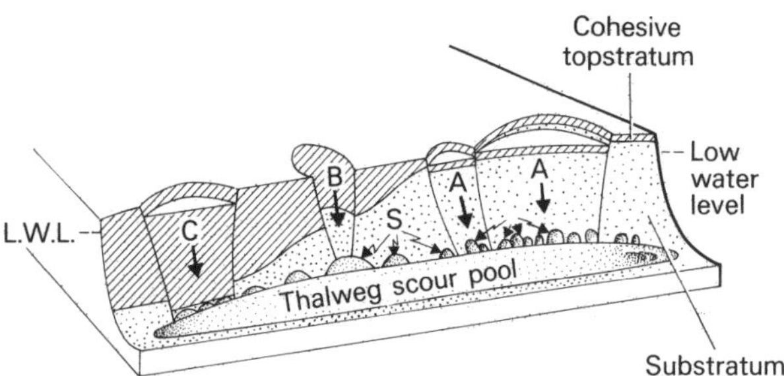

Fig. 3.21. Effect of cohesive topstratum thickness on bank failure. Accelerated scouring in the thalweg during high water stages is followed by subaqueous failures (either by shear or flow) in relatively non-cohesive substratum sands or gravels. For thin (A) and very thick (C) topstratums, subaqueous failures (S) are relatively numerous and small; they initiate thin upper bank failure by shear. Intermediate thickness topstratum (B) promotes small to large subaqueous failures (S) followed by bowl-shaped upper bank failure (after Turnbull *et al.*, 1966).

Fig. 3.22. Point bar of a pebbly meandering stream showing a scroll bar being driven onto the point bar surface at the downstream end. Flow is from right to left. Moffat Water, Southern Uplands, Scotland (photo: B.J. Bluck).

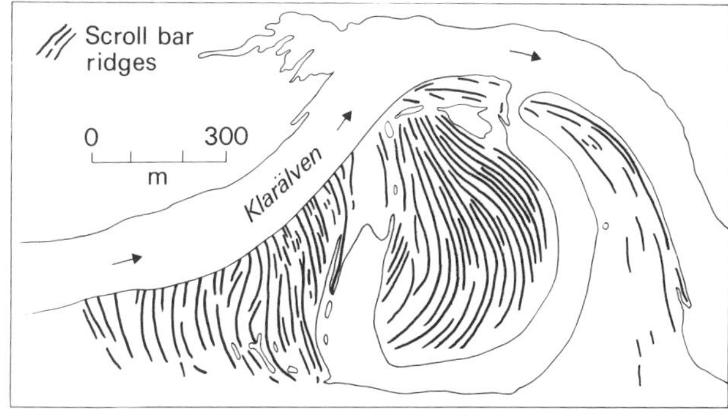

Fig. 3.23. Accretion topography of scroll bars on point bar bordering Klarälven, Sweden (after Sundborg, 1956).

bedforms. Scroll bars are ridges of sand which develop some distance down the point bar surface and are elongated more or less parallel to the contours of the surface. They have sharp downstream ends near the downstream limit of the point bar surface and they move up this surface until they are at bankfull level. They are then abandoned as a new one moves up behind (Fig. 3.22). They sometimes have slip faces directed towards the inner bank and they protect an area of still water which may accumulate fine sediments. The result is that the upper surfaces of point bars show a series of roughly concentric ridges and swales representing successive scroll bar accretions (Fig. 3.23) (Sundborg, 1956). These have been used to deduce erosional pathlines for individual meanders and thereby to establish the relative importance of down-valley and transverse migration (Hickin, 1974). As meanders grow there is a transition from transverse to down-valley migration.

The flow pattern around a meander bend is the key to understanding deposition on the point bar surface. The classical model developed most elaborately by Allen (1970a and b) and Bridge (1975) assumes bankfull discharge and a fully developed helicoidal flow around the bend. The diminishing velocities and depths and the upslope component of flow over the point bar surface cause bed shear stress to fall and the surface to operate as an elutriator, giving an upslope reduction of grain size (Fig. 3.24). Grain size distribution on the surface can be related to channel sinuosity.

The distribution of bedforms depends upon the depth and velocity of the flow and upon the available grain sizes. The precise distribution is therefore determined by channel size and sinuosity and by the grain size distribution of the available sediment, even if one only considers bankfull discharge. In a theoretical model, dunes should be the bedform on the lower part of the point bar and ripples or plane bed on the upper part. In many present-day rivers, dunes or sandwaves (which are not considered in the theoretical model) have been described from the deeper parts of meandering channels and ripples and plane beds from shallower areas (Lane and Eden, 1940; Sundborg, 1956; Harms, McKenzie and McCubbin, 1963; Frazier and Osanik, 1961; Bernard, Major, Parrott and LeBlanc, 1970).

(c) *The vertical sequence*. The model of point bar sedimentation is therefore fairly well established. Lateral

Fig. 3.24. The classical point bar model for a meandering stream (after Allen, 1964, 1970b).

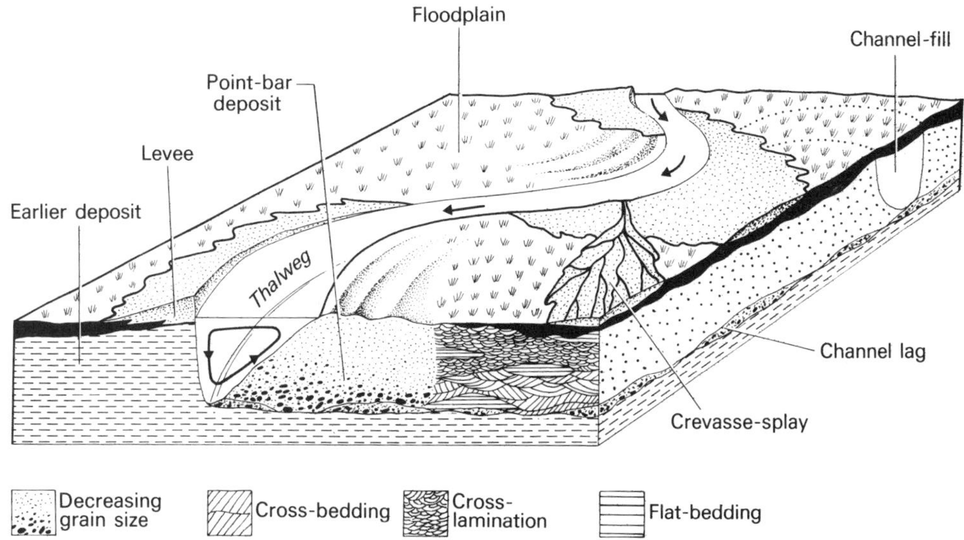

migration of the channel gives a tabular sand unit overlying a near-horizontal erosion surface with or without a lag conglomerate. The sands which form the bulk of the unit show an upward diminution of grain size and are cross-bedded with upwards reduction of set size. In the upper part cross-bedding gives way to ripple cross-lamination or parallel-lamination (Fig. 3.24) (Sundborg, 1956; Frazier and Osanik, 1961; McDowell, 1960; Bernard and Major, 1963). Tabular cross-bedded sets, significantly bigger than surrounding trough sets and with markedly divergent directions, may be produced by the upslope migration of scroll bars (Sundborg, 1956; Jackson, 1975a). The overall thickness of the tabular sand unit compares closely to the depth of the channel and the proportions of the various structures vary in relation to channel size and sinuosity (Allen, 1970a and b).

VARIATIONS AND COMPLICATIONS

The classical model is based upon the assumptions that bankfull discharge is the dominant control and that a fully developed helicoidal flow exists. Recent investigations have shown that many features of point bars are responses to flows other than bankfull discharge and that the flow pattern around meander bends may diverge very significantly from the helicoidal model.

(a) *Discharge variations.* In spite of the fact that flow in meandering streams tends to fluctuate less than in braided streams, discharge still varies widely and this may be reflected in the morphology of their channels and in their deposits. Carlston (1965) showed that meander wavelength is not absolutely related to bankfull discharge but to some lower figure. This is hardly surprising as bankfull discharge is relatively rare and discharges above and below this value tend to reduce flow velocities in the channel.

Some point bar surfaces have distinctly stepped profiles which can be described as 'two tier' (e.g. Harms, McKenzie and McCubbin, 1963; McGowen and Garner, 1970). These can be related to different dominant discharges.

On the dominantly sandy Beene point bar of the Red River, Louisiana, trough cross-bedding dominates the sequence above and below the level of the step. Silt, deposited during falling stage, mantles the floor of a shallow channel cut into the step and the higher deposits are somewhat coarser than those below (Fig. 3.25) (Harms, McKenzie and McCubbin, 1963).

Coarser, gravelly point bars may be more complex (McGowen and Garner, 1970; Bluck, 1971). High discharge

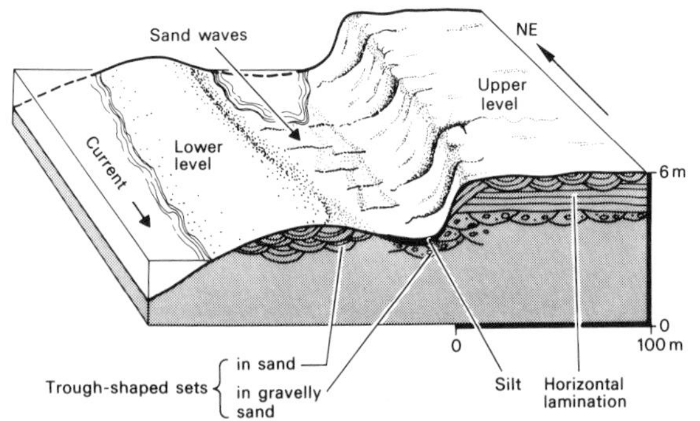

Fig. 3.25. Upstream part of Beene point bar showing the stepped profile and the associated internal structures (after Harms *et al.*, 1963).

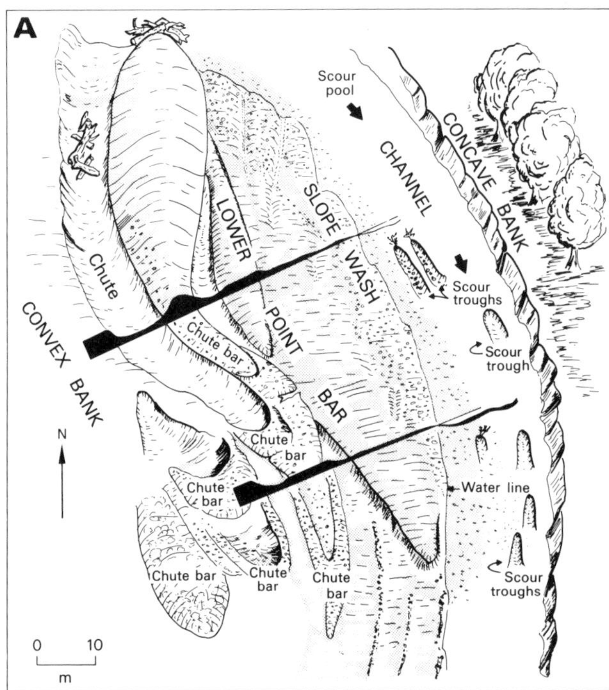

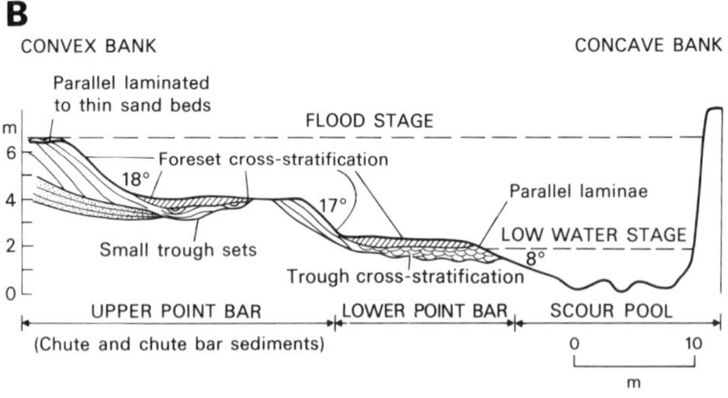

Fig. 3.26. Topographic features and internal structure of a coarse-grained point bar. (A) Plan, (B) cross-section (after McGowen and Garner, 1970).

flows cut across the point bar surfaces giving two major flow threads, one in the pool near the concave bank and one across the point bar surface. In the Colorado and Amite Rivers of Texas and Louisiana, channels ('chutes') are eroded into the upstream ends of the point bar surfaces and 'chute bars' are deposited at their downstream ends (Fig. 3.26) (McGowen and Garner, 1970). The surface of the lower part of the point bar is mainly a response to low discharge.

Chutes vary in size and shape and are normally floored by gravel. Their floors rise downstream until the flow is no longer confined and it can expand laterally to deposit the chute bar. Successive floods may partially fill the chutes with a succession of sediments arranged in small fining-upward sequences a few tens of centimetres thick and ending with a vegetated mud drape (Fig. 3.27). Each sequence represents a waning flood and may be eroded during subsequent floods.

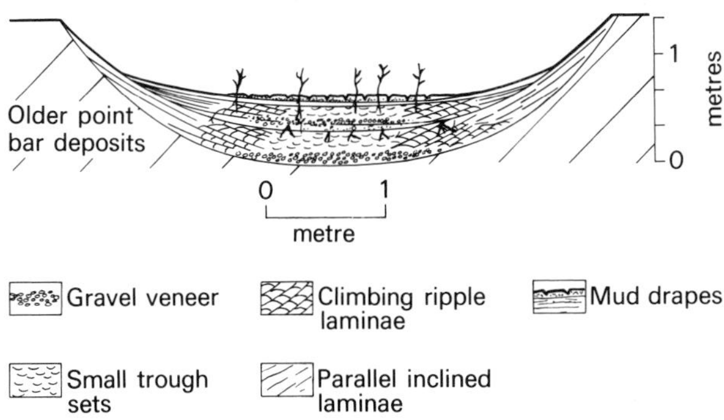

Fig. 3.27. Chute deposits of a coarse-grained point bar. Each flood episode deposits a thin, mud-draped, fining-upward sequence (after McGowen and Garner, 1970).

Chute bars occupy various levels on the upper point bar surface and develop both by lateral growth of the slip face and by vertical accretion of the top surface. Slip faces are convex downstream and may be 2 to 6 m high. In consequence, the vertical sequence will differ from that of the classical model in having one or more large-scale tabular sets of cross-bedding above a unit of trough cross-bedding. The large foresets will have a wider directional spread than the trough axes and will be overlain by horizontal sands or floodplain silts and sands (Fig. 3.26). If the upstream end of these point bars is preserved, the tabular sets might be replaced by a complex of small channels filled by small scale fining-upward sequences.

(b) *Complications in the flow pattern.* Following Bagnold (1960), Leeder and Bridges (1975) cast doubt upon the general applicability of a simple helicoidal flow pattern. At high curvatures, when r_m/w is less than about 2, large-scale flow separation occurs over the downstream end of the point bar surface (where r_m is the radius of curvature of the channel mid-line and w is the channel width). When this value is achieved in a developing meander, the growth direction changes from being dominantly transverse to dominantly down-valley. The effect of the vertical separation eddy on the distribution and orientation of bedforms on the point bar surface has not been investigated in detail but it seems that it will disrupt and probably prevent the development of the pattern predicted by the classical model.

A further complication is that, around any meander bend, the helicoidal flow is not immediately established, but develops gradually, so that in the upstream part of any meander bend the flow pattern is largely inherited from the next bend upstream with rotation in the opposite sense (Jackson, 1975a and b). In consequence, three major zones can be recognized around any bend, a 'transition zone' where the influence of the upstream bend prevails, a 'fully developed zone' where the local bend dominates and the classical model prevails and an 'intermediate zone' where one changes into the other. The extent of these zones varies with channel curvature and water stage. In the transition zone, velocity is highest near the inner bank, particularly at bankfull discharge, but the thalweg switches to the outer bank within the transition zone upstream of the point of switching of velocity asymmetry and sense of rotation of the helicoidal flow (Fig. 3.28).

These down-channel changes in flow pattern around a meander are reflected in the bedforms on the point bar surface and in the vertical sequence of sediments produced. In the fully developed zone, the classical model seems to apply with the addition of scroll bar sets in the upper part of the sequence. In the transition zone, the vertical grain-size transition is not clear and the sequence may even coarsen upwards. Similarly, the distribution of bedforms and sedimentary structures is much less well ordered, with ripples developing in deeper parts of the channel rather than dunes. Depending on the nature of the movement of a meander bend, different volumes of these different sequences will be preserved (Fig. 3.29).

3.5.3 **Channel cut-offs**

A freely meandering channel system is inherently unstable as differing rates of erosion in neighbouring meanders lead to periodic cut-offs of two main types: (a) chute cut-off and (b) neck cut-off (Fisk, 1947) (Fig. 3.30).

At flood stage streams tend to straighten the course of their main flow across meander bends by eroding chutes or deepening pre-existing swales. As they gradually take an increasing proportion of the discharge, activity in the main channel is reduced, with the result that it is gradually filled, first by bed load and later by silts and clays from suspension as the ends of the cut-off are plugged with bed load.

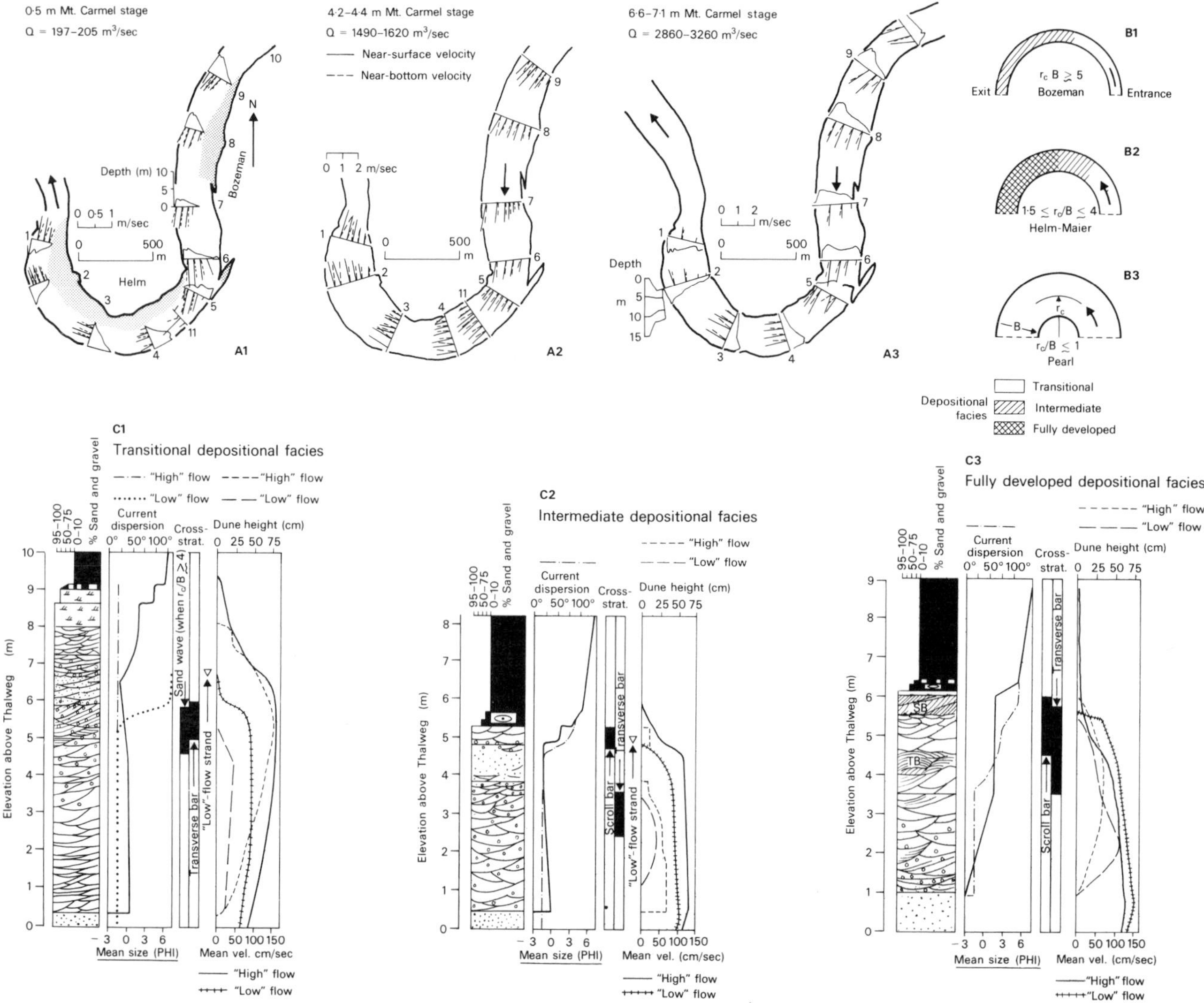

Fig. 3.28. The variation of depositional facies and their distribution on point bars in the Lower Wabash River, Illinois. A_1 to A_3: cross-channel profiles and velocity distribution around Helm Bend at increasing values of discharge. B_1 to B_3: Distribution of depositional facies with increasing channel curvature. C_1 to C_3: Depositional facies sequences from upstream to downstream on a point bar of suitable curvature (after Jackson, 1975b, 1976).

The concave banks of adjacent meanders may sometimes erode towards one another, narrowing the area of the point bar between them. If this neck is breached, the river will abandon the meander rather suddenly as the river rapidly shortens its course and steepens its slope. The abandoned channel will be rapidly plugged at both ends by bed material washed in by eddies and an oxbow lake will be created (Fig. 3.31). The lake will only receive sediment from suspension during flood and may exist as a lake for a long time. Bed forms, active on the point bar surface prior to cut-off, may be preserved below the suspended sediment (McDowell, 1960).

Both types of cut-off may be sites of dense vegetation and of accumulation of organic rich muds and silts, and form curved bodies of finer sediment bounded on the outer side by an

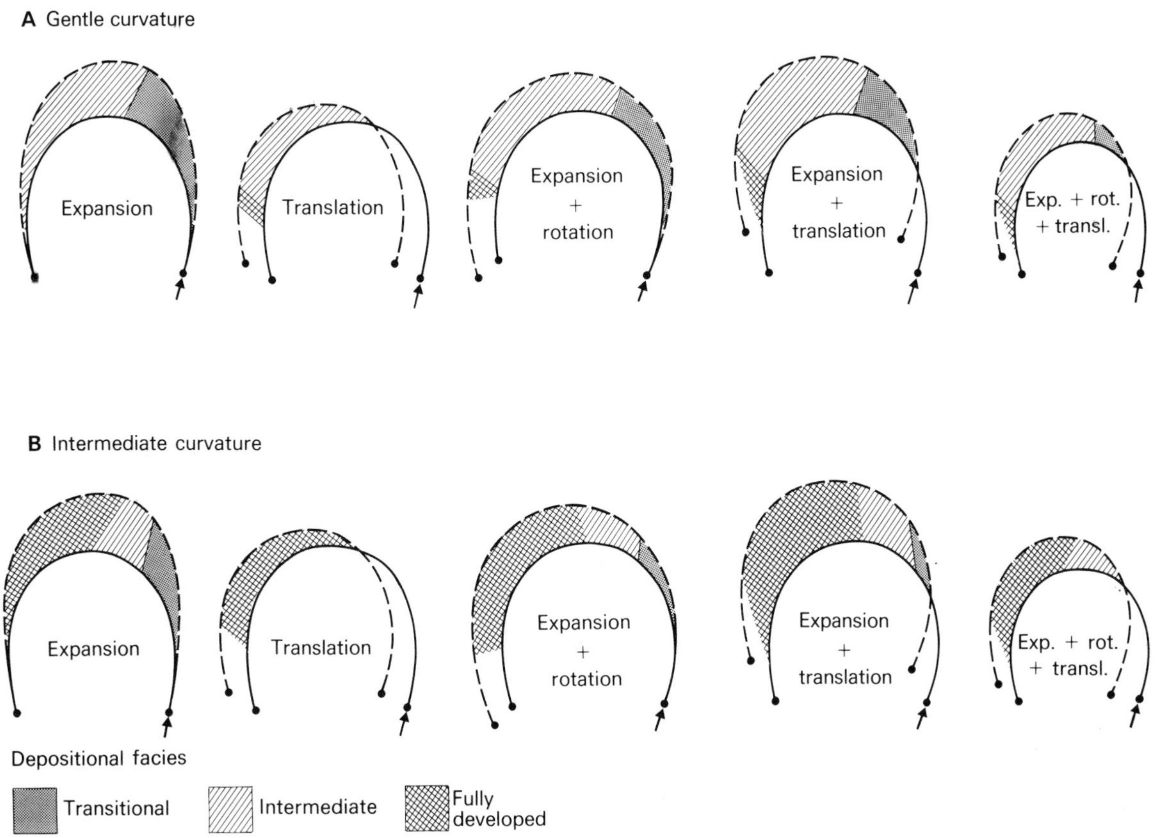

Fig. 3.29. Relative volumes of point bar facies sequences generated by different migrations of point bars (after Jackson, 1975a).

Fig. 3.30. Modes of channel shifting. (A) Chute cut-off. (B) Neck cut-off. (C) Development of new meander belt following avulsion (after Allen, 1965c).

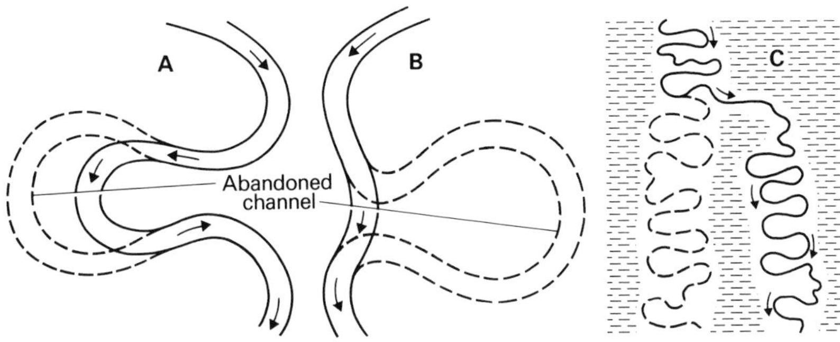

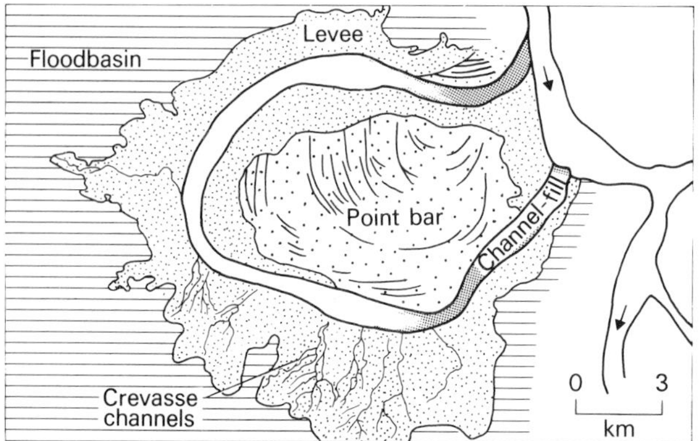

Fig. 3.31. Levee and crevasse topography associated with an oxbow lake caused by neck cut-off. False River cut-off channel, Mississippi (after Fisk, 1947).

erosion surface and on the inner side by an inclined surface of bedload sediments. They differ in the abruptness of the transition from bed load to suspended load sediment.

3.6 INTER-CHANNEL AREAS

So far we have confined our attention to the processes and products of channels and have said little about what goes on between or beyond them. The inter-channel areas, however, are normally much larger than those of the channels themselves and their deposits play an important role in the overall alluvial sequence. Unlike channels, inter-channel accumulations are strongly influenced by local climate. Two types of inter-channel area can be distinguished, (a) those influenced by an adjacent river (overbank) which in turn can be separated into proximal (levee) environments and distal (floodplain) environments and (b) those beyond the reach of river influence.

Though the inter-channel areas of different channel types have many features in common, meandering and low-sinuosity streams are more likely to have substantial inter-channel areas.

3.6.1 Overbank environments

LEVEES

Levees are ridges sloping away from the channel and are most conspicuous on the concave erosional banks of meanders (Fig. 3.31). They are submerged only at the highest floods. As flood water escapes from the channel on to the overbank area there is a fall-off in the level of turbulence and suspended sediment is deposited, the coarser sands and silts close to the channel, the finer sediments further out on the floodplain. Levees result from fall-out of the coarse components of the suspended load, and are commonly made up of an interbedded sequence of variable grain-size, reflecting fluctuating water stage. Plant growth may damp turbulence and accelerate sedimentation. The sedimentary structures are not well known, but ripple cross-lamination, particularly climbing ripple cross-lamination, and parallel-lamination, are common (e.g. Singh, 1972; Klimek, 1974b), though they may be destroyed by rootlets. Levees may be cut by crevasses giving concave upwards channels with concentric fills of silty and sandy laminae (Singh, 1972).

Levee deposits have a low preservation potential because they occur mainly on concave banks and therefore only escape erosion if the channel is cut off (Fig. 3.31).

FLOODPLAIN

Sedimentation and post-depositional changes on the floodplain depend on climate and on distance from the active channel. The floodplain is rarely inundated, the most frequent recurrence interval for over-bank flooding being between one and two years (Wolman and Leopold, 1957). Over-bank sedimentation rates are rather low, owing to the relatively high velocities of floodplain currents and low concentrations of suspended sediment at the flood peak. Most sedimentation is from suspension and deposits fine away from the channel. Only major floods deposit more than a centimetre or so of sediment and then only patchily. Vegetation localizes sedimentation and some areas may even be scoured during flood. Floodplain sediments dry out between floods and desiccation cracks and other features of sub-aerial exposure may develop.

In humid settings the floodplain may never fully dry out and remain a backswamp or even a lake. Vegetation then dominates and leads to the development of peats. Fluctuations of the water table and the growth of vegetation may produce soil profiles in the floodplain sediments.

In semi-arid settings vegetation is less abundant and so there is less organic matter in the sediment, less post-depositional disturbance by roots and less protection of the sediment surface from aeolian erosion and reworking. Desiccation cracking will be common in fine sediment and soils may begin to develop though they are more likely to form extensively in areas beyond river influence (Goudie, 1973). Similarly, the diagenetic processes leading to reddening may begin. With relatively small local streams, the distinction between channel and floodplain may be poorly defined (e.g. McKee *et al.*, 1967) and coarser sediment may be spread as a sheet beyond the normal channel. With larger rivers, the channel regime is often controlled by the climate of the source area rather than that prevailing in the floodplain. For example, the discharge of the lower reaches of the Nile owes little to the local climate.

Wind action is important in some areas with aeolian dunes

being built on the floodplain and finer material being winnowed out and redeposited (e.g. Higgins *et al.*, 1973). Large thicknesses of wind-blown silt may accumulate on floodplains and adjacent areas and form the bulk of the deposits (Lambrick, 1967). In highly ephemeral streams wind-blown dunes may block and divert channels and wind-worked and sorted sands may be reworked by subsequent stream activity.

CREVASSE SPLAYS

Coarser channel sediment is sometimes introduced to the floodplain by crevassing of a levee during flood. The deposits of crevasse splays occur as fans or tongues of sand elongated away from the river and originating at a crevasse cut in the levee. In the Brahmaputra (Coleman, 1969) the crevasse splays merge locally into an almost continuous sheet (Fig. 3.32).

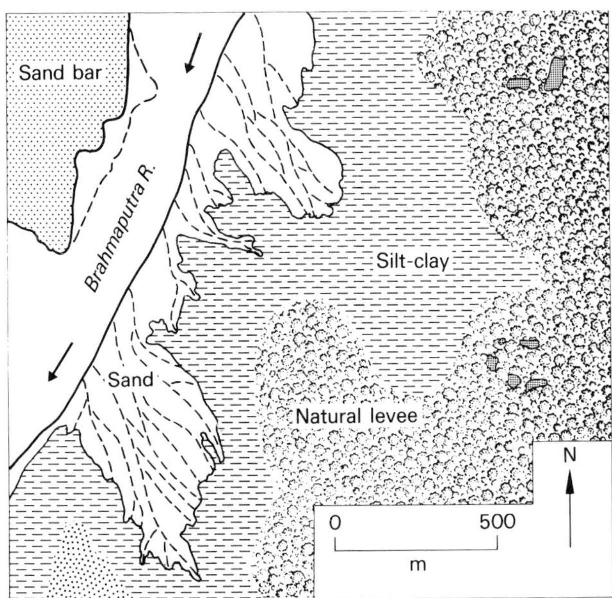

Fig. 3.32. Crevasse splays in the Brahmaputra (after Coleman, 1969).

The internal organization and geometry of crevasse splays are not well documented. Presumably beds of sand with basal erosion and internal evidence of waning flow are interbedded with floodplain silts and clays. Rapid sedimentation may be reflected in climbing ripple cross-lamination, though crevasse splays may be colonized by plants with resulting disturbance of depositional structures. The sands thin distally from the crevasse and have cross-lamination directions divergent from those of adjacent channel sands.

3.6.2 Areas beyond river influence

These areas encompass a whole range of environments both erosional and depositional. Here we confine our attention to those areas where alluvial sedimentation has been active in the past and is likely to return.

TERRACES

In many present-day alluvial settings terraces are important elements in the landscape and may be caused by lowering of base level, degradation of the hinterland or climatic change. Base level changes may be due to eustatic, isostatic or tectonic causes. Rapid climatic and vegetation changes, operating at a very local scale, have caused streams to be incised as 'arroyos' in alluvial plains of the south-western USA (Haynes, 1968; Cooke and Warren, 1973). The important processes are reworking by wind and *in situ* processes associated with soil formation, both strongly controlled by climate.

WIND ACTIVITY

Wind erosion and reworking will only be significant if vegetation cover is low, and are therefore more likely in arid and semi-arid settings or in high latitudes. Erosion generally winnows and removes finer particles, resulting in a deflation surface on the terrace top. With sufficient time, the coarse particles of the pavement may be shaped to ventifacts and may acquire a coat of desert varnish (Glennie, 1970). The material removed by wind may accumulate locally or be totally removed from the system. Sand is reworked into dunes (see Chapter 5); finer material, though forming less spectacular deposits, may be volumetrically as important as sand. It goes into suspension and is deposited locally or elsewhere as loess (Yaalon and Dan, 1974; Higgins, Ahmad and Brinkman, 1973; Lambrick, 1967). Wind not only deposits fine-grained sediment; it may also introduce material important for soil formation.

SOILS

Soil formation is the second main influence in interfluvial areas and again climate is the most important control. Despite their great importance, particularly in the investigation of problems of Quaternary geology, it is beyond the scope of this book to discuss soils in detail. Their study is a subject in its own right and many books present detailed descriptions and discussions (e.g. Bunting, 1967; Bridges, 1970; Hunt, 1972; Goudie, 1973; Birkeland, 1974). Only a few general points of potentially wider geological significance will be mentioned here.

Soils develop on both recently deposited sediments and poorly consolidated bedrock. Although the latter group may be sometimes important in the understanding of ancient

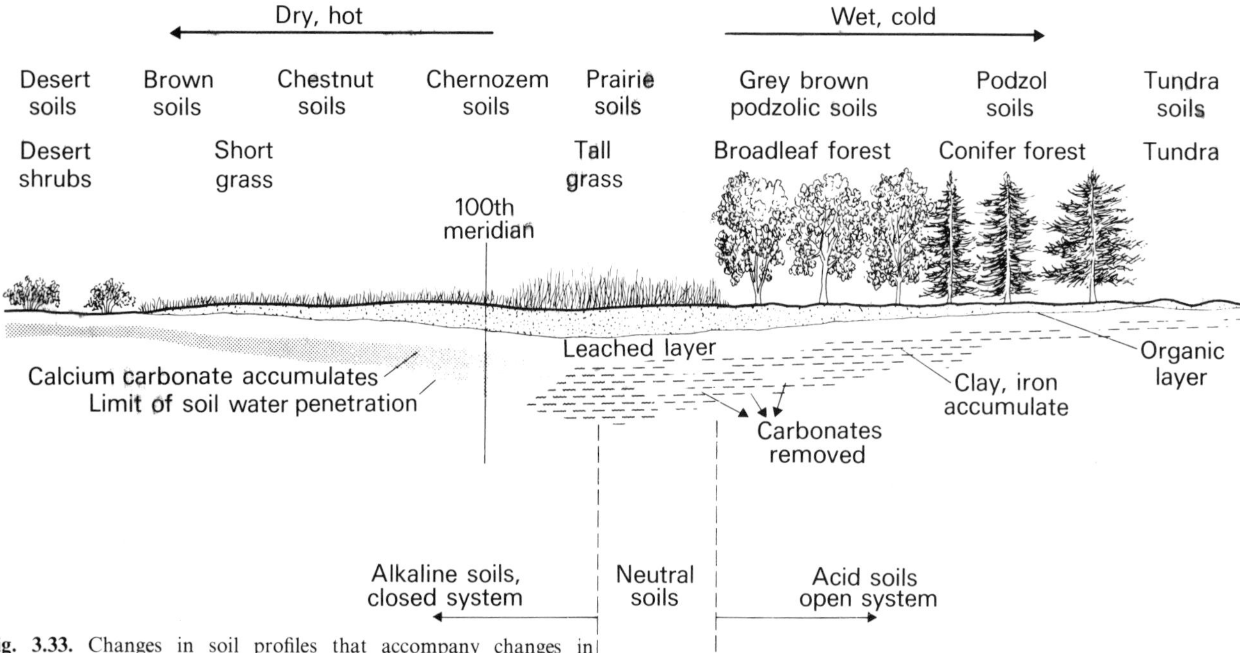

Fig. 3.33. Changes in soil profiles that accompany changes in vegetation and climate between the tundra in northern Canada and the deserts in south-western United States. At the 100th meridian the annual precipitation averages about 50 cm (after Hunt, 1972).

stratigraphical sequences, the former group is our main concern here. There have been many schemes for the classification of soils, attempting to relate the types to climatically controlled zonal systems (Fig. 3.33). While soils can undoubtedly be related to climate, it is important to realize, that, because most soil-forming processes act slowly, soil profiles of the present-day land surface may reflect not only the prevailing climate but also earlier regimes.

Most modern soils show the development of a vertical profile involving several horizons. The details of these horizons and their positions in the profiles are controlled by both the climate and the nature of the starting material. Most modern soils involve an organic-rich upper layer which grades down into mineral-dominated layers. The extent and importance of the organic layer is controlled by the relative rates of organic production and rotting which in turn depend upon climate and the level of the water table. While the preservation potential of the organic layer is small, except in swamps and peat bogs, it is important in influencing the rest of the soil profile. Rainwater percolating the organic layer is enriched in CO_2 and in various organic acids with a resultant decrease in the pH of the water. In addition, the roots of growing plants may extend into the lower mineral layers and both physically disturb the sediment and abstract material in solution from it.

In the underlying mineral layers, rainwater moves downwards. As this is usually acid, it removes alkali and alkali earth ions in solution and aids the breakdown of various silicate minerals to clays. Both the clays and the dissolved ions are transferred to lower levels in the profile where they may be deposited. In humid settings, however, the more soluble ions in solution may be removed from the system by transfer to the groundwater. The clays and the normally insoluble iron and aluminium which are mobilized by organic complexing are deposited in lower levels of the profile. In intense humid tropical settings, silica may also be leached from the soil and this leads to the concentration of the oxides of iron and aluminium giving laterites and bauxites. Such extremely leached soils often show cavernous and pisolitic textures. In cold humid settings, limonite may cement upper layers of the soil profile as a hardpan, precipitation being aided by organic activity.

In semi-arid and arid areas, the level of precipitation is too low to allow the development of an important organic layer and the consequent relatively high pH, combined with the low volume of percolating water, mean that solution rates are lower and that the water does not escape into the groundwater system but is retained within the soil profile and eventually lost through evaporation. The result is deposition of material from solution within the profile, under alkaline conditions. The depth at which material is deposited is directly related to the rainfall and to the permeability of the material.

The mineral most commonly accumulating in this way is

Table 3.1. Stages of the morphogenetic sequences in the development of caliche profiles and the youngest land surfaces on which the stages occur (after Gile, Peterson and Grossman, 1966)

Stage	Diagnostic carbonate morphology		Youngest, geomorphic surface on which stage of horizon occurs
	Gravelly soils	Non-gravelly soils	
I	Thin, discontinuous pebble coatings	Few filaments or faint coatings	Fillmore < 2,600 to 5,000 years
II	Continuous pebble coatings, some interpebble fillings	Few to common nodules	Leasburg > 5,000 years—latest Pleistocene
III	Many interpebble fillings |	Many nodules and internodular fillings	Picacho Late-Pleistocene
IV	Laminar horizon overlying plugged horizon	(increasing carbonate impregnation)	Picacho Late-Pleistocene
	(thickened laminar and plugged horizons) ↓		Jornada Mid- to Late-Pleistocene
		Laminar horizon overlying plugged horizon	La Mesa Mid-Pleistocene

calcite, though silica may also occur. Profiles which involve carbonate layers or nodules are called 'calcretes' or 'caliches', while those with secondary silica are 'silcretes'. Goudie (1973) has extensively reviewed calcrete formation and shown that while downwards movement of dissolved material is the main process, it is difficult to account for the supply of calcium carbonate to the sediment surface. Possible sources are carbonate-rich loess (Yaalon and Dan, 1974; Reeves, 1970) or leaf fall and plant drip. Areas down-wind of sites of gypsum precipitation have accelerated rates of caliche development presumably due to the introduction of calcium sulphate by wind and precipitation of carbonate by the common ion effect (Lattman and Lauffenburger, 1974). Rates of calcrete formation are variable but it seems that periods of the order of 10^3–10^5 years may be involved in the development of mature caliche profiles (Gardner, 1972).

Such profiles seem to develop through several stages. Isolated nodules and glaebules, possibly associated with roots, form first. They eventually coalesce into a honeycomb and give a continuous hardpan layer, commonly associated with a lamellar structure (Table 3.1) (Gile, Peterson and Grossman, 1966). The extent to which the profile has developed gives some measure of the time over which soil processes have operated. The carbonate layers may eventually deform into folds or pseudo-anticlines (Jennings and Sweeting, 1961; Reeves, 1970). This buckling may be associated with the development of patterned ground with a variety of morphological types commonly called 'gilgai' (e.g. Russell and Moore, 1972). Gilgai may, however, also result from seasonal drying and swelling of clays. The buckling and up-doming fractures the upper layers of caliches and the massive carbonate of mature profiles may be overlain by a brecciated layer (Gardner, 1972).

Silcretes seem also to be the product of semi-arid settings and occur in areas of very mature soil development. They are often associated with deeply weathered intermediate and basic igneous rocks but also develop more rapidly where pyroclastic sediments are abundant (Flach, Nettleton, Gile and Cady, 1969). Silica is mobilized in solution and reprecipitated locally though some may be introduced from greater distances. Precipitation may sometimes occur where upward moving silica-rich solutions meet downward percolating water rich in dissolved salts (Smale, 1973). In immature silcretes, small diffuse opal flocs coat grains but, with increased maturity, local centres of cementation gradually develop into more strongly indurated layers.

RED COLORATION

The origin of 'red beds' has been the subject of heated controversy with arguments centering on whether the pigment is detrital or diagenetic (see Glennie, 1970, for a concise discussion). Observations of alluvial fan and tidal flat sediments ranging in age from Pliocene to Recent in Baja California have demonstrated fairly clearly that, in modern sediments, reddening is essentially a diagenetic process (T.R. Walker, 1967). The change is possibly favoured by elevated temperatures, but a more important requirement is water. Biotite and hornblende break down to coat other grains with haematite. Simultaneously, clays are washed into lower levels of the soil profile. The processes are quite slow, the Recent sediments in California being grey, the Pleistocene ones yellow and only the Pliocene ones showing a full red coloration. The rate of reddening varies with lithology, clay-rich sediments altering more slowly due to lower permeability. Some desert dune sands seem to have reddened more rapidly than the Californian examples quoted (Glennie, 1970).

In more temperate climates there can still be some formation of calcium carbonate and ferruginous concretions but not to the same extent as in a semi-arid setting (Bernard and LeBlanc, 1965). According to T.R. Walker's (1967) conditions, reddening could also occur in temperate climates but the higher rates of organic productivity tend to make the soils richer in humus and thereby mask the process.

3.7 ANCIENT ALLUVIAL SEDIMENTS

Alluvial sedimentary rocks are generally recognized by the absence of marine fossils, the presence of red coloration, unidirectional palaeocurrents and channels, and evidence for emergence such as the presence of palaeosols and mudcracks. However, none of these features is diagnostic by itself since all may occur in other environments and in the Pre-Cambrian, where a marine fauna is lacking and soil processes were less active, it may be difficult to distinguish fluvial from shallow marine sediments.

However, once a broadly alluvial interpretation is established, it is possible to discuss more specific alluvial environments and to attempt to place the deposits in the continuum of river types seen at the present day. This is best done by recognizing that there are two main types of alluvial sequence; one dominated by sandstones and conglomerates with little fine-grained sediment and the other dominated by sandstones and siltstones with only rare conglomerates. The first type can be compared to the products found today in semi-arid fans, pebbly braided rivers and humid fans. The second type includes sediments found today associated with both low-sinuosity and high-sinuosity streams. These two broad facies types are, however, like the modern environments, intergradational and there are examples which fall close to the arbitrary boundary.

3.8 ANCIENT PEBBLY ALLUVIUM

The generation and preservation of thick sequences of conglomerates require substantial topographic relief and thus usually imply tectonic activity during or immediately prior to deposition.

Many alluvial fans are associated with normal faults and the development of grabens and half-grabens. Some of these are post-orogenic (e.g. Old Red Sandstones of S.W. Norway (Steel, 1976)), while others are associated with rifting or incipient rifting (e.g. New Red Sandstone, Hebrides, Scotland (Steel, 1974); Permian, E. Greenland (Collinson, 1972)). Other successions are not so directly related to fault lines but are the fills of basins flanking recently uplifted source areas (e.g. the Torridonian of N.W. Scotland (Williams, 1969), the Permian of S.W. England (Laming, 1966) and the Van Horn Sandstone of W. Texas (McGowen and Groat, 1971)). Such successions

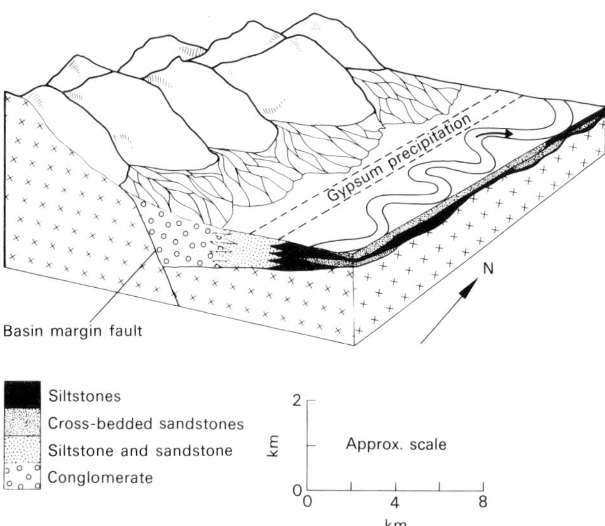

Fig. 3.34. The environment envisaged for deposition of the Røde Ø Conglomerate and associated sediments, Permian, East Greenland (after Collinson, 1972).

are often very extensive and are dominated by stream deposits in comparison with those in fault bound settings. This may be a function of bedrock type in the source area but will often be a result of a larger catchment area and lower slopes.

In the case of fans developed in response to active faulting, relief was often maintained and the position of the fan apex remained essentially fixed, while great thicknesses of sediment, often measured in kilometres, were deposited. The greatest thicknesses are usually close to the bounding fault where the facies reflect the proximity of the fan apex. Distal thinning varies in its rate between different fan deposits, the rate varying with size of fan and type of dominant process. Such thinning is usually accompanied by lateral facies changes, sometimes extending into the deposits of environments beyond the range of normal fan activity (Fig. 3.34).

3.8.1 Facies and their distribution

THE FACIES

Conglomerates make up a substantial proportion of the overall thickness, but sandstones and siltstones also occur. There are two main types of conglomerate, unstratified paraconglomerates with dispersed clasts and stratified orthoconglomerates, commonly with a clast framework and sometimes interbedded with sandstone. In some sequences this broad grouping can be further refined, though there is no scheme which commands universal support. Four conglomerate facies

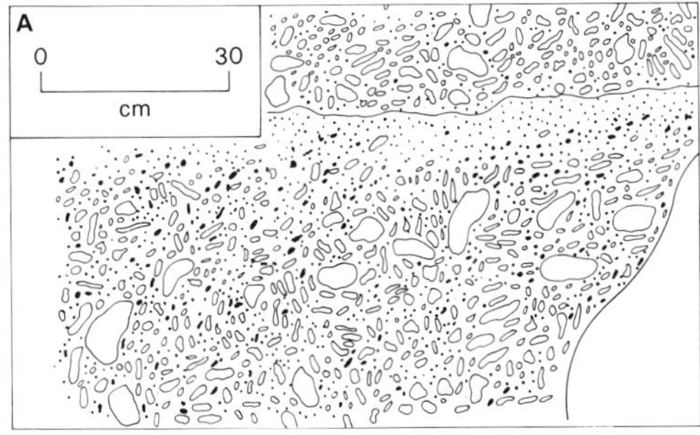

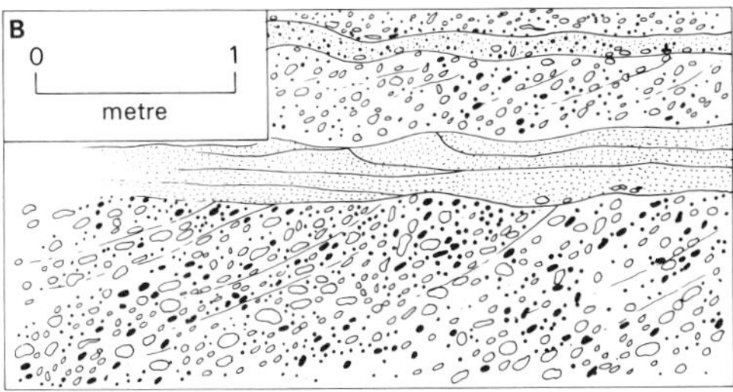

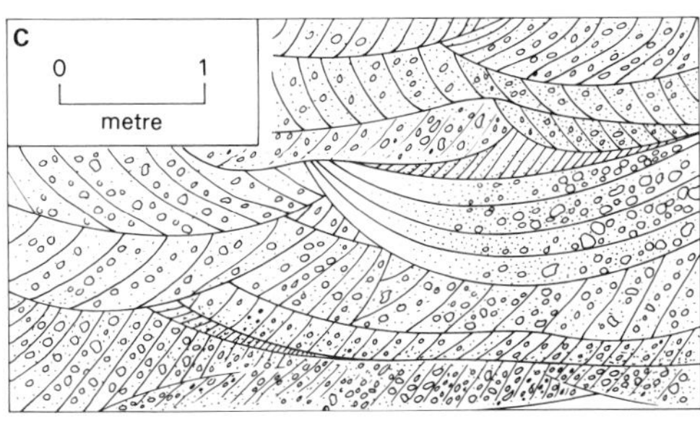

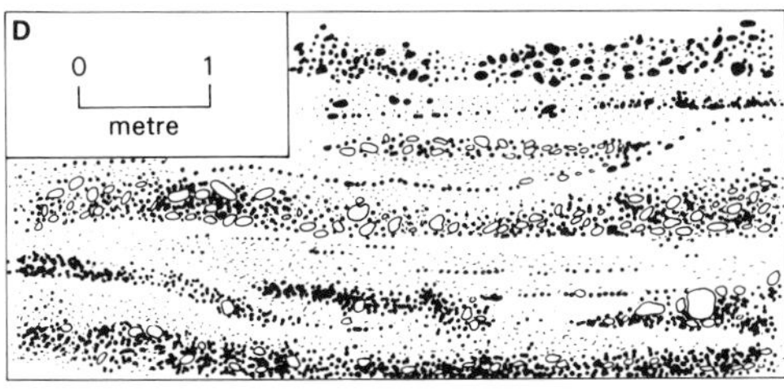

Fig. 3.35. Four conglomeratic facies recognized in Old Red Sandstone alluvial fan sediments, Firth of Clyde, Scotland. (A) is a paraconglomerate and is interpreted as a mudflow deposit; (B), (C) and (D) are all orthoconglomerates, (B) being interpreted as sheetflood and (C) and (D) as stream channel deposits (after Bluck, 1967).

were recognized in the Old Red Sandstone of the Firth of Clyde area (Fig. 3.35) (Bluck, 1967) and in the New Red Sandstone of the Hebrides (Steel, 1974), while in more humid stream-dominated fan deposits of the Van Horn Sandstone of West Texas, three broad facies were recognized (McGowen and Groat, 1971).

Paraconglomerates (Fig. 3.35A) are laterally extensive beds with non-erosive bases, interpreted as the products of high viscosity mudflows. They lack stratification and all sedimentary structures including imbrication, though the long axes of clasts may occasionally lie in the plane of the bedding and be aligned parallel to flow (Lindsay, 1968). Correlation between bed thickness and maximum particle diameter reflects a positive relationship between the competence and size of flows (Bluck, 1967). Paraconglomerate units are sometimes overlain by thin beds of sandstone showing parallel and low-angle bedding, which have been interpreted as a result of waning stage floods (Steel, 1974).

Orthoconglomerates are subdivided on the basis of their stratification. Bluck's scheme recognizes three types (Fig. 3.35B, C & D). Facies B consists of erosively based units which comprise a single cross-bedded set of rather sandy conglomerate sometimes overlain by an orthoconglomeratic lag and by smaller sets of cross-bedded sandstone. A correlation between maximum clast size and bed thickness suggests that the whole unit is the result of a single flow, interpreted as a sheetflood which deposited the lag gravel and the sandstone unit as the flood waned.

In Bluck's facies C and D, the bedding is more lenticular; there is no correlation between clast size and bed thickness and hence the facies are thought to result from continuous or multiple stream flows. Facies C is sandstone with scattered pebbles, cross-bedded in a variety of set sizes owing mainly to abundant erosion between sets. Framework gravels occur as lags and some units fine upwards. The facies is interpreted as the product of multiple reworking in stream channels. Facies D occurs in lenses up to 7 m wide associated with erosional scours, and interbedded with sandstone. The facies is again thought to be the product of streams.

In the more humid fan sequence of the Van Horn Sandstone, three facies are recognized, all involving orthoconglomerates (Fig. 3.36) (McGowen and Groat, 1971). These facies are somewhat more widely drawn than Bluck's facies and are more properly considered as facies associations. They are referred to proximal, mid-fan and distal settings and are highly intergradational. The proximal facies (Fig. 3.36A) consists mainly of thick framework conglomerates in units which have flat bases and convex upwards tops in sections transverse to the current direction. The units are elongate parallel to the current and are flanked by cross-bedded pebbly sandstones. Mid-fan facies (Fig. 3.36B) have less gravel and more scouring on the bases of beds, though some elongate gravel bodies persist. The

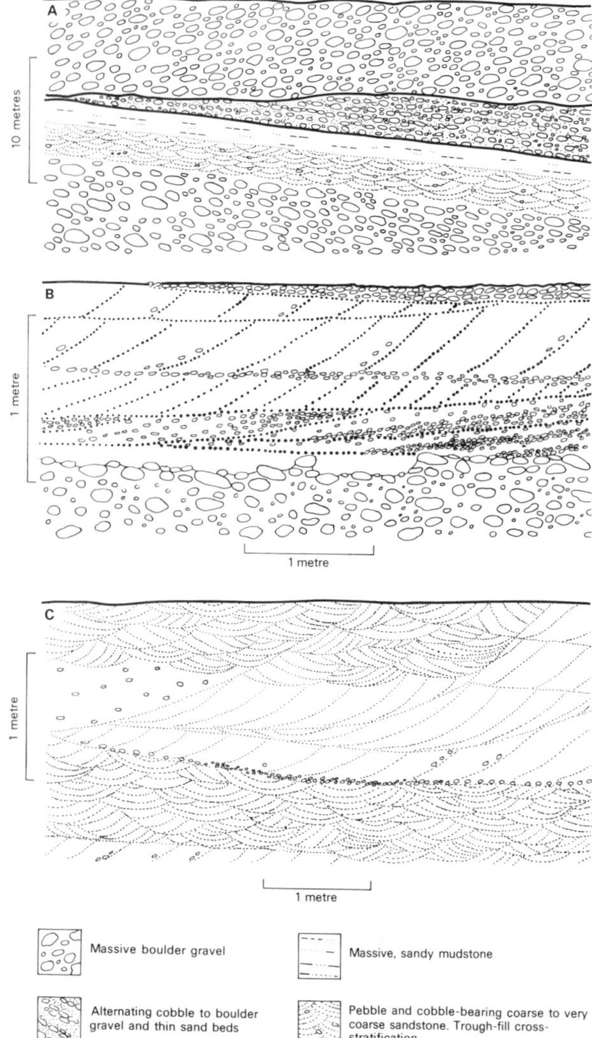

Fig. 3.36. Examples of the three facies associations recognized in the humid fan deposits of the Van Horn Sandstone, probably Pre-Cambrian, Texas. (A) is the proximal facies in which boulders may reach 1 m diameter, and where gravel is the dominant component. (B) is the mid-fan facies where conglomerates are interbedded with pebbly cross-bedded sandstones. (C) is the distal facies where tabular and trough cross-bedded sandstones dominate (after McGowen and Groat, 1971).

parallel bedded gravels are interpreted as longitudinal bars (cf. Rust, 1972a; N.D. Smith, 1974) and tabular cross-bedded sandstones flanking them as the product of transverse bars in the inter-bar channels (cf. N.D. Smith, 1974). In the distal facies (Fig. 3.36C) gravels are confined to thin beds and lenses

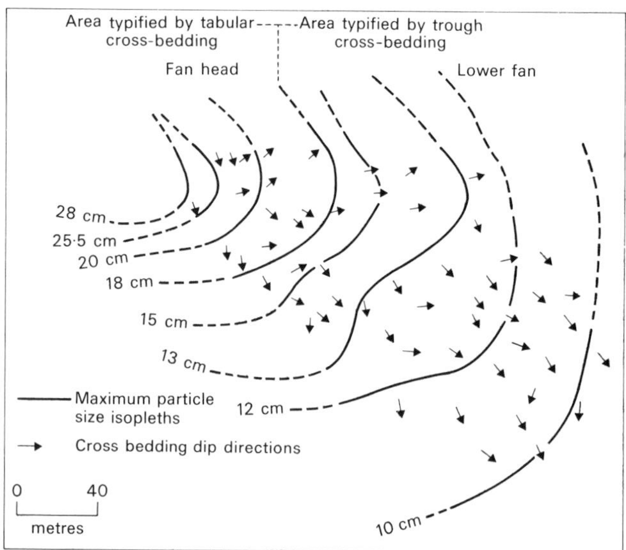

Fig. 3.37. A map of maximum particle size distribution in relation to palaeocurrent direction and cross-bedding type for a small alluvial fan in the Triassic of South Wales. The down-fan diminution in grain size is accompanied by a transition from tabular to trough cross-bedding and the palaeocurrents have a broadly radial pattern (after Bluck, 1965).

scattered in tabular and trough cross-bedded sandstones, as might be predicted from comparison with modern outwash areas (e.g. Boothroyd, 1972). Here the facies grade into those of sandy fluvial systems.

In the stream-dominated fan deposits of the Permian of S.W. England, two types of pebble imbrication were recognized in roughly parallel bedded conglomerates which were attributed to stream flows (Laming, 1966). 'Contact imbrication' occurs when the clast/matrix ratio is high and the inequant clasts are tightly stacked against one another. 'Isolate imbrication' is the product of low clast/matrix ratio with inclined clasts floating in the matrix.

In otherwise rather featureless gravels of the Carboniferous of the Intra-Sudetic Basin, Teisseyre (1975) attempted to use pebble imbrication to identify channels. Based on comparison with modern stream beds, he suggested that, within channels, pebble imbrication showed systematic changes in transverse profiles. In the centre of the channel imbricate clasts dip upstream while near the channel margins they dip with a component towards the centre line due to the influence of the banks on large-scale eddying.

The identification of channels in ancient alluvial fan deposits, particularly those of humid fans, can have great economic importance. Heavy minerals, particularly gold, occur commonly as intergranular fines in coarser orthoconglomerates. These tend to be more common in channels or in sites

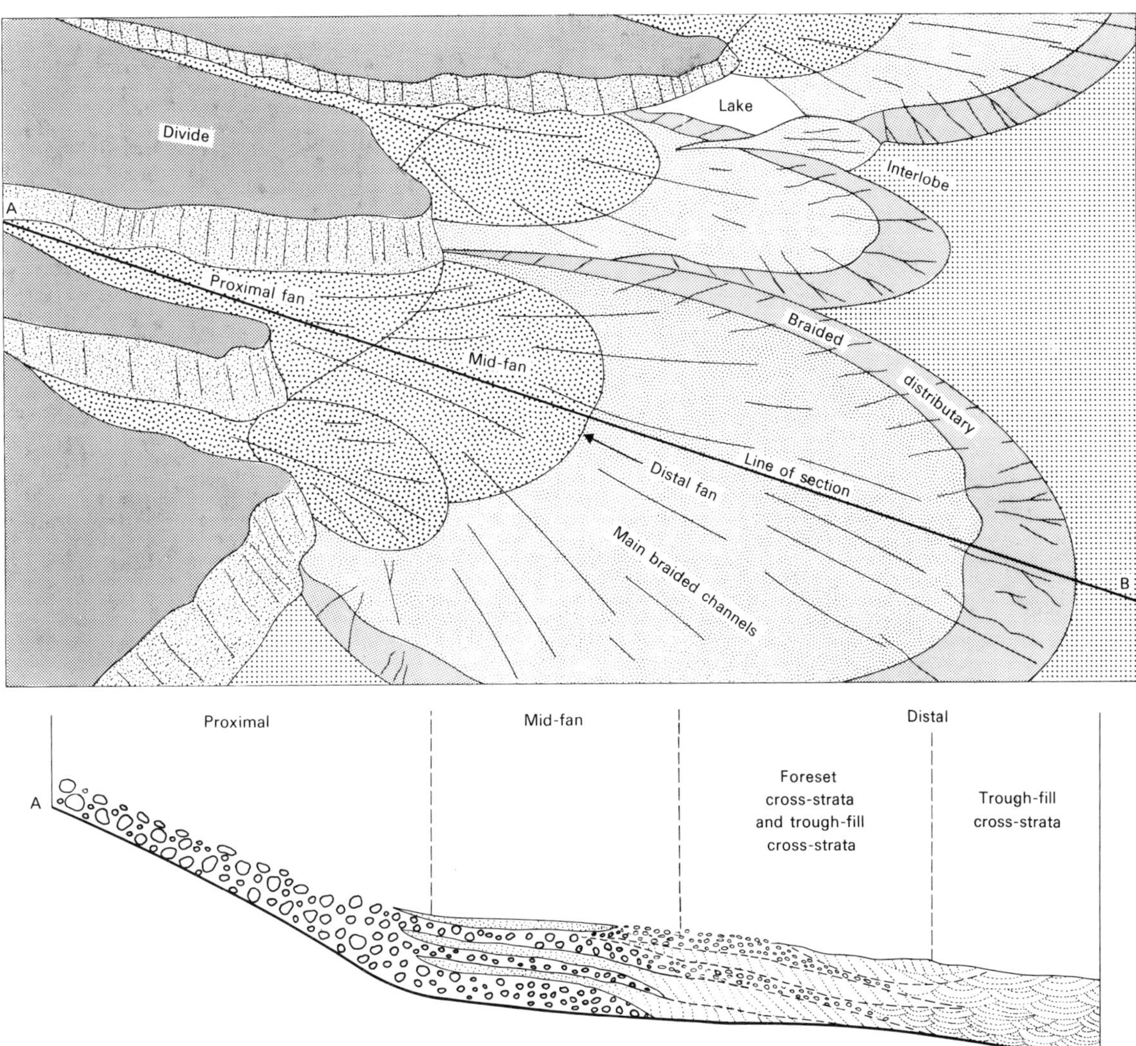

Fig. 3.38. Distribution of facies and environments in the humid deposits of the Van Horn Sandstone, Texas. The environmental and facies changes are all gradational and the diameters of the fans are of the order of 30–40 km (after McGowen and Groat, 1971).

of active reworking on the higher parts of the fan (e.g. Sestini, 1973).

LATERAL FACIES DISTRIBUTIONS

Within wedges of alluvial fan conglomerate there are lateral facies changes and a general down-current thinning, though the rates of change and of thinning will depend on the scale of the fan and the nature of the dominant processes. Bed thickness and the size of the largest clasts decrease distally (Fig. 3.37) (Bluck, 1967; Nilsen, 1969; Miall, 1970) as on present day fans (Bluck, 1964). In some small fan deposits in South Wales, maximum clast size falls off exponentially with distance (Bluck, 1965). On semi-arid fans there is often a distal decline in mudflow deposits and an increase in streamflood and channel deposits (Bluck, 1967; Nilsen, 1969; Steel, 1974;

Steel, Nicholson and Kalander, 1975) though Bluck (Fig. 3.39A) shows that distal mudflows may be interbedded with playa sediments deposited beyond the normal range of fan activity. The large travel distances of modern mudflows make this quite possible (e.g. Sharp and Nobles, 1953).

Some indication of the facies changes on humid fans are implicit in the facies recognized in the Van Horn Sandstone (McGowen and Groat, 1971). A distal fining is accompanied by a fall-off in the volume of framework conglomerates and an increase in the volume of cross-bedded pebbly sandstone (Fig. 3.38). The sizes of the cross-bedded sets diminishes distally and trough sets, the products of dunes, replace the tabular sets produced by transverse bars (Fig. 3.36, 3.38 and compare Fig. 3.37). Scouring and bed thickness diminish while muddy sands line small scours and sequences of small overlapping channels occur. These are interpreted as products of the braided distributaries of streams which emerge from internal drainage on to the fan surfaces. This pattern of lateral facies changes compares well with the model for a modern glacial outwash fan (Fig. 3.14) described by Boothroyd (1972) and Boothroyd and Ashley (1975).

LARGE-SCALE FACIES SEQUENCES

Fan successions frequently either broadly coarsen upward or fine upward on a variety of scales, the grain-size changes being reflected in the size of the largest clasts. The fining- or coarsening-upward sequences are commonly tens or hundreds of metres thick (Fig. 3.39) (Steel, 1974; Steel and Wilson, 1975). Fining-upwards sequences may result from waning sediment supply due to the wearing down of a source area without tectonic rejuvenation (Bluck, 1967; Nilsen, 1968; Deegan, 1973; Steel and Wilson, 1975). Bluck (1967) and Steel (1974) showed that these sequences coincided with an upwards facies change from mudflow domination to channel domination.

Coarsening-upward sequences and associated facies changes from channel domination to mudflow domination are usually less common. They are probably the result of progradation of fan lobes or of renewed or accelerated uplift and erosion of the source area. Some fans show a mixture of coarsening-upward and fining-upward sequences while others are dominated by fining-upward sequences and sudden

Fig. 3.39. Typical sections through the deposits of semi-arid alluvial fans from (A) the Old Red Sandstone of Scotland and (B and C) the New Red Sandstone of the Hebrides. Variations in maximum clast size through the sequence define coarsening- and fining-upward units and show the random interbedding of the facies (after Bluck, 1967 (A) and Steel, 1974 (B, C)).

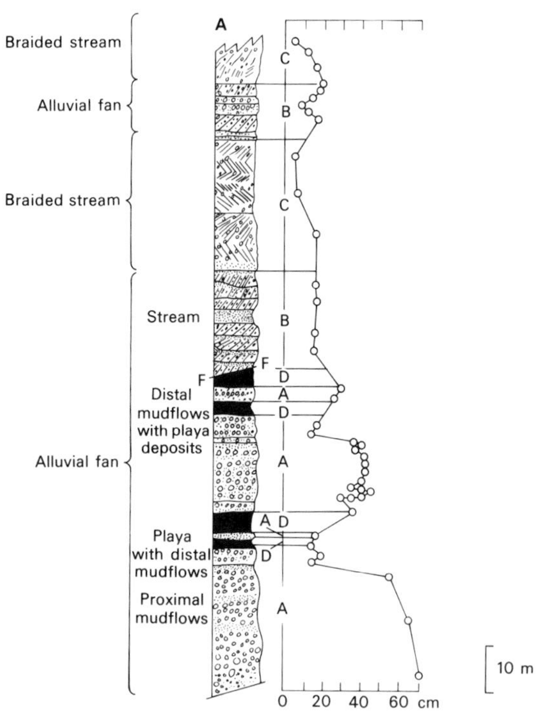

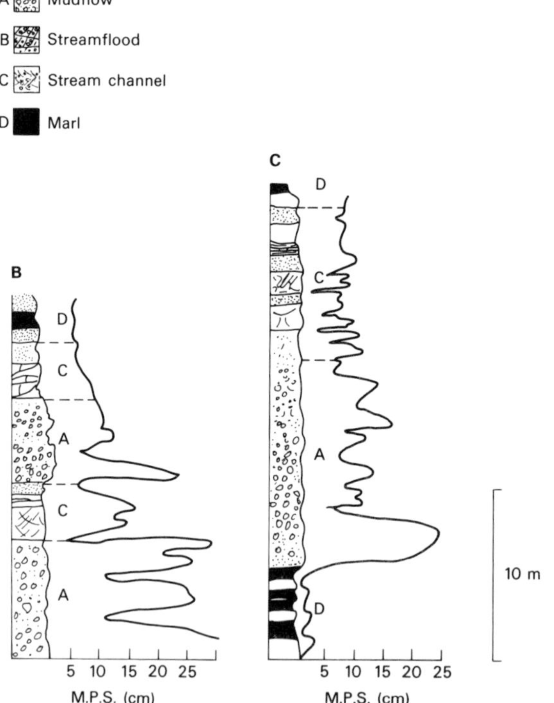

coarsening transitions. Successions dominated by fining-upward sequences may reflect highly episodic tectonic control while those with both fining and coarsening sequences were probably subjected to less violently fluctuating tectonics. It is also possible that some of the coarsening-upward and fining-upward may be the result of climatic or vegetation changes (Lustig, 1965; Croft, 1962).

PALAEOCURRENTS

Palaeocurrents from ancient fan sediments can be very valuable in building up a detailed picture of syn-sedimentary topography. Most fan sediments show fairly well-directed palaeocurrents (Nilsen, 1968, 1969; Collinson, 1972) though where mudflow deposits are abundant palaeocurrent readings can be difficult to obtain. It is sometimes possible to identity individual fans by radial palaeocurrent patterns (Fig. 3.37). In large Torridonian fans (Williams, 1969), a decrease in dispersion of current directions upwards through the sequence suggested that the fan head retreated. In addition, a sudden major change of direction at a particular level in the sequence suggests that an adjacent fan had taken over deposition at that point and that the bajada was composed of coalescing fans.

LOCAL FACIES RELATIONSHIPS IN WELL-STRATIFIED CONGLOMERATES

The sporadic nature of many of the processes and the unstable nature of the channel patterns on present-day fans and braided outwash areas means that sequences are likely to have a highly random and fragmentary preservation. Two small-scale facies patterns have, however, been recognized in Pleistocene gravels from Ontario and can be compared with present-day activity.

Braid bar pattern. Eynon and Walker (1974) described the internal structure of a large braid bar from gravel pit exposures. The bar is some 400 m long and in a section parallel to the current can be divided into a core, a bar front and an upstream section. Transverse to the current, it passes into channel deposits (Fig. 3.40).

The core is about 4 m thick and is composed of horizontally stratified, imbricated gravels, which accreted vertically on to a horizontal surface. The core is the eroded remnant of an earlier bar. The bar front gravels are cross-bedded in tabular sets up to 4 m thick which extend 150 m parallel to the current. The cross-bedding is complicated by reactivation surfaces like those seen in excavations of present-day transverse bars (Collinson, 1970b). They reflect fluctuating discharge or the migration of small superimposed bedforms on the bar top. Upstream of the core, small tabular sets of cross-bedding in a coarsening-upward sequence are thought to have accreted by dunes climbing the stoss-side during rising stage. Laterally the bar deposits interfinger with trough cross-bedded sands which

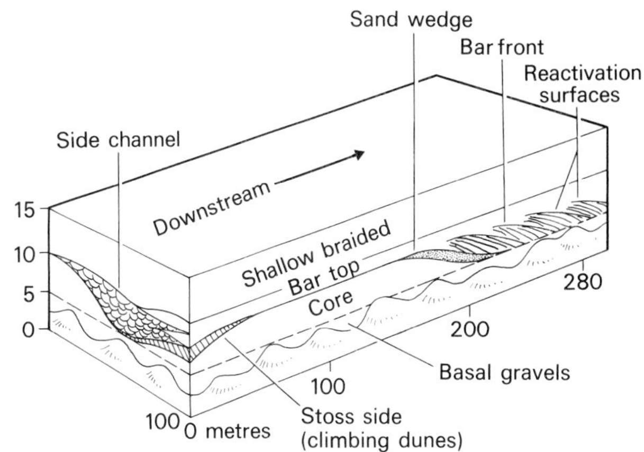

Fig. 3.40. Facies relationships of the braid pattern in Pleistocene outwash gravel, Ontario. The core of parallel-bedded imbricated gravels passes downstream into bar front cross-bedding and upstream into climbing dune cross-bedding. The lateral side channel is filled with trough cross-bedded sand (after Eynon and Walker, 1974).

are regarded as flanking channel deposits. The gravels do not extend to the base of the cross-bedded sands, however, suggesting that gravel moved over topographically higher areas rather than in the channels. This bar is unusual in comparison with what might be predicted from the behaviour of modern streams. The bar appears to have been very stable in comparison to the ephemeral nature of modern longitudinal bars. Most modern braided streams have the coarsest fraction

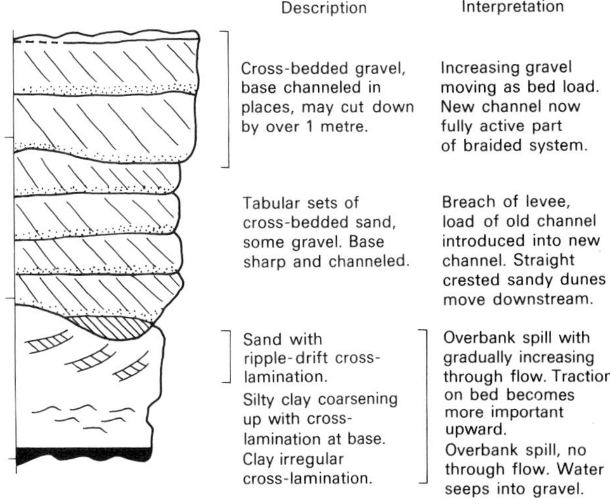

Fig. 3.41. The facies sequence associated with the re-establishment of flow in a channel from the Pleistocene outwash gravels of Ontario (after Costello and Walker, 1972).

of the bed in the channels rather than sand as seen here. *Channel reactivation pattern.* Costello and Walker (1972), working on a broadly coarsening-upward sequence of cross-bedded sands and gravels, described interbedded coarsening-upward units between 30 cm and 3 m thick. Basal massive and laminated clays mantle underlying sand. The clays pass gradationally upward into silts and ripple and climbing ripple cross-laminated sands which in turn are overlain by coarser, sometimes channel-bound cross-bedded sands. Cross-bedded gravels may erosively overlie the sands (Fig. 3.41). The sequence is thought to represent the gradual restoration of flow in a temporarily abandoned channel during rising water stage, as palaeocurrent directions within the sequence are all consistent. Large parts of these sequences may be removed by the erosion at the base of the gravel and so even the basal clay to silt coarsening is regarded as diagnostic of a braided stream by Costello and Walker. However, other situations, for example a chute on a pebbly point bar, could possibly give a similar sequence and therefore its value as a diagnostic criterion is not firmly established.

3.9 ANCIENT SANDY FLUVIAL SYSTEMS

3.9.1 Introduction

In most sandy fluvial sequences, two major facies associations can generally be distinguished. These are the 'coarse' and 'fine' members (Allen, 1965a), usually interpreted as laterally accreted channel deposits and vertically accreted inter-channel deposits respectively. Separation of the two associations, however, may be difficult especially when the coarse member grades vertically up into the fine member through the deposits of transitional environments such as levees.

The coarse member, with its more spectacular assemblage of sedimentary structures, has, until recently, received most attention and we now understand many aspects of its sedimentation. However, the same cannot be said for the fine member in spite of the fact that it is the key to the palaeoclimate.

3.9.2 Fine member (inter-channel) deposits

Fine member deposits can generally be separated into three primary depositional facies which may be modified by post-depositional *in situ* processes of pedogenesis and bioturbation.

SILTSTONES AND MUDSTONES

These are the most abundant facies and are generally laminated horizontally with a highly variable fissility. They may be homogenized by bioturbation and show evidence of subaerial exposure in the form of mudcracks, rainpits and footprints (e.g. Thompson, 1970).

The sediments may be further subdivided into red and grey beds, a colour distinction which also usually affects interbedded sediments. Red associations are more likely to show evidence of subaerial exposure while the grey beds are more likely to be disturbed by rootlets and to preserve organic matter as either comminuted debris or as coal.

The facies is interpreted as the product of suspended sediment fall-out on the floodplain or inter-channel area. Deposition may be from water due to overbank flooding or it may be of wind-blown loess when homogeneous silt, commonly coloured red, is abundant (Wills, 1970). The colour differences are attributed to the level of the water table in the depositional area. The red colour is related to oxidizing conditions and to periodic wetting (T.R. Walker, 1967) and in consequence, reflects a low water table. The grey colour is due to the maintenance of reducing conditions by a high water table, close to or even above the depositional surface as in a backswamp or lake.

SHARP-SIDED SANDSTONE BEDS

These commonly occur interbedded with siltstones and mudstones. They are usually thin, seldom more than a few tens of centimetres and exceptionally 1 m thick. They quite commonly wedge out laterally (e.g. Leeder, 1974) with convex upwards tops, though in other examples are parallel sided. They commonly look like turbidites, with sharp bases, solemarks, graded bedding and a Bouma (1962) sequence of internal structures with ripple drift cross-lamination particularly common. The occurrence of such beds in association with bird footprints and salt pseudomorphs led to arguments about the depth of deposition of 'flysch' (Mangin, 1962; De Raaf, 1964). However, the features are simply the products of episodic decelerating flows. In an alluvial context, therefore, they are interpreted as crevasse splay deposits particularly where laterally limited in extent (Allen, 1964; Leeder, 1974) or where more extensive and parallel sided, as distal sheet floods of a fan.

CROSS-BEDDED SANDSTONES

In some red, fine member sequences, there are cross-bedded sandstones whose grain size is more appropriate to the coarse member but which were deposited by inter-channel processes. Such units, which are often quite thick, may involve sets up to several metres thick. The sand is usually well sorted, lacks platy minerals and the grains may be well rounded. Such sandstones may be interpreted as the product of aeolian dunes migrating on the inter-channel areas particularly if their palaeocurrent directions are opposed to those of the associated waterlain sandstones (e.g. Laming, 1966). Further details of aeolian dunes are given in Chapter 5.

In other examples, thin units of cross-bedded sandstone with rounded and polished grains occur in inter-channel siltstones of reddened alluvial sequences (Thompson, 1970). The nature of the grains suggests aeolian activity but with thin beds usually only one set thick, it is difficult to know if the sands are the products of aeolian deposition or of aqueous reworking of aeolian sands on the floodplain.

PALAEOSOLS AND ASSOCIATED DEPOSITS

Inter-channel areas, being frequently subaerially exposed for long periods, may be sites of pedogenesis. In spite of the vast detailed knowledge of the structures and textures of modern soils, only a fraction of these features are preservable in the rock record because of the diagenetic alteration and compaction suffered on burial (Roeschmann, 1971). In Quaternary and Tertiary deposits, where recognition and interpretation of soils are of crucial stratigraphical importance, many of the original features are preserved and rather sophisticated interpretation is possible (e.g. Yaalon, 1971; Buurman and Jongmans, 1975). In the Mesozoic and older deposits, soil features are at present recognized only in a relatively crude way. The identification of fossil soils in the stratigraphic record must be based on both textural and compositional properties of the rock.

(a) *Nodules and cements* are commonly developed in soils as calcite, siderite and quartz. Calcite is particularly common as a nodular mineral in red sequences and it occurs as isolated nodules and as coalescing sheets or layers, sometimes with a horizontal lamellar structure (Fig. 3.42). Carbonate rich units range in thickness from a few centimetres to 2–3 m and are usually laterally continuous (Allen, 1974a; Burgess, 1961; Leeder, 1975). Sometimes a vertical sequence is seen with a zone of isolated nodules passing up into more continuous calcite. The more continuous layers may be buckled into gentle folds and on bedding planes the more isolated nodules may follow a polygonal pattern.

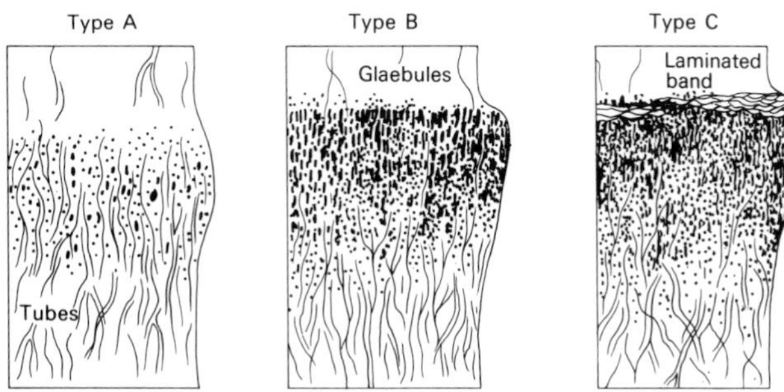

Fig. 3.42. Three inter-grading stages in the development of a vertical profile in pedogenic carbonate units of the Old Red Sandstone of Wales (after Allen, 1974a).

The carbonate-rich layers compare with caliche soils of modern semi-arid regions (Allen, 1974a; Leeder, 1975) and it is possible to use them as a measure of floodplain accretion rate, depending upon the stage of calcrete evolution which the sequence reaches (Leeder, 1975). Mature sequences with a lamellar upper part (Fig. 3.42C) require about 10,000 years to develop and they therefore imply that periods of this length occurred during which little or no sedimentation took place and the water table was deep. Such a sediment-starved inter-channel area may have been the result of channel incision and terrace development or of distant migration of the main river flow.

Not all carbonate layers in fine members are necessarily fossil caliches, however. In the Old Red Sandstone of Spitzbergen, laminated carbonate layers have ostracods and algal remains and are interpreted as the deposits of lakes (Friend and Moody-Stuart, 1970).

Siderite nodules, on the other hand, are commonly associated with grey fine members and in particular with rootlet-bearing beds below coal seams (Wilson, 1965). The nodules may be elongated normal to the bedding and are frequently associated with rootlets and plant fragments. They are interpreted as being due to precipitation from slightly reducing groundwater in a permanently saturated soil. In some examples, pyrite also occurs associated with the siderite but this commonly weathers to jarosite on exposure. Less common are brown seat-earths, containing nodules of sphaerosiderite. These commonly underlie thin coals and are thought to represent less saturated and partially emergent soils (Elliott, 1968).

Where the parent sediment is sand, silica cement occurs in two settings. In red beds, aeolian dune sandstones may show well developed syntaxial quartz or, more rarely, orthoclase overgrowths (Waugh, 1970). This is thought to be a relatively early cementation feature due to the precipitation of silica derived from the solution of quartz dust by alkaline ground water. The formation and evaporation of desert dew may have played an important role, giving an initial case-hardening to the dune followed by cementation of deeper levels (cf. Woolnough, 1930).

In grey beds, silica cementation results in quartz-rich seat-earths or ganisters. Here, the silica concentration is thought to be due to leaching in a water-logged profile.

(b) *Mottling and clay coatings.* The migration in solution of iron and manganese ions leads to the patchy accumulation of their oxides and hydroxides as grain coatings giving an overall mottled appearance to the sediment. Such mottling is a characteristic of gley soils where the movement of reducing pore waters is sluggish (Buurman, 1975). The unequivocal recognition of such soils is however very difficult.

At the microscopic level, the presence of clay coatings on sand-sized particles is a soil feature which is preserved in rocks as old as Palaeozoic (Teruggi and Andreis, 1971). Most of these are due to illuviation in the B-horizon of clays transported by downwards-moving soil water. Such coatings therefore reflect conditions of pore-water behaviour different from that associated with mottling.

(c) *Rootlets and Coal*. The penetration and disturbance of usually grey fine member sediments by rootlets is a diagnostic indication of an ancient soil. The rootlets penetrate the bedding at all angles and in the Carboniferous Coal Measures are usually preserved as thin carbon films (Wilson, 1965; Huddle and Patterson, 1961). Less commonly, major *Stigmaria* roots occur with smaller rootlets attached. These usually lie in a horizontal or subhorizontal plane reflecting the shallow rooting system of Carboniferous plants. The rootlets occur in host sediments of a wide range of grain-sizes from sandstone to mudstone. An upwards fining is quite commonly observed, possibly reflecting the decreasing energy due to the increasing density of plant growth (Wilson, 1965). The presence of abundant rootlets may lead to total destruction of original bedding, and intense slickensiding is caused by collapse and compaction of the roots (Huddle and Patterson, 1961).

Rootlet horizons are often overlain by coal seams of variable thickness which bear no relationship to the thickness of the underlying seat-earth. The coals are thought to be the product of *in situ* plant growth and decay with the seat-earth being the soil for only the earliest plants. Later plants probably rooted themselves in the rotting mat of earlier plant material. The environment had a high water table and may have been permanently submerged as an overbank back swamp.

(d) *Clay mineralogy*. The chemical reactivity of clay minerals makes them highly sensitive to pedogenic processes. By the same token, however, they are susceptible to diagenetic alteration and the more subtle properties may have a low preservation potential. Watts (1976) recognized palygorskite in a Permo-Triassic sequence as the product of a semi-arid soil. The most clearly developed trend is towards concentration of kaolinite in many fine member seat-earths at the expense of illite (Huddle and Patterson, 1961). This concentration, which is often accompanied by titanium enrichment, is interpreted as the result of *in situ* leaching in the soil below growing vegetation. However, the clay mineralogy of seat-earths is not only a function of pedogenesis and may also be due to the lithologies and weathering conditions in the source area (Wilson, 1965).

(e) *Soil sequences*. The above features of soils which may be preserved in the rock record occur together in various combinations. However, of the possible variations only extremes seem to have been commonly recognized. The red siltstones with carbonate concretions characteristic of semi-arid caliche are at one end of a spectrum which extends to grey palaeosols with rootlets, upward fining, ironstone nodules and overlying coal at the other end. There is much scope for extending recognition and interpretation of ancient soils into the central area of the spectrum.

3.9.3 **Coarse member (channel) deposits**

Five principal facies can be distinguished.

CONGLOMERATES

These generally occur as thin beds only a few clasts thick, associated with erosion surfaces. They are interpreted as channel lags. The clasts may be either extra-basinal or intra-basinal, derived from the interbedded inter-channel deposits. Thus the most common clasts may be of calcium carbonate or siderite concretions, large plant fragments or logs, depending on the nature of the inter-channel environment. Where coarse members dominate the sequence, the clasts may be the only indication of the inter-channel conditions. Bedding in the conglomerates is rare though clast imbrication occurs in places.

In addition to pebble lags, larger-scale breccias sometimes occur in the lowest parts of coarse members and involve blocks of bank material slumped into the channel without current reworking (e.g. Laury, 1971; Young, 1976).

CROSS-BEDDED SANDSTONES

These are by far the most common and often the only facies in coarse members. They encompass a wide range of grain-sizes and types of cross-bedding and occur at all levels within a coarse member. Trough cross-bedding is the most abundant type and occurs in sets up to about 3 m thick, though sets of less than 1 m are more usual. Trough cross-bedding is attributed to the migration of dunes with sinuous crest lines or a more three-dimensional form. Tabular sets range in size up to 40 m but again sets of up to 1 m are more common. Tabular sets result from the migration of sandwaves, transverse bars or, less commonly, scroll bars and chute bars. The exact origin can often only be deduced from the context and not from internal evidence. Both types of cross-bedding may be deformed by overturning of the foresets (e.g. Beuf *et al.*, 1971; Banks *et al.*, 1971; Mrakovich and Coogan, 1974). Internal evidence of water stage fluctuation in the form of reactivation surfaces (Sect. 3.4.2) and mud-draped foresets may occur. Exceptionally, bar fronts weather out in full relief showing gullying and terracing (Banks, 1973c) or the superimposition of ripples. Detailed interpretation depends upon the position of the cross-bedding in a sequence and upon the distribution and spatial relationships of palaeocurrents derived from it.

CROSS-LAMINATED SANDSTONES

These often form quite substantial parts of coarse members and may rarely form all of them. They usually occur in the upper parts of coarse members and involve finer and more micaceous sands than those of associated cross-bedding. The cross-lamination is normally of trough type and frequently shows ripple drift. When this is the case, the angle of climb may be used to estimate the instantaneous rate of vertical bed accretion (Allen, 1971a). Cross-lamination is interpreted as the product of migrating small-scale current ripples.

PARALLEL-LAMINATED SANDSTONES

These are normally of minor volumetric importance but occasionally form substantial thicknesses of coarse member deposits. They are usually fine-grained and primary current (parting) lineation is commonly found on bedding planes. They may occur at all levels within a coarse member though they are probably more common towards the top, particularly where cross-bedding is abundant. They are interpreted as the product of upper flow regime plane bed which develops beneath flow of high velocity or low depth.

EPSILON CROSS-BEDDING

A very important but not particularly common feature of coarse members is low-angled cross-bedding extending as a single set over the whole thickness of the member. This 'epsilon' cross-bedding (Allen, 1963) dips at right angles to the palaeocurrent direction derived from smaller scale structures within it, such as cross-lamination and small-scale cross-

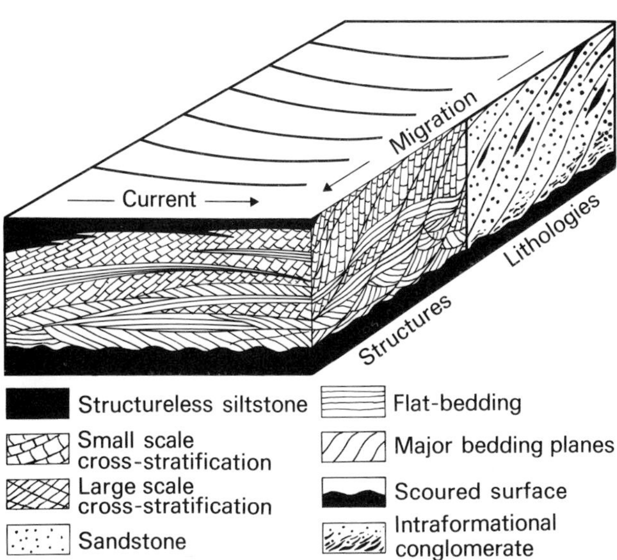

Fig. 3.43. Epsilon cross-bedding in the Old Red Sandstone. Major bedding surfaces dip at between 4° and 14°. Note the variability of the vertical sequences of structures. Large vertical exaggeration (after Allen, 1965b).

bedding (Fig. 3.43). The inclined bedding is defined by grain-size changes, generally in the siltstone to sandstone range. Coarse members with epsilon cross-bedding are seldom thicker than about 5 m (Allen, 1965b; Moody-Stuart, 1966; Allen and Friend, 1968; Nagtegaal, 1969). The cross-bedding is not always continuous in its development and may be broken by internal erosion surfaces (Allen and Friend, 1968; Beuf *et al.*, 1971).

The facies is interpreted as the product of lateral accretion on an inclined surface, usually taken to be a point bar surface. It has long been recognized in tidal creeks (Van Straaten, 1951). The relative rarity of this facies does, however, present a problem. If it is the product of point bar accretion, why is it not more common, reflecting the frequency of this process at the present day? While the orientation of exposures relative to the inclined surface may contribute to its scarcity, there is no reason to expect such surfaces to form at all unless there is a temporal fluctuation in conditions as well as the spatial variation implicit in the point bar model (see Sect. 3.5.2). Epsilon cross-bedding therefore probably reflects a rather fluctuating hydrological regime, and the occurrences of mud drapes and erosion surfaces associated with it seem to support this.

3.9.4 **Coarse member organization**

In order to determine the more detailed nature of the river channel responsible for a particular coarse member and thereby predict the behaviour and occurrence of the sandstone, several aspects of the coarse member need to be considered.

OVERALL BODY SHAPE

Coarse members vary greatly in their shape and size. They range from single channel units isolated in fine member deposits to composite units involving many channel units.

An appreciation of overall sandbody shape is important not only in distinguishing between types of river system but also in distinguishing the deposits of rivers from those of delta distributaries (e.g. Brown *et al.*, 1973). Distributary channel sands of fluvial dominated deltas are generally more elongate than the more tabular sand bodies of braided and meandering rivers.

Where coarse members are isolated in fine member sediments as in the Old Red Sandstone of Spitzbergen (Moody-Stuart, 1966) they always rest on a basal erosion surface. These may be flat and horizontal or they may be gently concave upwards (Fig. 3.44). Occasionally, more steeply inclined, sometimes vertical, channel margins occur. Concave-based lenticular channel bodies are commonly thought to be the product of low sinuosity streams which cut and then vertically filled their channels (Fig. 3.44) (Moody-Stuart, 1966; Kelling,

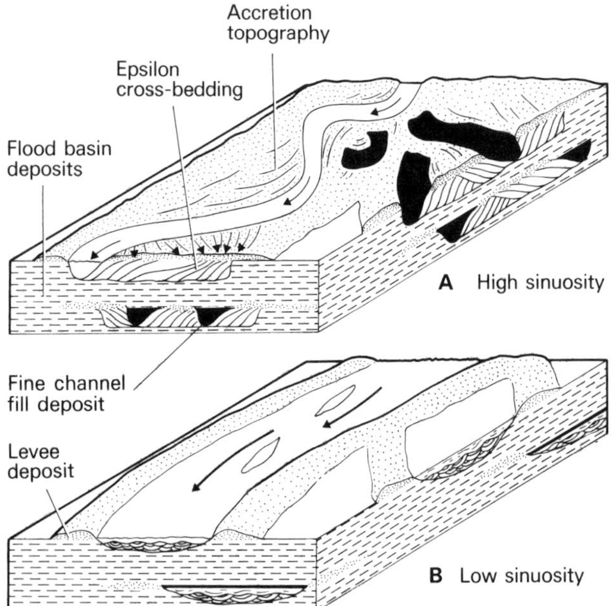

Fig. 3.44. Models illustrating the relationships between channel shape, internal structures and channel based on examples in the Old Red Sandstone of Spitzbergen (after Moody-Stuart, 1966).

1968; Thompson, 1970; Sykes, 1974). Flat-based sheets are commonly attributed to meandering streams where migration of the scour pool erodes to a fairly constant level (Allen, 1964; Moody-Stuart, 1966). Steep channel margins and epsilon cross-bedding are associated with the flat-based coarse members of the Spitzbergen Old Red Sandstone and support their interpretation as meandering stream deposits (Fig. 3.44A).

Composite units, comprising many channel fills, have commonly an overall sheet-like form. In the Westwater Canyon Member of the Morrison Formation (Jurassic) of New Mexico, sandstone forms a sheet 100 km wide, over 160 km long (parallel to the dominant palaeocurrent direction) and up to 60 m thick (Campbell, 1976). This sheet is built of a series of mutually erosive channel systems 1.5–34 km wide and 6–60 m deep (Fig. 3.45). Each system is in turn a composite of individual channels or channel fragments 30–350 m wide and 1–6 m deep. Channel systems are tabular in cross-section with rapid thinning at their edges whilst the individual channels have concave upwards bases. Channel edges are sinuous but non-meandering in plan, suggesting that the concave upwards base may reflect low sinuosity channels within a braided tract (channel system) involving multiple cut and fill as channels switch within the tract. The system compares in scale with that of the Kosi River fan (Sect. 3.2.3, Gole and Chitale, 1966).

Thus there seems no reason why low sinuosity streams should not migrate laterally and produce tabular sand bodies of wide lateral extent. Modern meandering streams, on the other hand, are laterally limited by erosion surfaces against inter-channel sediments or the clay plugs of abandoned loops, and steep channel margins might therefore be expected to be quite common in the deposits of meandering streams. They are, however, remarkably rare in some ancient sequences supposedly of meandering river origin, even taking into account the orientation of exposures relative to the palaeocurrent. The channel margin at Tugford, described by Allen (1964) is the only described example from the Anglo-Welsh Old Red Sandstone (Fig. 3.47C).

Fig. 3.45. Stratigraphic relationships of the channel systems which make up the sheet sandstone of the Westwater Canyon Member of the Morrison Formation (Jurassic), New Mexico. The section is roughly transverse to the dominant palaeocurrent direction (after Campbell, 1976).

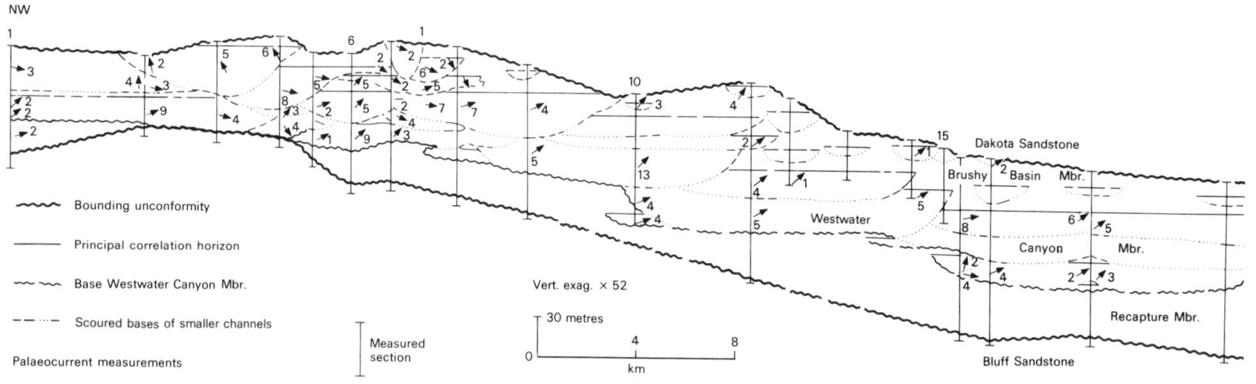

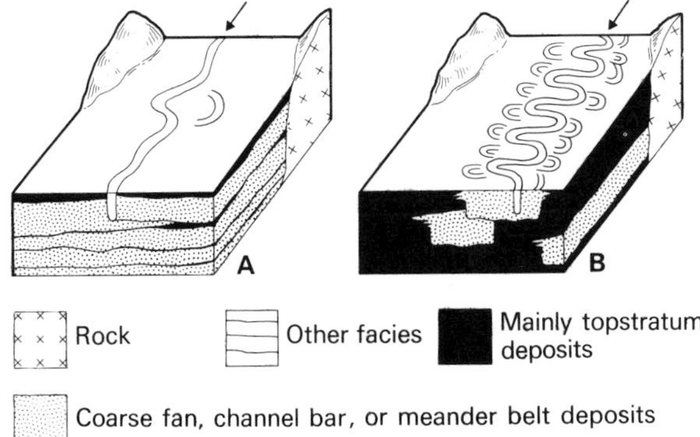

Fig. 3.46. Hypothetical models illustrating the broad facies relationships which might be produced by streams of (A) low-sinuosity and (B) high-sinuosity. The preservation of overbank sediments depends on the relationship between migration frequency and subsidence in both cases (after Allen, 1965c).

COARSE MEMBER/FINE MEMBER RATIO

Coarse members so dominate some sequences that little inter-channel sediment is preserved (e.g. Stokes, 1961; Selley, 1972; Campbell, 1976); in other cases channels are isolated in floodplain sediments (e.g. Moody-Stuart, 1966). The former situation is more likely on a sandy fan with low-sinuosity or braided channels. The latter is more likely on an alluvial plain, where high-sinuosity streams may have more associated overbank sediment than low-sinuosity streams (Thompson, 1970). This phenomenon is either related to the observed relationship between meandering and high suspended load (Schumm and Kahn, 1972) or to the suggested tendency for low-sinuosity streams to comb their floodplains (Fig. 3.46) (Allen, 1965c). The vital control in preserving overbank sediments in areas of laterally migrating channels is the relationship between migration frequency and overall subsidence. Multiple stacked erosion surfaces in coarse member sandstones may be due to scour and fill within a non-migrating channel and need not necessarily indicate frequent channel migration.

INTERNAL FACIES RELATIONSHIPS

The now classic fining-upward sequence was described by Bernard and Major (1963) and subsequently refined and extended by Allen (1964, 1965a, 1965b, 1970a, 1970b, 1974b), Visher (1965) and others. A sandstone overlying a horizontal erosion surface fines upwards and shows an upwards change

Fig. 3.47. Representative and standard coarse member sandstones from the Old Red Sandstone. (A, B) Welsh Borders, (C) Tugford, Clee Hills, (D) Mitcheldean, (E) Forest of Dean, (F) Clee Hills, (G) Spitzbergen. Note the variation of scale and of sequence. Where a range of thickness is given, the sequences are somewhat idealized standards (after Allen, 1965a).

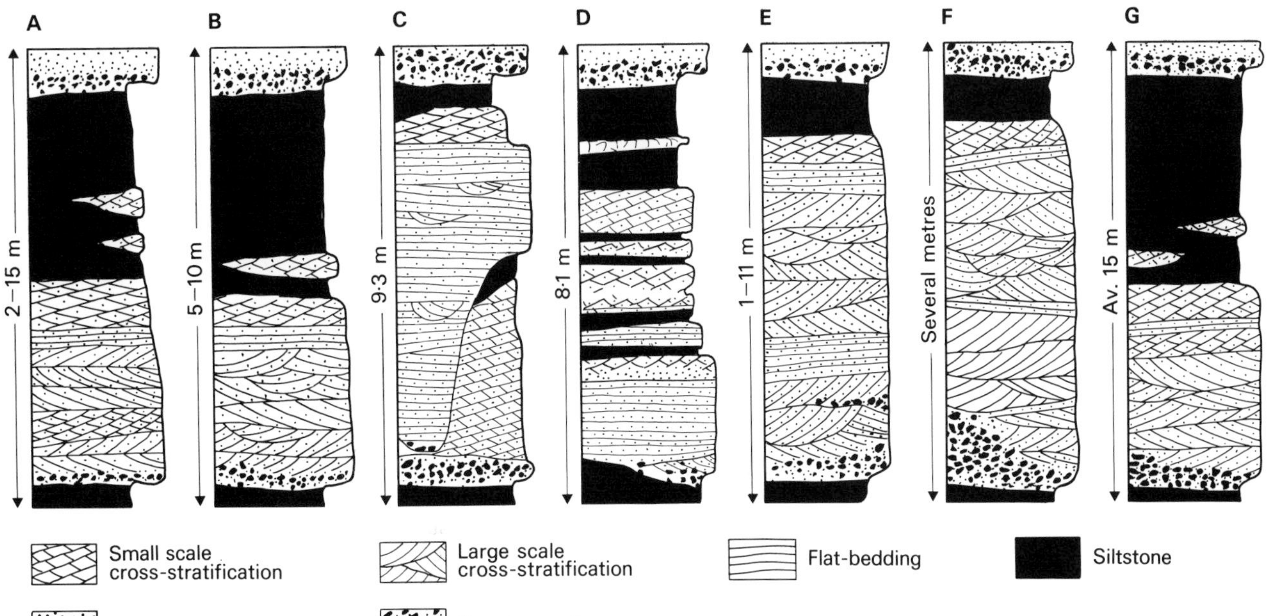

Fig. 3.48. Exposed scroll bars on the upper surface of channel sandbody in the Scalby Formation (Middle Jurassic), Burniston, Yorkshire. The individual point bar accretion units have erosive contacts with one another and restricted lateral extent.

from cross-bedding to parallel- and/or ripple cross-lamination before it grades into the overlying fine member. There may be a lag conglomerate at the base and the cross-bedding frequently shows an upwards decrease in set thickness. There is considerable variation in the lateral and vertical sequences and in the relative abundance of each facies (Fig. 3.47) (Allen, 1965a, 1974b). For example, parallel-laminated sandstone sometimes occurs low in the sequence (Fig. 3.47D).

A direct interpretation of the sequence indicates an upwards waning of stream power from an initially erosive phase. The influence of both velocity and depth on the stability of bed forms allows parallel-lamination to occur at a variety of levels. This interpretation of processes has been localized in the classic model as the product of point bar sedimentation with the thickness of the coarse member corresponding to the depth of the migrating channel. The vertical sequence is due to the spatial separation of the different bedforms on the point bar surface and the observed variations of sequence are attributed to differences in the channel slope and curvature (Fig. 3.24) (Allen, 1970a, 1970b).

In some coarse members, thinner than about 5 m, epsilon cross-bedding forms virtually the whole member. Moody-Stuart (1966) showed that epsilon cross-bedding occurred only in flat-based, steep-sided sandstone bodies while concave upwards channels had different fills (Fig. 3.44). Very rarely, exceptional exposures show extensive bedding planes on which scroll bar topography may be recognized on the tops of coarse members which show epsilon cross-bedding (Fig. 3.48) (Puigdefabregas, 1973; Nami, 1976). Where epsilon cross-bedding occurs, and particularly where it is associated with scroll bar topography and with restricted sandbody width delimited by steep channel margins, an interpretation as a meandering channel is justified. The occurrence of these features, however, raises the question of whether those coarse members without epsilon cross-bedding are meandering channel products as the classical model suggests. In some cases the evidence may not be seen because of the orientation of the exposure, the large scale of the surfaces or because the scour of the trough cross-bedding in larger channels has effectively obliterated the potential epsilon cross-bedding (Frazier & Osanik, 1961).

However, where channel margins are also absent, alternative interpretations must be considered. Many coarse members, especially those thinner than about 2 m, may not be channel deposits at all but be the products of sporadic sandy sheet floods of wide lateral extent. Such processes could well produce the commonly observed fining-upward sequences with their flat bases and intraformational conglomerates. Examination of some of the coarse member units which have been interpreted as meandering channel deposits shows that this could be the case. Allen's (1964) units at Lydney and Abergavenny (Fig. 3.49) and many of the coarse members from the Red Marls of Pembrokeshire (Allen, 1974b) could as easily be interpreted as flood deposits of crevasse splays or distal fans. There is a major problem in deciding which erosively based sheets are due to channel migration and which to episodic floods.

Several features might help to resolve the problem. Thickness may be a guide as it is difficult to imagine thick sands as the products of anything other than channels. No particular thickness will unambiguously discriminate, though a value of around 2 m seems of the right order. If major channel sandstones are present, the spatial and palaeocurrent relationships of thinner coarse members to these may provide useful information.

Concentration on the elegant point bar model has led to the development of few facies models for low sinuosity streams. Moody-Stuart (1966), Kelling (1968), Thompson (1970), Sykes (1974) and Campbell (1976) all record sequences which fine upwards, particularly in their upper parts and which lie above concave erosion surfaces atrributed to low-sinuosity channels (Fig. 3.50). The channel fills consist of massive, frequently conglomeratic sandstones at the base, overlain by cross-bedded sandstones with some parallel-lamination. Ripple cross-lamination at the top sometimes extends beyond the channel

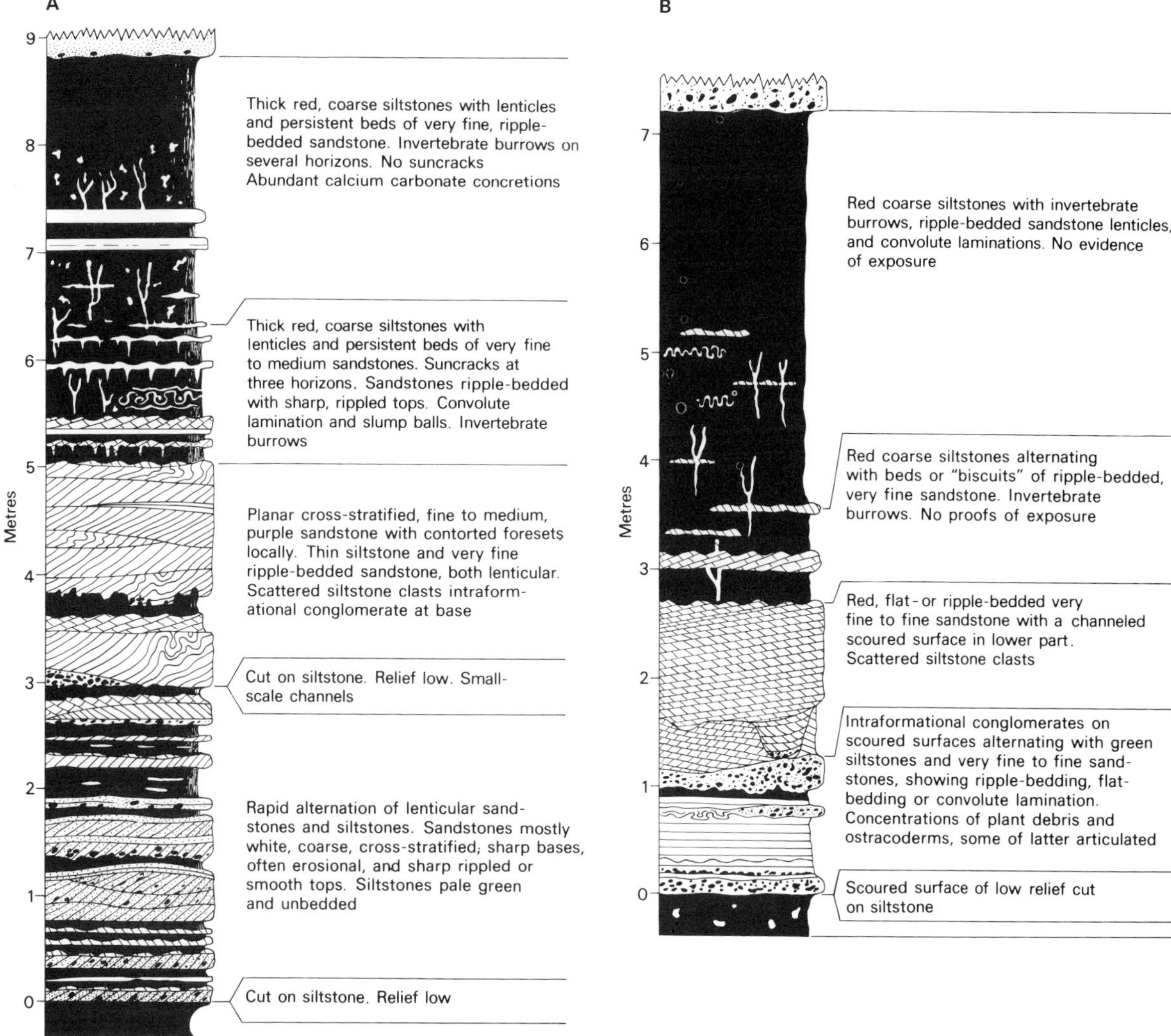

Fig. 3.49. Sequences in the Old Red Sandstone of the Anglo-Welsh region; (A) from Lydney, (B) from Abergavenny (after Allen, 1964).

margins and represents a lateral expansion of the flow prior to abandonment of the channel. Trough cross-bedding usually dominates with sets up to 1 m thick. Interbedded tabular sets seem to have no preferred level in the fill sequence. The fills are thought to represent a waning of the flow through time rather than a spatial segregation of bed forms within the channel as in the classical bar model.

In an attempt to set up a more complex facies model direct from the rock record, Cant and Walker (1976) recognized eight facies within the Devonian Battery Point Sandstone of Quebec and established a tentative model sequence which integrated facies and palaeocurrent information (Fig. 3.51). Above a scoured base, with intra-formational conglomerate, is a sequence dominated by trough and tabular cross-bedding. The lower part has poorly defined trough cross-bedding and this passes up into better-defined smaller troughs with a similar spread of palaeocurrent directions. Within this unit of small troughs, however, are larger tabular sets whose directions

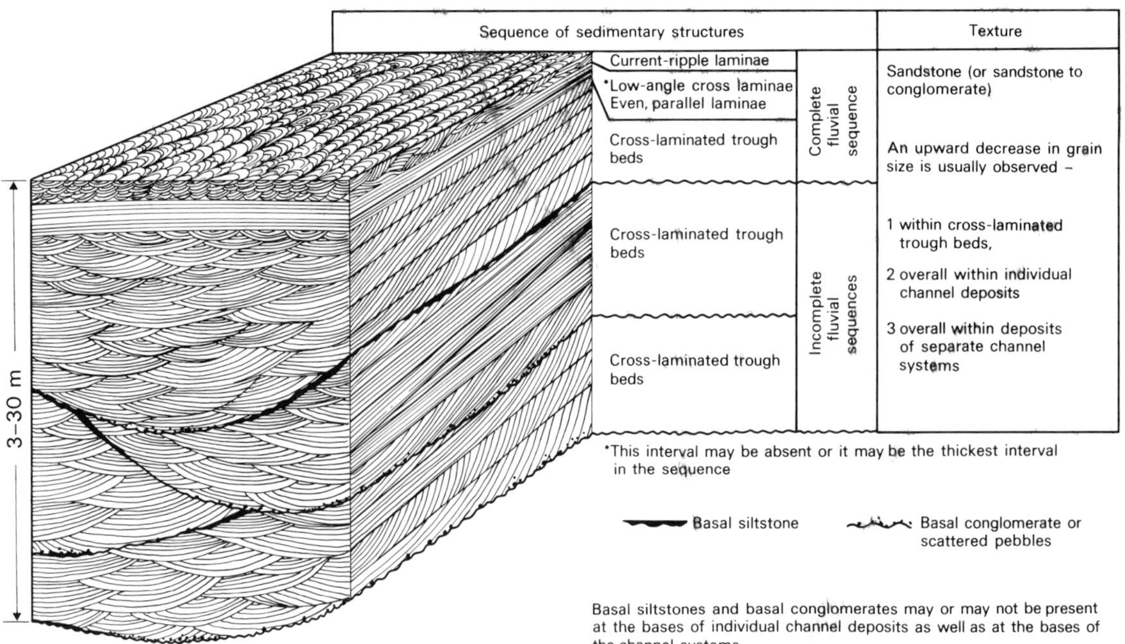

Fig. 3.50. The sequence of sedimentary structures and textures within a channel system of the Westwater Canyon Member of the Morrison Formation (Jurassic) of New Mexico. Note the concave upwards bases to the individual channels (after Campbell, 1976).

diverge by around 60° (maximum 90°) from those of the troughs and have a bimodal distribution about the trough mean direction. The top of the sequence consists of smaller-scale tabular sets and ripple cross-lamination.

The sequence is interpreted as a deposit of a low sinuosity stream, the large troughs being due to dune migration in the deeper parts of the channel and the tabular sets to transverse or cross-channel bars on higher areas. These bars moved with a strong transverse component as suggested by Smith (1972). The surrounding smaller troughs result from dunes moving in the inter-bar areas, burying the bars as they ceased to move. The smaller tabular sets result from dunes or small bars moving in the late stages of channel aggradation prior to abandonment.

This model, by combining facies analysis and palaeocurrent measurements, points the way in which future analysis must proceed. The model has only a small data base, however, and alternative interpretations are possible. Jackson (1976) has shown how, in the fully developed zone of meander bends, scroll bars on the point bar surface may give tabular sets, particularly in the upper part of the sequence, with a high divergence of direction to the surrounding troughs. Unless tabular sets can be shown to diverge in both rotational senses from the troughs *within the same channel unit,* Jackson's model could equally apply to the Battery Point Sandstone. Apparently such a situation does apply (Cant; personal communication) and the interpretation is therefore strengthened. Without the directional information, large tabular sets at high levels in trough-dominated sequences could also be due to chute bars (McGowen and Garner, 1970). These qualifications demonstrate clearly the need to erect facies models which combine facies description and sequence analysis very closely with palaeocurrent measurement. Only then is it possible to begin detailed interpretation either from first principles or by direct comparison with models derived from studies of modern rivers.

McCabe (1975, 1977), in erecting another facies model direct from the rock record, described large delta top channels in the Namurian of Northern England which are up to 40 m deep and probably more than 1 km wide (Fig. 3.52). They are filled with coarse, pebbly sandstone in four facies.

(a) Massive sandstone beds up to 2 m thick are apparently structureless and are irregularly though roughly horizontally bedded.

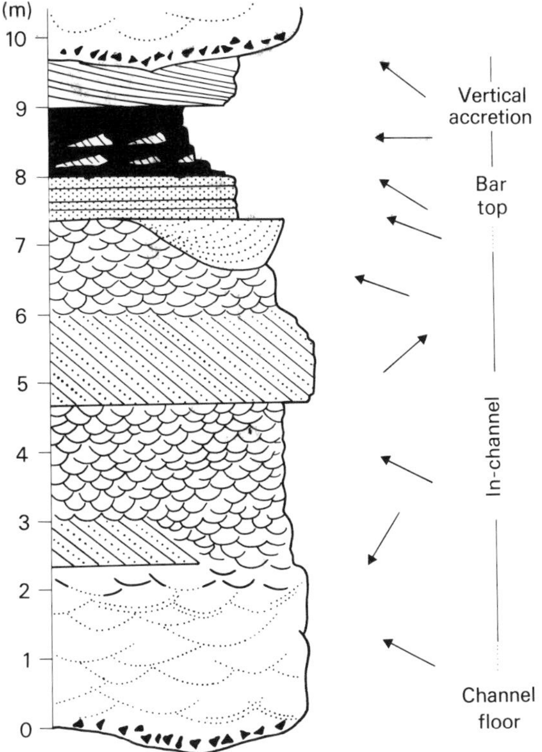

Fig. 3.51. Facies model for the Battery Point Sandstone (Devonian, Quebec) showing the relationship between vertical sequence of structures and their palaeocurrents. The sequence has been interpreted as the product of a low-sinuosity river (after Cant and Walker, 1976).

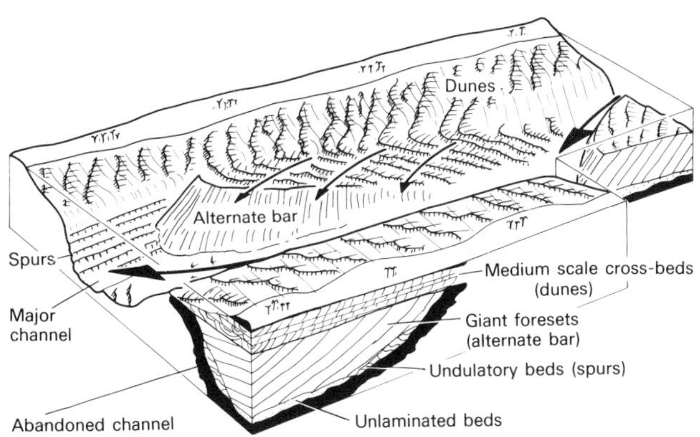

Fig. 3.52. Model for the large-scale delta-top, fluvial channels of the Namurian Kinderscout Grit of northern England. The channels are up to 40 m deep and about 1 km wide (after McCabe, 1975).

(b) Undulatory beds about 10 cm thick are rather clearly defined and undulate with a wavelength of between 10 and 20 m and a relief of about 1 m in sections normal to the independently determined palaeocurrent. The crests and troughs dip at low angles in either direction parallel to the palaeocurrent. Individual beds rise in height systematically from one crest to the next as they approach channel margins with an overall rise of 7 m being recorded across six undulations. The undulations are thought to be due to sand ridges aligned parallel to the current comparable with those of the Brahmaputra (Coleman, 1969).

(c) Large tabular sets of cross-bedding are up to 40 m thick, though normally less than 25 m, and extend horizontally for more than 1 km both parallel and perpendicular to the foreset dip direction. They were formerly thought to be Gilbert deltas (Collinson, 1968) (Chapter 6), but the discovery that they are channel-bound now suggests that they were formed in rivers, probably as alternate bars attached to channel sides and with major slip faces.

(d) Medium-scale cross-bedding is in sets normally less than 1 m thick.

The distribution of the facies is shown schematically in Fig. 3.52. The massive sandstone occupies the deepest part of the channel and undulatory bedding seems to occur near channel margins. Both are overlain by large-scale tabular cross-bedding. The dips of the large foresets diverge at around 40° from the inferred directions of elongation of associated sand ridges and have a component towards the centre of the channel. If the large foresets represent the slip faces of alternate bars, then the ridges may result from spiral vortices shed by the skewed slip faces. The massive beds could be due to the reattachment of strong separation eddies in the lee of the large bedforms which may inhibit sorting into beds. The medium-scale cross-bedding overlies the large-scale sets and generally has a direction close to that of the large-scale foresets. It is interpreted as the product of dunes which migrated on the tops of the alternate bars, and which fed sediment to their slip faces.

This model extends the range of fluvial possibilities beyond that which could reasonably have been predicted from our knowledge of present-day rivers.

PALAEOCURRENT DISTRIBUTION

The idea that a wide distribution of palaeocurrents is produced by meandering streams and a lower dispersion by low-sinuosity streams has been used to help interpret channel type (e.g. Kelling, 1968; Thompson, 1970).

Whilst this may have some truth when sampling is on a regional scale, the distribution of palaeocurrents within individual coarse members is seldom diagnostic of channel type. Channel bedforms are extremely complex in their behaviour (Sect. 3.4.2) and cross-bedding directions may indicate different patterns of bar movement rather than the channel type (Smith, 1972; Banks and Collinson, 1974). It is therefore important that sampling of palaeocurrents is done with careful regard to the type and position in the sequence of the sedimentary structures.

RÉSUMÉ

None of the approaches outlined above gives totally unambiguous interpretations of channel type in alluvial successions. Any interpretation must be based on a balanced appraisal of all types of data if they are available. Considerable refinement of the available models is still needed, particularly with regard to lateral variability of internal organization of sedimentary structures and its relationship to channel shape.

3.10 CHANGING ALLUVIAL SYSTEMS AND THEIR CONTROLS

In the earlier discussions of present-day rivers, little mention was made of the factors which determine where any particular channel lies in the continuum of channel patterns. The controls on river channel type have been covered extensively in the geological, geomorphological and engineering literature and detailed discussion here is inappropriate. However, in order to be at least aware of the influences which might have been operating during the deposition of those sequences where the nature of the river type appears to have been changing in time, it is necessary to outline briefly the effects of controlling factors.

3.10.1 Controlling variables and their effects

Data collected by geomorphologists and engineers both in the field and in the laboratory have led to the establishment of several empirical relationships between hydrological variables and channel morphology parameters, though often without any clear physical explanation of the relationships. The factors which appear to influence channel types most are (1) slope, (2) sediment type and discharge, and (3) water discharge and regime.

SLOPE

Plots of channel slope against bankfull discharge for rivers of different sizes show that meandering and braided rivers occur in fairly distinct fields (Leopold and Wolman, 1957; Ackers and Charlton, 1971; Schumm and Kahn, 1972). Whilst different sets of data show some variation in the positions of the discriminant lines between fields, it is clear that there is a transition from straight through meandering to braided with an increasing slope (Fig. 3.53).

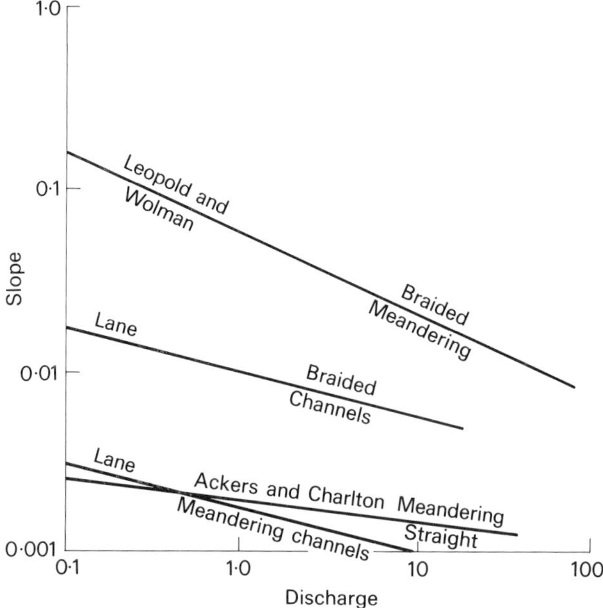

Fig. 3.53. Relationships between slope and discharge for various channel types showing discriminant lines between braided, meandering and straight channels (Leopold and Wolman; Ackers and Charlton) and typical lines for braided and meandering streams (Lane) (after Ackers and Charlton, 1971; Leopold and Wolman, 1957; Lane, 1957).

SEDIMENT TYPE AND DISCHARGE

The erodibility of bank sediment appears to exert an important control on channel type. With cohesive bank material, due usually to a high proportion of finer grained sediment, river channels tend to meander, whilst with highly erodible, non-cohesive bank material, channels tend to be braided (Schumm, 1963). Whilst the cohesive strength of the bank material, as expressed by the silt + clay percentage, is an important control, it has been argued that the nature of the sediment actually in transport is the critical factor, meandering being favoured by a high suspended sediment load (Schumm, 1968).

The amount of bed load sediment supplied to the river also influences channel type (Schumm, 1969). With an increase in bed load, channels widen and mid-channel bars form, leading to braiding. The nature and abundance of sediment load in a river depends on (1) the abundances of different types of bedrock in the source area, (2) the overall climate which influences the relative abundance of suspended load, and (3) the magnitude of sediment discharge which is influenced by relief, vegetation and run-off pattern. All these factors are intimately interrelated.

WATER DISCHARGE

The magnitude and distribution pattern of water discharge through a river channel both influence channel type. An increase in peak discharge values may cause a previously meandering river to widen and become braided, and braided streams commonly have more variable patterns of discharge than meandering streams. The discharge pattern, or river regime, can be related to the climatic setting (e.g. Beckinsale, 1969). The important aspects of regime appear to be the ratio of maximum to minimum discharge and the rates of change of discharge. The discharge pattern is not only a function of the pattern of precipitation in the source area but also depends upon the relief, bedrock type and vegetation cover of the catchment.

OVERALL CONTROLS

Slope, sediment discharge and water discharge are not independent variables. They are related in a complex network of links and feedback loops which involve tectonics and bedrock of the source, relief, vegetation and overall climatic setting. Their influences on channel types will therefore be far from clear cut and as more than one factor is likely to vary at once, particularly in response to climatic change, it will not always be possible to predict the extent to which a river will respond, and the best that can normally be done is to indicate in a qualitative way the likely direction of any change in channel pattern.

3.10.2 Examples of changes in channel pattern from geomorphology

Changes in channel type can take place over quite short intervals of time and may be deduced from historical records or read from relict features of earlier regimes on the land surface. The Cimarron River of Oklahoma and Kansas, for example, changed its channel pattern from meandering to braided with a dramatic increase of channel width in a matter of 30 years, triggered by a very large flood and developed by a period of drought which reduced floodplain vegetation. A later major flood appears to have initiated a period of floodplain restoration and a return to a narrower and more sinuous channel (Schumm and Lichty, 1963; Shelton and Noble, 1974).

Evidence of similar changes in channel type are preserved in the surface morphology of the Riverine plain of the Murrumbidgee River in New South Wales (Schumm, 1969). Here palaeochannels can be seen to have changed their width, depth and sinuosity in response to longer-term climatic changes which influenced not only the discharge but also the vegetation and sediment production.

3.10.3 Changes of channel type in vertical sequences

There is, in the geological literature, an increasing number of examples of sequences of alluvial sediments which show vertical changes of facies reflecting changing alluvial systems and therefore possibly changing hydrology, relief or tectonic activity.

In the Pre-Cambrian Torridon Group of north-west Scotland (Williams, 1966; Stewart, 1969) the lowest sediments, the Diabaig Formation, vary greatly in thickness and are confined to hollows in the underlying Lewisian basement. They are mainly fine sandstones and grey mudstones with mudcracks and shallow water ripple-marks. Breccias fringe and floor some of the basins. The formation seems to represent small playa or lacustrine basins, margined by small alluvial fans. Such basins represented local base levels and were filled at different times and at different topographic levels. The basin fills are overlain by the more widespread Applecross and Aultbea Formations which are composed of red sandstones and conglomerates several kilometres thick. These are interpreted, on the basis of palaeocurrent patterns, as the deposits of very large alluvial fans though the facies of individual sections indicate deposition by braided rivers, suggesting that the climate was humid. The lower part of the sequence is mainly conglomeratic with abundant channel morphologies preserved. These become less common upwards as the conglomerates give way to sandstones, presumably because the channels migrated more widely. The sequence has not been described in the detail

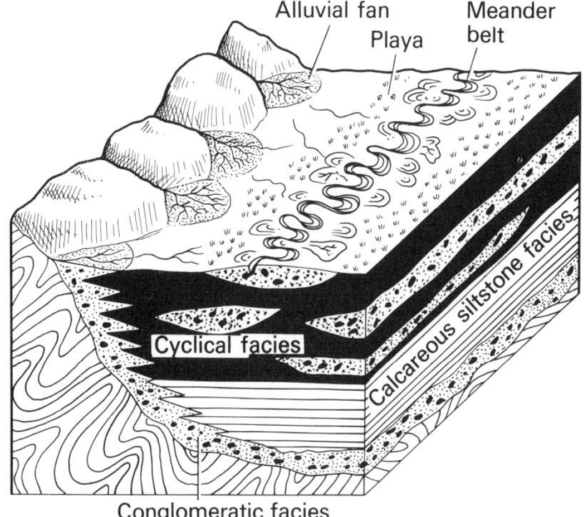

Fig. 3.54. Facies sequence and environmental interpretation of the conglomeratic and cyclical facies in the Old Red Sandstone of Anglesey, Wales (after Allen, 1965b). The calcareous siltstone facies is shown as representing a period of playa deposition prior to the arrival of the meander belt. Alternatively, it may have been deposited between the fans and the meander belt as the source area retreated.

necessary to reconstruct fully the channel types and determine their evolution. However, it appears to reflect a change from pebbly braided to sandy braided streams on fans supplied by a gradually degrading source area.

The Old Red Sandstone succession of Anglesey differs from that of the Torridonian in that the sandy fluvial system is separated from the pebbly alluvium by fine member deposits (Fig. 3.54) (Allen, 1965b). The basal conglomeratic facies, interpreted as alluvial fan deposits, passes upwards into siltstones with carbonate concretions and occasional thin conglomerates. These, in turn, are overlain by a thick sequence of cyclically interbedded sandstones and siltstones in which the sandstones commonly fine upward in units several metres thick. Epsilon cross-bedding occasionally occurs and the thicker siltstones have carbonate concretions. The calcareous facies can be interpreted as playa deposits or floodplain silts which gradually encroached on to the alluvial fan as relief subsided. Occasional sheetfloods still deposited conglomerates but accumulation was commonly slow or static, allowing caliche soil profiles to form. The cyclical facies are interpreted as floodplain and meandering channel sediments. They were deposited by streams flowing more or less parallel to the retreating scarp and unconnected with the drainage system which had earlier supplied the alluvial fans (cf. Fig. 3.34).

A more statistically based method of analysis, using changes in the variance of palaeocurrent directions and set thickness of cross-bedding, was used by Miall (1976) to deduce changes in channel type through a sequence of Cretaceous braided stream deposits in Arctic Canada. Within-channel variance of cross-bedding direction compared well with that of modern braided streams and a reduction of this parameter accompanied by an increase in set thickness was thought to reflect decreasing channel sinuosity. This, in turn, was related to tectonic rejuvenation of the source area.

These examples illustrate how alluvial facies and channel type may change through time and how a careful analysis of alluvial sediments can give an insight into the underlying controls upon deposition.

FURTHER READING

Leopold L.B., Wolman M.G. and Miller J.P. (1964) *Fluvial Processes in Geomorphology*. pp. 522. Freeman, San Francisco.

Chorley R.J. (Ed.) (1969) *Introduction to Fluvial Processes*. pp. 218. Methuen.

Goudie A. (1973) *Duricrusts in Tropical and Subtropical Landscapes*. pp. 174. Oxford University Press, London.

Gregory K.J. (Ed.) (1977) *River Channel Changes*. pp. 448. Wiley, Chichester.

Schumm S.A. (1977) *The Fluvial System*. pp. 338. Wiley, New York.

CHAPTER 4 Lakes

4.1 INTRODUCTION

A sedimentological account of lakes has a status similar to that of the hole in the doughnut. Ever since Gilbert's (1885) classic work on *The Topographic Features of Lake Shores,* lakes have played an important role in the development of ideas about deltaic, littoral, turbidity current and evaporitic processes, often acting as large-scale laboratories for the study of these processes. Consequently many aspects of lake sedimentation are covered in other chapters.

At the present day, lakes form about 1% of the earth's continental surface and are important sites of sediment accumulation since they intercept much of the sediment transported by rivers. In addition, particularly in basins of internal drainage, they may be the sites of chemical precipitation.

Being unaffected by the homogenizing effects of marine water, lake sediments, particularly in their chemistry, are sensitive to the influence of climate. Thus ancient lake sediments, particularly those of closed lakes, are valuable indicators of past climates and this feature will be emphasized in this chapter.

Views on the importance of lake deposits in the geological record have had a chequered history. Their importance was first recognized in the western USA, and many Old Red Sandstone and Triassic sequences in Europe were also interpreted as lake deposits. Some of these were subsequently reinterpreted as alluvial deposits when the importance of rivers as depositors of sediments was realized (Chapter 3). However, some of these reinterpretations are now seen to have gone a little too far and in some cases a further study may re-establish a lacustrine model.

Lake sediments are of considerable economic importance, particularly as sources of evaporitic minerals and of oil shales. In addition, it seems likely that some major iron ore bodies, particularly those associated with cherts in the Pre-Cambrian, may be lacustrine.

4.2 PRESENT-DAY LAKES

Present-day lakes are very variable in form, size and stability. The smallest lakes are those within environments such as coastal plains, alluvial plains, glaciated regions or deltas where lakes are an essential part of the environment and are formed by sedimentary damming, local erosion, solution, or a temporary cessation in sediment supply in a generally subsiding basin. These lakes are small, short-lived and rapidly filled. They are mainly discussed in other chapters. Similar temporary lakes occur within volcanic terrains.

Larger lakes are primarily tectonic in origin and they fall into two groups. (1) those which are formed in active tectonic areas either in extensional rift-valleys such as the East African and Baikal rifts or along strike-slip orogenic belts such as the Dead Sea; in these cases subsidence is rapid, sediment supply from nearby margins is frequently substantial, and the sedimentary fill is thick (perhaps 2 km in some East African lakes and 2–5 km for the Baikal rift) and rapidly deposited (see also Sect. 14.3.1). (2) those, such as Lake Chad and Lake Eyre, which are formed on long-lasting sags in cratonic areas; they persist for long periods of geological time, their margins fluctuate over hundreds of kilometres and clastic supply is relatively low.

Apart from clastic supply, sedimentation in lakes is affected by two principal factors: the chemistry of their waters and the extent of shoreline fluctuations. The salinity is governed both by the net water budget, which depends on the ratio of rainfall and inflow to outflow and evaporation, and by the ratio of precipitation to solution within the lake. The composition of lake water depends primarily on the bedrock lithology of the source area and local volcanicity. Lakes, such as Lake Chad, with no surface outflow, have shorelines which are extremely susceptible to fluctuations in water level, particularly if the climate changes. Others, such as the Great Lakes of North America and many large East African lakes, have a surface outflow which acts as a buffer against extreme fluctuations in lake level. Yet others, such as Lake Maracaibo, are connected to the sea which thus acts as a control on lake level, which, in turn, is affected by sea-level changes.

4.2.1 Lake water

Water properties and circulation patterns are important elements in the science of limnology and more extended discussions of many of the topics briefly reviewed here can be found in limnological textbooks (e.g. Hutchinson, 1957; Beadle, 1974).

WATER CHEMISTRY

In freshwater lakes in temperate climates and with a through drainage, the concentration of dissolved ions is usually small. Nevertheless, it plays a part in governing plant growth, thereby influencing levels of organic productivity and of dissolved oxygen in the water. High nutrient levels and high organic productivity may lead to rapid accumulation of organic debris on the lake floor and to oxygen depletion in the lower layers of lake water, a situation described as *eutrophic*. With low productivity the water column usually remains oxygenated, a situation described as *oligotrophic*. In tropical regions, even lakes with low levels of organic productivity may be reducing in their lower layers because of the strong thermal stratification (see below) (Beadle, 1974). River water entering a lake usually mixes eventually with the lake water but it may maintain its identity for some time before it is mixed. There may therefore be significant vertical and lateral variation of water chemistry within the lake, the pattern depending on how the waters circulate.

Saline lakes are those which normally contain water carrying more than $5^0/_{00}$ of dissolved solutes, the upper salinity tolerance of most freshwater aquatic organisms (Beadle, 1974, p. 264; Hardie, Smoot and Eugster, 1978). In lakes with internal drainage in semi-arid or arid settings, evaporite minerals may be precipitated because concentration of dissolved ions may equal or exceed those of sea water. $CaCO_3$ may also be precipitated at quite low concentrations if the pH is high. The composition of the dissolved material may be very different from that of the oceans and will depend on the nature of source rocks and local volcanic activity. Sulphate or nitrate levels may be particularly high in some cases and low in others (e.g. Dead Sea, Bentor, 1961). In the soda lakes of East Africa and South America, where particular types of volcanic activity occur, sodium, carbonate and halide ions may become highly concentrated and part of the dissolved material is supplied directly as hot brines from marginal and sub-lacustrine springs. A consequence of these ionic concentrations is the dramatically high pH which may exceed 10.

In such highly alkaline waters, the level of dissolved silica rises to more than 1,100 ppm (Eugster, 1970; Surdam and Eugster, 1976). These unusual brine compositions influence both the nature of the precipitated salts and the nature of chemical reactions on the lake floor between the brines and the clastic sediment particles. Increased salinities help to maintain the thermal density stratification of the lake waters (see below). This can lead to near-bottom conditions which are inhospitable to organisms and are possibly anoxic. Not all lakes with internal drainage show high ionic concentrations however. Lake Chad, for example, maintains low salinities apparently by the escape of more concentrated water as ground water through surrounding permeable soils (Dieleman and Ridder, 1964; Roche, 1970).

TEMPERATURE

The temperature of the water may vary greatly both between lakes and seasonally in the same lake. Most of the variation is in the surface layers, especially in deeper lakes. Many lakes exhibit density stratification, particularly where the lake is deep and there is a shallow sill. Where a thermocline develops, the colder lower layer remains relatively stagnant with most movement occurring in the upper warmer layer. In temperate latitudes density inversions can lead to seasonal mixing which oxygenates the bottom and allows the redistribution of nutrient salts. When the stratification is stable, as it commonly is in deep tropical lakes, the lower layer, the *hypolimnion,* will be anoxic, allowing preservation of organic matter on the lake floor (Fig. 4.1). This, in turn, leads to lower levels of productivity in the upper oxygenated layer, the *epilimnion,* because of the lack of redistribution of nutrients. Even in tropical lakes, however, the stability of thermal stratification varies with weather. Strong winds can agitate the upper layers and lead to mixing of the epilimnion with the upper part of the hypolimnion, with consequent lowering of the thermocline.

Temperature stratification greatly complicates the patterns of water movement in lakes so that each layer may have its own circulation pattern. Temperature-controlled density differences between lake and river waters are important in determining the distribution pattern of river-borne sediment in the lake. Both hyperpycnal (under) flow and hypopycnal (over) flow occur

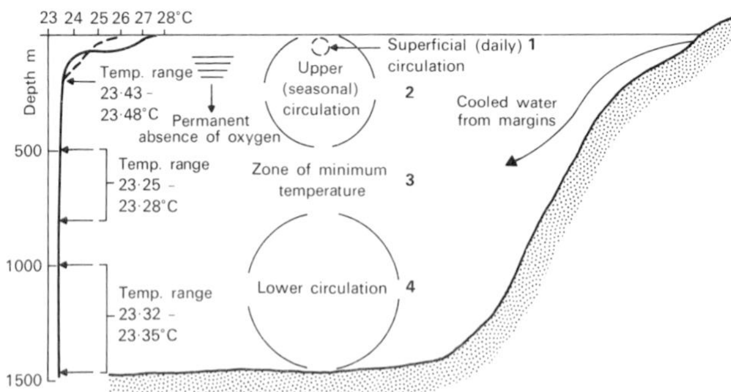

Fig. 4.1. Temperature profiles from Lake Tanganyika showing suggested regions of circulation and an 'epilimnion' down to about 50–80 m subject to daily circulation (1), a 'metalimnion' down to at least 200 m, subject to seasonal circulation (2) and a 'hypolimnion' which is free of oxygen and has a more or less uniform temperature (3 + 4) (after Beadle, 1974, Fig. 6.2).

(cf. Fig. 6.17) and, with density stratification, sediment-laden river water may even flow along the thermocline. River water dispersion commonly varies seasonally, sometimes showing overflow and sometimes underflow.

WATER LEVEL

Water level in lakes fluctuates on several scales (Fig. 4.2). In the short-term, seasonal factors control the discharge of tributary rivers and the level of evaporation. In the long-term, it may be the growth or decay of the outlet barrier or longer-term climatic changes which govern lake levels. Variations are greater in those lakes with internal drainage, since an outlet moderates the effects of changes in the water budget. The fluctuations in the area of Lake Chad during the present century from about 25,000 to 10,000 km^2 have been brought about by climatic changes. However, the size of the area is also a function of the very gentle topography of the lake basin. On at least three occasions in Pleistocene times, Lake Chad was over 300,000 km^2 in area due to longer-term changes of climate (Grove, 1970; Servant and Servant, 1970).

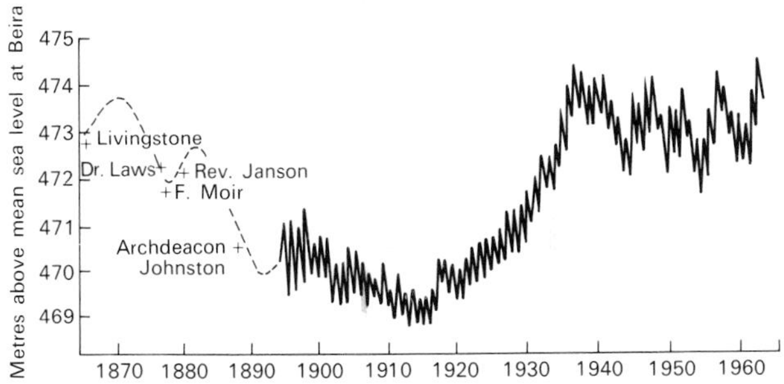

Fig. 4.2. Lake Malawi water levels: water level varies annually between 0.4 and 1.7 m and over a 20-year period has risen 6 m. After fall in water level the outlet was blocked by a barrier of sand, silt and vegetation which was not broken through until 1935. Since then, water level has not significantly fallen (after Beadle, 1974, Fig. 17.1).

WATER MOVEMENT

Lakes have complex patterns of water movement involving tides, waves and currents. Tides are only of significance in the very largest lakes and even there they only give rise to an oscillation in water level and never to strong currents.

Waves, on the other hand, transport sediment, assist in the mixing of lake water and cause large-scale water movements. At the shoreline coarser sediment is moved and reworked and fine sediment is moved off-shore (e.g. Norrman, 1964). Wind-driven waves develop beaches, spits and cliffs similar to those in shallow seas. The effects of waves on the shore will vary with wave fetch and the orientation of the shore with respect to the effective wind directions.

Waves are important in mixing the upper layers of lake water, thus countering the tendency of lake waters to become thermally or otherwise density stratified. Their effectiveness will depend on wave fetch and the strength and duration of winds. Deep tropical lakes show marked changes in the sharpness and depth of the thermocline in response to seasonal variations in wind strength (see Beadle, 1974, pp. 58–87, for a review).

Wind-driven currents and low atmospheric pressure cause large-scale wave-like oscillations of the water mass (Fig. 4.3). Such waves, termed *seiches,* result from the piling up of water at one end of the lake. Release of the force which causes the piling allows the water to propagate a large-scale wave-like motion along the length of the lake. Rebound from either end leads to oscillations of water level with a periodicity controlled by lake dimensions. In European lakes vertical amplitudes may be over a metre. As well as surface seiches, similar waves may occur internally by deformation of the thermocline.

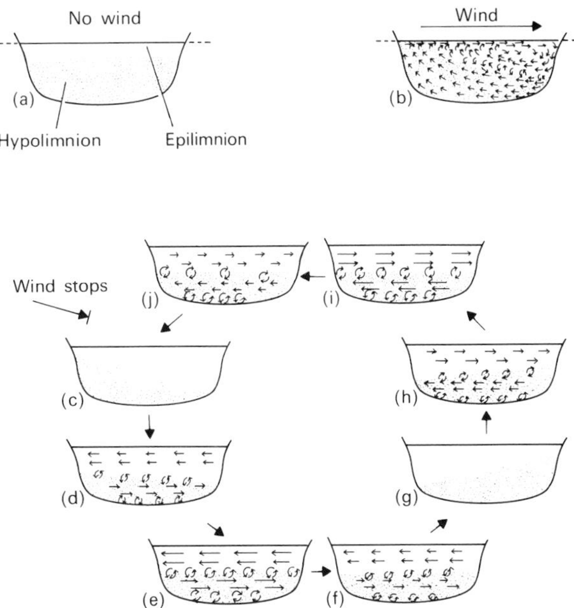

Fig. 4.3. Effect of a steady wind-driven current in a two-layered stratified lake. (a) No wind, (b) windward end of the lower layer is tipped upwards and its upper surface is eroded by the return current in upper layer, (c) wind stops and lower layer is reduced because part of it has been stirred into the upper layer, (d)–(j) one cycle of an internal oscillation wave (seiche) following cessation of wind to illustrate how opposing currents in the two layers cause turbulent mixing within layers, between them and along the bottom. Lengths of arrows and diameters of circles indicate the magnitude of velocities and the intensity of turbulence (after Beadle, 1974, Fig. 6.1).

Geostrophic effects may complicate the deformation by introducing a transverse component.

Current systems in lakes are very complex. Though they are seldom strong enough to transport sediment as bed load, they have very important influences on the dispersion of the finer sediment in suspension and upon the mixing patterns of the water masses. Currents are most commonly driven by wind drag, by atmospheric pressure inequalities, by the inertia of river inflow and by the simple through-flow of water from one end of the lake to the other. The last type is probably the only one of significance in small narrow unstratified lakes, but the others play a greater part with increasing lake size. Wind-driven currents are the most important in most large lakes. The shear of the wind on the water surface sets up a surface current and also causes the piling up of water at the down-wind side of the lake. Geostrophic effects deflect the water current to the right in the northern hemisphere and to the left in the southern.

These Ekman spirals are of great importance in determining lake circulation. The piling up of water in a lake leads to a compensating underflow either close to the lake floor or, in a stratified lake, at the thermocline. This flow will also develop an Ekman spiral and it is probable that shear at the thermocline may generate currents in the lower water layers. Very large lakes, such as the Great Lakes of North America, show large-scale gyral circulation patterns in the surface currents with components paralleling the shores.

Inertial flows from water introduced to the lake by rivers can occur at any level within the lake, depending upon the relative density of the two types of water, something which can change with the seasons. Such currents also show deflections to the side from geostrophic effects.

4.2.2 Sediment supply

Most of the clastic sediment deposited in lakes is transported there by rivers either in suspension or as bed load. Wind blown, ice rafted, and volcanic material may be locally important. In addition to the solids, material also arrives in solution though, except in lakes of internal drainage or with very high evaporation rates, most of this passes through the system without adding to the accumulating sediment. A certain amount of soluble material may be fixed by organisms growing in the lake and in surrounding swamps. These make a significant contribution both to the volume of accumulating sediment and to its oxidizing or reducing state. In marginal swamps, particularly in the tropics, water and bottom sediment conditions depend strongly on the type of vegetation. With open water and no high-growing subaerial plants, sub-aqueous photosynthesis is high and the water is well oxygenated. In swamps with high-growing plants such as papyrus, little light penetrates to the water surface. The lack of photosynthesis and the accumulation of falling canopy plants leads to peat accumulation in reducing bottom conditions (Beadle, 1974).

Not all sediment found on the floors of present-day lakes is a product of present processes. In glaciated regions, thick deposits of glacial and glacio-fluvial sediments may act as internal sources of sediment for reworking by lake processes. In other cases earlier deposits have been winnowed to a resistant pavement of lag gravels which can no longer be moved by lake currents or waves (e.g. Norrman, 1964).

The supply of sediment to lakes varies with river discharge which is often seasonally controlled. The seasonal contrast is particularly marked in high latitude lakes fed in whole or in part by glacial meltwaters. Here the extremely high load of summer months contrasts with the small and almost clear-water discharge of the winter. Supply to lakes may also fluctuate considerably in 'monsoonal' climatic settings.

4.2.3 Clastic sediment deposition

DEEPER WATER DEPOSITION

Clastic deposition in the deeper parts of lake basins is almost entirely from suspension. The particular nature of the distribution will depend upon the density stratification, if any, of the lake water and the density of the river supply (Fig. 6.17). Where the river water spreads out at the surface, coarser fractions of the suspended load will be deposited first and there will be a distal fining of sediment away from the river mouth in a way similar to that found in marine deltas. In saline lakes, flocculation of clays may accelerate settling, as in marine settings. The pattern of dispersion will be influenced by deflections of the inertial flow, by geostrophic effects and by surface currents. A coincidence of inertial flow direction and wind direction, as can happen in glacially-fed lakes with katabatic winds off the glacier, may lead to the strong development of Ekman spirals and to the asymmetrical deposition of sediment across the lake, creating a transverse slope to the lake floor.

Where water densities allow underflow of the river water, coarser sediment may be spread out over the lake floor. The density currents which operate in this way owe some of their excess density to turbidity but differ from episodic turbidity currents in being effective longer and in the main driving mechanism being temperature contrast. Clear-water underflows may also exist and be capable of reworking sediment. Such underflow systems may eventually lead to the development of a subaqueous fan in front of the river mouth with a system of channels, levees and lobes similar in many respects to submarine fans (e.g. Houbolt and Jonker, 1968). Slump-generated turbidity currents can also occur in lakes. Most described examples at the present day seem to be related to sites of dumping of industrial wastes. However, heavily laden rivers, feeding sediment to steep-sided lakes, may cause periodic oversteepening of the depositional surface leading to slumping.

The graded beds produced by river-generated underflow and slumping are, at least in temperate regions, introduced by irregular catastrophic floods and are not seasonally controlled. In Arctic regions, and in monsoonal climates, laminated lake floor sediments are more likely to be seasonal. Where glacially supplied these sediments are termed 'varves' or 'rhythmites' (Sect. 13.4.3). Fluctuations in supply may also be brought about by variations in phytoplankton, particularly diatom, productivity.

The nature of the bottom water influences the post-depositional history of the sediments. In lakes with reducing conditions at the bottom, sediment may have a high organic content and lamination is likely to be preserved, though escape of gases from the decay of organic matter can disturb lamination. Sulphide minerals may be precipitated due to the activity of sulphate-reducing bacteria. Oxygenated bottom waters may allow the existence of a benthic fauna and there may be disturbance of the lamination by burrowing organisms.

MARGINAL DEPOSITION

Lake marginal clastic sedimentation is concentrated around the mouths of rivers. Elsewhere beaches, spits and barriers may be formed by wave action, the processes and products differing little from their marine counterparts (see Chapter 7). The generally lower wave energy in lakes and the common fluctuation of water level are likely to produce sediments which are less well-sorted and less mature compositionally than on marine shorelines.

Deposition at the river mouth will depend upon the relative importance of suspended load and bed load, and upon the density relationships of river water and lake water. With overflow of river water (Fig. 6.17) and a dominant suspended sediment load, a delta similar to those forming in marine settings will develop. Such deltas are usually river-dominated on account of the relatively low levels of basinal energy in lakes. When bed load is important and it is rapidly dumped at the river mouth the delta slope may approach the angle of rest and a Gilbert-type delta (Fig. 6.1) may result.

With underflow of river water, delta formation will be inhibited as most of the sediment is transferred directly to the deeper parts of the lake floor (Houbolt and Jonker, 1968).

In areas of high topographic relief, alluvial fans may build directly into lakes. At the distal margin of the fan a steep slope develops, probably at the angle of rest, so that the overall sequence might be that of a Gilbert-type delta with steep foresets overlain by alluvial fan deposits. Lake fan deltas of this type are not well described, although the Laitaure Delta in Arctic Sweden may be regarded as a sandy example (Axelsson, 1967). If the slope into deep water is steep, little material is likely to be stored in a delta or fan at the shore and coarse debris may be carried into deep water by sliding and other mass flow mechanisms. If water level drops, shoreline features are abandoned to leave terraces at the lake margins. These may be erosional or depositional in origin and can include cliff lines, beach ridges, spits, delta lobes and a variety of small-scale terrace features all of very low preservation potential (e.g. Gilbert, 1885; Donovan and Archer, 1975). Not all lake margins are depositional and it is not uncommon for the deep water sediments to abut directly against an eroded bedrock margin.

4.2.4 Chemical and biological sediment deposition

The nature of sediments precipitated and deposited by chemical and biological activity is closely linked with the water chemistry, the drainage pattern and the climatic setting of the lake. Whilst chemical and biological sedimentation does take place in lakes with a through drainage, high rates of chemical sedimentation involving evaporites are most likely to occur in lakes of internal drainage, which preferentially occur in hot arid or semi-arid settings.

HIGH LATITUDE LAKES

Some Arctic or sub-Arctic lakes accumulate limonite in marginal marshy areas, particularly in positions close to mouths of inflowing streams. Such precipitation is thought to be due to the activity of bacteria (Harder, 1919) and is often in the form of pisoliths. It tends to become less abundant towards the central parts of the lakes. In some Arctic lakes, diatoms constitute an important component in the overall lake sediment.

Lakes which are frozen to the bottom in winter may have no effective sedimentation because the frozen bottom sediment is uplifted to the surface when the lake is inundated by meltwater in the spring.

TEMPERATE LAKES

Calcium carbonate may be precipitated if there are substantial amounts of calcareous rocks in the catchment area. In addition, certain macrophytic plants, which grow in marginal settings, precipitate coatings of low-Mg calcite on their leaves and stems and this is eventually added to the surface sediment at leaf fall or when the plant dies. More important, however, are the activities of blue-green algae which, by coating already existing clasts, cause both the precipitation and binding of calcium carbonate as grain coatings. With time these coated grains develop into oncolites. Those smaller than a certain size move in response to wave action and therefore grow in a roughly spherical form whilst larger ones are more stable and

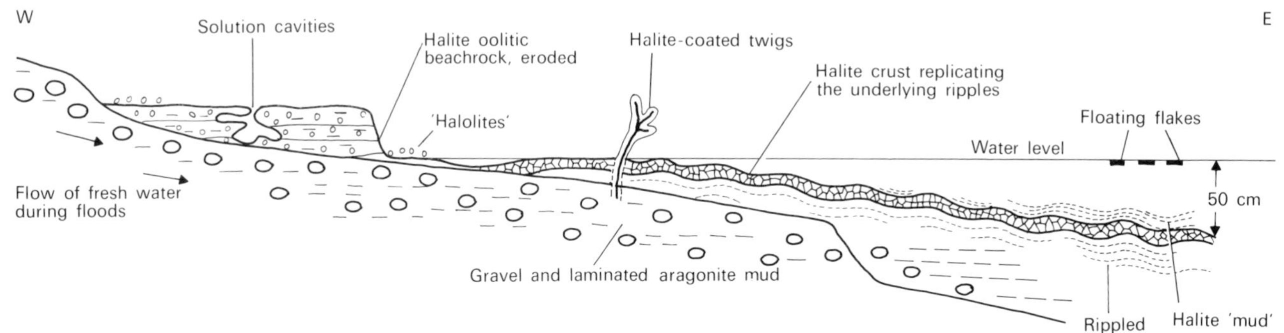

Fig. 4.4. Schematic section through the beach zone of a natural evaporitic pan within an artificially dammed part of the southern Dead Sea showing interrelations between environments of deposition and halite sedimentological features (not to scale) (after Weiler, Sass and Zak, 1974).

become disc-shaped. In Lake Constance, they occur most abundantly in a marginal belt between the high energy shoreline and deeper water where carbonate production is slower. The oncolites are associated with and closely overlie a crumbly carbonate sand which is probably of mixed origin, some grains being coated detrital grains and others crumbled oncolites (Schöttle and Müller, 1968). The fact that intact oncolites lie on the sand and not apparently within it suggests that they have a limited preservation potential. Yet another source of carbonates are molluscs which secrete aragonite shells and which are usually confined to the littoral zone. They may form the nuclei for oncolite growth.

On deeper lake floors, carbonates commonly occur as friable marls finely interlaminated with organic-rich layers. These develop seasonally, often as a result of seasonal precipitation by phytoplankton, and particularly where the water column is so stratified that the bottom layers are stagnant and reducing bottom conditions occur (Davis, 1900; Nipkow, 1928; Thiel, 1933). In many temperate lakes, centimetre-scale ferromanganese concretions occur, typically associated with sand and gravel substrates (Calvert and Price, 1977).

ARID LAKES

Arid lakes of internal drainage are commonly the sites of abundant chemical precipitation. Here we discuss only perennial saline lakes, that is those that persist for many years (tens, hundreds or even thousands) without drying up (Hardie, Smoot and Eugster, 1978). Playa lakes, or ephemeral saline lakes, are discussed in Chapter 5. Perennial saline lakes may be shallow, for example Great Salt Lake, Utah (about 12 m), or deep, for example the Dead Sea (about 400 m). Shallow lakes may rapidly turn into playas if the climate changes.

In perennial saline lakes, evaporation leads to concentration of the surface brine and the nucleation of saline minerals in the surface water. The concentrated brine and saline minerals sink toward the bottom of the lake and less dense, less concentrated water flows in over the brine. As evaporation continues, and assuming an increasingly arid climate in a non-alkaline lake, a sequence may be produced which gives (Hardie, Smoot and Eugster, 1978):

4 Halite + minor gypsum + traces of carbonates
3 Gypsum + minor carbonate (e.g. present Dead Sea)
2 Alkaline earth carbonates
1 Fresh water organisms (perhaps with marginal saline lagoons)

In the Dead Sea gypsum and aragonite are now being precipitated (Neev and Emery, 1967), but borehole evidence shows that, in early Holocene times, large thicknesses of rock salt and minor amounts of gypsum were deposited. At the present day, halite precipitation is confined to artificial evaporating basins at the southern end of the lake, where conditions simulate extremely arid natural conditions (Bentor, 1961; Weiler, Sass and Zak, 1974). Halite precipitates and accumulates in a variety of forms, which depend upon the level of agitation of the water (Fig. 4.4). In quiet off-shore areas, it accumulates as a crust by precipitation onto floor-nucleated crystals. Where wave action is more effective, surface-nucleated halite crystals occur and settle as a halite 'mud' which may be rippled. At the shoreface itself, the wave action maintains particle motion and causes them to grow as halite ooliths (Halolites) which accumulate as small beach ridges of low preservation potential (Weiler, Sass and Zak, 1974).

In the East African soda lakes, such as Lake Magadi, where volcanic activity leads to high concentrations of sodium carbonate and to the addition of abundant volcanic debris, complex and unusual precipitates and reaction products occur (e.g. Hay, 1968; Eugster, 1969; Surdam and Eugster, 1976). The most abundant precipitate, trona ($Na_2CO_3NaHCO_3.2H_2O$) occurs as a thick (40–60 m) layer beneath most of the lake floor. This sequence has some interbedded tuffs and clays which contain authigenic zeolites of which erionite ($NaKAl_2Si_7O_{18}.6H_2O$) is the most abundant, suggesting a complex series of reactions between fragments of volcanic glass and the lake brines. Fluorite is also common in the trona layer, reflecting reactions between the fluorine-rich brines and calcium in the sediment which is probably mainly present in detrital calcite (Surdam and Eugster, 1976). In the lower parts of the lake-floor sequence are layers rich in magadiite ($NaSi_7O_{13}(OH)_3.3H_2O$) and kenyaite ($NaSi_{11}O_{20.5}(OH)_4.3H_2O$). These are thought to result from annual periods of dilution of surface brines when the consequent lowering of pH caused the high level of dissolved silica to be partially deposited. Though present-day brines are saturated for both minerals, neither is being formed. This suggests that the magadiite and kenyaite deposits reflect a slightly more humid regime in the past.

Bedded and nodular cherts occur both below the magadiite layers of the lake basin and as their lateral equivalents around lake margins. Chert beds pass laterally into beds of magadiite over distances as small as 50 cm and are thought to result from the alteration, possibly by freshwater leaching, of the silicates. The reaction can possibly take place very early in post-depositional history as chert chips occur in some magadiite beds. Where subaerially exposed, around lake margins, the cherts may occur in association with large-scale polygonal surface patterns of cracks which show features of soft sediment extrusion. These features formed when the sediment was a paste of magadiite before its conversion to chert. Nodular chert commonly shows a highly characteristic surface pattern of small-scale cracking.

In lakes with a high sulphate concentration, such as Soda and Searles Lakes, California (Eugster and Smith, 1965), sodium sulphate may precipitate both on the lake floor and on the margins. At low temperatures, mirabilite ($Na_2SO_4.10H_2O$) is less soluble than halite and in winter it precipitates before halite in the Great Salt Lake, Utah. It dehydrates in the summer to thenardite (Na_2SO_4) in hot arid environments but remains as mirabilite in higher latitude lakes.

Internal drainage lakes are highly susceptible to quite small-scale changes of climate, particularly in the level of precipitation. Shallow ones are likely to dry out entirely from time to time and the chemical deposits of the perennial lakes may be interbedded with those of playa origin.

4.3 ANCIENT LAKE SEDIMENTS

4.3.1 General criteria of recognition

Apart from their relationships with other facies there are three groups of criteria which should help to identify ancient lake deposits: faunal, chemical and physical (see Feth, 1964; Picard and High, 1972 for review).

The first criterion is an absence of a marine fauna in deposits formed primarily from suspension combined, if possible, with the presence of a non-marine fauna such as gastropods, some types of bivalves, some types of ostracods, certain fish and vertebrate fossils. Because lakes are sensitive to frequent climatic changes, faunas are under high stress and usually therefore of low diversity. However, since similar faunas are found within fluvial channels and associated environments they are not by themselves diagnostic of permanent lakes and the distinction between the two is often difficult.

The chemistry of the deposits may be diagnostic because certain abnormal minerals or mineral assemblages can only have formed from waters whose chemistry has diverged greatly from that of normal sea water. In addition, fluctuations in the chemical composition and salinity of lake waters are several times more rapid and more substantial than any that have been postulated to have taken place in the oceans over geological time, and such fluctuations are also recorded in the sediments. In particular, annual varves are common.

Physical processes are in general similar to those found in marine basins. However, evidence for tidal currents should be absent and the shoreline facies should show evidence of very much reduced wave activity compared to marine shoreline deposits. Thus ancient lacustrine deltas tend to be elongate with well-developed mouth bar facies and repeated progradational sequences because the deltas were fluvial-dominated. Shoreline sediments are compositionally immature, with quartzose sandstone rare. Signs of emergence are common, reflecting the frequent, even annual, oscillations in lake level. Individual shoreline facies such as beaches, lagoons and barrier islands are never thick because the frequent longer-term oscillations in lake level lead to a rapid interdigitation of shoreline facies. In deeper water, in lakes where salinities are less than normal seawater, hyperpycnal flow (Fig. 6.17) may result in more or less continuous density currents flowing down the delta front and giving turbidites with less well-developed Bouma sequences than in marine environments.

In the following section examples are chosen to illustrate a range of climatic and tectonic settings. The most arid example is the Green River Formation. More humid lakes are represented by the East Berlin Formation, the Escuminac Formation and the Orcadian basin. In all the examples the climatic fluctuations are reflected in changing facies or shorelines; in addition, the Lockatong Formation, which is entirely fine-grained, shows alternations between through-drainage and closed-basin conditions.

The Green River Formation was formed in a broad, shallow, long-lasting depression. The others were shorter-lived and governed by fault-bounded margins.

4.3.2 Green River Formation (Eocene) of the Western USA

The Eocene sediments of Wyoming, Colorado and Utah comprise one of the largest and most extensively described sequences of ancient lake deposits. They have been studied in great detail not only because of their unusual and complicated nature but also because they contain some of the world's largest reserves of oil shale and trona (sodium carbonate). They are, however, only one of many ancient lake sequences which developed through the western United States between the Triassic and the Pleistocene (see Feth, 1964, for review). The Green River Formation itself occurs in several basins which were downwarped during the Eocene and which probably correspond to separate lake basins. The longevity of the lakes attests to a rather stable tectonic regime and the wide fluctuations in areal extent of the lake deposits through time can best be explained by long-term climatic changes.

LAKE GOSIUTE

The Green River Formation of the 'Lake Gosiute' area is underlain and overlain by dominantly fluviatile sequences (Bradley, 1964). It also interfingers with them around its margins with tongues up to several kilometres long (Fig. 4.5). It is divided into three lacustrine members, each of which reflects different extents of the lake water and different conditions in the lake. The control on these members seems to have been mainly climatic.

Until quite recently, the presence of thick extensive oil shales, in which delicate interlamination of kerogen and carbonate (dolomite and calcite) are preserved undisturbed, had been taken as evidence for a deep permanently stratified lake with a stagnant, anoxic hypolimnion. Similarly, the occurrences of trona beds were explained in terms of stratification of a deep lake by comparison with present-day East African lakes (e.g. Bradley and Eugster, 1969). In this interpretation the oil shale laminae were regarded as seasonal varves with the carbonate reflecting the dry season and the organic layer the wet season with strong algal productivity. However, this model has now been revised and present views envisage an essentially playa setting with shallow and rather ephemeral lakes (Eugster and Surdam, 1973; Eugster and Hardie, 1975; Surdam and Wolfbauer, 1975).

So far this revised model has been applied to the Tipton Shale and Wilkins Peak members. Its applicability to the Laney Shale Member has not been demonstrated but a shallow permanently oxygenated lake at this stage is suggested by the presence of well-preserved fossil catfish (Buchheim and Surdam, 1977).

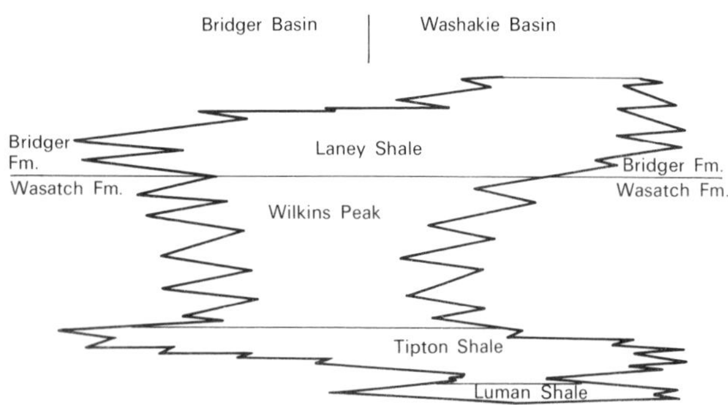

Fig. 4.5. Generalized stratigraphy of the Green River Formation in Bridger and Washakie (= Lake Gosiute) basins of Wyoming (after Surdam and Wolfbauer, 1975).

The facies

Six major lithofacies are recognized in the Wilkins Peak Member (Eugster and Hardie, 1975) and this scheme seems also broadly applicable to the Tipton Shale with the addition of stromatolitic limestone (Surdam and Wolfbauer, 1975).

Flat pebble conglomerate facies: Pebbles of torn-up dolomitic mud occur in beds up to 20 cm thick, which usually rest on mud-cracked mudstones of similar composition. The beds are traceable for distances of 30 km and therefore suggest a transgressive lag due to lake expansion over a mudflat of low relief.

Lime sandstone facies: Beds between 10 cm and 2 m thick can be traced for 35 km and comprise interbeds of wave-rippled calcarenites and mud-cracked dolomitic mudstone. Alternating periods of shallow agitated water and emergence are suggested. Radiating clusters of penecontemporaneous trona crystals of uncertain origin occur in the sands. The crystals may originally have been gaylussite ($Na_2Ca(CO_3)_2 . 5H_2O$), a common authigenic mineral in modern alkali lakes (e.g. Jones, 1965).

Mudstone facies: Laminated and thin-bedded dolomitic mudstones are characteristically strongly mudcracked, some cracks showing complex histories of opening and filling. They attest to desiccation which at times must have been so intense as virtually to brecciate the rock. An exposed playa mudflat, covered by occasional flood sheets which introduced the fine graded laminae as traction load, is envisaged. The thin beds perhaps represent deeper flood waters in playa depressions.

Oil shale facies: Two subfacies comprise the so-called 'oil shale'; organic-rich dolomitic laminites and oil shale breccias. The common occurrence of both mudcracks and breccias in all parts of the basin suggests accumulation of organic oozes in highly productive shallow water bodies which periodically dried out, causing preservation of very delicate algal, fungal and insect remains by heat fixation (Bradley, 1973). The breccias are thought to result from reworking of the desiccated ooze and the dolomitic interlaminae are interpreted as detrital silts washed in from surrounding mudflats during high water level. If this is the case, then the lamination is not a varving but a more random, flood-controlled effect.

Trona-halite facies: At least 42 beds of trona ($Na_2CO_3 . NaHCO_3 . 2H_2O$) most over 1 m thick and one 11 m thick are present in the Wilkins Peak Member (Culbertson, 1971). They are each areally widespread in the central part of the basin and commonly occur above a bed of oil shale. Halite is sometimes associated with the trona, either mixed with it or as a separate bed. Accumulation is thought to have compared with conditions in present-day soda lakes such as Lake Magadi (e.g. Eugster, 1970). Dolomite partings may reflect flood events or temporary expansions of the lake.

Siliciclastic sandstones facies: Several tongues of sandstone up to 10 m thick occur within the Wilkins Peak Member and they can be traced for distances of up to 100 km. The sandstones are well-sorted and commonly occur as mutually erosive, channel filling, fining-upward units, cross-bedded in their lower parts and ripple cross-laminated in their upper parts. They are interpreted as braided stream deposits.

Volcanic tuffs: Ashes are common either as distinct tuff beds or mixed with the carbonate mudstones. They do not contribute greatly to the general environmental model but the more persistent beds provide useful time lines in basin analysis. Several of them are strongly altered to zeolites such as analcime ($NaAlSi_2O_6 . H_2O$) or to authigenic feldspars. These authigenic minerals compare closely with the reaction products of tuffs in present-day alkali lakes (e.g. Hay, 1968; Eugster, 1969).

Stromatolitic limestones: Two units occur in the Tipton Shale Member and shows a range of stromatolitic growth forms from distinct individual algal heads through less domal types to rather planar types (Surdam and Wolfbauer, 1975). They can be traced continuously over long distances and are thought to reflect hypersaline conditions at the margin of the lake, where salinity inhibits predators (cf. Logan *et al.*, 1974).

Sequences (cycles) and environment

Within the Wilkins Peak Member, the facies succeed one another in sequences which show internal organization and have been called 'cycles' (Eugster and Hardie, 1975). Four basic types are recognized (Fig. 4.6). They are up to 5 m thick and can be traced for upwards of 20 km without significant thickness change. Individual components within cycles may, however, change laterally so that a type II may pass laterally into type I or type III. Type IV with a trona unit is confined to central areas of the basin.

The cycles are interpreted as transgressive–regressive couplets controlled by the short-term fluctuations of lake level in response to climatic change. The flat pebble conglomerates and lime sandstones are both transgressive units produced at the shore-line during different rates of expansion. The mudflats which produced the mudstone facies surrounded the central area of the basin and were normally subaerially exposed. They were only subjected to periodic flooding. They gradually extended their area into the centre of the basin during regressive phases while the oil shale, which accumulated in the topographically lower and therefore more permanently submerged area, contracted (Fig. 4.7).

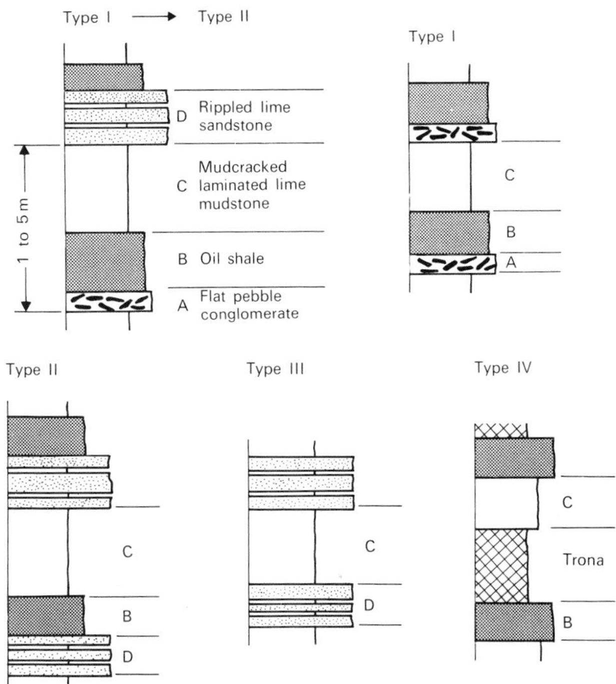

Fig. 4.6. Depositional cycles in the Wilkins Peak Member of the Green River Formation (after Eugster and Hardie, 1975).

The trona and halite deposits are thought to have precipitated from alkaline brines which resulted from the evaporative concentration of groundwater and spring water and the re-solution of efflorescent crusts. Precipitation of calcite and dolomite in more marginal areas led to high concentrations of alkali metal ions and increased pH to values where trona could precipitate. Seasonal variation led to flood and dry periods in the lake with resultant precipitation and re-solution of trona, with net accumulation (cf. Eugster, 1970). The climatic changes which led to these wet/dry cycles probably operated on a time scale of 20,000–50,000 years.

All of this expansion and contraction during the deposition of the Wilkins Peak Member took place, however, during what was, on a large view, a dry period. During deposition of the

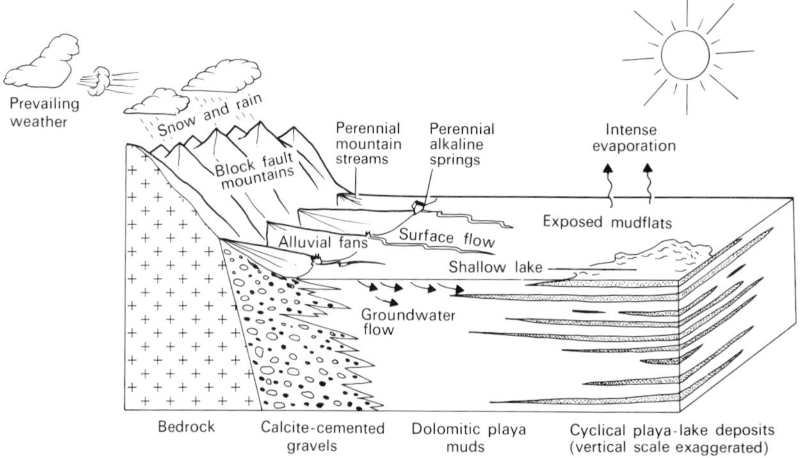

Fig. 4.7. Schematic block diagram showing general depositional framework envisaged for Wilkins Peak Member (after Eugster and Hardie, 1975).

Tipton Shale and Laney Shale Members the lake area was even more extensive, though again very shallow and largely explicable in terms of the playa model. Oil shale deposition was more extensive and persistent though still confined to central areas of the basin. The shorelines fluctuated, as demonstrated by the stromatolitic limestones which change their character as wave energy at the shoreline reduced in response to waning wave fetch (i.e. lake width) (Surdam and Wolfbauer, 1975). In these expanded lakes a broader facies pattern from marginal to central (lacustrine) can be drawn (Figs. 4.8 and 4.9).

Fig. 4.8. Schematic sections for lithofacies of marginal, mudflat and lacustrine environments of the Green River Formation (after Surdam and Wolfbauer, 1975).

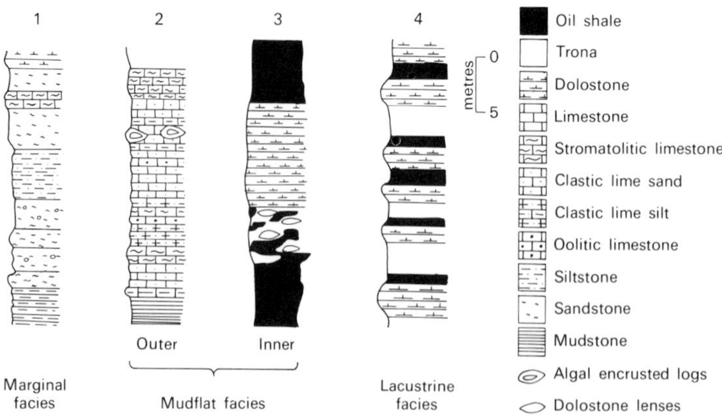

OTHER GREEN RIVER BASINS

Rocks assigned to the Green River Formation also occur in basins adjacent to the Bridger Basin. In the Piceance Creek Basin a similar pattern of sedimentation to that of the Wilkins Peak Member seems to operate (Lundell and Surdam, 1975). In the Uinta Basin, a cyclic sequence also occurs (Picard and High, 1968) and it seems likely that a similar interpretation may also apply there, though the original interpretation envisaged a central lake area with water up to 60 m deep.

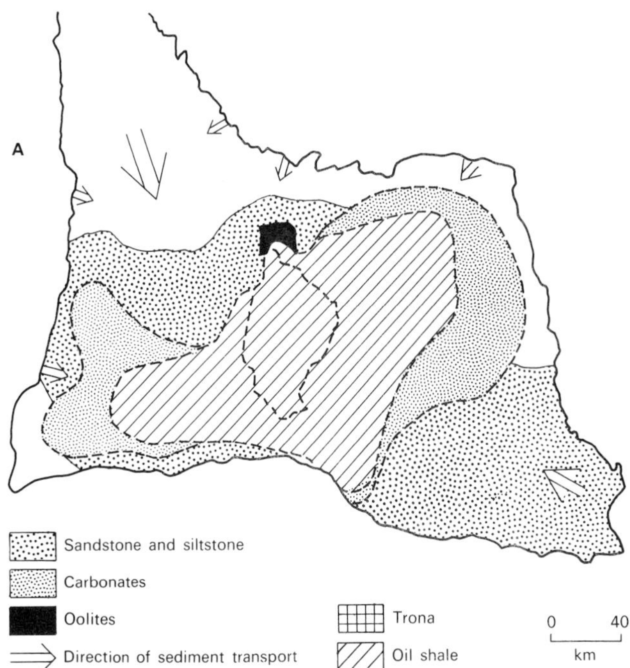

Fig. 4.9. Distribution of 'Lake Gosiute' lithofacies during (A) high stand in middle Tipton time and (B) low stand in middle Wilkins Peak time. Lacustrine facies represented by oil shale or trona; mud-flat facies by carbonates; marginal facies by sandstone and siltstone (after Surdam and Wolfbauer, 1975).

4.3.3 Triassic basins of Eastern North America

North-eastern USA and Eastern Canada were, in Triassic times, the site of intracontinental rifting and development of fault-controlled basins (Sect. 14.4.1). The basins, as preserved, are elongated roughly parallel to the present coast and appear to have operated as independent depositional basins (Klein, 1969). Three major basins are recognized on land; in New Jersey, in the Connecticut Valley and in Nova Scotia (Klein, 1962). Offshore, comparable basins may also exist, below the continental shelf (Fig. 14.14).

THE LOCKATONG FORMATION

The Lockatong Formation of the Newark Group of New Jersey is up to about 1,200 m thick and occupies a roughly central position in the basin. It wedges out laterally into alluvial fan deposits and is underlain and overlain by fluviatile sediments of the Stockton and Brunswick Formations. The Lockatong Formation itself consists entirely of fine-grained sediment, mainly in the clay and silt grade with some very sparse fine sandstones. These fine sediments are unusual in having a high content of the sodium-rich zeolite, analcime, particularly in the upper part of the formation, even though there are no volcanic rocks within the sequence.

Facies

Well-developed small-scale cycles or sequences, a few metres thick, characterize the formation (Van Houten, 1964). These vary in composition and have been divided into two main types, (a) 'detrital cycles' and (b) 'chemical cycles' (Figs. 4.10, 4.11). Within these cyclic sequences, five main rock types or facies have been recognized. These are highly intergradational and their recognition is based on quite subtle textural and structural differences.

1 The most abundant rock type or facies is tough, massive, medium to dark grey homogenous mudstone. Texture and composition differ between the detrital and the chemical cycles. In detrital cycles, wispy laminae of carbonaceous matter and of feldspathic siltstone occur and the rock has a hackly fracture. In the chemical cycles the mudstone may be calcitic or dolomitic and contain up to 7% sodium as albite or analcime, and the rock fractures in a more conchoidal pattern. Analcime and carbonate minerals may occur in crystalline patches giving a 'bird's-eye' fabric.

2 The second most abundant rock type is a dark grey-black carbonate-rich mudstone in which dolomite dominates. It is well-laminated ('varved') but disrupted by shrinkage cracks.
3 Thinly bedded calcareous siltstone with minor, very fine sandstone beds and lenses up to 1 m or so thick occurs mainly in the detrital cycles. Micaceous laminae, irregular ripples and cross-lamination and small-scale graded beds characterize the facies, which also has quite common trace fossils.
4 In the lowest part of the formation and in the south-west of the outcrop area, dark grey-red and grey-green mudstones, siltstones and minor very fine sandstones with ripples and cross-lamination occur. They are strongly disrupted by mudcracks and burrows.
5 A minor constituent of the sequence is black, silty calcareous shale with sub-lenticular lamination. Reddish-brown shales occur in the area of interfingering of the Lockatong and overlying fluviatile Brunswick formations.

Fossil content of the sediment is low, being confined to fishes, reptiles, ostracods and plants, all indicative of non-marine conditions.

Cycles

The small-scale cycles, which characterize the whole formation, are each a few metres thick and can be traced for a distance of 1 km or so in good exposure. Exposure is not good enough to establish their lateral extent but the implication is that they are widespread. The 'detrital cycles' which occur mainly in the lower part of the sequence are around 5 m thick and the 'chemical cycles' around 3 m thick. This small-scale cyclicity is superimposed upon larger patterns of fluctuation (Fig. 4.10).

The 'detrital cycles' comprise essentially coarsening-upward sequences with black pyritic mudstone passing up through interlaminated dolomitic mudstones into massive dolomitic mudstones, commonly with mudcracks and bioturbation (Fig. 4.11). There is no erosion between the sequences and each is interpreted as a small-scale regressive unit following a deepening of the lake. The upper parts of the sequences, with their evidence of frequent emergence, suggest increasingly exposed marginal mudflats at the expense of a shrinking lake area though with seasonal contraction and expansion. The presence of pyrite and the preservation of lamination in the lower parts of the sequence are thought to indicate reducing bottom conditions, probably induced by a thermally stratified, though shallow, water column.

The 'chemical cycles' are more common. Each shows an upwards transition from dark, laminated mudstone, with dolomite and calcite laminae and lenses, into more massive dolomite or analcime-rich mudstone with common micro-brecciation (Fig. 4.11). These cycles are also thought to result from a progressive reduction in lake area and general shallowing or emergence as clastic input is reduced. The onset

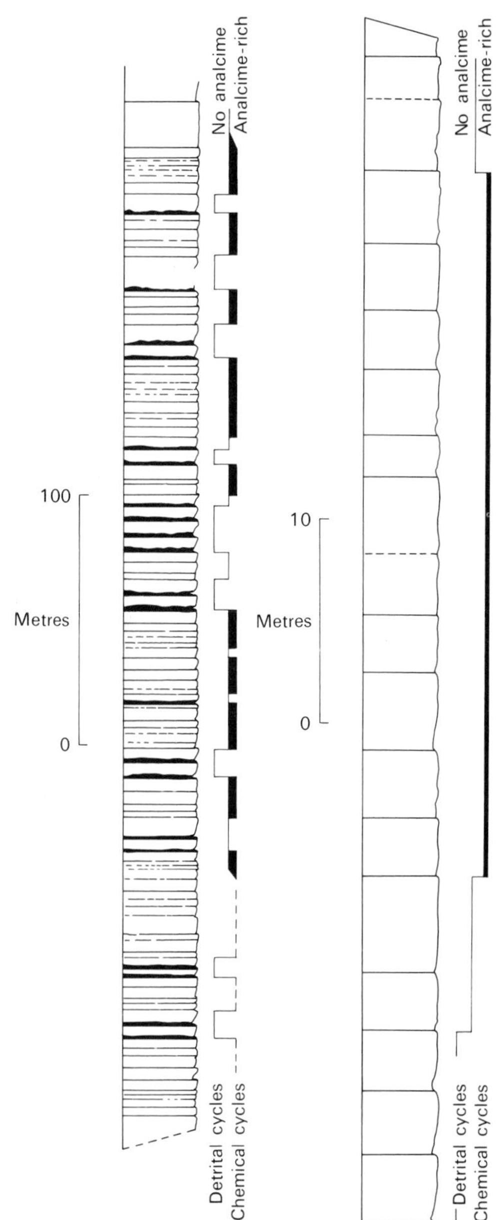

Fig. 4.10. Generalized stratigraphical sections of the Triassic Lockatong Formation showing both detrital and chemical short cycles (after Van Houten, 1964).

of the cyclic pattern may represent the initiation of an elongate lake by some form of damming of the drainage system which had deposited the Stockton Shale. The cyclicity itself is interpreted in terms of short-spaced climatic variation with

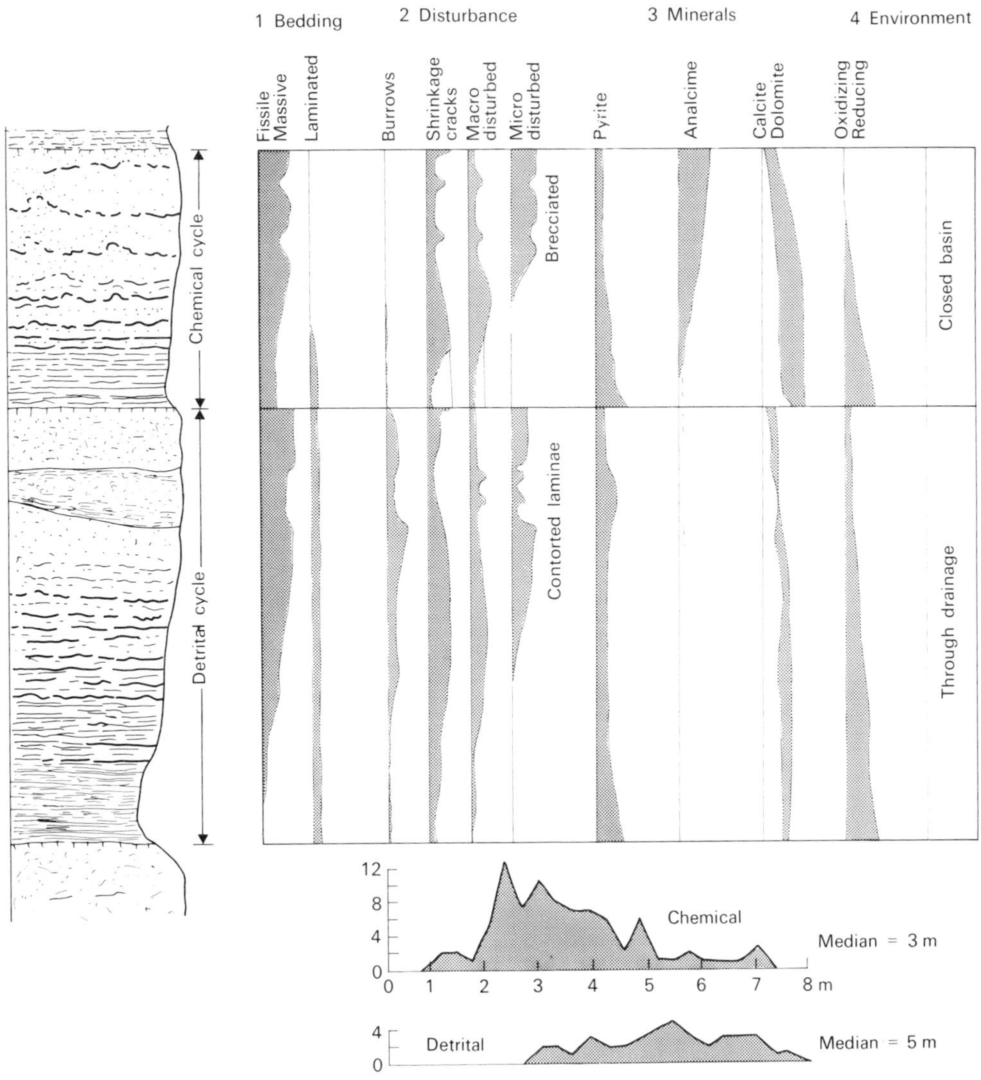

Fig. 4.11. Model of detrital and chemical short cycles in Triassic Lockatong Formation showing distribution and qualitative estimates of prevalence of sedimentary features and selected minerals (after Van Houten, 1964).

changing precipitation the dominant theme. It is tentatively suggested that the 21,000-year precession cycle might be the underlying control. Flocculation of colloidal clays in waters with a high cation concentration is thought to have produced the lack of lamination in the upper part. The analcime is of syngenetic or early diagenetic origin.

The two types of cycles are intergradational and the differences are interpreted as reflecting differences in the lake drainage. During deposition of the 'detrital cycles' the drainage is thought to have been characterized by through-flow during a relatively humid period, thereby ensuring that ionic concentrations in the lake water were kept low. The 'chemical cycles', on the other hand, are interpreted as reflecting a more arid period of internal drainage during which ionic concentrations rose. In particular, sodium concentration increased, due largely to weathering of a sodium-rich source area, thereby enabling authigenic analcime to precipitate (Van Houten, 1965). During the wetter phases which produced 'detrital cycles', the clastic input was characterized by a high content of albite.

THE EAST BERLIN FORMATION

In the area of the Connecticut Valley, active faulting on the eastern margin of a rift-valley basin (Hartford Basin) in Late Triassic and possibly Early Jurassic times allowed the development of alluvial fans and of more distal lakes, floodplains and rivers which flowed to the west (Sanders, 1968; Hubert *et al.*, 1976). The East Berlin Formation clearly shows the resultant facies change and two types of lacustrine sediments are recognized.

(1) Tabular units of red siltstone and fine sandstone a few tens of centimetres thick are interbedded with red mudstones. The sandstones show ripple cross-lamination and sometimes flaser-bedding, indicating currents or waves directed up or across the independently-defined palaeoslope. The sandstone and siltstone are interpreted as the deposits of small, shallow and short-lived floodplain lakes, though without clear comparison with a present-day analogue (Sanders, 1968; Hubert *et al.*, 1976).

(2) Grey and black beds occur in thicker more extensive cycles, which are thought to record perennial lakes developed during wetter periods. The cycles, which are separated from one another by floodplain red mudstones, are broadly symmetrical and have a central black pyritic mudstone unit (Fig. 4.12). This unit is confined to the central part of the basin for any particular cycle. The cycles are interpreted as recording the expansion and contraction of perennial lakes which on mapping evidence probably had dimensions of the order of at least 100 × 20 km, though circumstantial evidence suggests an area of 5,000 km^2. Slumped horizons can be used to estimate the direction of palaeoslope; and assumptions about the gradient needed for slumping, plus the areal dimensions lead to estimates of depth of the order of 80 m. Fish remains are common in the sequence and the presence of articulated skeletons and pyrite in the black mudstones suggests a thermally stratified water body with an anoxic hypolimnion. Fish kills may have occurred through overturn of the water column (cf. Beadle, 1974, p. 69). The dolomite/black shale interlaminated couplets are interpreted as reflecting short-term, possibly seasonal, fluctuations, with the carbonate layer recording a dry period and the carbon a wet period. This explanation contrasts with that offered for a similar facies in the Green River Formation (Sect. 4.3.2) where a playa model was envisaged. The absence of evaporites and flat pebble conglomerates are thought to warrant the deeper water model.

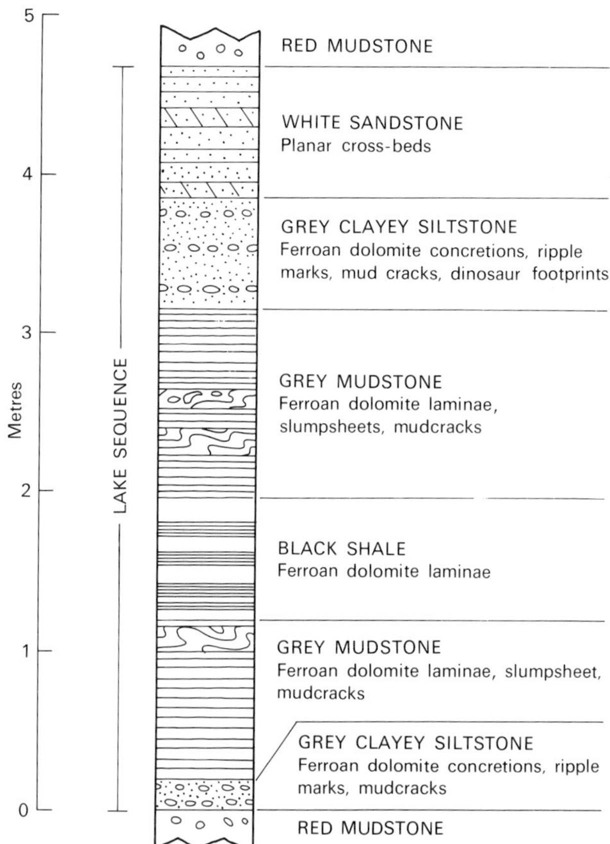

Fig. 4.12. Perennial lake sequence in the Jurassic East Berlin Formation of Connecticut. Black shale and grey mudstone contain fish, spores and pollen (after Hubert *et al.*, 1976).

4.3.4 Upper Palaeozoic Basins of the Maritime Provinces of Canada

In Devonian and Carboniferous times, the Maritime Provinces, in common with contiguous areas in northern parts of the British Isles, were subjected to a tectonic regime which led to a block and basin topography. Many of the basins were elongated rift valleys and had either sporadic or no marine connections (see also Sect. 8.4.3). In these basins lacustrine sediments interfinger with and over- or underlie alluvial sediments which commonly fringe the basin (e.g. Belt, 1965; Belt, Freshney and Read, 1967; Dineley and Williams, 1968).

The lake deposits themselves exhibit a variety of facies, many of which correspond with those already described. The sediments are dominantly fine-grained siltstones and mudstones with some sandstones.

In the Devonian Escuminac Formation of southern Quebec

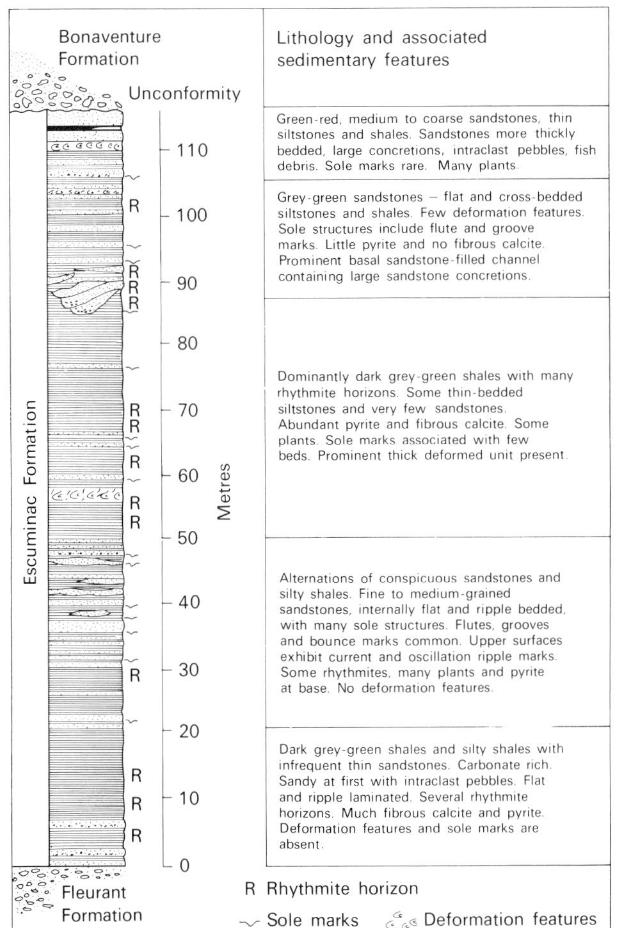

Fig. 4.13. Section through the Devonian Escuminac Formation of southern Quebec (after Dineley and Williams, 1968).

120 m of grey-green sandy siltstones and mudstones rest on an irregular surface of underlying alluvial fan conglomerates (Fig. 4.13). The fine sediments are commonly pyritic, carry a well-preserved fish fauna and have a well-developed fine lamination comparable with the 'rhythmites' of some glacial lakes (Sect. 13.4.3). Thus an anaerobic bottom, probably due to a stratified water column, is suggested. The fine sediments are interbedded with sandstone beds which are up to 60 cm thick and show good sole marks, grading, slump horizons and loading, comparing closely with classical distal turbidites. The lake therefore received pulses of coarser clastic sediment as turbidity currents. The upper part of the formation shows an increase in evidence of bottom currents, with thicker, sometimes cross-bedded, sandstones and plant fragments suggesting a gradual transition of the lake into a floodplain.

The Mississippian Albert Shale Formation (of the Horton Group) New Brunswick, represents a lake where clastic input was not distributed so widely and where chemical and organic components played a bigger part in the basin sediments (Greiner, 1962). Again, alluvial sediments underlie and overlie the formation. The Albert Shale Formation itself is about 1,300 m thick and shows a roughly symmetrical tripartite vertical sequence. The top and bottom members comprise siltstones, sandstones and some conglomerates. The sandstones commonly show ripples, mainly of wave origin and often with very shallow water features (double crests, etc.) The upper member is mainly calcareous and eventually passes into an evaporite unit of restricted distribution. Between the upper and lower members the Frederick Brook Member is the main lacustrine facies, comprising oil shale, calcareous shales and some thin sandstone beds. The well-laminated oil shales are highly bituminous and carry a well-preserved fish fauna. They constitute the major lithology in the central area but pass laterally into blocky mudstones with no fish.

The overall sequence is interpreted as the deposits of an elongate lake initiated as an alluvial valley and showing changes in its outlet through time. In the lower member, the shallow water was fresh, probably indicating through-flow. The Frederick Brook Member is interpreted as being more restricted and possibly stratified with an anaerobic bottom but the resemblance of the oil shales with those of the Green River Formation suggests that a playa interpretation may be possible. The abundant plant remains perhaps argue against this however. Subsequent resumption of clastic supply led to infilling and shallowing under a humid climate, as evidenced by abundant plant remains.

4.3.5 The Devonian Orcadian Basin of North-east Scotland

The Orcadian Basin of north-east Scotland was the site of thick accumulation of continental sediments through much of the Devonian. The Upper and Lower Devonian deposits are dominantly fluvial red beds but the Middle Devonian sediments are unlike the rest of the British Old Red Sandstone. The strata involved, the Caithness Flagstone Groups (Donovan, Foster and Westoll, 1974), are dominantly grey and grey-green fine-grained sediments. They are commonly strongly calcareous and have long been interpreted as lake sediments (Crampton and Carruthers, 1914), on the basis of their lithology and well-preserved fish fauna.

The Lower and Upper Caithness Flagstone Groups total some 4,000 m of sediment in which there is a broadly rhythmic gradational vertical variation in lithology. Around the lake margins there are distinctive lithologies and sharp lateral facies changes but within the body of the basin most of the lithologies are well laminated—or thinly parallel-bedded and appear to

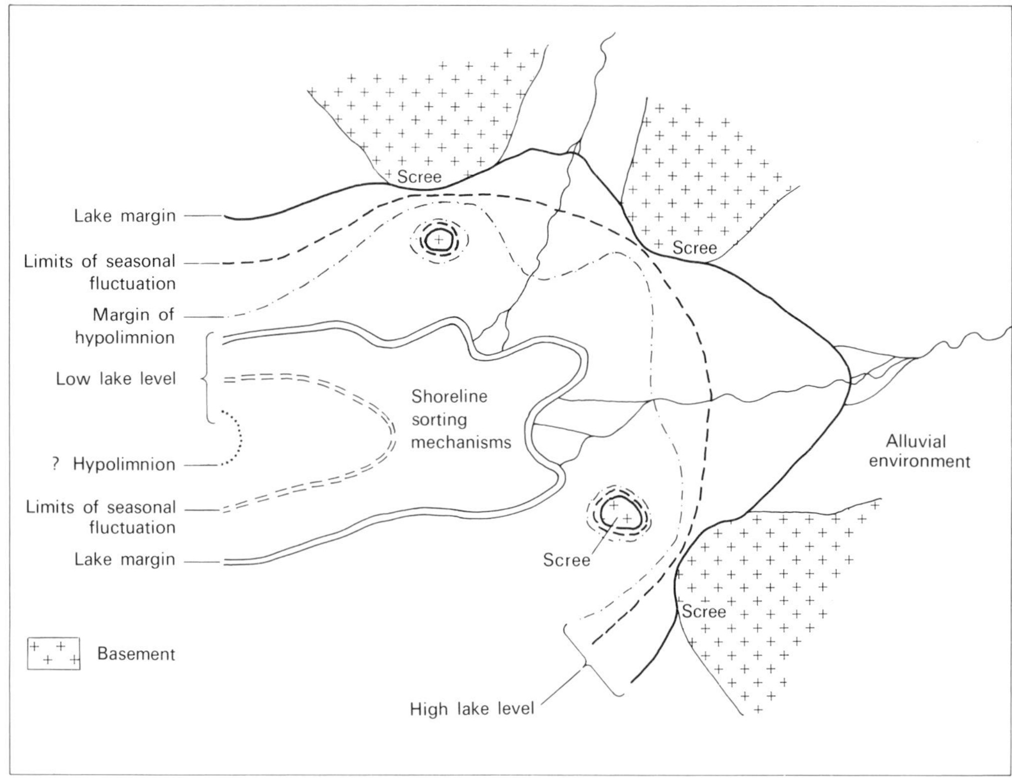

Fig. 4.14. Palaeogeographic model for the margin of the Devonian Orcadian Basin, Scotland, with lake margin separating an alluvial environment from that affected by shoreline sorting mechanisms (after Donovan, 1975).

form a highly gradational continuum of variation. At one extreme, they are laminites with alternating 0.1 to 1 mm thick laminae of microcrystalline carbonate and clastic material in variable proportions. When carbonate dominates, the clastic laminae may comprise only organic material and such sediments have been referred to as 'limestones' (e.g. Achanarras Limestone). Such carbonate-rich laminites are pyritic and rich in fossil fish. Towards the other extreme, carbonate is less important and grain-size and bed or laminae thickness increase with finer mudstone/siltstone interlaminations and subaqueous shrinkage cracks due to syneresis. With increasing grain-size, subaerial desiccation polygons are more abundant. This change is paralleled by increasing abundance of symmetrical ripples. Subaqueous shrinkage cracks associated with symmetrical ripples tend to have a preferred orientation parallel to the ripple crests (Donovan and Foster, 1972). At the coarsest extreme, discrete sandstone beds up to 5 cm thick are interbedded with desiccation cracked mudstones and siltstones. These have sharp bases and are laterally persistent and may have symmetrical or asymmetrical ripples on their top surface.

The fine carbonate-rich laminites are interpreted as the result of seasonally controlled deposition on the floor of a lake with a high though variable carbonate content. Algal blooms in summer lead to an increase in pH and to carbonate precipitation while wet periods (winter?) gave rise to clastic influx or to death of the algae (Rayner, 1963; Donovan, 1975; cf. Nipkow, 1928). The preservation of laminae and of fish and the presence of finely disseminated pyrite suggest anaerobic bottom conditions probably due to a stratified water column, while the occurrence of subaqueous shrinkage cracks would possibly be favoured by changing water chemistry if due to syneresis. The coarser lithologies with a more important clastic component, and with more evidence of wave action and

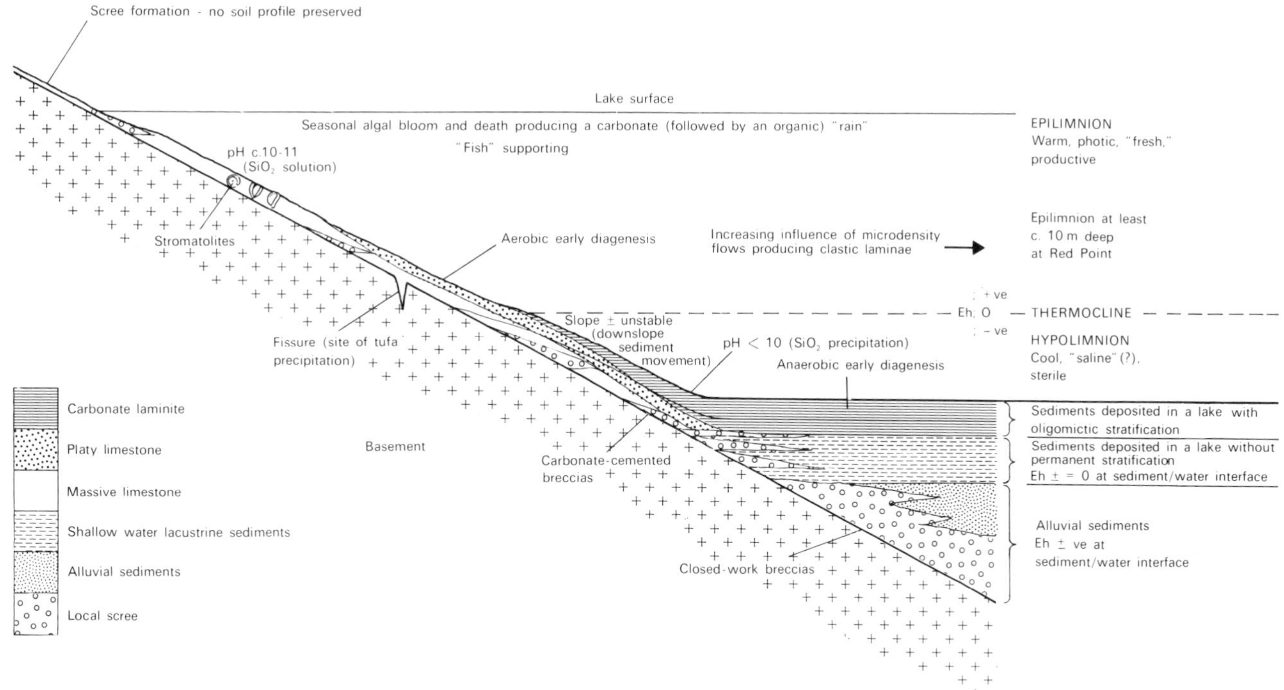

Fig. 4.15. Model for the margin of the Devonian Orcadian Basin, Scotland, illustrating vertical and lateral facies relationships and sedimentary processes which influence sedimentation of a *coincident* margin during lake transgression (after Donovan, 1975).

subaerial exposure, are the result of deposition around fluctuating lake margins. The variation of lithologies in a vertical sequence in a broadly rhythmic pattern is thought to reflect contraction and expansion of the lake due to climatic and/or tectonic causes.

The effects of expansion and contraction are most clearly seen at the lake margin, where periods when the lake margin coincided with the sedimentary basin margin can be distinguished from periods when it did not (Fig. 4.14) (Donovan, 1975). '*Non-coincident*' margins formed when lake level was low and marginal, clastic-dominated, flats developed which were periodically exposed, giving the subaerially cracked sequences. During episodes of '*coincident*' margin development, when water level was high, clastic supply was reduced as rivers alluviated their valleys and marginal topography was drowned. In the Orcadian Basin, lake margins have been recognized as unconformities between basement and lake sediments (Fig. 4.15). Above the unconformity the lake sediments are dominantly carbonate, with local breccias banked against the steeper topography (Fig. 4.16). Pebbles and earlier surfaces are commonly coated by stromatolitic carbonate while laminated tufa and granular carbonate fill cavities. The stromatolites are commonly dolomitized and show a range of growth forms from sheets to hemispheroids (Fannin, 1969; Donovan, 1973). The dolomitization does not affect the tufa and other interstitial carbonate and is therefore thought to be penecontemporaneous, reflecting very variable water chemistry.

Carbonates associated with the margin show distinct lateral facies changes and quite high depositional dips, due both to draping of the topography and to differential compaction (Donovan, 1975). At the margin, limestones are massive with birds-eye structure common while over a distance of a few tens of metres they pass laterally into carbonate-rich laminites. The lack of lamination in the marginal limestones is thought to reflect an oxygenated bottom where organic matter was not preserved. Some of the siltstone laminites associated with these margins show sole marks and grading. They are interpreted as small-scale turbidites deposited either off largely-starved river mouths or as a result of slumping off the marginal topographic

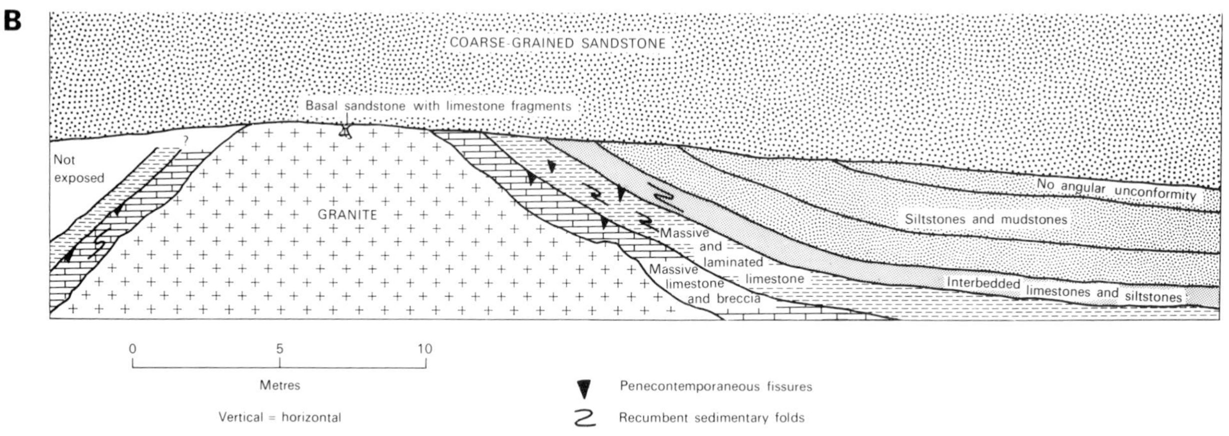

Fig. 4.16. Scale sections of basal facies relationships at the margin of the Devonian Orcadian Basin, Scotland (after Donovan, 1975).

slope. Some siltstones show penecontemporaneous recumbent folding due to slumping down the topographic slope, possibly aided by synchronous compaction.

4.3.6 Minor lake basins

Small lakes must have existed in the geological past both in their own right as small basins and as part of other, more widespread depositional environments such as alluvial floodplains or delta top areas. The deposits of such lakes have a low preservation potential but some examples are described.

The Bovey Tracey and Petrockstow Basins of Devon, England, are small elongate basins developed in Oligocene times in association with movement along a strike-slip fault (Freshney and Fenning, 1967). They accumulated thick sequences (800 m +) of kaolinitic muds and silts, much of which is exploited as ball clay for the ceramic industry. A certain amount of comminuted plant debris is preserved in the clays, and in the Petrockstow Basin seatearths, lignite beds and fining-upward units are quite common, suggesting that at times the surface of accumulation was a floodplain rather than a lake floor (Freshney, 1970). In the Bovey Tracey Basin, which is about 7–8 km wide, lacustrine laminated clays and silts occur in the central parts of the basin; alluvial sands and gravels are restricted to the margins. This basin is a smaller-scale and more humid version of the Ridge Basin of California with which comparison can be made (Sect. 14.3.3; Fig. 14.25).

In the Oligocene of the Isle of Wight broadly fining-upward units about 2 m thick pass up from irregular silts and muds at the base through muds with freshwater gastropods into a palaeosol (Daley, 1973a, 1973b). The sequence is interpreted as representing the inundation of floodplain areas by river water followed by a gradual silting process during which flow through the floodplain 'lake' became progressively more sluggish.

In the Permian Casper Formation of Wyoming and Colorado minor lenticular limestones within aeolian dune sandstones overlie low relief truncation surfaces in the cross-bedding (Hanley and Steidtmann, 1973; Pearson and Hanley, 1974). The limestones are up to 60 cm thick and several hundreds of metres in lateral extent. They contain ostracods and peloids; sparite fills desiccation cracks and inferred gas escape cavities. Disruption of the lamination is common. They are interpreted as the deposits of small lakes or ponds in interdune settings in an arid environment. The ponds are thought to have developed by local deflation of the dune sands to the water table, followed by a rise in the water table. The role of algae in precipitating the carbonate is not easily assessed but periodic drying out of the sediment surface is suggested by the desiccation cracks and the disrupted lamination which may represent reworked mudflakes.

4.3.7 **Conclusions**

It is difficult to draw general models because of the great variety of deposits and because modern research on ancient lakes is in its infancy. Many problems remain. The assumption that stromatolites, because of their restriction in modern marine environments to zones of hyper-salinity and their association in many ancient lake deposits with evaporites, mark periods or marginal zones of high salinity needs to be carefully re-examined in the light of recent discoveries of modern stromatolites in low-salinity lakes. In addition, it is curious that calcite marls and diatomites are apparently not as common in ancient lake sediments as in modern ones. It should, however, soon become possible to be more specific about water depth and to assess the importance of the deep stratified lake model compared to the playa one, as has been done for part of the Green River Formation.

Because of their value as indicators of palaeoclimate in continental regions and the increasing importance of this topic in understanding palaeo-oceanography, research on lakes, both modern and ancient, is certain to increase in importance.

FURTHER READING

HUTCHINSON G.E. (1957). *A Treatise on Limnology: Vol. 1, Geography, Physics and Chemistry*. pp. 1015. Wiley

BEADLE L.C. (1974). *The Inland Waters of Tropical Africa*. pp. 365. Longman.

MATTER A. and TUCKER M. (Eds.) (1978). *Modern and Ancient Lake Sediments. Spec. Publ. int. Ass. Sediment.* **2.**

LERMAN A. (Ed.) (in press). *Lakes: Physics, Chemistry and Geology*. Springer-Verlag, New York.

CHAPTER 5 Deserts

5.1 INTRODUCTION

The desert environment was one of the first to be used as a modern analogue for the detailed interpretation of ancient sedimentary facies (Walther, 1924; Wills, 1929; Shotton, 1937). Desert sandstones were also amongst the first to be studied for palaeocurrent directions (e.g. Brinkmann, 1933; Shotton, 1937; Reiche, 1938) and their recognition has been important in the reconstruction of palaeoclimates and palaeolatitudes (e.g. Köppen and Wegener, 1924). Studies of palaeomagnetism now offer a test of the reconstructed positions of continental masses in past geological times (e.g. Opdyke and Runcorn, 1960; Meyerhoff, 1970; Bigarella, 1972; Helwig, 1972; Robinson, 1973).

Economic pressures have also stimulated work on both recent and ancient desert sediments. Aeolian sandstones, which commonly have a high porosity, are important aquifers and the sites of hydrocarbon accumulation as, for example, in the gas fields of the North Sea (e.g. Butler, 1975; Van Veen, 1975). Ancient playa lake deposits are rich sources of evaporite minerals.

5.2 PRESENT-DAY DESERTS

5.2.1 Introduction and setting

Deserts are areas which are so dry that few forms of life can exist. Most areas of high aridity are hot during the day, though this is not always the case. High temperatures will tend to increase evaporation and it is the balance between precipitation and evaporation which finally determines the viability of plants. Low temperatures may, however, occur in some deserts, particularly those at high altitude such as in Chile and Turkestan, and those which extend to high latitudes. Even in hot deserts the temperature may have a high diurnal range, frequently falling below 0°C at night.

Arctic deserts are most likely to involve glacial and proglacial sediments (see Chapter 13). Hot deserts, which cover a much greater area of the present-day land surface, involve a variety of sub-environments such as aeolian sand seas (ergs), alluvial fans, ephemeral streams, playas and other lakes and areas of bare rock undergoing net erosion. Alluvial fans, ephemeral streams and lakes are dealt with in other chapters and they are only mentioned here where they contribute towards more general desert models. Areas of bare rock are frequently weathered and eroded into a variety of distinctive morphological features whose description is beyond the scope of this book and which have been extensively described elsewhere (e.g. Cooke and Warren, 1973).

Environments and processes which are particularly important in arid deserts include aeolian dune fields, inland sabkhas or playa lakes and the post-depositional development of a red coloration and of certain types of soil profile, particularly those involving duricrusts. The post-depositional processes have been dealt with in the 'Interfluvial' section of Chapter 3. Here we confine our attention to dune fields, inter-dune areas and inland sabkhas or playa lakes.

5.2.2 Desert climates

A world rainfall map shows that regions of high aridity occur as two belts either side of the equator centred around latitudes of 20–30° and associated with persistent high atmospheric pressures (Fig. 5.1). The centres of large continental masses are also arid. Departures from this pattern are quite common as shown by the elongated arid zones extending into high latitudes along the western seaboard of North and South America and by the monsoonal humid zones of south-east Asia. Such departures are controlled by oceanic and atmospheric circulation patterns and by local topographic features. Hence care has to be taken if the distribution of ancient desert sediments is used as an indicator of palaeolatitude, as the reconstruction of ancient oceanic circulations is not easy. In addition, the much larger continental masses inferred for Pre-Jurassic times might have contained very large deserts in their central parts.

Even the most arid deserts have occasional precipitation and this may be seasonal or sporadic. Australian deserts usually receive small amounts of seasonal precipitation, while the Sahara has a totally unpredictable rainfall pattern. Several years may pass without rain but when it occurs it tends to be in

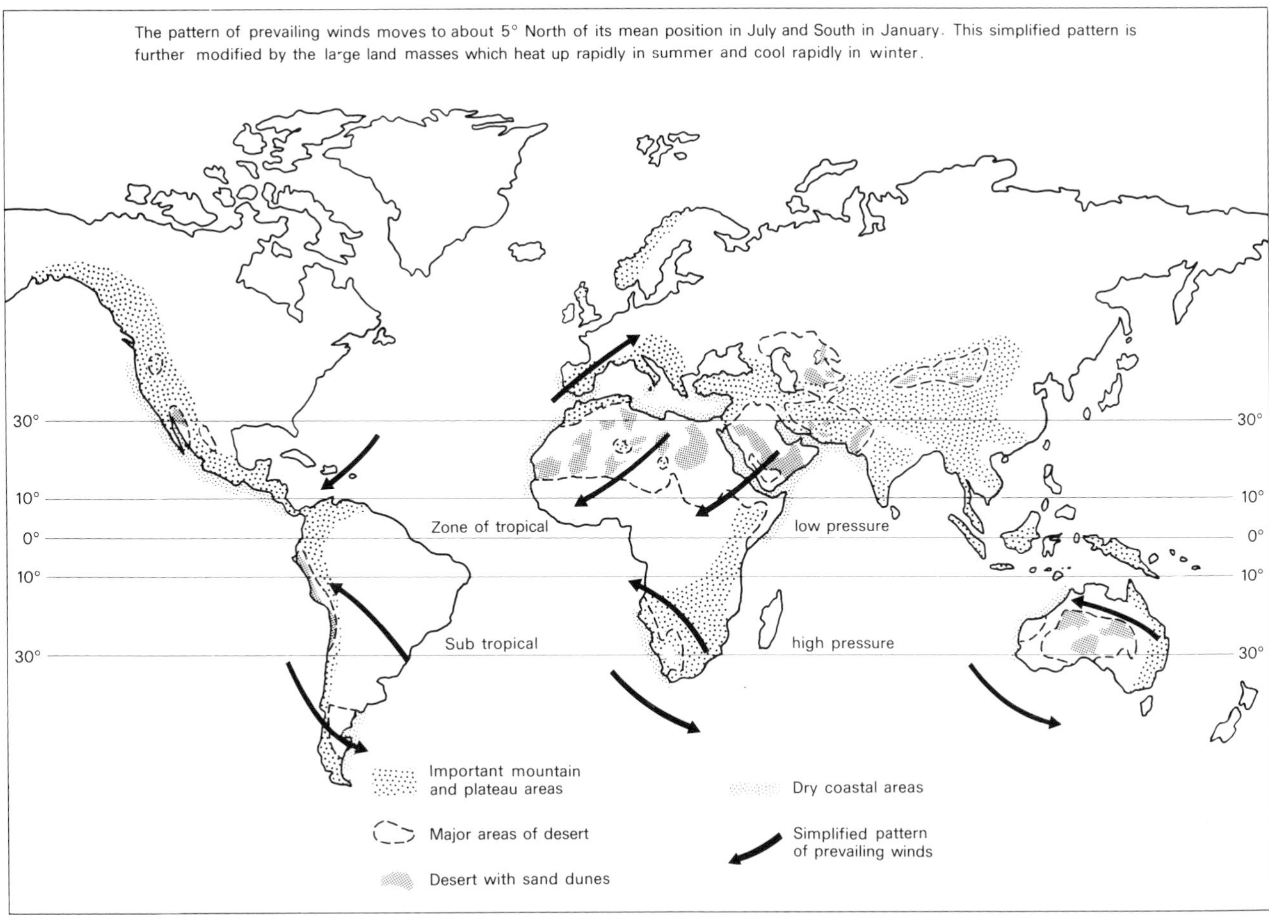

Fig. 5.1. Deserts of the world (after Glennie, 1970).

short, intense storms giving several centimetres of rain within a few hours. Because of the high permeability and low water table of many desert sediments, infiltration is considerable and surface streams do not flow far.

5.2.3 Tectonic setting of deserts

The occurrence of deserts is not directly related to tectonics. However, local tectonics control topography and thus the local climate and the development of a rain shadow. They also control the preservation of desert sediments.

Areas of thick accumulation of desert sediments occur both in active tectonic settings where fault-controlled basins are rapidly subsiding, as in South-western USA, and on stable cratonic areas, such as the Sahara. Many of the desert basins of both types have inland drainage. Such drainage patterns and the basinal sediments may be largely inherited from Pleistocene pluvial episodes when the climates of the present-day desert areas were more humid and run-off was less ephemeral (Peel, 1960). In the more tectonically active basins (e.g. SW USA; Iran; Danakil depression, Ethiopia), faulted margins commonly develop alluvial fans at the basin edges with other environments such as playa lakes and dune fields in the centre. Several of these desert graben have floors below sea level. In the intra-cratonic basins such as the Sahara or the Australian desert, the margins are much more diffuse. Fluvial processes are more evident at the margins while aeolian processes dominate the central regions. Large, generally intra-cratonic, areas dominated by wind-blown sand are called 'ergs' (Wilson, 1973) and they often seem to coincide in their extent with areas of pre-existing sandstone outcrop, suggesting that they owe their existence largely to an abundant supply of sand. Erg accumulation might, therefore, be the result of progressive recycling on a long time scale. Additionally, ergs normally occur in topographic lows where wind speeds decelerate by being able to expand vertically. Such topographic lows may also coincide with pre-existing sedimentary basins.

Wind-blown sand also accumulates in coastal regions forming dune fields elongated parallel to the coast. These are normally the product of strong onshore winds and thus are associated with coastal sediments, in particular those of beaches and coastal sabkhas. However, they have rather low preservation potential being susceptible to marine reworking during transgressions (see Chapter 7). Accordingly, they will not be described as it seems that their shape and internal structures do not differ significantly from those of ergs (Bigarella, 1972). The main differences are in the associated sediments and in the fact that coastal dune fields tend to have an overall linear trend.

5.2.4 **Sand-dominated deserts**

The mechanics of wind-transport of sand is beyond the scope of this book. Vaughan Cornish (1914), in an early exploration of this problem, emphasized the role of eddying but this approach has been superseded by the classic work of Bagnold (1941) which is still the clearest exposition of the subject. The most important difference between aeolian and aqueous transport is that, in aeolian transport, grain ballistics and intergranular collision are more important than fluid turbulence (Sharp, 1963, 1966).

The great sand seas or 'ergs' are the main sites of accumulation of wind-blown sand. They may be up to 5×10^5 km^2 in area and show a complex assemblage of bed forms; ripples, dunes and 'draas'. Ergs derive their sand by winnowing of the deposits of peripheral and internal streams and by attrition of older sandstones and other source rocks.

The sand supplied by ephemeral streams to the edges of ergs is poorly sorted unless it is derived from earlier aeolian sands. Wind is an efficient sorter of sediment because of its low density and viscosity and because it can only move sand-sized particles as bedload. The normal size range for wind-driven sand bedload is 0·1 to 1 mm. Coarser grains move less readily and finer material goes into suspension to be transported over longer distances and often out of the basin thereby contributing to the general deflation of the area. Sharp (1966) showed that in the Kelso dune field of California, the size distribution and sorting of the sand became established over a distance of some 15 km of saltation transport, though it took longer to achieve high rounding and polishing of the grains.

There have been many attempts to use grain-size statistics and grain shape properties as criteria for the recognition of aeolian sediments (e.g. Friedman, 1961). It seems, however, that the dominance of saltation load in aeolian and beach sediment populations means that they may often be confused (Moss, 1962, 1963). In addition to the rounding and sorting of grains by aeolian transport, there may also be a mineralogical sorting with a breakdown of minerals with cleavage and a survival of weakly cleaved minerals, in particular of quartz. Electron microscopy of desert aeolian sand grains shows that their surfaces have very subdued relief compared with those of aqueously abraded grains (Krinsley and Doornkamp, 1973). The smoothness is thought to result in part from intense physical attrition and in part from local, small-scale solution and precipitation during the diurnal temperature cycle.

Wilson (1972, 1973) showed that both grain size and thickness of the sand cover in ergs influence the scale and nature of the bed forms which develop (Fig. 5.2). In the thick accumulations of the Sahara and Arabia three scales of bed form are developed, the largest of which, draas, are large enough to carry superimposed dunes which in turn carry ripples of both aerodynamic and impact origin (Wilson, 1972). In areas of thinner sand accumulation, such as the deserts of Australia, draas are generally absent and dunes are the largest bed forms. All three scales show a wide variety of shape, considerably more complex than the range of shapes developed below flowing water. This complexity is partly because wind direction is more variable, due to its independence of topographic slope. Interference patterns caused by two or more effective wind directions are a common cause of complexity in dunes and draas, though ripples are generally small enough to change with the wind regime. Interpretation of wind direction based on dune morphology is not simple because of the time needed to remould substantially the largest forms. It is often difficult to say if a particular interference pattern is due to two or more currently active dominant wind directions or is the result of a longer-term change which has not yet obliterated the results of an earlier regime. Another problem is that because of the relatively high threshold velocities for sand entrainment by wind, grain movement is an intermittent process and the strong winds which actually move sands may have a different direction from the weaker winds which operate through a greater proportion of the time (Sharp, 1966).

RIPPLES

Three types of aeolian ripple occur either superimposed on larger forms or as features of interdune areas. The most common type is the impact ripple whose wavelength relates to the saltation path length of the moving grains (Bagnold, 1941). These ripples, with crest-lines transverse to the wind, develop in well-sorted coarse sands sometimes of granule grade, winnowed from the general sand population. They do not have avalanche faces and the highly sorted nature of the sand probably precludes their having well-developed internal lamination.

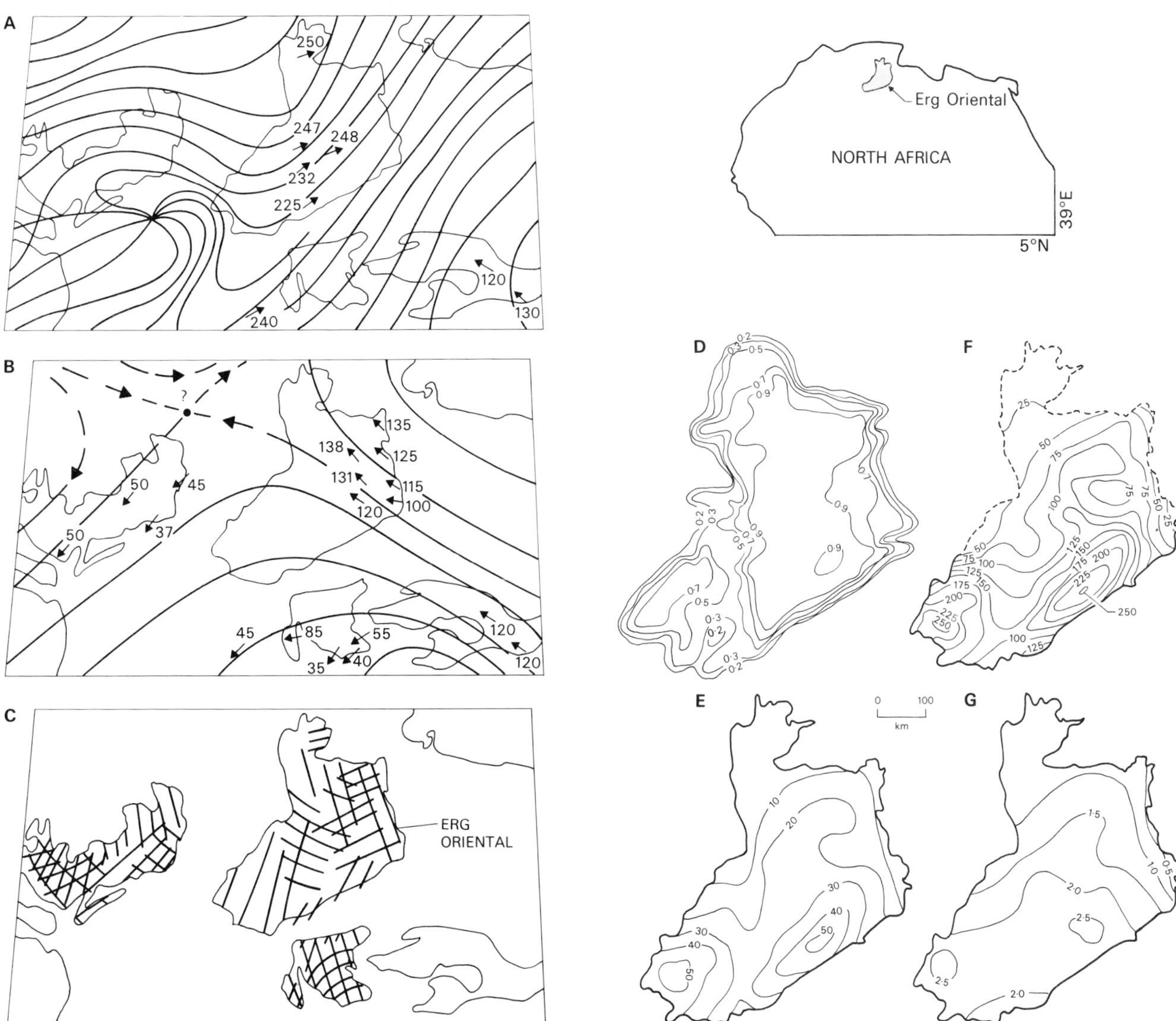

Fig. 5.2. The Erg Oriental, Algeria, showing its position in relation to regional patterns of sand transport and the relationships between sand cover and bed forms within it. A. Flow directions medium sand; B. Flow directions fine sand (arrows indicate directions taken from bed forms); C. Prominent draa trends (different trends are related to grain-size differences); D. Proportion of sand cover; E. Mean spread-out sand thickness (m); F. Mean draa height (m); and G. Mean draa wavelength (km) (after Wilson, 1973).

In addition, there is a less common ripple type which tends to have a strong longitudinal component to its shape and to be several times larger than the transverse impact ripples (Wilson, 1972). These so-called 'aerodynamic' or 'fluid drag' ripples owe their form to aerodynamic instability near the sand/air interface. Their internal structures are not described though they occur in rather fine sediment (Cooke and Warren, 1973).

The final form is the 'adhesion ripple' which develops on damp sediment and which is therefore more likely to be a feature of interdune areas. Moving sand grains adhere to the sediment surface in a rather irregular three-dimensional pattern sometimes described as warty rather than rippled (Reineck, 1955; Glennie, 1970). Grains are trapped on the up-wind sides of the ripples which may grow up to 5 cm high and

50 cm long. Glennie (1970) hints that relief of up to 30–40 cm may sometimes develop through adhesion. The distribution of such ripples is likely to vary through time, controlled as it is by the level of the water table. Little is known of the internal structures though Glennie (1970) shows a rather lenticular lamination in alleged adhesion ripple deposits.

DUNES AND DRAAS

It is possible to divide up the complex spectrum of dune and draa types into broad classes. The morphologies of the two size classes are sufficiently similar for them to be dealt with together. The distinction between draas and dunes is mainly one of size (Wilson, 1972) (Fig. 5.3). Dunes have wavelengths measured in 10's or 100's of metres and heights in metres, while draas have lengths of the order of kilometres and heights in 10's of metres. Rates of movement of both are related to their size with smaller forms able to move more quickly (Long and Sharp, 1964). Draas may move at rates of the order of 1 cm/year and may, because of the huge volume of sand involved, take a long time to develop (10^4–10^6 years). They are therefore likely to be to some extent relict features reflecting both ancient and prevailing effective wind regimes.

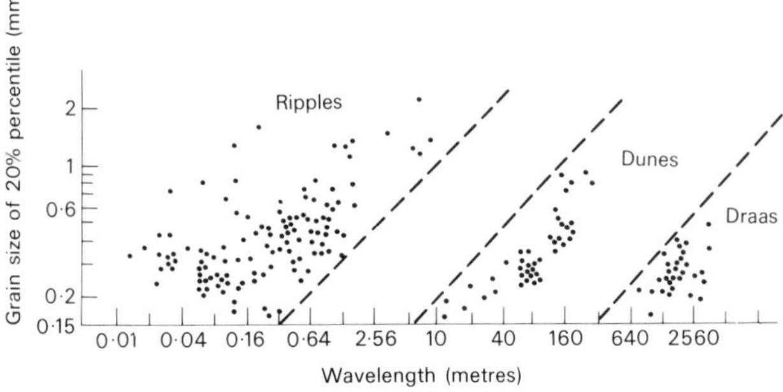

Fig. 5.3. Relationship of grain-size to wavelength for aeolian bed forms (after Wilson, 1972).

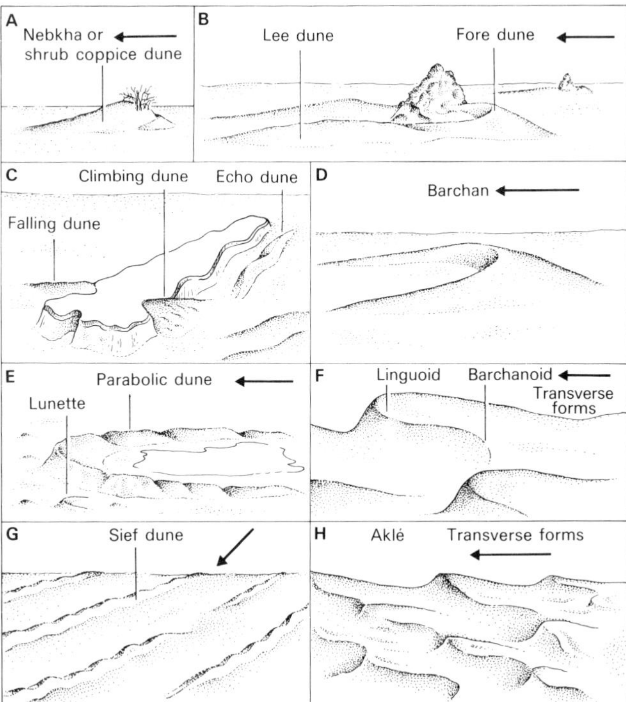

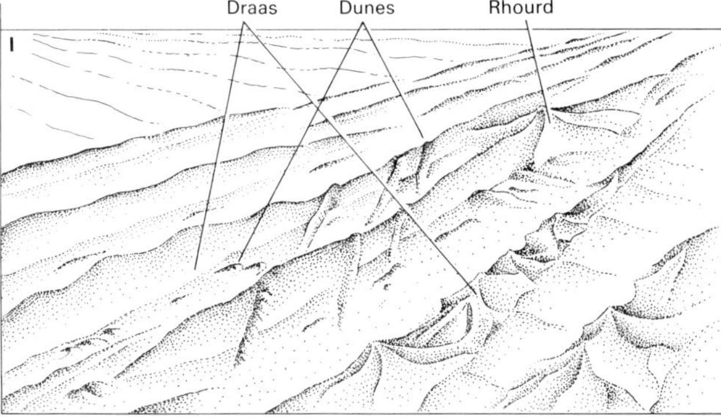

Fig. 5.4. Main types of aeolian dune and draa. The terms used in Figs. D–H may be applied to dunes as well as draas. Effective wind direction is indicated where appropriate. In Figs. C and I no one direction dominates. Figure I shows the size relationship between dunes and draas (after Cooke and Warren, 1973).

As with most types of natural bed form, any system of classification based on shape is, of necessity, arbitrary, but the following five classes cover the main types found. The first three are related in having their crest lines oriented more or less transverse to the dominant wind direction.

(a) *Long-crested transverse* forms have rather continuous crest lines oriented more or less transverse to the dominant wind direction and they seem to develop when there is only one effective wind direction (Fig. 5.4). The forms normally have avalanche faces on their lee sides. Larger draas sometimes have more gently inclined lee sides, upon which dunes are superimposed. Transverse forms often diverge from a straight-crested plan by showing curved (aklé) patterns made of alternating convex and concave downwind sectors. Transverse dunes and draas with continuous crestlines tend to be associated with areas of complete sand cover.

(b) *Barchans* are associated with areas of rather depleted sand supply where there is insufficient sand to cover completely the non-erodible pavement (Fig. 5.4). The degree of depletion controls the spacing of the isolated barchans, the less sand the

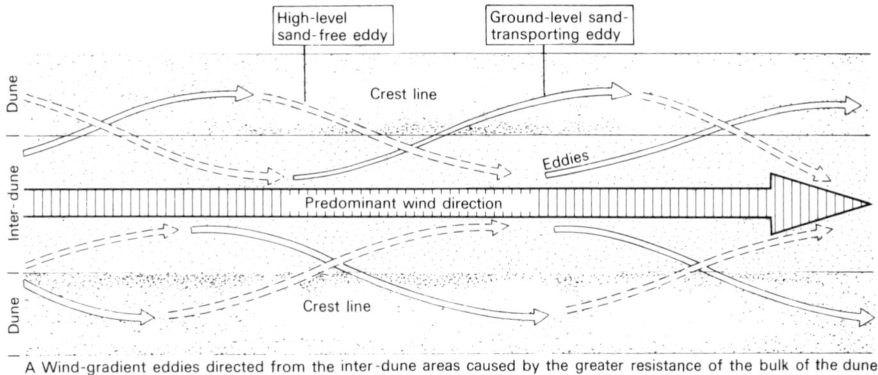

A Wind-gradient eddies directed from the inter-dune areas caused by the greater resistance of the bulk of the dune

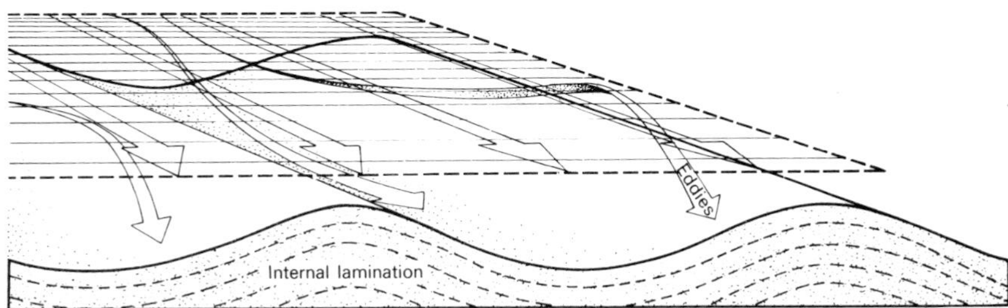

B Schematic representation of the spiral path followed by the wind eddies. They may give rise to crestal slip faces. A deviation of the wind from the predominant direction will accentuate the development of slip faces on one side of the dune

Fig. 5.5. Suggested relationship between the form of seifs and the large-scale structure of the near surface wind (after Glennie, 1970).

wider the spacing. The well-defined asymmetry and the slip faces of the barchans are due to their responding to a single effective wind direction. The highly concave downwind crest line and slip face die out downwind into projecting spurs which may be equally or unequally developed.

(c) *Parabolic* forms are the linguoid equivalents of barchans being isolated with lee faces convex downwind (Fig. 5.4). They are not as common as barchans but, like them, seem to be a response to depleted supply.

(d) *Longitudinal or seif* forms are elongated parallel to the effective wind direction (Fig. 5.4). They are remarkably parallel and laterally extensive, with lengths of the order of 10 km and exceptionally 100 km (Folk, 1971). They occur in groups over wide areas and commonly bifurcate in an up-wind direction (Folk, 1971). Crestlines may, in detail, be slightly sinuous and this pattern may carry with it the development of slip faces on alternate sides of the ridge. This form is thought to be associated with paired spiral vortices with axes parallel to the flow, the so-called Taylor-Görtler vortices (Fig. 5.5) (Folk, 1971; Wilson, 1972). Evidence of this secondary circulation pattern on large seifs occasionally comes from the pattern of superimposed dunes on their flanks.

(e) *Rhourds and other interference patterns* occur where effective wind patterns are variable and complex three-dimensional dunes develop (Fig. 5.4). The patterns are sometimes recognizable as the intersection of two or more linear trends but sometimes they break down into a series of more or less isolated peaks which, when developed at the scale of draas, are called 'rhourds'. Wilson (1973) illustrates a field of rhourds each over 200 m high which result from two dominant wind directions separated by an acute angle. Such peaks normally occur where there is sufficient sand to cover the whole surface, and the relatively flat areas between the rhourds often have a slower moving lag deposit of coarse sand. Many rhourds have complex patterns of slip faces and in many cases carry superimposed dunes. In other cases, three-dimensional dunes may be in the form of smooth domes with no well-developed slip faces (McKee, 1966).

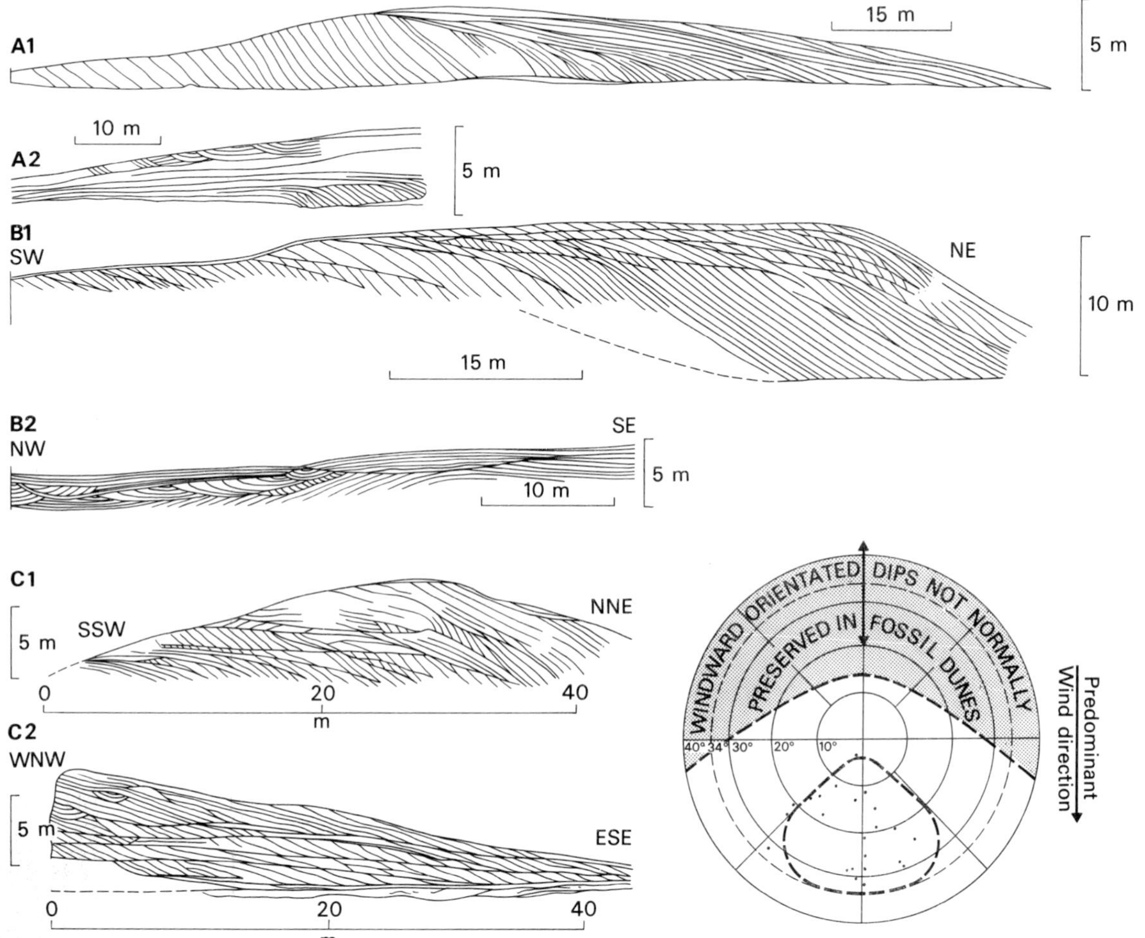

Fig. 5.6. A to C are of internal structures observed in excavations into the gypsum dunes of the White Sands National Monument, New Mexico (after McKee, 1966).
A. Dome-shaped dune: sections 1 and 2 are perpendicular.
B. Transverse dune: (1) parallel to the wind direction, (2) perpendicular to it through a lateral margin. Note the low-angle erosion surfaces in the upper part of the cross-bedded set.
C. Barchan dune: (1) parallel to the wind direction, (2) perpendicular to it through a lateral wing.
D. Hypothetical distribution of foreset orientation in barchan dunes (after Glennie, 1970).

INTERNAL STRUCTURES

Not a great deal is known of the internal structures of aeolian dunes and nothing directly about those of draas. Large-scale cross-bedding is the most abundant structure and the details of the cross-laminae are thought to differ somewhat from those of aqueous cross-bedding (Hunter, 1976, 1977). For aeolian sets of about 1 m height in fine to medium sand, the grain-flow (avalanche) cross-strata are inter-laminated with grain-fall deposits and wedge upwards from the base of the set, while the sub-aqueous grain flows are contiguous and extend over the height of the set. In addition, aeolian grain flows have much sharper lateral margins than subaqueous ones and are narrower for the same height of slip face. Whether these distinctions apply in larger sets is not clear.

In addition to normal cross-bedding, the laminae of dunes are sometimes deformed. This is generally attributed to the movement of partially cohesive sand masses on the slip face

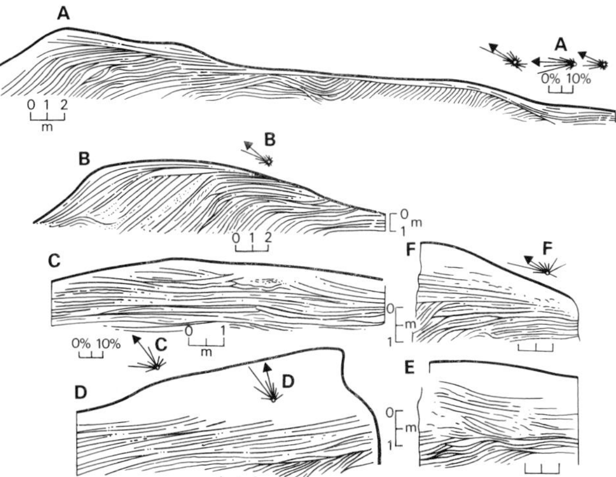

Fig. 5.7. Internal structures of coastal dunes, Parana, Brazil, with rose diagrams of foreset orientation. Note the similarity with the transverse dune from White Sands, New Mexico (Fig. 5.6, B(1)) (after Bigarella, Becker and Duarte, 1969).

after rain has fallen (Bigarella, 1972; McKee and Bigarella, 1972; McKee, Douglass and Rittenhouse, 1971).

At a larger scale, McKee (1966) recorded the internal structures of several different dune types in the White Sands of New Mexico (Fig. 5.6). These are composed of gypsum grains and an appreciation of how typical they are of dunes in general must await a more comprehensive study of other dune fields. Quartzitic coastal dunes seem, however, to have comparable internal structures (Fig. 5.7) (Bigarella, Becker and Duarte, 1969).

Of the types studied by McKee, transverse dunes had the simplest internal structure (Fig. 5.6B). Foresets, inclined at up to 35°, dip more or less parallel to the active slip face and are complicated by internal erosion surfaces which are roughly comparable to the reactivation surfaces of fluvial bars (Chapter 3, p. 29). These dip at low angles which range from near-horizontal near the dune crest to over 20° where they descend into the lower part of the set. The surfaces are probably products of reversed or transverse wind episodes, though the migration of smaller bed forms and the dune crest may produce similar effects.

The barchan dune had a similar structure though its internal erosion was more intense and irregular (Fig. 5.6C). A section perpendicular to the wind direction showed a similar pattern but with both foresets and erosion surfaces dipping at lower angles, apparently outwards from the core of the dune, which is not what might be intuitively imagined. The overall distribution of foreset azimuths is thought to be broad and unimodal (Fig. 5.6D) (Glennie, 1970).

A parabolic dune was also excavated by McKee. It showed large-scale cross-bedding directed down-wind with low-angle erosion surfaces separating bundles of foresets. In transverse sections the laminae curved convex-upwards. The dome-shaped dune (Fig. 5.6A) appeared to have evolved from a dune with a slip face, probably a straight crested transverse type, by gradual lowering of the slip face by successive erosional events.

In seif dunes the foresets dip away from the ridge axis, interdigitating in the axial zone (Fig. 5.8A and B) (Bagnold, 1941; Shotton, 1956; McKee and Tibbitts, 1964). This reflects the alternate growth of slip faces on the flanks of the dune ridge. Seif dunes may therefore produce a bimodal cross-bedding azimuth distribution symmetrical about the effective wind direction (Fig. 5.8C) (Shotton, 1956; Glennie, 1970). The situation will, however, be more complex in reality because of the migration of superimposed dunes on the flanks of the seifs.

INTERDUNE AREAS AND ROLE OF THE WATER TABLE

In those areas of aeolian dunes where sand is insufficient to cover the surface completely, the non-erodible surface may be simply scoured bedrock or it may be a deflation pavement covered with coarse-grained material. Such stony deserts, sometimes termed 'serir' or 'reg' in North Africa or 'gibber plains' in Australia, are covered with a poorly sorted assemblage of large grains up to boulder size derived either from the weathering of the underlying country rock or introduced by fluvial processes. The particles themselves are subjected both to wind abrasion and to solution and precipitation associated with the formation and evaporation of dew. The first process leads to the facetting of pebbles and cobbles to form ventifacts, while the second leads to the formation of desert varnish. It may sometimes prove difficult to distinguish wind polishing from the development of the varnish, and insolation splitting may in fact be responsible for some of the initial shapes rather than wind abrasion (Glennie, 1970). Excavations in interdune sediments show them to be rather poorly sorted with a roughly horizontal or low-angle bedding, probably reflecting an original sheetflood origin (McKee and Tibbitts, 1964).

In some deserts the water table may be quite close to the surface. This may be associated with topographic lows or it may be a feature of marginal areas where the desert passes into savannah. The presence of water at shallow depth influences the behaviour of wind-driven sand. Water-saturated sand is cohesive and therefore less erodible and the level of wetting will therefore act as a basement to dune movement. Erosion surfaces in aeolian dune deposits may therefore correspond to wetting levels. When developed on an extensive scale, such areas of damp pavement constitute inland sabkhas (see later) and may be the site of adhesion ripple development. A second effect of shallow water is to facilitate plant growth and this, in turn, influences the development and growth of dunes. In lightly vegetated areas, plants create local eddies and act as the

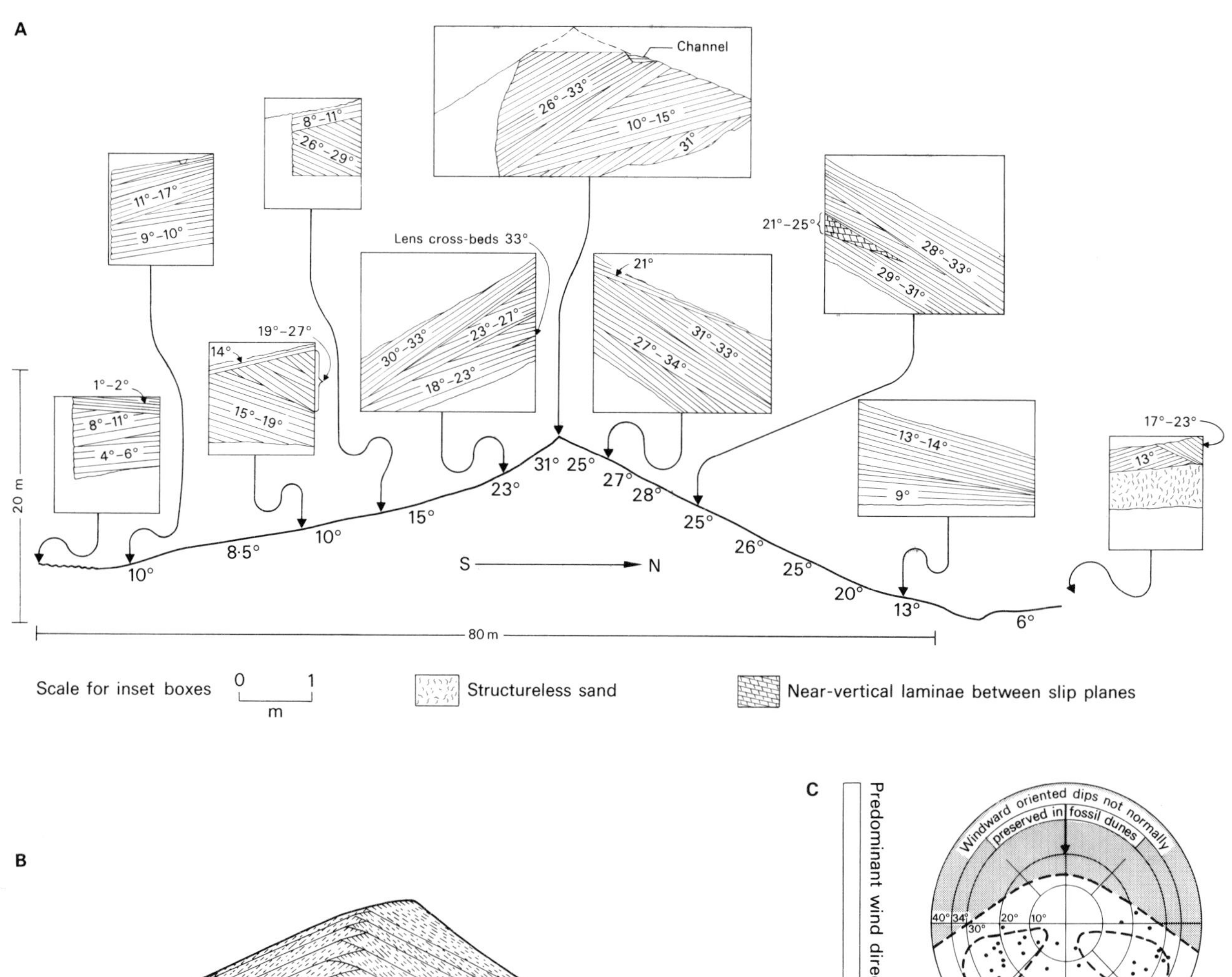

Fig. 5.8. Internal structure of a seif dune in transverse section. A. Small trenches dug into the surface of a Sahara seif dune (after McKee and Tibbitts, 1964; B. An idealized section of the likely overall structure (after Bagnold, 1941); C. Hypothetical distribution of foreset orientation in seifs (after Glennie, 1970).

nuclei to dunes. More intense plant growth fixes the sand and reduces movement though local 'blow-outs' may erode hollows and coppice dunes may occasionally develop (Fig. 5.4).

In semi-arid regions, dunes may develop somewhat sporadically because of long-term climatic changes with soil horizons and duricrusts forming in wetter periods (see Chapter 3, p. 39).

DISTRIBUTION OF SANDS AND BED FORMS WITHIN ERGS

At present it is very difficult to say how aeolian bed forms are distributed or how they relate to one another in space and time and it is almost impossible to develop any general large-scale model of aeolian desert basins. Wilson (1973) showed that a model erg, without bed forms and crossed by sand flow-lines, will be a convex lens tapering to the desert floor at the edges. Bed forms only occur where the sand flow is 'saturated', that is able to cover the substrate pavement. Many large bed forms within ergs are separated by pavement and therefore normally an alternation of saturated and unsaturated conditions, termed 'metasaturated', occurs (Wilson, 1973). Bed forms begin to form when metasaturation is achieved and they grow by trapping further sediment. They eventually coalesce when fully saturated. An erg may, therefore, have barchans or, more rarely, parabolic dunes around its margins while in the thicker central areas, transverse straight crested, longitudinal or three-dimensional forms may occur, the details depending on the prevailing wind. At a more local level, there may be a down-wind transition from longitudinal dunes, through a zone of coalescing barchans, to isolated barchans migrating on a non-erodible pavement (Norris and Norris, 1961).

5.2.5 Loess

Because wind winnows suspended load from bedload very efficiently, considerable volumes of silt-grade and finer material will be moved as dust-storms. Much of this finer material is transported beyond the desert basin and accumulates locally as blankets of finer material, termed 'loess'. In addition, finer material is dispersed further into a wide range of continental and marine environments by transport at high levels in the atmosphere and only rarely accumulates as recognizable deposits. Loess is usually associated with periglacial environments where ice grinding produces an abundance of silt-grade material (Smalley, 1966) and most of the world's loess deposits can be attributed to derivation from either modern or Pleistocene glaciated areas (Smalley and Vita-Finzi, 1968) (Sect. 13.3.5). There are, however, some areas with significant accumulations of loess which can only be derived from winnowing of hot arid desert regions where the grains are probably mainly produced as chips during the abrasive rounding of sand grains during bedload transport.

In Israel (Yaalon and Ginsbourg, 1966; Yaalon and Dan, 1974) desert loess accumulates mainly on desert fringes, where vegetation localizes deposition, and on hillsides leeward of the dominant wind direction. Not all the loess remains as a primary deposit, however, and much is reworked by water to accumulate in wadi beds and playas. There is a distal fining and thinning of loess deposits away from the source, and in desert fringe areas, where some rain falls, soils form upon it. A well-differentiated soil profile develops better in primary loess than in secondary reworked loess.

For the most part, loess blankets the existing morphology though sometimes it may be associated with a ridged topography of rather uncertain origin (Cooke and Warren, 1973). Loess areas sometimes pass rather sharply into the sand dune fields of adjacent deserts but large areas of loess around modern deserts are comparatively rare. Internally, loess deposits are structureless or faintly laminated and soil development may obliterate original textures.

5.2.6 Playas and inland sabkhas

The topographically lowest areas of many deserts are sometimes close to the water table and may contain ephemeral saline lakes. Such areas are termed playas in North America and inland sabkhas in the Middle East. They range in scale from a few square metres to 8,000 km^2, a size which Lake Eyre, Australia, reached at its maximum (Twidale, 1972); and they often lie below sea level. The water which feeds the lakes comes from two sources, surface run-off which comes from catastrophic storms and ground water which may be delivered by springs or by a general elevation of the water table relative to the sediment surface. Temporal climatic changes can have far-reaching consequences in these highly sensitive environments and many present-day playas were perennial lakes in the past (Stoertz and Eriksen, 1974). The present observed distribution of sediment types on and below the surfaces of playas may well be the product of complex recent climatic changes and their interpretation is therefore difficult (e.g. Death Valley, California) (Hunt, Robinson, Bowles and Washburn, 1966).

Many deserts occur as basins of inland drainage with lakes developed towards the centres. They may be fringed by dunes as in many passive, intracratonic basins or by alluvial fans in more tectonically active areas such as California (Fig. 5.9). Surface run-off may arrive via wadis, sometimes cutting through dune fields, or as the distal drainage of canyon-fan systems. The high permeability of many fan and erg sediments allows much of their discharge to be via the sub-surface. Flood flows of either type experience a sharp fall-off in their capacity and competence to transport clastic sediment on entering a lake and most deposition on the lake floor is from suspension, and therefore of dominantly fine-grained material. Thin graded beds may be produced by individual floods (Bull, 1972) though occasional major floods may introduce distal fan bed load sediment into the marginal playa sediments (Hooke, 1972).

As the lake bed dries, organic and inorganic agencies disrupt and disturb the bed. While the sediment is still wet, it is a good recipient of footprints, rainpits and other tracks, amongst which are the playa furrows produced by sliding, possibly wind-driven, boulders in certain Californian playas (Stanley, 1955). Such processes could introduce anomalously large boulders into an otherwise fine-grained environment. Later drying leads to the development of desiccation polygons on a variety of scales (Fig. 5.9). The polygonal sheets curl upwards and the cracks may be filled by wind-blown sand or by sand

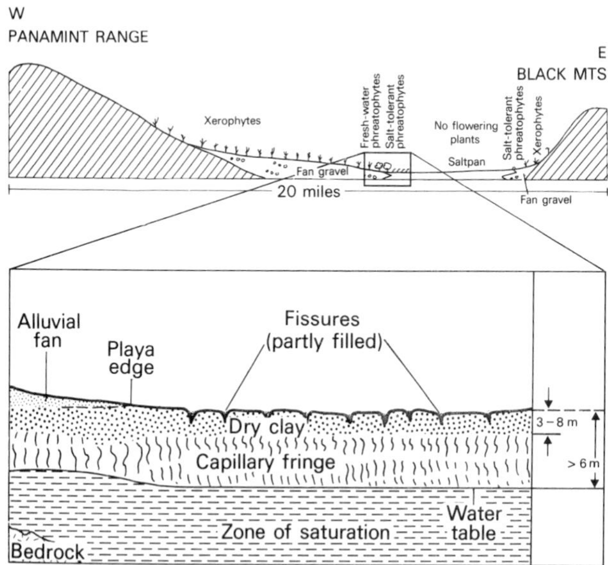

Fig. 5.9. Relationship between fringing alluvial fans, salt pans and the distribution of plants in Death Valley, California (after Hunt *et al.*, 1966). Enlargement shows a hypothetical cross-section of large-scale desiccation fissuring found in some playas and its relationship to the groundwater (after Neal, 1965).

Fig. 5.11. Salt ramparts between saucer-like polygons on the surface of the rough silty facies of the chloride zone of the Death Valley playa. The force of crystallization of the halite is thought to exploit weaknesses provided by desiccation cracks. In the background, an alluvial fan is building out on the playa surface (from Hunt, Robinson, Bowles and Washburn, 1966).

Fig. 5.10. In the sulphate zone of Death Valley playa, a layer of massive gypsum has a thin skin of anhydrite. It is underlain by calcitic sandy silt. Expanded section shows the salt profile in the carbonate zone where surface halite occurs as a blister-like crust above the gypsum of the sulphate layer. The calcite of the carbonate layer occurs as small euhedral crystals in a silty sand background (after Hunt, Robinson, Bowles and Washburn, 1966).

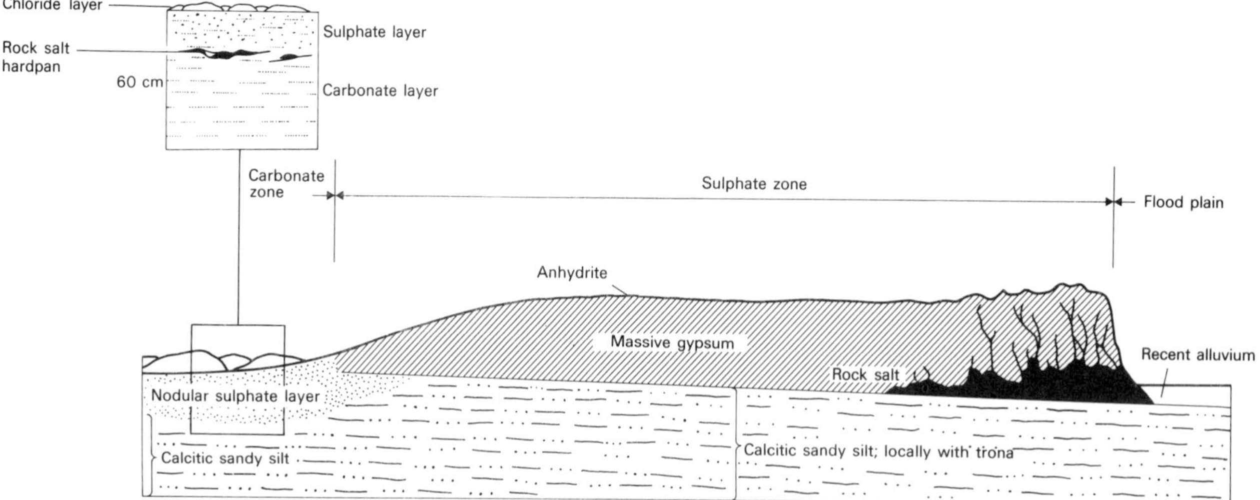

injected from below as sand dykes (Oomkens 1966; Glennie, 1970). The mud polygons may be eroded by later floods to give current-derived intra-formational breccias, though soft-sediment deformation alone may produce a 'diagenetic breccia' difficult to distinguish from current-formed breccias.

The high evaporation rates of many playas cause precipitation of salts from solution on or beneath the sediment surface. This is particularly important where ground water plays a major part in supply and where, in consequence, little clastic material is supplied by surface run-off. The nature of the older rocks of the catchment area will determine the type of salts precipitated in a playa and if older evaporites occur, deposition rates will be accelerated. The dominant mineral is usually halite though in some cases other species may be important, and in general playa evaporites comprise variable suites of minerals (e.g. Hunt, Robinson, Bowles and Washburn, 1966; Goudarzi, 1970).

It is sometimes possible to recognize distinct mineral facies and these may show a zonal arrangement within the basin, more soluble salts being deposited towards the centre (Hunt, Robinson, Bowles and Washburn, 1966; Amiel and Friedman, 1971). Local tectonics and hydrology will, however, complicate the picture. In Death Valley, for example, three main mineral facies are recognized each with sub-facies determined by clastic admixture or surface morphology. Carbonate, sulphate and chloride facies have a roughly concentric zonation with the chloride facies occupying by far the biggest area in the central part of the basin. Halite, however, is the dominant mineral in all areas and the carbonate and sulphate facies are defined by the presence of relatively minor amounts of calcite and gypsum. The carbonate facies occurs at the transition from distal fan to saltpan, the calcite occurring within the clastic sediments which fine into the basin. In silts of the carbonate facies, halite encrusts the sediment surface and a sulphate-rich layer with gypsum nodules lies between this and the carbonate-enriched sediments (Fig. 5.10). In the sulphate facies gypsum is precipitated at the surface either as a massive layer or as a crystal mush with an irregular microrelief (Fig. 5.10).

Four sub-facies are recognized in the chloride facies. The massive sub-facies has a layer of silt-free rock salt at the surface eroded into sharp grooves and pinnacles. The facies is thought to be the result of the evaporation of an earlier lake and not a product of prevailing conditions. The rough silty sub-facies involves a mixture of silt and halite up to 1 m thick deformed into spires or saucer-shaped slabs (Fig. 5.11). A thin silt layer caps the spires which are about 50 cm high and which may have developed from the weathering of earlier slabs. The smooth silty, sub-facies is the most extensive and consists of a smooth silt layer up to 15 cm thick which is underlain by up to 30 cm of salt and rests on clastic sediment. The silt layer consists mainly of small nodules of sulphates and borates, sometimes in the form of caliche layers, with a thin, hard crust with a carbonate and halite cement. The halite is thought to have formed at the top of the capillary fringe of the groundwater. The smooth sub-facies grades into the rough sub-facies where the surface is periodically submerged by flooding. The erosional sub-facies occurs where flooding has been intense and the silt has been removed to allow solution of the underlying halite. Where floods are seasonal, salt crusts are less well developed and a wet silty surface has either only a thin crust or a surface efflorescence during dry periods.

Desiccation polygons are commonly formed where evaporites precipitate. Subsequent precipitation of salts, generally halite, along the cracks builds ramparts between the polygonal areas (Glennie, 1970; Hunt, Robinson, Bowles and Washburn, 1966), presumably leading to considerable disruption of any bedding.

5.2.7 Overall desert models

Overall models of desert sedimentation depend on both the tectonics of the area, and the details of topography and climate. In consequence, models are variable but may possibly be categorized around two main types. In fault-bounded basins with tectonically active margins, such as Death Valley, California, alluvial fans may pass out into either playa lake or aeolian dune environments or into some combination of the two. Wind-blown dunes may mantle the lower parts of the alluvial fans, often migrating transverse to the slope of the fans. They may also migrate out over the playa. In more stable settings, sand seas (ergs) form most of the central parts of the basin and are fringed by alluvial or erosional environments. Local topographic lows may become the sites of inland sabkha development, but these are likely to be ephemeral. Over the ergs as a whole, wind directions may change systematically with a wheel-around effect and the wavelengths and heights of draas tend to increase towards central areas (Fig. 5.2) (Wilson, 1973).

A similar two-fold division has recently been proposed by Hardie, Smoot and Eugster (1978) who separate an *alluvial fan-ephemeral saline lake* model from an *ephemeral stream-floodplain-dune field-ephemeral saline lake* model.

5.3 ANCIENT DESERT SEDIMENTS

5.3.1 Introduction

The recognition of ancient desert sediments must rest on a variety of lines of evidence, none of which is unambiguous in isolation and which, even in combination, may leave room for doubt. The four 'criteria' most commonly applied, after a lack of fauna, are the presence of red coloration, aeolian dunes, evaporites and particular types of soil profile. Red coloration is dealt with in more detail in Chapter 3, but it is worth noting here that its post-depositional development requires some water. Aeolian dunes do not require desert conditions for their formation since they are a common feature of many coastlines and do not by themselves indicate that evaporation outweighed precipitation. Evaporites, on the other hand, do reflect this balance but may be of marine or continental origin. Caliche soils and silcretes indicate a low water table and are often the result of arid or semi-arid conditions. Of the four criteria, red coloration has traditionally been regarded as the most reliable, though there is now evidence that diagenetic reddening may also occur in moist tropical areas (e.g. Walker, 1974; Van Houten, 1972) and that red sediments occur in other environments, such as pelagic oceanic oozes. Its reliability is therefore doubtful and a diagnosis of ancient desert conditions must rely ideally on an appraisal of more than one feature.

5.3.2 Ancient aeolian sandstones

In spite of the many studies of palaeowind directions of aeolian sandstones (e.g. Baars, 1961; Stokes, 1961), few attempts have been made to distinguish ancient analogues of the various types of dune recognized in modern deserts. The recognition of sandstones as aeolian is usually based on a variety of criteria: medium to fine sand grade, good sorting, high roundness and sphericity, 'frosted' and polished grain surfaces, lack of clays and micas, lack of gravel, large-scale cross-bedding, well-defined lamination. While most of these criteria are valid, especially in combination, it is possible to be misled, particularly where aeolian sands have undergone local fluvial reworking or where alluvium or beach sediments have undergone minimal wind transport. In addition, large- and very large-scale cross-bedding can be the product of fluvial or shallow marine deposition. However, a coincidence of most of the criteria will generally lead to a fairly unambiguous diagnosis and thick blankets of well-sorted sandstone with large-scale cross-bedding with set thicknesses in metres are generally interpreted as aeolian. Medium- and small-scale aeolian cross-beds and flat-beds are rarely reported, however, except in close association with large-scale sets. This clearly raises the question of whether much aeolian cross-bedding is not being recognized as such.

Most attempts at a detailed interpretation of dune sandstones have relied heavily on the pattern of the orientation of dune foresets. Unimodal distributions have been attributed to transverse dunes or barchans (e.g. Shotton, 1956; Wright, 1956; Glennie, 1970, 1972) while tendencies to a bimodal distribution are explained as seif dunes (Shotton, 1956; Glennie, 1972). With subsurface information only (dipmeter or oriented core) this is probably the most powerful and possibly the only practical method. However, even with good surface exposure there has been little attempt to work out cross-bedding geometry in detail and thereby reconstruct dune type. Shotton (1937), in his pioneer study, recorded a spread of 160° in the cross-bedding directions from the Lower 'Bunter' Sandstone of the English Midlands. He explained this in terms of the movement of barchanoid dunes which he had already postulated. Glennie (1972, p. 106) suggested that while low-angle foreset dips might be associated with the flanks of seif dunes, horizontal laminae seem to be more commonly associated with barchans or transverse dunes. Sections in the cores of seif dunes might be expected to show bimodal cross-bedding directions while those of the flanks would be unimodal and might give a mistaken impression of the wind direction (Glennie, 1972). McKee (1945) suggested that the down-current concave curvature of foresets in the Coconino Sandstone was evidence for barchanoids. Clearly this

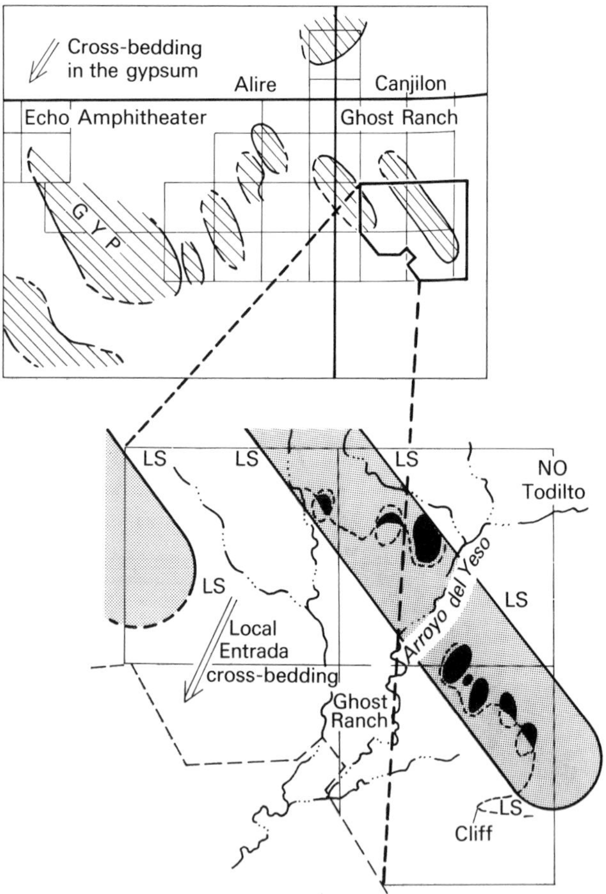

Fig. 5.12. Trends of (A) gypsum belts on a limestone substrate and (B) elongated dome-shaped dunes within one of the belts, and their relationship to the effective wind direction (Upper Jurassic Todilto Member of New Mexico) (after Tanner, 1965).

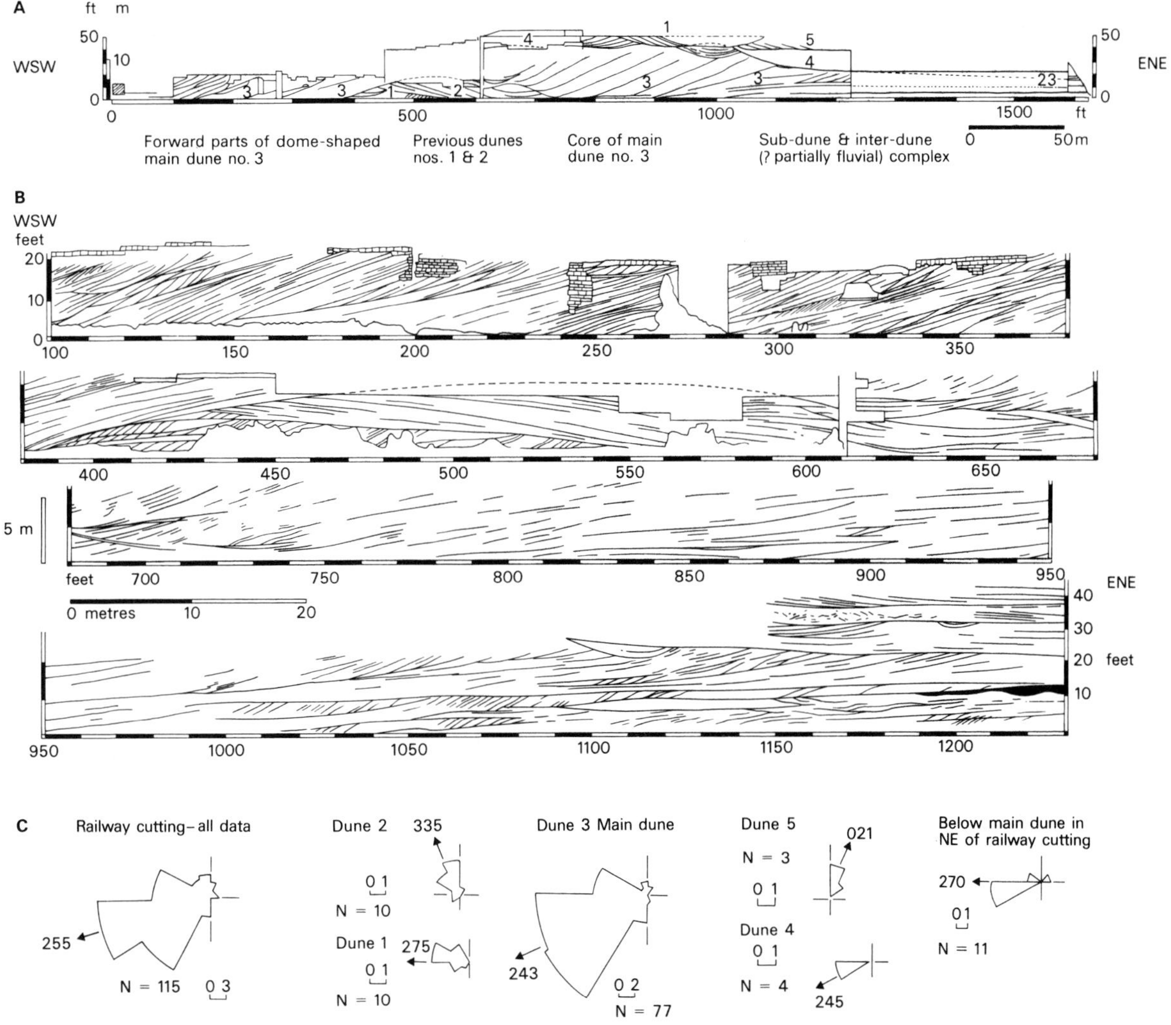

Fig. 5.13. Internal structures in the core of a supposed dome-shaped dune; Triassic, Cheshire, England. A. Shows the broad relationships of the main dune units; B. A more detailed section of the exposures in A; C. Foreset azimuth rose diagrams from each of the main units identified in A (after Thompson, 1969).

approach demands exceptional exposure to produce sound results.

In the Todilto Member of New Mexico, dome-shaped dunes of gypsum about 170 m wide, 350 m long and 30 m high were elongated more or less parallel to the foreset azimuth and had surfaces dipping at up to 40° (Tanner, 1965). Individual dunes rested on a limestone surface and were separated by tens or even hundreds of metres. The individual dunes are arranged in belts which are broadly transverse to the prevailing wind direction (Fig. 5.12).

A dome-shaped dune is also described from the Triassic of Cheshire, England (Fig. 5.13) (Thompson, 1969). Three-dimensional exposures in railway cuttings allow subdivision of the sequence into distinct units, almost entirely within cross-bedded, well-sorted, fine sandstones. The lowest unit comprises two small dunes eroded into a smooth convex-upwards morphology. The two dunes are separated by an erosion surface. The lower dune has medium-scale cross-bedding and the upper one low-angle lamination with a marked difference in their directions. The upper part of the higher set is eroded into a series of troughs, interpreted as blow-outs. Laterally equivalent to these lower dunes are interbedded siltstones and sandstones, aqueous deposits of interdune sabkhas or fluvial overbank ponds.

The main dune unit which overlies the aqueous deposits and the eroded dune remnant is at least 494 m long and 192 m wide. At the upwind end of exposure, the cross-bedded set is 2 m thick. It expands over a distance of some 60 m to a thickness of 15 m at the core of the structure. Here foresets are usually straight, but may sometimes be convex upwards. This and the azimuth distribution of foreset dips transverse to the main cross-bedding direction suggests a lee face that was convex in a down-wind direction, comparing with the structure of a modern dome-shaped dune (Fig. 5.6). The crest migrated down-wind as the dune grew in height. Small sets of internal cross-bedding and contorted layers locally complicate the foresets and an erosional trough at the top is interpreted as a blow-out. The toesets fill the blow-out troughs of the earlier dunes. Down-wind of the core, the set maintains its thickness with long, sometimes convex-upwards foresets though frequently broken by lower angle discontinuities comparable with the 'doubly cross, not to say furious', bedding of Reiche (1938) and with the reactivation surfaces found in aqueous cross-bedding. They compare also with features of modern transverse dunes (cf. Fig. 5.6), suggesting that the dune type gradually changed. On the flanks of the dune, some 200 m from the core, the cross-bedding is smaller, with abundant erosion surfaces. The foresets dip away from the core suggesting an overall radially outwards pattern on the fringes of the dome.

The significance of bounding erosion surfaces within aeolian sandstones has been explored by Brookfield (1977). Using the Cheshire Triassic example described above and the Permian Sandstones of Locharbriggs, Scotland, he recognized a three-fold hierarchy of bounding surfaces. The first-order surfaces are flat-lying or convex upwards, laterally extensive and can be widely spaced. They truncate both second-order surfaces and underlying dune cross-bedding. Second-order surfaces generally dip at moderate to low angles usually in a down-wind direction and separate sets of cross-bedding. Third-order surfaces correspond to reactivation surfaces, bounding bundles of foresets within the same set.

The first-order surfaces are thought to represent the boundaries between the migration products of successive draas across the area. Likely subsidence and bedform migration rates suggest that the bedforms must be on the scale of draas. This interpretation is at variance with that of Stokes (1968) who envisaged sand deflation to the water table to produce the surfaces. Water table fluctuations and dune migration were then thought to account for successive surfaces. Brookfield, however, pointed out that the water table in deserts is seldom horizontal and that there is no evidence of the interdune sabkha deposition that would occur if the water table was at the surface.

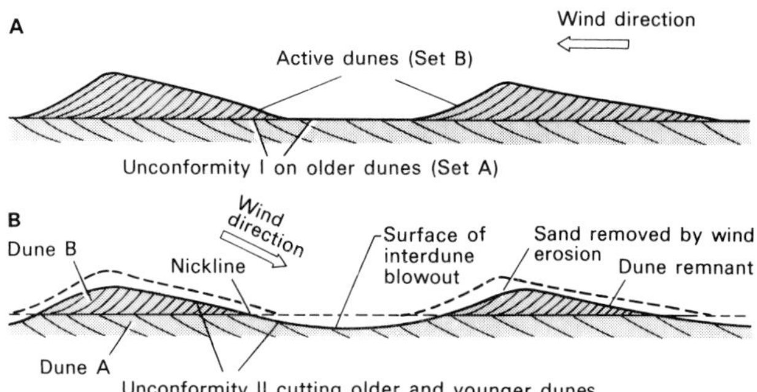

Fig. 5.14. Development of scoop-like erosion surfaces and nicklines within aeolian dune sands by the development of a blow-out due to a change in wind direction. A. Situation of earlier dominant wind; B. Situation after the change of wind pattern (after Walker and Harms, 1972).

Second-order surfaces are interpreted as the products of successive dune migrations over a larger (draa) form which possibly lacked its own slip face while third-order surfaces are attributed to short- or long-term changes in the pattern of air movement. Short-term changes are only likely to modify the upper parts of dunes; third-order surfaces which extend to the base of the set require longer-term changes.

In the Lyons Sandstone (Permian) of Colorado large tabular sets of cross-bedding up to about 12 m thick and several hundreds of metres in length are interpreted as aeolian dune deposits (Walker and Harms, 1972). The foresets commonly show wind ripples on their surfaces whose crest lines are parallel to the dip of the foresets, suggesting that transverse winds periodically reworked the slip faces. Some of the dune sets are eroded in their upper parts by concave upwards erosion surfaces interpreted as blow-outs developed between overlying dune units (Fig. 5.14) (Walker and Harms, 1972, Fig. 4). Animal tracks and small circular depressions interpreted as rain pits on foreset surfaces attest to periodic wetting of the dune by rain.

5.3.3 Ancient loess

Few ancient loess deposits older than the Pleistocene have been recognized. This is probably due to the rather featureless nature of such deposits and the consequent lack of positive criteria for their recognition. It is, however, quite likely that thick sequences of siltstones, often reddened, in sequences associated with alluvial or aeolian dune sandstones may be ancient loess. Wills (1970) suggested that the Keuper Marl (Triassic) of the British Midlands might be a loess and it is also possible that certain Old Red Sandstone 'marls', usually interpreted as floodplain deposits, may also be loess. Such deposits are often the parent sediments for soils, particularly caliches.

5.3.4 Ancient playa and inland sabkha deposits

Many present-day playas were once permanent lakes or arms of the sea and their present playa status may be temporary. Ancient playa deposits are also often intimately interbedded with the sediments of lakes or shallow seas or they may be associated with other types of typical desert sediments. Many inland evaporites are distinguished from marine evaporites by the common admixture of terrigenous impurities and by the associated sediments.

Some of the best documented examples come from the Permian and Triassic of North-western Europe. In the North Sea, the Rötliegendes contains a fine-grained clastic member, the Ten Boer Member, which is, in part, a monotonous succession of red mudstones with minor siltstones and sandstones and local nodular anhydrite (Glennie, 1972). The mudstones are commonly mudcracked and curled and frequently cut by sandstone dykes. Adhesion ripples are locally common and, where associated with nodular anhydrite, may be contorted. Thick sequences of adhesion ripple deposits require conditions of gradually rising water table such as might be found in more permanent sabkhas, while a lack of such ripples might indicate that the sabkha was ephemeral. Traced into the basin, the interbedded mudstones and siltstones of the Ten Boer Member pass into a thicker sequence of mudstones with abundant scattered and often euhedral halite, the so-called 'haselgebirge facies' (Kent and Walmsley, 1970). The facies lacks anhydrite and the halite may form up to 30% of the sequence.

In the Triassic halite deposits of Cheshire, England, halite also forms distinct layers up to several metres thick, often interbedded with clastic horizons. The halite layers internally show a variety of fabrics (Arthurton, 1973). Normally, crystals are elongated perpendicular to the bedding suggesting competitive upward growth into saturated brine from the floor of the evaporating water body. Crystallographic continuity is often maintained over several halite layers where clastic interlaminations are thin or absent. Sometimes the interlaminations or interbeds totally blanket the halite but where they are thin, the halite grows through the clastic layer which is incorporated as an inclusion in the halite. Bedding planes show polygonal patterns often of quite large dimensions associated with bedding festoons in vertical sections (Hollingworth, 1965). The patterns are due to either large-scale desiccation or pressure during halite growth comparable with the salt ramparts of modern playas (cf. Fig. 5.11) (Hunt, Robinson, Bowles and Washburn, 1966).

Haselgebirge and bedded halite facies are thought to represent different modes of halite precipitation. The haselgebirge facies is probably the product of displacive crystal growth in the vadose zone of flats bordering a substantial brine reservoir (D.B. Smith, 1971), while the bedded halite is the product of precipitation on the floor of a brine body which may have periodically totally dried out. Interbedded coarser clastic layers may be alluvial; interbedded silt or mud may be wind-blown desert loess (Wills, 1970). The two major halite facies may be related, with a dominance of bedded halite in the central parts of a basin and haselgebirge facies on the margins (D.B. Smith, 1971; Warrington, 1974). In this case the salt lake provided the reservoir for interstitial halite supply. The nature of the supply of brine to these Permian and Triassic environments is a subject of some controversy. It has been argued that at least episodic if not actually permanent connection with the sea is necessary to generate the volumes of salt involved (Evans, Wilson, Taylor and Price, 1968; Evans, 1970). However, whatever the source of the brines, a playa setting explains many of the features of the sediments.

5.3.5 Ancient desert alluvium

Alluvial sediments are described in Chapter 3 and little more need be added here. Around the margins of desert basins are alluvial fans or ephemeral streams. Interfluvial areas often develop soil profiles, particularly caliches, and red coloration. Where the alluvial deposits are associated with aeolian sands, the products of aeolian reworked alluvium and of fluvially reworked aeolian sands may be confused (e.g. Stokes, 1961). Even the presence of strong fluvial indicators such as mudflakes is not always unambiguous evidence. Sometimes it may only be possible to separate aeolian and fluvial deposits on the basis of palaeocurrent directions and a knowledge of regional palaeoslope, particularly when the available data is confined to cores.

5.3.6 Overall desert facies patterns

The facies assemblage and the relationships of facies associations, both lateral and vertical, are controlled to a large extent by the tectonic setting. In desert graben fills, alluvial fan deposits near the basin margin commonly pass distally into playa sediments with evaporites or into floodplain deposits of ephemeral or possibly more permanent streams with few dunes (e.g. Thompson, 1970; Collinson, 1972; Steel, 1974). In large desert basins with little active faulting and less well-defined margins, aeolian dune sandstones form broad lenticular sand bodies of thick lateral extent analogous to ergs (e.g. Baars, 1961). In Europe, the position of the Rötliegendes dune sandstones, overlying a gas source rock (Coal Measures) and underlying a Zechstein evaporite caprock, makes them the main reservoirs of the gas fields of Holland and the Southern North Sea. The aeolian dune sands occupy a transition zone

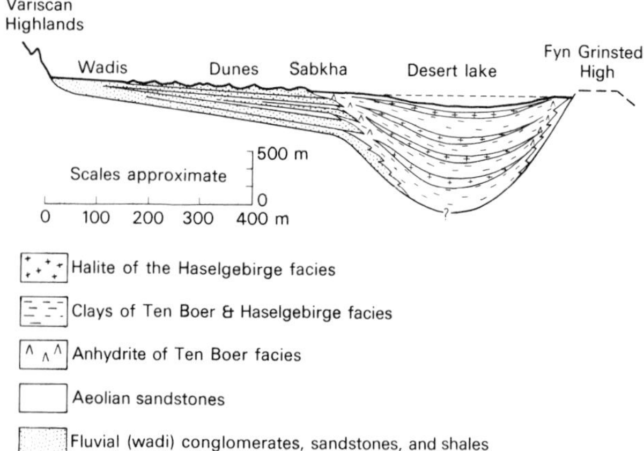

Fig. 5.15. The distribution of environments and facies in the Rötliegendes basin of the southern North Sea prior to the Zechstein transgression (after Glennie, 1972).

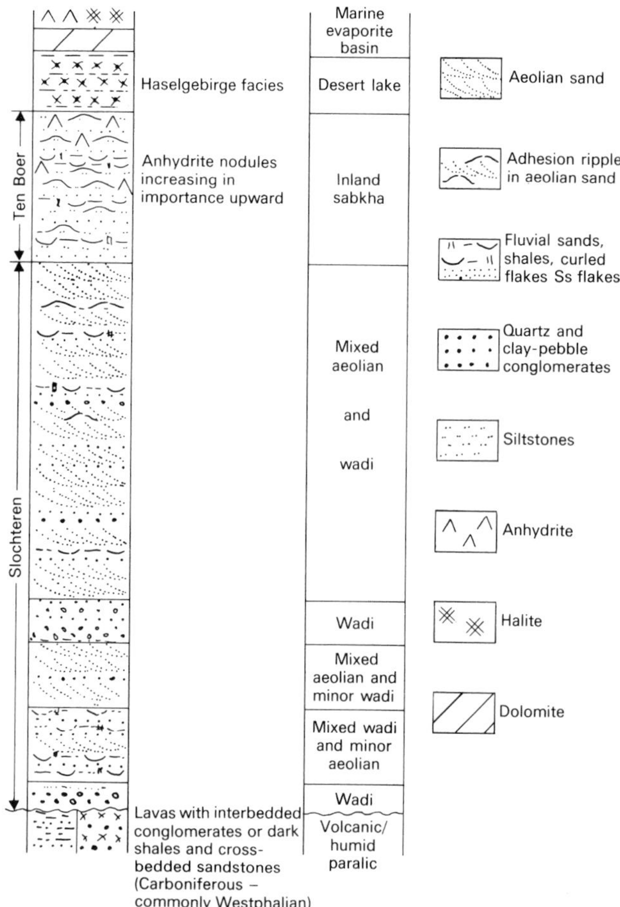

Fig. 5.16. Schematic section of the Upper Rötliegendes in north-west Europe from the south-central Rötliegendes basin. The section is composite and not all facies are present at any one place (after Glennie, 1972).

between fluvial and playa areas (Fig. 5.15) (Glennie, 1972; Marie, 1975; Van Veen, 1975). In the south, near to the rising sourcelands of clastic supply, wadi deposits and aeolian dune sandstones are randomly mixed, suggesting that the ephemeral streams periodically transected the dune fields (Fig. 5.16). Disparities of wind and water flow directions commonly occur. This sequence is banked up against the erosional relief of the basin margin. Upwards and distally, the dune sandstones become more important than the wadi deposits and localized interdune sabkhas are interspersed with the dune sandstones. Even further to the north, the aeolian sands gradually give way to inland sabkha sediments with anhydrite and accretion ripples suggesting a high water table. The deposits in turn pass into finer deposits with substantial halite, the deposits of an ephemeral lake. This transition takes place over a distance of some 200 km though similar changes can take place over much shorter distances given appropriate basin morphology (e.g. the Irish Sea Permo-Triassic; Colter and Barr, 1975).

The above examples are largely based on borehole records, but the onshore parts of the system are synthesized for the British area by D.B. Smith (1972) and described in detail by Laming (1966) for the Permian of Devon, England. In Devon, stream-dominated fan sediments, derived from a rising hinterland, are interbedded with aeolian sandstones. The large cross-bedded sets, up to 10 m thick and 60 m wide, are interpreted as the products of dunes of unspecified shape. The wind directions inferred from the cross-bedding have components across and up the alluvial palaeoslope. While this could be a purely fortuitous relationship, it is quite common in several interbedded alluvial/aeolian sequences and may reflect the present-day tendency for daytime winds to blow strongly towards upland areas (Glennie, 1972).

Following the period of tectonic activity, localized deposition, frequently in valleys or fault-bounded troughs, gave way to larger coalesced basins of deposition as topography degraded (D.B. Smith, 1972). Early marginal alluvial fan breccias and conglomerates diminished in extent as playas and aeolian dunes spread further afield.

FURTHER READING

Cooke R.U. and Warren A. (1973) *Geomorphology in Deserts*. pp. 374. Batsford, London.

Glennie K.W. (1970) *Desert Sedimentary Environments*. Developments in Sedimentology, **14**, pp. 222. Elsevier, Amsterdam.

CHAPTER 6 Deltas

6.1 INTRODUCTION

Deltas are discrete shoreline protuberances formed where rivers enter oceans, semi-enclosed seas, lakes or barrier-sheltered lagoons and supply sediment more rapidly than it can be redistributed by indigenous basinal processes. Generally, deltas are served by well-defined drainage systems which culminate in a trunk stream and supply sediment to a restricted area of the shoreline. Less organized or immature drainage systems produce numerous closely spaced rivers which induce uniform progradation of the entire coastal plain rather than point-concentrated progradation. At the river mouth, sediment-laden fluvial currents which have previously been confined between channel banks suddenly expand and decelerate on entering the standing water body. As a result the sediment load is dispersed and deposited, with coarse-grained bedload sediment tending to accumulate near the river mouth, whilst finer grained sediment is transported offshore in suspension to be deposited in deeper water areas. Basinal processes such as waves, tides and oceanic currents may assist in dispersion and also rework sediment deposited from the fluvial currents. Many of the characteristics of deltas stem from the results of this contest between fluvial and basinal processes.

Modern deltas often form extensive wetland areas of high biological productivity and are therefore important conservation areas (Gagliano and van Beek, 1973, 1975). For example, the recent acceleration in coastal erosion in the Nile delta brought about by completion of the Aswan dam has precipitated a major scientific enquiry in an attempt to preserve the fertile land and dwellings of the delta area (UNDP–UNESCO, 1976). Furthermore, as deltas are often major depositional centres they tend to produce exceptionally thick sedimentary successions. Ancient deltaic facies have been recognized throughout the geological column, and a large proportion of oil, gas and coal reserves are located in rocks of this type.

6.2 DEVELOPMENT OF DELTA STUDIES

Rigorous sedimentological studies of modern deltas commenced with Johnston's (1921, 1922) account of the Fraser River delta and classic work on the Mississippi delta (Trowbridge, 1930; Russell, 1936; Russell and Russell, 1939; Fisk, 1944, 1947). The exemplary nature of early research on the Mississippi delta caused it to be regarded as *the* delta model for a period, and this position was fortified by later work which provided further insight into this delta (Fisk, 1955, 1960; Coleman and Gagliano, 1964; Coleman, Gagliano and Webb, 1964). Other deltas were described at this time, but the prevailing attitude was largely concerned with demonstrating similarities between deltas. However, van Andel and Curray (1960) recognized the need for a critical comparison of modern deltas. Whilst noting basic similarities between deltas they

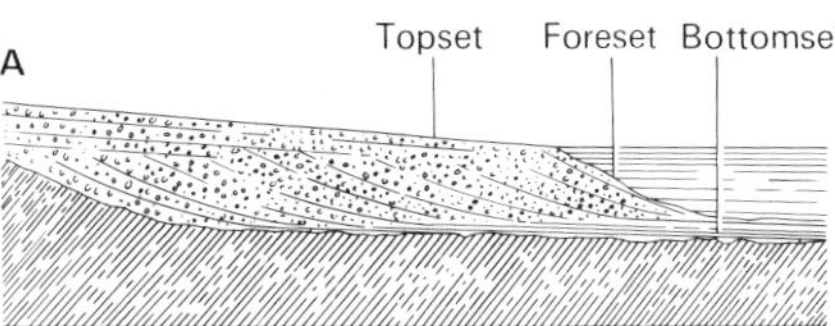

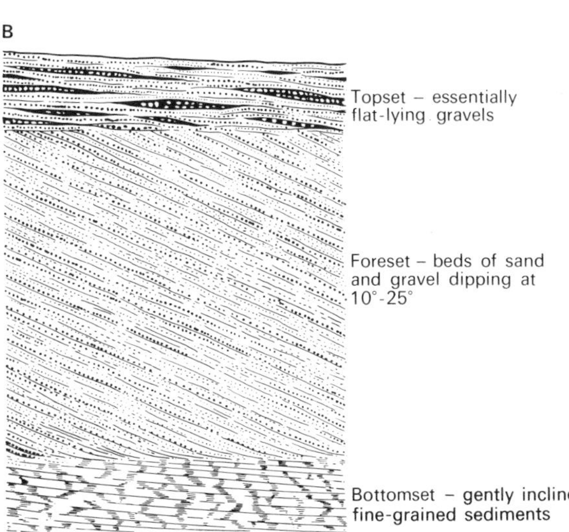

Fig. 6.1. (A) Section through a Pleistocene delta in Lake Bonneville; (B) Vertical facies sequence produced by delta progradation (after Gilbert, 1885; Barrell, 1912).

stressed that 'the striking variation in structure and lithology . . . as exemplified by the Rhône and Mississippi deltas, should not be underestimated. A study of comparative morphology and lithology of modern deltas appears highly desirable.' Subsequent publications on individual deltas concluded with a comparison of the described delta with other examples (e.g. Allen, 1965d; Van Andel, 1967), thus consolidating the trend towards the variability of deltas which currently constitutes the leitmotif of modern delta studies (Fisher *et al.*, 1969; Wright and Coleman, 1973; Coleman and Wright, 1975; Galloway, 1975).

Studies of deltaic facies commenced in ancient successions rather than modern deltas with Gilbert's (1885, 1890) descriptions of Pleistocene deltaic facies in Lake Bonneville. Glacial streams transporting coarse sediment produced a series of fan-shaped lacustrine deltas exposed by subsequent lake-level changes and channel dissection. The deltas have a three-fold structure which generated a distinctive vertical sequence of bedding types during progradation (Fig. 6.1). Barrell (1912, 1914) subsequently proposed criteria for the recognition of ancient deltaic deposits based on Gilbert's descriptions, and later applied these criteria to the Devonian Catskill Formation. The terms topset, foreset and bottomset were used to describe the delta structure, and the bedding, texture, colour and fauna of each component was discussed, thus initiating the facies approach in deltaic deposits.[1] Although Barrell stressed that not all deltas exhibited this Gilbert-type structure, the concept conditioned thinking on modern deltas for several decades, and the presence or absence of large-scale inclined foresets was considered an important criterion in the recognition of ancient deltaic successions.

In USA during the late 1940's outcrop and subsurface information began to be interpreted in terms of palaeo-environments and it was gradually realized that significant amounts of coal, oil and gas were located in ancient deltaic systems. As these studies were concerned with locating and tracing sandstone bodies they tended to concentrate on lateral facies relationships within well-defined stratigraphic intervals, thus permitting a deltaic interpretation to be offered with some degree of confidence (e.g. Pepper *et al.*, 1954; Busch, 1953; Nanz, 1954). A parallel development in USA and Europe was the recognition of coarsening-upward cycles or cyclothems which reflected a passage from marine facies upwards into terrestrial facies. The cycles were often attributed to delta progradation, and although this approach focussed attention on the vertical arrangement of facies it was not entirely beneficial to the development of delta studies. In many Carboniferous examples, where the approach was most eagerly applied, debates on the definition and genesis of the cycles often took precedence over analysis of the rocks. Controversy raged over the horizon at which cycles commenced and pleas were issued for more objective definitions of the cycles using statistical techniques. Discussions on the genesis of the cycles were often based on an idealized cycle rather than actual successions, and a variety of tectonic, climatic or sedimentological controls was invoked, none of which achieved a consensus. During the reign of this approach the sedimentary facies and their precise relationships were often neglected, with a tendency to view the succession as 'rocks that occurred rather than . . . sedimentary processes which happened' (Reading, 1971, p. 1410). In many cases it is only recently that the facies characteristics of these cycles have been scrutinized.

The economic importance of deltaic facies stimulated extensive borehole programmes in the Mississippi, Rhône and Niger deltas (Fisk *et al.*, 1954; Fisk, 1955, 1961; Oomkens, 1967, 1974; Weber, 1971). It became apparent from these studies that deltaic successions contain a wide variety of vertical facies sequences, with the type of sequence varying within a delta at different locations as well as between deltas. Current attitudes towards ancient deltaic successions stem largely from these studies. The sequence approach provides an alternative to idealized cycles and there has also been a greater tendency to discuss ancient deltaic successions in terms of different types of deltas, in harmony with the current emphasis on the variability of modern deltas.

6.3 A CONCEPTUAL FRAMEWORK FOR DELTAS

Comprehension of the variability of deltas, and the application of this information to ancient deltaic successions, requires a framework which summarizes interactions between variables which exert control on the development of deltas and indicates causally defined pathways between groups of variables. The framework adopted in this chapter regards delta regime as a general expression of the overall setting, and relates the regime of the delta to its morphology and facies pattern (Fig. 6.2).

The variables affecting delta systems stem from the characteristics of the hinterland and receiving basin. As the hinterland is responsible for supplying sediment, hinterland characteristics are largely reflected in the nature of the sediment and the fluvial regime which transports it. The receiving basin acts as a receptacle and perhaps the most important feature of a basin is the energy regime which contests the introduction of river-borne sediment. The basinal

[1] It is also noteworthy that Barrell referred to a deltaic cycle of sedimentation, but at this time the cycle was strictly Davisian in relating to the physiographic 'age' of the hinterland. High rates of sediment supply associated with a 'youthful' hinterland resulted in delta progradation, but as the hinterland matured and passed into 'old age' reduced sediment supply resulted in marine planation of the delta.

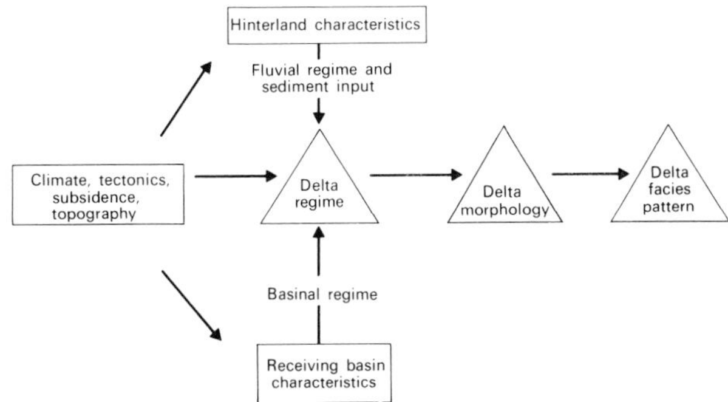

Fig. 6.2. Conceptual framework for the comparative study of deltas, applicable to modern deltas and ancient deltaic successions.

regime has dependent relationships with several major features of the basin such as shape, size, bathymetry and climatic setting, and can be considered to reflect these features. Interaction between the sediment-laden fluvial waters and basinal processes defines the delta regime which dictates the dispersal and eventual deposition of sediment in the delta area, and therefore occupies a focal position in this framework. An important feature which emerged from comparisons between modern deltas is a relationship between delta regime and morphology. Initially the classic birdfoot–lobate–arcuate–cuspate spectrum of delta types was related to increasing wave influence over fluvial processes (Bernard, 1965), and more recently this relationship has been extended by using data from a wide range of modern deltas, including those significantly influenced by tidal processes (Fisher *et al.*, 1969; Wright and Coleman, 1973; Coleman and Wright, 1975). Delta types therefore tend to be defined in terms of regime and illustrated by a characteristic morphology. The final link in the framework derives from comparison of the results of drilling programmes in the Mississippi, Rhône and Niger deltas which reveal that contrasts in overall facies patterns are related to differences in the regime and morphology of the deltas. Causative links therefore exist between delta regime, morphology and facies pattern.

This framework can also be applied to ancient deltaic successions, where studies are based on a partial record of the facies pattern gained from measured sections, subsurface cores or electrical logs, often widely scattered. Reconstruction of the facies pattern and analysis of the facies and sequences in terms of depositional processes can be used to reconstruct the regime and morphology of the former delta, permitting a regime-based delta type to be proposed. The importance of this is firstly that in thick basinal successions comprising a series of delta complexes it may be possible to detect temporal variations in delta type, reflecting evolution of the hinterland and/or receiving basin (Belt, 1975; Galloway, 1975). Secondly, as the nature of the palaeo-delta is inferred from a partial record of the facies pattern, it contains predictions on the remainder of the facies pattern in unexposed or unexplored areas. This is important in the exploration and development of hydrocarbons located in deltaic sandstone bodies as the postulated delta type provides a predictive model which can be tested by subsequent investigations.

6.3.1 **Hinterland and receiving basin characteristics**

The hinterland comprises the drainage basin and fluvial system where variables such as relief, geology, climate and tectonic behaviour interact to determine the fluvial regime and sediment supply which feed the delta (see Chapter 3 for discussion). With regard to deltas, important features include:

(1) The total amount of sediment supplied in relation to the reworking ability of the basinal processes.

(2) The calibre of sediment supply which influences the dispersion and deposition of sediment in the delta. Coarse-grained bedload sediment tends to be deposited in the immediate vicinity of the distributary mouth and either forms distributary mouth bars, or is reworked by wave and tidal processes into beach-barrier systems or tidal current ridge complexes. In contrast, finer grained suspended load sediment is generally transported offshore and dispersed with the aid of basinal processes over a wide area of the basin. Deposition produces an extensive mud-dominated platform in front of the delta which may be over-ridden by delta front sands as progradation continues.

(3) Fluctuations in discharge can be significant in determining the calibre of sediment supply. For example, rivers with erratic or 'flashy' regimes characterized by brief, episodic high discharge periods are more likely to supply coarse sediment to the delta than more stable regimes which tend to sort sediment prior to its reaching the delta.

(4) The timing of fluctuations in fluvial discharge relative to fluctuations in the basin energy regime also influences deposition in the delta area. If the maxima are in phase, basinal processes continually redistribute the river-borne sediment, but, if the maxima are out of phase, periods of virtually uncontested delta progradation alternate with periods of reworking by basinal processes (Wright and Coleman, 1973).

Characteristics of the receiving basin which influence the development of deltas include water salinity, basin shape, size and bathymetry, energy regime, and overall basin behaviour in terms of subsidence rates, tectonic activity and sea level fluctuations.

The relative density of river and basin waters is an important first-order control on the manner in which the sediment-laden river discharge is dispersed in the basin, and this is partly a function of the salinity of the basin waters (Bates, 1953). Where rivers enter freshwater basins there is either immediate mixing of the water bodies at the river mouth or the river discharge flows beneath the basin waters as a density current. In contrast, where rivers enter a saline basin the river discharge may extend into the basin as a buoyantly supported plume due to the higher density of seawater (see Sect. 6.5.2).

The basinal regime includes the effects of wave and wave-induced processes, tidal processes, and to a lesser extent semi-permanent currents, oceanic currents and wind effects which may temporarily raise or lower sea level. The type of basin is a prime control on the nature of the basinal regime. For example, at ocean-facing continental margins the full range of basinal processes affects the delta (e.g. the Niger delta; Allen, 1965d), whereas in semi-enclosed and enclosed seas wave energy is limited due to reduced fetch and tidal influence is minimal (e.g. the Danube, Ebro, Mississippi, Po and Rhône deltas). Deltas located in narrow elongate basins or gulfs connected to an ocean experience considerable tidal effects as tidal currents are amplified and may therefore transport considerable volumes of sediment (e.g. the Ganges–Brahmaputra delta). In smaller scale deltas prograding into barrier-sheltered lagoons or lakes, the influence of basinal processes is limited and the deltas are dominated by fluvial processes (Donaldson *et al.*, 1970; Kanes, 1970). Basin water depth and the presence or absence of a shelf-slope influence the basinal regime, particularly in terms of wave attenuation.

Finally, as deltas are topographically subdued areas at the margins of basins they are extremely sensitive to subsidence trends and sea level fluctuations in the basin. Delta sites may be affected by basement-related tectonics, as in the Ganges–Brahmaputra delta which is located in a downwarped basin with numerous active normal faults, and the Tertiary Niger delta which developed in a triple junction rift system (Morgan and McIntire, 1959; Coleman, 1969; Burke, Dessauvagie and Whiteman, 1971; Morgan, 1970). They may also be affected by sediment-induced or 'substrate' tectonics involving over-pressured shales which induce deep-seated lateral clay flowage, diapirism and faulting, as in the Mississippi and Niger deltas (Coleman *et al.*, 1974; Weber, 1971; Weber and Daukoru, 1975; see Sect. 6.8).

6.4 DELTA MODELS

In view of the variability of modern deltas a single delta model is no longer adequate. Instead a series of models is required and several schemes have recently been proposed (Fisher *et al.*, 1969; Coleman and Wright, 1975; Galloway, 1975).

On the basis of a qualitative comparison of modern deltas, Fisher *et al.* (1969) distinguish high-constructive deltas dominated by fluvial processes from high-destructive deltas dominated by basinal processes. Lobate and birdfoot types are recognized in the high-constructive class, whilst wave-dominated and tide-dominated types are distinguished in the high-destructive class (Fig. 6.3). Each type is illustrated by a characteristic morphology and facies pattern, described in terms of vertical sequences, areal facies distribution and sand body geometry. By virtue of stressing facies relationships the approach is directly applicable to ancient deltaic successions, but one disadvantage is that it concentrates on end-members of what is in reality a continuous spectrum. In addition, use of the term 'high destructive' is misleading in this context since all deltas are by definition constructive whilst active, and the term also confuses a class of deltas with a distinct phase of delta history which follows channel switching and delta abandonment – the so-called destructive phase (Scruton, 1960).

An alternative scheme involves analysis of statistical information from thirty-four present-day deltas using a wide range of parameters to illustrate the characteristics of the drainage basin, alluvial valley, delta plain and receiving basin (Coleman and Wright, 1975). Interaction of the variables defines a process setting which is unique to any delta, but multivariate analysis of this information produces six discrete delta models. Each model is defined by a setting in which the delta regime is pre-eminent, and illustrated by a sand distribution pattern which reflects the delta morphology (Fig. 6.4). The models are also described in terms of processes and morphology using representative modern deltas, and facies patterns are summarized by single, idealized vertical sequences. This scheme has an extremely broad data base in terms of the number of samples and the number of parameters, and an additional advantage is that initial description of the models is devoid of specific connotations associated with individual deltas. It is therefore an attractive scheme, but a major weakness is that idealized vertical sequences cannot summarize deltaic facies patterns which are ubiquitously characterized by extreme vertical and lateral variations.

The scheme adopted in this chapter is a modified version of a scheme proposed by Galloway (1975) which uses a ternary diagram to define general fields of fluvial-, wave- and tide-dominated deltas (Fig. 6.5). At present the positions of individual deltas are plotted qualitatively, but with information of the type currently being gathered from modern deltas a quantitative positioning may eventually be possible (e.g. Wright and Coleman, 1973; Coleman and Wright, 1975).

One point which should be borne in mind when using any regime-based scheme of delta models is that the regime of the delta plain is very often different from that of the delta front. For example, the delta front of the Rhône delta is wave-dominated, whilst the delta plain is largely fluvial-dominated by virtue of being sheltered from wave action by shoreline beach-barriers which enclose the delta plain. Also in areas of moderate to high tidal range, the upper delta plain is fluvial-dominated, the lower delta plain may be tide-dominated, and the delta front may be either tide- and/or wave-dominated. The

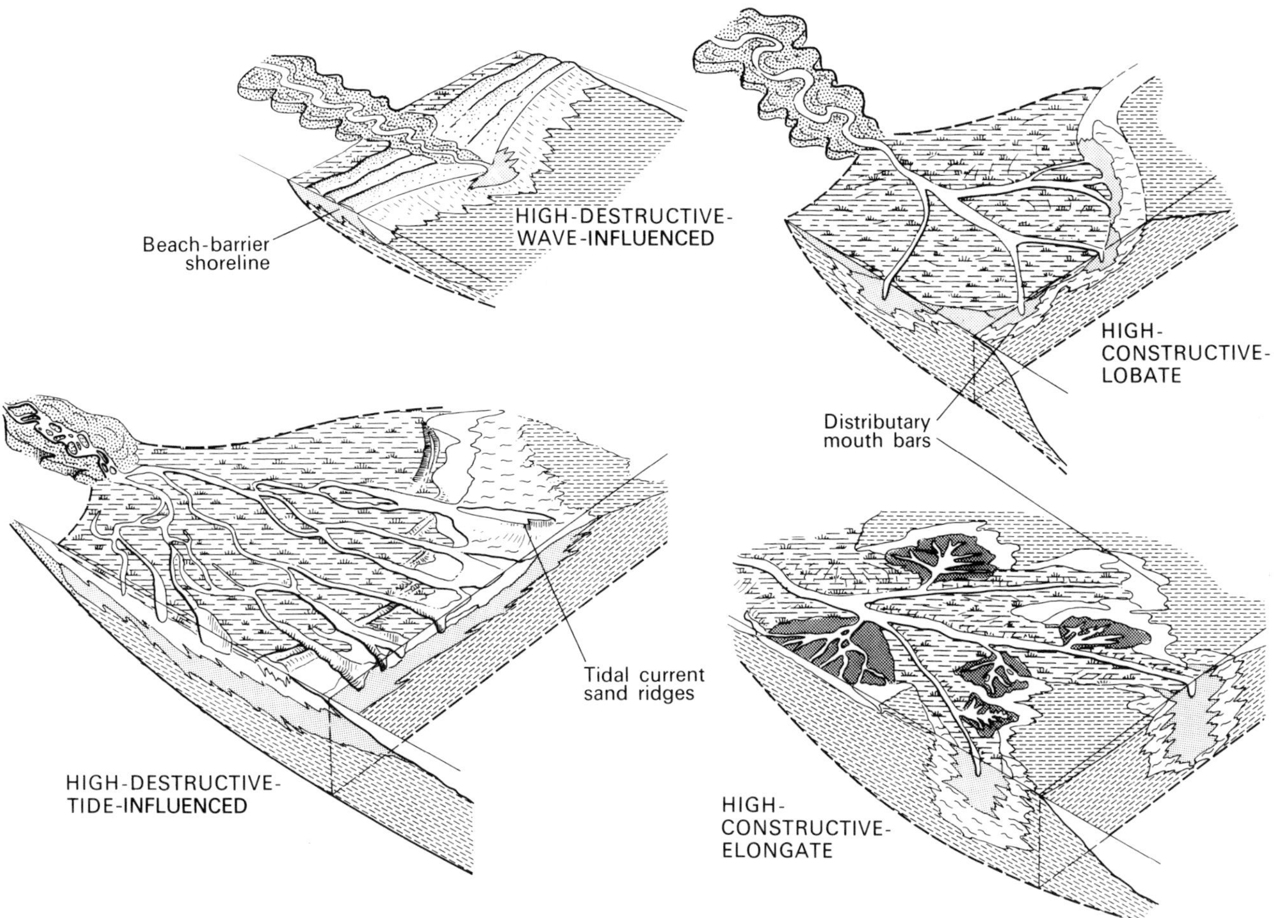

Fig. 6.3. High-constructive and high-destructive delta types as defined by Fisher *et al.* (1969).

regime of the delta front is used to define delta type.

It is also important to consider that the range of modern delta types is unlikely to be complete and it is probable that some ancient deltas had a subtly or radically different form. Where only a limited amount of delta is available, interpretation in terms of modern analogues is inevitable; but in well-exposed ancient deltaic successions it may be possible, and indeed necessary, to postulate a non-actualistic delta type.

6.5 FACIES ASSOCIATIONS IN MODERN DELTAS

Deltas comprise two basic components: the *delta front* which includes the shoreline and seaward-dipping profile which extends offshore; and the low-lying *delta plain* developed behind the delta front. As these components are often characterized by different regimes within, as well as between deltas, they are described separately.

6.5.1 The delta plain

Delta plains are extensive lowland areas which comprise active and abandoned distributary channels separated by shallow-water environments and emergent or near-emergent surfaces. Some deltas have only one channel (e.g. the São Francisco delta), but more commonly a series of distributary channels is spread across the delta plain, often diverging from the overall slope direction by 60° or more. These channels divide the total discharge of the alluvial system and supply it to the delta front.

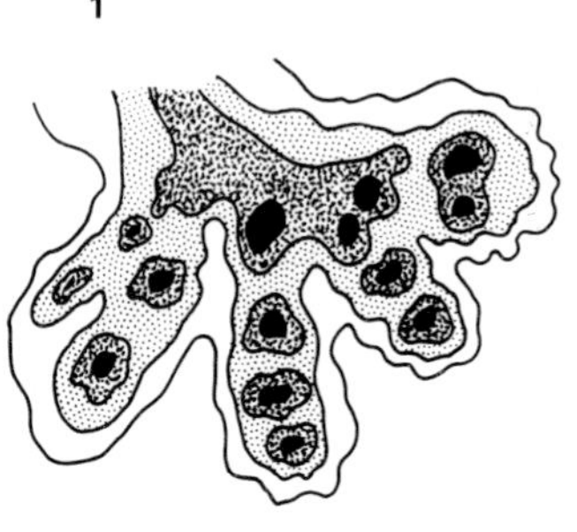

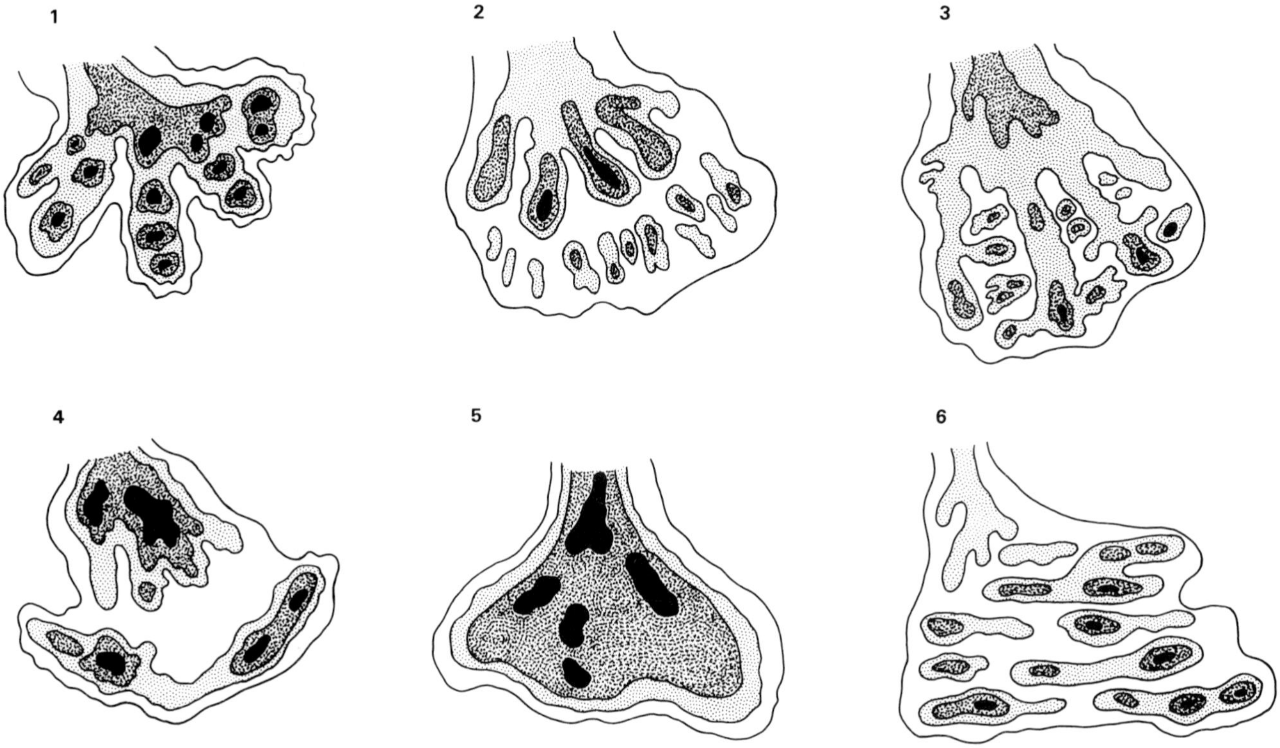

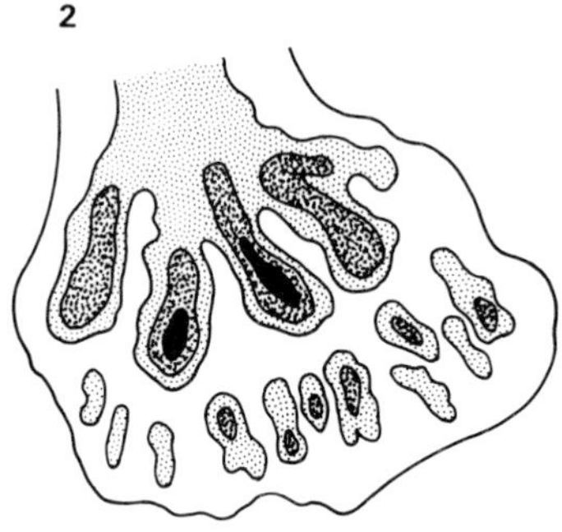

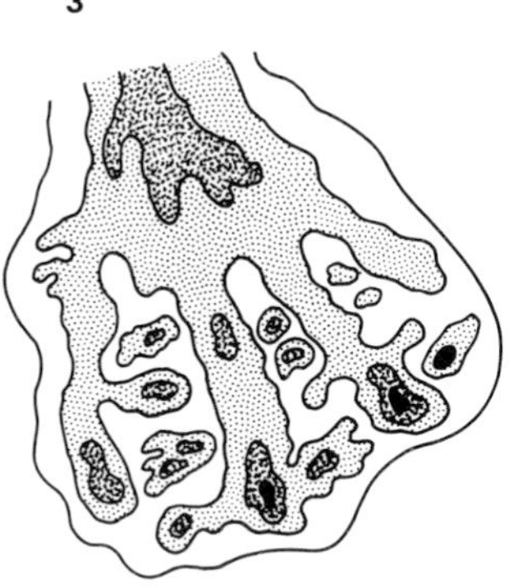

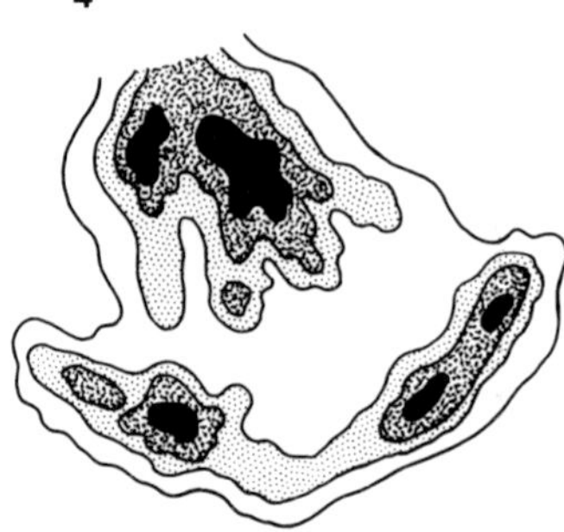

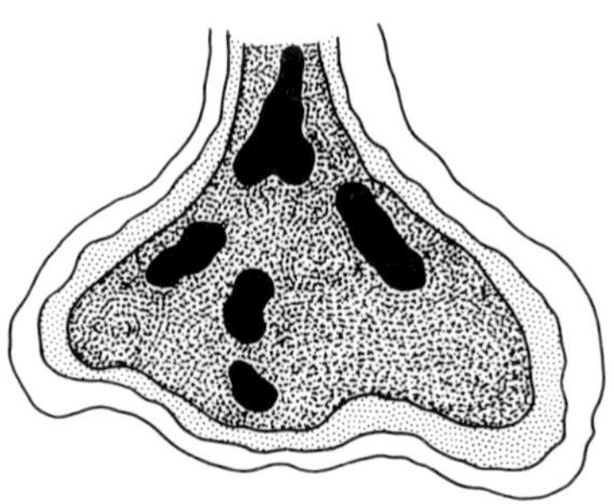

6

TYPE 1
Conditions: low wave energy, tidal range, and littoral drift, low offshore slope, fine-grained sediment load.
Characteristics: widespread, finger-like channel sands normal to the shoreline.
Example: modern Mississippi delta.

TYPE 2
Conditions: low wave energy, high tidal range, normally low littoral drift, narrow basin.
Characteristics: finger-like channel sands passing offshore into elongate, tidal current ridge sands.
Examples: Ord, Indus, Colorado, Ganges-Brahmaputra deltas.

TYPE 3
Conditions: intermediate wave energy, high tides, low littoral drift, shallow stable basin.
Characteristics: channel sands normal to shoreline, connected laterally by barrier–beach sands.
Examples: Burdekin, Irrawaddy and Mekong deltas.

TYPE 4
Conditions: intermediate wave energy, low offshore slope, low sediment yield.
Characteristics: coalesced channel and mouth bar sands fronted by offshore barrier islands.
Examples: Apalachicola and Brazos deltas.

TYPE 5
Conditions: high, persistent wave energy, low littoral drift, steep offshore slope.
Characteristics: sheet-like, laterally persistent barrier-beach sands with up-dip channel sands.
Examples: São Francisco and Grijalva deltas.

TYPE 6
Conditions: high wave energy, strong littoral drift, steep offshore slope.
Characteristics: multiple elongate barrier-beach sands aligned parallel to the shoreline with subdued channel sands.
Example: Senegal delta.

Fig. 6.4. Delta models based on multivariate analysis of parameters from a wide range of modern deltas and depicted by sand distribution patterns; increasing density of tone indicates increasing sand thickness (after Coleman and Wright, 1975).

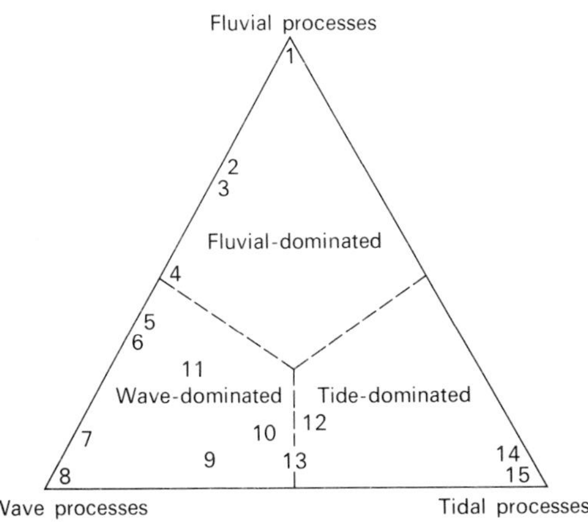

Fig. 6.5. Ternary diagram of delta types, based on the regime of the delta front area (modified after Galloway, 1975).

Between the channels is a varied assemblage of bays, floodplains, lakes, tidal flats, marshes, swamps and salinas which are extremely sensitive to climate. For example, in humid-tropical and sub-tropical settings luxuriant vegetation prevails over large areas of the delta plain as saline mangrove swamps, freshwater swamps or marshes (e.g. the Niger, Klang-Langat and Mississippi deltas). In contrast, delta plains in arid and semi-arid areas tend to be devoid of vegetation, and are characterized by calcretes (the Ebro delta) or salinas with gypsum and halite (the Nile delta). Alternatively, arid delta plains are dominated by aeolian dune fields, particularly in deltas with barrier-beach shorelines where sand is eroded from active and abandoned beach ridges (the São Francisco delta). Pingos, patterned ground and other cryogenic features occur in the delta plains of high latitude polar deltas, and tundra vegetation accumulates in shallow thaw ponds (e.g. the Mackenzie and Colville deltas; Mackay, 1963; Naidu and Mowatt, 1975).

Most delta plains are affected by fluvial or tidal processes, but only rarely by waves as wave-influenced deltas are characterized by beach-barrier shorelines which enclose and protect the delta plain. Waves are only effective in the open interdistributary bays of the Mississippi delta where they may rework part of the sediment supplied by fluvial processes.

FLUVIAL-DOMINATED DELTA PLAINS

Fluvial-dominated delta plains are either enclosed by beach ridges at the seaward end (e.g. the Rhône and Ebro deltas), pass downstream into a tide-dominated delta plain (e.g. the Niger and Mekong deltas), or are open at the seaward end and pass directly into the delta front (the Mississippi delta).

Fluvial distributary channels are characterized by unidirectional flow with periodic stage fluctuations, and are therefore similar to channels in strictly alluvial systems (see Chapter 3). High sinuosity patterns are common, but in certain arid or polar deltas with sporadic discharge and a high proportion of bedload the distributary channels are braided and anastomosing. Distributary channels in the Mississippi delta have a low sinuosity pattern and are not braided, even at low stage. However, a contrast with alluvial channels is that even in low energy basins the lower reaches of the distributary channels are influenced by basinal processes. For example, in the Mississippi delta flood tides and waves associated with strong onshore winds impound distributary discharge during low and normal river stages. Bedload transport is inhibited and fine-grained sediment may be deposited in the channel (Wright and Coleman, 1973, 1974). This sediment may be eroded during the next river flood, but some may persist to form drapes in the channel sequence. An additional feature which distinguishes fluvial distributary channels from strictly alluvial channels is the frequency of switching or avulsion. The abundance of abandoned distributary courses testifies to the frequency of this process which results from the creation of shorter, steeper courses as the delta progrades into the basin. During and after channel abandonment the effectiveness of basinal processes in the lower reaches of the former channels is enhanced, and in the Rhône and Ebro deltas abandoned channel mouths are sealed by wave-deposited beach sands (Kruit, 1955; Maldonado, 1975).

Facies and sequences of distributary channels resemble those of alluvial channels to a large extent. Cores in the Rhône and Niger deltas reveal erosive-based sequences with a basal lag, followed by a passage from trough cross-bedded sands upwards into ripple-laminated finer sands with silt and clay alternations, and finally into silts and clays pervaded by rootlets (Fig. 6.6). Some are composite or multi-storey sequences which either reflect repeated cut and fill within the channel, or minor fluctuations in channel location (Oomkens, 1967, 1974). The overall fining-upward of the channel sequence results either from lateral migration of the channel, or more commonly from channel abandonment, with the upper

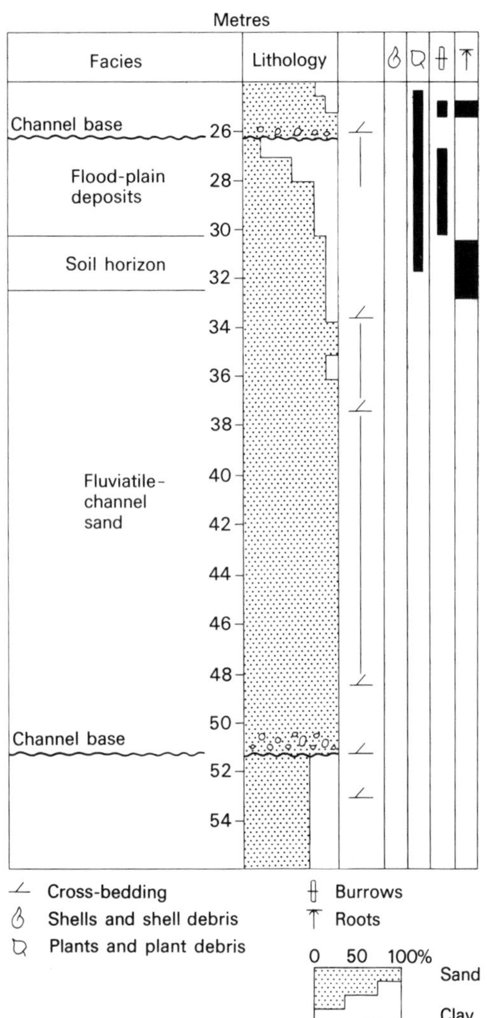

Fig. 6.6. Fluvial-distributary channel sequences from the Niger delta (after Oomkens, 1974).

fine member representing infilling of the channel by diminishing flow and perhaps later by overbank flooding from an adjacent active channel. In the lower reaches of the channels the introduction of sand by basinal processes during channel abandonment may suppress the fining-upward trend. Well-sorted, evenly-laminated sands with a marine fauna may terminate the channel sequence. An additional consequence of the switching behaviour of fluvial distributary channels is that they tend to be short-lived relative to upstream alluvial equivalents, causing the sand bodies to have low width to depth ratios. For example, in the Rhine and Rhône deltas the ratio decreases from 1,000 for the alluvial channels to 50 for distributary channels near the shoreline (Oomkens, 1974). This trend is also apparent in the Mississippi delta, although in this case it is partly related to a downstream change from high to low sinuosity channels.

Large-scale channel bank slumping can be an important feature of distributary channels as they often have fine-grained, cohesive bank materials. Scour during high river stage oversteepens the banks, inducing failure of the wetted

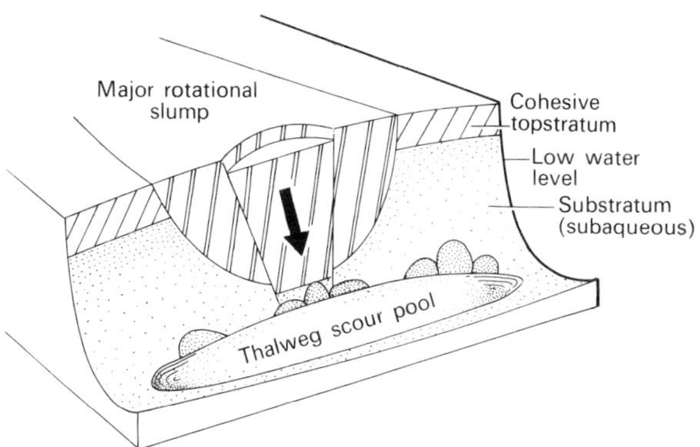

Fig. 6.7. Bank collapse by rotational slumping (after Turnbull *et al.*, 1966, and Laury, 1971).

sediments along rotational slump planes during low river stage (Fig. 6.7). Often the entire bank is slumped, and if the basal shear plane extends beneath the base of the channel slumped sediments may be preserved below the channel facies (Stanley *et al.*, 1966; Turnbull *et al.*, 1966; Laury, 1971).

Interdistributary areas of fluvial-dominated delta plains are generally enclosed, shallow water environments which are quiet or even stagnant, although in the open bays of the Mississippi delta waves induce mild agitation. This generally placid regime is frequently interrupted during flood periods as excess discharge is diverted from distributary channels into the bays. These flood-generated processes are the principal means of sediment supply to the interdistributary areas and morphological features which result from these processes include levees, various types of crevasse sands, and crevasse channels. Collectively, these features frequently fill large areas of the shallow bays and provide a platform for vegetation growth, gypsum and halite precipitation, or calcrete development, depending on the prevailing climate. The interdistributary areas tend to be dominated by a wide range of facies and sequences which reflect infilling by a variety of flood-generated processes (Elliott, 1974b; Fig. 6.8).

(1) Overbank flooding: involving sheet-flow of sediment-laden waters over the channel banks. Fine-grained, laminated sediment is deposited over the entire area, although frequently the laminations are destroyed by subsequent bioturbation. Coarser sediment is confined to the channel margins and contributes to levee development. As a result, levee facies comprise repeated alternations between thin, erosive-based sand beds representing sediment-laden flood incursions and silt-mud intervals deposited from suspension. The sands may be parallel- or current-ripple-laminated, whilst the finer sediments are frequently affected by rootlets, indicating repeated emergence or near emergence. Levee facies fine away from the channels, and lateral encroachment of the levees associated with channel migration or alluvial ridge build-up may therefore produce a coarsening-upward sequence characterized by increasing thickness of the sand beds upwards.

(2) Crevassing: in this process flood waters flow into the interdistributary area via small crevasse channels cut in the levee crest. There are two distinct mechanisms:

(a) Crevasse splay: a sudden incursion of sediment-laden waters which deposits sediment over a limited area on the lower flanks of the levees and the bay floor, producing locally wide levee aprons. The sediment may be deposited in numerous small, anastomosing streams, in which case the deposit comprises numerous small channel lenses each separated by a thin fine sediment drape. Alternatively the flow may develop into a density current and deposit an erosive-based lobe of sand which may either be a few centimetres thick or up to 1–2 m (Kruit, 1955; Arndorfer, 1973; Fig. 6.8). The thicker splay lobes often infill the bay and are overlain by facies reflecting emergence or near emergence.

(b) Minor mouth bar/crevasse channel couplets: in the Mississippi delta, couplets comprising semi-permanent crevasse channels and small-scale mouth bars are an important feature of the interdistributary areas (Fig. 6.9). Shallow crevasse channels bounded by subaerial levees flare at the mouth and deposit minor mouth bars which form localized shoal areas dipping gently into the bay (Coleman *et al.*, 1964). As a couplet progrades into the bay, proximal facies progressively overlie distal facies, and shallow borehole descriptions can be used to predict a series of vertical sequences (Fig. 6.8). Bioturbated muds and silts deposited on the bay floor pass upwards into interbedded silts and sands with multi-directional trough cross-lamination which is considered to reflect current and wave action on the mouth bar front. The upper part of the sequence is frequently removed by erosion at the base of the crevasse channel as progradation continues. Stage variations are particularly important in the crevasse channels as they may be temporarily abandoned at low river stage, resulting in complete cut-off of sediment supply until the next river flood period. Crevasse channel facies may therefore comprise sands with unidirectional, current-produced structures, together with numerous reactivation surfaces and fine-sediment drapes.

Close spacing of the couplets results in a laterally continuous front which advances into the bay, producing the sub-deltaic lobes of the Mississippi delta which have recently been documented in considerable detail in view of their importance as a possible means of rectifying land loss in the delta area (Coleman and Gagliano, 1964; Gagliano *et al.*, 1971; Gagliano and van Beek, 1975; Fig. 6.10). They develop from a crevasse break in the levee of a major distributary and have a life cycle of initiation, progradation and abandonment which spans

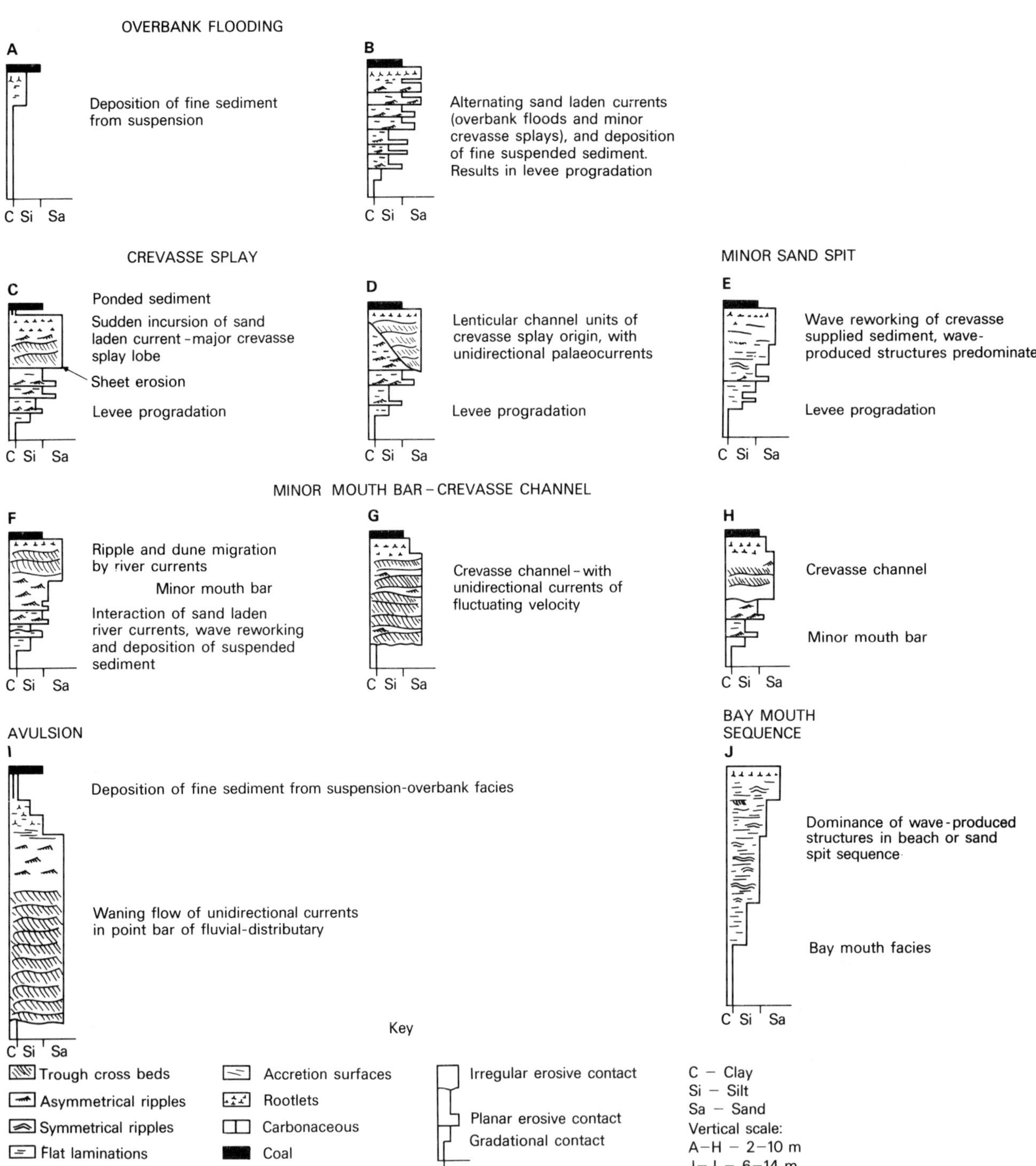

Fig. 6.8. Sequences produced in fluvial-dominated interdistributary areas (after Elliott, 1974b).

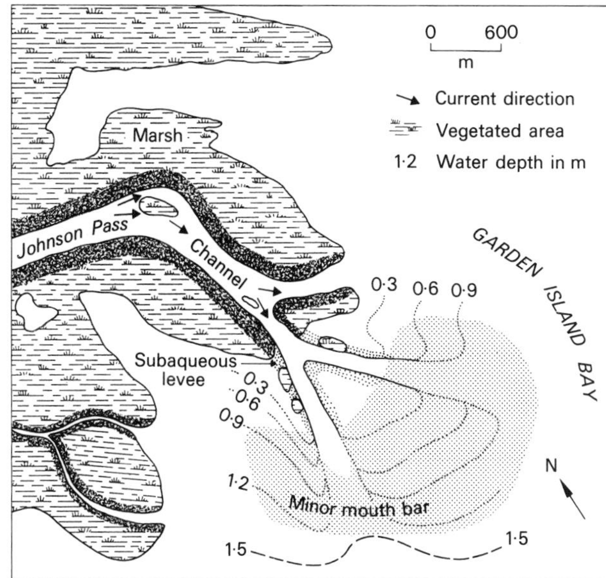

Fig. 6.9. A minor mouth-bar crevasse channel couplet in an interdistributary bay of the modern Mississippi delta (after Coleman, Gagliano and Webb, 1964).

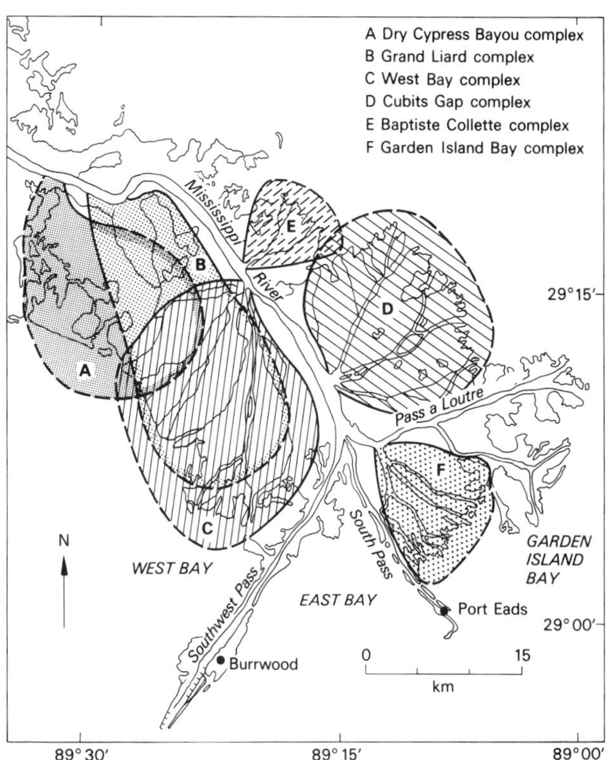

Fig. 6.10. Sub-deltas infilling interdistributary bays of the modern Mississippi delta (after Coleman and Gagliano, 1964).

100–150 years. Following an initial period of subaqueous crevassing during which the initial break is enlarged and made semi-permanent, numerous couplets prograde rapidly across the bay infilling large areas (Fig. 6.11). When infilling is accomplished or the crevasse point is healed, the sub-delta is abandoned. Compaction, subsidence and coastal erosion take over, producing large bays on the marsh surface. Oyster reefs form on distal levee ridges, and the front of the sub-delta is reworked by wave action. However, despite modification during abandonment the sub-deltas have a high preservation potential, with an average sediment retention figure of 70% (Gagliano *et al.*, 1971).

The spatial distribution of processes operating in a fluvial-dominated interdistributary area is determined by the distance from active distributary channels. Near-channel facies will be dominated by levee sequences and numerous crevasse splay lobes, whilst distal or central areas may comprise fine-grained bay floor sediments or crevasse channel/minor mouth bar couplets.

In the open bays of the Mississippi delta, waves rework the upper part of crevasse sands into minor sand spits, and sediment may be reworked directly from the distributary mouth as larger scale sand spits extending back into the interdistributary bay (Fisk *et al.*, 1954). These features are likely to produce small-scale and larger scale coarsening-upward sequences which terminate in wave-dominated sand units comprising well-sorted sands with flat lamination, wave ripples and perhaps low-angle accretion surfaces (Fig. 6.8).

TIDE-DOMINATED DELTA PLAINS

In areas of moderate to high tidal range, tidal currents enter the distributary channels during tidal flood stage, spill over the channel banks and inundate the adjacent interdistributary area. The tidal waters are then stored temporarily during tidal still-stand and subsequently released during the ebb stage. Tidal currents therefore predominate in the lower distributary courses, whilst the interdistributary areas assume the characteristics of intertidal flats.

Tidally influenced distributary channels have a low sinuosity, funnel-shaped form with a high width–depth ratio which contrasts considerably with the almost parallel-sided nature of fluvial distributary channels in areas of low tidal range. The properties of the tidal wave determine the rate at which the banks converge upstream, with standing tidal waves inducing an exponential rate of convergence whilst progressive tidal waves induce a linear rate (Wright *et al.*, 1973). In the Niger

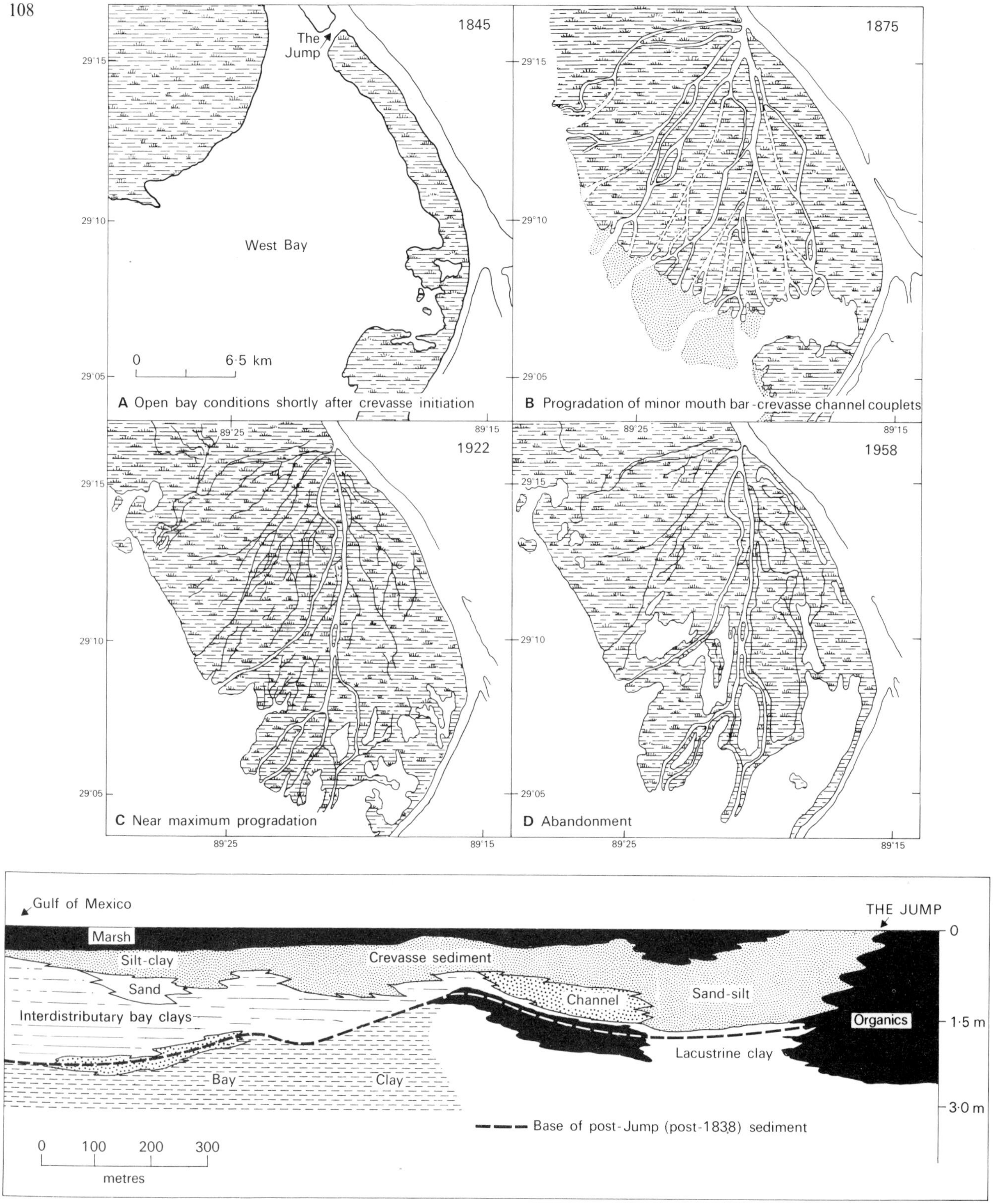

1845

The Jump

West Bay

0 6·5 km

A Open bay conditions shortly after crevasse initiation

1875

B Progradation of minor mouth bar-crevasse channel couplets

1922

C Near maximum progradation

1958

D Abandonment

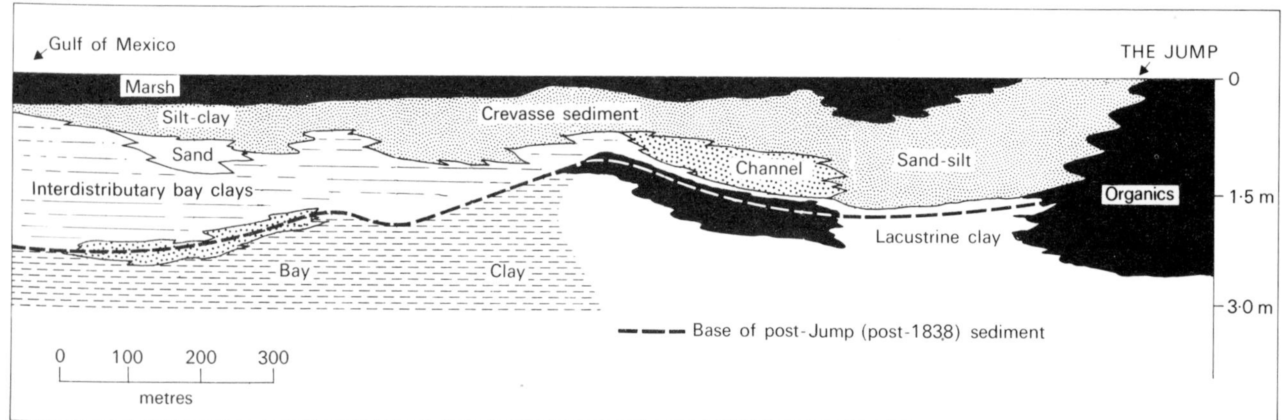

Fig. 6.11. Development of the West Bay subdelta and the resultant facies pattern (after Gagliano and van Beek, 1970).

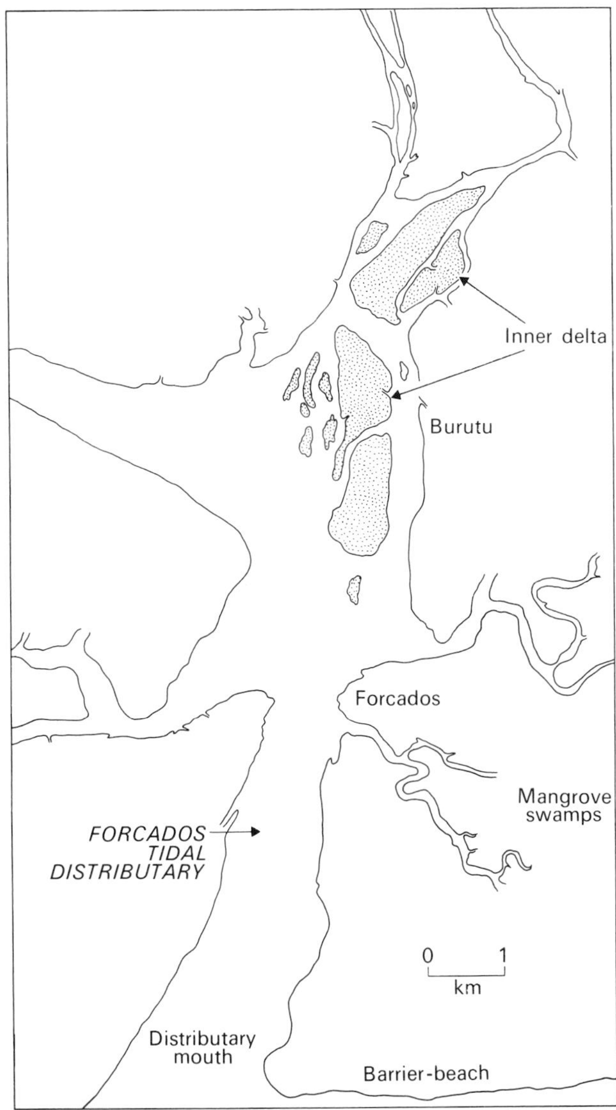

Fig. 6.12. Inner deltas located at the confluence of two tidally influenced distributary channels in the Niger delta (after NEDECO, 1961).

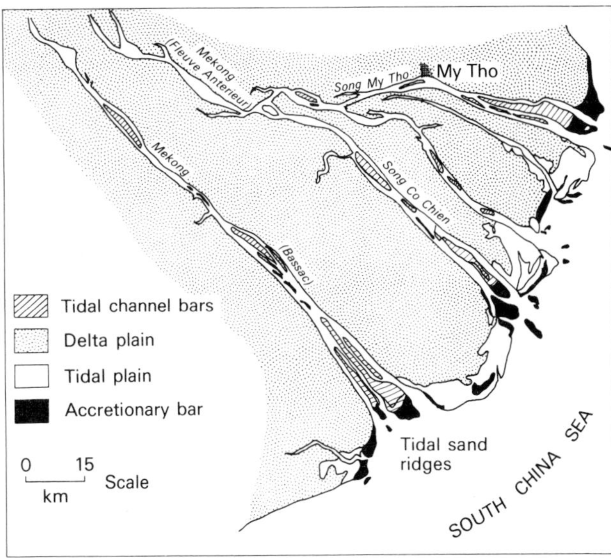

Fig. 6.13. Linear tidal channel sand ridges in the lower reaches of distributary channels in the Mekong delta (after Fisher *et al.*, 1969).

delta more than twenty tidal inlets ranging in depth from 9–15 m dissect the beach-barrier shoreline (NEDECO, 1961; Allen, 1965d). Dune bedforms predominate in the channels but in addition complex 'inner deltas' comprising a maze of sand bars and mudflats occur at confluences between distributary channels, upstream from their outlet to the sea (Fig. 6.12). The 'inner deltas' probably result from deposition of a large proportion of the sediment load as the river currents flare at the confluence and are impounded by tidal currents. Progradation of these features may produce sequences which reflect an interaction between fluvial and tidal processes and thereby resemble shoreline sequences, despite their position landward of the final distributary outlet. In the Mekong, Irrawaddy, Gulf of Papua, Ganges–Brahmaputra and other tidally-influenced deltas, the prevailing bedforms in the lower reaches of the distributary channels are linear sand ridges aligned parallel with the channel trend (Coleman, 1969; Fisher *et al.*, 1969; Coleman and Wright, 1975; Fig. 6.13). The ridges may be several kilometres long, a few hundred metres wide and between 10–20 m high, and reflect tidal current transport of sediment supplied by the river system. They compare favourably with the tidal current ridges of shallow shelf seas (see Chapter 9), but details of their morphology, long- and short-term behaviour, and facies characteristics have not been examined in deltaic settings.

Sequences from tidally influenced distributary channels in the Niger delta commence with a coarse, intraformational lag with a fragmented marine fauna, overlain by sands which exhibit a transition from decimetre-scale trough cross-bedding into centimetre-scale cross-lamination. The sand becomes finer upwards and there is also an increase in the clay content and the number of burrows (Fig. 6.14). Facies in the upper part of the channel sequences vary in accordance with channel position. Interior or delta plain distributaries pass upwards into rootlet-disturbed, organic-rich clays of the mangrove

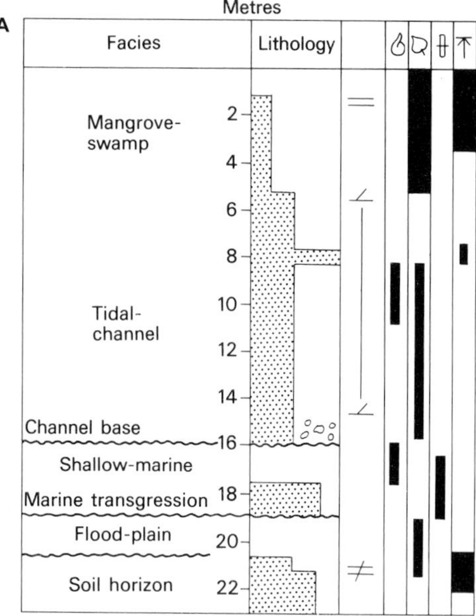

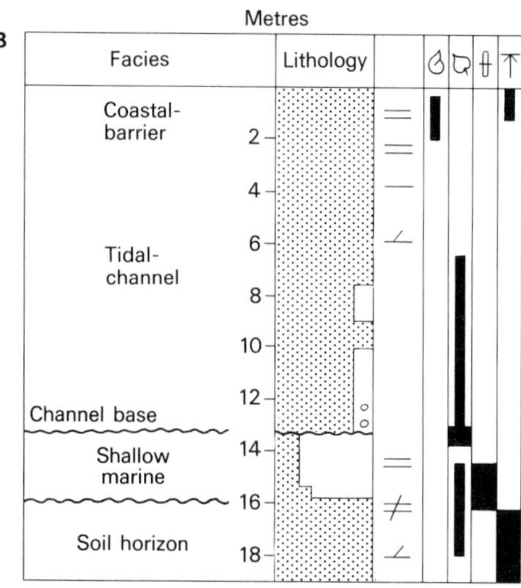

Fig. 6.14. Tidal distributary channel sequences from the Niger delta: (A) an inshore sequence terminating in mangrove swamp facies; (B) a shoreline sequence terminating in coastal barrier facies (after Oomkens, 1974); key as for Fig. 6.6.

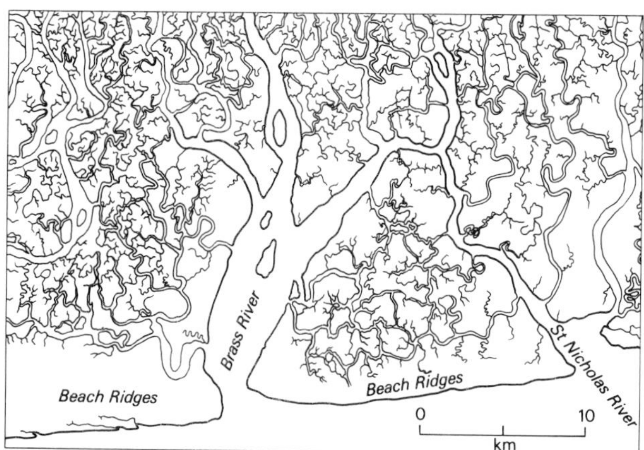

Fig. 6.15. Interdistributary area of the Niger delta comprising extensive mangrove swamps dissected by a maze of minor tidal channels (after Allen, 1965d).

swamp, whereas shoreline tidal channel sequences terminate in flat-laminated coastal barrier sands (Weber, 1971; Oomkens, 1974). Detailed observations in trenches cut through sub-Recent tidal distributary channels of the Rhine delta reveal the complexity of these sequences, in contrast to the limited observations possible in borehole cores (Oomkens and Terwindt, 1960; de Raaf and Boersma, 1971; Terwindt, 1971b). Although a general trend is occasionally discernible from trough cross beds with reversed palaeoflow directions upward into heterolithic facies comprising linsen and flaser bedding, this is frequently complicated by smaller scale fluctuations. Characteristic features include bimodality in flow direction and the frequency of small-scale facies variations in a vertical sense, both reflecting the fact that tidal currents fluctuate in direction and intensity on an extremely small time scale. Tidal distributary channels are less prone to switching and abandonment, but migrate continuously. Sand body shape and dimensions are therefore a function of channel form and the degree of lateral migration. Very little information is available at present, though in the former Rhine delta elongate tidal channel sands are considered to form complexes 20 km wide and 50 km long (Oomkens, 1974).

Interdistributary areas of tide-dominated delta plains include lagoons, minor tidal creeks and intertidal flats. During the tidal cycle the entire area is first inundated and then exposed, and is therefore sensitive to the prevailing climate. In the Niger delta, interdistributary areas are dominated by mangrove swamps (vegetated intertidal flats) dissected by tidal distributary channels and a complex pattern of meandering tidal creeks, each served by numerous small-scale dendritic drainage systems (Allen, 1965; Fig. 6.15). Sands are deposited by laterally migrating point bars in the tidal creeks and mangrove swamps develop on the surface left by the point bars. The entire delta plain probably comprises a sheet-like complex of small-scale, erosive-based sequences which pass upwards from point bar sands into the mangrove swamp facies, with

Fig. 6.16. Processes involved in the interaction between sediment-laden river waters and basin waters at the delta front (based on Wright and Coleman, 1974).

localized clay plugs representing infilled channels. The delta plain of the Colorado River delta at the head of the Gulf of California is also tide-dominated, but the arid climate causes the interdistributary areas to be desiccated mud- and sand-flats with localized salt pans, particularly near the supratidal limit (Meckel, 1975).

Tide-dominated delta plains therefore comprise tidally influenced (or dominated) distributary channel sequences and tidal flat facies which also reflect the prevailing climate. Despite the deltaic setting it is possible that there will be no evidence for fluvial processes in this association, except perhaps for a relatively abundant sediment supply in excess of that normally associated with non-deltaic tidally dominated areas. However, in a prograding succession this association will overlie a tide-dominated or wave-dominated shoreline sequence, and will itself be overlain by a fluvial-dominated upper delta plain association.

6.5.2 The delta front

This is the area in which sediment-laden fluvial currents enter the basin and are dispersed whilst interacting with basinal processes (Fig. 6.16). The radical change in hydraulic conditions which occurs at the distributary mouth causes the outflow to expand and decelerate, thus decreasing outflow competence and causing the sediment load to be deposited. Basinal processes either assist in the dispersion and eventual deposition of sediment, or rework and redistribute sediment deposited directly as a result of outflow dispersion.

Of prime importance in this part of the delta is the precise manner in which the outflow and basin waters mix at the distributary mouth. In an early example of the application of hydrodynamic principles to essentially geological problems, Bates (1953) contrasted situations in which the river waters

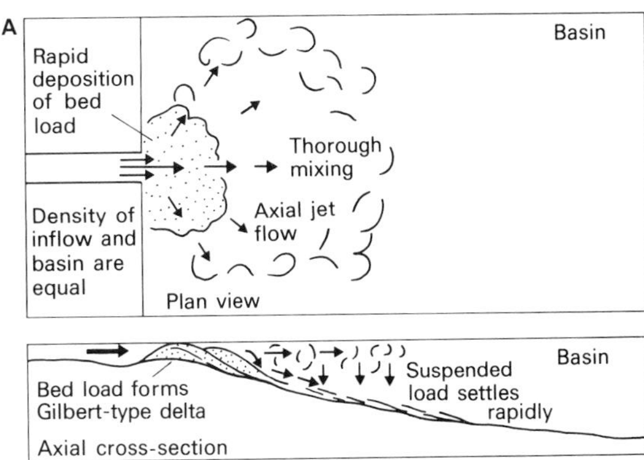

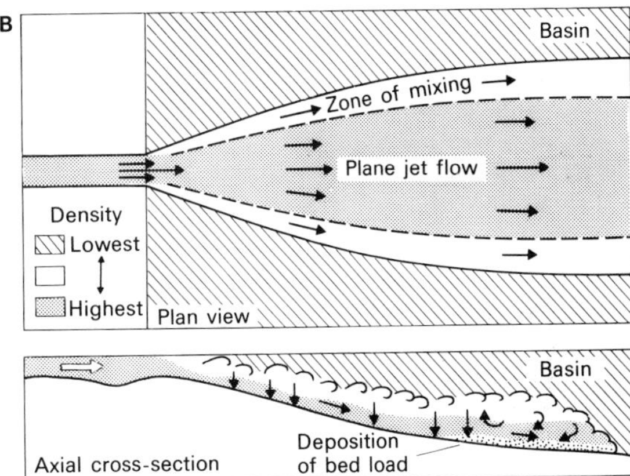

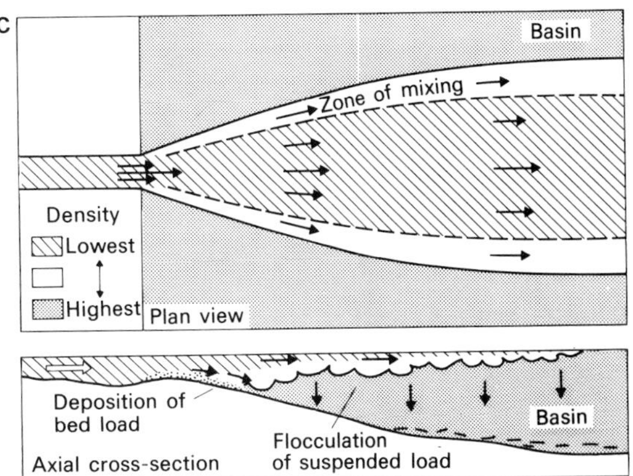

Fig. 6.17. Contrasted modes of interaction between sediment-laden river waters and basin waters, determined by the relative density of the water bodies: (A) homopycnal flow; (B) hyperpycnal flow; (C) hypopycnal flow (after Fisher *et al.*, 1969; originally from Bates, 1953).

were equally dense, more dense and less dense than the basin waters (homopycnal, hyperpycnal and hypopycnal flow; Fig. 6.17). If the water bodies are of equal density, immediate three-dimensional mixing occurs at the river mouth causing appreciable sediment deposition at this point. High density outflow tends to flow beneath the basin waters as density currents, causing sediment to by-pass the shoreline and thus restricting the development of a delta. If the outflow is less dense than the basin waters it enters the basin as a buoyantly supported surface jet or plume. This latter situation, hypopycnal flow, has been observed off the Mississippi and Po deltas (Scruton, 1956; Nelson, 1970) and is considered to operate wherever river outflow enters marine basins as sea-water is slightly denser than freshwater. The interaction between river outflow and basin waters in marine deltas therefore tends to be considered in terms of hypopycnal flow, but the possibility of other mechanisms operating, at least briefly, should not be neglected. Central to the idea of hypopycnal flow is the buoyancy of the outflow, but other important factors neglected until recently include inertial processes related to outflow velocity, and frictional processes which result from the outflow interacting with the sediment surface at the distributary mouth (Wright and Coleman, 1974). Differing combinations of buoyancy, inertial and frictional processes produce a series of hydrodynamic outflow dispersion models. For example, the ratio of inertial to buoyancy processes (represented by the densimetric Froude number F') defines two distinct models:

(1) Turbulent diffusion model: when inertial processes are dominant ($F' > 16.1$) the outflow is turbulent and assumes the form of a plane jet with turbulent eddies inducing exchange and mixing between the water bodies. Outflow dispersion results largely from turbulent diffusion which expands and dilutes the outflow, thus decreasing outflow competence. However, as most river mouths are characterized by shallow bars, frictional processes also operate in this model as the outflow has direct contact with the sediment surface.

(2) Buoyant outflow dispersion model: as F' decreases, strong density stratification is introduced into the water column causing the outflow to spread laterally as a buoyantly supported plume. In this situation turbulence is suppressed and dispersion of the outflow is brought about by the vertical entrainment of salt water across the density interface, at a rate determined by the amount that F' exceeds unity. Salt water entrainment not only causes the outflow to decelerate, but also increases outflow density, thus reducing buoyancy. Frictional processes do not operate in this model as the outflow extends over the receiving waters and therefore has no contact with the sediment surface.

Imposed on the primary outflow dispersion mechanism is the extent to which basinal processes either modify the mechanism, or redistribute sediment deposited as a result of outflow dispersion. Waves, wave-induced currents and tidal currents often assist in the dispersion of fine-grained suspended sediment and can also induce transport of considerable amounts of coarser bedload sediment.

Moderate wave action does not unduly interfere with operation of the outflow dispersion mechanism but may rework sediment deposited at the distributary mouth. Intense and persistent wave energy directly affects outflow dispersion and sediment is distributed according to the wave-induced circulation pattern. The effectiveness of wave processes in redistributing sediment supplied by rivers has recently been examined quantitatively using information from modern deltas (Wright and Coleman, 1973). Initially, the method involves calculating a long-term statistical average of wave power at the shoreline. Deep-water wave characteristics such as height, period and angle of incidence are collated, frequently having to be hind-casted from wind data using empirical relationships. However, as deltas tend to be fronted by broad, gently sloping subaqueous profiles the waves undergo significant refraction, shoaling and frictional attenuation as they approach the shoreline. Deep-water wave values are therefore transformed into nearshore values by computing these effects for any combination of wave conditions and bottom slope. Monthly weighted means of the total annual wave power are calculated in order to yield a long-term average of wave power. This parameter is then combined with river discharge data, providing a 'discharge effectiveness index' which describes the relative effectiveness of river discharge against wave reworking ability, and can be compared between deltas. Application of the method to the Mississippi, Danube, Ebro, Niger, São Francisco and Senegal deltas demonstrated a close correlation between delta front morphology and discharge effectiveness index.

The manner in which tidal processes operate in the delta front area has not been studied rigorously. Sediment transport in the lower part of distributary channels and distributary mouth areas may be influenced or dominated by the tidal current regime, depending largely on the tidal range and the effectiveness of tidal currents relative to sediment supply from the distributaries. In macro-tidal areas, tidal currents often confine direct fluvial discharge to the upper part of the delta plain, and sedimentation in the lower delta plain and delta front is then largely a response to the tidal current regime, except perhaps during major river floods. In meso-tidal areas the effects of tidal currents are confined to the distributary mouth area as the remainder of the delta front comprises wave-built beach-barriers. Primary outflow dispersion may still occur at the distributary mouth at high discharge periods, though the dispersion mechanism may be disrupted by tide-induced mixing of the water bodies.

There is a general tendency in the delta front for coarse sediment to be deposited at the distributary mouth, whilst finer sediment is transported further into the basin and deposited in deeper water offshore environments. Sediment deposition therefore constructs a seaward-dipping profile which slopes gently into the basin, generally at an angle of less than 2°, and fines progressively into the basin. The delta front progrades offshore in response to continued sediment supply and the net result is that former offshore areas are eventually overlain by the shoreline. In combination with the seaward-fining of the delta front this produces a relatively large-scale coarsening-upward sequence which reflects infilling of the receiving basin. The approach adopted in the following section involves describing different types of delta front in terms of processes and morphology, and assessing what is known of their facies characteristics.

A point worth emphasizing is that delta front progradation is rarely uniform and the facies patterns may not therefore be as orderly as portrayed in the limited descriptions from modern deltas. In addition to the vagaries of sediment-laden discharge entering a standing water body with its own regime, it must also be remembered that distributary channels rarely, if ever, divide the alluvial discharge equally. Sediment supply therefore varies from point to point around the delta front, and furthermore this sediment supply pattern is constantly changing as individual distributary channels wax or wane in response to avulsion and abandonment.

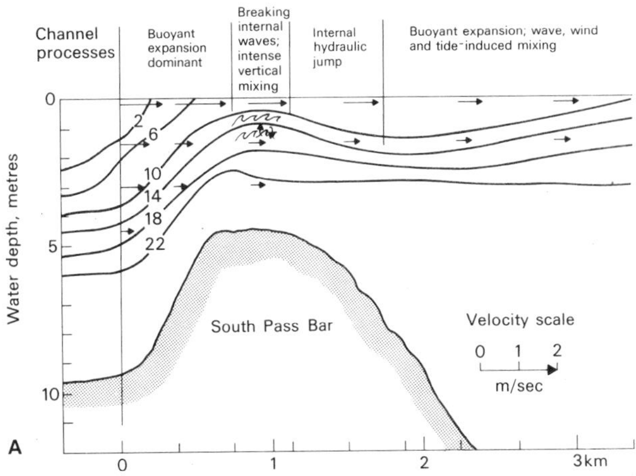

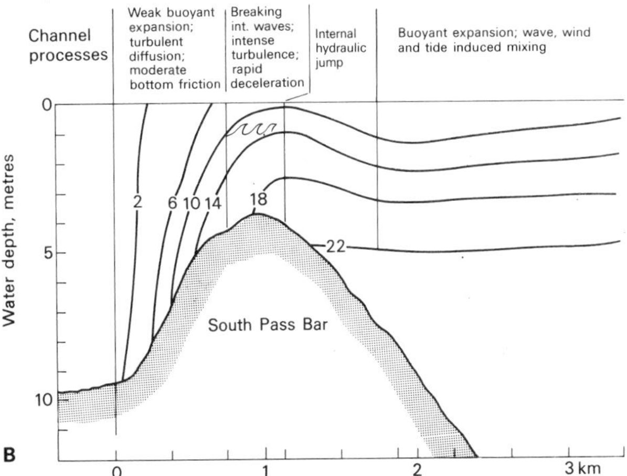

Fig. 6.18. Density and structure of river outflow at South Pass, Mississippi delta: (A) at low river stage and (B) at high river stage – both profiles taken during ebbing tide. A pronounced 'salt wedge' enters the lower reaches of the channel at low stage but is forced out during the next high stage period (after Wright and Coleman, 1974).

TYPE I: FLUVIAL-DOMINATED

The Mississippi delta is the only present-day marine delta in which delta front sedimentation is dominated by outflow dispersion characteristics with minimal interference from basinal processes. An extremely fine-grained sediment load comprising 98% clay and silt and 2% fine sand is supplied to the delta front and observations at one of the distributary mouths (South Pass) indicate that the dispersion of this sediment varies according to fluctuation in river stage (Wright and Coleman, 1974). During low and normal river stage the outflow velocity falls below a critical value and permits a low density 'salt wedge' to enter the lower reaches of the distributary channels. Under these conditions the buoyant outflow dispersion model operates, but at high river stage increased outflow velocities force the salt wedge out of the channel and the turbulent model prevails (Fig. 6.18).

Sediment deposition at distributary mouths constructs a series of discrete, isolated lunate mouth bars which protrude into the Gulf of Mexico (Fig. 6.19). In general these features comprise a variable *bar-back* area which includes minor channels, subaqueous levees and 'mid-channel bars'; a narrow *bar-crest* located a short distance offshore from the distributary mouth; and a *bar-front* which slopes offshore to the prodelta area (the term bar-front as used here incorporates the 'distal bar' of certain workers). In detail, there are two types of mouth bar in the Mississippi delta reflecting subtle variations in outflow dispersion (Coleman and Wright, 1975):

(1) Subaqueous jettied type, comprising a relatively deep distributary mouth which culminates in a narrow mouth bar connected to the channel by prominent subaqueous levees. This type is produced by a predominance of inertial processes supported by buoyancy processes during low stage (e.g. South Pass and South-West Pass).

(2) Middle ground type, in which the distributary channels are relatively shallow and bifurcate around a mid-channel

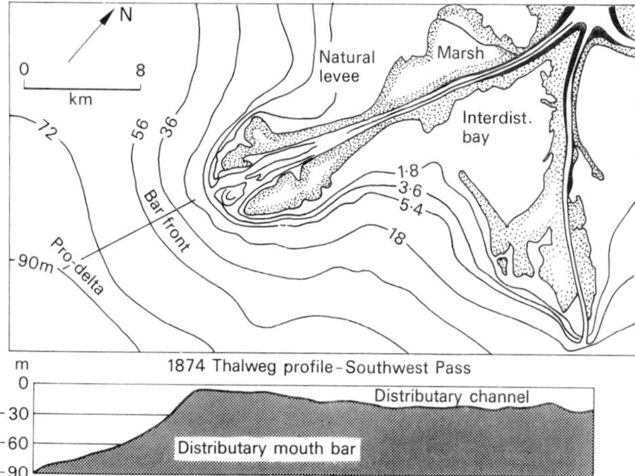

Fig. 6.19. Low sinuosity distributary channels and distributary mouth bars in the modern Mississippi delta (after Fisk *et al.*, 1954).

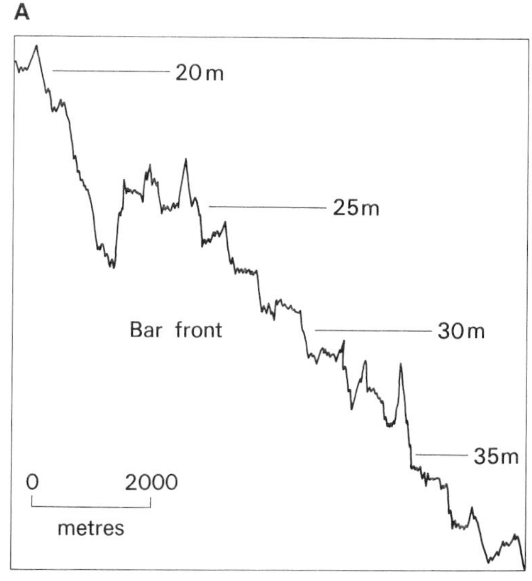

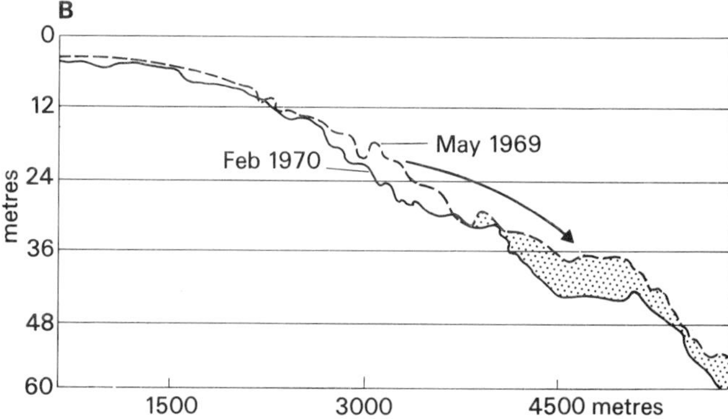

Fig. 6.20. (A) Irregular profile of a distributary mouth bar front in the Mississippi delta reflecting the presence of rotational slump planes; (B) offshore slumping associated with the slump planes, revealed by time-separated fathometer profiles (after Coleman *et al.*, 1974).

island at the mouth. Alternation between low and normal stage buoyancy processes and flood stage frictional processes favour the development of this type (e.g. the Pass a'Loutre and Main Pass complex).

Previously it was assumed that the bar front is relatively smooth, but fathometer profiles run off South Pass reveal frequent abrupt 'stairstep' changes in slope (Fig. 6.20). Seismic studies indicate that these irregularities are a surface expression of fault or slump planes along which large blocks of sediment are translated downslope. This process is a response to rapid sedimentation and oversteepening of the bar front, and is an integral part of progradation in the Mississippi delta (Coleman *et al.*, 1974). Normal progradation is occasionally interrupted by periods during which the upper bar front retreats whilst the lower bar front progrades by a corresponding amount, reflecting downslope translation of slump blocks along the fault or slump planes detected in seismic studies. Individual slump blocks average 90 m in width, 6 km in length and move downslope for distances in excess of 1.5 km. They are preserved intact and do not exhibit flowage structures, producing seemingly anomalous shallow water sand facies in deeper water mud-silt facies.

Observations at South Pass during the extreme river flood of 1973 demonstrated that the mouth bars advance in dune-like fashion, with the bar-crest and to some extent the bar-back being eroded, whilst the bar-front progrades and therefore has the highest preservation potential. The bar-crest *aggraded* rapidly during the flood with up to 3 m of sediment being deposited, then as the flood diminished this sediment was reworked by river currents and redeposited on the bar-front. Mouth bar *progradation* was therefore most marked immediately after peak flood, with the 10 m depth contour advancing 90–120 m within a few months of flood peak (Coleman *et al.*, 1974). Mouth bar progradation produces large-scale coarsening-upward sequences characterized by a thick mud-silt member at the base representing the prodelta and lower bar-front facies; repeated, small-scale interbedding of mud, silt and sand in the intermediate part reflecting the interaction between sediment-laden incursions from the distributaries, wave processes and deposition of sediment from suspension; and an upper sand-dominated member exhibiting current-produced structures (Fisk *et al.*, 1954; Fisk, 1955, 1961; Coleman and Wright, 1975; Fig. 6.21). However, mouth bars are three-dimensional features in which facies

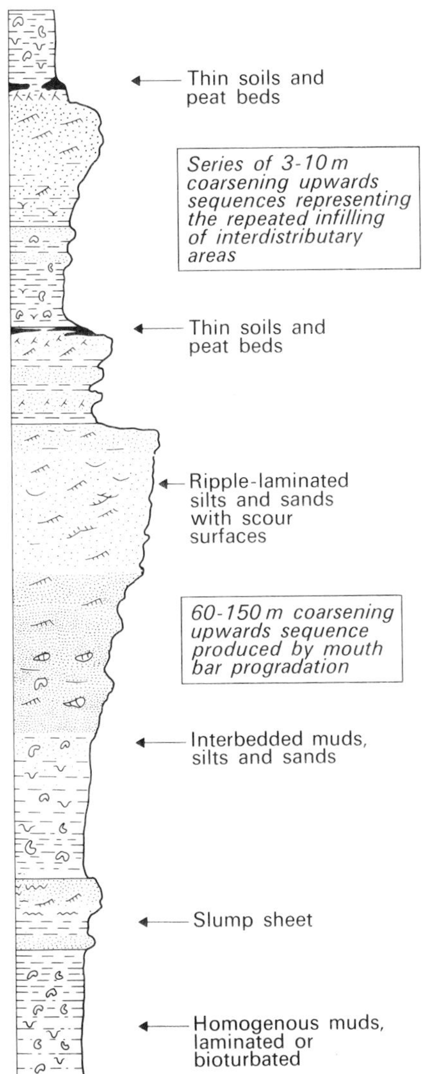

Fig. 6.21. Composite, idealized sequence produced by mouth bar progradation in the Mississippi delta (after Coleman and Wright, 1975).

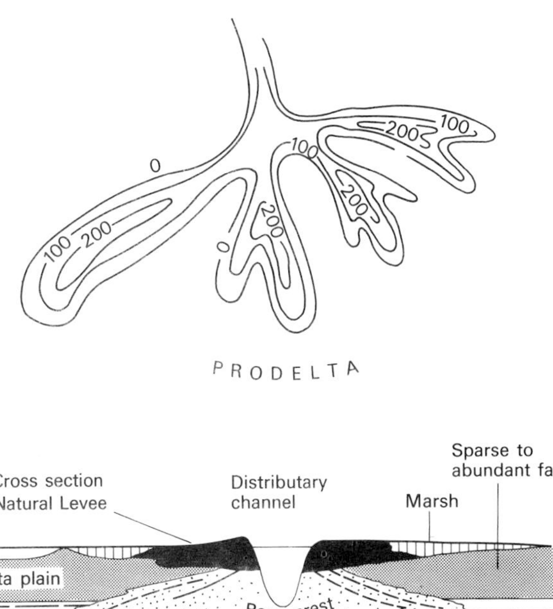

Fig. 6.22. Bar finger sands of the Mississippi delta (after Fisk, 1961).

characteristics are likely to vary markedly with respect to the position of the distributary outlet. Variations can be related to distance from the outlet in an axial-to-lateral sense and a proximal-to-distal sense. The proximal, axial part of the mouth bar will include a well-developed sand member which may be entirely absent laterally and distally. Furthermore, in the axial part of the mouth bar the distributary channel may erode the upper part of the sequence as progradation continues. Also, as processes vary across the mouth bar the facies details will vary. The facies characteristics of mouth bars cannot therefore be summarized by a single coarsening-upward sequence, but studies of the Mississippi mouth bars have so far neglected this fact.

As the introduction of sediment is virtually uncontested by basinal processes, progradation of individual distributary channel–mouth bar systems is unidirectional in a seaward direction. This has produced a series of radiating 'bar finger sands' which directly underlie the present distributaries and provide the birdfoot framework of the delta (Fisk, 1961; Fig. 6.22). Traditionally, they have been described as bi-convex, elongate sand bodies up to 30 km long, 5–8 km wide, and with a relatively uniform thickness of 70 m. More recently it has been demonstrated that this primary shape has been substantially modified by mud diapirism (Sect. 6.8.2).

A different type of fluvial-dominated delta front was formerly displayed by the small-scale Colorado River delta in East Matagorda bay, Texas (Kanes, 1970). The earlier lobe of this delta (pre-1930) was characterized by closely-spaced distributary channels and a continuous delta front composed of coalesced mouth bar sands. This important alternative to the classical birdfoot, fluvial-dominated form is also discernible in the pre-modern 'shoal water' lobes of the Mississippi delta (Fisk, 1955; Frazier, 1967), and ancient deltaic successions.

TYPE II: FLUVIAL-WAVE INTERACTION

In general this type of delta front is characterized by a relatively smooth cuspate or arcuate beach shoreline. Localized protuberances in the vicinity of the distributary mouth comprising subdued mouth bars flanked by beach ridge complexes reflect the fact that wave processes are partially capable of redistributing the river-borne sediment supply. Present-day examples occur in the Danube, Ebro, Nile and Rhône deltas, all of which are located in enclosed seas with moderate wave action but minimal tidal processes (microtidal). The modern Po delta has departed from its natural form as man has concentrated 60% of the discharge into one distributary – the Pila. This distributary mouth therefore has a prominent lunate mouth bar with a series of wave reworked sand bars at the crest, in some ways transitional between Types I and II (Nelson, 1970; Fig. 6.23).

The most thoroughly described example is the Rhône delta (Kruit, 1955; van Straaten, 1959, 1960). The delta front comprises laterally extensive beach ridges fronted by a relatively steep offshore slope (up to 2°) which descends to 50 m depth. Progradation is by beach ridge accretion and mouth bar progradation, and is most pronounced in the vicinity of the main distributary – the Grand Rhône. Elsewhere the beach zone of the delta front is thin, and in some places is retreating landwards (van Straaten, 1960; Fig. 6.24). One area of retreat occurs west of the Grand Rhône mouth where an earlier lobe of the delta is being reworked by wave action after abandonment. Shallow-water shoreline facies, including mouth bar facies, deposited during progradation may therefore be reworked after abandonment.

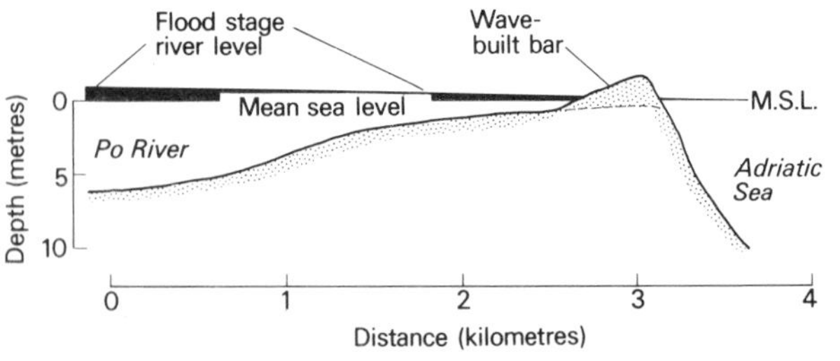

Fig. 6.23. Distributary mouth bar with wave-reworked fringe in the Po delta (after Nelson, 1970).

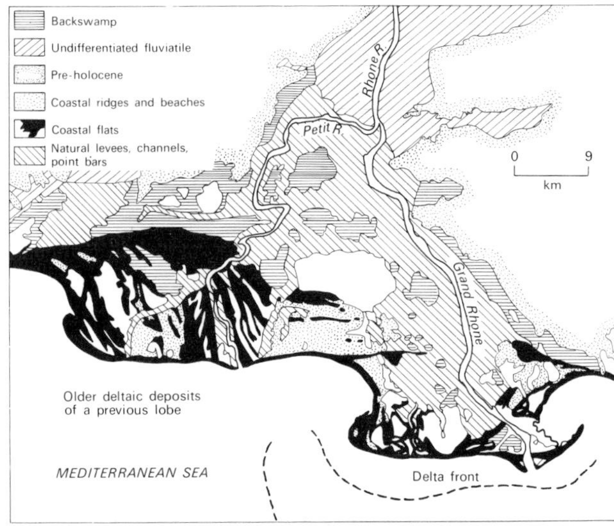

Fig. 6.24. Morphology of the Rhône delta (after van Andel and Curray, 1960).

Delta front coarsening-upward sequences have been described from the Rhône and Ebro deltas (Lajaaij and Kopstein, 1964; Oomkens, 1967, 1970; Maldonado, 1975). Bioturbated offshore clays pass upwards into finely laminated silts which gradually acquire discrete beds of silt and sand in the intermediate part of the sequence. Ripple-lamination is common at this level, but the sand member at the top of the sequence consists of well-sorted, horizontally-bedded sand (Fig. 6.25). Oomkens (1967) attempted to distinguish between

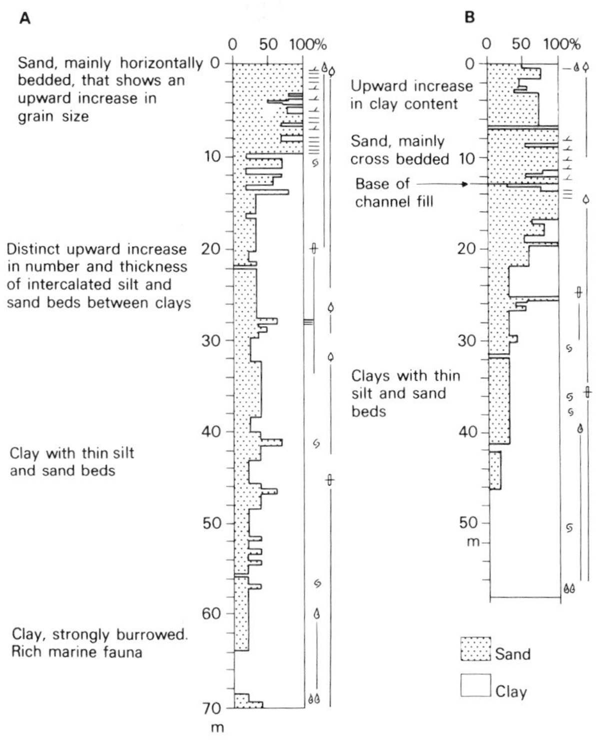

Fig. 6.25. Delta front coarsening-upward sequences from the Rhône delta: (A) coarsens upwards gradationally into a coastal barrier sand, whereas (B) is truncated at 13 m by an erosive-based distributary channel sequence (after Oomkens, 1967).

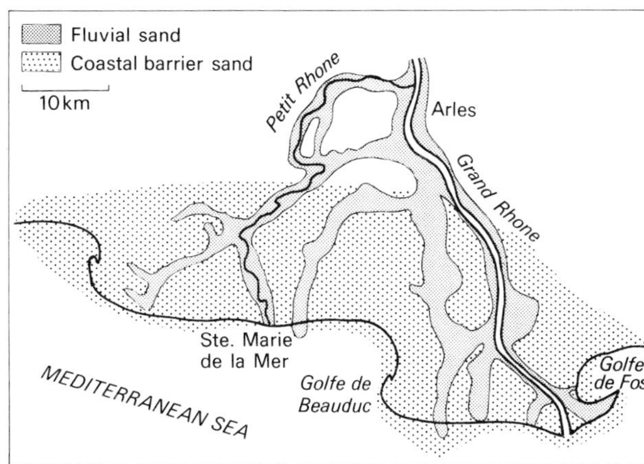

Fig. 6.26. Sand distribution pattern of the Rhône delta complex, illustrating a laterally extensive, slightly lobate coastal barrier sand cut locally by fluvial distributary channel sands (after Oomkens, 1967).

fluviomarine sequences produced by direct fluvial input of sediment near the active distributary mouth, and *holomarine* sequences in which sediment was supplied by longshore drift away from the distributary mouth. Lithologically, however, the sequences appear identical and can only be distinguished by their microfauna content which is more diverse and abundant in holomarine sequences. This suggests that all the delta front sequences terminate in a beach-barrier sand, and substantiates the view that shallow, proximal mouth bar sands are reworked after distributary channel abandonment.

The sand distribution pattern in the Rhône delta consists of a laterally extensive beach-barrier sand locally cut by linear distributary channel sands (Oomkens, 1967; Fig. 6.26). This pattern may resemble that produced by laterally coalescing distributary mouth bars in certain fluvial-dominated delta fronts, but points of difference include the lower number of distributary channel sands and the wave-dominated nature of the sheet sand in Type II.

TYPE III: WAVE-DOMINATED

In this type wave processes are capable of redistributing most of the sediment supplied to the delta front. It is therefore characterized by a regular beach shoreline with only a slight deflection at the distributary mouth and a relatively steep delta front slope. The formation of mouth bars is precluded and bathymetric contours parallel the shoreline. Progradation involves the entire delta front, rather than point-concentrated progradation, and is generally slow by comparison with other types. Abandoned beach-ridge complexes occur behind the active shoreline and in plan view the ridges are separated into discrete groups by discontinuities which reflect subtle changes in shoreline configuration induced by changes in the position of the distributary channels (e.g. the Grijalva delta, Psuty, 1966; Fig. 6.27).

No thorough studies of facies patterns in this type have been made, but a single, idealized coarsening-upward sequence has been described from the São Francisco delta front (Coleman and Wright, 1975). Bioturbated, fossiliferous muds at the base pass upwards into alternating mud, silt and sand beds with wave-induced scouring, grading and cross-lamination, and finally into a well-sorted sand with parallel and low-angle laminations representing a high-energy beach face. Aeolian sands succeed the delta front sequence, though the preservation potential of these sediments is uncertain.

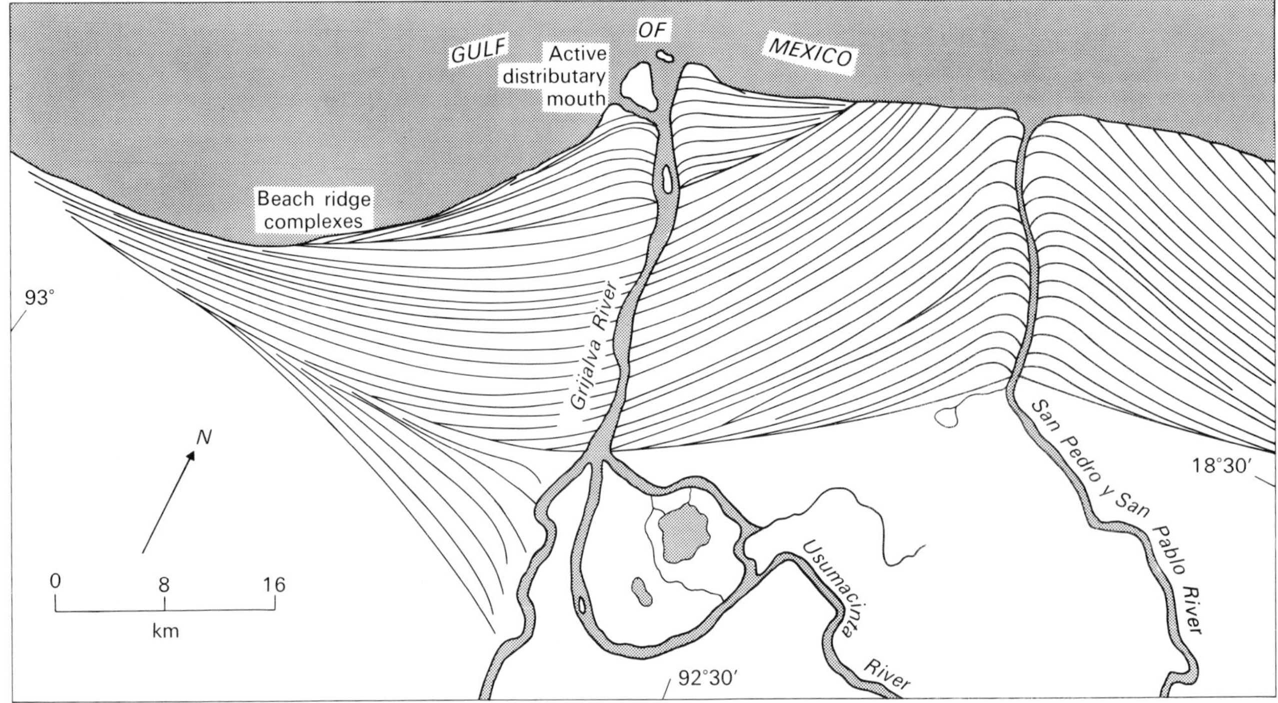

Fig. 6.27. Wave-dominated Grijalva delta (after Psuty, 1966).

This type may appear similar to Type II, particularly its vertical sequence, but minor differences may permit a distinction. For example, in Type III, all structures are wave-produced, and the fine member at the base of the sequence is often thinner due to wave distribution of suspended sediment beyond the delta area. Also, as progradation involves the entire delta front rather than being concentrated at certain points, the resultant sand body is likely to be a linear, sheet-like feature parallel to the shoreline, whereas in Type II it is broadly lobate. The facies of wave-dominated delta front areas may, however, be identical to those of non-deltaic, prograding beach-barriers and in this case the distinction may have to rest on regional considerations, in particular whether or not a distributary channel network is located behind the beach-barrier sands.

TYPE IV: FLUVIAL-WAVE-TIDE INTERACTION

In mesotidal settings, tidal currents frequently operate in conjunction with wave action at the delta front. Tidal effects are confined to distributary mouth areas whilst waves operate over the remainder of the delta front, and the shoreline therefore comprises wave-produced beaches or cheniers separated by tide-dominated distributary channels and mouth areas. Offshore the bathymetric contours and facies belts tend to parallel the shoreline, although there may be slight protrusions in the vicinity of distributary mouths. Examples of this type occur in the Burdekin, Irrawaddy, Mekong, Niger and Orinoco deltas, of which the best described is the Niger delta (Allen, 1965d; Fig. 6.28).

More than twenty tide-dominated distributary channels dissect the beach-barrier shoreline of the Niger delta and each distributary mouth has a shallow, sandy bar. These bars vary in shape from linear to arcuate and are deflected by longshore currents around the delta front (Fig. 6.29). They have been consistently described as river mouth bars, but river discharge is minimal at this point (NEDECO, 1961) and it seems more probable that they result from the expansion of tidal currents. The mouth bars and beach face descend to an inshore terrace rather than sloping uniformly offshore. This terrace, known as the 'delta front platform', occurs at 5–10 m water depth, is up to 20 km wide and has a distinct regime produced by the interaction of tidal currents, waves, longshore and semi-permanent currents (Allen, 1965). Beyond this platform the delta front slopes gently offshore into a low-energy environment mildly affected by waves, tidal currents and the Guinea Current which contours the prodelta slope. Detailed

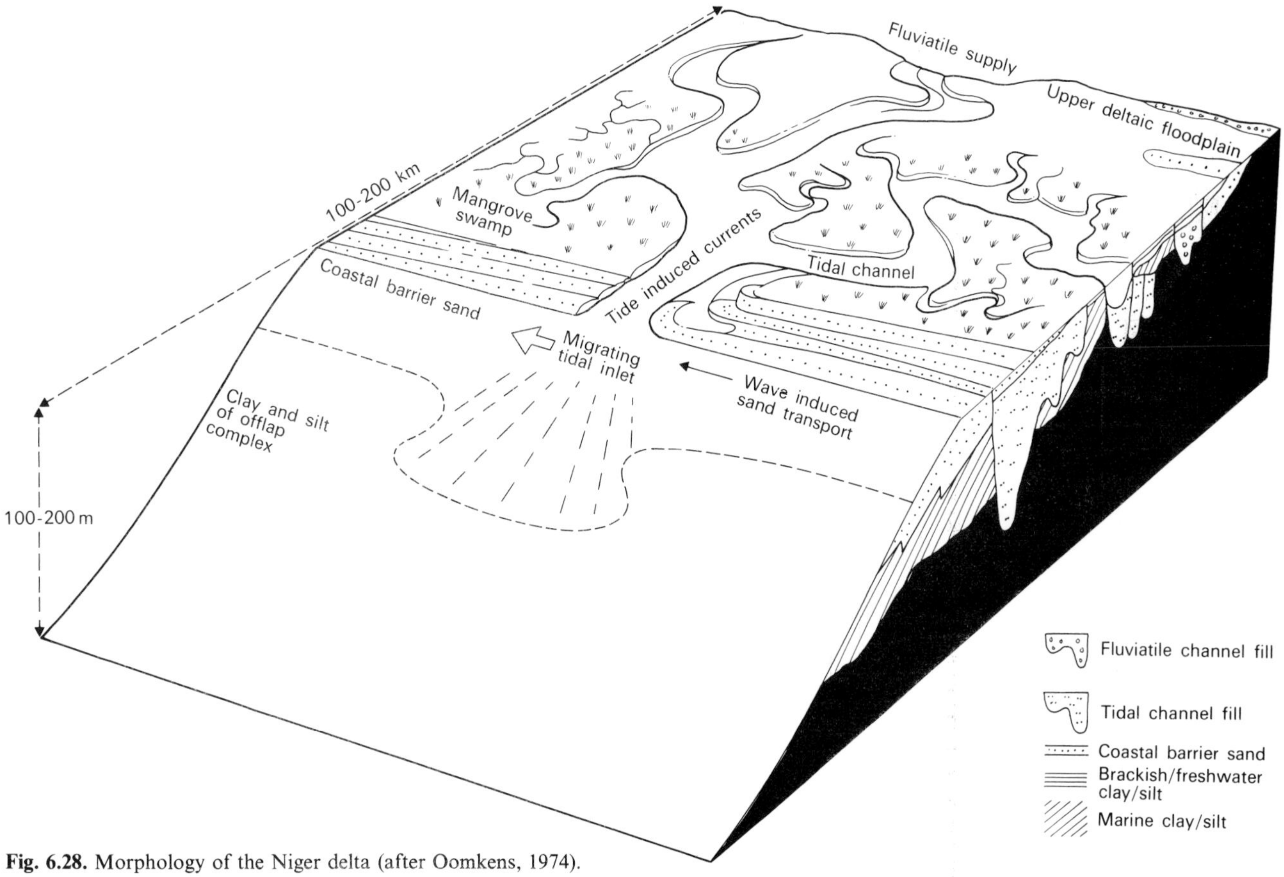

Fig. 6.28. Morphology of the Niger delta (after Oomkens, 1974).

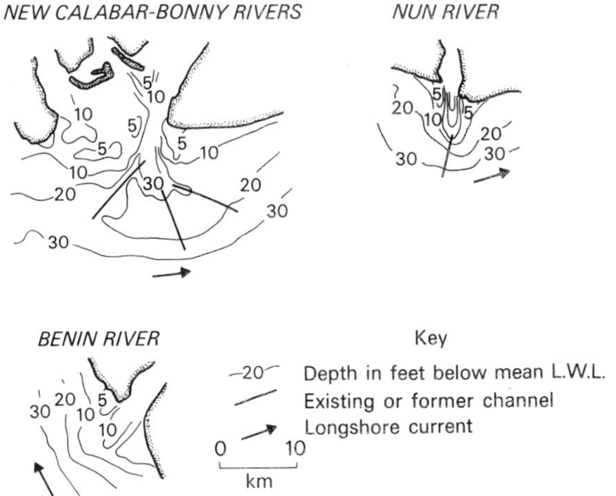

Fig. 6.29. Seaward-projecting bars at the mouth of tidally-influenced distributary channels in the Niger delta (after Allen, 1965d).

facies descriptions are available for the various subenvironments of the Niger delta front (Allen, 1965), and vertical facies sequences have also been described from cored boreholes (Weber, 1971; Oomkens, 1974). The sequences range in thickness from 10–30 m and commence with bioturbated clays with occasional silt and sand lenses which pass upwards into interbedded muds, silts and sands. Towards the top of a sequence there is either a gradational passage into well-sorted, parallel-laminated sands of the beach face, or the sequence is cut by a tidal channel sand (Fig. 6.30). Ideally this delta front would be represented by a sheet-like barrier-beach sand body frequently cut by linear tidal channel sand bodies normal to the shorelines and with up-dip and (?) down-dip extensions, but in fact sand body characteristics are extensively controlled by syn-sedimentary growth faults (Weber, 1971; Sect. 6.8.2)

In the Mekong and Irrawaddy deltas the shoreline consists of discontinuous chenier-like beach ridges rather than substantial beach-barriers (Fig. 6.31). Shoreline progradation in the Mekong delta has produced an area of abandoned beach

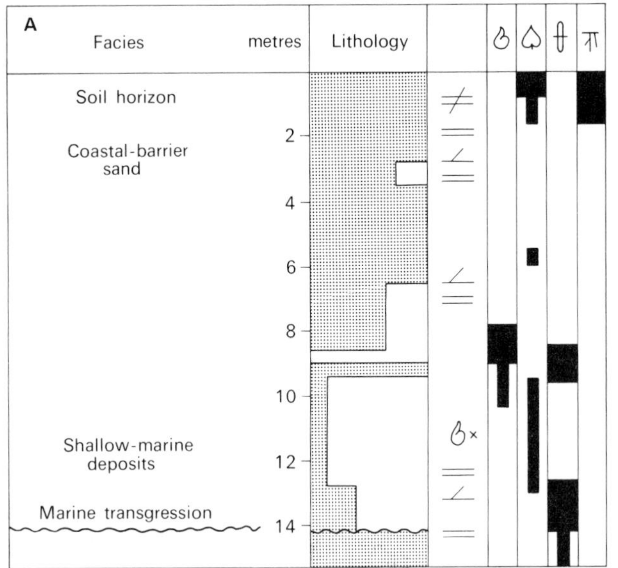

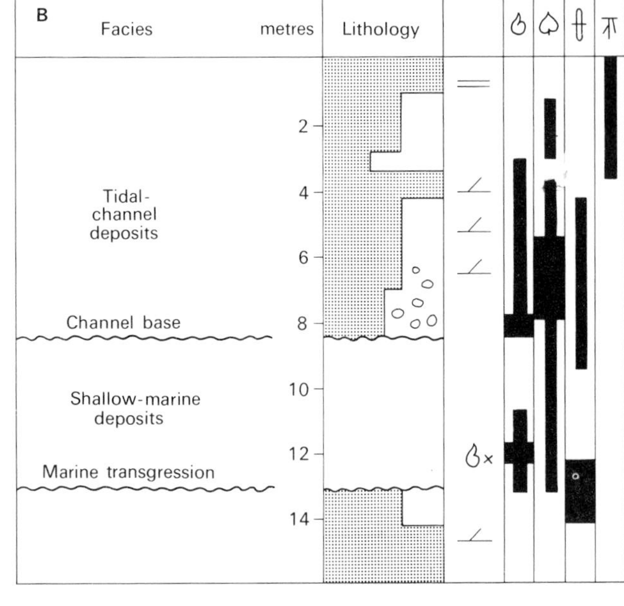

Fig. 6.30. Delta front sequences from the Niger delta: (A) terminates in the coastal barrier sand, whereas (B) has been cut by a tidal inlet (after Oomkens, 1974).

Fig. 6.31. The Irrawaddy delta (after Fisher *et al.*, 1969).

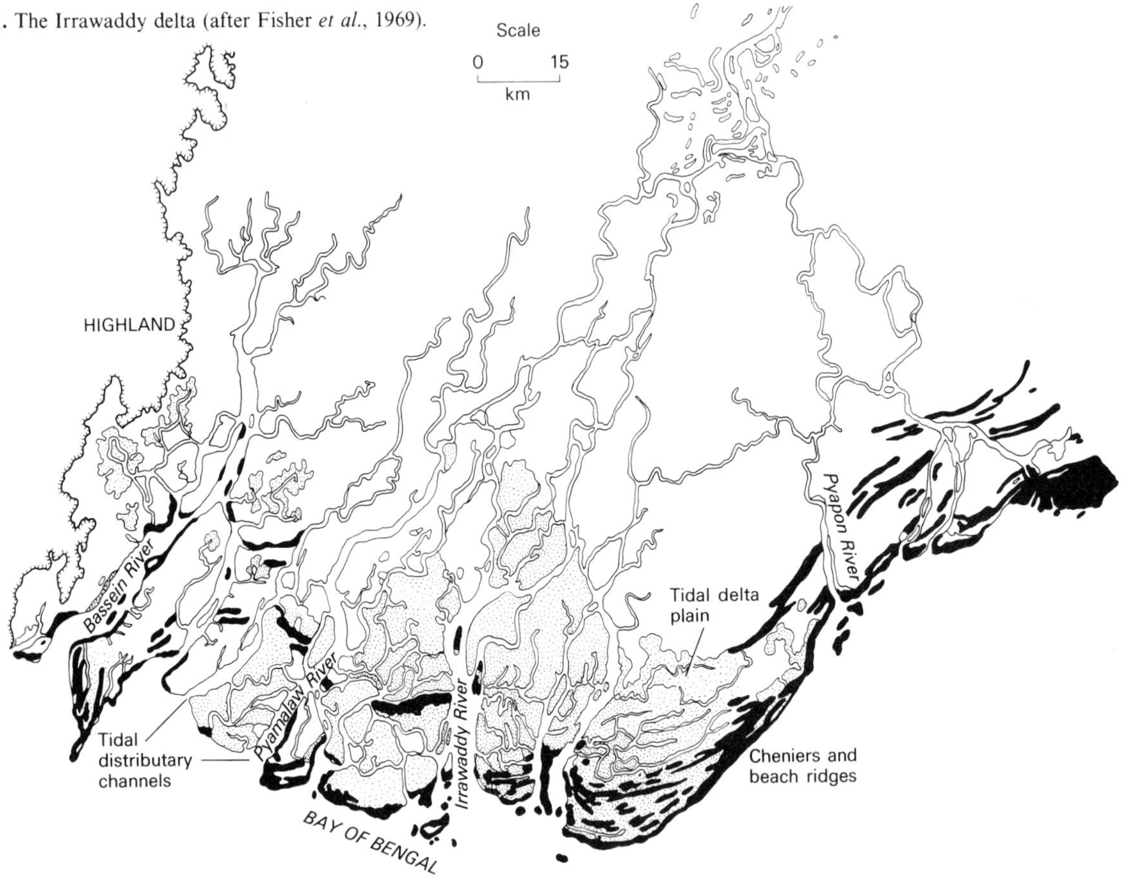

Scale
0 15
km
HIGHLAND
Bassein River
Pyamalaw River
Irrawaddy River
Pyapon River
Tidal delta plain
Tidal distributary channels
Cheniers and beach ridges
BAY OF BENGAL

ridges which extends inland for 56 km. Inland the ridges become progressively subdued and are eventually overlain by delta plain facies (Kolb and Dornbusch, 1975).

TYPE V: TIDE-DOMINATED

In macrotidal settings the shoreline and distributary mouth areas are an ill-defined maze of tidal current ridges, channels and islands which may extend a considerable distance offshore before giving way to the delta front slope. For example, in the Ganges–Brahmaputra delta a complex shoal area protrudes 95 km offshore (Coleman, 1969). The main features of this type of delta front are the tidal current ridges which radiate from the distributary mouths, as illustrated by the Gulf of Papua delta (Fig. 6.32). In the Ord River delta the ridges are on average 2 km long, 300 m wide and range in height from 10–22 m. Channels between the ridges contain numerous shoals and mid-channel islands covered by flood- and ebb-oriented bedforms (Coleman and Wright, 1975). Facies characteristics of the sequence produced by progradation of the Ord River delta front have been summarized in an idealized vertical sequence (Coleman and Wright, 1975; Fig. 6.33). The tidal current ridge sands at the top of the sequence comprise a complex of minor channels and bi-directional trough cross beds with occasional clay drapes, and, in terms of sand body characteristics, this type of delta front will probably produce relatively thick, elongate bodies aligned normal to the shoreline trend.

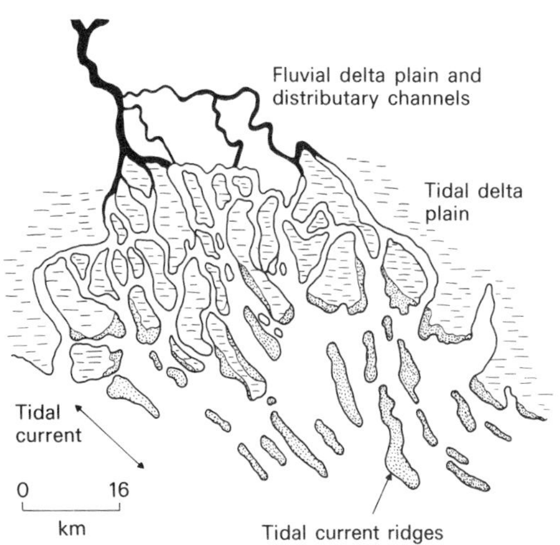

Fig. 6.32. The Gulf of Papua delta with extensive fields of tidal current sand ridges at the delta front (after Fisher *et al.*, 1969).

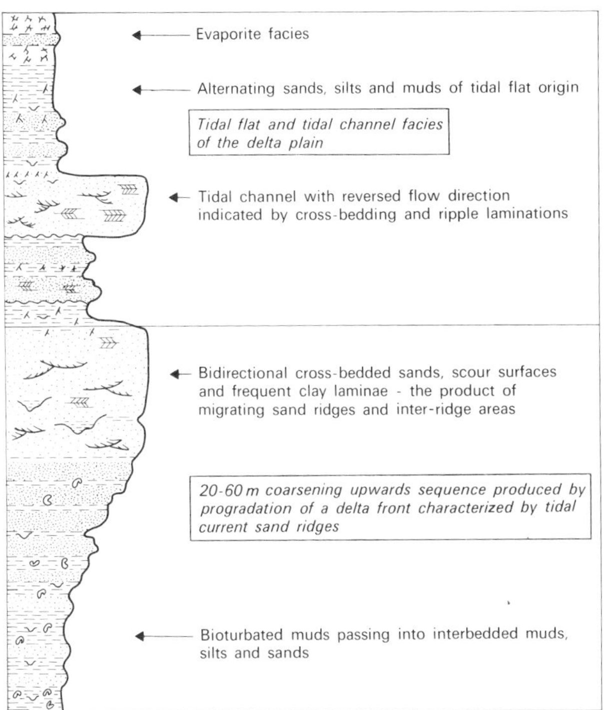

Fig. 6.33. Composite, idealized sequence through the tide-dominated Ord delta (after Coleman and Wright, 1975).

6.6 DELTA ABANDONMENT

Scruton (1960) stressed that deltas often have a two-fold history comprising a constructional phase during which the delta progrades, and a destructional or abandonment phase initiated by avulsion of the alluvial and/or distributary channel system supplying the delta. Although most sedimentation takes place during the constructional phase, consideration of the abandonment phase can greatly assist the interpretation of modern and ancient deltaic successions.

Alluvial- or distributary-channel switching most commonly results from over-extension of the channel system as the delta progrades into the basin. Shorter, steeper courses are generated, and if a crevasse breach is enlarged during a series of floods it may become a persistent feature of the channel network which gradually accepts an increasing proportion of the discharge of the parent stream until the latter is abandoned (Fisk, 1952a). In the present Mississippi delta the Atchafalaya River is diverting an increasing amount of discharge from the Mississippi River. From its point of bifurcation the Atchafalaya River flows only 227 km before reaching the Gulf of Mexico, whereas the Mississippi River flows 534 km. The

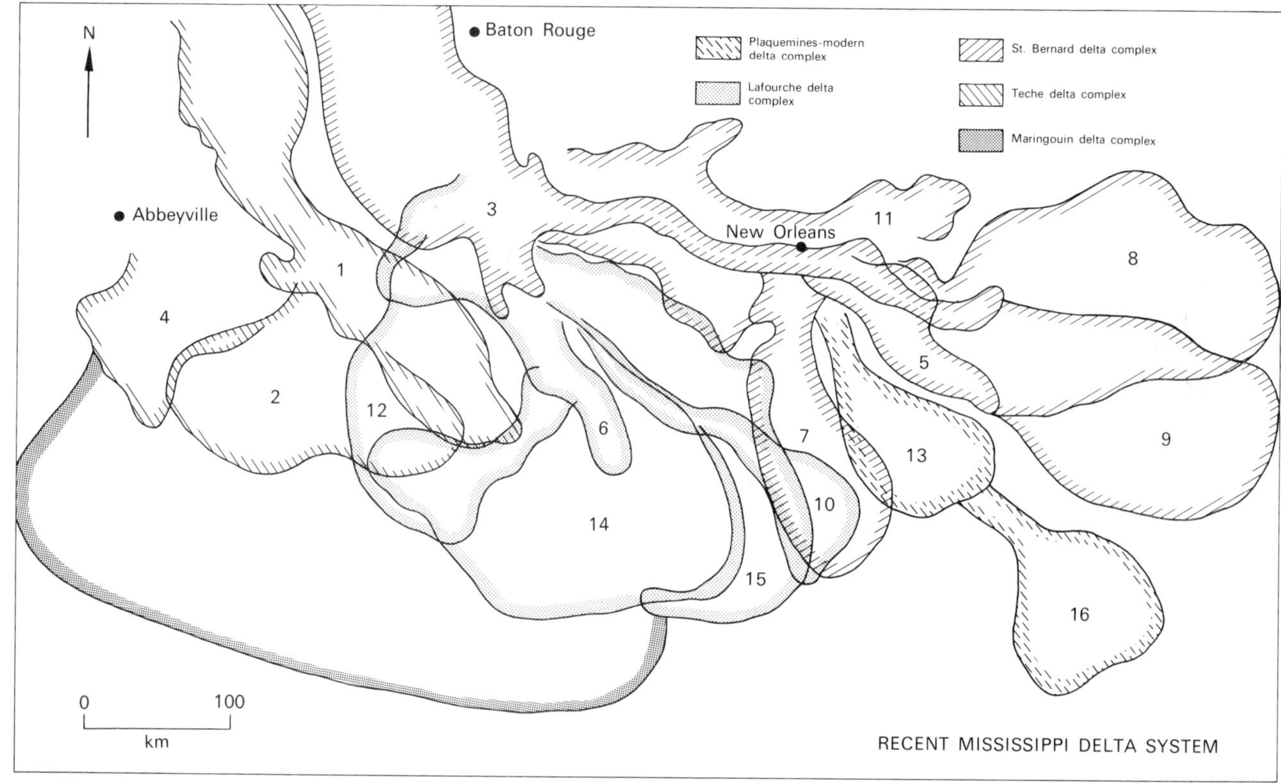

Fig. 6.34A. Areal distribution of delta complexes and lobes in the pre-modern and modern Mississippi delta (based on Frazier, 1967 and Fisher and McGowen, 1969).

gradient advantage of the Atchafalaya River is precipitating the next major abandonment phase of this delta complex, but man-made controls have so far limited its development.

Channel switching or avulsion causes the pre-existing delta or lobe to be abandoned and a new delta to be initiated elsewhere. Sediment supply to the formerly active delta diminishes and it therefore ceases to prograde. Basinal processes are enhanced in the absence of sediment input, causing at least local reworking, and compaction and subsidence to continue.

The present Mississippi delta was preceded by a series of 'shoal-water' deltas which prograded across the shallow shelf east and west of the modern delta site. Four major pre-modern delta complexes comprising fifteen lobes have been recognized (Frazier, 1967; Fig. 6.34). As these lobes have been successively abandoned during the last 6,000 years they are currently undergoing various stages of abandonment and a sequence of events can be demonstrated reflecting progressive changes during abandonment. The deltas are characterized by delta front sheet sands which extend over 800 sq km or more, with channel sand complexes up-dip (Fisk, 1955). Insight into the active form of the deltas can be gained from the extensive delta plain produced by the lobes and from the most recently abandoned Lafourche complex. Numerous distributaries traverse the delta plain, frequently bifurcating towards the shoreline, and it is considered that they produced a series of closely-spaced mouth bars which coalesced at the shoreline (Fig. 6.35). However, the Lafourche delta exhibits remnant beach-ridge complexes adjacent to the distributary mouth, and wave reworking may have been partially responsible for the sheet form of the delta front sand. Whichever applies, the net result is that the lobes subside uniformly rather than differentially due to the sheet-like sand distribution. Modification of the Lafourche complex is restricted to slight

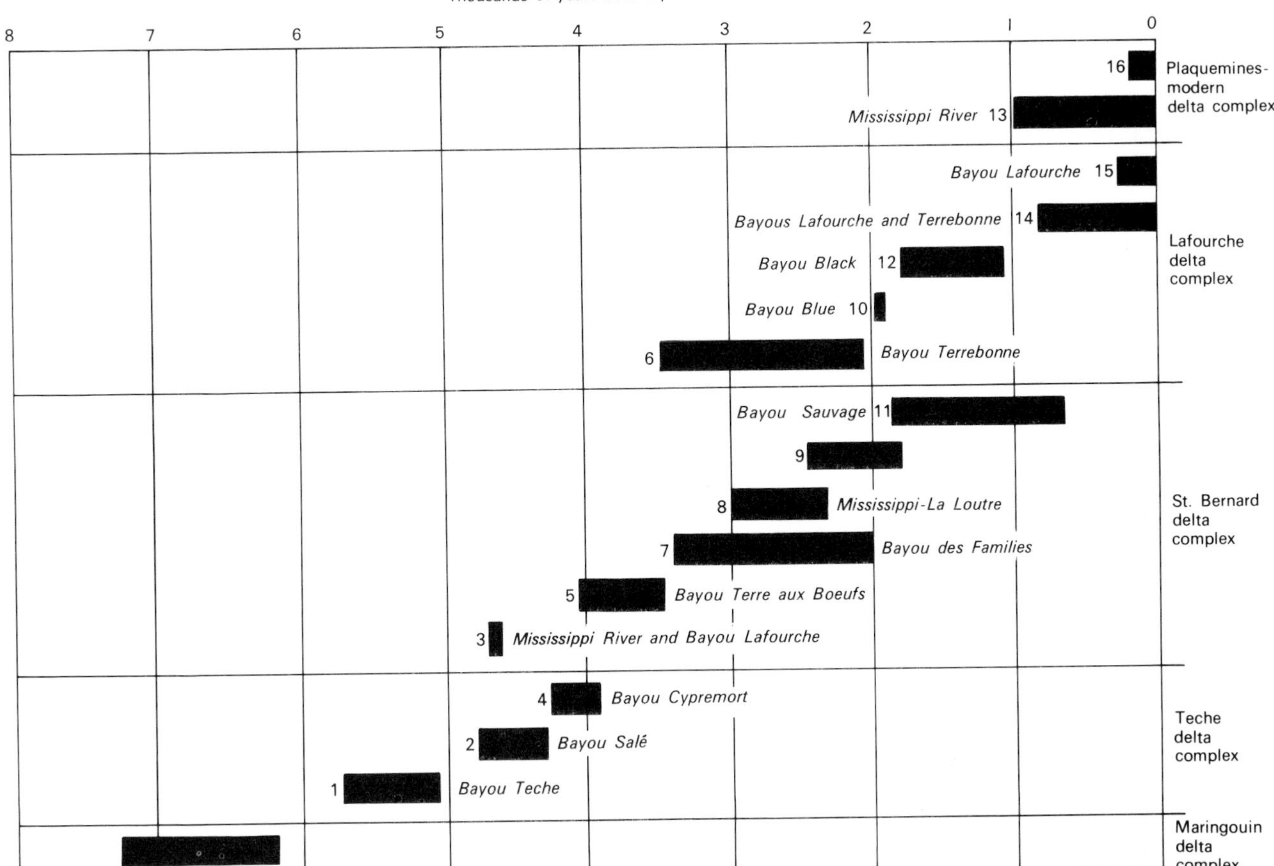

Fig. 6.34B. Chronology of delta complexes and lobes in the pre-modern and modern Mississippi delta (after Frazier, 1967).

smoothing of the shoreline, accompanied by lateral transport of the reworked sand to form a minor barrier island (Grand Isle). More advanced stages of abandonment are illustrated by the St. Bernard Complex which is characterized by a narrow, arcuate barrier island (the Chandeleur Islands) produced by wave reworking of the former delta shoreline (Fig. 6.36). This 'delta margin island' confines a shallow bay over the former delta plain in which fossiliferous clays, silts and sands are slowly accumulating. As subsidence continues the delta margin islands tend to migrate landwards as they are entirely dependent on the underlying abandoned lobe for sediment supply. Finally, in the older Teche and Maringouin complexes the former delta margin islands are marked by broad, submerged shoal areas several metres below sea level. Whilst these modifications are taking place in the vicinity of the former delta shoreline, the upstream areas are covered by peat blankets which may extend uninterrupted over several hundred square kilometres, and in offshore areas 'normal' or background sedimentation resumes. Most of the abandoned delta is therefore preserved, with only the former shoreline area being partially reworked. In this example, the abandonment phase produces a thin, laterally persistent unit which varies in facies across the abandoned delta but is generally distinguished by relatively slow rates of sedimentation. An upstream peat blanket passes laterally into fossiliferous clays, silts and sands of the protected bay (? restricted fauna), a thin sheet sand which is the transgressed remnant of a delta margin barrier island, and finally into a thin unit of offshore facies with a diverse and prolific marine fauna.

Radiocarbon dating of the peat blankets demonstrates that lobes within a complex are often contemporaneous, indicating both that separate distributary systems co-existed, and that certain complexes were contemporaneous. This suggests that at times the total discharge of the alluvial system was split

0 16
km
Distributary channels
Interdistributary lakes, marshes and swamps
Continuous delta front shoreline sand
Lower delta front - prodelta

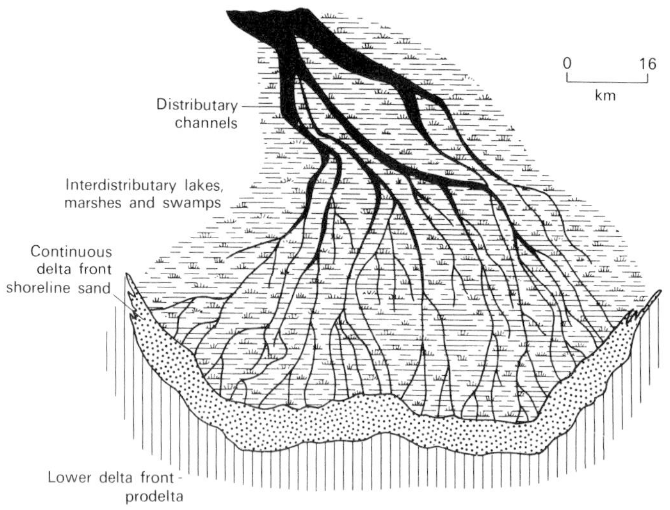

Fig. 6.35. Reconstruction of the abandoned Lafourche lobe of the pre-modern Mississippi delta (after Fisher *et al.*, 1969).

Fig. 6.36. The abandoned St. Bernard lobe of the pre-modern Mississippi delta illustrating post-abandonment modifications (after Coleman and Gagliano, 1964).

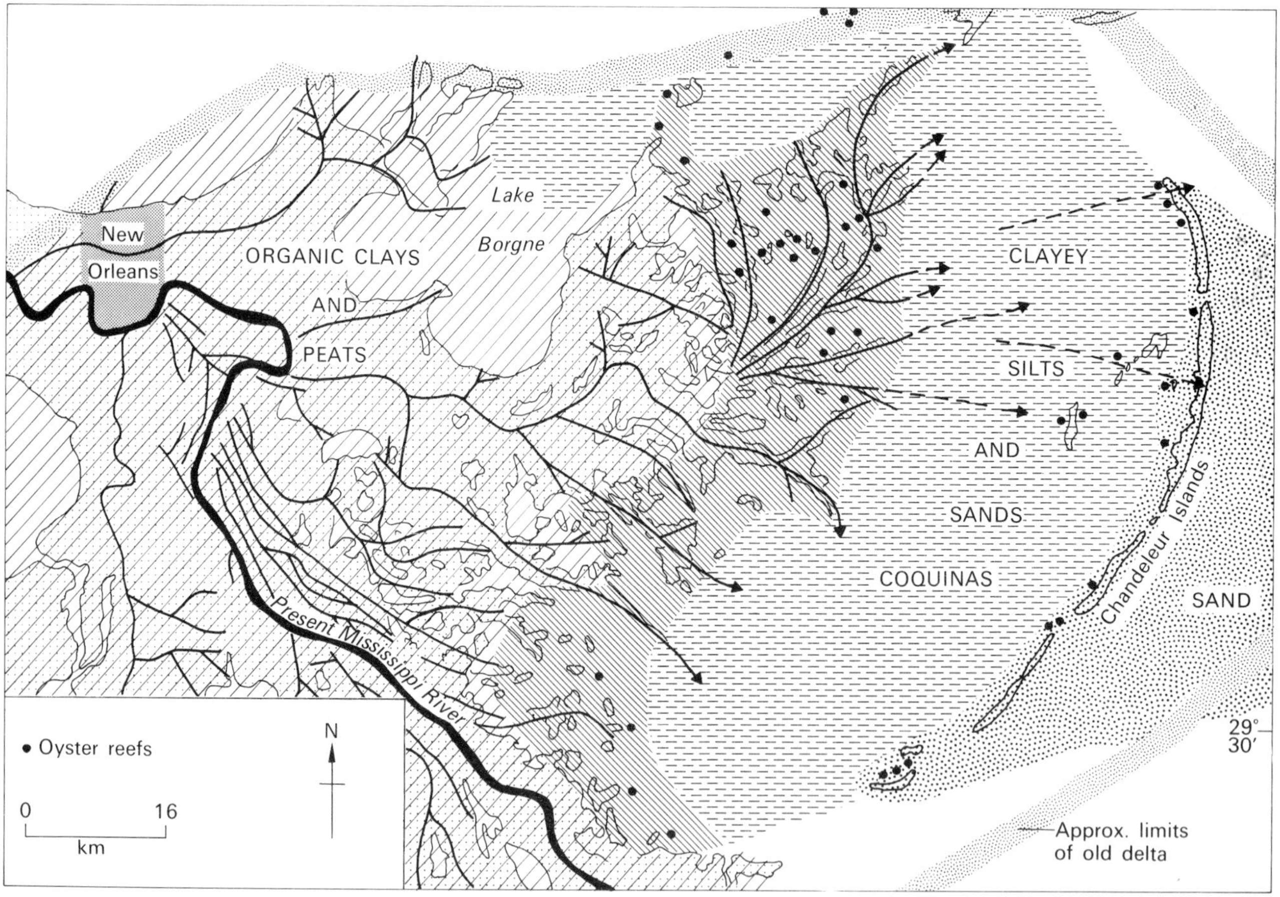

New Orleans
ORGANIC CLAYS AND PEATS
Lake Borgne
CLAYEY SILTS AND SANDS
COQUINAS
Chandeleur Islands
SAND
Present Mississippi River
• Oyster reefs
N
0 16
km
29° 30′
Approx. limits of old delta

between complexes (Frazier, 1967; Frazier and Osanik, 1969). This study also reveals that during the abandonment of one lobe another may be initiated, prograde to its full extent and commence its own abandonment phase. A thin abandonment facies horizon can therefore represent a far greater time interval than thick constructive facies associations.

As the concept of delta abandonment arose from the Mississippi delta its universality must be considered in view of the current emphasis on the variability of deltas. The initiation and abandonment of delta lobes or complexes is related to the frequency of channel avulsion. Rapid progradation rates in fluvial-dominated deltas produce significant shoreline protuberances and gradient advantages therefore abound. Avulsion occurs frequently and lobes proliferate. However, as wave effectiveness increases, progradation occurs over a broader area and the rate diminishes. Fewer gradient advantages are created and avulsion is therefore less frequent. In addition, after abandonment a greater proportion of the former delta is likely to be reworked by subsequent wave action. For example, in the Rhône delta only three lobes have formed during the same period in which the entire range of Mississippi delta lobes was produced (Fig. 6.37). If abandonment occurred in a wave-dominated delta reworking would probably obliterate the potential lobe and the entire area of the former delta would be transgressed. In tide-influenced deltas avulsion is likely to be confined to the fluvial channel reaches in the upper delta plain or alluvial valley as tidal channels tend to migrate continuously rather than switch direction. This is a response to the repeated, short-term and relatively small-scale fluctuations in regime, which contrast with the less frequent and more pronounced fluctuations in fluvial channel regimes which induce channel switching. Tide-influenced deltas often protrude significantly from the general shoreline trend and avulsion may therefore occur, but many of these deltas are located in narrow basins which restrict the development of discrete lobes.

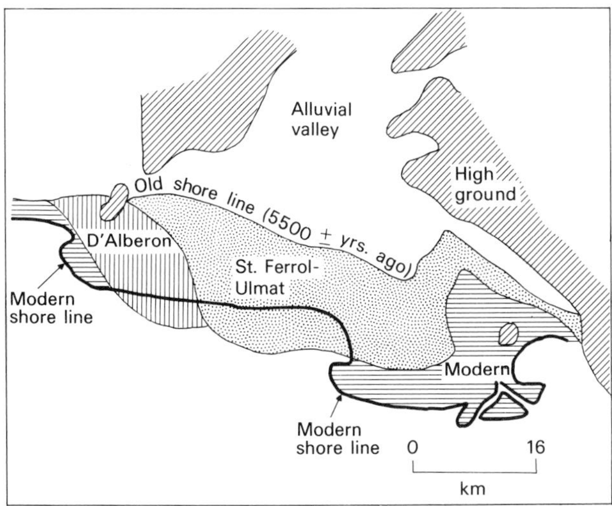

Fig. 6.37. Lobe development in the Rhône delta (after Scruton, 1960).

This brief comparison indicates that channel switching and the development of delta lobes is a preferential feature of fluvial-dominated deltas and to a lesser extent fluvial-wave interaction deltas. However, delta abandonment may also result from tectonically induced river capture. For example, in the Ganges–Brahmaputra delta, river capture results from contemporaneous basement faulting in combination with erratic major floods. Capture of the Hooghly river resulted in the abandonment of a large deltaic tract now occupied by a dense swamp area (the Sundarbans Jungle). In the event of tectonically induced abandonment the preservation of a delta lobe will depend largely on the subsidence rate and prevailing basin regime.

6.7 ANCIENT DELTAIC SUCCESSIONS

As deltas comprise an association of depositional environments ranging from fluvial-distributary channels to relatively deep-water offshore areas, recognition in the geological record requires the identification of a number of genetically related facies associations.

Three major facies associations characterize delta systems: the delta plain and delta front facies associations deposited whilst the delta is active, and the delta abandonment facies association.

6.7.1 Delta plain facies association

Fluvial-dominated examples of this facies association comprise large-scale fluvial-distributary channels, smaller scale crevasse channels, and a wide range of small-scale coarsening-upward sequences which reflect infilling of shallow water interdistributary areas (Ferm and Cavaroc, 1968; Collinson, 1969; Elliott, 1974b, 1975, 1976b; McBride *et al.*, 1975; Horne and Ferm, 1976).

The small-scale coarsening-upward sequences are on average 4–10 m thick and commence with finely laminated or bioturbated mudstones–siltstones deposited from suspension across the entire interdistributary area during river flood periods. Plant debris is often abundant, along with a brackish or freshwater fauna. This facies may constitute the entire bay-fill sequence, but more commonly the sequences terminate in a thin sandstone member. Facies details of the sandstones vary, depending on whether they reflect levee construction by overbank flooding, crevasse splay lobes, minor mouth bar-

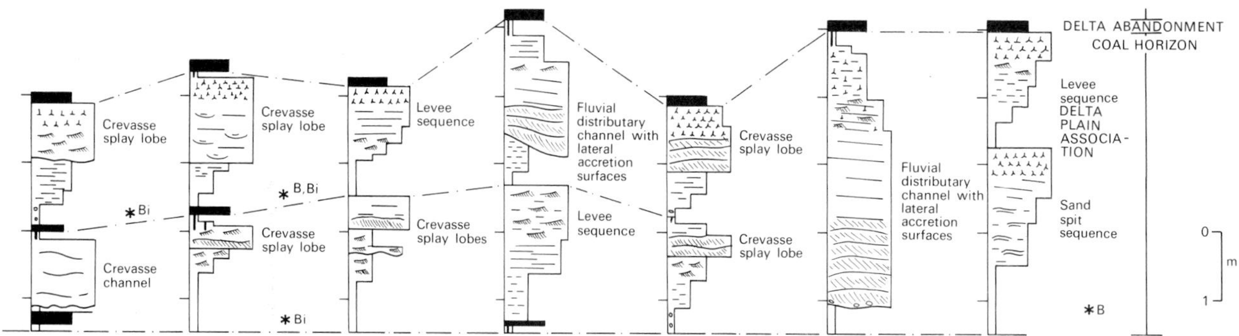

Fig. 6.38. Fluvial-dominated delta plain association from the Carboniferous of northern England (after Elliott, 1975).

crevasse channel couplets, wave reworked sand spits (Fig. 6.8). Each depositional process produces a relatively thin and impersistent sandstone body, but they may coalesce to produce thin sheet sandstones which infill large expanses of the interdistributary area (Fig. 6.38). Palaeosols frequently develop towards the top of the sequences, and coals accumulate under conducive climatic conditions.

Crevasse channels set in these sequences are generally 1–4 m thick and have a channel-fill which exhibits numerous reactivation surfaces, clay drapes and indications of temporary bedform emergence. These features reflect ephemeral flow in the crevasse channels resulting from healing or 'stranding' of the channel during periods of low river stage. Distributary channel sequences are larger scale and generally reflect more continuous discharge conditions, though still with stage fluctuations. The sequences are similar to those of fluvial channels, with channel-fill characteristics being dictated by channel processes and pattern. A high sinuosity pattern is inferred in one example from the presence of lateral accretion surfaces (Elliott, 1976), whereas in a separate example giant cross bed sets up to 40 m thick are interpreted as side-attached alternate bars in a low sinuosity channel (McCabe, 1977; see Chapter 3).

Channel abandonment often produces an overall fining upwards trend, with the fine member comprising ripple-laminated siltstones, occasional thin crevasse splay sandstones, plant-rich shales and palaeosol-coal units (Fig. 6.39).

Tide-dominated delta plain associations are likely to comprise tidal flat sequences, small-scale channel sandstone complexes produced by tidal creek systems, and larger-scale tidal-distributary channel sequences. However, although tidally influenced channel sequences have been inferred from bimodal current patterns in Cretaceous delta plain associations in the Western Interior, USA (van de Graaf, 1972; Hubert *et al.*, 1972), there are no descriptions of the entire association from an ancient delta system.

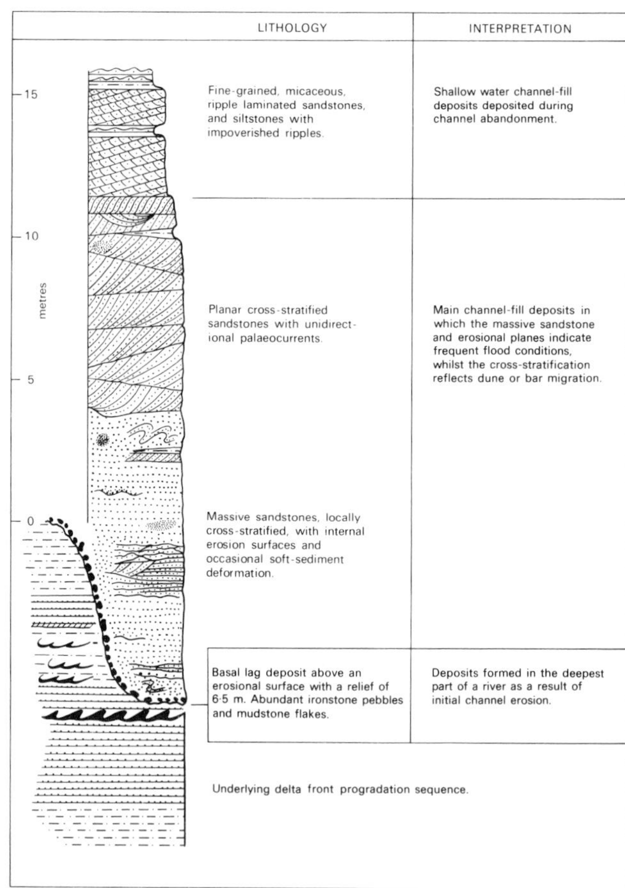

Fig. 6.39. Carboniferous fluvial-distributary channel sequence in south-west Wales (after Kelling and George, 1971).

6.7.2 Delta front facies association

The delta front is represented by a relatively large-scale coarsening-upward sequence which records a passage from fine-grained offshore or prodelta facies upwards into a shoreline facies which is usually sandstone-dominated. These sequences result from progradation of the delta front and may be truncated by fluvial- or tidal-distributary channel sequences as progradation continues. Although these sequences vary considerably in relation to the regime and morphology of the former delta front, so far only fluvial-dominated delta front sequences have been widely recognized in the geological record (Fisher *et al.*, 1969; Kelling and George, 1971; Brown *et al.*, 1973; Erxleben, 1975; Elliott, 1976a; Horne and Ferm, 1976).

Fluvial-dominated sequences commence with a thick, uniform interval of mudstones or fine- to medium-grained siltstones deposited from suspension at the base of the delta front and beyond (Fig. 6.40). This facies may appear massive, but more commonly exhibits diffuse banding defined by slight variations in grain size which reflect fluctuations in the supply of suspended sediment. Bioturbation may disrupt this banding and marine faunas occur, but faunal density and diversity is generally low due to the almost continuous fall-out of sediment from suspension, and the possibility of water salinity being lowered slightly by freshwater input. Plant debris occurs in this facies and is presumably an additional consequence of sediment input being direct from the distributaries. In several examples of this facies, thin erosive-based coarse siltstone–fine sandstone beds occur within the mudstones–siltstones (de Raaf *et al.*, 1965; McBride *et al.*, 1975). Structures in these beds indicate that they are the product of waning currents, and they often resemble thin, relatively fine-grained turbidites. It is probable that sediment-laden discharge issuing from the distributary mouth during flood periods was periodically denser than the basin waters and therefore entered the basin as bottom-hugging currents which developed into turbidity currents. In extreme cases, turbidity currents and related

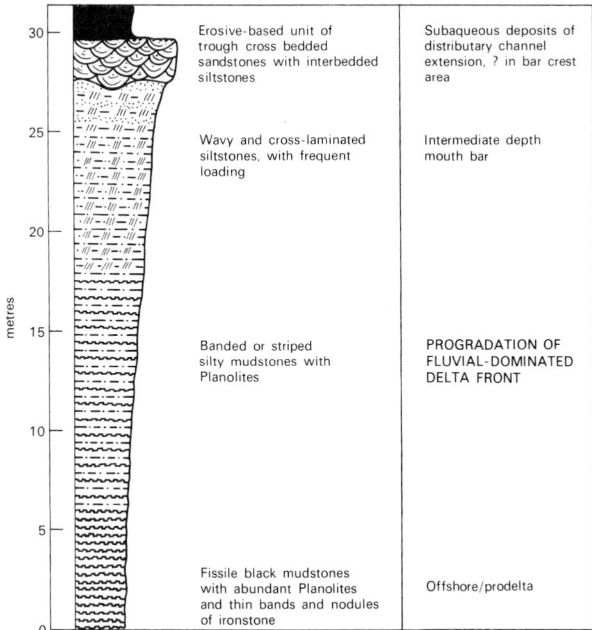

Fig. 6.40. Salient features of a fluvial-dominated delta front sequence; considerable departures from this idealized summary occur in the majority of cases, in particular the facies of the intermediate and upper members are extremely variable and sediment-induced deformation may occur at any level (modified after Kelling and George, 1971).

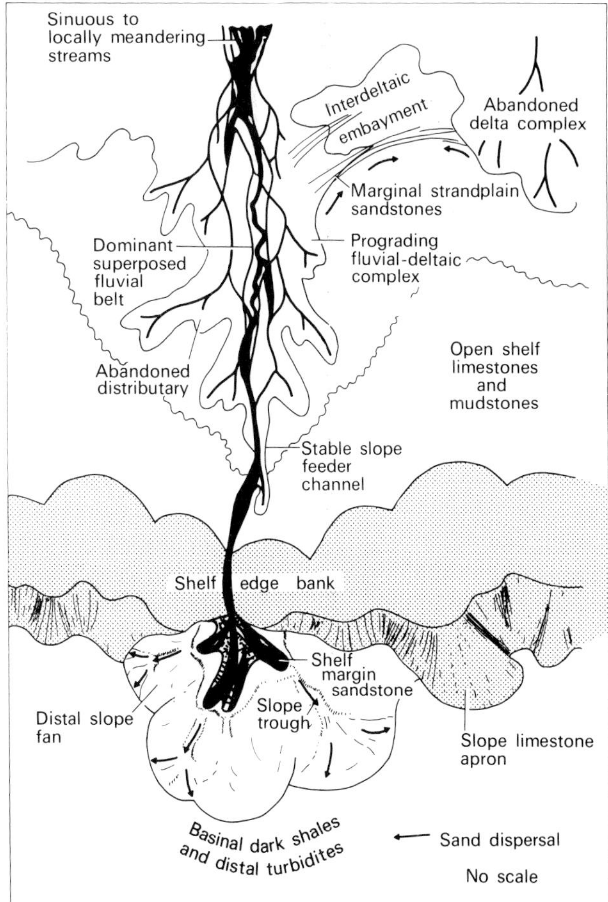

Fig. 6.41. Delta system directly supplying a slope-submarine fan system, reconstructed from the Carboniferous Cisco Group in East Texas, USA (after Galloway and Brown, 1973).

processes construct substantial submarine fans beyond the delta front (Walker, 1966; Collinson, 1969, 1970a; Galloway and Brown, 1973; Fig. 6.41).

The intermediate parts of these sequences comprise mudstone–siltstone background sediment into which coarser siltstone and sandstone beds are repeatedly intercalated. Initially the background sediment is a direct continuation from below, but as the sequence is ascended thin siltstone laminae, small-scale ripple laminations and ripple form sets appear. This evidence of agitation in the background sediment is often considered to reflect the inception of wave base in the sequence. The coarser beds represent flood-generated sediment incursions from the distributary mouth. Various types of beds are deposited in this intermediate depth area of the delta front, determined by the amount of sediment supplied during a flood, the precise manner in which it is dispersed, the rate of flood rise and wane, and the level of basinal processes at the time. In most cases the coarse beds have planar erosive bases and exhibit waning flow sequences involving a passage from parallel lamination upwards into asymmetrical ripple laminations. These beds are considered to result from sediment-laden traction currents flowing down the delta front from the distributary mouth. Upper surfaces of these beds are often sharply defined and may exhibit straight-crested symmetrical ripple marks reflecting post-flood wave reworking of the upper few centimetres. Other beds have gradational bases and indicate increasing flow velocity and sediment transport upwards, perhaps related to a different mechanism of outflow dispersion and more gradual flood rise. Towards the top of the sequence the coarse beds become thicker and dominate the facies. Individual sandstone beds are laterally continuous, but lenticular units representing minor subaqueous extensions of the distributary channel may occur at the top. A wide range of structures reflect high rates of sediment transport and deposition, whilst the background sediment is confined to thin intervals between sandstone beds.

In wave-influenced sequences the fine member at the base of the sequence is generally less well-developed as suspended sediment is widely distributed beyond the immediate delta area (Fisher *et al.*, 1969). The intermediate and upper parts of the sequence may comprise well-sorted sandstones with a predominance of symmetrical ripples, wave-produced parallel-lamination and symmetrically filled scours, reflecting partial or complete wave reworking of the river-supplied sediment. In the Cretaceous Cody Shale–Parkman Sandstone delta succession of the Western Interior, USA, the upper part of the large-scale delta front coarsening-upward sequence is dominated by flat laminated sandstones with occasional sets of trough cross-bedding of variable direction. Interpretation of this facies as a wave-produced beach-barrier sandstone is further supported by resistivity data which reveals low-angle, seaward-dipping surfaces in the sandstone member reflecting former positions of

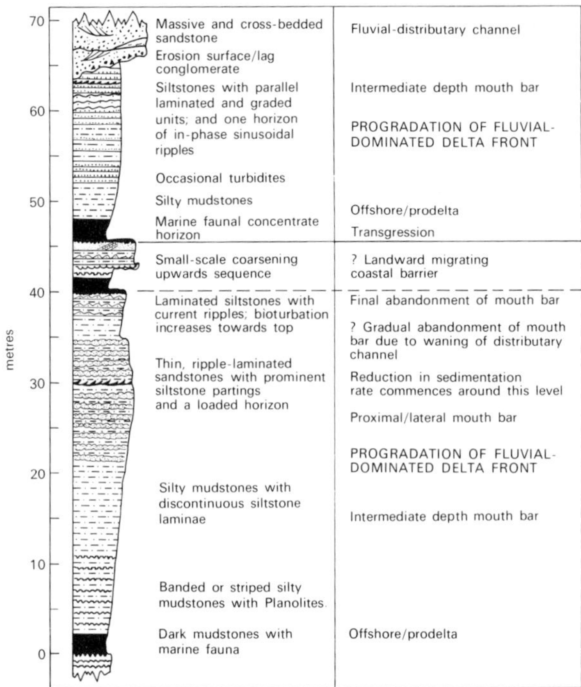

Fig. 6.42. Variations in fluvial-dominated delta front sequences: in the first sequence (0–40 m) the upper part is characterized by a reduction in the rate of sedimentation, possibly related to gradual waning of the distributary channel; in the second sequence (45–70 m) the fluvial-distributary channel cuts into the delta front sequence (modified after Kelling and George, 1971).

the beach face (Asquith, 1970, 1974; Hubert *et al.*, 1972).

Tidally-influenced delta front sequences have not been described from the geological record. Recognition is likely to depend on the characteristics of the sandstone member at the top of the sequence and to a lesser extent the intermediate part of the sequence. Features which would support an interpretation along these lines include bimodal sets of trough cross-bedding and ripple-lamination, frequent reactivation surfaces and clay drapes, and complex interfingering between tidal current ridge facies and between-ridge channel facies.

So far delta front sequences have been discussed as relatively uniform, coarsening-upward sequences, but several factors may give rise to departures from the expected sequence. Proximity to the distributary mouth in an axial-lateral sense is important, and related to this is the spacing of distributary channels. Away from the distributary mouth the sand member may be thin or even absent, but conversely this area may be influenced by two distributary mouths if the channels are closely spaced. These complications are compounded by the tendency of individual distributaries to wax or wane with time, or migrate laterally, both of which cause sedimentation to vary continuously at a given site (Fig. 6.42).

6.7.3 **Delta abandonment facies association**

In general, abandonment facies comprise thin but laterally persistent marker horizons composed of facies which reflect relatively slow rates of sedimentation. Until recently studies of ancient deltaic successions neglected these facies and concentrated on the manifestly obvious progradational facies of the delta plain and delta front. However, where abandonment facies are developed they permit greater refinement in the interpretation of ancient deltaic successions (Fisher *et al.*, 1969; Elliott, 1974a). Their importance is four-fold:

(1) as distinctive marker horizons they permit correlation in successions which are otherwise characterized by lateral impersistence of facies;

(2) the horizons only develop in the area of the abandoned delta (or lobe) and therefore define its areal extent;

(3) the horizon dates delta (or lobe) abandonment and is therefore significant in reconstructing delta history;

(4) as the conditions of the abandonment period vary between deltas in terms of the energy level of basinal processes and the subsidence rates as dictated by sand distribution and overall mud–sand ratio, the abandonment facies may differ and reflect delta type (Fisher *et al.*, 1969).

In fluvial-dominated delta systems in the Tertiary of the Gulf Coast, abandonment facies marker horizons are extremely thin in comparison with progradational facies. Distal parts of the abandoned deltas have a veneer of well sorted, fine-grained bioturbated sandstones, shell debris and marine mudstones, whilst upstream or proximal areas are characterized by laterally extensive lignites traceable over several thousand square kilometres (Fisher and McGowen, 1967). Comparable facies associations occur in Carboniferous fluvial-dominated deltas (Ferm, 1970; Elliott, 1974a; Erxleben, 1975; Fig. 6.43). In one example a 20 cm coal extends uniformly across underlying facies variations in the upstream area and represents a former peat blanket. Distally this passes into a thin interval of bioturbated marine sandstone and coquinoid limestones. The sandstone facies is interpreted as the transgressed remnant of a delta margin island (Chandeleur-type), whilst the limestone represents a return to normal marine sedimentation in the absence of clastic input (Elliott, 1974a; see Fig. 6.49).

In wave and/or tidally influenced deltas, basinal processes are simply enhanced after delta abandonment and there is less of a contrast between progradational and abandonment facies, at least in the delta front area. A relatively thick transgressive sheet of marine sandstones, siltstones or mudstones gradually extends across the entire area of the former delta. In the Cody Shale–Parkman Sandstone delta for example, 4–18 m of

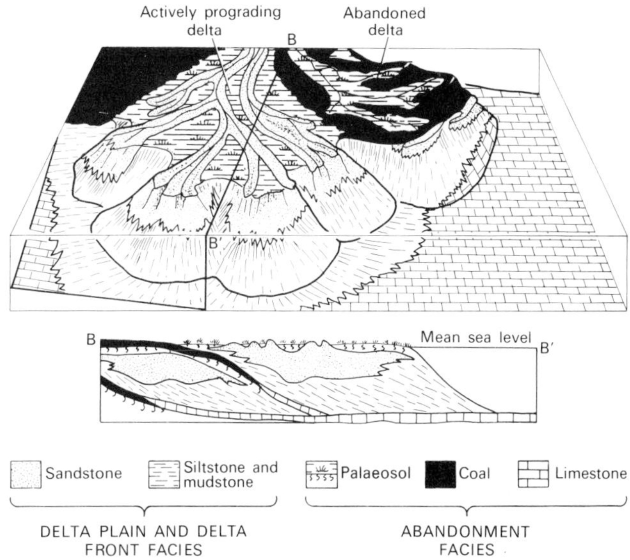

Fig. 6.43. Palaeosol-coal-limestone abandonment facies association enveloping delta front and delta plain facies of Carboniferous deltas in the United States (after Ferm, 1970).

marine sandstone transgress across the delta plain facies association (Hubert *et al.*, 1972). However, in the delta front sequence it may be difficult to define the transgressive abandonment facies.

6.7.4 **Recognition of delta type in the geological record**

Early attempts at distinguishing different types of deltas in the geological record are largely confined to discriminating between lobate and birdfoot Mississippi types. However, as the range of deltas was presumably comparable, if not greater, in the geological past it follows that a more diverse range of delta types should be recognizable in the ancient record.

Various approaches can assist in the elucidation of palaeodelta type:

(1) Vertical facies sequences provide the basis for most interpretations as information of this nature is available in most studies at an early stage in the investigation. The nature of delta front sequences is particularly important in distinguishing different regime-defined deltas.

(2) Lateral facies relationships on a regional scale permit the reconstruction of proximal–distal and axial–lateral changes. Abandonment facies marker horizons are particularly useful in this respect.

(3) Sand isopach data may indicate the palaeo-delta type, though often only at a late stage in the investigation. Lateral correlations are a prerequisite of this approach, and vertical

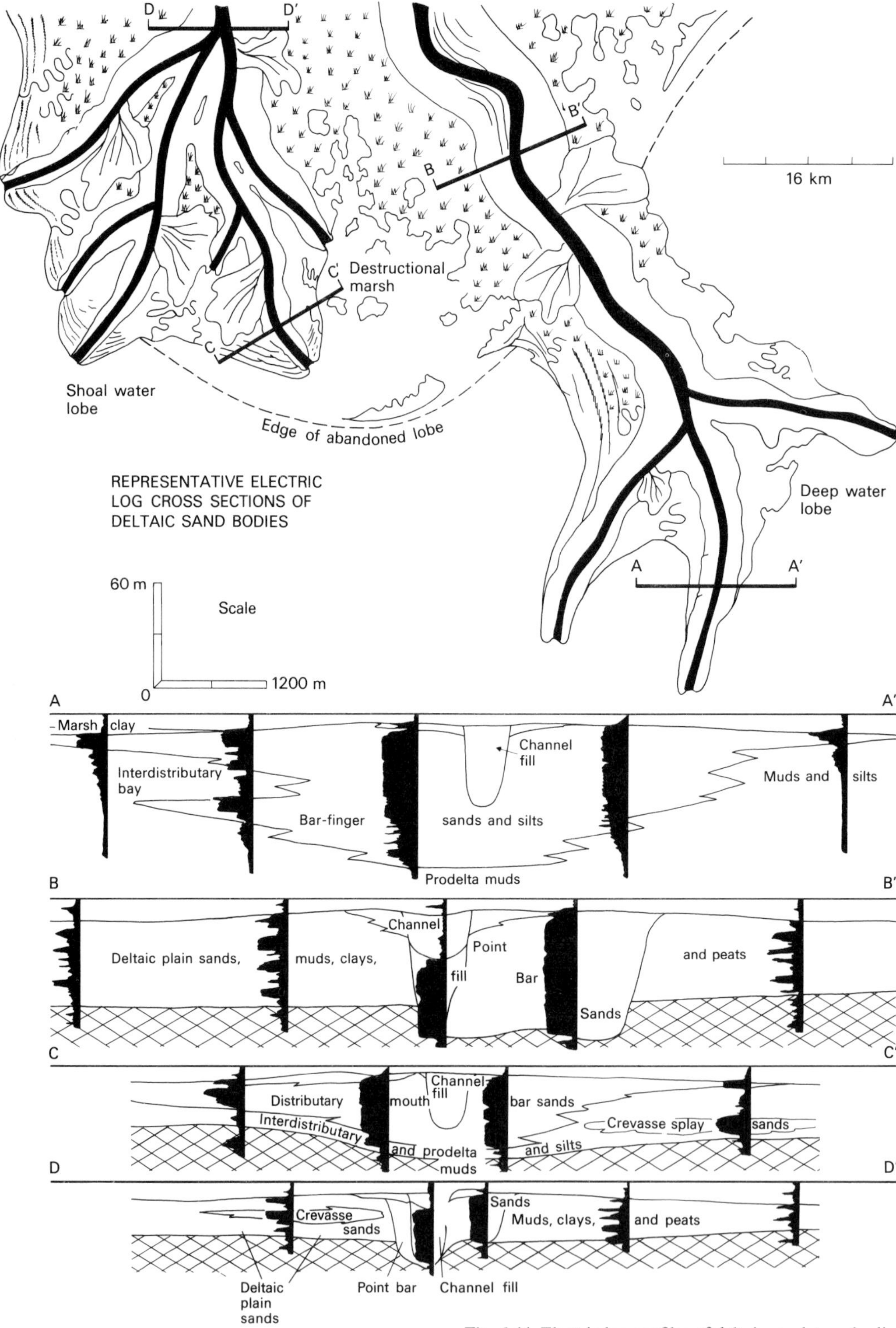

Fig. 6.44. Electric log profiles of deltaic sandstone bodies based on the Tertiary fluvial-dominated Holly Springs delta system in the Gulf Coast, USA (after Galloway, 1968).

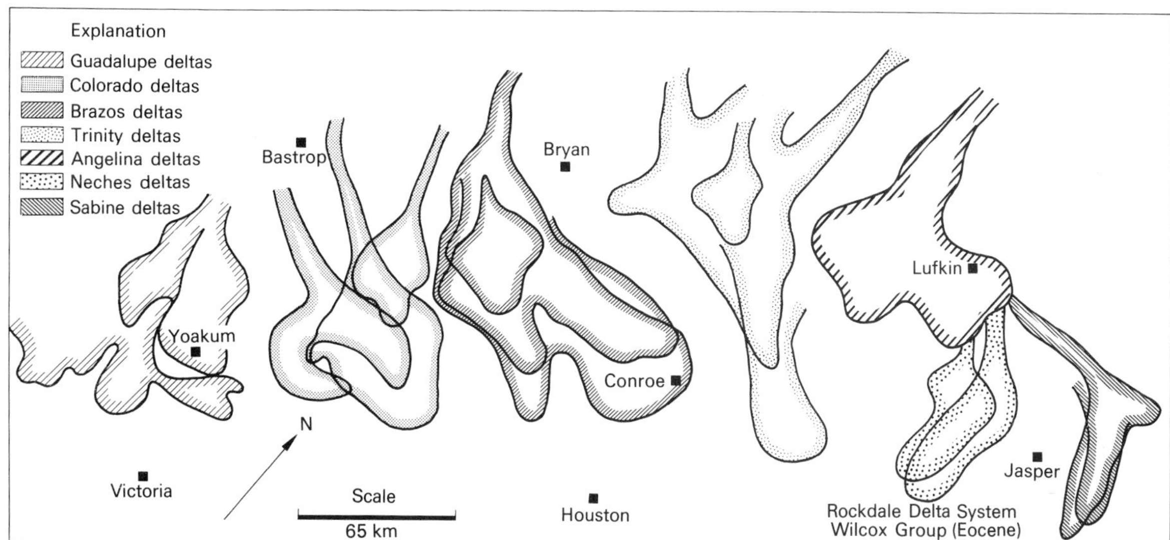

Fig. 6.45. Delta complexes and lobes in the Tertiary fluvial-dominated Rockdale delta system in the Gulf Coast, USA (after Fisher and McGowen, 1969).

sequence data greatly improves the quality of the interpretation.

(4) Abandonment facies marker horizons may reflect the sand distribution and indirectly indicate the delta type. This can be particularly important in distinguishing birdfoot and lobate fluvial-dominated deltas.

Other data such as palaeocurrents and compositional petrography should be integrated with these lines of approach to complete the analysis, but in both cases the data should be related to the palaeo-environmental interpretation of the facies and sequences.

TERTIARY DELTAS IN THE GULF COAST

The presence of hydrocarbons in Tertiary successions on the northern flank of the Gulf Coast basin has stimulated investigations of these successions and resulted in a wealth of sub-surface information to supplement that gained from surface exposures. Time-equivalent strata have been studied over extensive areas, revealing large-scale facies patterns of depositional systems. The Tertiary record is dominated by deltaic systems recognized using vertical sequences and sandstone body characteristics derived largely from electric logs (Fisher, 1969; Fisher *et al.*, 1969; Fig. 6.44).

In the Eocene *Lower Wilcox Group* the fluvial-dominated Rockdale delta system extends over 6,000 sq. km and has a maximum thickness of 1,500 m (Fisher and McGowen, 1967). Abandonment facies marker horizons and sand isopach maps define three major delta complexes comprising sixteen lobes which range from 500–5,000 sq. km in lateral extent and 90–450 m in thickness (Fig. 6.45). The lobes are either rounded, lobate features considered analogous to pre-modern shoal-water Mississippi delta lobes, or elongate narrow forms reminiscent of the modern birdfoot Mississippi delta. Fluvial-dominance is suggested by details of the delta front and delta plain facies and the marked thickness contrast between progradational and abandonment facies. On a regional scale the delta system passes upstream into a major fluvial system, and laterally into strandplain and beach-barrier systems. The latter is considered characteristic of fluvial-dominated deltas in this setting and further emphasizes the analogy with the Holocene Gulf Coast. Similar examples of fluvial-dominated delta systems in this area include the Holly Springs delta system (Galloway, 1968), the Yegua Formation (Fisher, 1969) and the Fayette delta system (Fisher *et al.*, 1969).

Facies patterns in the *Upper Wilcox Group* differ radically from the Lower Wilcox Group (Fisher, 1969). Electric log profiles permit distributary mouth bar sequences to be distinguished from coastal barrier sequences in the delta front. The mouth bar sequences are only locally developed and are flanked by extensive areas of coastal barrier sequences arranged in a cuspate or chevron pattern. Maximum sand development is parallel to depositional strike, supplemented by a series of slightly lobate protrusions tied to linear sand

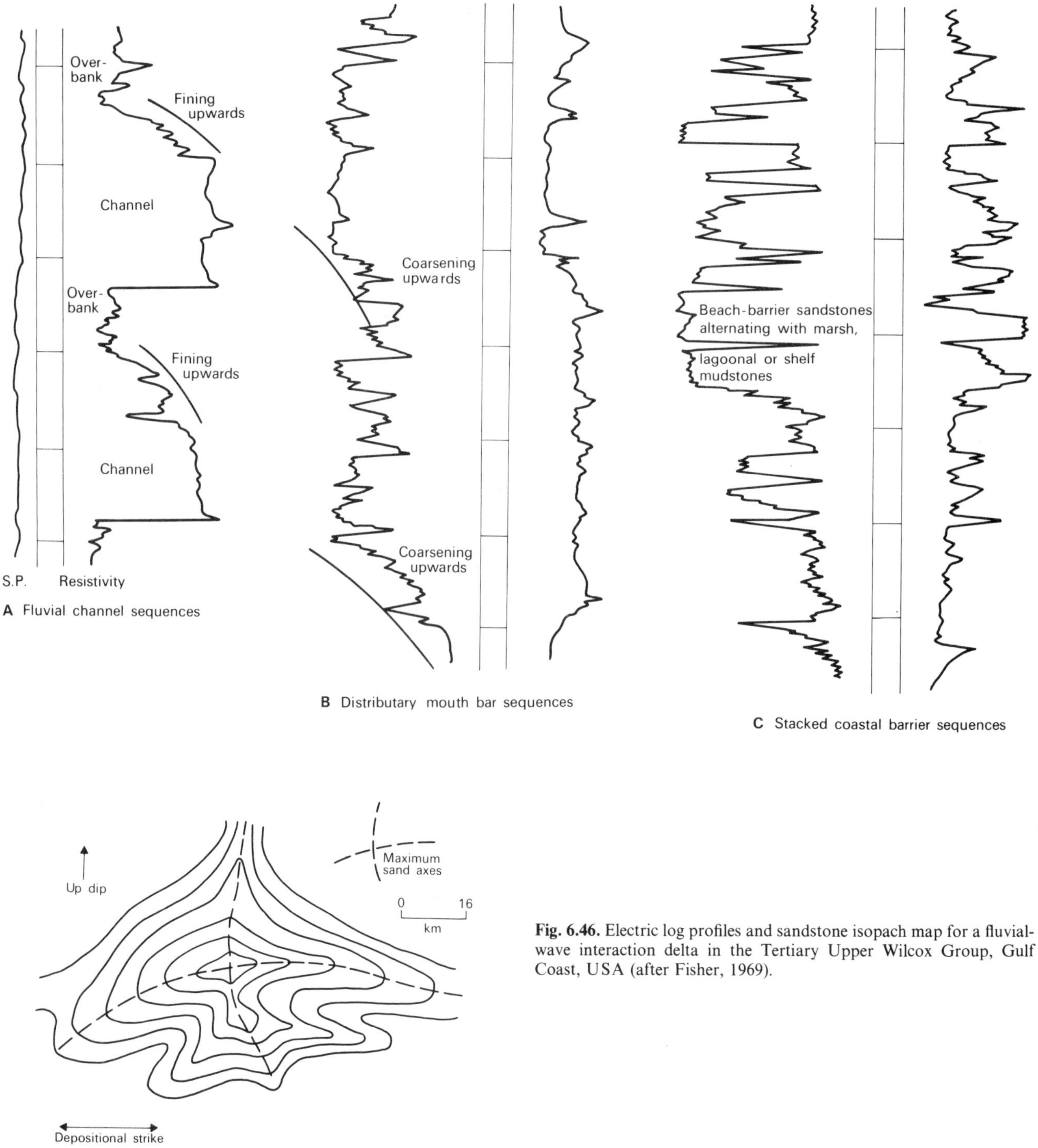

Over-
bank
Fining
upwards
Channel
Over-
bank
Fining
upwards
Channel
S.P.
Resistivity
A Fluvial channel sequences
Coarsening
upwards
Coarsening
upwards
B Distributary mouth bar sequences
Beach-barrier sandstones
alternating with marsh,
lagoonal or shelf
mudstones
C Stacked coastal barrier sequences
Up dip
Maximum
sand axes
0
16
km
Depositional strike

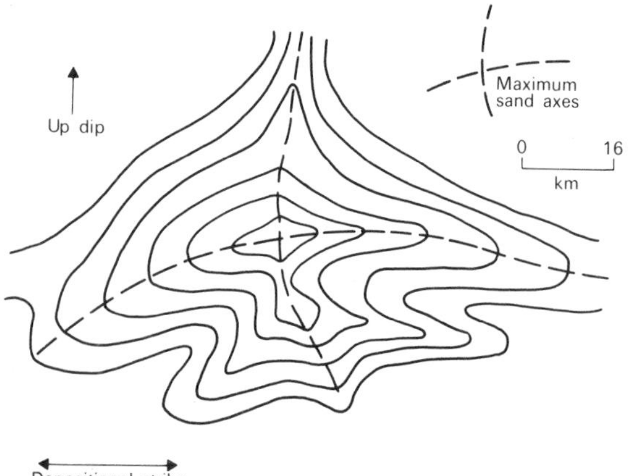

Fig. 6.46. Electric log profiles and sandstone isopach map for a fluvial-wave interaction delta in the Tertiary Upper Wilcox Group, Gulf Coast, USA (after Fisher, 1969).

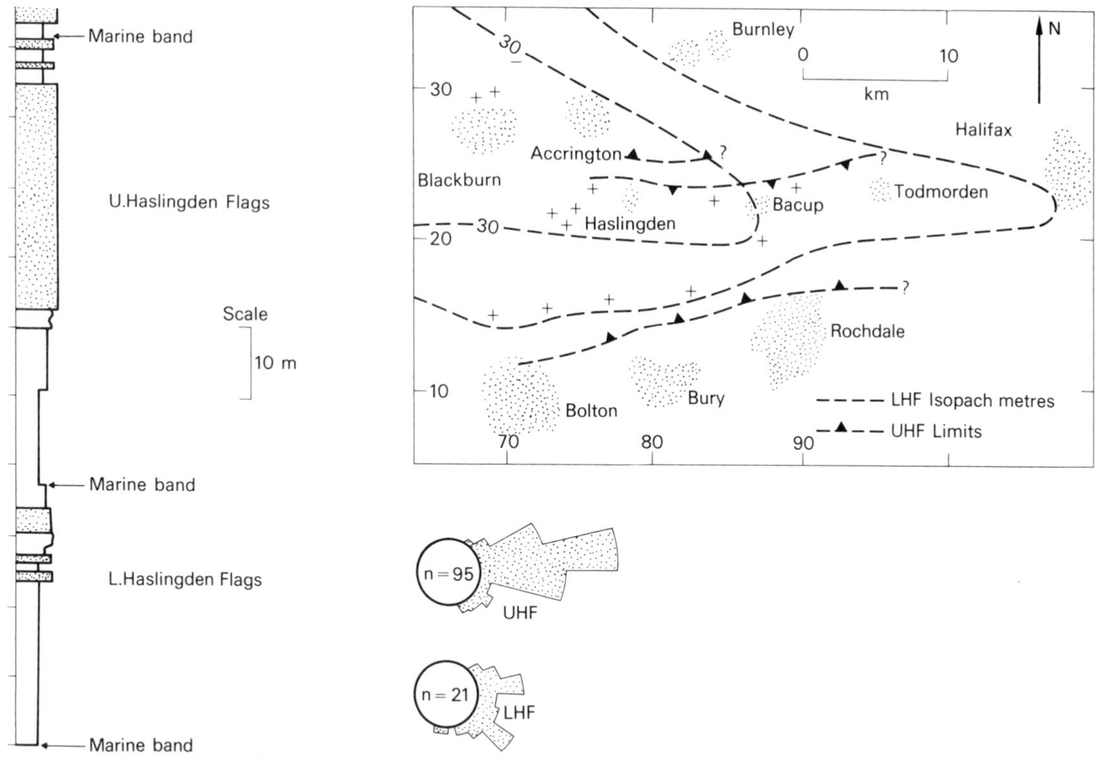

Fig. 6.47. Vertical sequence, isopachs and palaeocurrents for the Lower and Upper Haslingden Flags in the Upper Carboniferous of northern England; interpreted as the repeated progradation of bar finger sands in a fluvial-dominated delta (after Collinson and Banks, 1975).

maxima which represent distributary channel sands (Fig. 6.46). The sequences and sand distribution pattern are interpreted in terms of a wave-dominated delta, but the localized preservation of mouth bar sequences and the pattern of the coastal barrier sands suggest a closer resemblance to fluvial-wave interaction deltas. Other examples of wave-influenced deltas in these successions include the Middle Vicksburg delta system of Texas (Gregory, 1966) and the Frio delta system (Boyd and Dyer, 1964; Fisher, 1969).

CARBONIFEROUS DELTAS IN EUROPE AND THE UNITED STATES

During the Upper Carboniferous copious amounts of clastic sediment were supplied to the margins of many basins in Europe and the United States, creating major river deltas in which many of the coal seams accumulated. So far, only fluvial-dominated deltas have been recognized, but there appears to be a wide range within this class.

The *Haslingden Flags delta system* developed in the confined Central Pennine basin in northern England and is represented by two large-scale delta front coarsening-upward sequences (Collinson and Banks, 1975; Fig. 6.47). Facies characteristics of the sequences suggest that they reflect a fluvial-dominated delta front, with sedimentary structures being current-produced throughout and minimal evidence for basinal processes. The sandstone members at the top of the sequences are linear lobes up to 30 m thick and 8 km wide, aligned parallel to palaeoflow. They are interpreted as bar finger sands of a fluvial-dominated elongate delta and isopachs for the lower example diminish to zero in the palaeocurrent direction, suggesting that the precise form of the bar finger has been preserved after abandonment. The upper bar finger sand exhibits large-scale units of inclined bedding up to 27 m thick dipping perpendicular to the palaeocurrents indicated by ripple lamination within the units. These surfaces are regarded as lateral accretion surfaces formed at the sides of mid-channel bars and sub-channels in the axial mouth bar. The depth of the

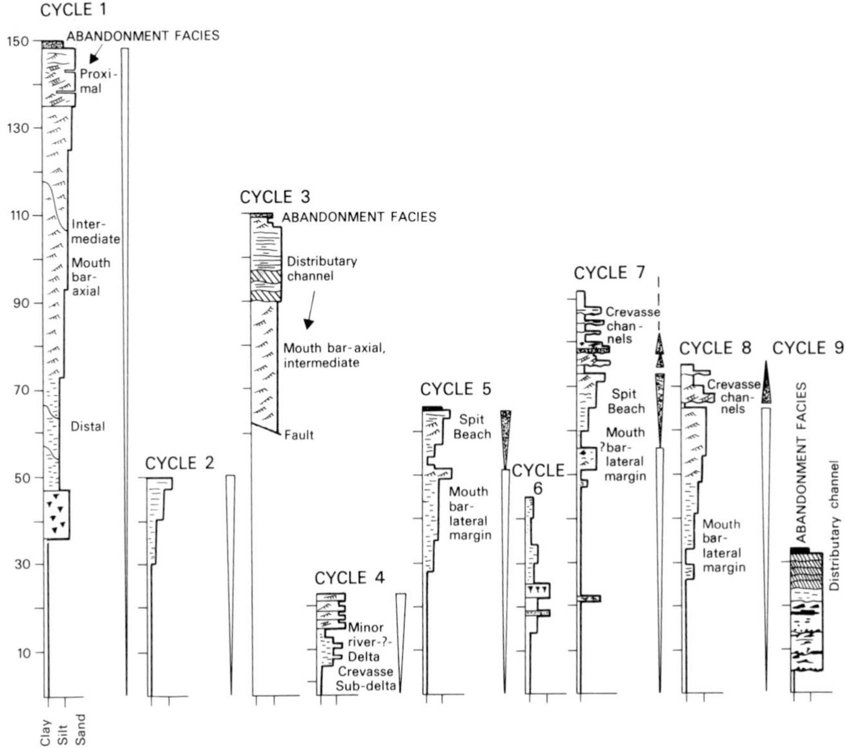

Fig. 6.48. Interpretation of Carboniferous cycles in the Bideford Group, north Devon, in terms of differing positions in a bar finger system of a fluvial-dominated delta (after Elliott, 1976a).

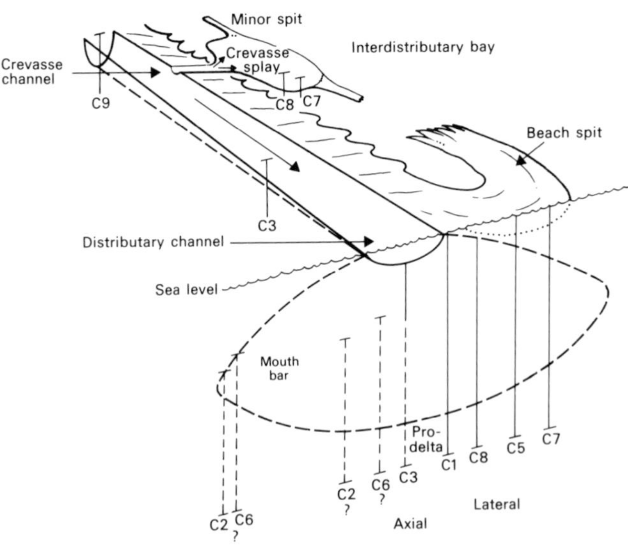

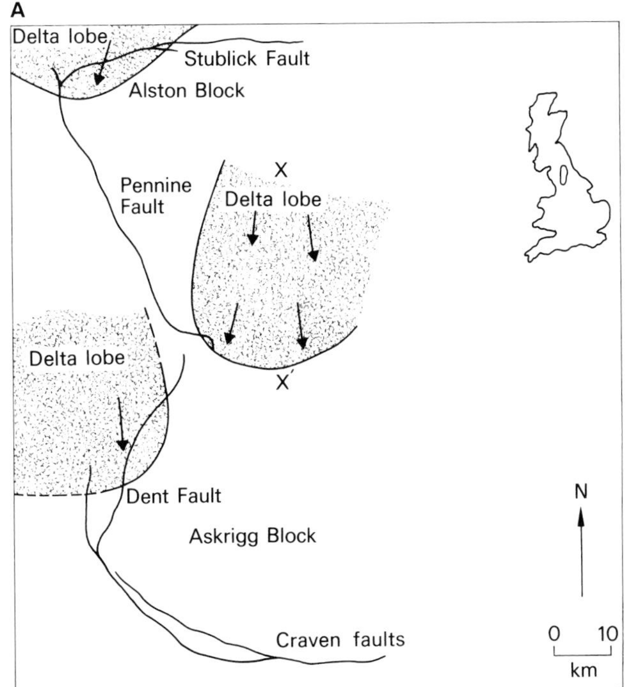

Fig. 6.49. Fluvial-dominated delta lobes from a Carboniferous Yoredale cyclothem in northern England: (A) plan view of three lobes; (B) longitudinal section through the central lobe (after Elliott, 1975).

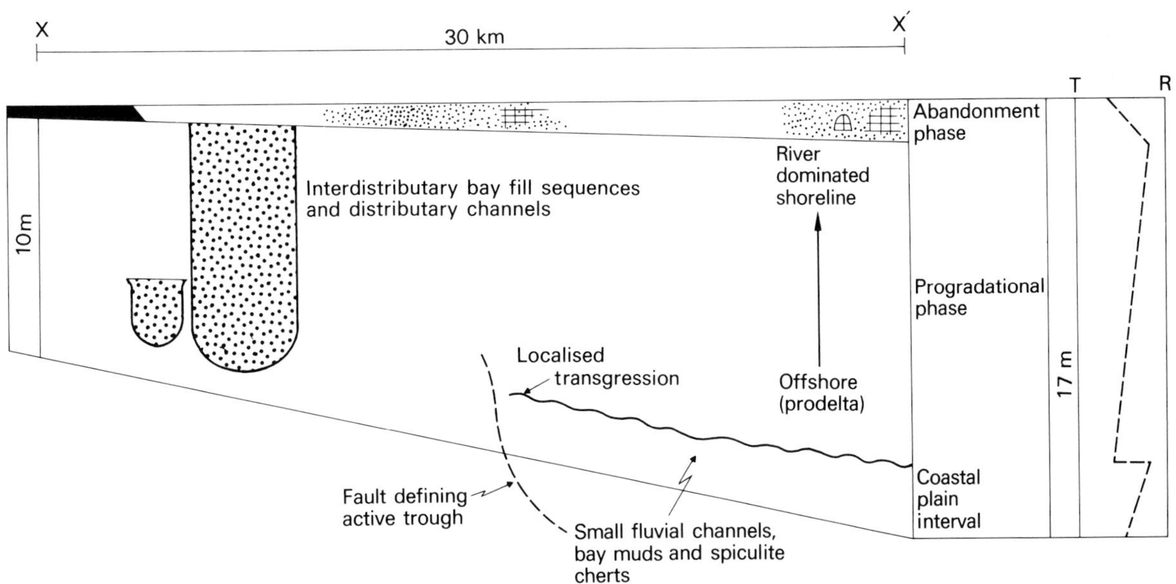

surfaces suggests that the distributary channels did not shoal appreciably towards the mouth, unlike those of the modern Mississippi delta, and this may reflect a contrast in the outflow dispersion mechanism arising from reduced water salinity in the confined basin.

The *Bideford Group* deltas in the Westphalian of north Devon have also been interpreted as fluvial-dominated, elongate deltas, but this is inferred solely from vertical sequences (de Raaf *et al.*, 1965; Elliott, 1976a) (see also Sect. 2.1.2; Figs. 2.1, 2.2). The succession comprises nine coarsening-upward 'cycles', each representing the progradation of a delta into a moderately deep basin. Thick progradational facies can be distinguished from thin abandonment facies horizons, and delta front and delta plain facies can be differentiated in the progradational facies (Fig. 6.48). The delta front is represented by large-scale (50–150 m) coarsening upwards sequences and two distributary channel sandstones (20–26 m thick). The delta plain facies includes small-scale interdistributary-fill coarsening-upward sequences and small-scale crevasse channels. Abandonment facies are represented by thin horizons of bioturbated siltstone, and, in one case, a thick horizon of impure coal. Comparison of the cycles reveals that (1) the upper sandstone member is frequently absent; (2) abandonment facies only occur where there is a substantial sandstone member at the top of the cycle; and (3) delta plain sequences are preferentially developed above mudstone–siltstone dominated delta front sequences devoid of a significant sandstone member. These contrasts are explicable in terms of differing locations in an elongate, birdfoot delta. The frequent absence of the upper sandstone member suggests that it was impersistently developed in individual deltas as bar finger sands. Lateral margins of the bar fingers provided shallow platforms on which delta plain facies were deposited, thus explaining the preferential development of delta plain facies. After abandonment the bar fingers were shallow, elevated areas where deposition was slow and abandonment facies accumulated. The adjacent mud-silt dominated areas subsided more rapidly and experienced only a brief cessation in deposition.

A delta system in the *Yoredale Series* in northern England reveals three delta lobes separated by inter-deltaic clastic embayments (Elliott, 1974a, 1975; Fig. 6.49). The study is based on vertical and lateral facies relationships, and lobe recognition is facilitated by abandonment facies marker horizons. The most complete lobe can be traced over 700 sq. km and is exposed in a series of cross-sectional traverses. The upstream cross-section is dominated by two small-scale (4.5 m) coarsening-upward sequences which reflect the repeated infilling of shallow interdistributary bays by levees, crevasse splay lobes and minor beach spits. Small-scale crevasse channels and larger-scale distributary channels locally dissect these sequences, and one of the distributary channels is represented by a 1.5 km wide sandstone body with large-scale lateral accretion surfaces produced by the lateral migration of point bars (see Fig. 6.38). The downstream cross-section is represented by a single 9–17 m coarsening-upward sequence which records the progradation of a fluvial-dominated, mouth bar shoreline, and variations in palaeocurrents and sandstone composition across the section suggest the presence of two laterally coalescing mouth bars. Downcurrent changes in sequence types are parallelled by facies changes in the abandonment phase marker horizon from coal to marine sandstones and limestones (see Sect. 6.7.3).

The *Namurian Kinderscout Grit* delta system is one of two major incursions of clastic sediment into the Central Pennine basin (Reading, 1964; Walker, 1966; Collinson, 1969; Fig. 6.50). It comprises a passage from basinal muds into a distal turbidite apron, a submarine fan complex, a delta front slope and finally the delta plain, with a total thickness of 700 m. The delta front is represented by a 60 m coarsening-upward sequence dominated by mudstone and siltstone, but with steep-sided, sand-filled channels at various levels. Towards the base of the sequence the channels are filled by turbidites, whereas towards the top they exhibit the preferred sequence: erosive base → massive bedded sandstone → horizontally bedded sandstone → cross bedded sandstone. It appears that the delta front received relatively fine-grained sediment from suspension, whilst sand largely by-passed the delta front and was discharged directly down the delta front in subaqueous channels. These channels were initially characterized by sediment-laden traction currents, but developed into turbidity current channels lower on the slope and supplied sediment directly to the submarine fan. This implies that buoyancy processes did not operate at the distributary mouth, possibly due to reduced salinity levels in the confined basin. Delta plain facies include bay mudstones and siltstones, crevasse splay sandstones and small-scale (4 m) coarsening-upward sequences capped by erosive-based minor channel sequences, possibly reflecting minor mouth bar–crevasse channel couplets. Fluvial channels with large-scale alternate bars occur in the delta plain facies (see Chapter 3). Finally, a condensed horizon with a prolific marine fauna (the Butterly Marine Band) reflects abandonment of the delta. Although the delta is fluvial-dominated it contrasts markedly with known examples. This absence of a modern analogue is significant as it indicates that in well-documented areas it is sometimes necessary to invoke a non-actualistic model.

In the Carboniferous of the United States relatively small-scale fluvial-dominated delta systems are often intimately associated with carbonate shelves (Ferm, 1970; Brown *et al.*, 1973; Galloway and Brown, 1973; Donaldson, 1974; Erxleben, 1975; Horne and Ferm, 1976). The *Cisco Group* deltaic system in north-central Texas repeatedly prograded into a stable cratonic basin with a shelf-slope-basin floor form and

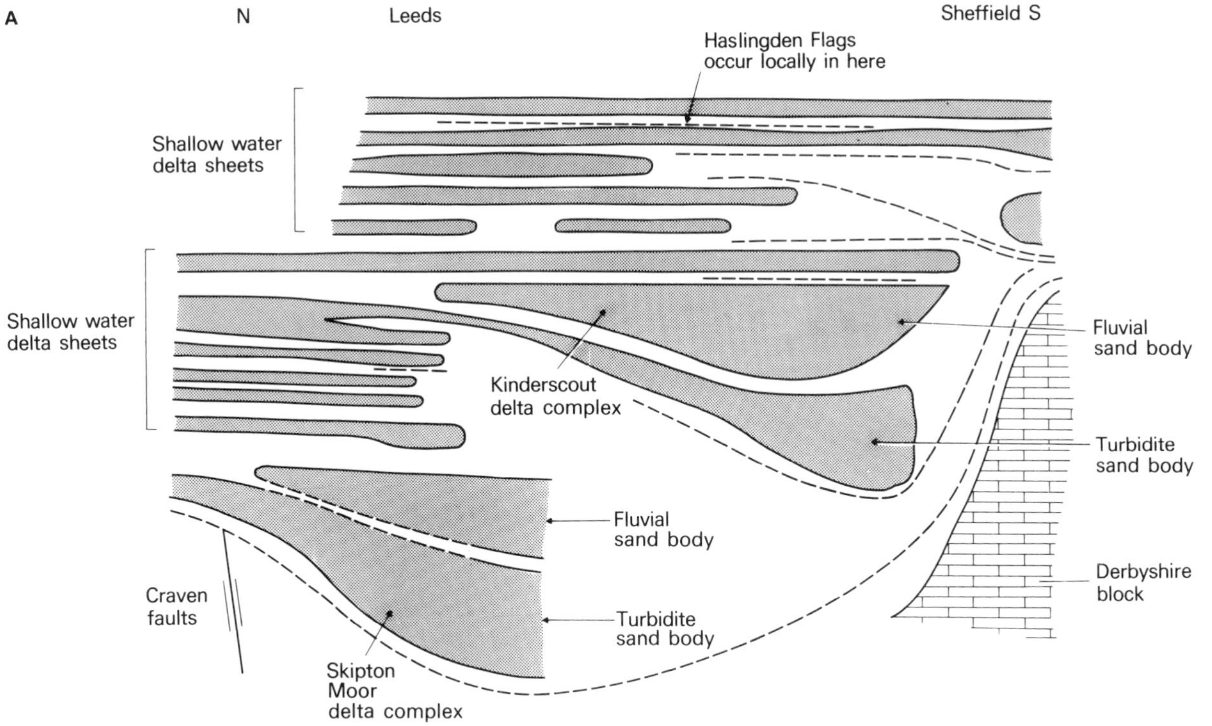

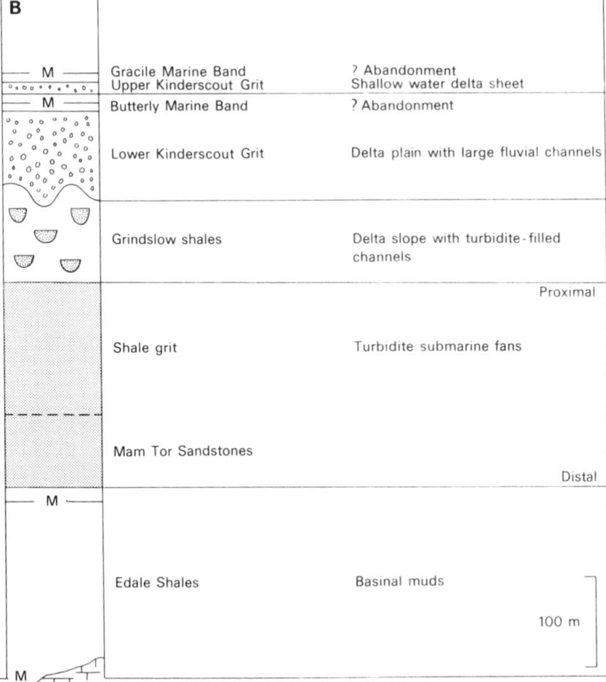

Fig. 6.50. (A) Delta systems infilling the Carboniferous Central Pennine Basin in northern England; (B) sedimentary succession of the Kinderscout delta system; and (C) depositional systems of the Kinderscout delta (after Reading, 1964; Walker, 1966; and Collinson, 1969).

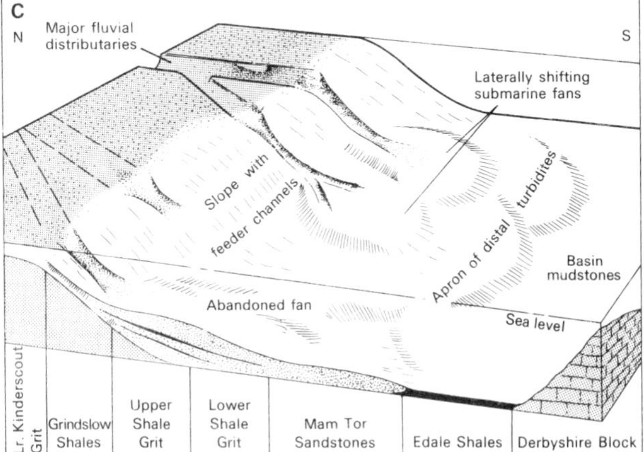

carbonate accumulations at the shelf edge (Galloway and Brown, 1973; Fig. 6.41). Elongate and lobate fluvial-dominated deltas occur in 10–15 main progradational episodes and are the only delta types developed in the basin. The principal depositional site was, however, the basin slope which comprises a series of clastic wedges deposited by resedimentation processes which commenced at the shelf edge and propagated down the slope.

6.8 SEDIMENT-INDUCED DEFORMATION

The theme of the chapter so far has been that deltaic facies patterns are controlled largely by the regime and morphology of the delta, but several studies have demonstrated that this primary facies pattern can be significantly modified by synsedimentary deformation operating on a wide range of scales. The deformation may be related to basement tectonics, as in the Ganges–Brahmaputra delta, but there is also a discrete class of synsedimentary deformational processes related solely to sedimentary factors. In the depocentre of the Mississippi delta these processes have produced a diverse range of deformational features including mud diapirs, rotational slumps, fault-defined grabens, surface mud flows and deep-seated faults (Fig. 6.51). Several of these features have also been described from the Fraser, Magdalena, Niger and Orinoco deltas, and the larger scale features have been detected seismically in several subsurface studies of ancient deltaic successions.

6.8.1 Deformational processes

In the Mississippi delta, surface slumping results from oversteepening of the delta front slope as mentioned earlier (see 6.5.2), but many deformational features involving mass movement are attributed either to the effects of exceptional waves and sediment degassing, or to deep-seated clay flowage.

Under extreme wave conditions, for example during hurricanes, wave-induced bottom pressures may exceed sediment shear strength and initiate surface mass movement. This effect is enhanced if high concentrations of methane gas exist in the sediment as this substantially reduces its shear strength, and may cause the sediments to degassify when subjected to high bottom pressures. Surface slumps and mudflows may form in areas with minimal slopes as a result of these processes (Coleman *et al.*, 1974).

However, deep-seated clay flowage is the most important process responsible for generating synsedimentary deformation in deltaic areas, as demonstrated in the Mississippi and Niger delta depocentres (Coleman *et al.*, 1974; Weber and Daukoru, 1975). This process involves the development of overpressured conditions during the

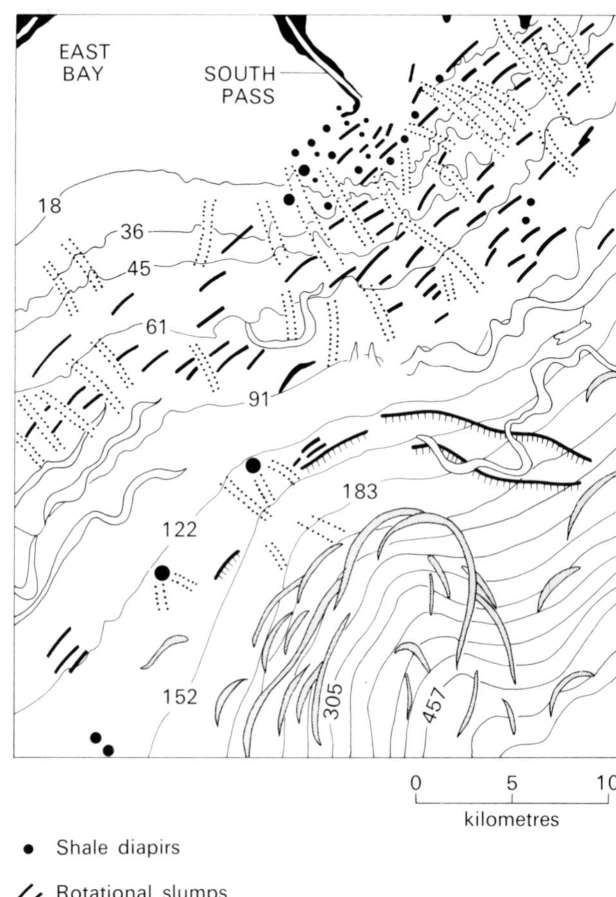

Fig. 6.51. Sediment-induced deformational features in the vicinity of South Pass in the modern Mississippi delta (after Roberts *et al.*, 1976).

compactional history of rapidly deposited clays. In the initial stages of compaction water is easily expelled, but as compaction continues the rate of water expulsion decreases as permeability falls. Eventually the orderly expulsion of water is prohibited and high pore-water pressures develop in the clays. At this point compaction ceases and the clays are then overpressured and undercompacted. Methane generation from bacterial decomposition of organic matter may also contribute

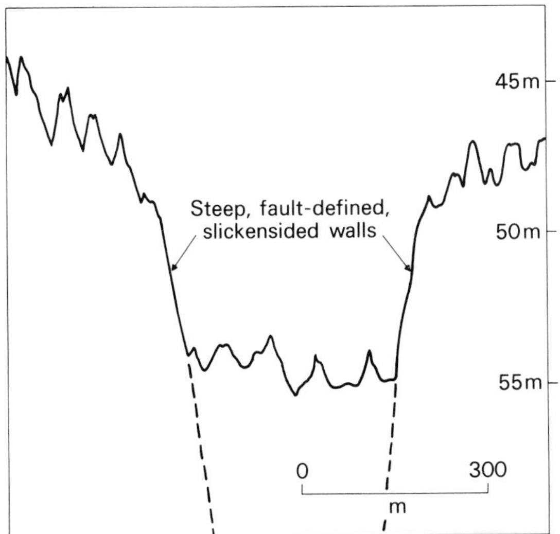

Fig. 6.52. Fathometer profile across a delta front gully in the Mississippi delta, illustrating the graben-like form with steep walls and flat bottom (after Coleman *et al.*, 1974).

to the development of overpressured conditions (Hedberg, 1974). In addition to being out of pressure equilibrium with their surroundings, overpressured clays have a low viscosity and are therefore unstable and potentially mobile. Continued loading in the depocentre can induce slow, but continuous deep-seated mass movement which produces deformational features in the depocentre (donor area) and the offshore area in front of the delta (recipient area).

6.8.2 Deformational features

DELTA FRONT GULLIES

The upper part of the Mississippi delta front is dissected by innumerable shallow gullies which were originally attributed to the downslope mass movement of sediment at the surface (Shepard, 1955). However, shallow seismic profiles, coring and scuba diving have indicated that this mechanism is untenable, and they are now interpreted as a response to deep-seated clay flowage (Coleman *et al.*, 1974).

The gullies are prolifically developed near distributary mouths, but also occur in the intervening areas. They are initiated at water depths of 70–80 m and grow landwards and seawards over a period of approximately two years. Those formed near the distributary mouth are infilled within 4–6 years, but offshore gullies and those located between distributary mouths persist for fifteen years or more. Gullies extend from shallow depths (7–10 m) down to 100 m water depth in some cases, and range in size from 3–25 m deep, 570–700 m wide and 8–10 km long. They are generally steep-sided (up to 21°), but flat-bottomed, with v-shaped profiles being rare. Coring reveals that the sediments are undisturbed across the gully floor, but are slickensided along the side walls, suggesting that the gullies are downfaulted grabens defined by normal faults at the gully sides (Fig. 6.52). The tensional stress required to generate the grabens results from lateral clay flowage from beneath the depocentre.

DIAPIRS AND SHALE RIDGES

Diapirism has been well documented in the Mississippi delta as 'mudlumps' frequently emerge near the distributary mouths

Fig. 6.53. Diapiric mudlumps in the Mississippi delta with high-angle reverse faults in the diapir crests and exceptional thicknesses of distributary mouth bar between the diapirs (after Morgan *et al.*, 1968).

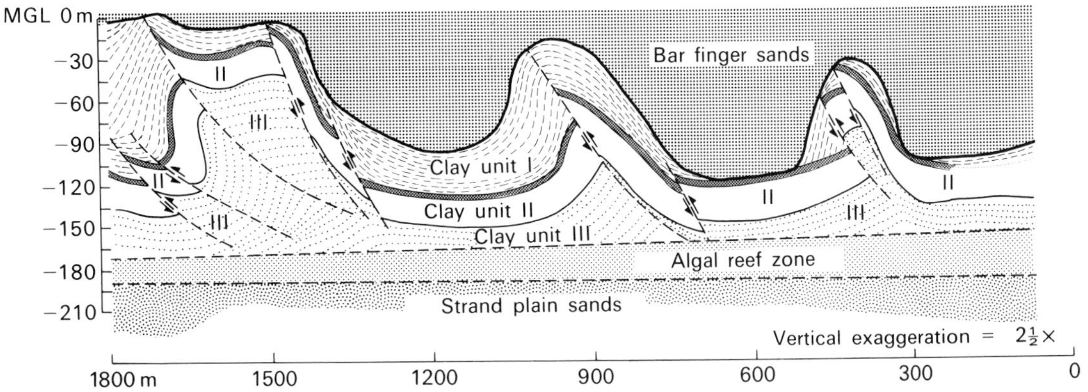

and form temporary islands amenable to direct observation (Morgan, 1961; Morgan *et al.*, 1968). Isopach maps reveal that these strongly diapiric mudlumps have substantially modified the bar finger sands. Instead of being uniform, linear bodies 13–20 m thick as originally described (Fisk, 1961), they comprise a series of discrete sand pods up to 100 m in thickness separated by areas of minimal sand thickness (Coleman *et al.*, 1974; Fig. 6.53). Surface exposures of the mudlumps reveal steeply dipping, stratified delta front sediments with numerous small anticlines, *en echelon* normal faults, reverse faults, radial faults and thrusts. Other features include small mud cones formed by extrusion of methane-rich muds from fault planes, and planation horizons produced by wave erosion of the exposed mudlump. Clays involved in the diapirism exhibit intense brecciation and later fractures (Morgan *et al.*, 1963; Coleman *et al.*, 1974).

The mudlumps are considered to be thin spines superimposed on linear shale folds or ridges, with large-scale high-angle reverse faults in the mudlump crests producing most of the uplift. There is a close relationship between diapiric activity and distributary mouth bar sedimentation with the appearance of mudlumps invariably coinciding with rapid

Fig. 6.54. Plan view of growth faults in the Niger delta illustrating their lateral impersistence, slightly curved form concave to basin, and their general parallelism with the delta front (after Weber and Daukoru, 1975).

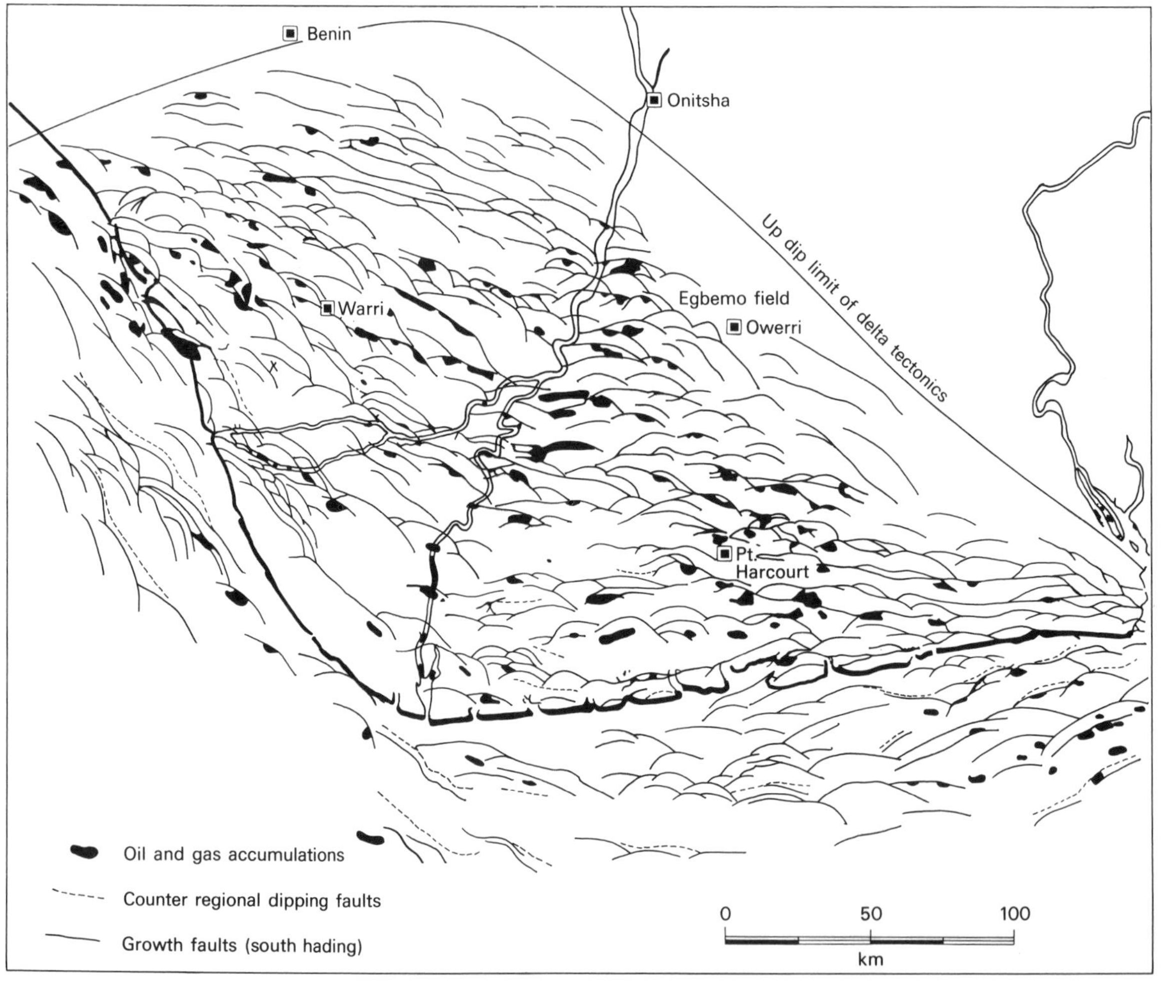

sedimentation during river flood periods, and the site of mudlump activity migrating seawards in concert with mouth bar progradation (Morgan *et al.*, 1968).

All other examples of diapirs or shale ridges in the Fraser, Magdalena, Niger and Orinoco are submerged and occur considerable distances in front of the delta (Mathews and Shepard, 1962; Shepard *et al.*, 1968; Shepard, 1973a; Weber and Daukoru, 1975; Nota, 1958). In the Niger delta, for example, shale ridges occur along the continental slope in front of the extensively growth-faulted depocentre. These offshore shale ridges are probably a surface expression of overpressured clays which flowed away from the depocentre, in which case the Mississippi mudlumps may reflect the re-activation of pre-existing offshore shale ridges by direct *in situ* loading as the delta progrades onto the continental slope.

GROWTH FAULTS

These form a discrete class of synsedimentary faults defined by characteristics which reflect their genesis. They commonly form in delta depocentres, and their initiation, development and decay is intimately related to sedimentation in the depocentre.

The faults parallel the shoreline, with active faults situated in the vicinity of the shoreline, incipient faults in front of the shoreline and a zone of decaying faults behind the shoreline. Individual faults often have curved traces concave to the basin and are of limited lateral persistence, though they frequently merge to form extensive fault lines (Fig. 6.55). In cross-section the faults have a concave profile which flattens with depth, passing into bedding plane faults. The faults are normal with downthrow consistently into the basin, and an essential characteristic is that the amount of downthrow varies from almost zero at the top of the fault to a maximum at some mid-point in depth, and finally decreases as the fault plane flattens (Fig. 6.55). In large-scale fault systems maximum throws in excess of 1,000 m have been noted (Busch, 1975). The sedimentary succession is appreciably thicker on the downthrow side and also contains a significantly higher sand proportion. Oligocene growth faults in Texas defined discrete basins which exerted considerable control on gross facies patterns (Walters, 1959) and in the Niger delta growth faults separate thick, lens-shaped sedimentary units (Short and Stauble, 1967; Weber, 1971). Rotation of the downthrown strata along the curved fault planes often produces broad anticlines (rollovers), frequently accompanied by smaller, antithetic fault systems. As the rollovers occur in thick, sand-dominated successions they often form excellent hydrocarbon reservoirs (Weber, 1971; Busch, 1975). Numerous workers have inferred the presence of positive elements at depth on the upthrown side of the faults, and improved seismic techniques have recently led to the recognition of linear masses of overpressured shale at these sites (Bruce, 1973; Weber and Daukoru, 1975).

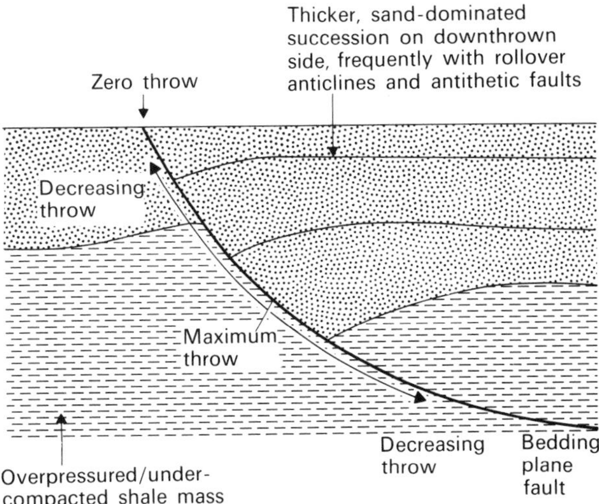

Fig. 6.55. Salient features of growth faults in a section normal to fault strike, i.e. normal to the shoreline (based on Ocamb, 1961; Weber, 1971; and Bruce, 1973).

Numerous mechanisms have been proposed to account for growth faults, many of which are now untenable. Recently, Bruce (1973) proposed a mechanism which stressed the role of overpressured shale masses at depth on the upthrown side, and incorporated earlier ideas on deep-seated substrate movement (Walters, 1959; Ocamb, 1961) and differential compaction (Carver, 1968). In this mechanism, differential compaction in the depocentre produces subdued shale masses in the vicinity of the gross sand-to-mud transition. Growth faults form on the seaward side of a shale mass and once developed define a subsiding depocentre and are therefore perpetuated (growth stage) as sedimentation continues. As the depocentre migrates basinwards the faults decay, resulting in an upwards diminution of fault throw. New faults are therefore initiated progressively basinward and higher in the stratigraphic succession, as observed in numerous examples.

SEDIMENT-INDUCED GRAVITY SLIDING AND OVERTHRUST FAULTING

It has been argued that the overpressured conditions which result in mud diapirism are also conducive to synsedimentary gravity sliding (Chapman, 1973, p. 163, 1974). Large-scale mud diapirism can deform successions into steep, narrow anticlines separated by broad gentle synclines aligned parallel to depositional strike. If a sufficient slope exists, sheets of sediment between diapir crests may slide downslope. The slope value required is relatively low if interstitial fluid pressures approach the total overburden pressure, and in this situation sliding takes place on surfaces within the overpressured clays. Overthrusting may occur at the leading edge of the slide sheet, extending the range of sediment-induced deformation still further. Possible examples of this type of deformation occur in Tertiary depocentres in Borneo, New Zealand and Trinidad.

6.8.3 Sediment-induced deformational features in exposed deltaic successions

Certain of the deformational features described above have been identified in exposed deltaic successions. For example, growth faults have recently been recognized in the Cretaceous of Colorado, the Triassic of Svalbard, and the Carboniferous of the United Kingdom and USA (Weimer, 1973; Edwards, 1976a; Chisholm, 1977; Brown *et al.*, 1973; Erxleben, 1975). In Svalbard, small-scale growth faults affecting delta front coarsening-upward sequences are exposed almost normal to fault strike. Towards the top they are steeply dipping normal faults, but they flatten with depth and become bedding plane faults in the underlying shales. Erosion surfaces occur in the sandstones on the upthrown side, whilst on the downthrown side there is a greater thickness of sandstone and the beds exhibit local tilting and rollover anticlines. In the Cretaceous Rencôncavo basin (Brazil), delta front facies reflect a predominance of sediment-gravity flow processes induced partly by growth faulting and the oversteepening of slopes by diapirism (Klein *et al.*, 1972).

Although sediment-induced deformation is by no means an ubiquitous process in deltaic successions it is possible that examples of the larger-scale features have been overlooked or mis-interpreted in exposed areas. Where deltaic successions include, or are underlain by, thick intervals of fine-grained sediments, deformational features can be expected.

FURTHER READING

Shepard F.P., Phleger F.B. and Van Andel Tj.H. (Eds.) (1960) *Recent Sediments, northwest Gulf of Mexico*. pp. 394, Am. Ass. Petrol. Geol., Tulsa.

Fisher W.L., Brown L.F., Scott A.J. and McGowen J.H. (1969 *Delta Systems in the exploration for oil and gas*. pp. 78 + 168 figures and references, Bur. econ. Geol., Univ. Texas, Austin.

Morgan J.P. and Shaver R.H. (Eds.) (1970) *Deltaic Sedimentation Modern and Ancient*. pp. 312, Spec. Publ. Soc. econ. Paleont. Miner., **15,** Tulsa.

Broussard M.L. (Ed.) (1975) *Deltas, models for exploration*. pp. 555, Houston geol. Soc., Houston.

CHAPTER 7 Clastic Shorelines

7.1 INTRODUCTION

Numerous attempts have been made to classify the wide range of present-day non-deltaic shorelines using one or more of the main controls on coastal development such as sediment supply, hydrodynamic setting, climate, sea level history, tectonic setting and the pre-existing structure of the depositional area to categorize the shoreline types (Johnson, 1919; Shepard, 1963; Cotton, 1952; Armstrong-Price, 1955; Davies, 1964; Inman and Nordstrom, 1971). A scheme which is currently gaining acceptance as a sedimentological classification of shorelines stresses the importance of tidal range and proposes three divisions: microtidal, <2 m; mesotidal, 2–4 m; and macrotidal, >4 m (Davies, 1964; Hayes, 1976; Hayes and Kana, 1976). The global distribution of these ranges can be mapped (Fig. 7.1), but more significantly the development of shorelines can be considered in terms of these ranges (Fig. 7.2). The coast of western Europe illustrates the relationship between tidal range and coastal morphology, with elongate barrier islands in microtidal areas off Holland and north-west Denmark gradually decreasing in length in mesotidal areas due to dissection by tidal inlets, and finally passing into the estuary–tidal flat complex of the macrotidal German Bay (Fig. 7.3). Although this chapter uses conventional morphological terms to describe the different types of shoreline, Davies's thesis is adopted.

7.2 MODERN BEACHES AND BARRIER ISLANDS

Beaches and barrier islands are long, narrow sand accumulations which occur within deltas (see Chapter 6), along depositional strike from deltas, or in an oceanic or lacustrine context devoid of any connection with deltas. Both systems are aligned parallel to the shoreline, but beaches are attached to the land, whereas barrier islands are separated from it by a shallow lagoon and are also often dissected by tidal inlets. Conditions conducive to the formation of beach and barrier island systems include: (1) a steady supply of sand to the shoreline, either directly by river input or by longshore drift; (2) a hydrodynamic setting characterized by low, moderate or

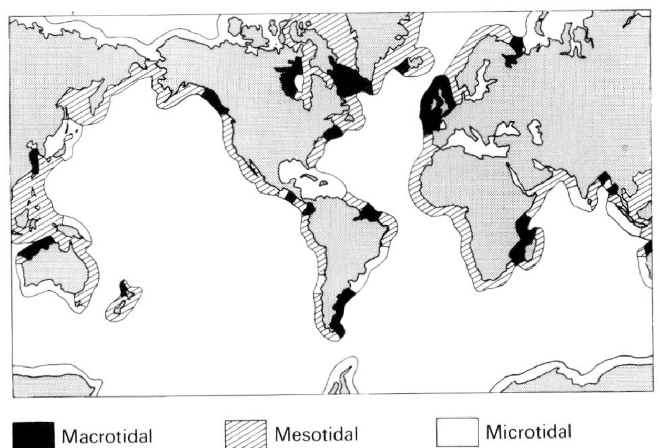

Fig. 7.1. Distribution of microtidal (<2 m), mesotidal (2–4 m) and macrotidal (>4 m) ranges (after Davies, 1964).

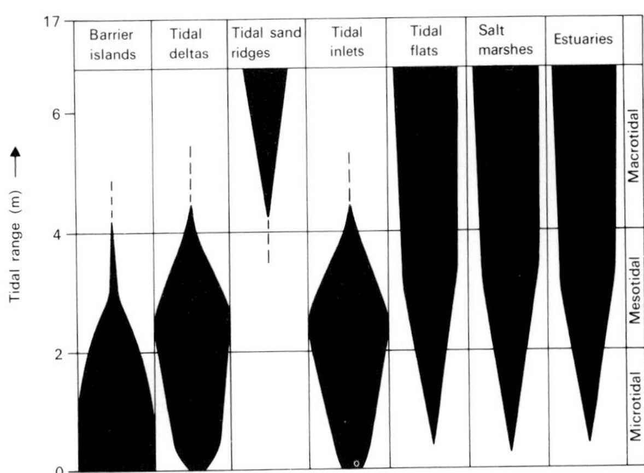

Fig. 7.2. Relationship between tidal range and coastal morphology (modified after Hayes, 1976).

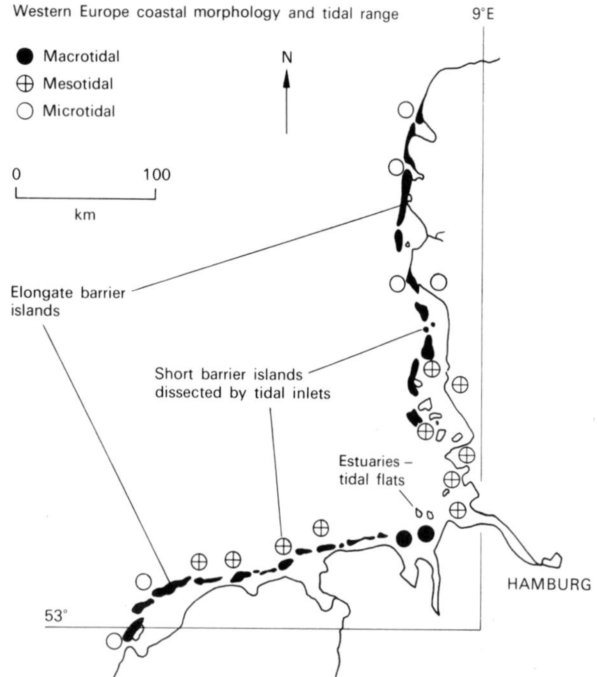

Fig. 7.3. Relationship between tidal range and coastal morphology in western Europe (modified after Hayes, 1976).

high wave energy, but limited tidal range (microtidal–mesotidal); and (3) a moderately stable, low gradient coastal plain. Estimates of the present extent of these systems range from 5,710 km to 32,038 km depending on the definition adopted, but are sufficient to substantiate their importance (Berryhill *et al.*, 1969; Cromwell, 1971). East Coast and Gulf Coast examples in the United States are the most intensively studied (LeBlanc and Hodson, 1959; Shepard, 1960; Bernard, LeBlanc and Major, 1962; Hoyt and Weimer, 1963; Kraft, 1971), but other well-documented examples occur in north Holland, west Africa and Australia (Allen, 1965e, van Straaten, 1965; Hails, 1968; de Jong, 1971; Bird, 1973).

7.2.1 **Wave processes and sediment transport**

Beaches and barrier islands are constructed by wave processes, and during the last few decades these processes have been intensively studied by direct observation, experimentation and theoretical procedures. Discussions have centred on three main points: (1) waves are transformed and induce sediment transport as they approach the shoreline; (2) waves generate nearshore currents which are also capable of transporting sediment; and (3) temporal fluctuations in wave regime result from alternations between fairweather and storm conditions.

WAVE TRANSFORMATION

In shelf areas, wave-induced orbital motions produce oscillatory flow which may impinge on the sea bed and initiate sediment transport. The circular orbits of deep water waves become elliptical as the water shallows and give rise to a straight line to-and-fro motion in the direction of wave travel (Sect. 9.4.3). As waves continue to approach the shoreline, interference with the sea bed becomes more pronounced and eventually the waves break (Shepard and Inman, 1950; Ingle, 1966; Madsen, 1976). A series of hydrodynamic zones are associated with this transformation (Fig. 7.4), commencing with the *wave build-up* or *shoaling wave zone* in which the wave alters from an essentially sinusoidal or trochoidal form into a solitary form. Each solitary wave produces a landward-directed bottom surge which may transport sediment as bedload, and there is also a weaker, seaward-directed return flow

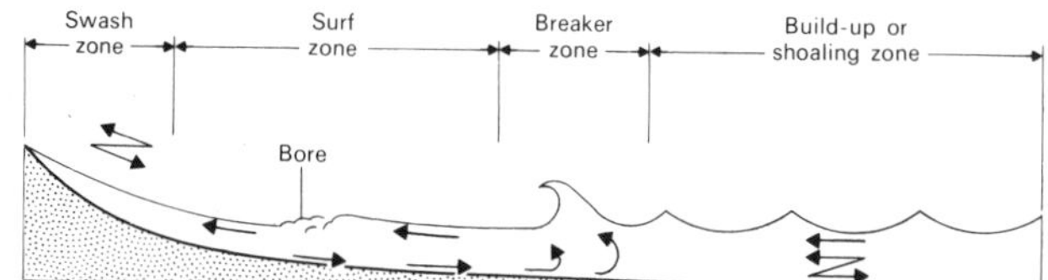

Fig. 7.4. Nearshore wave transformation zones (modified after Shepard and Inman, 1950; Ingle, 1966).

from the shoreline. As the wave travels landwards it increases in height, steepens and eventually breaks, thus initiating the *breaker zone*. Spilling, plunging, collapsing and surging types of breaking waves are recognized, with the development of each type determined by beach slope and wave steepness (Galvin, 1968). High energy conditions in the breaker zone cause coarse grains to saltate in a series of elliptical paths parallel to the shoreline, whilst finer sediment is temporarily suspended (Ingle, 1966; Fig. 7.5). Breaking of the wave in turn generates the *surf zone* in which a shallow, high velocity translation wave or bore is projected up the beach face. Coarse sediment is transported landwards as bedload, and the development of short duration suspension clouds or 'sand fountains' reaches a maximum in the inner part of this zone (Brenninkmeyer, 1976). The landward-directed flow is complemented by a seaward-directed return flow, but this may be concentrated into rip currents. Finally, in the *swash zone* at the landward limit of wave penetration, each wave produces a very shallow, high velocity, landward-directed swash flow which is followed immediately by an even shallower, seaward-directed backwash flow. Conditions are variable, but the overall trend during a swash-backwash event is of decreasing flow power during the swash period, a brief still-stand during which water may percolate into the bed, and finally

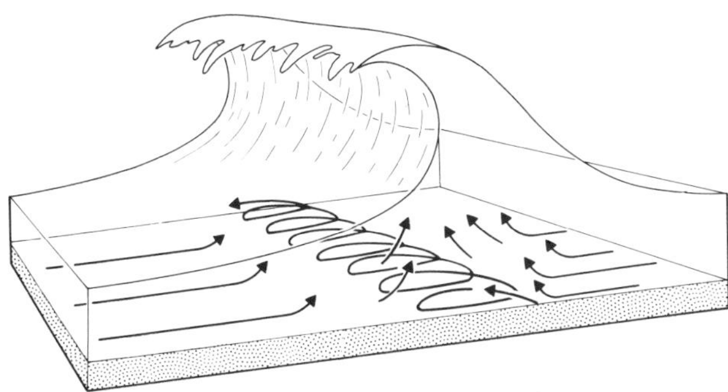

Fig. 7.5. Sediment transport associated with a breaking wave: coarse-grained sediment moves as bedload in a series of elliptical paths parallel to the coast, whilst finer sediment is suspended (modified after Ingle, 1966).

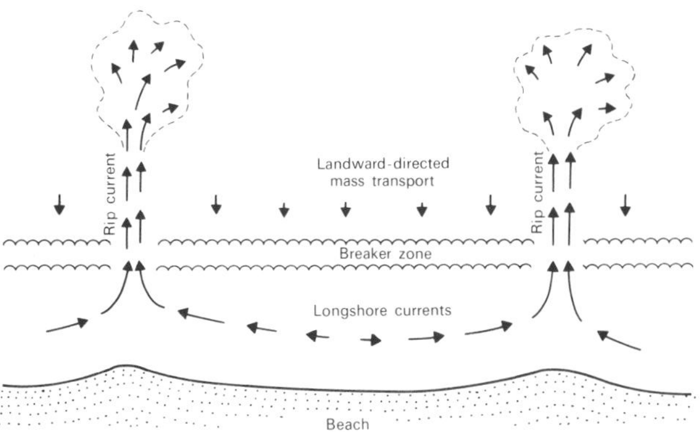

Fig. 7.6. Wave-induced nearshore circulation system (after Shepard and Inman, 1950).

accelerating flow during the backwash period. Bedload transport is predominant and Froude numbers for these flows often attain values of 0.8 and greater (Waddell, 1976; Wright, 1976). Where waves approach the shoreline obliquely, sediment is transported alongshore in a 'saw-tooth' manner as the swash flow is directed across the beach face in the direction of wave travel whilst the backwash flow is normal to the beach face.

The distribution of these hydrodynamic zones varies with the slope and structure of the beach face. Steeply sloping beaches rarely possess a surf zone as the waves break close to the shoreline, whereas on gentle beach faces the waves break a considerable distance offshore and generate a broad surf zone. On beach faces with more than one nearshore bar, hydrodynamic zones may be repeated as waves break on the outer bar, re-form, and break again on the inner bar (Davidson-Arnott and Greenwood, 1976; see Sect. 7.2.3). In addition, the zones migrate across the beach face as the tide rises and falls, with higher tidal ranges inducing a greater degree of spreading.

WAVE-INDUCED NEARSHORE CURRENTS

Two wave-induced nearshore current systems have been recognized: a cell-like circulation system of rip currents and associated small-scale longshore currents; and a larger-scale longshore current system generated by waves approaching the shoreline obliquely (Shepard and Inman, 1950; Ingle, 1966; Komar and Inman, 1970; Komar, 1976a, 1976b, 1976c).

The cell-like circulation system comprises a series of narrow, high-velocity rip currents directed seawards from the surf zone and linked laterally by longshore currents. Water lost from the surf zone by the rip currents is replaced by landward-directed mass transport of water, thus completing the circulation system (Fig. 7.6). Rip currents have been observed to operate in water depths of up to 5 m and often attain appreciable velocities. Besides eroding shallow channels and generally deflating the upper beach face, they also deposit aprons of sediment farther offshore.

The larger-scale longshore currents flow parallel to the shoreline in the surf and breaker zones. Interaction of the currents with surf zone processes results in unidirectional sediment transport along the shoreline, witnessed by sediment accumulation on the upstream side of groynes designed to arrest this transport. The net tendency is to progressively straighten the shoreline as protrusions refract waves and therefore concentrate energy, whereas adjacent embayments dissipate energy and act as sediment traps. Maximum development of longshore currents is favoured by high wave energy and a low angle of wave approach.

TEMPORAL VARIATIONS IN WAVE REGIME: FAIRWEATHER VERSUS STORM PERIODS

Nearshore wave-induced sediment transport can be considered in terms of an *onshore–offshore component* which results from wave transformation supplemented by nearshore currents, and a *coast-parallel component* which results from the interaction between longshore currents and surf zone processes (Swift, 1975a, 1976a). Superimposed on these process-response elements are temporal variations in wave regime between fairweather and storm conditions. During fairweather periods, waves tend to be relatively low amplitude, long period swells which induce only a moderate amount of sediment transport in relatively shallow waters. However, during storms higher energy, short period waves lower the wave base and transport considerable volumes of sediment. The nature of onshore–offshore sediment transport varies considerably between these periods. During fairweather periods, wave-induced bottom surge is directed landwards and sediment is deposited on the upper beach face.

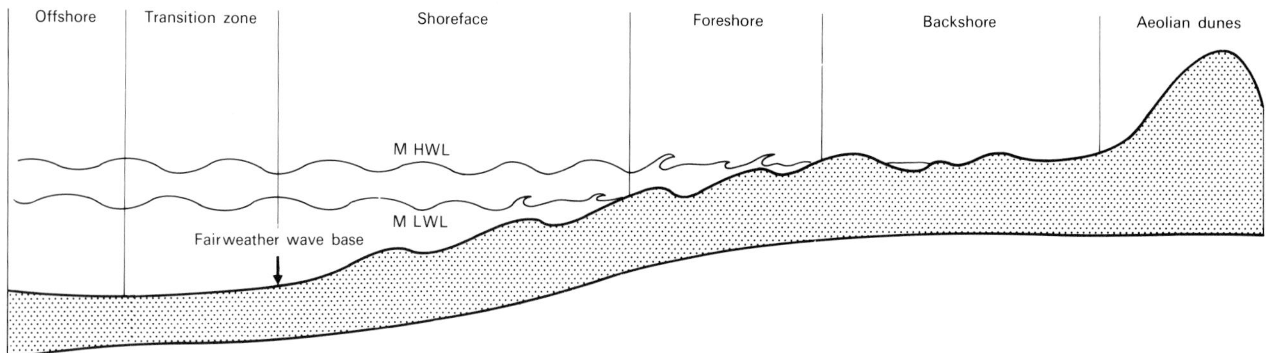

Fig. 7.7. Beach face sub-environments.

As the nearshore cell circulation system is weak, sediment loss by rip currents is low and the beach face tends to aggrade. However, during storms the efficiency of the landward-directed bottom surge is decreased as wave steepness increases. Sediment supply to the beach face is therefore diminished, and in addition, there is considerable sediment loss from enhanced rip currents. Beach profile studies demonstrate this repeated alternation between fairweather aggradation and storm period deflation which is often termed the beach cycle (Sonu and van Beek, 1971).

Other effects of major storms include the landward retreat or local breaching of the aeolian dune ridge, and in barrier islands washover fans advancing into the lagoon (Hayes, 1967a; McGowen and Scott, 1976). The effects of storms are therefore differential, with some areas experiencing substantial deposition whilst others are eroded.

7.2.2 The beach face

This area comprises a series of sub-environments which are aligned parallel to the shoreline and superimposed on a general seaward fining of sediment. From landward to seaward the sub-environments are: aeolian dunes→backshore→foreshore→shoreface, with the latter passing transitionally into the offshore or shelf area (Fig. 7.7).

AEOLIAN SAND DUNES

Aeolian sand dunes form complex ridges which cap the beach face above mean tide level. They generally result from wind reworking of sand emplaced on the upper beach by storm waves and attain a height of several metres. More extensive complexes of larger-scale dunes form in areas of significant sand supply and wind regime (van Straaten, 1961; Bigarella, Becker and Duarte, 1969; Orme, 1973). One system extends along the entire Pacific-facing coast of Washington, Oregon and California, and in this example vegetation-free dunes with transverse and oblique ridge patterns are distinguished from precipitation ridges and parabolic dunes which develop partly in response to vegetation (Cooper, 1958, 1967).

Dunes are well developed in arid, semi-arid and temperate

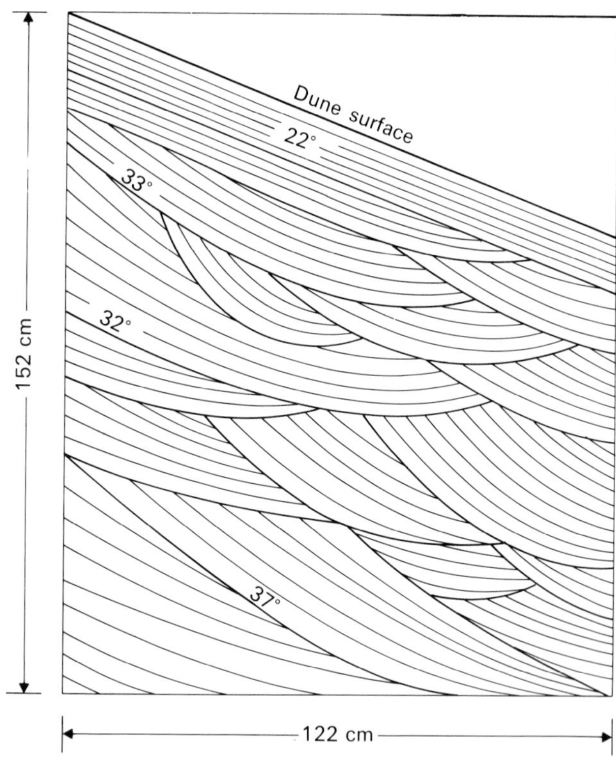

Fig. 7.8. Internal structures of coastal dunes on Mustang Island, Texas (after McKee, 1957).

areas, but are poorly developed in tropical and sub-tropical areas where the combined effects of dense vegetation, relatively low wind velocities and dampening of the sand limit their formation (e.g. West African coastal areas; Allen, 1965e). Coastal aeolian dunes are generally composed of well sorted, marine derived sands characterized by sets of steeply dipping cross-bedding of variable direction (Fig. 7.8; see Chapter 5). However, the primary characteristics are often modified or obliterated by fluctuating ground water levels, plant activity and soil formation (McBride and Hayes, 1962; Hails and Hoyt, 1969; Bigarella *et al.*, 1969).

BACKSHORE–FORESHORE

The *backshore* is the supratidal part of the beach which is only inundated during storms, whereas the *foreshore* is the intertidal area. A low ridge or 'berm' generally separates these sub-environments, but they may be considered together as the upper part of the beach face. Sediment emplaced on the backshore during storms is partly reworked by wind action during fairweather periods, producing a veneer of ripples and dunes, or localized deflation surfaces. Foreshore areas are generally characterized by swash zone processes, although on low gradient or mesotidal beaches they may also experience surf zone processes. Surface bedforms often reflect high flow power conditions, with a predominance of current lineated plane beds, rhomboid ripples and antidunes (Wunderlich, 1972; Hayes and Kana, 1976), though during lower wave energy periods the area may be covered by straight-crested symmetrical or asymmetrical ripples. These bedforms are often superimposed on low amplitude, swash-generated bars, referred to collectively as rhythmic topography (Dolan and Ferm, 1968; Dolan, 1971). The most common type is probably 'ridge and runnel topography' which comprises a series of coast-parallel, asymmetrical ridges separated by shallow troughs or runnels 100–200 m wide. This topography is created by swash-backwash processes during fairweather periods when sediment is continually being added to the beach face. The ridges migrate landwards and infill the runnel immediately up-beach, but during storms the topography is planed off as the beach face is deflated. As storm conditions wane the ridges re-form and the cycle is renewed (Davis and Fox, 1972; Davis, Fox, Hayes and Boothroyd, 1972). It appears that the development of this topography is favoured by moderate wave energy conditions acting on flat, fine-grained, mesotidal beaches with abundant sediment supply. Other types of coast-parallel rhythmic topography include beach cusps spaced uniformly along the high water mark, and sets of crescentic bars and inner bars formed lower on the foreshore. Offshore from Florida, wave-induced bars are oriented perpendicular or at a high angle to the shoreline. Maximum relief seldom exceeds 1 m, but they range from 300–2,000 m in length and are considered characteristic of low wave energy, gently sloping beach faces (Niedorada and Tanner, 1970; Niedorada, 1972). An additional feature of foreshore areas is the small-scale braided rivulets and channels which locally rework beach sediment (Clifton *et al.*, 1973).

Despite the diverse range of sedimentary structures visible at various times on foreshore areas, the types of internal structures in foreshore facies are limited as sediments deposited during and immediately after high wave energy storm periods tend to have the highest preservation potential. The predominant structure is parallel lamination dipping gently seawards at 2°–3° (Thompson, 1937; Hoyt and Weimer, 1963). These laminations comprise distinctive two-fold couplets with a thickness of 1–2 cm or less (Clifton, 1969). Each couplet commences with a basal fine-grained and/or heavy mineral layer, overlain by a coarser grained and/or light mineral layer. In plan view the laminae are irregular ellipses several tens of metres long parallel to the shoreline, but rarely more than 10 m wide normal to the shoreline. Subtle truncation planes reflecting brief periods of erosion often divide the laminations into discrete sets. The laminations are attributed to grain segregation under conditions of plane bed sediment transport during swash-backwash flow. High angle, landward-dipping laminations recognized in foreshore facies are considered to result from the migration of foreshore ridges (Hoyt, 1962; Davis *et al.*, 1972). Dips of up to 30° have been observed though swash-backwash flow often reduces this angle, and a distinctive vertical sequence may be formed by ridge migration (Fig. 7.9). Finally, it has recently been argued that antidune

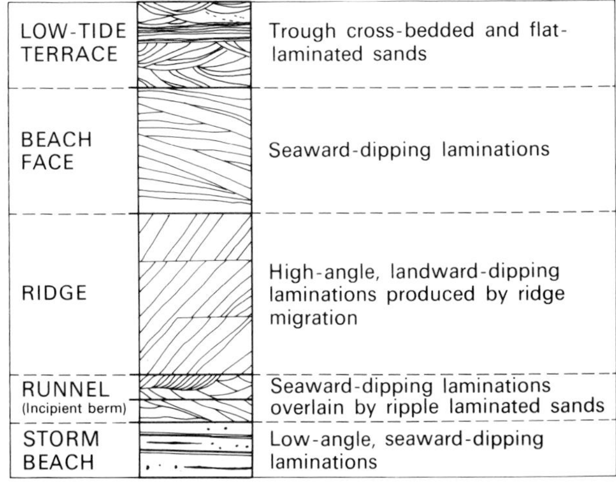

Fig. 7.9. Idealized vertical sequence produced by the landward migration of a foreshore ridge; no vertical scale given, but the probable thickness ranges from 20–80 cm (modified after Davis, Fox, Hayes and Boothroyd, 1972).

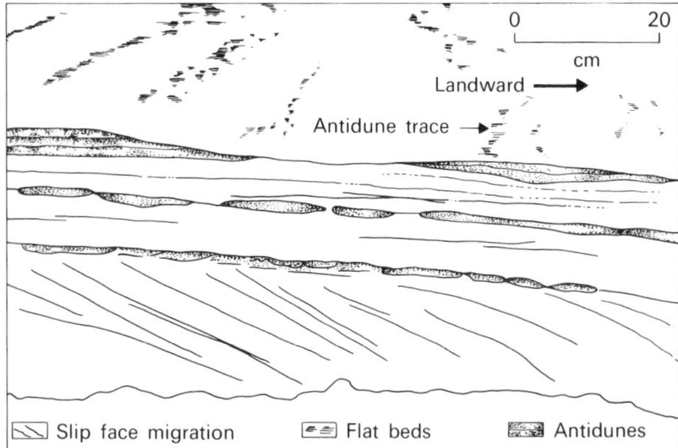

Fig. 7.10. Foreshore facies composed of high-angle, landward-dipping laminations produced by migration of a foreshore ridge, overlain by flat laminated sands with three sets of low-angle antidune cross-laminations; antidune bedforms occur on the surface (after Hayes and Kana, 1976).

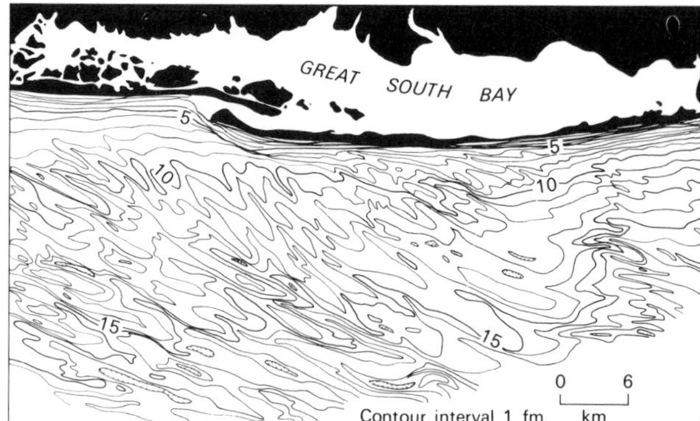

Fig. 7.11. Shoreface-attached linear shoals off Fire Island, Long Island, New York (after Duane, Field, Meisburger, Swift and Williams, 1972).

laminations may also be significant in foreshore facies (Wunderlich, 1972; Hayes and Kana, 1976). Trenches cut through low amplitude, high wavelength, quasi-symmetrical bedforms interpreted as antidunes reveal thin, lenticular sets of low-angle laminations which generally dip up-current (Fig. 7.10). Single sets of these laminations may be interspersed through the foreshore facies, and multiple sets may form if sediment deposition rates are high.

SHOREFACE

The shoreface is the sub-tidal part of the beach face, commencing at mean low tide level and terminating at the fairweather wave base. The upper shoreface is relatively steep (1 : 10), but the slope decreases offshore until at the seaward extremity in water depths of 12–20 m it has a value of 1 : 200 (Swift, 1976). In physical terms the shoreface is an area of transition between offshore or shelf processes and nearshore processes such as the breaking of waves. The upper shoreface is dominated by wave-driven flow associated with shoaling waves and to a lesser extent rip currents, whereas the lower shoreface may also be influenced by tidal currents and storm-induced coastal flows of the shelf regime (Swift, 1975a, 1976a).

The shoreface areas of several barrier islands along the East Coast, USA are characterized by shoreface-attached linear shoals trending obliquely to the shoreline (Fig. 7.11; Sect. 9.6.3). In contrast, the shoreface zone off Cape Cod exhibits 30 or more bars which parallel the shoreline (Nilsson, 1973), whereas other examples have no major topographic features but are ornamented by a fluctuating assemblage of bedforms (e.g. the Oregon coast; Clifton, Hunter and Phillips, 1971).

Facies characteristics of the shoreface are generally superimposed on a progressive seaward fining of sediment. Originally this was explained by the null-line hypothesis which stated that sediment of a given grain size occupied a position on the shoreface determined by the balance between a downslope-directed gravity force and an upslope-directed fluid force imparted by shoaling waves (Johnson and Eagleson, 1966). This is now regarded as an oversimplification, and the seaward fining of sediment is attributed primarily to sorting in plumes of suspended sediment supplied to the shoreface by rip currents, with other factors such as the landward increase in wave-induced shear stress supporting the trend (Cook, 1969; Cook and Gorsline, 1972; Swift, 1976a). Departures from this seaward fining trend occur in several present-day shoreface areas where the shelf regime impinges on the lower shoreface area during coastal retreat, causing the lower shoreface facies to be characterized by relatively coarse-grained sediments with high energy bedforms and internal structures (Swift, 1976).

Facies details across the shoreface are determined largely by the prevailing hydrodynamic regime in which wave action is pre-eminent, though biogenic processes are also important, and shoreface facies often reflect an interplay between these processes. In high wave energy settings primary structures predominate as biogenic structures are not developed or have a low preservation potential, whereas in low wave energy settings biogenic reworking may be so thorough that the sediment becomes homogenous, poorly sorted and virtually structureless. Shoreface areas of intermediate wave energy are characterized by a distinctive facies which reflects a repeated alternation between physical and biogenic processes related to storm and fairweather periods. In Sapelo Island, USA, this facies comprises alternating horizons of parallel laminated and bioturbated sand (Howard, 1971; Howard and Reineck, 1972b), and shallow cores taken between 5–21 m water depth off Long

Island, New York, reveal similar units with the preferred sequence: basal gravel→evenly laminated sand→bioturbated or wave ripple-laminated sand (Sanders and Kumar, 1975a; Kumar and Sanders, 1976). It is considered that storm period waves scour the shoreface, temporarily suspend the sediment, and redeposit it as laminated sands as the storm wanes. Then, during fairweather periods, burrowing organisms and/or lower energy waves rework the upper part of the laminated sand. Identical beds, termed 'sublittoral sheet sands' or 'storm lag deposits', have been described in ancient shelf facies associations (see Sect. 9.11.2 and Fig. 9.50).

7.2.3 Facies profiles of the beach face

Several studies have recently described the facies characteristics of present-day beach profiles, employing a variety of techniques including box cores, shallow borings, lacquer peels and X-ray radiography of samples. Study areas include the Pacific-facing Oregon coast, the Atlantic-facing coast off Sapelo Island, Georgia; the Gulf of Gaeta in the north Tyrrhenian Sea; Kouchibouguac Bay, New Brunswick; and Padre Island in the Gulf of Mexico (Clifton *et al.*, 1971; Howard and Reineck, 1972b; Reineck and Singh, 1971; Davidson-Arnott and Greenwood, 1976; Hill and Hunter, 1976). The wave regime controls the structural configuration and facies characteristics of the beach face, and although quantitative data are lacking it appears that the Oregon coast is a high wave energy area, whereas the others are intermediate to low wave energy areas, though Padre Island is frequently affected by major storms or hurricanes.

HIGH WAVE ENERGY BEACH FACE PROFILES

The Oregon coast stands apart as a very high wave energy setting (Clifton *et al.*, 1971). Although observations for this study were made during *fairweather* conditions, wave energy was still appreciable with breaking waves 1.5–3.0 m high recurring every 8–12 seconds. Landward- and seaward-directed wave surges produced during transformation are more important than longshore-, rip- or tidal currents, though the tidal range is not stated. The beach face fines seawards from gravel and coarse sand into fine sand, and is devoid of any major morphological features. However, various bedforms occur in zones which are approximately coincident with wave transformation zones (Fig. 7.12). Landward-directed currents produced by shoaling solitary waves are mainly responsible for producing the asymmetrical ripple, lunate megaripple and outer planar zones, whereas the inner rough and inner planar zones reflect a predominance of seaward-directed return flows in the inner part of the surf zone and the swash-backwash zone. To some extent these hydrodynamic-bedform zones produce distinctive facies belts, but as the zones migrate landwards or seawards as conditions fluctuate they generally give rise to complex assemblages of structures. For example, box cores in the asymmetrical ripple zone reveal small-scale cross-lamination superimposed on larger scale scour-and-fill structures presumably generated by earlier storm waves (Fig. 7.13).

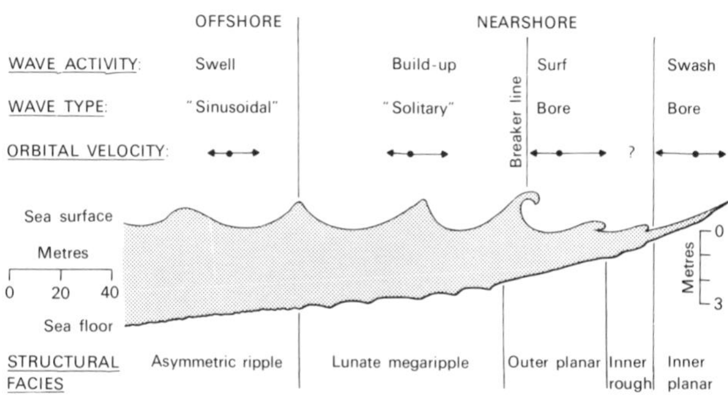

Fig. 7.12. Relationship of bedforms to wave characteristics on the high wave energy beach face of the Oregon coast (after Clifton, Hunter and Phillips, 1971).

INTERMEDIATE TO LOW WAVE ENERGY BEACH FACE PROFILES

Kouchibouguac Bay, New Brunswick, is a microtidal area characterized by a series of barrier islands and spits composed mainly of fine sand (Davidson-Arnott and Greenwood, 1974, 1976). Mean wave height during storms is 1–2 m with periodicities of 4–8 seconds, and there are well-developed longshore and rip currents. Beach faces exhibit a continuous outer crescentic bar and a discontinuous system of inner bars, and although sediment on the outer bar is slightly finer grained and more bioturbated, facies belts are to a large extent repeated across the beach face as waves break on the outer bar, re-form and break again on the inner bar (Fig. 7.14). The facies can be summarized in terms of four sub-environments related to the bars: (1) the seaward slopes – composed of small-scale, landward-directed ripple lamination interbedded with seaward-dipping parallel lamination; (2) the bar crests – in which parallel laminations are interbedded with sets of small- to medium-scale trough cross-bedding produced by lunate megaripples; (3) the landward slopes – characterized by low-angle, landward-dipping laminations and sets of trough cross-bedding on the outer bar, and high-angle, landward-dipping foresets up to 1 m high on the inner bar; (4) the troughs – composed of finer-grained sediment with ripple laminations produced by longshore currents parallel to the shoreline. Rip currents produce channels which locally dissect the inner bars and deposit seaward-directed planar cross-bedding in relatively

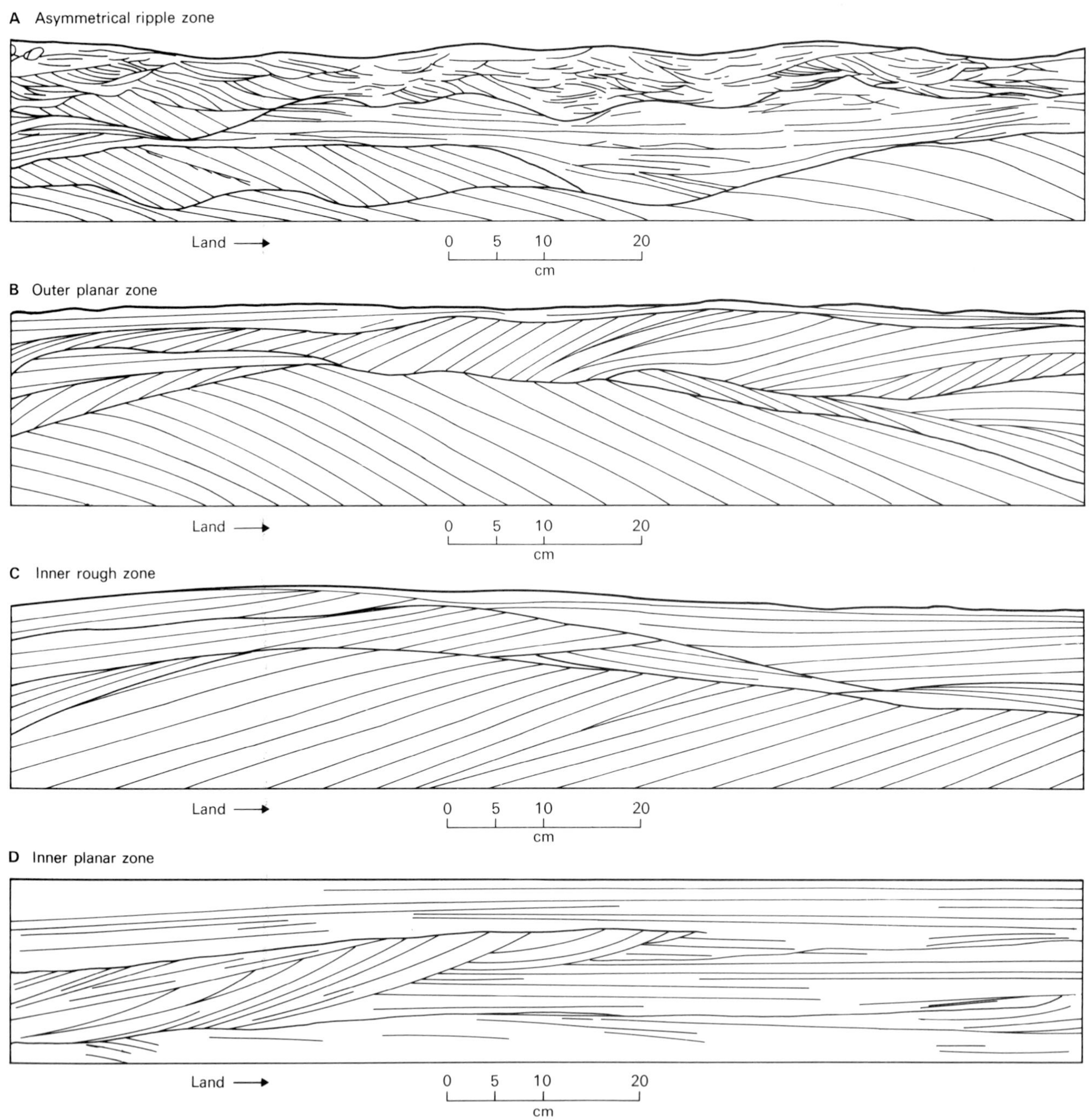

Fig. 7.13. Assemblages of internal structures across the Oregon coast beach face: (A) asymmetric ripple zone – small-scale ripple laminations superimposed on seaward- and landward-dipping foresets reflecting larger bedforms generated by storm waves; (B) outer planar zone – parallel laminations superimposed on seaward- and landward-dipping foresets; (C) inner rough zone – predominantly seaward-dipping foresets; (D) inner planar zone – swash-formed parallel laminations with inter-bedded seaward-dipping foresets of the inner rough zone (after Clifton, Hunter and Phillips, 1971).

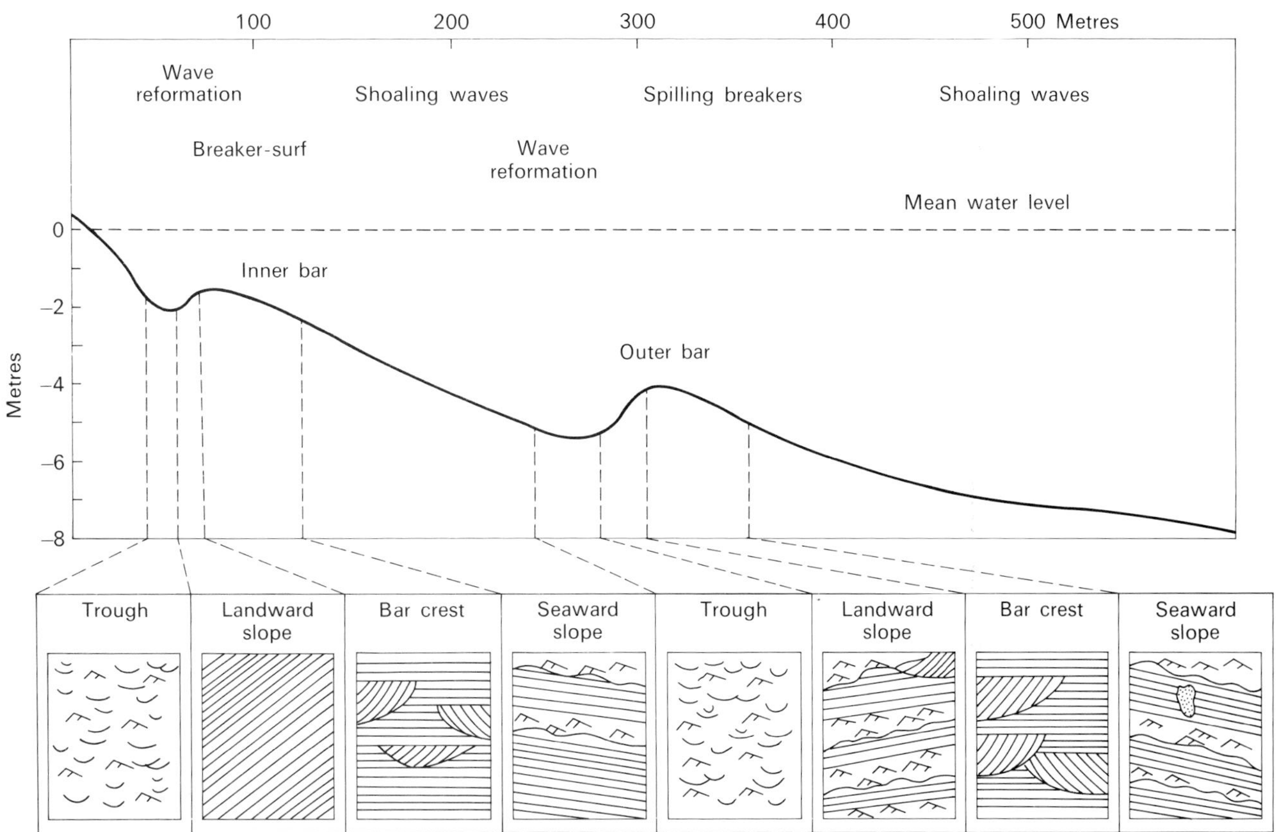

Fig. 7.14. Wave transformation and facies zones for the Kouchibouguac Bay beach profile, illustrating a certain amount of repetition across the inner and outer bar systems (after Davidson-Arnott and Greenwood, 1976).

coarse sand. These sets presumably occur in a channel form, but may fan out lower down the seaward slope of the inner bar.

Sapelo Island is one of a series of barrier islands along the mesotidal Georgia coastline (Howard, Frey and Reineck, 1972). Average wave height is 20–30 cm, and although major storms affect the area they are relatively infrequent and waves only occasionally exceed 1.7 m. The island is in fact a Pleistocene remnant bounded by tidal inlets and estuary mouths to the north and south, but it is fronted by present-day beaches separated from the main island by a narrow strip of salt marsh (Fig. 7.15). The beach face dips gently seawards at less than 2° and is characterized by ridge and runnel topography. The offshore area is composed of relict Pleistocene sands and the profile therefore commences at 10 m water depth with bioturbated muddy sands devoid of laminations which pass landwards into a sand facies composed of alternating horizons of parallel lamination and bioturbation (Howard and Reineck, 1972a, 1972b, Fig. 7.16). This change reflects an increase in the importance of wave energy in a landward direction, and this trend continues as ripple laminated sands eventually pass into the foreshore facies composed of low-angle, seaward-dipping laminations and higher-angle, landward-dipping laminations. Bioturbation is a sensitive indicator of conditions in this profile as it diminishes towards the shore as wave energy increases, and also exhibits conspicuous zonation across the profile (Frey and Howard, 1969; Frey and Mayou, 1971; Hertweck, 1972). This is particularly apparent in the form and configuration of decapod burrows. Irregular, ramifying burrows of *Callianassa atlantica* in the shoreface area give way to vertical burrows of *Callianassa major* in the foreshore, whereas the backshore exhibits inclined J-, U- or Y-shaped burrows of the ghost crab

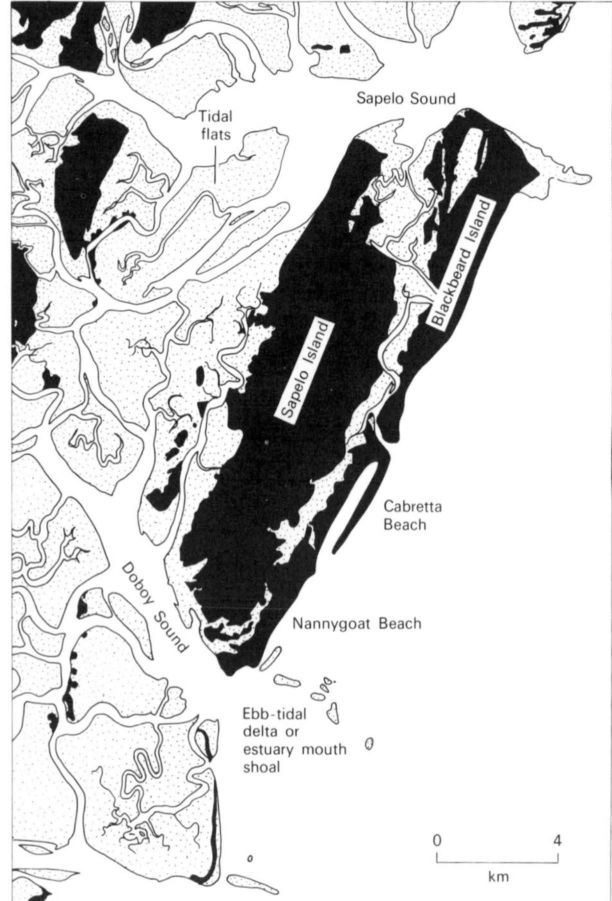

Fig. 7.15. Sapelo Island, Georgia, USA: a Pleistocene barrier island fronted by present-day beaches and bounded to the north and south by major tidal inlets which drain extensive tidal flats behind the island (after Howard, Frey and Reineck, 1972; Hertweck, 1972).

Ocypode quadrata. Burrow form reflects the behavioural responses of the animals to the prevailing conditions and provides a useful adjunct to sediment grain size and structures in facies definition.

Padre Island, Texas, is a major barrier island located in the microtidal Gulf of Mexico (Dickinson, 1971; Hill and Hunter, 1976). Mean wave heights are restricted to 0.3–1.0 m but are sufficient to generate a nearshore circulation system and longshore currents. Rhythmic topography developed along the foreshore and upper shoreface produces a combination of parallel lamination, small-scale ripple lamination and medium-scale cross-bedding (Fig. 7.17). However, in all parts of the beach face except the upper foreshore swash-backwash zone, there is an interplay between sediment deposition immediately after storms and intense biogenic reworking during low wave energy fairweather periods. Plots of hurricane paths demonstrate that very little of the Texas shoreline has escaped the direct passage of at least one hurricane this century, and the importance of these events is enhanced if one considers the total area affected by each hurricane (Hayes, 1967). In this area major storms and hurricanes are therefore relatively common events affecting nearshore sedimentation (McGowen and Scott, 1976).

SUMMARY

Direct comparison of the results of these studies is hindered by differences in emphasis and terminology, and in addition there are real differences between the study areas in hydrodynamic setting, morphology, sediment grain size and climate. However, the following points can be made:

(1) Foreshore facies are broadly similar between study areas despite obvious contrasts in wave regime.

(2) Shoreface facies differ according to the wave regime, with high wave energy areas being dominated by physical sedimentary structures, whilst intermediate to low wave energy areas record an alternation between physical and biogenic structures.

(3) The sequence of bedforms generated on the beach face by waves can be considered in a manner analogous to the flow regime concept, with bedform type being related to flow conditions and sediment grain size (Clifton *et al.,* 1971; Clifton, 1976; Davidson-Arnott and Greenwood, 1976). A model erected by Clifton (1976) recognizes four main fields: no sediment movement→symmetrical bedforms→asymmetrical bedforms→flat bed, and describes the development of these fields in terms of maximum orbital bottom velocity, velocity asymmetry, median grain size and wave period.

(4) Each profile is characterized by a seaward fining of sediment, and offshore progradation of the beach face will therefore produce a coarsening-upward sequence (Fig. 7.18). The basic motif will involve a passage from fine-grained offshore facies into well-sorted sands of the foreshore/backshore, but the sequences will differ in detail and these differences will reflect the physical characteristics of the beach face.

7.2.4 Tidal inlets

Tidal inlets act as passageways for tidal waters in barrier island systems, with tidal flood waters entering via the inlet, being stored temporarily during tidal still-stand, and subsequent draining via the inlets as the tibe ebbs. The extent to which tidal inlets are developed and maintained in a barrier island is related to the tidal range, and microtidal and mesotidal barrier

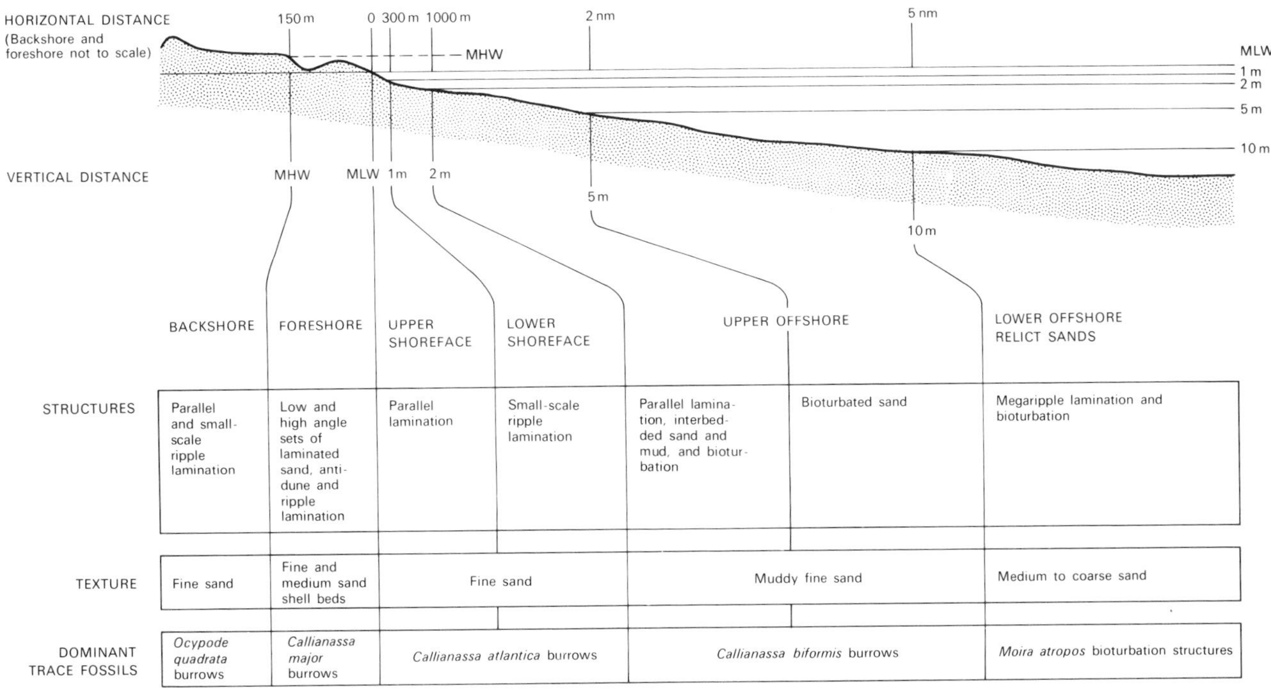

Fig. 7.16. Facies characteristics across the Sapelo Island beach profile (modified after Howard and Reineck, 1972b).

islands can be distinguished on this basis (Hayes and Kana, 1976). However, it is important in present-day examples to distinguish between true tidal inlets and drowned river valley estuaries which are not developed in response to tidal range. Barrier islands in microtidal settings have few, if any, tidal inlets. In the Gulf of Mexico a few narrow inlets are maintained but stretches of barrier island more than 100 km long are frequently devoid of tidal inlets (e.g. Padre Island; Fig. 7.17). Breaches cut during storms may act as tidal inlets for a brief period after the storm (El-Ashry and Wanless, 1965), but they are soon healed by sediment deposition as tidal currents are insufficient to maintain permanent channels. In mesotidal settings tidal inlets are more numerous and repeatedly dissect the barrier island into short segments.

Three types of tidal inlet are recognized: overlap, symmetrical and offset (Fig. 7.19), with the development of each type related to longshore drift characteristics (Byrne, Bullock and Tyler, 1976; Goldsmith, Byrne, Sallenger and Drucker, 1976). In general, they range from several hundred metres to a few kilometres in width and may be up to 20 m deep. The side of the inlet which is upstream relative to the longshore drift direction tends to be a depositional bank, whereas the opposite bank is erosive. Dunes and sandwaves generally cover the sub-tidal part of the depositional bank, and are overlain by a series of wave-induced recurved spits attached to the channel margin. Although tidal inlets form only a part of barrier island systems, even in mesotidal settings, they are often extremely significant in terms of the overall facies pattern as they tend to migrate laterally in the direction of longshore drift (Shepard, 1960; Hoyt and Henry, 1967). During migration, previously deposited barrier sands are eroded, and a substantial wedge of sediment is deposited on the updrift margin. The extent to which inlet deposits are developed depends on: (1) the frequency of inlets along the barrier island; (2) the rate and consistency of direction of inlet migration; (3) the nature of barrier migration, whether regressive, transgressive or stationary; and (4) the rate of barrier migration relative to the rate of inlet migration. Lateral migration rates of present-day tidal inlets are variable, but can be extremely high. The updrift depositional margin of Doboy Sound, Sapelo Island, has migrated 1 km in approximately 4,000 years and produced an inlet sand body 1 km long parallel to the shoreline and 10–13 km wide normal to the shoreline (Hoyt and Henry, 1967; Figs. 7.15 and 7.20). However, the western tip of Fire Island inlet has migrated 8 km in 115 years at an average rate of 64 m per year, and a rate of

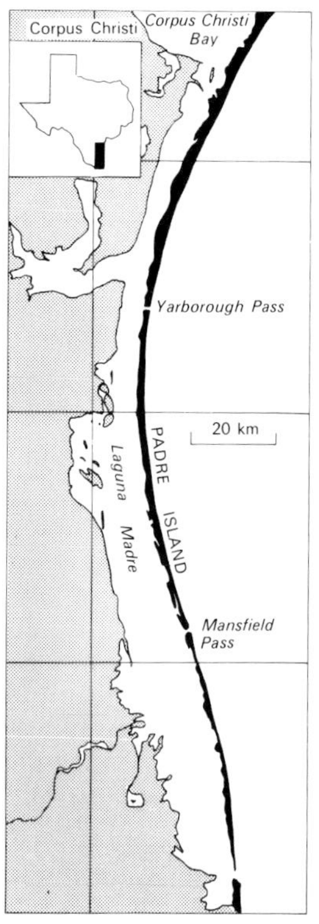

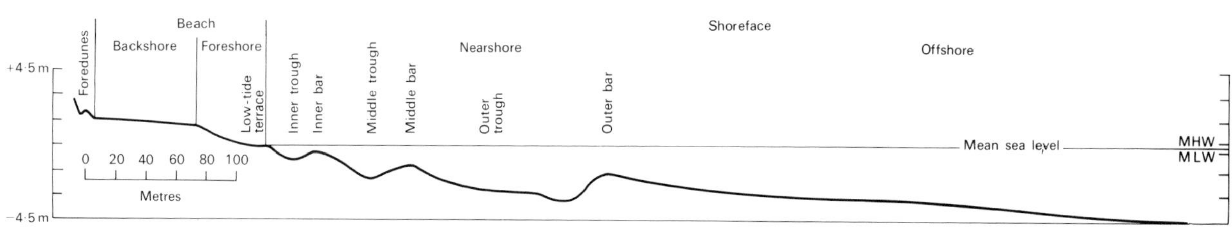

Fig. 7.17. Map and beach face profile of Padre Island, Texas (after Dickinson, 1971; Hill and Hunter, 1976).

90 m per year has been estimated for the inlet associated with Nauset Spit off Massachusetts (Kumar and Sanders, 1974; Hine, 1976). As the inlets scour well below sea level their deposits have a high preservation potential, and if high rates of migration are sustained in a constant direction, inlet deposits may constitute the entire depositional record of a barrier island system.

FACIES ASSOCIATIONS OF TIDAL INLETS

Important features of tidal inlet facies noted in early studies include a basal erosion surface floored by a shell gravel lag with an extremely mixed faunal assemblage, major lateral accretion surfaces dipping into the channel reflecting former positions of the depositional bank, and large-scale sets of cross-bedding separated by thin silt and clay drapes (Shepard, 1960; Hoyt and Henry, 1967). A recent account of Fire Island inlet on the south side of Long Island, New York, provides the first rigorous description of tidal inlet facies (Kumar and Sanders, 1974). This overlap inlet is 1 km wide and up to 10 m deep (see Figs. 7.11 and 7.19). Tidal currents scour the channel base and generate asymmetrical sand-waves which migrate in the ebb direction and are modified, but not reversed, by flood-tidal currents. In shallower parts of the channel these bedforms are absent and plane bed conditions prevail. The channel is capped by a subaqueous spit which resembles a Gilbert-type micro-delta, and a subaerial spit at the inlet margin. The deposits of this inlet comprise five distinctive facies corresponding to

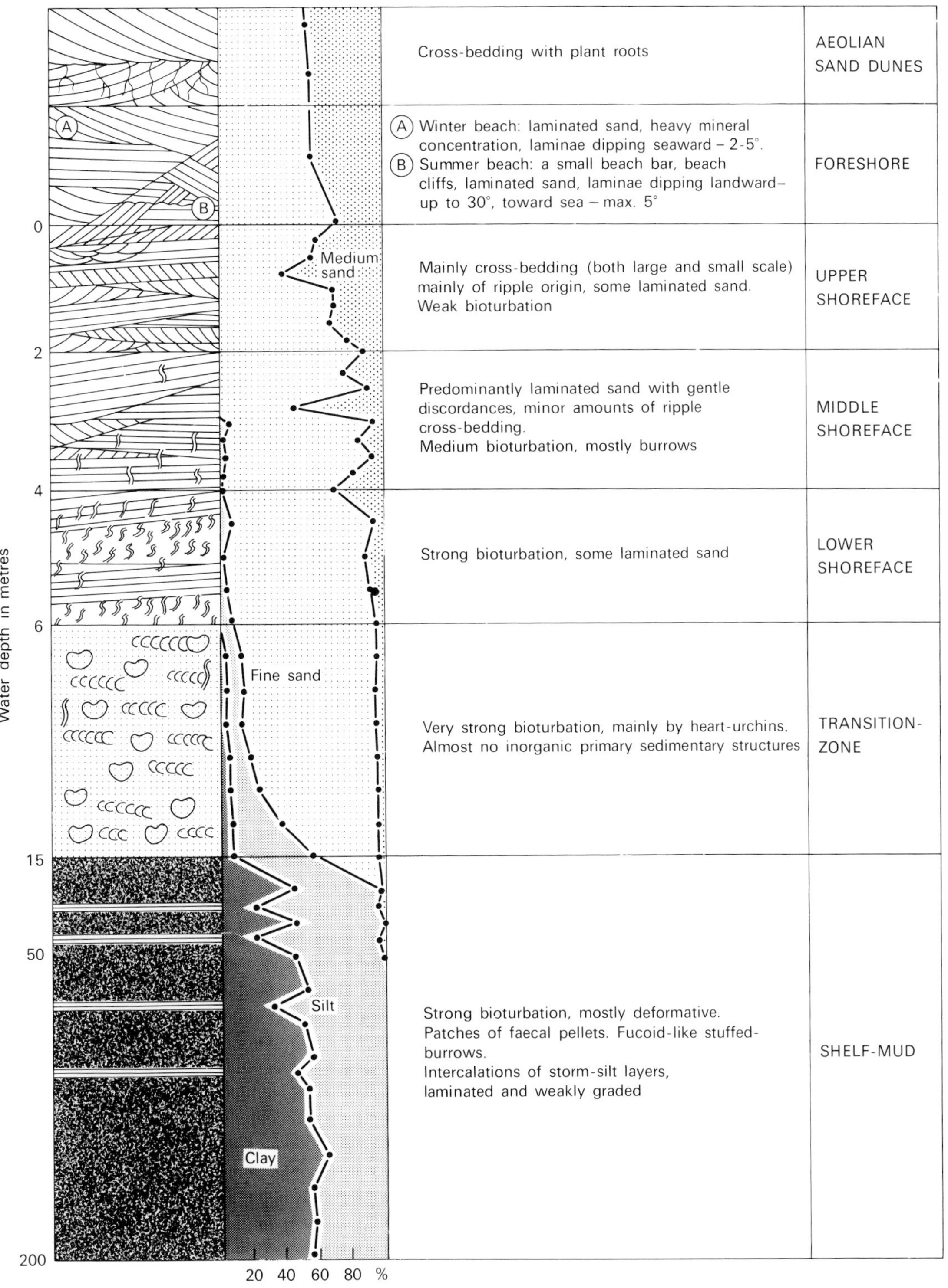

Fig. 7.18. An idealized summary of the coarsening-upward sequence which would develop by progradation of the low wave energy beach face in the Gulf of Gaeta, Italy; illustrating an upwards decrease in bioturbation structures and corresponding increase in wave-induced structures (after Reineck and Singh, 1971, 1973).

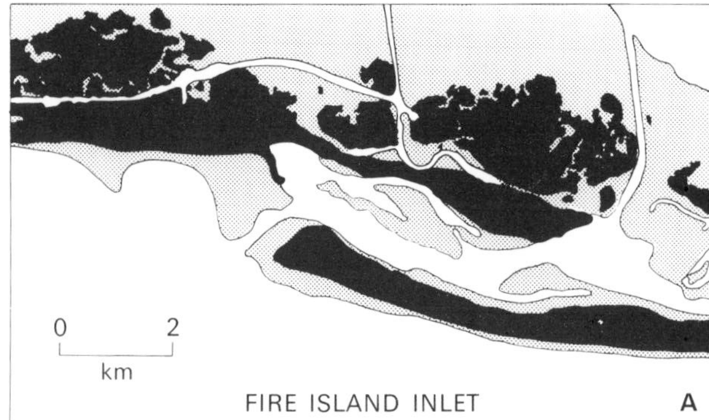

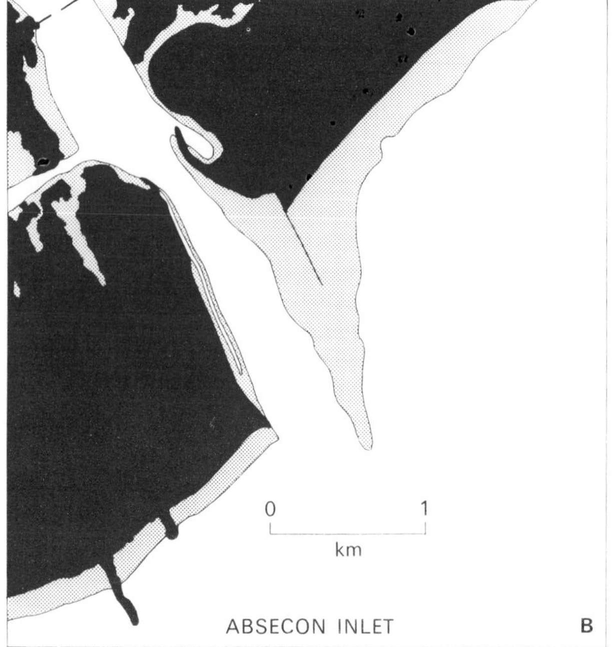

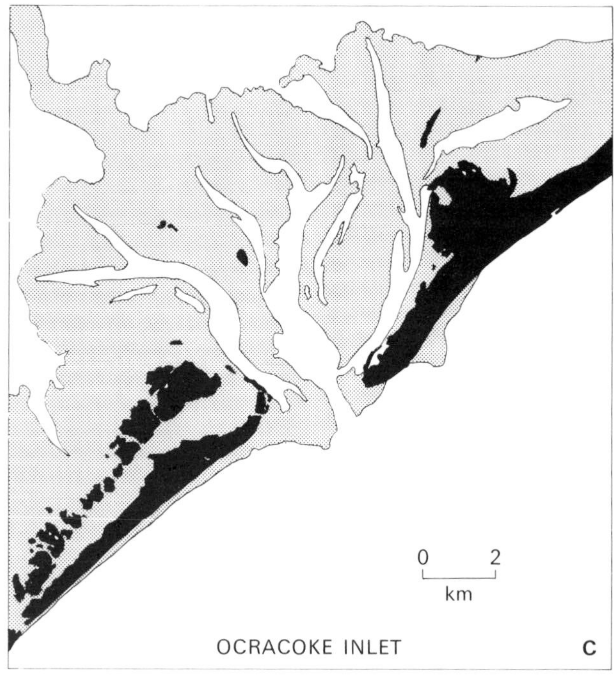

Fig. 7.19. Types of tidal inlet: (a) overlap; (b) offset; (c) symmetrical; note also the well-developed ebb and flood tidal deltas associated with the symmetrical inlet (after Swift, 1976a).

channel floor, deep channel, shallow channel, spit platform and subaerial spit sub-environments (Fig. 7.21). Facies in the lower part of the sequence reflect a predominance of tidal processes, whereas the upper part records an interplay between tidal and wave processes. A comparable, though idealized, sequence based on studies of tidal inlets along the Massachusetts shoreline recognizes four facies (Hayes and Kana, 1976): (1) coarse sand-shell debris channel facies with large-scale, bi-directional planar cross-bedding and interbedded medium-scale trough cross-bedding; (2) medium-grained sand dominated by small- to medium-scale trough cross-bedding; (3) flat-laminated sands with variable dip directions; and (4)

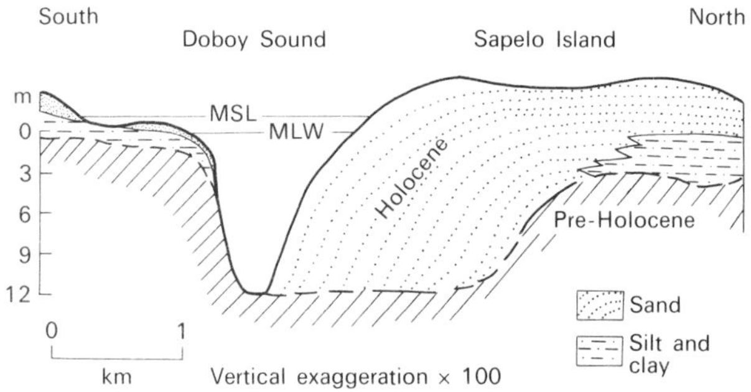

Fig. 7.20. A laterally accreting sediment wedge deposited on the updrift margin of Doboy Sound, Sapelo Island (after Hoyt and Henry, 1967); see Fig. 7.15 for location.

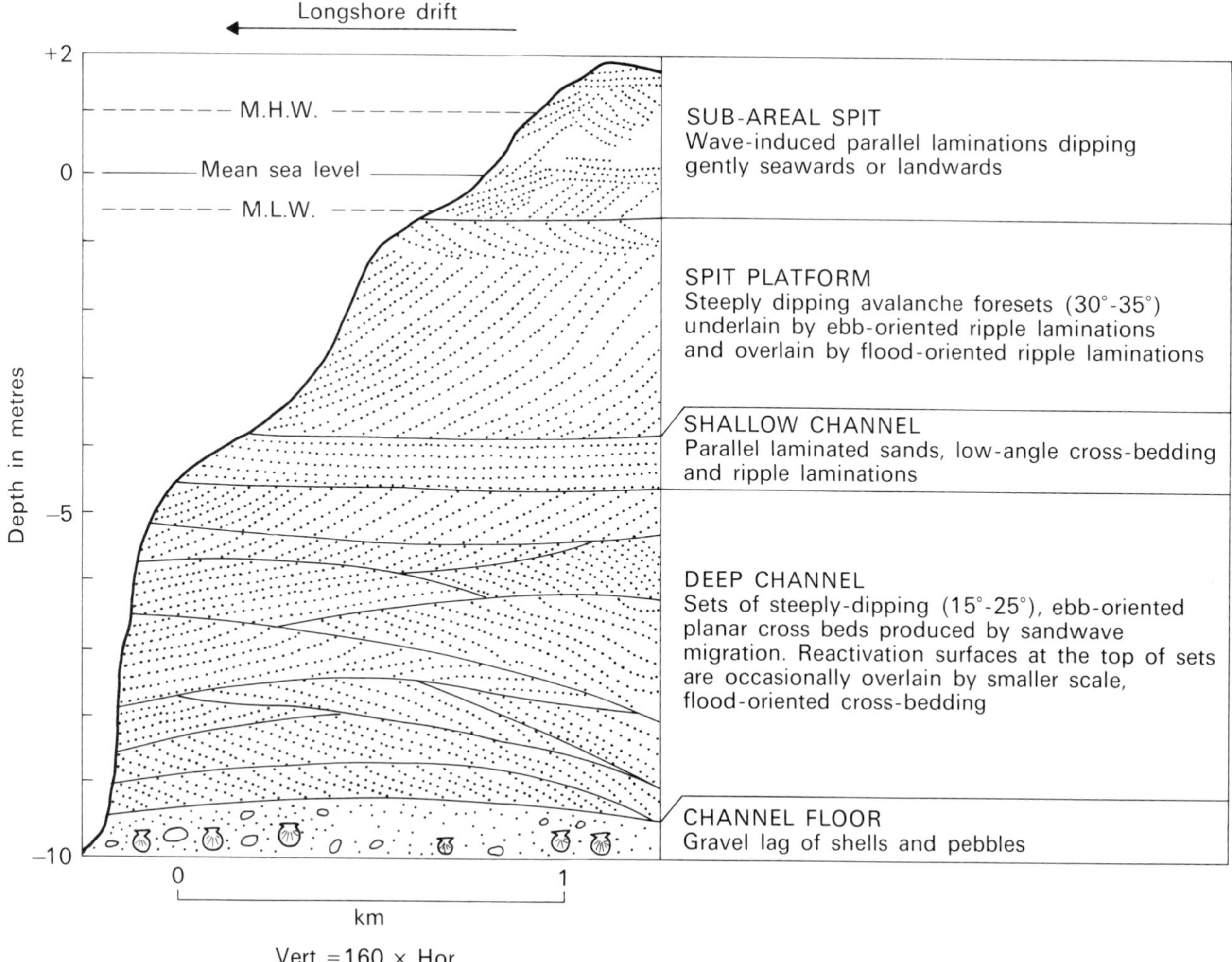

Fig. 7.21. Vertical profile and facies characteristics of a tidal inlet in Fire Island, Long Island, New York (after Kumar and Sanders, 1974); see Fig. 7.19a for a plan view of the inlet.

fine-grained, well-sorted aeolian dune sands with trough cross-bedding.

TIDAL DELTAS

Ebb-tidal deltas form at the seaward mouth of tidal inlets, whereas flood-tidal deltas form at the landward mouth and are best regarded as part of the lagoonal environment (see later). Ebb-tidal deltas result from an interaction between tidal currents and waves, and form broad sand accumulations which may coat the beach face if they migrate laterally with the inlet. They conform to a general pattern in which the main ebb channel is flanked laterally by levee-like bars and fronted by a terminal lobe and swash bars (Oertal, 1972, 1976; Hayes and Kana, 1976; Fig. 7.22). Tidal processes predominate in the ebb channel and the central part of the lobe, whilst wave processes are most effective on the flanks. Areas of seaward-directed dunes and sandwaves are therefore flanked by landward migrating swash bars (Hine, 1976). Alternative terms for ebb-tidal deltas include inlet-associated bars and estuary entrance shoals, and as they project onto the shelf they are also discussed in Chapter 9 (Sect. 9.6.3).

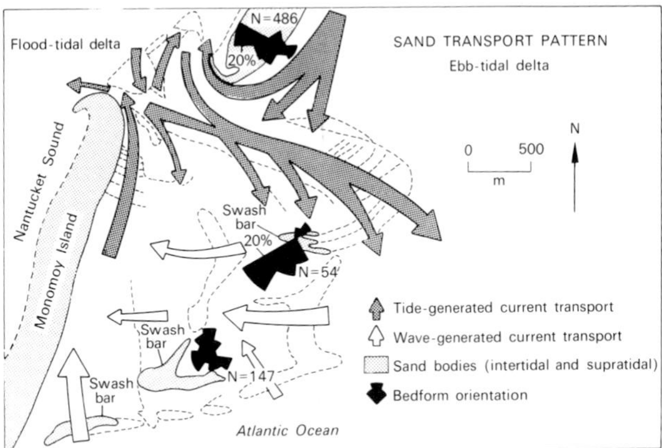

Fig. 7.22. Sediment transport in an ebb tidal delta, illustrating that the central part is dominated by tidal currents, whereas the lateral margin is dominated by wave processes (after Hine, 1976).

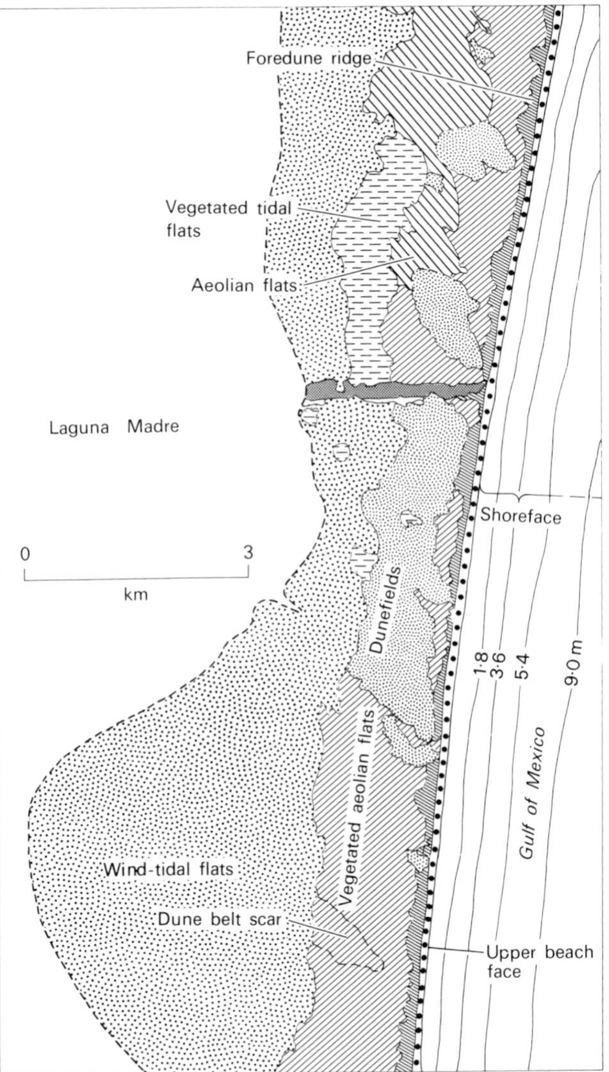

Fig. 7.23. Extensive back-barrier flats composed of washover sands partially reworked into aeolian dunes, Padre Island, Texas (after Dickinson, 1971).

7.2.5 **Lagoons**

Lagoons are extremely variable shallow water environments located behind barrier islands. This variability is in part attributable to the prevailing climate, as illustrated by lagoons in the Gulf Coast which range from semi-arid in the south-west to humid in the north-east (Rusnak, 1960; Shepard and Moore, 1960), and there are also marked differences between lagoons in microtidal and mesotidal settings. In *microtidal lagoons* abnormal salinities may prevail – either brackish or hypersaline – as the water body has limited communication with the open sea due to the paucity of tidal inlets. Also, as there are no passageways to accommodate landward-directed storm surges, the barrier islands are frequently inundated or breached during storms (Hayes and Kana, 1976). Washover fans are therefore common and they often coalesce to produce an irregular apron of sand termed the back-barrier flat which projects into the lagoon (Fig. 7.23). In *mesotidal lagoons* salinity values are normal due to continual exchange between lagoonal and open sea waters via tidal inlets. Flood-tidal deltas are important sites of sediment deposition at the lagoonal mouths of inlets, and extensive areas of the lagoon may comprise intertidal flats and salt marshes dissected by a network of tidal creeks (Allen, 1965e; Phleger, 1965; Frey and Howard, 1969; Fig. 7.24). Washover fans may occur locally, but are not developed to the same degree as in microtidal lagoons.

LAGOONAL FACIES

In general, lagoons accumulate fine-grained sediment deposited from suspension, but minor depositional systems set into this background such as washover fans, flood-tidal deltas, intertidal flats, tidal channels and small-scale river deltas extend the range of lagoonal facies.

The clays and fine silts deposited in quiet, open water areas may be finely laminated, but more commonly they are structureless due to bioturbation. In humid and sub-humid areas these fine-grained sediments are often rich in organic matter, including plant debris washed in by rivers. In contrast,

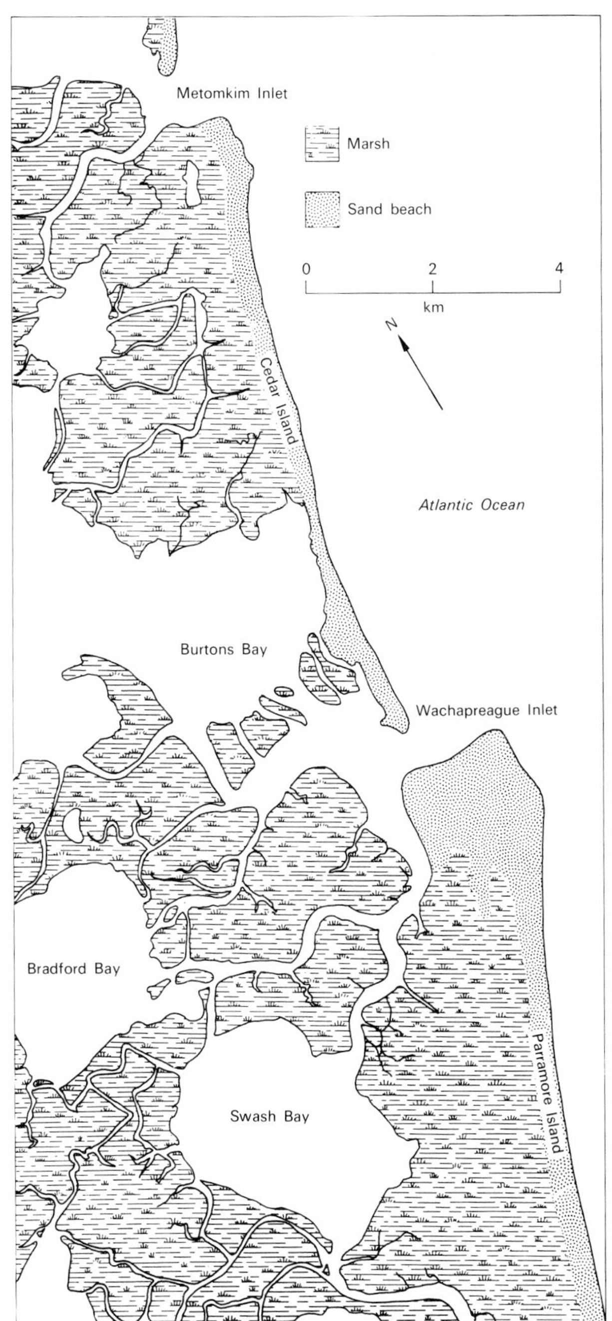

Fig. 7.24. Mesotidal lagoons behind barrier islands along the Virginia shoreline (after Morton and Donaldson, 1973).

the equivalent facies in arid and semi-arid lagoons have a lower organic content and may contain evidence of periodic desiccation. Mudflats which infill over 200 sq. km of the Laguna Madre are exposed for prolonged periods between major storms and subjected to wind action which produces mud-cracked surfaces, deflation surfaces and dunes formed of aggregated clay pellets. Fine laminae of granular gypsum form just below the sediment surface, and algal mats which coat the surface are also incorporated as sedimentation continues (Fisk, 1959; Rusnak, 1960). Coarse silt/fine sand laminae and ripple form sets are produced in open water facies by slight wave agitation, and the distal fringes of washover units may produce thin, sharply defined sand beds with parallel- or current-ripple-lamination. Oolites, coated shell fragments, and micro-coquinas accumulate locally on wave exposed landward shorelines of Laguna Madre (Rusnak, 1960).

The abundance and diversity of lagoonal faunas varies according to salinity values between lagoons, and also within lagoons both temporally and spatially. Abnormal salinity values result in low diversity assemblages with high numbers of individuals, whereas normal salinity conditions in the vicinity of tidal inlets favour high diversity assemblages in which open marine and lagoonal species may be mixed (Parker, 1960; Phleger, 1960; Rusnak, 1960).

Washover fans result from sediment being eroded from the seaward side of the barrier and transferred into the lagoon

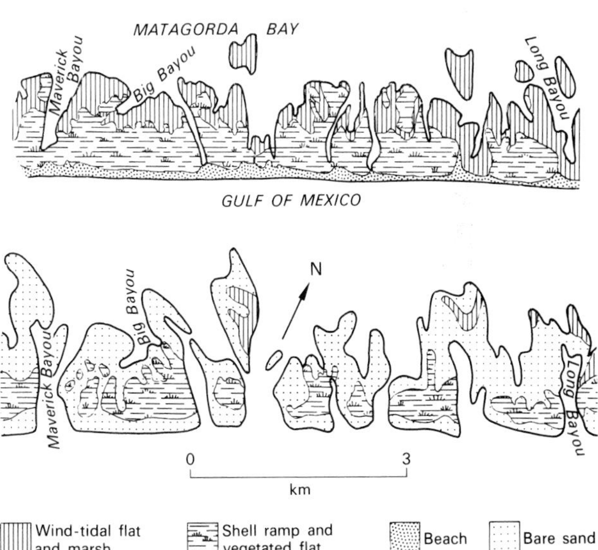

Fig. 7.25. The breaching of Matagorda Peninsula by Hurricane Carla in 1961, illustrating the development of numerous washover channels; note the way in which several earlier channels have been re-excavated (after McGowen and Scott, 1976).

during storm periods. In some cases the sediment-laden waters simply overtop the barrier crest in a sheet-like fashion, but generally the crest is breached at selected points and sediment is transferred in washover channels (Fig. 7.25). Most channels are cut above normal sea level and are therefore stranded after the storm, but channels cut below normal sea level persist for a short period after the storm and may be re-opened during subsequent storms (Hayes, 1967). On reaching the back-barrier, the channelized washover flow expands into a sheet flow which deposits a lobe-shaped sand unit. The depositional product of individual washover events is a thin, sheet-like sand bed with a planar erosive base. For example, at one point Hurricane Carla deposited an 8.5 cm washover sand unit over a back-barrier algal mat in Padre Island (Hayes, 1967a). Large-scale rhomboid bedforms have been described from the Texas Gulf Coast (Morton, 1978). Sedimentary structures described from washover fans include parallel to sub-parallel laminations in the main part of the unit, and medium-scale landward-dipping foresets at the distal margin where the fan leaves the back-barrier and enters the standing water body of the lagoon (Schwartz, 1975; Fig. 7.26). Shortly after deposition the washover sands may be bioturbated whilst the sediment is still moist. In present-day examples in Georgia fiddler crabs (*Uca pugilator*) and ghost crabs (*Ocypode quadrata*) produce distinct burrow forms, thin pelletoidal layers, and localized areas of intense mottling at the fan margins where moist conditions persist (Frey and Mayou, 1971).

Many washover fans are composite bodies, either because co-existing fans overlap and merge, or more commonly because a major breach through the barrier island is re-opened during successive storms. A composite fan on St. Joseph Island attains a maximum thickness of 1.25 m and extends over 7 sq. km (Fig. 7.27). In this example individual washover units exhibit the sequence: erosion surface – shell-rich basal layer with a mixed assemblage – sands with parallel, ripple or antidune laminations, locally disrupted by rootlets (Andrews, 1970). Subtle scour surfaces and thin intervals of wind reworked sand separate individual washover units (Schwartz, 1975).

Flood-tidal deltas form at the lagoonal inlet mouth in response to flow expansion. Initially, sheet flow conditions deposit a series of overlapping sediment fans at the inlet

Fig. 7.26. Sedimentary structures in washover fan sands (after Schwartz, 1975).

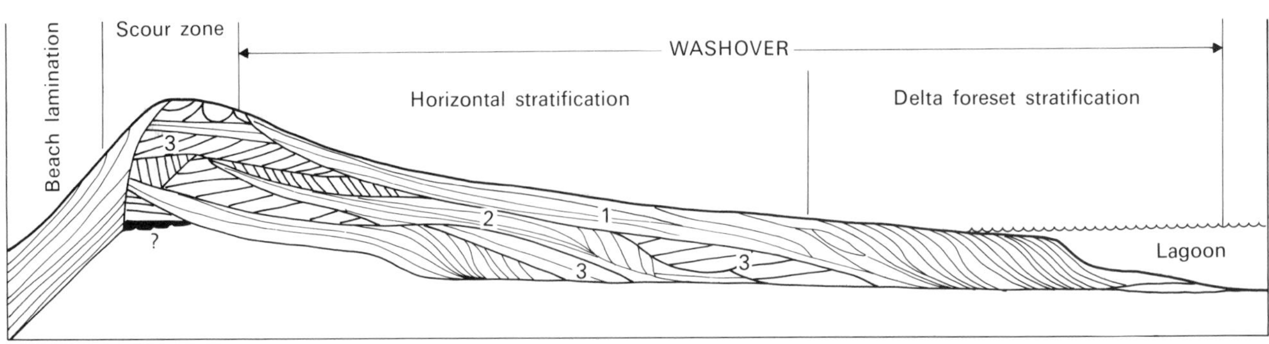

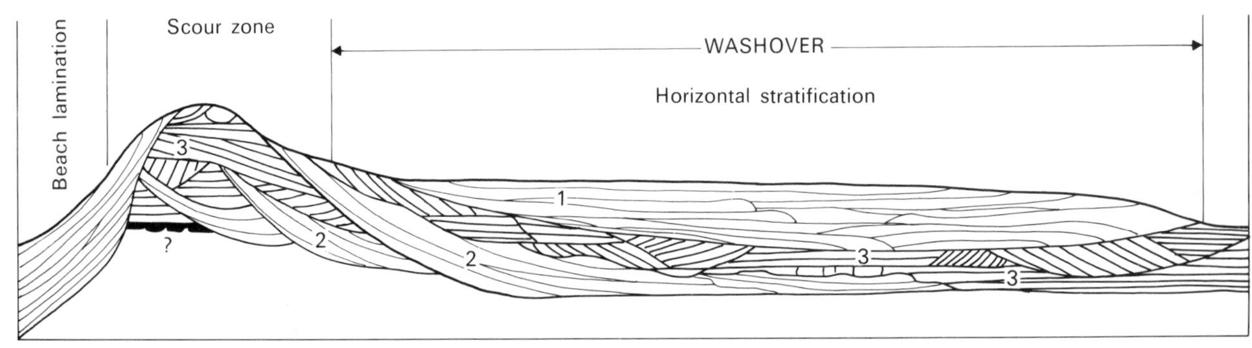

Legend 1 Newly deposited washover 2 Old washover 3 Aeolian deposits

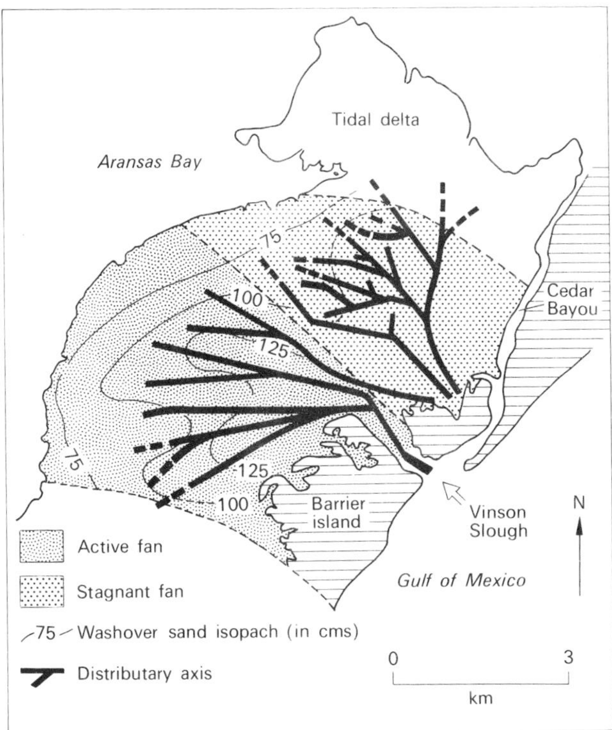

Fig. 7.27. A composite washover fan on St. Joseph Island, Texas (after Andrews, 1970).

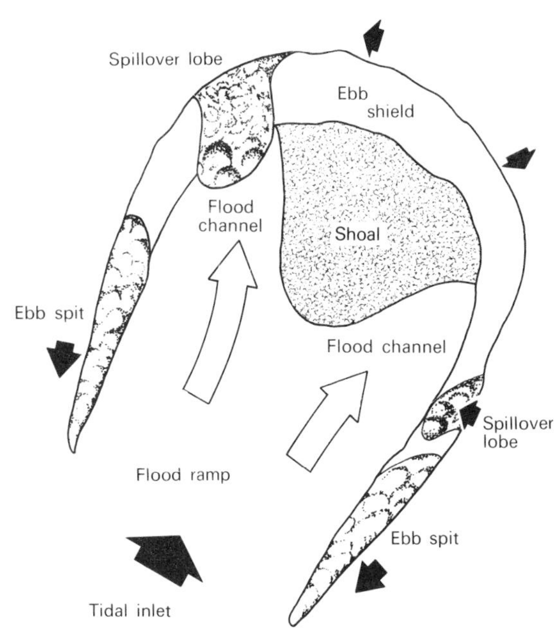

Fig. 7.28. Typical morphology of a flood-tidal delta (after Hayes, 1976).

mouth, but with time flow is concentrated into channels stabilized by levees (Morton and Donaldson, 1973). A flood-tidal delta recently initiated in Chatham Harbour, Massachusetts, illustrates the early stage of development (Hine, 1976). At present it comprises two distinct complexes of coalescing spill-over lobes covered by sandwaves, lunate and cuspate megaripples, and ripples which are predominantly landward-directed, although minor topographic features often divert the currents and cause bedform directions to spread about the mean. The mature morphology of flood-tidal deltas comprises a ramp dissected by flood-tidal channels and fringed by a series of ebb-produced spits, shields and spill-over lobes (Hayes, 1976; Hayes and Kana, 1976; Fig. 7.28). Sedimentation rates in flood-tidal deltas are often high, and if the deltas migrate laterally in concert with inlet migration they may form a significant proportion of the lagoonal facies.

The facies patterns of small-scale river deltas which occasionally form at the landward margin of the lagoon resemble those of fluvial-dominated deltas (Donaldson, Martin and Kanes, 1970; Kanes, 1970; see Chapter 6), and lagoonal intertidal flats and tidal channels produce facies associations comparable with those in non-lagoonal areas (Frey and Howard, 1969; see Sects. 7.5.1, 7.6).

7.2.6 Beach and barrier island migration

The overall facies pattern of a beach or barrier island is determined by its long-term response to sea level fluctuations, subsidence/emergence rates of the depositional area, and variation in sediment supply (Dickinson, Berryhill and Holmes, 1972). Under conditions of continued sediment supply, stable sea level and low to moderate subsidence rates facies prograde offshore, whereas a reduction in sediment supply, a rise in sea level or a high rate of subsidence induces landward migration of the facies. Alternatively, the system may remain stationary, as argued for Padre Island, Texas, which is considered to be thickening in place (Dickinson, 1971).

PROGRADATIONAL SYSTEMS

Progradational beaches and barrier islands usually exhibit striking ridge and swale patterns behind the active shoreline (Bernard *et al.*, 1959; Curray and Moore, 1964; van Straaten, 1965; Stapor, 1975; Wilkinson, 1975; Fairbridge and Hillaire-Marcel, 1977). Each ridge records a former position of the beach face and the ridge pattern therefore reflects the

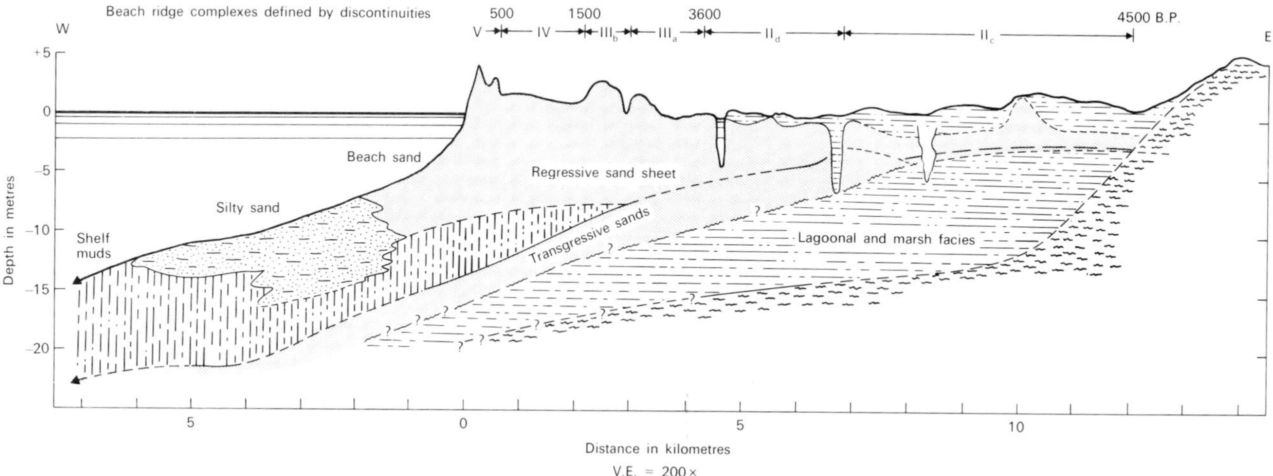

Fig. 7.29. Cross-section through the beach complex of Costa de Nayarit, Mexico, illustrating an extensive sheet sand produced by beach progradation, overlying shelf muds and silts and an earlier trangressive sand (after Curray, Emmel and Crampton, 1969).

progradational development of the sytem. The coastal plain of Costa de Nayarit, Mexico, extends over 225 km with an average width of 5 km and exhibits approximately 280 beach ridges (Curray, Emmel and Crampton, 1969). The ridges range in height from 1–2 m, are spaced 15–200 m apart, and can be traced laterally for 50 km or more. They are considered to form by the landward migration and eventual emergence of breaker bars which isolate the former beach face and create a depression between the ridges which subsequently becomes a shallow water, marshy area. In plan view, extensive discontinuities truncate the beach ridges and divide the coastal plain into five sets of beach ridges. The discontinuities reflect periodic changes in sediment supply or nearshore processes which induce re-alignment of the coastline. Under conditions of stabilized sea level which commenced 4,750–3,600 B.P., offshore progradation of this coastal plain has produced a laterally extensive sand sheet which overlies the shelf facies (Fig. 7.29). Boreholes through Galveston Island, Texas, reveal details of a coarsening-upward sequence from lower shoreface into foreshore and aeolian dune facies, and demonstrate that progradation produced a lenticular sand body 5–15 m thick aligned parallel to the shoreline (Fig. 7.30). Excavated coastal barrier deposits in Holland display a similar sequence and also exhibit prominent major bedding surfaces dipping gently seawards. These surfaces are interpreted as former beach face surfaces and are presumably an internal reflection of beach ridges (van Straaten, 1965).

The details of the progradational sequence will vary according to the wave regime and beach face characteristics, and the presence of laterally migrating tidal inlets will substantially modify the facies pattern produced by prograding barrier islands (see Sect. 7.2.4).

TRANSGRESSIVE SYSTEMS

Most major present-day barrier islands migrated landwards across the shelf during the Holocene transgression, and some are currently continuing this trend. The most widely accepted mechanism of landward retreat involves *shoreface retreat* in which, under conditions of rising sea level, sediment is eroded from the upper shoreface and emplaced either in the lower shoreface–offshore area, or in the lagoon as a series of washover fans (Bruun, 1962; Fig. 7.31). As the upper shoreface or breaker-surf zone passes across the former barrier it erodes the underlying facies deposited during earlier stages of landward barrier migration, and the transgressive sequence therefore includes a planar erosion surface at the base of the lower shoreface facies. Where subsidence rates are high relative to the rate of shoreface erosion a relatively thick and complete transgressive sequence will be preserved, but where subsidence rates are low the sequence will be incomplete (Fischer, 1961; Swift, 1968; Dillon, 1970).

In the Middle Atlantic Bight (East Coast, USA), barrier migration across the shelf left a thin, discontinuous sand sheet with the sequence: back-barrier facies→planar erosion surface with a gravel or shell lag→inner shelf facies. The passage of the upper shoreface across the area produces the erosion surface and accounts for the absence of barrier facies. Brief pauses in

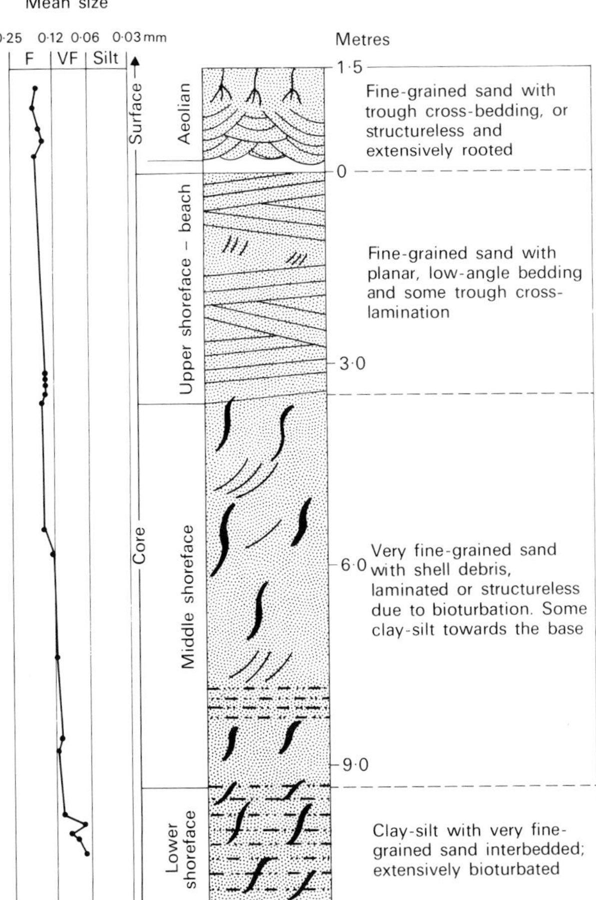

Fig. 7.30. Vertical sequence and sand body produced by seaward migration of Galveston Island, Texas (after Bernard, LeBlanc and Major, 1962).

barrier migration are reflected in a series of terraces up to 10 m high spaced across the shelf every 10–20 km. During a pause the barriers build upwards and perhaps prograde briefly until the transgression resumes (see Fig. 9.20). The ridges are therefore planed-off remnants of still-stand barriers and are probably composed of abnormal thickness of lower shoreface facies (Stahl, Kozcan and Swift, 1974; Swift, 1975a). Comparable sequences recording infilling of the lagoon by washover sands and subsequent truncation by the shoreface erosion surface occur in the Holocene succession of the Rhône delta (Oomkens, 1967, 1970). In this example a coarse-grained bioturbated lag overlies the erosion surface, but the sediment rapidly becomes fine-grained (Fig. 7.32). An almost complete vertical transgressive sequence is preserved in a barrier island along the Delaware coastline which is currently retreating landwards (Kraft, 1971; Kraft, Biggs and Halsey, 1973; Fig. 7.33). However, this sequence has accumulated prior to shoreface erosion, and will only be preserved in this state if landward migration of the barrier ceases.

An alternative mechanism of landward barrier migration, termed *in-place drowning*, has recently been revived to explain facies patterns off Fire Island, New York (Sanders and Kumar, 1975a, 1975b). In this mechanism, as sea level rises the barrier remains in place until the wave breaker zone reaches the top of the barrier. At this point the breaker zone jumps landward to the inner margin of the lagoon, thus drowning the barrier (Fig. 7.34). Under these conditions the breaker zone skips across the shelf rather than combing the entire area. Support for the operation of this mechanism off Fire Island involves a 600–800 m wide 'shoestring sand' up to 10 m thick aligned parallel to the present shoreline for 32 km. Borings on the landward side of this sand body reveal grey and brown back-barrier sands with peat beds, whilst on the seaward side laminated and bioturbated moderate wave energy shoreface sands prevail. The sand body is therefore interpreted as a barrier island preserved by in-place drowning.

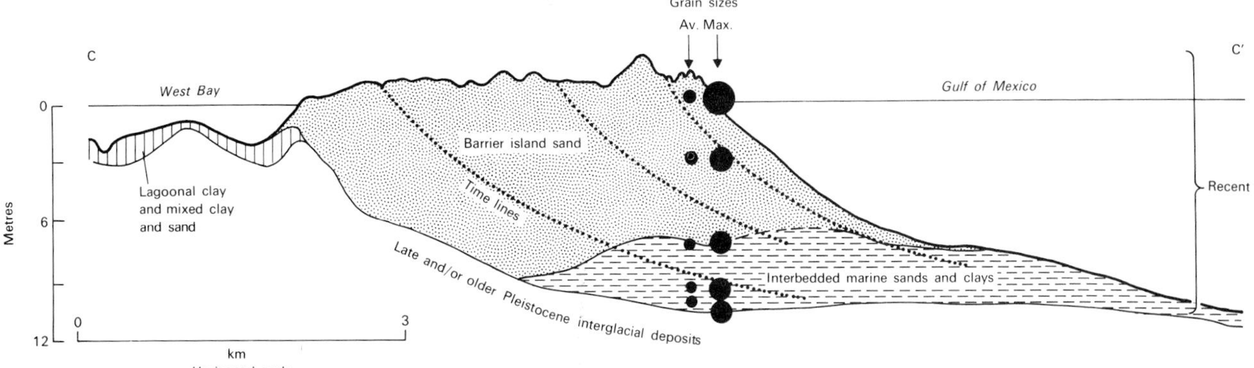

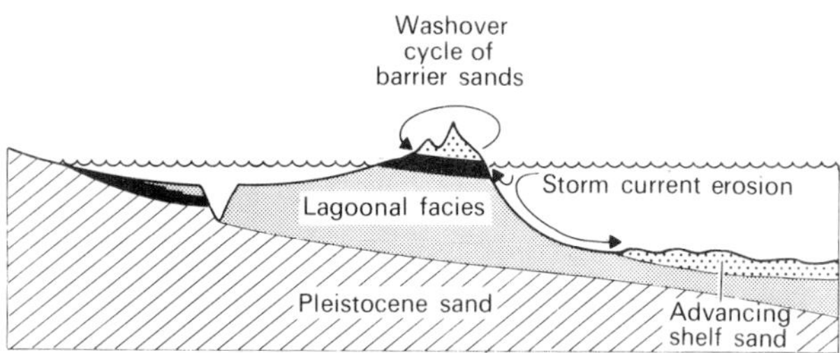

Fig. 7.31. Mechanism of landward barrier migration by shoreface erosion (after Swift, 1975a, originally after Fischer, 1961; Brunn, 1962).

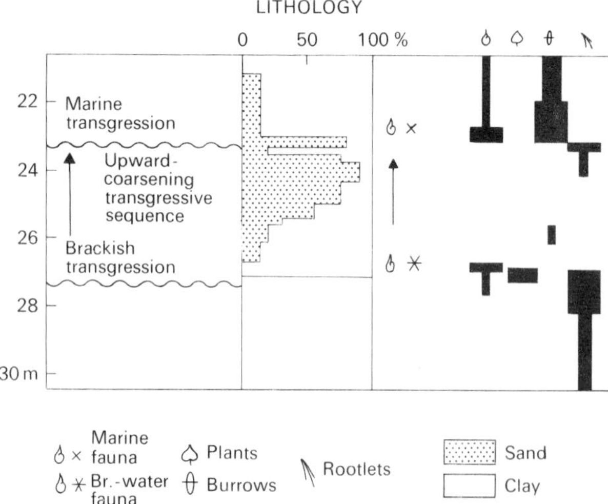

Fig. 7.32. Vertical sequence produced by landward migration of a barrier island – lagoon system illustrating the intra-sequential erosion surface (marine transgression plane) produced by shoreface erosion (after Oomkens, 1967).

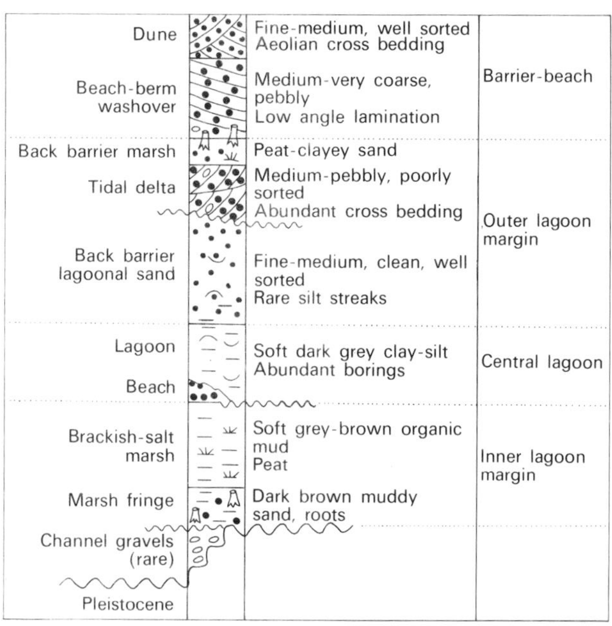

Fig. 7.33. Transgressive sequence deposited prior to shoreface erosion (after Kraft, 1971; Kraft, Biggs and Halsey, 1973).

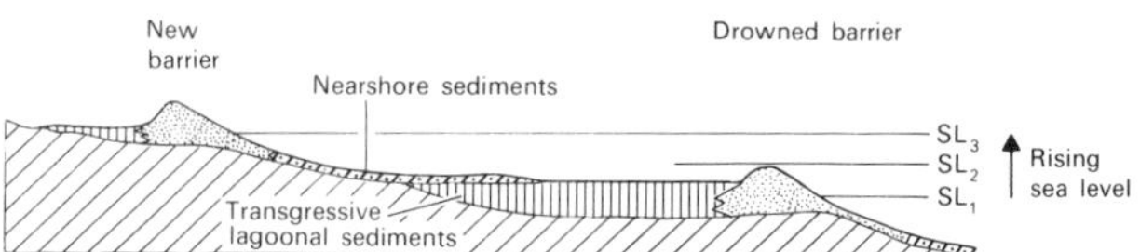

Fig. 7.34. The in-place drowning mechanism of landward barrier migration in which the wave breaker zone jumps across the shoreline to the inner lagoonal margin after surmounting the beach-barrier, thus drowning the beach-barrier (after Sandars and Kumar, 1975a).

Kumar and Sanders (1974) assert that inlet migration is only significant in barrier islands which are migrating landwards, since a lagoonal area must be maintained to accommodate the tidal waters. They imply that inlets will not be maintained during barrier progradation, but this view is oversimplified since lagoons can be maintained in prograding mesotidal barriers by tidal channels and creeks eroding the back of the barrier. The role of inlet migration during the landward migration of barrier islands is not clear, though it may be significant since the rate of inlet migration is generally many times greater than the rate of barrier retreat. If the barrier migrates by shoreface retreat a thin sheet of inlet deposits may be preserved below the shoreface erosion surface, but if the in-place drowning mechanism operates, broad channel sand bodies aligned parallel to the shoreline may be preserved intact.

7.2.7 The genesis of barrier islands

Debate on the genesis of barrier islands spans more than 125 years and involves three main rival theories. De Beaumont (1845) suggested that they result from the upward building and eventual emergence of offshore bars; whereas Gilbert (1885) argued that they result from the breaching and detachment of coast-parallel spits. At a later date, Hoyt (1967) reactivated the debate by proposing that shoreline-attached beaches are drowned to form barrier islands – the submergence hypothesis. Points for and against each theory have been discussed at length (see Schwartz, 1973, for compilation), but when applying any theory to major present-day barrier islands it must be stressed that they are considered to have originated on the outer shelf and subsequently migrated landwards to their present position.

Evidence relating to the origin of these barriers should be located on the outer shelf but has been largely destroyed during landward migration of the barriers (Otvos, 1970; Pierce and Colquhoun, 1970; Swift, 1975a; Field and Duane, 1976). There is therefore little direct evidence for the origin of these barriers, but Swift (1975a) argues that barriers along the central part of the East Coast, USA, probably formed by the submergence of mainland beaches since the coastal morphology at the low stand of sea level was conducive to this mechanism.

7.3 ANCIENT BEACH AND BARRIER ISLAND FACIES

The first criteria used in recognizing ancient beach and barrier island deposits concerned the shape, orientation and stratigraphic relationships of sandstone bodies produced in progradational systems. Many early examples were recognized in sub-surface studies as laterally extensive sandstone bodies aligned parallel to the inferred palaeo-shoreline. Subsequently it was argued that facies deposited in these systems were petrographically distinctive in comprising extremely well-sorted orthoquartzites which reflected intense winnowing and grain attrition by wave action (Ferm, 1962). Then, as the bedforms and internal structures of modern systems became known this information was applied to ancient analogues (Berg and Davies, 1968; Davies, Ethridge and Berg, 1971). All three lines of evidence are currently in use, and it is now possible to distinguish progradational sequences produced by beach faces of differing wave energy, and transgressive sequences. However, the implications of tidal inlet migration in ancient barrier island facies have hardly been considered so far.

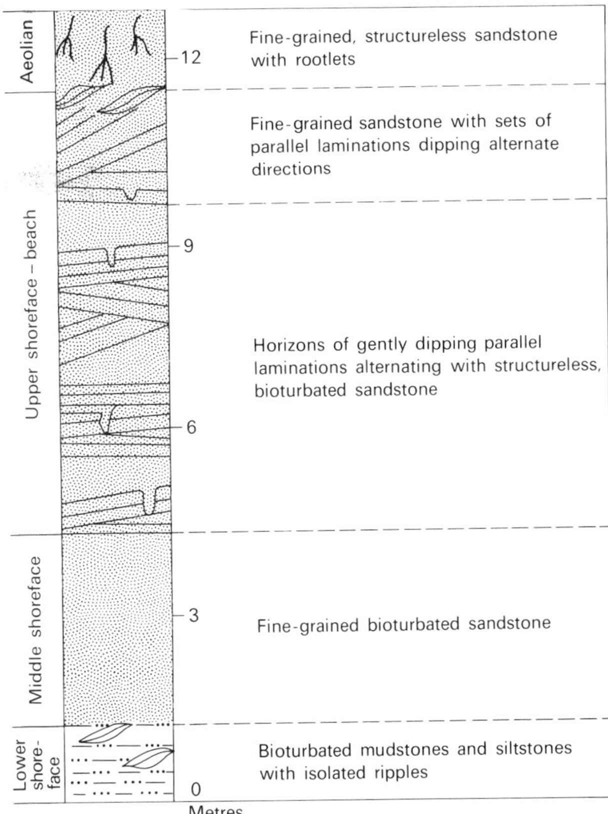

Fig. 7.35. Coarsening-upward sequence produced by barrier island progradation in the Lower Cretaceous Muddy Sandstone (after Davies, Ethridge and Berg, 1971).

7.3.1 Facies associations in progradational systems

Several progradational beach and barrier island sequences have been recognized in surface and subsurface studies in the Cretaceous epeiric seaway of the Western Interior, USA. The Lower Cretaceous Muddy Sandstone is an important oil reservoir which is only 6.5 m thick on average but is extremely persistent along depositional strike (Berg and Davies, 1968; Davies *et al.*, 1971). Facies patterns are dominated by a gradational coarsening-upward sequence which is considered to reflect a passage through the shoreface into foreshore and aeolian dune environments (Fig. 7.35). The high level of bioturbation in the lower and middle shoreface suggests that low wave energy conditions prevailed, with sediment emplaced during storms being completely reworked. Adjacent lagoonal facies generally comprise irregularly inter-bedded mudstones, siltstones and sandstones with occasional lagoon-fill sequences involving washover fan sandstones (Fig. 7.36). Upper Cretaceous rocks in Utah reveal a passage from highly bioturbated offshore siltstones upwards into parallel laminated and bioturbated sandstones identical to the facies observed off Sapelo Island and Fire Island. This shoreface facies is overlain by ripple-laminated and trough cross-bedded sandstones deposited in the wave breaker-zone, and parallel-laminated sandstones produced by swash-backwash flow on the foreshore (Howard, 1971, 1972). The Upper Cretaceous Gallup Sandstone in north-western New Mexico reveals a lateral passage through the seaward face of an ancient beach complex (Campbell, 1971). The facies are arranged in a series of offlapping imbricate units 1–6 m thick and 0.5–2.0 m wide, with each unit reflecting an episode of beach face progradation (Fig. 7.37). Collectively they form a sediment body 320 km long and 160 km wide which is comparable to the present-day Costa de Nayarit beach ridge coastal plain.

In the Carboniferous of the Pocahontas Basin (Virginia, USA), orthoquartzitic sandstone bodies produced by barrier

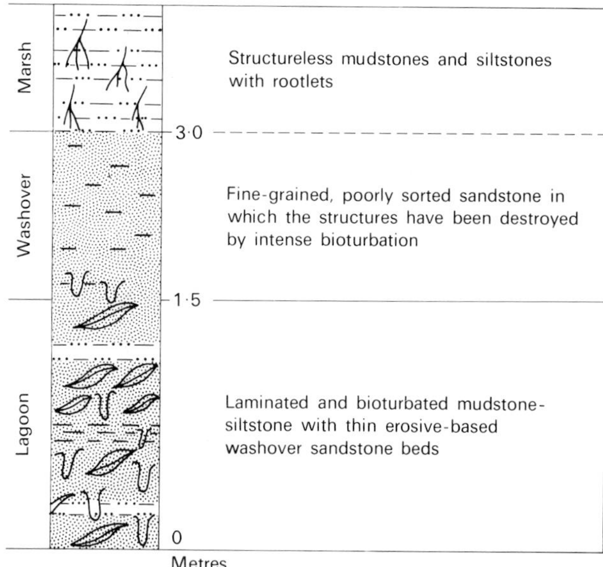

Fig. 7.36. Lagoon-fill sequence involving a washover sand unit in the Lower Cretaceous Muddy Sandstone (after Davies, Ethridge and Berg, 1971).

island migration are an important component of the succession, and judging from the descriptions of tidal inlets and related facies it seems probable that the barrier islands were active in a mesotidal setting (Horne and Ferm, 1976; Hobday and Horne, 1977). The sandstone bodies are 11–26 m thick, 2–8 km wide and several tens of kilometres long parallel to the palaeo-shoreline. Progradation of the beach face is represented by coarsening-upward sequences similar to those described above, but these sequences are often truncated by channel sandstones interpreted as tidal inlets. One section examined in detail reveals the complexities of a tidal inlet, flood- and ebb-tidal delta association (Hobday and Horne, 1977). The association is dominated by multi-storey channel sandstones in which the basal erosion surface is overlain by trough and planar cross-bedded standstones in which there is an upward decrease in set size, and upward increase in palaeocurrent dispersion. Reactivation surfaces, alternations between small-

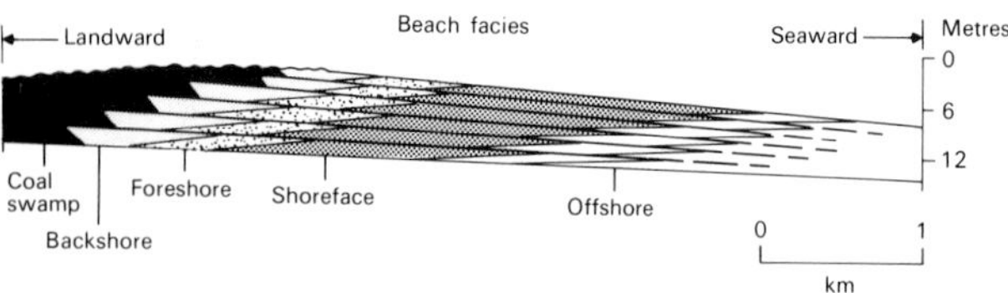

Fig. 7.37. Lateral facies relationships in a prograding beach face in the Upper Cretaceous Gallup Sandstone (after Campbell, 1971).

and medium-scale cross-bed sets and the variable palaeocurrent pattern testify to fluctuations in sediment transport due to reversing tidal currents. Occasionally, siltstones occur at the top of the channels, but in most cases this facies is cut out by erosion at the base of the overlying channel. These sequences are considered to result from the lateral migration of tidal inlets. Ebb-tidal deltas are represented by parallel-bedded sandstones which dip gently seawards and are characterized internally by parallel- and wavy-lamination. Sets of cross-bedding intercalated within the bedded sandstones are oriented perpendicular to the dip of the bedding surfaces, and there are occasional low-angle discordances in bedding. Symmetrical and asymmetrical ripples with rills occur on bedding surfaces and provide evidence of low stage run-off and emergence. Swash-backwash action on the front of the ebb-tidal delta is considered to have produced the seaward-dipping bedding surfaces, with the cross-bed sets being produced by longshore currents. The flood-tidal delta consists of wedge-shaped sandstone units 1.6–5.3 m thick separated by depressions containing laminated siltstones. Each sandstone unit includes high-angle, landward-dipping planar cross-bedding reflecting flood-oriented sand waves, and low-angle seaward-dipping parallel-lamination produced by wave swash bars during tidal ebb.

In the Carboniferous of northern England a barrier island sandstone body can be traced laterally over several hundred square kilometres using surface exposures (Elliott, 1975). Over most of the area a 10–12 m coarsening-upward sequence with low-angle beach face accretion surfaces reflects seaward progradation of an intermediate wave energy beach face (Fig. 7.38). However, at the southern end of the sandstone body, offshore mudstones and siltstones are erosively overlain by laminated and bioturbated sandstones identical with the facies observed off Sapelo Island and Fire Island (see Sect. 7.2.2.). The erosion surface at the base of this facies implies that the barrier was migrating landwards, though the thickness of the shoreface sands (7.5 m) suggests that landward migration was followed by a period of barrier still-stand. Support for the idea of landward migration is provided by the lagoon facies in this area which include washover sandstone units characterized by low-angle cross-bedding and flat-lamination. It appears that for a period the tip of the barrier was migrating landwards by shoreface erosion and washover fan sedimentation (? due to limited sediment supply), whilst the central part of the barrier was prograding seawards. The absence of tidal inlet deposits implies that microtidal conditions prevailed in the area.

The majority of beach and barrier island facies recognized so far are composed of fine- to medium-grained sandstones, but a coarser-grained beach progradational sequence directly analogous to the high wave energy beach face of the Oregon coast has been described from the Quaternary of California (Clifton *et al.*, 1971; Fig. 7.39).

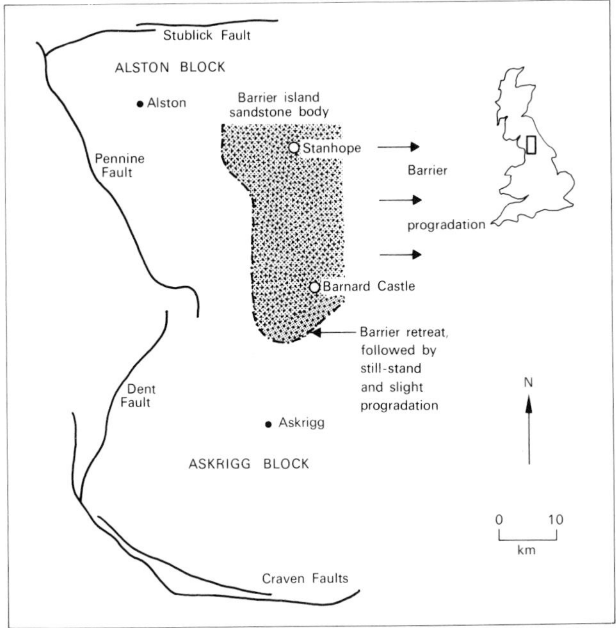

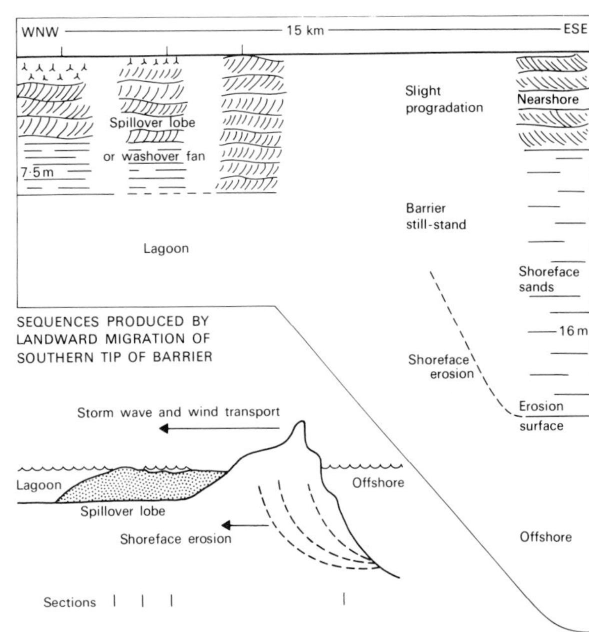

Fig. 7.38. Location map and vertical sequences of a barrier island sandstone body from the Upper Carboniferous of northern England (after Elliott, 1975).

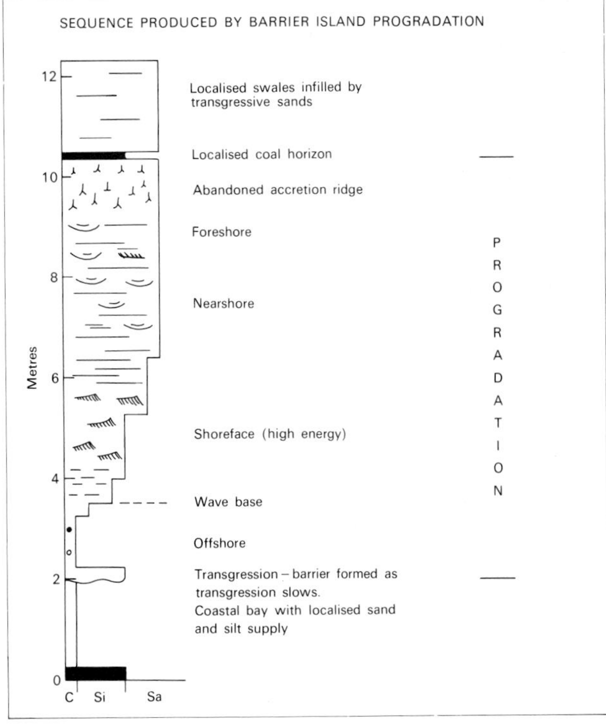

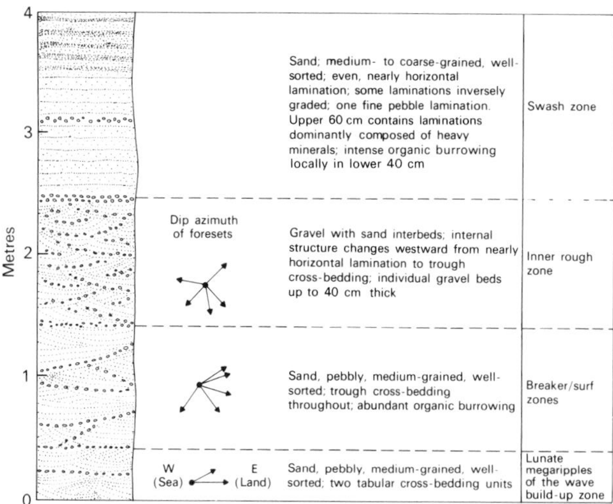

Fig. 7.39. Coarse-grained, high wave energy upper beach face facies from Quaternary of California (after Clifton, Hunter and Phillips, 1971).

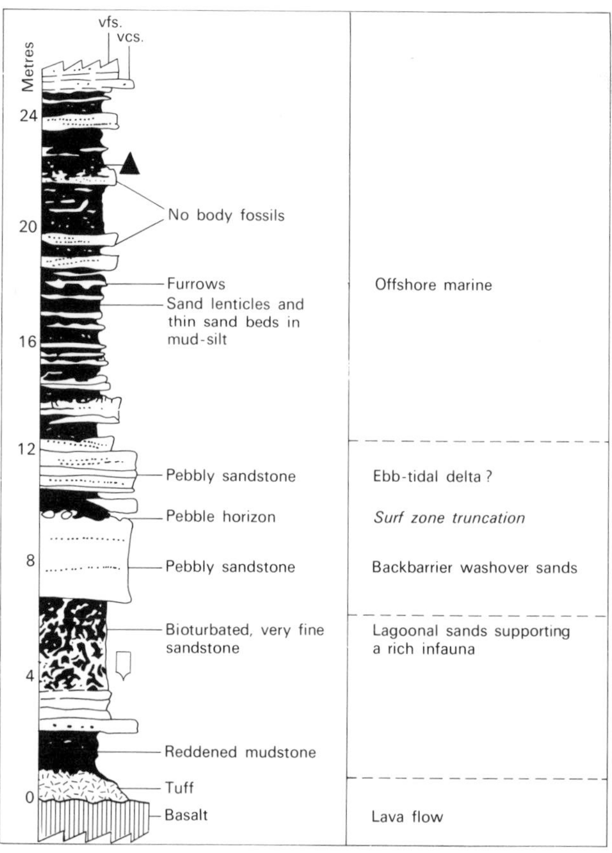

Fig. 7.40. Vertical sequence formed by the landward migration of a barrier island in the Lower Silurian of south-west Wales (after Bridges, 1976).

7.3.2 Facies associations in transgressive systems

Very few transgressive barriers have been recognized in the geological record so far, possibly because they tend to produce relatively thin sequences. However, several examples occur in the Lower Silurian of south-west Wales which was deposited during a eustatic rise in sea level (Bridges, 1976). The sequences commence with lagoonal facies which often transgress across the subaerial surface of lava flows (Fig. 7.40). Lagoonal facies consist mainly of intensely bioturbated, fine-grained sediments with a low diversity fauna. Thin sandstone beds set in this facies are interpreted as washover fan or flood-tidal delta deposits, and 1–3 m thick fining-upward units are considered to reflect the infilling of washover channels. Sandstones and conglomerates 0.15–8.0 m thick which abruptly overlie the lagoonal facies are characterized by sub-horizontal laminations, large symmetrical ripples and concentrated lenses of granules and pebbles which are considered to indicate wave action. Some of these facies are interpreted as proximal washover fans and others as remnants of transgressed barriers. Reworked pebble lags at the top of these sandstones reflect the passage of the surf zone across the area, and these surfaces are overlain by bioturbated mudstones and sandstones with a diverse, open marine fauna. Four transgressive sequences are recognized but each differs, reflecting the morphological complexity of the former shoreline.

In the Pre-Cambrian Ekkerøy Formation in Finnmark, north Norway, a thin (3.0–3.5 m) wave-dominated succession is attributed to the landward migration of a beach (Johnson, 1975). The succession can be traced laterally for 6 km and exhibits the general sequence: planar erosion surface→coarse to medium-grained sandstones with seaward-dipping parallel lamination and isolated sets of cross-bedding, produced by swash-backwash action on the beach face→medium-grained sandstone with predominantly landward-dipping planar and trough cross-bedding reflecting the landward migration of dunes→fine- to medium-grained sandstone with asymmetrical and symmetrical wave ripples. The basal erosion surface, the upwards decrease in wave energy and sediment grain size, and the lateral uniformity of the sequence suggest that it formed as a result of the landward migration of a beach during a rise in sea level. Furthermore, the facies bear a close resemblance to those of the Oregon Coast, suggesting that the beach face was characterized by high wave energy conditions.

A lagoon-fill sequence related to landward barrier migration during a transgression occurs in the Pleistocene of Zululand (Hobday and Orme, 1974). The sequence commences with lagoonal mudstones and sandstones overlain by a lignite bed deposited at the outer (seaward) lagoonal margin. A 15 m sandstone member with sets of steeply dipping cross-bedding up to 9 m thick directly overlies the lignite bed. The sandstone is interpreted as a major washover fan, with the large-scale foresets reflecting exceptionally high wave energy conditions.

7.4 CHENIERS

Cheniers are isolated sand or shell-debris ridges set in a marshy coastal plain. The classic examples occur downdrift from the Mississippi delta where individual cheniers are up to 3 m high, 300–1,000 m wide, and 50 km long (Russell and Howe, 1935; Price, 1955; Byrne, LeRoy and Riley, 1959; Hoyt, 1969). They are generally slightly curved, with smooth seaward margins but ragged landward margins due to washovers. In cross-section they are bi-convex and generally become more subdued inland. The cheniers bifurcate and fan towards estuary mouths, forming slightly cuspate features (Fig. 7.41). In this respect chenier complexes appear superficially similar to certain wave-influenced deltas, but they are distinguished by the alternation of chenier ridges with intervening mud-silt marsh areas, and the *upstream* deflection of chenier ridges at the estuary channel margins. Away from the estuaries coalesced cheniers form composite bodies up to 15 km wide.

Conditions conducive to the formation of cheniers include low wave energy, low tidal range, effective longshore currents and a variable supply of predominantly fine-grained sediment. During periods of high sediment supply longshore currents are incapable of sorting the sediment and deposition results in poorly-sorted mudflats which prograde seawards. However, if

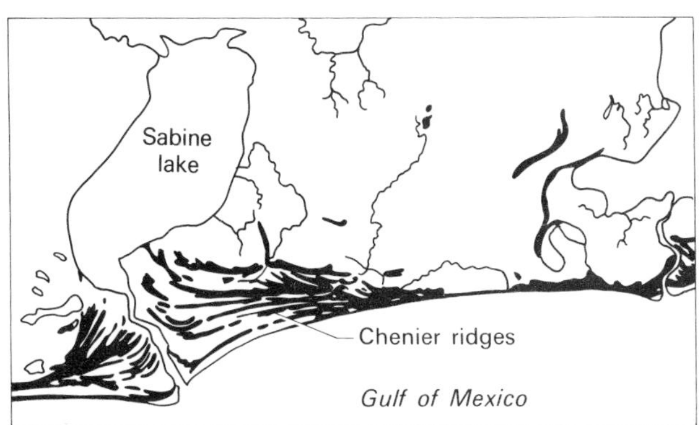

Fig. 7.41. The cuspate arrangement of chenier ridges around estuaries (modified after Gould and McFarlan, 1959).

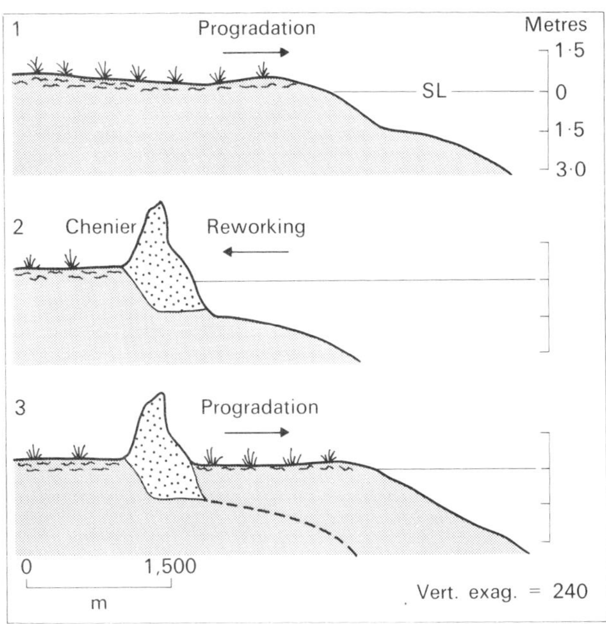

Fig. 7.42. Chenier formation by means of alternating mud flat progradation and wave reworking (after Hoyt, 1969).

sediment supply diminishes the longshore currents and waves erode and winnow the mudflat sediments into chenier ridges (Fig. 7.42). The shoreline therefore progrades in a step-wise fashion, and this is reflected in three distinct types of shoreline which exist in the chenier complex (Beall, 1968). Mudflat shorelines protected by inner shoreface breaker bars prevail during periods of high sediment supply, but are replaced during the early stages of chenier formation by a transitional shoreline composed of eroded mudflat and marsh sediments. Wave reworking in periods of reduced sediment supply produces a beach shoreline with breaker bars and well developed washover deposits which eventually constitutes the chenier ridge. The fan-shaped form of chenier complexes around estuaries is attributed to 'dynamic diversion' caused by tidal flows cutting across the longshore currents and forming an obstacle to sediment transport (Todd, 1968). Periods of chenier formation are considered to relate to switching of the Mississippi delta: when the active delta was located in the east longshore sediment supply was diminished and cheniers formed; when the delta was located farther west increased sediment supply resulted in mudflat progradation (Gould and McFarlan, 1959).

Cheniers and mudflats are also developed along the coastline of Surinam and Guyana, fed in this case predominantly from the Amazon mouth (Brouwer, 1953; Vann, 1959). Chenier ridges composed of approximately equal proportions of sand

and shell-debris occur in an area of tidal flats north of the Thames estuary (Greensmith and Tucker, 1968, 1975). The cheniers are asymmetrical, lenticular bodies which are crudely layered and cross-bedded. They project 0.5–3.0 m above the adjacent tidal flat and are up to 25 m wide. Supply to the chenier system is considered to be from immediately offshore rather than alongshore, and the cheniers reflect a complex history of shoreline progradation and retreat related to sea level changes.

Concerning chenier facies, the Louisiana coastal plain cheniers produce thin, but persistent, sand bodies set in fine-grained marsh-mudflat facies (Fisk, 1959). Boreholes reveal that chenier sands transitionally overlie gulf-bottom sands and silty clays, forming small-scale coarsening-upward sequences capped by a soil horizon and marsh facies. Since the end of the Holocene transgression the chenier plain has prograded seawards by 15 km, producing a substantial sediment wedge (Byrne *et al.*, 1959). The most reliable criterion for recognizing cheniers in the ancient record is the presence of extremely linear sandstone bodies isolated in mudstone-siltstone-marsh facies. For this reason it seems, ancient examples are confined to several subsurface studies which recognize them as elongate, bi-convex, shoestring sandstones (Fisk, 1955; Byrne *et al.*, 1959; Fisher, Proctor, Galloway and Nagle, 1970).

7.5 MODERN ESTUARIES

An estuary is a physically complex environment defined by Pritchard (1967) as 'a semi-enclosed coastal body of water which has a free connection with the open sea and within which sea water is measurably diluted with fresh water derived from land drainage'. Present-day examples occur in association with tidal flats, barrier islands and river deltas, and fall into two distinct categories: those which are simply the tidally influenced lower stretches of rivers, and those formed by the drowning of river or glacial valleys during the Holocene transgression. Tidal processes are of the utmost importance, and two points concerning physical processes in estuaries deserve emphasis. Firstly, as fresh water and salt water are of differing density, a variety of circulation patterns develops depending on the manner in which the water bodies mix (Pritchard, 1955; Cameron and Pritchard, 1963). In virtually tideless areas, the less dense fresh water extends over a salt wedge as a distinct layer and the water column is highly stratified in a vertical sense. However, as tidal processes become more significant they induce turbulent mixing of the water bodies until eventually the water column is homogenous. Secondly, the properties of the tidal regime and the manner in which these properties change along the estuary are crucial to sediment transport and ultimately facies characteristics (Postma, 1967; Wright, Coleman and Thom, 1973, 1976; Boothroyd and Hubbard, 1976). Important properties of the tidal regime include the maximum flood- and ebb-tidal velocities at different points in the estuary, the difference between flood- and ebb-velocity maxima, and the time span during which a given velocity is maintained. The single most important feature of the tidal regime is the time/velocity relationship of ebb- and flood-tidal flows. Tidal currents are regarded as symmetrical if slack water periods occur at low and high tide, and maximum flood- and ebb-tidal velocities occur at the mid-point of flood and ebb periods. More commonly, however, tidal currents are asymmetrical as slack water does not occur at low and high tide, and maximum velocities are offset from the mid-points. Maximum ebb currents often occur late in the ebb period near low water and are therefore concentrated in deep ebb channels, whereas flood currents are offset towards high tide and affect topographically higher areas.

Hayes (1976) distinguishes microtidal, mesotidal and macrotidal estuaries, but uses an extremely broad definition of estuaries which results in the microtidal estuaries being lagoons whereas the mesotidal estuaries are indistinguishable from tidal inlets. Only the macrotidal estuaries accord with Pritchard's definition of an estuary. Nevertheless, tidal range is an important control on estuarine sedimentation and the following account is a modified version of Hayes' scheme.

MICROTIDAL, MESOTIDAL AND MACROTIDAL ESTUARIES

This section strives to concentrate on estuaries created by tidal waters entering the lower parts of river channels and influencing sedimentation as this type of estuary exhibits a relationship with tidal range, being preferentially developed in areas of high tidal range (see Fig. 7.1). Drowned river- or glacial-valley estuaries developed as a result of the Holocene transgression are therefore omitted as they develop indiscriminately with respect to tidal range, though it has to be admitted that there is a 'grey' area composed of estuaries initially developed by drowning but now adjusted to the present hydraulic setting.

In microtidal areas the development of estuaries is often inhibited by the ability of river discharge to prevent tidal waters from entering the channel. However, low discharge rivers which empty into microtidal seas may develop estuarine characteristics, and tidal waters may enter low to moderate discharge rivers during low river stage and temporarily create estuarine conditions. These microtidal estuaries are characterized by a highly stratified circulation pattern with a well-developed salt wedge, and the distributary channels of certain river deltas are low stage, ephemeral examples (Sect. 6.5.1).

In mesotidal areas, tidal processes are more effective and

may significantly influence sediment transport in the lower parts of river courses. The estuaries are characterized by a partially or completely mixed circulation system, and comprise a broad funnel-shaped mouth which passes upstream into an area of tidally influenced meandering channels before reaching the fluvial-dominated river channel. The presence of the latter is the main distinguishing feature between mesotidal estuaries and tidal inlets. Prominent depositional features include flood- and ebb-tidal deltas, transverse bars and spill-over lobes, ornamented by a diverse assemblage of bedforms. Discrete flood- and ebb-tidal channels dissect these features and frequently exhibit sandwaves with superimposed mega-ripples (Boothroyd and Hubbard, 1976).

Macrotidal estuaries are also characterized by a funnel-shaped form headed by a tidally influenced meandering channel, but in this case the predominant depositional features in the main part of the estuary are linear tidal sand ridges (Fig. 7.43). Tidal currents induce complete mixing of the water bodies and the circulation system is therefore vertically homogenous. The Ord River, western Australia, occurs in a deltaic setting, but illustrates the main features of this type of estuary (Wright *et al.*, 1973, 1976). Mean tidal range is 4.3 m (spring 5.9 m) and the tidal prism exceeds river discharge throughout the lower 65 km of the channel. Suspended sediments are deposited on intertidal flats at the margins of the estuary, whilst bedload sediments are moulded into quasi-symmetrical tidal sand ridges 10–22 m high, up to 2,000 m long and 300 m wide. Smaller scale bedforms superimposed on the crests of the sand ridges often have opposed directions on the different parts of the same ridge, suggesting that the ridges form in areas of bedload convergence (Sect. 9.5.3). At the estuary

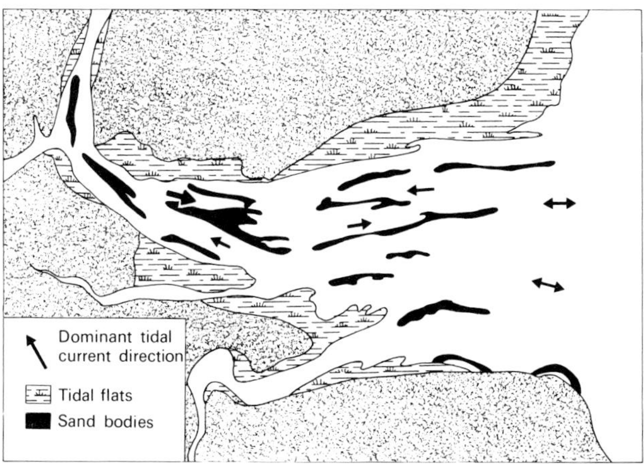

Fig. 7.43. Generalized form of a macrotidal estuary, characterized by a funnel-shaped channel, with elongate tidal current ridges in the centre of the channel flanked by tidal flats at the channel margins (after Hayes, 1976).

mouth, flood- and ebb-currents are of approximately equal magnitude and sediment transport is therefore bi-directional. Part of this bi-directional transport is accomplished by the reversal of medium- and small-scale bedforms, but of greater significance is the fact that flood- and ebb-dominated bedform migration takes place along closely spaced but separate courses. Farther upstream, though still in the funnel-shaped part of the estuary, the tidal wave becomes asymmetrical and the velocity of flood-tidal currents increases relative to the ebb currents. Strongly asymmetrical, flood-oriented megaripples 30–100 cm high retain their orientation during ebb flow, but smaller scale ripples are often reversed. In the upstream area of meandering channels, flood dominance becomes more pronounced and point bars are covered by flood-oriented medium-scale bedforms.

7.5.1 Estuarine facies

There are few comprehensive descriptions of modern estuarine facies, but with the aid of box cores, relief casts and X-ray radiographs, sedimentary facies have recently been described in mesotidal estuaries along the East Coast, USA (Howard and Frey, 1975a, 1975b). Ossabaw Sound dissects a series of barrier islands and comprises a complex of major bedforms and flood- and ebb-tidal channels which extends 3 km offshore before terminating in an ebb-tidal delta (Greer, 1975). Seaward progradation of the inlet shoal produces a coarsening-upward sequence in which bioturbated shelf sands are overlain by well-sorted sands with trough cross-bedding, ripple lamination and parallel-lamination, produced by an interaction between tidal and wave processes. This sequence may be truncated by the estuary channel as progradation continues and the erosion surface is overlain by a variable fining-upwards sequence with the general trend: lag deposit→tidally-dominated channel facies→tidal flat facies (Fig. 7.44). The tidally influenced meandering stretch of Ogeechee River, Ossabaw Sound, is characterized by point bars dominated by trough cross-bedded coarse sand and ripple-laminated finer sands in the downstream area, and interbedded fine sands and muds with wavy and flaser bedding in the upstream area (Howard, Remmer and Jewitt, 1975; Fig. 7.45). These channels migrate laterally and are considered to dominate the depositional record beneath extensive salt marches bordering the channels. If progradation continues beyond the point depicted in Fig. 7.44, the upper part of the estuarine sequence may be reworked by smaller-scale, meandering estuarine channels.

Excavations in one of the major Dutch estuaries, the Haringvliet, provided an unparalleled opportunity to study sub-tidal estuarine facies in the overall setting of the Rhine delta (Oomkens and Terwindt, 1960; de Raaf and Boersma, 1971; Terwindt, 1971b; see Sect. 6.5.1). Three facies are recognized: (1) coarse-grained sand with opposed sets of large-

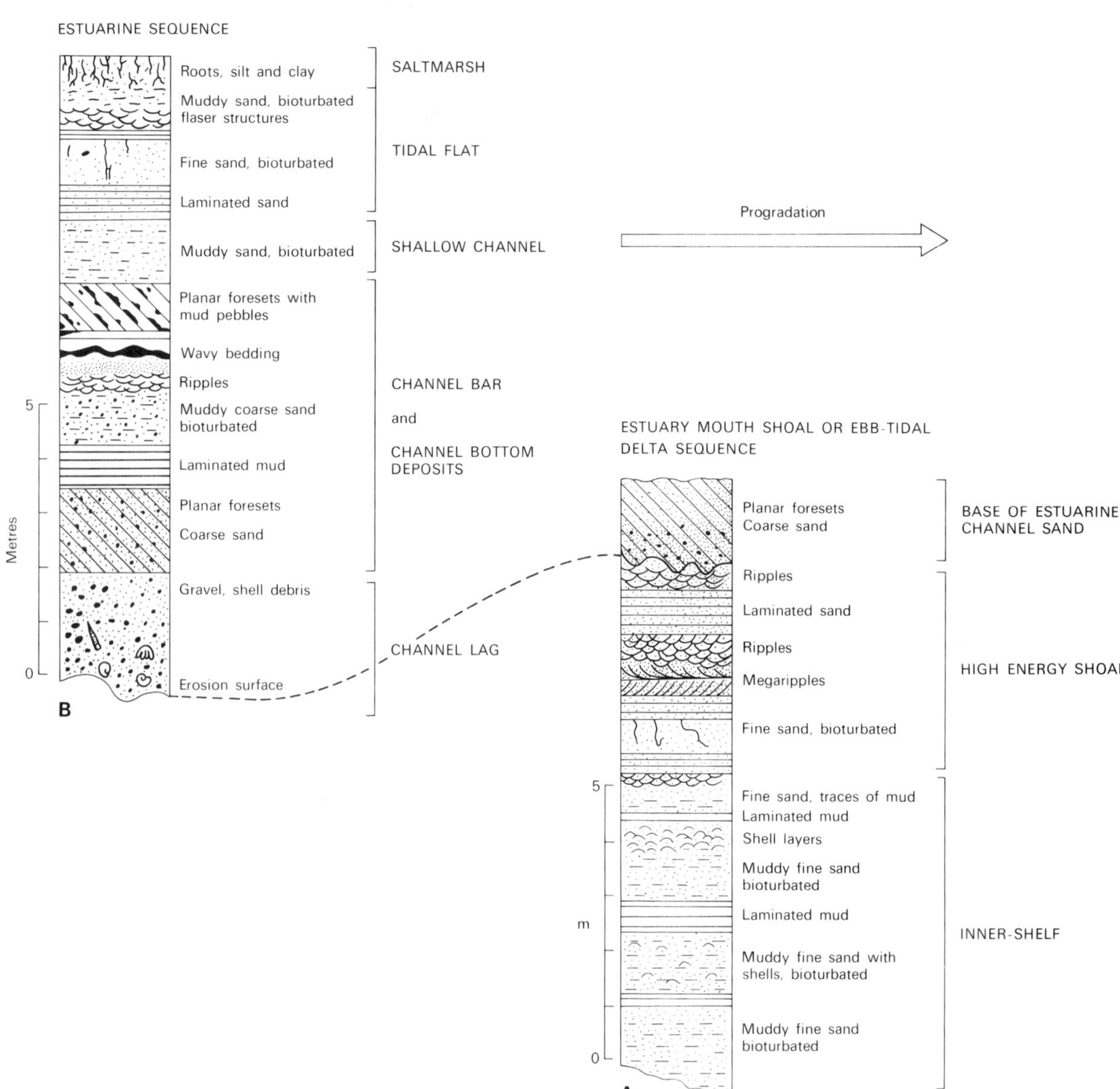

Fig. 7.44. Sequence produced by the progradation of an ebb tidal delta (A) and estuarine channel (B); Ossabaw Sound, Georgia Coast, USA (after Greer, 1975).

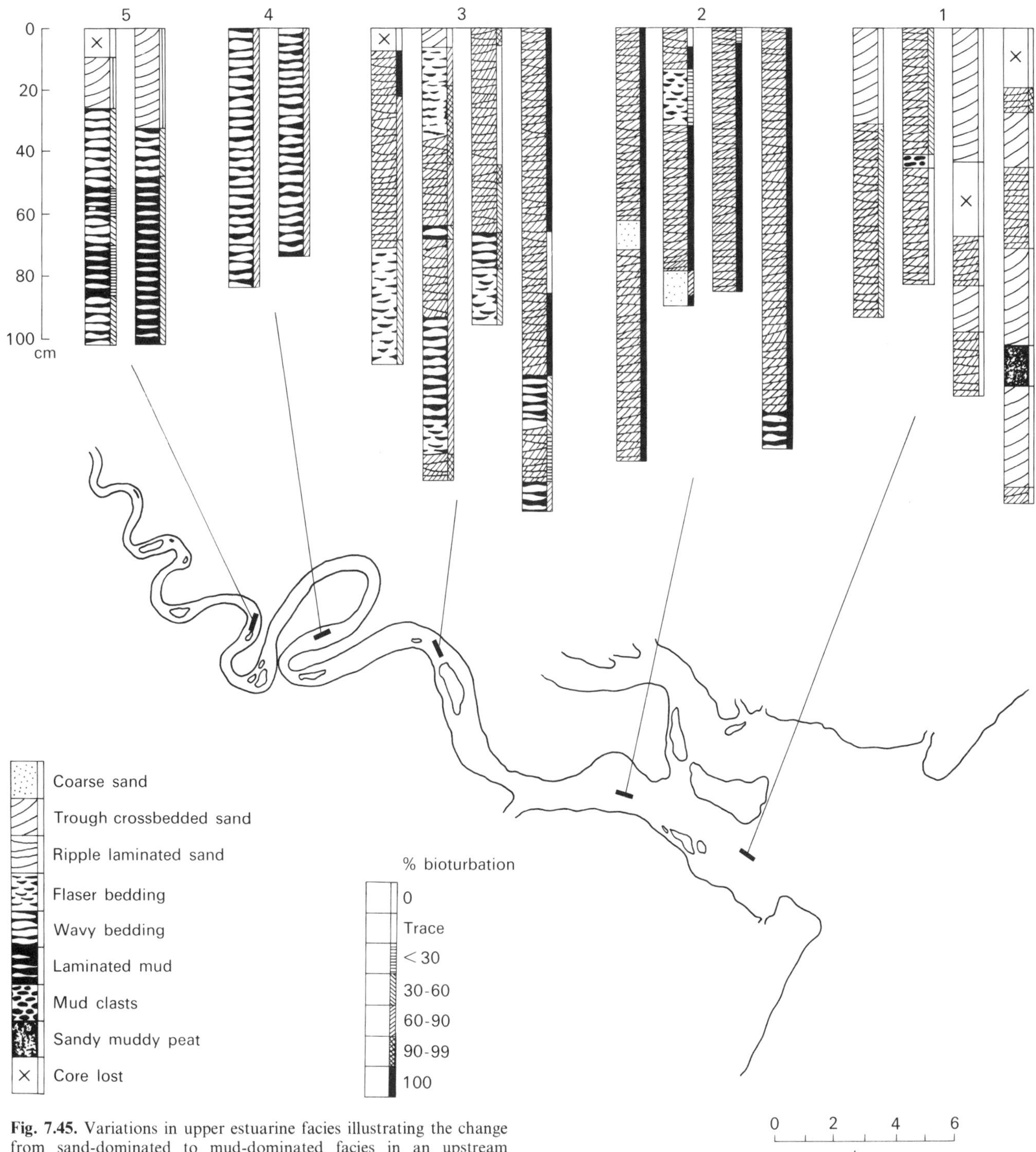

Fig. 7.45. Variations in upper estuarine facies illustrating the change from sand-dominated to mud-dominated facies in an upstream direction (after Howard, Elders and Heinbokel, 1975).

scale trough or planar cross-bedding in which the foresets are occasionally draped by thin silt layers; (2) medium- to fine-grained sand exhibiting flat-lamination, ripple-lamination and flaser bedding; (3) fine-grained sands and clays with lenticular bedding produced by ripple- and parallel-laminated sand beds being draped by thin mud layers (Terwindt, 1971b). Marked lateral and vertical facies changes characterize the facies pattern and reflect the variable conditions imparted by tidal processes. Directional bimodality and repeated small-scale changes from bedload transport to deposition of fine sediment from suspension are the most diagnostic features to emerge from these studies, though the former is not an ubiquitous feature.

7.6 MODERN TIDAL FLATS

Tidal flats occur in low wave energy, mesotidal and macrotidal settings where they dominate extensive stretches of shoreline, or form within coastal embayments, lagoons, estuaries and tidally-influenced deltas. Intensively studied examples occur in north Germany and Holland, the Wash (eastern England), the Bay of Fundy and the Gulf of California (van Straaten, 1954, 1961; Klein, 1963, 1967b; Evans, 1965, 1975; Reineck, 1963, 1967; Thompson, 1968, 1975).

In general, tidal flats comprise almost featureless plains dissected by a network of tidal channels and creeks. During flood period, tidal waters enter the channels, overtop the channel banks and inundate the adjacent flats. Following a period of still-stand the tidal waters drain via the channels and re-expose the flats. The *supratidal zone* lies above mean high tide level and is therefore sensitive to climate. In temperate areas salt marshes cover this area and accumulate interlaminated clays and silts in which the laminae are extensively disrupted by bioturbation, rootlets and nodule development (Reineck, 1967). In arid to semi-arid areas, desiccated mudflats with evaporites prevail. Bioturbation is minimal, but the growth of gypsum and halite crystals, supplemented by shrinkage of the muds, disrupts the laminations and produces a porous mud-evaporite mixture (Thompson, 1968, 1975). The *intertidal zone* comprises smooth, seaward-dipping flats dissected by large- and small-scale tidal channels. In general, intertidal flats range from mud-dominated near the high-water mark to sand-dominated near the low-water mark (van Straaten, 1954, 1961) though there are often departures from this overall trend. In the Wash embayment, for example, the range of subenvironments from high to low tide level is: salt marsh→higher mud flats→inner sand flats→*Arenicola* sand flats→low mud flats→lower sand flats (Evans, 1965; Fig. 7.46). Relatively weak tidal currents and waves interact on sand-dominated intertidal flats to produce extensive areas of asymmetrical and symmetrical

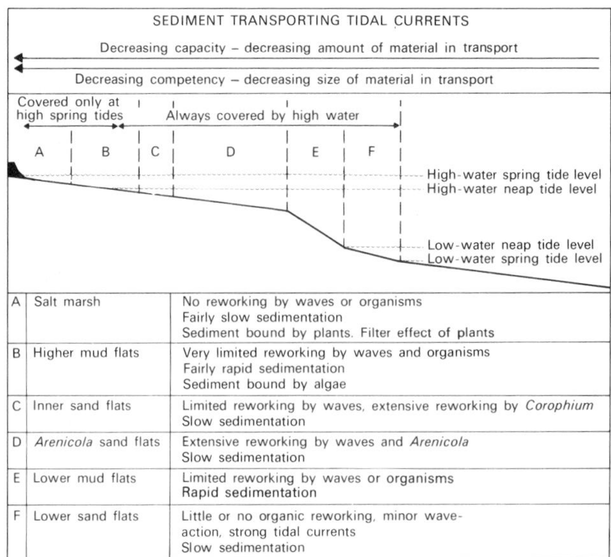

A	Salt marsh	No reworking by waves or organisms Fairly slow sedimentation Sediment bound by plants. Filter effect of plants
B	Higher mud flats	Very limited reworking by waves and organisms Fairly rapid sedimentation Sediment bound by algae
C	Inner sand flats	Limited reworking by waves, extensive reworking by *Corophium* Slow sedimentation
D	*Arenicola* sand flats	Extensive reworking by waves and *Arenicola* Slow sedimentation
E	Lower mud flats	Limited reworking by waves or organisms Rapid sedimentation
F	Lower sand flats	Little or no organic reworking, minor wave-action, strong tidal currents Slow sedimentation

Fig. 7.46. Facies belts in tidal flats of the Wash embayment, eastern England (after Evans, 1965).

ripples, often with complex interference patterns. Intertidal flat facies are dominated by interlaminated clays, silts and sands exhibiting prolific flaser, wavy and lenticular bedding (Fig. 7.47). These facies reflect constantly fluctuating but relatively low energy conditions, with brief periods of sand and coarse silt bedload transport by tidal currents and waves alternating with fine sediment deposition from suspension (Reineck and Wunderlich, 1968). Faunal zonation and fluctuating levels of faunal diversity across the intertidal flats affect the type of bioturbation in the sediments, and in general the intensity of bioturbation is inversely related to the rate of sediment deposition.

The tidal channels often have highly sinuous channel patterns with laterally migrating point bars on the inner depositional bank. In the Solway Firth (Scotland), point bars in gently curved meanders have a uniform sigmoidal profile, whereas tightly curved meanders have two-tier point bar profiles with a platform produced by flow separation around the meander bend (Bridges and Leeder, 1976). Channel migration is most pronounced during late ebb-tidal flow and often occurs at a rate of several tens of metres per year. Wetted channel bank sediments are exposed during low water periods and blocks of sediment slump into the channel along rotational slip planes. This process contributes substantially to channel bank erosion and can also produce rotated blocks of sediment which are preserved in point bar facies. Relatively coarse sands with shell debris and abundant mud clasts floor the channels and form a basal lag to the channel sequence. The point bar is

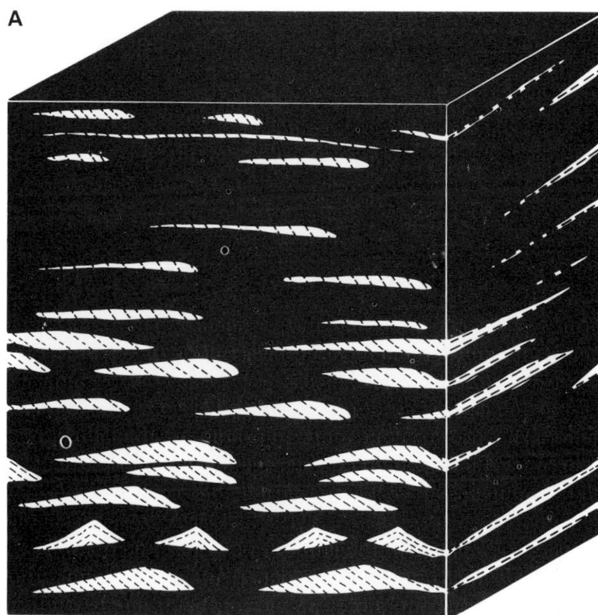

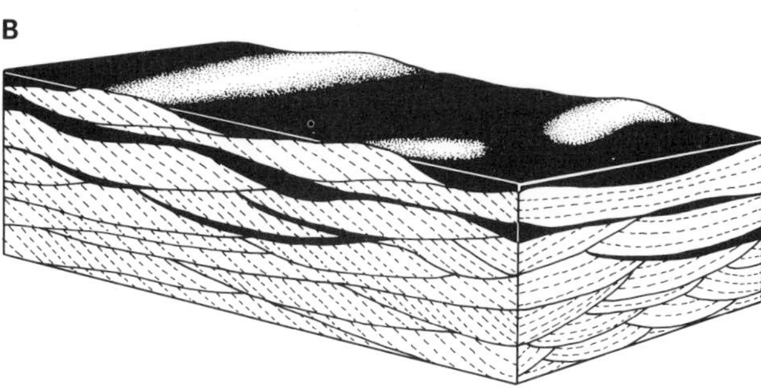

Fig. 7.47. (A) Lenticular bedding with thin sand lenses and isolated ripple form sets in mud-silt; (B) flaser bedding composed of mud-draped ripples (after Reineck and Wunderlich, 1968; Reineck and Singh, 1973).

Fig. 7.48. Lateral accretion bedding in alternating sands and mud-silts developed by point bar migration in meandering tidal channels (after Reineck, 1967).

composed of thin, interlaminated clay-silt and sand beds characterized by lateral accretion bedding dipping gently into the channel (van Straaten, 1954; Reineck, 1958; Fig. 7.48). Sections through point bar facies often reveal inclined erosion surfaces which are considered to reflect scouring of the point bar surface during exceptional discharge conditions (Bridges and Leeder, 1976). Depending on channel spacing and the rate and direction of channel migration, a large part of the intertidal flats may be underlain by a sheet of point bar sediments with the sequence: basal lag→interlaminated clay→silt and sand with lateral accretion surfaces→intertidal flat facies. The intertidal flats may pass into the offshore area with no major features, but in sand-dominated areas such as the German Bay and Bay of Fundy the subtidal zone may include a complex of channels, bars and shoals (Sect. 9.5). In the German Bay the channels are deep, broad, funnel-shaped extensions of estuaries with sand-wave, megaripple and ripple bedforms (Reineck, 1963; Reineck and Singh, 1973). Sand shoals or bars exist at the mouths of the channels and between the channels, creating a sand-dominated depositional province which extends over several hundred square kilometres (Fig. 7.49). Medium- and large-scale cross-bedding are common features of the channel facies, whereas the shoals and bars are characterized by finer-grained sands dominated by ripple lamination and wave-induced flat lamination.

Progradation of tidal flats tends to produce a fining-upward sequence which reflects a transition from low tide level sand flats, upward into high tide level mudflats, and eventually into supratidal flats. This sequence may be cut at any level, and in some cases completely replaced, by erosive-based tidal channel sequences (Reineck, 1967, 1972; Mackenzie, 1972; Evans, 1975). However, tidal flats in the Gulf of California are dominated by fine-grained sediments and the facies changes from subtidal to supratidal therefore occur within a silt-clay sequence with no fining-upward trend (Thompson, 1968, 1975).

7.7 ANCIENT ESTUARINE AND TIDAL FLAT FACIES ASSOCIATIONS

Criteria for the recognition of tidal processes in ancient sedimentary successions discussed in connection with tidally influenced shallow marine facies (Sect. 9.10) apply equally to estuarine and tidal flat facies (de Raaf and Boersma, 1971; Klein, 1971). They include: (1) a close spatial and temporal association of current-formed structures indicating bi-

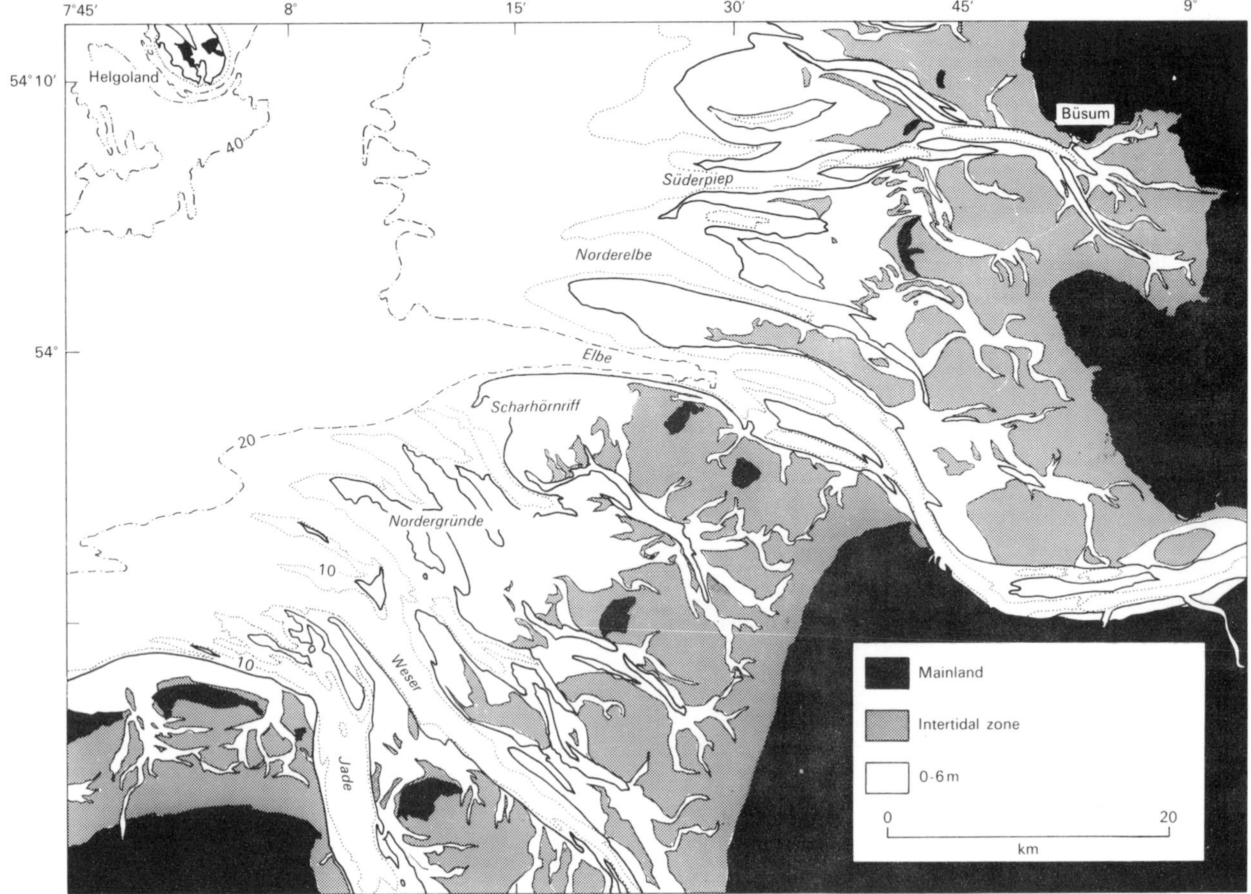

Fig. 7.49. Intertidal and subtidal flats dissected by estuaries, German Bay, North Sea (after Reineck and Singh, 1973).

directional flow; (2) an abundance of reactivation surfaces; and (3) the presence of facies which reflect small-scale, repeated alternations in sediment transport conditions. Additional criteria which apply only to intertidal settings include indications of repeated emergence and the presence of surface run-off features formed as the tide falls (Klein, 1971).

There are relatively few examples of ancient estuarine facies associations. In Holland, several Pleistocene sequences are interpreted as subtidal estuarine channels (de Raaf and Boersma, 1971). In general, the sequences comprise a basal erosion surface overlain by a thin, intraformational conglomerate and trough cross-bedded sands with bimodal palaeocurrents. Cross-bed foresets are draped by clay laminae and interlaminated clay-silts, suggesting that bedform migration was intermittent in response to tidal current fluctuations. Finer grained linsen- and flaser-bedded facies succeed the channel sands and also exhibit bimodal palaeocurrents. In this example, discrimination between estuarine and tidal inlet interpretations is tentatively based on the seaward-directed palaeocurrent mode being attributed to fluvial outflow rather than ebb-tidal flow, and is supported by the close proximity of fluviatile facies.

Lateral and vertical facies relationships in part of the Tertiary Lower Bagshot Beds of southern England have also been interpreted as an estuarine association (Bosence, 1976). Large channel-like scours with thick intraformational conglomerates grade upward into flat-bedded and planar cross-bedded sands, with silt drapes on some of the foresets. Individual channel-fills have unidirectional palaeocurrents, but the overall directions are bimodal. Intercalated between the channel sands are erosive-based sheet-like units of interlaminated silts and fine ripple-laminated sands which frequently exhibit flaser and lenticular bedding. Bioturbation is particularly prominent below erosion surfaces and includes *Ophiomorpha nodosa* which is considered to indicate inshore marine conditions. The channel sands are interpreted as the relatively deep parts of estuarine channels, whereas the sheet-like units of linsen and flaser facies represent laterally accreting point bar facies higher on the inner depositional bank.

In contrast, numerous examples of tidal flat facies associations have been recognized in successions ranging in age from Pre-Cambrian to Recent (Klein, 1970a, 1971; Kuipers, 1971; Mackenzie, 1972, 1975; Sellwood, 1972b, 1975; Johnson, 1975; Rust, 1977; Tankard and Hobday, 1977). In the majority of examples recognition is based on an assemblage of structures and facies considered to indicate tidal processes, and a fining-upward sequence which reflects tidal flat progradation. In the Lower Jurassic of Bornholm (Sellwood, 1972, 1975), the sequence commences with fine to medium-grained sandstones with bimodal planar and tabular cross-bedding characterized by reactivation surfaces and clay drapes (? subtidal-low sand flat). These sandstones pass vertically and laterally into a facies composed of thin, laterally persistent sandstone beds with symmetrical and asymmetrical ripple forms draped by clays, giving rise to flaser, wavy and lenticular bedding (higher intertidal flats). The sequence terminates with a series of thin coal horizons separated by units of lenticular and flaser bedding which are permeated, and in some cases homogenized, by rootlets (supratidal marsh). In the Pre-Cambrian Ekkerøy Formation (Finnmark, North Norway) a 3 m fining-upward sequence with a lateral extent of at least 6 km is considered to reflect progradation of a tidal channel–tidal flat association (Johnson, 1975). The tidal channel comprises an erosive-based conglomeratic lag overlain by coarse-grained sandstones with planar, trough, sigmoidal and complex low-angle sets of cross-bedding. Palaeocurrents are bimodal, with the inferred ebb direction being dominant and producing the low angle and sigmoidal sets, whereas the flood direction produced mainly planar sets of cross-bedding. The overlying tidal flat facies include fine- to medium-grained, ripple laminated sandstones, and thin, tidal current and storm-generated sandstone units interlaminated with silty mudstones. Palaeocurrents in the intertidal flat facies are more variable, but are dominated by the flood-tidal component.

FURTHER READING

GINSBURG R.N. (Ed.) (1975) *Tidal Deposits. A casebook of Recent examples and fossil counterparts*. pp. 428, Springer-Verlag, Berlin.

DAVIS R.A. and ETHINGTON R.L. (Eds.) (1976) *Beach and Nearshore Sedimentation*. pp. 187, Spec. Publ. Soc. econ. Paleont. Miner. **24,** Tulsa.

HAYES M.O. and KANA T.W. (Eds.) (1976) *Terrigenous clastic depositional environments – some modern examples*. pp. I-131, II-184, Tech. Rept. **11**-CRD, Coastal Res. Div., Univ. South Carolina.

KOMAR P.D. (1976) *Beach Processes and Sedimentation*. pp. 429, Prentice-Hall, New Jersey.

STANLEY D.J. and SWIFT D.J.P. (Eds.) (1976) *Marine Sediment Transport and Environmental Management*. pp. 602, John Wiley, New York.

CHAPTER 8 Arid Shorelines and Evaporites

8.1 INTRODUCTION

Coastal environments in which evaporite minerals are formed are essentially areas of low clastic input with high net evaporation, yielding carbonates and evaporites. The coastlines are characterized by lines of islands which shelter hypersaline lagoons with reduced tidal ranges. A closely related suite of facies is produced ranging from subtidal to supratidal, as sediment build-up produces a regressive facies sequence.

Study of modern environments has centred on the Trucial States of the Arabian Gulf with its subtidal carbonates, algal mats, and sakhha sulphates and Baja California with its fine-grained clastics and halite. The Shark Bay area of Western Australia and the Bahama Banks (see Chapter 10) also have some similar facies.

The discussion of ancient models immediately leads to problems. Numerous ancient analogues of the sabkha environment have been described, but the origin of ancient evaporite deposits such as the Miocene of the Mediterranean or the Devonian Elk Point Basin of Canada (saline giants) is still hotly debated. Study of these latter areas leads us into the controversy over a deep versus shallow-water origin for evaporites.

8.1.1 History of research

It is surprising that sedimentological research into modern examples of these facies only began in the early 1960s, with the chance discovery of anhydrite in a Trucial Coast salt-flat (Curtis *et al.,* 1963). Since then the Trucial Coast and Qatar Peninsula (Fig. 8.1) have become the model area for sabkha evaporites, thanks to the efforts of Shearman, Evans, Kinsman, and their associates and to the work carried out by Shell Research. The work in these two areas has been placed in the wider context of the whole Gulf in the book edited by Purser (1973b). All this work has been dominantly carbonate-oriented and little is yet known about the interesting north-west coast of the Arabian Gulf, where carbonate sediments grade into clastics, and coastal sabkha facies are produced in quartz-sand sediments. Comparable sediments occur in the Ranns of Kutch, India (Glennie and Evans, 1976).

Unlike the Trucial Coast, where halite is found only as an ephemeral mineral, the tidal flats of both coasts of Baja California contain halite-bearing salinas. Work by Holser (1966) and Phleger (1969), and a general sedimentary study of the Colorado delta area and the tidal flats of the Gulf of California coast (Thompson, 1968), led Shearman (1970) to study a halite-bearing salina (Salina Ometepec) on the tidal flats near the Colorado delta.

8.2 THE TRUCIAL COAST, ARABIAN GULF

8.2.1 General setting of the Abu Dhabi region

The region is dominated by a wave and wind (Shamal) approach from the north-north-west and has a tidal range of 2.10 m at the coastal barrier, reduced to 1.20 m or less at the back of the coastal lagoons. Monthly average air temperatures range from 47 to 12°C. Lagoon waters vary from 40 to 15°C and exposed sediment surfaces from 50+ to 15°C. The open Gulf waters range in salinity from 40 to 45‰, whilst the lagoons reach 70‰ in their coastal reaches. Rainfall is sporadic, but torrential when it occurs, averaging 4–6 cm/year. Net evaporation rates are high in the open Gulf, being about 150 cm/year.

The bathymetry of the area is typical of the southern shore of the Arabian Gulf (Fig. 8.1). In front of the coastal barrier the sea-floor shelves from 15 to 5 metres. It is characterized by bioclastic carbonate sands with a rich variety of skeletal fragments (including bivalves, perforate foraminifera, gastropods, coral/algal grains) and a varying, largely carbonate, mud-content (Evans *et al.,* 1973; Wagner and van der Togt, 1973).

A traverse inland, from in front of the barrier islands, gives a succession of sedimentary environments (Fig. 8.2B) which will yield a preferred regressive vertical sequence.

8.2.2 The coral/algal facies

Small coral/algal thickets, of insufficient relief to be called reef masses, are developed in front of some barrier islands

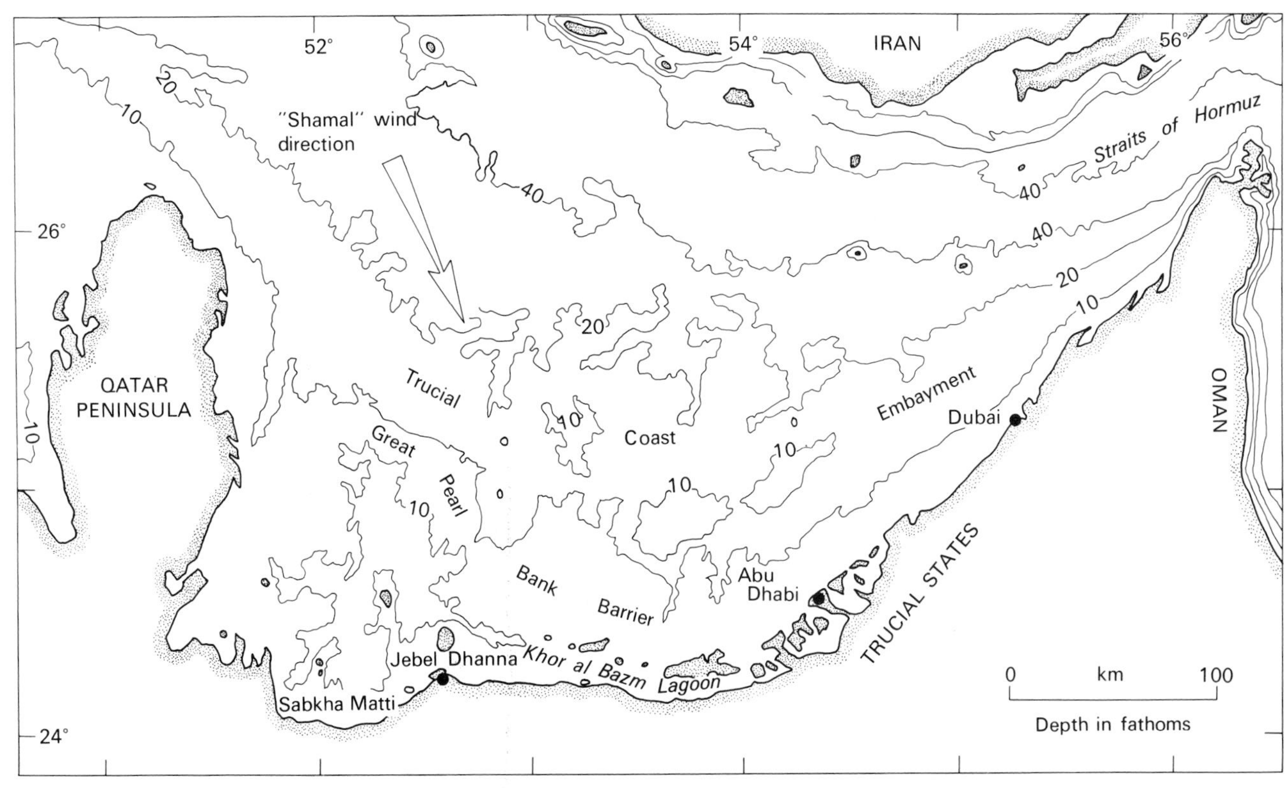

Fig. 8.1. Map of the Trucial Coast embayment showing morphology, channels, terraces and algal-mat/mangrove areas.

(Fig. 8.2). They are sparse patches containing the echinoid *Echinometra malthei,* which is very abundant and a few coral genera (mainly *Acropora, Porites* with *Platygyra, Cyphastrea* and *Stylophora*). Much of the coral is dead and coated with the calcareous algae *Lithothamnion, Lithophyllum* and *Goniolithon.* The corals tolerate conditions which are more extreme (salinity of 40–45‰, water temperature of 30–40°C) than those in which corals are usually found (Kinsman, 1964).

8.2.3 **Tidal channels and deltas – oolitic sands**

The barrier islands along the Abu Dhabi coast are separated by channels several km wide and up to 7–10 m deep, which are flanked by deltas of oolitic sand (Fig. 8.3 and 8.4). The deltas occupy the shelving area where sediment grains are agitated by opposing onshore waves and offshore ebb-tidal currents. The coarsest pure oolitic sands occur along the edges of the channel and form brilliant white levees, which are exposed at low tide. Oolitic grain size (and the thickness of the oolitic cortex) decreases away from these levees, until in depths of water of about 2 m there is an abrupt change to a mixed ooid-pellet-bioclastic sand. This mixed sand occupies parts of the channels and forms the beach and aeolian dunes which are up to 20 m high in places. The shape of the delta is controlled by the ebb-flow tidal currents, and the seawards-orientated spill-over lobes are visible. Near the periphery of the deltas there are bars, 100 m wide and 1–2 m high, which have steeper landward-facing slopes, probably produced by the onshore waves and surface currents. These peripheral areas of the delta may sometimes be stabilized by seagrass (*Halodule*). Loreau and Purser (1973) suggest that ooids are kept near the levees by the opposing onshore-wave and ebb-tidal currents. Some grains fall into the channels where they are mixed with skeletal debris and contribute to the prograding delta front. Other grains which reach the periphery of the delta may be swept onto the beach to contribute to the lateral accretion of the barrier islands, to be returned to a delta or to be blown into the aeolian dunes.

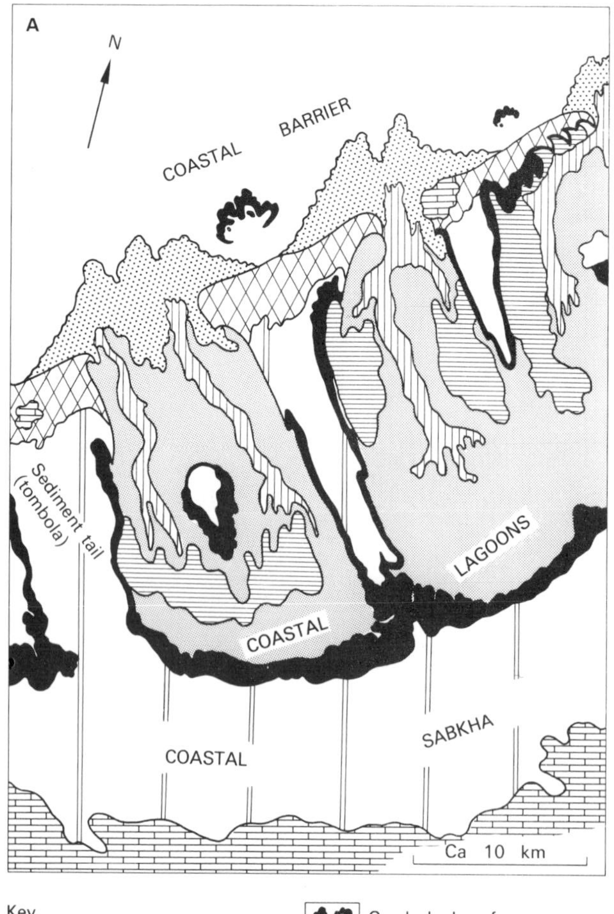

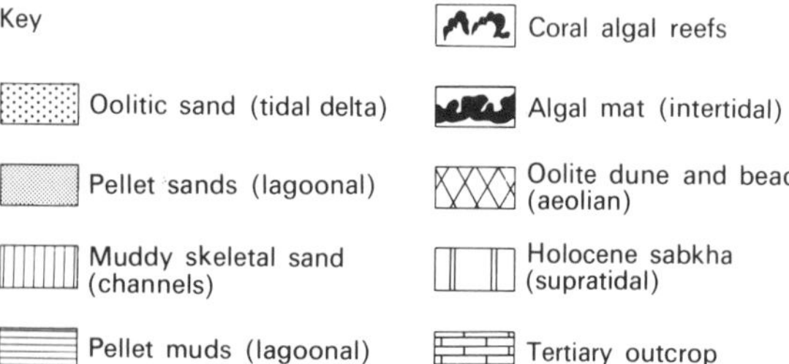

Evans *et al.* (1973) give species lists for the live and dead foraminifera and ostracodes from this and the succeeding sediment types.

8.2.4 The lagoons – subtidal to lower intertidal pellet mud

In addition to small coral/algal reefs the lagoon contains three main sedimentary environments (Fig. 8.3).

The main channel is either excavating Pleistocene limestone or contains coarse bioclastic sands or gravels. Near the islands and tidal deltas, ooids are common.

The subtidal facies is characterized by grey muddy sands rich in peneroplid foraminifera and bivalves, stabilized in places by the seagrasses *Halophila* spp. and *Halodule urinervis*. Bush (1970) thought that the aragonite which constitutes the mud fraction of these sediments is precipitated by indirect organo-physical and direct physico-chemical means from the lagoon waters. Coralline algae are not found in the lagoons (compare with the Bahamas–Florida region, Chapter 10). Kinsman and Park (1976) describe small isolate domed stromatolites up to 6 cm high by 8 cm in diameter in waters up to 3 m deep. These gelatinous masses show a millimetre-scale layering and are dominated by the blue-green alga *Schizothrix sublittoralis*.

The lagoon terraces occur around lagoon-islands which have Pleistocene centres and join up to form a wide shallow terrace at the rear of the lagoon that passes under the algal mats. These terraces consist of Recent lithified sediments overlain by a thin veneer of pellet and bioclastic sediment which may abrade the surface of the lithified sediment to produce a Recent hardground. This lithified surface (Fig. 8.5) is partly intertidal and partly subtidal.

Towards the back of the lagoon the sediment is largely composed of faecal pellets produced by the action of grazing cerithid gastropods on the aragonitic mud. This facies extends from the subtidal to the lower intertidal, finally giving way to the algal mats (Fig. 8.6). Again the pellet mud sediment is

Fig. 8.2. A: Schematic sketch map of the Abu Dhabi region, showing distribution of main sedimentary facies (after Purser and Evans, 1973). B: Diagrammatic profile across the inner shelf to the mountains (after Kinsman, 1966).

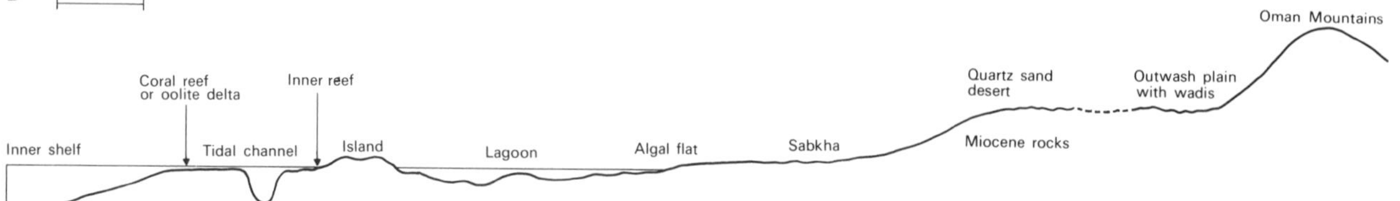

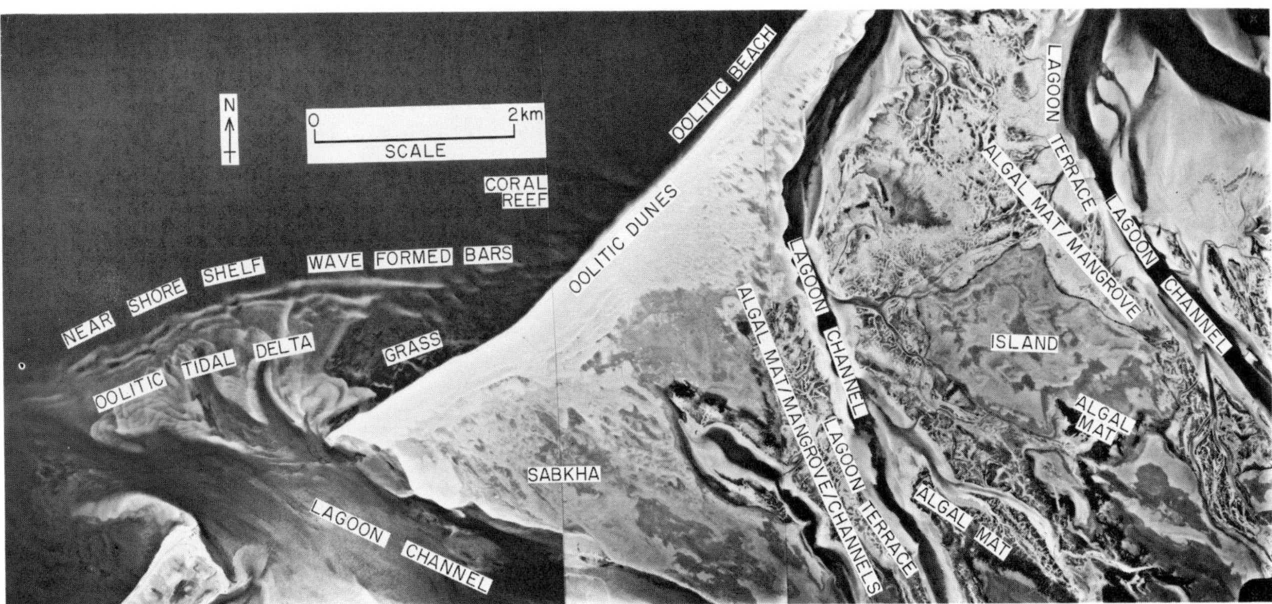

Fig. 8.3. Aerial photo-mosaic of the area around Sadiyat island, showing details of the oolitic tidal delta and the lagoonal complex of channels, terraces and algal-mat/mangrove areas.

present as a thin veneer on top of a bored, lithified surface (Fig. 8.5) which can be traced both under the subtidal sediments and inland under the sabkha. On breaking through this layer, several 3–5 cm thick bored layers, separated by loose sediment, are seen. The layers are all made of the same pellet mud and differ only in the presence or absence of aragonitic needles and high-magnesium calcite cement. Near Qatar similar layers are undergoing submarine cementation at the present time (Shinn, 1969). The continuing cementation has caused the development of submarine anticlines, 'tepees' and large polygonal fracture systems (10–40 m across) as the cemented layers expand, possibly due to the forces of crystallization (Shinn, 1969, Fig. 17; and Evamy, 1973, p. 329). These features have not been observed in the Abu Dhabi region.

8.2.5 The upper intertidal zone – the algal mats

From about the middle intertidal zone landwards, the length of time of exposure makes the environment too hostile for grazing cerithids, and stable substrates are colonized by blue-green algae. Below this position the algae try to colonize the sediment surfaces, but fight a losing battle with the gastropods (Fig. 8.6). Seen from the air this algal-mat zone has a black

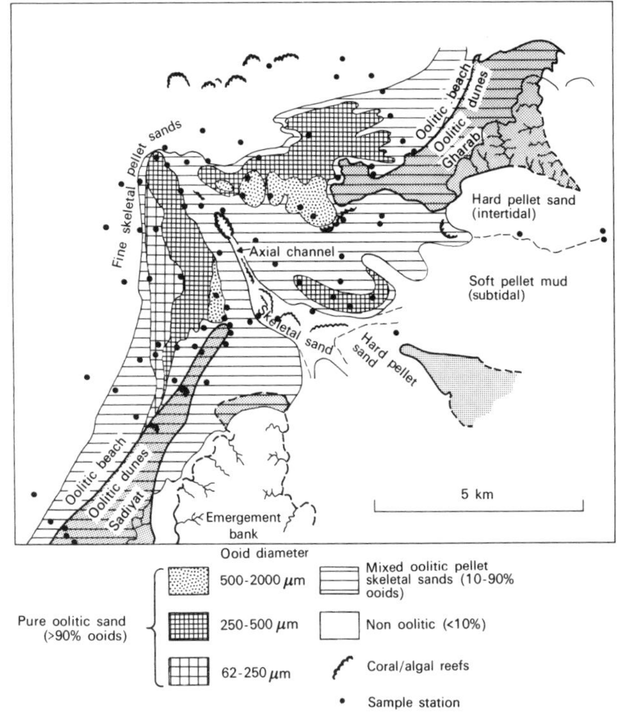

Fig. 8.4. Map of an oolitic tidal delta (from Loreau and Purser, 1973).

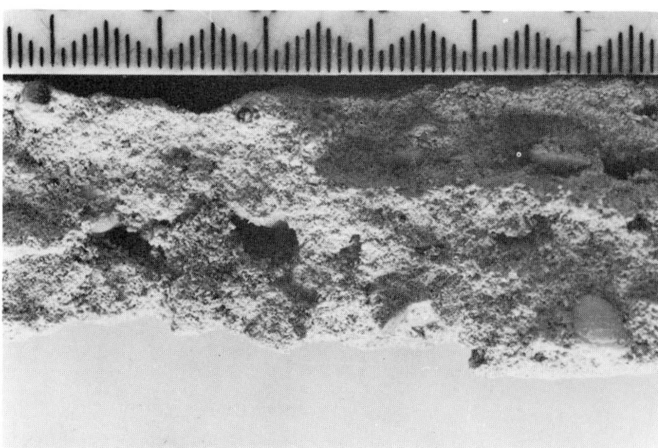

Fig. 8.5. Lithified subtidal sediment from the Abu Dhabi lagoon. Scale in mm.

Fig. 8.7. Aerial photograph of the algal mat zone, Abu Dhabi coast. Tracks of footsteps give scale (provided by R.K. Park).

colour and is patterned with an irregular meshwork of channels and polygon-mat covered ponds (Fig. 8.7).

This algal belt may be up to 1–2 km wide in areas of low slope (about 1 : 3000, with a spring tide range of 120 cm) and is characterized by algal mats or Recent stromatolites. Sediments are laminated on a millimetre scale, consisting of alternate algal-rich and sediment-rich layers (Fig. 8.8), though detailed study shows that individual laminae have a complex structure (Park, 1976). The blue/green mucilaginous algae trap and bind sedimentary particles on their leathery surface, but do not secrete calcareous skeletons. The control of periodicity in algal laminations has been much debated. Park (1976) gives a good review, from the proponents of annual lamination, or storm control to diurnal lamination, in a variety of geographical and geological environments. Whilst Monty (1967) and Gebelein (1969) have recorded diurnal growth in subtidal algal mats, they both stressed the point that intertidal mats, subjected to irregular wetting, are likely to show irregular growth patterns. Park (1976, 1977) believes that the thick sediment layers are formed during storms which drive a wedge of sediment-rich water on shore. Though 5–8 severe storms occur each year on the Trucial Coast, each may not be represented by a sediment layer. He estimates that Abu Dhabi algal mats accrete at about 2–5 mm per year. Compaction and loss of fluids from the sediment could produce a 70–80% volume reduction, giving an apparent growth rate of about 0.7 mm/year in the 4,000 year old buried algal mats.

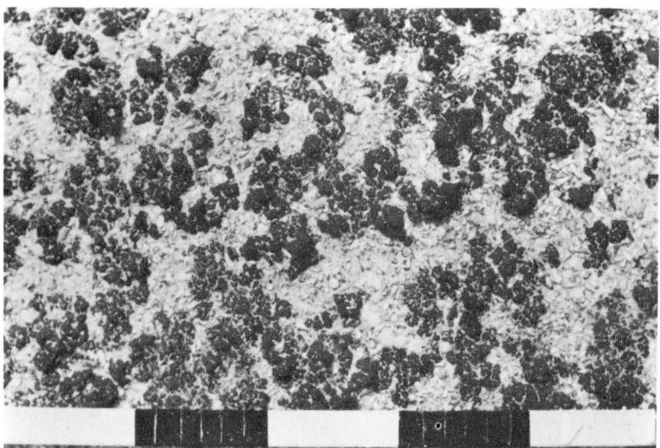

Fig. 8.6. Sediment surface in the middle intertidal zone, Abu Dhabi lagoon. Pustular blue–green algal mat is competing with the grazing cerithid gastropods. White bar on scale is 10 cm.

In their study of an algal mat zone just west of Abu Dhabi, at Khor al Bazm (Fig. 8.1) Kendall and Skipwith (1968) defined four zones within the algal belt in terms of surface morphology of the algal mat. These zones reflected frequency and duration of subaerial exposure and the salinity of tidal waters. A more detailed study of the coastal belt just south-east of Abu Dhabi (Park, 1973; Kinsman and Park, 1976) defined six major mat morphologies, dominantly controlled by the flooding or wetting frequency of the tidal or storm-driven lagoon waters (Figs. 8.9 and 8.10). Golubic (1976) gives detailed descriptions of the algal assemblages characterizing all the mat forms.

(1) *Pustular mat* is commonly the initial colonizer of the mid and lower intertidal areas (Fig. 8.6). Where high salinities limit the activity of grazing gastropods it may occur down to −65 cm. It is interesting to think that in Pre-Cambrian times, before the advent of grazing gastropods, many of these mat forms could have established themselves in the subtidal zone (Garrett, 1970). This mat form is dominated by the coccoid

Fig. 8.8. Smooth mat, Abu Dhabi coast.
(A) From the centre of a polygon
(B) From a polygon rim
White bar on scale is 10 cm.

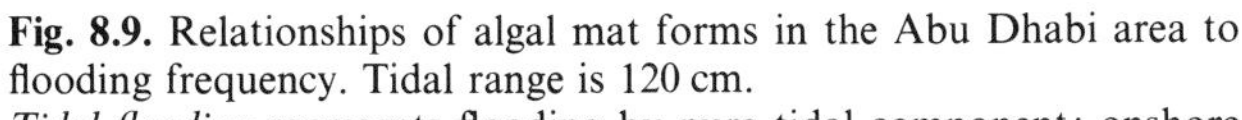

Fig. 8.9. Relationships of algal mat forms in the Abu Dhabi area to flooding frequency. Tidal range is 120 cm.
Tidal flooding represents flooding by pure tidal component; onshore winds increase the flooding frequency for any location, as does the presence of intertidal depressions.
Flooding frequency is indicated as number of flood episodes per year.
L and HWST are low and high water spring tide levels (from Park, 1973).

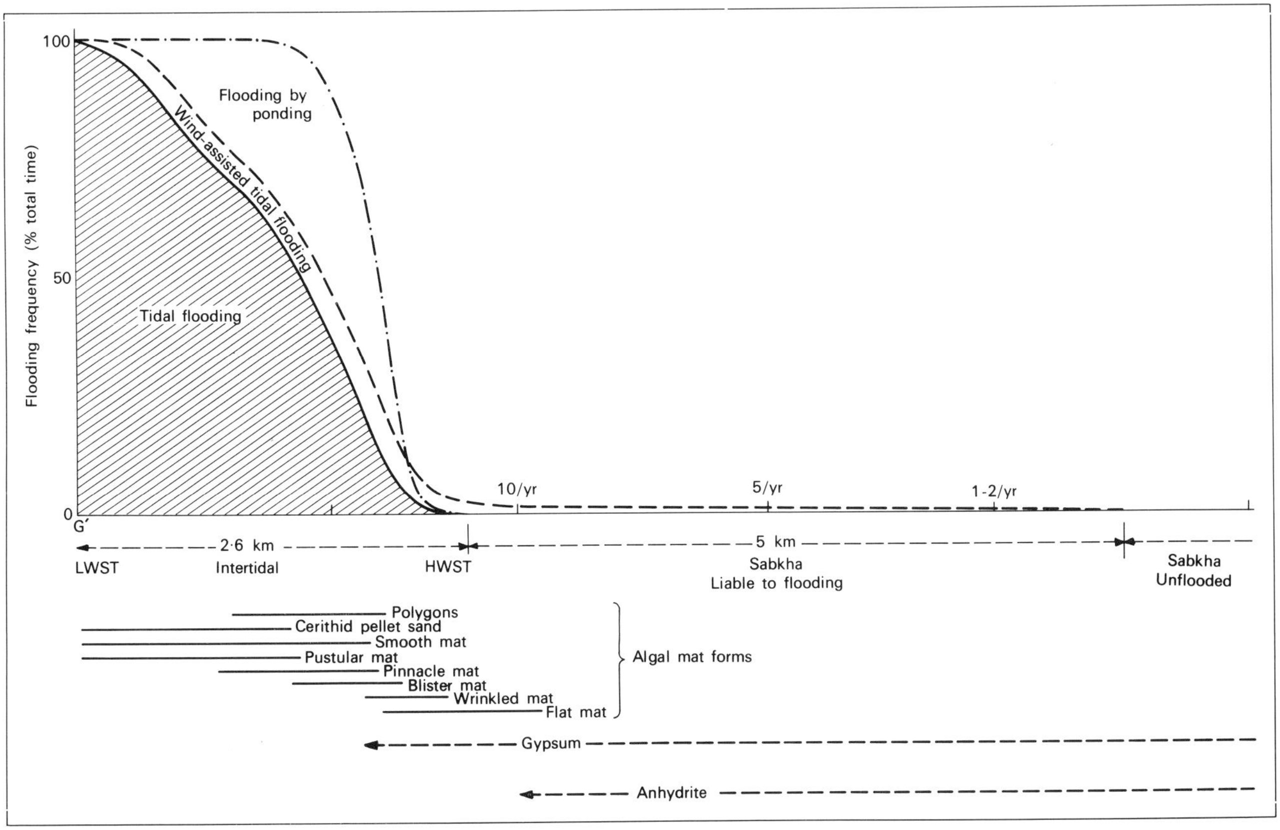

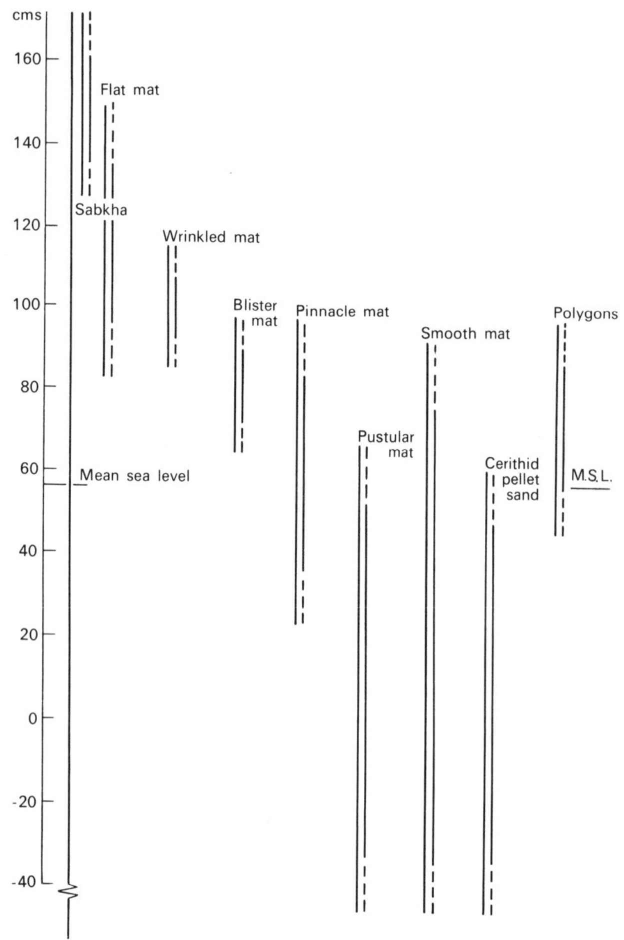

Fig. 8.10. Vertical range of mat forms in the Abu Dhabi area with respect to mean sea level. The total vertical range of each algal mat form is indicated by the solid line to the left; the solid part of the right hand line for each form indicates the vertical range over which the form covers more than 30% of the substrate surface (from Park, 1973).

alga *Endophysalis magna* and produces a clotted sediment with little or no organic peaty-fraction. The preservation potential of this form is low.

(2) *Smooth mat* is the most common mat form (Fig. 8.8) generally from +15 to +90 cm above LWST, but in high-salinity areas it may occur down to −40 cm. It is best developed in pool and channel sites, covers 30–60% of the typical algal belt and is commonly broken into polygons (lily-pads) by desiccation shrinkage. It is dominated by the filamentous alga *Microcoleus chthonoplastes*. Profiles may grade from carbonate-sediment rich (sediment layer up to 1 cm, organic layer less than 0.5 mm) at the base to almost pure algal peats at the surface. This mat form has the highest preservation potential.

(3) *Pinnacle mat* develops as a surface modification of smooth mat. The pinnacles are mainly formed of the large filamentous alga *Lyngbya aestuarii.* Though covering 30–40% of the algal belt they only occasionally acquire sediment laminae, in which case they may be preserved as sedimentary features.

(4) *Blister mat* has a leathery, domed and blistered appearance and is often broken into polygons which are infilled with gypsum discs.

(5) *Wrinkle mat* is a thin, surficial mat form, often underlain by a gypsum crystal mush. This mat is underlain and disrupted by gypsum crystals which are being precipitated from the concentrated pore-water brines (see next section).

(6) *Flat mat* is a thin, discontinuous mat. This and the previous mat form are only flooded by the highest tides (Fig. 8.10) or by storms, and the algae are struggling to survive.

COMPARISON OF MAT FORMS WITH OTHER AREAS

Unfortunately the nomenclature of Recent stromatolites has not become standardized, but the work in the Abu Dhabi area can be compared with that of Logan *et al.* (1974) in Hamelin Pool, S.E. Shark Bay, W. Australia (Table 8.1).

To proceed further with a comparison of mat zones between the Trucial Coast and Shark Bay is difficult, because in the former area LWST is used as a datum, where in the latter area PLWL is used (prevailing low water level – the lowest water level under prevailing wind conditions, Logan *et al.,* 1974, p. 144). However, the sequence of mat forms in Shark Bay is certainly similar (Fig. 8.11) though the smooth mats develop a wide variety of structures (shaving brush forms, calyx structures, inclined and confluent mat columns – see Logan *et al.,* 1974, Figs. 16–18).

Comparison of mat forms with quartz-rich stromatolites – Recent marine stromatolites have been described from the Florida, Bahamas region (Chapter 10), from the Trucial Coast and Shark Bay (see above). In all these cases carbonate grains are being trapped and bound by the algal sheaths. Stromatolitic facies have now also been recognized in quartz-rich sediments from Mauritania, West Africa (Schwarz *et al.,* 1975) and Ceylon (Gunatilaka, 1975). The sequence of mat forms obtained is very similar (Table 8.1), but Schwarz *et al.* believe that the preservation potential of the Mauritania mats is very low. Lack of cementation, together with the intense burrowing activity of crabs and the grazing of fish (producing a contoured mat surface) are all combining to destroy the stromatolites.

Table 8.1 A comparison of algal mat nomenclatures

Vertical position on coast		Kinsman and Park (1976) Abu Dhabi lagoon, Trucial Coast	Kendall and Skipwith (1968) Khor al Bazm, Trucial Coast	Logan *et al.* (1974) Shark Bay, W. Australia	Schwarz *et al.* (1975) Mauritania, W. Africa	Gunatilaka (1975) N.W. Ceylon
SUBTIDAL		Domal	—	Colloform		
Lower/middle	INTERTIDAL	Pustular	Cinder	Pustular	Knoll and cuspate	Flat
Middle/upper		Smooth	Polygon	Smooth, gelatinous	Flat	Blistered and crinkled
		Pinnacle	—	Tufted	—	
Extreme upper		Blister		Blister	Gas-domed	
		Wrinkle	Flat and crinkle	—		
		Flat		Film	Crinkle	Rounded zone
SUPRATIDAL						

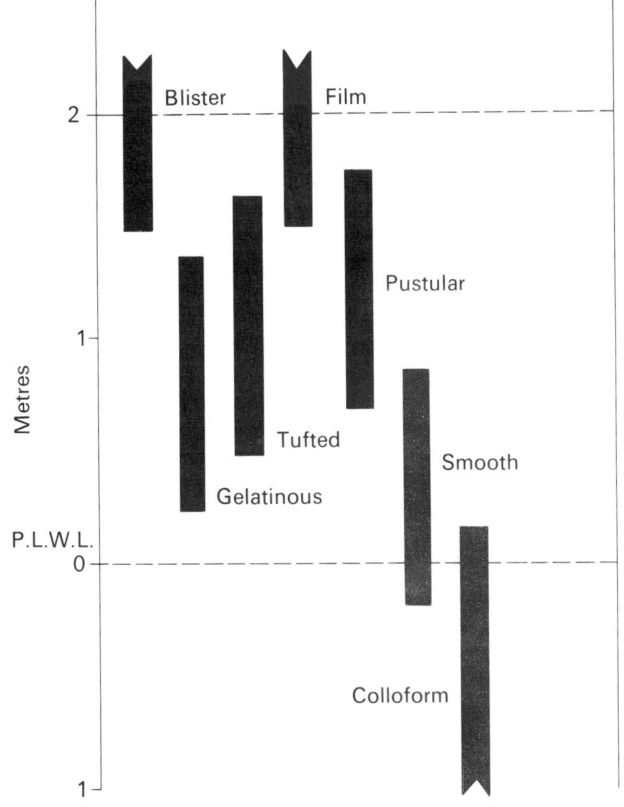

Fig. 8.11. Zonation of algal-mat types in relation to tidal levels in Shark Bay, W. Australia. P.L.W.L. – prevailing low water level (from Logan *et al.*, 1974).

PRESERVATION POTENTIAL OF MAT FORMS

Kinsman and Park (1976) and Park (1976, 1977) show, by a study of buried algal mat sections (radio-carbon dates up to 4,000 B.P.), that few surface mat forms are preserved. Firstly the top 20–30 cm of algal mat are destroyed by the precipitation of gypsum crystals. This leaves a mat thickness of 50 cm or less. In buried sections only smooth mat, with or without polygons (Fig. 8.8), is preserved. Algal filaments, cells or pigments are rarely identifiable in sections over 100 years old, but occasional sheaths of *Lyngbya* have been found in 3,000–4,000 year old sections. Park (1976) believes that the original diurnal (micro-) lamination of the algal mats becomes unidentifiable in buried sections more than 300–400 years old. This leaves us with the characteristic stromatolite millimetre-scale laminations which probably represent a limited number of storm periods (Fig. 8.12).

In areas where the slope of the intertidal zone is steeper than 1:1000 algal mats do not develop. Here patches of lithified sediment may have thin algal veneers, or muddy sediments may contain a variety of sometimes-lithified crab burrows and support the mangrove *Avicenna marina*. Low beach ridges of pelleted, or skeletal (dominantly cerithid gastropods) sands or gravels may be found. Though the beach sediments may be totally bioturbated by the crabs, the beach ridge is often well laminated, or cross-bedded shorewards, and topped with small dunes.

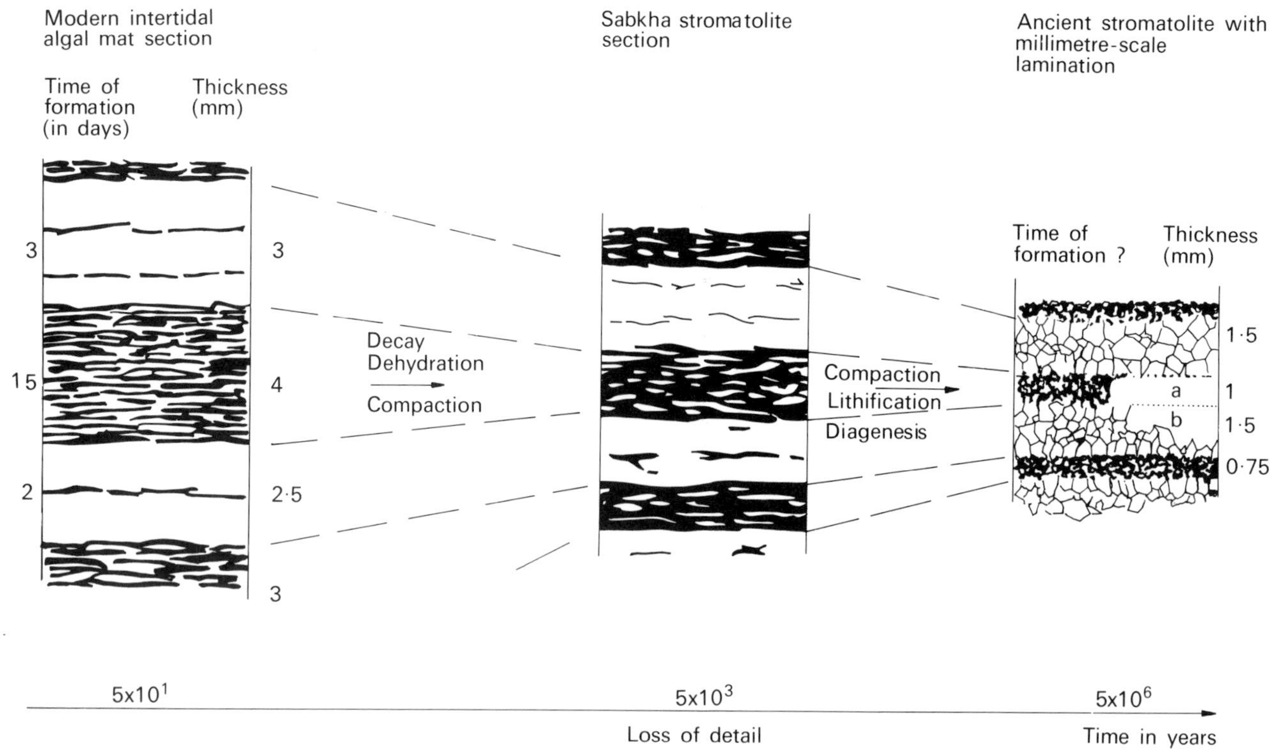

Fig. 8.12. Schematic diagram illustrating the possible consequences of compaction upon a normal algal mat accumulation, and its possible fossil equivalent. (a) Micrite/microspar: commonly dolomite: (b) Spar: almost invariably calcite (from Park, 1976).

Fig. 8.13. Section in the sabkha, south east of Abu Dhabi Island. The buried algal mat is just visible below the white 'cottage cheese' anhydrite. Scale is in cm.

8.2.6 The supratidal zone – the sabkha

The term sabkha is the Arabic for salt flat and is given a variety of phonetic renderings in the literature – sabkha, sebkha, sabkhah, sabkhat, sebkhat, sebjet. Kinsman (1969b, p. 832) proposed that all workers standardize on *sabkha*.

The characteristic brown sediment of the supratidal zone consists of aeolian carbonate and quartz grains, which have accumulated by adhesion to the moist sabkha surface. The salt flat has a slope averaging 1:3000 and is a deflation surface whose elevation is controlled by the local ground-water level.

DIAGENETIC MINERALS

In this regime of strong net evaporation the pore waters become highly concentrated and are drawn towards the surface, causing the precipitation of diagenetic minerals. The most important of these is anhydrite (Fig. 8.13) which is present in a variety of forms.

The lower continuous mass of white anhydrite (with a cottage cheese texture), is now accepted as partially originating by secondary replacement of gypsum which was originally present in the upper parts of the algal mat section (Butler,

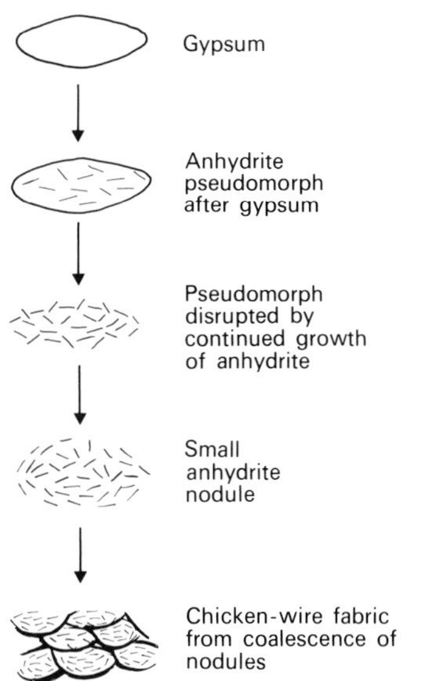

Fig. 8.14. The replacement of gypsum by anhydrite, leading to a chicken-wire fabric. Based on a sketch by D. Shearman.

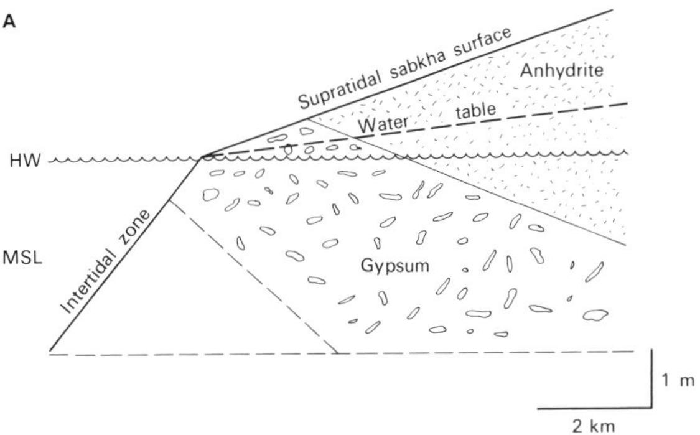

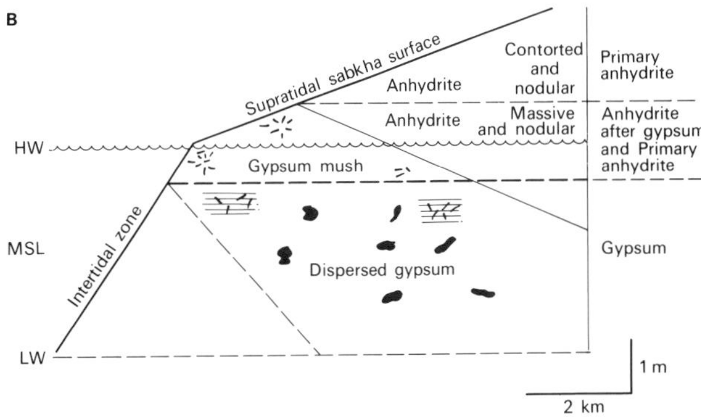

Fig. 8.15. A. A cartoon of gypsum and anhydrite distribution within the sabkha environment.
B. A more detailed description of the form of the sulphates (from Kinsman, 1966).

1970). The original gypsum crystal is pseudomorphed by a felted mass of anhydrite laths. Subsequent precipitation of more anhydrite disrupts this pseudomorph and produces a nodule. Nodules coalesce and produce a thick layer of anhydrite. Sediment trapped between the growing nodules may produce a chicken-wire fabric (Fig. 8.14). Above this zone the characteristic contorted layers (preferred to 'enterolithic structure' – since they are layers not tubes) and anhydrite nodules are developed. These also consist of felted masses of anhydrite needles, but Kinsman (1974) stresses that this anhydrite is of primary origin and explains the chemical conditions required for its formation (Fig. 8.15).

A variety of other diagenetic minerals are formed in this environment by reaction between the carbonate fraction of the sediments and the pore-water brines (Kinsman, 1966).

Halite is present as an ephemeral white surface crust which marks the area of recent tidal flooding, when onshore winds may drive a wedge of water up several kilometres onto the sabkha.

Gypsum is present in the upper algal mat zones, in places (such as old channels) making up to 50% of the top metre of sediment. It is precipitated from the pore waters as discoid crystals, flattened normal to the c-axis, and up to 5 cm long. These crystals tend to be fairly clear. In contrast, the large disc-shaped gypsum crystals which grow in the upper parts of the sabkha are full of included grains.

Dolomite first appears at the junction between the algal mats and the sabkha, and increases in amount inland. By a distance of 10 km inland the whole of the top metre of sediment may be replaced by fine-grained (2–3 mm), calcium-rich, disordered (proto-) dolomite. The sabkha pore-waters acquire a high mg^{2+}/Ca^{2+} ratio by the precipitation of gypsum, which begins when concentrations reach five times that of sea water. Subsequently, when mMg^{2+}/mCa^{2+} values reach between 5 and 10 (Kinsman, 1966), the fine-grained aragonite in the sediment begins to be replaced by dolomite. This dolomite is the same as that described from other areas of recent sedimentation (Shinn *et al.*, 1965).

Celestite singly and in sheath-like aggregates is also formed as the strontium-rich aragonite (7,000 ppm Sr) is replaced by dolomite (less than 1,000 ppm Sr).

In areas of the sabkha where the sediment is more quartz rich, or lacks the fine-grained aragonite mud, dolomite does not

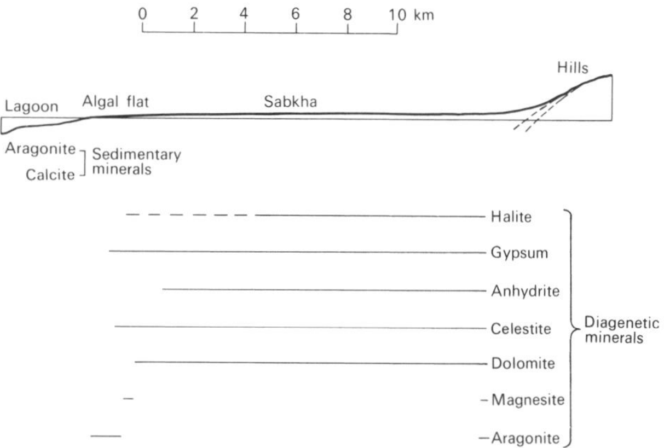

Fig. 8.16. Sabkha profile indicating approximate relative positions at which the early diagenetic minerals are developed (from Kinsman, 1966).

form. Here the mMg^{2+}/mCa^{2+} ratio of the pore waters may rise above 10, and huntite and magnesite may form (Kinsman, 1967).

THE SABKHA BRINES

Analysis of these brines has been important in understanding the diagenetic mineralogy of these sediments (Kinsman, 1966; Butler, 1969; Patterson and Kinsman, 1977). Bush (1970) was also able to show that the typical sabkha brine has a composition similar to the fluid inclusions found in lead-zinc deposits of the Mississippi Valley type (Table 8.2). This is particularly interesting in view of the known association of this type of ore body with carbonate-evaporite rocks (see Park and MacDiarmid, 1975, Chapter 15).

By a careful programme of levelling and sampling across the sabkha, Kinsman's group have shown that the water table slopes gently seawards across the coastal sabkha, at a depth of one metre or less. For these brines the chloride content increases rapidly both from the landward and seaward margins of the sabkha to a chloride plateau of 4.7 m Cl^-/kg (Fig. 8.17). Lagoon waters at 70‰ salinity are about 1 m Cl^-/kg. The seaward flow of these brines is maintained by input from the continental ground-water system. Sporadic recharge occurs in the seaward part of the sabkha by the infiltration of sea water which is driven onto the sabkha by storms. Concentration of these brines occurs because of the high net evaporation. However, Kinsman (1976) believes that annual net evaporation rates may only be ca. 6 cm/year from the sabkha brines (compared to 150 cm/year in the open Gulf). He found that average day-time relative humidities are 35–45%, but at night they exceed 90%. Consequently the sabkha absorbs moisture during the night and takes some part of the day to boil this off.

Clearly these sabkha brines contain a mixture of marine and non-marine waters. Patterson and Kinsman (1977) show that mCl^-/mBr^- and mK^+/mBr^- ratios are more reliable indicators of brine origin than are mNa^+/mCl^- and mCl^-/mK^+ ratios (Table 8.3 and Fig. 8.18).

Since the maximum Holocene transgression (about 5,000 years B.P.) the sabkha sediments have filled in 5–10 km of originally marine lagoon. Patterson and Kinsman (*op. cit.*) show that whilst the seaward margin of the chloride plateau (Fig. 8.17) has moved seawards at the same pace as the sediment offlap rate (i.e. 5–10 km/5,000 years), the landward margin has only moved 2–3 km. This has produced a gradual widening of the belt of high (greater than 4 m Cl^-/kg) salinity brines. Since the long-term preservation of anhydrite requires brines of 3.8 m Cl^-/kg (Kinsman, 1974, Fig. 8), though some rehydration of anhydrite to gypsum has occurred in the extreme landward area of the sabkha, the belt of nodular anhydrite will increase in width with time.

Table 8.2 Brine compositions in ppm, taken from Bush (1970, p. B142)

Ion	Sea water	Sabkha brine	Fluid inclusions (Tri-state, Mississippi Valley)
Cl^-	19,370	168,800	124,600
SO_4^{2-}	2,712	200	3,300
Ca^{2+}	412	20,700	18,000
Mg^{2+}	1,290	14,400	2,400
K^+	399	3,800	2,700
Na^+	10,760	58,900	57,100

Table 8.3 Ion ratios which characterize sabkha brines (data from Patterson and Kinsman, 1977).

	Marine	Mixed marine/ continental	Continental
mCl^-/mBr^-	1,000	1,000–5,000	5,000
mK^+/mBr^-	25	25–150	150

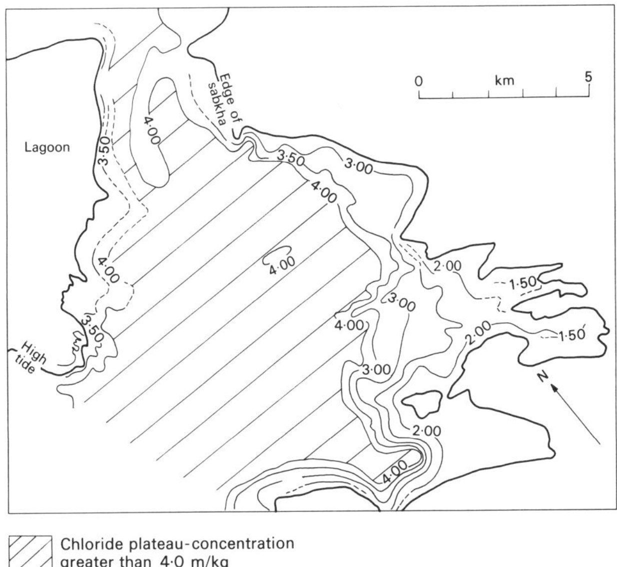

Fig. 8.17. Chloride concentration (in mCl⁻/kg) of sabkha brines inland of Abu Dhabi Island (from Patterson and Kinsman, 1977).

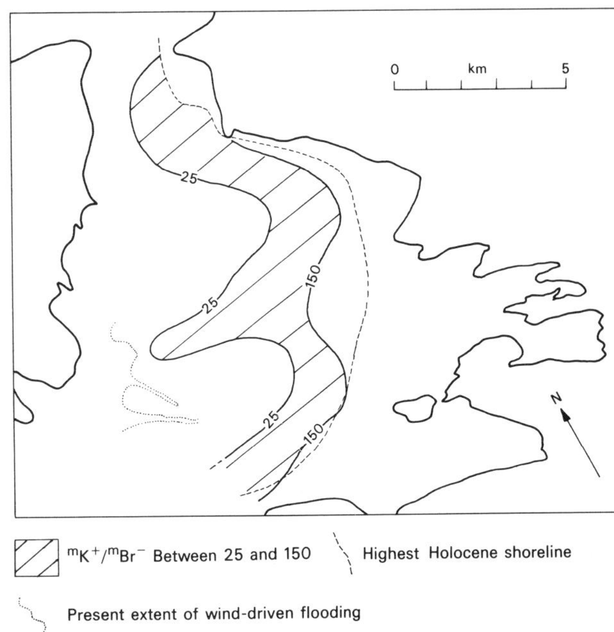

Fig. 8.18. Ratio of mK^+/mBr^- of sabkha brines. Locality as in Fig. 8.17 (from Patterson and Kinsman, 1977).

8.2.7 **Terrestrial environments**

Buttes of Pleistocene limestone, which probably stood above the level of the Holocene transgression, still stand above the sabkha as wind-fretted islands. At two localities, Scholle and Kinsman (1974) discovered aragonitic and high-magnesium calcite caliche, consisting of laminated, lacy and pisolitic carbonate crusts. Purser and Loreau (1973) describe a range of similar aragonitic crusts from east of Qatar. All the authors suggest that these crusts with metastable mineralogies, and marine or brine associated chemistry and isotopic composition, formed in the vadose environment. However, vadose (and phreatic) pore waters in this area have such unusually high salinities that cements of marine character develop with vadose textures. Thus, *vadose* will not always imply fresh water and Scholle and Kinsman (1974, p. 914) suggest that pisolites in the Permian Reef Complex, New Mexico (Dunham, 1969) could have been produced in this way.

Moving inland from the sabkha onto the Miocene-based hinterland our traverse continues across quartz sand, aeolian dune fields, inland sabkha and an outwash plain (see Glennie, 1970) into the Oman mountains. Thus this area gives a lateral sequence of sedimentary facies from shallow-marine carbonates, through evaporites to continental sandstones.

8.2.8 **A stratigraphic cross-section (Fig. 8.19)**

The sequence of surface environments described above is mirrored in the sedimentary sequence seen in the dominantly regressive wedge of diachronous sedimentary facies which underlie the sabkha.

The Holocene transgression flooded an earlier continental sabkha and reached the highest Holocene shoreline (Fig. 8.19). This older surface is underlain by units 7 and 8 which unconformably overlie the Miocene basement. At that time sedimentation rates were too slow to keep pace with the transgression and only 10 cm of intertidal sediments are preserved (units 6 and 5). Subsequent regression has been caused mainly by sediment offlap (about 2 m/year), together with a fall in sea level of ca. 120 cm.

The main regressive sedimentary facies comprise the subtidal, intertidal (lower and upper) and supratidal sediments (units 4, 3, 2 and 1) characterized by muds, algal mats and sulphate-bearing sands. A pit dug anywhere in the present sabkha will reveal this sequence of sediments (Fig. 8.20).

It is this facies sequence which Shearman (1966) so excitingly first compared with known ancient successions.

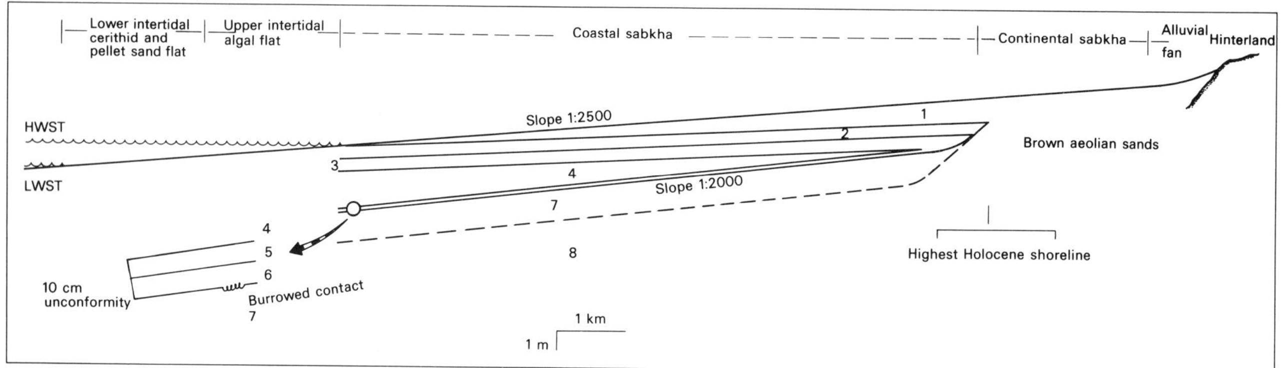

1 Supratidal - detrital sands and evaporites
2 Upper intertidal - algal facies or skeletal gravels
3 Lower intertidal - cerithid and pellet sands
4 Subtidal - lagoonal muddy skeletal sands
5 Lower intertidal - cerithid and pellet sands
6 Upper intertidal - algal facies
7 Grey aeolian sands
8 Brown aeolian sands

Fig. 8.19. Composite NW–SE stratigraphic section across the Abu Dhabi coastal-continental sabkha belt. (HWST and LWST – high- and low-water spring tide levels, respectively). The stratigraphic section, from the surface downward, is as follows:
(1) *Supratidal Facies*: 0–100 cm, a wedge of detrital sediments thickening inland from the shoreline. Dominated by anhydrite in inland areas: may be magnesite-rich at base.
(2) *Upper Intertidal Facies*: 60 cm: commonly an *algal facies* but coarse skeletal sands or gravels are equivalent sediments in high energy zones; anhydrite may be present in upper part; gypsum often abundant; intensely dolomitized in some areas. Full regressive development.
(3) *Lower Intertidal Facies*: 60 cm: cream, muddy, pellet sands with abundant cerithids; large gypsum crystals common; fairly intensively dolomitized in some areas. Full regressive development.
(4) *Subtidal Facies*: 0–300 cm: grey-brown, peneroplid foraminifera and bivalve-rich muddy sands; thickens seaward. Minor dolomitization.
(5) *Lower Intertidal Facies*: 2–5 cm: Cream, pellet sands with abundant cerithids. Attenuated transgressive development.
(6) *Upper Intertidal Facies*: 2–5 cm: *algal facies* in some places, otherwise mixed pellet and detrital sands. Attenuated transgressive development.
Unconformity: generally fairly sharp: sometimes burrowed.
(7) *Grey Aeolian Sands*: 50–150 cm: grey, cross-bedded aeolian sands: original iron oxide films around grains now present as iron sulphide minerals; some gypsum.
(8) *Brown Aeolian Sands*: thickness unknown (probably about 5–10 m): brown, cross-bedded aeolian sands with some gypsum.
Unconformity.
Miocene Rocks underlie this succession
(from Patterson and Kinsman, 1977).

8.3 BAJA CALIFORNIA, MEXICO

8.3.1 Introduction

This region is characterized by a climate that is slightly more arid than that in the Arabian Gulf. Mean maximum temperatures vary from 41° to 20°C, whilst mean minima are 25°C and 6°C respectively. Rainfall averages 8 cm per year (mostly in December to February and in thunderstorms in August and September). Net evaporation is quoted at 290–180 cm per year (Thompson, 1968). Tidal ranges in the area vary from 1.5 m (neap) to 8.5 m (maximum spring).

8.3.2 Salina Ometepec

A slight topographic depression near the landward margin of the tidal flats is occupied by Salina Ometepec (Fig. 8.21). Here the surface is underlain by up to 25 cm of halite rock, which lies on the clayey supratidal flats. Maximum or storm spring-tides flood the salina, and first gypsum and then halite form during the drying-out of this trapped brine body. Shearman (1970) recognized two stages in this process. During the first, the *brine-pan phase,* halite crystals nucleate at the brine surface. Small pyramidal hopper crystals form, hang together and aggregate into rafts. Some of these rafts sink to the floor of the pool, others blow across the pan to bank up on the lee side. When the pan has dried out, up to 5 cm of halite rafts will have accumulated on the pan floor. The second *dry salt-pan phase* then begins as the moisture content of the rafts dries and causes an efflorescent halite growth of tiny cubes to cover the surface. Individual rafts begin to lose their definition as redistribution of the halite occurs. The next flooding of the pan dissolves some of the halite, but gradually a layered halite rock is developed with gypsum layers within it. In thin section the

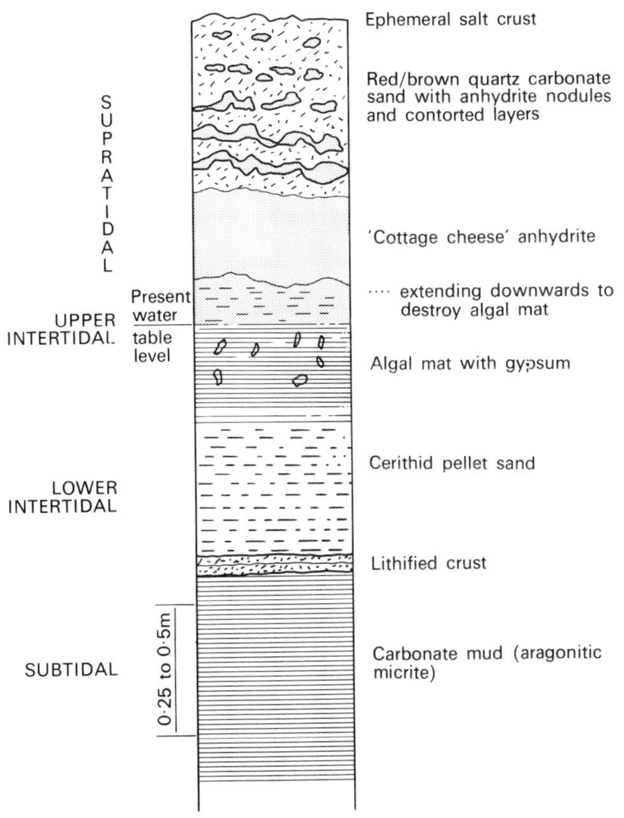

Fig. 8.20. Graphic log of the regressive sabkha sequence in Abu Dhabi.

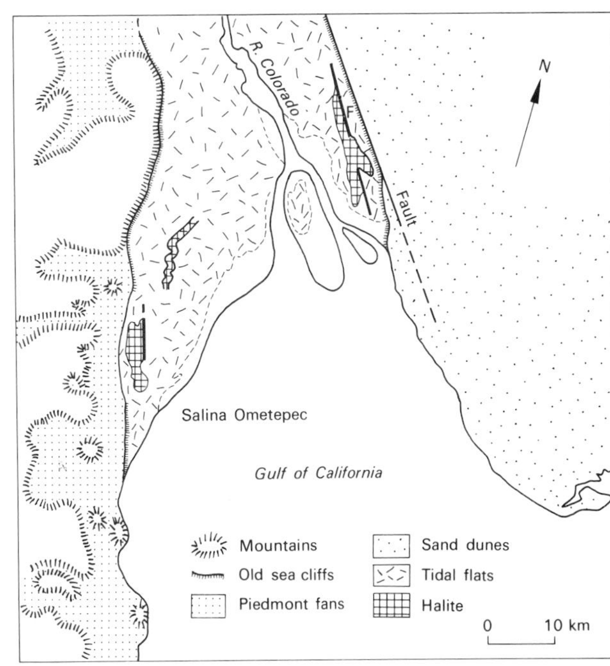

Fig. 8.21. Tracing from space photograph showing location of halite flats, Baja California, Mexico (from Shearman, 1970).

halite has brine inclusions (trapped during rapid crystal growth) arranged in a chevron pattern. Inter-chevron cavities are initially void, where halite dissolution has occurred. Secondary precipitation of clear halite occurs within these voids from the trapped brines. If the voids were completely filled with clear halite the rock would be identical to the *chevron halite* described from ancient saline giants (e.g. Wardlaw and Schwerdtner, 1966). The upward-pointing chevrons of the halite rock suggest growth upwards from a surface by competing halite crystals (Fig. 8.22). Arthurton (1973) was able to produce a similar texture, by growing halite on the floor of shallow bowls. Shearman (1970) suggests that the laminated nature of this rock reflects the successive flooding episodes, not annual cycles as suggested in the past for laminated halite deposits.

This work on a modern salina is the only description of halite deposits forming at the present day under marine influence (except for the work on the Gulf of Karabogaz, off the Caspian Sea, described by Strakhov, 1970, Part II, Chapter 1, but this description lacks details of the facies). This begins to reveal a paradox. A good model exists for the sabkha-type sediments, but these are not volumetrically important in the ancient. Also, no thick modern sequence of sabkha-type deposits has been found. This could be because of their sensitivity to sea-level changes. The many giant gypsum/halite deposits of the geological record are left without a modern analogue, certainly none of a similar scale (see Sect. 8.5).

8.4 ANCIENT SABKHAS

8.4.1 Introduction

Once Shearman (1966) had compared the facies sequence of the Trucial Coast sabkha environment with the succession in the Lower Purbeck (uppermost Jurassic) of the Warlingham borehole (just south of London), many similar ancient subtidal to supratidal sabkha sequences were described, for example in the Visean of Eire (West *et al.*, 1968), the Upper Devonian of western Canada (Fuller and Porter, 1969), the middle Carboniferous of the Maritime Provinces, Canada (Schenk, 1969), the Upper Jurassic of the Arabian Gulf (Wood and

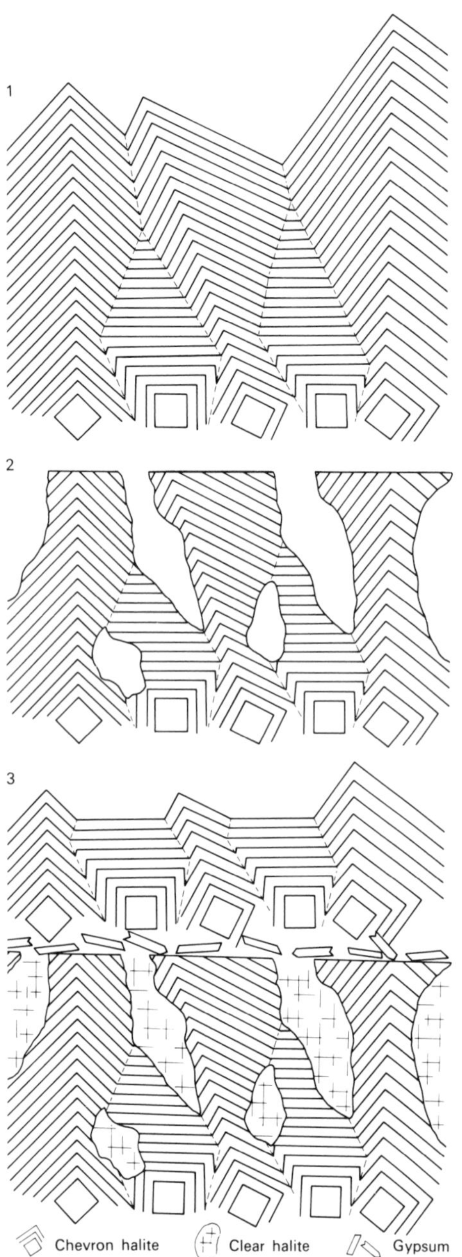

Fig. 8.22. Schematic diagram showing mode of development of layered halite rock (after Shearman, 1970).

Wolfe, 1969), the Lower Eocene of Jamaica (Holliday, 1971), the Upper Permian of Cumberland, England (Arthurton and Hemingway, 1972) and the late Pre-Cambrian of Kitwe, Zambia (Clemmey, 1974).

All these examples are dominated by carbonate–sulphate evaporite facies, some passing laterally into red-beds. In the Stettler Formation (Fuller and Porter, op. cit.) there is also a thin (100 m) development of halite within this succession. As Selley (1976a) suggests, this could easily form within a sabkha if local subsidence caused the water table of the concentrated brines to be above the surface. Thus a brine pool would form where halite would precipitate on evaporation, the brines being already depleted in calcium sulphate (see p. 188). This sort of halite development is a minor part of the succession and is much less important than that described in Sect. 5.3.4.

8.4.2 The Lower Purbeck of southern England

The Purbeck sections have been chosen for fuller descriptions for a number of reasons. Firstly, they were the first rocks to be

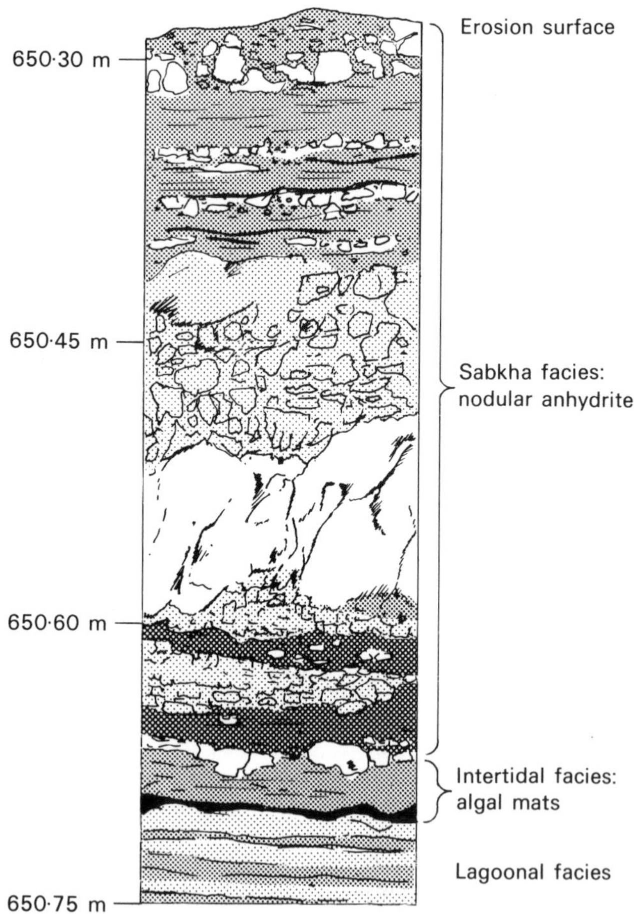

Fig. 8.23. A complete sabkha cycle in the Lower Purbeck Beds from the Warlingham borehole (from Shearman, 1966).

compared with the modern sabkha (Shearman, 1966; but also see Evans *et al.*, in discussion of West, 1964, p. 327) (Fig. 8.23). Secondly, the similarities in small-scale and large-scale fabric and petrography between the lower Purbeck rocks and the Recent sabkha are most striking.

THE LOWER PURBECK OF SUSSEX

Outcrops occur in a series of *en echelon* inliers in the Weald around Brightling (Howitt, 1964) (Fig. 8.24). The lowest Purbeck beds, containing gypsum seams, lie with a sharp contact on the calcareous sandstones of the Portland beds (Anderson and Bazley, 1971) and are only seen in the many boreholes sunk in the area. These gypsum seams are worked in a number of mines (Northcott and Highley, 1975).

A typical borehole (Holliday and Shepard-Thorne, 1974) shows a succession of sabkha cycles with laminated pelmicrites, algal-laminated pelmicrites with birds-eye fabric (see Shinn, 1968a), and nodular and chicken-wire anhydrites. As in most ancient gypsum/anhydrite deposits a complex diagenetic history was unravelled by Holliday and Shepard-Thorne (1974), which is similar to that proposed by West (1964, 1965) from work at Swanage, Dorset (Fig. 8.25).

Five diagenetic stages are recognized:

Stage A – interstitial displacive growth of discoid gypsum crystals soon after completion of sedimentation – such as takes place in algal mats and buried-channel muds of the Trucial Coast (see p. 187 above).

Stage B – replacement of gypsum within the unconsolidated sediment by aphanitic anhydrite (as described by Butler (1970) in the sabkha – see p. 186 above) to form 'cottage cheese'. The soft felted mass of anhydrite laths does not preserve the pseudomorphs. Depending on the amount of interstitial material a net-texture (more matrix) or a macro-cell, chicken-wire texture (less matrix) can develop.

Stage C – further anhydrite (primary) may be precipitated as described by Kinsman (1974) for the sabkha (see p. 187 above). In the Purbeck rocks these anhydrite laths tend to be coarser (up to 1 mm long) than the anhydrite of stage B (0.1 mm long).

Stages D and E – burial to depths of more than 600 metres and with temperatures of more than about 60–80°C causes any remaining gypsum to dehydrate. Subsequent uplift allows gypsum to form by hydration of some (though not usually all) of the anhydrite, and porphyroblastic gypsum (Stage D) and/or alabastrine gypsum (Stage E) may form.

In addition, at various stages sulphates are often replaced or pseudomorphed by calcite and/or amorphous silica. Folk and Pittman (1971) for example believed that such chalcedony, if length-slow, is a 'testament to vanished evaporites'. Replacement by celestine is also common.

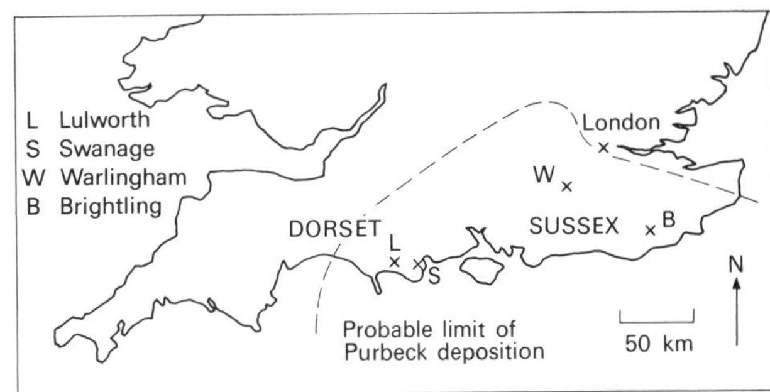

Fig. 8.24. Map of Purbeck exposures and boreholes in southern England mentioned in text.

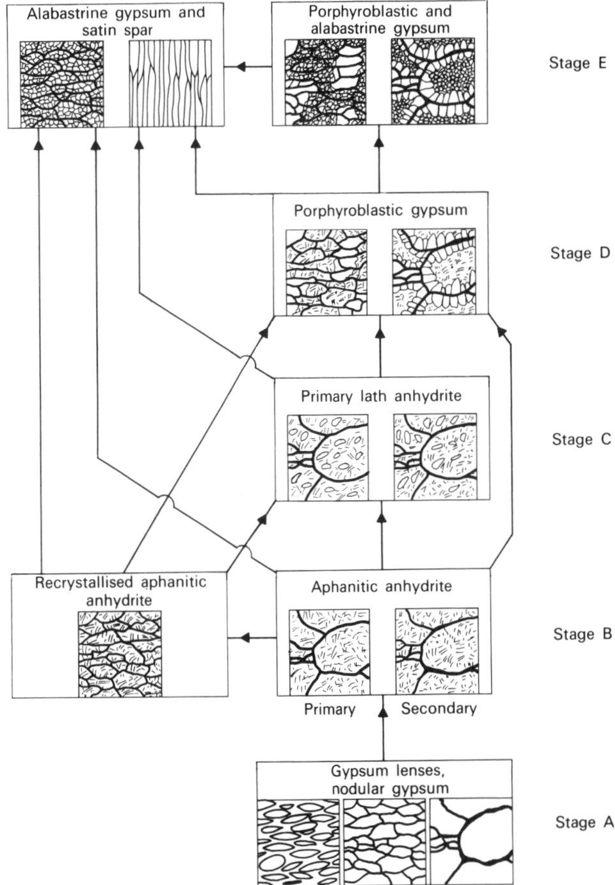

Fig. 8.25. Diagenetic fabrics in the Lower Purbeck calcium sulphate rocks of southern England.

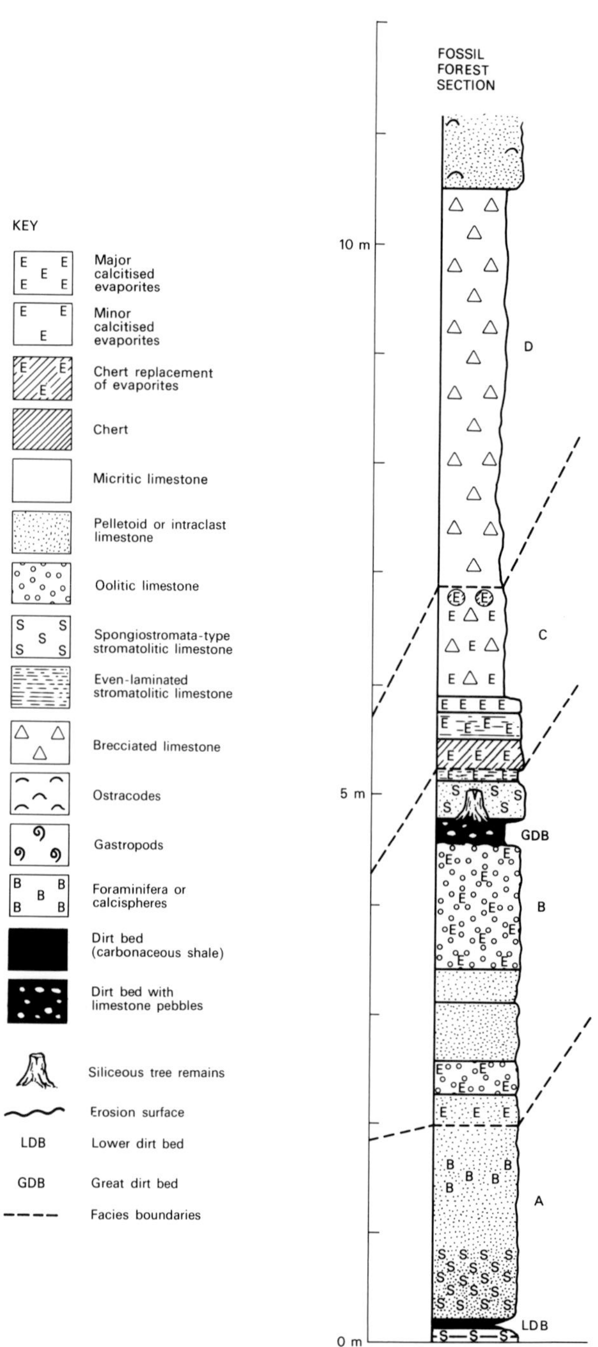

Fig. 8.26. Graphic log of the sequence of facies associations in the basal Purbeck at Lulworth, Dorset (after West, 1975, Fig. 2).

Fig. 8.27. Stromatolite with cross-cutting pseudomorphs after gypsum, Facies B; Lulworth, Dorset.

Fig. 8.28. Sabkha facies, with contorted layers of calcitized sulphate, Lulworth, Dorset.

THE LOWER PURBECK OF DORSET

The careful petrographic and sedimentological work of West (1964, 1965, 1975) describes these beautiful and much-visited sections along the Dorset coast (Fig. 8.24).

The old emotive names for these lower Purbeck beds were:

Cypris Freestones
Broken Beds
At the base: Caps and Dirt Beds

The base of the Purbeck passes down, without a sharp break, into the Portland shallow-marine oolitic and shelly limestones (Sect. 10.2.5; Townson, 1975).

The two lower Purbeck-age units are correlated with the gypsum-bearing horizons of Sussex (Anderson and Bazley, 1971, plate VI) described above. West (1975) has erected a facies model for the lower Purbeck of Dorset in which he replaces the three formations listed above by four sedimentological facies-associations (Fig. 8.26).

West believes that the basal Purbeck began with moderately hypersaline, subtidal and intertidal limestone facies (A), followed by a more hypersaline, intertidal facies with gypsum (B). The overlying main evaporite facies (C) is totally replaced, but probably formed in very hypersaline intertidal to supratidal conditions. The top facies (D) reflects a reversion to less hypersaline, intertidal to subtidal conditions.

Facies A – Limestone with Foraminifera – consists of thin-bedded pellet limestones, containing the Spongiostroma-type stromatolites. These occur in beds from 0.3–5 m thick showing rapid lateral variations (Brown, 1963). The algal heads built layers 5–10 cm thick, containing millimetre-scale laminations. Algal genera are described by Pugh (1968). The general lack of fauna and the pellet limestones suggest hypersaline (50–70‰?) subtidal conditions like those of the Trucial Coast lagoons and the Spongiostroma-type stromatolites may compare with the subtidal domal forms described by Kinsman and Park (1976; see Sect. 8.2).

Facies B – Limestones with Stromatolites and some replaced gypsum – contains both algal mats and Spongiostroma-type stromatolites which are much larger than in facies A and form metre-high mounds with a tree-trunk centre – as developed at Lulworth in the classic Fossil Forest horizon. The algal mats contain gypsum crystals, now replaced by calcite, of similar form and cross-cutting the algal laminae (Fig. 8.27) as those seen in the Trucial Coast.

Facies C – Calcitized Evaporites – is an unfossiliferous sequence of porous, coarsely crystalline (in places sac-

Fig. 8.29. Pelsparite with casts of halite crystals, Facies D, Lulworth, Dorset.

charoidal) limestones which are mostly secondary replacements of anhydrite-bearing rocks. They contain celestite and nodules of chert with pseudomorphs after anhydrite.

This facies also largely contains brecciated limestones – called the Broken Beds, though in some places they extend down into facies B or up into facies D. West (1975) shows that the presence of pseudomorphs after anhydrite in the Broken Beds means that brecciation occurred after diagenetic stage C (Fig. 8.25) and believes that it was caused by tectonic activity.

This facies could represent supratidal conditions, since the calcitized fabrics strongly resemble sabkha anhydrite fabrics, but dolomitized sediments are not present, and West (1975) believes the environment may have been less arid than the Trucial Coast and may have had some intertidal influences. Clear sabkha facies, but again calcitized, are only seen much higher in the Purbeck, near the Cinder Bed (Fig. 8.28).

Facies D – Limestones with Ostracods – is relatively free of gypsum, rich in ostracodes and consists of bio-, intra-, and pelsparites often with ripple-marked surfaces. The pelsparites contain good casts of halite crystals (Fig. 8.29).

The limestone types and the abundant ostracods suggest dominantly subtidal conditions, with occasional drying out to produce halite crystals. Dirt beds – or carbonaceous shales occur in both Facies A and B. West (1975) believes that they are weathering horizons from a semi-arid climate and therefore represent periods of emergence.

Facies model – a typical Dorest Purbeck cycle is:

3. Gypsum, anhydrite (both calcitized) and limestone
2. Stromatolitic and pellet limestones
1. Dirt bed, with trees, on an erosion surface.

Each cycle of sedimentation began with a transgression of this emergent surface and ended with sediment build-up again causing exposure. The inferred palaeogeography for two such cycles is shown in Fig. 8.30.

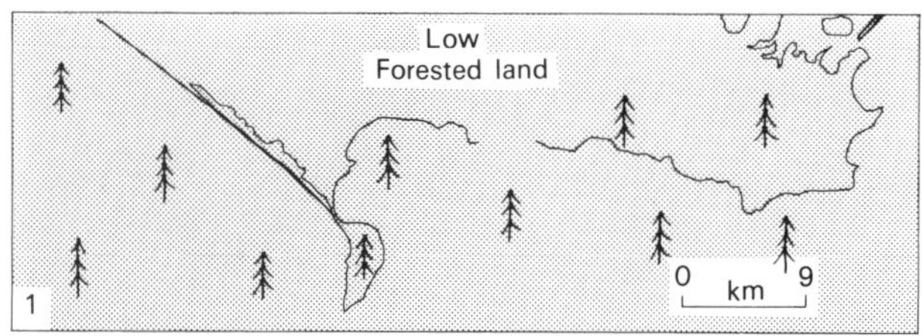

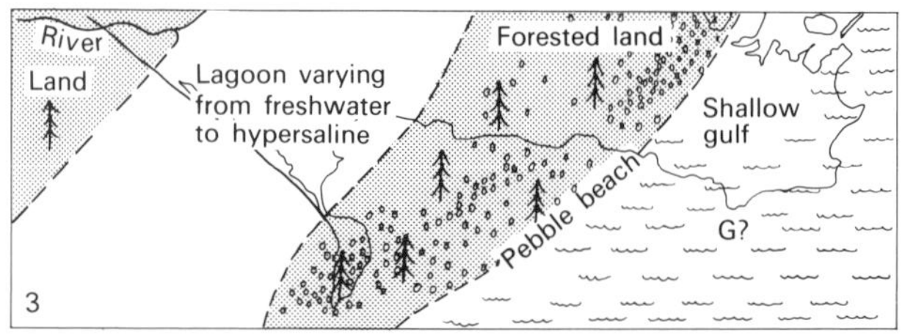

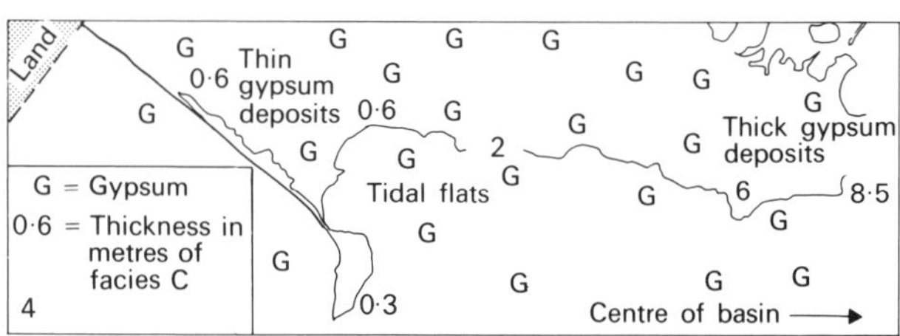

Fig. 8.30. Palaeogeography of the Lower Purbeck in Dorset. 1 – at the time of deposition of the Lower Dirt Bed. 2 – during the deposition of the limestones between the Lower and Great Dirt Bed. 3 – during the formation of the Great Dirt Bed. 4 – during deposition of the main evaporites (facies C) (from West, 1975).

These cycles differ from the regressive sequences of the Trucial Coast because they lack good subtidal carbonate muds and they contain fossil soil horizons. It is probable that the Purbeck environment was not as arid as that of the Trucial Coast (West, 1975).

To conclude, we see that the Lower Purbeck of Dorset has affinities with a modern arid shoreline. Our knowledge of the latter (algal mats, sabkhas etc.) helps us greatly in interpreting individual facies, but the actual sequence of facies associations in West's (1975) model is somewhat different.

NEW BRUNSWICK PLATFORM
FUNDY
BASIN
a
N
MEGUMA PLATFORM
0 50 km
Quebec
New Brunswick
P.E.I.
Maine
Nova Scotia

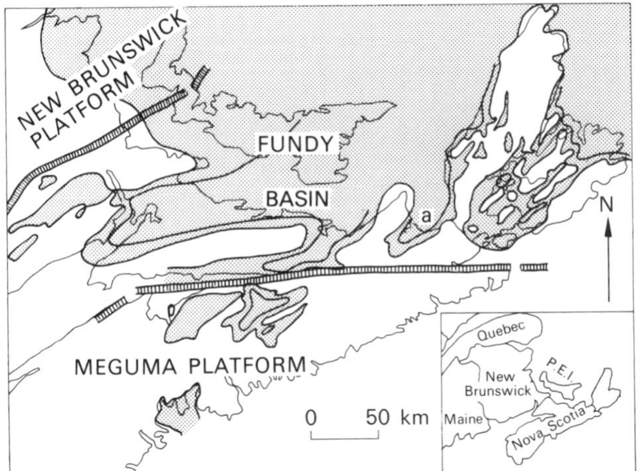

Fig. 8.31. The Fundy basin. a is the Antigonish-Mabou sub-basin. Basins are stippled; highs are clear (from Schenk, 1969).

8.4.3 The Middle Carboniferous of the Maritime Provinces, Canada

The Late Devonian Acadian orogeny produced a complex rift-valley system, the Fundy Basin, in the Maritime Provinces (Fig. 8.31) (see also Chapter 4.3.4). Within this basin, Middle Carboniferous Windsor Group sedimentation occurred in a series of linear sub-basins separating horst blocks which provided coarse detritus. Schenk (1969) compares one of these sub-basins, the Antigonish–Mabou sub-basin (Fig. 8.31) with the present-day Trucial Coast. The beauty of this comparison is that the Windsor Group shows all facies from coarse alluvial fans, coming off the horst blocks, to marine oolite-shoals and reefs and the full classic carbonate-sulphate-red bed facies sequence is seen (Fig. 8.32).

Facies I – consists of fossiliferous bituminous biomicrite believed to have been deposited in quiet-water normal-marine conditions on an inner-shelf area.

Facies II–VI – form the sequence from reefoid masses and oolite-shoals (biosparites and oosparites), through lagoonal muds (dolomitic biomicrite) and intertidal stromatolites to a supratidal sabkha (dolomitic sulphates with felted anhydrite and chicken-wire structure).

Facies VIII – is the dominant Windsor Group facies association consisting of red silty clays, red sands and coarse-grained pebbly feldspathic sandstones with festoon-type cross-stratification. Schenk (1969) interprets these as alluvial fans supplied by the horst blocks.

Facies VII – a blue-black lutite can underlie all these facies, and may represent an initial transgressive unit over which the main regressive sediment-wedge developed.

Sedimentary cycles – Schenk (1969) was able to recognize a number of carbonate–evaporite-red bed cycles within the 1,000 m of Windsor Group sediments. Each began with a rapid transgression followed by a slower regression. He believes that fluctuations in supply of terrigenous detritus into a continuously subsiding basin could be the mechanism controlling the cyclic sedimentation.

It is unfortunate that the Recent sabkha sequences have not been forming long enough for such a repetitive sediment column to be produced.

Fig. 8.32. Facies models of the Windsor Group, showing a carbonate–sulphate–red bed facies sequence. See text for descriptions (from Schenk, 1969).

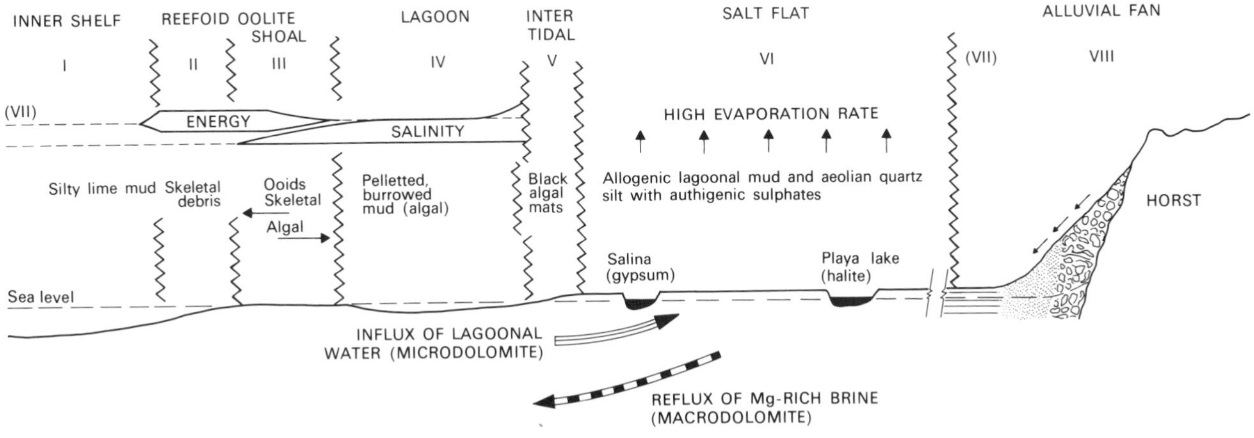

INNER SHELF
REEFOID OOLITE SHOAL
LAGOON
INTER TIDAL
SALT FLAT
ALLUVIAL FAN
I
II
III
IV
V
VI
(VII)
VIII
(VII)
ENERGY
SALINITY
HIGH EVAPORATION RATE
Silty lime mud
Skeletal debris
Ooids Skeletal
Algal
Pelletted, burrowed mud (algal)
Black algal mats
Allogenic lagoonal mud and aeolian quartz silt with authigenic sulphates
HORST
Salina (gypsum)
Playa lake (halite)
Sea level
INFLUX OF LAGOONAL WATER (MICRODOLOMITE)
REFLUX OF Mg-RICH BRINE (MACRODOLOMITE)

8.5 ANCIENT SALT DEPOSITS (SALINE GIANTS)

8.5.1 Introduction

The study of evaporites leaves us with a paradox. We can easily recognize ancient sabkhas from our study of modern environments. But there are many large (both in area and in thickness) evaporite basins dominated by halite deposits, for which we have no modern equivalent. Three such examples (saline giants) will be considered here, because they illustrate the concept that the presence of an evaporite bed does not immediately imply one specific sedimentary environment. This is partly because we have no definitive model for evaporite formation, but also because of the ease with which evaporite minerals undergo diagenetic change.

Various aspects of the 'deep-water' versus 'shallow-water' model will be discussed. An excellent review of the development of ideas on these models was presented by Hsü (1972). Braitsch (1971) describes the physico-chemical aspects of evaporite formation, based on the results of the large and active school of German workers (see Eugster; 1973, for a review of the early work by Van't Hoff). Borchert and Muir (1964) described evaporites in the days when 'deep-water' models were *de rigeur* and Cramer (1969) gave a comprehensive bibliography of evaporite formations.

8.5.2 The Upper Elk Point (Middle Devonian) Basin

This large, well studied basin (Fig. 8.33) is of great economic interest because the carbonates within it contain a number of large oil fields and sulphur deposits. They are also associated with Mississippi Valley-type mineralization (e.g. Jackson and Beales, 1967). In addition the evaporite rocks are rich in potash. This basin also highlights the arguments about evaporite origin. For example, succeeding papers in one issue of the *American Association of Petroleum Geologists Bulletin* advocate a shallow-water (Fuller and Porter, 1969) and deep-water (Klingspor, 1969) origin for the same formations.

The stratigraphic nomenclature of the Elk Point Basin is complex and confusing within the Elk Point Group in that the main formations are named differently in different states:

N.W.T.	ALBERTA	SASKAT-CHEWAN	MAIN LITHOLOGIES
Watt Mtn.	Watt Mtn.	2nd Red Beds	Clastics
Presqu'ile (Barrier Reef)	Muskeg	Prairie	Evaporites
Keg River	Keg River	Winnipegosis	Carbonates

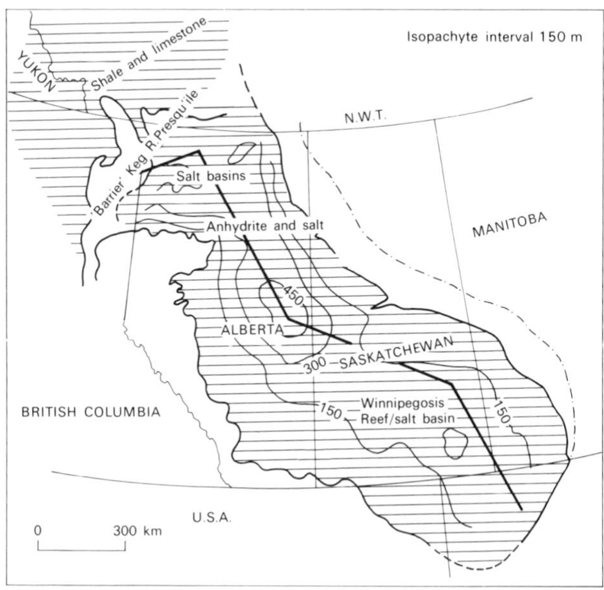

Fig. 8.33. Location map for the Upper Elk Point (Middle Devonian) basin. The north-eastern boundary of the basin is an erosional edge. The reconstructed basin edge is shown by a dashed line. Line of section (Fig. 8.34) is marked.

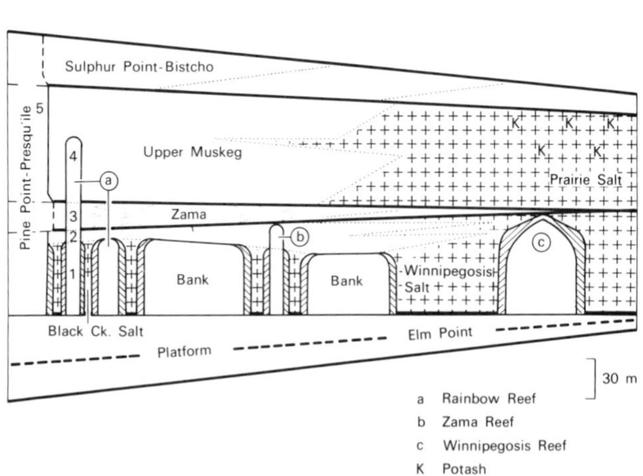

Fig. 8.34. Diagrammatic NW–SE cross-section showing main facies and stratigraphic relations in Middle Devonian Upper Elk Point basin. Numbers indicate main stages of reef growth: (1) Black Creek stage; (2) unnamed (three-anhydrite) stage; (3) Zama stage; (4) high Rainbow stage; and (5) top Muskeg–Presqu'ile stage. Anhydrite flanks first-stage reefs and banks at edge of marginal shelf (represented by oblique hatching), and laminite floors the intervening basins. Section has an average horizontal compression of about 5,000 times. Line of section shown on Fig. 8.33. (From Fuller and Porter, 1969).

Table 8.4 Stratigraphic subdivisions of the Winnipegosis and Prairie Formations, Elk Point Group, Saskatchewan (from Shearman and Fuller, 1969)

Formation Name	Lithological Subdivision	Sedimentary Stage
DAWSON BAY		
PRAIRIE	Salt Potash Salt Zama Dolomite and Evaporite	Deposits of the Prairie Salt Basin
	Anhydrite and Salt Salt Laminates	Deposits of the Winnipegosis Salt Basin
WINNIPEGOSIS	Reef Platform	
ASHERN	Basal shale	

The main rock types in the Saskatchewan area are shown in Table 8.4 and the stratigraphic relationships are summarized in the cross-section given in Fig. 8.34.

Shearman and Fuller (1969) and Fuller and Porter (1969) believe that Winnipegosis time began with typical warm-marine conditions and the formation of large upstanding reef masses. A very profound lowering of sea-level by a hundred or so metres then emptied the Elk Point Basin, producing arid intertidal conditions in which algal mats (laminites) colonized the inter-reef areas; sabkha deposits (flanking anhydrite) covered the reef margins. This fall in sea-level was caused by a damming of the northern barrier reef, allowing the water to evaporate away. Petrographic analysis shows that the laminites, now calcitized, had a complex diagenetic history. Original carbonate/organic matter layers were invaded by growing anhydrite. Slight oscillations in the water level caused these sediments to be drowned intermittently. Such flooding by relatively dilute waters calcitized the uppermost anhydrite layers, resulting in a calcite/anhydrite (organic matter) laminite. Shearman and Fuller show that the nodules in the anhydrite beds, which form enterolithic layers or chicken-wire structures, contain a felted mass of anhydrite laths, identical to those formed in the present-day sabkhas (see Sect. 8.2). This anhydrite is interbedded with dolomitic carbonates. They also point to the karstic features in the reef tops as evidence of exposure in the basin.

The Winnipegosis anhydrites vary from centimetres to 30 metres thick and are interpreted as a succession of sabkha-type sequences. Presumably, slight transgressions or irregularities in subsidence rate caused the breaks in succession. The overlying salt deposits (up to 200 m thick) are also laminated, and

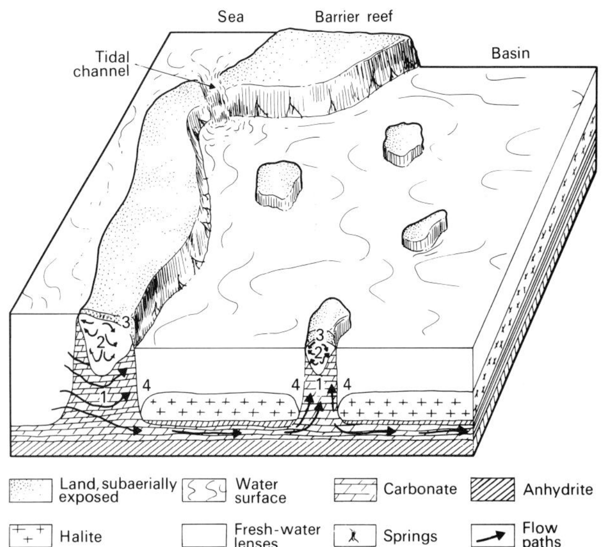

Fig. 8.35. Block diagram showing suggested conditions in late Winnipegosis and Prairie times. 1 – normal sea water; 2 – fresh-water lens; 3 – vadose or sub-aerial zone; 4 – normal sea-water flow (causing solution of salt). (From Maiklem, 1971).

contain *chevron halite* (see Sect. 8.3; Wardlaw and Schwerdtner, 1966). Following Shearman (1970) these deposits would be interpreted as salina deposits which gradually filled up the basin. Potash deposits also formed as the final precipitates in more localized bitterns. It seems likely that brines were supplied from the north-west over the Keg River–Presqu'ile Barrier.

Klingspor (1969) describes and figures the Muskeg Formation and shows that it contains similar laminites, evaporites and banded salt deposits. He dismisses the similarity of the laminites to algal mats as coincidence and proposes that they formed in a quiet-water deep basin under strong evaporite conditions. The development of brines able to precipitate halite under these conditions is fully explained in Schmalz (1969) and below (Sect. 8.5.4). The Open University (1976), in a well illustrated account of the Elk Point Basin, similarly sweep away the idea of an intertidal origin for the laminites, claiming that their persistence over several kilometres proves a quiet-water deep-water origin. However, it is worth noting that present-day algal mats can be traced over several kilometres (see Sect. 8.2 and Park, 1977).

Maiklem (1971) believed that reef formation (in places forming masses up to 200 metres high) ended in the Elk Point Basin when sea-level dropped about 30 metres. He believed that blocking of the barrier reef led to 'evaporative downdraw' within the basin. The difference in water levels outside and inside the basin (Fig. 8.35) would force normal marine waters through the reef masses, causing diagenetic changes (cementation, dolomitization) within them, but also maintaining a supply of salts to the basin.

Overall, the evidence in this basin seems to point to salina and sabkha conditions for the formation of the upper Elk Point evaporites, but the proposals made by Hsü (1972) for the Messinian (uppermost Miocene) of the Mediterranean may also apply (see Sect. 8.5.5 below).

8.5.3 **The Zechstein (Upper Permian) Basin of Europe and the North Sea**

THE GERMAN ZECHSTEIN

The Zechstein salt deposits of Germany, which are up to 1,200 m thick, have long been mined and studied. Indeed the first mining of potash deposits in the world was in Stassfurt in 1862. German geologists have made painstaking and detailed studies of these salt deposits. Figures 8.36 and 8.37 show the form and sedimentary sequence of the Zechstein basin.

Four main evaporite cycles are recognized (Z1 to Z4) and an ideal cycle would be:

Top: Anhydrite (retrogressive)
Potash salt
Halite
Anhydrite (progressive)
Dolomite
Limestone
Base: Siltstone, clay etc.

The German workers (e.g. Lotze, 1938; Richter-Bernburg, 1955) believe the basin contained about 1,000 m depth of sea water, and had a shallow bar in the north-west over which less saline sea-water flowed into the basin. Each of the four main cycles (a small fifth one Z5 is now also recognized, Smith, 1973) represents a damming of this entrance, causing an increase in evaporation to give first sulphates and then halides. The cycle ended with a breaching of the bar, causing dilution of the brines (see below for a model for deep-water evaporite formation).

The halite and potash beds are often rhythmically banded on a centimetre scale with halite-rich and clay–dolomite–anhydrite-rich laminae. These have been interpreted as varves, or annual layers (Richter-Bernburg, 1955). During the hot, dry summer seasons an increase in evaporation caused precipitation of halite; in the cooler, warmer, winter seasons an increased inflow of river waters dilutes the brines, brings in detritus and produces the anhydrite/detritus-rich layers. German workers use the bromine content of the salt deposits as a direct reflection of the salinity of the brines from which the salt precipitated. Kühn (1952) showed that the summer laminae were richer in bromine (and hence of higher salinity) than the winter laminae. Richter-Bernburg (1955) counted these annual 'varves', related their thickness to the eleven-year sunspot cycle, and concluded that the Zechstein salts formed in 500,000 years. He noted that the Zechstein stage probably lasted for 20–30 million years, and believed (rather ahead of his time) that much of this period was represented by non-deposition. For some of the salt beds he concluded that the rate of deposition must have been about 6 cm per year. This gave weight to his belief in a deep-water basin, since this rate is far in excess of the possible subsidence rate: the basin therefore must have been deep *before* deposition commenced. All this interpretation hinges, of course, on the correct interpretation of the varves as annual. Shearman's (1970) studies of salina salts (see Sect. 8.3) suggest that these laminae represent major flooding and evaporation events (perhaps occurring every few years), thus giving a longer, more reasonable, time for Zechstein salt formation. However, Shearman assumes a shallow-water origin for such laminated salt deposits.

The recent discovery of redeposited anhydrite beds, interpreted as turbidites, and of mass-flow breccias and slumps in the German Zechstein is, however, powerful evidence for a deep basinal origin of *part* of this formation (Schlanger and Bolz, 1977). Some compromise, including both shallow and deep environments, is thus probably applicable to the Zechstein.

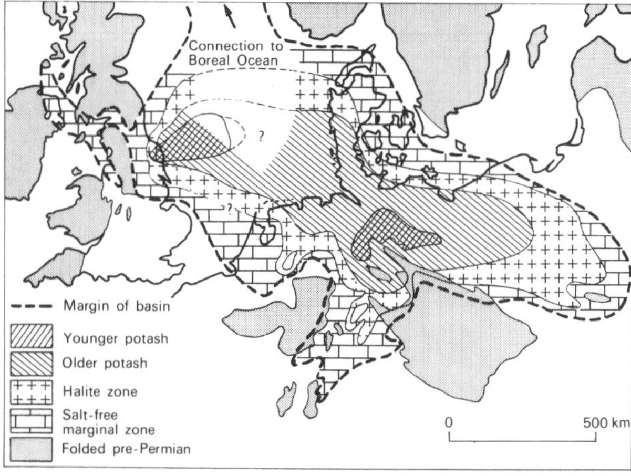

Fig. 8.36. The Zechstein Basin (after Borchert and Muir, 1964).

Fig. 8.37. (opposite page) The Zechstein succession in Germany (after Richter-Bernburg, 1955). The top horizons of Zechstein 4 – the Grenzanhydrit – would now be assigned to Zechstein 5. The arrows summarize the changes in net evaporation.

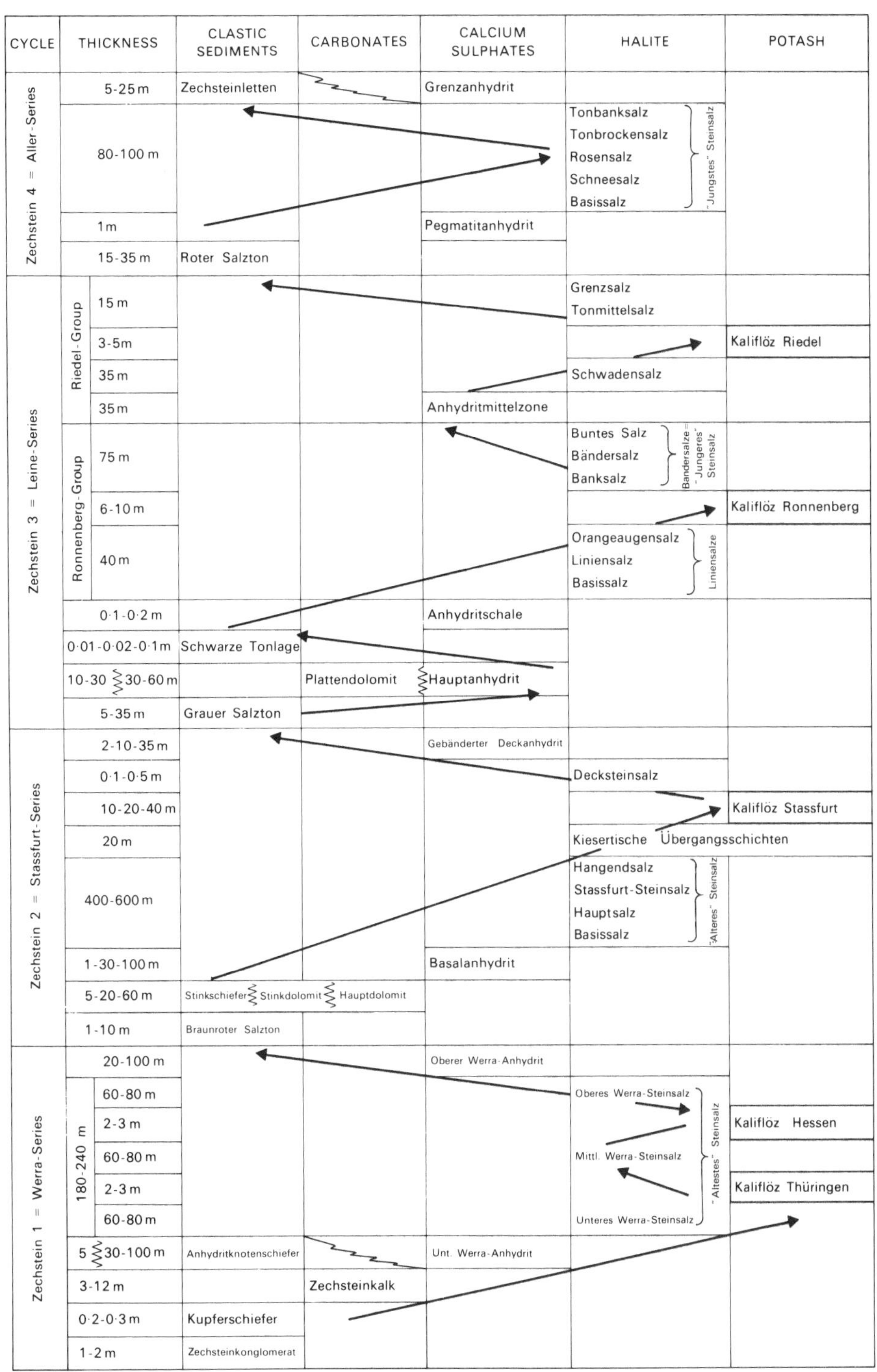
CYCLE
THICKNESS
CLASTIC SEDIMENTS
CARBONATES
CALCIUM SULPHATES
HALITE
POTASH
Zechstein 4 = Aller-Series
5-25 m
Zechsteinletten
Grenzanhydrit
80-100 m
Tonbanksalz
Tonbrockensalz
Rosensalz
Schneesalz
Basissalz
"Jungstes" Steinsalz
1 m
Pegmatitanhydrit
15-35 m
Roter Salzton
Zechstein 3 = Leine-Series
Riedel-Group
15 m
Grenzsalz
Tonmittelsalz
3-5m
Kaliflöz Riedel
35 m
Schwadensalz
35 m
Anhydritmittelzone
Ronnenberg-Group
75 m
Buntes Salz
Bändersalz
Banksalz
Bandersalze = "Jungeres" Steinsalz
6-10 m
Kaliflöz Ronnenberg
40 m
Orangeaugensalz
Liniensalz
Basissalz
Liniensalze
0·1-0·2 m
Anhydritschale
0·01-0·02-0·1 m
Schwarze Tonlage
10-30 30-60 m
Plattendolomit
Hauptanhydrit
5-35 m
Grauer Salzton
Zechstein 2 = Stassfurt-Series
2-10-35 m
Gebänderter Deckanhydrit
0·1-0·5 m
Decksteinsalz
10-20-40 m
Kaliflöz Stassfurt
20 m
Kiesertische Übergangsschichten
400-600 m
Hangendsalz
Stassfurt-Steinsalz
Hauptsalz
Basissalz
"Alteres" Steinsalz
1-30-100 m
Basalanhydrit
5-20-60 m
Stinkschiefer
Stinkdolomit
Hauptdolomit
1-10 m
Braunroter Salzton
Zechstein 1 = Werra-Series
20-100 m
Oberer Werra-Anhydrit
180-240 m
60-80 m
Oberes Werra-Steinsalz
2-3 m
Kaliflöz Hessen
60-80 m
Mittl. Werra-Steinsalz
2-3 m
Kaliflöz Thüringen
60-80 m
Unteres Werra-Steinsalz
"Altestes" Steinsalz
5 30-100 m
Anhydritknotenschiefer
Unt. Werra-Anhydrit
3-12 m
Zechsteinkalk
0·2-0·3 m
Kupferschiefer
1-2 m
Zechsteinkonglomerat

THE NORTH SEA ZECHSTEIN

The edge of the Zechstein basin reaches the shores of north-east England (Stewart, 1963) and the potash beds of Z2 are now mined in Yorkshire. Exploration for North Sea oil and gas over the past fifteen years has enabled correlations to be made across the basin (Fig. 8.38) and has led to new stratigraphic proposals for the Permian of the whole British Isles (Smith *et al.,* 1974). Unfortunately less has been learnt about the sedimentology of these strata because of halokinetic disruption and partly because little detailed work has been published for proprietary reasons. However, Taylor and Colter (1975) describe the members of typical Zechstein cycles and suggest a model for the basin. They recognize two parts of the basin; the western platform and the eastern (central Zechstein) basin, as shown in Figs. 8.39 and 8.41.

The following succession is found in the second Zechstein cycle (Z2).

Carbonates – Taylor and Colter (1975) consider that the Upper Werra Anhydrite is a platform equivalent of the basin carbonates of Z2 (compare with Fig. 8.38). The Upper Werra is dolomitic and contains crinkly laminations of anhydrite which are very reminiscent of recent algal mats (cf. laminites of the Elk Point basin). These anhydrite beds pass laterally into dark, bituminous laminated carbonates – the Stinkkalk (Fig. 8.40B). The platform anhydrites are overlain by oolitic dolomitic sands of shallow-water origin. These pass laterally into burrowed, pelleted ostracod-rich muds and then into foetid dark limestones (Fig. 8.40D).

Evaporites – 50 to 100 m of anhydrite, polyhalite and halite is overlain by 300–600 m of halite, which contains 10 m of potash-rich beds 10 m from the top. The evaporites grade from anhydrite (sabkhas?) to halite and thicken basinwards (Fig. 8.40F). Thickness and sedimentary structures are not easily determined, since this thick halite unit is the main cause of Zechstein halokinetics.

The Z2 cycles are capped with a 'regressive' anhydrite unit (Deckanhydrit) which Taylor and Colter (1975) interpret as a

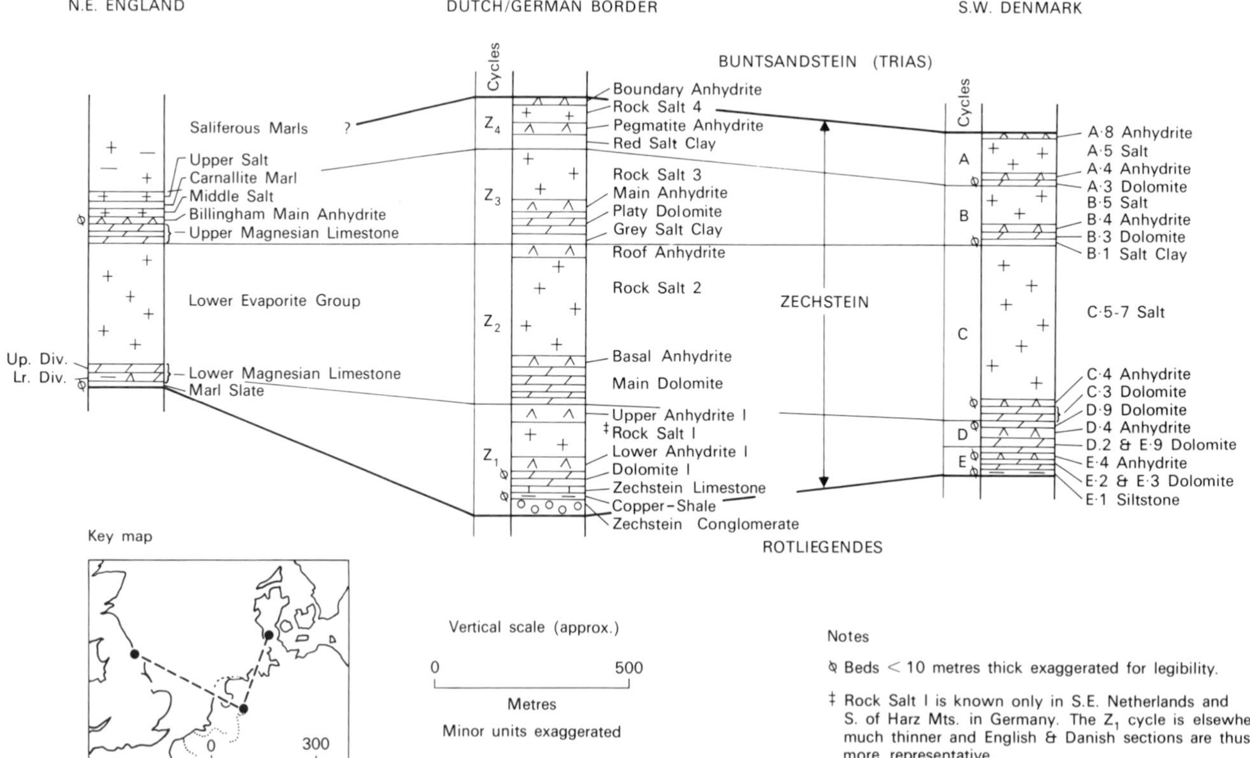

Fig. 8.38. The Zechstein successions bordering the North Sea area (from Brunstrom and Walmsley, 1969).

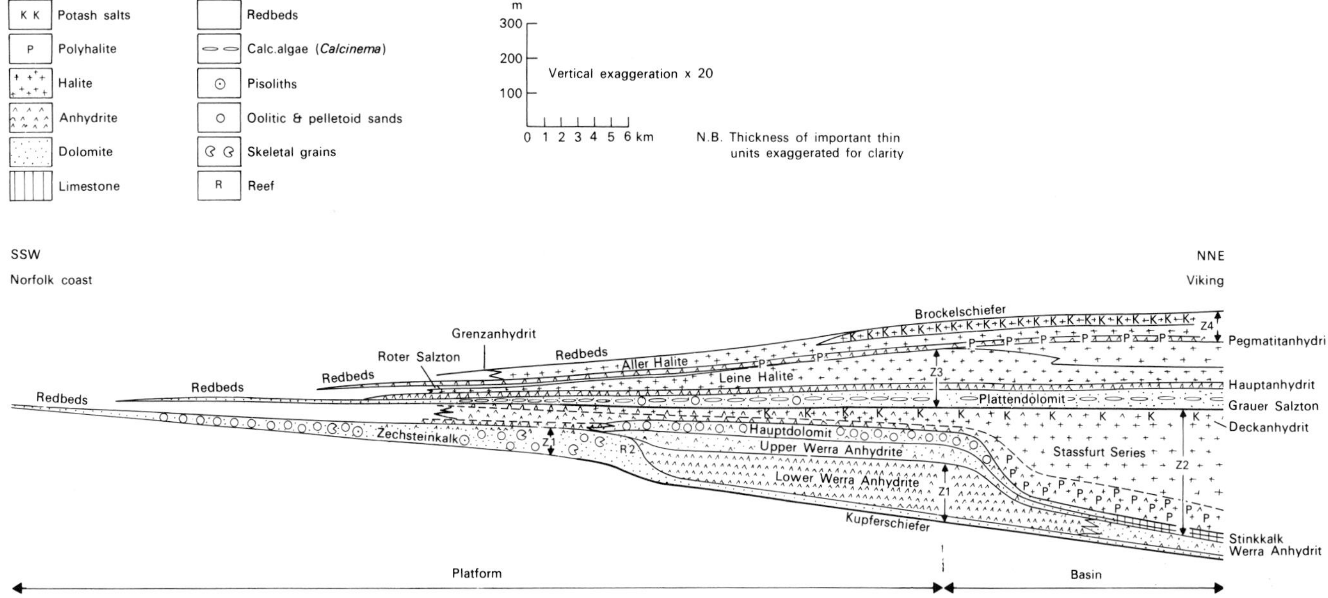

Fig. 8.39. Generalized cross-section in the southern North Sea Basin, from the Norfolk coast towards the Viking Field (from Taylor and Colter, 1975).

Fig. 8.40. Isopach and facies maps for the second Zechstein cycle in the southern North Sea (from Taylor and Colter, 1975).

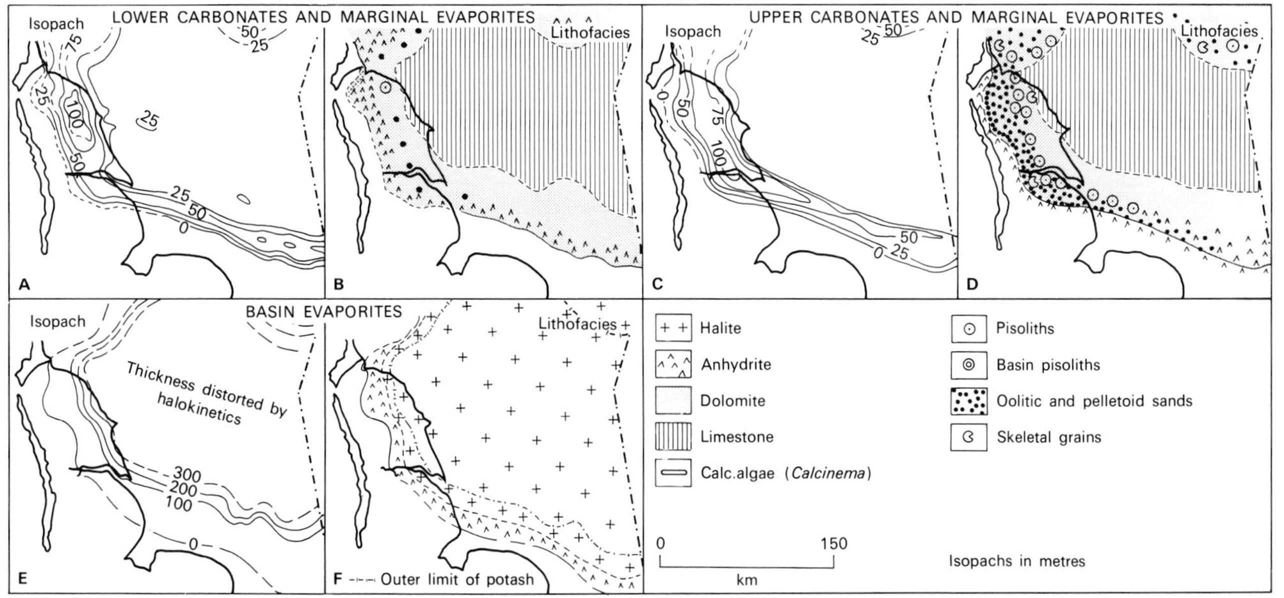

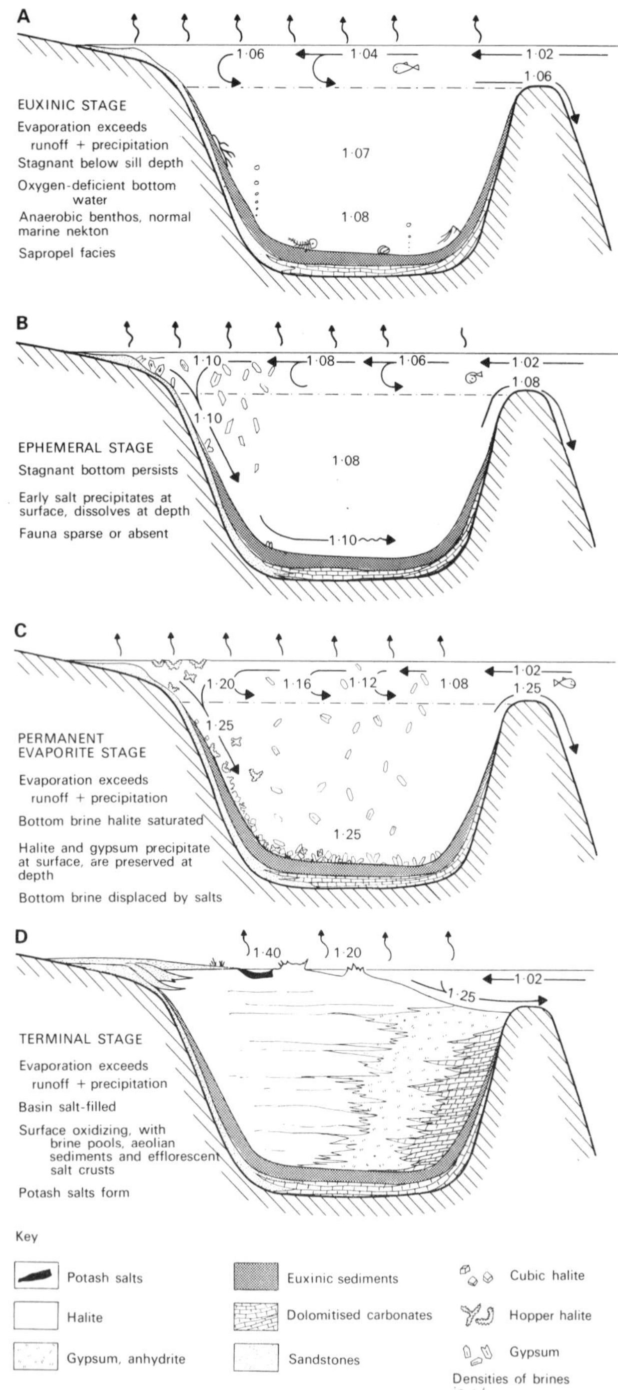

Fig. 8.41. A model for deep-water evaporite deposition. Four stages in the infilling of such a basin are shown (after Schmalz, 1969).

sedimented solution–residue from the underlying halite. This thin unit (about 1 m) represents the third-cycle transgression.

Taylor and Colter see the North Sea Zechstein as an adaptation of the classical barred basin model (see below). In a tideless, almost current-free, inland sea, carbonate sediments formed prolifically on the shallow margins of the basin, rapidly building out a wedge of sediment. As subsidence of this basin continued, communication with the northern ocean was restricted, causing a drop in sea level. This exposed the basin margins, allowing sabkhas to develop before the basin waters became sufficiently concentrated to precipitate evaporites. As the major restriction to circulation was maintained, evaporites formed in the basin, until potash deposits were laid down in the last stages before circulation was restored.

8.5.4 Deep-water evaporite models

Barred-basin models have been popular since the 1870s (see the review by Hsü, 1972) when Ochsenius realized that if the Zechstein basin was 720 m deep and full of sea water, and was completely isolated, it would only produce 12 m of salt deposit on desiccation. Since there are 490 m of salt at Stassfurt this necessarily implied the opening and closing of a flood-gate forty-one times. This, he thought, defied belief and consequently proposed that a continuous supply of sea water occurred across a shallow barrier. Innumerable variations on

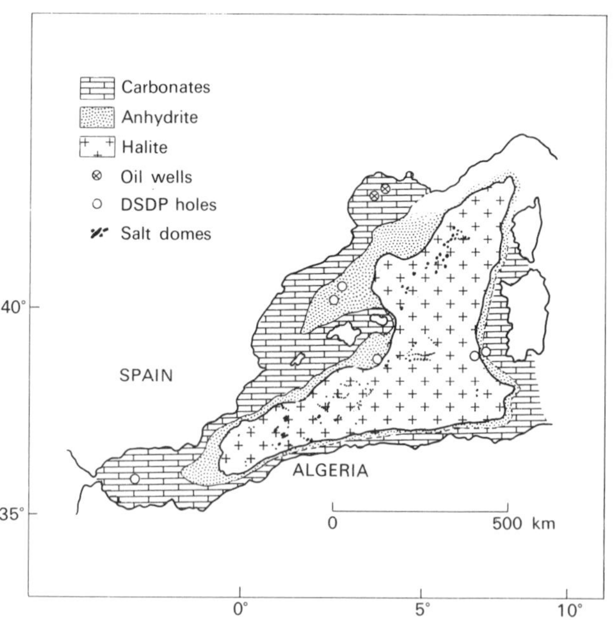

Fig. 8.42. The distribution of Messinian evaporites in the Balearic region (from Hsü, 1972).

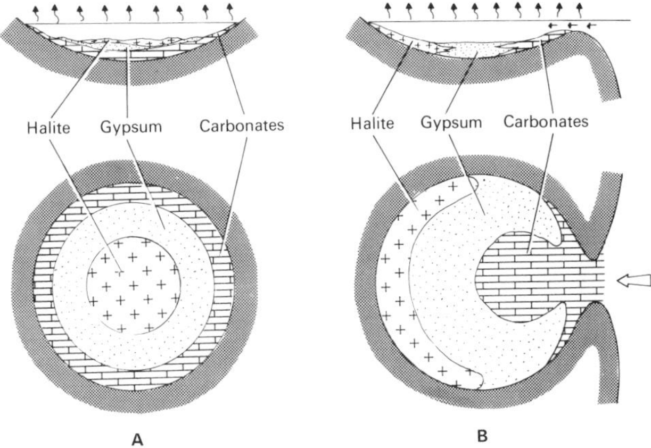

Fig. 8.43. Theoretical facies distribution in a desiccating basin (A) and in a barred basin (B) (from Hsü, 1972).

this theme have occupied the intervening century. Recently Schmalz (1969) described a succession of states for such a model (Fig. 8.41). In a basin where evaporation is in excess of total water input (that is run-off and precipitation) a stable water mass of denser brine gradually builds up below the sill (Fig. 8.41A). The precipitation of evaporites begins in the shallower, more concentrated waters of the basin margins (Fig. 8.41B) which act as a supply area to the deeper parts of the basin, in which the brine density gradually builds up. The terminal stage (Fig. 8.41D) could finish in other ways. If the barrier becomes blocked, potash 'salinas' could form and gradually become overlain by aeolian and outwash clastics. Alternatively, if evaporation became less than run-off and precipitation, evaporite formation would be disrupted, a new euxinic clastic layer might form, and the cycle could start again when net evaporation resumed.

There are a number of semi-enclosed basins at the present day which could fit this model: the Mediterranean, the Red Sea and the Arabian Gulf, but none at present has sufficient net evaporation (largely due to run-off from large river systems) to maintain evaporite formation. To those who can accept the idea that there may be some ancient environments for which we have no present analogue, Schmalz's (1969) model is highly plausible.

8.5.5 The Mediterranean Messinian Basin

The lands surrounding the Mediterranean have long been known to possess Upper Miocene evaporites; and from geophysical evidence of diapiric structures it was suspected that they also underlay the floor of the Mediterranean itself. Leg XIII of the Deep Sea Drilling Project encountered a 500 m

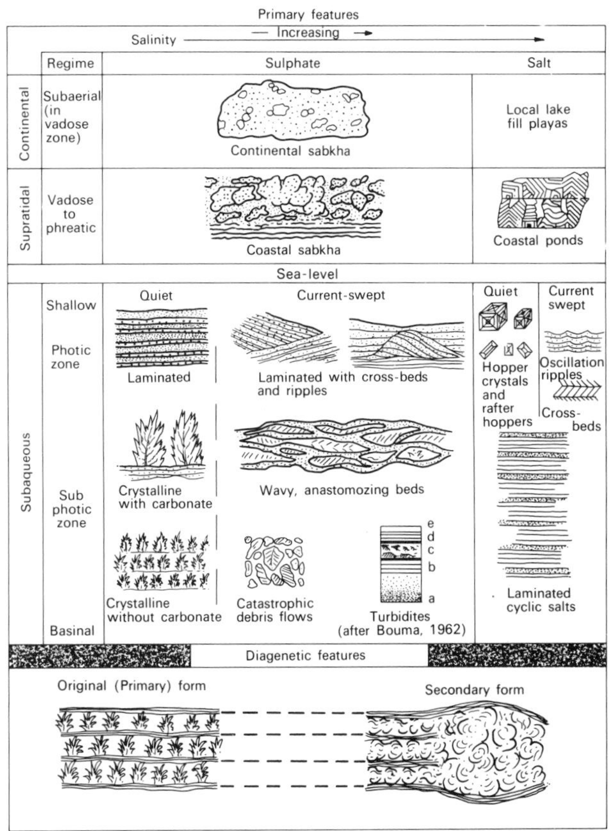

Fig. 8.44. The form of evaporite deposits found in the Messinian of the Sicilian Basin (from Schreiber *et al.*, 1976).

thick succession of evaporites (Ryan and Hsü *et al.*, 1973). These were shown to be of Late Miocene, Messinian (6.5 to 5.0 m.y. B.P.) age. Hsü (1973a) noted that the anhydrite layers were enterolithic, bearing a strong resemblance to sabkha deposits. They formed in a basin whose topography was very similar to the present form of the Mediterranean (Fig. 8.42) and were found to be stratigraphically associated with deep-water clastic sediments. Yet Hsü found desiccation cracks within the evaporite formations – clear evidence of exposure. To explain those apparently conflicting observations he made a revolutionary proposal, which has found much support and, he claims (Hsü *et al.*, 1977), is reinforced by the second DSDP visit to the area. Hsü (1972) suggested that the Messinian evaporites formed in a desiccating deep basin, whose floor was several kilometres below global sea level. This proposal seems to fit the observations very well. For example one would expect a *bull's eye* distribution of evaporites in such a drying-out basin. Figure 8.42 shows this to be the case in the

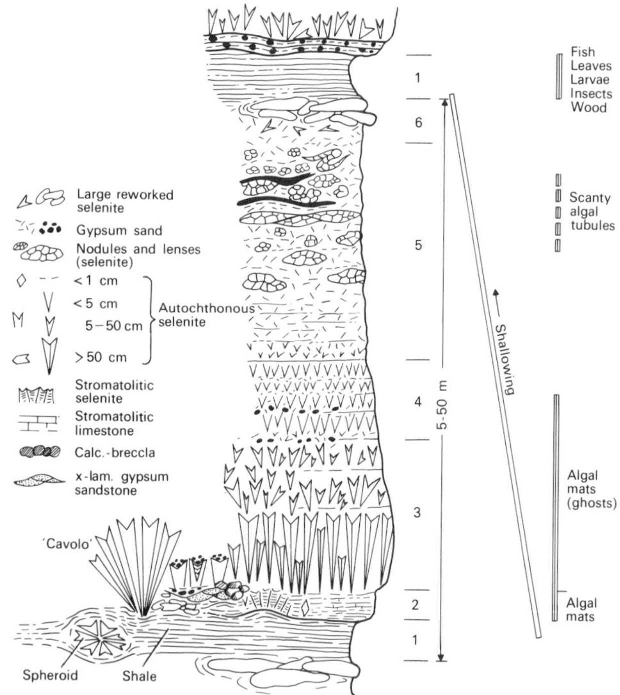

Fig. 8.45. The sedimentary cycle found in Messinian gypsum beds of the Northern Apennines. See text for details of the facies (from Vai and Ricci Lucchi, 1977).

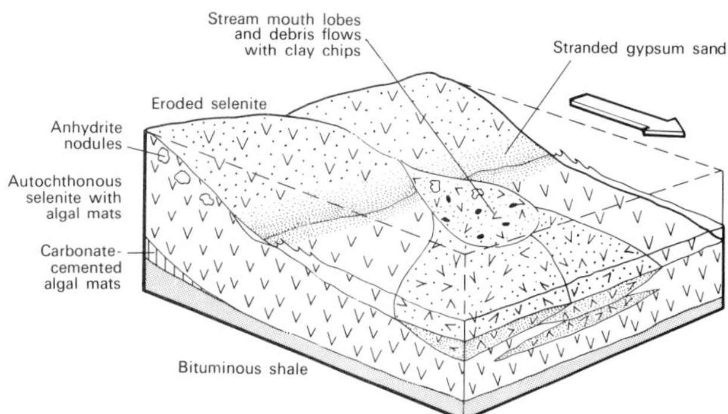

Fig. 8.46. The environment of deposition of gypsum beds in the Messinian of the Northern Apennines (from Vai and Ricci Lucchi, 1977).

Mediterranean. The traditional deep-water deep-basin model should have a *tear-drop* pattern of evaporite facies imposed by the positioning of the sea-water source (Fig. 8.43). Hsü observes that the Zechstein deposits (see Sect. 8.5.3) have a 'multi-centred bull's-eye' distribution and the Elk Point Basin (Sect. 8.5.2) has an 'elongate bull's-eye' pattern. The Mediterranean model may, therefore, be applicable in these two cases.

These fascinating proposals have led to many publications on the Messinian evaporites on land (Drooger, 1973). The deposits in the Sicilian Basin are particularly interesting in this respect: Schreiber *et al.* (1976) describe the varied forms of gypsum and anhydrite that are found in environments that vary from continental to basinal, and suggest that their morphology can be diagnostic of their sedimentary environment (Fig. 8.44).

Vai and Ricci Lucchi (1977) describe massive (up to 35 m thick) gypsum beds found in the Messinian of the Northern Apennines. These remarkable sediments (Fig. 8.45) seem to have formed on the margin of a basin, and contain reworked intraformational (cannibalistic) debris-flow beds of gypsum.

The facies in Fig. 8.45 begin with a bituminous shale (1) totally lacking macrobenthos whose environment of deposition is uncertain. This is overlain by intertidal stromatolitic sediment rich in selenite (facies 2). Facies 3 and 4 are interpreted as forming in the same environment, but are characterized by very large (10–100 cm) swallow-tail gypsum crystals. These large crystals look very similar to those found growing in the Abu Dhabi sabkha at the water-table level (see Sect. 8.2). Facies 5 and 6 consist of nodular and chaotic selenite crystals which Vai and Ricci Lucchi (1977) believe were transported into the basin by fluvial action and moved forward intermittently as sluggish debris-flows. They believe that the whole cycle represents a regressive sequence at the edge of a desiccating basin (Fig. 8.46).

8.6 CONCLUSION

The study of evaporites is at an interesting stage. Sabkha, salina and basin-edge facies are now recognizable in the ancient record through their good modern analogues. But are the thick salt deposits at the basin centres the result of deep-water deep-basin or desiccating deep-basin deposition? Although the latter is no longer an outrageous hypothesis its applicability to 'saline giants' other than the Mediterranean Messinian still needs to be confirmed.

FURTHER READING

Kirkland D.W. and Evans R. (Eds.) (1973) *Marine Evaporites: origin, diagenesis and geochemistry*. pp. 426. Dowden, Hutchinson & Ross, Stroudsburg.

Purser B.H. (Ed.) (1973) *The Persian Gulf. Holocene carbonate sedimentation and diagenesis in a shallow epicontinental sea*. pp. 471. Springer-Verlag, Berlin.

CHAPTER 9 Shallow Siliciclastic Seas

9.1 INTRODUCTION

Shallow seas lie between those parts of the sea dominated by nearshore processes and those parts dominated by oceanic processes. They have moderate depths (ca. 10–200 m) and comprise two main types (Shaw, 1964; Heckel, 1972): (1) marginal or pericontinental seas as exemplified by present-day continental shelves, and (2) epeiric or epicontinental seas which extend over continental regions to form shallow, partially enclosed basins such as the Baltic Sea, Hudson Bay, North Sea and Yellow Sea.

There have been relatively few sedimentological studies of ancient offshore shallow marine siliciclastic deposits, resulting in a scarcity of ancient shallow marine models. This can be partially attributed to the rather slow development of modern shelf studies caused by inaccessibility and the extreme variability of the physical and biological processes in this complex environment. In addition it partly results from the atypical nature of modern shelves compared with ancient shallow seas since, following the Holocene transgression, they all show some degree of disequilibrium. Thus while modern shelves provide valuable transgressive models, they are not directly applicable to ancient regressive or equilibrium situations.

Nevertheless the modern shelf environment provides a basis for reconstructing the physical and biological processes operating in ancient shallow seas. A previous review of this complex subject by Heckel (1972) emphasized the organic aspects of these deposits (see also Chapter 10). This chapter, however, concentrates on the physical processes of sedimentation in modern shelf seas and utilizes these observations in the reconstruction of palaeohydrodynamic regimes in ancient shallow seas.

9.1.1 Historical development

Ideas in the sedimentology of modern continental shelves and ancient siliciclastic seas have, for the most part, followed diverse courses.

For many years the continental shelf was considered to be an equilibrium surface whose surficial sediments *should* show a progressive decrease in grain size when traced offshore (Johnson, 1919). However, Shepard (1932) demonstrated by bottom sampling that in reality most shelves are covered by a complex mosaic of sediments. This complexity is partly due to the presence of relict sediments (Emery, 1952, 1968) which were originally deposited in various shallow water and terrestrial environments and subsequently drowned during the Holocene transgression. Bottom sediment sampling dominated shelf studies into the 1950s with the greatest upsurge in information accompanying World War II in response to the need to interpret acoustic data used to detect submarines (Emery, 1976). This was because acoustic reflectivity and reverberation of sound is controlled by the type of shelf substrate, particularly sand, mud and rock bottoms.

The process-response phase of shelf studies was initiated by Van Veen (1935, 1936) in the southern North Sea and continued in the seas around the British Isles (Stride, 1963). It began in N.W. Europe for three main reasons. First it has long been known that powerful tidal currents are responsible for the present-day movement of sand banks in these and other tide-dominated seas. Secondly, the regular acquisition of hydrographic data coupled with the movement of major sand bodies is critical to the safety and economy of some of the world's busiest shipping lanes. Finally, the improvement of acoustic methods enabled large areas of the continental shelf to be mapped by side-scan sonar techniques which could be integrated with the detailed hydrographic data. During the 1960s and early 1970s rapid advances were made in the importance of tidal currents in sediment transport and development of bedforms (Stride, 1963; Reineck, 1963; Belderson and Stride, 1966; Houbolt, 1968; Kenyon, 1970; Kenyon and Stride, 1970; McCave, 1971a, b, and c; Terwindt, 1971a). It was also demonstrated that wave agitation affected most shelf bottoms during storms, enabling sediment transport rates to be increased significantly (Hadley, 1964; Draper, 1967; Johnson and Stride, 1969).

The initial overemphasis on relict sediments delayed the application of process-response models to other types of shelves, notably the wind/wave-dominated type. However, the results of detailed studies on several North American shelves from the late 1960s onwards have shown that present-day processes can overprint pre-Holocene deposits through *in situ* reworking (Swift, 1969a and b, 1970; Swift, Stanley and Curray, 1971). Furthermore, since most North American shelves are wind/wave dominated, emphasis has been placed on seasonally-controlled shelf hydraulic regimes, with their alternation of storm and fair weather periods (Swift, 1969a; Swift *et al.*, 1972a, 1973; Komar *et al.*, 1972; Kulm *et al.*, 1975).

Thus, the greatest significance of the 1970s shelf studies is the increased appreciation of modern processes and products, which are more directly applicable to the interpretation of ancient deposits.

Environmental reconstruction of ancient shallow marine siliciclastic deposits has developed from two lines of study: (1) palaeoecological, and (2) sedimentological. In palaeoecology emphasis is placed on the reconstruction of substrate conditions, life habitats and palaeocommunities by interpretation of body fossils, trace fossils and their assemblages (e.g. Ladd, 1957; Hedgpeth, 1957; Ager, 1963; Imbrie and Newell, 1964; Schäfer, 1972; Scott and West, 1976).

In sedimentology, early emphasis was on sand body geometry, textural trends and palaeocurrent directions but, with the scarcity of modern analogues, there were few attempts to relate observed products to known processes. However, in the 1970s processes and products were related in studies of cross-bedded sand deposits whose structures and palaeocurrent patterns were thought to have been formed in tidal shelf environments (e.g. McCave, 1968; Narayan, 1971; De Raaf and Boersma, 1971; Swett, Klein and Smit, 1971). These studies were the direct result of work on modern tidal shelf seas in the 1960s; thereby illustrating how work on a modern environment can give impetus to ancient studies.

The storm versus fair weather concept (Swift, 1969a) was first applied to ancient deposits by Hobday and Reading (1972) while Banks (1973a) and Brenner and Davies (1973) integrated both concepts of tidal shelf currents and superimposed storm activity in high energy shallow marine sandstones, a feature emphasized by later workers (e.g. Anderton, 1976; Johnson, 1977a). In ancient seas with only weak currents, however, both Goldring and Bridges (1973) and Banks (1973b) noted the importance of storm events and particularly the high preservation potential of their deposits.

Thus, there is now an upsurge in the sedimentological analysis of ancient shallow marine deposits concurrent with an increase in publications of modern shelf studies (Emery, 1976, p. 582). Because the deposits of ancient shallow marine siliciclastic environments are so poorly known, and because they are potentially of high economic importance, they should prove a particularly valuable line of research in the near future.

9.2 MODERN SILICICLASTIC SHELF MODELS

Emery (1952, 1968) recognized six main types of shelf sediment: (1) detrital (sediment now being supplied to the shelf), (2) biogenic (shell debris, faecal pellets etc.), (3) residual (*in situ* weathering of rock outcrops), (4) authigenic (e.g. glauconite, phosphorite etc.), (5) volcanic, and (6) relict (remnant from an earlier environment and now in disequilibrium). Emery considered that 70% of present-day continental shelves are covered by relict sediment, while the remainder consists of modern sediments occurring mainly as a thin strip of coastal sands and as mud belts adjacent to large deltas and river mouths (Emery, 1968). However, more recent studies suggest that the proportion of relict sediment is considerably smaller (generally <50%) because large areas of Arctic shelves are covered by recent sediments (Creager and Sternberg, 1972, p. 350, their Fig. 25; McManus, 1975).

A dynamic model of shelf sedimentation was proposed by Curray (1964, 1965) and Swift (1969a, 1970) who looked at patterns of shelf sediment distribution with regard to processes. They recognized three main shelf facies: (1) *shelf relict sand blanket* comprising pre-Holocene deposits in disequilibrium with present-day processes, (2) *nearshore modern sand prism* comprising shoreline beaches, barriers and shoreface, and including a seaward thinning nearshore sand zone, and (3) *modern shelf mud blanket* consisting of fine-grained sediment which has bypassed the nearshore zone and been deposited on various parts of the shelf. Growing awareness of the importance of present-day shelf processes led to the differentiation between true '*relict sediment*', which should denote unreworked sediment, and '*palimpsest sediment*', which is a reworked sediment possessing aspects of both its present and former environments (Swift, Stanley and Curray, 1971). This includes many of those continental shelves experiencing intense wave and current activity which are reworking Pleistocene deposits, such as the tidal seas of N.W. Europe (Sect. 9.5), the tidally influenced parts of the Northwest Atlantic shelf (Stewart and Jordan, 1964; Smith, 1969), and the storm-dominated ridge and swale topographies of the Southern and Middle Atlantic Bights (Sect. 9.6.3).

These earlier views have been extended by Swift (1974 and 1976b) in a 'transgressive–regressive' model, which incorporates the physical processes operating within the nearshore-inner shelf zone, the rate of sea level fluctuation, and the nature and rate of sediment supply.

9.3 GEOLOGICAL CONTROLS OF SHELF SEDIMENTATION

There are several partially interdependent factors which influence the nature of sedimentary facies on present-day siliciclastic continental shelves: (1) rate and type of sediment supply, (2) type and intensity of the shelf hydraulic regime, (3) sea level fluctuations, (4) climate, (5) animal-sediment interactions, and (6) chemical factors.

9.3.1 Rate and type of sediment supply

The type and rate of sediment supplied directly from continental regions to the adjacent shelf is largely determined by the degree to which river mouths and estuaries have readjusted to the Holocene transgression. Direct sediment supply is negligible, except at the mouths of the largest rivers, and many estuaries actually receive sediment *from* the shelf.

Sediment which does bypass river mouths is overwhelmingly dominated by fine-grained suspended sediment, consisting mainly of mud. Its transport, deposition and accumulation are mainly governed by the rate and concentration of external sediment supply, and the type and intensity of the shelf hydraulic regime, particularly near-bed wave activity (McCave, 1970, 1971a and 1972). The most favourable sites of mud accumulation are therefore adjacent to major river mouths, such as the Amazon, Mississippi and Ganges–Brahmaputra, where mud blankets frequently extend across the full width of the shelves (Fig. 9.1). However, there are only twelve major rivers transporting significant quantities of sediment on to continental shelves (Holeman, 1968), including the Ganges–Brahmaputra and Yellow rivers which together carry 20% of the total (Drake, 1976).

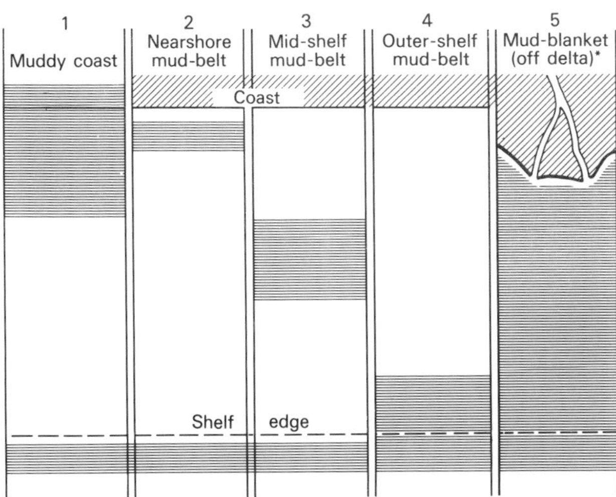

Fig. 9.1. Schematic representation of five situations of shelf mud accumulation (from McCave, 1972).

Modern shelf sands are mostly relict. However, occasionally the shelf receives a direct external sand supply such as through river mouths during floods (Drake *et al.*, 1972) and from beaches during storms, particularly by storm surge ebb currents (Hayes, 1967a; Reineck and Singh, 1972). The most likely long-term supply of sand is from tide-dominated deltas where near-equilibrium conditions enable interchange of sediment and water between inshore and shelf environments (cf. Coleman *et al.*, 1970; Coleman and Wright, 1975).

9.3.2 Type and intensity of the shelf hydraulic regime

The physical processes operating in shelf seas are summarized in Sect. 9.4 and discussed in further detail in relation to their products in Sect. 9.5 and 9.6. At this stage, however, it suffices to contrast deposits of tide-dominated and wind/wave-dominated hydraulic regimes (cf. Sect. 9.5 and 9.6).

Wind/wave- or storm-dominated shelves are controlled by seasonal fluctuations in wave and current intensity with active sediment transport restricted to intermittent winter storms. Apart from drowned barrier shorelines, such as eastern USA, fine-grained sediments and small-scale ripple bedforms are predominant.

Tide-dominated shelf seas are swept daily by powerful bottom currents enabling a wide range of bedforms to develop and sand to be transported in much greater quantities than in non-tidal seas.

9.3.3 Sea-level fluctuations

Sea-level fluctuations determine water depth which influences such features as hydraulic energy at the seabed, and river mouth equilibration which controls sediment supply offshore.

Fluctuations in the rate and extent of transgressions and regressions in the shoreline environment, controlled by the rate of deposition and the rate of relative sea-level change, will significantly influence the spatial and temporal distribution of facies on the shelf (Curray, 1964).

The impact of sea-level fluctuations on shelf sediments is emphatically illustrated by reference to the Holocene transgression, which has produced an extensive relict sediment cover and the characteristic shelf physiography of modern shelves (Curray, 1969).

9.3.4 Climate

Climate controls shelf sediments mainly by its effects on the hinterland. It determines the type and rate of weathering and erosion, thereby affecting the type of sediment available for transport. It also determines the mode of transport (water, wind or ice) which in turn affects the rate of supply of sediment to the receiving basin. On a global scale these factors have produced a broad latitudinal zonation of shelf sediment types (Hayes, 1967b; McManus, 1970; Senin, 1975).

Temperature and precipitation are the dominant factors controlling the type of sediment on the shelf (Fig. 9.2). Their most extreme values coincide with the most marked latitudinal variations in shelf sediments (McManus, 1970). In polar

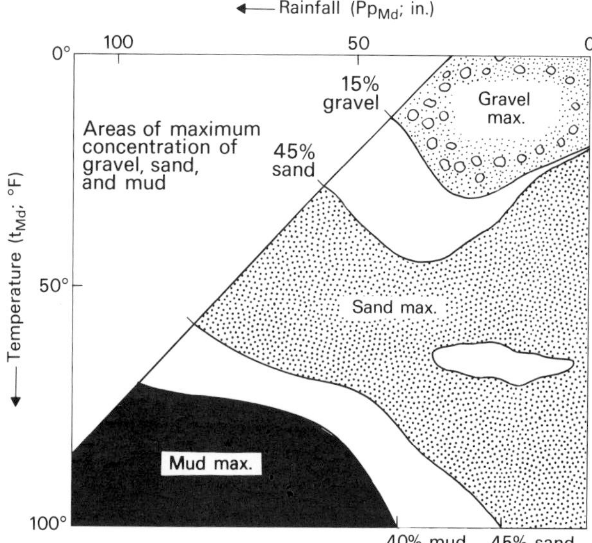

Fig. 9.2. Areas of maximum concentration of gravel, sand and mud on the inner continental shelf compared to temperature and rainfall of adjoining coastal regions (from Hayes, 1967b).

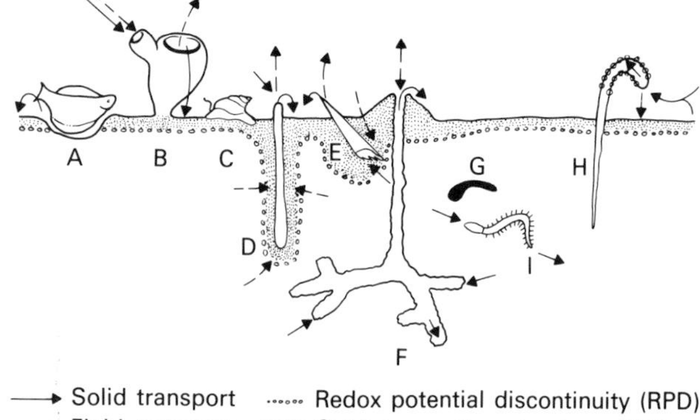

Fig. 9.3. Some methods of organism modification of the benthic boundary layer. A. Surface dweller (fish) disturbing surface sediment; B. Epifaunal suspension feeder converting suspended solids into deposit faeces; C. Epifaunal deposit feeder (gastropod) disturbing the surface, forming a mucus trail and increasing particle size by faecal deposition; D. Infaunal suspension feeder (polychaete) circulating the interstitial water; E. Infaunal deposit feeder (polychaete) transporting sediment upward and water downward; F. Burrower (crustacean) transporting sediment upward and horizontally; G. Animal with mineral hard parts (bivalve) converting dissolved ions into sedimentary particles; H. Tubiculous animal (polychaete) concentrating specific components of the sediment; and I. Burrower (polychaete) reworking the sediment (from Webb *et al.*, 1976).

climates, for example, mud deposits contain a paucity of clay minerals, gravel may be extensive adjacent to actively eroding and depositing land ice, quartz grains display glaciated surface textures, and chlorite is characteristic of the fine-grained sediment. In a tropical rainy climate, however, mud is abundant and contains a high proportion of clay minerals, kaolinite is frequently abundant near small rivers, and quartz is a dominant constituent of coarse-grained sediments. In extremely hot, dry climates sand deposits contain quartz grains with aeolian surface textures. Most intermediate climates do not give rise to distinctive sediment types.

Marked seasonal variations also frequently control the main periods of shelf sedimentation. In polar climates sedimentation rates are greatest during brief ice-free periods when direct deposition occurs from ice-melt and increased river discharge. Sediment transport rates also fluctuate widely in tropical rainy climates between wet and dry seasons.

9.3.5 **Animal-sediment interactions**

Shallow marine sediments are continually modified by biological and physicochemical processes, which are most active in the upper 10–15 cm of the substrate (Fig. 9.3). In the marine environment the benthic biomass reaches a maximum on continental shelves, ranging from 150–500 g/m^2, whereas on abyssal plains it falls to 1 g/m^2 (Menzies *et al.*, 1973).

Many morphological and physiological adaptations of benthic animals appear to be related to the physical properties of their sediment substrate, resulting in animal-sediment associations and communities in which similar groups or species consistently occupy a similar substrate (Petersen, 1913; Thorson, 1957). However, species distribution is also partially determined by whether the sediment surface is responding to the shelf hydraulic regime (recent) or whether it is relict (Webb *et al.*, 1976; Fig. 9.4).

On mobile substrates with active bedload transport of sand both benthic microfauna and macrofauna are characterized by low population density, relatively few species and an absence of epifauna. Finer grained and less mobile sand deposits possess greater substrate stability and 'hardness' (Richards and Parks, 1976), resulting in an increasing number of species, greater population density and a predominance of suspension feeding fauna. As sediment size decreases the proportion of suspension feeders also decreases and the proportion of deposit feeders increases (Saila, 1976).

Benthic organisms also have an important impact on substrate stability and sediment cohesiveness, which can lead to either substrate stabilization, or destabilization. Substrate stabilization results from sediment binding of organisms due to the secretion of organic films over the sediment surface providing greater adhesion, growth of dense populations of

SEDIMENT TYPE	SEDIMENT ORIGIN			
	Recent (hydrodynamic equilibrium)		**Relict** (hydrodynamic disequilibrium)	
	Physical Substrate Conditions	Biogenic Activity	Physical Substrate Conditions	Biogenic Activity
GRAVEL	Highly unstable	Almost no fauna	Stable pore spaces filled with smaller grains.	Well-developed epifauna & interstitial infauna.
MEDIUM-COARSE SAND	Mobile substrate due to tidal currents & waves. Bedload transport in the form of dunes, sand waves & sand ridges. Active water exchange.	Rich interstitial & meiofauna. Poorly developed macrofauna. No epifauna.	Immobile substrate. No sedimentation.	Low diversity of meiofauna. Well-developed macrofauna. Suspension feeders & deposit feeders common.
FINE SAND	Firm substrate due to packing by wave & tidal action. Small-scale wave & current ripples. Active sediment transport on a small-scale. Small interstitial voids.	Poor interstitial fauna. Difficult to burrow, but suitable organisms may be common. Suspension feeders dominate.	Soft & loosely packed. Readily transported sediment.	Interstitial fauna poorly understood. Suspension feeders dominate.
SAND-CLAY	Fluctuating intensity of tidal & wave activity (fair weather & storm conditions). Alternating sand-clay layers (storm sand layers). High primary productivity & large organic matter content, the latter decreasing with depth.	All feeding types present. Micro-, meio- & macrofauna abundant. Populations may be large but subject to great cyclical changes. Species diversity increases with depth. Secondary production highest among sediment types.	Not affected by tidal currents. Homogenised sand-clay sediment forming 60% of shelf sediments. Fewer environmental fluctuations & lower primary production than in Recent sand-clay.	Probably all feeding types present. Reduced organic matter reduces population sizes. Species diversity high due to lack of dominance & greater predictability of environment. Highly bioturbated giving increased structural complexity of environment.
MUD	Water content high. High suspended sediment concentrations &/or low wave effectiveness.	Mainly deposit feeders. No interstitial species; micro- & meiofauna in superficial layer. Burrows unstable.	Hard substrate, including peat & former lagoonal/estuarine muds.	Mainly suspension feeders. No interstitial fauna; micro- & meiofauna in superficial layers. Often burrowing species with many permanent burrows.

Fig. 9.4. Relationship between organism activity and substrate type as determined by whether the sediment is in hydrodynamic equilibrium (recent) or disequilibrium (relict) (from Webb *et al.*, 1976).

vertical tubes, marine algae, kelps or vascular plants, and by chemical cementation such as the formation of calcareous crusts by algae and invertebrates.

Substrate destabilization is apparently more widespread. It results from biological and physiochemical mechanisms such as particle agglomeration and flocculation, mechanical disturbance (burrowing etc.) organic trapping and re-suspension of sediment and the production of faecal pellets. These processes occur most effectively in conjunction with the feeding activities of epifauna and infauna. Particle agglomeration and flocculation causes fine-grained suspended sediment, such as fine silt and clay, to form relatively large aggregates with settling velocities similar to those of coarse silt and fine sand (e.g. Johnson, R.G., 1974). This process increases suspended sediment concentrations near the bed and also reduces both bed cohesiveness and its resistance to resuspension (Rhoads, 1972; Stanley *et al.*, 1972). Similarly, intense bioturbation by deposit feeding organisms in the upper 10 cm of the substrate significantly modifies the physical properties of muds by producing a granular surface texture 5–10 mm thick (Rhoads, 1972). The ensuing increased pore space prevents compaction, resulting in a thixotropic sediment containing 50–60% water by weight. Sediment is also disturbed by the grazing and dwelling activities of animals, such as fish, which aid erosion and maintain a higher concentration of sediment in suspension.

Suspension feeders can also significantly increase sedimentation rates by extracting solids from sea water and ejecting them as faecal pellets with relatively large settling velocities. In the Dutch Wadden Sea, for example, Verwey (1952) suggests that mussels deposit approximately 150,000 metric tons of sediment per year.

Some of the most complete studies of animal-sediment relationships in the modern shallow marine environment have been accomplished by German sedimentologists and biologists, notably in three main areas: Heligoland Bay in the southern North Sea (Schäfer, 1972), Gulf of Gaeta in the Tyrrhenian

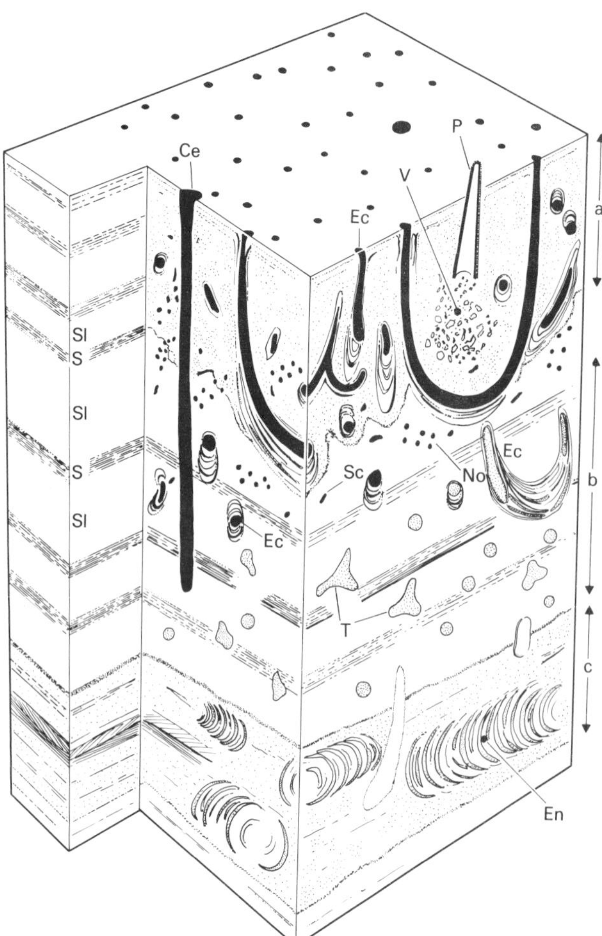

Fig. 9.5. Schematic block diagram of bioturbation structures based on box core samples from muddy sediments, Heligoland region of the southern North Sea. Primary bedding (left) comprises laminated (?storm) sand layers alternating with muds. Bioturbation structures in the three layers are: (a) bioturbation by *Echiurus*, (b) *Thalassinoides* burrows with a lower degree of bioturbation than in (a), and (c) thick sand layer with bioturbation structures of *Echinocardium cordatum* (En). Other bioturbation structures are: multi-walled U-shaped burrows of *Echiurus* (Ec); shell nests (V) by *Pectinaria* (P); burrows of *Cerianthus lloydii* (Ce) and *Notomastus latericeus* (No) (from Reineck and Singh, 1973, after Reineck *et al.*, 1967).

Sea, Italy (Dörjes, 1971; Gadow, 1971) and Sapelo Island, USA and its adjacent shelf (Dörjes, 1972; Hertweck, 1972). For example, in Heligoland Bay there is a transitional seaward zonation of sedimentary and biofacies (Schäfer, 1972). Here the characteristics of each facies are determined by the interaction of physical and biogenic processes, with the activity of burrowing benthic organisms recorded in a variety of bioturbation features (Fig. 9.5).

9.3.6 Chemical factors

There are several major chemical components which occur in shallow marine sediments consisting of (1) biologically and non-biologically produced carbonates, (2) alumino-silicate minerals, (3) quartz, (4) iron and manganese hydroxide, (5) biogenic silica, and (6) biologically produced organic matter (Hatcher and Segar, 1976).

Several coastal and shelf processes have direct and indirect effects on seawater chemistry and the chemical characteristics of shelf sediments (Fig. 9.6). These processes partially control the precipitation of authigenic minerals, such as chamosite, glauconite and phosphorite, which in some areas are characteristic of the shallow marine environment. Chamosite and glauconite are now found in areas of low detrital sedimentation, while their distribution appears related to water depth and temperature (Porrenga, 1967). Chamosite occurs in warmer waters usually between 10 and 170 m deep, whereas glauconite is found in cooler water ranging from 10 to 2,000 m deep. Phosphate rich waters are commonly found in zones of coastal upwelling resulting in the direct precipitation of calcium phosphate as nodules or laminae, or replacement of calcium carbonate. Upwelling frequently occurs along the west coasts of continents resulting in excessive concentrations of phytoplankton accompanied by phosphate enrichment (Bromley, 1967).

Chemical precipitation can also cause sediment cementation and adhesion which may have an important effect on the substrate by increasing its stability and reducing its erodibility (Hatcher and Segar, 1976).

9.4 PHYSICAL PROCESSES (GENERAL)

Currents and waves on the continental shelf are mainly generated by meteorological (local winds and waves), tidal and other forces such as global atmospheric circulation systems controlled by solar radiation. These forces generate four main types of current: (1) oceanic circulation (semi-permanent) currents, (2) tidal currents, (3) meteorological currents, and (4) density currents (Fig. 9.7). Seismically-induced waves, notably tsunamis, occur on shelves adjacent to seismically active regions (Coleman, 1968).

9.4.1 Oceanic currents

Oceanic circulation patterns are the result of differential temperature balance between high and low latitudes, manifested by a flow of heat from the equator to polar regions.

COASTAL/ SHELF PROCESSES	INFLUENCE ON SHELF SEDIMENTS
1. Estuarine (freshwater-seawater) Interface	Important changes in equilibria between dissolved & solid chemical species.
2. Mechanisms of Scavenging from Solution.	Removal of elements & compounds from seawater through their incorporation into shelf sediments by 3 main mechanisms: a. direct precipitation, b. adsorption, & c. biological processes. Suspended particles are particularly active due to their large surface area compared to their mass & direct derivation from the coastal zone resulting in their disequilibrium with seawater.
3. Seawater Composition & Residence Time	Many dissolved components carried in suspension are not in equilibrium with solid phases. Reactions occur between dissolved & solid phases causing a net loss of elements from seawater & addition to sediments.
4. Seasonal Changes	a. Sediment supply partly controlled by river discharge, b. More sunlight increases photosynthesis thereby producing more organic matter. The latter extract chemical species from seawater, c. Chemical factors are influenced by water temperature, & d. Fluctuating intensity of the shelf hydraulic regime affects chemical & biological processes which are most active during less intense periods.
5. Sediment Transport Processes	Chemical composition of shelf sediments determined by both: a. physical sorting, b. chemical changes during transport. This induces dissolution and reprecipitation.
6. Tidal Effects	a. Controls the frequency & intensity of sediment transport processes in tide-dominated seas, & b. Influences nearshore zone by forming a highly productive organo-chemical area in tidal flats. Constant exchange during tidal cycle between inshore tidal flats & offshore shelf waters.
7. Benthic Communities	Organisms use various sediment components as food sources. Biomass & excreta constitute an important component in shelf sediments. Estimated that 75% of shelf benthic fauna depends on detritus & associated organisms for a food source (Riley & Chester 1971).
8. Post-Depositional Chemical Changes	Precipitation of authigenic minerals characteristic of the shallow marine environment: a. Chamosite & glauconite (iron minerals). Include 4 main facies depending on the dominance of oxide, silicate, carbonate or sulphide (Porrenga 1967). Depth restrictions roughly temperature dependant (chamosite warmer-water; glauconite cooler-water), & b. Phosphorite typically occurs in shallow, marine waters (30-300m depth, Bromley 1967). Phosphate enrichment in zones of coastal upwelling increasing the proportion of phytoplankton.

Fig. 9.6. Influence of coastal and shelf processes on the chemical characteristics of shelf sediments (mainly from Hatcher and Segar, 1976).

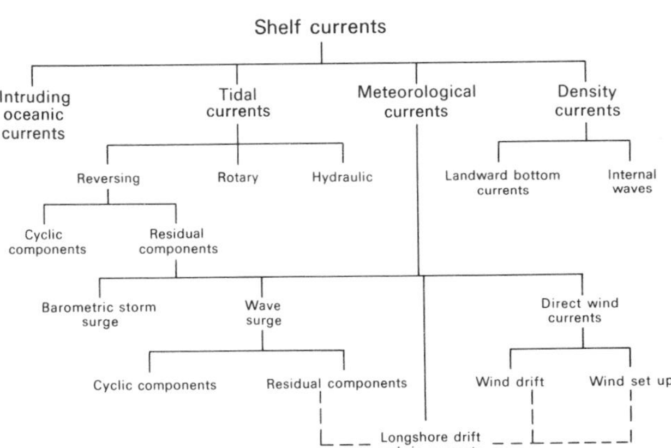

Fig. 9.7. Summary of the main physical processes influencing shelf hydraulic regimes (from Swift *et al.*, 1971).

The resulting currents mainly consist of a relatively shallow, wind-driven component which roughly corresponds to global wind patterns, and a deep water thermohaline component driven mainly by temperature and salinity variations in the stratified ocean waters. The boundary between these components is a zone of rapid temperature and density changes between the warm, relatively light surface waters and the cool, relatively dense deeper waters, known as the thermocline. However, in many cases this distinction is obscure, particularly in the case of major currents (e.g. Gulf Stream), due to vertical water movements. Although the major currents lie on the oceanward side of the shelf edge, they induce active interchange between oceanic and shelf waters. This is frequently manifested by large eddies spinning off the main current on to the outer continental shelf, but also, as with the Gulf Stream, it results from lateral migration accompanying large-scale meandering of the currents. The Florida Current exhibits seasonal variations in shelf-ocean water interchange (Niiler, 1975). During the summer the main current encroaches on to the shelf, whereas during winter it moves offshore accompanied by the oceanward sinking of colder and denser shelf water.

Current velocities are variable, ranging from a few cm/sec to more than 200 cm/sec (Von Arx, 1962; Groen, 1967) but sediment transport capabilities are largely unknown. The weaker currents are capable of transporting suspended sediment whereas the stronger currents are thought to be responsible for current erosion of the Blake Plateau (Swift, 1969b) and for transporting sand as bedload in megaripples on the outer Saharan continental shelf (Newton, Seibold and Werner, 1973).

9.4.2 Tidal currents

Tides are the result of gravitational attraction mainly by the moon, and to a lesser extent the sun, on the surface waters of the Earth. The theoretical equilibrium tide, which assumes a continuous ocean surface in equilibrium with these gravitational attractions, is made extremely complex in reality by such features as the motion of the Moon around the Earth and the Earth–Moon system around the Sun, by the distribution of the continents and by local variations induced by basin physiography. The resulting sea level fluctuations are the product of these forces occurring twice daily (semidiurnal tides), once daily (diurnal tides) or in various combinations (mixed tides). Superimposed on these daily variations are longer term fluctuations which influence the size of the tides and strength of the tidal currents. Most important are the fortnightly inequalities leading to *spring tides* when gravitational effects of the sun and moon are additive, and to *neap tides* when they are in opposition.

Tidal currents on continental shelves are propagated as waves generated in deep ocean basins, a feature known as *co-oscillation*. Thus enclosed seas or those with only a small oceanic connection are usually tideless or only weakly tidal. The magnitude of the tides is dependent on the natural

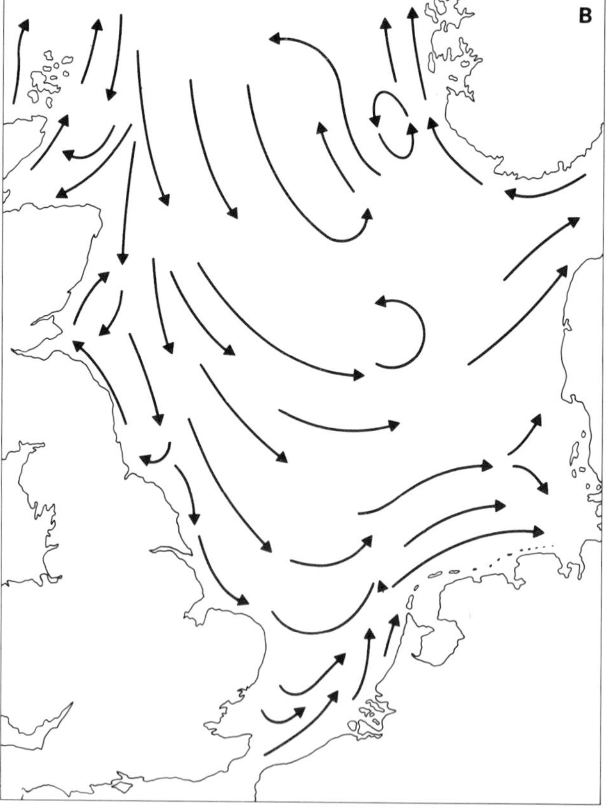

Fig. 9.8. Summary of the tidal regime in the North Sea: A. The amphidromic tidal system illustrating the circulation of tidal waves around amphidromic points. The co-tidal lines (continuous) show the time of high water in 'lunar hours' and the co-range lines (dotted) show the average tidal range, and B. Average surface currents (cf. sand transport paths, Fig. 9.12) (from Houbolt, 1968; Harvey, 1976).

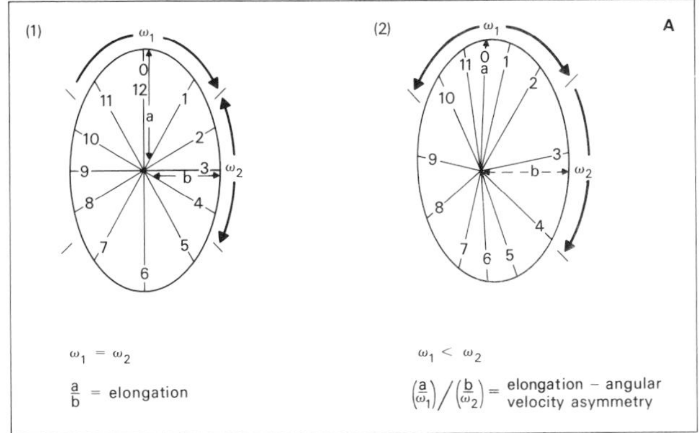

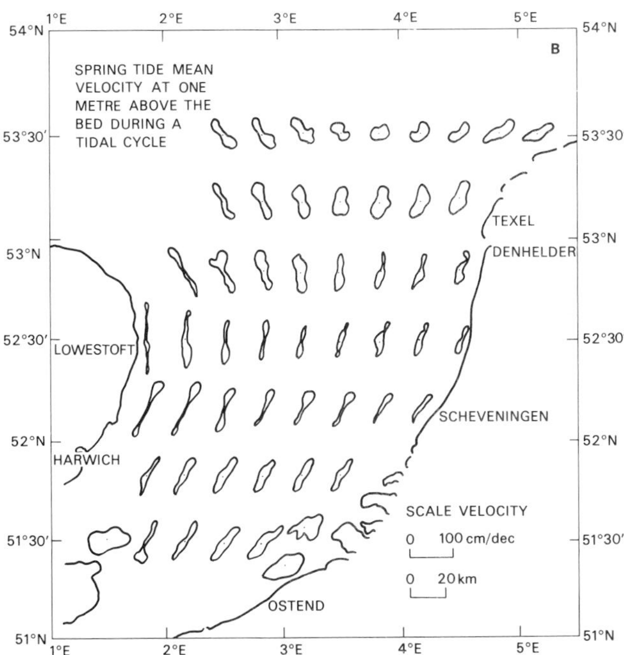

Fig. 9.9. Examples of tidal ellipse asymmetry illustrating the reversing and partly rotary nature of tidal currents at a point: A. Simple elongation with angular velocity symmetry (1), and complex elongation with angular velocity asymmetry (2). Position of lunar hours indicates a clockwise rotating tide; B. A range of tidal current ellipses based on velocities at 1 m above the bed (from McCave, 1971b).

oscillation period of the basin, determined by its physiography and mean water depth, and is greatest when this oscillation period corresponds to that of the principal tide-producing force. In the open sea and wide bays the Coriolis effect causes the tidal currents to change direction constantly, with water particles following an elliptical path in a horizontal plane. In more restricted areas such as partially enclosed shallow shelf seas, however, the shoaling effect of the sea bed and constraints imposed by basin configuration produce elliptical or rectilinear reversing currents (Defant, 1958).

Some of these phenomena are exemplified in the partially enclosed North Sea, where a progressive tidal wave is propagated from the Atlantic Ocean parallel to the shoreline. This results in an *amphidromic system* in which the time of high water and the tidal wave moves in a rotary, anticlockwise path around a point of constant sea level, an *amphidromic point* (Fig. 9.8). The tidal currents at any one point, however, are essentially rectilinear (Fig. 9.9).

Highest values of tidal range accompanied by strong tidal current velocities occur where the natural period of resonance of a basin is closely similar or equal to that of the dominant tide producing force. This phenomenon of *resonant amplification* is best illustrated by the Bay of Fundy whose resonance period is 12.58 h which closely corresponds to the lunar semidiurnal period of 12.42 h. This results in the world's largest tidal range of approximately 15 m accompanied by powerful (1–2 m/sec) tidal currents.

9.4.3 Meteorological currents

Circulation systems in shallow shelf seas are either dominated or strongly influenced by meteorological forces, notably in the form of wave- and wind-induced processes. They are the dominant source of energy on open shelves where tidal influence is negligible. Meteorological forces manifest themselves by the transfer of energy through direct wind stress and fluctuations in barometric pressure. These forces induce four main types of water movement: (1) wind-driven currents, (2) oscillatory and wave-drift currents, (3) storm-surge, and (4) nearshore wave-induced longshore and rip currents.

Wind-driven currents are the result of direct wind shear stress on the water surface with the resulting currents deepening with time as energy is transferred through turbulent mixing. These currents can therefore be considered as the direct product of atmospheric circulation systems, which operate over several spatial and temporal scales (Mooers, 1976): (1) tens of kilometres and a time span of a few hours for diurnal land–sea breeze systems; (2) hundreds of kilometres operating over a few days for mesoscale disturbances such as atmospheric warm and cold fronts, extratropical cyclones and anticyclones; and (3) thousands of kilometres persisting for several months for large-scale features such as midlatitude high pressure cells and other large-scale pressure systems. These conditions will generate waves and currents in shelf waters which vary strongly in magnitude, intensity and periodicity.

Surface currents generated by wind stress deviate from the wind direction in response to the Coriolis effect, while at depth this deviation is intensified (the Ekman spiral effect). Powerful currents, for example, accompany winter storms in which nearbed current velocities are commonly greater than 25 cm/sec with maximum recorded values around 80 cm/sec at depths of 50–80 m (Smith and Hopkins, 1972). The dominant currents are usually undirectional but raw data indicate a broad spread. Intermittency in this type of shelf current can be correlated with fluctuations in wind speed (Fig. 9.17).

Wave processes are the product of an orbital motion of water associated with surface wave motions. In deep water, the water particles move in almost circular orbits, with the orbital diameter equalling wave height at the surface and decreasing progressively with depth (Fig. 9.10A). When waves move into shallow water the circular orbits of the water particles at the surface become progressively more elliptical with increasing depth (Fig. 9.10B). At the sea bed this elliptical motion is replaced by a straight-line to-and-fro motion of the water, in which the orbital velocity varies with the surface wave motions. Variations in the strength of the two directions of orbital wave motion frequently result in unidirectional flow causing a preferred direction of water and sediment transport (Madsen, 1976).

Sediment transport on all continental shelves is significantly influenced by the interaction of waves and unidirectional currents. Wave action can place sediment in suspension so that even weak unidirectional currents are capable of transporting sediment (Komar, 1976d).

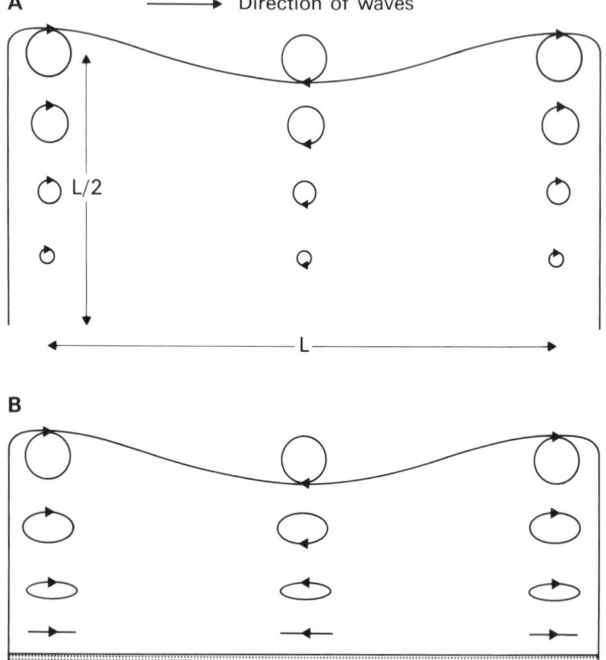

Fig. 9.10. Particle motion in A. a deep-water wave, and B. a shallow-water wave (from Harvey, 1976).

Storm-surge is caused by a marked reduction of barometric pressure and/or high wind stress. This may produce an abnormally high water level at the coastline followed by a drastic lowering of water level. Intense wave agitation accompanies these surges, but it is the seaward returning storm-surge ebb current that is the more significant transporting agent on the shelf, resulting in redeposition of nearshore sediments offshore (Hayes, 1967a; Gadow and Reineck, 1969; Reineck and Singh, 1972). A model of storm-surge currents induced by a hurricane predicts current velocities on the shelf of the order of 1 m/sec (Forristall, 1974).

9.4.4 Density currents

Variations in temperature, salinity and concentration of suspended sediment lead to differential densities of sea water. These generate density currents which occur as layers in the water column.

Density stratification may occur at the mouths of rivers where a plume of suspended sediment is carried offshore by fresh water flowing above a denser water wedge (Fig. 6.17). In the Middle Atlantic Bight the seaward flow of lighter surface water causes a shoreward flow of bottom waters (Bumpus, 1973).

9.5 TIDE-DOMINATED SHELF SEDIMENTATION

Several present-day continental shelves experience semi-diurnal tides with a large tidal range (>3–4 m) and maximum surface current speeds (mean Springs) ranging from 60 to >100 cm/sec. These conditions characterize many partially enclosed seas and blind gulfs such as the seas of N.W. Europe, Gulf of Korea, Persian Gulf, Gulf of California, Malacca Straits, Taiwan Straits and many others (e.g. Off, 1963). Typical products of tidal currents in all these seas are bedforms such as large-scale linear sand ridges and sand waves, which have developed mainly from the *in situ* reworking of pre-Holocene sand and gravel deposits.

Although tidal currents are bidirectional, rectilinear or rotary they develop essentially unidirectional sediment transport paths because: tidal current ebb and flood velocities are usually unequal in maximum strength and duration; ebb and flood currents may follow mutually exclusive transport paths; the lag effect associated with a rotating tide delays the entrainment of sediment; and a single tidal current direction may be enhanced by other currents, such as wind-driven currents. The interaction of these processes is exemplified by the world's most intensively studied tidal seas, namely those of N.W. Europe, whose hydraulic regimes are in partial equilibrium with bedforms and sediment transport paths (Kenyon and Stride, 1970).

9.5.1 Sedimentary facies along tidal current transport paths

Distinctive bedforms and sedimentary facies characterize tidal current transport paths whose trend may be defined by: (1) direction of maximum near-surface tidal current velocities, (2) elongation of the tidal current ellipse, (3) facing direction of sand wave lee slopes, (4) trend of sand ribbons, (5) direction of decreasing grain size, and (6) direction of decreasing near-surface tidal current velocities (Stride, 1963; Belderson and Stride, 1966; Kenyon and Stride, 1970; McCave, 1971b, 1972).

When completely developed these transport paths contain the following depositional zones: (1) sand ribbons, (2) sand waves, (3) sand patches, and (4) mud (e.g. Fig. 9.34B).

(1) *Sand Ribbons* are longitudinal bedforms developed parallel to the maximum current velocities in response to a presumed helical flow structure of the currents. They consist of up to 15 km long ribbons or strips of sand up to 200 m wide and not greater than 1 m thick, with intervening strips of the underlying immobile gravel (Kenyon, 1970). Typical conditions for their formation include maximum near-surface current velocities in excess of 100 cm/sec, a paucity of sand and water depths generally between 20 and 100 m (maximum 150 m). Kenyon (1970) distinguishes four main types of sand ribbon based on their external morphology (Fig. 9.11). Sand ribbon types A, B and C are restricted to zones situated immediately downcurrent of a zone of erosion. Type D, however, occurs between sand waves in the sand-deficient parts of sand wave zones.

(2) *Sand Waves* are large-scale, transverse bedforms, generally with straight crests and well defined avalanche faces. They are characteristic of many modern tidal shelves but have been most extensively studied in the southern North Sea (Stride, 1963; 1970; Houbolt, 1968; McCave, 1971b; Terwindt, 1971a; Langhorne, 1973). Tidal sand waves are generally greater than 1.50 m high with wavelengths greater than 30 m. However, they are usually considerably larger, varying between 3 and 15 m high with wavelengths between 150 and 500 m.

The asymmetry of these bedforms, which is frequently in the direction of maximum near-surface tidal current velocities, has been regarded as indicating a response to present-day tidal currents (Stride, 1963; Houbolt, 1968). This idea is strongly supported by the fact that facing directions of sand waves in the southern North Sea are in agreement with calculated sand transport data, and they occur where there is adequate current velocity (mean spring tide >0.60 m/sec), low to moderate wave activity, and pronounced asymmetry of the tidal current ellipse (McCave, 1971b). However, accurate migration rates of the southern North Sea sand waves have never been obtained and

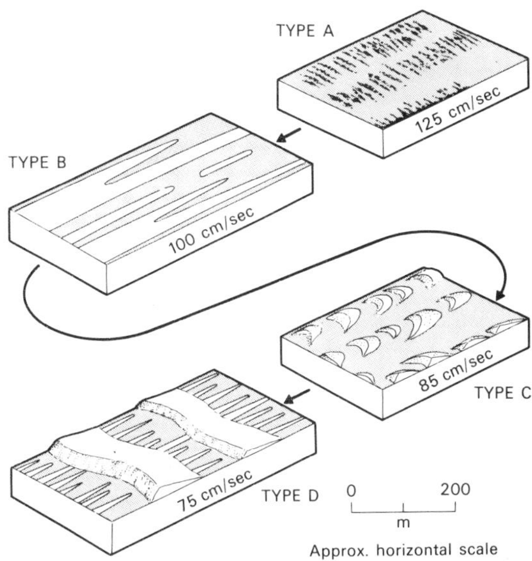

Fig. 9.11. Morphology of the main types of sand ribbon found in European tidal seas, and the typical maximum near-surface current velocity associated with each type (from Kenyon, 1970).

Terwindt (1971a) questions the relationship between sand wave asymmetry and present-day hydraulic conditions, because sand waves are occasionally symmetrical or reverse their asymmetry along the crest of a single sand wave. In addition megaripples migrate obliquely to sand wave crests and mutually opposing migration directions of megaripples on the lee and stoss sides of single sand waves occur. Similar observations have led Langhorne (1973) to invoke a secondary flow regime around sand waves.

The complexity of these surface morphological features of sand waves should be recorded in their internal structures, but at the moment data are restricted to the shallow cores (30–70 cm) of Reineck (1963) and Houbolt (1968) revealing variably dipping cross-bedding.

(3) *Sand Patches* form extensive flat sheets of well-sorted, fine sand, which may grade vertically and laterally into poorly sorted muddy sand. Although relatively extensive, this facies has never received the same attention as other facies. It is probably covered in ripples and supports a varied infauna (Belderson and Stride, 1966). The resulting facies will therefore reflect the interaction of bedload rippling, deposition from suspension of fine grained sediment and reworking by benthic organisms.

(4) *Mud zones* are usually located at the ends of tidal current transport paths (Stride, 1963; McCave, 1971c). Suspended sediment is moved parallel with or laterally across paths of bedload transport by both the general tidal circulation patterns

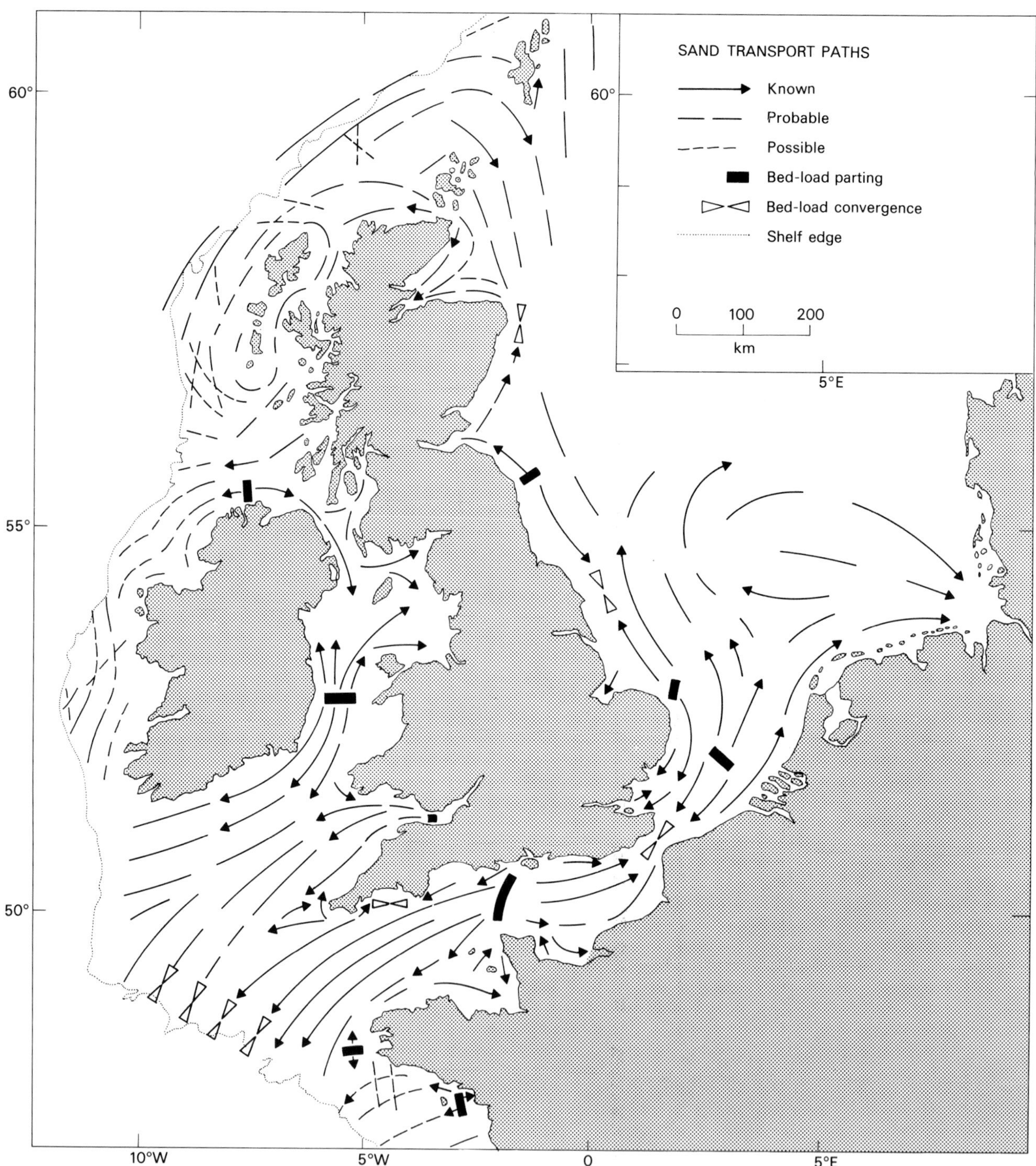

Fig. 9.12. Distribution of the dominant sand transport paths on the N.W. European continental shelf (after Stride, 1965; Kenyon and Stride, 1970).

and wind-driven currents (Stride, 1963). Its deposition is controlled by tidal current boundary shear stress, wave effectiveness and suspended sediment concentration (McCave, 1970, 1971a and c). Mud accumulates in a wide variety of situations but because wave activity has such a dominant control most of the extensive mud areas are in moderately deep water (> 30 m). In the Heligoland region of the German Bight a combination of moderate suspended sediment concentration and low wave effectiveness allows mud to accumulate in depths of 20–40 m at rates of 15.5 cm/100 years despite maximum near surface tidal current velocities of 60 cm/sec, or 30 cm/sec at 1 m above the bed (Reineck *et al.*, 1967; McCave, 1970). On the other hand, in depths of over 100 m, the sea bed of St. George's Channel is swept clear of mud by tidal currents in excess of 100 cm/sec. Similarly the absence of mud from large areas of the northern North Sea is attributable to a low rate of deposition combined with sufficiently frequent disturbance by waves (McCave, 1971c).

9.5.2 Sediment dispersal patterns

The spatial distribution of facies suggests that the dominant sediment dispersal and bedform patterns are parallel to present-day tidal currents, and that these are mainly parallel to the coastline (Fig. 9.12). These general patterns appear to be controlled by the tidal currents and are only rarely affected by local conditions such as basin physiography, bed roughness, sand availability, or exposure to storm waves (Kenyon and Stride, 1970).

The most completely developed transport paths are about 400 km long, but many are incomplete, and others converge or diverge. Zones of bedload convergence mark sites of deposition and may be represented by sand wave fields with 'sand hills' such as those off the east coast of England (Dingle, 1965). Zones of bedload parting are more common, and are usually areas of erosion.

Deposition may occur in zones of bedload parting when there is an influx of sand from the shoreline zone due to a combination of the tidal lag effect and the sense of rotation of the tidal current vectors (Stride, 1974). Stride's conclusion that sand is not entirely derived from reworked Holocene material but includes material from the shoreline has important implications concerning mechanisms of sand dispersal and supply in ancient shelf seas (Sect. 9.13).

Although transport paths for suspended fine-grained sediment may differ from bedload paths, being influenced by both wind-drift and general circulation patterns, mud appears to accumulate preferentially at the ends of these transport paths since they frequently coincide with areas of low density and frequency of wave activity (McCave, 1970, 1971a and c).

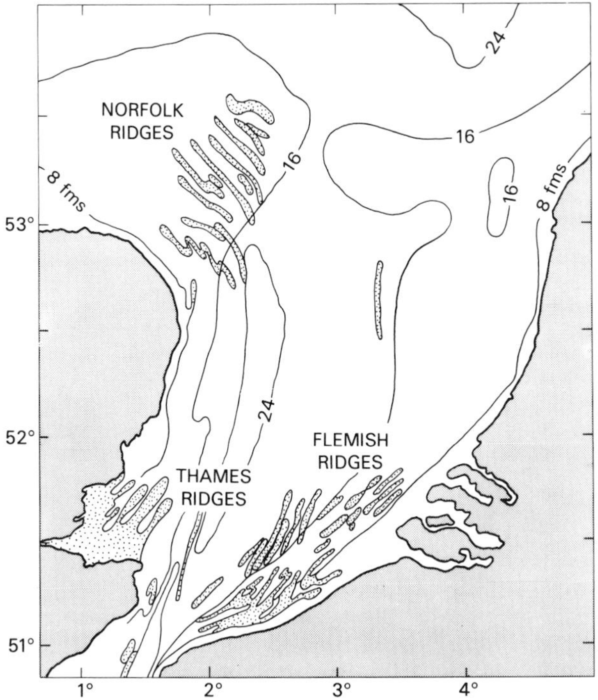

Fig. 9.13. The main fields of tidal sand ridges in the southern North Sea (after Houbolt, 1968).

9.5.3 Tidal sand ridges

Tidal sand ridges are large-scale, linear bedforms whose long axes are oriented parallel to the direction of strongest tidal currents, and are characteristic of many tidal seas. Inshore they are associated with tide-dominated deltas (Chapter 6) and estuaries (Chapter 7). In the offshore environment they do not occupy an established position within tidal current transport paths; their distribution is more closely related to the transgressive history of adjacent shorelines and inshore regions (Swift, 1974).

In the southern North Sea (Fig. 9.13) sand ridges are 10–40 m high, 1–2 km wide, up to 60 km long and their crests spaced at 4–12 km. Water depth between the banks is 30–50 m while their crests lie at depths of only 3–13 m. The ridges commonly carry dunes facing towards the ridge crests and are asymmetric to the north-east, with the steeper face inclined up to 4° or 7° (Caston, 1972). This steeper slope faces the direction of lateral migration as demonstrated by large-scale internal bedding planes (Fig. 9.14) (Houbolt, 1968). They are composed of well-sorted, medium-grained sand with fragmented shell debris, and shallow cores reveal abundant cross-bedding. Grain size distribution across the ridges is relatively uniform, but usually a basal lag conglomerate occurs consisting of

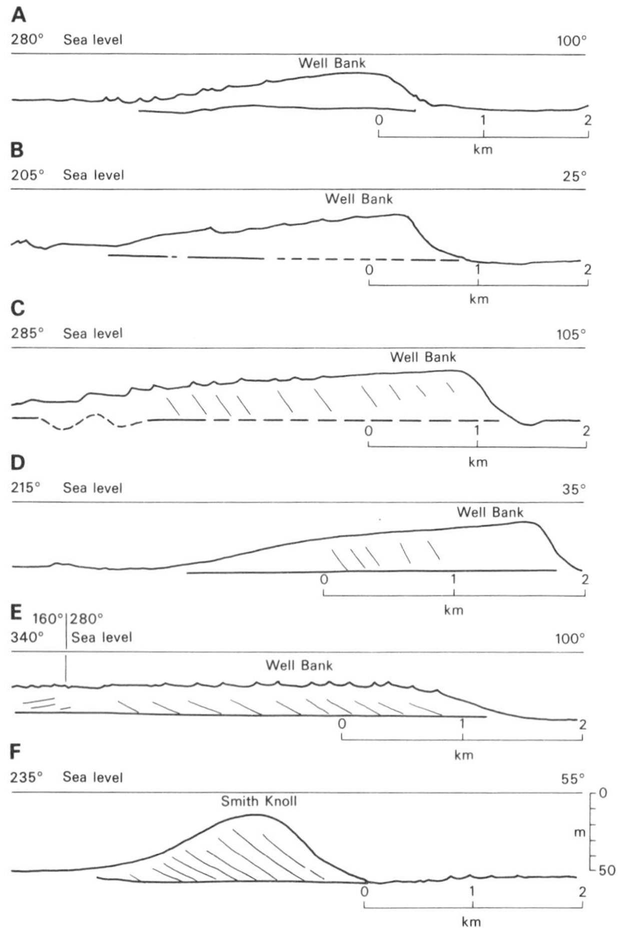

Fig. 9.14. Interpretation of sparker profiles over Well Bank and Smith Knoll (Norfolk ridges) indicating the large-scale, low-angle accretion surfaces (from Houbolt, 1968).

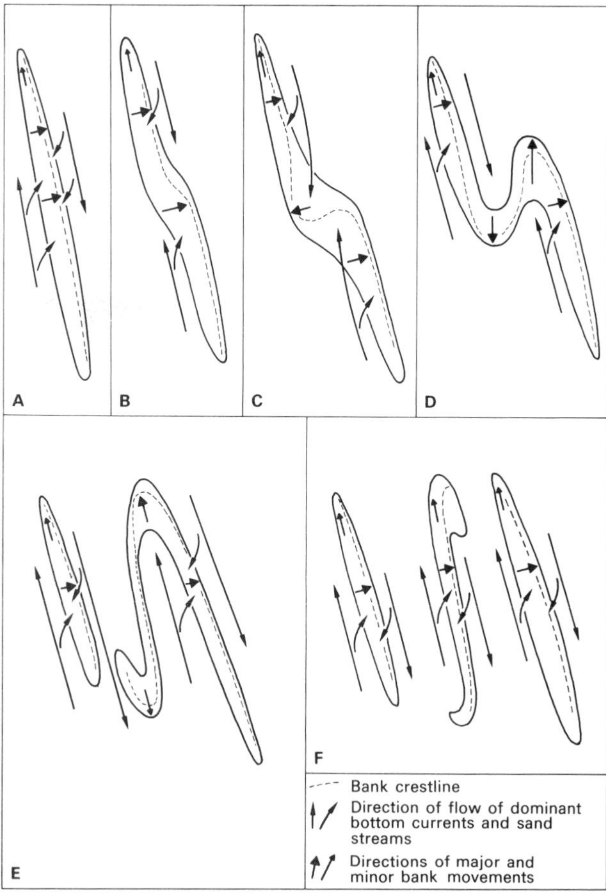

Fig. 9.15. Model for the growth and development of linear tidal sand ridges. A linear sand ridge is built between two mutually evasive ebb and flood tidal channels (stage A). Inequality of the secondary cross-shoal currents leads to the destruction of the straight crest-line (stages B and C). The resulting double curve develops into an incipient pair of ebb and flood channels (stage D). The channels continue to lengthen resulting in parallelism of the centre ridge with those adjacent to it (stage E). The initial cycle is thus complete but can continue with three ridges instead of one. This sequence is based on the inner ridges (Haisborough Sand-Hearty Knoll system) of the Norfolk Banks (from Caston, 1972).

pebbles, coarse shells and *Sabellaria* worm-tube aggregations. This probably reflects continued erosion of the trough floor, which is thereby kept clean of fines. Elsewhere, however, there is an upward increase in grain-size (Meckel, 1975) probably due to wave winnowing of fines from the more frequently reworked sands of the crest and upper flanks (cf. Stubblefield *et al.*, 1975).

Tidal sand ridges were either initiated by present-day hydraulic conditions, or are a partial relict feature of an earlier stand of sea-level but responding to modern conditions (Swift, 1975b). Their world-wide association with areas now experiencing strong tidal currents led Off (1963) to conclude a tidal genesis. In the southern North Sea they occur at the base of the Holocene transgression and are interpreted as tidally reworked sands from former glacial outwash fans and fluviodeltaic deposits (Houbolt, 1968). Since sand is now being transported around the ridges Houbolt considered them to be in dynamic equilibrium with present-day tidal conditions and proposed a spiral current hypothesis for their formation. This hypothesis envisages ridge construction by tidal currents flowing obliquely across the gentle stoss side towards the ridge crests and depositing sand on the steeper slope mainly by avalanching.

The existence of across-bank transport was confirmed by sand waves supposed to be moving in opposite directions across both bar flanks and converging at the crest (Caston, 1972). Ridge asymmetry is considered indicative of the predominance of one of these sand streams, but this may also partially reflect a lag-effect in the entrainment and settling of sediment during the tidal cycle (Stride, 1974).

While this demonstrates present-day ridge activity, it is probable that they were initiated during a period of lower sea-level. Swift (1975), for example, suggests that the two types of sand ridge found in modern tidal seas, namely those in estuaries and those adjacent to linear shorelines, were initiated as channel-shoal couplets formed either inshore or on the shoreface and were subsequently maintained offshore. Thus, during transgression, offshore tidal currents flowed parallel to currents in the inshore/nearshore environment, enabling the channel-shoal couplets to be maintained and enlarged offshore.

Thus the Norfolk Banks are interpreted as a system of mutually evasive ebb and flood tidal channels which were initiated on a sandy shoreface by coast-parallel tidal currents (Robinson, 1966) and were subsequently developed offshore during transgression (Swift, 1975). During this period of detachment further growth may have occurred through erosion initiated by across-shoal tidal currents (Caston, 1972; Fig. 9.15), one stage of which may be seen in the parabolic shapes of the Haisborough-Winterton ridge system (Fig. 9.13).

A similar method of growth and detachment is also thought to occur in estuaries, while several other examples of offshore sand ridges may also owe their initiation to a period of lower sea-level (e.g. Jansen, 1976; Bouysse *et al.*, 1976). A broadly similar mechanism has been proposed for morphologically similar sand ridges developed in non-tidal, storm-dominated seas (e.g. Duane *et al.*, 1972; Swift *et al.*, 1972b, 1973; Sect. 9.6.3).

9.6 STORM-DOMINATED (WIND- AND WAVE-DRIVEN) SHELF SEDIMENTATION

9.6.1 Introduction

In many shelf seas, including most pericontinental seas, tidal currents are less important than wind- and wave-induced currents. Tidal range is small. It rarely exceeds 2–3 m and maximum tidal current velocities are usually less than 30 cm/sec. On the other hand meteorological factors dominate the hydraulic regime which is characterized by a strong seasonal aspect. The most intense wave action occurs on those shelves which face the prevailing westerly winds and are open to oceanic waves (e.g. Bering Sea, Washington–Oregon shelf, N.E. Atlantic shelf), whereas less intense conditions occur on shelves in the lee of prevailing winds (e.g. eastern USA). Even less intense conditions characterize partially enclosed seas, such as the Gulf of Mexico, which experience lower and more sporadic wave activity partly due to the restricted wave fetch.

While many of these shelves are dominated by relict sediments others, such as the Oregon, Moroccan and Niger shelves, are moving from disequilibrium towards equilibrium conditions. The Southeast Bering shelf, however, is in an advanced stage of textural equilibrium in response to a storm-dominated hydraulic regime (Sharma, Naidu and Hood, 1972; Sharma, 1975). The surficial sedimentary cover consists mainly of fine- to medium-grained sand with a progressive seaward decrease in grain-size and degree of sediment sorting, both trends closely paralleling the bathymetric contours (Fig. 9.16). The main source of energy is wave action which is most intense during winter storms and is accompanied by coast-parallel, residual currents with maximum near surface velocities of 50–70 cm/sec. The facies distributions are therefore interpreted as reflecting a progressive decrease in energy across the shelf in response to increasing water depth and decreasing intensity of wave disturbances.

On the Gulf of Mexico shelf, textural grading takes place by mixing of the Holocene basal transgressive sands and modern muds (Swift *et al.*, 1971). The externally derived modern shelf muds are reworked into the basal transgressive sand facies by periodic wave surges and benthic organisms, producing a facies with irregular laminations and abundant mottling (Moore and Scrutton, 1957). As distance from the shoreline increases the well-sorted transgressive sands, with their unimodal grain-size distributions, give way progressively into less well-sorted sediments with bimodal and polymodal grain size distributions (Curray, 1960).

One type of bed likely to be deposited and preserved on wind- and wave-dominated shelves is that resulting from storm events, but descriptions of modern examples are scarce (Hayes, 1967a; Swift *et al.*, 1972a; Kumar and Sanders, 1976). The main effect of storms is that waves rework the shelf bottom well below fair weather wave base (cf. Komar *et al.*, 1972), while unidirectional bottom currents, such as storm surge ebb and wind-driven currents, enhance the more constant tidal and oceanic currents. The main factors influencing the type of bed deposited during a storm are: (1) energy level of the hydraulic regime, (2) type of sediment available, (3) direction of the storm-generated currents with respect to the shoreline and/or local shelf sediment sources, and (4) distance from the shoreline combined with water depth. The preservation potential of any storm-deposited bed depends on the subsequent hydrodynamic history of an area, but is generally high since the energy level of fair weather periods is lower than that of storms.

The remainder of this section looks in more detail at the two most intensively studied examples of storm-dominated shelf

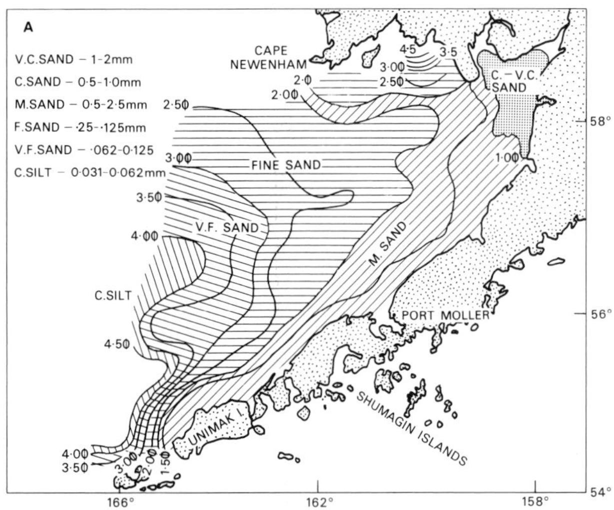

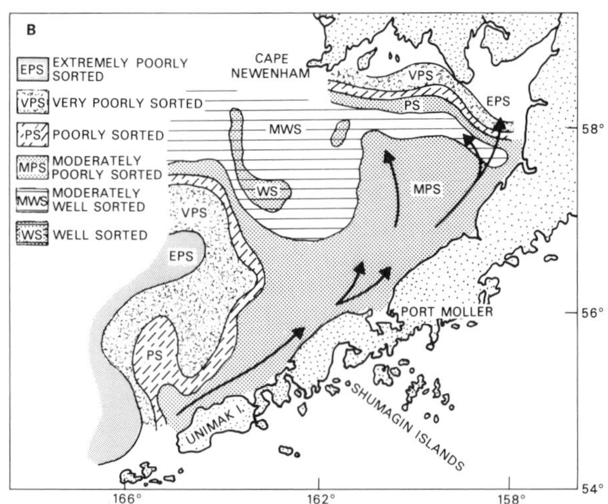

Fig. 9.16. A modern texturally graded shelf as exemplified by Bristol Bay, southern Bering Sea; A. grain size distribution, and B. variation of sediment sorting and the general pattern of shelf currents (from Sharma *et al.*, 1972).

sedimentation, the Oregon–Washington shelf and the Atlantic shelf of the USA.

9.6.2 Storm-dominated (wind- and wave-driven) sedimentation on the Oregon–Washington shelf

On the Oregon–Washington shelf the distribution of sedimentary facies is controlled by several factors (Kulm *et al.*, 1975), particularly river discharge, sediment supply, upwelling, and physical and organic reworking.

HYDRAULIC REGIME

This is dominated by meteorologically-induced currents and their seasonal variations but tidal and oceanic currents are of secondary importance.

Seasonally fluctuating semi-permanent currents are represented by the California and Davidson Currents. The latter has the greater direct influence on the shelf (Smith and Hopkins, 1972). It migrates onto the shelf particularly during winter when the nearbed current flows northwards, but reverses during summer. Although current velocities of a few centimetres per second are incapable of eroding the sea bed, they can transport suspended sediment and may be enhanced by northward-flowing wind-drift currents during winter.

Mixed and semidiurnal tides, with a tidal range of 2–3 m, develop rotary tidal currents which affect other bottom currents but are relatively weak on their own. Tidal currents on the middle and outer shelves have mean velocities of only 10 cm/sec, ranging from 0 to 30 cm/sec (Harlett and Kulm, 1973). However, on the inner shelf mean current velocities may be up to 30 cm/sec and are frequently supported by wave surge.

The dominant source of energy is provided by meteorologically-induced currents, including both wave-drift (wave surge) and direct wind-driven currents, which exhibit the strongest seasonal variability. During summer currents are generally only capable of reworking the sediment surface on the inner shelf. The coastal zone experiences upwelling as surface waters move offshore in response to the northerly winds (Smith, R.L., 1974; Halpern, 1976). The middle and outer shelf areas are dominated by biogenic activity and the deposition from suspension of silts and muds.

During winter the intensity of physical processes increases dramatically when strong winds and oceanic storm waves move across the shelf. An important meteorological feature is an elongate low-pressure system in the Gulf of Alaska which generates exceptionally strong southerly winds blowing approximately parallel to the Oregon–Washington coast. Direct wind stress generates mainly unidirectional currents flowing to the northwest at the bed (Fig. 9.17). Current velocities are frequently of 40 cm/sec and even exceed 80 cm/sec (Smith and Hopkins, 1972). At these velocities the currents erode and transport sand and silt both as bedload and in suspension (Sternberg and Larsen, 1976), and move it across the shelf towards the shelf edge.

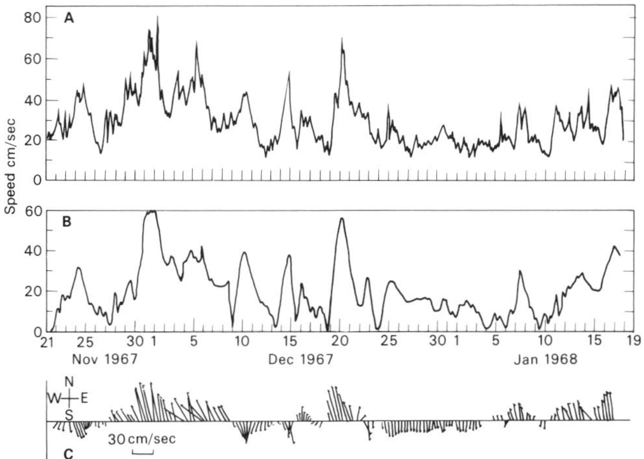

Fig. 9.17. Current speed and direction recorded during winter on the Washington–Oregon shelf. The raw data A. have been averaged over 24 hours B. The peak current speeds are related to storms when the strongest currents flowed to the northwest C. (from Smith and Hopkins, 1972).

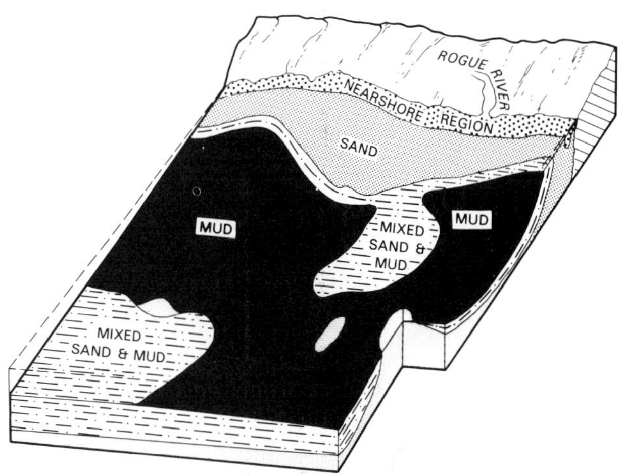

Fig. 9.18. Schematic representation of sediment facies distribution on the southern Oregon shelf (from Kulm *et al.*, 1975).

The effectiveness of these currents may be significantly increased by waves initiated by storms in the north Pacific and propagated across the shelf (Komar *et al.*, 1972; Sternberg and Larsen, 1976). In response a series of north–south (coast-parallel), symmetrical wave ripples are formed out to water depths of 204 m (Komar *et al.*, 1972). Moderate storms and forerunners to major storm waves are believed to ripple the shelf bottom to depths of up to 100 m, while average summer waves are restricted to depths of less than 85 m. Thus winter storms generate a downwelling situation in which unidirectional bottom currents flow slightly obliquely offshore or parallel to the shoreline, while waves migrate eastwards and trend approximately parallel to the shoreline.

SEDIMENTARY FACIES

On the Oregon shelf there are three main facies: sand, mixed sand and mud, and mud (Kulm *et al.*, 1975). The distribution of these facies is largely controlled by sediment supply, shelf hydraulic regime and burrowing benthic organisms. Although there is no systematic depth dependency of the facies, in general sands occur on the inner shelf and shoreline (Clifton, Hunter and Phillips, 1971), muds occur on the middle and outer shelf, and mixed sands and muds are located either between areas of sand and mud or on the outer shelf (Fig. 9.18).

(1) *Sand Facies* extends offshore into water depths of 50–100 m and consists of relict sands and gravels, modern very fine sands and a mixture of the two. The modern detrital sands are composed mainly of quartz, feldspar and rock fragments with fresh surface textures in contrast to the typical iron oxide coatings of relict sands (Emery, 1968). Minor constituents include diatoms, Radiolarians, Foraminifera and glauconite. The latter is most common on the outer edge of the shelf and on topographic highs.

Primary sedimentary structures are largely obliterated through intense bioturbation, but stratification is occasionally preserved. This mainly consists of horizontal lamination which leads Kulm *et al.* (1975) to invoke dominantly upper flow regime conditions, although this is at variance with the observation of Komar *et al.* (1972) who suggest that lower flow regime conditions are more important. The horizontal lamination may reflect deposition of sand from suspension during storms (cf. Reineck and Singh, 1972), while the rarity of unidirectional cross-lamination may partially reflect the more prolonged periods of wave reworking which obliterate evidence of the shorter-lived currents.

Due to the effective trapping of coarse sediment in estuary mouths, only very fine sand (3–4 ϕ) is transported directly in suspension on to the shelf. Modal grain size is the most sensitive parameter for analysing sand dispersal in these fine grained sediments, and this has been used on the northern and central Oregon shelf to indicate a new equilibrium dispersal pattern (Kulm *et al.*, 1975). Here sediments finer than 2.75 ϕ gradually decrease in grain-size with increasing water depth. Only near the Columbia River mouth is this pattern disturbed by the increased proportion of mud and silt.

(2) *Mixed Sand and Mud Facies* is the product of organic reworking of the sand and mud facies, the bioturbation having completely obliterated any primary sedimentary structures. On

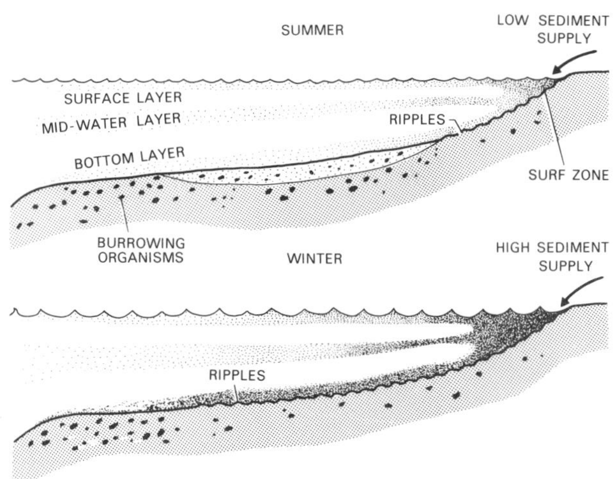

Fig. 9.19. Seasonal sedimentation pattern on the Oregon continental shelf. Turbid layers are represented by stippling (from Kulm *et al.*, 1975).

the outer shelf it consists of reworked modern muds and relict transgressive sands (Fig. 9.18).

(3) *Mud Facies* has a patchy distribution but predominates on the middle shelf and on parts of the outer shelf, fluctuating in thickness from an average of 10 cm to over 40 cm.

Mud is the most common type of sediment being introduced to the Oregon shelf through major river mouths. The suspended sediment is concentrated above three surfaces (Fig. 9.19): (1) seasonal thermocline (surface layer), (2) permanent pycnocline (mid-water layer), and (3) bottom (bottom layer). All show a marked seasonal variation (Harlett and Kulm, 1973; Kulm *et al.*, 1975).

The surface layer moves seaward from the coast as a plume, with decreasing intensity offshore. The plumes are most dense during maximum river discharge, which is during spring and summer. The mid-water layer corresponds to the level of the permanent thermocline (Harlett and Kulm, 1973). It becomes thicker seaward, but diffuse near the shelf edge, and it also migrates vertically in response to changing temperatures. Suspended sediment is supplied to it from the surface layer through viscous settling. In addition, biogenic material is introduced during upwelling when the denser upwelled water sinks to the mid-water layer in the middle part of the shelf.

The bottom layer receives sediment both from the overlying layers and from a mud substrate, where waves and currents resuspend the bottom sediment. Areas of high concentration may develop into low-density flows giving velocities of 15–20 cm/sec (Komar *et al.*, 1974). The thickest mud layers accumulate close to the major points of supply and where wave activity is only moderate. The most important source is the Columbia River with its annual suspended sediment load of approximately 11,000,000 m^3 (Kulm *et al.*, 1975). However, the maximum rate of mud accumulation is only 6 cm/1,000 years, this slow rate being presumably due to resuspension and offshore transport during winter storms. In summer, when waves are less capable of resuspending bottom sediment, the mud facies is most extensive and previous sand substrates are frequently draped by mud (Fig. 9.19).

9.6.3 Storm-dominated (wind-driven) sedimentation on the N.W. Atlantic shelf

The Atlantic shelf of eastern North America is also storm-dominated but contrasts with the Oregon–Washington shelf in the following respects: (1) The inner shelf is characterized by a complex sand ridge topography (Fig. 9.20), which reflects the successive detachment and reworking of a barrier shoreline during the Holocene transgression (Field and Duane, 1976). (2) The surface structures and textures of this sand sheet are partly the result of *in situ* reworking by the modern hydraulic regime, and there is little direct terrigenous input. (3) The wave regime is less intense because the shelf lies in the lee of the continent with respect to the prevailing westerly winds. (4) The continental margin is tectonically inactive and is thus flanked by a hinterland of low relief and a low-lying coastal plain protected by extensive barrier islands.

HYDRAULIC REGIME

At the coast the tidal range is frequently only 1 to 2 m, and offshore maximum tidal current velocities are usually less than 20 cm/sec. The dominant features of the shelf hydraulic regime are the meteorologically-induced currents and their seasonal fluctuations.

During summer, shelf waters are relatively calm and develop a density stratification accompanied by a slow, southerly moving, coast-parallel drift in response to the direction of the prevailing winds and freshwater runoff (Harrison *et al.*, 1967). Wave intensities are insufficient to resuspend fine sediment so that suspended sediment concentrations remain low, except near estuaries and tidal inlets, and the sea bed is undisturbed, except for the shoreface (less than 15 m deep) and the crests of some sand ridges (Swift *et al.*, 1972a).

In winter, water stratification is destroyed by falling water temperature and increased turbulence from direct wind stress. Successive mid-latitude lows migrate approximately parallel to the coast and bring intense storms which generate strong northeasterly winds (Beardsley and Butman, 1974). Surface waters are driven onshore by direct wind stress resulting in a build-up of water in the coastal region which is locally enhanced by a convergence of flow paths. The ensuing pressure head may be relieved by downwelling resulting in an oblique seaward-

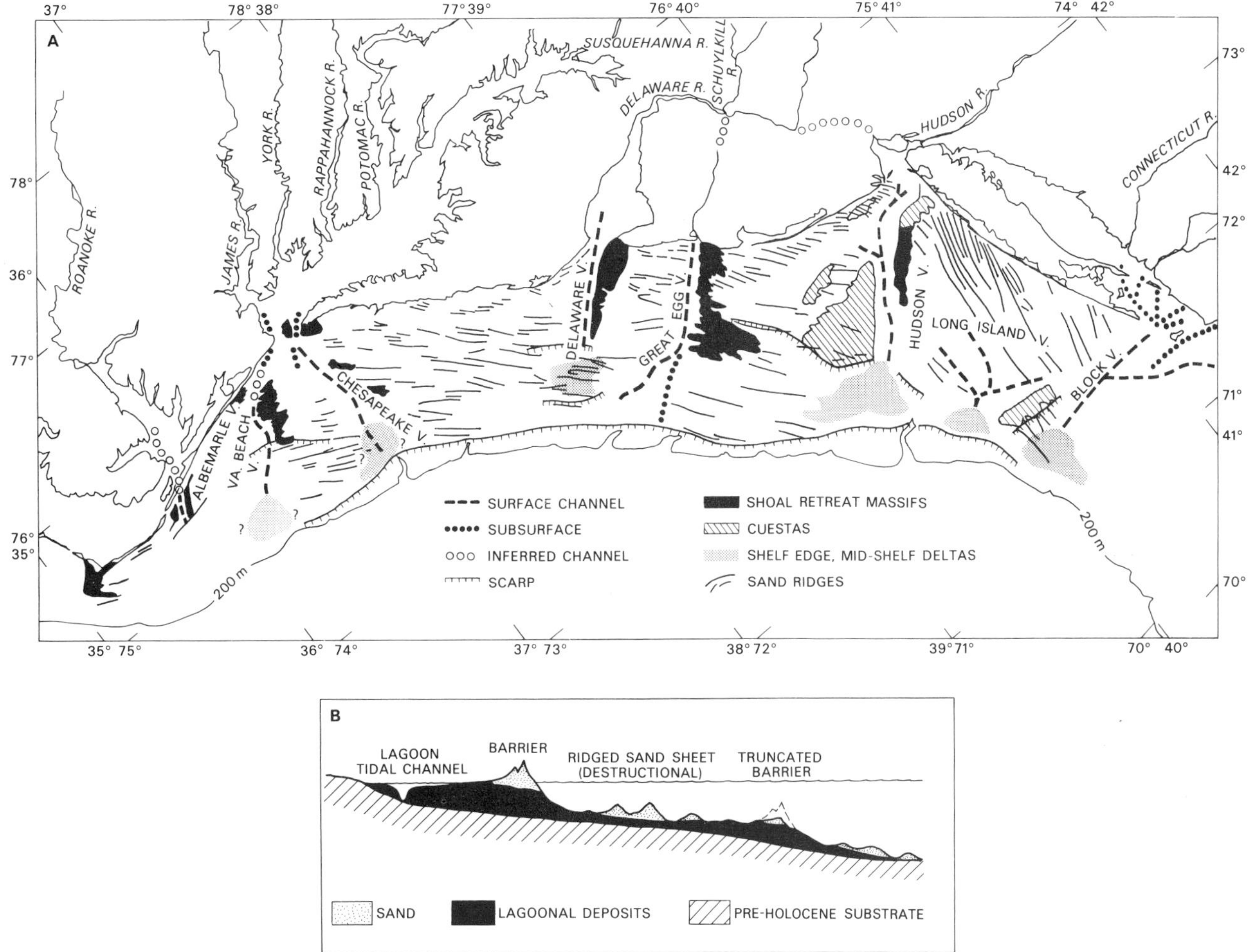

Fig. 9.20. A. The main morphological features of the Middle Atlantic Bight, eastern USA shelf, and B. Schematic profile across this shelf illustrating the transgressing barrier and the progessively abandoned inner shelf ridged sand sheet overlying lagoonal and pre-Holocene deposits (from Swift *et al.*, 1973; Swift, 1974).

directed bottom current (Swift *et al.*, 1973). Wind-drift currents are considered to be the dominant component although this also contains a subordinate wave-drift component consisting of oscillatory wave surge (Swift, 1972). Bottom currents move at velocities of 2–5% of the surface wind speeds (Weigel, 1964). Thus a northeasterly wind of 12.5 m/sec has generated a southerly moving bottom current of 20 cm/sec, and bottom currents exceeding 30 cm/sec have also been recorded (Duane *et al.*, 1972; Swift *et al.*, 1973). The efficiency of these bottom currents is increased by the simultaneous occurrence of moderate to strong wave activity.

RELICT COMPONENT OF THE SURFICIAL SAND SHEET

The surficial sands of the inner shelf and shoreface are either reworked from the Holocene substrate, derived laterally from an adjacent shelf, or left stranded by barrier and headland erosion and retreat (Field and Duane, 1976). The morphology of this sand sheet, therefore, preserves both the primary (first-order) morphological features of the retreating shoreline, such as former fluvial-estuarine and barrier island systems, and superimposed (second-order) features related to modern conditions (Swift *et al.*, 1973).

The first-order features consist of zones trending across the shelf normal to the shoreline. They include: (1) former river valleys and estuary mouths, and (2) former zones of littoral drift convergence which occur at headlands. Thus in the Middle Atlantic Bight relict channels extend across the shelf frequently connecting present-day rivers with canyon heads (Fig. 9.20). These zones are frequently the sites of former estuary-mouth sand bars (inlet-associated shoals), or former headland sand bodies (cape-associated shoals). The retreat paths of these two types of sand bar or 'shoal retreat massifs' (Swift *et al.*, 1973) represent the thickest zones of the surficial sand sheet.

The smaller-scale, second-order, features are represented by ridges and swales (Swift *et al.*, 1972a). They mark the retreat path of the shoreface, and are composed of large elongate sand ridges (Fig. 9.20). They occupy the large plateau-like areas, between the first-order features, where the surficial sand sheet is thinnest (0–10 m). They are, therefore, mainly found adjacent to linear barrier coastlines but because they are responding to the hydraulic regime they also develop on top of the first-order shoal retreat massifs (Swift *et al.*, 1972b; 1973).

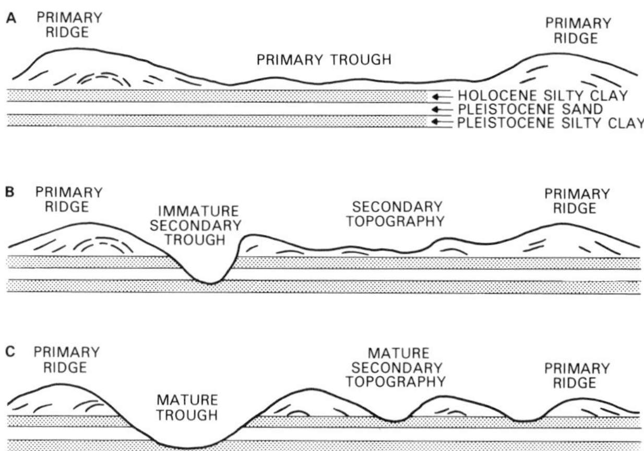

Fig. 9.21. Evolution of storm-generated linear sand ridges, based on the central New Jersey shelf. A. Incipient ridges are formed during shoreface retreat and erosion by storm-generated (wind-driven) currents. Sand continues to be eroded from the troughs and deposited on the ridges as water depth increases during the Holocene transgression. B. Intense storm erosion locally penetrates the early Holocene lagoonal clays and forms a secondary immature trough. C. As downcutting in the secondary trough decreases, lateral erosion increases to form a mature trough. Secondary ridges result from similarly eroded troughs situated up-current (from Stubblefield and Swift, 1976).

MORPHOLOGY AND PROCESSES OF THE INNER SHELF SAND BARS

The surficial sand sheet consists of three main types of sand bar:

(1) *Linear Sand Bars* are 3 to 10 m high, 1 to 2 km wide, several tens of kilometres long, and include two main types: those which are attached to the shoreface and those which are isolated (Duane *et al.*, 1972). Bar crests trend obliquely to the coast forming an acute angle of intersection of approximately 22°, which generally opens to the north (Fig. 9.20). They occur adjacent to barrier islands and form the ridge and swale topography of the inner shelf, but they also continue into deeper water. Cores and seismic profiles reveal a sharp contact between the sand bars and the underlying muddy sediments of Holocene lagoons (Fig. 9.21).

Duane *et al.* (1972) propose a hydraulic model for the initiation, growth and maintenance of these sand bars (Fig. 9.22), which is based on observations at False Cape and Bethany Beach. Shoreface-connected ridges evolve to isolated types primarily because of a rise in sea level leading to shoreface retreat accompanied by the intense, southerly moving, storm-generated currents and wave surge. The ridges are isolated by headward erosion at the base of the shoreface and grow because of control of sediment transport by presumed helical flow cells in the storm currents. The latter are considered to cause bottom-current convergence at the crests and divergence in the troughs (Fig. 9.23). The ridges are, therefore, sites of aggradation while the troughs are sites of erosion. As sea level rises the ridges become progressively detached and eventually 'relict' in the sense that they were formed under hydrodynamic conditions which were different to those they now experience. Once formed, however, the present shelf hydraulic regime appears to maintain the sand ridges since they are not destroyed by wave action and occasionally show signs of lateral migration.

The sand bars appear to respond to seasonal variations in the hydraulic regime. As a result of the oscillatory wave surge of the fair weather summer regime, wave ripples occur on the crest and upper flanks of the bars (Swift *et al.*, 1972a). During this period fines are moved from upper parts of the bars and deposited in the troughs. This process produces the characteristic coarsening-upward sequence and eventually winnows and degrades the bars so that a secondary lag is formed on their crests (Stubblefield *et al.*, 1975).

Products of the intense winter storm regime are less easily studied. However, the storm-generated, southerly flowing unidirectional currents appear to be the main process in maintaining the northeast–southwest trending sand ridges. Evidence of strong and fluctuating bottom currents is provided by longitudinal current lineations, linguoid current ripples, cross-bedded units within cores, fining-upward units, occasional examples of sand waves (McKinney *et al.*, 1974; Stubblefield *et al.*, 1975; Hunt *et al.*, 1977) and the variably oriented internal bedding planes recognized in seismic profiles

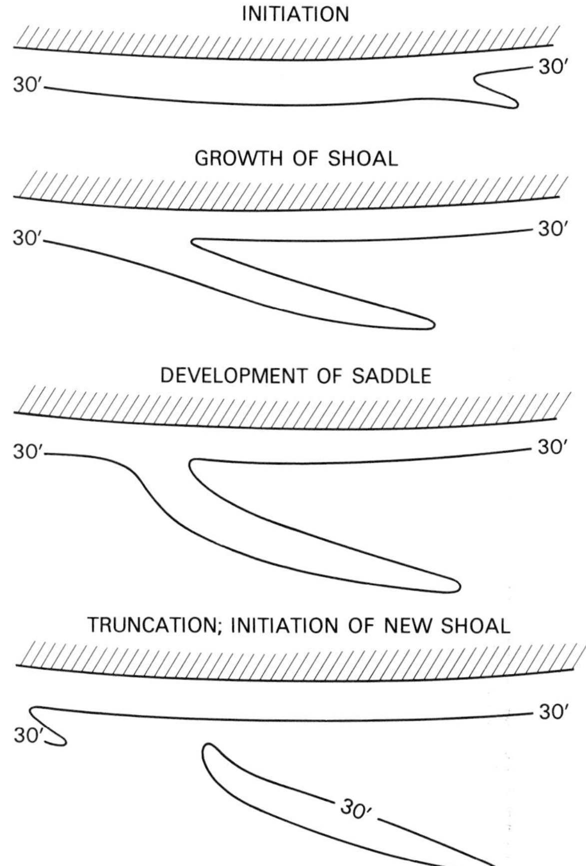

Fig. 9.22. Model for the development of offshore linear sand ridges by the progressive detachment of shoreface-connected ridges (from Duane *et al.*, 1972).

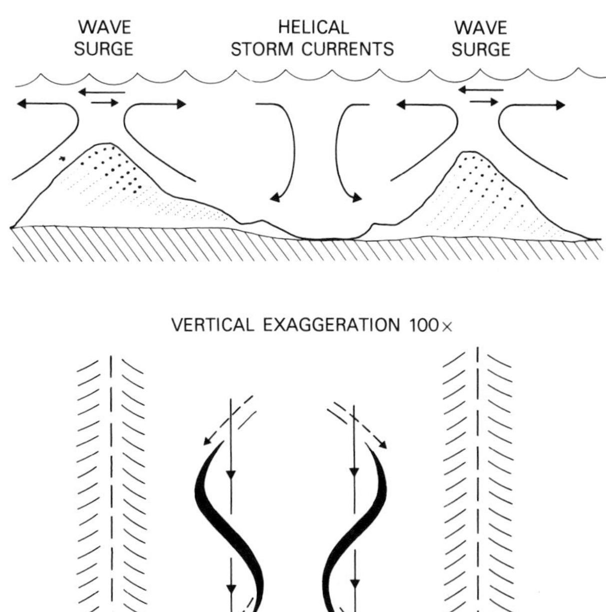

Fig. 9.23. Schematic representation of the inferred secondary flow currents (helical flow structure) and storm-induced wave-surge, believed to be associated with the storm-generated currents and linear sand ridges on the eastern USA shelf (from Swift *et al.*, 1973).

(Swift *et al.*, 1973). Furthermore, textural differentiation of the surficial sediments from the crests, flanks and troughs of bars on the Central New Jersey shelf led Stubblefield *et al.* (1975) to propose a three-fold hydraulic process model consisting of intense storm, transitional and fair weather periods.

(2) *Inlet-Associated Bars* (also called 'estuary entrance shoals' and 'ebb tide deltas') consist of seaward convex bars at the mouths of major estuaries and tidal inlets. These bars extend seaward for several kilometres and have relatively smooth arcuate fronts. Small bars are usually bisected by a single channel and flanked by swash bars, swash platforms and chutes (Oertel, 1972). Wider estuary mouths are filled by sand banks which are themselves dissected by several interdigitating, mutually evasive ebb and flood tidal channels, particularly when sediment is abundant (Ludwick, 1970; Swift *et al.*, 1972b).

They are important depositional sites where sands, transported laterally as littoral drift, may either be deposited on the seaward sides of the bars or transported up into the estuary or inlet.

The bars have complex hydraulic systems in which tidal currents are dominant inshore, giving way offshore to the storm-dominated shelf regime. The bar crests are characterized by a wide range of bedforms with tidal currents forming dunes and wave-induced currents forming swash bars and swash platforms (Oertel, 1972; Oertel and Howard, 1972). As the storm-generated currents gain dominance, the sand bars may be destroyed or progressively re-orientated into the direction of the shelf currents to form linear sand bars (Fig. 9.24).

(3) *Cape-Associated Bars* consist of sand bars which are adjacent to prominent headlands, and possess a characteristic seaward convex 'hammer-head' shape (Fig. 9.25). The headlands act as shields to the high rate of longshore sediment supply and their characteristic shape is fashioned by a combination of wave refraction and the strong southerly moving, storm-generated currents (Swift *et al.*, 1972b). They extend offshore, transverse to the shelf, as transgression proceeds, with the shelf hydraulic regime developing superimposed linear sand ridges and troughs. The super-

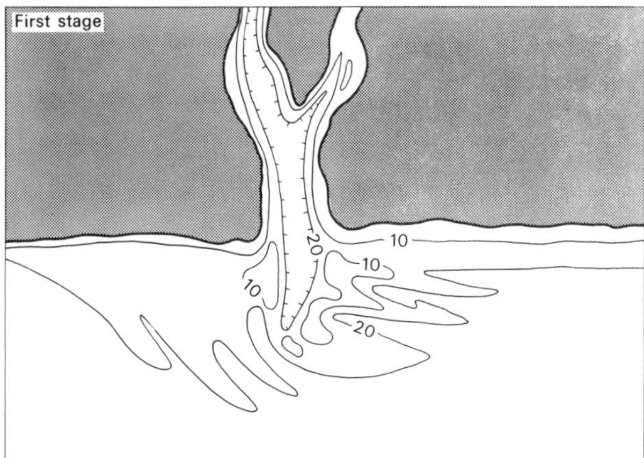

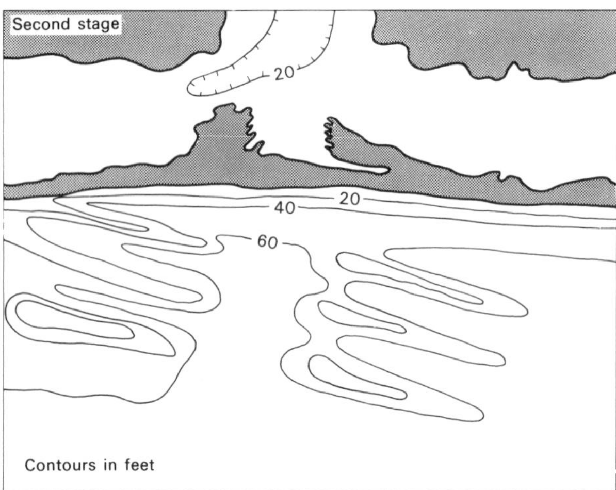

Fig. 9.24. Schematic model for the evolution of inlet-associated bars into offshore linear sand ridges, based on the eastern USA shelf. In the *first stage* (e.g. Delaware Bay and Chesapeake Bay) there is an offshore transition from tide-dominated inshore channels and bars, through seaward flaring into storm-dominated linear sand ridges. The orientation of the latter gradually swings into the path of the storm-generated currents. In the *second stage* (e.g. Virginia Beach Ridges) coastal retreat during transgression has left a shelf transverse sand body (shoal retreat massifs of Fig. 9.20 and 9.25) which is reworked with superimposed linear sand ridges formed by the unidirectional storm-generated currents (from Swift *et al.*, 1972b).

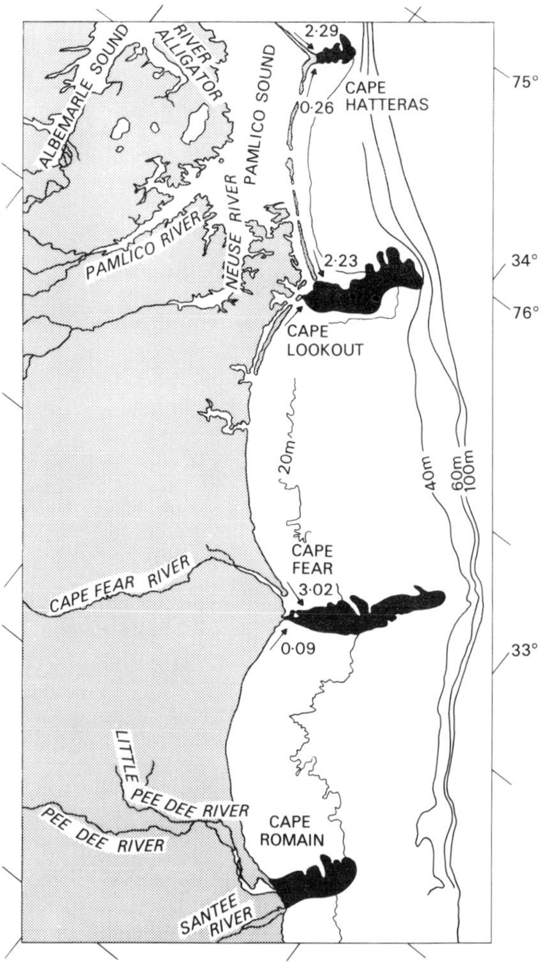

Fig. 9.25. Cape-associated sand bars, across part of the eastern USA, forming at sites of littoral drift convergence (sand transport rates in yd/yr $\times 10^{-3}$) (from Langfelder *et al.*, 1968, in Swift, 1976a).

imposed ridges are convex seaward and have their long axes normal to wave approach. The troughs usually shallow to the south, and are thought to have been cut by storm-generated currents (Swift *et al.*, 1973). In addition, they occasionally carry sand waves whose lee slopes face south, parallel to ridge elongation (Hunt *et al.*, 1977).

9.7 COMPARISON OF TIDAL AND STORM-GENERATED SAND RIDGES

Despite the contrasting hydraulic regimes of tidal and storm-generated linear sand ridges, they have similar morphologies and origins. Both are believed to originate as inshore channel-shoal couplets which during transgression become detached from the shoreline (Swift, 1975b). In both cases their maintenance on the shelf appears to be largely controlled partly by the parallelism of the strongest shelf currents with the initial trend of the channel-shoal couplets, and by convergent flows over the shoals.

These similarities raise the question as to whether these

contrasting sand bodies could be distinguished if preserved in the geological record. Both are frequently characterized by a coarsening-upward sequence due to the winnowing of fines from the ridge crests and their redeposition in the quieter, deeper waters of the troughs. However, where bottom currents actively scour the troughs a coarser lag deposit overlies a planar erosion surface.

The reversing character of tidal currents should be the critical difference, but in cores or at outcrop only one tidal current direction may be recorded (Johnson, 1977a). Although both types are cross-bedded it is only in tidal seas that dunes and sand waves are common on the ridge flanks. In storm-dominated seas sand waves are only locally abundant and may be transitory features, because of the seasonal aspect of the hydraulic regime. Furthermore, suspension deposition may be equal or even more important than bedload transport on storm-dominated ridges.

This leads to the suggestion that extensively cross-bedded sandstone bodies are probably tide-dominated (see Sect. 9.10) and sandstone bodies exhibiting a broader range of structures and depositional units may be storm-dominated. However, more information is desirable on the internal structure of modern sand ridges, particularly those that are storm-dominated.

9.8 CRITERIA FOR RECOGNIZING ANCIENT SHALLOW MARINE SILICICLASTIC DEPOSITS

The most reliable criteria for recognizing ancient shallow marine deposits are features controlled by the salinity and depth of sea water. Thus marine body fossils, trace fossils, certain minerals and geochemical parameters may be characteristic. Sedimentological data such as texture, sedimentary structures, lithofacies associations, sand body geometry and palaeocurrent patterns, are usually not in themselves diagnostic.

Invertebrate fossils provide the most reliable means of distinguishing marine from non-marine environments because they are principally controlled by salinity and salinity variations. In addition many marine organisms appear to have occupied similar salinity ranges through time with several ancient groups having modern descendants or relatives now inhabiting open marine environments (Fig. 9.26A). The sublittoral (or level bottom) environment is identified by stenohaline species which occupy relatively narrow, normal marine salinity ranges. In fossil assemblages this includes most corals, cephalopods, articulate brachiopods, echinoderms, bryozoa and certain calcareous foraminifera. In contrast,

Fig. 9.26. Modern distribution of major fossilizable non-vertebrate groups relative to A. salinity, and B. water depth (from Heckel, 1972).

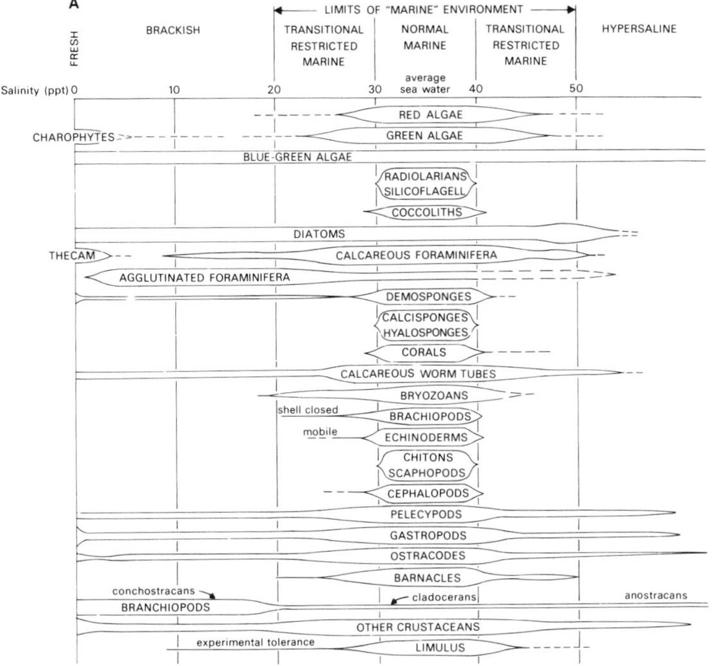

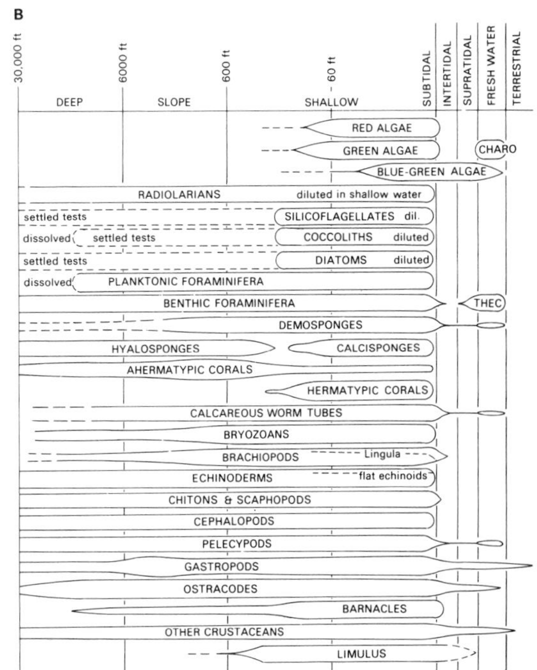

euryhaline species, occupying broad salinity ranges in both brackish and hypersaline waters, are recognized in fossil assemblages by some types of ostracod, bivalve and gastropod. In addition species diversity is low.

Water depth is more difficult to ascertain since some groups show little direct depth dependence (Fig. 9.26B) while other predominantly shallow water forms may have been transported to deeper water. In most field occurrences differentiation may be made between 'nearshore' and 'offshore' rather than actual water depth. However, since many fossil communities are largely depth-dependent, bathymetric zonation of sublittoral environments is sometimes possible (Ziegler, Cocks and Bambach, 1968).

Detailed reconstruction of the sublittoral environment is achieved through the palaeoecological analysis of marine benthic faunas and their relationship to the substrate (biostratinomy, Seilacher, 1973). Marine palaeoecologists concentrate on benthic faunas because they live in or on the bottom and are therefore likely to be preserved in, or close to, their life habitat, with the minimum of post-mortem transport.

The main physical parameters influencing marine benthic organisms are temperature variation, dissolved oxygen, light intensity, salinity, hydraulic energy and substrate. Most of these parameters are indirectly related to water depth. However, only the substrate is directly available for the palaeoecologist to study but this provides abundant information about ancient siliciclastic sea beds, such as rates of sedimentation, hydraulic energy, grain size, sorting, degree of consolidation, stability and content of organic matter. Although a close relationship exists between substrate and benthic animal communities in modern seas (Thorson, 1957) the same relationship is not widely documented in rocks, apparently due to a lack of direct fauna-sediment studies. However, Fürsich (1976, 1977) has provided a comprehensive reconstruction of Corallian (Jurassic) sublittoral communities through detailed studies of fauna-substrate relationships (Fig. 9.27).

The degree of reconstruction of faunal communities is limited by several geological factors which affect preservation and the fossil record. For example, shell preservability is affected by biological, mechanical and chemical destruction which occur both on and beneath the substrate. Extensive chemical dissolution can occur during initial burial and subsequent diagenesis. Post-mortem transport may have been extensive in many ancient current- and storm-swept shallow seas thereby leading to shell abrasion and mixing of benthic communities. However, the absence of species may not indicate environmental changes but may reflect chance perhaps related to variations of larval transport or the transient, short-lived nature of many populations. Thus a uniform environment does not necessarily support an evenly distributed population nor preserve an evenly distributed fossil assemblage.

Several texts discuss in detail these and other principles of palaeoecology (e.g. Ladd, 1957; Ager, 1963; Imbrie and Newell, 1964; Raup and Stanley, 1971; Scott and West, 1976).

Trace fossils are particularly important in sedimentary rocks since they provide an *in situ* record of animal activities within or upon the substrate (Frey, 1975). Although the structures produced by benthic organisms do not necessarily distinguish particular animals they preserve the reaction of individuals to the substrate. Thus the trace fossils which characterize the sublittoral or shelf environment mainly record the activities of suspension feeders whereas in deeper water or lower-energy environments the proportion of elaborate sediment feeders increases. This forms the basis of Seilacher's (1967) bathymetric zonation of trace fossil communities (Fig. 9.28) of which the *Cruziana* facies characterizes the sublittoral environment, particularly where it experiences moderate to high-energy conditions. However, since the sublittoral environment is subjected to a wide range of physical energy levels and diverse substrate types there are probably several trace fossil communities. Furthermore, water depth is only one of several other environmental factors controlling biogenic activity in the substrate. Food availability, which is independent of bathymetry, is an important factor (Frey, 1971). Osgood (1970) also doubts the existence of a sharp distinction between *Cruziana* facies and *Nereites* facies and notes that *Zoophycos* occupies transitional zones in both deep and shallow water deposits.

As a criterion for inferring a shallow marine environment trace fossils assume paramount importance in deposits devoid of body fossils. Many Cambrian and Ordovician shallow marine sandstone deposits lack body fossils, but have rich trace fossil assemblages which confirm a shallow marine environment (e.g. Banks, 1973a; Baldwin, 1977). The occurrence of *Chondrites* in the black Hünsruck Shale of Germany also demonstrates the presence of an infauna indicating that it did not accumulate in a hydrogen sulphide-rich, euxinic environment devoid of animals (Richter, 1931, quoted in Häntzschel, 1975, p. 35). In addition, associations of trace fossils and body fossils may enable differentiation between fresh water, brackish water and open marine deposits in some Upper Carboniferous deposits (Seilacher, 1964).

Certain *authigenic minerals* are either largely or entirely restricted to the marine environment and can therefore be used as indicators of marine sediments. In the shallow marine environment the most characteristic authigenic minerals are various iron silicates, such as glauconite and chamosite, and some phosphates.

Phosphates mainly occur in modern shallow marine and pelagic deposits, but only rarely in continental ones. They are particularly characteristic of areas of slow clastic sedimentation, frequently occurring on topographic highs, and in areas

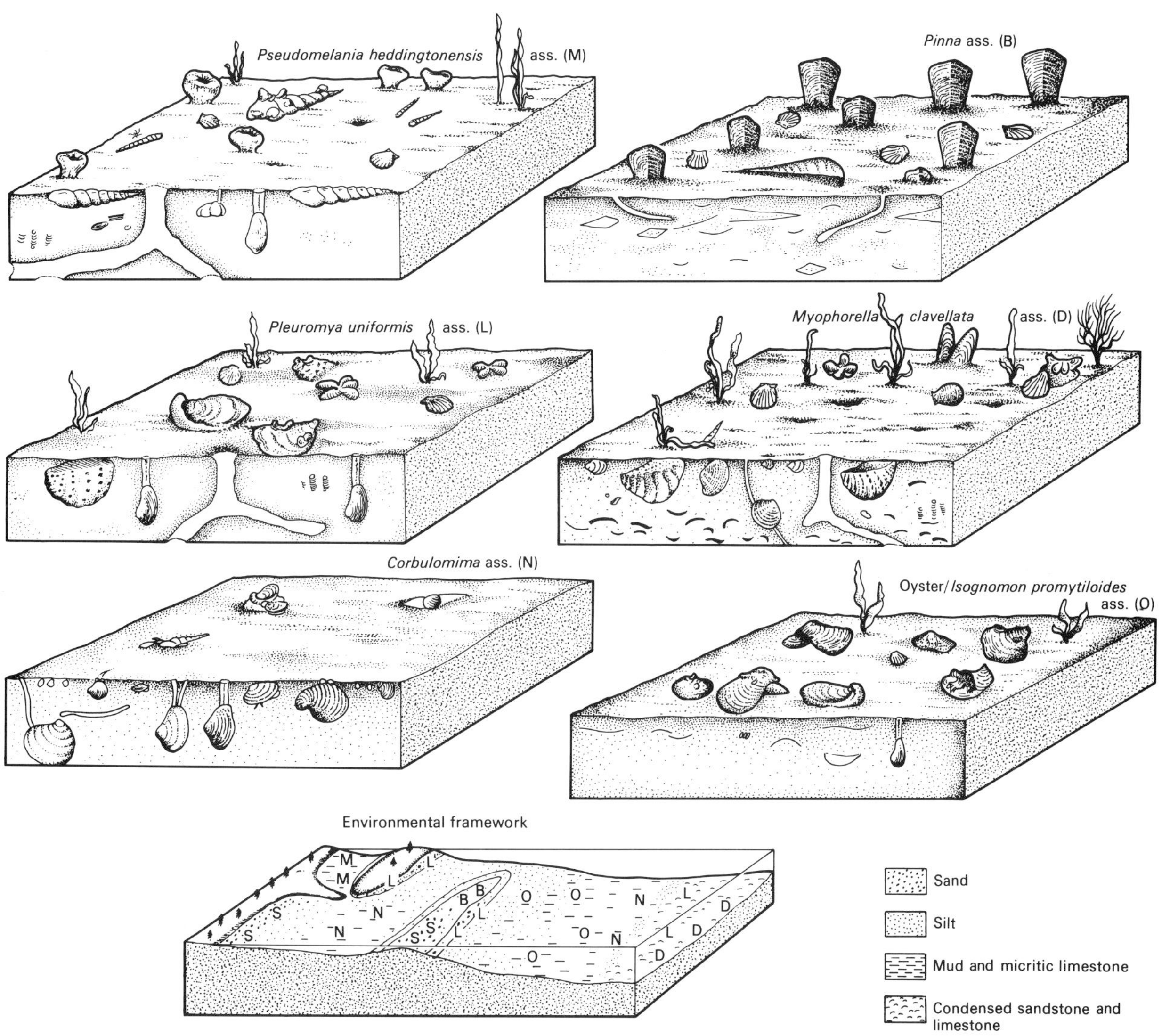

Fig. 9.27. Reconstruction of the Corallian (Jurassic) sublittoral environment (central block), in southern England and northern France, in terms of sediment type and benthic faunal associations. Attempted reconstructions of six faunal associations illustrate some fauna-substrate relationships and the inferred life habitats of some of the characteristic marine benthic faunas (from Fürsich, 1977).

of coastal upwelling (Sect. 9.3.6). Most ancient occurrences, ranging from the Pre-Cambrian onwards, are demonstrably shallow marine by their association with shelf-dwelling calcareous organisms, cross-bedding, abrasion features on phosphoclasts, remnants of reef-building algae and lateral interfingering with shallow water clastic sediments (Blatt *et al.*, 1972, pp. 546–550). In the Phosphoria Formation (USA) phosphates reach their maximum development in an outer shelf to deep basin transition (McKelvey, 1967). They are associated with oolitic and pisolitic limestones and calcareous sandstones, which interfinger basinward with cherts and muddy cherts while landward they pass into continental red beds. Phosphatic nodules also characterize horizons of slow or non-deposition in shallow marine clastic deposits (e.g. Casey and Gallois, 1973; Spearing, 1976).

Glauconite is restricted to the marine environment forming

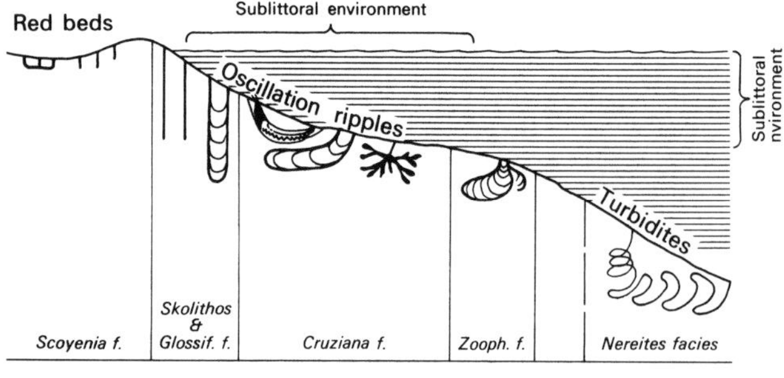

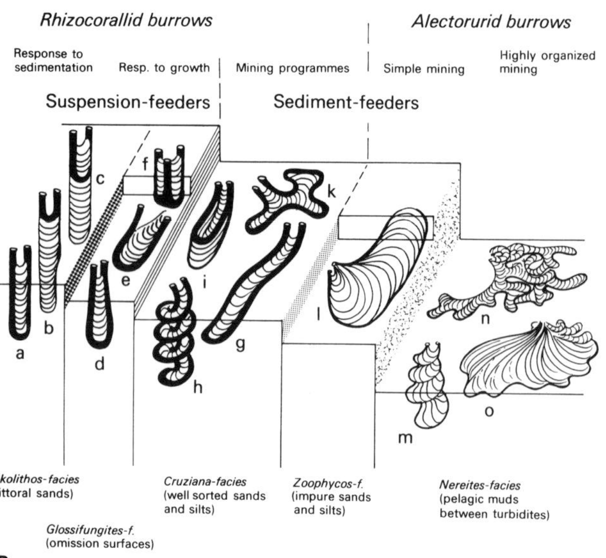

Fig. 9.28. Types and distribution of trace fossil communities in the sublittoral environment: A. generalized bathymetric distribution of the major trace fossil communities, and B. bathymetric zonation of fossil spreite burrows indicating the predominance of suspension feeders in the shallow water high-energy zone, which are gradually replaced in the deeper water lower energy environments by increasingly elaborate sediment feeders (from Seilacher, 1967).

by several possible processes such as direct precipitation from sea water, alteration of detrital phyllosilicate minerals such as illite and biotite, but mainly by the alteration of organic matter particularly faecal pellets (Burst, 1968a and b). This probably accounts for the rarity of glauconite in Proterozoic shallow marine deposits in contrast to its rich occurrence in certain Mesozoic and younger deposits (e.g. Spearing, 1976; Selley, 1976b). Glauconite accumulates mainly by the transport and deposition of granules and pellets with other clastic material, and is a common constituent of shallow marine 'greensands'.

Chamosite is also largely shallow marine and is the dominant primary iron silicate in Phanerozoic ironstones. It is mainly associated with clastic sediments, limonite (or haematite) oolites, or siderite deposits.

Trace elements which are salinity-dependent provide a means of differentiating between marine and non-marine sediments. Palaeosalinity estimates have been obtained from boron–gallium and boron–lithium ratios in illite and mudstones (Degens *et al.,* 1957; Walker and Price, 1963; Reynolds, 1965), calcium–iron ratios in sedimentary phosphates (Nelson, 1967) and carbon–oxygen isotope ratios in limestone, fossil shells and sandstone deposits (Keith and Degens, 1959). The merits and limitations of these techniques are beyond the scope of this study but have been thoroughly discussed elsewhere (e.g. Eager and Spears, 1966; Müller, 1969; Guber, 1969; Harder, 1970; Ernst, 1970).

Textural maturity is a characteristic of shallow marine sands due to long periods of transport with winnowing of fines. This probably mainly results from extensive reworking in the high-energy coastal zone. Certainly many modern shelf sand deposits have undergone relatively little reworking during the last 7,000 years. However, wave and current activity may have been more intensive in ancient shallow, high-energy epeiric seas enabling some *in situ* winnowing. The textural maturity of shallow marine sands is emphasized in some ancient shelf successions by limestone–quartz–arenite associations where there is a marked deficiency of fine-grained sediment (e.g. Swett and Smit, 1972; Pettijohn, Potter and Siever, 1973, pp. 492 and 544–572). Reduction of textural maturity accompanies a decrease in physical energy conditions and also results from biogenically and/or authigenically introduced clays. Thus a nearshore to outer shelf transition may be recognized by a progressive truncation of the coarse sand fraction, increased bioturbation and a decrease in sorting (e.g. Swift, Heron and Dill, 1969).

Sedimentary structures are rarely environmentally diagnostic but can be indicative of specific processes. Thus bidirectionally orientated, current-formed sedimentary structures have long been considered indicative of tidal currents. Similarly, wave ripples and wave ripple cross-lamination are indicative of wave action, but this is not restricted to a shallow marine environment.

Although not diagnostic there are several *lithofacies assemblages* and *vertical sequences* which are characteristic of shallow marine deposits (Sect. 9.10 and 9.11). In particular there are some noticeable dissimilarities between the lithofacies sequences of offshore shallow marine deposits and those of the better known inshore, shoreline and continental deposits. In the absence of salinity indicators a combination of other sedimentological data, such as palaeocurrent patterns, texture, sand body geometry and trend and lithofacies associations can provide a means of differentiating between open marine and other shallow water environments (e.g. Hobday and Reading, 1972; Anderton, 1976).

A complete spectrum of *palaeocurrent patterns* has been described from shallow marine deposits, but some recurring patterns have been noted (e.g. Klein, 1967a; Selley, 1968). Thus, while there is no pattern diagnostic of a shallow marine environment, palaeocurrents may provide strong supplementary evidence when combined with textural variations, sand body geometry and lithofacies associations (Sect. 9.10 and 9.11).

9.9 SUMMARY OF SHALLOW MARINE SILICICLASTIC LITHOFACIES

Shallow marine siliciclastic deposits can be divided into lithofacies on the basis of their grain size, texture, sand–mud ratio, bed thickness, sedimentary structures, degree of bioturbation and trace fossil and body fossil content (Fig. 9.29). An increase in the relative abundance of faunal content necessitates a biofacies classification which will either complement or supersede the lithofacies scheme (e.g. Sutton, Bowen and McAlester, 1970; see also Chapter 10).

The present classification is partly developed from that of Boersma (in Ginsburg, 1975). The three major facies represent a continuum of physical energy conditions.

Sandstone Facies (*S*) is determined initially by its bulk sand content (90–100%) and subdivided on the basis of the dominant sedimentary structure into three subfacies: (1) cross-bedded (Sa), (2) parallel-laminated (Sb), and (3) cross-laminated (Sc).

Parallel lamination usually reflects upper flow regime conditions, possibly combined with deposition from suspension, caused by either unidirectional currents or wave action. Cross-bedding and cross-lamination reflect lower flow regime conditions during the migration of dunes (megaripples) and small-scale ripples (current, wave or combined-flow varieties) respectively (e.g. Simons *et al.*, 1965; Harms and Fahnestock, 1965; Harms, 1975).

The mobility of the substrate in these high-energy deposits generally prevents faunal colonization and fossil remains are usually scarce apart from comminuted shell debris.

Heterolithic Facies (*H*) consists of thinner bedded deposits with a variable sand content (10–90%) which enables subdivision into three subfacies: (1) sand-dominated (Ha, 75–90% sand), (2) mixed (Hb, 50–75% sand), and (3) mud-dominated (Hc, 10–50% sand).

The sand-dominated subfacies (Ha) is usually characterized by relatively thick (5–20 cm) parallel-sided, laterally persistent sheet sandstones separated by variously ripple laminated sandstones. The thicker sheet sandstones, which superficially resemble turbidites, also characterize the mixed and mud-dominated subfacies where they are thinner (3–10 cm). Mud layers in the mixed subfacies usually occur as continuous and discontinuous drapes forming flaser and wavy bedding (Reineck and Wunderlich, 1968; Reineck and Singh, 1973, pp. 97–102). Linsen bedding characterizes the mud-dominated subfacies. It forms when discontinuous sand patches and starved ripples are transported across muddy substrates.

The variability of this facies reflects fluctuations in the intensity and periodicity of hydrodynamic conditions and in the supply of sediment. Both bedload and suspension sedimentation are active, giving rise to a wide range of sedimentary structures. High-energy events are recorded by the thicker, laterally persistent sheet sandstones (Sect. 9.11). The type of ripple lamination in the flaser-wavy-linsen bedded units provides a means of distinguishing between the relative importance of wave and current processes (see Sect. 9.11).

Fossils frequently occur as rich concentrates along the base of the sheet sandstones where they form transported assemblages (e.g. Goldring and Bridges, 1973; Brenner and Davies, 1973; Bridges, 1975). The upward increase in the intensity of bioturbation in the sheet sandstones (laminated to burrowed) records the recolonization of the substrate during normal conditions following periods of exceptional high energy (e.g. Goldring and Bridges, 1973; Howard, 1972; see also shoreface sands, Chapter 7). This facies is commonly associated with rich trace fossil assemblages (e.g. *Cruziana* facies of Seilacher, 1964, 1967) with the alternate deposition of sand and mud being particularly conducive to the preservation of trace fossils.

Mud Facies (*M*) consists either of muds with thin intercalations of silt, sand, shell beds (coquinas) or bioclastic limestone (Ma) or of homogenous muds with scattered faunal remains (Mb). Thin (1–2 cm) graded beds reflect periodic deposition from suspension. Faunal remains provide the greatest information for reconstruction of muddy shelf environments, particularly where shells have undergone little or no transport. The orientation, fragmentation and sorting of shell beds provide information on hydrodynamic conditions in these low-energy sediments (e.g. Brenner and Davies, 1973; Seilacher, 1973; Bridges, 1975).

Classification of shallow marine lithofacies in terms of the dominant physical process has to be approached with caution. This is due to the difficulty of interpreting sedimentary structures and lithofacies in terms of a specific process or group of processes operating in the shallow marine environment. It is relatively easy, for example, to relate a sedimentary structure to a general process (e.g. cross-bedding represents megaripple migration in response to a unidirectional current) but it is considerably more difficult to specify the process which generated the current. In the shallow marine environment the latter may have been tidal currents, oceanic currents, storm-generated (wind-driven) currents, barometric storm surges,

FACIES	SUBFACIES	TYPICAL LOG	INTERNAL STRUCTURE	SAND CONTENT	BED OR SET THICKNESS	INFERRED PROCESSES & NOTES
SANDSTONE FACIES	Sa Cross-Bedded		Tabular } Cross-bedding; Trough } Cross-bedding	90-100%	ca 10-200 cm	Cross-beds variable in type and set thickness. Represents dunes/megaripples (trough sets) and sand waves (tabular sets).
	Sb Flat Bedded		Parallel and low-angle lamination		variable	Wave- or current-formed lamination associated with high-energy conditions.
	Sc Cross-laminated		Cross-lamination		1-5 cm	Cross-lamination. Varies in relation to ripple type, notably current, combined-flow and wave ripples.
HETEROLITHIC FACIES	Ha Sand Dominated		Parallel lamination	75-90%	5-20 cm (max 200 cm)	Alternations of parallel and cross-laminated sheet sandstones. Thicker sheet sandstones may form 20-90% of this subfacies. Amalgamation may be common. Sand deposited from suspension & as bedload. Variable reworking by current and wave ripples. Sheet sandstones commonly inferred to be the product of intense storm conditions. May contain transported shell debris. Bioturbation increases in the finer grained intercalations.
			Parallel to cross-lamination		5-20 cm (max 200 cm)	
			Low-angle and trough lamination		5-20 cm (max 50 cm)	
			Isolated tabular cross-bedding		5-20 cm (max 50 cm)	
			Sandy flaser bedding		1-5 cm	
	Hb Mixed		Parallel lamination	50-75%	1-10 cm	Mainly ripple laminated sandstones & mudstones with subordinate parallel laminated sheet sandstones (10-50%). Variable types of cross-lamination in response to current, combined-flow and wave ripples. Storm and fair weather increments may be recognized as above. Upper part of sheet sandstones bioturbated.
			Parallel to cross-lamination		1-10 cm	
			Low-angle lamination		1-10 cm	
			Flaser-wavy bedding		1-3 cm	
	Hc Mud Dominated		Parallel lamination	10-50%	1-5 cm	Mainly linsen bedding with rare sheet sandstones (5-10%). Sand lenses formed by current or wave processes. Sandstone interbeds deposited from suspension during storms. Suspension deposition of muds predominant fair weather process. Latter commonly intensively bioturbated.
			Parallel to cross-lamination		1-5 cm	
			Linsen bedding		1-3 cm	
MUD FACIES	Ma		Graded sand &/or shell-rich layers	0-10%	0.1-2 cm	Mainly muds with thin sand interbeds and sand and silt streaks. Deposition entirely from suspension. Wave and current activity only accompany rare storms. Intensive bioturbation, *in situ* or slightly transported benthic faunas.
	Mb		Mud		< 0.5 cm	

Fig. 9.29. Facies scheme for siliciclastic sediments with particular reference to the main types of lithofacies in the sublittoral environment.

wave-generated currents (e.g. wave surge, longshore currents, rip currents) and tectonically-induced surges such as tsunamis (Sect. 9.4).

It is possible, nevertheless, to make significant predictions concerning the type of shallow marine palaeohydrodynamic conditions by combining lithofacies and sedimentary structures with other sedimentological factors: e.g. type of facies alternation, sequences, spatial distribution of facies, textures, palaeocurrent patterns, sand body geometry, shape and trend, palaeogeography and general position within a sedimentary sequence.

9.10 TIDE-DOMINATED FACIES: RECOGNITION AND MODELS

Ancient, high-energy, tide-dominated shallow marine sandstones have first to be interpreted as offshore shallow marine, and then tidal action has to be recognized as the dominant source of energy.

Shallow marine sandstones can be identified by the following features: (1) open marine macro- and/or micro-fauna usually occurring either as concentrations of shells, or as dispersed and comminuted shell debris, (2) assemblages and/or communities of trace fossils and characteristic bioturbation, (3) interfingering of sandstones with fully marine shales, (4) glauconitic and/or phosphatic accessory mineralogy, (5) high textural and mineralogical maturity of the sandstones, and (6) laterally extensive, low-relief erosion surfaces and an absence of deep channelling.

Accepting an offshore shallow marine environment, tidal currents can be inferred from the following: (1) bidirectional current-formed sedimentary structures and bimodal palaeocurrent patterns with the two modes approximately 180° apart, reflecting flow reversals in rectilinear tidal currents, (2) multimodal palaeocurrent patterns reflecting either temporal fluctuations in direction or the rotary nature of tidal currents, possibly with superimposed storm activity, and (3) the abundance of cross-bedding reflecting dunes and sand waves which today are mainly found in tidal seas.

9.10.1 Sedimentary structures due to tidal currents

Bidirectional current-formed sedimentary structures are the most diagnostic feature of tidal currents and may be found in the form of *herringbone cross-bedding* in which superimposed sets of oppositely dipping cross-strata are thought to preserve both ebb and flood tidal flows. However, the two tidal current directions may also be recorded by complicated features, particularly in larger structures. In the Lower Cretaceous Lower Greensand of Southern England large-scale avalanche sets, about 3–6 m thick, alternate with thinner cross-bedded sets

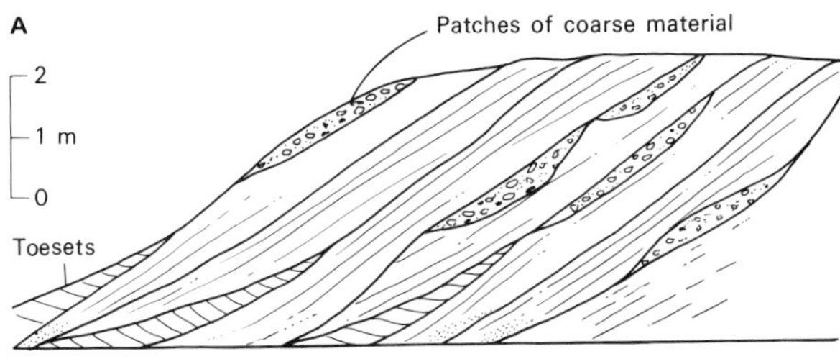

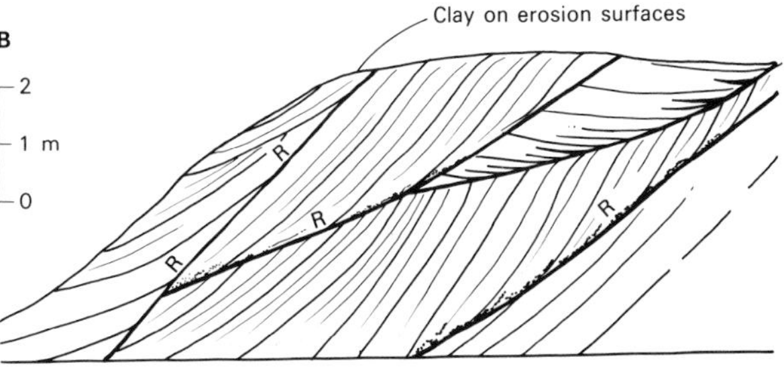

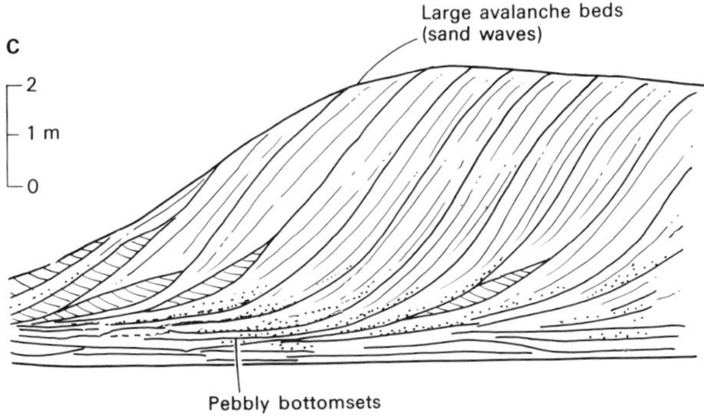

Fig. 9.30. Large-scale cross-bedding related to tidal sand waves in the Lower Cretaceous Woburn Sands (Upper Sands), Leighton Buzzard, Southern England. Associated structures include reversely dipping sets, reactivation surfaces with mud drapes and backflow toesets (from Bentley, 1970).

(10–50 cm thick) which frequently dip up and also down the larger-scale surfaces (Fig. 9.30; De Raaf and Boersma, 1971). The thicker sets are interpreted as sand waves which record the direction of the strongest tidal currents. However, the reversing currents were sufficiently powerful to modify the sand wave lee slopes and to generate up-slope migrating dunes (Fig. 9.30).

The degree of bidirectionality of the dip directions of cross-bedding is variable on both outcrop and regional scales. It is made difficult to interpret because it encompasses spatial variations of sand transport and energy conditions spanning considerable time intervals (several hundreds to thousands of years). Thus two oppositely directed palaeocurrent modes are unlikely to represent small-scale time fluctuations (days or months) of two reversing tidal currents but instead much longer-term fluctuations (several tens or hundreds of years). This and other factors (Sect. 9.10.2) help to explain why unidirectional cross-bedding is more common at outcrop (e.g. McCave, 1973; Anderton, 1976).

To demonstrate tidal activity in cross-bedded sandstones where one direction is overwhelmingly dominant is more difficult. In some circumstances the weaker reversing currents form secondary structures, which may support such an interpretation. For example, oppositely dipping small-scale current-ripple cross-lamination occasionally forms both within and between thicker cross-bedded units. In finer-grained, tidal, heterolithic deposits cross-lamination formed by small-scale current-ripples is more susceptible to preserving current reversals at outcrop.

Reactivation surfaces are common in tidal sands and have been interpreted as the result of modification of bedform morphology by the weaker reversing tidal current (Klein, 1970a). However, this structure may also form both in rivers, as a product of fluctuations in discharge (Collinson, 1970b), and in flumes, due to random fluctuations in the migration of current ripples under uniform, unidirectional flow (Allen, 1973).

A number of other features are commonly associated with offshore, shallow marine sand deposits. They are not in themselves diagnostic of tidal activity, but may be indicative of other longer term processes in the shallow marine environment.

Mud Drapes are a common feature in tidal sand deposits (e.g. Reineck, 1963; Houbolt, 1968; Narayan, 1971; de Raaf and Boersma, 1971). Several workers have thought that they form by fall-out of fine-grained suspended sediment during the slack water period of the tidal cycle (e.g. Reineck and Wunderlich, 1968; Klein, 1970b). However, assuming an abnormally high suspended sediment concentration (e.g. 20 mg l^{-1}) and settling velocity (e.g. $3 \times 10^{-2} \text{ cm sec}^{-1}$) it would take at least 6 days of continuous mud deposition for a 2 mm thick mud layer (300 mg cm^{-2}) to form (McCave, 1970). If continuous mud deposition is restricted to a 2 hour slack water period in each tidal cycle 18 days would be required. Thus mud drapes in offshore tidal sand deposits are not the result of diurnal and semi-diurnal tidal fluctuations. Instead they are more likely to be due to a combination of abnormally high suspended sediment concentrations, low current velocities and low wave intensity over a longer period (McCave, 1970). Suitable conditions may immediately follow a storm when suspended concentrations are high combined with a decrease in wave and current activity. Similar conditions are inferred for a waning flow sequence which terminates in a silt layer draping small-scale current ripples on larger foresets in the Lower Greensand (Narayan, 1971). These fine-grained layers may be intensively bioturbated sediment, supporting the suggestion of long periods of negligible sand transport (Narayan, 1971; Nio, 1976).

Erosion Surfaces of broad lateral extent, low relief, and lacking deep channelling contrast with those formed in estuaries, barrier inlets and rivers. They may be accompanied by shell concentrations, intensive bioturbation and phosphatic or glauconitic mineralization, as in the Lower Greensand (Narayan, 1971; Casey and Gallois, 1973). These surfaces indicate that erosion was extensive and/or that sand transport was negligible for long periods.

Large-scale, low-angle erosion surfaces in the Eocene Roda Sandstone Formation (Spanish Pyrenees) are interpreted as the product of current erosion. Planar surfaces, associated with an *in situ* open marine fauna, such as colonial corals and sponges, and intense bioturbation, reflect a complete cessation of sand transport (Nio, 1976).

Three types of major erosion surface are distinguished by Anderton (1976) in the late Pre-Cambrian Jura Quartzite (Fig. 9.31). Morphologically they range from shallow channels (up to 2 m deep) to laterally persistent planar surfaces of very low relief (0.2 m). Occasionally they are draped by layers of fine-grained sediment or by thin (one-grain thick) layers of granules and pebbles. In the Jura Quartzite the cross-bedded

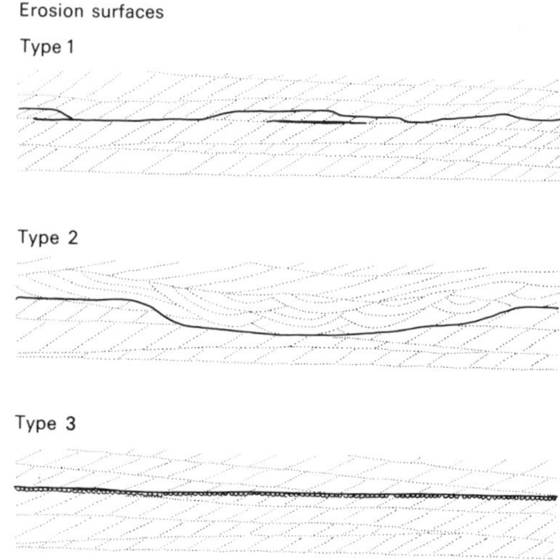

Fig. 9.31. Three types of erosion surface in the late Pre-Cambrian Jura Quartzite, an inferred tidal shelf sandstone deposit in the Scottish Dalradian; Type 1 irregular to undulating with low relief (decimetres to metre) and occasionally floored by a thin siltstone layer (black); Type 2 channelled erosion surfaces up to 2 m deep; and Type 3 laterally persistent and planar (relief never >20 cm) and occasionally mantled by a one-grain thick layer of granules or pebbles (from Anderton, 1976).

sandstones are mainly interpreted as the product of dune migration by tidal currents, whereas the intercalated erosion surfaces are regarded as due to periodic storms, which were sufficiently powerful to prevent sand accumulation and to form winnowed pebble layers. The surfaces overlain by mud drapes record a considerable delay prior to renewed sand transport, a feature possibly comparable to the burrowed and phosphatized surfaces of younger deposits.

Where these erosion surfaces occur in demonstrably tidal deposits several conditions may be envisaged for their formation: (1) reduced sand supply, (2) lateral migration of the main sediment transport paths, (3) lateral migration of sand ridges across scoured troughs, and (4) fluctuations in tidal current velocity due to (a) long term fluctuations (e.g. spring-neap cycles, winter storms), (b) increased water depth (transgression), (c) widespread reduction in tidal current activity due to changing basin configuration (cf. Johnson and Belderson, 1969), and (d) increased tidal current velocities possibly when fair weather currents were enhanced by storms (e.g. Johnson and Stride, 1969).

9.10.2 **Palaeocurrent patterns: tidal versus wind-driven circulation patterns**

Palaeocurrents enable reconstruction of sediment transport in ancient shallow marine cross-bedded sandstones. However, it is frequently uncertain whether dunes and sand waves are exclusively indicative of tidal currents (McCave, 1973; Anderton, 1976), or are also the product of unidirectional, oceanic and /or wind-driven (storm-induced) currents (Hobday and Reading, 1972; Dott, 1974).

Bedload movement of sand and the development of dunes, sand waves and linear sand ridges, is today mainly restricted to tide-dominated seas (cf. Sect. 9.5 and 9.6). In some ancient shallow marine deposits the bidirectional, current-formed sedimentary structures are easily interpreted as tidal (e.g. De Raaf and Boersma, 1971; Banks, 1973a). In other cases the regional palaeocurrent pattern is unidirectional or a single mode is overwhelmingly dominant (e.g. McCave, 1973; Anderton, 1976). However, even a few isolated palaeocurrent reversals may enable tidal currents to be invoked.

Palaeocurrent patterns with a single mode or one dominant mode can be preserved in tidal seas because the strengths of the ebb and flood currents differ. Single tidal current directions may be enhanced due to the basin shape; the main ebb and flood currents frequently follow mutually exclusive paths several hundred metres to kilometres apart and one tidal current direction may be enhanced by superimposed storm-generated currents. In the case of present-day large-scale bedforms, such as sand waves and sand ridges, there is a far greater preservation potential of those structures formed either by the stronger tide or on accretionary lee-slopes. Thus tidal currents may have been responsible for some ancient cross-bedded, shallow marine sandstones, even if only unidirectional palaeocurrent patterns are seen.

However, contrary to some earlier assumptions, tidal currents are not the sole cause of bedload transport of sand on continental shelves. Bedforms resulting from unidirectional flows develop on the eastern USA shelf due to storm-generated wind-driven currents (Hunt *et al.,* 1977; Sect. 9.6.3) and on the Saharan shelf due to oceanic currents (Newton *et al.,* 1973). Although these examples are apparently rare today it is possible that given a sufficient sand supply, currents like the Gulf Stream could generate more widespread megaripples, and these conditions may have existed in some ancient shallow seas.

The inference that wind-driven currents (both storm-generated and semi-permanent) may have dominated circulation patterns in shallow epeiric seas has been particularly emphasized by Dott (e.g. Dott and Batten, 1971, p. 225). Their inferred pre-eminence stems from his conclusions that in shallow epeiric seas tidal currents will always be of secondary importance to wind-driven currents. This is because their extreme shallowness (usually less than 100 m deep) and lateral extent should reduce tidal action to a minimum through a dampening of the tidal wave by energy loss due to frictional drag with the sea bed (Heckel, 1972, p. 231). This situation has certainly not materialized in some of our best known modern epeiric seas, namely those bordering North-west Europe, where tidal conditions are dominant. Similarly we have seen extensive, ancient shallow epeiric seas with strong tidal activity (e.g. Swett *et al.,* 1971; Banks, 1973a). Thus it cannot be inferred that most ancient epeiric seas necessarily had negligible tidal circulations.

It is possible, however, that unimodal palaeocurrent patterns in cross-bedded, shallow marine sandstones reflect unidirectional wind-driven currents. Some Upper Cambrian, cross-bedded, shallow marine sandstones of the Great Lakes region, North America, have been interpreted as the product of a trade-wind-dominated marine circulation system, which transported sand uniformly to the south and south-west (Michelson and Dott, 1973; Dott, 1974). This idea receives added support from the supposed close position of the equator during Upper Cambrian times (Fig. 9.32). The total dismissal of tidal currents, however, remains open to question. Raw palaeocurrent data reveal considerable local variability with occasional bimodal patterns with the two modes 180° apart (e.g. Michelson and Dott, 1973, p. 787). Statistical refinement of the raw data, however, smooths-out the 'odd' directions and emphasizes the dominant direction. As stressed earlier (Chapter 2) a statistically insignificant observation may be geologically significant. Thus some palaeocurrent data interpreted as wind-driven are similar to those interpreted elsewhere as tide-dominated (cf. Narayan, 1971; Anderton, 1976).

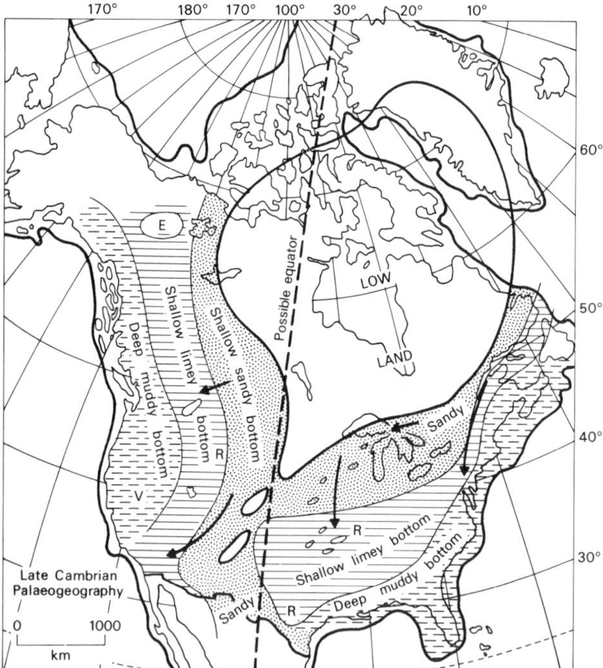

Fig. 9.32. Late Cambrian palaeogeography of North America illustrating broad lithofacies distributions and dominant sediment transport paths (from Dott and Batten, 1971).

Supporting evidence for tidal conditions can be looked for in related shoreline deposits where intertidal sequences may allow estimation of palaeo-tidal range (Klein, 1971). In this Upper Cambrian example there is evidence of tidal flat sedimentation in associated inferred shoreline deposits (Lochman-Balk, 1971). In addition tidal activity has been emphasized in other Cambrian sediments around the North Atlantic (Swett and Smit, 1972).

In conclusion, bidirectional large-scale current-formed structures are positive evidence of tidal currents. Unidirectional palaeocurrent patterns are inconclusive. The extent to which bidirectional tidal currents and unidirectional wind-driven currents may interact in shallow seas seems infinite, and at this stage impossible to decipher in ancient deposits.

Palaeocurrents and other sedimentological and palaeoecological data are combined in three facies models of inferred tidally-dominated, shallow marine sandstones: (1) blanket sandstones including various undifferentiated sand bodies, (2) sand wave deposits, and (3) linear sand bar deposits.

9.10.3 Blanket sandstones

Typical blanket sandstones are extensively cross-bedded and texturally and mineralogically mature. They are usually of considerable lateral extent (several hundred kilometres) and are abundant in the late Pre-Cambrian and Lower Palaeozoic (e.g. Pryor and Amaral, 1971; Swett and Smit, 1972; Banks, 1973a; Johnson, 1977b), but occasionally occur in Mesozoic and younger systems (e.g. Narayan, 1971). The late Pre-Cambrian Jura Quartzite is a representative example.

The *Jura Quartzite* (Anderton, 1976) is excellently exposed in islands off western Scotland (Fig. 9.33) and it is believed to have been mainly deposited in a shallow marine environment. Several important sedimentological trends can be followed along strike for at least 80 km. Sandstones predominate in the south (Islay and Jura) but are replaced northwards by finer-grained sediments, accompanied by a rapid reduction in the number and thickness of sandstone beds in Ardmucknish Bay. In addition the dominant palaeocurrent modes are also directed towards the north and north-east and there is a northward decrease in minimum thickness from ca. 5 km on Jura to ca. 0.5 km in Ardmucknish Bay (Fig. 9.33).

LITHOFACIES IN THE JURA QUARTZITE

Three main lithofacies are distinguished on the basis of grain size and sedimentary structures: (1) coarse facies (sandstone facies Sa), (2) fine facies (heterolithic facies Hb and Hc, mudstone facies M), and (3) coarse/fine facies alternations (summarized facies nomenclature from Fig. 9.29).

Coarse facies (*Sa*) consists mainly of medium to coarse grained sandstone with subordinate layers of very coarse sandstone to pebbly gravel, and rare siltstone lenses. Cross-bedding is the dominant internal structure, represented mainly by tabular sets (average 0.1 to 2 m thick, maximum 4.5 m thick), trough cosets (up to 0.5 m thick), climbing tabular cosets (up to 0.25 m thick) and low-angle sets (up to 0.5 m thick). Individual sets and cosets are bounded by erosional surfaces which range from being flat and planar to gently undulatory with low-relief channels (Fig. 9.31). Secondary structures include various soft-sediment deformation features in which foresets are recumbently deformed, buckled or convoluted.

The cross-bedding is a product of migrating dunes and sand waves. The dominant sediment transport direction was to the north and north-east but secondary palaeocurrent modes directed towards the south confirm the periodic bidirectionality of the bottom currents, thereby supporting the hypothesis of a tide-dominated current system. The vertical alternation of cross-bedding and erosion surfaces records marked fluctuations in the strength of these bottom currents. Thus the migration of bedforms is inferred to be a product mainly of normal tidal currents, which were periodically replaced by even higher energy events causing widespread erosion. The most likely interpretation for this type of sequence is the alternation between fair weather and storm conditions.

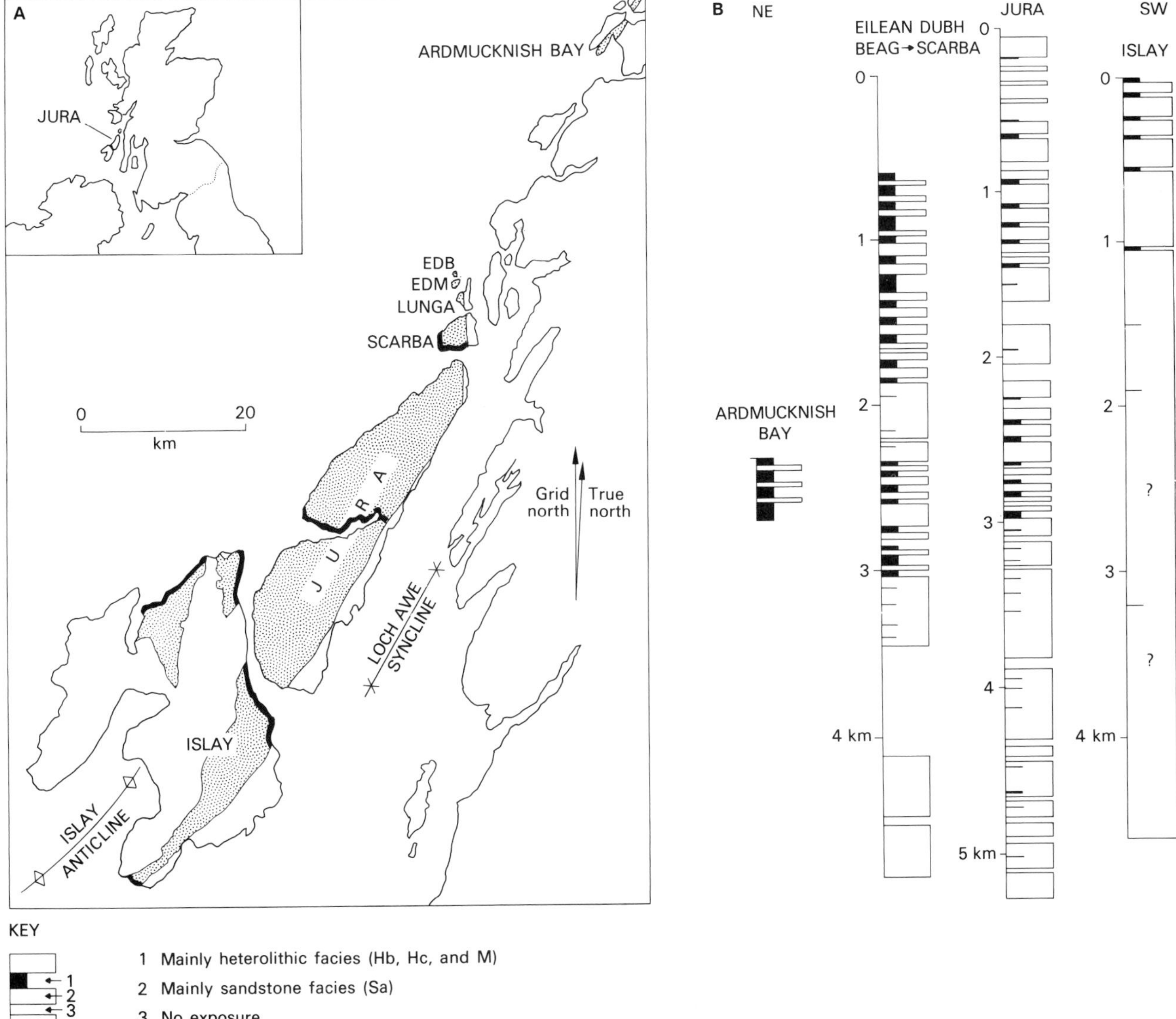

Fig. 9.33. A. Distribution of the late Pre-Cambrian Jura Quartzite in southwest Argyllshire, Scotland, and B. generalized vertical sections illustrating the major lithofacies distributions. Note the northeastward thinning and fining (from Anderton, 1976).

Fine facies (*H and M*) consists of very fine- to medium-grained sandstone interbedded with siltstone and mudstone, including both sandstone-dominated lithologies (Ha–Hb facies) and mudstone-dominated lithologies (Hc–M facies).

The sandstone beds are 1 to 50 cm thick and laterally persistent (several tens to hundreds of metres). They have sharp, planar to slightly erosive bases and planar, non-erosive tops, and their internal structures are parallel-lamination, current-ripple cross-lamination including occasional climbing cosets, grading and loading. There are distinct upward changes in stratification types, the most common being parallel-lamination to cross-lamination. The interbedded siltstones and mudstones form units up to 1 m thick, which may possess thin sandy laminae.

The sandstone beds record relatively high-energy events which alternated with periods of lower energy when fine grained layers were deposited. The sandstone beds display several features comparable to turbidites, because they form from currents of decreasing strength as indicated by their erosive bases and non-erosive tops, parallel-lamination to cross-lamination and grading. In the shallow marine environment they are probably the product of storms when large volumes of suspended sediment are transported offshore in suspension by decelerating currents. This interpretation has also been applied to other similar types of heterolithic facies found in shallow marine deposits (e.g. Banks, 1973a; Goldring and Bridges, 1973).

Coarse/fine facies alternations comprise decimetre to metre thick alternations of the previous two facies. The coarser sandstones are medium- to coarse-grained, form beds 0.1 to 1 m thick and are cross-bedded or graded. They have erosive and occasionally channelled bases and non-erosive tops. The sandstone bodies include two types: (1) *tabular sand bodies* 0.2 to 1 m thick and composed of solitary tabular sets or climbing tabular cosets; (2) *shoestring sand bodies* infilling elongate channels 5 to 30 cm deep and 0.4 to 1.3 m wide, which parallel the palaeocurrent direction and show graded bedding, parallel- and cross-lamination, and climbing tabular cosets. They resemble erosional furrow and gutter casts described from other shallow marine deposits (e.g. Bridges, 1972).

This facies alternation is the product of marked fluctuations between normal sedimentation and periods of much greater energy.

These three facies are interpreted as reflecting the interaction of fair weather and storm conditions superimposed on a tide-dominated hydraulic regime.

RECONSTRUCTION OF THE JURA QUARTZITE SHELF SEA

Vertical facies alternations are random, consisting of erosively based sand bodies (coarse facies) alternating with finer grained sediments (fine facies). A lateral facies pattern is inferred with coarse facies deposited in the south while fine facies accumulated in the north. This inferred facies distribution forms the basis of an idealized Jura Quartzite shelf model, which suggests that: (1) the dominant northward sediment transport path was paralleled by decreases in grain size and mean tidal current velocities, and (2) sediment dispersal was largely controlled by tidal currents in the coarse facies with the importance of storm processes, as a depositor of sediment, increasing downcurrent.

Anderton (1976) suggests that different bedforms will develop along a tidal current transport path (Fig. 9.34A). His reconstruction is based partly on interpretations of the lithofacies but also on observations from modern tidal current transport paths (Fig. 9.34B). Four hydraulic regimes are envisaged: fair weather, moderate storm, intense storm and post-storm conditions.

During fair weather conditions there is a downcurrent decrease in grain size accompanied by four main depositional zones: (a) winnowed gravel, (b) dunes and sand waves, (c) current-rippled sands, and (d) mud belt. Bedform migration is relatively slow, reaching a maximum during spring tides and a minimum during neap tides. Cross-bedded and cross-laminated sands are the main preserved features.

During moderate storms increased sediment transport rates cause the fair weather zones to migrate downcurrent. Dunes are partially eroded in the proximal zone while distally a thin sand layer thinning and fining downcurrent may be deposited.

During intense storms, occurring simultaneously with spring tides, maximum rates of sediment transport occur as currents are enhanced by storm surge and/or wind-driven currents. In the proximal zones erosion produces shallow channels, planar erosion surfaces and winnowed pebble lags. Downcurrent zones experience rapidly migrating bedforms, including climbing dunes, with thinner storm sand layers deposited distally. The ensuing sediment sheet has a high chance of preservation.

Post-storm conditions represent a transitional period between the height of a storm and the return of fair weather conditions. Prior to the re-establishment of fair weather bedforms there may occur a period of laterally extensive mud deposition, if suspended sediment concentrations were sufficiently high. In the Jura Quartzite coarse facies, mud and silt frequently drape planar erosion surfaces.

9.10.4 **An example of ancient sand waves**

Ancient sand waves are preserved as relatively thick (ca. 1–2 m) tabular, avalanche sets (e.g. Nio, 1976). However, since the lee faces of modern sand waves are frequently less than the angle of repose they may also be preserved as low-angle (ca. 10°) foresets. These large-scale structures may form part of blanket sandstone deposits (e.g. Jerzyiewicz, 1968; Pryor and Amaral, 1971; Anderton, 1976). In the Eocene Roda Sandstone Formation of the southern Pyrenees, however, sand wave structures occur in several separate sandstone bodies (Nio, 1976).

The Roda sandstones occupy a total thickness of up to 160 m in the middle of a 400–600 m succession of neritic marls. They are assigned to two laterally interfingering environments on the basis of broad macro- and micro-faunal associations (Fig. 9.35): (1) open marine, and (2) brackish water/estuarine. Only deposits of the former environment are considered here.

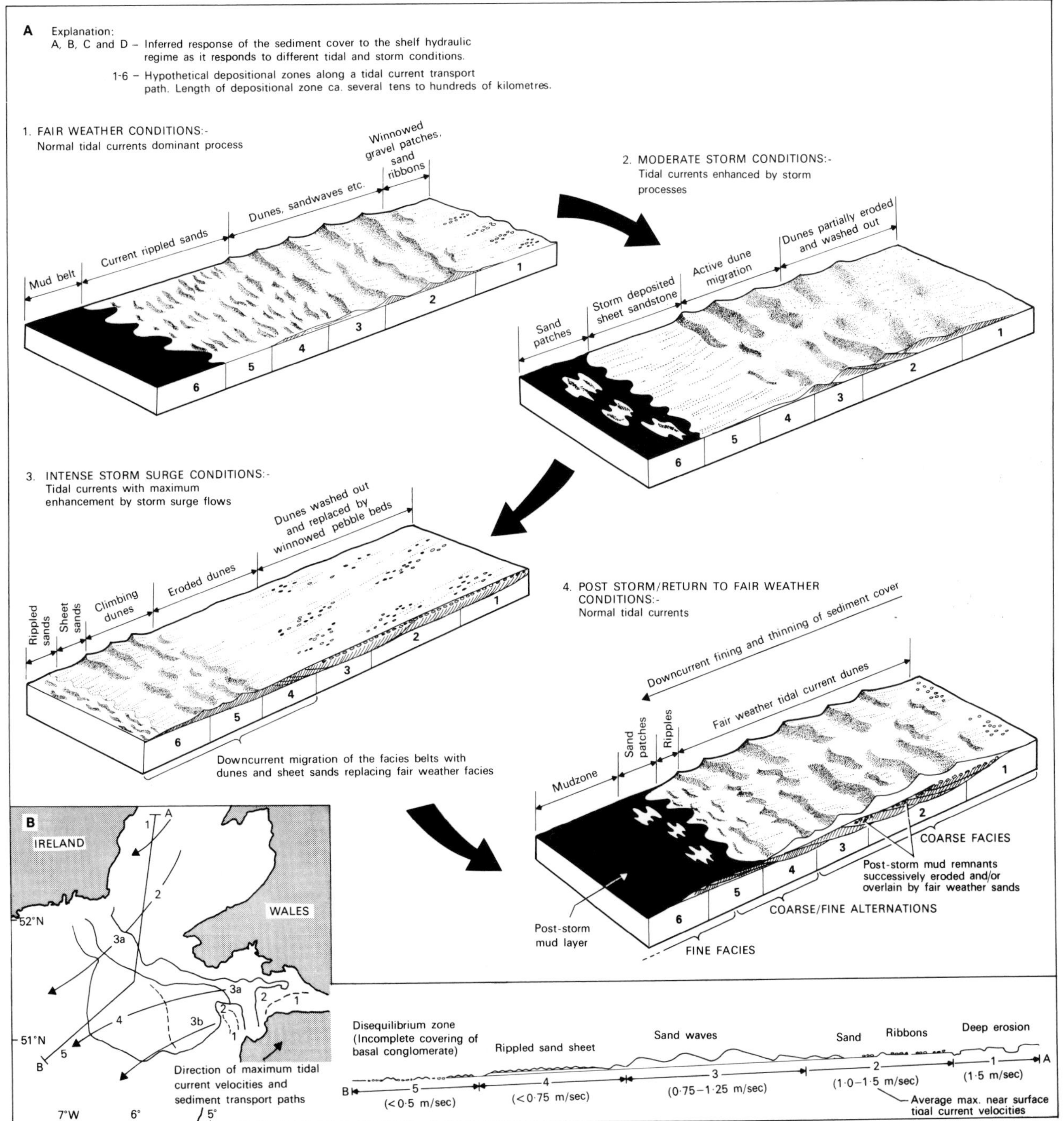

Fig. 9.34. A. Hypothetical reconstruction of the Jura Quartzite shelf sea as developed under different tidal and storm conditions. See text for explanation (after Anderton, 1976); and B. Possible modern analogue as exemplified by the tidal current transport paths in the Celtic Sea (from Belderson and Stride, 1966).

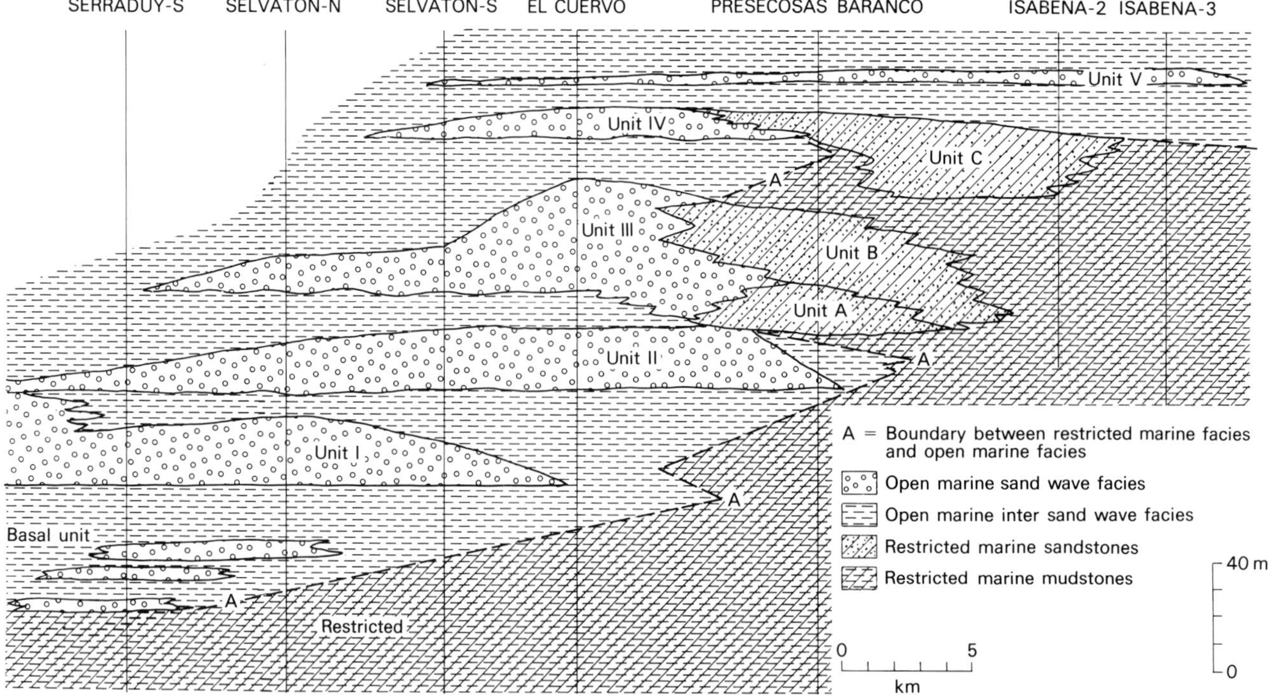

Fig. 9.35. Lithofacies distributions in the Eocene Roda Sandstone Formation, southern Pyrenees, Spain. Units I–V represent open marine sand wave facies separated by richly fossiliferous inter-sand wave facies (marls). These deposits interfinger to the south with restricted marine deposits (sparsely fossiliferous).

LITHOFACIES OF THE OPEN MARINE PART OF THE RODA SANDSTONE FORMATION

Five major sandstone bodies each 10–30 m thick (termed 'sand wave facies') are separated by 5–30 m thick intervals of finer-grained, richly fossiliferous marls, bioturbated silty sandstones and bioclastic limestones (termed 'inter-sand wave facies'). The sandstones contain abundant marine shell debris and are surrounded by marls with rich, open marine faunas. When traced downcurrent each sandstone body reveals a recurring sequence of sedimentary structures, which enables five subfacies to be distinguished (Fig. 9.36).

Initial sand wave subfacies consists of coarse-grained, cross-bedded sandstone, usually comprising a unit 6 to 10 m thick. Tabular cross-bed sets (0.6–2 m thick), occasionally with low-angle foresets (ca. 10–15°), are the dominant structure, and indicate unidirectional sediment transport to the south-west.

Sand wave (sensu stricto) subfacies comprises solitary, tabular cross-bedded sets, 8 to 20 m thick (Fig. 9.37), which show internal erosion surfaces, reactivation surfaces and slightly sinuous foreset traces. Foresets dip to the south-west, parallel to sand body elongation.

Proximal slope subfacies consists of tabular cross-bedded sandstones (sets 0.5–3 m thick) separated by large-scale, low-angle (8–15°) erosion surfaces. The erosion surfaces are separated mainly by tabular sets dipping down the slopes and more rarely by sets dipping up the slopes. Palaeocurrent directions are more variable than in the previous subfacies but a south-westerly mode is always dominant. Bioturbation is more common, consisting of simple vertical and horizontal burrows, and is most evident in association with major erosion surfaces.

Distal slope subfacies is characterized by a progressive downcurrent decrease in maximum grain size and increase in the proportion of fine-grained sediment, mainly as mud drapes, and increased bioturbation. Cross-bedded sets are thinner (ca. 5–30 cm) and the low-angle surfaces are more gently inclined (<10°) and more closely spaced. This subfacies terminates downcurrent in finer-grained, inter-sand wave facies.

Abandonment subfacies comprises a thin (ca. 10–50 cm), richly fossiliferous bed which drapes the whole sand body surface. The most common type is a foraminiferal bioclastic limestone bed but occasionally this is replaced by small, *in situ* colonial corals associated with other open marine shelly faunas (e.g. echinoderms, bivalves and calcareous foraminifera).

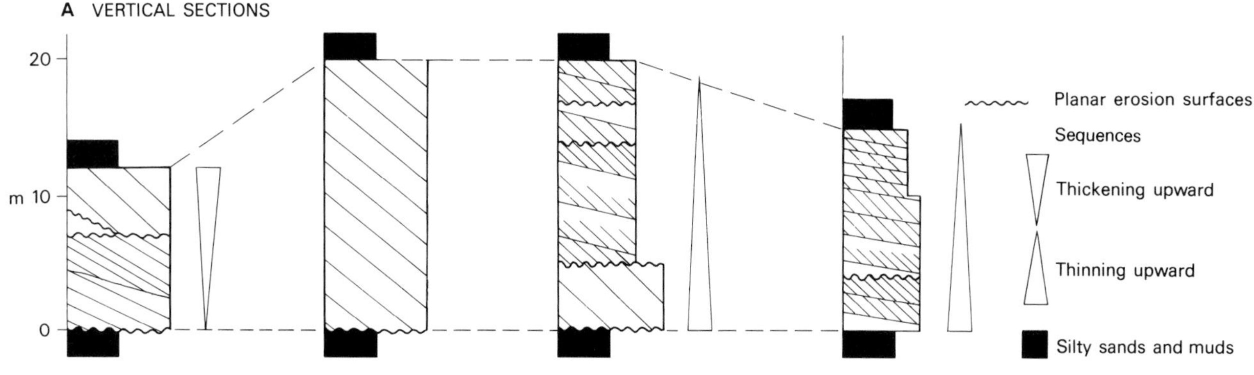

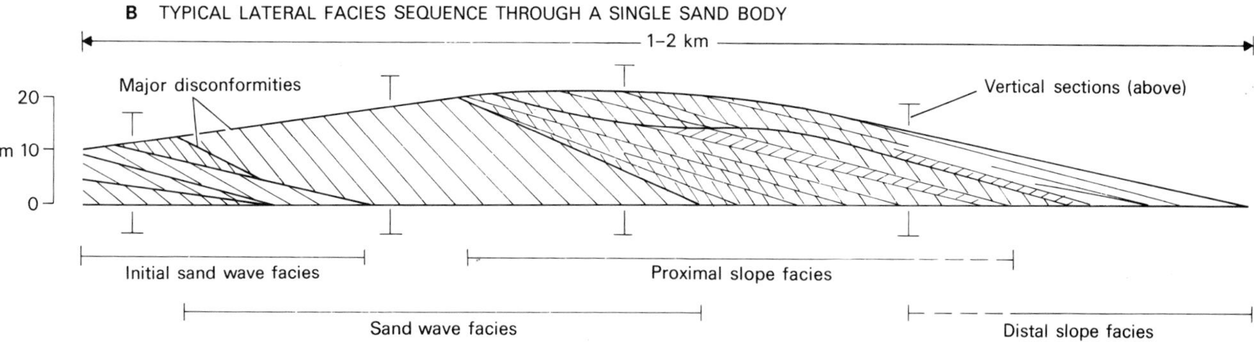

Fig. 9.36. Schematic vertical and lateral sections through an idealized sand body from Units I–V (Fig. 9.35); A. typical vertical sections showing thickening- and thinning-upward sequences and major erosion surfaces, and B. lateral sequence of sedimentary structures and their facies subdivision.

A MODEL FOR THE DEVELOPMENT OF A SINGLE SAND BODY

In every exposed sandstone body these five subfacies are located in a predictable lateral facies sequence (Fig. 9.36). Using an idealized sequence, therefore, it is possible to propose a process model which attempts to explain this recurring facies sequence in terms of the aggradational and degradational history of individual sand bodies (Fig. 9.38).

Initial sand wave phase: coarse sands are introduced into a muddy marine environment, the sand body growing from isolated sand patches into a continuous sand sheet, whose surface is covered by megaripples and small sand waves migrating to the south-west. The megaripples appear to build vertically and downcurrent into larger bedforms (*sand waves sensu stricto*).

Sand wave phase: large, straight to slightly sinuous crested sand waves migrate to the south-west by a process of avalanching at the time of maximum energy.

Slope sedimentation phase: sand wave crests are degraded and eroded by progressive planing-off of the sand wave topography. Currents are more variable in strength, direction and periodicity as indicated by the oppositely dipping cross-bedding, erosion surfaces, mud drapes and burrowed horizons.

Abandonment phase: active sand transport ceases; there is faunal colonization including colonial corals, concentration of shell debris, especially forams, and draping by fine-grained sediment. Bottom currents either suddenly cease or switch elsewhere. This phase is followed by a return to a muddy, low-energy substrate supporting numerous benthic organisms.

The model proposed is independent of any particular type of marine current. However, occasional evidence of bidirectional currents, the size of the sand waves and the palaeogeographic setting of a partially-enclosed, funnel-shaped sea suggest that tidal currents were probably the main source of energy.

This sand wave complex appears to have developed in response to an abundant sand supply and strong tidal currents.

Fig. 9.37. Large-scale cross-bedded sandstone of the sand wave subfacies, overlain by smaller cross-bedded sets of the proximal slope subfacies, Eocene Roda Sandstone Formation, southern Pyrenees, Spain.

Fig. 9.38. Process reconstruction of the structural development of a typical Roda Sandstone sand wave body, illustrating the four major phases of development.

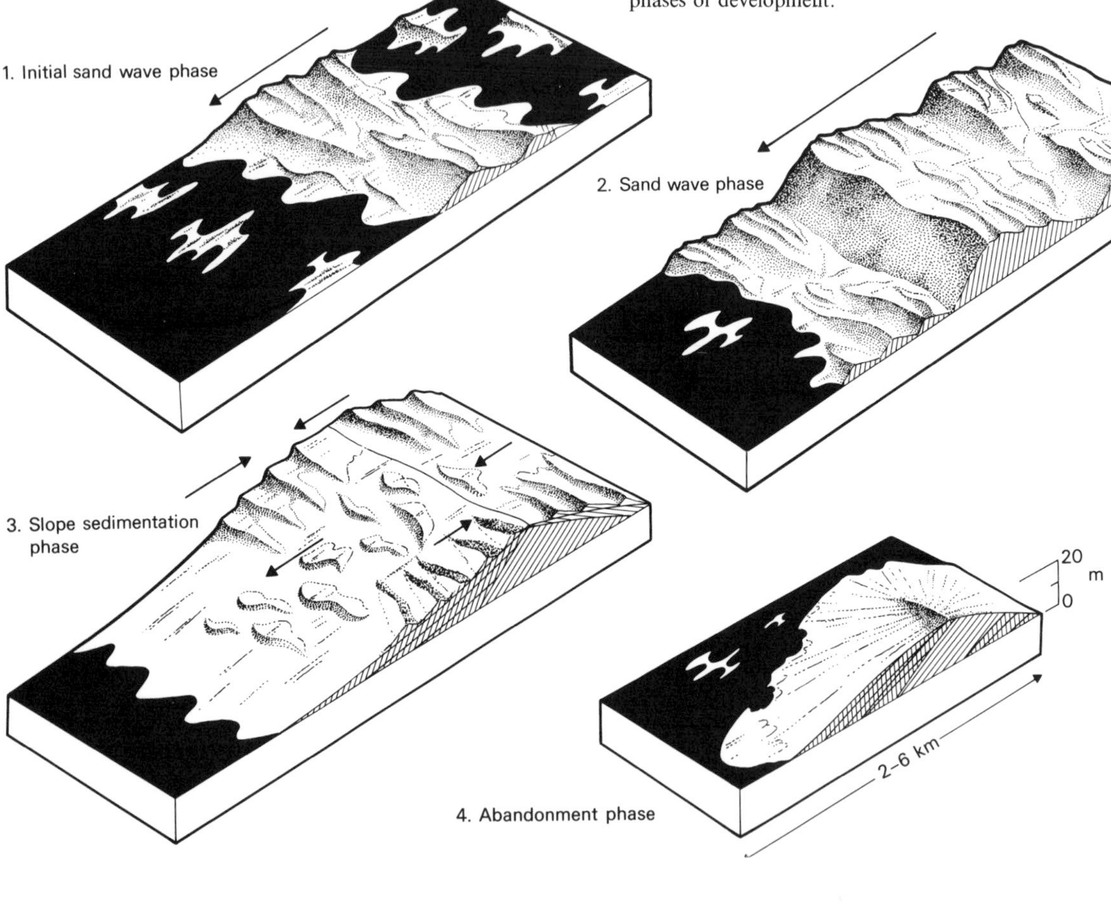

1. Initial sand wave phase

2. Sand wave phase

3. Slope sedimentation phase

20 m

0

2–6 km

4. Abandonment phase

B

Fig. 9.39. A. An example of a late Pre-Cambrian inferred tidal sand bar deposit from Finnmark, North Norway. The major elements are: (1) westerly inclined bar-flank accretion surfaces (dipping to the left), (2) north-westward directed (into the page) trough cross-bedded sandstones between the major accretion surfaces, and (3) coarsening-upward sequence; and B. part of a mutually evasive ebb and flood tidal channel complex which dissects the bar crests, and supports a tidal current hydraulic regime. Numbers adjacent to annotated vertical sections refer to the lithofacies described by Johnson, 1977a (after Hobday and Reading, 1972; Johnson, 1977a).

An additional factor may be that a marine transgression is a causal factor of this and other sand wave deposits (Nio, 1976).

9.10.5 **Linear sand bar deposits**

In ancient deposits linear sand bars are distinguished either by the parallelism of sandstone body length and dominant palaeocurrent direction (e.g. Berg, 1975) or by the preservation of large-scale inclined surfaces normal to the palaeocurrent direction and reflecting the bar flanks (Fig. 9.39; Hobday and Reading, 1972; Johnson, 1977a). The best documented examples occur as isolated, elongate sand bodies with marine muds, as in the Mesozoic of western North America. The dimensions of these ancient sand bar deposits, their internal sequence which commonly coarsens upward, and their palaeocurrent patterns are similar to both tidal current and storm-generated modern linear sand ridges (Fig. 9.40).

UPPER JURASSIC SAND BAR DEPOSITS, WESTERN INTERIOR, USA

During the widespread Upper Jurassic regression, mudstones coarsened upwards into a laterally extensive sheet of cross-bedded fine-grained orthoquartzitic sandstone (Brenner and Davies, 1973, 1974). The mudstones, with interbedded bioclastic limestones and coquinoid sandstones, indicate a low-energy, shallow marine shelf environment. The sandstones are also considered to be shallow marine by virtue of their textural and mineralogical maturity, glauconitic composition, locally intense bioturbation and concentrations of fragmented bivalves (coquinoid sandstones). Furthermore, the absence of both typical beach structures and evidence of emergence suggests sandstone deposition in a subtidal environment, while multidirectional palaeocurrent trends are considered indicative of tidal currents and superimposed storm processes.

The sandstone sheet comprises several coalesced sandstone bodies which interfinger laterally with finer-grained muddy sandstones and mudstones. Preservation of the sandstone bodies shows them to be 200 m to 2 km wide, up to 21 m high, and at least 1 to 5 km long (Fig. 9.41). Thus, they have a primary linear sand body morphology.

The coarsening-upward sequence is interpreted as due to the migration of linear sand bars and their finer grained interbar troughs over a lower-energy muddy shelf environment (Fig. 9.42). Within each of the three main environments of sand bar, interbar trough and shelf muds, coquinoid sandstones occur. These are thought to have been formed

Examples	MODERN TIDAL SAND RIDGES Sand bar/Sandstone body geometry			Internal organization		Current/palaeo-current patterns	Crest line orientation with respect to shoreline	Additional remarks
	height/thickness	width	length	cu=coarsen upwards	fu=fine upwards			
General world-wide examples (Off 1963)	6-30 m	2-11 km	2-80 km	usually cu			usually parallel	Relatively uniform morphology of a wide variety of sand ridges.
Southern North Sea (Houbolt 1968)	10-40 m	2-4 km	60 km	none or fu		bidirectional but with 1 mode dominant either side of ridge crest	parallel	Flat erosive base with a basal lag conglomerate in the troughs.
Gulf of California (Meckel 1975)	6-9 m	2-4 km	15 km	cu		unidirectional parallel to bar crest	parallel to gulf axis	Cross-beds dip inshore in response to dominant flood tide.
Malacca Straits (Coleman & Wright 1975)	up to 14 m	?	?	fu (?)		unidirectional (dominant current) (inferred)	parallel	Flat erosive base due to strong currents in the troughs.
Timor Sea/Ord River delta (Coleman & Wright 1975)	10-22 m	0.3 km	2 km	cu		?	parallel	Associated with tidally-influenced river delta.
	MODERN STORM-GENERATED LINEAR SAND RIDGES							
Atlantic shelf bordering the eastern U.S.A. (Swift *et al.* 1972 a and b, 1973, 1975; Duane *et al.* 1972; Stubblefield *et al.* 1975)	4-10 m	2-4 km	20-60 km	cu		unidirectional	oblique to subparallel	Sand ridges constructed by unidirectional storm-generated (wind-driven) currents. Differences in internal structure between storm-generated and tidal sand ridges is uncertain.
	ANCIENT OFFSHORE LINEAR SAND BARS							
Gallup sandstone, U. Cretaceous, New Mexico (Campbell 1971)	6 m	3 km	6-64 km	cu		unidirectional parallel to bar crest	parallel	Coalesced sand bars produce multiple sand bodies. Seaward imbricate pattern of sand bodies.
Shannon sandstone, U. Cretaceous, Wyoming (Spearing 1975)	15-22 m	30-50 km	50-100 km	cu		unidirectional parallel to bar crest	parallel	Two elongate sand bodies
Sussex sandstone, U. Cretaceous, Wyoming (Berg 1975)	12 m	8-48 km	32-160 km	cu		?	parallel	Several elongate, isolated sand bodies
Viking Formation, L. Cretaceous, S. Saskatchewan (Evans 1970)	3-10 m	11-22 km	113 km	cu		unidirectional parallel to bar crest	oblique	Imbricate overlapping of sand bodies separated by bentonite layers
U. Jurassic, Wyoming and Montana (Brenner & Davies 1973, 1974)	5-15 m	0.2-2 km	>5 km	cu		multidirectional distribution	parallel	Cross-shoal channels dissect bar crest at high angles
U. Dakkavarre Formation, Late Pre-cambrian, Norway (Johnson 1977a)	4-10 m	>3 km	?	cu		unidirectional parallel to bar crest	parallel	Internal large-scale accretion surfaces reflect bar flanks. Coalesced sand bodies

Fig. 9.40. Summary of the main morphological characteristics of modern tidal and storm-generated linear sand ridges, and comparison with the geometries of ancient sublittoral sandstone bodies.

during storms and consist of three types: channel lags, storm lags and swell lags.

Channel lags comprise variably comminuted shell debris consisting mainly of the thick-shelled *Ostrea* with subordinate amounts of smaller oysters, thin-shelled bivalves, crinoids and brachiopods. They are restricted to the sandstone facies (bar and interbar troughs) where they form cross-bedded units 1 to 4 m thick which include lag deposits at the bases of 4 m deep channels which obliquely dissect the bar crests (Fig. 9.42).

Storm lags are more laterally extensive and consist of similar fragmented shell debris which forms sheet-like beds (3–30 cm thick) with planar erosive bases and non-erosive or gradational tops.

Swell lags consist of whole, or nearly-whole, thin shelled pecten-like bivalves (e.g. *Camptonectes*). The shells form convex-upward flat sheets usually less than 20 cm thick but forming units up to 3 m thick. This type of bed is restricted to the shelf mud facies.

These three storm deposits vary according to the fair weather energy levels of contrasting subenvironments. The highest-energy deposits are the channel lags which are interpreted as storm-surge channel-fills formed when storm activity, coupled with increased tidal current velocities, induced cross-shoal currents to dissect the bar crest. Alternatively the channels may be an integral part of the sand bar environment (e.g. Robinson, 1966). These channels concentrate the highest flows and therefore may preserve coarser sediment irrespective of the speed of filling (cf. Johnson, 1977a and Fig. 9.39B). However, Brenner and Davies (*op cit*) suggest that the channels were filled rapidly as the storm waned, preserving a coarse lag concentration of shell debris. Storm lags are thought to result from the reworking of the sea floor over extensive areas by

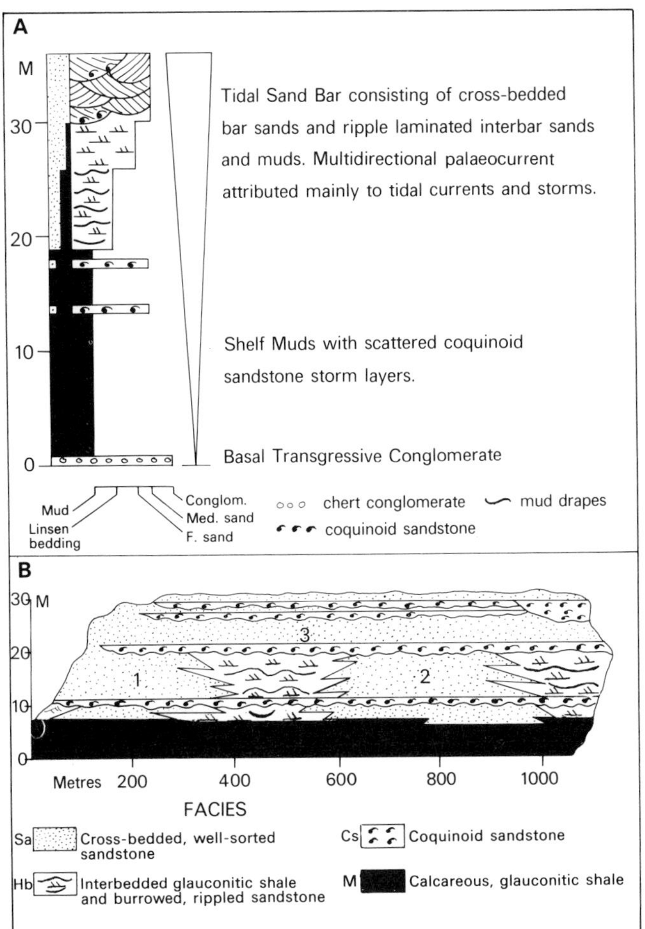

Fig. 9.41. Facies relationships in Oxfordian (Upper Jurassic) sand bar deposits, Western Interior, USA: A. typical coarsening-upward sequence reflecting sand bar progradation across a muddy shelf, and B. lateral facies relationships in these linear sand bodies; bars 1 and 2 are viewed normal to sand body elongation, bar 3 is parallel to sand body elongation (from Brenner and Davies, 1973).

storm waves and associated currents, but may also form at the mouth of storm surge channels. Swell lags are thought to represent the weaker impact of storms in the lower-energy and deep-water muddy shelf, the shells remaining unbroken by storm waves.

LINEAR SANDSTONE BODIES FROM THE CRETACEOUS OF WESTERN NORTH AMERICA

During the Cretaceous of western North America north–south trending seaways appear to have been particularly conducive to the generation of strong marine currents (tidal, oceanic or storm-generated) leading to the development of coast- and current-parallel, elongate shelf sand bodies (Figs. 9.40 and 9.45).

Most of these sandstone bodies: (1) consist of well-sorted, glauconitic and quartzose sands, (2) are cross-bedded, (3) coarsen upwards, (4) have palaeocurrent modes parallel to sand body elongation, (5) are between 10–30 m thick, 2–60 km wide and up to 160 km long, (6) have planar bases and convex-upward tops, and (7) are either surrounded by or interfinger with marine muds. They frequently form oil reservoirs since they may be both surrounded and capped by potentially hydrocarbon bearing, impermeable muds (e.g. Sussex Sandstone, Berg, 1975, Fig. 9.43).

The Lower Cretaceous Shannon and Sussex Sandstones have coarsening-upward sequences which have been interpreted as the result of downcurrent and lateral migration of linear sand ridges (Figs. 9.43 and 9.44; Berg, 1975; Spearing, 1976). In the Shannon Sandstone, sediment transport was consistently towards the south-west, parallel to sand body elongation, whereas wave ripple crests were mainly normal to this trend (Fig. 9.45). Herringbone sets within the cross-bedded sandstones may be indicative of tidal reversals but the unidirectional palaeocurrent pattern is inconclusive and could also represent semi-permanent and/or storm-enhanced current systems. Indeed the relationship between current direction and wave ripple trend is particularly suggestive of storm-enhanced currents.

9.11 WAVE- AND STORM-DOMINATED FACIES: RECOGNITION AND MODELS

In seas where tidal and other persistent bottom currents are absent or of secondary importance the main sources of energy are meteorologically induced, notably by waves and by storms. How, then, can the products of these processes be recognized and how do they differ from those of tidal current-dominated seas?

9.11.1 Wave-formed sedimentary structures

The most frequently recognized forms of wave action in ancient deposits are symmetrical and asymmetrical wave ripples. The latter are distinguished from current-formed ripples by their morphology (Reineck and Singh, 1973, pp. 24–28; Harms, 1975), in particular their straight and bifurcating crest lines. However, since bedding planes are not always exposed it is important to distinguish wave- from current-formed cross-lamination.

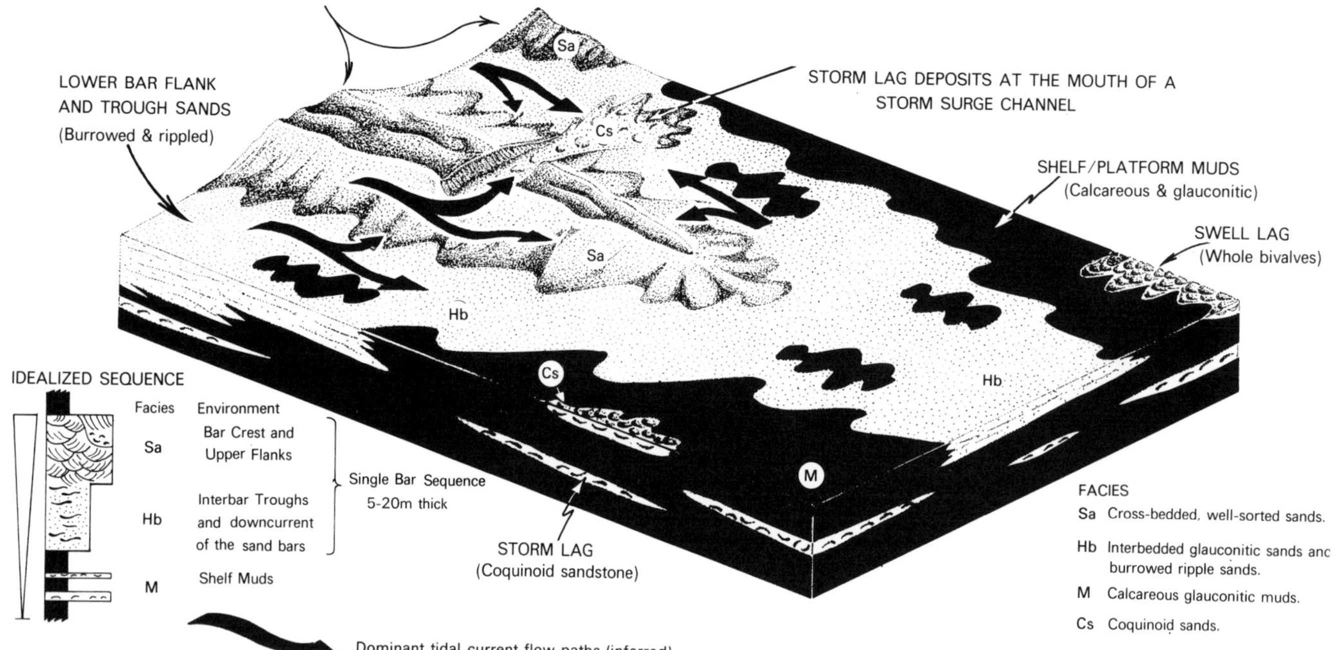

Fig. 9.42. Reconstruction of the Oxfordian sublittoral environment during progradation of the linear tidal sand bars (after Brenner and Davies, 1973, 1974).

Fig. 9.43. Description and interpretation of the Upper Cretaceous Sussex Sandstone, Wyoming, USA. The coarsening-upward sequence and linear sand body geometry (Fig. 9.44) are interpreted as the product of prograding elongate tidal sand bars (after Berg, 1975).

Units of Berg,1975	Thickness	Description	Interpretation of the Facies	Interpretation of the Sequence
1	1.5-2.5 m	Bioturbated muds with reworked sands & granules. Horizontal bedding.	Shelf muds with storm deposited sand layers.	Sudden cessation of active sand transport. Shelf muds blanket the linear sand bars.
2	0.30-1.0 m	Pebbly sandstone with chert & mudstone pebbles. Chert pebbles concentrated in top layer.	Wave reworked sands. Related to wave processes concentrated on bar crest.	Gradual increase in wave & current activity in response to a prograding linear tidal (?) sand body. Elongate geometry of several sand bodies parallel to shoreline & to current transport path suggests they may represent linear tidal (?) sand ridges which formed topographic highs. Coarsening upward sequence reflects both progradation & preferential reworking of bar crest & flanks.
3	1.5-5.5 m	Cross-bedded & flat bedded sandstone Cross-bed sets ca. 5-20 cm thick. Numerous mudfakes & occasional mud-drapes. Rare cross-laminated sands.	SE (?) migration of dunes in response to tidal currents, possibly enhanced by storms. Abundant penecontemporaneous erosion of mud. Sands deposited on upper bar flanks.	
4	1.8-5.0 m	Cross laminated fine grained sandstone with numerous mud-drapes. Occasional mudflakes & cross-bed sets. Minor bioturbation.	Deposition by current ripples in response to tidal (?) currents. Currents of fluctuating strength. Abundant fine grained suspended sediment. Deposition on lower bar flanks & troughs.	
5	0.9-4.0 m	Muds with rippled sandstone lenses & sandstone interbeds 1-5 cm thick. Minor bioturbation.	Suspension deposition of muds alternating with periodic sand influxes (? distal storm layers). Deposition downcurrent of the sand bars.	
		Massive marine muds but with little bioturbation.	Shelf muds	

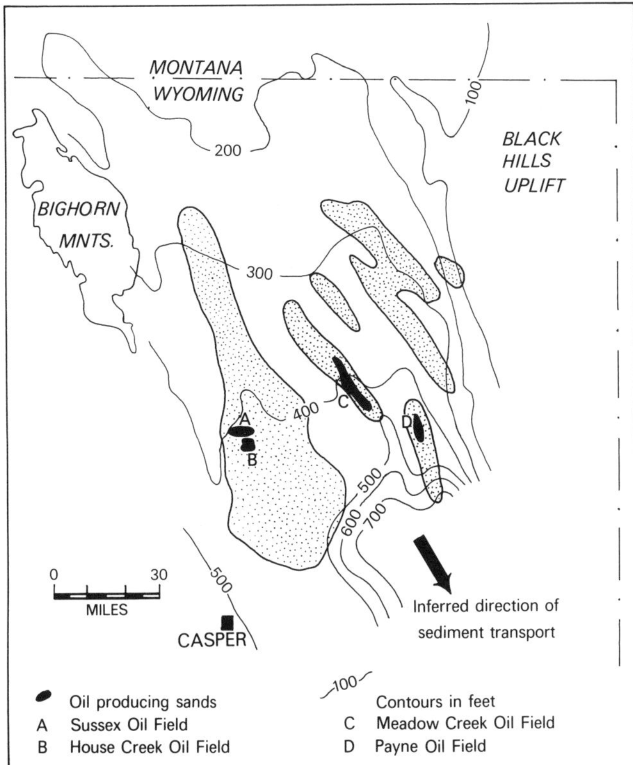

Fig. 9.44. Subsurface geometry of the Sussex Sandstone with superimposed isopachs (in feet) of the 'Sussex' time-rock unit. Sand body elongation is inferred as being parallel to the dominant palaeocurrent direction. This trend is the same as that recorded at outcrop for the underlying Shannon Sandstone which was deposited in a similar sublittoral environment (Fig. 9.45) (from Berg, 1975).

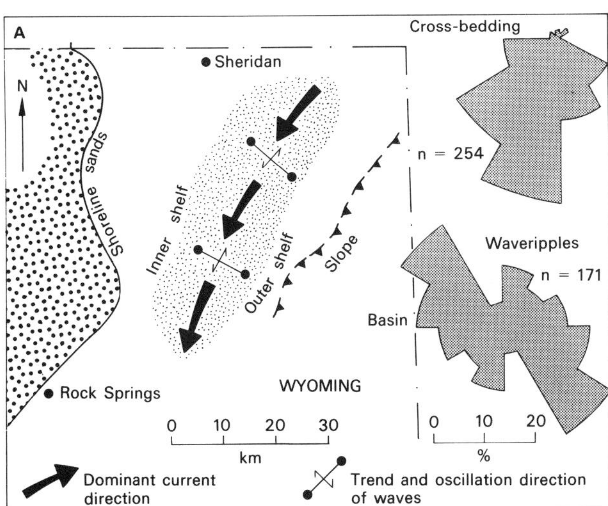

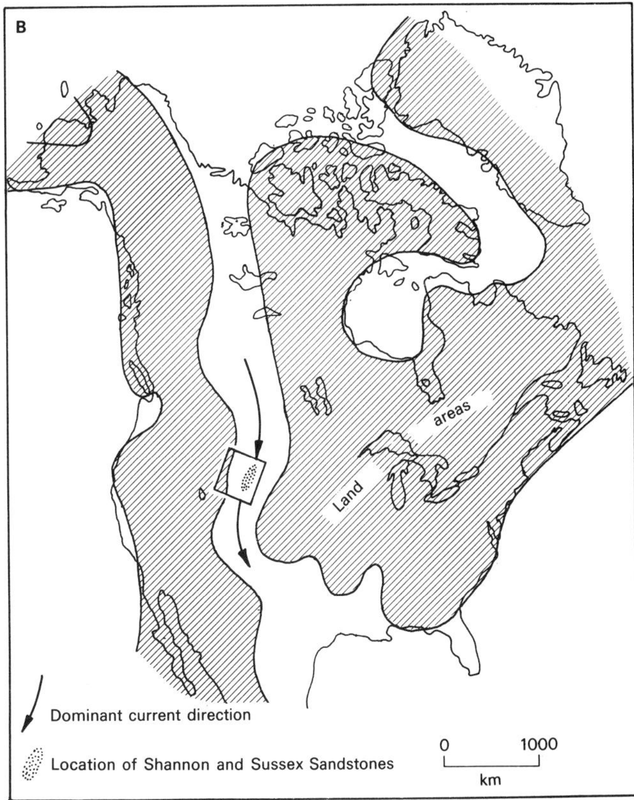

Fig. 9.45. Palaeogeographic reconstruction of the shelf environment during deposition of the Upper Cretaceous Shannon Sandstone: A. detailed palaeogeography in Wyoming illustrating the relationship between sand body elongation and palaeocurrent patterns (from Spearing, 1975, 1976) and B. general location of the Shannon and Sussex sandstones in the elongate Upper Cretaceous (early Campanian) seaway of western North America (from Williams and Stelck, 1975).

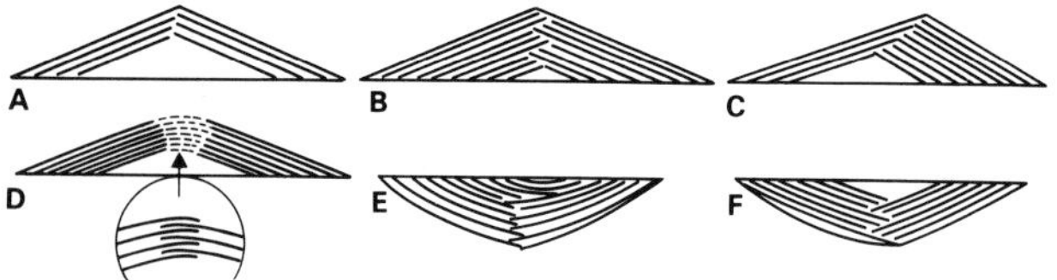

Fig. 9.46. Types of chevron structures developed in wave-generated ripples (from Reineck and Singh, 1973, after Boersma, 1970).

A characteristic group of structures produced by symmetrical wave ripples are chevron structures, consisting of superimposed, oppositely dipping laminations (Fig. 9.46). They are produced by equal or near-equal oscillatory wave velocities in which the ripples grow mainly through vertical accretion. In many cases, however, wave ripples migrate in one preferred direction thereby producing unidirectional cross-lamination (e.g. Newton, 1968). Boersma (1970; see also de Raaf *et al.*, 1977) notes several differences from current-ripple cross-lamination. Wave ripple cross-lamination has (1) a less trough-like shape, (2) an irregular and undulating lower bounding surface, (3) bundle-wise upbuilding of foreset laminae, (4) swollen lenticular sets, and (5) offshooting and draping foresets (Fig. 9.47; terminology of de Raaf *et al.*, *op. cit.*). The internal cross-lamination may also be preserved as a series of almost asymmetrical lenses representing the preserved troughs between ripple crests (Fig. 9.48). Foresets may dip in opposite directions and may superficially resemble trough sets, although the straight-crested form allows distinction from current ripples (Harms, 1975).

Wave ripple cross-lamination has been described from a variety of heterolithic facies, including linsen and flaser bedding, from the Lower Carboniferous of southern Ireland (de Raaf *et al.*, 1977). Here the cross-lamination is mainly unidirectional with the dominant mode interpreted as directed towards the shoreline. The rare examples of oppositely dipping sets are thought to result from seaward flowing currents possibly caused by rip currents, rather than from tidal reversals for which there is no complementary evidence.

Difficulties in recognizing wave-formed cross-lamination and distinguishing it from current-formed varieties are increased where current and wave processes interact on the sea bed. This is illustrated in some shallow water, mainly subtidal, heterolithic facies in the late Pre-Cambrian Stangenes Formation in north Norway (Baldwin and Johnson, 1977). Here north and south flowing (?tidal) currents generated sinuous-crested to linguoid current ripples which migrated in opposite directions parallel to the shoreline. Superimposed on these current ripples are north–south trending wave ripples, indicating east–west oscillatory motion, which modified and

Fig. 9.47. General features diagnostic of wave-ripple cross-lamination (from de Raaf *et al.*, 1977, after Boersma, 1970).

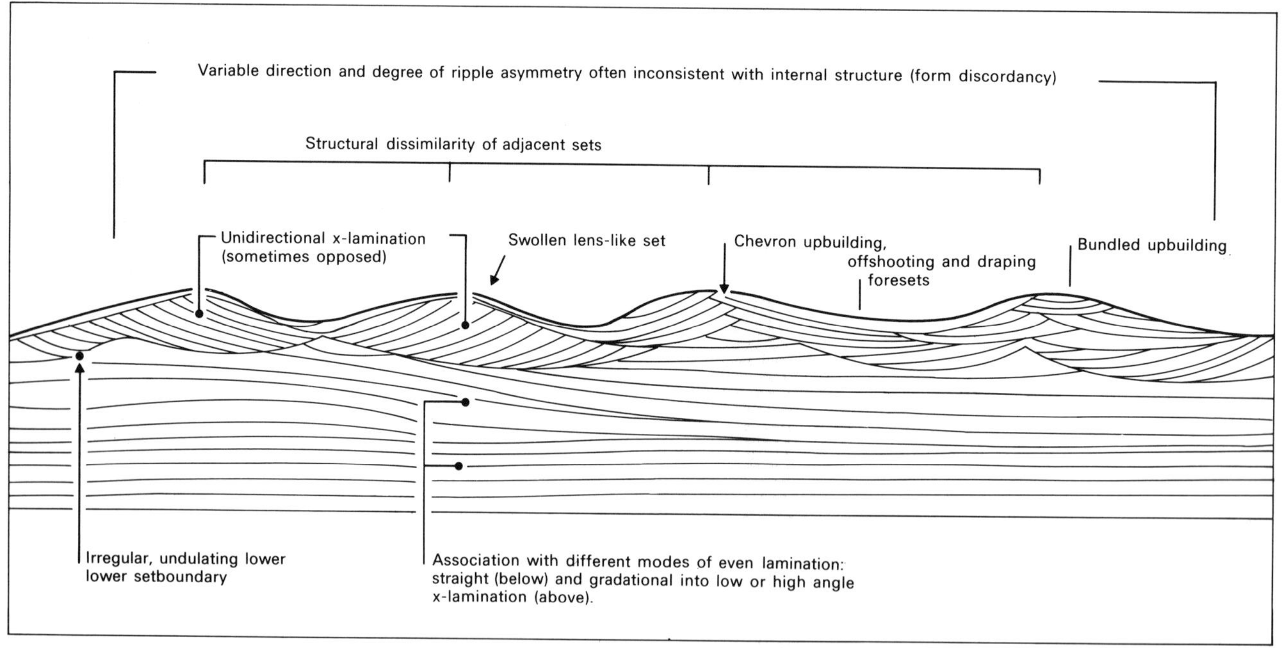

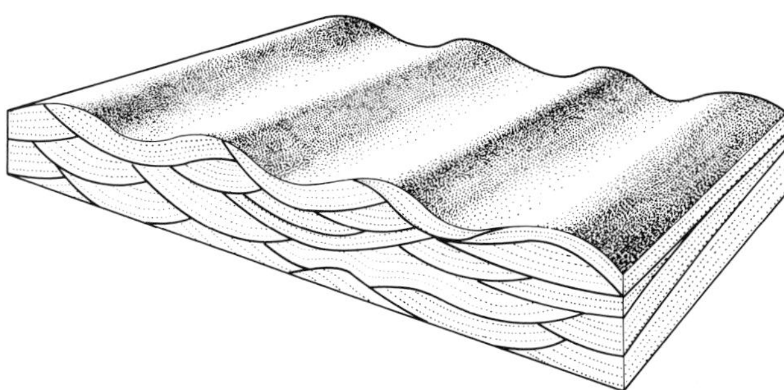

Fig. 9.48. Wave ripple cross-lamination developed in asymmetrical wave ripples. Characteristic features include irregular lower bounding surface, interfingering of lenses, cross-lamination of both lee and stoss sides of the wave ripples, predominance of trough lamination and foreset drapes (from Reineck and Singh, 1973, after Boersma, 1970).

occasionally obliterated the current-ripple morphology (Fig. 9.49).

9.11.2 Sublittoral sheet sandstones

The most characteristic storm deposits in the sublittoral environment are well-sorted, fine- to medium-grained, sheet sandstones (Fig. 9.50) usually interbedded with mudstones and siltstones which are frequently fossiliferous or bioturbated. These sandstones have been called 'sublittoral sheet sands' by Goldring and Bridges (1973), and 'storm lag' deposits by Brenner and Davies (1973).

The sandstones are usually 5 to 30 cm thick (max. ca. 2 m), laterally persistent for several tens of metres to kilometres with sharp, planar to gently undulating erosive bases and non-erosive tops. As bed thickness increases the basal erosion surface tends to become more visibly erosive and locally channelled to depths of ca. 30–50 cm. The base of the bed may possess sole marks and be lined with intraformational conglomerate, pebble layers or broken shell fragments. Some erosion surfaces exhibit delicate tool marks along the walls of large, elongate grooves or furrows (e.g. Bridges, 1975; Anderton, 1976). The upper boundary may be sharp (planar or wave rippled), gradational or bioturbated.

The dominant internal structure is parallel-lamination, which may display primary current lineation. This is commonly associated with low-angle (5–8°) cross-stratification infilling broad shallow undulations or troughs several metres broad but only a few tens of centimetres deep (hummocky cross-stratification of Harms, 1975). Other common associations of sedimentary structures include: cross-lamination, climbing cross-lamination, upward transitions from parallel- to cross-lamination, grading and occasional load structures.

These 'turbidite-like' sandstone beds are interpreted as having been deposited from waning sediment-laden currents

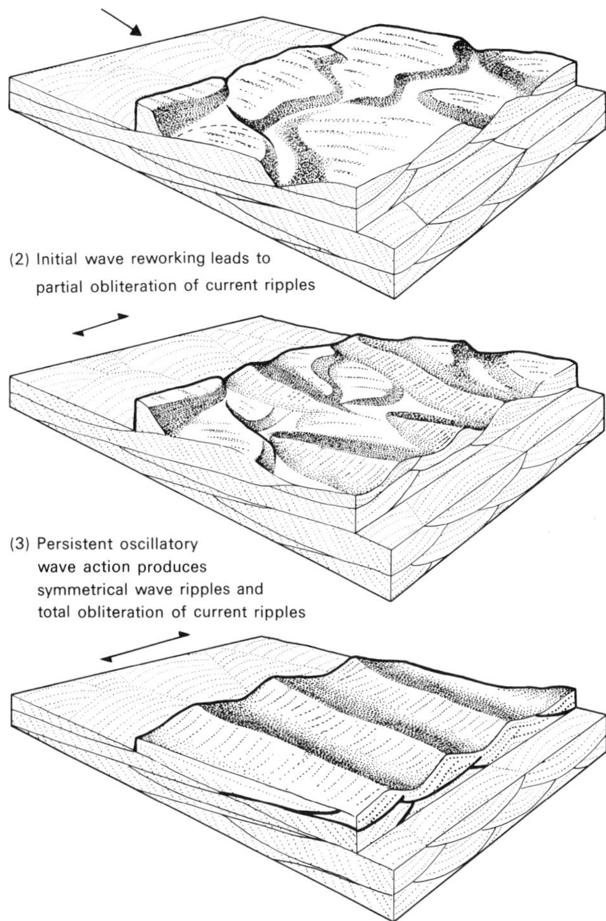

Fig. 9.49. Interaction of small-scale wave and current ripples in a late Pre-Cambrian sublittoral environment (Stangenes Formation, Finnmark, north Norway). North and south migrating linguoid and sinuous-crested current ripples preserve varying degrees of reworking by east–west oscillating waves. Thus wave ripple marks trend normal to the crests of the current ripples with the latter showing varying degrees of modification as shown by the sequence $1 \rightarrow 2 \rightarrow 3$.

(Fig. 9.50). Each bed is thought to record a single storm event comprising: (1) initial storm erosion, (2) deposition, (3) post-storm reworking.

In non-tidal seas *initial erosion* is due to oscillatory wave reworking and/or storm-generated current erosion, which entrain sediment and place it in suspension. Shells and gravel may be reworked to form a winnowed lag over the basal erosion surface. *Deposition* occurs as the storm decreases in intensity with parallel-lamination forming as sand falls out

TYPICAL BED	BED THICKNESS	INTERNAL STRUCTURES	PROCESS INTERPRETATION	ENVIRONMENTAL INTERPRETATION	High energy shelf (sand dom.)	Low energy shelf (mud dom.)
	20 – 150 cm.	Cross-bedding (trough, tabular and climbing). Discontinuity or reactivation surfaces.	Dune migration in response to storm-enhanced tidal currents. Fluctuations in flow power.	Extension of dune field down the tidal current transport path when intense storm conditions enhance normal tidal currents.	PROXIMAL STORM DEPOSITS	ABSENT
	20 – 150 cm.	Parallel lamination with parting lineation.	Upper flow regime plane bed with sediment movement conditions. Sand deposited as bed load, but introduced mainly from suspension currents commonly of decreasing strength.	Combination of (1) exceptionally high energy conditions associated with intense wave action and storm-generated currents, and (2) abundant sand availability in shoreline and subtidal zones. Forms a facies belt immediately down-current/offshore from extensive tidal dune fields. Thick sheet sandstones also characterize upper shoreface deposits (chapter 7).		
	20 – 150 cm.	Low-angle trough cross-bedding with parting lineation on foresets.	Transitional flow regime period or washed-out dune phase. Combination of bed load and suspension deposition of sand. Currents of decreasing strength.			
	ca. 20 cm.	Graded units occasionally passing upwards into parallel lamination.	Deposition from suspension by a current of decreasing strength.			
	5 – 10 cm.	Graded: massive or parallel laminated.	Deposition from suspension by a decelerating current.	Depending on the overall sequence this could represent more distal storm deposits to the thicker beds or they may represent less intense energy conditions, depending on strength of tidal, storm and wave processes.	INTERMEDIATE STORM DEPOSITS	PROXIMAL STORM DEPOSITS
	5 – 10 cm.	Parallel or low-angle lamination with wave ripples on top.	Upper flow regime conditions, followed by wave reworking.			
	5 – 10 cm.	Parallel to wave-ripple lamination.	Deposition by a decelerating current, followed by progressive oscillatory wave action.			
	5 – 10 cm.	Parallel to current-ripple lamination.	Deposition by a decelerating current: plane bed with movement to ripples.			
	5 – 10 cm.	Current ripple cross-lamination, occasionally climbing.	Lower flow regime current ripple phase.			
	0.5 – 3 cm.	Graded to parallel lamination.	Deposition from suspension.	Interpretation depends on position within the sequence but characterizes the distal parts of tidal current transport paths and low energy shelf deposits.	DISTAL STORM DEPOSITS	DISTAL STORM DEPOSITS
	0.5 – 3 cm.	Parallel lamination.	Deposition from suspension.			
	0.3 – 3 cm.	Current ripple cross-lamination (continuous layers).	Migration of current ripples with moderate sand supply.			
	0.3 – 3 cm.	Current ripple cross-lamination (discontinuous lenses).	Migration of current ripples with deficient sand supply.			
	1 – 5 mm.	Flat sand layers.	Sand incursions from suspension.			

Fig. 9.50. Summary of the internal stratification sequences and their interpretation from some typical sublittoral, inferred storm-generated, sheet sandstones.

from suspension. High bed shear stress persists at the sediment/water interface, sweeping the sand out as flat sheets with the development of primary current lineation. The low-angle trough cross-stratification can also be attributed to high-energy conditions, possibly comparable to the 'washed-out dune' phase of Simons *et al.* (1965). Waning current conditions are demonstrated by graded beds and by the frequent occurrence of lower flow regime structures, notably current and wave ripple cross-lamination, immediately above the parallel-laminated or cross-stratified division. *Post-storm reworking* is indicated by wave rippling or bioturbation.

9.11.3 **Sublittoral sheet sandstone facies associations**

These sandstone beds occur in three main associations: (1) shoreface-shoreline association, (2) tide-dominated shelf association, and (3) non-tidal (storm- and wave-dominated) open shelf association.

The shoreface-shoreline association forms an important part of shoreface deposits in prograding beach barrier sequences (see Sects. 7.2.3, 7.2.6). In the tide-dominated shelf association sheet sandstones characterize those heterolithic facies which are interbedded with cross-bedded, tidal sand bodies (see Sect. 9.10). The non-tidal, storm- and wave-dominated open shelf association is significantly affected neither by a nearby shoreline, nor by persistent background marine currents. In those which are wave-dominated sheet sandstones may be associated with wave-built bars (de Raaf *et al.*, 1977). In those which are storm-dominated sheet sandstones are frequently randomly interbedded with open marine, occasionally richly fossiliferous mudstones (e.g. Bridges, 1975; Kelling and Mullin, 1975) and may form coarsening-upward sequences indicative of a prograding shelf (Banks, 1973b).

The latter two examples seem to illustrate how two types of meteorologically dominated hydraulic regime can be reconstructed (1) wave-dominated model proposed for the Lower Carboniferous of southern Ireland, and (2) storm-surge ebb model proposed for the late Pre-Cambrian Innerelv Member of north Norway.

9.11.4 **Wave-dominated facies model**

A shallow marine clastic sequence dominated by heterolithic lithologies in the Lower Carboniferous of southern Ireland has been divided into four facies on the basis of grain size and sand–mud ratio (de Raaf *et al.*, 1977): (1) streaked muds (M), (2) lenticular beds (Hb–Hc), (3) parallel- and cross-laminated sandstones (Ha), and (4) large-scale structured sandstones (Sb). The general character of these facies can be obtained from Fig. 9.29 using the abbreviations.

Streaked muds (M) consist of mudstones with thin silty and sandy intercalations in which the latter are mainly parallel-laminated, undulatory or more rarely cross-laminated (Fig. 9.51, 1–4). Increasing sand content and higher energy conditions characterize the *lenticular beds (Hb–Hc)* which comprise sandstone layers with continuous and discontinuous mudstone drapes (flaser bedding). Unidirectional cross-lamination with form-discordant lenses is the dominant structure with subordinate low-angle and undulatory even lamination (Fig. 9.51, 5–8). All structures appear to be wave-generated. *Parallel- and cross-laminated sandstones (Ha)* are fine- to medium-grained (Fig. 9.51, 9–12). Cross-laminated sandstones include both wave- and current-formed varieties including climbing cosets, and the various laminations are frequently arranged in microsequences reflecting increasing and decreasing energy conditions. The *large-scale structured sandstones (Sb)* represent the highest energy deposits in this continuum. Parallel and low-angle bedding are the dominant structures with subordinate cross-bedding. The lack of emergent features indicates that sedimentation was in a sublittoral environment.

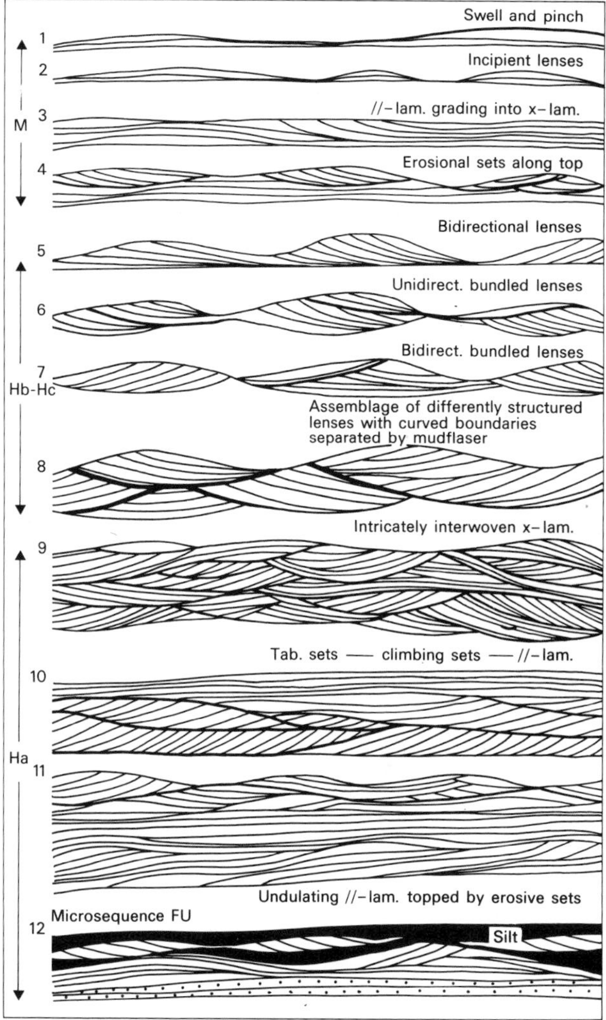

Fig. 9.51. Some characteristic wave-dominated structures from sublittoral lithofacies in the Lower Carboniferous, County Cork, southern Ireland (from de Raaf *et al.*, 1977).

The main arguments for interpreting these deposits as wave-dominated are: (1) the types of lamination are more typical of wave processes than of unidirectional currents, and (2) unidirectional palaeocurrent patterns, which, based on regional palaeogeography, are directed onshore and are interpreted as representing the direction of wave surge.

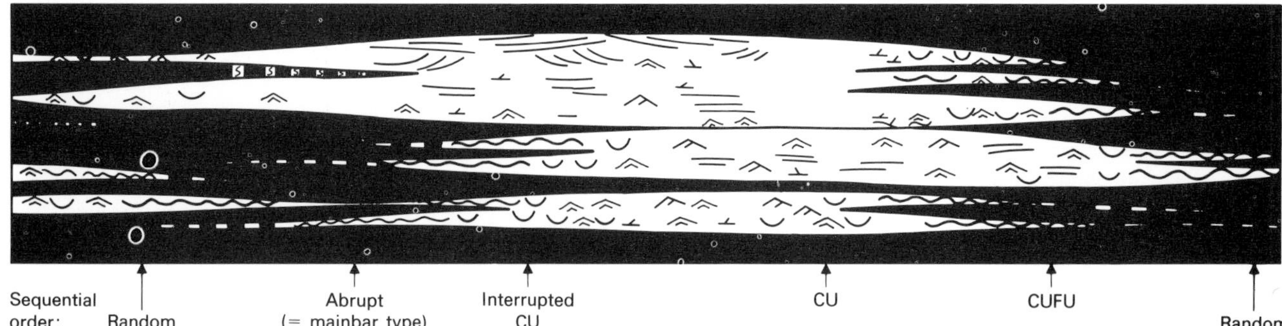

Fig. 9.52. Relationship between the types of vertical sequence and three inferred types of wave-generated sand bar: incipient, submerged and locally emergent types. The main types of facies alternation are: random, abrupt, coarsening-upward (CU), interrupted CU, multiple coarsening/fining-upward (CUFU) (from de Raaf *et al.*, 1977).

The various types of sandstone lamination can therefore be interpreted in terms of fluctuating wave intensity. Since the latter is largely determined by water depth, the four facies are probably depth related. The highest-energy sand layers in each facies can be equated with increased wave agitation accompanying storms, with several types of bed similar to the sheet sandstones described earlier (Sect. 9.11.2). The occasional unidirectional current structures may also be related to storm- or wave-induced currents.

These facies are arranged into various sequences, most of which coarsen upward. The coarsening-upward sequences reflect progradation in a sublittoral environment, and terminate with a sand body deposited in a very shallow, sublittoral environment. The discontinuous geometry of the sandstones suggests that they formed part of elongate sand bars, perhaps topographically elevated. The sand bodies wedge out in various directions with thinner sandstones represented by different types of sequence (Fig. 9.52). The consistent palaeocurrent directions indicate that the sand bodies were longshore bars whose long axes paralleled the shoreline. The different types of sequence may reflect both particular stages of sand body development and lateral variations within a single sand body.

9.11.5 Storm surge ebb facies model

The late Pre-Cambrian Innerelv Member from north Norway comprises 300 m of mudstones, siltstones and very fine sandstones. It has been divided by Banks (1973b) into five gradationally linked facies usually arranged into coarsening-upward sequences (Fig. 9.53). The facies sequence is believed to reflect shallowing in a sublittoral environment, which never became a shoreline.

The sequence of facies 1 to 5 represents a continuum of gradually increasing energy conditions passing from mainly siltstones and mudstones with subordinate sandstone beds (facies 1 and 2), through alternating sheet sandstones and siltstones (facies 3 and 4) into channel-filled lenses of fine sandstones, siltstones and mudstones (facies 5). However, within each there was an alternation between relatively high-energy conditions when sediment was mainly deposited by decelerating suspension-laden currents, and relatively low-energy conditions when waves reworked fine-grained sediment as it was deposited from suspension.

Since the five facies are considered to be laterally equivalent, these two contrasting hydraulic regimes can be correlated between the facies. Highest-energy conditions were represented by unidirectional currents, which were most probably directed offshore and responsible for the sheet sandstones. The sedimentary structures and waning flow sequences suggest that storms were responsible for these conditions. In addition, the unidirectional palaeocurrent pattern precludes tidal activity, and the intermittent character of the currents argues against oceanic currents and for storm activity. Following modern analogues (Hayes, 1967a; Gadow and Reineck, 1969) storm surge ebb currents were considered the most likely seaward-flowing currents. The environment of deposition in this case is therefore a low-energy coastline adjacent to a fine-grained and generally low-energy seaward-fining shelf in which sediment transport was largely restricted to periodic meteorological disturbances.

Fig. 9.53. Storm surge ebb facies model based on the late Pre-Cambrian Innerelv Member (after Banks, 1973b).

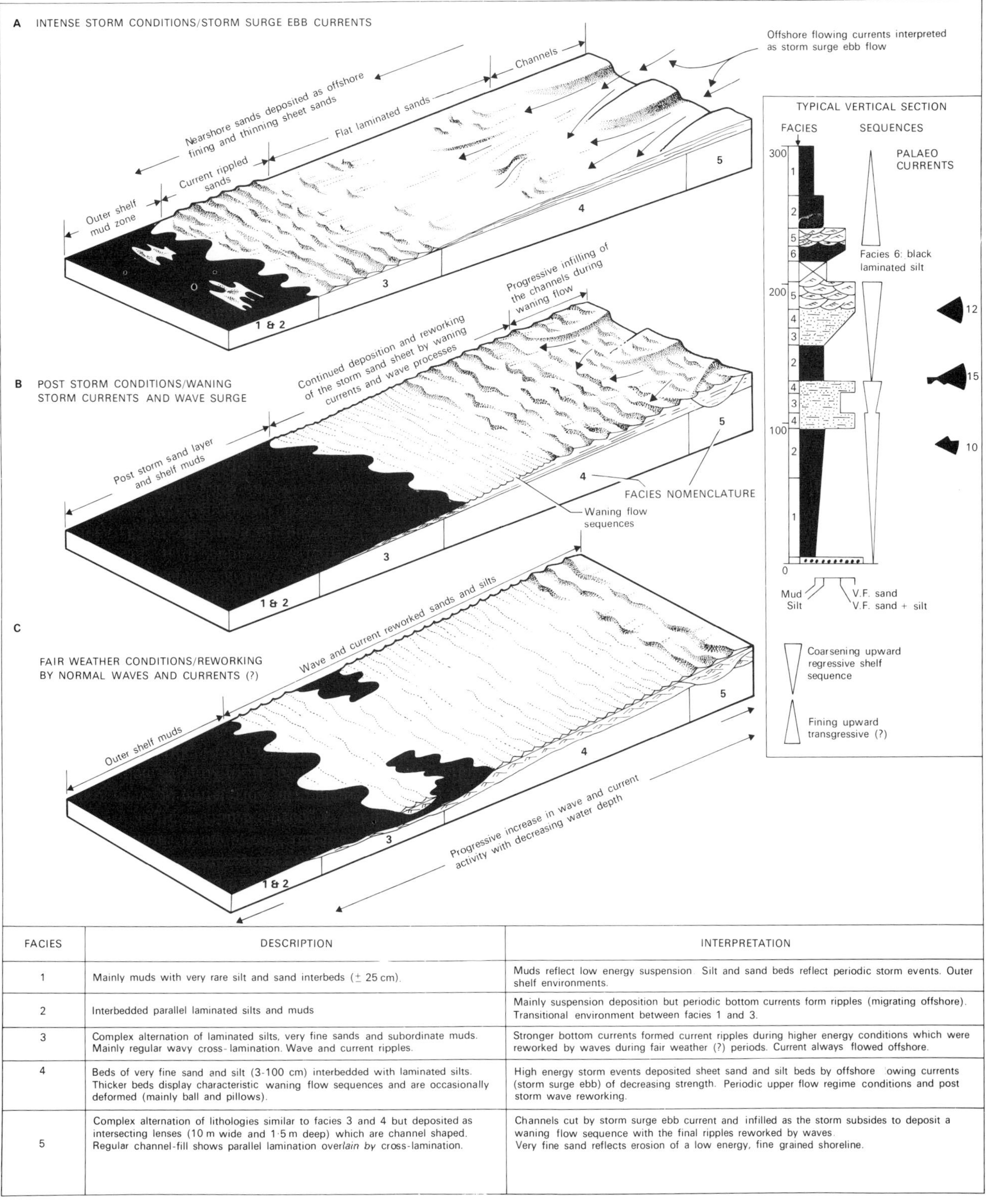

FACIES	DESCRIPTION	INTERPRETATION
1	Mainly muds with very rare silt and sand interbeds (± 25 cm).	Muds reflect low energy suspension. Silt and sand beds reflect periodic storm events. Outer shelf environments.
2	Interbedded parallel laminated silts and muds	Mainly suspension deposition but periodic bottom currents form ripples (migrating offshore). Transitional environment between facies 1 and 3.
3	Complex alternation of laminated silts, very fine sands and subordinate muds. Mainly regular wavy cross-lamination. Wave and current ripples.	Stronger bottom currents formed current ripples during higher energy conditions which were reworked by waves during fair weather (?) periods. Current always flowed offshore.
4	Beds of very fine sand and silt (3-100 cm) interbedded with laminated silts. Thicker beds display characteristic waning flow sequences and are occasionally deformed (mainly ball and pillows).	High energy storm events deposited sheet sand and silt beds by offshore flowing currents (storm surge ebb) of decreasing strength. Periodic upper flow regime conditions and post storm wave reworking.
5	Complex alternation of lithologies similar to facies 3 and 4 but deposited as intersecting lenses (10 m wide and 1·5 m deep) which are channel shaped. Regular channel-fill shows parallel lamination *overlain by* cross-lamination.	Channels cut by storm surge ebb current and infilled as the storm subsides to deposit a waning flow sequence with the final ripples reworked by waves. Very fine sand reflects erosion of a low energy, fine grained shoreline.

9.12 ANCIENT MUDDY SHELF DEPOSITS

Muddy shelf sediments have mostly been studied from their palaeoecological aspects, and strictly sedimentological information – excluding the clay mineralogy – is often sparse. If modern environmental analogues are valid, thick, mud-dominated shelf deposits probably accumulated most rapidly in areas of low wave and current agitation (Sect. 9.3.1), particularly where suspended sediment concentrations were high.

Some shelf mudstone sequences can be fitted into a broad environmental context by reference to their location within the general facies succession. For example, in prograding shoreline deposits shelf mudstones form the lower part of coarsening-upward sequences; in deltaic sequences fossiliferous shelf mudstones are gradationally overlain by unfossiliferous prodelta mudstones and siltstones (Chapter 6), while in other shorelines they pass transitionally upward into shoreface deposits (Chapter 7). However, thick shelf mudstone sequences often show subtle internal variations that are only interpretable in the light of detailed palaeontological analysis (e.g. Sellwood, 1972; Duff, 1975; Fürsich, 1977).

During much of Jurassic times a large portion of north-west Europe was covered by a shelf sea (Fig. 9.54) in which mud was the dominant sediment (Hallam and Sellwood, 1976). This epeiric basin provided striking examples of various mud facies which may be models for other basins that have not been studied in as much detail. During the Lower Jurassic three major mud facies may be recognized: 'Normal', 'Restricted' and 'Bituminous' (Morris, 1977) (Fig. 9.55). Swimming and floating organisms (e.g. ammonites, belemnites, fish and coccolithophorids) are found in all three mud facies, but the benthic organisms were strongly facies controlled.

Normal mud facies consists of claystones and mudstones with a diverse benthic assemblage of burrowing and epifaunal organisms. The burrowers are represented by both trace-fossils (e.g. *Chondrites, Rhizocorallium,* and *Thalassinoides*) and body-fossils, the latter being dominated by bivalves. The burrowers include both suspension-feeding and deposit-feeding groups.

The restricted mud facies contains fewer burrowers of all types and fewer epifaunal organisms. The benthos is mostly dominated by infaunal deposit-feeding bivalves such as protobranchs and specialized epifaunal bivalves such as *Posidonia* (*Bositra*) and *Inoceramus.* Trace fossils are few, mainly consisting of thin horizontal burrows, attributed to deposit-feeders (Sellwood, 1972). *Chondrites* and other trace-fossils are rare.

The bituminous mud facies contains few benthic genera –

Fig. 9.54. Distributions of shelf facies in the Jurassic epeiric basin of the north-west European region (after Hallam and Sellwood, 1976).

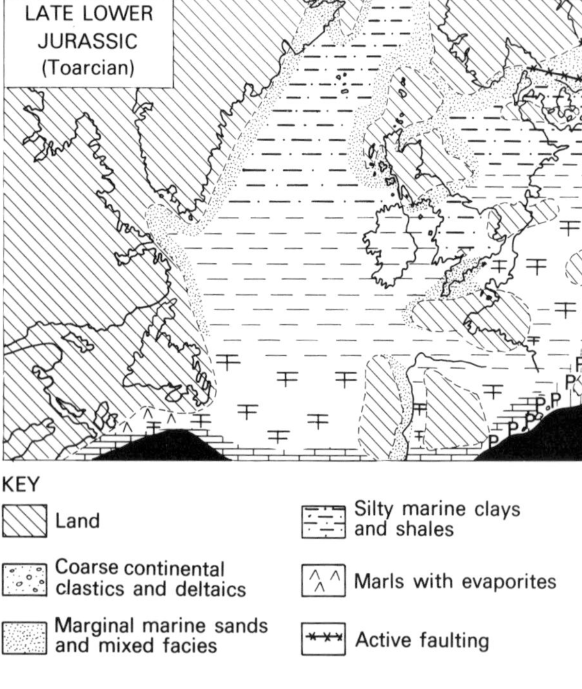

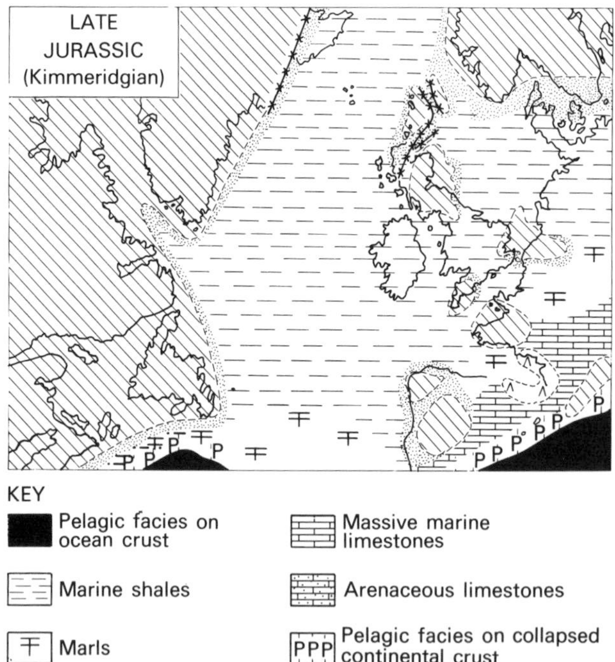

FACIES	BIVALVE GROUPS	TRACE FOSSILS	BENTHIC FORAMS.	CONCRETIONS	ENVIRONMENTAL INTERPRETATION
NORMAL SHALE	EPIFAUNAL AND INFAUNAL SUSPENSION-FEEDERS, INFAUNAL DEPOSIT-FEEDERS	*CHONDRITES,* HORIZONTAL BURROWS	COMMON	SIDERITIC AND CALCAREOUS	Sea water Abundant O_2; Mild oxidizing conditions; Reducing conditions; Sediment
RESTRICTED SHALE	DOMINANT INFAUNAL DEPOSIT-FEEDERS	FEW HORIZONTAL BURROWS	RARE	CALCAREOUS	Sea water Abundant O_2; Reducing conditions; Sediment
BITUMINOUS SHALE	DOMINANT EPIFAUNAL SUSPENSION-FEEDERS	NONE	NONE	PYRITIC CALCAREOUS	'Soupy' layer; Sea water Abundant O_2; Abundant H_2S; Reducing conditions; Sediment

Fig. 9.55. A classification of shale facies and criteria for their recognition (after Morris, 1977).

except the specialized epifaunal groups mentioned above. These are commonly found adhering to the upper surfaces of larger particles such as ammonites and fossil driftwood which provided a suitable substrate. The sediment is mostly unbioturbated and retains well-developed laminae that are rich in kerogen, thus providing the bituminous content of the sediment. In vertical sequences 'normal', 'restricted' and 'bituminous' facies are often arranged in symmetrical cycles on both large (tens of metres) and small (decimetre) scales with the 'normal' passing upwards into 'restricted' and then 'bituminous' shales. It seems that periodically the basin became progressively more restricted environmentally before subsequent amelioration.

Bituminous shales probably formed in tranquil basins with a restricted oxygen budget and reflect a combination of oxygen deficiency and poor substrate conditions at the sea-bed. Some of the bituminous shale facies developed over large areas encompassing most of the north-west European epeiric basin (e.g. during late Lower Jurassic and late Upper Jurassic times). They provided ideal hydrocarbon source-rocks during their later burial (Hallam and Sellwood, 1976).

More detailed accounts of the palaeoecological studies on these Jurassic mud facies can be obtained from Hallam (1975), Duff (1975), Sellwood (1972a) and Sellwood (1978).

Mud-dominated shelves have developed in many places and at different times, notably during the Cretaceous in both Europe and North America (e.g. Kauffman, 1967, 1974; Simpson, 1975) and in the Lower Palaeozoic (e.g. Ziegler, Cocks and McKerrow, 1968; Ziegler, Newall, Halbeck and Bambach, 1971; Cocks and McKerrow, 1978) but as yet most of the studies on these sediments have been mainly palaeontological.

9.13 SILICICLASTIC SEDIMENT SUPPLY TO SHALLOW SHELF SEAS

The type of sediment available is *the* most important single control of shallow marine siliciclastic facies. Having summarized the processes and products of the shallow marine environment this chapter is concluded with a brief discussion on how sediment, in particular sand, reaches the shelf. As Curray (1975) emphasizes '(the) seaward transport of sand beyond the shore zone is one of the very important problems in sedimentary geology today. Can sand leak out to the continental shelf, or are we misinterpreting the environments of many ancient sand bodies?'.

Three main factors controlling the type and rate of sediment supplied to shallow seas are terrestrial supply, sediment storage in the coastal zone and sediment transfer to the offshore zone.

Terrestrial supply is determined by the amount and type of sediment available in the hinterland, and the rate of supply to the shoreline.

Sediment storage involves the type of sediment trapping and degree of reworking in the shoreline zone. This is determined by the interaction of fluvial and marine processes, which is manifested in river mouth (deltaic) and shoreline configuration (see Chapters 6 and 7).

Sediment transfer from inshore and shoreline zones to the shelf environment is controlled by the interaction of nearshore and inner shelf hydraulic processes. These are determined by fluvial activity, tidal range, wave intensity, storm frequency, sea level fluctuations (transgressing or prograding shoreline) and the type of shelf hydraulic regime (tide-dominated, wave- or storm-dominated, etc.)

The most efficient direct (external) supply of sand from terrestrial environments to the shelf is through tidal river mouths (tide-dominated deltas) where inshore and inner shelf hydraulic regimes are in equilibrium. This is exemplified today be several major tide-dominated river deltas with flared mouths (e.g. Ganges–Brahmaputra, Klang–Langat, Colorado). Transgressions provide the most important mechanism for placing large quantities of sand on the shelf. Thus major sand bodies deposited during lower sea level stands (e.g. beach, barrier, fluvial and deltaic sand deposits) are drowned during rising sea level and reworked *in situ* by shelf processes. When coastal erosion, resulting from delta switching or transgression and shoreface retreat, is accompanied by strong shelf currents sand is moved offshore. In addition, intense storms are capable of extensive coastal erosion, with storm surge ebb currents particularly effective in transporting shoreline sands offshore. However, the daily persistence of tidal currents makes them the most efficient of all shelf currents in transporting sand across shallow shelves.

The prerequisites for the accumulation of many blanket sandstone deposits (Sect. 9.10.3) were an abundant sand supply, fluctuating sea level and strong tidal currents. Numerous episodes of shoreline progradation and retreat coupled with extensive coastal erosion would ensure a continuous supply of sand. The inferred pre-eminence of tidal activity as the major source of energy in these basins would presumably cause most major fluvial incursions to develop into tide-dominated deltas or tidal estuaries. This mechanism of transporting sand directly onto the inner shelf has been postulated for inferred tidal sand ridge deposits in a regressive, tide-dominated deltaic sequence from the Cretaceous Colorado Group in Canada (Simpson, 1975).

Sand may also be directly supplied to the shelf from rocky coasts or protruding rock islands surrounded by a high-energy sea. The latter is well exemplified at Baraboo, Wisconsin, where, during the late Cambrian, 40–100 m high quartzite islands supplied a range of clastic detritus to the adjacent shelf (Dott, 1974).

Any fine-grained sediment released to these extensive, high-energy shallow seas was presumably successively winnowed and deposited either in protected inshore environments or in deeper offshore regions.

Sedimentation of muddy shelf deposits is controlled by the concentration of fine-grained suspended sediment entering shelf seas and by the intensity of shelf hydraulic processes, particularly the strength and frequency of wave action.

FURTHER READING

Swift D.J.P., Duane D.B. and Pilkey O.H. (Eds) (1972) *Shelf Sediment Transport: Process and Pattern.* 656 pp. Dowden, Hutchinson & Ross.

Stanley D.J. and Swift D.J.P. (Eds) (1976) *Marine Sediment Transport and Environmental Management.* 602 pp. John Wiley & Sons.

CHAPTER 10 Shallow-water Carbonate Environments

10.1 INTRODUCTION

10.1.1 Historical background to research

Although both Darwin and Lyell appreciated the significance of organic activity in the generation of modern and ancient carbonate sediments, the first major attack on the problems of limestone formation and diagenesis was by Sorby. His work began in 1851 with a paper on a chertified limestone from the Upper Jurassic and continued for 53 years. Because so many limestones were made up of organisms, he set out to study the structure and mineral composition of skeletal materials. He was also acutely aware of the diagenetic problems associated with the interpretation of shelly faunas after the selective loss of the aragonitic elements. Ideas and deduction based upon a wealth of observations were rigorously tested by experiment. He considered the problem of ooid formation and concretion growth and he studied thin sections of most British limestones.

Although Sorby pointed out the problems that others could work on, like dolomitization, there seems to have been little immediate follow-up in Britain. Folk (1973) attributes this to the apparently uninspiring mineralogy of sediments and the preoccupation of palaeontologists and stratigraphers with the discovery of new species and formations. However, in France, Cayeux started to publish a steady stream of work that partly stemmed from his studies on the French Chalk. This work, partly interrupted by the First World War when he worked on ironstones, culminated with the publication of his 'Magnum Opus' on the sedimentary rocks of France (Cayeux, 1935).

The Royal Society financed an expedition to Funafuti Atoll to test Darwin's theory of atoll formation and under the direction of Sollas, a hole was cored to a depth of 1,114′ (340 m). In a memorable volume (Bonney, 1904), which includes Sorby's last paper, Cullis (1904) published a milestone paper on the petrographic treatment of limestones. Staining techniques were extensively used and his observations on cement development were brilliantly illustrated. Cullis noted an increase in cementation and the disappearance of aragonite with depth where calcite and dolomite dominated. Dolomite and aragonite were seen to be mutually exclusive.

The analysis of modern carbonate environments began with the expedition of the father and son Agassiz to the Bahamas (Agassiz, 1894, 1896). Subsequently Vaughan (1910) accompanied by Field began pioneering work on the Cenozoic history of Florida.

With the few exceptions already given, the first two decades of this century saw disappointingly few advances in diagenetic and environmental studies. This was partly because limestone study in general fell into the hands of palaeontologists eager for fossils but oblivious of their matrix.

Major advances were to come as a result of the International Expedition to the Bahamas (in 1930) which led to an analysis of algal-binding and a general review of carbonate production on the Banks (Black, 1933a and b). Field and Hess (1933) gave the results of the borehole put down through the Banks.

Between the Wars and after the Second World War the discovery of major carbonate oil reservoirs, particularly in the Middle East, gave a new impetus to carbonate research within oil companies (particularly the Shell Development Co.). At first little was published but the mid- to late-fifties became the 'renaissance' years in the study of carbonates with a rebirth of petrography.

The new ideas and attitudes came from Illing (1954), Newell *et al.* (1951), Ginsburg (1956, 1964), and Cloud (1955) who were revising carbonate petrography and laying the foundation of facies analyses.

Undoubtedly one of the papers that had an enormous impact at the start of this 'new age' was that by Ginsburg (1957) which dispelled the need to bury limestone in order to cement it. Vadose zone cementation suddenly became fashionable and continued to dominate ideas on early cementation until submarine cementation was demonstrated by Ginsburg *et al.* (1968) from the Bermudan reefs and by Shinn (1969) and Taylor and Illing (1969) from the Persian Gulf.

The upsurge in the facies approach to carbonates stemmed largely from the needs of oil reservoir analysis and progress continued with the studies of Purdy (1961, 1963a and b) in the Bahamas, with Ginsburg's group in Florida and later with the Koninklijke/Shell and Illing Groups in the Persian Gulf.

Shearman and his co-workers from Imperial College brought a new word to the geological vocabulary – 'sabkha'. New vistas of research were opened into the chemistry of diagenesis and particularly dolomite, gypsum and anhydrite formation and to some workers all ancient evaporites were slotted into sabkha models. Only recently have they been seriously questioned in the light of deep-sea drilling, particularly in the western Mediterranean (Sect. 8.5.5).

In the early fifties too, geological and biological expeditions were sent to Pacific atolls (e.g. Ladd *et al.*, 1950; and Emery *et al.*, 1954). Soon after, each atoll was annihilated in a nuclear blast.

The great advances in the study of modern environments were essential to provide working facies models, but into the fifties there had been no substantial improvement in the petrographic approach to diagenesis since the work of Cullis and Sorby. Improvements in staining techniques were achieved (Friedman, 1959; Evamy and Shearman, 1965, 1969 and many others) and by the late fifties a lead was being taken on the petrographic front by Bathurst (1958, 1959, 1964) culminating in 1971 with the publication of his masterly textbook on carbonates and their diagenesis.

While the late fifties witnessed many advances in the facies approach there was still no satisfactory classification. Grabau (1904, 1913) introduced the twofold grain-size terms *calcirudite, calcarenite* and *calcilutite* and these terms are still in common usage. Pettijohn (1949) simply subdivided limestones into *allochthonous, autochthonous, biohermal* and *biostromal* but the subtleties of differentiating between limestone types were not possible before Folk's (1959) paper on petrographic classification was published. This elegant approach, inspired by Krynine's treatment of sandstones, employed the relationships between textural maturity, allochems, spar and micrite and provided both an accurate description and an indication of the energy of deposition. The Folk classification was later restated together with that of Leighton and Pendexter (1962) in the milestone volume edited by Ham (1962). This volume was largely based on oil company work and also contained the other major classification in common contemporary usage, that of Dunham (1962) whose prime concern was the nature of the grain-support.

The sixties and seventies have seen extension, consolidation and application of the works cited and the derivatives that they have inspired. Text-book summaries range now across the whole spectrum of research from the purely chemical (including Berner, 1971; and Lipmann, 1973) to more general compilations (Bathurst, 1975; Purser, 1973a; Milliman, 1974; Wilson, 1975). Recent developments have included the recognition of environment-specific cement fabrics (James *et al.*, 1976); the reappraisal of stromatolites (Walter, 1976), the extension of petrographic analyses with the advent of cathodoluminescence and the scanning electron microscope, and the advancement of chemical approaches to diagenetic problems. In his presidential address to the Geological Society of London in 1879 Sorby admitted that despite working for nearly thirty years on various questions 'essential to the proper elucidation of his subject' he felt 'painfully conscious how much still remains to be learned'. This is still true.

10.1.2 **Major controls on carbonate production and distribution**

The dominance of carbonate deposits upon certain sectors of the world's continental shelves is, today, directly related to two major factors: relative lack of clastic deposition and high organic productivity. All modern carbonates occur in areas that are not generally receiving large amounts of silicate detritus and the bulk of carbonate material on modern shelves is ultimately of organic origin – either directly as skeletal material or indirectly as a precipitated by-product of organic activity.

The rate of organic productivity in the marine environment is controlled by many variables but in general there is a progressive increase from the higher to lower latitudes as solar illumination increases. Productivity in the equatorial and subtropical belts is also promoted by oceanic upwelling which is particularly strong along the western borders of continents in the northern hemisphere and their eastern borders in the southern, and this process re-circulates nutrients that would otherwise have remained on the ocean bed (see Chapter 11).

Shelf carbonates are not, however, restricted to latitudes between 30°N and 30°S (Chave, 1967; Lees and Buller, 1972). For example, many thousands of square kilometres of virtually pure shelf carbonates are currently accumulating off southern Australia, between latitudes 32° and 40°S (Conolly and Van der Borch, 1967; Wass *et al.*, 1970), and in smaller, discrete, patches elsewhere in the world including areas as far north as western Ireland and western Scotland.

Temperature combines with other factors like salinity, CO_2 balance, water depth, nature of local current regimes, light penetration, effective day length, nature of substrates and turbidity in controlling shelf carbonate sedimentation. However, Lees (1975) makes a good case for regarding temperature and salinity as being the prime controls on a global scale. The chemistry of carbonate equilibria is discussed by Bathurst (1975, p. 242), Berner (1971), Berner and Morse (1974), Berner and Wilde (1972) and many others. Lees and Buller (1972) reviewed the global distribution of grain types in the sand and coarser fractions in carbonate sediments deposited in <100 m of water and found that they could be grouped into a few major grain associations. They distinguished two skeletal-grain associations (Fig. 10.1): (1) the *foramol* association typical of 'temperate-water carbonates' which included the debris of benthic foraminifera, molluscs, barnacles, bryozoans and calcareous red algae as the dominant

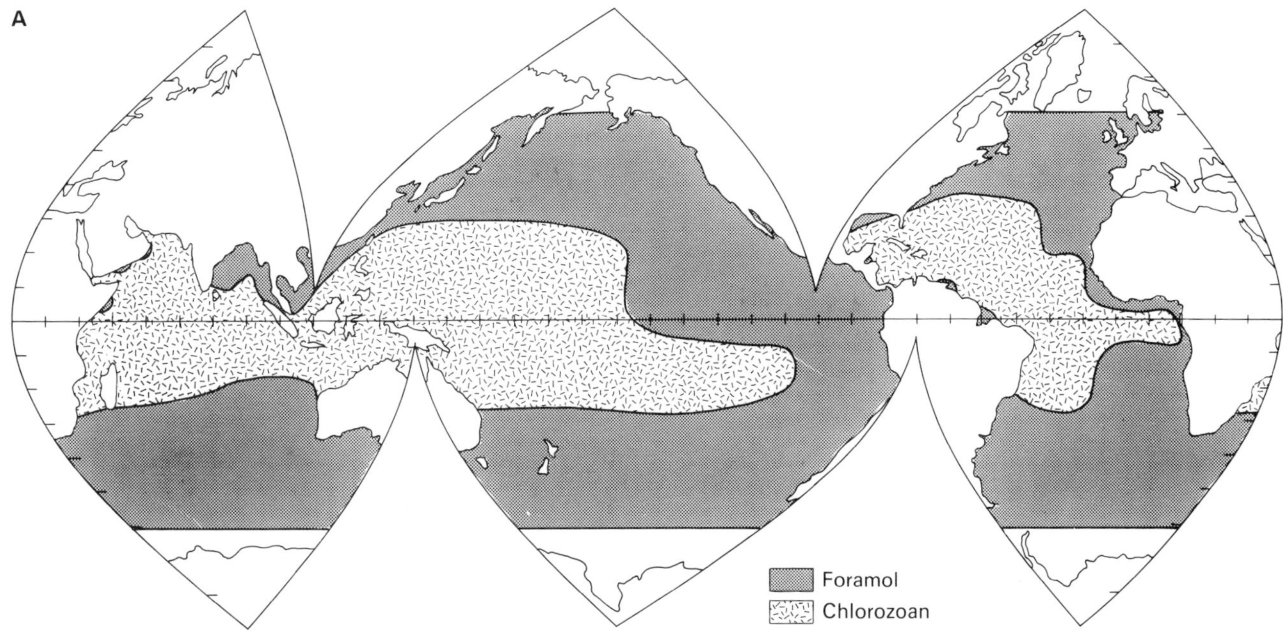

Fig. 10.1A. Predicted distributions of skeletal grain associations in shallow-water (0–100 m) carbonate sediments of present-day seas and oceans between 60°N and 60°S. The distributions illustrate potential not fact. Although they only concern shallow-water areas, distributions are extended across the oceans' areas because shoal areas occur within oceans (after Lees, 1975).

Fig. 10.1B. Predicted distributions of non-skeletal grain associations in shallow-water (0–100 m) carbonate sediments of present-day seas and oceans between 60°N and 60°S. Same restrictions apply as in Fig. 10.1A (after Lees, 1975).

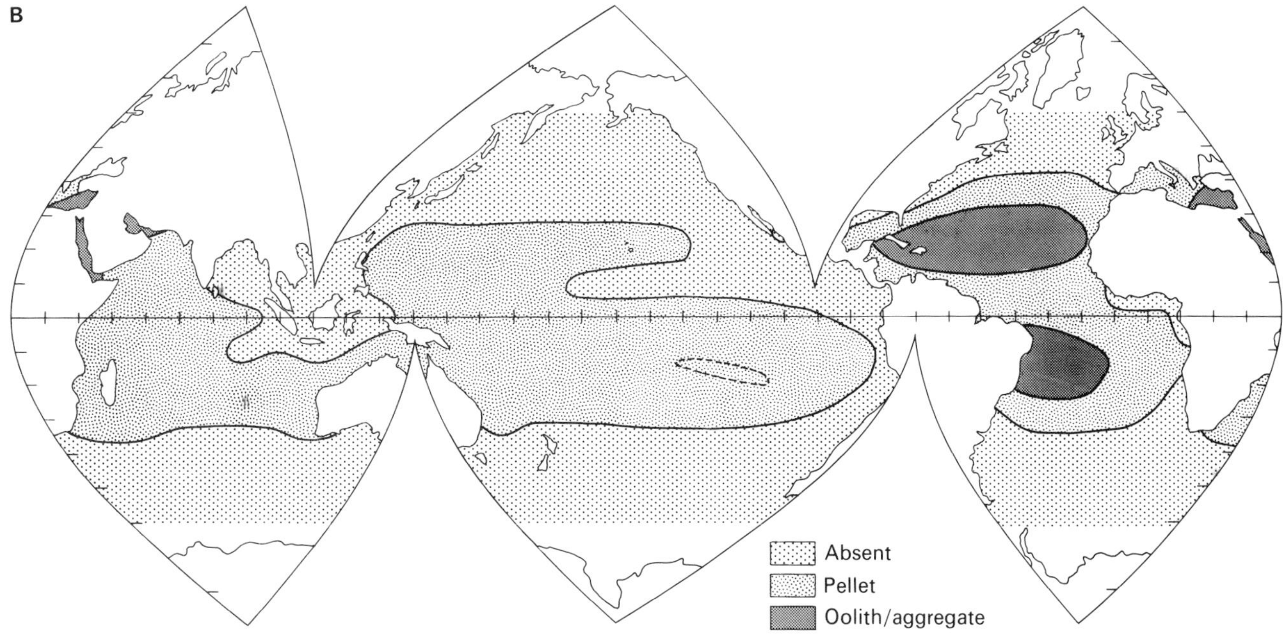

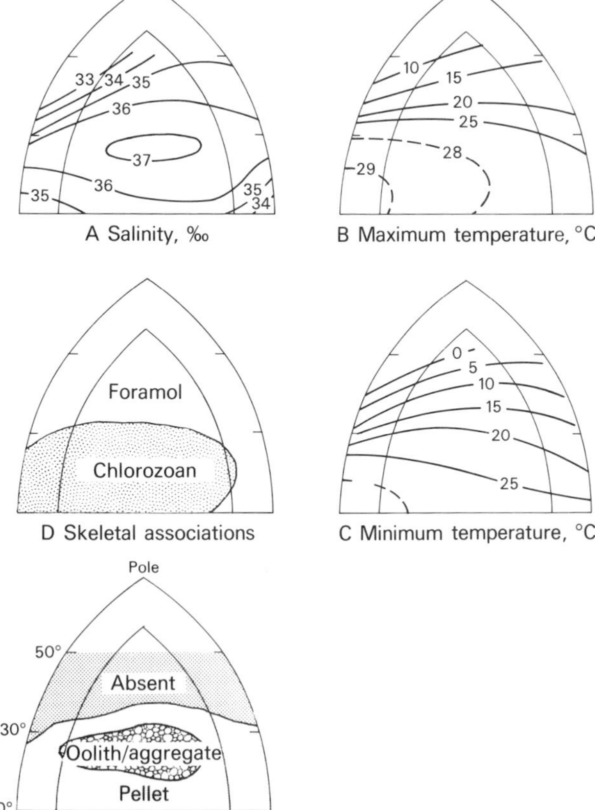

Fig. 10.2. Predicted distributions of skeletal and non-skeletal grain associations in shallow-water (0–100 m) carbonate sediments in an ideal ocean (N. Hemisphere). Predictions based on applications of the Salinity Temperature Annual Range diagrams (Figs. 10.3 and 10.4). Marginal continental shelves are shown diagrammatically (after Lees, 1975).

components. Other groups such as echinoderms, ostracods and sponge spicules are accessories. (2) the *chlorozoan* association, typical of 'warm water', which contains many of the *foramol* components but includes hermatypic corals and calcareous green algae. It lacks barnacles and has fewer bryozoans.

Recent non-skeletal grains are also distributed in a systematic way (Fig. 10.2) there being three associations based on grain-types present: (1) non-skeletal grains absent; (2) only pellets present; (3) ooliths and/or aggregates present either with or without pellets. The oolith/aggregate association (type 3) is essentially restricted to *chlorozoan* areas whereas the pellet association (type 2) extends further into regions occupied by the *foramol* assemblage. Over the rest of their range the *foramol* sediments contain no non-skeletal grains (type 1).

The *chlorozoan*/non-skeletal grain associations are mostly restricted to within 30° of the equator and temperature appears to be a prime controlling factor. In support of this, the *chlorozoan* association is seen to occur only where the minimum near-surface temperature exceeds 14°–15°C while the *foramol* association tolerates much lower temperatures. However, the latter is also found in areas where the minimum exceeds 15°C suggesting that some other factor is involved (Figs. 10.3 and 10.4).

Presence or absence of non-skeletal grain associations *appears* to be explained by water temperature, with non-skeletal grains occurring only where minima exceed 15°C and means exceed 18°C. The type of carbonate grain is not wholly determined by temperature but both pellet associations and oolith/aggregate associations are inhibited by minimum temperatures below about 15°C. However, salinity and temperature may compensate each other because the *chlorozoan* association is inhibited at high temperatures if the salinity falls below a certain value ($\sim 31^0/_{00}$), and yet it develops at relatively low temperatures where salinity is sufficiently high. This compensation also appears as an effect in controlling non-skeletal associations since the oolith/aggregate association depends on salinity. In environments with extreme ranges of salinity an association containing calcareous green algae but lacking corals is found and is termed *chloralgal* by Lees (1975).

Most of the supposed present-day sites of active oolith and aggregate formation are concentrated near the tropics. The equatorial belt appears to contain none. The zones of maximum salinity in oceanic waters lie at about 25°N and 25°S and Lees suspects that this correlation between grain-type and salinity is directly related. Both tropical belts normally have annual precipitation rates that are less than evaporation rates, whereas in the equatorial belt rainfall is greater than evaporation.

So far we have considered, on a global scale, organism/grain-type associations. Much of the contemporary carbonate accumulating in shelf environments consists of mud ($<60\ \mu$) the origin of which has long been a thorny problem (see Bathurst, 1975). In the Great Bahama Banks a possible source is provided by the disintegration of calcareous green algae but Bathurst (1975, p. 279) concludes that this process cannot account for all the mud. In areas like the Trucial Coast this mechanism is impossible because of the rarity of calcareous green algae (Kinsman, 1969a). It is probably necessary, in both areas, to invoke the direct precipitation of aragonitic mud from sea-water, possibly as a by-product of organic activity. In other areas (e.g. Belize and South Florida) the disintegration of calcareous skeletal material may provide the only required source (Matthews, 1966; Scholle and Kling, 1972; Stockman *et al.*, 1967). Reviewing all these data, Lees (1975) believed that mud was dominantly derived from skeletal breakdown of *foramol* components in this association while in

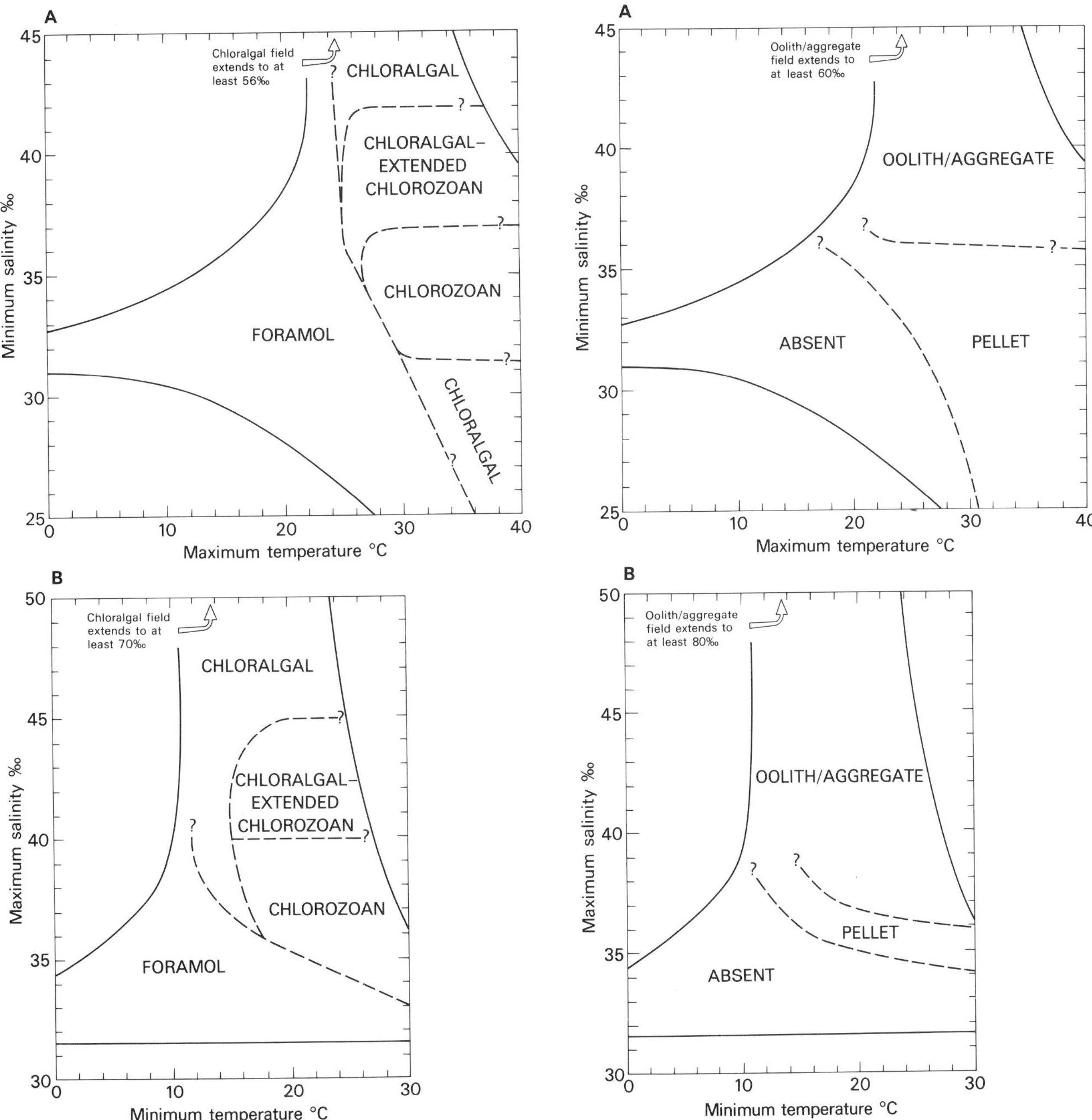

Fig. 10.3. Salinity Temperature Annual Ranges diagram pair for skeletal grain associations in modern shelf carbonate sediments. The dotted lines indicate the approximate limits of near-surface salinity/temperature combinations existing in present-day shelf seas and lagoons (after Lees, 1975).

Fig. 10.4. Salinity Temperature Annual Ranges diagram pair for non-skeletal grain associations in modern shelf carbonate sediments. The dotted lines indicate the approximate limits of near-surface salinity/temperature combinations existing in present-day shelf areas and lagoons according to the data used to construct the diagrams (after Lees, 1975).

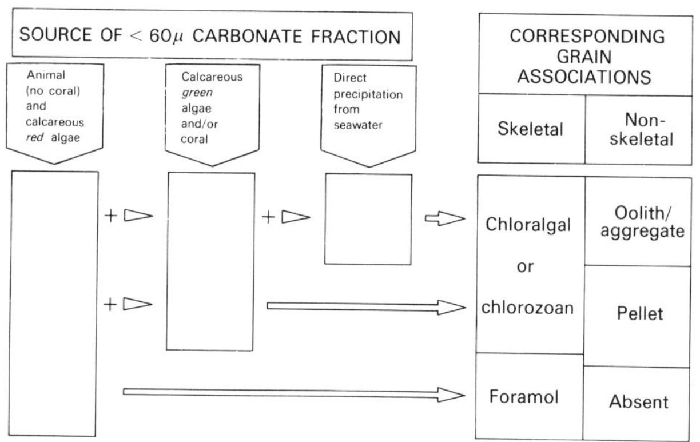

Fig. 10.5. Possible relationships between the various types of carbonate mud (<60 μ) and the grain associations (after Lees, 1975).

the *chlorozoan* association material derived from the *foramol* element is vastly outweighed by that derived from corals and green algae. He also concluded that inorganic precipitation was only likely in areas falling clearly within the oolith/aggregate field (Fig. 10.5).

10.2 SUBTROPICAL CARBONATE SHELVES

10.2.1 General settings

In the subtropical and dominantly chlorozoan belt carbonate shelves fall into two major categories:

1. Protected shelf lagoons ('Rimmed Shelves' of Ginsburg and James, 1974) of the Bahamian, Florida-, Belize-, Batabano- (Fig. 10.6) and Great Barrier Reef types.
2. Open Shelves (open deeply submerged inclined shelves) which include Yucatan, Western Florida, Western North Atlantic, the eastern Gulf of Mexico (Fig. 10.6), and North Australia.

The shelf lagoons consist of a shallow sea-floor ($\sim$10 m average depth) enclosed within topographic barriers which have formed as coral and coralgal reefs, islands and shoals. The shelf-margin is usually precipitous, rising from abyssal depths. The presence of fringing barriers and the shallow depth of the platform surface combine to produce low energy environments within the shelf-lagoon. Major wave effectiveness is limited to the outer margin so that mud is dominant over most of the platform. The main physical factors operating within the shelf lagoon are currents generated by winds and tides and these processes are strongly influenced by the topographic irregularities of the sea-floor (Purdy, 1963).

The open shelves (of Ginsburg and James, 1974) are inclined seawards toward a shelf-break at 140–230 m, and since there

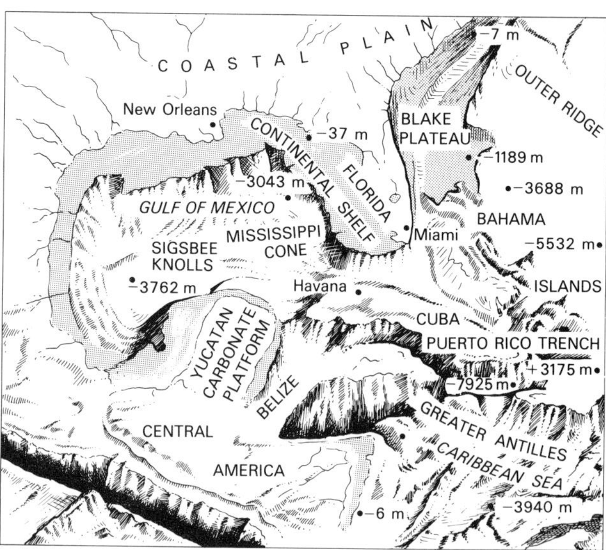

Fig. 10.6. Locality map for the Gulf of Mexico, Caribbean and Bahamian areas showing the general physiography.

are no physical barriers the wave-zone is spread over the shelf-floor and oceanic and tidal currents are also active. Such shelves may be high energy environments with an abundance of coarse-grained detritus. The coarse-grained debris includes 'clean' calcarenite with the finer-grained carbonate being confined, in the main, to the deeper (low energy) outer margins where pelagic sedimentation becomes significant.

Carbonate shelves, like their terrigenous–clastic counterparts, have been profoundly influenced by Quaternary sea-level fluctuations. The last glacial stage (Wisconsian or Devensian, 20,000 yrs B.P.) resulted in an erosional unconformity that is now overlain by unconsolidated sediments. Most 'Recent' sediment on modern carbonate shelves began to accumulate during transgression so that the subsequent morphology, facies distributions and the siting of reefs are being strongly influenced by the nature of the shelf surface (Purdy, 1974a). The Pleistocene transgressive histories of shelf lagoons contrasted markedly with those of 'open shelves'. The shelf lagoons of the Bahamas, for example, did not become inundated until the final phases of trangsgression about 5,000 yrs B.P. Flooding of the platform surface was a relatively rapid affair and subsequent facies distributions within this, and ancient, shelf lagoon areas are likely to have been strongly influenced by only subtle changes in sea-level and by the karst morphology that had developed on the platform.

In the more steeply inclined Yucatan and western Florida shelves inundation began about 20,000 yrs B.P. at the shelf edge and there was a marked shift in successive 'depth controlled' environments with rising sea-level. Ideally, this would have provided a vertical sequence commencing in shallow and passing up into deeper-water sediments.

Non-skeletal grains form in shallow water on these shelves (Sect. 10.1) and consist of ooids, peloids and aggregates. Since carbonate production is mostly dependent upon organic activity, evolutionary changes and extinctions make analogies increasingly tenuous before the Cenozoic. Equally dangerous would be direct comparisons between the narrow continental margins of the present and the vast epeiric seaways of the past. The very rapid sea-level fluctuations of the Pleistocene may not have occurred during most of the Phanerozoic. Finally, it has even been suggested (Sandberg, 1975) that the seas and oceans may have suffered fundamental changes in their Mg/Ca ratio as recently as the Mid-Cenozoic. However, geochemical models for the evolution of the ocean (e.g. Lafon and MacKenzie, 1974) suggest a constancy of composition for the last 1.2×10^9 yrs.

Bearing these warnings in mind we will now briefly review some of the main characteristics of modern subtropical shelves and attempt to reconstruct some ancient examples that may be comparable.

10.2.2 Major environments, sub-environments and facies in 'warm-water' carbonate systems

Five major environmental zones can conveniently be recognized: Supratidal Zone; Shore Zone; Marine Platform; Reef Belt and Shelf Slope and within each of these zones many interrelated sub-environments occur. By comparison with *wholly* clastic sequences, the problems of generating facies-models from non-clastic sediments are much greater. In addition to the physical process/product relationship we also have the additional complexities of diagenesis and organism response to take into account.

SUPRATIDAL ZONE

This zone, which lies above the reach of the highest normal spring tides, may be many kilometres in width with a gently undulating topography. Processes acting in the supratidal zone are strongly influenced by climate and particularly by rainfall.

In regions with high seasonal rainfall, like Florida and the Caribbean, dry seasonal evaporation may lead to the temporary formation of evaporites, but these are soon removed during rainy periods. Marshes develop in the supratidal zone (Fig. 10.7) and these are frequently flushed by fresh-water while hollows on the marsh surface may remain as temporary pools of fresh or brackish water. Marsh surfaces are extensively colonized by blue-green algae that form the diet of abundant grazers (e.g. gastropods). The algal crusts grow upon carbonate muds and pelletal silts that are laminated on a centimetre scale. These have been deposited during wind-tide and storm inundations of the supralittoral area. The sediments are normally penetrated by the roots of grasses and mangroves accompanied by the discrete burrows of worms and land-crabs. Gas-heave vugs resulting from the decay of organic material selectively follow the laminations and in the consolidated sediments of the vadose zone, such vugs (or 'birdseyes') have a high preservation potential (Shinn, 1968b). Small marine shells and sand-grade particles may be blown from the shore zone inland by the wind.

In arid regions (see Chapter 8 for main discussion), evaporites form in the supratidal zone (termed the "sabkha" in the Persian Gulf) (Sect. 8.2.6). Rooted vegetation is either impoverished or not developed and abundant burrowers are unable to survive. Thus the vuggy, laminated, lime mud sediment is mostly disturbed only by the expansive growth of minerals like anydrite and gypsum. After wind-tide inundation, ponded sea water on the flats evaporates to dryness and temporarily produces small deposits of halite and other salts. In the arid areas, wind plays a significant role in redistributing sediment. In both arid and humid areas, where strong evaporation leads to the refluxing of sea-water through the sediment, the marsh may be one of the sites of aragonite replacement by dolomite during early diagenesis.

SHORE ZONE

The shore zone often consists of a complex of sub-environments including intertidal flats, channels, levees, 'ponds' and beach-ridges (Figs. 10.7, 10.8 and 10.9).

The extent of the intertidal flats depends upon the tidal range and the slopes of accretion surfaces. In arid environments different types of algal mat are often developed in well defined zones (Sect. 8.2.5). In more humid regions and in the lower intertidal zones of arid regions, although blue-green algae grow abundantly, the laminated sediments that they trap are often thoroughly bioturbated by invertebrates and rooted halophytes. In the Bahamas the intertidal flats are of smaller extent than the 'ponds', channels and supralittoral marsh. Their sediment consists mostly of pellets, mainly between 40 μ and 80 μ and other larger pellets. As well as burrows, the laminated intertidal flat sediments are penetrated by mangrove roots. Most of the sediment on the flats is derived from the adjacent marine area and is brought on to the flats during storm-surges. In the Persian Gulf, the lower intertidal zone consists of pelletal lime muds with algal-grazing cerithids and burrowing crabs. In some places black mangroves live. In the upper intertidal zone the muddy sediment brought in by storms is trapped by well-developed and mostly undisturbed algal mats that may have interstitial gypsum crystals.

Channels consist of complex tributary and distributary systems which are the main pathways for tidal current exchange and of water-movement during storm surges.

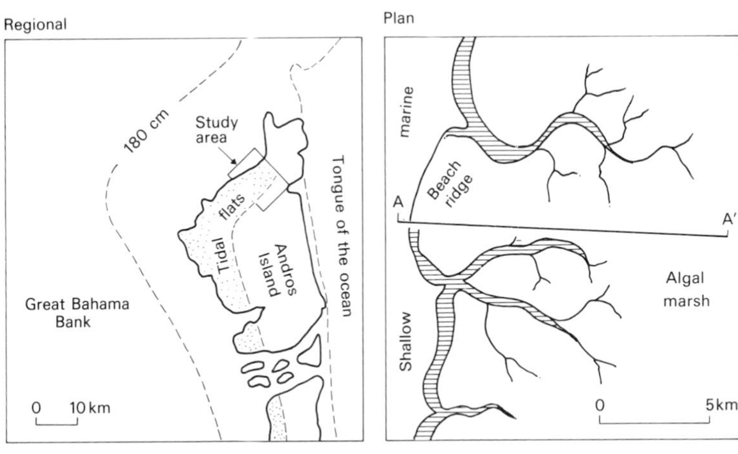

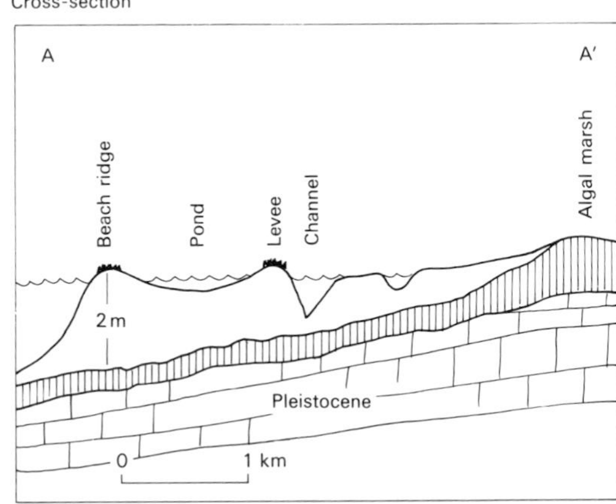

Fig. 10.7A. Generalized plan and cross section of an Andros tidal-flat (after Ginsburg and Hardie, 1975).

Current velocities within them can vary greatly and the sediment ranges in type from shell coquina to carbonate mud. Bars within the tidal channel systems may have rippled sediments with shells, lithified mudlumps and pellets. Ripple-marks are seldom preserved because of the abundant burrowing organisms. Channel migration leads to the incorporation of dolomitic and mudclast conglomeratic sediments from the levee areas (see later) into the channel-fill and the overall sequence resulting from channel migration generally fines upward (Fig. 10.8). In humid areas like the Bahamas, the channels carry a great deal of fresh water during the rainy season. Channels normally contain a more diverse fauna and flora than the adjacent flats and in the Bahamas this includes peneroplid forams, cerithids, crustacean and annelid

Fig. 10.7B. Schematic block diagram of an Andros tidal flat (after Ginsburg and Hardie, 1975).

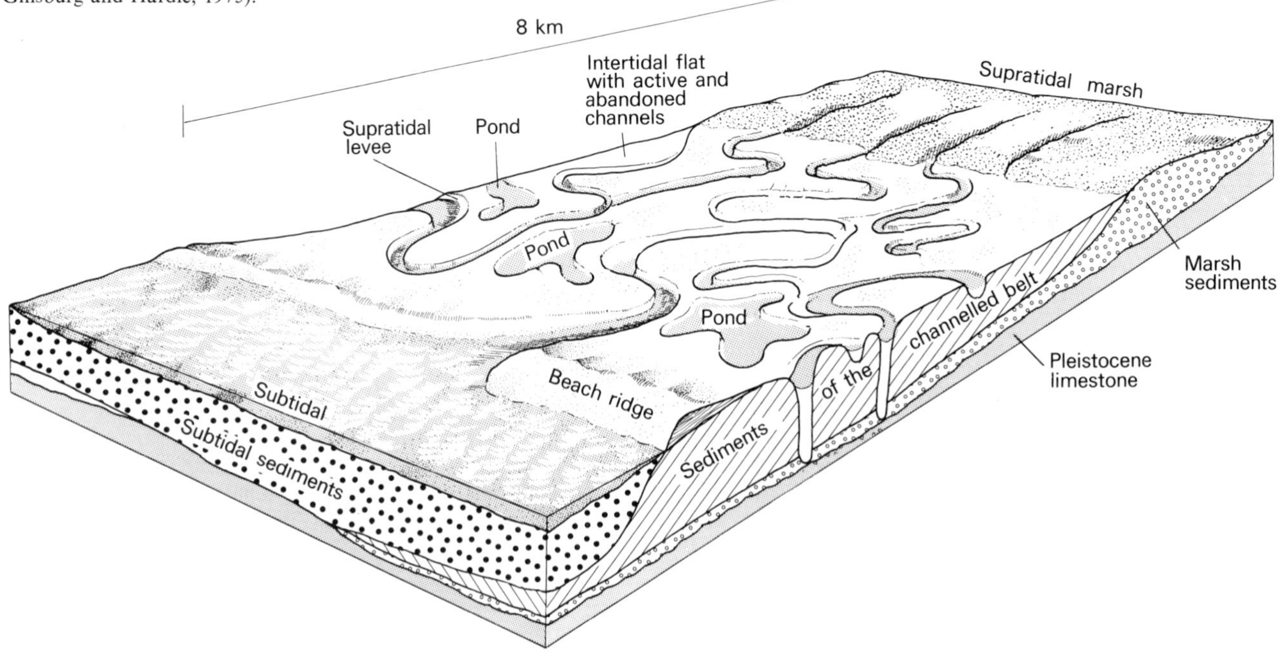

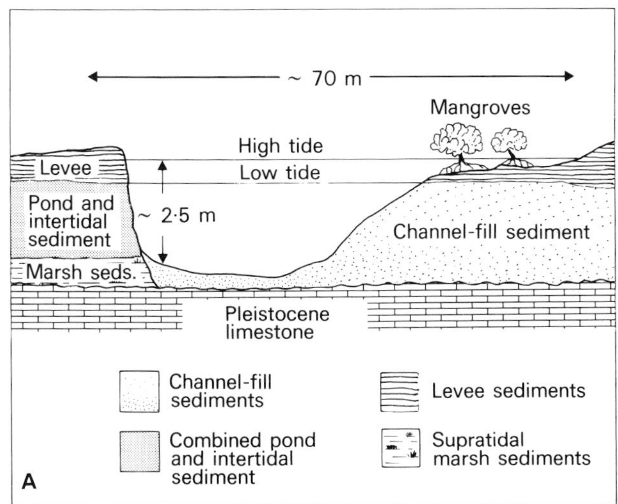

Fig. 10.8A. Interpretative cross section of a tidal channel. Erosion of left hand bank produces a lag (after Shinn *et al.*, 1969).

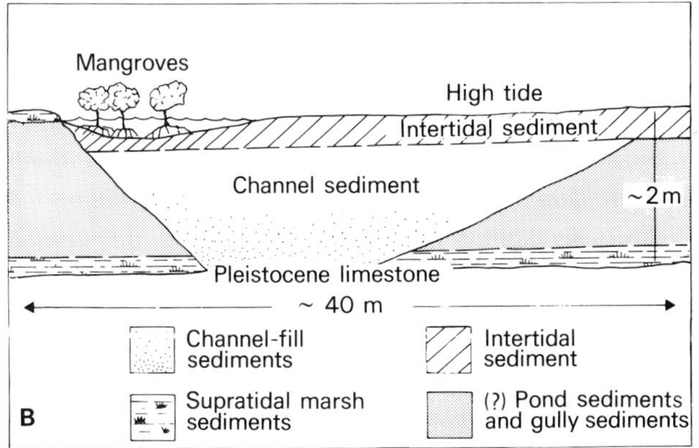

Fig. 10.8B. Interpretative cross section of an abandoned tidal channel (after Shinn *et al.*, 1969).

burrowers and sometimes patches of sponges. On the arid Trucial Coast, coarse lags of coral and shell debris are thickly encrusted with algae and bryozoans and infested by algal/fungal borers.

The outer banks of the channel meanders are often constructed to heights of 30 cm or more above the normal high-tide level and form prominent levees. The sediment in the levees is coarser grained than that on the flats and in the Bahamas consists of sand-grade pellets. The sediments are often thinly laminated and contain up to 10% dolomite that has grown *in situ* at the expense of aragonite. The sediments are sometimes graded with well developed algal films on the exposed substrates. Birdseye vugs locally demarcate the

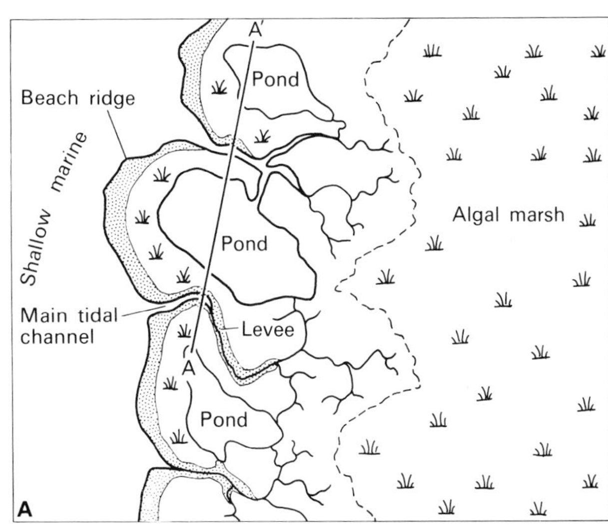

Fig. 10.9A. Major physiographic-hydrographic subdivisions at the shore zone in the lee of Andros (after Shinn *et al.*, 1969).

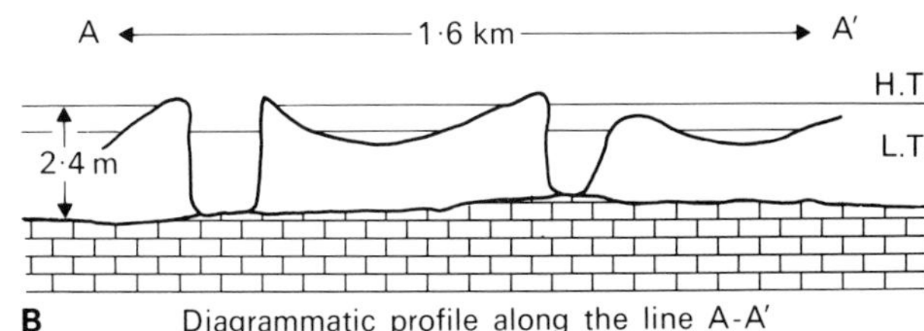

Fig. 10.9B. Schematic plan of two channel systems showing relationship of ponds, intertidal flats, and levees. Cross section A–A' shows shallow nature of ponds and relationship to normal high tide (HT) and normal low tide (LT) (after Shinn *et al.*, 1969).

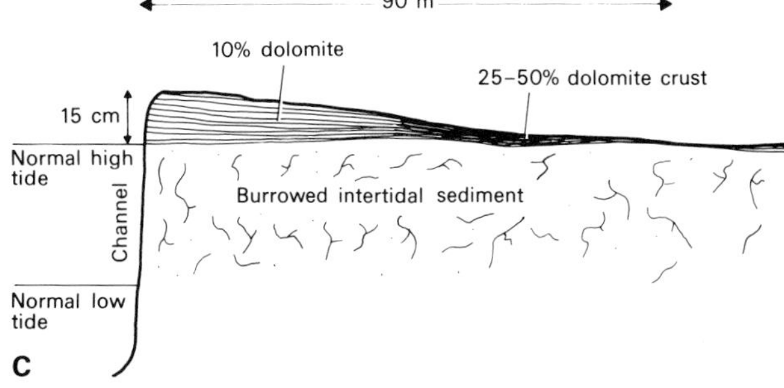

Fig. 10.9C. Schematic cross section of a tidal levee. Dolomitic sediment (generally crusts) is thinner but more concentrated on the low flanks of levees where the sedimentary surface is just cms. above normal high-tide level. Sedimentation rate is lower on the flanks, thus dolomitized sediment is less diluted by the addition of nondolomitic sediment (after Shinn *et al.*, 1969).

individual laminations which are best preserved in the thickest and highest parts on the levee (which are those closest to the channel). Because the levee crests are frequently exposed, few animals are present and the laminations have a high preservation potential.

In the Bahamas much of the interchannel area is occupied by 'ponds' of salt water. These 'ponds' are dammed by the channel levees (Figs. 10.7 and 10.9) and contain the finest sediment on the entire Banks. Very few distinct pellets are visible and the sediment has been almost wholly homogenized by bioturbation. The fauna of the 'ponds' is similar to that of the channels. These areas may become brackish or wholly fresh for weeks at a time during the wet season and then 'blooms' of fresh-water organisms may occur.

Beach-ridges up to 2 m high extend all along the tidal flat belt in the Bahamas (Figs. 10.7 and 10.9). Their steepest slopes (1°–5°) face the seaward side and they merge with the tidal flat behind. Their sediment consists of laminated pellets and fine sand-grade skeletal fragments. The skeletal material may be ripple cross-laminated and generally the grain-size of sediments decreases in the landward direction. The well-defined graded laminations are picked out by birdseye vugs and thus resemble the levee and marsh sediments. Preservation potential is high because the emergent ridges with their widely varying temperatures and salinities (day/night, and seasonal contrasts; and dry and rainy phases) produce an environment inhospitable to most burrowers (except land-crabs). Although trees (e.g. black mangrove) are anchored on the higher parts, the ridges are not generally colonized and laminae are undisturbed. Cementation tends to be early here too.

In the arid Trucial Coast, beaches exposed to the strong north-westerly winds and the 2 m spring tides are composed of coarse sand with aeolian quartz, ooids, shell debris and intraclasts. Internal bedding surfaces are parallel to the beach surface and the ridges are extensively colonized by land-crabs that produce discrete burrows.

MARINE PLATFORM

Depending upon the overall energy regime, the platform may be dominated by either lime-sand or lime-mud. If the shelf is broad and shallow or fringed by reefs and/or sand shoals then the marine platform is a restricted marine environment suffering strong seasonal and diurnal variations in temperature and/or salinity. In the low-energy marine platforms of the Bahamas the sediments are mostly pelleted muds that have been intensely bioturbated by *Callianassa* and other invertebrates. Abundant green algae (*Penicillus* and *Halimeda*) appear to be capable of supplying many of the aragonite needles in the lime-muds (Stockman *et al.*, 1967) as well as providing sand-grade particles which also come from molluscs, foraminiferans and echinoderms. Large areas are stabilized by marine grasses. On the Trucial Coast and in Florida Keys much of the marine platform is blanketed with sand-grade material and both regions support sparse patch-coral populations. However, the Persian Gulf contrasts with the Caribbean as it lacks the green algae that are so characteristic there.

In regions of higher energy, and particularly towards the outer margins of the platform, sand-belts occur with sand-waves and sand-bars of various types (Ball, 1967). It is within the general region of the sand-belts that many of the coated grains (including ooids) develop. Many are subsequently redistributed across the platform, particularly by storm processes (Fig. 10.10).

Ooids range in size from 0.3–1.0 mm and consist of concentrically laminated aragonitic layers up to 40 μ thick. They are composed of aragonite needles 2 μ long and mucilaginous organic material that is of either fungal or bacterial origin. The needles are arranged randomly in the tangential surface of the grain. Most grains have a highly polished surface that probably results from abrasion.

Following modern experimental work (Weyl, 1967) it seems that ideal conditions for ooid growth are (1) surrounding sea water supersaturated with respect to aragonite, (2) the presence of suitable nuclei, (3) agitation of grains, (4) their maintenance in a growth-promoting environment. Growth of ooids probably follows the cyclic pattern of: (a) rapid growth leading to increased stability of the coating, (b) a steady state phase as the coat either wholly or partially dissolves, (c) a resting phase with the ordering of the remaining overgrowth and (d) erosion and renewed rapid growth. The old idea of particle adhesion providing a 'snowball-like' growth probably only operates in the growth of algal oncolites. The organic material within ooids may even assist in the precipitation process as proteins have been found within the aragonite needles. Loreau and Purser (1973) have suggested that the 1–2 μ aragonite needles grow with an original haphazard or radial orientation while the ooid is at rest in relatively protected micro-environments. Tangential ordering results secondarily as the ooid interacts with other grains during transport, the aragonite needles being compacted to form a dense tangential fabric. (These arguments are developed in Bathurst, 1975, p. 303 and p. 551).

In the Trucial Coast rippled oolites lie at the seaward entrances to the tidal channels. By contrast, in the Bahamas the main growth areas are the marine sand belts that lie parallel to the major slope-break near the platform margin, in water of 3–5 m. In both cases, the sand-bodies are surfaced by ripples and mega-ripples. In the Bahamian case (Ball, 1967) spillover lobes form during storms and the largest spillovers are directed on to the platform (Fig. 10.10). Ooids form today in water of less than 5 m while hard peloids and rounded polygenetic grains of micrite form at less than 10 m (Ginsburg and James, 1974).

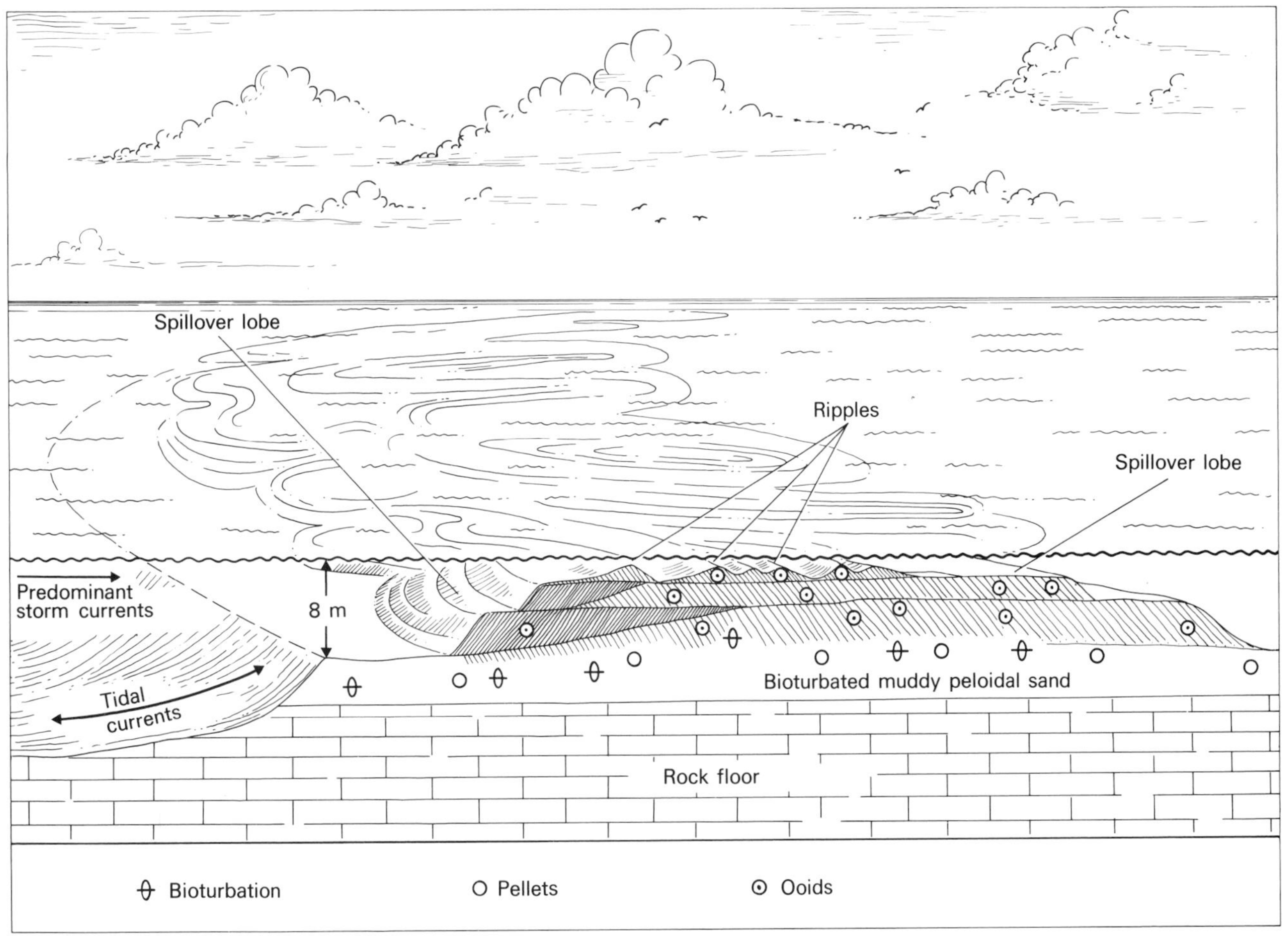

Fig. 10.10. Schematic block diagram of an oolitic sand belt – based on Cat Cay (after Ball, 1967).

Mobile sand belts usually have impoverished benthonic faunas because of the instability of the substrate. However, most of the major sand-bodies only undergo rapid migration during storms and at other times are covered by thin films of 'sub-tidal algal mat' (Bathurst, 1967). The regions of more active sand movement grade into marginal areas where mobility is less and colonization by grasses has occurred. Here, the more stable substrates support platform benthic communities, with abundant burrowers. Ooids and coated grains that have been driven on to the platform during storms consitute 'interior sand blankets' which themselves grade into the more muddy sediments of the lowest energy environments where structures have invariably been destroyed by bioturbation. Regions on the platform typified by short periods of bottom-agitation followed by longer periods of bottom-stability allow the development of cemented aggregates termed 'lumps' by Illing (1954) and grapestone by Purdy (1963a) who followed Sorby's lead. These composite grains range in size from 0.5–2.5 mm and consist of shell fragments, ooids, and pellets cemented by microspar. Algal cells living between the loose grains may induce precipitation of the early cement. In the Bahamas, the bottom topography in the grapestone facies (of Purdy, 1963b) consists of banks and mounds up to 30 cm high bound by marine grass, and cones of sediment a few centimetres high at the openings of crustacean burrows. Much of the bottom is grass covered.

Larger, plant-stabilized mounds (several tens of metres long and over 2 m elevation) may also occur in the absence of vigorous wave action. Such mounds may show well defined zonations of plants and animals (Turmel and Swanson, 1969,

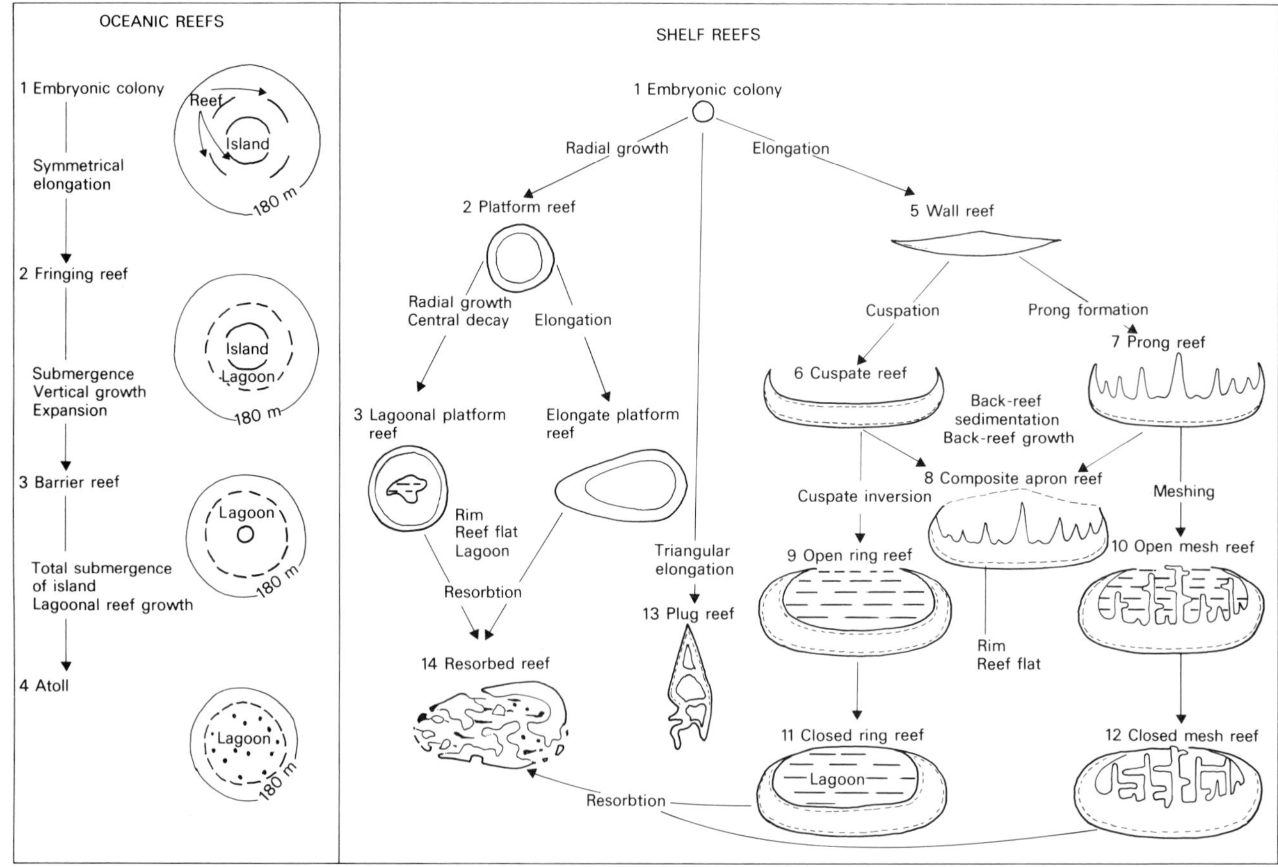

Fig. 10.11. Classification of reefs (after Maxwell, 1968).

1976). In other regions (e.g. Gulf of Batabano, Cuba) skeletal sand forms major spreads on the platform surface and particularly towards platform-margins, where higher wave and current energies prevent lime mud deposition.

REEFS AND ATOLLS

The three major types of contemporary carbonate build-up are barrier reef, fringing reef and atoll (Fig. 10.11; for fuller consideration of ancient equivalents see Sect. 10.4.2). However, shelf-reefs growing on submerged continental margins are not simple structures but comprise giant reef complexes which include fringing, barrier and atoll forms (Maxwell, 1968; Purdy, 1974a). The presence or absence of a reef-belt is controlled by a number of factors such as upwelling and steepness of slope. The distribution of reefs is often of fundamental importance in determining the energy input into the shelf and the ensuing hydraulic regime controls the distributions of shelf facies.

Barrier reefs are separated from their adjacent land areas by several kms of lagoon (tens of kms in the case of the Great Barrier Reef). At the present time, the largest is the Great Barrier Reef of Australia. This aggregation of reefs, which varies from 16–320 km in width, is nearly 2,000 km long. Unlike many other present-day reefs, it is primarily a coral reef with calcareous algae playing a secondary role (Ladd, 1971). However, Hill (1974) reports that both coralline algae and bryozoa are important frame-binders so Ladd's claim may be exaggerated. In other areas, the frames of individual reefs are composed of corals welded by red algae and in some cases calcareous algae are more important than the corals.

Fringing reefs are in continuity with the adjacent shore zone (at low tide). They consist of the coral-algal association and some may have a windward marginal algal ridge.

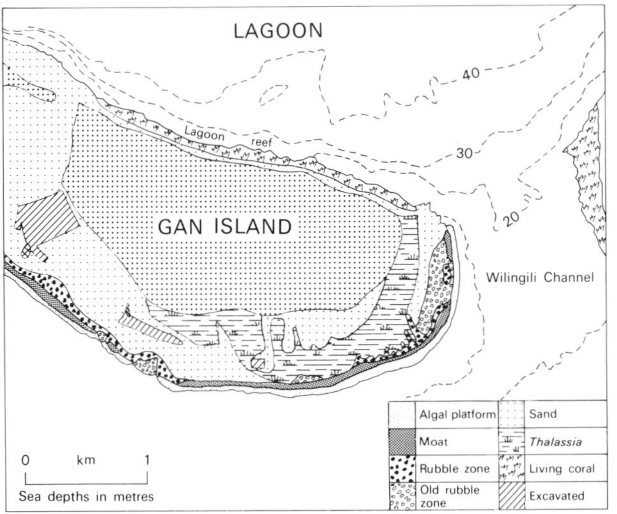

Fig. 10.12A. Gan Island, Addu Atoll, its surrounding reefs and sediment distribution (after Spencer-Davies *et al.*, 1971).

Atolls (oceanic reefs) are areas of shallow-water carbonate deposition in the open ocean. Most have been built on volcanic cores and typically form annular reefs at or near the surface of the sea. At their windward seaward margins they often have a low wave-resistant ridge composed of calcareous algae (e.g. *Lithothamnium*) and the seaward slope is often cut by a series of regularly-spaced grooves (chutes) separated by a series of spurs. The buttress-like spurs are the results of reef growth. Similar features are found in non-atoll shelf-reefs too. The maximum zone of coral growth is in water down to 20 m and many atolls show a striking zonation of facies parallel to the reef front (Figs. 10.12 and 10.13; Spencer-Davies *et al.*, 1971). Leeward edges of atolls usually lack the algal ridge while the lagoons may show a gradation from reef-derived coral-algal rubble close to the reef to foraminiferal lime muds in the more distal parts of the lagoon. Atolls in the Caribbean are considered 'pale images' of their Pacific counterparts (Milliman, 1969) both in terms of their scale and in the diversities of their biota. They also lack true algal ridges.

Fig. 10.12B. Diagram showing coral zonation on the lagoon reef, Gan. A chute formation on the reef slope is shown (after Spencer Davies *et al.*, 1971).

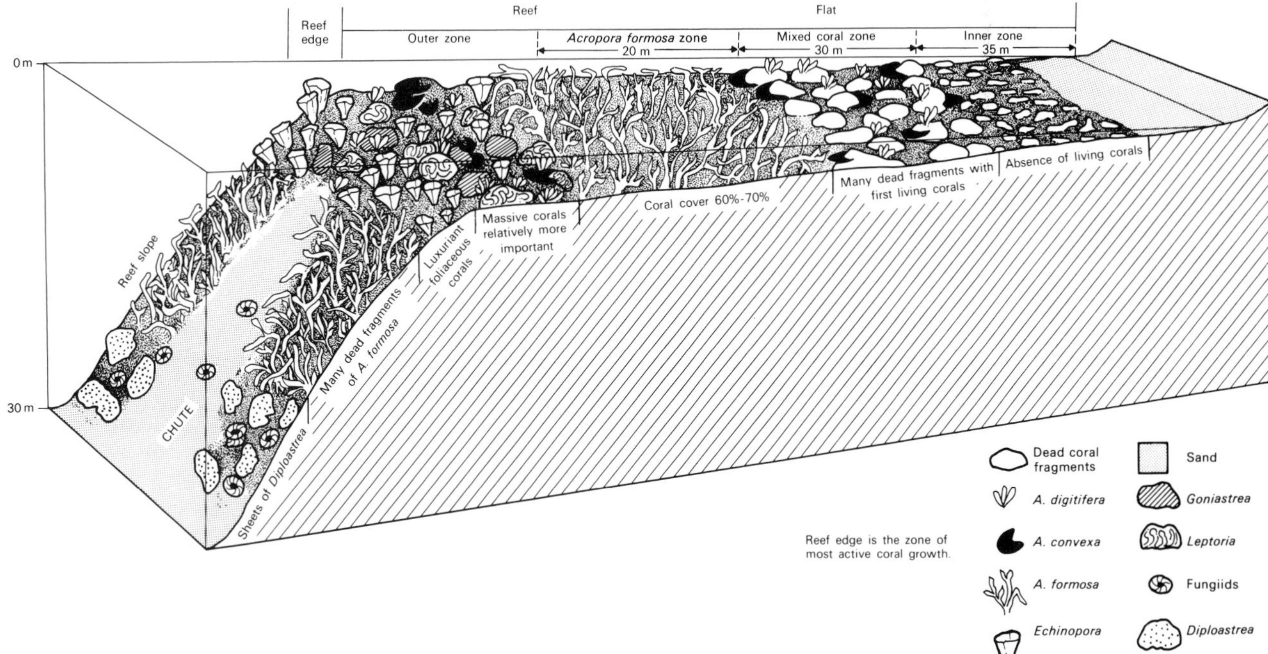

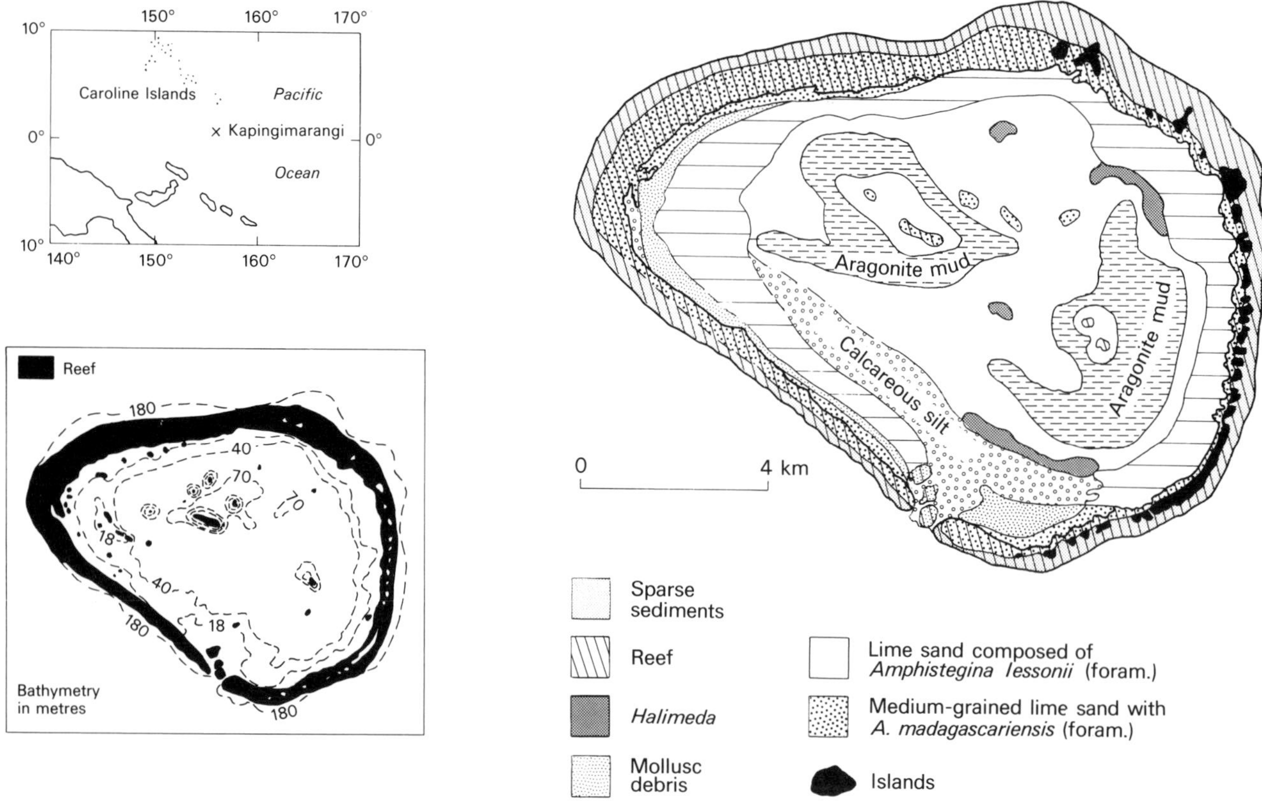

Fig. 10.13. Kapingimarangi Atoll, its location, bathymetry and sediment distribution (after McKee *et al.*, 1959). Horizontal lines indicate area with coral debris.

SHELF SLOPE

Modern carbonate shelf slopes adjacent to oceans are dominated by pelagic sediments (Ginsburg and James, 1974) with planktonic foraminiferans and nannoplankton often being the dominant components. Mass-flows lead to sediment transfer on modern slopes and calcareous mass-flow deposits derived from the shallower zones probably characterized ancient slope areas also. These features are also discussed in Chapter 12.

GENERAL FACIES SCHEME

In the Ancient, slopes can seldom be shown to have been continental slopes. Intracratonic (epeiric) carbonate facies have left the most extensive record upon the continents. Epeiric basins would have been separated from adjacent epeiric platforms by slopes of varying steepness depending on the structural controls and Wilson (1975, p. 350) (Fig. 10.14) has attempted to portray the expected facies distributions from an Ancient epeiric seaway in nine standard belts. As he says, no one Ancient or Recent example necessarily includes all nine belts and as we shall see, factors such as tectonic regime, steepness of slopes, energy levels in adjacent basins and overall climatic regime are merely a few of the variables that can affect facies development and distributions. However, conceptual models of this sort are necessary in view of the absence, at the present time, of analogues of the vast epeiric seas that lay over cratonic hinterlands during much of the Palaeozoic and Mesozoic.

Figures 10.15 A–G simplify as cartoons the facies distributions from some of the examples cited later and these models are introduced here, along with Wilson's (1975, p. 350) conceptual model, as bases for comparison and contrast.

The four pelagic facies belts of the epeiric 'slope' and epeiric 'basinal' environments are left until Chapter 11.

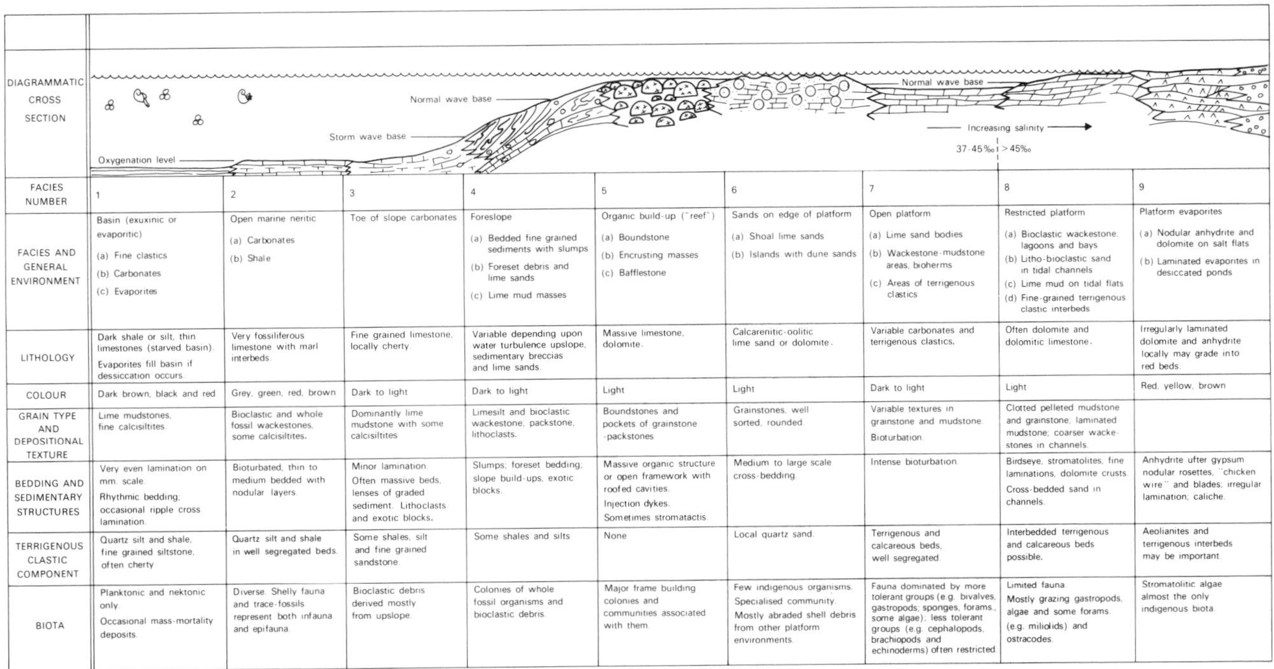

DIAGRAMMATIC CROSS SECTION									
FACIES NUMBER	1	2	3	4	5	6	7	8	9
FACIES AND GENERAL ENVIRONMENT	Basin (exuxinic or evaporitic) (a) Fine clastics (b) Carbonates (c) Evaporites	Open marine neritic (a) Carbonates (b) Shale	Toe of slope carbonates	Foreslope (a) Bedded fine grained sediments with slumps (b) Foreset debris and lime sands (c) Lime mud masses	Organic build-up ("reef") (a) Boundstone (b) Encrusting masses (c) Bafflestone	Sands on edge of platform (a) Shoal lime sands (b) Islands with dune sands	Open platform (a) Lime sand bodies (b) Wackestone-mudstone areas, bioherms (c) Areas of terrigenous clastics	Restricted platform (a) Bioclastic wackestone: lagoons and bays (b) Litho-bioclastic sand in tidal channels (c) Lime mud on tidal flats (d) Fine-grained terrigenous clastic interbeds	Platform evaporites (a) Nodular anhydrite and dolomite on salt flats (b) Laminated evaporites in desiccated ponds
LITHOLOGY	Dark shale or silt, thin limestones (starved basin). Evaporites fill basin if dessiccation occurs.	Very fossiliferous limestone with marl interbeds.	Fine grained limestone, locally cherty.	Variable depending upon water turbulence upslope, sedimentary breccias and lime sands.	Massive limestone, dolomite.	Calcarenitic-oolitic lime sand or dolomite.	Variable carbonates and terrigenous clastics.	Often dolomite and dolomitic limestone.	Irregularly laminated dolomite and anhydrite locally may grade into red beds.
COLOUR	Dark brown, black and red	Grey, green, red, brown	Dark to light	Dark to light	Light	Light	Dark to light	Light	Red, yellow, brown
GRAIN TYPE AND DEPOSITIONAL TEXTURE	Lime mudstones, fine calcisiltites.	Bioclastic and whole fossil wackestones, some calcisiltites.	Dominantly lime mudstone with some calcisiltites	Limesilt and bioclastic wackestone, packstone, lithoclasts.	Boundstones and pockets of grainstone -packstones	Grainstones, well sorted, rounded.	Variable textures in grainstone and mudstone. Bioturbation.	Clotted pelleted mudstone and grainstone; laminated mudstone; coarser wacke-stones in channels.	
BEDDING AND SEDIMENTARY STRUCTURES	Very even lamination on mm. scale. Rhythmic bedding; occasional ripple cross lamination.	Bioturbated, thin to medium bedded with nodular layers.	Minor lamination. Often massive beds, lenses of graded sediment. Lithoclasts and exotic blocks.	Slumps; foreset bedding; slope build-ups, exotic blocks.	Massive organic structure or open framework with roofed cavities. Injection dykes. Sometimes stromatactis.	Medium to large scale cross-bedding.	Intense bioturbation.	Birdseye, stromatolites, fine laminations, dolomite crusts. Cross-bedded sand in channels.	Anhydrite ufter gypsum nodular rosettes, "chicken wire" and blades; irregular lamination; caliche.
TERRIGENOUS CLASTIC COMPONENT	Quartz silt and shale, fine grained siltstone, often cherty	Quartz silt and shale in well segregated beds.	Some shales, silt and fine grained sandstone.	Some shales and silts	None	Local quartz sand.	Terrigenous and calcareous beds, well segregated.	Interbedded terrigenous and calcareous beds possible.	Aeolianites and terrigenous interbeds may be important.
BIOTA	Planktonic and nektonic only. Occasional mass-mortality deposits.	Diverse. Shelly fauna and trace-fossils represent both infauna and epifauna.	Bioclastic debris derived mostly from upslope.	Colonies of whole fossil organisms and bioclastic debris.	Major frame building colonies and communities associated with them.	Few indigenous organisms. Specialised community. Mostly abraded shell debris from other platform environments.	Fauna dominated by more tolerant groups (e.g. bivalves, gastropods; sponges, forams., some algae); less tolerant groups (e.g. cephalopods, brachiopods and echinoderms) often restricted.	Limited fauna. Mostly grazing gastropods, algae and some forams (e.g. miliolids) and ostracodes.	Stromatolitic algae almost the only indigenous biota.

Fig. 10.14. The scheme of standard facies belts proposed by Wilson (1970, 1974, 1975).

Wilson (*1975*) *recognizes:*

(1) A facies that accumulates in deep intracratonic and marginal cratonic basins at depths below the zone of oxygen depletion. It consists of interbedded organic, argillaceous and pelagic carbonate laminae. Laminations are undisturbed by bioturbation and constitute a letal-pantostrate biofacies (cf. Schäfer, 1962) with a high preservation potential of organisms entombed there.

(2) If the water is well circulated (and thus oxygenated), basinal lithofacies consists of burrowed muddy limestones and interbedded marls accumulate. Here, depths are mostly below normal wave-base but the bottom is within the reach of intermittent storms.

(3) At the toes of platform-margin slopes (Fig. 10.14), pelagic deposits are interbedded with mass-flow carbonates derived from the adjacent platform. These constitute basin margin successions.

(4) Foreslope deposits mostly consist of debris derived from the platform. Because of the slopes, bedding is often disordered by slumping. However, organic build-ups are locally present. Micrites, calcarenites and breccias are all possible lithologies in this situation.

(5) Organic buildups are of variable character depending on a number of factors including the hydraulic energy of the adjacent basin, steepness of slopes, rates of productivity and the type of community constructing the build-up complexes. They produce massive skeletal limestones.

(6) Winnowed platform-edge sands consisting of shoals, beaches, tidal bars or barrier islands form in water to depths of about 10 m but tend to contain poor benthonic communities because of their high environmental stresses (high energy and unstable bottom conditions). Ooid, peloid and skeletal calcarenites are the major rock-types.

(7) Open marine platform environments are found in open lagoons and bays (shelf lagoon) behind the outer platform edge. They form in very shallow water down to a few tens of metres. Environmental stresses are often quite high because of strong seasonality. Shallow waters restrict tidal and wave processes thus limiting circulation. Autochthonous micrites are interbedded with calcarenites derived from the sand belt, mostly as a result of storm processes.

(8) Restricted platform environments include sediments deposited in 'ponds', lagoons and tidal flats. Alternating periods of exposure and flooding by either fresh or hypersaline water may be coupled with strong seasonal or even diurnal temperature changes. All these factors combine with generally poor circulation to produce very high-stress environments and restrict colonization to relatively low-diversity organism

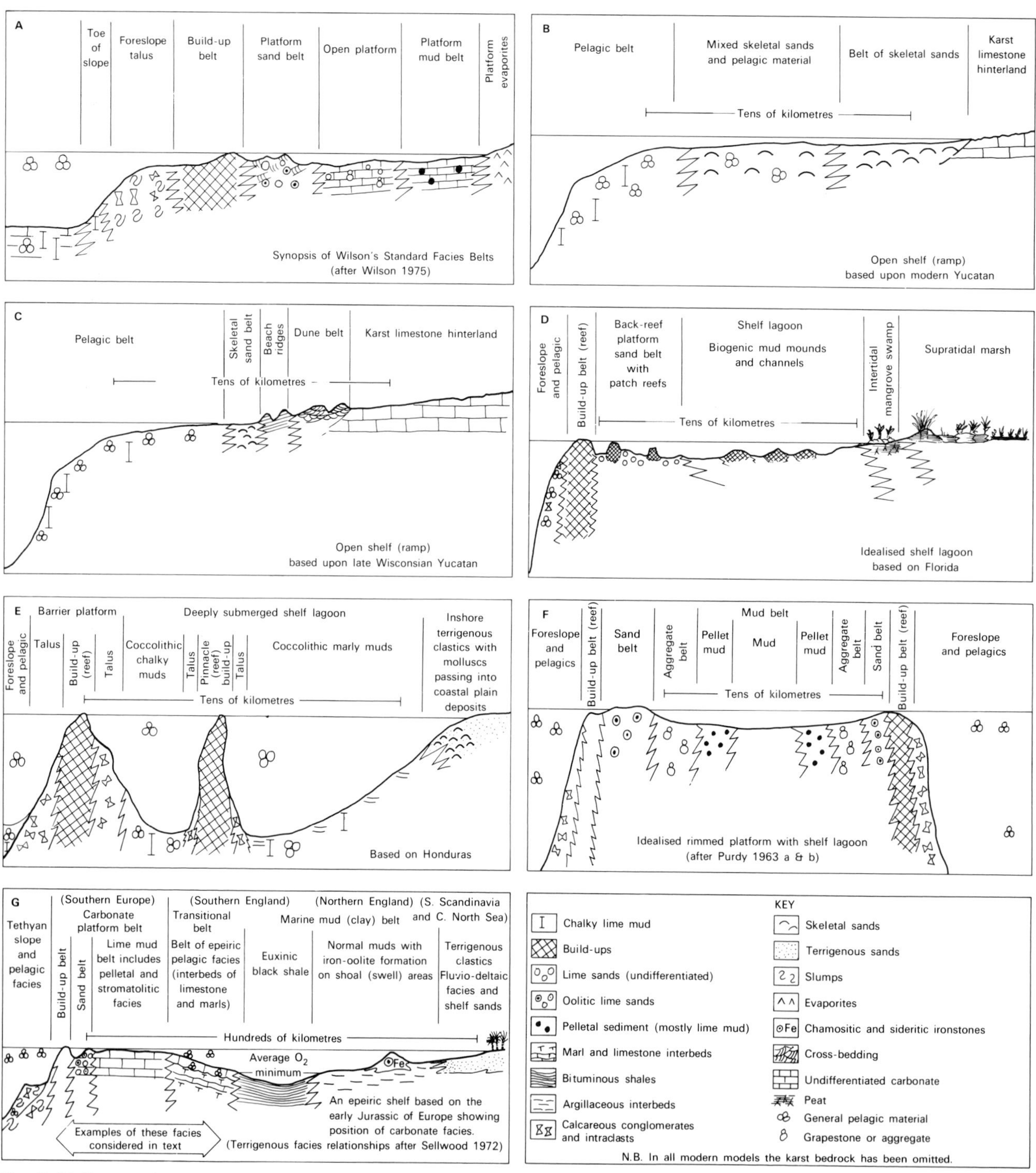

Fig. 10.15. Comparative models of modern and ancient facies belts.

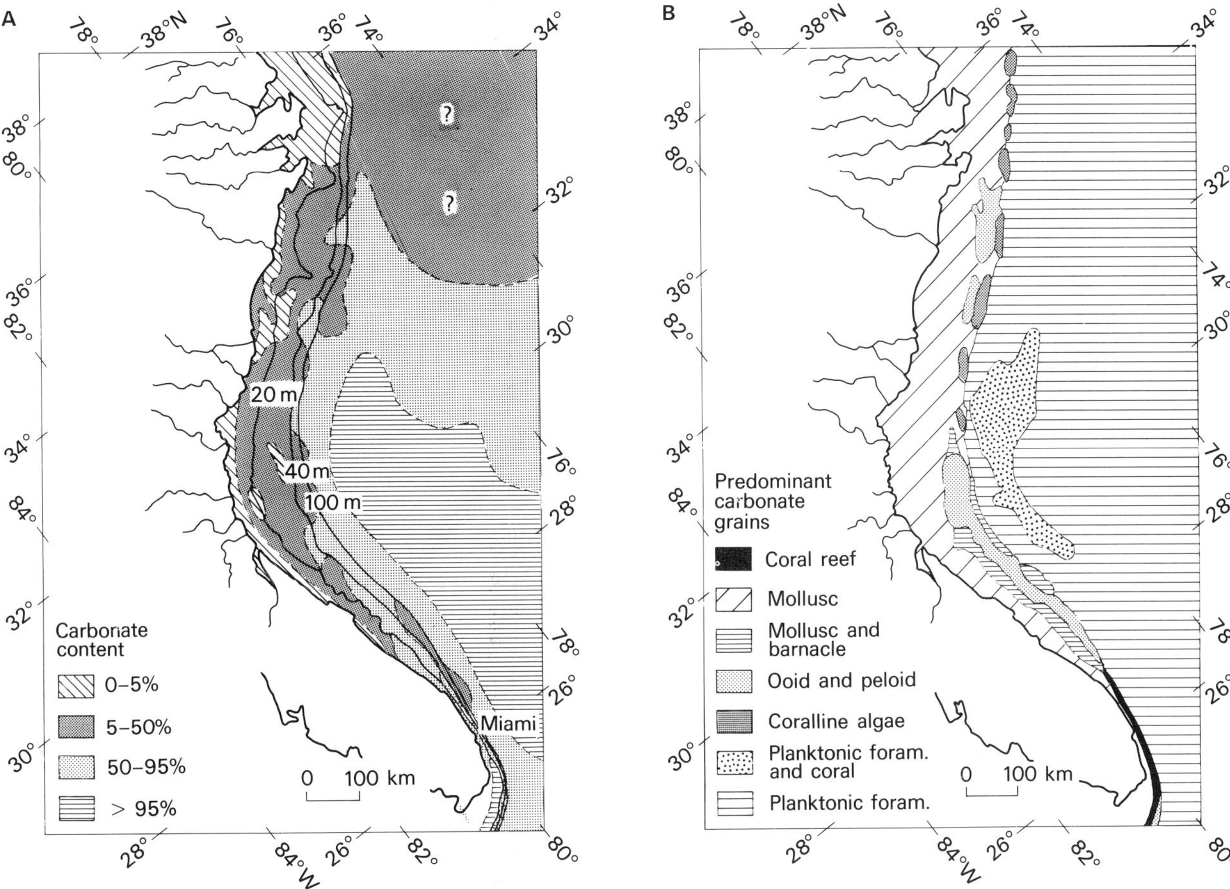

Fig. 10.16A. Bathymetry and carbonate content on the Atlantic Shelf of North America (after Milliman *et al.*, 1972; Milliman, 1972).

Fig. 10.16B. Distribution of predominant grain types on the Atlantic Shelf of North America (after Milliman *et al.*, 1972; Milliman, 1972).

communities. With exposure there is often early cementation. Coarse sediment is deposited either during storm surges or as a result of wind action.

(9) Platform evaporites form in supratidal environments in arid climates where the local water budget is one of net evaporation (see Chapter 8). High temperatures are either constant or seasonal. Flooding by marine waters is normally sporadic and results from storm-surges. Gypsum and anhydrite form both diagenetically and as primary precipitates while dolomitization of existing aragonite may often occur.

Having reviewed the more general aspects of shallow-water carbonate facies we will now consider some specific areas of carbonate deposition. We must bear in mind, however, that all modern shelves have suffered a recent, and very major, transgression. Ancient limestone successions mostly represent progradational sequences. Thus we are forced to use modern facies analogues in a predictive way. Predictions are, perhaps, justifiable if recurring facies associations can be seen in the geological record. Facies trends can be successfully predicted in ancient rocks, particularly in oil exploration, and successful predictions lend support to the models constructed from these principles.

10.2.3 Examples of open shelves

Modern open shelves that are dominated by carbonates are relatively narrow and seldom exceed 500 km from shore to shelf-break. According to Ginsburg and James (1974) they fall into two major categories:

(1) those with appreciable terrigenous clastics and with the

dominant carbonate grains occurring in distinct zones that parallel the shoreline; and
(2) those that are only slightly affected by clastics.

The western North Atlantic Shelf and the eastern Gulf of Mexico Shelf fall into category (1) while that of Yucatan falls into category (2).

CLASTIC-INFLUENCED OPEN SHELVES

The transgressive Atlantic shelf of southern N. America (Fig. 10.16)

This shelf spans the transition zone from temperate to subtropical seas. However, it is convenient to include it here as a model of mega-facies development. The sediments are mostly autochthonous and represent reworked Holocene deposits, the shelf having been affected by rapid transgression accompanied by downdrift and shoreface bypassing (Emery, 1968; Milliman *et al.*, 1972; Pilkey, 1964; Swift, 1974). There is a clear decrease in the amount of carbonate from south to north which is mostly a response to Pleistocene climatic change.

Following the Wisconsian glacial stage, rising sea-level allowed the deposition of shallow-water carbonates near the shelf-margin. These consisted of ooid sands and calcarenites composed of coralline algae. With continuing post-Wisconsian transgression, molluscan debris accumulated on the inner shelf. Benthic communities were at first represented by shallow-water species that were replaced by deeper-water types as the depth increased. Deepening of the water halted ooid formation at the shelf-margin. Subsequently, as water levels approached their present positions, reefs and patches of branched corals began to grow along the more southerly margin of the shelf and pelagic facies formed on the slope.

The eastern Gulf of Mexico

This shelf is 130 km wide from shore to the 70 m shelf-break and lies west of low-lying peninsular Florida where Tertiary and Pleistocene limestone karsts form the catchment for short streams. The slope is smooth with no relief (slope of 0.4 m/km increasing to 1.6 m/km at the shelf-break) and the zonation of major grain-types parallels the shore and shelf margin (Fig. 10.17).

Most of the sediment is sand-sized with little mud (Fig. 10.17). Irregular patches of shell-hash and coquina are scattered on the central part of the shelf and lime-mud occurs only on the outer shelf. There is a gradation from terrigenous sands on the inshore third of the shelf to carbonates on the remaining offshore portion and carbonate sand grains in these sediments occur in bands of varying width that are parallel to both the shore and the shelf-break. Molluscs provide the dominant carbonate grains in the inner shelf but at 60 m depth

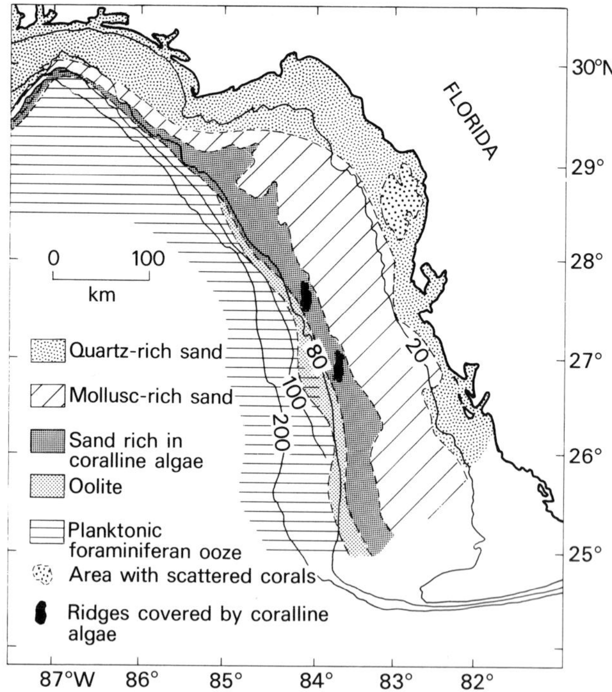

Fig. 10.17. Bathymetry and facies distribution, Eastern Gulf of Mexico Depths are in metres (after Ginsburg and James, 1974).

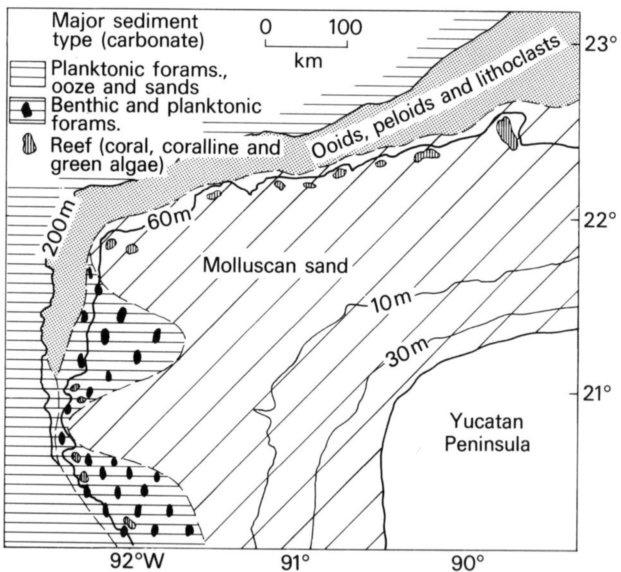

Fig. 10.18. Bathymetry and distribution of major carbonate sediment types, Yucatan Shelf, Mexico (after Logan *et al.*, 1969).

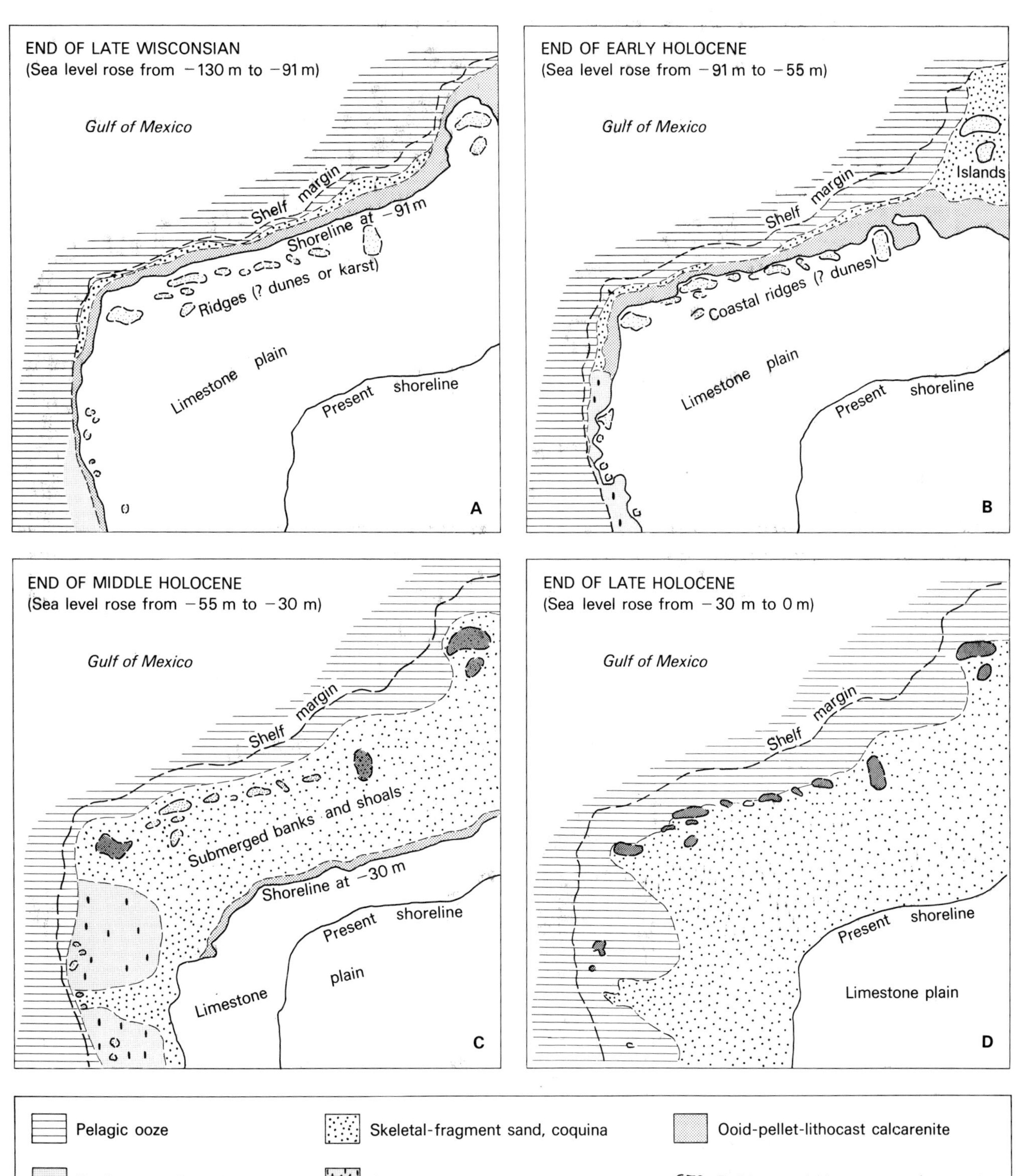

Fig. 10.19A–D. Facies development during the transgressive history of the Yucatan Shelf (after Logan *et al.*, 1969).

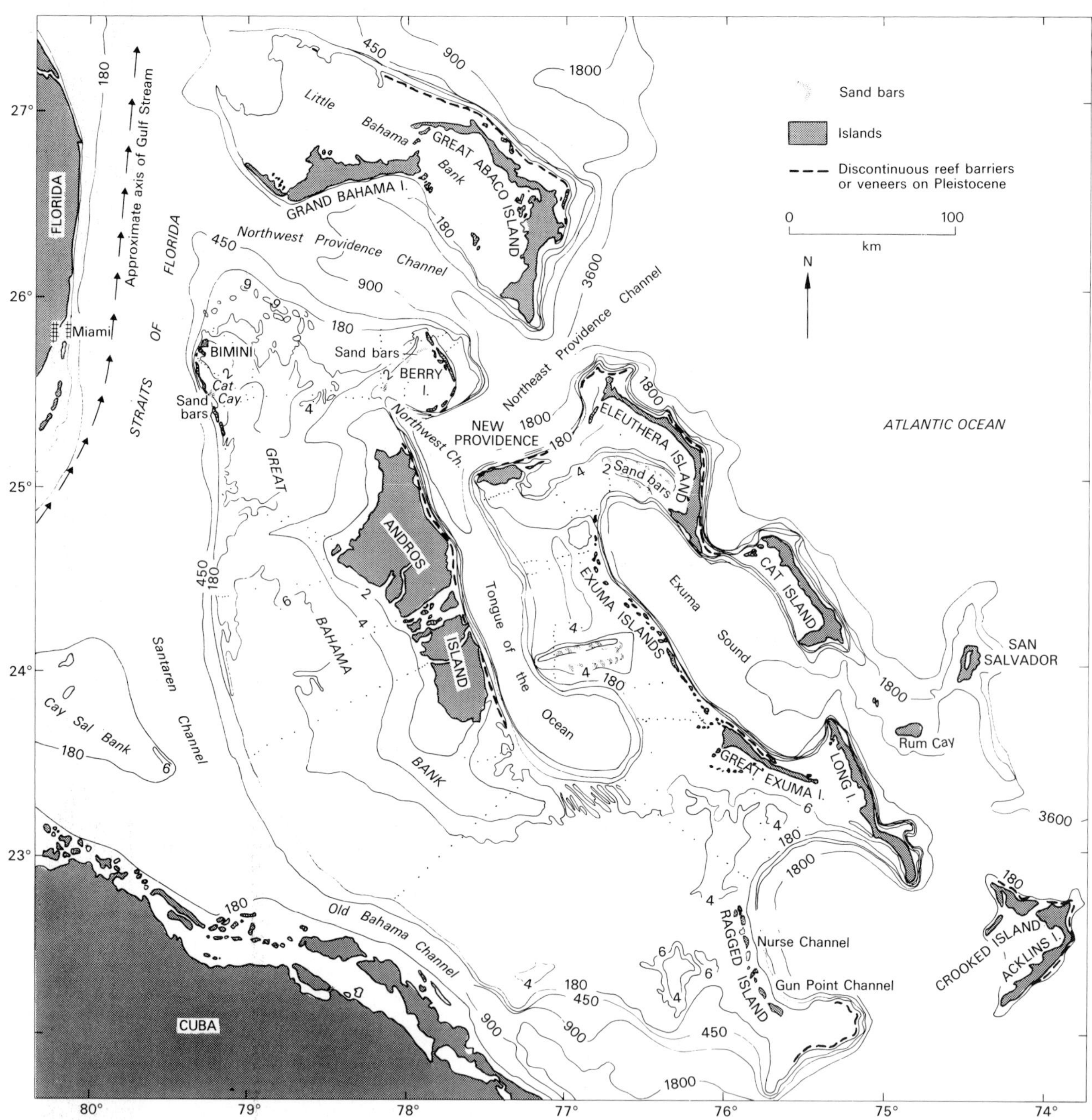

Fig. 10.20A. General bathymetry of the Bahamian–Florida area. Depths are in metres. (after many sources including Multer, 1971; Bathurst, 1971).

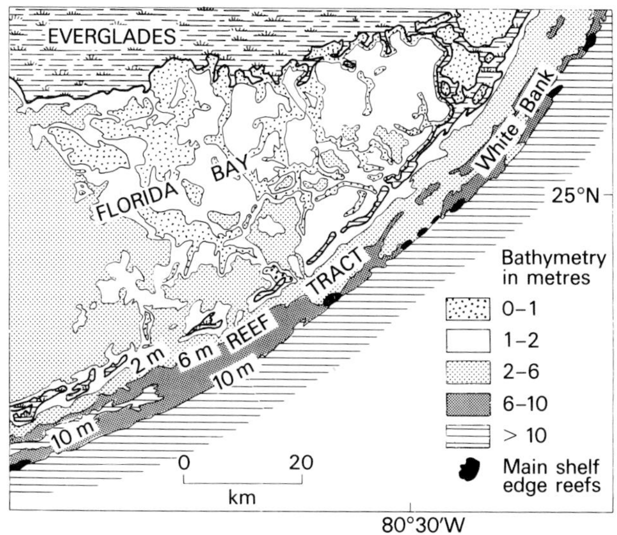

Fig. 10.20B. Bathymetry of Florida Bay (after Ginsburg, 1956).

molluscan quartzose sands are replaced by carbonate sands with coralline algae. Between 80–100 m the bottom is floored by relict (autochthonous) ooid sand containing pelagic and benthonic forams which dominate the deeper-water sequences. Landward of the shelf margin, local carbonate build-ups occur as ridges rising 15 m or so from depths of about 40 m. These are covered by branched and massive corals, bryozoans and the green alga *Halimeda*.

Holocene sediments seldom exceed 1 m in thickness and most are autochthonous (in the sense of Swift, 1974). The ooid sands now found at 80–100 m originally formed in water that was ~5 m deep in early Holocene times and are thus analogous to those on the Atlantic Shelf of southern N. America. Molluscan sands and the coralline algal sands are probably younger than both the ooid and *Halimeda* sands.

CLASTIC-FREE OPEN SHELVES

Yucatan Shelf

This region lies between 20° and 23°N in the Gulf of Mexico (Fig. 10.18). The Yucatan peninsula consists of a low, undulating karstic plain that has no surface drainage. Soil-filled depressions support dense rain-forest (Logan *et al.*, 1969). The shelf area, extending over some 34,000 km^2, has a thin veneer of modern deposits overlying mainly relict sediments.

The shelf platform is crossed by three terraces that were probably cut during the Holocene rise of sea-level (Logan *et al.*, 1969). The platform's width varies from 160 km to 290 km from the shore to the sharp shelf-break at depths ranging from 82–315 m. There is no guarding rim of reef barriers although pinnacle reefs and knolls rise from the 60 m isobath and divide the Inner Shelf (0–60 m) from the Outer Shelf (60–210 m) of Logan *et al.* (1969). The shelf is thus open to the influence of ocean waves and the Yucatan Current.

Tidal ranges are low: normal spring ranges are about 30 cm. Winds from the SE, E and S are dominant but the fetch is limited. Those from the N and NE provide higher energies on the shelf because their fetch is larger and bottom drag slight. Hurricanes affect the area but as yet their effects are largely unknown. Annual water temperatures range from 24°–30°C, but cold water (17°–18°C) upwelling off the eastern side possibly explains the absence of coral reefs there.

Playa lagoons and swampy mudflats with sedges and mangroves are present on the seaward margin of the karstic hinterland and the shore is marked by a series of coastal dune-ridges parallel to the wave-dominated shell beach. The sediments veneering the inner shelf are mollusc-rich skeletal sands and coquinas. The outer shelf beyond the reef pinnacles (Fig. 10.18) is carpeted by relict ooids, peloids and intraclasts from the underlying limestone. From 90 m down to the shelf-break these sands are increasingly diluted by the 'rain' of modern planktonic pelagic foraminiferan tests, so that beyond the shelf-break the pelagic facies is predominant.

During a late Wisconsian (Devensian) low stand of sea-level, when the shoreline lay at about −130 m (Fig. 10.19A), a series of coastal ridges (?dunes) was constructed while pelagic sediments accumulated just offshore (Fig. 10.19A). With rising sea-levels an onlapping sequence developed. During the low sea-level stands, the karst margin plain extended over the area which is now the inner shelf, with a solution rim and conical karst developing close to the present shelf margin (Fig. 10.19A–D, cf. Purdy, 1974a).

With the subsequent rise in sea-level the shoreline advanced and biostromes developed on these solutional karst remnants. Although ooid production no longer occurs in the area, during the late Wisconsian and early Holocene ooid-pellet calcarenites were generated to the seaward of the conical karst rim. After the Flandrian transgression, a static sea-level resulted in the modern facies pattern (Fig. 10.18) with pelagic deposition beyond the 50 m isobath and some winnowing of sediment down to depths of about 110 m. This situation has been complicated by the biogenic mixing of the contemporary sediments with older, relict sands.

On the Inner Shelf the skeletal-fragment sand and coquina is mostly generated from the living mollusc communities on the sea-floor and wave action seems to be the dominant agent of sediment dispersal, but with some additional influence from oceanic currents. At or near to the shelf-break, vigorous reefs and banks grow on karstic pedestals.

10.2.4 Examples of rimmed shelves

BAHAMAS/FLORIDA

The Bahama Platform and the South Florida Shelf are the vestiges of a once continuous platform which, in late Mesozoic times, incorporated most of peninsular Florida and extended northwards as far as the now subsided Blake Plateau. Although Florida is underlain by continental basement rocks, controversy still exists as to the nature of the Bahamian basement. Most workers agree that the Bahama province is not part of the Atlantic shelf of North America; it is thought by some to be underlain by (?Jurassic) oceanic crust (e.g. Dietz *et al.*, 1973; Sheridan, 1974). More detailed environmental reviews are given in Shinn *et al.* (1969), Bathurst (1971) and Multer (1971).

BAHAMA PLATFORM

The Bahama Platform lies between about 22°N and 28°N and comprises a group of barely submerged carbonate plateaus with intervening deeply submerged embayments (e.g. Tongue of the Ocean). The Platform lies off the Atlantic coast of Florida and enjoys a humid subtropical climate. The area of shallow marine carbonate deposition encompasses some 96,000 km^2 and is bounded on all sides by steep marginal slopes (often exceeding 40°) that terminate in water depths of hundreds or even thousands of metres (Fig. 10.20A). On the Great Bahama Bank mean monthly surface water temperatures range from 22° to 31°C. In most places, turbulence is generally great enough to ensure a relatively uniform vertical distribution of temperature. Water depths over the Platform seldom exceed 15 m and are normally less than 6 m. Dominant wind directions are from the east, and eastern margins of the Platform are rimmed by islands of Pleistocene limestone composed of cemented aeolian dunes. Where islands form a continuous barrier circulation is restricted, salinities are raised and lime mud accumulates. However, where marginal barriers are discontinuous, current velocities are higher and mud-free lime sands are formed.

Water circulation over the Great Bahama Bank is governed by a combination of tidal and wave-action. Tides are semi-diurnal but strongly influenced by both wind direction and velocity. The mean spring range is 41 cm with an extreme range of 95 cm (Ginsburg and Hardie, 1975). However, water levels may be raised locally by more than 3 m at the Bank margin during a hurricane.

Tidal water moves on and off the Banks with a radial pattern, the flood currents being slightly stronger than those of the ebb. Near the open sea tidal currents of 25 cm/sec are common but in the channels velocities in excess of 1 m/sec have been recorded and in general there is a decrease in tidal influence inward from the Bank margin so that mid-Bank ranges are negligible.

The Bahama Bank lies in the trade wind belt and from March until August the prevailing wind is from the east or southeast. In winter the winds are from the exactly opposite quadrant. Andros Island provides an effective barrier to free water circulation on the Bank, where tidal action is also damped, and evaporation allows hypersaline water to accumulate in the lee of Andros (Figs. 10.21 and 10.22). However, the high rainfall (100–150 cm) prevents salinities in excess of 38‰ from being developed over most of the Bank and evaporite minerals are not preserved. On the other hand, the persistence of the established salinity pattern is promoted by: (a) the small tidal exchange, (b) the shelter provided by Andros and (c) the opposing directions of the summer and winter winds. All of these factors combine to prevent large-scale water movement, but excess Bank water probably escapes south-eastwards.

Only about 3.5 m of unconsolidated Recent (post-Flandrian) sediment veneers the Pleistocene bedrock. Five main facies are recognized (Purdy, 1963a and b) which, under ideal circumstances, should perhaps have formed five concentric bands parallel to the Bank's margin (Purdy, 1963b). The 'patchlike' pattern has been controlled, to a large extent, by the underlying karst topography.

The facies and their distributions (Fig. 10.21C) have been mapped by Purdy (1963a and b) and the communities (Fig. 10.21D) by Newell *et al.* (1959) and Coogan (in Multer, 1971). The communities are controlled primarily by both substrate and turbulence and there is a strong correlation between community and facies (Fig. 10.21D).

We have already considered some of the factors controlling the composition and distribution of the facies (Sect. 10.2.2), but the community/facies relationships deserve closer attention and are summarized in Table 10.1.

Dolomitic crusts containing between 80% and 10% of dolomite form in the supratidal zone. Individual dolomite crystals range in size up to 2 μ (Shinn *et al.*, 1969) and their small size makes their origin difficult to interpret. They probably result from a combination of primary precipitation from the interstitial sea water and from replacement of original aragonite. Some of the dolomite may, however, be detrital. The highest percentages of dolomite and the hardest dolomitic crusts occur only a few cms above the normal high-tide levels (Shinn *et al.*, 1969). Relatively pure, very high calcium dolomite is not observed. If high concentrations of dolomite represent the most complete stage of replacement, then the relationship suggests that dolomite quality improves as replacement proceeds. The high-concentration dolomites may, however, represent a residuum from the natural leaching of calcite, aragonite and very high-calcium dolomite and the

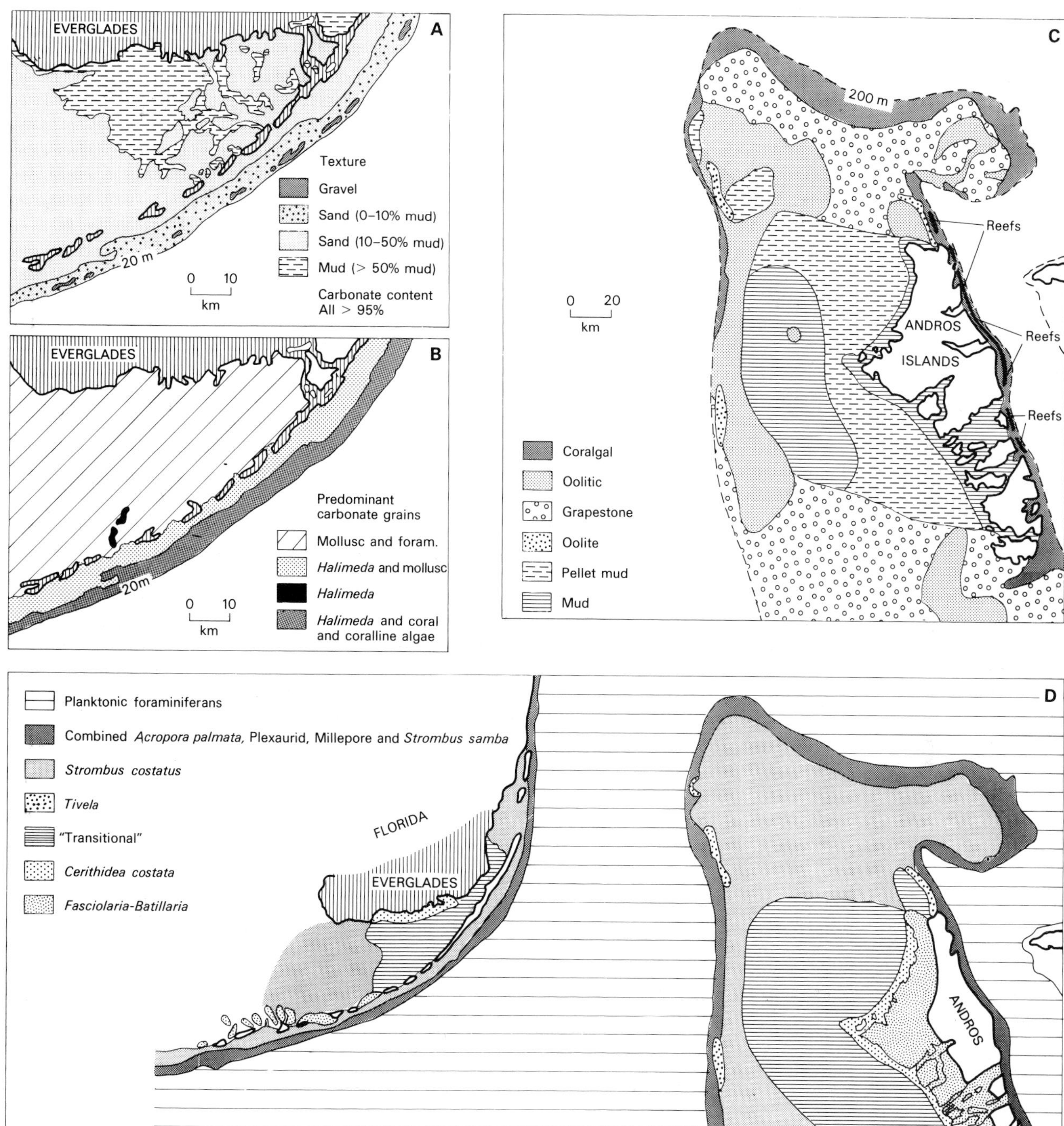

Fig. 10.21A. Sediment texture on the South Florida Shelf (after Swinchatt, 1965; Ginsburg and James, 1974).

Fig. 10.21B. Predominant carbonate grains upon the South Florida Shelf (after Ginsburg, 1956).

Fig. 10.21C. Sediment distribution upon Great Bahama Banks (after Purdy, 1963A & B).

Fig. 10.21D. Biofacies of the Florida and Great Bahama areas (after Coogan *in* Multer, 1971).

Table 10.1 Lithofacies, habitats and communities of the Great Bahama Bank and (where appropriate) of Florida. Compiled from Bathurst (1971) and Coogan (in Multer, 1971)

LITHOFACIES	HABITAT	COMMUNITY (BIOFACIES)
Coralgal	Reef (outer reef of shelf margin) High diversity community including about 30 coral species; coral frames bound by coralline algae. Niches in frame colonized by molluscs, echinoids, foraminiferans, hydrocorallines, annelids, alcyonarians and fish, Spur and groove development. Optimum growth conditions extend from 1 m to 50 m depth. Comprises the windward reefs in both Florida and the Bahamas.	*Aeropora palmata*
	Rock pavement Occurs in back-reef areas, local patch reefs attached to rock bottom covered by blanket of ephemeral lime-sand. Corals (e.g. *Montastrea* and *Diploria*) common, with gorgonaceans and plexaurid sea-whips. Biota dominated by strongly cemented and encrusting species. Also found in Florida immediately behind the reef flat.	Plexaurid
	Inshore rocky shoreline Areally restricted and strongly zoned. Dominated by cemented or closely attached biota and includes green algae, coralline algae, sponges, barnacles, chitons, gastropods, bivalves and echinoids. Rather variable depending on tidal range and degree of exposure (a chloralgol association).	*Littorina
	Rock ledges and prominences Subtidal rock ledges along exposed shorelines. Transitional with Plexaurid community. Many molluscs from the reef and rocky shoreline are found here. Corals include rock pavement species plus the hydrozoan *Millepora*. Mostly attached and encrusting species here.	*Millepora*
	Subtidal unstable sand Found on the outer platform of the Bank margin and in the immediately back-reef area of the Bahamas and Florida. Lime-sand only partially stabilized by marine grasses. Bottom is rippled and there is much sediment movement, providing a high-stress habitat. The conch *Strombus*, burrowing bivalves and sand-dollars are the typical fauna.	*Strombus samba*
Oolitic and Grapestone	Vegetation-stabilized sand Is the most widespread habitat and contains the most diverse biota. The *community* develops in the sheltered waters of the back-reef and open lagoonal areas adjacent to the Bank edge in both the Bahamas and Florida. In the Bahamas ooids are not forming in this lithofacies. However there is about 89% of non-skeletal sand grains in the Bahamian oolitic facies and about 83% in the grapestone facies. These grains are composed of faecal pellets, mud aggregates, grapestone, cryptocrystalline grains and ooids. Ooids account for 67% of the grains in the oolitic facies but only comprise about 15% in the grapestone facies. Green and red algae are common, molluscs are abundant (especially burrowers). Stabilization is either by algae or grass. Ooids are not forming in Florida, but the community develops under physical conditions that are comparable to those in the Bahamas.	*Strombus costatus*
Oolite	Intertidal, bank-edge unstable oolite Contemporary oolite is forming in the Bahamas and the facies contains 90% of ooids. These sand shoals provide extremely mobile and grass-free habitats that are localized to actively growing intertidal oolite bars. The community is not found in Florida. Almost devoid of biota apart from the active burrowing clam *Tivela*.	*Tivela abaconis*
Mud and Pellet Mud	Muddy-sand with normal to hypersalinities Found in areas away from the shelf edge and transitional to the muddy shorelines and tidal flats. The community is transitional between the diverse *S. costatus* and that of the euryhaline mangrove association. It is found in the sluggish hypersaline water in the lee of Andros (Fig. 10.21D). *Didemnum* is a tunicate; other members of the biota include green algae, grasses, the bryozoan *Schizoporella* and one coral (*Manicina*) plus a few echinoids and the mollusc *Pitar*. It is difficult to reconcile the Floridan and Bahamian faunal lists and the community (biofacies) has been termed '*transitional*'.	*Didemnum*
	Subtidal variable salinity, muddy bottom Occurs nearshore with a low-diversity salinity-tolerant biota. Only two molluscs – *Ceritidea* and *Pseudocyrena* – are present accompanied by the non-calcareous alga *Batophora* and miliolid and peneroplid foraminiferans. Found in Bahamian areas receiving rainwater run-off, and in Florida close to the Everglades.	*Cerithidea costata*
	Intertidal and supratidal mangrove association Muddy intertidal shorelines and supratidal flats are colonized by the red and black mangrove (*Rhizophora mangle* and *Avicenna nitida*). Sheltered marshes, mud flats and lagoonal shores of western Andros and mud islands in Florida support this community. The sediments are stromatolitic and rich in the grazing gastropods *Fasciolaria* and *Batillaria*.	*Fasciolaria–Batillaria*

* Indicates communities too restricted to be shown on Fig. 10.21D.

pitted texture of many exposed crusts indicates that some leaching has occurred (Shinn *et al.*, 1969).

SOUTH FLORIDA SHELF

The southern end of the Florida peninsula is surrounded by a shelf area stretching in a broad arc southwards and westwards for some 360 km beyond Miami. About 10 km shorewards of the break, the shelf is fringed by an arcuate chain of low islands (the Florida Keys) which are composed of Pleistocene limestones. The Keys attain heights of about 3.5 m and separate the very shallow area of Florida Bay from the more exposed reef tract (Fig. 10.20B). From the reef tract, the sea-floor slopes away gently seaward to the Pourtales Terrace (200–300 m) which is itself bounded to seaward by a precipitous slope that plunges into Florida Strait (800–1,000 m).

The reef tract exhibits a complex of actively growing coral reefs, sand shoals, grass-covered banks and elongate depressions. Hurricanes may pound the reef tract with 3–5 m waves that cause wholesale disruption and redistribution of coral heads and skeletal debris. In the Upper Florida Keys (towards Miami) there are two discontinuous but parallel ridges: an outer ridge of shelf-margin reefs, coral rubble and skeletal sand: and an inner ridge of skeletal sand with scattered patch reefs on its seaward side. Between the Keys and the 9 m isobath the temperature ranges between 15° and 35°C while salinities vary between 32‰ and 38‰.

The reefs, like those of the Bahamas, have coral frames bound by red algae. Their seaward slopes are marked by spur and groove topography (Shinn, 1963). Coral fragments accompanied by algal debris dominate the sandy sediment that veneers the Pleistocene bedrock between individual reefs.

In the back-reef lagoon, immediately behind the reef flat, *Halimeda* and mollusc fragments are most conspicuous and in this relatively sheltered environment a rock pavement community is developed (cf. Table 10.1). Most of the back-reef lagoon is divisible into: (1) a seaward zone consisting of a limestone pavement, either bare or with patch-reefs and a veneer of lime-sand and (2) a landward zone where the grass *Thalassia* forms an almost continuous carpet. In the seaward zone there is a modified Plexaurid-*Millepora* community; in the landward zone, a diverse assemblage of molluscs, green algae and corallines is stabilized by vegetation (cf. Table 10.1).

Plant-stabilized mud mounds (e.g. Rodriguez Bank) up to 5.5 m high developed over originally mud-filled hollows whose surfaces were stabilized by *Thalassia*. They accreted as sediment was baffled by the plant cover and now form mangrove-capped areas rising to 30 cm above mean low water. Communities are strikingly zoned around these mounds with *Thalassia* and codiacians dominating the inner zone which is succeeded outwards by narrow zones with *Porites* and

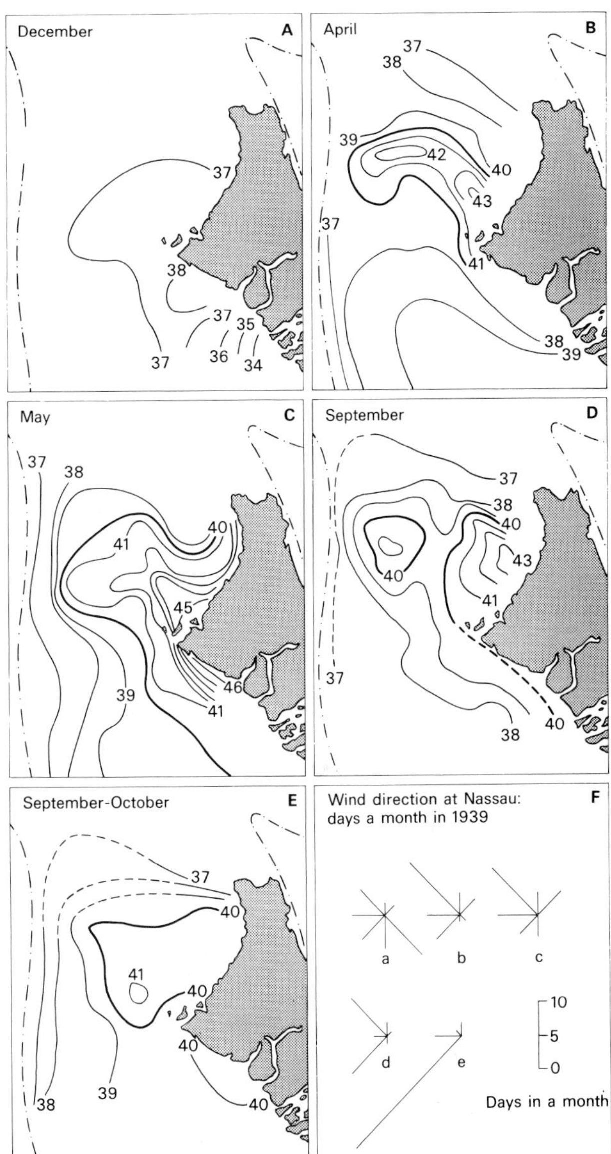

Fig. 10.22. Isohalines (parts per mille) and wind directions for the Great Bahama Bank (from Bathurst, 1971).

Goniolithon.

Florida Bay is a broad and very shallow area extending from the coastal mangrove swamps at the margins of the Everglades to the Keys (Figs. 10.20B, and 10.21). There is a large fresh-water input from the Everglades and salinities in the Bay can range from 10–55‰ with local peak values of 70‰ having been recorded. Water temperatures can range seasonally from

15–40°C although the deeper channels may only range from 20°C to 30°C. It is, therefore, a high-stress environment with a restricted biota.

The northern part of the Bay is protected by a continuous island barrier. The Bay itself contains a network of grass-covered mud-banks that may rise 2–3 m above adjacent hollows to within 30 cm of the water surface. Tidal damping by friction with the shallow bottom totally inhibits tidal exchange in the north.

In the southern part of the Bay, the island barrier is discontinuous allowing tidal exchange which provides stronger currents. Consequently, winnowing of the sediment leaves some parts of the Bay veneered by only a thin (30 cm) layer of muddy skeletal sand.

In the Bay, the habitat mostly comprises the 'muddy, normal to hypersaline transitional' type (a modified *Didemnum* community of Table 10.1). Inshore, however, where strongly fluctuating salinities occur, the *Cerithidea costata* community (of Table 10.1) is developed on muddy bottoms and is itself transitional with the intertidal and supratidal mangrove association (*Fasciolaria-Batillaria* community of Table 10.1).

Supratidal dolomite comparable with that from the Bahamas is also developed in the Florida Keys (Shinn, 1968b). Selective dolomitization of the more permeable sediment results from the migration of magnesium-enriched brines which are formed by evaporation.

THE GREAT BARRIER REEF, AUSTRALIA

The Great Barrier Reef of Australia represents the largest aggregation of reefs in the world. The Reef runs for some 1,900 km along the continental shelf of Queensland from about 9°S to 24°S (Hill, 1974) and comprises some 2,500 separate reefs, most of which cover no more than a few square km.

The Queensland Shelf is fringed by the complex of shelf edge reefs which guard a deep interior basin with relatively open circulation in which a widespread zone of terrigenous clastic sediments is developed. The shelf is narrowest at its north end (Fig. 10.23), widening and deepening southwards. In the north, the shelf edge is abrupt and steep while in the south both the landward and seaward parts of the shelf are separated by an embayment and the outer shelf edge slopes gently down into the Coral Sea Platform.

The tropical climate is humid with little evaporation (Hill, 1974). Three major rivers discharge between 20° and 25°S but the northern 1,000 km is entirely free of major stream discharge. Salinities over the reef average 34.7‰ with maximum values of about 35.5‰. The reefs lie in the surface water layer which extends down to about 100 m and water temperatures range seasonally from 21–29°C. There is little vertical gradient in the surface water layer because of the intense turbulence caused by tidal and trade-wind induced currents. Locally, the tidal range is about 10 m and in most places there are strong tidal currents, particularly towards the shelf margin.

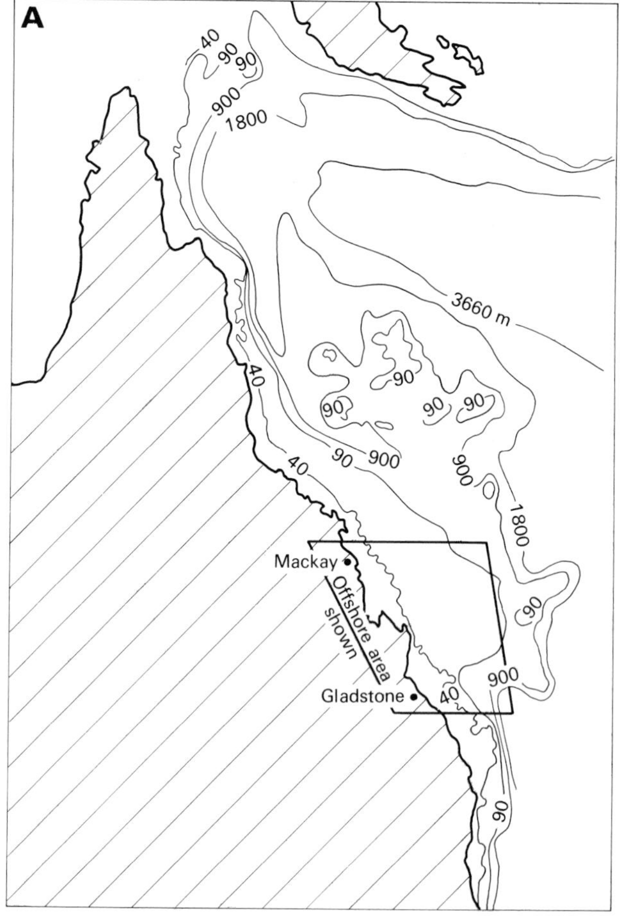

Fig. 10.23. The Great Barrier Reef, E. Australia.
Fig. 10.23A. Location of the Great Barrier Reef Complex.

The reefs consist mostly of calcareous skeletal detritus and skeletons. Corals are the dominant organisms with hydrocorallines, coralline algae, molluscs, foraminiferans, echinoderms and bryozoans accounting for most of the rest. Corals and hydrocorallines form the building-bricks of the reef which are then bound together by coralline algae and bryozoans (Hill, 1974). Other organisms help to fill in the crevices between the framework. Organism diversities are high. There are, for example, 350 species of coral and 146 species of echinoderms.

Reef shapes depend upon the shape of the substrate upon which they were initiated and upon the direction and force of prevailing waves and currents. In the north, linear reefs are

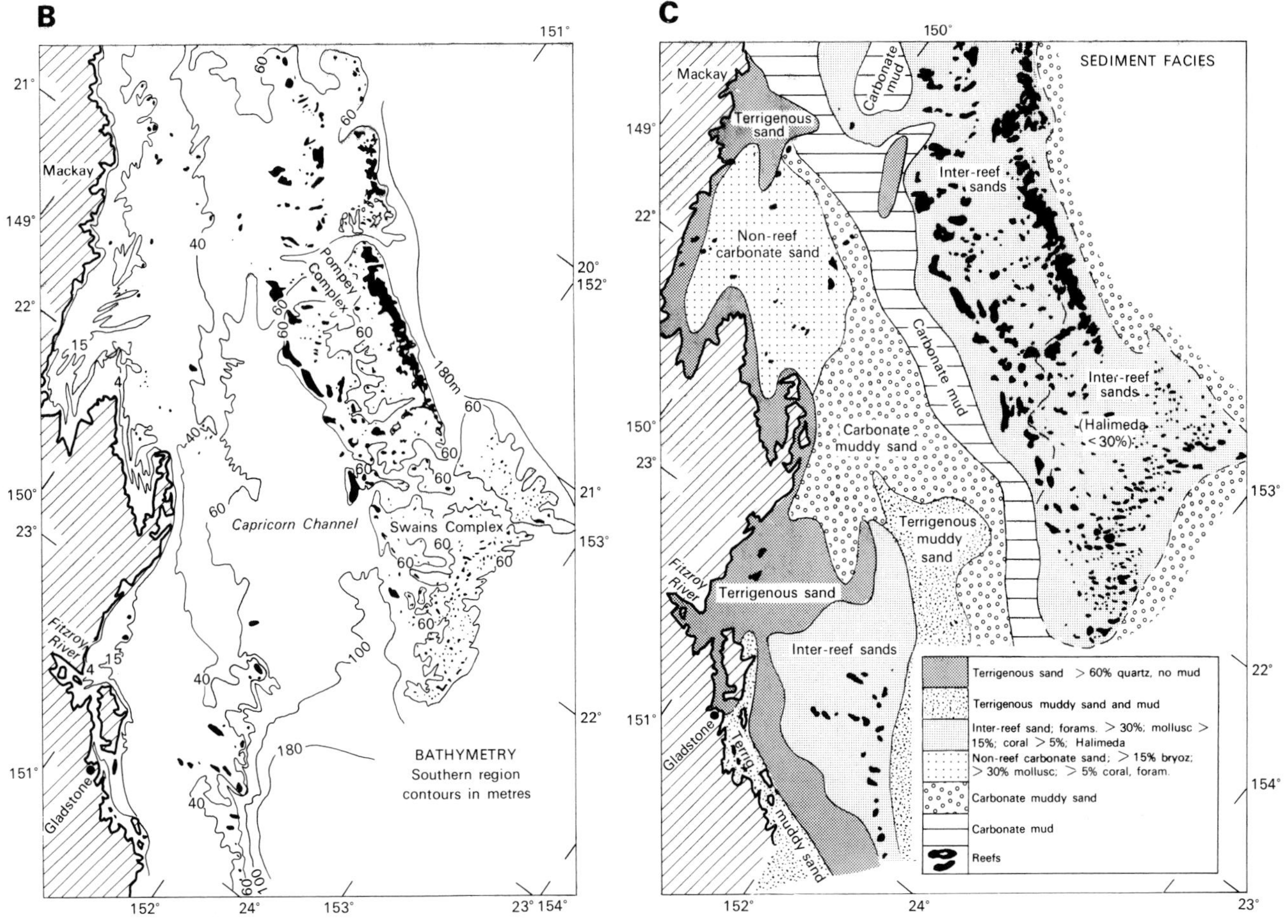

Fig. 10.23B. Bathymetry of the Southern Region around Swains Complex (after Maxwell, 1968; Maxwell and Swinchatt, 1970).

Fig. 10.23C. Sedimentary facies of the Southern Region around Swains Complex (after Maxwell, 1968; Maxwell and Swinchatt, 1970).

found that become less abundant southwards. The central parts of the shelf have fewer marginal reefs and this change in style coincides with the change in character of the shelf edge from precipitous to gentle. The southern sector has numerous elongate and ring reefs (see Fig. 10.11). Zonation is a notable feature of reefs. The outer reef platform is relatively barren and consists of rippled sands with relatively few corals. The slope from the platform to the growing reef rises at about 30°. The top 2 m of this slope is marked by the reef front which is the zone of most active coral growth and where spur and groove develops. Behind the reef front is the algal rim, a zone of up to 500 m width that is exposed at low tide, and dominated by red algae (e.g. *Lithothamnium*). Where the reef is swept clear of shingle and boulders a reef flat develops behind the algal rim. This zone is also exposed at low tide and consists of three main sub-zones: an area of living corals and coral pools, a dead coral zone and a sand flat.

The surface sediments present a fairly uniform pattern of nearshore terrigenous sands and muds, a central zone of mixed terrigenous and carbonate muddy sands and an outer fringe of reefs. In the south, where the shelf is broadest, terrigenous material rarely extends more than half-way across. However, in the north even the inter-reef sediments at the shelf edge may contain up to 40% of terrigenous material. Maxwell (1968) considers that much of the sand and gravel of the inner shelf is relict and that muds are the only terrigenously derived materials to be currently accumulating.

Lime-mud, oolites and cemented carbonate grains are unknown from the area. Reef growth of the type currently developing probably commenced in the Pleistocene and Purdy

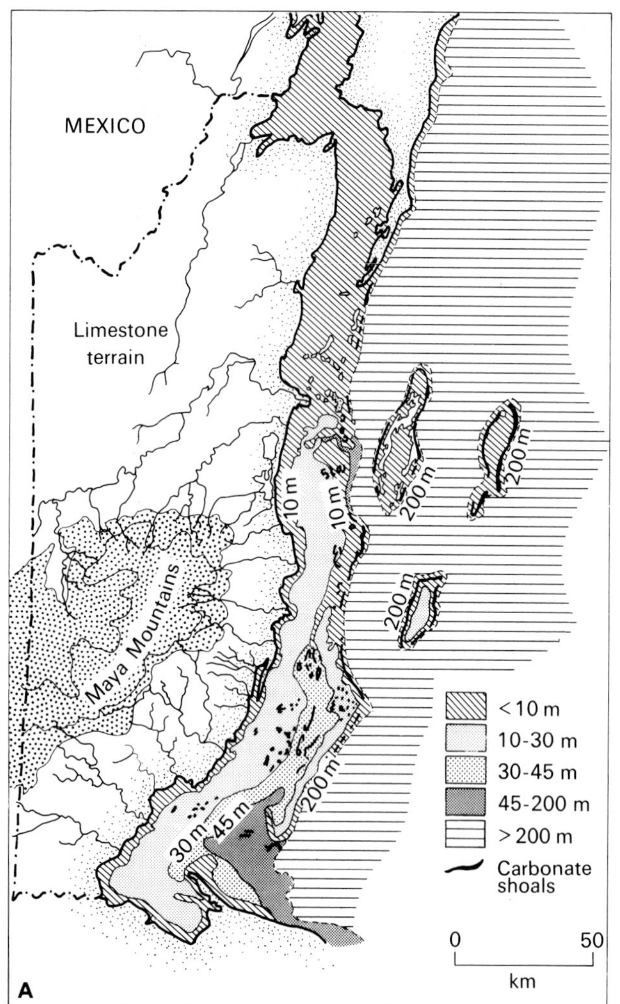

Fig. 10.24A. The shelf of Guatemala and Honduras, bathymetry (after Purdy, 1974; Ginsburg and James, 1974).

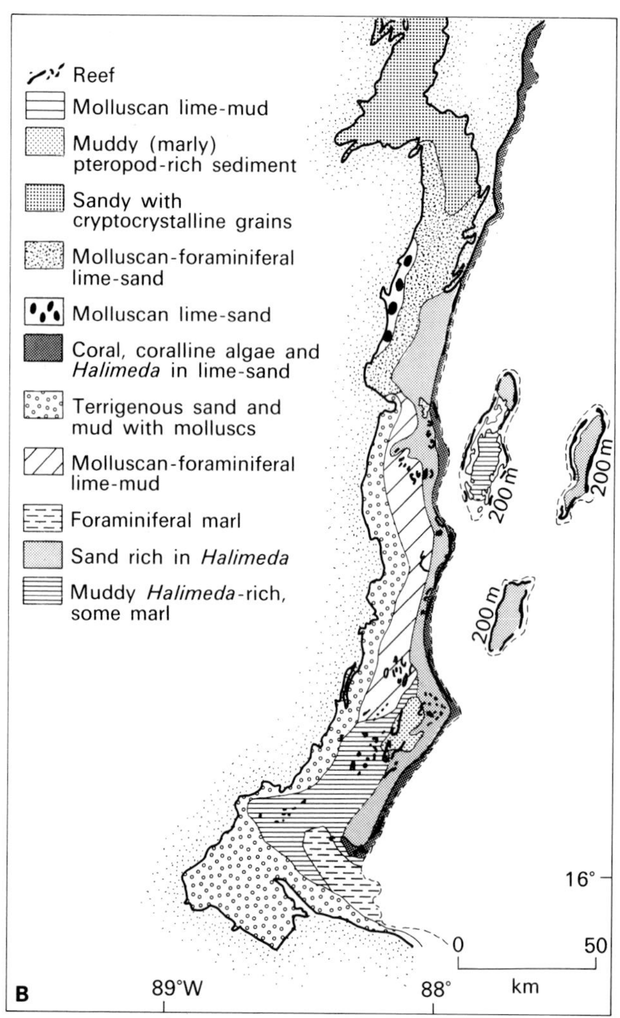

Fig. 10.24B. Facies distribution on the shelf of Guatemala and Honduras (after Purdy, 1974; Ginsburg and James, 1974).

(1974a) has presented a strong case for believing that the original pre-reef substrate was a karst, the modern reefs being initiated on antecedent karst-induced elevations.

THE SHELF OF GUATEMALA AND HONDURAS (Fig. 10.24a and b)

On a small scale this provides a range of facies reminiscent of the more extensive carbonate shelves of the geological past with near-shore clastics, mid-shelf muds and a shelf-break reef barrier (Matthews, 1966; Kornicker and Bryant, 1969; Cebulski, 1969; Scholle and Kling, 1972; and Ginsburg and James, 1974). The region lies between 15° and 20°N and consists of a shelf 480 km long and from <2–29 km wide. The depositional province consists of a narrow, deep, back-reef lagoon in which fine-grained coccolithic carbonate muds are accumulating. Temperatures are uniformly high with winter temperatures in the 25.6°–26.7°C range in the mid-shelf surface waters, and salinities of 36‰ (although they are much lower near the coast). The shelf is wind-dominated with strong easterlies blowing throughout the year except in November and December when they veer to the north and often become gale-force. Occasional hurricanes also pass over the area. Tides on the other hand are unimportant. Maximum ranges are in the order of 1 m while semi-diurnal tides are about 30 cm.

Three major depositional areas can be recognized on the shelf: the Barrier Platform, the Shelf Lagoon and an inshore area of terrigenous influence.

The Barrier Platform, with its precipitous seaward face, is 1–10 km wide and has depths below 3 m. It is essentially a barrier-reef belt which effectively restricts large-scale water movement from the open ocean on to the shelf. However, the barrier is not complete and circulation is not totally cut off. Faunal diversities are high in both mud- and sand-grade sediment producers and terrigenous clastic sediments are lacking. The reefs consist of corals and coralline algae accompanied by abundant *Millepora*. The growing reef spurs are often surrounded by a rubble of cemented coral branches. In the north there are no reefs, but to the south reefs and lagoon atolls are abundant, each fringed by a windward crescent of coral growth and a leeward sand belt.

Behind, the barrier platform drops away into the deep shelf lagoon. Depths increase from north to south within the shelf lagoon, reaching nearly 60 m near the southern end. Pinnacle reefs sometimes rise from depths as great as 43 m. The sediments of the shelf lagoon are dominantly pelleted lime muds but transitional calcarenitic facies occur close to the barrier while contamination by up to 50% montmorillonite occurs on the mainland side.

Much of the lime mud is derived from the breakdown of larger carbonate skeletons (Matthews, 1966), but some (~20%) is derived from coccoliths living in the surface waters of the lagoon. Eight species inhabit the lagoon, comprising the normal offshore assemblage for the area, although the lagoon occasionally becomes brackish. The shelf lagoon environment is more stressful than that of the open ocean. The higher-stress environments of the shelf lagoon are reflected in the subordination of planktonic foraminifera, and pteropods and even coccoliths disappear in the northern basin of the lagoon where the stresses are greatest. *Orbulina, Globigerina, Globorotalia* and pteropods occur in most lagoon sediments but decrease in abundance inshore. Coccoliths are low Mg-calcite and, unlike Bahamian lagoons, there is a significant addition of low Mg-calcite from these coccolith sources. It should be noted that the Honduran muds consist of large amounts of aragonite and high-magnesian calcite in addition to the coccolithic calcite. Recent oceanic Globigerinid oozes normally contain around 10% coccoliths compared with 18–21% for Cretaceous chalks (Scholle and Kling, 1972). Thus, Cretaceous chalks are more comparable with the Honduran 'lagoonal' muds rather than the modern open-sea oozes (see Chapter 11). Much of the sediment is pelleted and contains small amounts of glauconite while foraminiferan tests contain iron oxides and aragonite cements.

From the lagoon towards the shore the proportion of terrigenous sediment increases. It consists mostly of clay-grade material derived from the many coastal rivers, particularly in the southern shelf lagoon where the shore consists of quartz sands that pass into calcareous montmorillonitic muds about 2 km into the lagoon. The Mayan Mountains (Fig. 10.24A) are drained by short vigorous streams which produce locally low salinities within the lagoon. To the north along the shelf the hinterland is low-lying and the coast consists of swamps with lagoons and mangrove marshes. However, wave action within the lagoon has sufficient energy to generate a series of barriers and spits that often have sandy outer beaches.

In Honduras Bay and near the Honduras–Guatemala coasts the sediments contain less than 50% $CaCO_3$. The fauna is dominated by molluscs while barnacles and scaphopods are increasingly important (a *foramol* assemblage promoted by salinity conditions).

Sand is not usually transported within the shelf area at depths greater than 7 m except during major storms and hurricanes. However, fragments of *Halimeda* and coralline algae have been reported at depths below 230 m. *Halimeda* plates have a very low bulk density (see Purdy, 1974b, p. 833) and their widespread distribution probably results from wave action, although storms or even tectonically induced slumping has also been invoked to explain their distribution.

10.2.5 **Ancient analogues of subtropical carbonate shelves**

GENERAL REVIEW

Regional environmental models from the geological record involving ancient limestone complexes are dominated by rimmed-shelf examples many of which are associated with evaporites (see Wilson, 1975 for reviews and Carbonate Buildups Sect. 10.4.2). However, there are examples which lack a reef-belt equivalent and where regional facies distributions fit an open-shelf model. In the Upper Jurassic, which the Smackover Formation from Texas to Alabama and the British Portland Group each represent regressive sequences from essentially open shelves. Both are, however, associated with evaporites in the regressive phase. In the Palaeozoic the Carboniferous Limestone (Dinantian/Mississippian) in SW Britain may also represent the deposits of an open shelf environment.

TETHYAN MESOZOIC CARBONATE PLATFORM FACIES; AND PALAEOZOIC CARBONATE FACIES OF THE CENTRAL APPALACHIAN SHELF

Tethyan carbonate platform facies

During the Mesozoic, thick sequences of massive platform limestones and dolomites formed in the Tethyan region from

Central America, through Mediterranean Europe into the Middle East. These sediments compare, in a general way, with modern and ancient Bahamian-type platform deposits.

Triassic

In the Late Triassic, a thickness of 1.0 to 1.5 km of cyclically arranged lagoonal, intertidal and supratidal dolomitic limestones accumulated over a 20 km wide belt in the Northern Limestone Alps between Lofer and Vienna (Fischer, 1964, 1975). To the south, this belt was bounded by a rim of reefs that defined the southern edge of the Dachstein bank (Fig. 10.25, and Sect. 10.4.2).

The base of each cycle (Fig. 10.25A) is marked by a weathered and solution-riddled surface that represents a phase of exposure. This surface is overlain by red or green argillaceous sediment containing limestone cobbles and is interpreted as a fossil soil. Cavities in the underlying limestone are often partially filled with this material. Above is a sequence of dolomitic, birdseye limestones (loferites) consisting of cream and light-grey micritic limestones with abundant fenestral (or birdseye) pores that have become filled by geopetal mud and sparry cement. Algal lamination is either flat or crinkled, and prism-cracks are often well-displayed. Laminated loferites grade into massive loferites that show 'clotted structures' consisting of vague pelletal structures comparable with peds from modern soils (Fischer, 1975). These loferites are interpreted as intertidal and supratidal deposits.

Above the loferites, a massive calcarenite (1–20 m thick) with a rich marine fauna is developed. Megalodontid bivalves are very conspicuous while sponges, corals, bryozoans, brachiopods, echinoids and a variety of other molluscs are all found. The fauna becomes progressively impoverished as the sequence is traced northward into the dolomitic ultra-back-reef facies (Hauptdolomit, Fig. 10.25B). Photic conditions are indicated by the presence of algae: rhodophytes, codiacians and dasycladaceans.

Evaporites and evaporite pseudomorphs are not associated with these limestones, suggesting that the climate was humid. The megalodontid calcarenites probably accumulated in a vast lagoon, not more than a few metres deep (Fischer, 1964, 1975). Normal salinities probably existed towards the reef belt (Sect. 10.4.2) but became progressively more saline into the backreef (Hauptdolomit).

To explain the cyclicity, Fischer favoured regional oscillations in relative sea-level and proposed amplitudes of 5 m with a periodicity of 50,000 yrs. These oscillations, he believed, represented either interruptions of regional subsidence patterns by episodes of uplift, or eustatic changes. Such variations would have produced a broadening lagoon during transgressive phases accompanied by landward migration of the intertidal fringe, while regression would have involved a seaward retreat of the intertidal belt and a narrowing of the lagoon. Exposure of lagoonal sediments during regressive phases led to the development of karst surfaces.

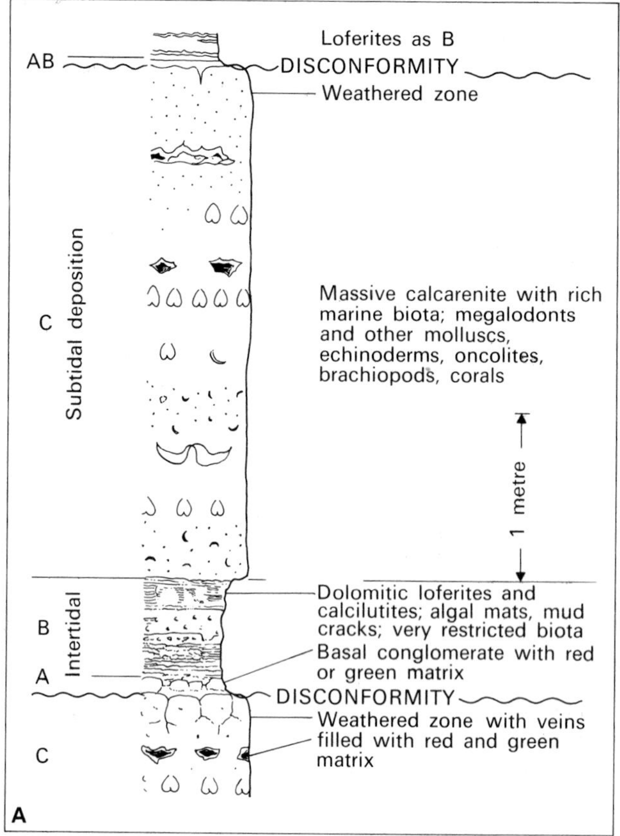

Fig. 10.25A. An idealized Lofer Cyclothem (after Fischer, 1964, 1975).

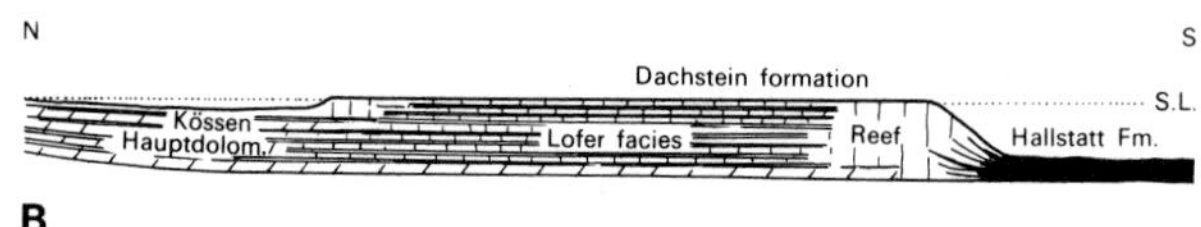

Fig. 10.25B. Diagrammatic restoration of Late Triassic facies in the Northern Limestone Alps. Cycles shown in 10.25A occur in the Lofer back-reef facies (after Fischer, 1964).

Fischer discounted a steady-state tidal marsh model on the grounds that:
(1) it would not explain the occurrence of emergent horizons and soil formation; (2) units produced in such a model should be very lenticular and (3) the marine fauna in the calcarenites is too rich for the calcarenitic sequence to have accumulated in tidal creeks. Of these arguments, the first is probably the strongest because a steady-state model, just involving

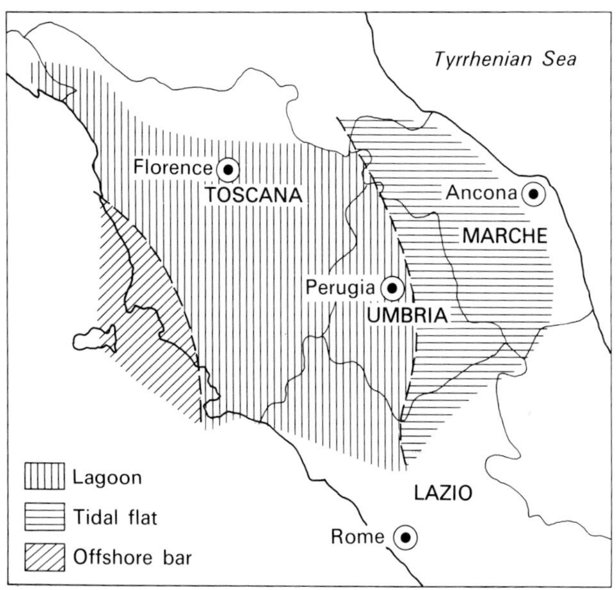

Fig. 10.26A. Locality map and facies distributions in the Calcare Massiccio of Italy (after Colacicchi *et al.*, 1975).

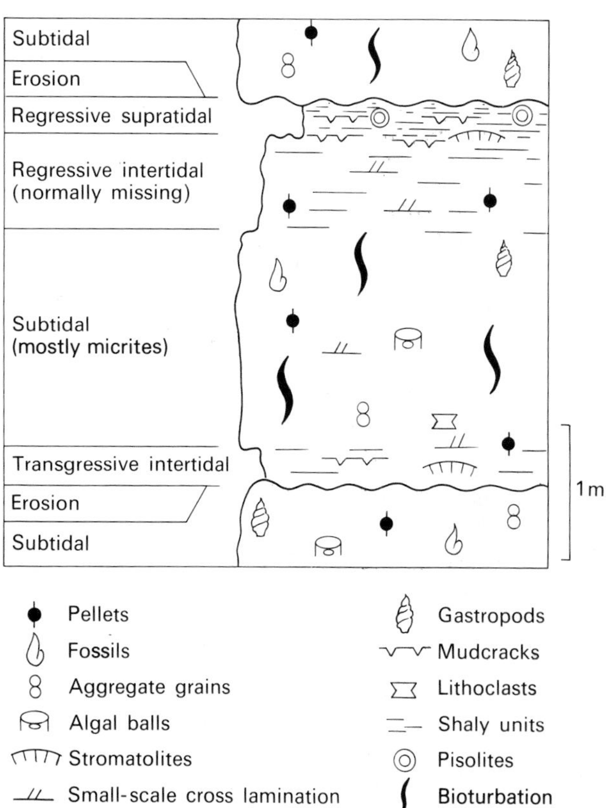

Fig. 10.26B. Idealized cycle for the Calcare Massiccio; Umbria-Marche area of Italy (after Colacicchi *et al.*, 1975).

subsidence and sedimentation, should lead to a regressive sequence whereas each Lofer cycle (Fig. 10.25) is a transgressive sequence.

Jurassic

In parts of the Italian Apennines (Fig. 10.26) the widespread Calcare Massiccio (Lower Lias) also exhibits remarkably cyclic facies sequences in a formation that crops out over a wide area. In the east (Umbria-Marche), cyclic sequences (Fig. 10.26B) commence in white micrites with peloids, oncolites and bioclasts that are interpreted as lagoonal facies by Colacicchi *et al.* (1975). Some of the peloids were probably bound together to form grapestone facies. These micrites are cut through by subtidal channel-fill sequences of grainstones containing oncolites, ooids, intraclasts, peloids and bioclasts with low-angle cross-lamination. Some of the oolitic and bioclastic lime-sands display spar-filled cavities that resemble the 'keystone vugs' of Dunham (1970) from recent beaches. Incompletely bioturbated pelletal micrites with stromatolitic laminations and desiccation cracks accumulated as intertidal-flat sediments (Fig. 10.26) while an association of laminated and dolomitic micrites with desiccation cracks and birdseyes represents tidal-channel levee deposits. Finally, beds with vadose pisolites and others cut by cavities filled with red-brown clay represent supratidal and karst environments (Bernoulli and Wagner, 1971). These essentially regressive cycles conform to a steady-state tidal marsh model.

In Western Umbria and Tuscany (Fig. 10.26) the Calcare Massiccio is mostly represented by bioturbated oncolitic and peloidal micrites. There are no signs of desiccation and the beds represent a shelf-lagoon environment. Unlike the Lofer facies, the Calcare Massiccio is not seen in association with reefs but structures and facies compare closely with those from the modern Bahama Banks (Colacicchi *et al.*, 1975; Boccaletti and Manetti, 1972; Passeri and Pialli, 1972).

Late Jurassic–Early Cretaceous sediments recently dredged from the collapsed Mesozoic platform of Gentry Bank in the Bahamian Chain included oolitic, oncolitic and peloidal sediments mostly with sparitic cements (D'Argenio *et al.*, 1974). Small shallow-water benthic foraminifera (miliolids and textularids) were present and cyanophyte algal debris also. During the Cretaceous, thick carbonate platform sequences formed synchronously through Sicily, S. Italy, Yugoslavia and Greece, and developed facies identical to those dredged from the Gentry Seamount (D'Argenio *et al.*, 1974).

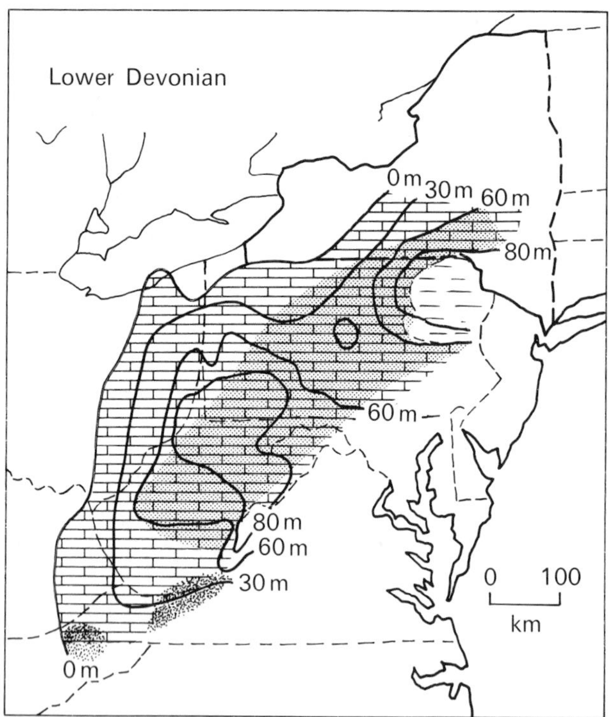

Fig. 10.27A. Generalized facies and thicknesses for the Lower Devonian (Helderberg) in the Central Appalachians (after Laporte, 1971, 1975). Dashes indicate marine shales.

Extensive carbonate platforms were established along the Tethyan margins by the Late Triassic. During the Early Jurassic many, but not all, of these broke up into a series of blocks that underwent differential subsidence throughout much of the remaining Mesozoic Era. Only in the areas mentioned above did carbonate platforms persist through into the Cretaceous (Bernoulli and Jenkyns, 1974). Collapsed platforms became the sites of pelagic deposition (see Chapter 11).

Palaeozoic carbonate facies of the Central Appalachian Shelf (Fig. 10.27)

Within the Cambrian to Devonian carbonate units of the Central Appalachians (Laporte, 1971, 1975) tidal flat, shallow subtidal, deep subtidal and carbonate buildup environments are recurring depositional themes. The major lithological, palaeontological and stratigraphical characteristics of the environmental suites are given in Table 10.2 (p. 302). Through time, the faunas changed but early Devonian facies and community reconstructions (Figs. 10.28A–D) provide examples of the various organism-sediment relationships (Walker and Laporte, 1970).

There are many other examples of ancient rimmed-shelves in the literature (see Wilson (1975); Ginsburg (1975); Ziegler *et al.* (1974) and Sect. 10.4.2).

SMACKOVER FORMATION (OXFORDIAN), SOUTHERN USA

The Smackover Formation (Fig. 10.29) occurs in the subsurface between Texas and Alabama in an arc parallel to the margin of the Gulf of Mexico (Bishop, 1968; Wilson, 1975, p. 294). In some places within this belt the Smackover is represented by more than 300 m of offlapping carbonates. Grainstones provide the best reservoir facies for a large number of oil fields although some production also comes from dolomitized mudstones. In Arkansas production is primarily from lime-grainstones with preserved primary porosities of 20–25% while in eastern Texas the reservoirs consist of dolomitized grainstones where porosities sometimes exceed 30%.

Fig. 10.27B. A restored section in the Helderberg Group, Lower Devonian, New York (after Laporte, 1971, 1975).

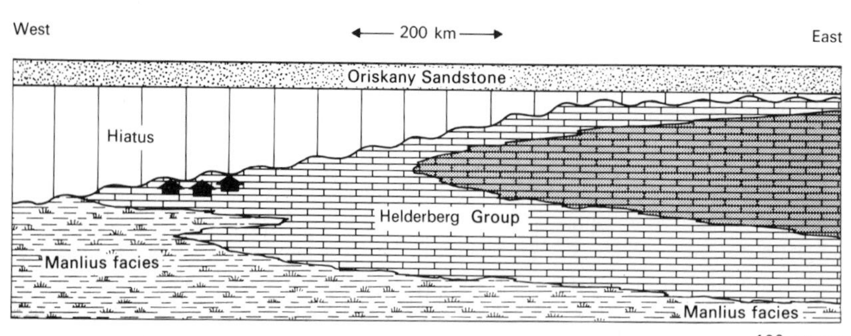

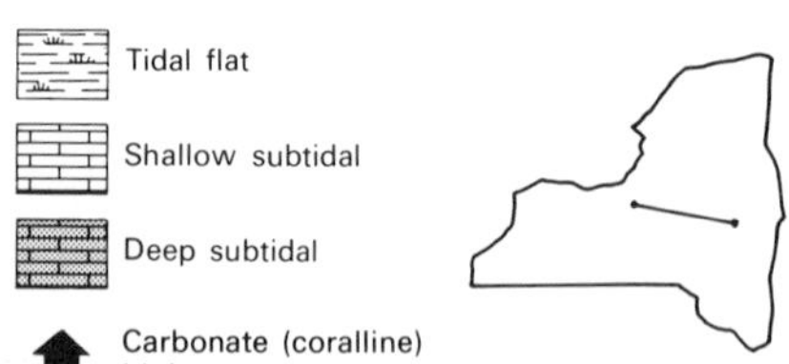

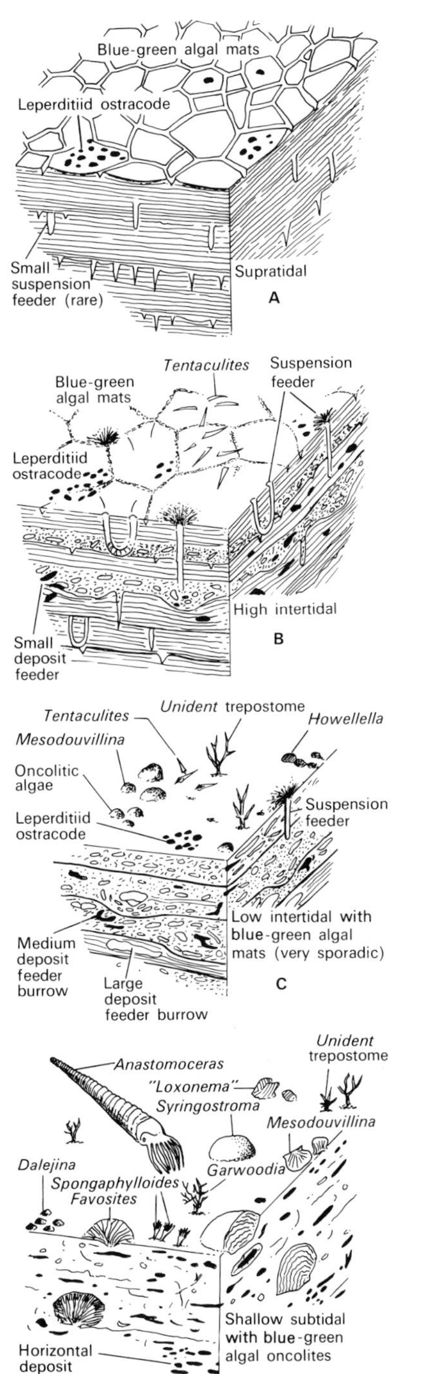

Fig. 10.28. Biofacies reconstructions of environments represented in the Manlius facies (after Laporte, 1975).

A vertical sequence (Fig. 10.29B) shows deeper water dense micro-laminated kerogenic micrites with undisturbed foraminiferal laminations. With increasing ooid concentrations these micrites pass up into oolitic grainstones which are in turn overlain gradationally by the anhydritic limestones and anhydrites of the Buckner Member.

The whole sequence has been interpreted as a progradational wedge that built southwards after an initially rapid transgression. At any one time during progradation a supratidal and back shelf zone of red-beds and evaporites passed seawards into a turbulent shelf zone where ooids formed and grainstones were deposited. Offshore, there was a rapid passage out through slope and deeper water lime-mudstones and, in the deepest water, sapropelic lime-muds (Fig. 10.29C).

The back shelf and shore zones are represented by a highly variable sequence of red-beds that commonly contain nodules and laminations of white to red anhydrite. White and pink sandstone interbeds up to 1.5 m thick also occur, often with anhydritic cements. Updip, in the north, these sediments rest directly upon Palaeozoic floor. To the south there is often a transition downwards from red-beds and evaporites into grain-supported limestone. In the transitional facies, pellets are a major component while ooids, if present, are mostly recrystallized; bioclasts of shell and algal debris are absent.

The grainstone facies contains oolitic-coated pellets that are often well-sorted. Coarser grained oncolite and pisolite-bearing packstones also occur with occasional codiacian and dasycladacean algal remains providing evidence of strong photic conditions. These sediments are largely devoid of terrigenous clastics and probably formed on a shoal at or near the prograding shelf margin. Low faunal diversities reflect both

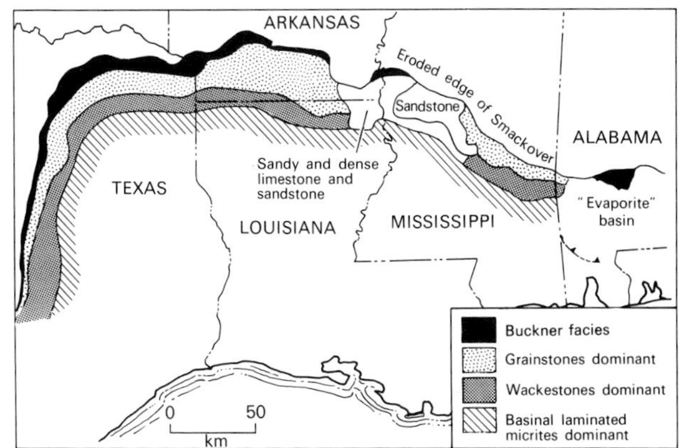

Fig. 10.29A. Distribution of major environments during the deposition of the Upper Smackover from East Texas to Alabama (after Bishop, 1968).

the high stress environments of the oolite shoals and the raised salinities of the shelf.

The slope facies consists of massively bedded peloidal wackestones and mudstones with a characteristic foraminiferal assemblage. Derived algally-coated grains occur, but pellets are once more the dominant allochem. Remains of the planktonic crinoid *Saccocoma* are locally common and the mudstones have been intensely bioturbated ('disturbed mud' of Bishop, 1968).

The lack of reefs and the close juxtaposition of the various facies provide the main evidence for an open shelf environment. Thus, some comparison with Yucatan and the Atlantic shelf of N. America is possible. However, the facies distributions on both the Atlantic and Yucatan shelves were achieved during transgression in a humid area. The evaporites associated with the Smackover indicate an arid environment comparable with the modern Persian Gulf.

THE LATE JURASSIC (KIMMERIDGE FORMATION, PORTLAND GROUP AND PURBECK FORMATION) OF SOUTHERN ENGLAND

Like the Smackover Formation, these rocks were laid down as a regressive sequence (Fig. 10.30A). They have recently been

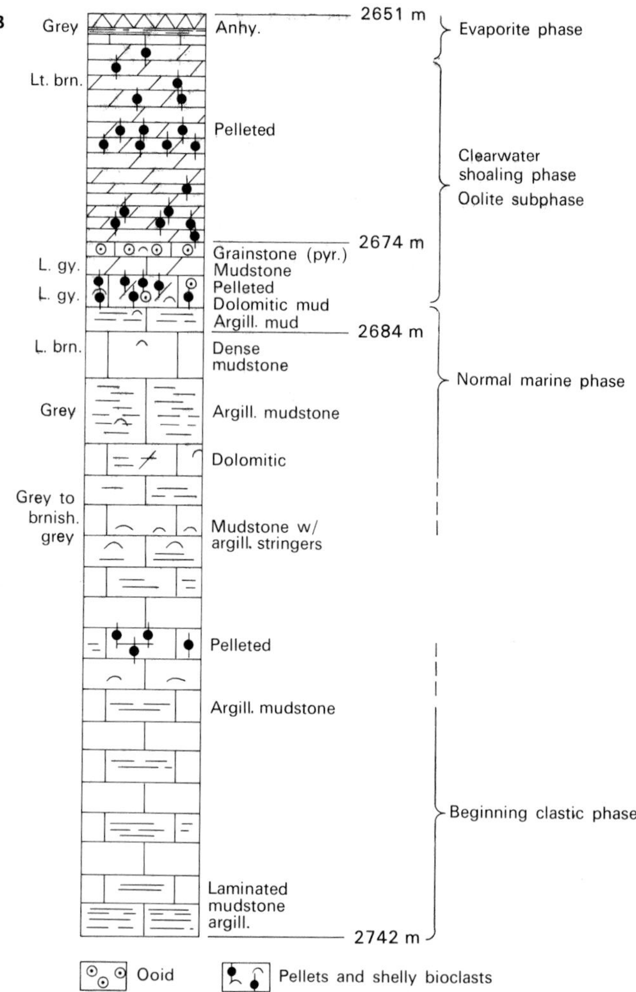

Fig. 10.29B. Vertical section through the Smackover Formation in East Texas to illustrate the upward shoaling sequence (after Wilson, 1975).

Fig. 10.29C. Facies model to illustrate the progradational nature of the Smackover and associated facies (after R.C. Vernon, *pers. comm.*, 1977).

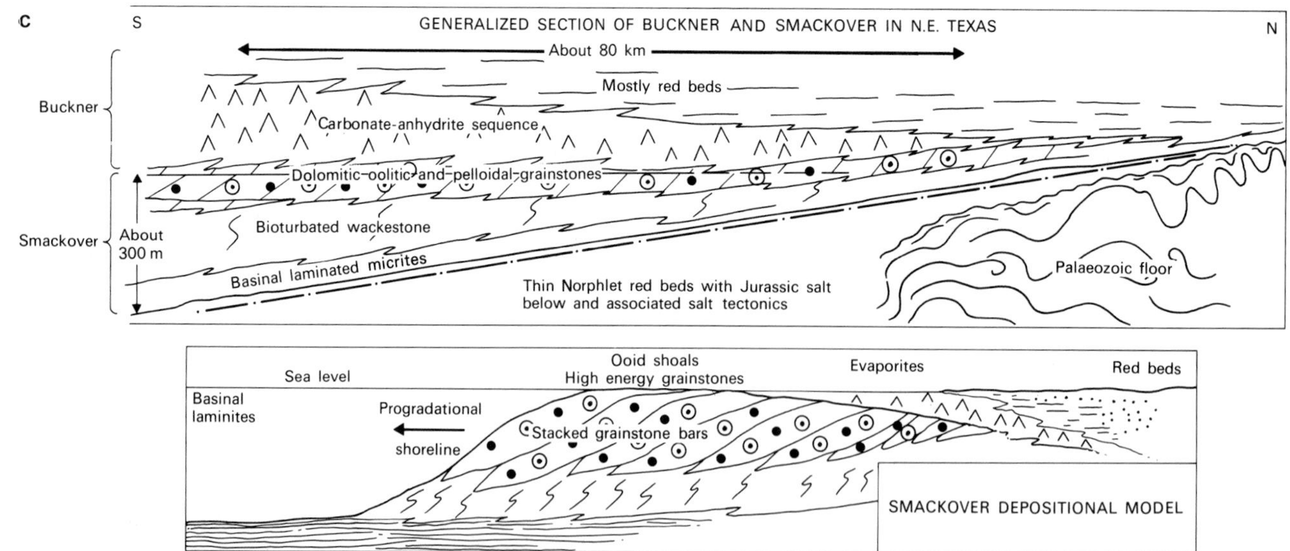

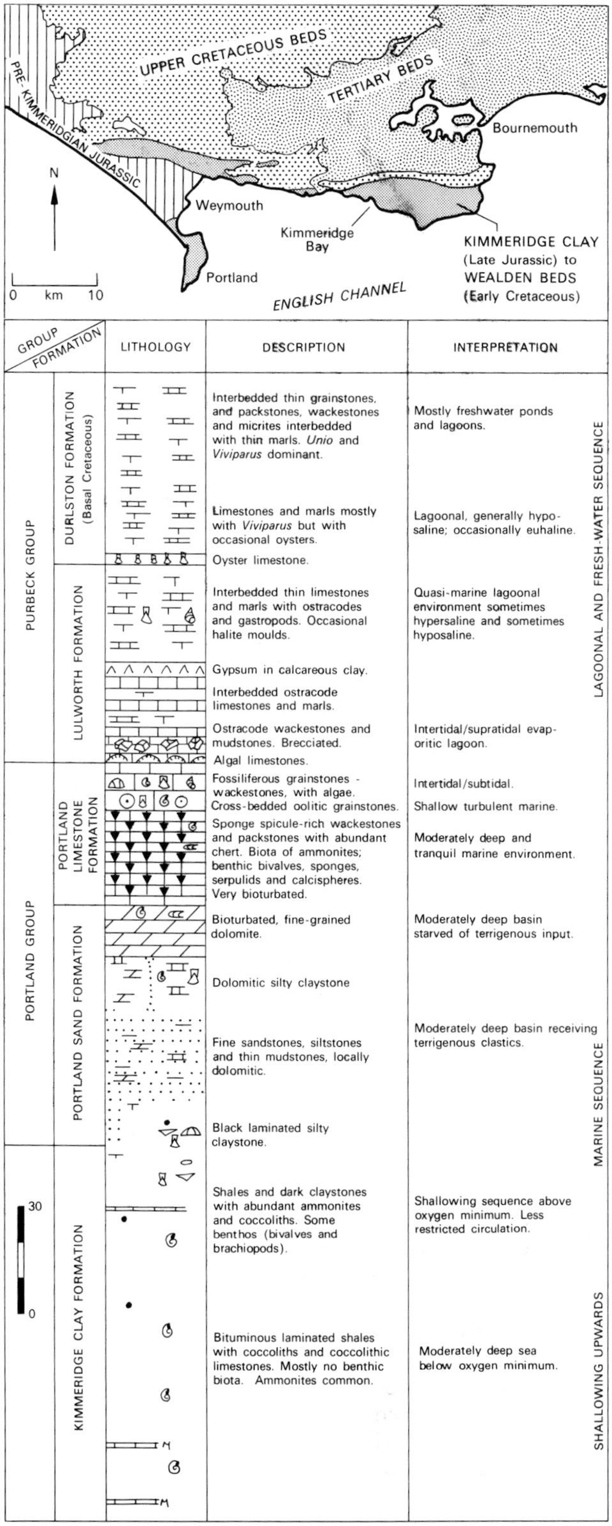

UPPER CRETACEOUS BEDS
TERTIARY BEDS
PRE KIMMERIDGIAN JURASSIC
Bournemouth
Weymouth
Kimmeridge Bay
Portland
ENGLISH CHANNEL
KIMMERIDGE CLAY (Late Jurassic) to WEALDEN BEDS (Early Cretaceous)
N
0 km 10
GROUP
FORMATION
LITHOLOGY
DESCRIPTION
INTERPRETATION
PURBECK GROUP
DURLSTON FORMATION (Basal Cretaceous)
Interbedded thin grainstones, and packstones, wackestones and micrites interbedded with thin marls. Unio and Viviparus dominant.
Mostly freshwater ponds and lagoons.
Limestones and marls mostly with Viviparus but with occasional oysters.
Lagoonal, generally hypo-saline; occasionally euhaline.
Oyster limestone.
LULWORTH FORMATION
Interbedded thin limestones and marls with ostracodes and gastropods. Occasional halite moulds.
Quasi-marine lagoonal environment sometimes hypersaline and sometimes hyposaline.
Gypsum in calcareous clay.
Interbedded ostracode limestones and marls.
Ostracode wackestones and mudstones. Brecciated.
Intertidal/supratidal evaporitic lagoon.
Algal limestones.
LAGOONAL AND FRESH-WATER SEQUENCE
PORTLAND GROUP
PORTLAND LIMESTONE FORMATION
Fossiliferous grainstones - wackestones, with algae.
Intertidal/subtidal.
Cross-bedded oolitic grainstones.
Shallow turbulent marine.
Sponge spicule-rich wackestones and packstones with abundant chert. Biota of ammonites; benthic bivalves, sponges, serpulids and calcispheres. Very bioturbated.
Moderately deep and tranquil marine environment.
PORTLAND SAND FORMATION
Bioturbated, fine-grained dolomite.
Moderately deep basin starved of terrigenous input.
Dolomitic silty claystone
Fine sandstones, siltstones and thin mudstones, locally dolomitic.
Moderately deep basin receiving terrigenous clastics.
Black laminated silty claystone.
MARINE SEQUENCE
KIMMERIDGE CLAY FORMATION
Shales and dark claystones with abundant ammonites and coccoliths. Some benthos (bivalves and brachiopods).
Shallowing sequence above oxygen minimum. Less restricted circulation.
Bituminous laminated shales with coccoliths and coccolithic limestones. Mostly no benthic biota. Ammonites common.
Moderately deep sea below oxygen minimum.
SHALLOWING UPWARDS
30
0

newly interpreted by Townson (1975) and West (1975). The Portland carbonates grade down into the deeper-water shale sequence of the Kimmeridge Clay which includes sapropels and coccolith laminations. The Portland is overlain by the evaporite-bearing limestone sequence known as the Purbeck Group (Sect. 8.4.2).

The lower part of the Portland consists of dolomitic claystones that accumulated in a wave-free marine environment. Dolomite formation was possibly the result of an *in situ* replacement of lime-mud reacting with refluxed brines (Townson, 1975). The middle part of the Portland consists of cherty sponge-rich micritic limestones in which redistribution of sponge spicule silica accounts for the silification in these beds. The upper part of the Portland consists of cherty limestones that pass up into shallow-water cross-bedded grainstones. Ooids in the grainstones may have formed on shoals that developed over elevated 'swell' areas situated over structurally or halokinetically controlled elevations on the sea-floor (Fig. 10.30B). The marine grainstones of the upper sequence are overlain by stromatolitic and evaporite-bearing micrites with abundant exposure features (Figs. 10.30A and 10.31).

The model generated by Townson (Figs. 10.31A and B) accounts both for the vertical sequence and the lateral distribution of facies in the area. Although the general sequence conforms well with that expected from an open shelf there are complications to the story. The later evolution of the area appears to have involved ooid formation on an offshore shoal while the evaporite facies probably developed shoreward of an algal 'barrier' (Figs. 10.31A and 10.31B).

GENERAL FACIES DISTRIBUTIONS DURING THE DEPOSITION OF SOME OF THE HIGHEST PORTLAND GROUP
CHERTY MICRITIC BEDS
OOLITIC GRAINSTONES
Lime mud ± spicules
Medium ooid sand
% ooids decrease and biochems increase
SWELL
Medium biochem sand
CHERTY MICRITIC BEDS
Shelly lime mud and sponge spicules
?Muddy fine biochem sand?
LOWER PART OF OOLITIC GRAINSTONES

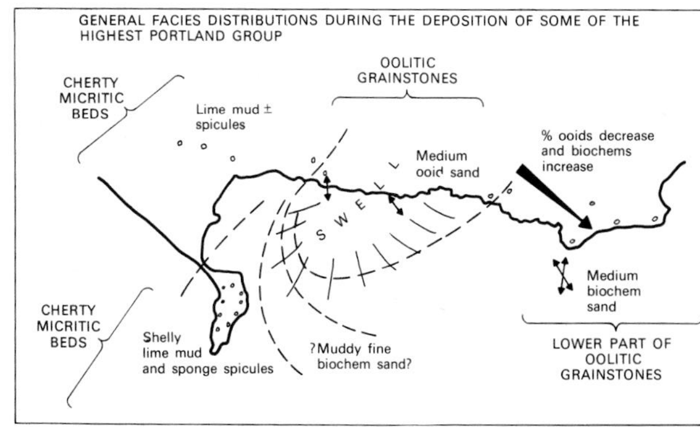

Fig. 10.30B. Map of facies distributions during Portland Group deposition in Dorset (after Townson, 1971).

Fig. 10.30A. Vertical sequence through the latest Jurassic of Dorset, S. England, illustrating the shoaling upward sequence.

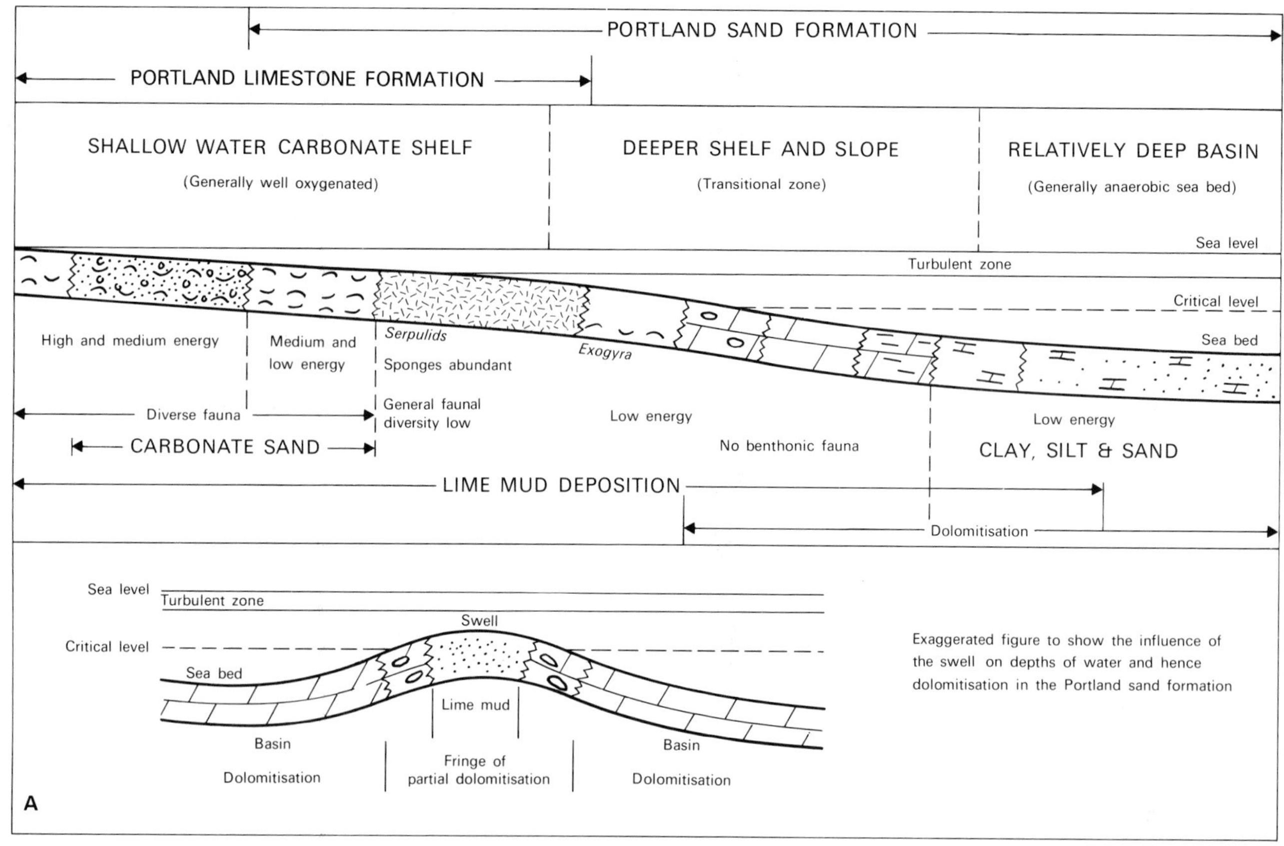

Fig. 10.31A. General model for environments of deposition of the Portland Group of Dorset (after Townson, 1975).

EARLY CARBONIFEROUS (MISSISSIPPIAN) SHELF OF SOUTHERN BRITAIN

The early Carboniferous transgression generated vast carbonate shelf deposits in North America and Europe. Within the carbonate succession of Southern Britain, six major facies associations are recognized (Ramsbottom, 1973) which often occur in a cyclic arrangement (Fig. 10.32). Although oversimplified, the facies associations at particular times tended to occur within more-or-less linear belts that were parallel to the major basinal trends (Fig. 10.32). Through much of the early Carboniferous, the main belt of ooid accumulation lay close to the line of the Variscan Front in South Wales and the Mendip region (Fig. 10.32). In this region, facies belts are narrow and, as with the Smackover, there are no major fringing 'reefs'.

Radiolarian and goniatite-bearing mudstones were deposited below wave-base in a deep marine environment to the south (Fig. 10.32). The bioclastic limestones and calcareous mudstones are probably facies analogues to the molluscan shell sediments that formed offshore in Yucatan seaward of the ooid zone (Fig. 10.19).

The oosparites and occasional oomicrites with their low density faunas, by analogy with modern ooids, probably formed in <5 m of water. The oolites are locally capped by stromatolites with desiccation cracks and show vadose features indicating that ooid ridges were periodically exposed.

North of the oolite belt, a further belt of micrites was developed with stromatolites and dolomite. The Carboniferous stromatolitic micrites with their associated dolomitization are comparable with the lagoonal micrites of the Trucial Coast (Murray and Wright, 1971; Chapter 8) where the lagoon is protected by oolite barriers (Fig. 10.33). In the Bristol area

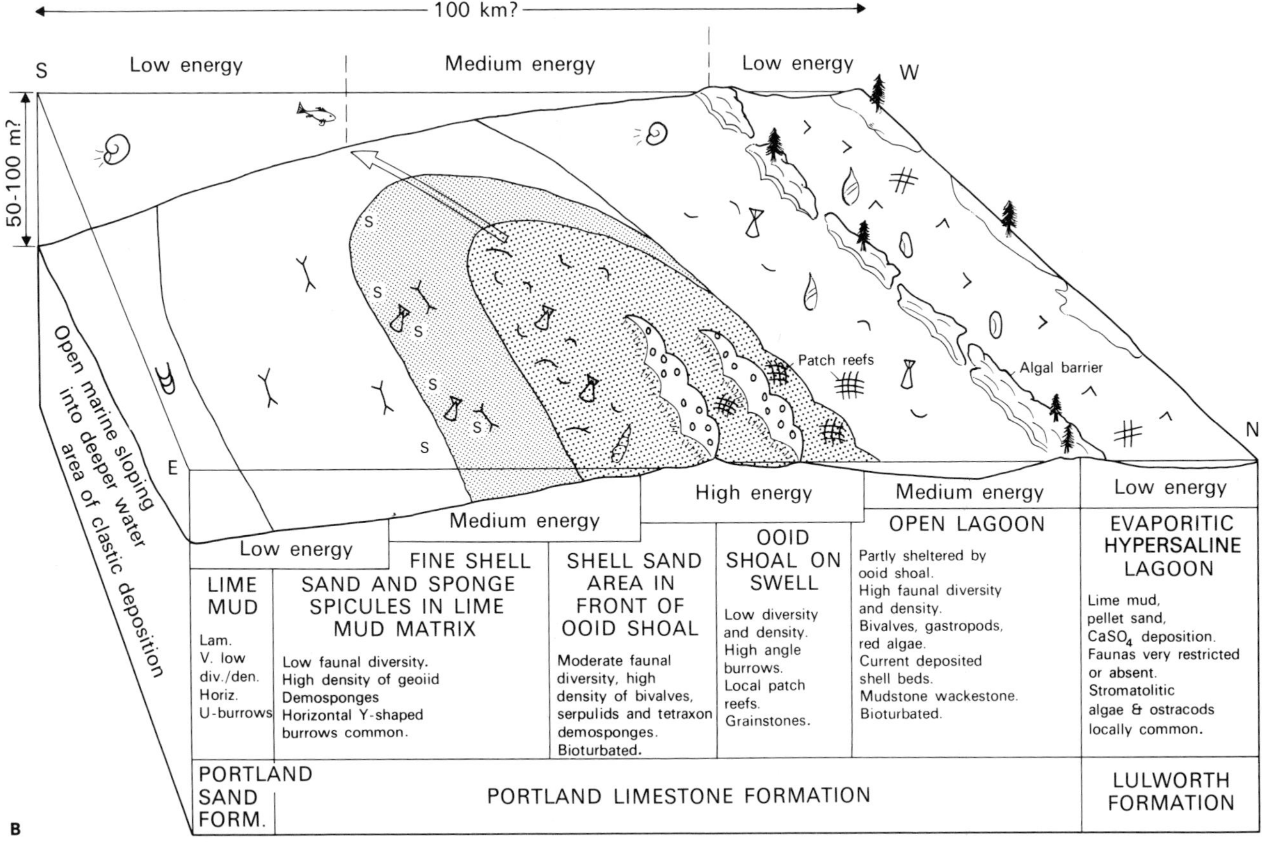

Fig. 10.31B. Generalized model for the deposition of latest Jurassic facies in southern England (after Townson, 1975).

they consisted of two major types:
(1) Crinkle Mat-type laterally linked hemispheroids (LLH) of Logan *et al.* (1964) similar to those that form in the intertidal zone of the modern Persian Gulf (Kendal and Skipworth, 1969a and b; Sect. 8.2.5) and
(2) Columnar '*Collenia-Cryptozoon*' type growth forms up to 730 mm thick and 30 mm in diameter that compare with the columnar stacked hemispheroid (SH) stromatolites from the low intertidal zone of Shark Bay (Hagan and Logan, 1975). In Shark Bay the height of the columnar stromatolites approximates to the tidal range (Hagan and Logan, 1975).

Northward, and locally along the outcrop in southern Wales, terrigenous sand wedges were derived from the southern margin of St George's Land (Fig. 10.32). The position of the oolite belt in the S. Wales–Mendip region was possibly structurally controlled.

Ramsbottom (1973) has suggested a depth zonation as the cause of sediment and community relations in the Lower Carboniferous of Britain (Fig. 10.32) and although there are many problems concerning a simple depth relationship the general facies trends conform reasonably well with an open shelf model.

Discussion

Each of the Ancient open shelf analogues cited above displays facies associations that lacked barriers and organic buildups and have sequences that commenced deposition in deep water and ended in shallow water. The shallowest phases were often evaporitic. In interpreting carbonate sequences there is always the risk that the interpretation will be forced into an arbitrary 'box' (i.e. open shelf). Ancient shelf carbonates cannot be neatly boxed but we have selected three examples which appear to show the sequence expected.

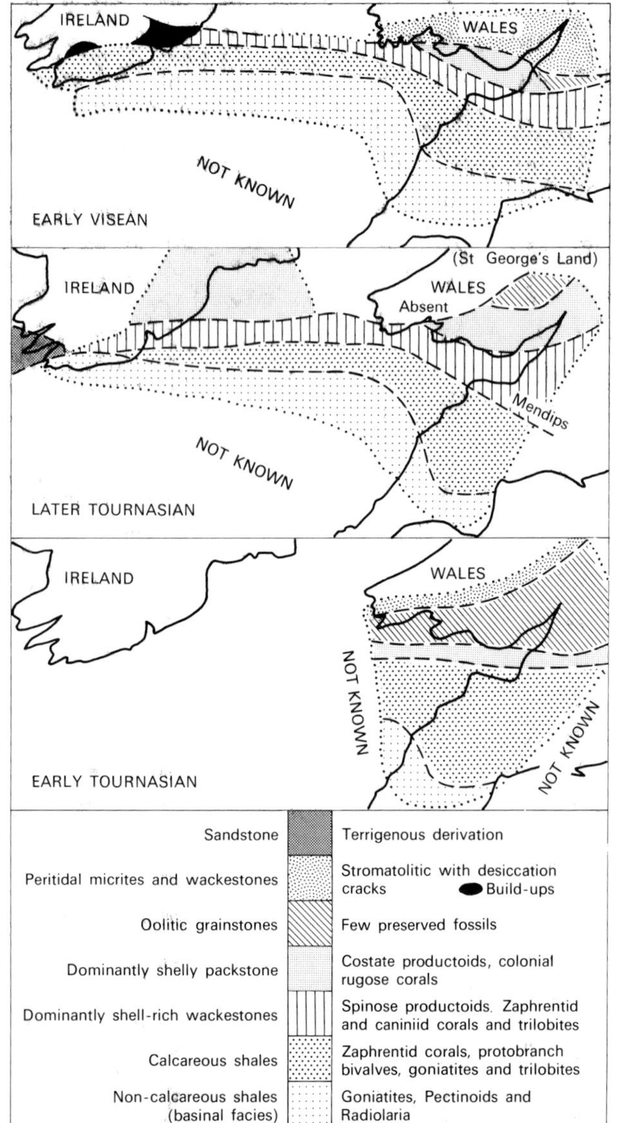

Fig. 10.32. Facies distributions and facies sequences in the Early Carboniferous of western Britain and southern Ireland (after Ramsbottom, 1970, 1973). Key illustrates cyclic sequence.

10.3 EXAMPLES OF TEMPERATE WATER CARBONATE SHELVES

Until recently little was known about carbonate sediments from shallow temperate waters. However, it is now evident that they are locally abundant (Lees, 1975). Specific examples of modern case studies include Boillot *et al.* (1971), and Lees *et al.* (1969). The following modern examples fall into the *foramol* association of Lees (1975) which contain foraminifera–mollusc–barnacle faunal assemblages and lack oolites and aggregates.

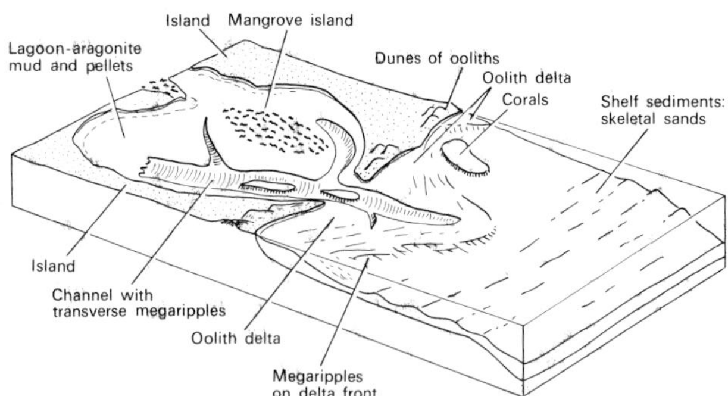

Fig. 10.33. Reconstruction of the modern carbonate environments of the Trucial Coast, Persian Gulf (after Murray and Wright, 1971).

10.3.1 Mannin Bay, western Ireland

The coast of Connemara (Figs. 10.34 and 10.35), in western Ireland, is low-lying and there are no major rivers flowing from a hinterland of durable igneous and metamorphic rocks. Most of the terrigenous clastic sediments are derived from the erosion of Pleistocene glacial sediments. Offshore a variety of marine calcareous sediments is currently forming (Lees *et al.*, 1969; Bosence, 1976). One area around Mannin Bay illustrates the major depositional characteristics of the carbonates. Air temperatures in the region normally range from 5°C (Feb mean min) to 15°C (August mean max), and surface waters range from 7.2–18.5°C. Dominant winds are onshore from the SW (230°) and mean tidal ranges are 4.3 m to 1.9 m (mean neaps). Salinities vary little offshore (from 33.5‰–35‰) although fresh water run-off reduces salinity locally inshore. The proportion of terrigenous clastic material in Mannin Bay varies from zero to 96%. In the central parts of the Bay less than 26% is usual although clastic-dominated samples have been obtained offshore. The bathymetry (Fig. 10.35A) consists of a series of drowned glacial channels. The four major sediment groups are given in Fig. 10.35B and consist of:
(1) muds; (2) a coastal fringe of sands; (3) deposits of the calcareous alga *Lithothamnium* and (4) a variable association of sands, shells and gravels occurring patchily on a rocky substrate.

Muds: are restricted to the bay-heads and to areas with water depths less than 10 m. The sediments consist of black gelatinous muds interrupted by irregular patches of gravel. Filamentous mats of algae and, in summer, dense forests of seaweed cover the surface but carbonate contents seldom

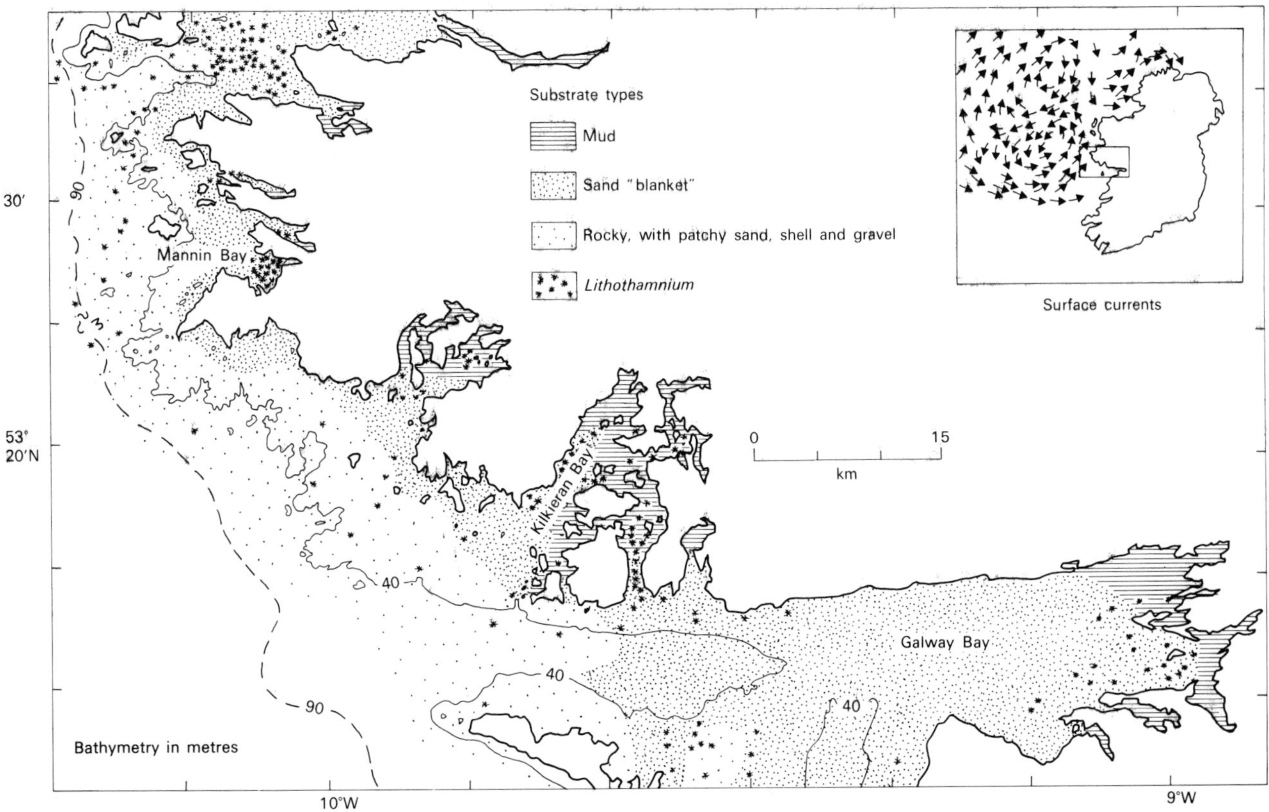

Fig. 10.34. The coast of Connemara showing localities mentioned in text, bathymetry and general sediment distribution (after Lees *et al.*, 1969).

exceed 40%. Their major biogenic constituents are benthonic molluscs and forams, with molluscan material increasing toward the more variable salinity bay-heads.

Sands: occur as a blanket, up to 5 km wide, containing up to 80% carbonate. Grain sizes range from fine to coarse and surfaces are locally rippled although burrowers are locally abundant. The carbonate fraction of the finer sands is composed of debris from molluscs, forams, echinoderms, bryozoans, ostracods and sponges. In proximity to the *Lithothamnium* facies, *Lithothamnium* material and debris from grazing gastropods increase. The more open parts of the bay contain very coarse sand composed of large proportions of *Mytilus* and barnacles derived from offshore islands and rocky shoals.

Lithothamnium facies: consists of 'gravels' comprising rhodoliths of the unattached calcareous algae *Lithothamnium* and *Phymatolithon*. They live both in quiet and exposed waters and are restricted by light levels to depths of less than 16 m. Rhodoliths vary in shape from sphaeroidal through ellipsoidal to discoidal. Dense branching of the algal thalli develops in response to apical damage incurred during rolling and thus densely branched forms are found in exposed areas while open forms with less branching occur in quiet areas. Locally the algae produce 30 cm high autochthonous banks with diverse faunas. The development of banks is primarily controlled by wave-induced currents and the algae are broken down to form a mobile algal gravel which supports a poor fauna (Bosence, 1976).

Hard Substrates: Rocky outcrops usually carry heavy growths of seaweed and in the littoral zone abundant barnacles and *Mytilus*.

Two types of hardened contemporary sediments ('beach rock') have been recognized: (1) lithified inter-tidal *Lithothamnium* rock in which the cement consists of micritic low-Mg calcite within a frame of *Lithothamnium* fragments. These are still in high-Mg calcite although heavily micritized.

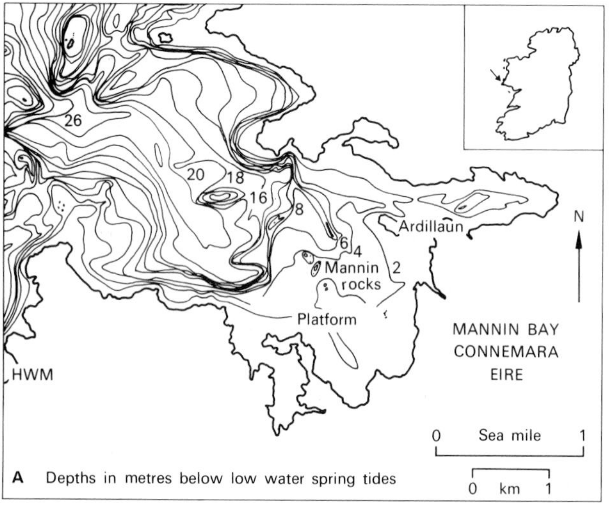

Fig. 10.35A. Bathymetry of Mannin Bay (after Bosence, 1976).

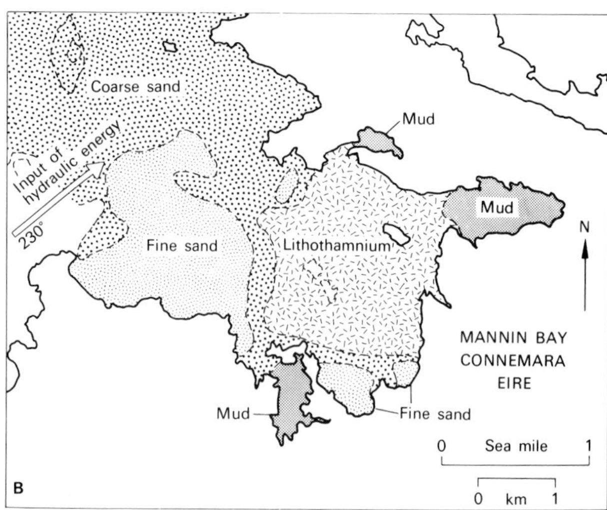

Fig. 10.35B. Lithofacies in Mannin Bay and the direction of main energy input (after Bosence, 1976).

(2) Patches of lithified mollusc-shell sand that occur in the upper intertidal zone.

The distribution of the sedimentary facies relates to the direction from which maximum energies are derived ($\sim 230^{\circ}$). In detail, the growth-forms of the rhodoliths also relate to the environmental energy which decreases shorewards (Bosence, 1976).

10.3.2 Ancient examples of possible temperate water carbonates

As the published work on temperate shelf carbonates is so limited it is to be expected that few ancient analogues have been described. However, possible comparisons do exist and over coming years many more may follow.

UPPER CRETACEOUS (IVÖ KLACK), SOUTHERN SWEDEN (Fig. 10.36A and B)

The Upper Cretaceous Campanian rocks exposed at Ivö Klack in Southern Sweden (Surlyk and Christianson, 1974) represent transgressive carbonates resting upon deeply weathered Pre-Cambrian bedrock that formed an archipelago of low islands. On one of these islands a rocky shoreline fauna developed which, though dominated by bivalves (especially oysters), includes the most northerly known Cretaceous reef-corals and rudists. Densely-branched red algal rhodoliths similar to those described by Bosence (1976) from the more exposed parts of Mannin Bay are present also, although here they are accompanied by green algae.

Surlyk and Christianson believe that a subtropical climate is indicated because of the presence of seven species of presumed hermatypic corals. However, corals are subordinate to a modified *foramol* assemblage of barnacles, molluscs, calcareous algae and bryozoa. This appears to represent an odd mixture of chlorozoan and foramol, putting the area at the limit of these associations (Lees, pers. comm. 1977).

Added support for a subtropical interpretation came from some early oxygen isotope data of Lowenstam and Epstein (1954) indicating a minimum temperature of 17°C. However, the fauna is sufficiently dissimilar from that of a modern subtropical assemblage to warrant further consideration. On palaeomagnetic grounds, Ivö Klack would have been at about 35°–40°N in Upper Cretaceous time (Van der Voo and French, 1974; Smith *et al.*, 1973) but similarities to the western Ireland carbonates allow at least a general comparison. Their respective settings of high organic productivity and low terrigenous run-off seem to have been identical although strict climatic comparison is more doubtful.

MIOCENE OF THE HOLY CROSS MOUNTAINS, POLAND

Lithothamnium calcarenites make a substantial contribution to some Late Miocene sediments on the southern margin of the Holy Cross Mountains (Radwanski, 1968, 1969, 1973). The Late Miocene transgression produced a 'Dalmatian-type' coast in this region (Fig. 10.37) with rocky cliffs of Mesozoic limestone in the west and a more gently shelving terrigenous-

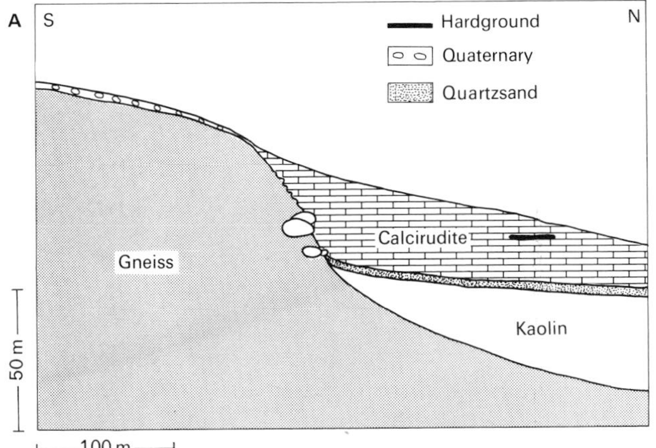

Fig. 10.36A. Distribution of facies at Ivö Klack, a schematic diagram showing the distribution of boulders mapped by Surlyk and Christianson, 1974 (after Surlyk and Christianson, 1974).

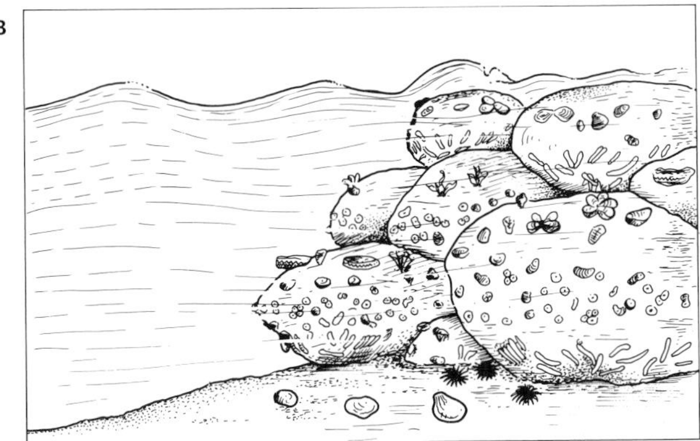

Fig. 10.36B. Reconstruction of biofacies at Ivö Klack showing the cementing fauna of the Cretaceous rocky coast attached to the boulders.

clastic dominated shoreline in the east where a series of rhodolithic limestones occurs above quartz sands and ahermatypic corals, while in the west similar limestones succeed clays and marls. The coralline-algal sand facies includes both *in situ* material and cross-bedded units of derived algal rhodoliths.

The fauna is dominated by bivalves and gastropods with bryozoans, barnacles and a variety of other groups. The whole assemblage of fauna and sediments appears to be comparable with those ot Mannin Bay (Sect. 10.3.1). On the basis of faunal evidence, Bałuk (1971) believed that the Miocene climate was subtropical although the palaeolatitude would have been about 40°N (Van der Voo and French, 1974; Smith *et al.,* 1973). Colonial corals are absent and the faunal assemblage as a whole compares far more closely with the *foramol* association of Lees and Buller (1972) than with a normal subtropical *chlorozoan* assemblage.

OTHER ANCIENT EXAMPLES?

Apart from the two possible examples cited above, where are there others? *Lithothamnium* limestones are fairly common in the late Tertiary around the Mediterranean. They are well exposed in the late Miocene–Pliocene of the Balearics (personal observations) and have been described from Malta (Pedley, 1976; Pedley and Waugh, 1976; Pedley *et al.,* 1976) but no comprehensive accounts exist.

Many ancient tillite-bearing sequences are associated with carbonates (including large amounts of dolomite), for example the Varangian of Norway and Scotland (Sect. 13.5.4). Buller (1969) suggested that certain Permian limestones from Tasmania that are also interbedded with glacial sequences would be likely places to search.

As modern temperate-water carbonate environments become better documented further examples of ancient equivalents may be recognized.

10.4 CARBONATE BUILDUPS THROUGH TIME

10.4.1 Modern buildups

A carbonate buildup is a body of carbonate rock possessing topographic relief above coeval substrates. Such buildups (autochthonous mounds) may be termed reefs if they display evidence or potential for maintaining growth in the wave zone (Heckel, 1974). Each buildup owes its origin to the growth of particular suites of organism communities that provide either a framework of skeletal material for the buildup or a mechanism for selectively baffling out sediment. The modern coral reef is, ecologically, a steady-state 'oasis' of organic productivity with high population densities, intense carbonate metabolism and complex food-chains generally surrounded by waters of relatively low mineral and plankton content (Stoddart, 1969). Through Phanerozoic time, many groups of organisms have either constructed or significantly contributed to buildups (Fig. 10.38; Heckel, 1974; Copper, 1974). Barriers comprising complex chains of linear buildups are recognizable in the geological record because they separate slope and basinal sequences from coeval shallow-water deposits.

In modern reefs, organisms in their original growth position may only account for 10% or less of the total reef mass (Ladd, 1971). The remainder is provided by organisms inhabiting the intra-reef frame and by material derived from abrasion and predation of the frame itself. In modern sub-tropical reefs, the

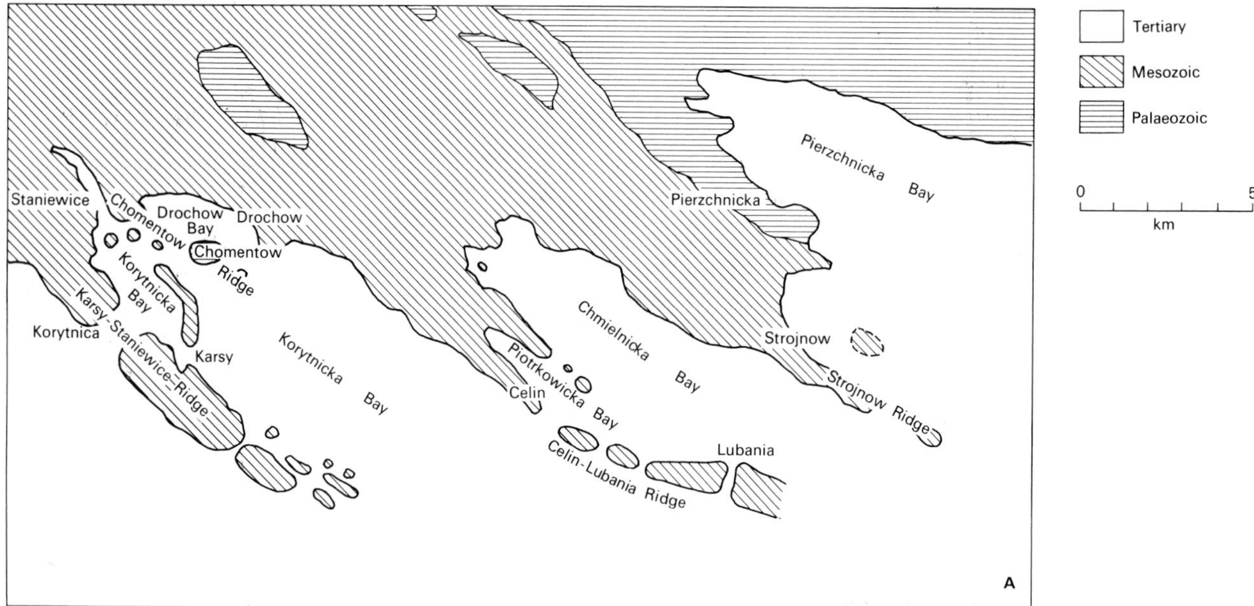

Fig. 10.37A. Locality map of a small part of the Holy Cross area of Poland (after Radwanski, 1968).

Fig. 10.37B. Schematic cross sections through the area to show facies relationships. These sections are extremely generalized and are not drawn to scale (after Radwanski, 1968).

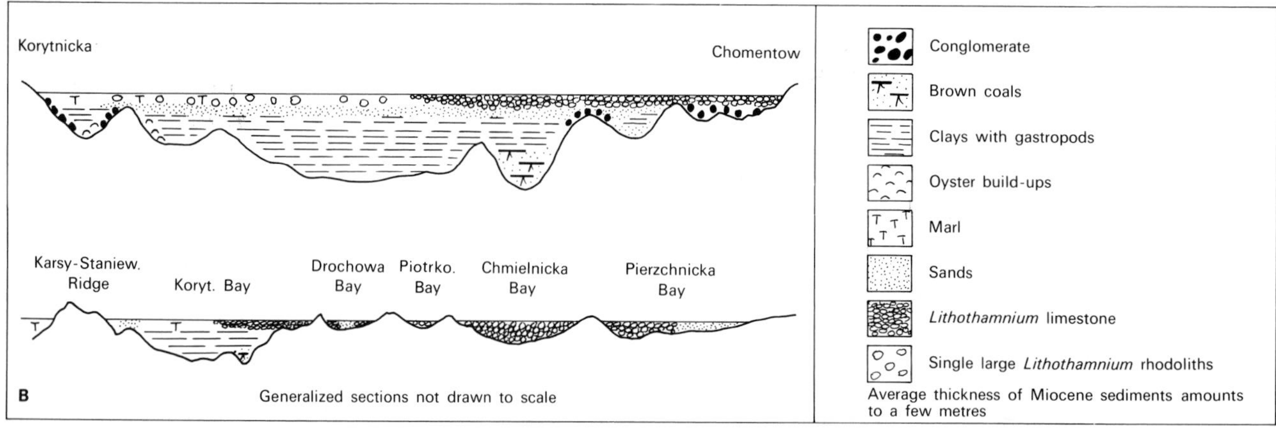

Fig. 10.38. The possible ecological distribution of dominant organisms in skeletal buildups through time (after Heckel, 1974).

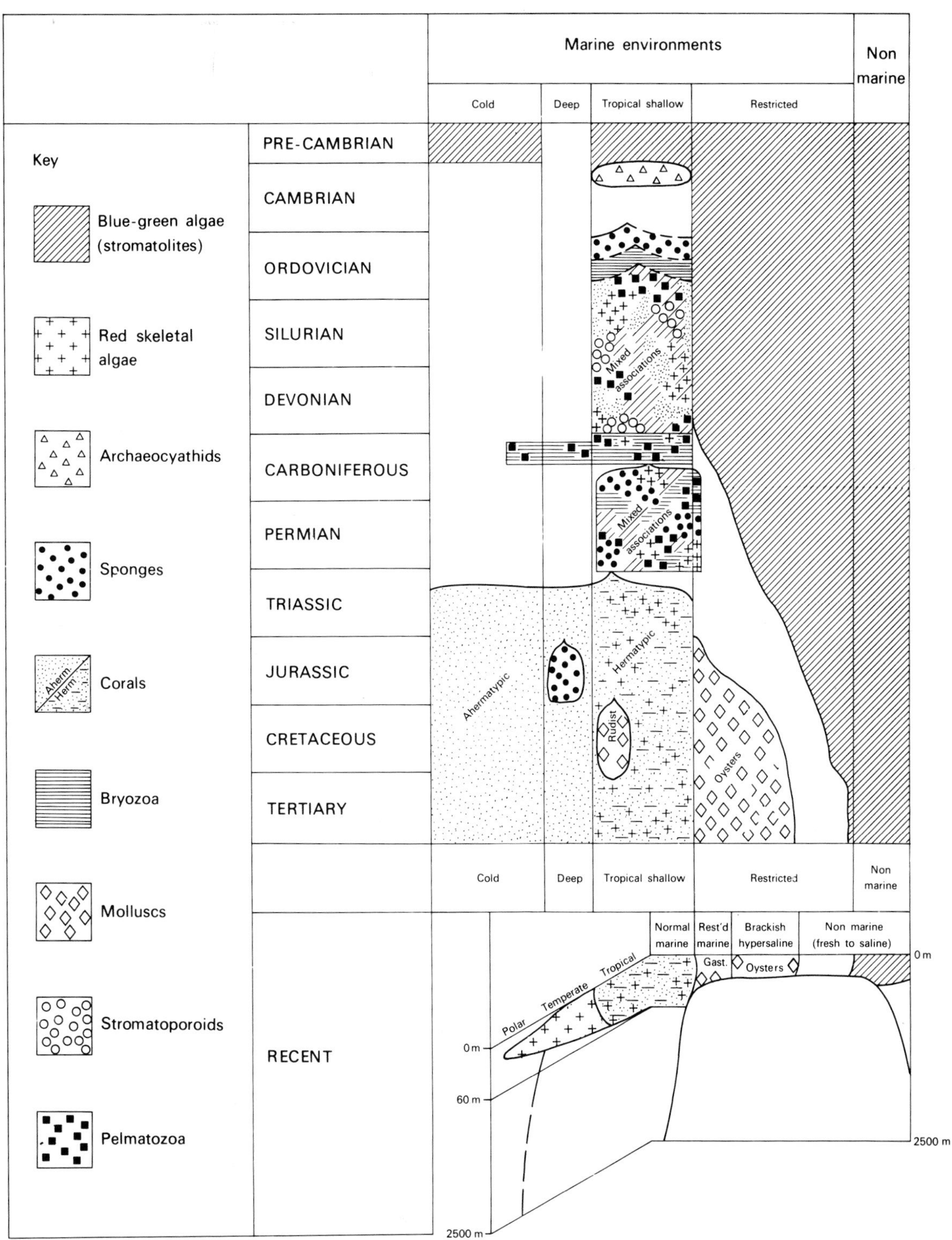

Marine environments
Non marine
Cold
Deep
Tropical shallow
Restricted
Key
Blue-green algae (stromatolites)
Red skeletal algae
Archaeocyathids
Sponges
Corals
Aherm
Herm
Bryozoa
Molluscs
Stromatoporoids
Pelmatozoa
PRE-CAMBRIAN
CAMBRIAN
ORDOVICIAN
SILURIAN
DEVONIAN
CARBONIFEROUS
PERMIAN
TRIASSIC
JURASSIC
CRETACEOUS
TERTIARY
RECENT
Mixed associations
Ahermatypic
Hermatypic
Rudist
Oysters
Normal marine
Rest'd marine
Brackish hypersaline
Non marine (fresh to saline)
Gast.
Polar
Temperate
Tropical
0 m
60 m
2500 m

frame is formed by hermatypic corals (which have dinoflagellate symbionts (zooxanthellae) living within their tissues) often welded into a rigid frame by red algae.

The main ecological requirements for modern hermatypic coral growth are: (1) Shallow water (down to 100 m), (2) Warm water (18°–36°C), (3) Normal salinities (27–40‰), (4) Fairly strong sunlight, (5) Abundant nutrients to supply zooplanktonic prey for the polyps, (6) Stable substrates for attachment (Ginsburg, 1972). Lacking osmoregulatory organs, coral polyps are extremely vulnerable to rain and flood waters, especially at low tides. There are many hazards in closely modelling ancient coralliferous buildups upon their Recent counterparts (Teichert, 1958a; Heckel, 1974; Moore and Bullis, 1960; Stoddart, 1969, and many others). Buildups with ahermatypic coral-frames grow in deep and shallow waters of low turbidity outside the areas of hermatypic dominance (see summary in Heckel, 1974). Sometimes ahermatypic buildups form deeply submerged but living barriers between present-day shelf and slope facies (e.g. off the French coast). Other organisms capable of producing modern buildups in shallow-water include vermetid gastropods, sabellariid worms and coralline algae but they are insignificant by comparison with the coral-dominated reefs.

The coelenterates have not always dominated buildup communities (Fig. 10.38). Perhaps their fluctuating fortunes have been in part resource based. Corals are carnivores but there is no doubt that the zooxanthellae translate photosynthate to the animals. The precise relationship between algae and polyps is a subject of continuing controversy (Stoddart, 1969; Wethey and Porter, 1976), but it is well known that corals reject their symbionts when starved of zooplankton. Zooxanthellae probably play a major role in removing coral wastes and in promoting carbonate precipitation. Through geological time, it is possible that coral-dominated communities have grown best at or near ocean-fringes where zooplankton populations were kept high as a result of upwelling. Epeiric basins have mostly contained buildups dominated by other groups.

The gross morphology of many modern subtropical reefs appears to be related to the underlying karst topography with

Table 10.2 Lithologic, palaeontologic, and stratigraphic characteristics of the tidal flat, shallow subtidal, deep subtidal, and organic buildup facies suites of the Cambrian through Devonian carbonates in the central Appalachians (after Laporte, 1971). (see p. 290).

FACIES CHARACTERISTICS \ FACIES SUITES	Tidal flat	Shallow subtidal	Deep subtidal	Organic buildups
Mud cracks and birdseyes	typical	—	—	—
Scour and fill w/pebble cgls	typical	—	—	—
Laminations	typical	—	—	—
Early dolomite	typical	—	—	—
Sparite/micrite	variable	high–low	low	variable
X-stratification	small-cale	medium-scale	—	sometimes present
Burrow-mottling	rare	common	abundant	rare
Oolites	—	often present	—	—
Bedding	thin-medium	medium-thick	thick, massive	unbedded, massive
Quartz and clay			sometimes abundant	
Algal structures	stromatolites	oncolites	—	typical in C–O
Burrows	vertical	vertical and horizontal	horizontal, abundant	rare
Fossil abundance	low	very high	variable	very high
Fossil diversity	low	medium	usually high	medium-high
Major taxa	trilobites and/or ostracodes	calc. algae, pelmatozoa, brachs, and ectoprocts	brachs, trilobites, ectoprocts, and pelmatozoa	tabulate and rugose corals, stromatoporoids
Vertical facies variations	sharp and frequent	transitional and common	very gradual and infrequent	complex
Areal facies variations	outcrop scale	relatively persistent	basinal scale	outcrop scale to several kms
Facies strike	variable	parallel to basin margin	parallel to basin axis	variable

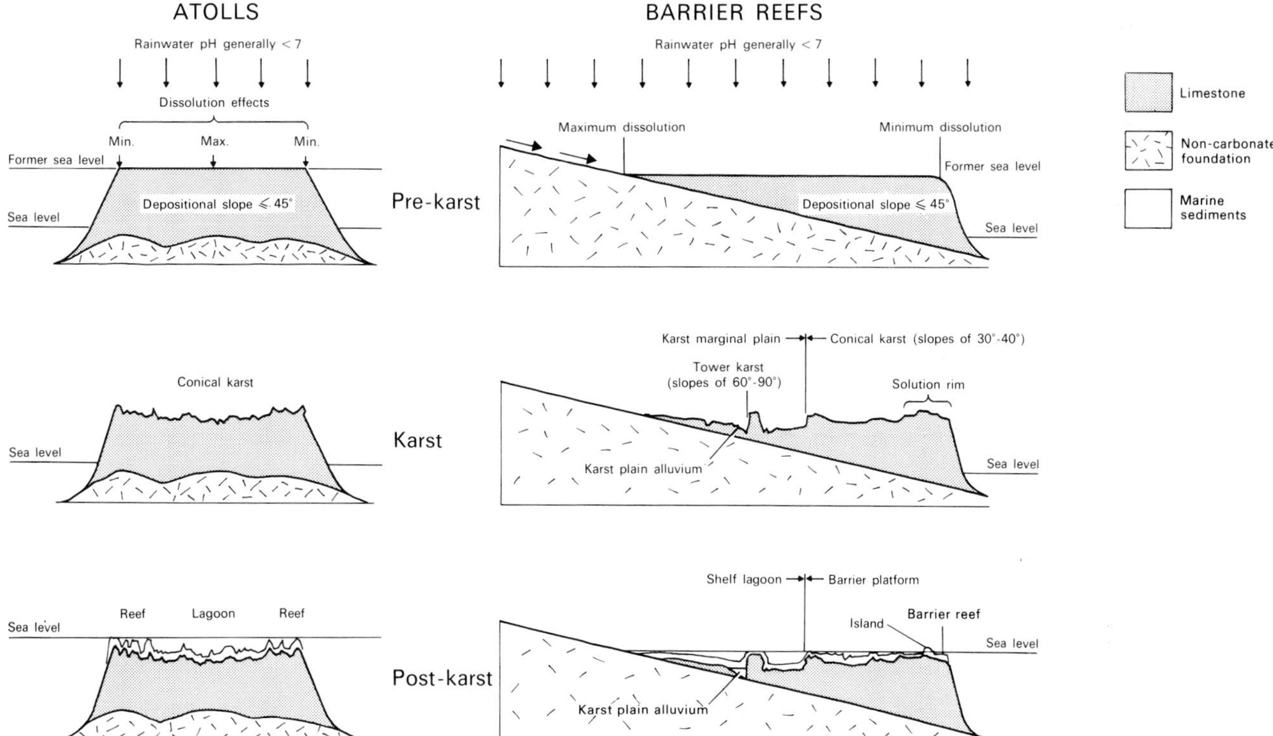

Fig. 10.39. The evolution of atolls and barrier reefs according to the antecedent karst theory. Both sequences start with subaerial exposure and end with the results of deposition on a drowned karst topography (after Purdy, 1974a).

reefs growing in many areas as veneers on drowned karsts (Purdy, 1974a, and Fig. 10.39). In the Persian Gulf reef communities grow upon the crests of offshore salt domes (Purser, 1973a).

The three main morphological elements of a reef comprise the fore-reef, reef-flat, and back-reef. The fore-reef consists of reef-derived talus associated with a biota that is distinct but gradational with that of the reef-flat at shallow depths (e.g. Figs. 10.40, 10.12). The back-reef is mostly a calm shallow water area with greatly diminished organism diversities by comparison with the reef belt. This restriction in biota results from the generally higher variability in the environment and the increased stresses there.

A comprehensive review of buildups through time is impossible here, but references in Heckel (1974) and Wilson (1975) will provide a basis for further reading. To complement the review in Fig. 10.38 examples from the literature have been selected to illustrate the evolution of major buildup types through time.

10.4.2 **Ancient analogues**

CENOZOIC

Buildup communities have evolved through time and in general the most ancient communities provide the least reliable comparisons to those of the Recent. Coral/algal buildups probably stand close comparison with their modern counterparts as far back as the Eocene, and deep drilling on atolls has penetrated Eocene facies closely comparable to modern equivalents.

MESOZOIC

Coral/algal buildups grew in the Tethyan belt from Mexico through Europe to Japan during the Cretaceous, but in the Mid to Late Cretaceous rudist bivalves dominated buildup construction through this region. In the Pacific, a drowned

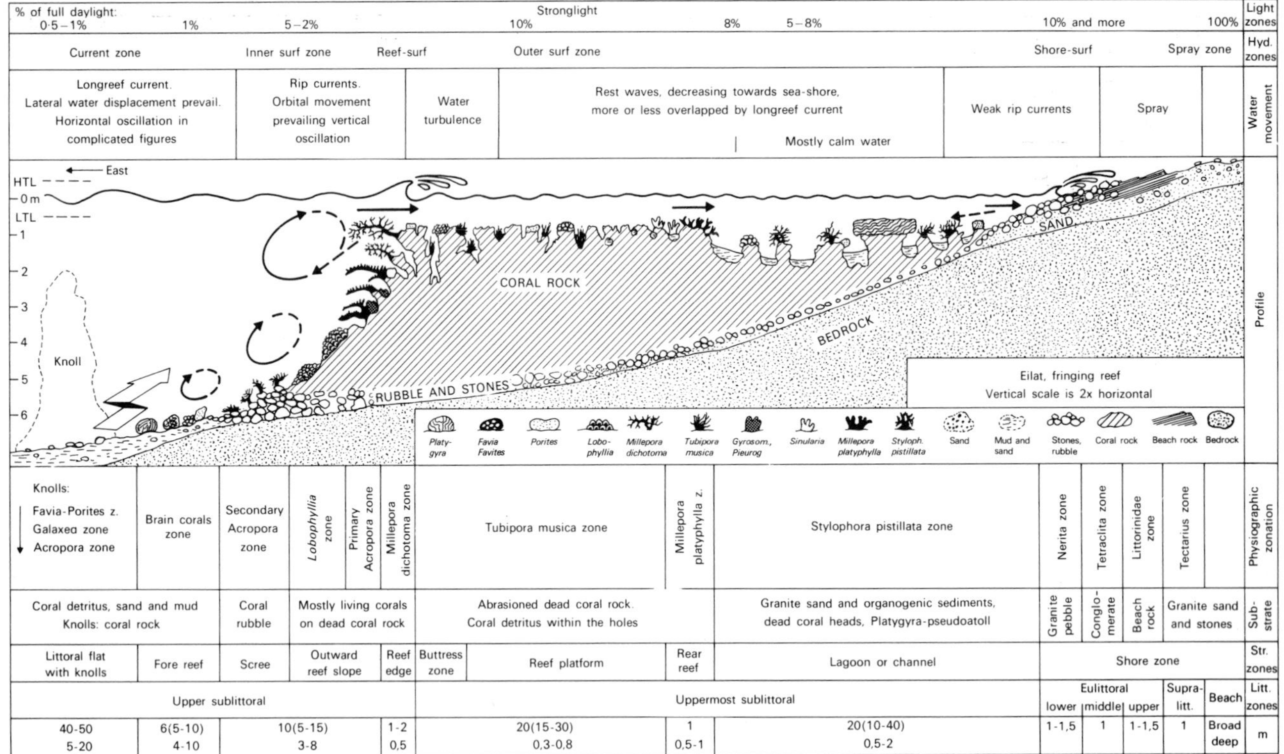

Fig. 10.40A. A profile through the northern fringing reef of Eilat showing the main ecological zones and their controlling influences (after Mergner, 1971).

Cretaceous atoll is rimmed by fossil rudists (Ladd, 1971). Rudist buildups provided a rim to the ancestral Gulf of Mexico through much of the Cretaceous and associated platform and basinal facies have been well mapped because of the large oil fields associated with the reef belt (Figs. 10.41A and B).

Rudists were almost certainly suspension-feeders (Skelton, 1976). Thus their feeding requirements differed from those of corals and direct ecological comparisons between Cretaceous rudistid and modern coralgal reefs cannot be made. It is probable that at least some rudistids, like modern *Tridacna*, were symbiotic with zooxanthellae.

In the Caribbean (Kauffman and Sohl, 1974) and elsewhere the rudists were distributed over shallow shelf environments that had very low sedimentation rates, with the rudist frameworks being dominant everywhere except in the shallowest subtidal zones where coral patches grew. Rudistid communities developed best on calcarenitic or coarser substrates and, although accessory organisms like oysters, corals, algae, stromatoporoids, hydrozoans and bryozoans were commonly present in the most mature frameworks, rudists were dominant. Competitive restriction and/or secretory activity of the rudist mantle may have caused this dominance which led to the lack of an effective binding agent for many of the frames. Sometimes rudist/algal associations are found (Wilson, 1975, p. 323) and in these cases radiolitid rudists were generally associated with red algae in an outer-shelf position (Fig. 10.42). Codiacians occurred with caprinid rudists over a wide range of environments while requienids and blue-green algae grew together in some extreme 'back-reef' positions. The absence of associated ammonites, echinoids and brachiopods may also be attributed to the ability of rudists to grow in fluctuating temperature, oxygen and salinity regimes.

During the Late Jurassic, in Bavaria and elsewhere, buildups were composed of flat and cup-shaped sponges and algae. They occur in a micritic matrix associated with bryozoans, brachiopods and serpulids. Ammonites, bivalves and belemnites also occur. These associations formed in quiet water environments that may have been relatively deep but photic. These structures and associated sediments are reviewed in Gwinner (1976) and Barthel (1970).

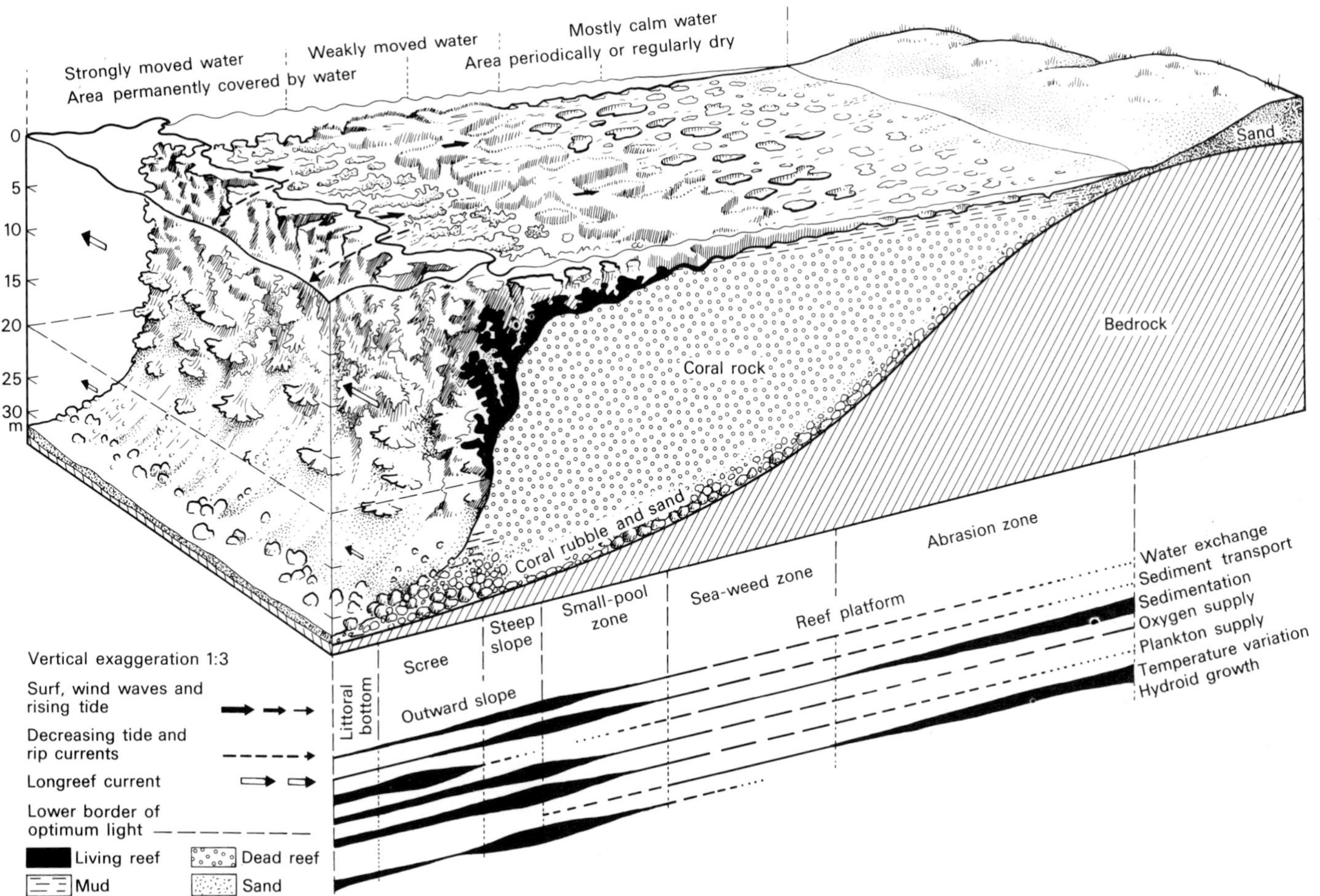

Fig. 10.40B. Block diagram of the northern fringing reef of Eilat illustrating the reef structure, zonation and major ecological conditions within the reef (after Mergner, 1971).

Late Triassic barrier buildup complexes developed in Germany and Austria (Fig. 10.43; from Zankl, 1971). They separated shallow-water carbonates from deeper-water calcilutites into which beds of buildup-derived talus were emplaced. These Dachstein buildups consist of dome-like patches of calcisponges and hexacorals (75%) that are bound by solenoporoid red algae, hydrocorallines, cyclostome bryozoans, encrusting forams and stromatolites. The immediate back-reef sequence consists of coated and well rounded calcarenite grains with abundant dasycladacean green algae and megalodont bivalves. Shelfwards these sediments grade into stromatolitic and birdseye laminates of the inter- and supra-tidal mudflats (see Sect. 10.2.5).

LATE PALAEOZOIC

While the Triassic and later buildups have a fairly familiar aspect, the situation is very different in the Palaeozoic. Before the great end-Palaeozoic extinctions Late Palaeozoic buildup communities often included fenestellid bryozoans, productid brachiopods and rugose tetracorals.

Of the Permian buildup complexes, that of Capitan is best known. It consists of a plexus of barrier buildups that formed on the margin of the Delaware basin of Texas and New Mexico (Fig. 10.44). The actual barrier buildup itself ('reef') contains encrusted upright calcisponges bound by problematic hydrocorallines (Tubiphytes), stromatolitic algae, bryozoans

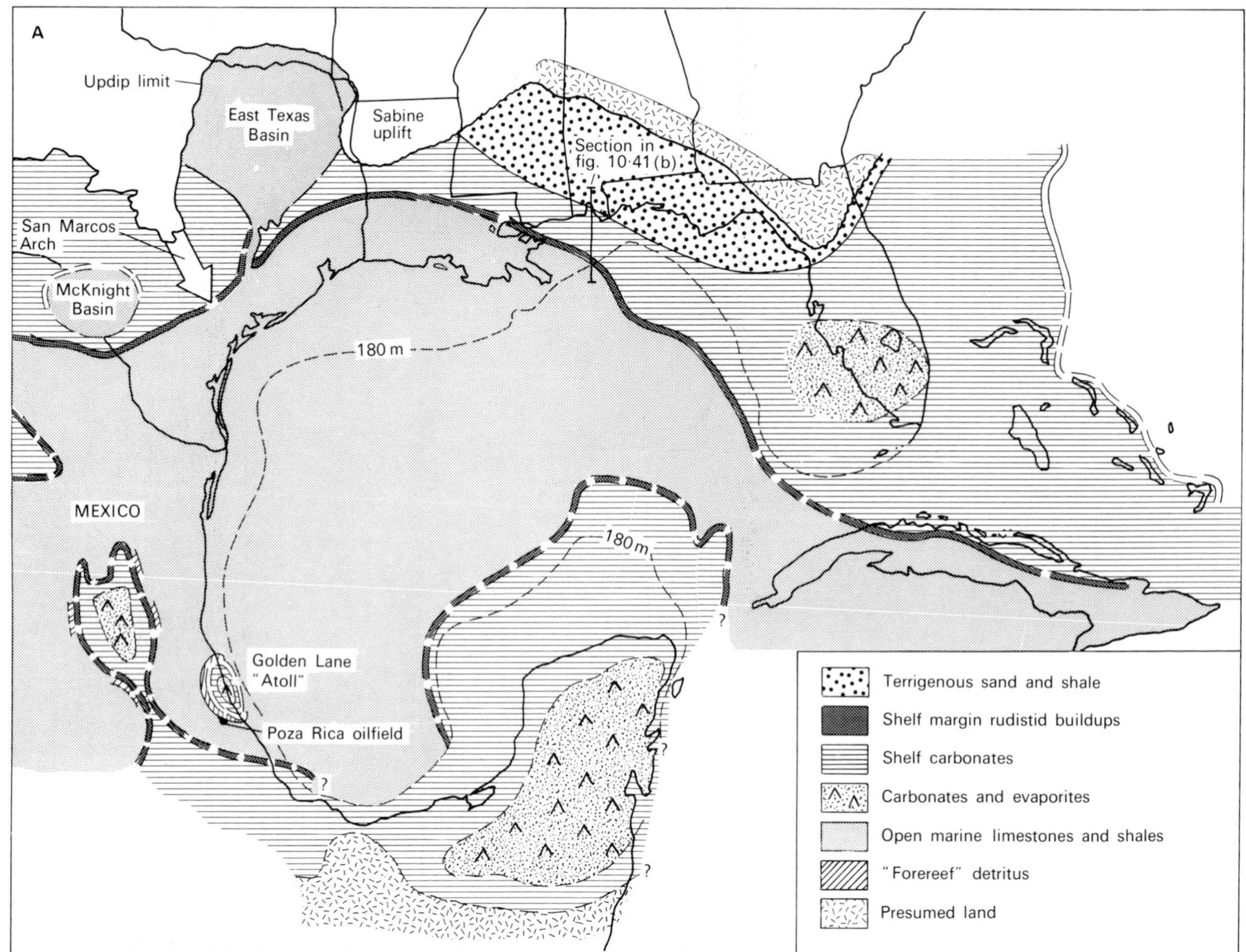

Fig. 10.41A. Major facies distributions in the Early Cretaceous of the Gulf of Mexico region showing the distribution of the rudist buildup trend (after Bryant *et al.*, 1970; with many additions from R.C. Vernon *pers. comm.*, 1977).

and solenoporoid red algae intergrown with cyclostome bryozoa. Large amounts of other skeletal debris are present including pelmatozoa, brachiopods and foraminiferans. Only about 3% of the fossils are in place. The basinal side of the buildup complex consists of lime mud with interbedded coarse and fine skeletal debris and clasts of buildup material. This consolidated 'fore-reef' sediment also occurs as clasts within the finer sediment at the foot of the 'fore-reef' slope. Shelfwards, winnowed skeletal sands with both dasycladacean and codiacian algae, oncolites, molluscs and forams occur although this material and the buildup lithologies are extensively dolomitized. Vadose pisolites are also present and the calcarenites pass laterally into laminites and evaporites (Newell *et al.*, 1953; Dunham, 1972; Wilson, 1975, p. 217 et seq.). Dunham (1972) tried to show that marked vertical movements of sea-level had affected the shelf-margin and that during times of low-stands, intense vadose zone diagenesis was initiated and slide conglomerates developed on the buildup slope. He attributed the sea-level variations to waxing and waning glaciations on Gondwanaland. However, several environmental models could be fitted to the Capitan complex (Fig. 10.44) with either interpretation A or C being the best fits.

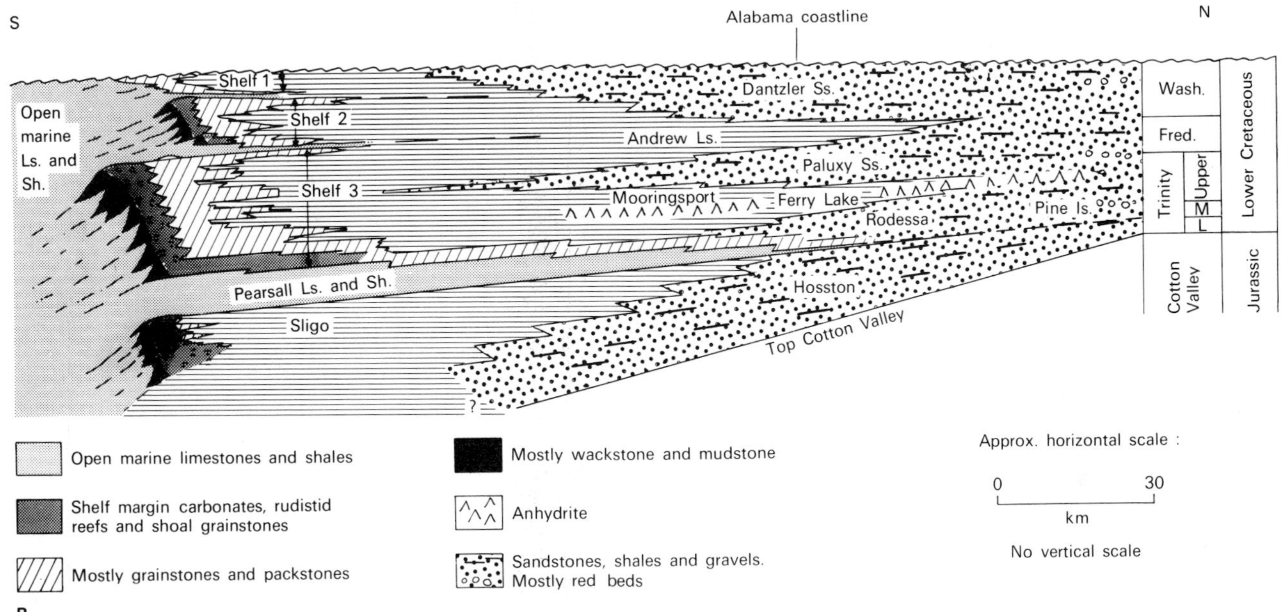

Fig. 10.41B. Schematic regional dip section along the line shown in 10.41A. Section provided by R.C. Vernon, Shell Oil Co., Houston.

Fig. 10.42. Generalized biofacies across the margin of the Golden Lane 'Atoll', Mexico, shown in Fig. 10.41A (modified from Wilson, 1975).

MICROFACIES	Light to dark. Homogenous or laminated micrite. Occasionally calcarenitic. Pelagic in origin.	Dark micrite with thick to thin rhythmic bedding and slump structures. Pelagic and microbioclastic.	Coarse lithoclastic-bioclastic limestones. Boulders in micritic matrix	Rudistid knolls with creamy, shelly, thick-bedded limestone. Micritic matrix.		Oolitic-bioclastic grainstone.	Light micrite with thin to medium beds. Cyclically arranged miliolid grainstone, bioturbated wackestone to laminated fenestral micrite. Dolomite crusts.	Anhydrite facies within Golden Lane Bank.
BIOTA	Planktonic microfossils. Ammonites, Globigerinids, Tintinnids, Coccolithophorids.	Planktonic microfossils. Ammonites, Globigerinids, Tintinnids, Micropeloids, Coccolithophorids.	Mixed but mainly debris from upslope.	Caprinids, Radiolitids, Colonial corals, Stromatoporoids, encrusting algae, boring bivalves, benthonic forams.	Dominated by caprinids with requienids and the oyster-like *Chondrodonta*, Miliolids and gastropods.	Caprinid debris, radiolitids, stromatoporoids, solenoporoid red algae, Codiacian algae, dasyclad algae and miliolids and gastropods.	Restricted stromatolites, biostromes of requienids and *Gryphaea* when marly, dasycladaceans, *Dicyclina*, Miliolids.	Miliolids only
ENVIRONMENTAL INTERPRETATION		TAMABRA FACIES		Outer knolls	Inner knolls	Interreef sands; TANINUL FACIES	Increasing salinity → EL ABRA FACIES	
EXTENT	BATHYAL-TAMAULIPAS FACIES many km wide	TOE OF SLOPE (TAMAULIPAS FACIES) ~ 2 km wide	FOREREEF up to 15 km wide	SHELF MARGIN KNOLLS AND PATCH REEFS a few km wide			INNER BANK ~ 100 km across	

N
S
Tens of kilometres
BASIN WITH BITUMINOUS MARLS
Minor reef knolls
Shallow water mud banks with inter- and supratidal flats
SAND FLATS
REEF BELT
Reef knolls
Reef detritus
Basinal marls
MARLS
Sea level
Kössen Beds
Hauptdolomit
Cyclically arranged loferites and calcarenites
Lofer facies
Back reef
Reef belt
Fore reef
Zlambach Marl
Hallstatt Limestone
DACHSTEIN LIMESTONE

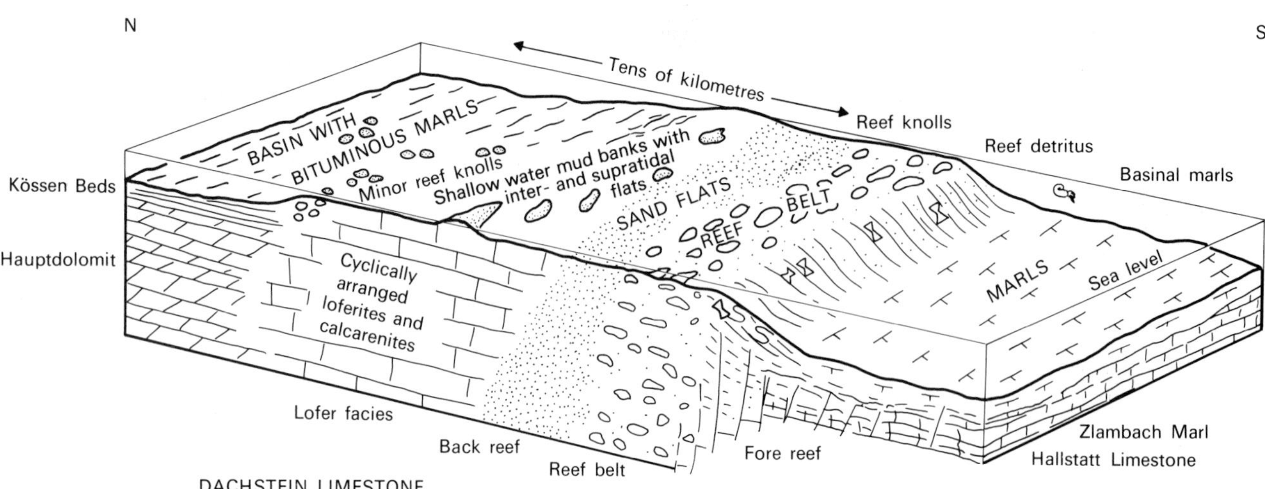

Fig. 10.43. Palaeogeographic reconstruction of the depositional environments during the Late Rhaetian in the Northern Limestone Alps of Austria. Compare with Fig. 10.25B (after Zankl, 1971).

Fig. 10.44. Three alternative models of the Permian Reef Complex, West Texas and New Mexico (after Dunham, 1972).

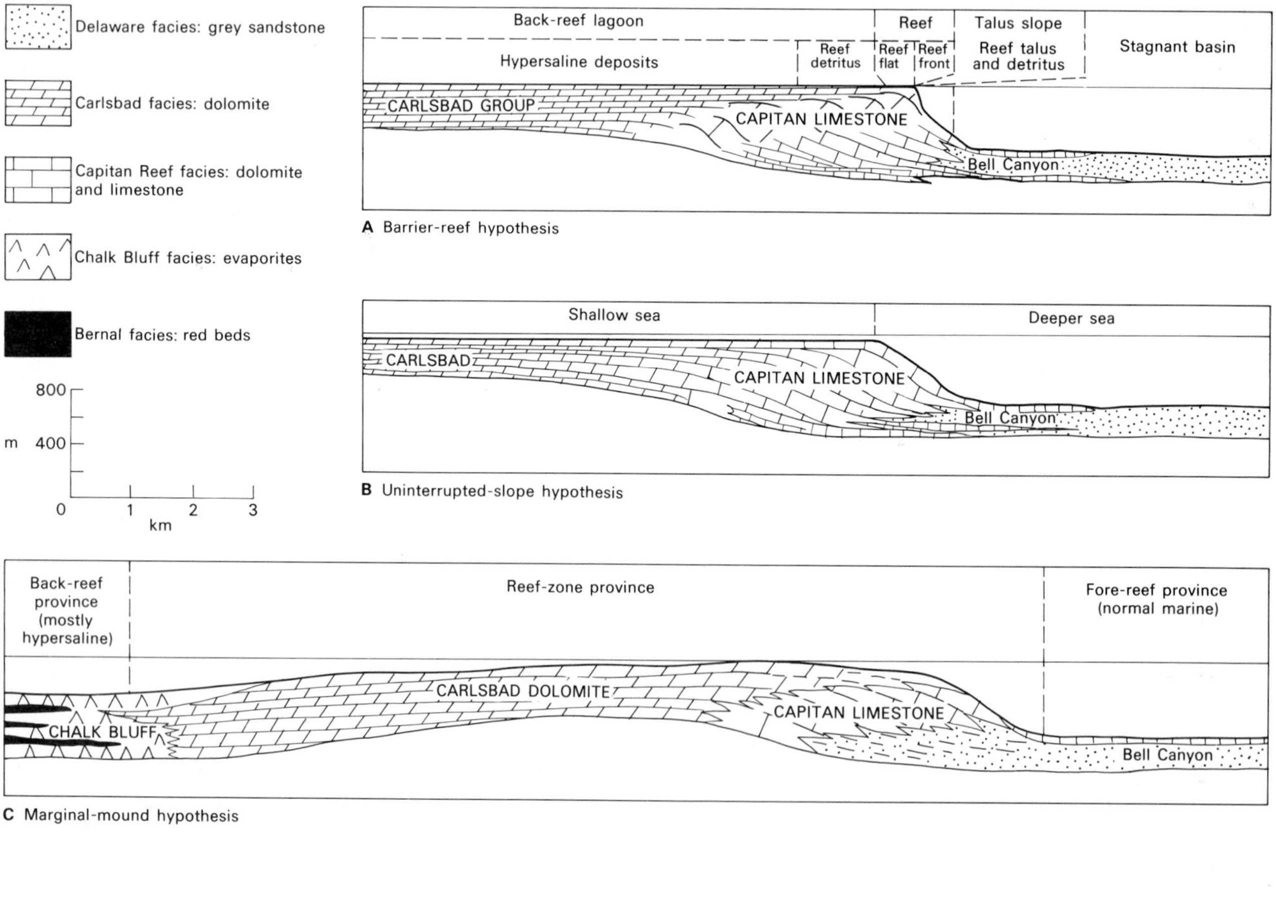

Delaware facies: grey sandstone
Carlsbad facies: dolomite
Capitan Reef facies: dolomite and limestone
Chalk Bluff facies: evaporites
Bernal facies: red beds
800
m 400
0
1
2
3
km
Back-reef lagoon
Reef
Talus slope
Hypersaline deposits
Reef detritus
Reef flat
Reef front
Reef talus and detritus
Stagnant basin
CARLSBAD GROUP
CAPITAN LIMESTONE
Bell Canyon
A Barrier-reef hypothesis
Shallow sea
Deeper sea
CARLSBAD
CAPITAN LIMESTONE
Bell Canyon
B Uninterrupted-slope hypothesis
Back-reef province (mostly hypersaline)
Reef-zone province
Fore-reef province (normal marine)
CARLSBAD DOLOMITE
CHALK BLUFF
CAPITAN LIMESTONE
Bell Canyon
C Marginal-mound hypothesis

Evidence provided by Harms (1974) suggests that the basin was density stratified and received dense saline shelf waters through channels cutting the surrounding banks.

In the Early Carboniferous, buildups were mostly Waulsortian-type mounds that sometimes formed major mound complexes – as in Ireland (Lees, 1961, 1964). Sometimes mounds rose more than 100 m within shale-dominated basins. The mounds consist of a core of micrite with variable amounts of cavity-filling and neomorphic carbonate spar often showing a complete spectrum from a mud-rich facies with scattered fenestellids to a spar-rich facies with a meshwork of fenestellids. The main recognizable skeletal constituents are fenestellid bryozoans with crinoids and other invertebrate bioclasts scattered throughout. Depositional accretion surfaces within the mounds dip at slopes of up to 50° and are depicted by shell-fragment layers and Stromatactis.

In Ireland the Waulsortian mounds were widely distributed over the shelf but were more abundant in a broad zone separating the deeper water mud-belt to the south from the lagoonal facies to the north (Lees, 1961). One of the many problems concerning micritic mounds of Waulsortian type involves the maintenance of relief and steep slopes, particularly when there is apparently no organic/skeletal frame to the structure. Complicated hypotheses have been evolved to account for the generation of such buildups (Fig. 10.45, from Wilson, 1975) but studies of lime-mud and sand banks in Shark Bay (Davies, 1970) and Bahamas/Florida (Ginsburg and Lowenstam, 1958) have shown that provided a suitable baffling agent is present (in modern areas this is normally marine grass) then lime-mud banks develop as largely self-propagating systems. In the Waulsortian mounds the interacting organisms (partly preserved and partly diagenetically lost) may have constructed community mounds comparable in a general way to those of Shark Bay. The presence of associated slumped flank beds provides evidence of periodically mobile slopes but the general absence of current scours and winnowed cross-beds suggests that the mounds formed in a low-energy and possibly deep-shelf position.

The Mid to Late Devonian witnessed an explosive and almost world-wide development of coral/stromatoporoid buildups. Through the Silurian and Early Devonian, tabulate corals were dominant, but rugose tetracorals then became increasingly important.

The Givetian–Frasnian reefs of western Canada (Jamieson, 1971) and of the Canadian Rockies (Nobel, 1970) provide some excellent examples of both rigid frame and open skeleton construction buildups that were controlled in their distribution by subtle shifts in both sea-level and bottom topography. Particular communities were confined to specific environments. Massive stromatoporoids, crustose coralline algae and colonial corals were mainly restricted to the 'reef' and 'fore-reef', while organisms with more delicate branching skeletons,

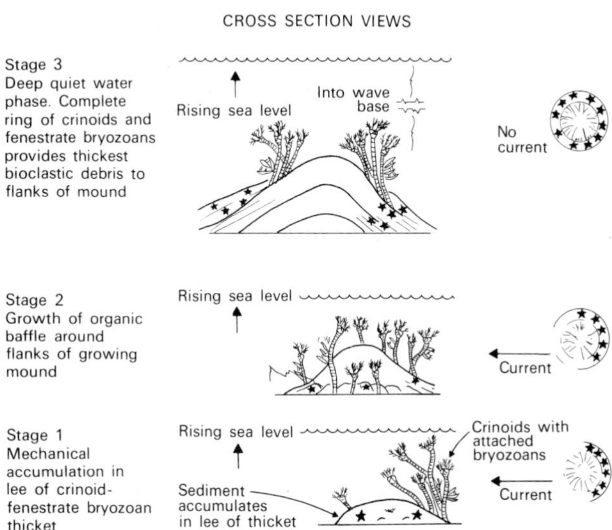

Fig. 10.45. A possible model of Waulsortian mound development involving progressive colonization of a sediment mound that formed initially in the lee of a crinoid-bryozoan thicket (after Wilson, 1975).

like *Amphipora,* dominated the presumed lagoonal environments. An environmental model was constructed by Jamieson (1971) for the Alexandra Reef-complex in which she recognized deep-water fore-reef, reef, protected shelf lagoon and semi-emergent coastal mudflats (Fig. 10.46). Algae contributed greatly to carbonate formation and in Jamieson's view, many of them resembled modern reef-building algae 'to a striking degree'. They also showed distribution patterns similar to those of algae in modern seas.

A complex of patch-reefs a few metres to several tens of metres across constructed a partial barrier. In the 'back-reef' area behind this barrier 'beehive-shaped' knolls occurred. These were up to 2 m high and were dominated by *Amphipora,* crustose coralline algae and small corals. Mud-mounds were also locally developed. In Australia (Playford and Cockbain, 1969), W. Europe (Krebs, 1974) and many other parts of the world (Heckel, 1974) comparable buildups were extensively developed.

EARLY PALAEOZOIC

In the Silurian, important buildup complexes grew in Gotland, Illinois–Indiana and England. In the USA, buildups were best developed in the east towards the margin of the central craton (Fig. 10.47). In addition, the three cratonic basins of Michigan, Illinois and Texas–New Mexico were each rimmed by buildups. In Illinois the buildups occurred in irregular clusters where rates of clastic deposition were low (Figs. 10.47A and B;

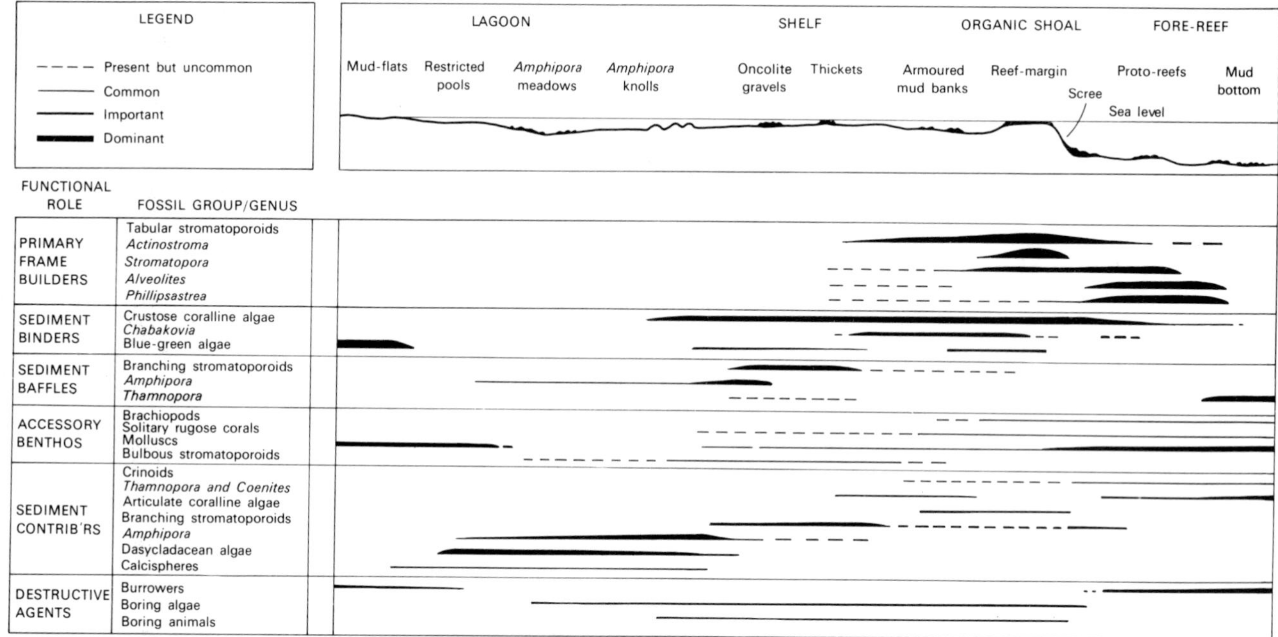

Fig. 10.46. Reconstructed profile of the Alexandra Reef-complex, Northwest Territories, Canada (after Jamieson, 1971).

Fig. 10.47A. Middle Silurian buildups and carbonate facies of the American Mid-west with an inset showing general Silurian facies in North America (after Shell Oil Co., 1975; Lowenstam, 1950; Mesolella *et al.*, 1974; Wilson, 1975).

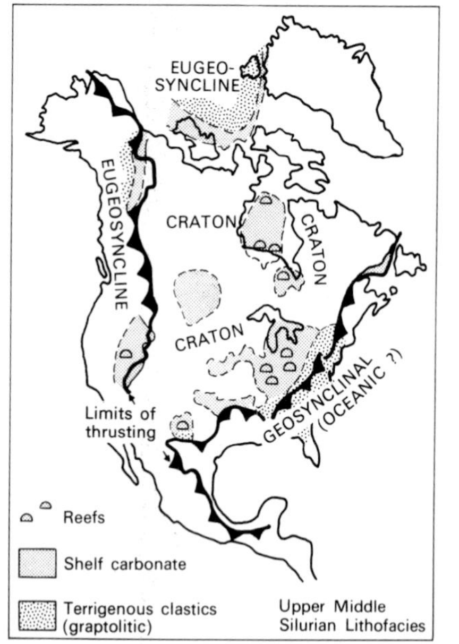

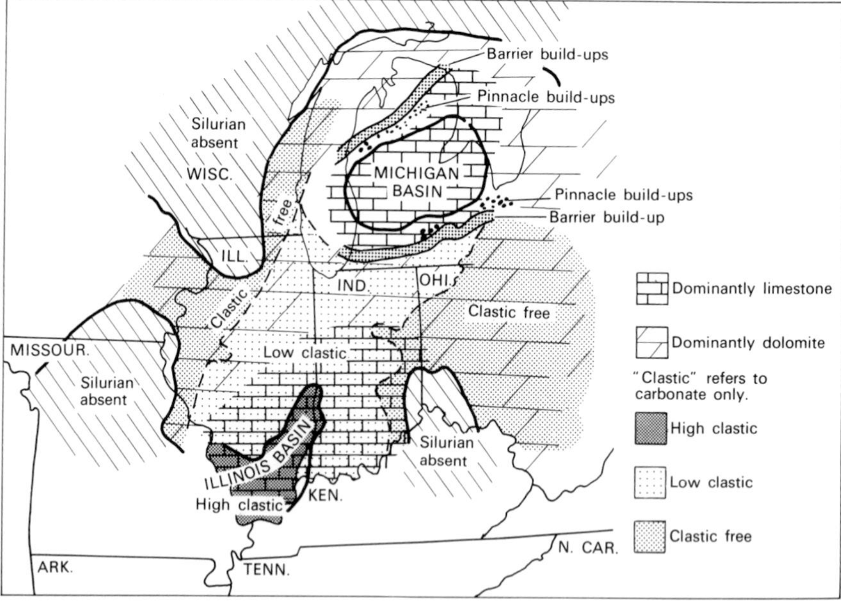

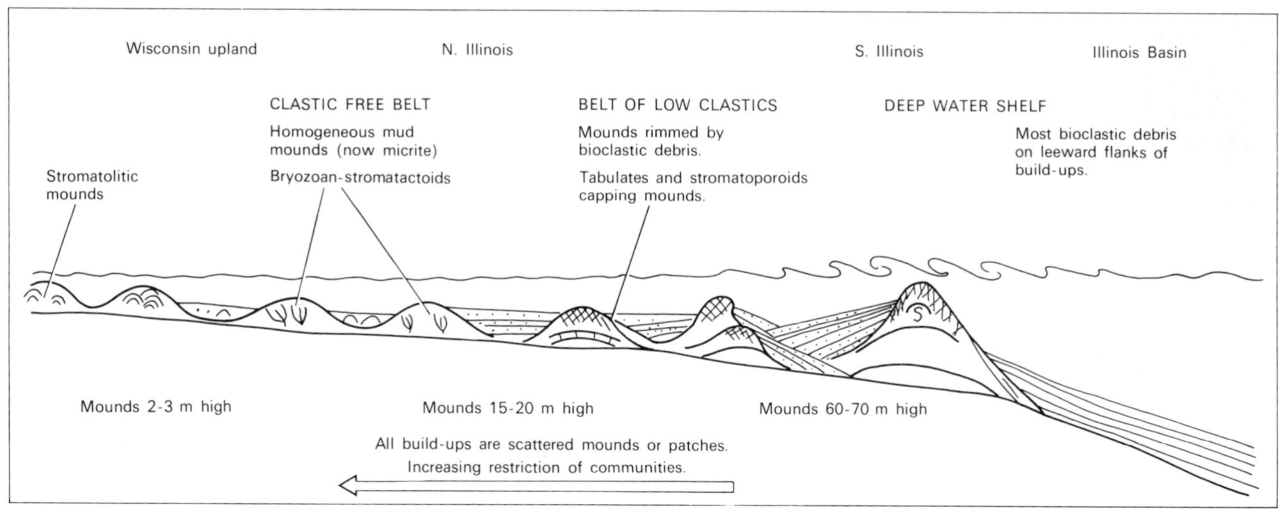

Fig. 10.47B. Generalized facies model across the Middle Silurian shelf of the American Mid-West (after Wilson, 1975).

Shaver, 1974; Wilson, 1975, p. 103). Shaver (1974) recognized five buildup types: (1) Algal spongiostrome stromatolitic mounds, (2) Bryozoan-mud mounds with stromatactoid structures, (3) Crinoid-tabulate coral mud mounds, (4) Wave-resistant framework reefs, (5) 'Pinnacle reefs'.

The wave-resistant frameworks achieved heights of 30–70 m and grew on top of crinoidal wackestones containing some tabulate corals, a few stromatoporoids, sponges and trilobites. Upwards, there is a passage from mudstones and wackestones containing hydrozoans and Stromatactis into the true framework association. The framework contains abundant stromatoporoids, clumps of tabulate corals, and the debris of crinoids, brachiopods and molluscs. Fauna changed as the frame was built into the wave zone (Fig. 10.48); sponges disappeared while larger tabulate coral-clumps accompanied the stromatoporoids (Textoris and Carozzi, 1964). In addition to the corals, there were also surprisingly large populations of other carnivores (especially nautiloids) providing a striking analogy to the trophic structure of modern reefs. Upward growth of the frame was completed when it reached the wave zone, and extensive crinoid-rich flank beds then developed by lateral accretion. Preferential growth of the buildup in one particular direction has been interpreted as growth in response to the prevailing wind. Complexes of similar buildups also formed a barrier around the sediment-starved Michigan basin (Fig. 10.47) but additionally, within the basin, 'Pinnacle reefs' formed where biogenic construction kept pace with subsidence.

The first large Gotland buildups (reviewed in Manten, 1971) occurred in the Wenlockian and resemble some of the middle-stage (unit 3 in Fig. 10.48) development phases in the American examples (Wilson, 1975) with stromatoporoids as the dominant frame builders. The latest Gotland buildups (Holmhällar-type of Manten, 1971) are Ludlovian and consist of large flattened but irregular masses composed almost exclusively of stromatoporoids and algae. Crinoid debris dominates the flank deposits. Wilson (1975, p. 117) suggests that the essential differences between the Gotland and American buildups were caused by a combination of factors: (1) the Gotland shelf was of probably normal marine salinities with good circulation, whereas the American situation was more restricted, (2) clay was more abundant in Gotland, (3) the smaller size of the Gotland buildups resulted from either shallower water depths and/or slower subsidence rates. In Britain the Wenlock and Woolhope Limestones of the Late Silurian resemble the Gotland buildups (Scoffin, 1971).

Ordovician buildups were less well developed than those of later times. However, several new groups of skeletal organisms evolved, particularly in the Mid-Ordovician. These included bryozoans, lithistid sponges and tabulate corals (Fig. 10.38), while the red (solonoporoid) algae, although present from the Cambrian, became both abundant and diverse (Heckel, 1974; Wray, 1971, 1977).

In the Cambrian, buildups are even more difficult to understand. Vaguely linear complexes of stromatolitic mounds

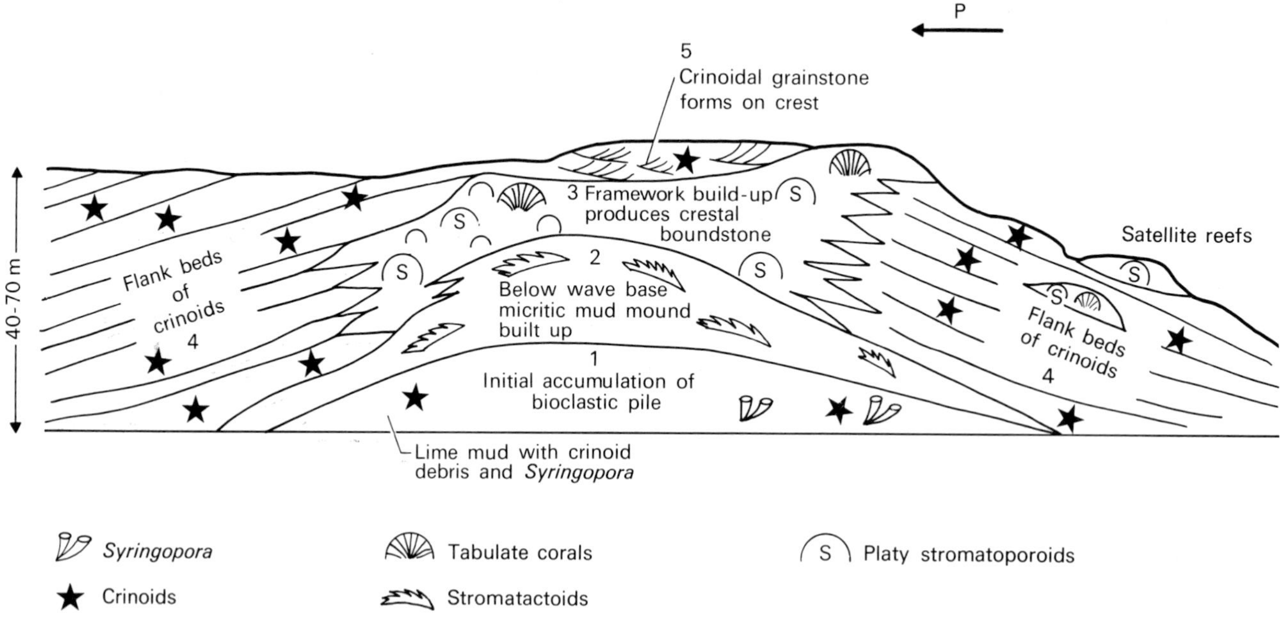

Fig. 10.48. Buildup development in the 'low clastics' carbonate belt of the Mid-Silurian shelf of the American Mid-west (modified from Wilson, 1975).

many kms in length and up to 1 m in thickness are known from the Upper Cambrian of Tennessee (Oder and Bumgarner, 1961) and some structures from the Upper Cambrian of New York State have even been termed 'barriers' in the literature (Goldring, 1938).

The archaeocyathids, members of an extinct and possibly sponge-like phylum, also produced buildups in N. America, Europe, Africa, Australia and Antarctica during the Early Cambrian (e.g. James and Kobluk, 1978). The cup-like archaeocyathid organisms were bound together by the enigmatic *Renalcis* and *Epiphyton* (both possible algal structures). Low mounds up to 2 m high, but with diameters of some 30 m or more, are known and Russians working in the Siberian fold belt (reported in Hill, 1972) have proposed a depth zonation model for some 'algal'-archaeocyathid associations. Recently, Brasier (1976) has suggested that some of the assumed 'algal' structures were vadose 'coniatolites' suggesting contemporary exposure of some archaeocyathids.

PRE-CAMBRIAN

Here, interpretations of buildup associations become wildly speculative. Hoffman (1974) has reviewed the place of stromatolites through geological time and it is apparent that stromatolite preservation potential was much higher prior to the diversification of grazers and burrowers. In the Proterozoic of NW Canada, Hoffman recognized facies associations analogous to those produced on later platform margins (Fig. 10.49). Branching columnar stromatolites produced buildups up to 20 m thick separated from each other by channels filled with intraclast grainstones. In some places such

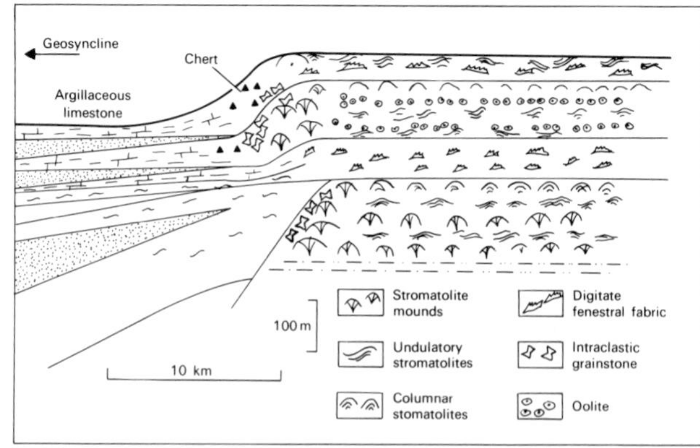

Fig. 10.49. Facies distribution across a Middle Pre-Cambrian carbonate platform. Coronation geosyncline, north-western Canadian Shield, Northwest Territories, Canada (after Hoffman, 1974).

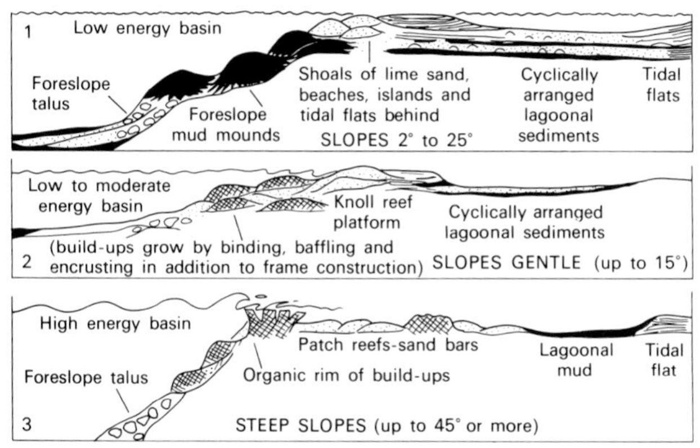

Fig. 10.50. Three types of carbonate shelf margin: (1) Foreslope mud mounds fringing a low energy basin; (2) Knoll reef platform fringing a moderate to low energy basin; (3) Organic reef rim fringing a high energy basin (after Wilson, 1975).

buildup complexes formed until well into the Ordovician, particularly along structural hinge-lines separating basins from platforms (Hoffman, op cit.).

SUMMARY

The various types of buildup seen in the geological record prior to the Cenozoic should not be too closely compared with modern reefs that are dominated by hermatypic corals. The various types reviewed here may fit into major shelf models in the way postulated by Wilson (1975, p. 360, Fig. 10.50) but in applying such models, the evolutionary background and trophic requirements of the organisms involved must be considered.

FURTHER READING

BATHURST R.G.C. (1971) *Carbonate Sediments and their Diagenesis*. 620 pp. Elsevier, Amsterdam.

WILSON J.L. (1975) *Carbonate Facies in Geological History*. 471 pp. Springer-Verlag, Berlin.

CHAPTER 11 Pelagic Environments

11.1 HISTORICAL INTRODUCTION

11.1.1 Pelagic sediments in the oceans

Any examination of pelagic sediments requires a journey back through time to the voyage of H.M.S. *Challenger*. This vessel plied the seas between the years 1872 and 1876 and made the first systematic study of the sedimentary nature of the ocean floor. On board the *Challenger* was a Canadian-born Scots-educated oceanographer named John Murray and it was through his tireless effort and enthusiasm that the pioneer descriptions of deep-sea deposits became works of lasting scientific value (e.g. Watson, 1967/68).

A number of preliminary papers were published in the 1870s but it was not till more than a decade or so later that the final comprehensive account of oceanic sediments appeared. This volume, the most seminal work in geological oceanography ever published, bears the simple title 'Deep Sea Deposits'; it was co-authored by Murray and a Belgian priest, the Abbé Renard, and published by Her Majesty's Stationery Office in 1891. In this volume the most important types of deep-sea deposit were painstakingly documented.

The scientific impact of the '*Challenger*' Expedition was enormous. In Murray's own words 'The results of the "*Challenger*" Expedition . . . became the starting point for all subsequent observations' (Murray and Hjort, 1912). A number of other oceanographic cruises, involving European and American scientists, followed this pioneer expedition. From 1877–1880, under the direction of Alexander Agassiz, the U.S. Coast Survey steamer *Blake* explored the Caribbean, the Gulf of Mexico, and the coasts of Florida. From 1890–1898 the Austrian steamer *Pola* investigated the bottom of the Mediterranean and Red Seas; the German Deep Sea Expedition (1898–1899), utilizing the steamer *Valdivia*, undertook collection of material from the Indian Ocean, the Atlantic and the Antarctic. The reports of these expeditions, however, did but largely echo the results of the *Challenger*.

Around the turn of the century the U.S.S. *Albatross*, under Alexander Agassiz, carried out scientific investigations in the Pacific; and in 1910 the Norwegian vessel *Michael Sars* carried Sir John Murray – on what was to be his last major voyage – to the North Atlantic; results from this expedition, led jointly by Johan Hjört, appeared in 1912.

After the First World War the German ship *Meteor* carried out investigations in the Atlantic (1925–1927), samples from which were described by Correns (1939). Around the same time the American ship *Carnegie* was collecting material from the floor of the Pacific; descriptions of this material were not published till near the end of the Second World War (Revelle, 1944). In both these studies the new tool of X-ray diffractometry greatly aided investigation of the mineralogy of fine-grained material. The war itself acted as a stimulant to submarine research; both the Germans and Americans, for example, produced bottom-sediment charts (Shepard, 1948). During 1947–1948 the Swedish Deep Sea Expedition took place, and the relatively new technique of piston coring yielded abundant samples described by Arrhenius (1952).

Subsequent years saw a pronounced shift in oceanographic studies to the United States, research primarily being sponsored by Woods Hole Oceanographic Institution, Lamont-Doherty Geological Observatory and Scripps Institution of Oceanography. A considerable amount of surficial sediment-coring was undertaken by scientists operating out of these institutions. As the techniques of oceanic investigation became more sophisticated there was a steady issue of papers on deep-sea sedimentation.

The evolution of deep-sea drilling techniques using shipboard computer-controlled dynamic positioning opened up a new world of marine geological exploration. The American vessel *Glomar Challenger*, which sailed on its maiden leg in July 1968, has given us tantalizing glimpses of oceanic stratigraphy as far back as the Jurassic. Since 1969 a growing stack of weighty turquoise volumes – the Initial Reports of the Deep Sea Drilling Project – have occupied the shelves of many libraries: lavishly illustrated, these volumes bear witness to the influence of the scanning electron microscope as a powerful tool in studying the micro- and ultra-structure of pelagic sediments. With the participation of France, Germany, Japan, the United Kingdom and the USSR the Deep Sea Drilling Project (DSDP) has now been transmuted into the International Phase of Ocean Drilling (IPOD). We must wait to judge the full scientific impact of this international venture.

11.1.2 Pelagic sediments on land

Interest in the terrestrial record of pelagic sedimentary rocks dates back to the time of the *Challenger* Expedition (e.g. Jenkyns and Hsü, 1974). Geologists working in Scotland, the Caribbean, the Alps and elsewhere, publishing in the late nineteenth century, variously claimed recognition of deep-sea deposits on land. At his request many of these rocks were sent to John Murray for examination. Neumayr (1887), for example, recorded the fact that Murray considered certain Mesozoic Alpine rocks to be more closely comparable with deep-sea deposits than any other material he had examined. Murray himself claimed that the Tertiary *Globigerina* Limestones from Malta had 'formed in deep water (300 to 1,000 fathoms), at some distance from a continental shore, but still within the influence of detrital matter brought down by rivers' (Murray, 1890). However, in the *Challenger* report itself, Murray and Renard (1891) wrote that only a few 'doubtful exceptions' existed to prove the rule that the pelagic deposits of the oceans could nowhere be found on the continents. Walther (1897) remarked that Murray considered only the Maltese Limestone to be a true deep-sea deposit. This change of viewpoint seems to have been forced on Murray by the prevailing Anglo-American belief in the permanency of continents and ocean basins. This doctrine, which can be found in the work of Louis Agassiz (1869), had clearly attained a stranglehold on many scientists by the end of the nineteenth century. A few occurrences of deep-sea deposits, however, showed unusual tenacity in the pages of geological literature. The Tertiary deposits of Barbados described by Jukes-Browne and Harrison (1892) are among these, as are the red radiolarian clays from Timor, Borneo and Rotti (Molengraaf, 1915, 1922). These East Indian deposits contain sharks' teeth and manganese nodules and were specifically compared by Molengraaf to certain of the *Challenger* samples. In the Alps, Steinmann (1905, 1925) professed the view, in opposition to Murray, that the Mesozoic radiolarites were analogues of Recent siliceous oozes and he noted also the common association of the cherts with mafic and ultramafic igneous rocks.

Reaction to these ideas in North America was, at best, non-committal. Grabau (1913), in discussing the radiolarian ooze of Barbados, and certain Triassic and Jurassic deposits in the Alps, commented: 'Whether these deposits will eventually prove to have such an (abyssal) origin, or whether they too, may not be of shallow-water origin, must for the present remain undecided.' Twenhofel (1926) remarked that 'Malta, Barbadoes and Christmas Island in the East Indies are said by Walther to possess true deep-sea oozes of Tertiary age' (cf. Walther, 1911). Twenhofel was, however, inclined to view the Mesozoic red clays of Borneo, Timor and Rotti as true deep-sea deposits and indeed considered them the most important occurrences known. He also briefly referred to the radiolarian cherts of the Franciscan Formation and followed Davis (1918) in accepting a shallow-water depositional environment. Davis' detailed account leant heavily on the close stratigraphical association of the cherts with the Franciscan Sandstone which was viewed as a shallow-water deposit. These interpretations were, of course, made without the benefit of the turbidity-current hypothesis and belong to an era of misunderstood bathymetric criteria.

In Great Britain writers of text books between the wars adopted perhaps somewhat less ambiguous positions. Marr (1929), for example, wrote that, 'this Barbados Earth approaches more closely in character to the modern abyssal ooze than any other of the deposits of the past times which have hitherto been described.' Hatch, Rastall and Black (1938) opined that 'a few examples, such as those of Barbados and Timor, appear to be true abyssal deposits, comparable with the modern radiolarian oozes.'

This uneasy situation, whereby the exact bathymetric interpretation of fossil red clays and radiolarites remained uncertain, lasted until the sixties. Generally, continental workers favoured a deep-sea, but not necessarily oceanic, interpretation for certain Mesozoic Alpine rocks (e.g. Gignoux, 1936; Trümpy, 1960). Geochemical and mineralogical work on both the Barbados Earth and the Timor red clays confirmed that they were comparable with Recent oceanic deposits (El Wakeel and Riley, 1961; El Wakeel, 1964); yet the tectonic implications of these occurrences were not, at that time, fully explored.

Finally, with the advent of plate tectonics, and interpretation of ophiolite assemblages, the old ideas on the presence of oceanic sediments on land have been vindicated. Separation of true oceanic deposits, originally laid down on ocean crust, from similar, if not identical pelagic sediments laid down at the edges or in the interiors of continents, has also clarified the issue. Palaeo-oceanography is no longer excluded from the domain of the field geologist.

11.2 DEFINITIONS AND CLASSIFICATIONS

'Pelagic sediments' have been variously defined and the term has now reached the stage (in common with words like 'greywacke' and 'flysch') when any author can claim, like Humpty Dumpty, that it means exactly what he wants it to mean – neither more nor less. Early definitions are offered by Murray and Renard: 'Pelagic deposits are situated at a considerable distance from land, and for the most part in the greatest depths of the ocean. It is only in exceptional circumstances that sandy or other particles immediately derived from the land make up any considerable portion of these deposits. The most characteristic minerals are derived from volcanic eruptions, floating pumice, or are of secondary

Table 11.1 Classification of deep-sea pelagic and terrigenous sediments according to Murray and Renard (1891). This classification has the virtue of extreme simplicity

Terrigenous deposits	Shore formations, Blue mud, Green mud and sand, Red mud,	Found in inland seas and along the shores of continents.
	Coral mud and sand, Coralline mud and sand, Volcanic mud and sand,	Found about oceanic islands and along the shores of continents.
Pelagic deposits	Red clay, Globigerina ooze, Pteropod ooze, Diatom ooze, Radiolarian ooze,	Found in the abysmal regions of the ocean basins.

origin formed *in situ*. The remains of pelagic organisms that have fallen from the surface form the principal part of many of these deposits, as indicated by the names: Pteropod, Globigerina, Diatom and Radiolarian Oozes. In some of the deeper regions of the ocean these organic oozes are replaced by Red Clays, formed for the most part by the disintegration of rocks and minerals *in situ*' (Murray and Renard, 1891). Their classification is illustrated in Table 11.1.

Twenhofel (1926) stated simply that: 'The pelagic sediments consist of the hard parts of organisms which lived in the upper lighted waters, or on the bottom'. It is clear that he visualized pelagic sediments purely as biologically derived material. Shepard (1948) in his first edition of *Submarine Geology* followed Revelle (1944) in restricting pelagic deposits to material 'of red, brown, yellow or white color (indicative of "slow" deposition) and to sediments which have less than about 20% of allogenic minerals and rock particles less than 5 microns in diameter.' Kuenen (1950a) offered: 'The pelagic (or eupelagic) sediments are characterized by the absence of terrestrial mineral grains larger than the colloidal fraction. The most common constituents are clay minerals and remains of planktonic unicellular organisms'. Kuenen also described 'hemipelagic sediments' at some length, including under this term 'blue, red, yellow, green, coral, calcareous and volcanic muds', deposits that were constituted by 'mainly pelagic and but little terrigenous matter'. The term 'hemipelagic', as understood by Kuenen in 1950, had certainly been current for several decades.

Revelle, Bramlette, Arrhenius and Goldberg (1955) in a discussion of Pacific pelagic sediments began their paper by announcing: 'we are here considering sediments derived in major part from particles which settled out or were precipitated from the water column far from continents and thus without much terrigenous material above clay size. Such sediments generally accumulate slowly and commonly have colors indicating marked oxidation.' This description clearly owes a lot to Murray and Renard. According to Bramlette (1961), who followed a similar definition to Revelle, Bramlette, Arrhenius and Goldberg, 'the most distinctive chemical and other characteristics of pelagic deposits . . . seem related primarily to their slow accumulation and resulting approach to chemical equilibrium.' Arrhenius (1963) preferred to define pelagic sediments on the basis of a maximum value of the depositional rate of a terrigenous component – a rate in the order of a few mm per thousand years. 'Within the basin accumulating pelagic sediments, the terrigenous deposition rate appears to vary not much more than one order of magnitude (5×10^{-5} to 5×10^{-4} cm/year), whereas values much higher and varying by several orders of magnitude are characteristic of sediments fringing the continents.' Publishing in the same year Shepard (1963), in the second edition of *Submarine Geology*, commented that older classifications of deep-sea

Table 11.2 Classification of deep-sea pelagic sediments according to Berger (1974). This classification shows the influence of the Deep Sea Drilling Project particularly by including various lithified facies

I. Pelagic deposits (oozes and clays)
   $<25\%$ of fraction >5 µm is of terrigenous, volcanogenic, and/or neritic origin.
   Median grain size <5 µm (excepting authigenic minerals and pelagic organisms).
   A. Pelagic clays. $CaCo_3$ and siliceous fossils $<30\%$.
      (1) $CaCO_3$ 1–10%. (Slightly) calcareous clay.
      (2) $CaCO_3$ 10–30%. Very calcareous (or marl) clay.
      (3) Siliceous fossils 1–10%. (Slightly) siliceous clay.
      (4) Siliceous fossils 10–30%. Very siliceous clay.
   B. Oozes. $CaCO_3$ or siliceous fossils $>30\%$.
      (1) $CaCO_3$ $>30\%$. $<\frac{2}{3}$ $CaCO_3$: marl ooze. $>\frac{2}{3}$ $CaCO_3$: chalk ooze.
      (2) $CaCO_3$ $<30\%$. $>30\%$ siliceous fossils: diatom or radiolarian ooze.

II. Hemipelagic deposits (muds)
   $>25\%$ of fraction >5 µm is of terrigenous, volcanogenic, and/or neritic origin.
   Median grain size >5 µm (excepting authigenic minerals and pelagic organisms).
   A. Calcareous muds. $CaCO_3$ $>30\%$.
      (1) $<\frac{2}{3}$ $CaCO_3$: marl mud. $>\frac{2}{3}$ $CaCO_3$: chalk mud.
      (2) Skeletal $CaCO_3$ $>30\%$: foram ~, nanno ~, coquina ~.
   B. Terrigenous muds. $CaCO_3$ $<30\%$. Quartz, feldspar, mica dominant. Prefixes: quartzose, arkosic, micaceous.
   C. Volcanogenic muds. $CaCO_3$ $<30\%$. Ash, palagonite, etc., dominant.

III. Pelagic and/or hemipelagic deposits.
      (1) Dolomite-sapropelite cycles.
      (2) Black (carbonaceous) clay and mud: sapropelites.
      (3) Silicified claystones and mudstones: chert.
      (4) Limestone.

deposits had not taken into account the significance of fine-grained turbidites and ice-rafted debris. He then added: 'Another serious difficulty in the older classification of the deep-sea sediments is in defining *pelagic sediments*. Concerning the meaning of this name many heated debates have been held. Therefore, without hope that any broad agreement can be reached on the definition, pelagic sediments will be considered here as those sediments of the deep part of the open ocean that settled out of the overlying water at a considerable distance from land and in the absence of appreciable currents so that the particles are either predominantly clays or their alteration products or consist of some type of skeletal material from plants or animals. The clays may be derived from the land either by water or by wind or may come from volcanic dust or meteorites'. In the latest (1973b) edition of this book Shepard essentially adopted the same stance towards pelagic deposits as he did a decade earlier. Interestingly enough in both editions, coral-reef debris, coral sands and coral muds are included as pelagic deposits.

In a recent article by Berger (1974) another classification, including hemipelagic deposits, is set up (Table 11.2). It is apparent that no shallow-water material is included as true pelagic sediment.

As the above references illustrate various problems exist concerning the definition and classification of pelagic deposits. Dictionaries generally offer 'of or pertaining to the open sea' as a working definition. This is simple enough; however, mid-Pacific atolls 'pertain to the open sea', yet their deposits may be variously supra-, inter- and subtidal. Feeding moderate to great depth into the definition does not help as shallow-water

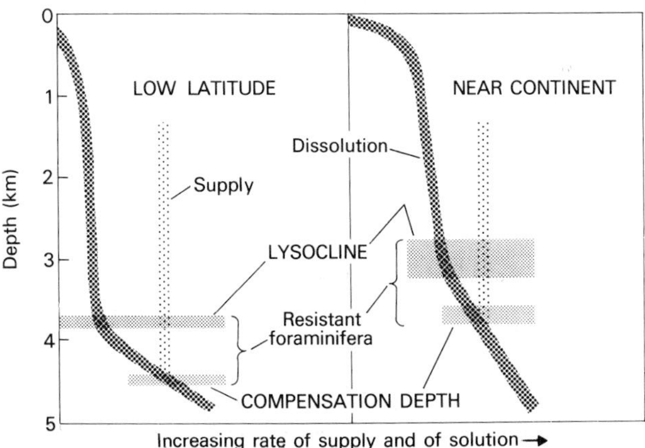

Fig. 11.1. Model of relationship between pelagic sedimentation (supply) and carbonate dissolution in the southeast Pacific, illustrating the significance of the lysocline and calcite compensation depth (after Berger, 1970). Explanations for the differences in dissolution curves from near-continent and low-latitude regions are offered by Berger (1971) and Berger and Winterer (1974).

foraminiferal-nannofossil oozes occur in certain elevated oceanic areas. That 'pelagic' refers to both organisms and sediments is also part and parcel of the problem. Twenhofel (1926), for example, with his simple equation between phyto- and zooplankton and deep-sea deposits necessarily excluded 'red clay' as a pelagic sediment. Yet 'red clay' pertains to the open sea, even though it is partly of terrigenous origin. Rather than becoming involved in what is to some extent a semantic wrangle, *pelagic* is here used in a qualitiative and descriptive sense to be essentially synonymous with 'of the open sea' but excluding reefs and reef-associated settings. In this chapter, various open-sea environments – divided on the basis of tectonics, topography and geography – are outlined and their sedimentary record sketched; such sequences when found in the ancient, may, with due appreciation of palaeotectonics and palaeogeography, merit interpretation as pelagic deposits.

11.3 PELAGIC SEDIMENTS IN THE OCEANS

11.3.1 Introduction to pelagic sedimentation

Two major controls on pelagic sedimentation are the calcite compensation depth (CCD) and the fertility of the surface waters. The CCD, below which calcite does not accumulate on the ocean floor, is simply the level where the rate of solution of biogenic carbonate is balanced by the rate of supply; the level of increased solution rate, or lysocline, may be shallower (Fig. 11.1) or deeper than this (Berger, 1970; Milliman, 1975). That the accumulation of calcite, and aragonite, was depth dependent was known by Murray from his experience on H.M.S. *Challenger* (Fig. 11.2). Thus above the CCD there may accumulate calcareous oozes, essentially constituted by the low-magnesian calcitic skeleta of Foraminifera and nannofossils which, in shallower depths, contain admixtures of aragonitic pteropods. Below this, radiolarian and diatom oozes and red or brown clays may form. Sedimentary rates are given in Table 11.3.

Biogenous deposits will only accumulate rapidly below areas of high fertility where upwelling brings nutrient-rich waters to the surface, but silica will only tend to become a quantitatively important component in sediments near or below the calcite

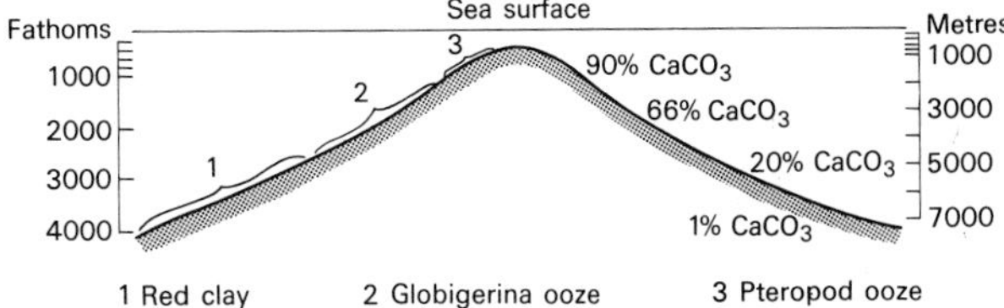

Fig. 11.2. Distribution of pelagic sediments in relation to depth (after Murray and Hjört, 1912).

Table 11.3 Rates of accumulation of Recent and sub-Recent pelagic facies (after Berger, 1974)

Facies	Area	mm/10^3 years
Calcareous ooze	North Atlantic (40–50°N)	35–60
	North Atlantic (5–20°N)	40–14
	Equatorial Atlantic	20–40
	Caribbean	~28
	Equatorial Pacific	5–18
	Eastern Equatorial Pacific	~30
	East Pacific Rise (0–20°S)	20–40
	East Pacific Rise (~30°S)	3–10
	East Pacific Rise (40–50°S)	10–60
Siliceous ooze	Equatorial Pacific	2–5
	Antarctic (Indian Ocean)	2–10
Red clay	North and Equatorial Atlantic	2–7
	South Atlantic	2–3
	Northern North Pacific (muddy)	10–15
	Central North Pacifc	1–2
	Tropical North Pacific	0–1

compensation depth. Siliceous oozes are found in equatorial zones, the subarctic and subantarctic and in certain continental-margin regions. Transfer to the sea floor of small biogenic particles such as diatoms and coccoliths is effected by the sinking of faecal pellets ejected by predatory organisms; however, in transit and on the sea floor, soluble skeleta may be dissolved leaving only the more robust forms as a sedimentary record. In barren regions of the oceans, red clays, chiefly derived from aeolian, volcanic and cosmic sources, accumulate by default. In sub-polar regions ice-rafted debris becomes a significant additive, and all pelagic sediments if traced towards continental margins contain increasing quantities of re-deposited shallow-water material. These and other sedimentary processes are ably reviewed by Ramsay (1973), Berger and Winterer (1974), Berger (1976), Davies and Gorsline (1976) and Windom (1976). For a global sedimentary distribution map see Fig. 11.3. Marine ferromanganese deposits, commonly found in pelagic environments, are exhaustively described in Glasby (1977).

The above remarks give an essentially static view of pelagic sedimentation. Clearly, however, during Pleistocene glacial advances ice-rafted debris penetrated further into oceanic realms (e.g. Connolly and Ewing, 1965; Keany, Ledbetter, Watkins and Huang, 1976). Deep Sea Drilling also reveals a mobile palaeo-oceanographic picture that departs significantly from that of the present. For example, synchronous variations in Cenozoic sedimentary rates, of probable climatic origin, are now established for the major oceans (Davies, Hay, Southam and Worsley, 1977).

The equatorial zones are characterized by high productivity such that the rate of supply of biogenic carbonate (and silica) is relatively great; thus the CCD is depressed in this zone (Fig. 11.4). Enhanced accumulation of carbonate leads to an equatorial sediment bulge. Coupling this effect with plate motion leads to the concept of 'plate stratigraphy', best illustrated in the case of the Pacific Ocean (Berger and Winterer, 1974). Northward movement of the Pacific Plate displaces the sediment bulge which can be located as an expanded sequence by Deep Sea Drilling in areas north of the equator (Fig. 11.5). Indeed the age and position of the bulge may be used to give an estimate of the rate of sea-floor spreading.

The CCD changes not only in space but also in time. If deep-sea solution is considered as a mechanism of returning to the World Ocean calcium carbonate which has been removed by organisms, to maintain a steady state, temporal variation of the CCD becomes a possible factor (Berger, 1971). For example, a spread of particularly successful calcareous planktonic organisms may suddenly remove unprecedented amounts of $CaCO_3$ from marine waters; transgressions increasing the volume of shelf-sea carbonates will also subtract $CaCO_3$ from the ocean. This depletion could then be balanced by rise of the CCD causing solution of increased amounts of pelagic carbonate. Cenozoic excursions of the CCD, documented by Berger and Winterer (1974) and van Andel (1975), are illustrated in Fig. 11.6.

Although it was formerly assumed that the oceanic stratigraphic column was complete we know now that this is not so. Abyssal currents have been recognized for some time but only since the advent of Deep Sea Drilling has the importance of sea-floor erosion been fully appreciated: pelagic sections in all oceans, in a variety of topographic settings, are punctuated by unconformities of differing temporal extent (e.g. Hollister, Ewing *et al.,* 1972; Rona, 1973; Davies, Weser, Luyendyk and Kidd, 1975; Kennett and Watkins, 1976; van Andel, Heath and Moore, 1976).

These unconformities are apparently formed by erosive flow of density-driven bottom waters derived from polar ice-caps and by concomitant increased shallower-water circulation; as such these breaks are climatically controlled (e.g. Johnson, D.A., 1974; Lonsdale, 1976). The paths of these currents are a function not only of oceanic topography but also of continental disposition; thus continental drift will affect the location of unconformities (e.g. Kennett, Burns, Andrews, Churkin, Davies, Dumitrica, Edwards, Galehouse, Packham and van der Lingen, 1972; Berggren and Hollister, 1977). Such currents, if dependent on the presence of polar ice caps, would clearly be subdued in times of equable climate; in this case

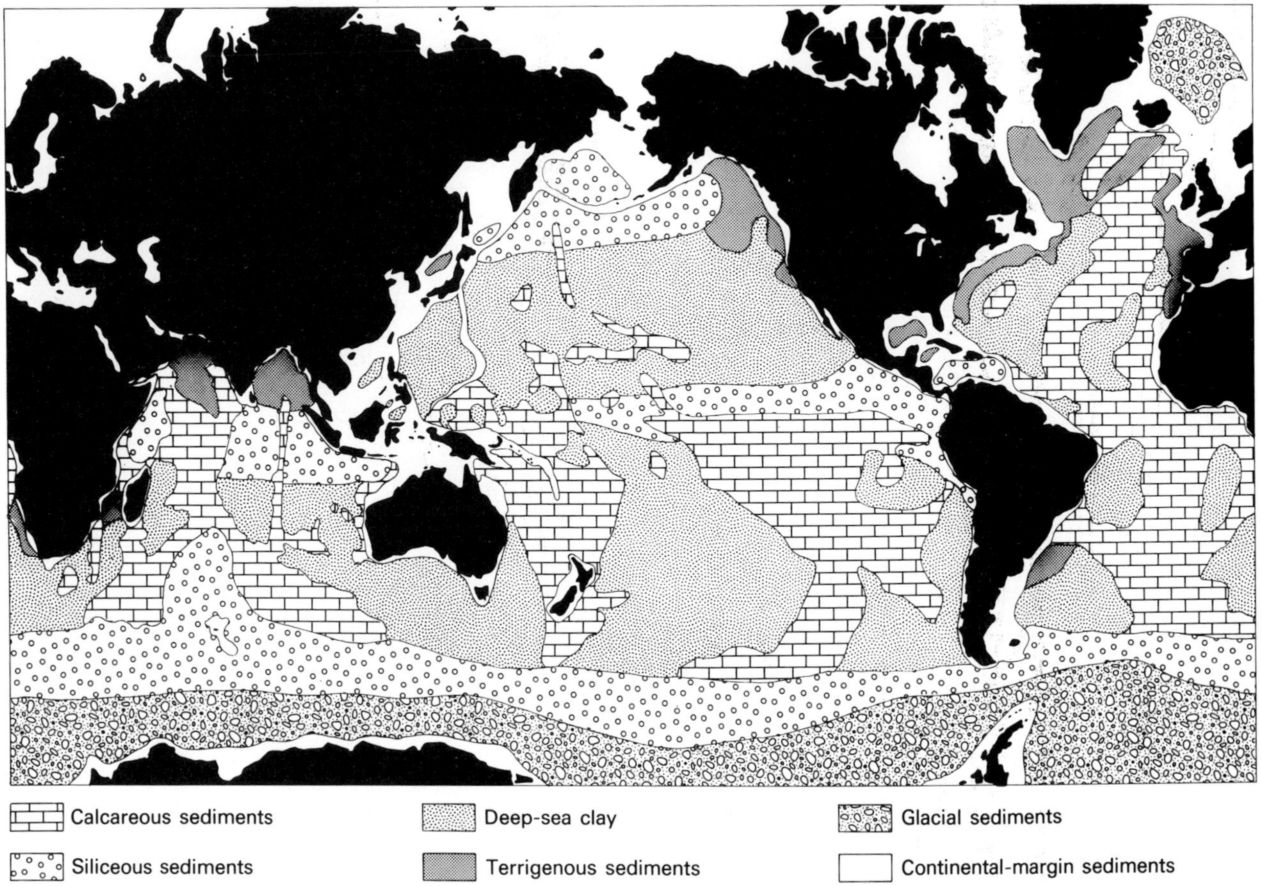

Calcareous sediments
Deep-sea clay
Glacial sediments
Siliceous sediments
Terrigenous sediments
Continental-margin sediments

Fig. 11.3. Global distribution of principal types of pelagic sediment on the ocean floors (after Davies and Gorsline, 1976).

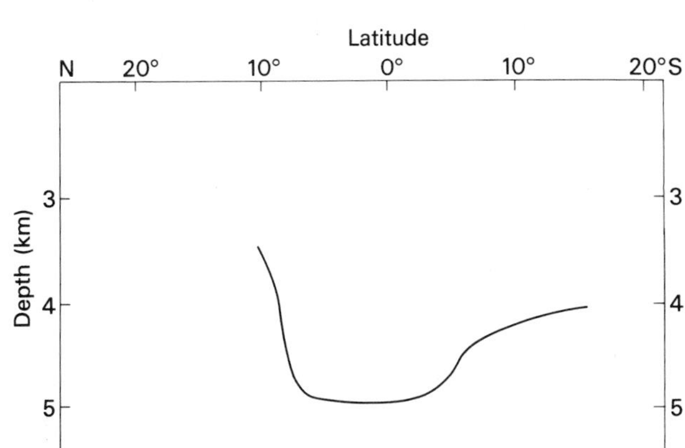

Fig. 11.4. Configuration of the present calcite compensation depth in the eastern tropical Pacific between 100°W and 150°W (after Berger and Winterer, 1974).

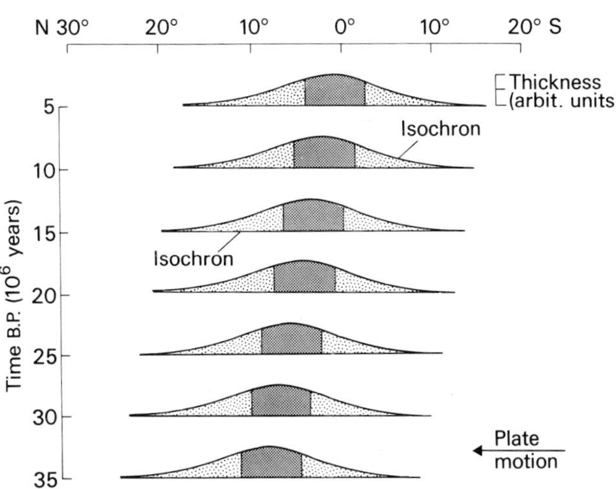

Fig. 11.5. Schematic model of the displacement of the equatorial sediment bulge by northward movement of the Pacific Plate (after Berger and Winterer, 1974).

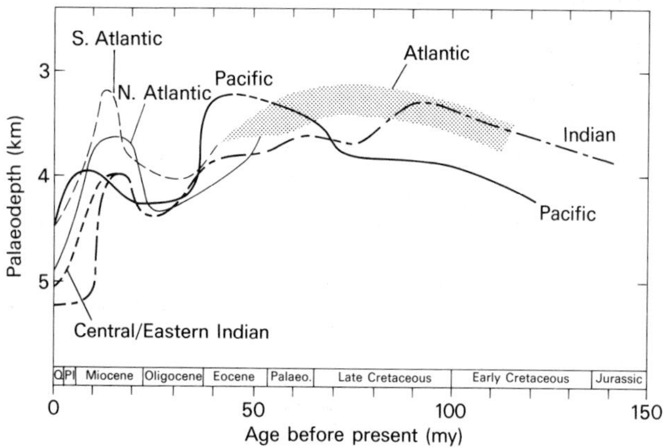

Fig. 11.6. Proposed temporal variations in the calcite compensation depth for the major oceans (after van Andel, 1975). Data points from Atlantic Deep Sea Drilling Sites produce the diffuse Cretaceous and Eocene zone shown in the figure. The techniques for generating palaeo-CCD curves are dealt with by Berger and von Rad (1972) and Berger and Winterer (1974).

oceanic mixing processes may have been so severely curtailed that the waters would have become stratified. Minimal recycling of cold, oxygenated bottom waters, coupled with heightened organic productivity and consequent increased oxidation of phytoplankton tests, would favour the formation of anoxic waters. The existence of such waters during the mid- to Late Cretaceous is demonstrated by the presence of organic-rich black shales on aseismic ridges, oceanic plateaus, and in deep basins throughout the world ocean and as intercalations in coeval pelagic sediments on land (see Sect. 11.3.3, 11.3.6, 11.4.2, 11.4.4, 11.4.5). The formation of such organic-rich levels may be interpreted as a function of the intersection of an expanded oxygen-minimum layer with bottom sediments (Fig. 11.7). Such an extended level of anoxic water was probably favoured not only by the mild climate of the Late Cretaceous but also by the transgression which stimulated production of organic carbon in newly created marginal and shelf seas (Schlanger and Jenkyns, 1976; Fischer and Arthur, 1977).

In the following pages various pelagic environments are outlined and their sedimentary record sketched. In order to place these environments in palaeo-oceanographic perspective considerable reference is made to the results of the Deep Sea Drilling Project. Major submarine topographic features referred to in the text are illustrated in Fig. 11.8 to 11.10.

11.3.2 Spreading ridges

Spreading ridges are loci of active seismicity, volcanism and high heat flow. Some, like the Mid-Atlantic Ridge, are rifted and characterized by differential relief of kilometres scale and exposure of a variety of mafic and ultramafic rocks partly in the form of breccias and basaltic sands; others, like the East Pacific Rise, are topographically more subdued with surfaces dominated by extrusives (van Andel and Bowin, 1968) (Fig. 11.11). Even in axial zones young ridge basalts show surficial alteration to palagonite with the formation, in cracks and vesicles, of K-rich smectitic clays (Melson and Thomson, 1973; Hekinian and Hoffert, 1975). Although locally sediment-starved, ridges may be partly covered, on both high and low ground, by sediments that possess a unique and distinctive chemistry. Perhaps the best studied of these so-called active-ridge sediments are those found on the fast-spreading (16 cm/year total rate, Rea, 1976) slightly rifted East Pacific Rise (Boström, 1973). These deposits are predominantly dark brown in colour although exhibiting lighter shades where locally diluted with carbonate. Darkest-coloured sediment is concentrated on the ridge crest itself. The pigment-carrying

Fig. 11.7. Model of a stratified ocean with well-developed and expanded oxygen-minimum layer. As illustrated here, stagnant bottom-water exists in restricted ocean basins (after Schlanger and Jenkyns, 1976).

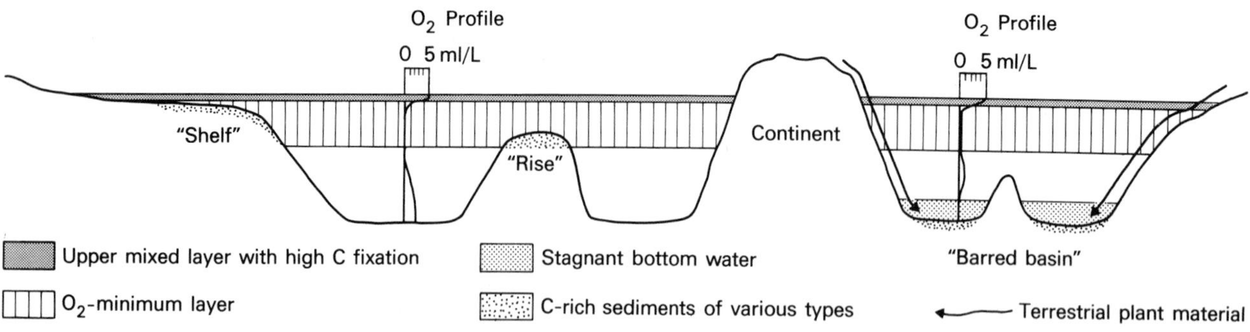

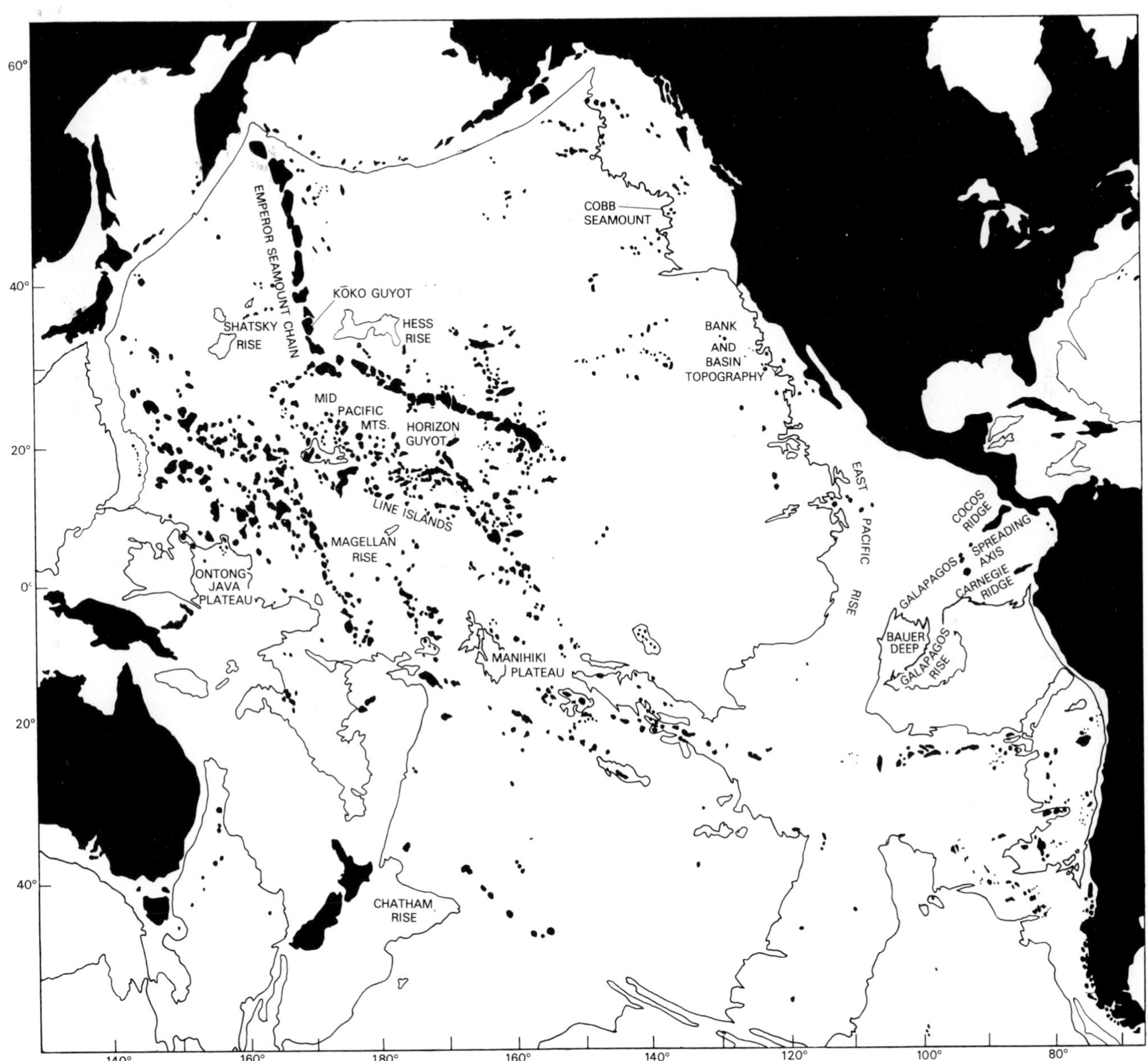

Fig. 11.8. Map of the Pacific Ocean, illustrating topographic features mentioned in the text. Submarine elevated areas picked out in black. Bathymetric contour at 4 km. (after Chase *et al.* *Topography of the Oceans*).

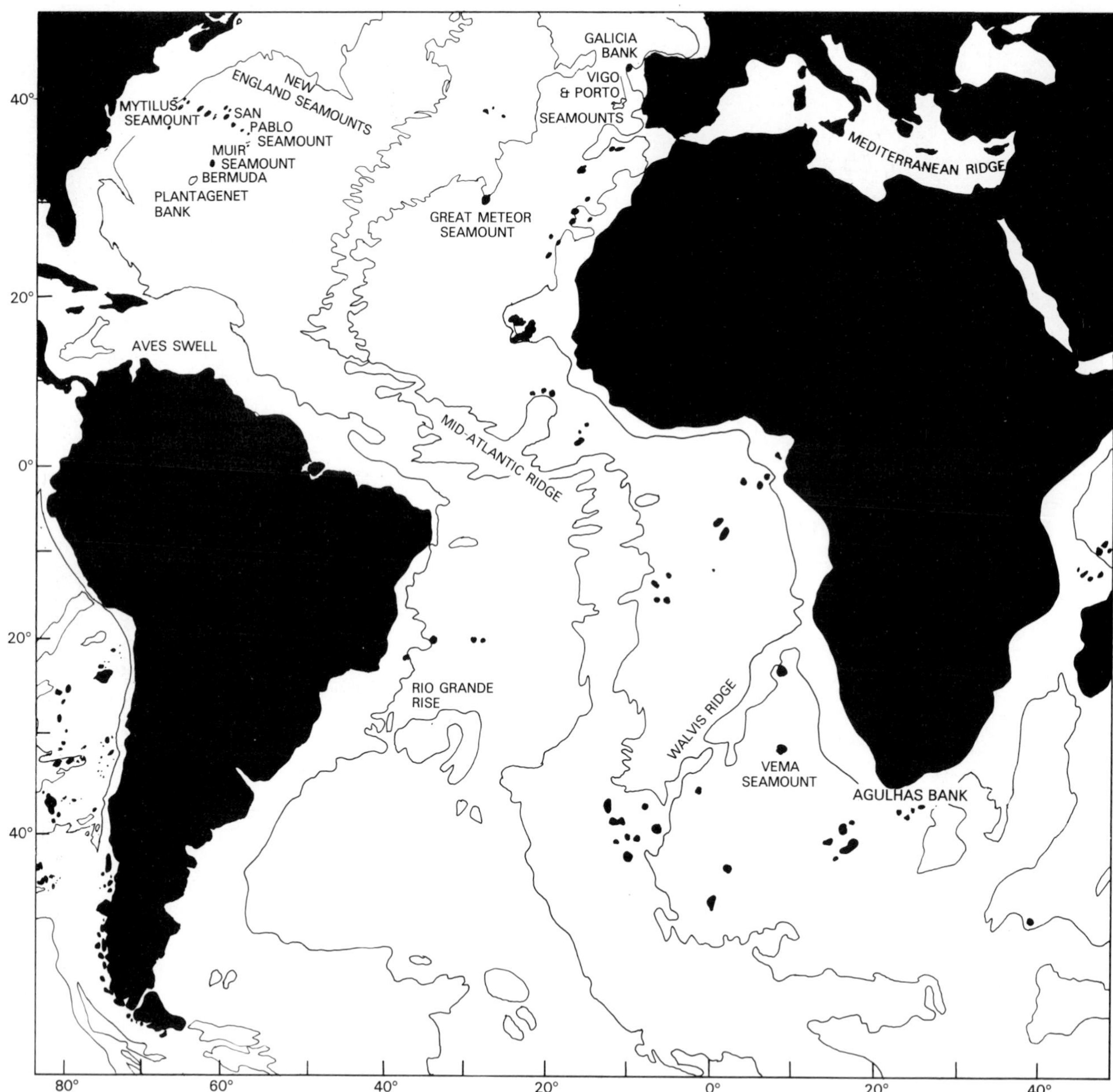

Fig. 11.9. Map of the Atlantic Ocean, illustrating topographic features mentioned in the text. Submarine elevated areas picked out in black. Bathymetric contour at 4 km. (after Chase *et al. Topography of the Oceans*).

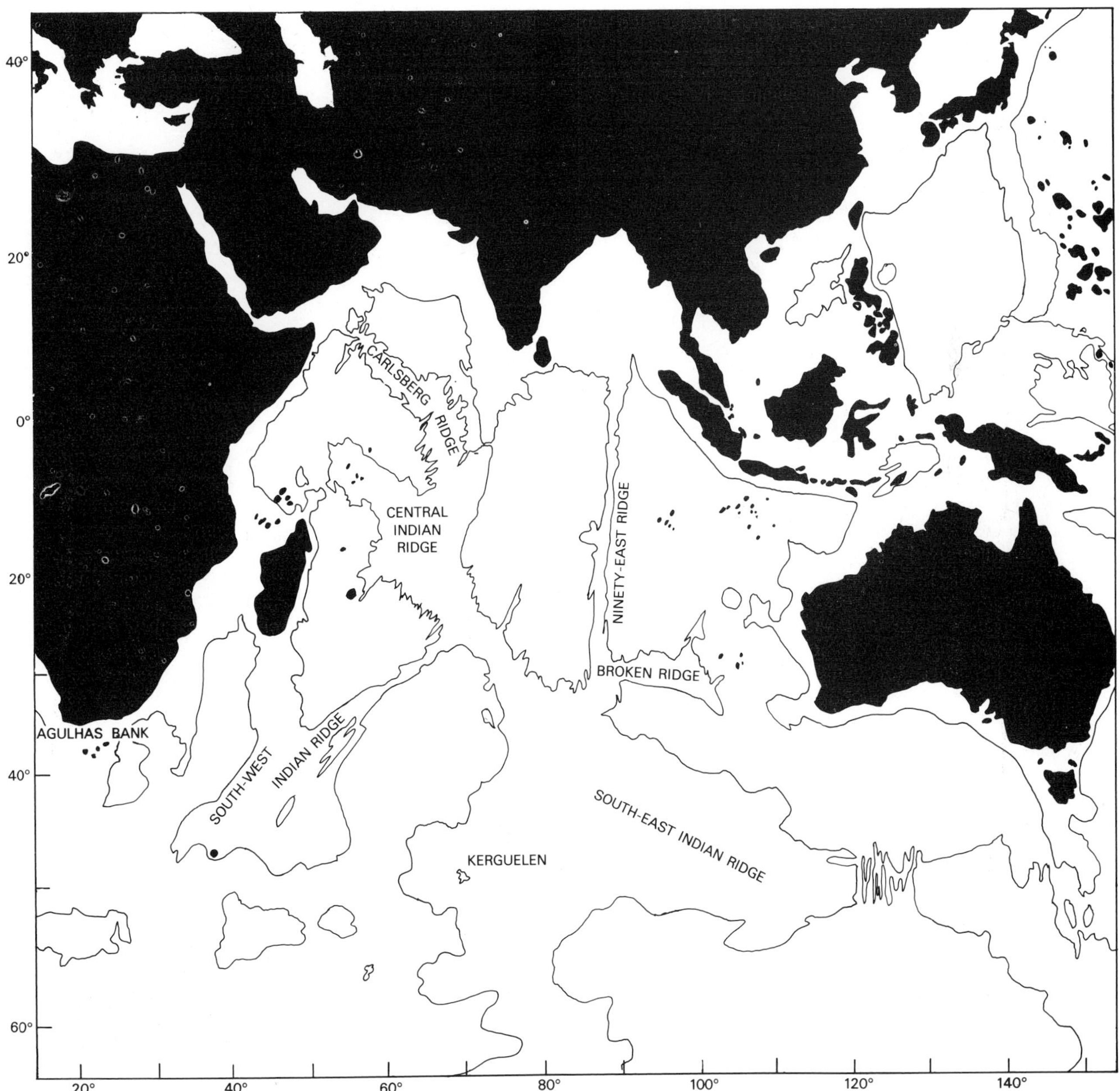

Fig. 11.10. Map of the Indian Ocean, illustrating topographic features mentioned in the text. Submarine elevated areas picked out in black. Bathymetric contour at 4 km. (after Chase *et al. Topography of the Oceans*).

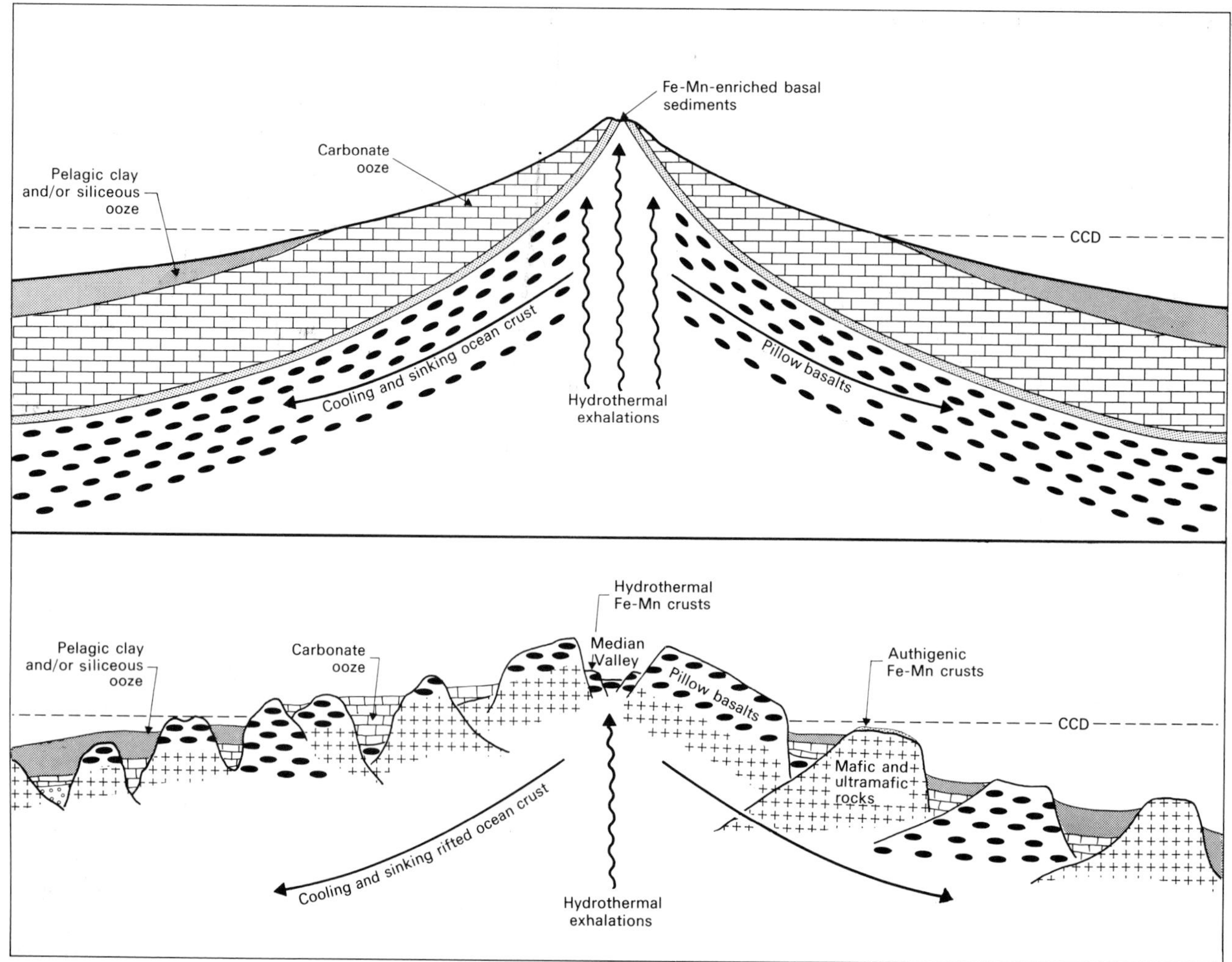

Fig. 11.11. Above: Sediment distribution on a fast-spreading ridge of East-Pacific-Rise type. Below: Sediment distribution on a slow-spreading rifted ridge of Atlantic type. Note the enhanced development of Fe–Mn enriched basal sediments on the fast-spreading ridge: mineralized crusts on the rifted ridge change from hydrothermal to authigenic as they move away from the vents in the median valley. Note furthermore the difference in igneous rock-sediment relationships between the two types of ridge and the difference in sediment geometry. The CCD controls the change in facies down the ridge. Not to scale (modified from Garrison, 1974; Davies and Gorsline, 1976).

material is usually described as colloidal X-ray-amorphous iron–manganese oxide–hydroxides; these take the form of roughly spherical globules of micron scale which, in fact, comprise goethite, iron-rich smectite and various manganese oxide-hydroxides (Dymond, Corliss, Heath, Field, Dasch and Veeh, 1973). High accumulation rates of these sediments are apparently directly related to the rates of sea-floor spreading so that the East Pacific Rise is characterized by a thick development of this particular facies relative to other ocean ridges (Boström, 1973).

Since new ocean crust moves down and away from spreading loci and is gradually covered by a blanket of pelagic sediments, it follows that ancient active-ridge sediments should be revealed as a basal layer overlying basalt in Deep Sea Drilling sites (Fig. 11.11). Such material, variously described as 'basal ferruginous sediments', 'iron-rich basal sediments' has now been cored from the Atlantic, Pacific and Indian Oceans at varying distances from spreading ridges (e.g. Horowitz and Cronan, 1976; Cronan, 1976; Cronan, Damiani, Kinsman and Thiede, 1974) and presumably occurs on tholeiitic basalt in much of the world ocean.

The chemistry of these active ridge sediments is dissimilar to average deep-sea clays in that the former are abnormally poor in Al, Ti and SiO_2 but enriched in Fe, Mn together with a host of other metals (Cu, Pb, Zn, Ni, Co, Cr, V, Cd, U, Hg) plus As and B.

The origin of active-ridge sediments is controversial. Some authors (e.g. Boström, 1973) have favoured a mantle-derived magmatic source for the enriched elements, suggesting passage up from depth in CO_2-rich fluids. The alternative model is that proposed by Corliss (1971) and later developed by Dymond, Corliss, Heath, Field, Dasch and Veeh (1973) and Bonatti (1975), whereby the debouching metal-rich fluids are assumed to derive from basalt–sea-water interaction. It is suggested that sea-water penetrates into and circulates within newly extruded basalt in the ridge zone such that the igneous material is leached and a dilute metal-rich fluid produced which is finally extruded as a thermal spring (Fig. 11.12). Within the ridge system itself these fluids probably carry the metals as chloride complexes which, on exhalation in cold oxygenated sea-water, are rapidly precipitated as tiny globules of iron–manganese oxide–hydroxide. Although some of the trace elements may be derived from leaching of the lava, some are added post-depositionally by 'scavenging' from surrounding sea-water: iron–manganese oxide–hydroxides are efficient agents of sorption from dilute solution (e.g. Goldberg, 1954; Krauskopf, 1956).

Support may be mustered for both these hypotheses. Boström, for example, points to the fact that continental carbonatite magmas are commonly enriched in Mn, Fe, Ba, P and rare-earth elements and that a CO_2-rich phase in ocean ridges would be a likely purveyor of these trace materials. Other authors (e.g. Corliss, 1971) have demonstrated that the slowly cooled interiors of ocean basalts relative to quenched flow margins are depleted in Mn, Fe, Co, rare-earths and other elements. Furthermore the chemical signatures of active-ridge sediments, in terms of rare-earth abundances and strontium isotopes, clearly suggest an origin from sea-water. In some elegant laboratory experiments Hajash (1975) and Seyfried and Bischoff (1977), have shown that heated basalt and sea-water will react such that Fe, Mn, Cu, Zn, Ba and other metals are leached out. As is commonly the case, these two proposed sources for active-ridge sediments – primary magmatic, secondary leaching – may represent end members of a spectrum: nature may have chosen the middle ground between them.

As well as metal-rich sediments, oxide–hydroxide crusts, also ascribed to hydrothermal sources, occur around volcanic vents and on the basaltic talus of spreading ridges; an Fe-rich Mn-poor deposit was sampled from the East Pacific Rise (Bonatti and Joensuu, 1966; Bonatti, Kramer and Rydell, 1972), and Mn-rich, Fe-poor varieties have been collected from the median valley of the Mid-Atlantic Ridge and from near the Galapagos spreading axis (Scott, Scott, Rona, Butler and Nalwalk, 1974; Moore and Vogt, 1976). In these latter cases Fe may have been separated out *within* the oceanic basalts as a sulphide phase such that the debouching hydrothermal fluids were preferentially enriched in Mn (Fig. 11.12). The chemistry of the precipitate changes with distance from the thermally active ridge zone; this is interpreted as reflecting decline of hydrothermally supplied metals and increase in those of hydrogenous or authigenic origin (Scott, Malpas, Rona and Udintsev, 1976). A proportionality between thickness of these Fe–Mn crusts and distance from the spreading axis has also been established for the East Pacific Rise, the Carlsberg Ridge in the Indian Ocean and the Mid-Atlantic Ridge near 45°N (Menard, 1960a; Laughton, 1967; Aumento, 1969). This effect

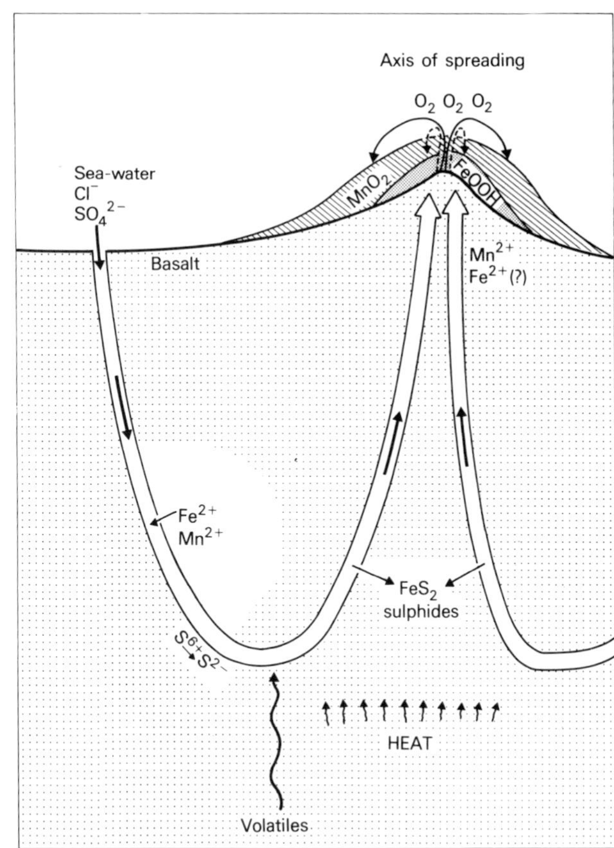

Fig. 11.12. Model illustrating basalt–seawater interaction, hydrothermal circulation and metallogenesis at oceanic spreading centres. Sulphides are assumed to precipitate out within the igneous basement. Exhaled fluids are converted to oxide-hydroxides. Fe-rich varieties, being easily oxidized, occur near the hydrothermal vent, Mn-rich species are precipitated further afield (after Bonatti, 1975).

is presumably related to the time an accreting basaltic surface spends in contact with sea-water.

Away from the crestal portions of ridges biogenic sediments assume some importance, particularly in depressions. On the Mid-Atlantic ridge biogenic sediments are dominated by carbonate derived from planktonic Foraminifera, pteropods and calcareous nannofossils (Emery and Uchupi, 1972). Locally they are lithified by high-magnesian calcite cements (Bartlett and Greggs, 1970). Lithified low-magnesian calcite oozes and aragonite cements also occur in certain fractures and small pockets in basalts from this area (Garrison, Hein and Anderson, 1973; Hekinian and Hoffert, 1975). The clay component of the sediments is chiefly illite; fresh-water diatoms have also been recorded, testifying to an aeolian source for some fine-grained material (Folger, 1970). The Mid-Atlantic Ridge, between 22 and 23°N, has a crestal region with only a patchy sediment cover, while valleys and ponds, beginning some 75–100 km from the axis, are characterized by thick fills (Figs. 11.11, 14.12); adjacent hills also have a sediment cover (van Andel and Komar, 1969). The sediments in ponds are calcareous, and developed as tan oozes interbedded with layers of foraminiferal sands that are graded, current-laminated and, in rare instances, cross-bedded. Age determinations on the Foraminifera show that they include Quaternary and Late Tertiary forms. These features clearly suggest that the sands are redeposited, apparently by turbidity currents that rebounded off the pond walls and returned through their own tails. Redeposited material may also include ferromagnesian minerals and breccias of serpentinized peridotite set in low-Mg calcite and aragonite cements. The igneous constituents were apparently derived from upthrust ultramafics exposed in fracture zones (Bonatti, Honnorez and Gartner, 1973; Bonatti, Emiliani, Ferrara, Honnorez and Rydell, 1974).

In another area of the Mid-Atlantic Ridge, van Andel, Rea, von Herzen and Hoskins (1973) have shown that the amount of redeposition decreases with distance from the ridge crest until the sedimentary rate approaches the background level of pelagic sedimentation. Finally as the ridge flank and its overlying sediment dive below the calcite compensation depth, the facies deposited will change from carbonates to red clays and siliceous oozes (Fig. 11.11). This change is abrupt; on the East Pacific Rise there is a 100-fold increase in the degree of carbonate dissolution between depths of 3,870 and 3,950 metres (Broecker and Broecker, 1974).

11.3.3 Aseismic volcanic structures

A variety of topographic features occurs in the oceans which are unrelated to present-day sea-floor spreading. They are, however, all built of anomalously thick piles of volcanic rock. Such features include aseismic ridges, seamounts and plateaus. To some extent it is artificial to separate them since a string of seamounts may merit the term 'ridge', and such a ridge may terminate in a plateau. Their origins, although not properly understood, may be related to the former presence of hot spots or 'leaky' transform faults that resulted in profuse outpourings of basalt for a limited period of time (e.g. Jackson, Silver and Dalrymple, 1972; Sclater and Fisher, 1974; Winterer, Lonsdale, Matthews and Rosendahl, 1974; Moberly and Larson, 1975). Since these topographic features commonly rise above the CCD they usually carry a veneer, dome or thick package of carbonate sediment.

ASEISMIC RIDGES

Aseismic ridges are long linear volcanic features that rise a kilometre or two above the ocean-floor; they are present in all oceans. On some of these ridges Deep Sea Drilling has revealed an early history of probable subaerial exposure (e.g. Late Cretaceous Ninety-East Ridge in the Indian Ocean, von der Borch and Sclater *et al.,* 1974). Brown coal and possible lagoonal or shelf replacement dolomites actually occur atop the Ninety-East Ridge (Cook, 1974; von der Borch and Trueman, 1974); other volcanic-dominated sediments contain abundant bivalves and gastropods, plus bryozoans, echinoderm fragments, solitary corals and even a terebratulid brachiopod (Davies and Luyendyk *et al.,* 1974). *Inoceramus* is locally abundant. Foraminifera of shelf aspect also occur in these near-basal sections.

Early shallow-water history is also established for the Rio Grande Rise and Walvis Ridge in the South Atlantic where basal facies (?Campanian and ?Aptian respectively) are developed as biosparites rich in bivalves, echinoderms and red algae (Maxwell *et al.,* 1970; Bolli and Ryan *et al.,* 1975). Other non-spreading ridges (e.g. Carnegie Ridge, Cocos Ridge, eastern equatorial Pacific) have an entirely pelagic sedimentary history (van Andel and Heath *et al.,* 1973). However, those ridges which possess basal shallow-water deposits all bear an overlying succession of pelagic character, generally nannofossil-foraminiferal chalks and oozes locally containing siliceous microfossils and chert. Such sedimentary piles, typically a few hundred metres thick, are usually cut by hiatuses. Organic-rich layers are recorded from the upper Albian of the Walvis Ridge (Bolli and Ryan *et al.,* 1975).

The surface structure of non-spreading ridges is commonly irregular and sediment may preferentially occur in local ponds. Abyssal dunes, both transverse and barchan type, have been recorded from the north flank of the Carnegie Ridge where they rest on a ferromanganese substrate exposed by erosion down through calcareous ooze (Lonsdale and Malfait, 1974). Erosional furrows are also cut into the sedimentary mantle of Cocos Ridge (Heezen and Rawson, 1977).

Although the pelagic deposits that form on spreading and

aseismic ridges are similar there are important differences in their sedimentary successions. Spreading ridges lack any record of shoal-water deposition, being typically generated at depths around 2,700 metres in major oceans (Sclater, Anderson and Bell, 1971). Aseismic ridges, on the other hand, locally 'grew' into waters shallow enough to develop shoal-water carbonates before abandonment of the volcanic centre and resulting subsidence. Such a thermal motor, however, being relatively short-lived, did not produce the continuous metal-rich hydrothermal discharge characteristic of spreading ridges.

VOLCANIC SEAMOUNTS

Volcanic seamounts and guyots, essentially conical features that rise up from the ocean deeps, occur throughout the ocean basins but are particularly widespread in the Pacific and, to a lesser extent, the Atlantic. The sedimentary cover on certain guyots in the Mid-Pacific Mountains exceeds 100 m in thickness (Karig, Peterson and Shor, 1970). In many cases, however, the rocks exposed on the surface of seamounts include volcanic material and relict sediment, suggesting that erosion or non-deposition have been active for millennia. Rippled sands and other bed-forms, discussed below, also testify to the influence of active water currents (Lonsdale, Normark and Newman, 1972; Johnson and Lonsdale, 1976). Around these topographic features there are sedimentary aprons of derived material (Sect. 11.3.4).

The earliest sediments on seamounts, which may be interbedded with flows, are dominated by volcanic constituents and their alteration products; smectitic clays, zeolites, and iron oxyhydroxides have, for example, been dredged from Pacific seamounts and guyots (Lonsdale, Normark and Newman, 1972; Piper, Veeh, Bertrand and Chase, 1975). Deep Sea Drilling on the Emperor Seamount Chain and the Line Islands, also in the Pacific, revealed basal volcaniclastic sandstones usually showing grading and cross-lamination, and claystones vividly coloured in red and (at the Line Island site), blue and green (Natland, 1973; Jenkyns and Hardy, 1976). Minerals in these deposits include smectitic clays, illite, plus traces of pyroxene, feldspar and very rare anatase. Chemically such deposits are distinct from the spreading-ridge sediments described above and from normal Pacific pelagic clays. The basal seamount sediments show enrichment only in Fe and Ti relative to average Pacific clays; their trace-element levels are generally lower. Relative to spreading-ridge sediments the seamount deposits are enriched only in Ti. The distinctive chemistry of the multicoloured claystones and associated volcanic sandstones seems to result essentially from *in situ* weathering of basalt; hydrothermal solutions have played a strictly limited role.

That many western Pacific guyots, now lying at depths of a kilometre or two, were formerly near sea-level is shown by their post-volcanic capping of limestones containing rudists, corals, Bryozoa, coralline algae, echinoids, bivalves and agglutinating Foraminifera, i.e. a typical reef fauna (Matthews, Heezen, Catalano, Coogan, Tharp, Natland and Rawson, 1973). This fauna is dated as mid-Cretaceous. Overlying the reef material are pelagic chalks and oozes. The change in facies is attributed to Albian–Cenomanian eustatic rise in sea level and consequent drowning of the guyots. More recent changes from reefal to pelagic carbonates are manifested on Kōko Guyot, in the Emperor Seamount chain (Matter and Gardner, 1975). In the Atlantic, Mytilus Seamount has yielded Foraminifera and molluscs of possible Late Cretaceous age and calcareous algae of ?Eocene age (Ziegler, in Emery and Uchupi, 1972); this seamount now lies some 2 km below the surface and is capped by ice-rafted debris and foraminiferal-nannofossil oozes and scoured by tidal currents (Pratt, 1968; Johnson and Lonsdale, 1976). Generally, basal reefal, and/or volcaniclastic sediments pass up into calcitic or perhaps, as on part of the Bermuda Pedestal, calcitic and aragonitic pelagic oozes (Chen, 1964). Seamount flanks, however, may preserve a post-volcanic record of clay or siliceous ooze.

The Tertiary succession of many seamounts is dominated by foraminiferal-nannofossil oozes. If, at a certain time, the seamount lay beneath a particularly fertile part of the ocean the record of that age may be substantial (e.g. Lonsdale, Normark and Newman, 1972). The Eocene phase of high silica manufacture is commonly recorded by chert levels from seamounts in both the Pacific and Atlantic (Lonsdale, Normark and Newman, 1972; Heezen, Matthews, Catalano, Natland, Coogan, Tharp and Rawson, 1973; Schlanger and Jackson *et al.*, 1976; Tucholke and Vogt *et al.*, 1975) (cf. Fig. 11.14).

Somewhat problematic is the record of unconformities within the sediments of certain seamounts, particularly around the flanks. Heezen, Matthews, Catalano, Natland, Coogan, Tharp and Rawson (1973) have related these gaps in the pelagic record to eustatic fall in sea-level bringing the sediment cap within reach of near-surface water currents. Changes in the patterns of deep ocean currents, or slumping down the sides of seamounts, may also be responsible.

The non-depositional conditions that may characterize the flanks and at times the summits of seamounts do generate a specific type of environment where chemical reactions, dependent on the long-standing proximity of the sediment–water interface, can come to fruition. Such reactions include formation of a hard lithified substrate by precipitation of a submarine cement, commonly high-magnesian calcite (e.g. Fischer and Garrison, 1967; Bartlett and Greggs, 1970; Milliman, 1971), replacement of calcareous material by phosphate (e.g. Marlowe, 1971; Heezen, Matthews, Catalano, Natland, Coogan, Tharp and Rawson, 1973) and formation of ferromanganese nodules and crusts (e.g. Mero, 1965; Aumento, Lawrence and Plant, 1968) which, in areas adjacent to continents, may be mixed with glauconite (Palmer, 1964).

The lithified zones examined by Bartlett and Greggs (1970) from Atlantic seamounts form lithified/non-lithified couplets and are interbedded with ferromanganese deposits. Lithification is usually accomplished by precipitation both of Mg-calcite rim cement within skeletal tests and as micritic matrix which may be, as on Great Meteor Seamount, developed in pelletal fabric (von Rad, 1974). These limestones are in isotopic equilibrium with enclosing sea water, as are some generally older low-Mg calcite limestones dredged from deeper waters in the same area: it seems probable therefore that the initial Mg-rich cement may invert to a low-Mg variety within the deep marine environment. All these processes are a trifle enigmatic given that water at these depths is apparently undersaturated with respect to all forms of calcium carbonate (Milliman, 1971).

Phosphatized limestones from seamounts are generally relict. Those recorded from Aves Swell in the Caribbean are dated as Eocene–Oligocene; some Pleistocene–Lower Holocene material is also present (Marlowe, 1971). Western Pacific guyots also contain phosphatic foraminiferal-nannofossil chalks and rudistid limestones; this material ranges in age from Cretaceous (pelagic and reefal facies) to Eocene (pelagic facies only). In both these cases the phosphatic material, termed carbonate apatite by Marlowe, francolite by Heezen, Matthews, Catalano, Natland, Coogan, Tharp and Rawson has variously replaced walls of body fossils, microfossils and coccolith-rich matrix. Iron–manganese oxides are intimately associated with the phosphatized material. The origin of the phosphates is probably related to the former presence of abundant phosphate-rich plankton entering the sediments during a time of very low sedimentary rate.

Ferromanganese crusts and nodules occur on many seamounts (e.g. Mero, 1965; Heezen and Hollister, 1971). San Pablo Seamount, for example, which is part of the New England chain, carries a thick pavement on its flanks and top (Aumento, Lawrence and Plant, 1968). This crust has a laminated internal structure, possibly of bacterial 'stromatolitic' origin (cf. Hofmann, 1969). Other organic residents in nodules and crusts include sessile Foraminifera; serpulids, corals, bryozoans and sponges occur as surficial encrustations (Wendt, 1974). Ferromanganese deposits from seamounts may also bear a distinctive chemical signature (e.g. Cronan and Tooms, 1969; Piper, 1974). These authors suggest that high Co, and possibly Ba, Pb, V and the rare earths Yb and Lu generally characterize nodules from seamounts; deeper-water nodules tend to be richer in Ni and Cu. These variations may relate to oxygenation levels in the water, sedimentary rate, proximity or otherwise to basaltic lava, and diagenetic recycling of certain elements.

Since seamounts are commonly current-scoured, fine nannofossil ooze may be winnowed away and coarser calcarenites left behind as a lag deposit; this material generally comprises foraminiferal tests and, if depths are sufficiently shallow, aragonitic pteropods. Thus rippled pteropod-globigerinid sands are recorded from Muir Seamount in the Atlantic (Pratt, 1968). Submarine foraminiferal dunes formed by accelerated tidal currents are described from Horizon Guyot; these sinuously crested structures, about 1 m tall, are highly asymmetric and have a wavelength of around 30 m (Lonsdale, Normark and Newman, 1972).

Horizon Guyot is now some 2 km deep. Shallower seamounts, however, may rise to within the photic zone, and thus provide a favourable milieu for light-dependent organisms. Examples of photic seamounts are Cobb Seamount, a Pacific volcanic feature that rises to within 34 m of the surface, Plantagenet Bank, an Atlantic seamount lying 55 m below sea level and Vema Seamount, off South Africa, whose summit plateau is some 120 m deep (Budinger, 1967; Gross, 1965; Simson and Heydorn, 1965). The flora and fauna from the shallower parts of Cobb Seamount include algae-encrusted basalt pebbles, plus bivalves, gastropods, brachiopods and polychaetes; the bivalve *Mytilus* is particularly abundant. On Plantagenet Bank the bottom sediments comprise fragments and nodules of calcareous algae, Foraminifera, corals, echinoids, green algae and Bryozoa. Dolomite is present in the sedimentary section at 20 m below the sediment–water interface. Vema Seamount which, like Cobb Seamount and Plantagenet Bank, is within reach of scuba divers, is also covered by algal biscuits or rhodolites (c.f. Bosellini and Ginsburg, 1971). The fauna includes deep-water corals, sponges, hydroids, echinoderms, bivalves, gastropods, ascidians and lobsters (Berrisford, 1969). In all the above cases it is likely that the sediments contain appreciable quantities of calcareous nannofossils.

OCEANIC PLATEAUS

Oceanic plateaus are vast (hundreds of thousands km^2) areas of relatively thick elevated ocean crust that commonly rise to within 2–3 km of the surface above an abyssal sea floor that lies several kilometres deeper. Iceland, the Galapagos volcanic platform and the Azores platform may represent rough modern analogues of this situation. In the Pacific these features (e.g. Ontong Java, Manihiki Plateaus, Shatsky, Hess and Magellan Rises) total about 2% of the Ocean Basin (Moberly and Larson, 1975). In the Indian Ocean the Kerguelen Plateau and Broken Ridge are further examples.

The Pacific plateaus are intriguing in that they are covered by extremely thick sedimentary blankets (Fig. 11.13). On Magellan Rise the 1,170 m-thick Jurassic to Recent succession is almost entirely constituted by calcium carbonate derived from nannofossils and Foraminifera. The sediments rest on

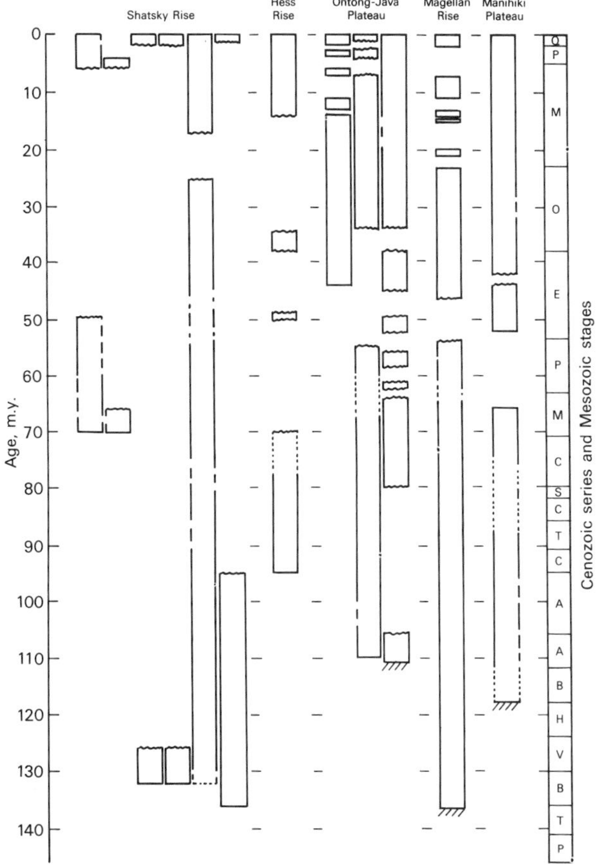

Fig. 11.13. Known stratigraphical record of central and western Pacific oceanic plateaus. Wavy-lined gaps in columns are from known unconformities or condensed sections. Straight-lined gaps show incomplete coring which may or may not include unconformities. Dashed outlines of columns indicate stratigraphical uncertainties. Basement, where reached, is basalt. Maximum sedimentary thicknesses recorded are 640 m (Shatsky Rise, incomplete section), 350 m (Hess Rise, incomplete section), 1,260 m (Ontong Java Plateau), 1,170 m (Magellan Rise), 910 m (Manihiki Plateau) (after Moberly and Larson, 1975 and DSDP sources).

altered and brecciated basalt (Winterer, Ewing *et al.,* 1973). The youngest sediments are predominantly oozes; older material tends to be more lithified but there is no exact depth-of-burial/lithification dependence (Schlanger and Douglas, 1974). Two angular unconformities occur in the section: one basal Tertiary, the other Miocene. Quartzose and porcelanitic cherts are locally developed in Cretaceous and Eocene strata, presumably attesting to former times of abundant biogenic silica manufacture; although recognizable remains of Radiolaria or diatoms are scarce it is assumed they have been diagenetically rearranged to form siliceous nodules (e.g. Wise and Weaver, 1974). Planktonic Foraminifera, either with original calcitic tests or replaced by silica, are often recognizable within these cherts (Lancelot, 1973). The basal levels of the Magellan Rise section yielded two aptychi, clearly indicating that ammonites were living on and around the plateau in latest Jurassic and Early Cretaceous time (Renz, 1973).

Other plateaus possess a similar stratigraphy to this, although they differ in detail. For example, six unconformities are tentatively recorded from Hess Rise and the Ontong Java Plateau (Moberly and Larson, 1975; Andrews, Packham *et al.,* 1975). Also notable is the presence, in the Cenomanian-Turonian of Hess and Shatsky Rises and the Barremian-Aptian of Shatsky Rise and the Manihiki Plateau, of black organic-rich bituminous horizons, implying relatively anoxic bottom conditions at these times (c.f. Fig. 11.7). The 1,260 m-thick succession on the Ontong Java Plateau contains basal ashy limestones and cherts whose igneous components were probably introduced by turbidity currents and reworked by moving bottom waters, possibly under tidal influence (Klein, 1975); Recent superficial oozes have undergone considerable soft-sediment deformation, a process aided by carbonate dissolution on the deeper parts of the edifice (Berger and Johnson, 1976). The Manihiki Plateau bears, at least locally, a particularly thick basal section of igneous derivation. Overlying highly vesicular basalt are more than 250 m of derived greenish-black breccias and graded volcaniclastics that pass upwards into limestones, chalks and claystones locally containing chert, and oozes: the total section is some 910 m thick (Jenkyns, 1976). Native copper, which occurs as blebs and strands within the volcaniclastics, is probably derived from post-depositional injections of cupriferous fluids into the sediment column; these fluids were perhaps of related origin to those debouching at spreading ridges. At the transition zone between the volcaniclastics and overlying calcareous deposits abundant benthonic molluscs are present (Kauffman, 1976). These include an older 'shelf-depth' bivalve and gastropod fauna and a younger bivalve-dominated assemblage of deeper-water bathyal character: *Inoceramus* is common in this group. The molluscs therefore testify to the post-eruptive subsidence of the plateau.

All plateaus clearly show a general history of subsidence offset somewhat by the counteracting accumulation of thick sedimentary sections. However, the sediments and fauna of Broken Ridge, Indian Ocean, point to at least one phase of uplift punctuating the overall deepening of the depositional zone (Davies, Luyendyk *et al.,* 1974). We might ask why topographic eminences of this sort come to support the thickest section of pelagic sediments in the South Pacific – as is the case with the Manihiki Plateau (Winterer, Lonsdale, Matthews and Rosendahl, 1974). The main reason seems to be simply that the Plateau lies under the equatorial high productivity zone. Thus supply of nannofossil and foraminiferal carbonate is abundant and, since the Plateau surface lies above the calcite compensation depth, accumulation is rapid. Other plateaus such as Hess Rise, which now lies to the north of the equator, possess thick Upper Cretaceous sections laid down when the plateau was temporarily under equatorial regions before continued northward movement of the Pacific Plate pushed it into less fertile regions (Sect. 11.3.1) (Moberly and Larson, 1975).

The net effect on plateaus has been one of successful deposition; nevertheless erosional and solutional influences have affected the sedimentary piles to produce unconformities (Sect. 11.3.1). Such conditions appear to pertain locally on the surface of the Manihiki Plateau at the present time since ferromanganese nodules – witnesses of non-depositional environments – cover parts of the surface (Heezen and Hollister, 1971).

11.3.4 Deep ocean basins

In the deep parts of the open ocean are the areas of abyssal plains and abyssal hills. Above the CCD, sediments are calcareous; below it they comprise red clay, radiolarian and diatomaceous ooze. Around topographic highs, redeposited volcanogenic, shallow-water and pelagic material accumulates as sedimentary aprons (e.g. Menard, 1964; Lonsdale, 1975). Near continental margins terrigenous material is emplaced by turbidity currents – where it may be sandwiched between red clays – to build the level abyssal-plain topography (see Chapter 10).

Calcareous units in this deep environment may evince a cyclic repetition of pale lime-rich and darker brown siliceous and clay-bearing material. Described initially by Arrhenius (1952) from Pleistocene cores these cycles are now known to extend back to the Upper Eocene (Hays *et al.*, 1972); their origin may be related to climatic variation that controlled biogenic productivity, input of detrital clay, or deep-sea dissolution or all of these (Berger, 1974; Gardner, 1975). In the Indian and Pacific Oceans the Pleistocene carbonate-rich layers apparently correlate with glacial periods, although the reverse seems to be true in the Atlantic. Furthermore, the Pacific cycles are clearly diachronous. The exact origin, therefore, of these cyclic alternations is not completely understood.

The red clay of the barren ocean comprises a variety of clay minerals, chiefly illite and montmorillonite, plus lesser and local amounts of kaolinite and chlorite together with a certain amount of X-ray-amorphous iron–manganese oxide–hydroxide, authigenic zeolites such as phillipsite and clinoptilolite, local amounts of palygorskite, and cosmic spherules (e.g. Arrhenius, 1963; Griffin, Windom and Goldberg, 1968; Berger, 1974). Associated detrital material includes feldspars, pyroxenes and quartz, the first two of which are assumed to derive from volcanic intra-oceanic sources, whilst the latter is probably aeolian. Certain oceanic clay minerals, particularly illite and to a lesser extent kaolinite, are also assumed to derive as fallout from high-altitude jet streams; the aeolian contribution to deep-sea clay is probably between 10 and 30%.

The Sahara Desert is a well-documented supplier of clay to the Atlantic deeps; material deriving from African dust storms can be traced to the Caribbean. The Indian and North Pacific Oceans probably derive their aeolian clays from the Asian mainland; in the South Pacific Australia is a likely source (Windom, 1975). Montmorillonite, however, is generated within the ocean basin by submarine weathering of volcanics; chlorite is detrital, being derived from low-grade metamorphic terrains.

The cosmic constituents of red clays, recognized by Murray and Renard (1884), include not only black magnetic spherules of nickel–iron but also chondrules of olivine and pyroxene (Arrhenius, 1963). There appear to have been marked changes in the input of metallic spherules during Cenozoic time; although this phenomenon might be explicable in terms of changing sedimentary rates (Sect. 11.3.1).

Red clays form the most common substrate for ferromanganese nodules (Murray and Renard, 1891; Mero, 1965) and the two occur together over much of the deep North Pacific floor. Such nodules tend to be more Mn-rich than those on calcareous substrates. Populations of nodules, distinct in terms of chemistry, mineralogy and morphology, occur in different parts of the oceans (Fig. 11.15) (e.g. Cronan and Tooms, 1969; Cronan and Moorby, 1976; Meylan, 1976). Whales' earbones and sharks' teeth are accessories in most areas (Murray and Renard, 1891).

Siliceous oozes, derived from Radiolaria, diatoms, silicoflagellates and sponge spicules, presently occur in three major areas; a global southern belt, a North Pacific zone including the back-arc basins, and a near-equatorial belt which is better developed in the Pacific and Indian Oceans, than in the Atlantic (Fig. 11.3). In the northern and southern areas diatoms are particularly important; in the equatorial zones Radiolaria locally dominate (Lisitzin, 1971). In the Eocene, however, under different palaeo-oceanographic conditions, the Atlantic equatorial belt was extensive, and may have been connected with the Pacific through central America (Fig. 11.14) (Ramsay, 1973).

Older deposits differ from those exposed at the surface. Siliceous frustules commonly dissolve within the sediment; only some 2% of the original standing crop survive as part of the sedimentary record (Heath, 1974). At depth a grain-size increase is evident, caused by the low-temperature solution of opaline tests and precipitation of aggregated lussatite (= opal-CT) 'lepispheres' (micron-scale spherules) which ultimately

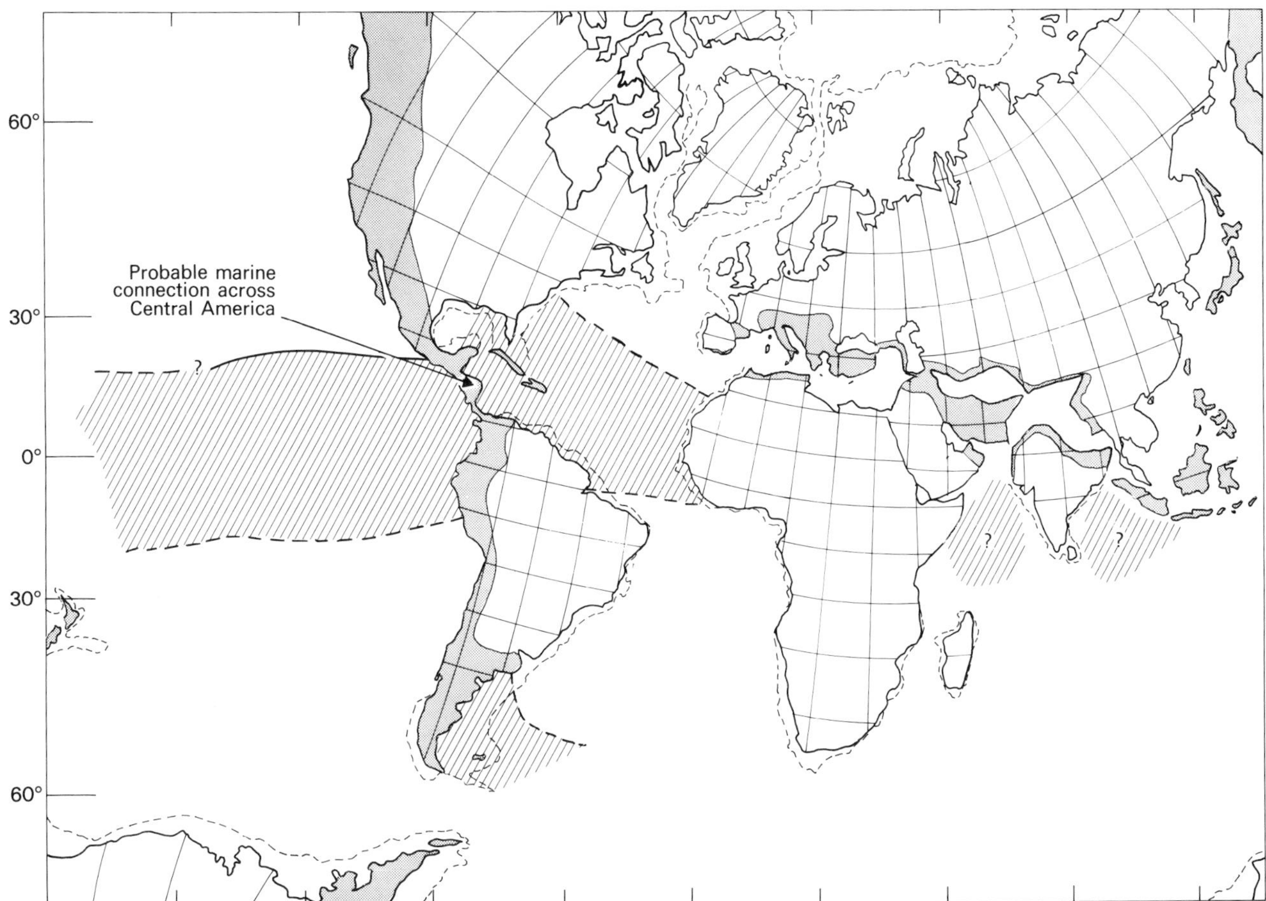

Fig. 11.14. Postulated global silica belts during the Eocene. Stippled areas on continents are Tertiary orogenic belts (after Ramsay, 1973).

recrystallize to quartzose chert, processes catalyzed by the presence of calcium carbonate; lussatitic porcelanites are thus particularly common in clay-rich and lime-poor matrices (Jones and Segnit, 1971; Lancelot, 1973; Wise and Weaver, 1974; Kastner, Keene and Gieskes, 1977).

The mineralogical nature of equatorial Pacific sediments has changed in the last 50 million years. From middle Eocene to late Miocene times the sediments were dominated by montmorillonite-rich, quartz- and pyroxene-poor detritus; subsequently, during late Miocene or early Pliocene times, chlorite, kaolinite and pyroxene were supplied in greater quantities to the western Pacific sediments. Finally, from the end of the Pliocene to the present day, quartz and illite-rich detritus dominated. The pyroxene is probably the signature of active marginal Pacific volcanism (Heath, 1969).

Given that the crust of the deep ocean basins was created at more shallow ridges, generally above the calcite compensation depth, it follows that the total stratigraphy of the deep-sea floor will show red clay or siliceous oozes underlain by calcareous ooze, chalk or limestone, in turn underlain by metal-rich spreading-ridge sediments (Fig. 11.11). Thus the deep-sea floor will be covered by a moderately thick sedimentary blanket and ridge-derived topography will have been smoothed over. Nonetheless, abyssal hills may punctuate the low-relief flat plains, as they do in most of the Pacific and parts of the Atlantic (e.g. Menard, 1964; Emery and Uchupi, 1972). These abyssal hills, with relief commonly in the range 50–250 m, are characterized, at least in the Pacific, by complicated patterns of deposition and erosion (Johnson and Johnson, 1970; Moore, 1970): although thicker sediments occur in the valleys, their surface layers are commonly of Tertiary age, whereas Quaternary deposits occur on the top

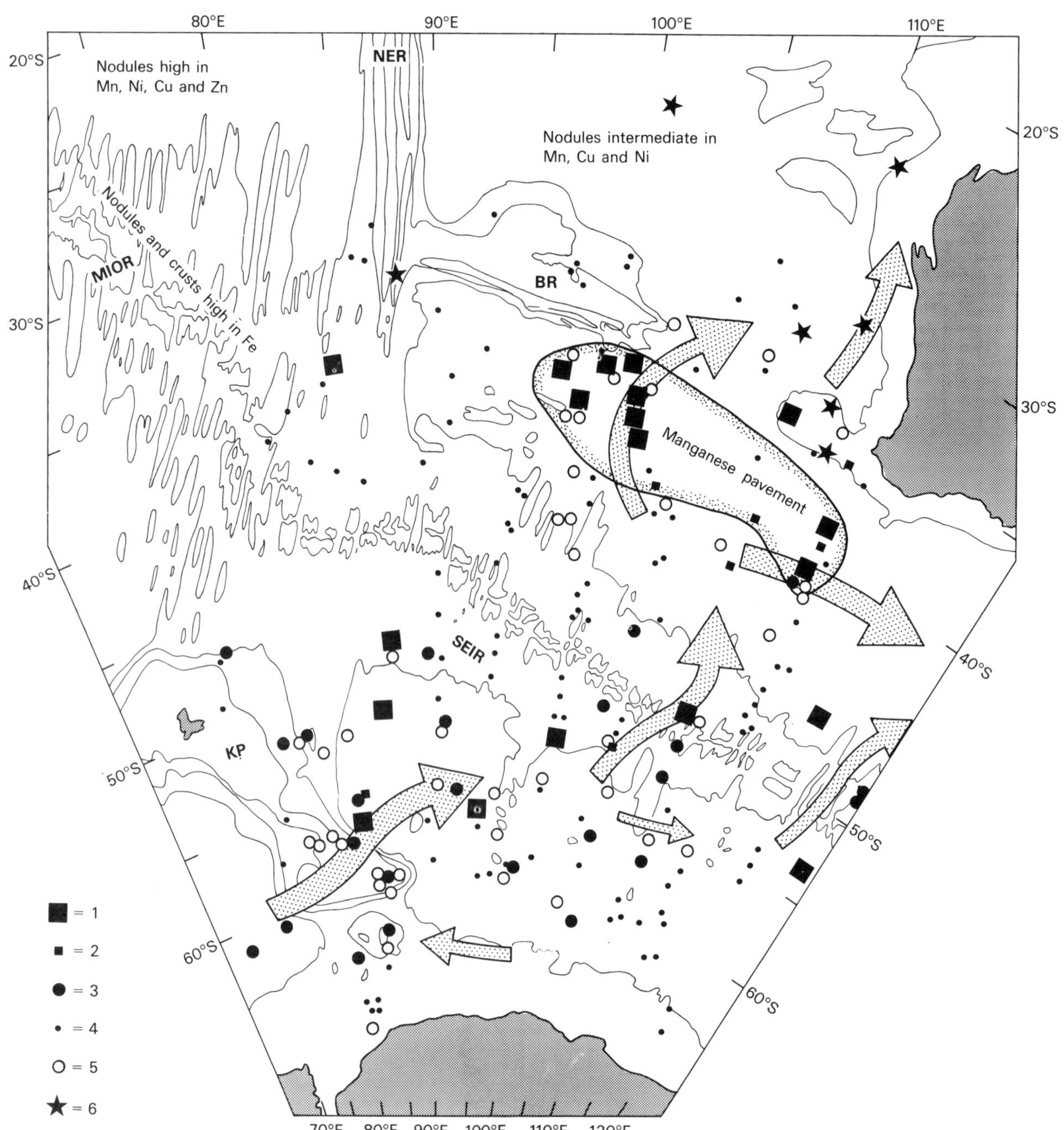

Fig. 11.15. Map illustrating proposed sea-floor currents and their relationship with ferromanganese nodule distribution in the south-east Indian Ocean. Bathymetric contours 2,000, 3,000 and 4,000 m. Symbols marking sites of interpreted observations based on bottom photographs or sediment core analyses are explained as follows: (1) manganese nodule pavement to abundant manganese nodules; (2) common to scattered manganese nodules; (3) strong bottom current activity (including ripple marks, scour, and distinct lineations); (4) core or photograph showing minimal evidence of current activity; (5) piston cores featuring disconformities and (6) site of Deep Sea Drilling Project core featuring major disconformity. The foot-shaped area shows the inferred limits of the Southeast Indian Ocean Manganese Pavement. Large arrows denote inferred paths and directions of major bottom-water flow in the region (after Kennett and Watkins, 1975). Data on chemical composition of Fe–Mn nodules and crusts from Cronan and Moorby (1976). NER – Ninety-east Ridge; MIOR – Mid-Indian Ocean Ridge; BR – Broken Ridge; SEIR – South-east Indian Ridge; KP – Kerguelen Plateau.

and particularly the flanks of the hills, where they are associated with ferromanganese nodules. These sedimentary relations may be related to Late Tertiary step faulting on the sides of the abyssal hills, and/or differential erosion.

As noted in Sect. 11.3.1 parts of the deep-sea floor are and were passageways for abyssal currents. The oceanic sedimentary record is cut by numerous stratigraphical gaps, and Tertiary sediments are locally exposed on the sea floor. Particularly significant are the apparently world-wide hiatuses during the Palaeocene and Oligocene, which are not necessarily limited to the deep ocean (Rona, 1973; Kennett, Houtz, Andrews, Edwards, Gostin, Hajos, Hampton, Jenkins, Margolis, Ovenshine and Perch-Nielsen, 1975). Ripples, scours, furrows and lineations mark the paths of Recent currents; particularly striking, however, is the relationship between erosion surfaces and Fe–Mn nodules (Kennett and Watkins, 1975; Pautot and Melguen, 1976). This is well illustrated in the case of the South Tasman Sea and the south-east Indian Ocean (Fig. 11.15); the nodule field of the latter area encompasses some 10^6 km^2. Nodule growth was clearly favoured by a non-depositional environment.

One area in the Pacific, the so-called Bauer Deep, is of particular interest. This area, located between the active East Pacific Rise and the abandoned Galapagos Rise, is a locus of high heat flow and characterized by depths greater than 4 km, apparently below the calcite compensation depth (Sect. 11.3.2). Surficial sediments are extremely poor in carbonate but enriched in iron, manganese, and other elements (McMurty and Burnett, 1975; Sayles, Ku and Bowker, 1975). These metal-rich deposits occur throughout the depositional column which includes much nannofossil ooze (Yeats, Hart *et al.*, 1976). The Bauer Deep is the most extensive area of metal-rich deposits not associated with a spreading ridge, although other examples, possibly linked to fracture zones, are known (Bischoff and Rosenbauer, 1977). The metalliferous sediments of the Bauer Deep, chiefly iron-rich smectites and Fe–Mn oxide–hydroxides, are chemically very similar to those on the East Pacific Rise, but locally contain enhanced quantities of SiO_2, Ni, Co and Cu. Major elements, isotope ratios and rare-earth element abundances all suggest an origin via seawater. Thus they apparently form in a similar way to spreading-ridge sediments. Indeed, according to Lonsdale (1976), the metal-rich sediments in the Bauer Deep are not generated *in situ* but transported from the East Pacific Rise by deep-sea currents.

11.3.5 **Small pelagic basins**

The oceanography of small pelagic basins may be distinct from large oceans in terms of salinity, productivity, water mixing and other factors, such that distinctive sedimentary environments are produced. In young rifted oceans the crust will, of course, be relatively shallow which may favour deposition of carbonate. Oceans like the Red Sea and Gulf of California, however, have very different sedimentary records and these are distinct again from those of back-arc basins of Japan Sea type. Post-collision basins of the Mediterranean type are also characterized by a distinctive sedimentary record.

The Red Sea and the proto-central and South Atlantic have an early record of evaporite deposition, apparently partly on oceanic crust (e.g. Kinsman, 1975). The Miocene to Pleistocene post-evaporitic sediments of the Red Sea include dark grey dolomitic silty claystones that pass up into grey claystones containing a certain amount of nannofossils, overlain by grey silty clay, nannofossil chalk and ooze. Only the top sedimentary layers, 100–200 m thick, can justly be called pelagic; locally, Foraminifera and pteropods make up more than 50% of the sediment whereas siliceous microfossils are rare (Stoffers and Ross, 1975). Detrital material includes quartz, mica, amphibole and a range of clay minerals dominated by montmorillonite and palygorskite. In the Pleistocene of the Red Sea there are lithified layers cemented either by aragonite, commonly as radiating crystals in recrystallized pteropods, or high-magnesium calcite (Gevirtz and Friedman, 1966; Milliman, 1971). Pteropods also occur as moulds produced by sub-surface dissolution (Friedman, 1965). The cemented high-magnesian micritic clasts, which contain Foraminifera and pteropods, seem to be partly derived from the reconstitution of low-magnesian calcitic nannofossils. Some magnesian–calcite lutite, apparently a primary precipitate, also occurs in surficial sediments. Formation of these lithified carbonate layers is probably linked to conditions of former elevated salinity: such a situation will of course only be possible in a restricted ocean. The Red Sea also contains spectacular metal-rich deposits of hydrothermal origin (Degens and Ross, 1969).

The Mediterranean Sea, which may or may not be partly or fully oceanic, is a post- or syn-collision basin and includes a number of Recent sedimentary types (Fig. 11.16), many of which have a strong terrigenous component, particularly near the mouths of large rivers (Emelyanov, 1972). Volcaniclastics and trace-element-rich iron-oxide deposits occur near eruptive centres (Bonatti, Honnorez, Joensuu and Rydell, 1972). On certain volcanic seamounts there are ferromanganese crusts associated with carbonate sands rich in cosmic spherules (Selli, 1970; Del Monte, Giovanelli, Nanni and Tagliazucca, 1976). Calcareous biogenic sediments are widely distributed in areas not influenced by terrigenous input; such deposits are essentially *Globigerina*-nannofossil oozes with variable amounts of pteropods. A considerable amount of the carbonate in Mediterranean deep-sea sediments is an inorganic high-magnesian calcite precipitate, probably formed during periods of elevated temperature and salinity (Milliman and Müller, 1973; Sartori, 1974). Of particular interest are the high-magnesium calcite concretions found between Crete and

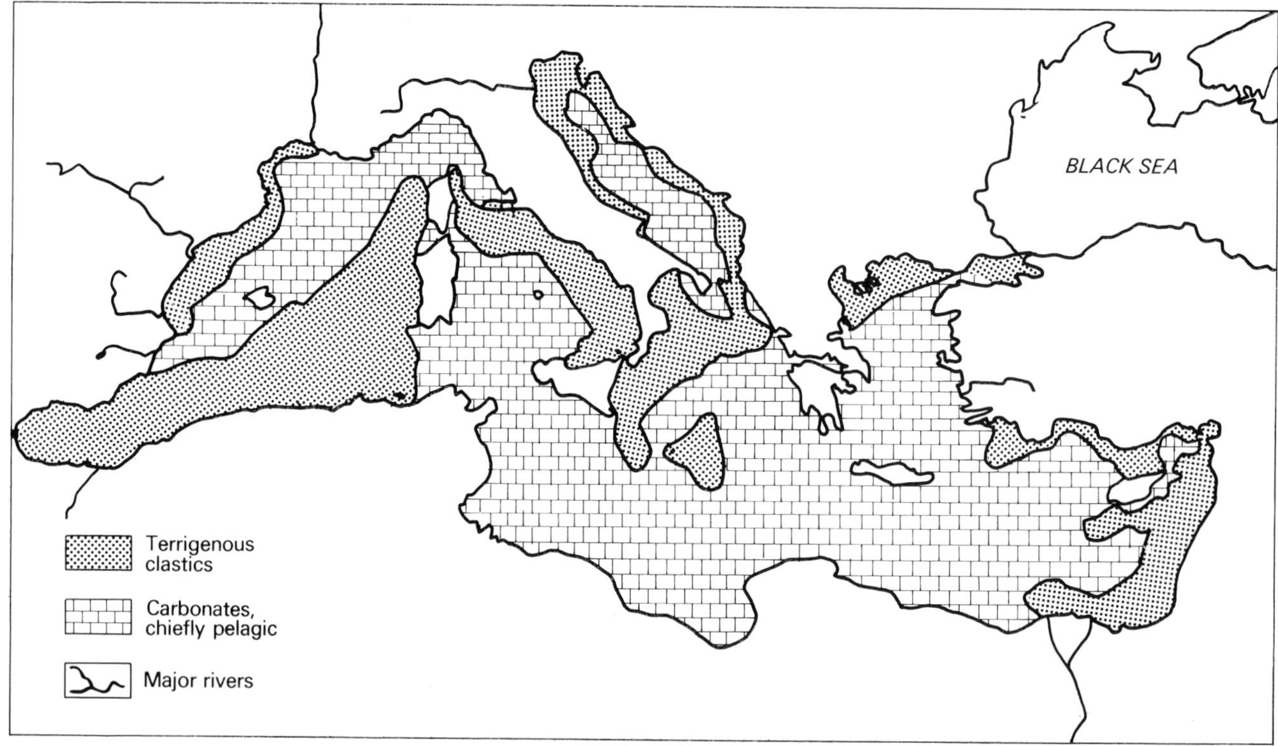

Fig. 11.16. Sedimentary distribution map of the Mediterranean Sea (simplified after Emelyanov, 1972).

Africa. Described in detail by Müller and Fabricius (1974) from the 2–3 km deep Mediterranean Ridge, these concretions are of centimetre scale, may be soft or hard and possess a micritic matrix entombing globigerinid Foraminifera, pteropods and heteropods. As is the case with the Red Sea 'lithic crusts', coccoliths are locally made over to a high-magnesian calcite precipitate.

The Pliocene–Quaternary history of the area has been dominated by 'normal' deposition of marly nannofossil foraminiferal oozes, punctuated in the eastern Mediterranean by episodes of stagnation witnessed by sapropelic layers (Nesteroff, Wezel and Pautot, 1973; Hsü, Montadert *et al.*, 1975). Both the Foraminifera and the sapropels are locally redeposited. Cosmic spherules are relatively abundant in the Pleistocene of the Mediterranean Ridge (Cita, Bigioggero and Ferrario, 1975). Below the Pliocene sequence, the basal part of which is a high-carbonate pelagic foraminiferal/nannofossil deposit, Messinnian gypsum, anhydrite, halite and dolomite are present, testifying to the Late Miocene Mediterranean salinity crisis (Hsü, Montadert, Bernoulli, Cita, Erickson, Garrison, Kidd, Mèlierés, Müller and Wright, 1977). Below this lie older marls and oozes, commonly rich in nannoplankton and Foraminifera.

Apart from the evaporite deposits the general sedimentary pattern in the central parts of the basin has been one of calcareous pelagic deposition with a fine-grained terrigenous overlay. This kind of stratigraphic record may be expected in a small post-orogenic basin, still seismically active, that is surrounded by mountain chains from which clastic material is continuously supplied.

The Gulf of California is, like the Red Sea, a young ocean (see Sect. 14.3.1); however, its situation on the west side of a major continent renders it a favourable site for upwelling due to the removal of surface waters away from the coast by offshore winds (Lisitzin, 1971). Diatoms and to a lesser extent Radiolaria dominate in the central part of the Gulf (Fig. 11.17); clay-rich sediments are important on its eastern side where rivers draining the Mexican mainland debouch into the sea (Calvert, 1966a). Glauconite occurs in many non-depositional areas (van Andel, 1964). Laminated diatomaceous facies, manifested by alternate diatom-rich and clay-rich laminae, are confined to the poorly oxygenated basin slopes

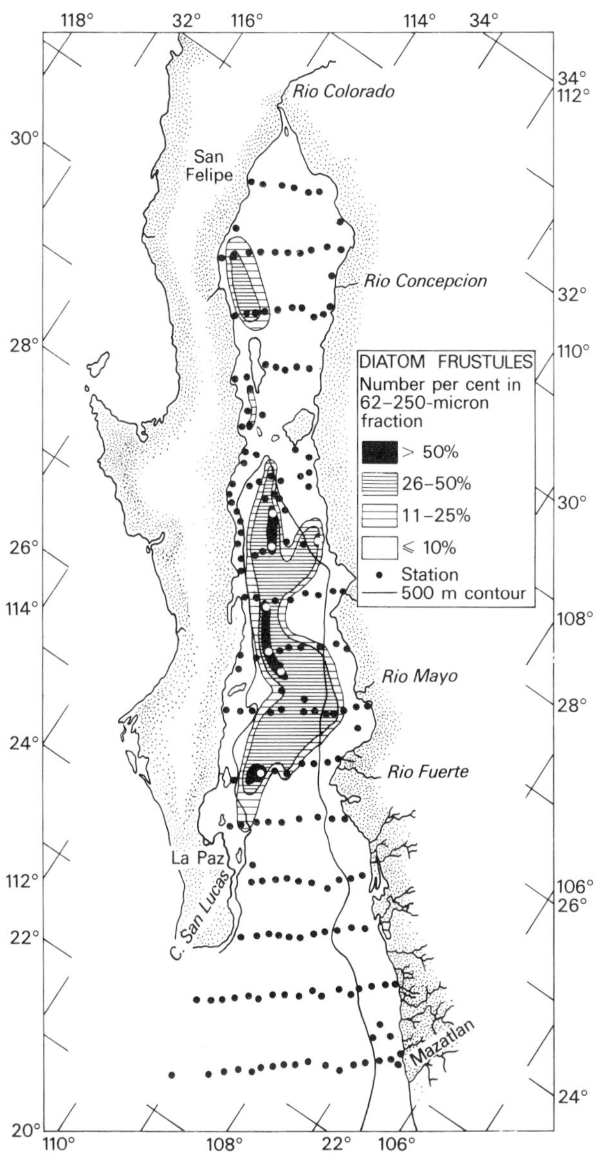

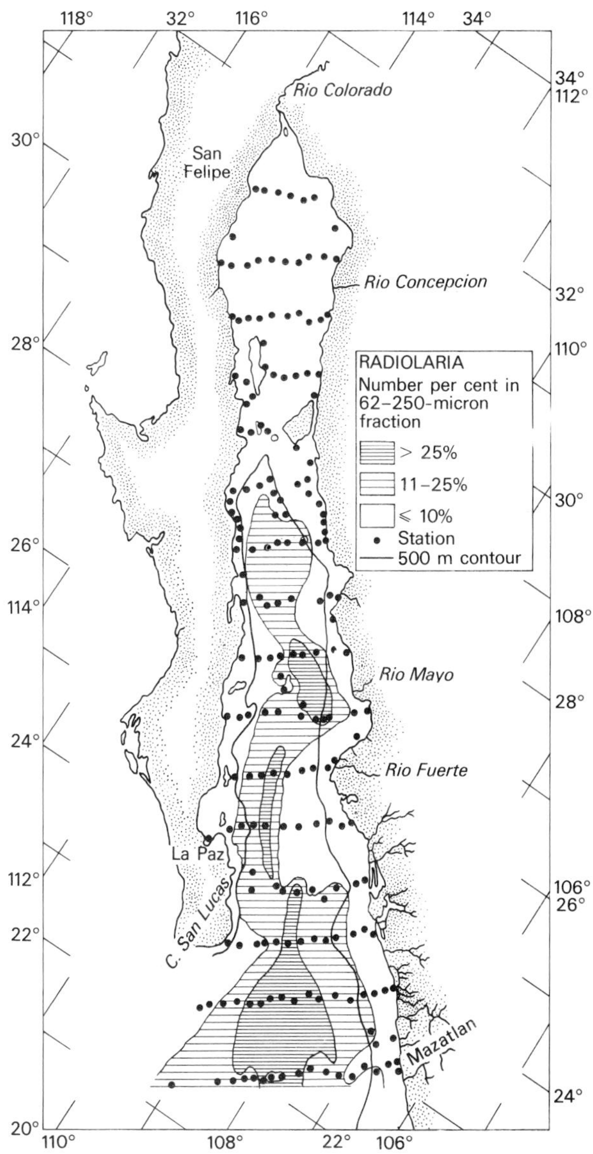

Fig. 11.17. Diatom and radiolarian distribution in sediments of the Gulf of California (after Calvert, 1966a).

where bioturbation is eliminated: depths here range from 300 to 1,400 metres. As diatom productivity is reasonably constant throughout the year the laminae are attributed to seasonal influx of detrital material. Sedimentary rates are between 4.7 and 5.4 cm/10 years (Calvert, 1966b).

The North Pacific back-arc basins are also loci of intense silica manufacture, particularly the Bering Sea, Sea of Okhotsk and, to a lesser extent, the Sea of Japan (Lisitzin, 1971). In those areas bordered by active island arcs volcanic products are important sedimentary constituents, commonly developed as turbidite aprons and horizons of volcanic ash (e.g. Karig, Ingle *et al.*, 1975). Metal-rich deposits occur in some hydrothermal areas (Zelenov, 1964; Ferguson and Lambert, 1972). Most of these marginal basins were apparently produced in the Late

Oligocene to Early Miocene (Dott, 1969) and many of them have a record of diatomaceous sedimentation that extends back to this period. Massive and laminated diatomaceous sediments, associated with small and local amounts of nannofossil and *Globigerina* carbonate, and interbedded with terrigenous and volcanogenic turbidites, have been recorded from the Miocene and Pliocene of the Sea of Japan and Bering Sea (Scholl and Creager, 1973; Karig, Ingle *et al.*, 1975). Locally the diatomite has been converted to chert (Garrison, Rowland, Horan and Moore, 1975).

The sedimentary record of ancient small ocean basins may be established by looking at the sediments of the proto-Atlantic. Drilling on both leg 11 and leg 41 of the Deep Sea Drilling Project revealed basal Kimmeridgian–Oxfordian sediments comprising red slightly nodular pelagic marly limestones, commonly involved in slump structures and containing turbidites entirely composed of pelagic material: Radiolaria, planktonic crinoids (*Saccocoma*), thin-shelled bivalves and nannofossils (Hollister and Ewing *et al.*, 1972; Lancelot and Seibold *et al.*, 1975). Overlying these red marly limestones are white to grey limestone sequences of Tithonian to Neocomian age; their basal portions are locally slumped and chert nodules occur at some levels. Ammonites and aptychi are recorded; the microfauna includes calpionellids, Radiolaria, sponge spicules and *Stomiosphaera*. Above these pelagic facies, Lower Cretaceous black clays, rich in organic matter (up to 14.8%), occur: these may indicate stagnation of the whole proto-central Atlantic at this time (Berger and von Rad, 1972). Such widespread stagnation was clearly favoured by both a tendency towards global oceanic anoxia and the restricted nature of the young ocean basin (Sect. 11.3.1; Fig. 11.7).

11.3.6 **Continental-margin seamounts, banks, plateaus and basins**

PELAGIC SETTINGS

Many continental margins possess a submarine topography of seamounts, banks and plateaus interspersed with deeper basins. These structural features are generally non-volcanic and are considered to rest on attenuated continental crust with perhaps some overlap on to oceanic basement. Some areas of this type are developed as carbonate platforms (e.g. Bahama Banks) and have a considerable record of shallow-water sedimentation dating back to the initiation of ocean-basin formation; in other, more basinal regions, pelagic sedimentation prevails. The development of pelagic sedimentation in continental-margin settings is dependent on a virtual absence of terrigenous material due to primary lack of supply and/or the presence of an intervening clastic trap.

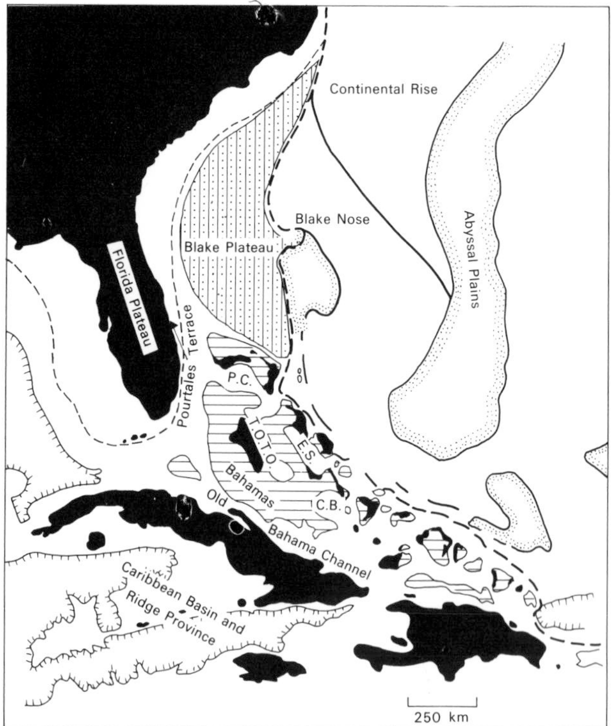

Fig. 11.18. Physiographic map of the continental margin, western central Atlantic. PC – Providence Channel; TOTO – Tongue of the Ocean; ES – Exuma Sound; CB – Columbus Basin (after Emery and Uchupi, 1972).

In the western central Atlantic region one of the most prominent submarine elevated structures is the Blake Plateau (Fig. 11.18) which has an area of some 228,000 km^2 and an average depth about 850 m (Pratt and Heezen, 1964). A Deep Sea Drilling core on the Blake Nose demonstrated conclusively that this part of the Blake Plateau lay near mean sea level until early Cretaceous time (?Barremian) when it abruptly subsided: birdseye limestones of tidal-flat origin are overlain by red pelagic carbonates containing goethitic pisolites and crusts. Subsidence may have taken place somewhat later in more southerly parts of the Blake Plateau (Sheridan, 1974). The red biomicrites of the Blake Nose site contain fragments of corals, codiacian algae, sponges, bryozoans and belemnites plus benthonic Foraminifera, echinoderm skeleta, ostracods, spicules, Radiolaria and abundant coccoliths (Enos and Freeman, in press). Nannofossils actually occur within the cortices of the ferruginous pisolites. Dredging at the northern end of the Plateau has recovered an upper Campanian–lower Maastrichtian fauna of ammonites, bivalves and gastropods from outcrop on the sea floor (Pratt, 1971). The higher parts of the Blake Plateau section are characterized by a regional unconformity between the Campanian and Albian. Cherty horizons are developed in the Upper Eocene and Palaeocene; otherwise most of the section is nannofossil ooze. Due to the erosional effect of the Gulf Stream the amount of deposition in later Tertiary time has not been great and the sequence is condensed (Benson, Sheridan *et al.*, 1976). As with Pacific seamounts, carbonate ooze has been phosphatized; later submarine erosion has left the mineralized material (francolite) as a lag deposit of phosphatic slabs. The age of fossils found in the phosphorite embraces the period Cretaceous to Pliocene. Evidence for replacement is clear and many of the nodules have undergone several phases of phosphatization. Glauconite is present within certain nodules and also occurs as discrete grains on the western part of the Plateau, where it may be relict (Pratt, 1971).

Covering much of the phosphatic layer is a continuous ferromanganese pavement, possibly embracing an area of 5,000 km^2, which passes into a realm of ferromanganese nodules to the south and east and phosphate nodules to the west (Fig. 11.19). This ferromanganese pavement lies at a depth of some 400 m in the north and 800 m to the south; the mineralized crust itself is about 7 cm thick and contains traces of encrusting organisms such as Foraminifera *within* the ferromanganiferous material, plus serpulids, bryozoa, sponges and deep-water corals on the outer surface. Encrusting organisms also occur on the undersides of the crust, testifying to excavation of sediment. Similar faunal traces, plus fungal borings, occur within and without the ferromanganese nodules on the Plateau (Wendt, 1974).

The ferromanganese nodules are unlike those from deep oceans in that they are rich in carbonate, locally in the form of aragonite veinlets, poor in silica and more humbly endowed with manganese and trace-metals. Mineralogically the nodules comprise goethite and todokorite but much of the material is X-ray amorphous. The nodules have formed by both primary accretion and by partial replacement of a phosphatic precursor.

Granitic boulders that occur locally on the Plateau are attributed to tree-rafting from the Guyana Shield (Pratt, 1971).

The sediments on the Plateau include *Globigerina*-nannofossil oozes which in areas of high current velocity have been winnowed into rippled foraminiferal sands, locally containing pteropods and heteropods (Fig. 11.19). Some fine-grained calcilutite, probably derived from the Bahama Banks, occurs in southerly regions. Under the axis of the Gulf Stream there are ahermatypic coral banks up to 20 m high; coral detritus is also abundant in this region. In some areas the *Globigerina*-pteropod sands have been moulded into submarine dunes or sand-waves on which slabs of ferromanganese material rest (Hawkins, 1969). Submarine lithification has been active in these non-depositional environments such that cemented foraminiferal–mollusc–coral calcarenites, commonly manganese-stained, have been produced (Milliman, 1971; Pratt, 1971): the sand waves themselves are also locally lithified. The cement is a high-magnesian calcite (Gevirtz and Friedman, 1966). It is clear from this description that the facies on the Plateau closely resemble those on volcanic seamounts (Sect. 11.3.3).

In basinal parts of the Blake–Bahama region the background sedimentation is represented by foraminiferal–nannofossil oozes: these pelagic deposits accumulate in the channels and gulfs that transect the Bahama Banks (Tongue of the Ocean, Providence Channel, Exuma Sound, Columbus Basin; Fig. 11.18). At the base of the Little Bahama Bank, possibly extending into the Providence Channel, is a series of so-called lithoherms or steep-sided mounds, up to 50 m high, built of sequentially layered soft and lithified pelagic carbonate containing some shallow-water grains. These structures, of obscure origin, occur in water depths of 600–800 metres (Neumann, Kofoed and Keller, 1977). Interrupting the pelagic stratigraphy of the deep channels are discrete beds of skeletal biosparite variously containing ooliths, pellets, corals, bryozoans, calcareous algae, ostracods, echinoderms, molluscs and Foraminifera, all of obvious shallow-water derivation (Bornhold and Pilkey, 1971; Emery and Uchupi, 1972). These beds, commonly graded, are interpreted as turbidites derived from the edge of the adjacent Bahama Banks; their high-Mg calcite and aragonitic mineralogy further distinguishes them from the low-Mg calcitic pelagic ooze. This deep-water environment, periodically affected by influxes of derived carbonate-platform material, existed at least as far back as the Late Cretaceous (Paulus, 1972).

South of the Blake Plateau, lying seaward of the Florida Keys, is the Pourtales Terrace (Fig. 11.18). This feature, a submerged edge of the Florida carbonate platform, is covered by water some 200–400 m deep. Eocene–Pliocene facies are shallow-water carbonates cut by karstic solution holes; Mio-Pliocene facies are lag deposits of rounded and polished sea-cow bones, sharks' teeth, coprolites and irregularly shaped phosphatized limestone cobbles (Gorsline and Milligan, 1963). This lag material rests on a bored ferromanganiferous hardground containing conglomerates, breccias and fills of pelagic carbonate in cavities. The complete section on the Terrace records the drowning of part of a carbonate platform and the formation of a condensed sequence under the influence of the swift Florida Current (Gomberg, 1973).

On the eastern side of the Atlantic off the Iberian coast are three non-volcanic seamounts (Black, Hill, Laughton and Matthews, 1964; Dupeuble, Rehault, Auzietre, Dunand and Pastouret, 1976). On all of these seamounts Upper Jurassic deposits are developed as platform carbonates. On Vigo Seamount, however, the topmost Tithonian is represented by

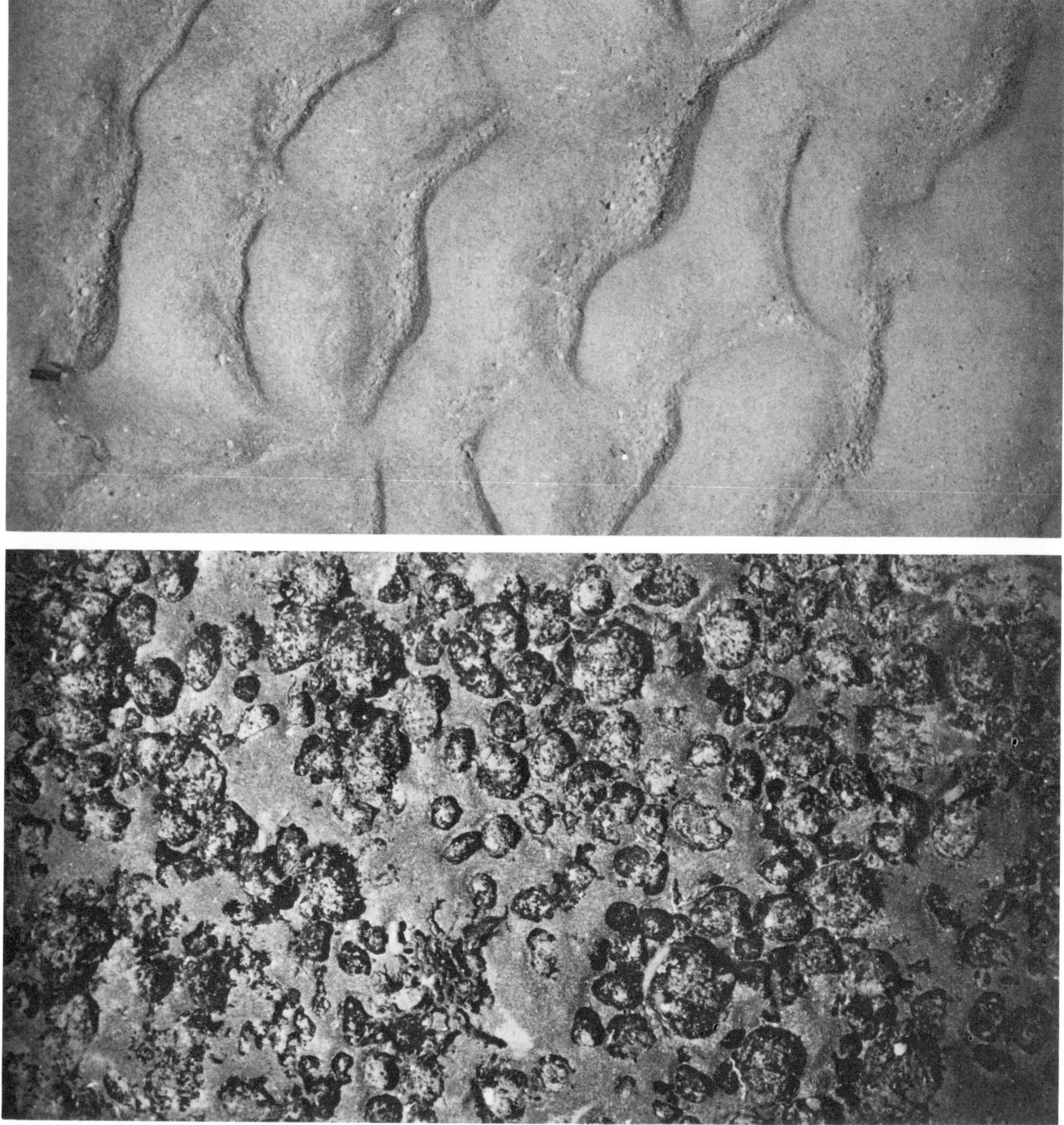

Fig. 11.19. Bottom photographs of two areas of the Blake Plateau. Above: Globigerinid-pteropod sands moulded into asymmetrical ripples, 30° 51′N, 78° 55′W, Depth 805 m. Below: Ferromanganese nodules partly covered by globigerinid-pteropod sands, 30° 58′N, 78° 17′W, Depth 883 m. Photographs courtesy of D.A. Johnson.

calpionellid-bearing biomicrites; similar facies, of Berriasian age, have been dredged from Porto Seamount. It is probable that a Bahamian-type situation existed in this area until the latest Jurassic when disintegration and subsidence of the carbonate platform allowed the invasion of pelagic conditions. Lower Cretaceous rudistid limestones from Galicia Bank suggest that shoal-water conditions persisted longer in this more northerly outpost; pelagic deposition probably commenced in the Maastrichtian. The Tertiary sections on all three features are dominated by foraminiferal–nannofossil oozes and chalks. On Vigo seamount such material is locally impregnated with iron oxy-hydroxides. Recent deposits include rippled foraminiferal shell-hash sands which, on Galicia Bank, attain the status of small dunes. This bank is a favoured habitat for crinoids. Ice-rafted pebbles and boulders occur atop these two seamounts.

HEMIPELAGIC SETTINGS

Another shallow-water area that has had at times a record of pelagic sedimentation is the seaward part of the Agulhas Bank, off South Africa, where Miocene to Recent foraminiferal–bryozoan nannofossil limestones occur (Siesser, 1972). These limestones are locally associated with glauconite and a phosphorite (carbonate fluorapatite) that has apparently formed by replacement of the pelagic–carbonate precursor (Parker and Siesser, 1972; Dingle, 1974). As well as phosphatized sediments conglomerates also occur, the components of which are phosphatized limestone pebbles set in a matrix of glauconite, quartz sand, microfossils and carbonate–fluorapatite cement (Parker, 1975).

A similar environmental setting is that of Chatham Rise, an elevated area of probable continental crust, mostly shallower than 500 m, situated some 800 km east of South Island, New Zealand (Norris, 1964). Sediments include abundant Recent glauconite formed by replacement of faecal pellets, or as internal casts of Foraminifera including *Globigerina,* or as indeterminate grains. Phosphorite nodules produced by replacement of foraminiferal ooze also occur; most of this is relict Miocene material, although some may possibly be Recent.

Another environmentally similar area includes the bank and basin topography off western Mexico and Southern California (Emery, 1960b). Sediments on these banks, where rates of detrital sedimentation are relatively slow, contain glauconite and phosphorite and generally comprise foraminiferal (benthonic and planktonic) and molluscan sands associated with bryozoan and echinoid remains. Glauconite has formed by replacement of clays, volcanic particles, skeletal carbonates and faecal pellets as well as by cavity-filling and accretion; a certain amount of this mineral is a Miocene remnant (Pratt, 1963). Much of the phosphorite is of similar age; some is younger (D'Anglejean, 1967). In deeper areas, such as the Santa Barbara Basin, conditions are locally anaerobic such that dark-coloured diatomaceous/silt–clay varves are formed; these deposits contain abundant planktonic Foraminifera, the aragonitic skeleta of pteropods and fish scales (Hülsemann and Emery, 1961; Berger and Soutar, 1970; Soutar and Isaacs, 1974). In the shallower (<480 m) oxygenated zone the sediment is bioturbated and coloured a homogenous green; benthonic Foraminifera and bivalves such as *Cardita, Lucina* and *Lucinoma* are abundant, as are gastropods and echinoids. Pteropods are absent. Clearly, in the more oxygenated environment, dissolution of calcareous skeleta is a significant phenomenon.

In all the above examples it may be argued that proximity to the coastline renders the environment less than fully pelagic; this is reflected perhaps in the abundance of benthonic Foraminifera and the presence of minerals whose genesis depends on high organic contents and/or specific land-derived detrital minerals. Glauconite, for example, apparently forms by the interaction between kaolinite-type clays – which are spatially linked to land-masses – and Fe^{2+} and K^{+} ions (Giresse and Odin, 1973). Replacement and primary phosphorites are formed from anoxic interstitial waters that contain P supplied by the dissolution of the zoo- and phytoplankton that contain this element in their protoplasm. Upwelling of nutrient-rich waters to promote plankton productivity – a common phenomenon along certain continental margins – is thus a critical parameter for the formation of phosphorite (e.g. Gulbrandsen, 1969; Tooms, Summerhayes and Cronan, 1969; Burnett, 1977).

11.4 PELAGIC SEDIMENTS ON LAND

11.4.1 Introduction

Pelagic sediments occur on all continents throughout the Phanerozoic column, and are manifested in a variety of facies. Some occur as parts of thrust-sheets in mountain chains; others are essentially undeformed. In this section oceanic and continental-margin facies are described, as well as deposits of small pelagic basins and epeiric seas.

Ophiolite assemblages, generally comprising variably serpentinized periodotite and dunite, gabbros, sheeted dyke complexes and pillow lavas, are generally interpreted as fragments of old ocean crust and mantle (e.g. Moores and Vine, 1969; Bailey, Blake and Jones, 1970; Dewey and Bird, 1971; Coombs, Landis, Norris, Sinton, Borns and Craw, 1976); thus the sediments that stratigraphically overlie the pillows can be interpreted as oceanic. In many instances, however, actual contact between the extrusives and sediment is not seen and the original relationship must be inferred; extreme complications arise when blocks of mafic and ultramafic igneous material occur chaotically in oceanic sediments, which may include flysch, as part of an olistostrome or melange.

In some cases the post-emplacement history of an ophiolite complex may have been characterized by an episode of shallow-water sedimentation such that neritic or littoral carbonates overlie the oceanic igneous and sedimentary rocks. If exposure is poor such shoal-water facies may be mistakenly considered as an autochthonous sedimentary cover of the ophiolite and a shallow depth for the formation of the oceanic crust has to be postulated (cf. Abbate, Bortolotti and Passerini, 1972). Thus a variety of tectonic and sedimentary factors may conspire to render original depositional relationships of oceanic sediments somewhat obscure. In areas where ultramafics are apparently missing yet abundant pillow lavas occur interbedded with, and overlain by, pelagic sediment, geochemical criteria may be used to pinpoint oceanic derivation of the flows even if they are considerably altered (e.g. Pearce and Cann, 1971, 1973).

Regional palaeogeographic studies on pelagic rocks that possess a complete and laterally extensive field record may reveal that they were laid down in restricted gulfs or embayments. Such a palaeogeographic configuration of small pelagic basins may also have pertained in the case of some of the ophiolite-linked (i.e. oceanic) Mesozoic and Palaeozoic deposits, but lack of outcrop does not allow such sophisticated restorations. In the Tertiary examples described here, the nature of the basement is not always clear; in some cases at least it is demonstrably continental.

Given that the sediments that overlie ophiolites are oceanic, the deposits of tectonically adjacent nappes may, in some cases, have been laid down in areas that were or later became continental-margins, floored by continental crust. In the absence of such juxtaposed oceanic rocks the style of deformation and type of metamorphism may forcibly suggest the imprint of tectonic and thermal processes unique to continental edges. Finally, the palaeotectonic and palaeogeographic picture that emerges from a study of the facies may only be compatible with a continental-margin setting.

Recognition of epeiric or epicontinental pelagic facies relies on an appreciation of the cratonic nature of the underlying continental basement; because of the stable nature of cratons the sedimentary cover will remain largely undeformed unless involved in subsequent episodes of rifting and compression.

11.4.2 **Pelagic sediments with inferred oceanic basement**

CAMBRO–ORDOVICIAN BLACK SHALES AND MULTI-COLOURED CHERTS OF THE PROTO-ATLANTIC (IAPETUS) OCEAN: NEWFOUNDLAND AND SCOTLAND

Cambro–Ordovician ophiolites occur in western Newfoundland and, in mediocre exposure, between Girvan and Ballantrae, Southern Uplands, Scotland. In Newfoundland the pillow lavas and hyaloclastites that form the upper part of the ophiolite stratigraphy are overlain by and interbedded with sediments variously described as black shales and cherts (i.e. the Cambrian *Bay of Islands Complex*, Stevens, 1970) and red argillite, red and yellow chert and lenticular manganiferous argillites that pass up into volcanogenic sediments (e.g. Ordovician *Betts and Tilt Cove ophiolite*, Dewey and Bird, 1971). The *Cambellton Sequence*, assumed to be Cambrian to Early Ordovician in age, contains metamorphosed pillow lavas capped by dark-coloured clays and siliceous argillites that are moderately thick-bedded and possess faint laminar structure; about 30 m above the sedimentary-igneous contact a manganiferous zone containing rhodonite and rhodochrosite (max. Mn = 9%) is locally developed (Kay, 1975). Higher sediments are graded terrigenous clastics.

The pillow lavas, tuffs and agglomerates of the Ballantrae ophiolite, which locally contain interstitial limestone and siliceous mudstone, are capped by black shales and red and green laminated cherts (Peach and Horne, 1899). The black shales have yielded an Arenigian graptolite fauna; the cherts contain abundant recrystallized Radiolaria which are locally preserved complete with radiating spines in the more clay-rich or mineralized zones (Hinde, 1890).

All the sediments described above are interpreted as having been laid down on a spreading ridge of a small ocean or oceans (Dewey, 1974; Mattinson, 1975): the near-basal manganiferous layers in the pelagic argillites of the *Cambellton Sequence* may thus be analogous to Recent oceanic hydrothermal deposits (see Sect. 11.3.2). Subsidence of these pelagic sequences down the ridge flank eventually resulted in a bathymetry deep enough to allow invasion of turbidity currents carrying continent-derived material.

Similar Lower Palaeozoic deposits, interpreted as oceanic, have been described from the Urals, the Tasman 'geosyncline' and the inner zone of the Cordilleran Foldbelt, eastern U.S.A. (Hamilton, 1970; Oversby, 1971; Churkin, 1974; Stewart and Poole, 1974).

PERMIAN OCEANIC SEQUENCE FROM NEW ZEALAND: VOLCANICLASTICS AND MOLLUSCAN CARBONATES

In the Permian ophiolite complex of the Dun Mountain Belt, South Island, New Zealand, tholeiitic pillow lavas are overlain, locally with unfaulted contact, by up to 500 m of red and green volcanic breccias and sandstones, plus younger black argillites, green cherts and impure limestones (basal units of *Maitai* and *Bryneira Groups*) (Waterhouse, 1964; Landis, 1974; Coombs, Landis, Norris, Sinton, Borns and Craw, 1976). Certain of the volcaniclastic layers are slumped; others are graded, containing pebbles of igneous material at the base, and passing up through grits and sandstones to limestones. These volcanogenic layers, which may be cut by clastic dykes, comprise fragments of spilite, dolerite and granophyre with argillite, chert and greywacke and grains of quartz, feldspar, clinopyroxene, epidote, chlorite, muscovite and clay minerals. The overlying limestones, commonly grey or pink, are locally graded and cross-bedded; they comprise micrite and the bivalve *Atomodesma* plus its prismatic fragments. This total mixed carbonate–volcaniclastic sequence changes thickness dramatically (circa 30 m–1,170 m) across the outcrop.

The graded volcanogenic conglomerates may be interpreted as redeposited ridge-derived material possibly shed from fault scarps; the graded *Atomodesma* limestones presumably represent those parts of the pelagic cover that have flowed downslope as turbidity currents. Redeposition into volcanically floored ponds is also suggested by the great lateral variation in thickness of the volcanic/carbonate formation (cf. Fig. 11.11). Implicit in this interpretation is the idea that *Atomodesma*, from which all the limestone apparently derives, must have flourished in an open-marine habitat: some support for this comes from the fact that the bivalve is probably ancestral to the inoceramids (Kauffman and Runnegar, 1975), whose Cretaceous representatives are a common constituent of Deep Sea Drilling cores.

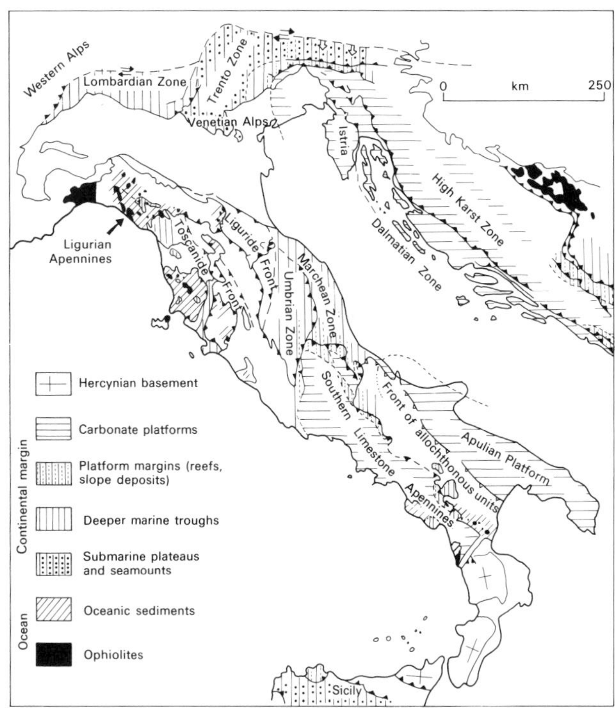

Fig. 11.20. Map of the Alps and Italy showing principal facies domains of the mid-to-late Mesozoic. Ophiolites of the Ligurian Apennines represent fragments of a Tethyan Ocean with its accompanying sediments. Carbonate platforms, submarine plateaus, seamounts and basins represent parts of continental margins, floored by continental crust (modified, with additions, from Bernoulli, 1972). Compare Fig. 11.18.

MESOZOIC OCEAN-RIDGE SEQUENCE FROM THE LIGURIAN APENNINES, ITALY: RED MANGANIFEROUS CHERTS, WHITE LIMESTONES AND BLACK SHALES

In the Ligurian Apennines (Fig. 11.20) lightly metamorphosed pillow lavas are capped by a series of green, brown and red radiolarian cherts (*Diaspri*) that vary in thickness from 0 to 350 m across the outcrop (Decandia and Elter, 1972; Elter, 1972). Red cherts are common near the base of the sequence where they may be interbedded with flows, and commonly contain mafic igneous material either disseminated in the siliceous matrix or as discrete graded units of volcanogenic sandstone and arkose (Abbate, Bortolotti and Passerini, 1972). Locally, sandwiched between the pilllow lavas, are conglomerates comprising fragments of spilite, foliated gabbro, serpentinite and radiolarian siltstones (Fig. 11.21). Graded units devoid of igneous material also occur in the cherts; these beds may possess load-and-scour structures, cross-, undulose-, and parallel-lamination and their basal layers are densely packed with Radiolaria (Garrison, 1974). Locally the cherts have yielded fragments of silicified wood (Abbate, Bortolotti and Passerini, 1972). Manganese deposits occur sporadically near the basalt–radiolarite contact (e.g. Bonatti, Zerbi, Kay and Rydell, 1976). These deposits, both massive and banded, have thicknesses of a few metres or more and extended several hundred metres laterally; they are concordant with the cherts and indeed the two may be interbedded. The mineralogy is dominated by braunite; the chemistry is Mn-rich (max. Mn = 47.62%) and Fe-poor (Fe < 1%). Trace metals, particularly Ba, and to a lesser extent Ni, Co, Cu, Cr and Zn are also present; and indeed may also be concentrated in the non-mineralized cherts (Thurston, in Spooner and Fyfe, 1973). Above the basal levels the radiolarites take on their more typical aspect of centimetre-scale quartzose chert beds interleaved with paper-thin dark red argillites or thicker siliceous mudstones. Frequency of Radiolaria is highly variable; and the skeleta are more recrystallized in the lighter-coloured than in the darker red cherts (Franzini, Gratziu and Schiaffino, 1968; Thurston, 1972).

Towards the top, the radiolarites become less obviously cherty before they interbed with and finally give way to a white thinly bedded micrite containing calpionellids (*Calcari a Calpionelle*); in some places, where radiolarites are absent, this calcareous unit rests directly on ophiolitic basement. The *Calcari a Calpionelle*, which contain grey marly interbeds,

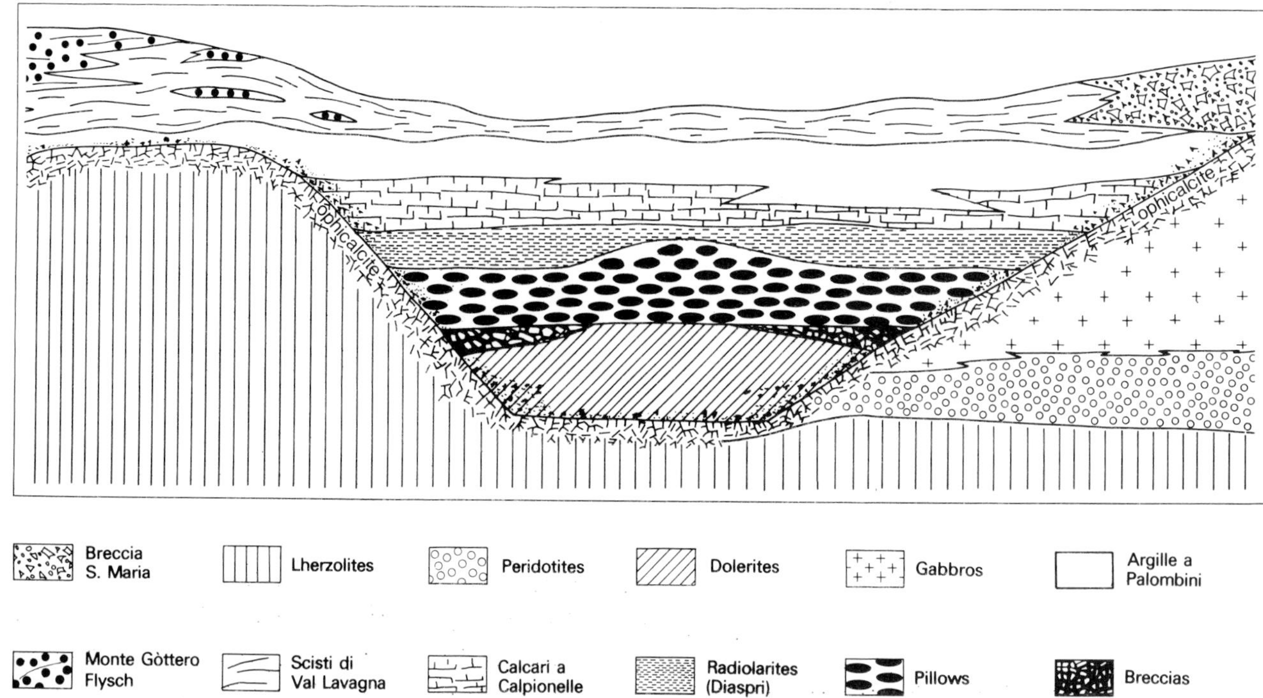

Fig. 11.21. Schematic section of pelagic sediment-igneous rock relationships in the Ligurian Apennines. No exact scale is implied, but horizontal distance envisaged is in the area of kilometres; vertical thicknesses of sedimentary units are typically a few hundred metres. Rapid lateral changes of facies and thickness are characteristic of this terrain. Ophicalcite signifies a network of calcite-filled veins cutting the lherzolites, peridotites and gabbros (modified from Elter, 1972).

range in thickness from 0 to 230 metres, and contain green tuffaceous layers and the odd blue–black chert nodule. Flute and groove casts occur on the soles of certain parallel-laminated beds; rare graded beds contain derived ooliths and fragments of oolitic and crinoidal limestones (Decandia and Elter, 1972; Andri and Fanucci, 1975). Additional fauna of the *Calcari a Calpionelle* includes carbonate-replaced Radiolaria and rare sponge spicules set in a matrix of grossly recrystallized nannofossils. The calpionellids indicate a Berriasian age for this formation.

The *Calcari a Calpionelle* are overlain by and pass laterally into the *Argille a palombini,* a somewhat thicker series of dark grey argillites, marls and partly chertified limestones that are locally graded and bear sole marks. This formation may rest directly on the *Diaspri.* The fauna includes radiolarians and rare Foraminifera, the latter suggesting an Albian age for the top of the formation.

The *Argille a palombini* are in turn capped by the *Scisti di Val Lavagna* (Albian to Cenomanian), a series of black marls and calcilutites, some 100–1,000 metres thick, that in higher reaches contain a terrigenous clastic component. Small globigerinid Foraminifera occur rarely. Locally the *Scisti di Val Lavagna* contain olistostromes and olistoliths of ophiolitic rocks. As the terrigenous clastic component increases, in both lateral and vertical directions, the formation gradually takes on the aspect of flysch.

As implied above, the *Diaspri,* the *Calcari a Calpionelle* and the *Argille a palombini* may all rest depositionally on the ophiolites. Furthermore, the nature of the igneous base varies laterally; the radiolarian cherts, for example, rest atop pillow lavas, gabbros and serpentinized ultramafics within a 25 km^2 area (Spooner, in Garrison, 1974; Barrett and Spooner, 1977). In some areas a system of veins, filled with white radiaxial fibrous and equant sparry calcite, talc, and red and green calcareous sediment, cuts through the serpentinites and gabbros (Abbate, Bortolotti and Passerini, 1972; Folk and McBride, 1976a). Such carbonate-rich brecciated levels are termed *ophicalcites* and may be capped by both igneous and sedimentary rocks. Inferred sedimentary-igneous relations are illustrated in Fig. 11.21.

The variety of igneous-sediment contacts indicates clearly that considerable syn-sedimentary faulting took place, which locally exposed ultramafics on the sea floor. Dramatic lateral changes in sedimentary thickness also point to accumulation of pelagic material in local ponds. The *ophicalcites* have parallels from the Mid-Atlantic Ridge as do the redeposited volcanogenic sandstones and breccias and the apparently hydrothermal manganese deposits in the radiolarian cherts (see Sect. 11.3.2) (Bonatti, Emiliani, Ferrara, Honnorez and Rydell, 1974; Garrison, 1974; Bonatti, Zerbi, Kay and Rydell, 1976; Barrett and Spooner, 1977). These features suggest a slow-spreading rifted ridge of Atlantic type (cf. Fig. 11.11). The trace-element enrichment of cherts, the manganese deposits, and the nature of the ophiolitic rocks themselves suggest that sea-water circulation was occurring within this ancient ridge (Spooner and Fyfe, 1973).

After the sedimentary regime had changed from siliceous to calcareous (see Sect. 11.4.2), redeposition processes down the ridge flanks were clearly still active: the paucity of nannofossils in the limestones has been ascribed to their destruction during transport although it may be a function of geothermal or tectonic processes. Derived oolitic materials suggest that parts of the oceanic slab had, by Berriasian time, sunk to a position sufficiently basinal to be reached by material from a marginal carbonate platform. The *Argille a palombini* have also been interpreted as a largely resedimented formation where the marly intervals represent the background sediment and where the more calcareous units are derived, presumably from higher up the ridge (Andri and Fanucci, 1975). The *Scisti di Val Lavagna* may record the greatest bathymetry attained in the ocean; they were deposited, probably near the calcite compensation depth, during an episode of oceanic anoxia (see Sect. 11.3.1).

Apart from the radiolarian cherts, all of the above-mentioned sediments have parallels in coeval material cored in the central Atlantic (Sect. 11.3.5), thus reinforcing the interpretation of the Ligurian Ocean as an Atlantic-type feature (Bernoulli and Jenkyns, 1974). Similar sections recorded from the Vourinos ophiolite, Greece (Pichon and Lys, 1976) and the Western Alps of France (Lemoine, 1972) may be relics of the same ocean.

THE TROODOS MASSIF OF CYPRUS: CRETACEOUS OCEAN FLOOR AND ACTIVE-RIDGE SEDIMENTS

The uplifted Troodos Massif of Cyprus is perhaps one of the best documented ophiolites in the world and its sedimentary cover is now admirably described (Fig. 11.22) (e.g. Moores and Vine, 1971; Robertson and Hudson, 1973, 1974; Robertson, 1975, 1976; Robertson and Fleet, 1976). Immediately overlying and occasionally interbedded with the upper pillow lavas – the youngest extrusives of Troodos – are local accumulations of chestnut-coloured fine-grained carbonate-free mudstones termed *umbers*; they are rich in Fe, Mn and a host of other

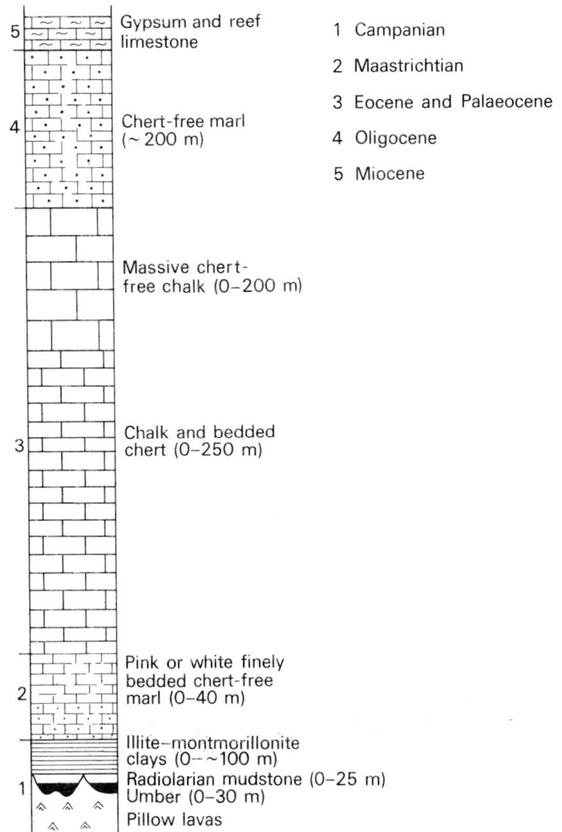

Fig. 11.22. Composite section of Upper Cretaceous and Tertiary oceanic sediments of Cyprus with capping of Miocene gypsum. Igneous base is the Troodos ophiolite (after Robertson and Hudson, 1974).

metals including rare earths (Robertson and Hudson, 1973; Robertson and Fleet, 1976). Such sediments sit in hollows, ponds and graben in the lava surface and their thickness varies accordingly; in fault-controlled depressions the thicker sections (10–15 m) show evidence of post-depositional slumping. The underlying lavas, which are locally highly vesicular, are veined and brecciated, containing iron oxides, smectitic clays and calcite. Other types of basal sediment, the so-called *ochres*, typically Fe-rich and Mn-poor, are usually associated with the massive sulphides of the ophiolitic complex.

The *umbers*, which themselves are locally silicified, pass up into Campanian radiolarian cherts and mudstones which are lime-free and, like the *umbers*, restricted to hollows in the lava surface; direct contacts between radiolarites and lavas occur locally. The radiolarian rocks, pink to pale grey, are regularly bedded and finely laminated; they reach thicknesses of up to 35 m but are generally thinner. Mineralogically they comprise both quartz and lussatite (see Sect. 11.3.4). The radiolarians show variable preservation; most are replaced by chalcedonic quartz, others remain as isotropic silica, similar to the matrix. Locally, the radiolarian rocks are overlain by bentonitic (illite–montmorillonite) clays, of Maastrichtian age, 0–100 m thick, that tend to fill the upper parts of hollows that are floored with umbers and radiolarites.

Above this the succession varies: intercalations of allochthonous masses are present in some sections, elsewhere pelagic chalks, composed of nannofossils and pelagic Foraminifera, follow the radiolarian rocks or rest directly on the upper pillow lavas. The lower chalks, also Maastrichtian, are discontinuous, but the higher levels (Palaeocene–Miocene) are more widespread. The Maastrichtian chalks contain the micron-sized bladed spherulites of lussatite termed lepispheres (see Sect. 11.3.4). The Palaeocene–Lower Eocene section is richly endowed with nodular and bedded chert layers. The bedded varieties, plus the overlying chalk, generally occur in couplets characterized by a graded tuffaceous base passing up into finely laminated chalk; the basal granular levels are preferentially silicified. Capping Oligocene and Lower Miocene unconsolidated chalks and marls are reef limestones and gypsum.

The pelagic sediments of the Troodos Massif may be readily compared with Recent oceanic deposits. Most diagnostic are the *umbers* which compare remarkably well with the less calcareous varieties of active ridge sediments (see Sect. 11.3.1) and a comparable origin is inferred (Robertson and Hudson, 1973) (cf. Fig. 11.12). The rare-earth element distribution in inter-lava sediments supports this interpretation by evincing a distinctive sea-water signature, characterized by a negative cerium anomaly (Robertson and Fleet, 1976). The *ochres* may partly derive from the same source but *in situ* submarine oxidation, erosion and deposition of massive sulphides have also been important (Robertson, 1976). The presence of *umbers,* taken together with the extrusive nature of the igneous basement, suggests that the Troodos Massif may be best compared with the fast-spreading, slightly rifted and hydrothermally vigorous East Pacific Rise (Fig. 11.11).

The radiolarian rocks record a time of high fertility in the Troodos Ocean when an abundant siliceous fauna flourished; the clay sequence that follows is interpreted as a hemipelagic product from both continental and island-arc sources (Robertson and Hudson, 1974; Robertson and Fleet, 1976). The overlying chalks also attest to fertile surface waters in which coccoliths and planktonic Foraminifera had largely replaced radiolarians. The change from siliceous to carbonate facies, paralleling that in the Ligurian Apennines, may be the result of uplift above the calcite compensation depth (cf. Fig. 11.11). The Palaeocene–Lower Eocene chalks were evidently laid down around topographic highs and were later redeposited: the more porous basal levels of the turbidites became favoured loci of chertification. Topography was apparently more subdued when the Lower Miocene chalks were deposited. Subsequent history of the massif has involved uplift leading ultimately to emergence.

Comparable ophiolites and *umbers,* probably pertaining to the same ocean as the Troodos, outcrop in north-west Syria (Parrot and Delaune-Mayere, 1974).

TIMOR: CRETACEOUS RED CLAYS, SHARKS' TEETH AND MANGANESE NODULES

The fossil red clays from western Timor are a classic example of oceanic sediments exposed on land (see Sect. 11.1.2) (Molengraaf, 1915, 1922; El Wakeel and Riley, 1961; Audley-Charles, 1965). The deposits comprise yellow, red and brown clays scattered with ferromanganese nodules and micro-nodules plus Upper Cretaceous elasmobranch teeth and fish bones. Their structural context shows that they are rafts in a Miocene olistostrome (Audley-Charles, 1972). The red clays contain abundant ill-defined radiolarian tests plus fragments of volcanic rocks, serpentinite pieces, feldspars and quartz; the clay fraction comprises illite, chlorite and montmorillonite. The black to brownish-black ferromanganese nodules, generally of ellipsoidal to spheroidal form, are of centimetre scale, concentrically laminated and possess a tubercular and finely granulated outer surface. Both the clays and Fe–Mn nodules are chemically analogous to their Recent oceanic counterparts; and they have been interpreted as deposits formed on an abyssal plain to the north of Late Cretaceous Timor which, at that time, may have constituted the northern continental margin of Australia (Audley-Charles, 1972). During the Late Cretaceous, Australia and Antarctica were joined and the circumpolar current would have been deflected north of Australia (Kennett, Burns, Andrews, Churkin, Davies, Dumitrica, Edwards, Galehouse, Packham and van der Lingen, 1972). This erosional environment would have favoured growth of the nodules (Fig. 11.23), as is postulated for the current-scoured South Tasman Sea and Indian Ocean manganese pavements today (see Sect. 11.3.4) (Glasby, 1978).

On the island of Roti, adjacent to Timor, comparable red clays and ferromanganese nodules occur, tentatively dated as Jurassic (Molengraaf, 1915; Jenkyns, 1977). The same environmental model applies.

FRANCISCAN FORMATION: RADIOLARITES, LIMESTONES AND MUDSTONES

In western California the ophiolite-bearing *Franciscan* outcrops. Variously described as a 'Series, Formation, Group,

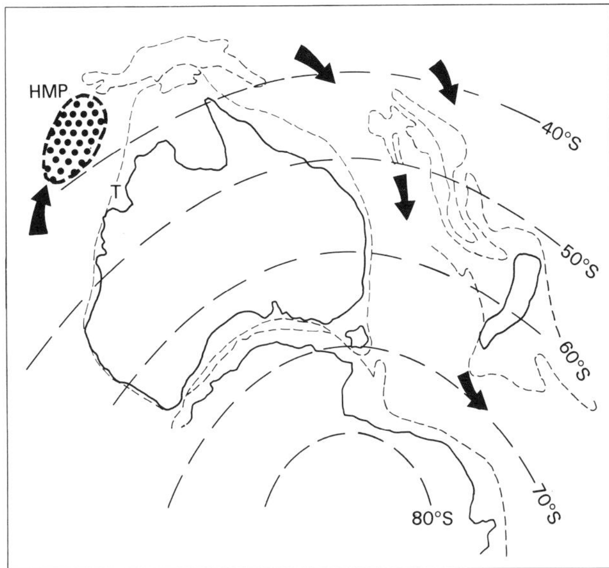

Fig. 11.23. Late Cretaceous–Early Tertiary reconstruction of Australasia with suggested directions of 'circumpolar' current (arrows) which could have favoured growth of a manganese pavement (cf. Fig. 11.15) to the north of Timor, which at that time constituted the northern continental margin of Australia. Later tectonic movements emplaced the red clays and Fe–Mn nodules on to this margin; subsequent subsidence isolated Timor from the mainland. T – Timor; HMP – Hypothetical manganese pavement (adapted from Kennett, Burns, Andrews, Churkin, Davies, Dumitrica, Edwards, Galehouse, Packham and van der Lingen, 1972).

Assemblage, melange and complex' it comprises blocks of mafics and ultramafics, blueschists, pillow lavas and various sedimentary rocks (Fig. 11.24) (e.g. Bailey, Irwin and Jones, 1964; Hsü, 1971; Blake and Jones, 1974). Small amounts of pelagic carbonates occur locally within pillow lavas (Hopson, pers. comm., 1976), but most prominent are red and brownish-red radiolarites of Tithonian–Neocomian age, resting immediately on or interbedded with basalt; these cherts and paper-thin argillites are essentially identical with their Ligurian counterparts (Davis, 1918; Pessagno, 1973).

Commonly associated with the radiolarites as part and parcel of a melange are sheared and locally metamorphosed dark mudstones and greywackes (Bailey, Irwin and Jones, 1964). The bivalves *Buchia* and *Inoceramus* and ammonites point to Late Jurassic and Early Cretaceous ages for this material. In some areas radiolarian cherts and greywackes are interbedded (Davis, 1918; Blake and Jones, 1974).

Limestones are of minor volumetric importance; they divide into two basic types: one, the so-called *Laytonville Limestone*, is a red micritic facies of probable Cenomanian–Turonian age; the other, the *Calera Limestone*, is represented by both white

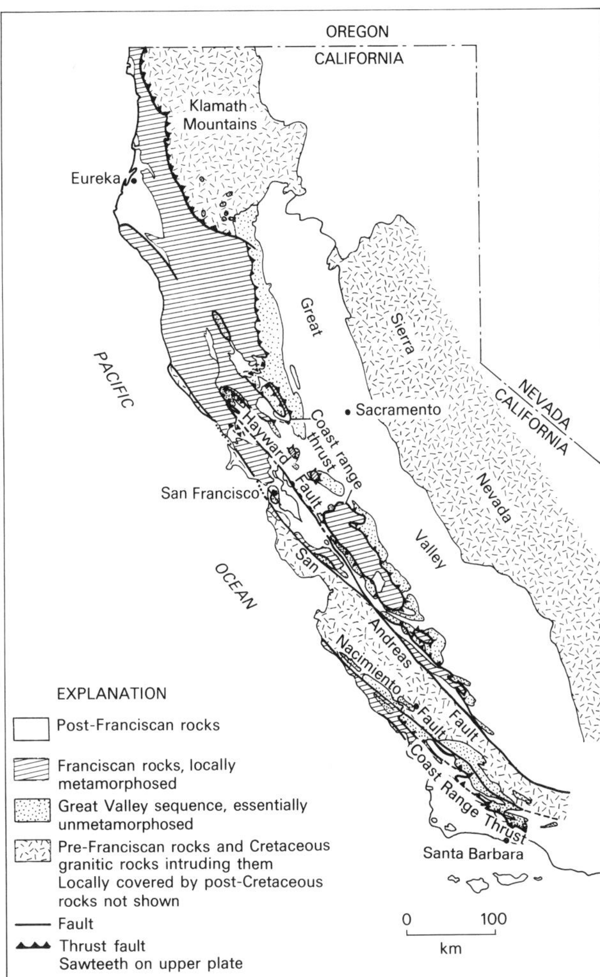

Fig. 11.24. Geological map of western California, showing the distribution of and tectonic relationship between the Franciscan and the Great Valley Sequence (simplified after Bailey, Blake and Jones, 1970).

and black varieties (Bailey, Irwin and Jones, 1964; Wachs and Hein, 1974, 1975). The *Laytonville Limestone*, found north of San Francisco, occurs as tectonic inclusions in shale and resting on and between pillows of non-vesicular basalt; the facies grade from massive to thinly laminated and contain planktonic Foraminifera, and Radiolaria preserved in interbedded lenses of red chert. The micritic matrix is almost entirely composed of calcareous nannofossils (Garrison and Bailey, 1967).

The *Calera Limestone*, tentatively dated as Lower Cenomanian, may be divided into a lower, black bituminous unit and an upper light-coloured unit. The lower unit, which may contain >2% organic matter, lacks planktonic Foraminifera but contains radiolarian moulds and partly recrystallized nannofossils. The upper unit, which is less recrystallized, contains planktonic Foraminifera, plus echinoderm fragments and radiolarians. Planktonic Foraminifera occur as graded beds at some levels; benthonic Foraminifera are present in the more micritic zones. The *Calera Limestone* also includes isolated pods of shallow-water carbonate facies that contain pisoliths, ooliths, pellets, fragments of coral, echinoderms, molluscs and coralline algae (Wachs and Hein, 1974).

The depositional environment of minor inter-pillow nannofossil carbonates and the overlying Tithonian radiolarian cherts may be interpreted as that of an oceanic ridge with the basal siliceous sediments undergoing limited redeposition down the flanks to produce parallel-laminated deposits (e.g. Chipping, 1971). Accumulation of these siliceous sediments on an abyssal plain is also possible, given sufficient distance from the continental margin (Hsü, 1971).

The Franciscan Limestones need to be interpreted in light of the fact that coeval sediments are dominated by terrigenous clastics. This suggests some kind of topographic eminence as a depositional locus. Since the *Laytonville Limestone* is associated with essentially non-vesicular basalt Matthews and Wachs (1973) suggested formation in water depths greater than 3.5 km. The depositional environment of the *Calera Limestone* is more intriguing given the occurrence of apparent shallow-water varieties, although these may conceivably be redeposited. Formation of the *Calera Limestone* at modest depth is, however, indicated by the highly vesicular nature of associated volcanic rocks. The lower organic-rich unit of the *Calera Limestone* may be interpreted as the deposit of an isolated enclosed basin perched on a ridge (Wachs and Hein, 1975). Alternatively this deposit may be but one of many victims of a global anoxic event, independent of local basin geometry (see Sect. 11.3.1). The upper unit of the *Calera Limestone* is also interpreted as being deposited on some kind of topographic high: parts of this feature were, however, basinal as indicated by the presence of redeposited planktonic foraminiferal layers.

The exact nature of the submarine high is open to dispute. Possibilities include seamounts, remnant arcs, aseismic ridges and oceanic plateaus. Certainly those aseismic structures whose sedimentary record includes shallow-water carbonates overlain by foraminiferal–nannofossil oozes (see Sect. 11.3.3) are viable models.

CALIFORNIA COAST RANGE OPHIOLITE: INTER-PILLOW LIMESTONES, RADIOLARITES AND BLACK SHALES

To the east of the Franciscan terrain, and lying structurally upon it, is the *California Coast Range Ophiolite*, of Callovian

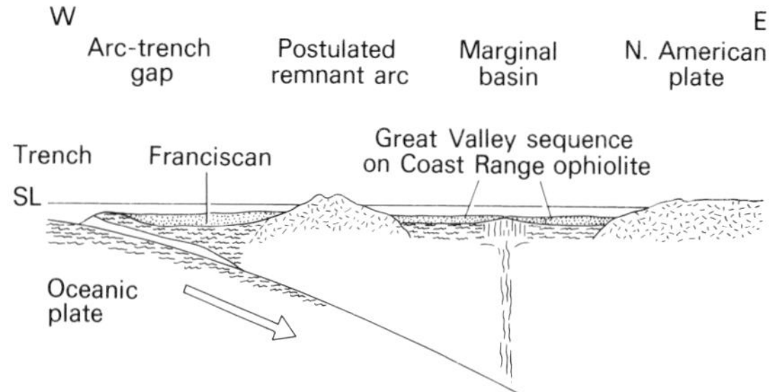

Fig. 11.25. Postulated Late Jurassic palaeotectonic setting of Franciscan sediments, in an arc-trench gap, and the Great Valley Sequence in a marginal basin, off western North America. Not to scale (after Blake and Jones, 1974).

age, that constitutes the base of the *Great Valley Sequence* (Bailey, Blake and Jones, 1970; Hopson, Frano, Pessagno and Mattinson, 1975). Locally the pillow lavas contain interstitial grey limestones composed of recrystallized nannofossil ooze. In some place the flows are overlain by red and black radiolarian cherts and mudstones, of comparable age and facies to the Franciscan although richer in graded and laminated bands of volcanic ash (Jones, Bailey and Imlay, 1969; Pessagno, 1973). Younger sediments are dominantly quartzo-feldspathic clastics and commonly show graded bedding and sole marks.

The difference in age between the ophiolite itself (Callovian) and the oldest dated sediments (Tithonian) suggests that the spreading centre was a zone of erosion for several million years (Hopson, Frano, Pessagno and Mattinson, 1975). Some nannofossil carbonate may have trickled between the basalt pillows but siliceous ooze and redeposited volcanic ash were the first major deposits. As the sea floor became topographically subdued the area was covered first with fine-grained mudstones and later with a blanket of turbidites. A palaeoenvironmental model for both the basal *Great Valley Sequence* and *Franciscan* is illustrated in Fig. 11.25.

TERTIARY OF OLYMPIC PENINSULA, WASHINGTON STATE, USA: PELAGIC LIMESTONES FROM SEAMOUNTS?

The *Crescent Formation,* exposed in the eastern part of the Olympic Peninsula, Washington State, USA, contains a remarkably thick section of metamorphosed Lower and Middle Eocene Hawaiian-type tholeiites (Glassley, 1974; Cady, 1975). These extrusives are apparently stratigraphically underlain by and pass laterally into graded terrigenous clastics. Associated with the volcanics are widespread small-scale manganese deposits, some of which have textures similar to marine ferromanganese nodules. The chemistry of the deposits is Mn-rich and Fe-poor (Sorem and Gunn, 1967).

The volcanics occur as two separate centres, a 15-km-thick

pile to the east and south-east and a thinner (circa 5 km) pile to the north with the clastic sediments between them. In the southern area the lower 10 km of basalt, characterized by pillow structures, contain intercalations of red pelagic limestone (Garrison, 1972, 1973). The limestones occur within and between pillows, as inclusions in flows, and as components of variously sized breccias. The bedded varieties are generally bright red or mottled, laminated and very fine-grained: planktonic Foraminifera are abundant in a fine-grained matrix comprising more or less recrystallized nannofossils. Some limestones, typically green, greenish-grey and yellow-pink, have been caught up in the lava and only rarely do these yield recognizable outlines of foraminiferal tests. In other cases the inter-pillow limestones appear as finely laminated internal sediments that commonly contain admixtures of volcanic debris; void-filling radiaxial fibrous calcite commonly rims the edges of the former voids and may be interspersed with internal sediment. Pelagic limestones also occur within the lavas in fractures, vesicles and even the cores of pillows.

In the higher levels of the volcanic pile, where columnar joints and scoria are characteristic, interbeds of grey limestone containing middle Eocene benthonic Foraminifera are present (Cady, 1975). Above the *Crescent Formation* middle Eocene to middle Miocene terrigenous clastics are interbedded with andesitic flows, tuffs and breccias.

Glassley (1974) interpreted the Crescent Formation as the product of ocean-ridge volcanism; the great thickness of lavas was attributed to tectonic replication: the manganese deposits could thus be viewed as hydrothermal deposits like those from spreading ridges (see Sect. 11.3.2). However, the geometry of the two volcanic centres led Cady (1975) to postulate that the *Crescent Formation* was composed of two seamount-like features that formed on the ancient Pacific Ocean and were later inserted on to the western margin of North America. The expression 'seamount-like' is used advisedly since the intertonguing of the volcanic pile with coeval redeposited terrigenous clastics suggests that the features never attained distinct bathymetric profiles. Since these volcanic mounds were apparently stratigraphically underlain by turbidites of the basal *Crescent Formation,* these features must have grown near the edge of North America; however, to begin with, elevation was apparently sufficient to shield the growing lava piles from clastic influences so that foraminiferal–nannofossil oozes could be laid down (Fig. 11.26). Some sediments were forcefully injected by lavas resulting in recrystallization of nannofossils; others filtered passively into inter-pillow voids which were also host to submarine carbonate cements. Structures in the volcanic rocks suggest that the upper parts of the pile were extruded from deep through shallow-water to locally terrestrial environments, an interpretation in harmony with upward change from planktonic to benthonic foraminiferal faunas in the included limestones.

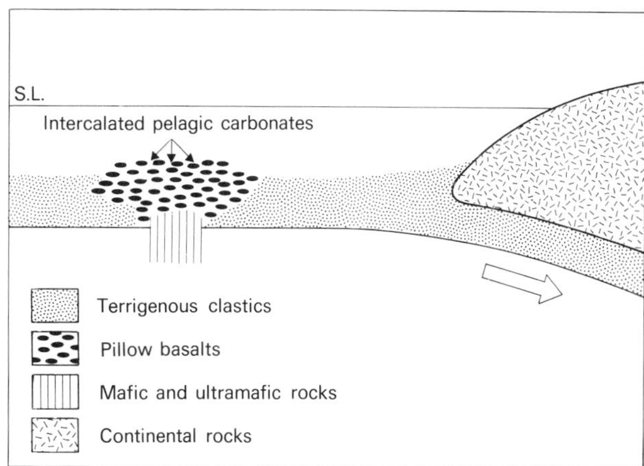

Fig. 11.26. Postulated Eocene palaeotectonic setting of pelagic carbonates of the Crescent Formation, Olympic Peninsula, Washington, USA. A 'seamount-like' feature, surrounded by terrigenous clastics, is coated with a veneer of nannofossil ooze. Not to scale (modified from Cady, 1975; Garrison, 1975).

BARBADOS: OCEANIC SEDIMENTS FROM AN ISLAND-ARC COMPLEX

The oceanic deposits of Barbados have a historical pedigree dating back to their description by Jukes-Browne and Harrison (1892) (see Sect. 11.1.2). These sediments overlie the *Scotland Group,* a folded series of mid-Eocene sandstones and shales that resemble flysch (Senn, 1940), and are in turn capped by coralline limestones. The middle Eocene to uppermost Oligocene *Oceanic Group* is constituted by a thick (? > 1,600 m) sequence of highly variable pelagic sediments including foraminiferal, radiolarian and nannoplankton clays and marls, radiolarites, spiculites, diatomites, brown clays and volcanic ash beds (Saunders, 1965; Lohmann, 1973). Small fish teeth and cosmic spherules are common (Hunter and Parkin, 1961). The radiolarian clays are composed of silica and illite with minor montmorillonite (El Wakeel, 1964).

The close similarity of the Barbados deposits to Recent oceanic sediments in terms of composition, structure, and mineralogy renders a deep-sea interpretation viable (Jukes-Browne and Harrison, 1892; Senn, 1940; El Wakeel, 1964). However, according to Beckmann (1953), the Foraminifera in these rocks suggests depositional depths of around 1,000–1,500 metres. Against this may be set the fact that Deep Sea Drilling has revealed closely comparable coeval sediments at depths of more than 5.5 kilometres (Bader *et al.,* 1970). Using this and other evidence Westbrook (1975) has suggested a 4- to 5-km post-Early Pliocene uplift of the Barbados Ridge, of which the island itself is a part. In this geotectonic context a model may be generated for the depositional environment of the oceanic

A MIDDLE EOCENE

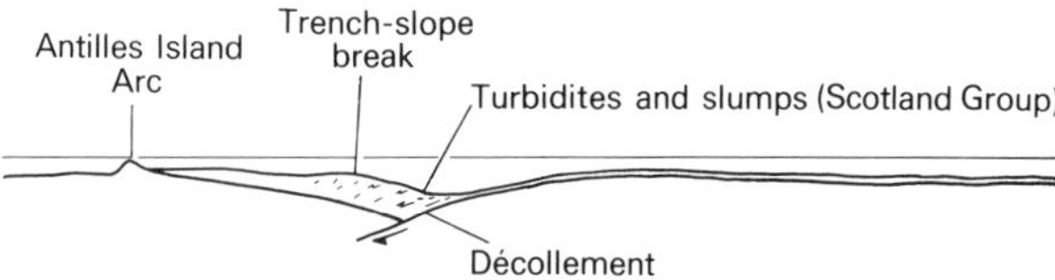

B LATE EOCENE-MIOCENE

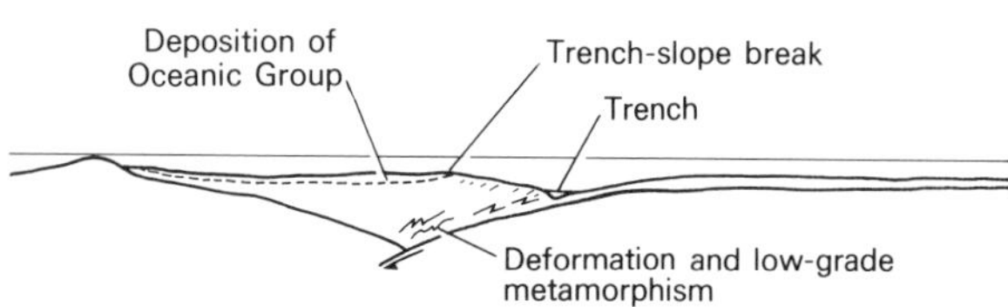

C PRESENT

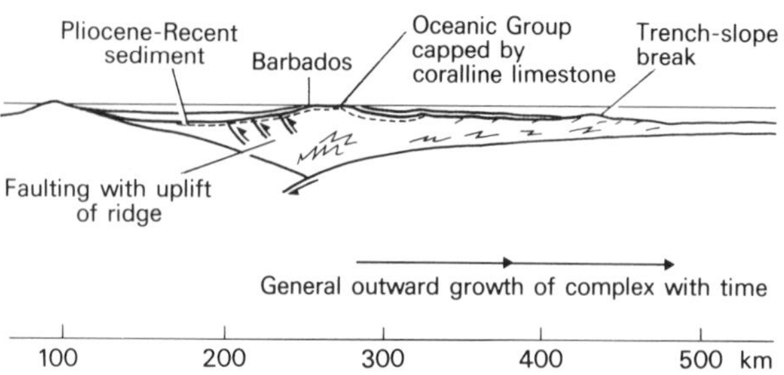

Fig. 11.27. Sequential palaeotectonic diagram illustrating the probable evolution of the Barbados Ridge and the Antilles Island Arc. (a) The Scotland Group is deposited by turbidity currents and gravity sliding. (b) The Oceanic Group is laid down on a sedimentary ridge behind the trench-slope break. (c) Uplift of the Barbados Ridge into shallow water allows development of coralline limestones and finally results in the exposure of the Oceanic Group on land. Vertical exaggeration 2:1 (after Westbrook, 1975).

deposits in the broader setting of the Lesser Antilles Island Arc (Fig. 11.27).

Pelagic sediments are recorded from various Pacific island arcs and arc-related islands (Mitchell, 1970; van Deventer and Postuma, 1973; Garrison, Schlanger and Wachs, 1975). Clearly the vertical uplift that can take place on and around such structures is the motive force behind the subaerial exposure of deep-sea deposits in these tectonic settings.

PALAEO-OCEANOGRAPHIC PERSPECTIVE

In many of the afore-mentioned examples assumed oceanic-ridge basalts are capped by radiolarian cherts, a stratigraphical relationship dissimilar from the present-day where spreading axes protrude above the calcite compensation depth and hence play host to a cover of biogenic carbonate (e.g. Garrison, 1974; Jenkyns and Hsü, 1974; Bosellini and Winterer, 1975; Hsü, 1976). Subsidence of Recent ridges results in a facies change from carbonate to red clays and siliceous oozes (Fig. 11.11). In the Ligurian Apennines and on the Troodos Massif the sequence is doubly contrary in that basal carbonate-free radiolarites and umbers are overlain by limestones and chalks. However, the fact that inter-pillow limestones occur in some ophiolites (e.g. Ballantrae, Franciscan, California Coast Ranges) and that ophicalcites are present in the Ligurian ultramafics points to initial elevation of the spreading centre above the calcite compensation depth. Following entrapment of carbonate within the pillow basalts the environment may have remained essentially non-depositional for millions of years, as is suggested for the *California Coast Range Ophiolite* (see Sect. 11.4.2). When deposition recommenced the oceanic slab may have sunk below the CCD. This, however, is only a partial explanation: the record of radiolarian chert on ophiolites has no parallel in any known deep-sea stratigraphy. Clearly these ancient oceans were loci of particularly vigorous radiolarian manufacture, linked presumably to upwelling and high fertility.

The ophiolitic record itself may, however, be biased in that these igneous complexes probably represent fragments of crust from relatively *small* oceans (e.g. Dewey and Bird, 1971; Gass, Smith and Vine, 1975; Mattinson, 1975). Thus, for Recent parallels, we should perhaps look not at mature oceans but smaller basins such as the Red Sea and Gulf of California: such restricted oceans, if characterized by high productivity, tend to have shallow calcite compensation levels (Berger and Winterer, 1974; Bosellini and Winterer, 1976). Thus if Mesozoic radiolarians occupied the environmental niche that diatoms do at present, oceans comparable with the Gulf of California may have allowed deposition of carbonate-free radiolarian ooze at depths of 3 km or less (Jenkyns and Winterer, in prep.). This depth approaches very closely that of a mature spreading ridge (Sclater, Anderson and Bell, 1971) and could therefore account for the presence of radiolarian ooze as the first major deposit on ophiolitic basement.

As the basin widened the CCD may have been depressed to levels more typical of mature oceans, allowing deposition of carbonate at depths down to 4 km or so. In the case of the Ligurian sequence such a depression would have been aided by the rapid increase in abundance of calcareous plankton in latest Jurassic time; the significance of the younger Cyprus sequence in this respect is, however, more problematic.

Of the other sediments described above those from the Olympic Peninsula, Barbados and Timor match more closely the present deep-sea record and are probably parts – albeit marginal parts – of large oceans. The Permian sequence from

New Zealand may in turn be regarded as problematic in that there is so little chert present; possibly the Dun Mountain ophiolite was also part of a large ocean.

11.4.3 **Deposits of small pelagic basins**

NEOGENE DIATOMITES FROM THE PACIFIC RIM

Around the margins of the Pacific Ocean there are abundant Miocene and Pliocene diatomaceous deposits. The best documented example of these is the Miocene *Monterey Formation* of California (Bramlette, 1946). The rocks are light coloured and vary from quartzose chert, through porcelanite to soft opaline diatomites. Many porcelanites are constituted by lussatite that characteristically occurs as micron-sized lepispheres (see Sect. 11.3.4) (Oehler, 1975). Generally the more lithified varieties occur towards the base of the section where they are typically developed as centimetre-scale beds separated by paper-thin partings (Murata and Larson, 1975). The total thickness of the Formation varies considerably from place to place but locally extends to around 3 kilometres. It is stratigraphically bracketed by and may pass laterally into terrigenous clastics and shales, locally calcareous. Pelletal phosphorites and phosphatic mudstones are most common near the base (Smith, 1968). A rhythmicity is imposed on some sections by the repetitive presence of graded centimetre-thick sandstones. In the purer diatomites a very fine millimetre lamination is visible; such beds, generally a metre or so in thickness, alternate with more clay-rich diatomaceous units that are several times thicker. Calcareous beds, partly dolomitic, occur at some levels as do carbonate concretions. Rhyolitic tuffaceous layers, usually altered to bentonites, are common throughout the sequence. Parts of the *Monterey Formation* are bituminous and contain recognizable organic matter.

Diatoms are clearly the most abundant fossils but other siliceous biota such as radiolarians, silicoflagellates and sponge spicules are locally significant. Benthonic and planktonic Foraminifera and rare pectinid bivalves also occur. Fish skeletons and scales may often be found on bedding planes; more rarely the bones of whales and sea cows and the remains of birds and land mammals come to light. Leaves of land plants have also been recorded.

The *Monterey Formation* is interpreted as being deposited in irregular basins, locally deep, and separated by shallow, marine sills or by land areas. Only around the margin of these basins were clastics and land-derived faunas deposited in important amounts; in deeper central regions sedimentation was essentially a function of siliceous phyto- and zooplankton manufacture. The general paucity of terrigenous material in these land-linked basins is strange; it may be related to arid and/or bevelled surfaces or a geography that deflected major rivers away from the depositional site. Certainly some clastics were introduced, possibly by volcanically and seismically triggered turbidity currents, to parts of the basins where they accumulated as thin graded layers. Benthonic Foraminifera may also have been redeposited in this way. The finer millimetre laminations were interpreted by Bramlette (1946) as a product of annual climatic events governing input of fine-grained detritus; this suggests sedimentary rates of 1 cm/70 years. This very same explanation was adopted by Calvert (1966b) to explain the more rapidly formed varves of Recent diatomites in the Gulf of California; and indeed this small ocean is an excellent environmental model for the *Monterey Formation* (see Sect. 11.3.5). Clearly the Neogene basins were sites of high organic productivity dependent on upwelling nutrient-rich waters derived from the lower levels of the Pacific Ocean.

Enhanced production of planktonic organic matter probably caused removal of considerable quantities of oxygen from the Monterey seas, thus shifting much of the water masses towards anoxia: within those parts of the basin that intersected the oxygen-minimum zone (cf. Fig. 11.7) benthonic organisms were necessarily excluded and primary sedimentary lamination was preserved. Any organic matter entering the burial stage would also escape oxidation and could ultimately end up as bituminous matter. In such an anoxic environment periodic mass mortality of fish might take place, thus accounting for their numerous skeleta in the sediment. The palaeobathymetry of the Monterey basins is difficult to assess and is certainly variable; depths from 60–70 m to 2,000 m have been suggested on the basis of Foraminifera (Smith, 1968; Ingle, 1973).

Diatomaceous deposits very similar to the Monterey Formation of California occur around the north and central Pacific (Oregon, USA, Soviet Far East, Japan and South Korea) and, in the southern hemisphere, outcrop in Chile and Peru (Orr, 1972; Ingle, 1973, 1975; Garrison, 1975). The Japanese deposits are perhaps the best documented and are apparently in every way similar to the *Monterey Formation*: features in common include variable but great thickness (locally 1 km), calcareous concretionary zones, and basal phosphatic levels. The Japanese deposits also show the same progressive increase of porcelanite and chert towards the base of the sections (Garrison, 1975).

The environmental setting of these circum-Pacific Neogene diatomites was apparently similar to that of the *Monterey Formation*. Some of the depositional sites may have been parts or wholes of back-arc basins comparable with the present-day Japan Sea; others may have formed by simple rifting; still others may have been – as is suggested for the Monterey Basin – related to transform faulting (Dott, 1969; Crowell, 1974b; Garrison, 1975).

MIOCENE DIATOMITES FROM THE MEDITERRANEAN REGION

Aquitanian–Burdigalian and Messinian diatomites occur in many of the lands bordering the Mediterranean. Detailed descriptions of the older facies are furnished by Filipesco (1931), Colom (1952) and Campisi (1962) from Romania, Spain and Sicily respectively. Messinian facies from Algeria, Sicily and north Italy are documented by Anderson (1933), Ogniben (1957) and Sturani and Sampò (1973).

The Aquitanian–Burdigalian deposits contain appreciable quantities of lithified porcelanite and chert, and are stratigraphically associated with terrigenous clastics, locally glauconitic, and skeletal limestones. The Messinian examples are usually light, porous, extremely friable and essentially constituted by opaline silica; typically diatom-rich beds of a metre or so are interlayered with darker grey to brown calcareous and dolomitic shales and mudstones of comparable or lesser thickness. Below these diatomaceous facies are pelagic clays; above limestone and gypsum follow. The total section of diatomaceous and associated rocks varies from next to nothing to a few tens of metres (Messinian of north Italy and Sicily) to a few hundred metres (Messinian of Algeria, Aquitanian–Burdigalian of Spain). The diatom-rich units commonly possess a millimetre lamination produced by alternations of carbonate and siliceous material; rarely they are homogenous. Interbedded marls are themselves partly laminated and partly massive. Phosphatic concretions occur locally in the Messinian of Italy and Sicily. Most diatomites are rich in organic matter, including hydrocarbons, although this material is oxidized at outcrop. Apart from the normal Messinian diatomite-marl rhythm, volcanic horizons can also occur; these are abundant in the Algerian deposits but rather rare in Sicily.

Fig. 11.28. Pelagic fish of the scombrid family from Messinian diatomite. Scale bar = 2 cm. Monte Capodarso, central Sicily.

The most obvious macrofossils, at least in the Messinian deposits, are pelagic fish and their debris, usually selectively concentrated at certain levels (Fig. 11.28). In the Algerian deposits, bivalves such as *Pecten, Cardita, Arca, Modiola* can be common (see Sect. 11.3.6) and gastropods, echinoids, bryozoa, brachiopods and ostracods also occur. Bivalves also occur in the Spanish deposits. In most diatomites careful search may reveal the trace of the odd leaf or wood fragment. Worm burrows are not uncommon. The microscopic biota of the rocks includes, apart from diatoms, Radiolaria, silicoflagellates, sponge spicules, plus Foraminifera, both pelagic and benthonic, rare pteropods and heteropods, and abundant coccoliths and discoasters. The diatoms themselves are of both pelagic and benthonic/neritic types (Colom, 1952; Baudrimont and Degiovanni, 1974).

Interpretation of Mediterranean diatomites is similar to those in the circum-Pacific regions. Environmental settings envisaged are narrow basins, gulfs and bays, with limited connection to the open sea (e.g. Burdigalian North Betic Strait, Fig. 11.29).

Palaeobathymetry is controversial but seems to be most reliably established in the Messinian examples where the fish faunas were taken by Sturani and Sampò (1973) – on the basis of comparisons with the same species in the Recent – to suggest depths around 400–500 metres. The exceptional state of preservation of these fossils also suggests bathymetric levels below zones of wave and current agitation. Lack of oxygen in the bottom water clearly prevented bacterial decay of the fish and furthermore ensured that the millimetre lamination of the diatomites was preserved. Interpretation of this lamination as a varve enabled Ogniben (1957) to estimate a sedimentary rate in the range 1 cm per 20–100 years for these facies. The massive diatomites, which have suffered bioturbation, were clearly laid down in more oxygenated waters. Exact environmental interpretation of these Messinian deposits must, however, also be viewed within the framework of a deep desiccated Mediterranean (Hsü, Montadert, Bernoulli, Cita, Erickson, Garrison, Kidd, Mèliéres, Müller and Wright, 1977).

As with the *Monterey Formation* active upwelling of nutrient-rich water, in this case from the Miocene Atlantic, must have produced rampant diatom populations in the Mediterranean gulfs; occasionally plankton blooms resulted in mass mortality of fish, thus supplying phosphates to the sediment.

MIOCENE *GLOBIGERINA LIMESTONE* AND NODULAR PHOSPHATES FROM MALTA

On the island of Malta the Aquitanian through Langhian is represented by the so-called *Globigerina Limestone* (see Sect. 11.1.2) underlain by a top-Oligocene coralline facies (Murray, 1890; Pedley, House and Waugh, 1976). The pelagic

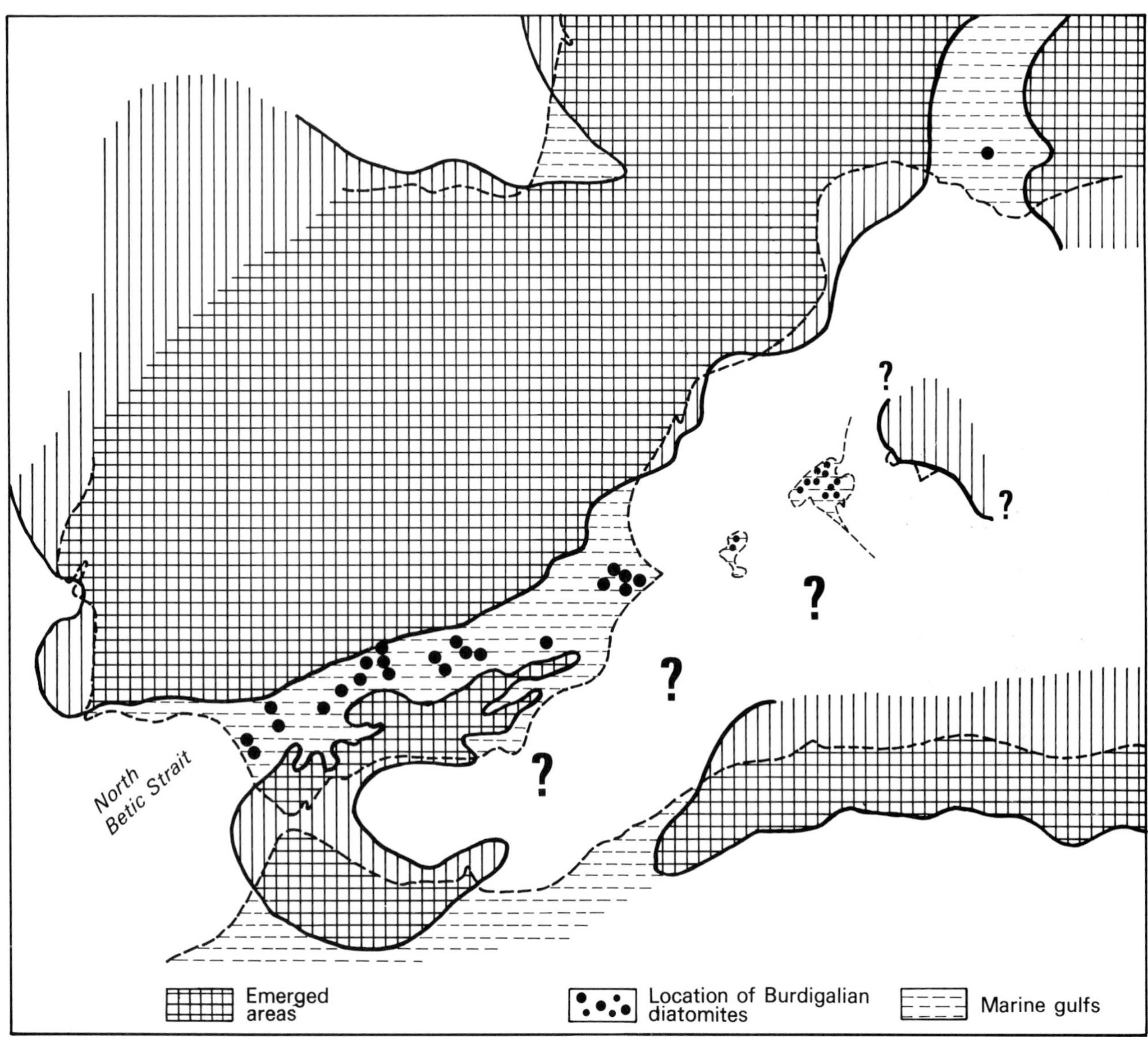

Fig. 11.29. Palaeogeographic sketch of the North Betic Strait during the Burdigalian. Dots indicate exposures of diatomaceous rocks (after Colom, 1952).

limestones are massively bedded grey and yellow globigerinid biomicrites that range in thickness, across the island, from about 20 to 200 metres. Echinoderms, molluscs, particularly *Pecten,* ostracods and pteropods occur in this formation; there is also ample evidence of bioturbation. Calcareous nannofossils are locally abundant. The detrital fraction of these rocks includes clay minerals, plus quartz, glauconite and various minerals of igneous derivation, all commonly less than 0.1 mm in diameter.

Within the sequence there are two major and several minor nodular phosphatic beds, a few centimetres to more than 1 metre in thickness. These brown-coloured layers of aggregated concretions contain echinoids as an *in situ* fauna, *Thalassinoides* burrows and derived phosphatized casts of molluscs and corals, plus sharks' teeth, turtle, crocodile, whale and seal bones. Nodules of khaki chert occur locally. Overlying the *Globigerina* Limestone are pale grey globigerinid marls which comprise the so-called *Blue Clay*.

Murray (1890) suggested formation of the *Globigerina* Limestone in relatively deep water, in the hundreds- to thousands-of-metres range. This view ignored the stratigraphical context of a formation that overlies a deposit containing coral-rich patch reefs and red algae. It seems more likely, following Pedley, House and Waugh (1976), that the pelagic limestone was laid down on a submarine plateau sheltered from clastics, where condensed phosphatic deposits could be formed and where depths were a few tens to a few hundred metres.

PALAEO-OCEANOGRAPHIC PERSPECTIVE

The global spread of Miocene diatomites, given that they occur not only around the Pacific, the Mediterranean but also along the coastal plains of the Western Atlantic (Wise and Weaver, 1973), suggests that conditions particularly suitable for diatom manufacture were realized during this part of Tertiary time. In many, but not all cases these deposits are associated with volcanics (e.g. Taliaferro, 1933); however, Calvert (1966a) showed clearly that for the present-day Gulf of California there is no need to appeal to igneous sources of silica as a nutrient source. Abundant diatom populations are simply related to the upwelling of fertile waters, and such conditions pertain in many small basins.

Partly the abundance of Miocene diatomaceous sediments may have been a tectonic phenomenon, in that these restricted basins had to be produced in the first place. However, the parallel abundance of Miocene phosphate (see Sect. 11.4.3) relative to its meagre Recent record, must also point to this part of the Tertiary as a time of peculiarly vigorous upwelling. Coupled with this was a probable anoxic tendency of much of the world ocean at this time (Fischer and Arthur, 1977). Given that upwelling is generally linked to the distribution of winds and ocean currents, it is probable that the trigger for the production of Miocene diatomites was climatic.

Palaeo-oceanographic conditions in the Pliocene, although locally advantageous to diatom manufacture, may be regarded as more like those of the present day.

11.4.4 **Continental-margin facies**

LOWER PALAEOZOIC BLACK SHALES, DIRTY LIMESTONES AND CHERTS FROM EUROPE AND NORTH AMERICA

The pelagic or hemipelagic black graptolitic shales that occur in Cambrian and, more particularly, in Ordovician and Silurian strata are widespread in the Caledonian, Hercynian and Appalachian orogenic belts of Europe and North America. Silurian facies from Yorkshire and Westmorland, England may be divided into two types (Rickards, 1964): the first contains graptolites to the virtual exclusion of all other forms of life except those of planktonic or pseudoplanktonic lifestyle; the second, stratigraphically higher, type is bioturbated and contains a normal benthos as well as graptolites. These British graptolitic mudstones comprise clay- to silt-grade quartz, altered feldspars, mica, iron minerals, carbonates and clay minerals. Local variants include green and red mudstones, levels with mud and silt laminae and graded limestones (Piper, 1975; Ziegler and McKerrow, 1975). Rickards (1964) and Piper (1975) also distinguish a lamination composed of banded layers rich in carbonaceous material and pyrite. Organic carbon levels locally rise to 3.68% in these graptolitic facies. Middle Cambrian to lowermost Ordovician black graptolitic shales from the Oslo region contain similar sedimentary structures to those from England; organic carbon contents rise to around 10% in Upper Cambrian facies but decrease in the Lower Ordovician. These rocks have a mineralogy and chemistry that suggests continental derivation from the Baltic Shield; higher Ordovician sediments were apparently supplied by an active island arc (Björlykke, 1974).

Redeposition processes were clearly important during the accumulation of most black-shale facies and an environment comparable to a Recent abyssal plain is possible (Piper, 1975). The organic-rich and benthos-poor nature of certain of these facies may reflect a time of oceanic anoxia when abundant planktonic algal remains could be entombed in the bottom sediments (see Sect. 11.3.1).

Pelagic limestones are locally associated with graptolitic shales and also constitute an important facies in their own right. An example is the Cambrian *Conestoga Limestone* and its correlatives, exposed in the Appalachians of Pennsylvania and Vermont (Rodgers, 1968): this is a thin-bedded dark argillaceous facies, lacking fossils and considerably deformed. It contains carbonate breccias derived from an adjacent shallow-water carbonate platform. A similar palaeogeographical configuration may be generated for the Upper Cambrian and Lower Ordovician strata of eastern Nevada where the deeper pelagic facies are developed as dark laminated calcareous mudstones locally rich in sponge spicules, bearing trilobites, and containing beds of redeposited shoal-water material (Cook and Taylor, 1977). The dark mudstones are assigned to a continental-slope setting (Fig. 11.30), an environmental model that could also hold for similar Silurian–Lower Devonian graptolitic facies from the same area (Matti, Murphy and Finney, 1975).

Similar deposits are developed in the Ordovician of the *Ouachita System* of West Texas, Arkansas and Oklahoma (McBride, 1970). In the overlying formation (Silurian to

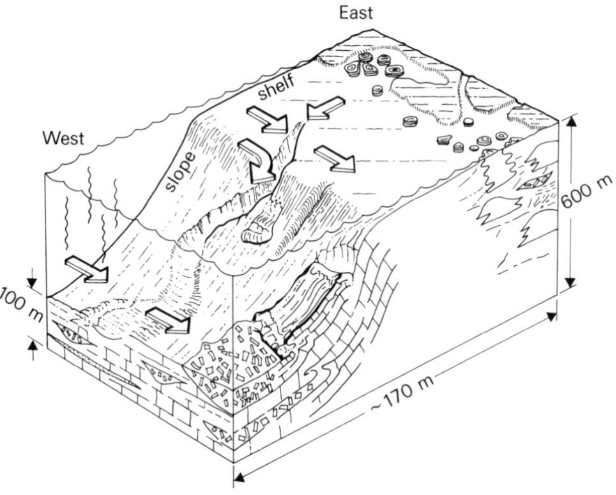

Fig. 11.30. Model of Late Cambrian continental margin in Nevada, USA. Shallow-water carbonates and patch reefs of the shelf pass westward to a slope, incised by canyons; slumps and beds of redeposited shoal-water carbonates and dark spicular mudstones are characteristic of this zone. Arrows indicate major directions of sediment transport (after Cook and Taylor, 1977).

Fig. 11.31. Late Devonian–Early Carboniferous palaeotectonic and palaeogeographical reconstruction of the Ouachita Basin; the Caballos and Arkansas Novaculite are thus interpreted as being deposited along an ancient continental margin of North America. Key to letters: North America, MC – Mexico City, A – Austin, Texas, B – Birmingham, Alabama, NY – New York City. Laurasia, L – London. Gondwana, D – Dakar, B – Bogota (after Lowe, 1975).

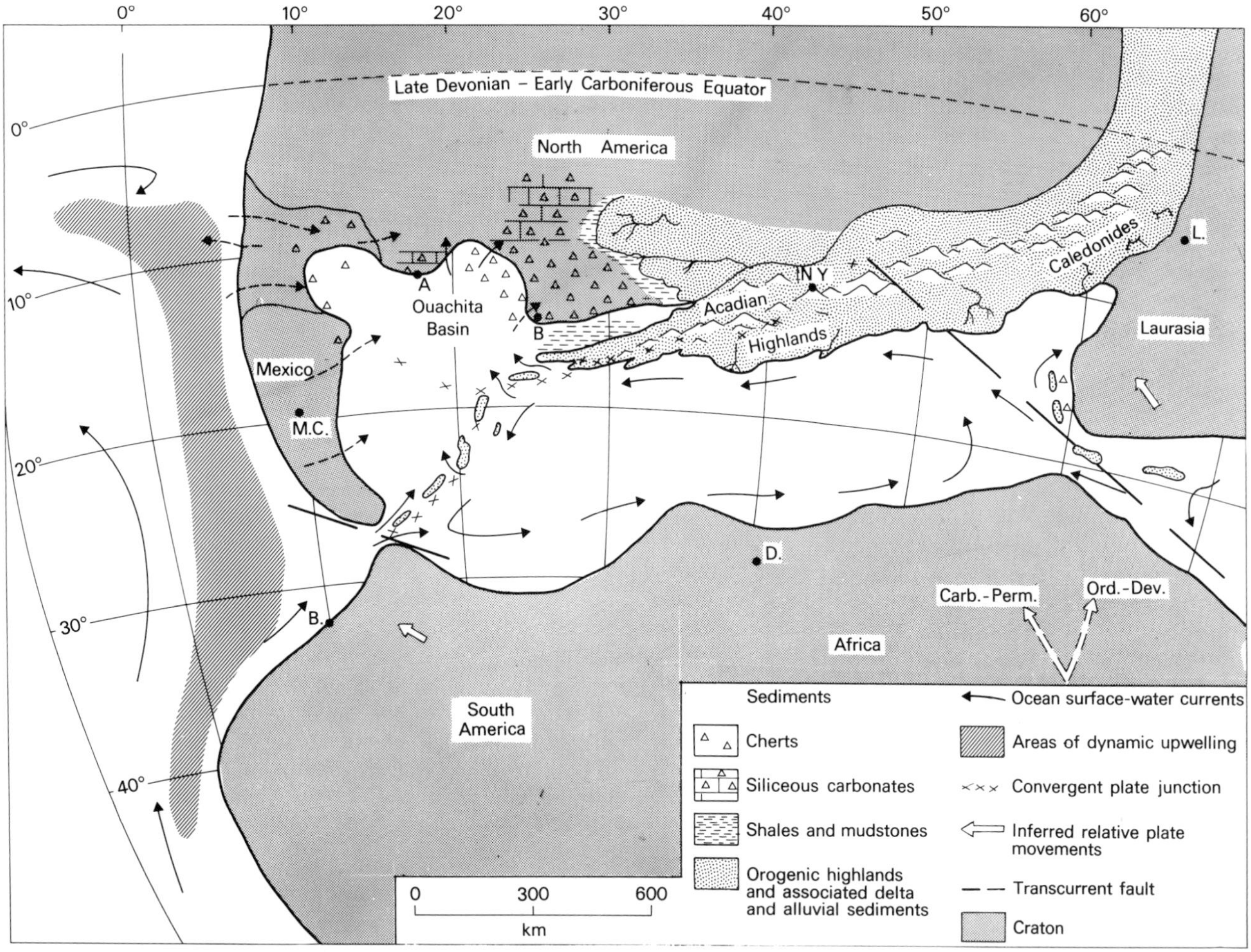

Carboniferous) chert becomes the dominant rock (circa 90%) with subsidiary calcarenite, shale, sandstones and pebble conglomerates (*Caballos* and *Arkansas Novaculite*: McBride and Thomson, 1970; Lowe, 1975, 1976; Folk and McBride, 1976b). The cherts, which include white, green, tan, grey, black, brown, blue and red varieties, are – as in many siliceous sequences – cyclically interbedded with clay. It is the pure milky white variety (novaculite) that gives its name to the formation; other types owe their spectral pigments to clay minerals, iron and manganese. The cherts themselves, which are locally fractured and brecciated, were probably derived, via solution–precipitation, from opaline radiolarians and sponge spicules; the clay is interpreted as a terrigenous product that settled through the water column. The clastics associated with this sequence also exhibit symptoms of redeposition from shallow-water areas. This dominantly siliceous section, unlike the underlying calcareous sequence, was deposited under oxidizing conditions: both formations may be interpreted as deep-water deposits (McBride and Thomson, 1970; Folk and McBride, 1976b). Estimated average sedimentary rates for the *Caballos Novaculite* are a few mm per 10^3 years, comparable to Recent radiolarian oozes. In the *Arkansas Novaculite* possible seasonal varves, comparable to those in the Gulf of California (see Sect. 11.3.5), have been recognized; these yield the vastly higher sedimentary rate of 1.25 cm/10 years (Lowe, 1976). These two contrasting figures may only be reconciled if there are stratigraphic lacunae in the section. A palaeogeographical sketch is given in Fig. 11.31.

PALAEOZOIC RED, GREEN AND GREY LIMESTONES, SHALES AND CHERTS FROM EUROPE AND NORTH AFRICA

In the Cantabrian Mountains of north-west Spain, Hercynian nappes contain Lower and Middle Cambrian red and green nodular biomicrites and green shales, of probable pelagic origin; these rocks overlie shallow-water platform carbonates and together they comprise the so-called *Lancara Formation* (Zamarreño, 1972). Containing a fauna of trilobites, brachiopods, echinoderms, and locally glauconitic, the red and green nodular facies strongly resemble the Devonian–Carboniferous *Cephalopodenkalk* and *Griotte* and the Jurassic *Adneter Kalk* and *Ammonitico Rosso* (see Sect. 11.4.4).

Similar facies of Devonian–Carboniferous age are distributed in a zone extending from South-west England through central Europe to North Africa, south of the 'Old Red Sandstone' landmass. Two main lithological types are recognizable, typically overlying shallow-water carbonates: (1) a so-called *Schwellen*-facies, comprising stratigraphically condensed limestones rich in goniatites (*Cephalopodenkalk*) and (2) a so-called *Becken*-facies of stratigraphically expanded silty shales locally rich in ostracods (e.g. Tucker, 1973a, 1974). The condensed facies are commonly developed as red and grey biomicrites to biosparites containing, apart from ammonoid moulds, thin-shelled bivalves, gastropods, ostracods, conodonts, Radiolaria, Bryozoa, trilobites, brachiopods, solitary corals and stromatoporoids, fish remains and isolated crinoid ossicles (Bandel, 1974). Burrows are also present. Locally styliolinid shells and their enclosing calcareous cement entirely constitute the limestone (Tucker and Kendall, 1973). Distinctive sedimentary features include hardgrounds, ferromanganese and phosphatic nodules, calcite-filled cracks and neptunian dykes (Krebs, 1972; Szulczewski, 1971, 1973; Tucker, 1973b). Foraminiferal-red algal associations occur at some levels. The discontinuity surfaces include planar varieties that cut through both shells and cavity-filling calcite, and those with irregular centimetre-scale relief, commonly coated with ferromanganese oxides and colonized by sessile Foraminifera, more rarely by corals, stromatoporoids and crinoids. The bed-parallel cracks (sheet cracks) are a few centimetres high, up to 3 metres long, and are filled with internal sediment, radiaxial fibrous and equant calcite. Neptunian dykes, of more vertical orientation, are common within the pelagic facies; these vary in width from 1–15 cm, may penetrate to a depth of 50 or so centimetres, and are filled with fine-grained laminated carbonate. Neptunian dykes penetrating the underlying reef and platform facies may be considerably larger and may show multiple intrusions. The ferromanganese nodules, discussed in detail by Tucker (1973b), are developed as incrustations around skeletal calcite and limestone clasts; nodules are richer in iron than they are in manganese (average $Fe = 10.5\%$; average $Mn = 2\%$).

Lateral facies variants of the limestone described above are nodular carbonates (locally termed *Griotte*; Fig. 11.32), and shales with calcareous nodules; features associated with hardground formation are absent whereas soft-sediment deformation and syn-sedimentary brecciation are common.

The stratigraphically expanded *Becken*-facies are developed as red to black shales, in some places calcareous, in others carbonaceous and locally possessing silty laminae. Faunally these beds are impoverished; they are dominated, in the Givetian and Frasnian, by styliolinids and in the Famennian by ostracods. Graded terrigenous clastics, tuffs and shoal-water carbonates, commonly echinodermal, occur intercalated in these shales; pebbly conglomerates and slump blocks of *Schwellen* material are also found (Szulczewski, 1968; Tucker, 1969; van Straaten and Tucker, 1972; Bandel, 1974).

Lower Carboniferous rocks include rare pelagic *Schwellen* facies comparable to those of the Devonian (e.g. Szulczewski, 1973; Bandel, 1974; Uffenorde, 1976) and widespread benthos-poor grey and black shales associated with radiolarian cherts that are locally manganiferous (Zakowa, 1970; Meischner, 1971; Waters, 1970). The cherts themselves may be graded

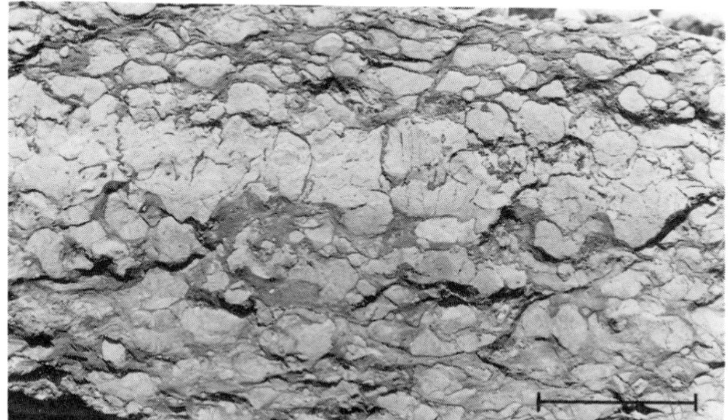

Fig. 11.32. Devonian nodular pelagic limestone (*Griotte*): the calcareous kernels are wrapped in anastomosing marl seams but the bedding is clear. Such nodular structures, common in certain Palaeozoic pelagic facies, are developed also in the Triassic and Jurassic of the Tethyan region (*Ammonitico Ross* facies) and in the Cretaceous European *Chalk*. Mont Peyroux, Montagne Noire, France. Scale bar = 5 cm. Photo courtesy of M.E. Tucker.

(Swarbrick, 1967). Volcanic rocks are a common accessory. In some areas graded shallow-water carbonate fossils and grains occur as intercalated beds (e.g. Franke, Eder and Engel, 1975).

The stratigraphically expanded shales and condensed cephalopod limestones from the Devonian are interpreted as deposits of *Becken* (basins) and *Schwelle* (swells) respectively: the Cambrian example from Spain falls into this latter category. Depositional rates for the condensed facies were a few mm/10^3 years. The *Schwelle* were variously constituted by stable blocks of reef and platform carbonates, volcanic lava piles and uplifted basement. On these submarine elevations accumulation of carbonate was probably hindered by current agitation, and submarine solution/lithification and precipitation of Fe–Mn oxyhydroxides (glauconite in the Cambrian example) were successfuly promoted (see Sect. 11.3.3, 11.3.6). Syn-sedimentary lithification accounts for many features of the *Schwellen* facies: the formation of various hardgrounds, for example, the submarine-cemented sheet cracks and the neptunian dykes, which could only develop by fracture of a consolidated substrate. Interpretation of depositional depths is possible for those sequences containing foraminiferal-red algal nodules, the growth of which was necessarily constrained by the photic zone (circa 200 m).

On the slopes of these various rises the calcareous nodular facies (cf. Sect. 11.3.5 for a Recent analogue) locally moved downhill to produce slump-folds and breccias: marginal faulting is also indicated by the neptunian dykes in the *Schwellen*-facies and the presence of blocks of the same in basinal shales. Clay-rich sediments were apparently selectively fractionated into the deeper water which was also periodically favoured with influxes of turbidites from carbonate and terrigenous-clastic shelves. Tucker (1973a) estimates depths of around 1,000 m for these basinal facies but there are no reliable bathymetric handholds.

The Early Carboniferous saw a smoothing out of topographic differences with a few stable limestone blocks remaining as high ground where condensed pelagic carbonates could accumulate (e.g. Szulczewski, 1973; Uffenorde, 1976). In deeper water, black shales and radiolarian cherts could point to waters whose bottom levels were anoxic (cf. Fig. 11.7) and whose higher levels were fertile. Intra-basinal redeposition of siliceous oozes took place locally, but the bulk of the resedimented material was shed from carbonate-platforms. Depths of these basins may have been comparable to those in the deeper parts of the Devonian seas.

TRIASSIC-JURASSIC RED, GREY, AND WHITE LIMESTONES, MANGANESE NODULES, PELAGIC 'OOLITES' AND CHERTS FROM THE TETHYAN REGION

Tethyan Triassic and Jurassic pelagic facies, although faunally different, are otherwise generally comparable: in both cases such sediments overlie and laterally interdigitate with platform carbonates (e.g. Bernoulli and Jenkyns, 1974). These Mesozoic pelagic rocks bear an uncanny resemblance to their European Palaeozoic counterparts and again a basic division into stratigraphically condensed and expanded facies may be made (Fig. 11.33).

The condensed facies is typically developed as a metre or so of red biomicrite containing a fauna of ammonites, belemnites, gastropods, thin-shelled bivalves, brachiopods, rare single corals, globigerinid Foraminifera, Radiolaria, sponge spicules, conodonts (in the Triassic only), ostracods and pelagic and benthonic echinoderm fragments. Burrows occur in some horizons. Problematic nannofossils are rare in the Triassic (e.g. Zankl, 1971; Di Nocera and Scandone, 1977) but are locally abundant in the Jurassic (Fischer, Honjo and Garrison, 1967; Garrison and Fischer, 1969; Bernoulli and Jenkyns, 1970; Jenkyns, 1971a). The biomicrites may pass laterally into lenses of ammonite-, pelagic bivalve- and brachiopod-rich crinoidal biosparites (e.g. Jenkyns, 1971b; Sturani, 1971).

Ammonites in the condensed micrites are commonly closely packed together in layers; in such instances they are typically present as moulds or partial moulds that are highly corroded and bored, particularly on the former upper surface of the shell; corrosion on both surfaces is more rare (Fig. 11.34) (Wendt, 1970; Schlager, 1974). Such partial etching usually goes hand in hand with encrustations by Fe–Mn oxyhydroxides which also fill algal/fungal borings that infest many fossil fragments. Such mineral coatings also occur as thick crusts and pavements on hardgrounds. In the Austrian Triassic mineralized hardgrounds occur in dense succession where they

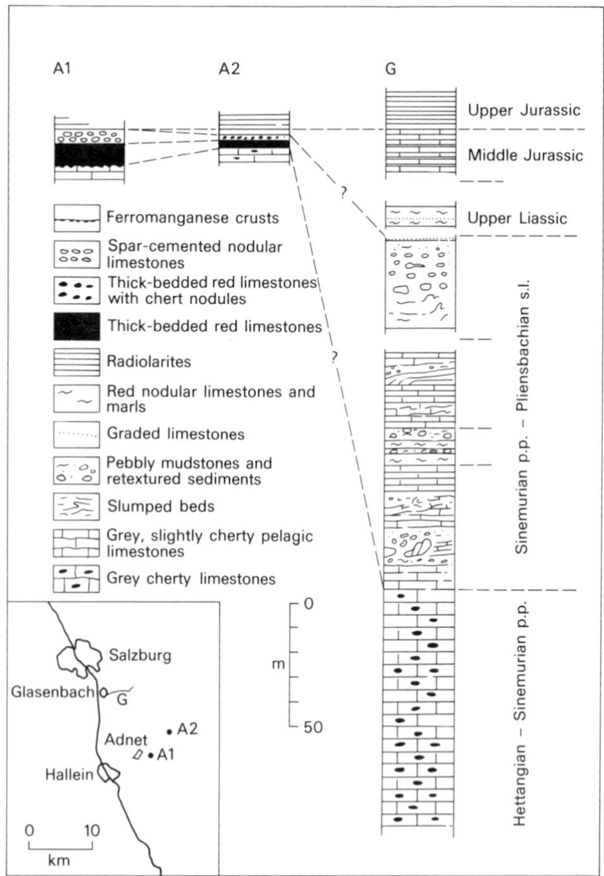

Fig. 11.33. Columnar sections of an expanded Jurassic sequence (Glasenbach) and a coeval condensed sequence (Adnet) in the Eastern Alps of Austria. The condensed sequences are rich in fauna and impregnated with Fe–Mn oxide-hydroxides; the expanded sequences contain graded beds and show evidence of hard- and soft-sediment deformation (after Bernoulli and Jenkyns, 1970).

are colonized by foraminiferal pillars or 'reefs' (Fig. 11.35) (Wendt, 1969). Only in the Jurassic are ferromanganese nodules well developed, particularly in western Sicily, Austria, the Venetian Alps of North Italy, and Hungary (Jenkyns, 1970; Germann, 1971; Sturani, 1971; Galácz and Vörös, 1972; review in Jenkyns, 1977).

The west Sicilian nodules, which are carbonate rich, are usually several centimetres in diameter, concentrically laminated and usually formed around corroded ammonites, calcareous intraclasts or lava fragments; some lack obvious nuclei. Chemically, the nodules may be Mn-rich (circa 40%), Fe-rich (circa 50%) or contain roughly balanced amounts of the two elements; in this latter case enrichment of minor

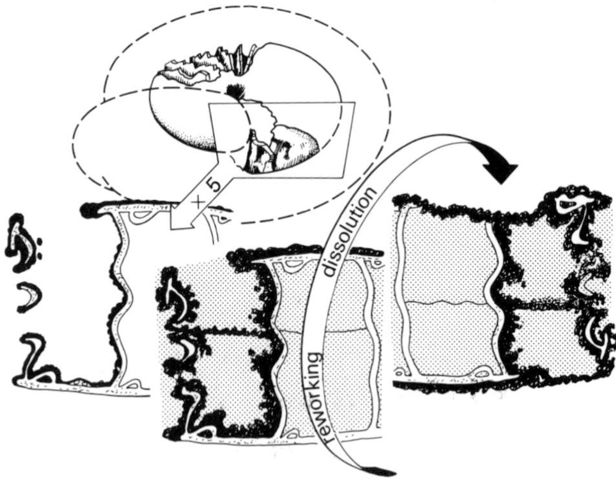

Fig. 11.34. Diagram illustrating corrosion and Fe–Mn mineralization patterns in ammonite shells and their relationship to reworking (after Wendt, 1970).

elements (Ni, Co, Cr, V, Ti) is common. Minerals identified in these nodules include calcite, goethite, haematite and posssibly todorokite. Nodules typically possess a colloform microfabric and may enclose sessile Foraminifera and serpulids; mineralized algal borings are also common (see Sect. 11.3.6). The structurally similar Austrian nodules, in which pyrolusite has been identified, are chemically very variable with Mn-contents ranging from 0.06 to 23.9% and Fe-contents ranging from 0.1 to 17.6%. Ni, Co, Cu, Zn, Pb and Cr show variable levels of enrichment.

In a few localities in Sicily the Fe–Mn nodules are associated with a pisolitic ironstone, rich in ammonites, belemnites, gastropods, brachiopods, crinoid ossicles and sharks' teeth. The pisoliths, rich in Fe and Mn, are also well endowed with trace metals; their mineralogy includes calcite, goethite, haematite and chamosite. Traces of boring algae are visible in many pisoliths. Of particular significance is the presence of fragments of sanidine trachyte as the cores of certain pisoliths.

Associated with the ferromanganese nodules in Jurassic condensed sequences from western Sicily, the Austrian and the Venetian Alps are calcareous stromatolites, typically developed as linked hemispheres or as cupolas on ammonites (Jenkyns, 1971a; Krystyn, 1971; Sturani, 1971). A Callovian (Middle Jurassic) condensed layer in the Villány Mountains of Hungary, which contains chamositic and limonitic ooliths, manifests a profusion of stromatolitic forms including laterally linked undulose clumps, pillars and columns as well as subspherical oncolites which have usually formed around belemnites and ammonites (Radwański and Szulczewski, 1965). Inverted stromatolitic growths are common. In the

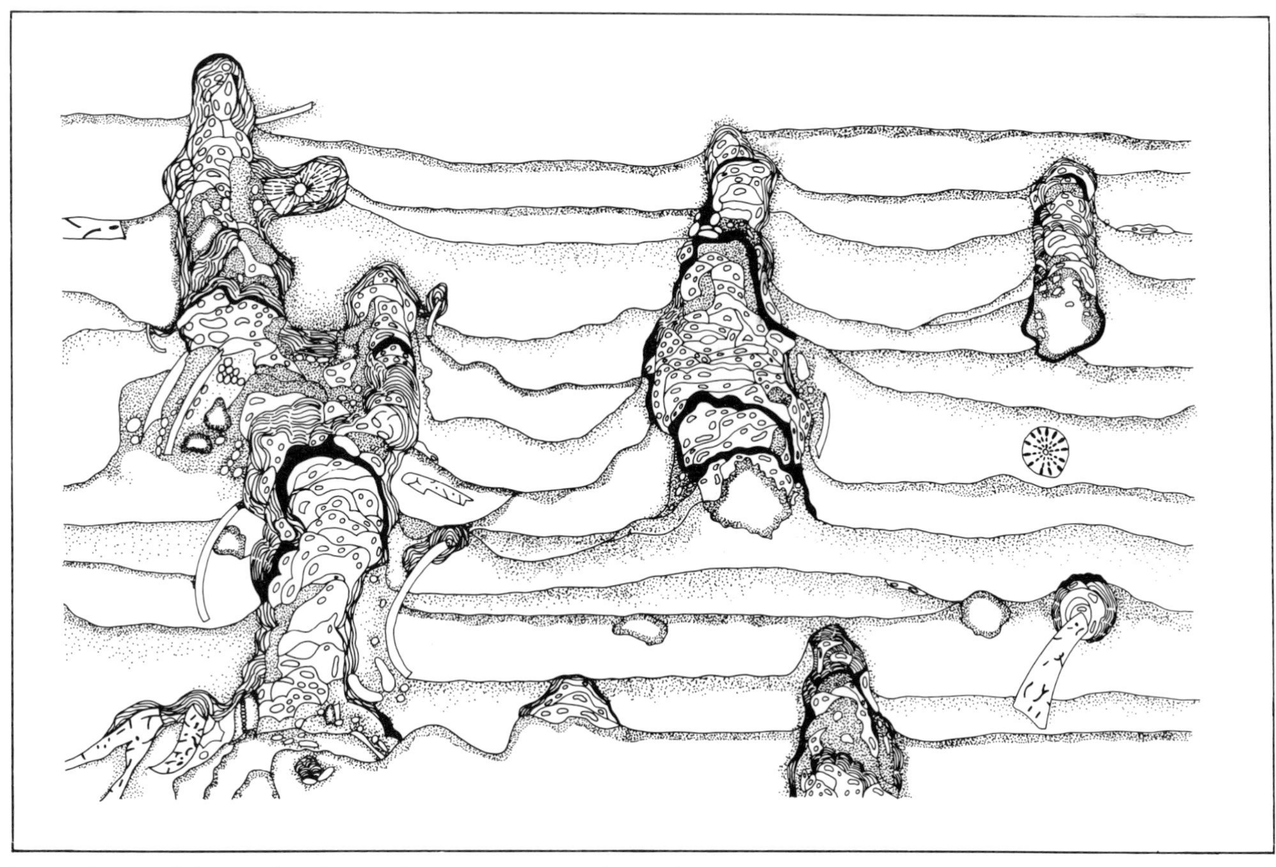

Fig. 11.35. Reef-like sessile foraminiferal colonies (chiefly *Tolypammina*), mineralized by colloform Fe–Mn oxide-hydroxides, and spatially linked to a dense succession of hardgrounds. Vertical distance shown is 1 cm. Triassic Hallstätter Kalk, Feuerkogel, Eastern Alps, Austria (after Wendt, 1969).

interstices of the algal clumps coarser sediment is concentrated; both the stromatolites and their matrix contain calcareous nannofossils (Bernoulli and Jenkyns, 1974). Condensed Triassic facies from Yugoslavia also display stromatolitic cupolas on ammonites (Wendt, 1973).

Neptunian dykes and sills are characteristic of Triassic and Jurassic condensed sequences, occurring both within the pelagic facies themselves and cutting through underlying platform carbonates (Fig. 11.36) (Schlager, 1969; Wendt, 1971). The bed-parallel fissures are commonly rich in small brachiopods, ammonites and other molluscs whose shells generally lack the corrosion phenomena seen in the normal stratigraphic succession (Wendt, 1971; Schlager, 1974). The successions in the sills are also very much reduced in thickness and may display sedimentary features typical of condensation; such sediments locally pass laterally into cements of radiaxial fibrous calcite that fill the whole of the fissure. The sub-vertical dykes, commonly showing multiple intrusions, are faunally very sparse.

The red biomicrites of the condensed facies typically pass up into a more nodular and marly lithology (cf. Fig. 11.32), included under the terms *Hallstätter Kalk* (Triassic) and *Adneter Kalk* or *Ammonitico Rosso* (Jurassic). These facies are characterized by the presence of lime-rich micritic nodules of centimetre scale set in a darker red marly matrix; spar-cemented nodular varieties occur locally (e.g. Fig. 11.33). In western Sicily the basal levels of the Jurassic *Ammonitico Rosso* contain corroded ammonites bearing stromatolitic cupolas, but such features are very rare (Jenkyns, 1974). Faunas in these nodular limestones are similar to those in condensed sequences except that hard-substrate dwellers are generally absent, and calcareous nannofossils are very scarce. In the Bathonian of the Pieniny Klippen Belt of Poland such facies contain rounded fragments of igneous, metamorphic and sedimentary rocks

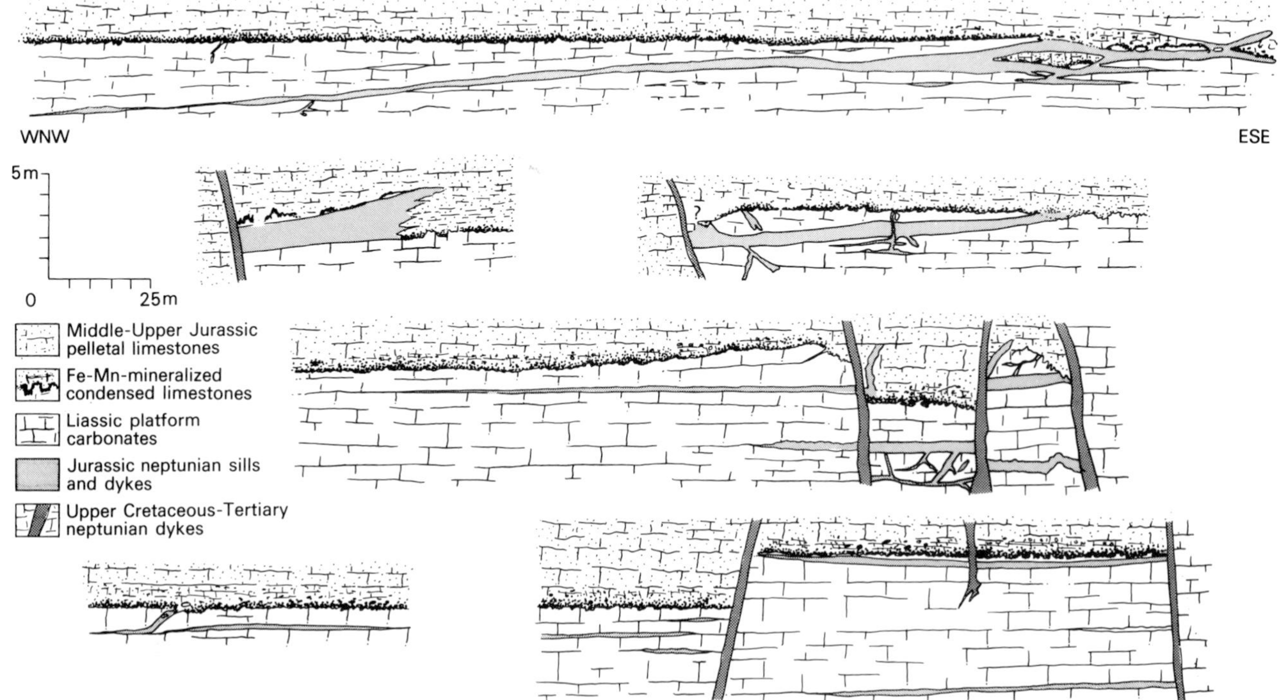

Fig. 11.36. Sketch sections showing change from Lower Jurassic carbonate-platform facies to Middle Jurassic condensed facies with Fe–Mn nodules and crusts. Sections are cut by numerous neptunian dykes and sills fed from above. Rocca Busambra, western Sicily (after Wendt, 1971).

(Birkenmayer, Gasiorowski and Wieser, 1960).

Overlying the nodular carbonates are red and grey cherty micrites and radiolarian cherts a few metres or tens of metres thick; these are rich in Radiolaria with, in Jurassic examples, solution-welded nannofossils supplying any carbonate matrix (Garrison and Fischer, 1969). Capping these siliceous facies in Jurassic successions are thicker white calpionellid-bearing micritic limestones of similar age and facies to the oceanic deposits of the Ligurian Apennines (see Sect. 11.4.2). The continental-margin facies, locally termed *Biancone* or *Maiolica,* are, however, particularly rich in calcareous nannofossils. The odd blue-black chert nodule occurs in these fine-grained limestones.

In some parts of the Tethyan region, specifically Polish Carpathians, southern Hungary, Franco–Italian Alps, and western Sicily, stratigraphically condensed facies pass up into a different rock-type; this is a grey-, cream- or rose-coloured micro-oncolitic stromatolitic facies locally attaining a thickness of 300 metres (Kaszap, 1963; Jenkyns, 1972; Bernoulli and Jenkyns, 1974). The fauna is typically pelagic, including calpionellids; and the free-swimming crinoid *Saccocoma,* globigerinid Foraminifera and thin-shelled bivalve fragments all act as nuclei to the concentrically laminated or massive cortex of the micro-oncolites. This cortex is partly composed of coccoliths and their debris.

The successions described above, and various local variations on the same theme, constitute distinct genetic types characterized by the presence of condensed sequences that typically rest directly on platform carbonates. Other sections of the same age-span, however, are considerably thicker, possibly extending up to 4 km, as in the Lower Jurassic of the Southern Alps on the Swiss–Italian frontier (Bernoulli, 1964). Such very thick sections are developed as grey burrow-mottled limestone-marl interbeds, rich in Radiolaria and sponge spicules and locally chertified; Triassic equivalents are known from Austria and elsewhere (e.g. Schlager, 1969; Bernoulli and Jenkyns, 1974). In the Lower Jurassic of Hungary, the Carpathians and the Eastern Alps such facies contain organic-rich laminae and Fe- and Mn-carbonates. Overlying the Jurassic grey limestones are more condensed red nodular and marly limestones similar to the *Ammonitico Rosso* described above (see Sect. 11.4.4) but considerably richer in clay.

Fig. 11.37. Slump-rubble, pebbly mudstone breccias and disrupted beds of red nodular and marly limestone. Lower Jurassic expanded (basinal) sequence. Kammerköhr Alm, Tirol, Eastern Alps, Austria (from Bernoulli and Jenkyns, 1970).

Throughout such grey and red sequences there is evidence of bedding disturbance in the form of slump-folded complexes, massive pebbly mudstone breccias (Fig. 11.37), and graded and laminated limestone beds that locally bear flutes on their soles. In the Austrian Glasenbach section (Fig. 11.33) and elsewhere these graded beds show an evolution up the sequence from white crinoid–echinoid biosparites in the grey limestone-marl inter-beds, then pink crinoidal biosparites at the base of the red nodular limestones and marls, red mixed crinoidal–pelagic bivalve biosparites/biomicrites higher up and finally red pelagic bivalve biomicrites/biosparites with only very rare crinoids at the top of the red limestones and marls (Bernoulli and Jenkyns, 1970). Whole lithified blocks of limestone, from centimetre to decametre scale, also occur interbedded in the sequence: many of these, such as red crinoidal biomicrites, biomicrites with corroded ammonite shells, and ferromanganese-encrusted ammonites, are typical of condensed facies. Calcareous nannofossils are locally abundant in the grey limestone-marl interbeds and in the graded and laminated units.

Above the red nodular and marly limestone green and red radiolarites, a few tens of metres thick, follow; in one Austrian locality such rocks contain spectacular breccias, slump-folds and graded and laminated deposits. The derived material includes older reef and pelagic limestones (Schlager and Schlager, 1973). Overlying the radiolarites are thick (circa 250–350 m) white Jurassic–Cretaceous micritic limestones (*Biancone, Maiolica*) that locally contain interbedded grey shales which can be bituminous (Bitterli, 1965). Ammonites are found infrequently although their aptychi are more common; Radiolaria, sponge spicules and calpionellids constitute the microfauna. Burrow traces are present. The micritic matrix is formed by a welded mosaic of partly recrystallized nannofossils (Garrison and Fischer, 1969; Bernoulli, 1972). Parallel- and cross-laminated pelagic limestones, manifesting crude grading of Radiolaria, occur at some levels; beds rich in pellets, echinoderm fragments, bivalves, ostracods, Bryozoa, calcareous algae, oncolites and ooids may also be present, locally mixed with planktonic fauna. These horizons with shoal-water carbonate elements are commonly graded and may bear flute casts on their soles; secondary chert is a common accessory.

The interpretation of these Triassic and Jurassic pelagic sequences, which parallels that of the similar Palaeozoic facies described previously, is illustrated in Fig. 11.38. Sedimentary rates are given in Table 11.4. On the submarine highs or seamounts current activity is indicated by overturned ammonites and stromatolites, the even concentric growth of the Fe–Mn nodules, which locally contain reworked lava fragments, and the fact that many sediments occur in neptunian dykes and sills while they are absent from the normal stratigraphic succession (Jenkyns, 1971a). These sills are intriguing in that they were often filled slowly and apparently supplied a habitat peculiarly favourable for small animals (Wendt, 1971). The crinoidal, ammonite and bivalve biosparite lenses may be viewed as sand-waves formed perhaps by tidal currents, thus paralleling similar features described from Recent aseismic ridges, seamounts and marginal plateaus (see Sect. 11.3.3, 11.3.6). In this largely non-depositional setting the familiar processes of carbonate dissolution/precipitation and encrustation by Fe–Mn oxyhydroxides could all take place, much as they do on Recent seamounts (see Sect. 11.3.3, 11.3.6). The spar-cemented

Table 11.4 Proposed ranges in rates of accumulation for Jurassic to Cretaceous pelagic sediments of the Tethyan continental margins. (After Schlager, 1974)

Facies	Area	mm/10^3 years
Condensed Cephalopod Limestone	Austrian Alps	0.5–1.5
	Italian Apennines	<1–6.5
Nannofossil Limestone (*Maiolica Biancone*)	Austrian Alps	17–51
	Italian Apennines	8–10
Radiolarite	Austrian Alps	3–4
	Italian Appennines	3–9

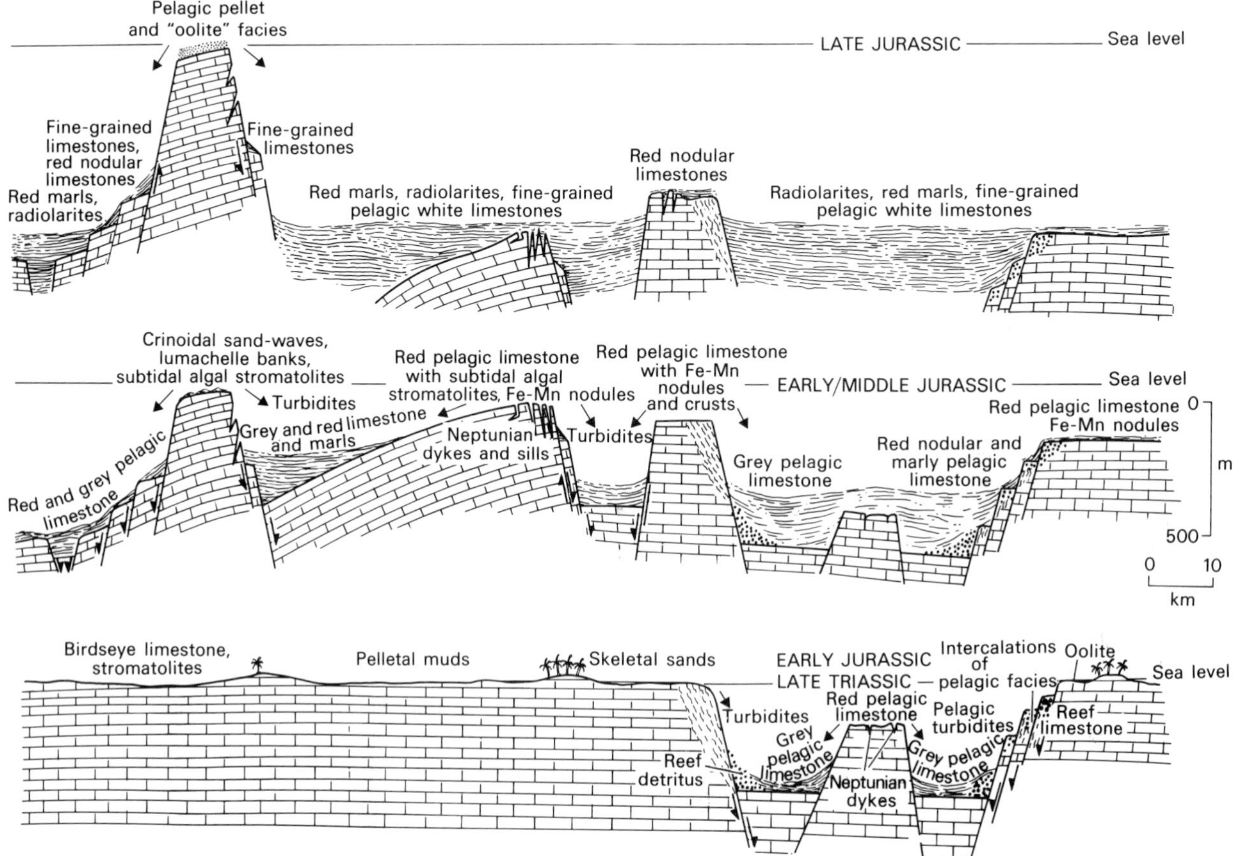

Fig. 11.38. Sketch of the palaeogeographic evolution of the Tethyan continental margin during the Triassic and Jurassic. Block-faulting and differential subsidence affected most (but not all) carbonate platforms and gave rise to a seamount-and-basin topography on which coeval condensed and expanded sequences were deposited. A general evening-out and deepening of the sea floor is indicated for the Late Jurassic except for those seamounts that were accumulating pelagic 'oolites'. Some seamounts and plateaus (e.g. Trento Plateau, Fig. 11.20) persisted through much of the Cretaceous (after Bernoulli and Jenkyns, 1974).

nodular limestones probably record precipitation of a submarine cement around exhumed and transported nodules (e.g. Garrison and Fischer, 1969; Bernoulli and Jenkyns, 1974). The Pb-rich chemistry of the Fe–Mn nodules from the Austrian Jurassic and the enrichment in Co, Ba, and V plus paucity of Cu in the west Sicilian examples points to a relatively shallow environment of formation (see Sect. 11.3.3), a conclusion supported by the presence of thallophyte borings and algal stromatolites indicating photic depths (circa 200 m max.) for certain condensed sequences. Shallow photic depths are also indicated for the Upper Jurassic micro-oncolitic facies which may have been generated in highly turbulent water (Jenkyns, 1972).

Syn-sedimentary faulting – locally accompanied by submarine volcanism – on and around the seamounts, is forcibly indicated by the presence of sediment- and carbonate cement-filled submarine fissures; pebbly mudstones, breccias, turbidites and displaced lithified blocks in basinal sequences may also testify to tectonic activity. The evolution of turbidites in the Glasenbach section is taken as reflecting a gradual sinking of a seamount source, on which crinoid ossicles were gradually replaced by pelagic bivalve fragments as the sand-sized components. These Jurassic submarine horsts (Fig. 11.38), essentially built of platform carbonates, may be directly compared with the Blake Plateau, the Pourtales Terrace and the non-magnetic seamounts off the Iberian Coast (see Sect. 11.3.6). The presence of apparently tree-rafted pebbles on the Blake Plateau (see Sect. 11.3.6) lends credence to the idea of a similar mode of transport for the exotic rocks in the Bathonian red nodular limestones from Poland (Birkenmayer, Gasiorowski and Wieser, 1960).

In the deeper basins, conditions in the Early Jurassic were locally stagnant giving rise to organic-rich deposits and Fe–Mn carbonates, as in the present Baltic Sea (Hartmann, 1964) and in certain Cretaceous settings (Fig. 11.7). The change from grey limestone—marl interbeds to red nodular limestones up-section must reflect a decrease in the amount of buried organic matter on the sea floor, resulting either from higher oxygen levels in the bottom water or lower sedimentary rates. The change from red nodular limestones to radiolarites in both seamount and basin sequences indicates less pronounced bottom relief by the end of the Jurassic, except in those areas where rejuvenation of submarine topography resulted in exposure of older sediments and their subsequent basinward transport to and accumulation in a submarine fan (Schlager and Schlager, 1973). Depths of several kilometres, presumably below the calcite compensation level, have been estimated for these siliceous facies (e.g. Garrison and Fischer, 1969; Bernoulli and Jenkyns, 1974; Hsü, 1976). The problem in interpreting the change from radiolarites to the white nannofossil limestones has been discussed above (see Sect. 11.4.2). The Late Jurassic–Early Cretaceous coccolith ooze apparently acted as an efficient sedimentary blanket by ironing out much, but not all, of the remaining deep sea-bottom topography. Local persistence of fringing carbonate platforms is indicated by the derived beds of shallow-water material. Anoxic bottom waters in the Early Cretaceous (see Sect. 11.3.1) favoured formation of bituminous intercalations.

Many of the Mesozoic facies described above from the Alpine–Mediterranean region occur, as exotic blocks, in the Tibetan Himalaya and in Timor (Wanner, 1931; Heim and Gansser, 1939); the same palaeogeographic models may be applied.

CRETACEOUS TO TERTIARY OF THE VENETIAN ALPS, NORTH ITALY

In parts of the Venetian Alps, north Italy (Fig. 11.19) the Lower Cretaceous is represented by the so-called *Biancone* (see Sect. 11.4.4) which is developed as a slightly cherty white to rose nannofossil micrite containing calcite-replaced radiolarians and calpionellids (Castellarin, 1970). Maximum thickness of this facies is about 40 m but it is generally much less; and where the deposit thins almost to nothing it is replaced by a 15 cm zone of brown ferruginous phosphatic crusts, angular cherty fragments and rounded micritic clasts. Small quartz and glauconite grains occur as accessories. Within the calcareous parts of this so-called hardground, cosmic spherules have been found (Castellarin, del Monte and Frascari, 1971). The mineralized crusts, which contain detectable quantities of Ni and Co, have commonly replaced the enclosing micrite and associated calcareous clasts. The time interval embraced by the hardground includes the Aptian and parts of the Barremian and Albian stages.

Overlying the *Biancone* is the so-called *Scaglia Rossa,* typically developed as a pink to red slightly nodular marly limestone rich in calcareous nannoflora and planktonic Foraminifera. Trace fossils are abundant and include *Thalassinoides, Chondrites* and in the higher, more marly parts of the succession, *Zoophycos* (Massari and Medizza, 1973). In the lower Turonian and Albian there are bituminous levels rich in uranium (Fuganti, 1966). Rounded pebbles of crystalline rocks, quartzite and other material occur sporadically.

The *Scaglia Rossa* is also characterized by stratigraphic lacunae, lenticular condensed zones, or hardgrounds that bear spectacular brown Ni–Co-bearing ferruginous-phosphatic crusts locally containing volcanic fragments, sharks' teeth and rhynchonellid brachiopods (Malaroda, 1962; Massari and Medizza, 1973). Two main hardgrounds are distinguished; one embraces all or part of the Campanian–Maastrichtian interval and bears a modest amount of mineralization; the other, at the Cretaceous–Tertiary boundary, involves the topmost Maastrichtian, the Palaeocene and Lower Eocene. This latter discontinuity zone shows evidence of abundant and complex authigenic mineralization. The burrows, for example, which generally contain fills of younger pelagic and neritic sediment and/or goethitic and phosphatic material, may show partly replaced walls; and breccias of reworked phosphatic–goethite crusts and limestone clasts may themselves bear a mineralized overprint. The ferruginous–phosphatic material is commonly developed in a colloform structure and can, like Devonian and Triassic–Jurassic hardground mineralization, contain sessile Foraminifera (see Sect. 11.4.4). Cosmic spherules have also been found in these condensed facies (Castellarin, del Monte and Frascari, 1971).

Using criteria discussed previously (see Sect. 11.4.4) these condensed and mineralized Cretaceous pelagic sequences are interpreted as being laid down, in photic depths, on the top of a fault-bounded seamount: the so-called Trento Plateau (Fig. 11.20; cf. Fig. 11.38) (Castellarin, 1970). The high content of cosmic spherules is consistent with this view since such magnetic particles are concentrated on Recent seamounts where depositional rates are slow (see Sect. 11.3.5). The Ni and Co content of the Fe–P crusts probably stems from diagenetic remobilization of the extra-terrestrial constituents. Deposition of the phosphatic material itself must have been linked to upwelling of nutrient-rich water along the margins of the Plateau; direct comparisons may be made with the Miocene history of the Blake Plateau, the Pourtales Terrace and the Agulhas Bank (see Sect. 11.3.6). Thus the pebbles of crystalline rocks in the *Scaglia Rossa* may have been tree-rafted (see Sect. 11.3.6). The presence of bituminous shales in the Albian and lower Turonian on the Trento Plateau suggests that the edifice was at times in contact with anoxic waters (cf. Fig. 11.7).

It is pertinent to note that coeval Cretaceous facies in Italy include deep-water nannofossil carbonates and marls, as well as carbonate-platform deposits (Fig. 11.20).

CRETACEOUS OF NORTH-EAST MEXICO

In north-east Mexico two major facies are represented in the Lower to mid Cretaceous: massive platform carbonates, locally rich in rudists, and thin-bedded dark bituminous cherty limestones. This latter lithology, which is rich in pelagic microfossils such as calcispheres, globigerinids, calcified Radiolaria and calpionellids, laterally passes into the coeval shallow-water carbonates. Around the margins of the two facies domains, breccias and graded units of shoal-water material are intercalated in the pelagic limestones. In more distal regions the cherty micrites are locally pelletal and show millimetre laminations (Enos, 1974, 1977; Wilson, 1975). The pelagic sequences may be interpreted as deposits formed in deep-water anoxic channels transecting carbonate platforms (Fig. 11.39), in every way similar to those that cross the Bahama Banks (Fig. 11.18).

Similar Cretaceous deep-water pelagic limestones, locally bituminous, occur in Venezuela and Trinidad (Hedberg, 1937; Kugler, 1953).

PALAEO-OCEANOGRAPHIC PERSPECTIVE

One of the problems to be faced with respect to the above-mentioned continental-margin facies is the origin of the pelagic carbonate in the limestones. The source of the micritic component is particularly puzzling in the case of Palaeozoic sequences where there is very little evidence for the existence of a calcareous nannoflora, although problematic records do exist from the Permian and Carboniferous (Pirini-Radrizzani, 1971; Gartner and Gentile, 1972). Equally meagre but more convincing are Triassic occurrences (e.g. Fischer, Honjo and Garrison, 1967; Zankl, 1971; di Nocera and Scandone, 1977), and certain Lower Jurassic beds are very rich in nannofossils (e.g. Garrison and Fischer, 1969; Bernoulli and Jenkyns, 1970). However, the local variation in preservation of such Jurassic examples, even within coeval facies, suggests that sedimentary factors and diagenesis can destroy coccoliths or at least render them unrecognizable; nowhere is this more striking than in the nodular *Ammonitico Rosso* facies which, regardless of age, is usually devoid of nannofossils even if they are abundant in beds stratigraphically above and below (Jenkyns, 1974). Tectonic influences may also cause recrystallization (Bramlette, 1958).

Thus there may have been an abundant Palaeozoic nannoplankton that has left negligible trace because of the

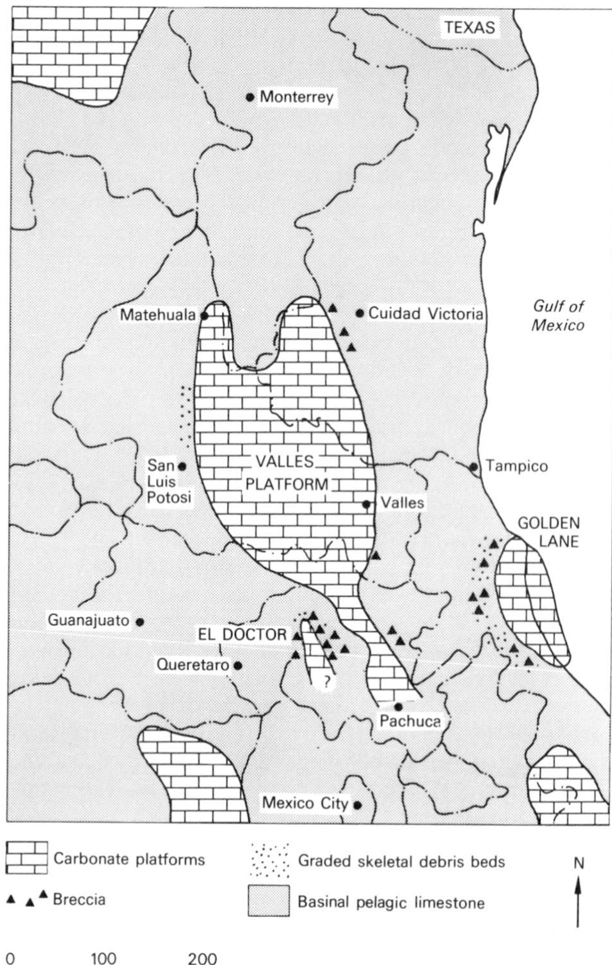

Fig. 11.39. Carbonate facies of the mid-Cretaceous of Mexico showing pattern of carbonate platforms and deep pelagic basins (cf. Fig. 11.18) (after Enos, 1974).

destructive action of various sedimentary, diagenetic and tectonic influences – influences that are inevitable on continental margins involved in orogenesis. It is, however, possible that many primitive nannofossils were incompletely calcified, such that on death they disintegrated into their component crystals leaving no recognizable skeleton but contributing significantly to the sediment (Jenkyns, 1971a). From the Late Jurassic (Tithonian) times on, after the explosive spread of the calcareous nannoflora, a large phytoplankton population was always present to supply abundant low-magnesium calcite to the pelagic environment (e.g. Garrison and Fischer, 1969).

Looking back, however, at the examples described above one notices that all depositional sites of pelagic limestones were spatially related to carbonate platforms (cf. Wilson, 1969). Recent examples of such shallow-water complexes are prolific producers of aragonite and high-Mg calcite and some of this material can be washed into deeper water; the pelagic basins that transect the Bahama Banks, for example, contain considerable quantities of silt-sized aragonite as well as calcareous nannofossils (e.g. Pilkey and Rucker, 1966; Neumann and Land, 1975). The metastable aragonite disappears as it becomes buried in the sedimentary column (e.g. Hollister, Ewing *et al.,* 1972). Part of the apparently pelagic carbonate laid down on ancient seamounts, slopes and basins may therefore have been wafted through the water-column from similar shallow-water sites (see Sect. 11.3.6). Finally, it is possible that a certain amount of micrite may derive from the macerated skeleta of macrofossils and from inorganic precipitation of high-Mg calcite, as in the Mediterranean and Red Sea (see Sect. 11.3.5).

Another problem, specifically concerned with the condensed pelagic facies, is the control on the nature of mineralization on hardgrounds. Fe–Mn oxyhydroxides characterize parts of the Devonian, the Triassic and Jurassic; phosphates are typical of other parts of the Devonian and the Cretaceous. Changes in mineralization may also be observed in the oceans: the Miocene mineralization on the Blake Plateau is phosphatic, while Recent precipitates are ferromanganiferous (see Sect. 11.3.6). Clearly the Fe–Mn deposits are formed in oxidizing conditions whereas phosphate generation is apparently favoured by a reducing environment (e.g. Gulbrandsen, 1969). The abundance of Cretaceous phosphate may therefore be related to the anoxic tendencies of the world ocean during parts of that time (see Sect. 11.3.1) (Fischer and Arthur, 1977); and the same conditions may have pertained during some of the Devonian. Local upwelling of phosphate-rich bottom waters along many ancient continental margins was another critical parameter.

11.4.5 **Deposits of epeiric seas**

CAMBRO-ORDOVICIAN OF THE BALTIC SHIELD

In the Baltic Shield there are several places where flat-lying Lower Palaeozoic sediments of probable pelagic origin are developed (Fig. 11.40). The Upper Cambrian comprises soft black shales rich in organic matter and pyrite; intercalated are lenses and beds of bituminous limestones, locally rich in trilobites limited to one or two species (Lindström, 1971). Sparse brachiopods may also be found. There is no trace of bioturbation. The fine component comprises clay minerals, quartz and feldspar. This Upper Cambrian facies, a mere 16

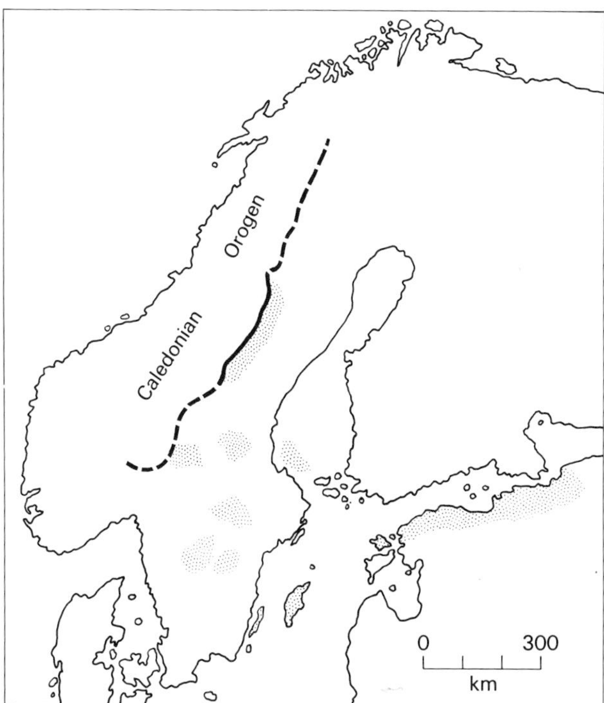

Fig. 11.40. Location of erosional remnants of pelagic Ordovician cephalopod limestone (stippled) on the Baltic Shield (after Lindström, 1974).

metres thick, persists locally into the Tremadocian where it contains dendroid graptolites, conodonts and a generally more abundant fauna than the beds below. Locally, at this level, there are some sandy intercalations. Overlying these rocks is a series of cephalopod limestones, generally less than 50 m thick, that extend into the Caradocian where they occur interbedded with bentonites. These facies vary in colour from light greyish-green to red and reddish-brown and are generally constituted by well-bedded fine-grained limestones, interleaved with more marly horizons. Locally there are nodular horizons where the calcareous kernels occur in a more marly matrix. Limestone beds can wedge out or pass laterally into marls (Jaanusson, 1955, 1960, 1972). This succession is particularly well-known on account of the presence of discontinuity surfaces and small limestone folds (Fig. 11.41) in the Lower Ordovician (Jaanusson, 1961; Lindström, 1963). The folds are convex both upwards and downwards, generally occur singly, and are invariably underlain by an undeformed bed of marl. The wavelength of the folds is commonly around 0.5 metres or less, and their amplitude is usually about 15 cm; crestal regions may be perforated by boring organisms and can show an apparently corroded surface. The discontinuity surfaces or hardgrounds

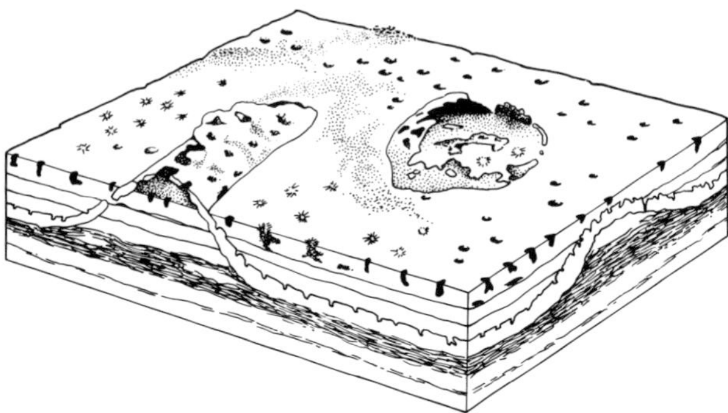

Fig. 11.41. Reconstruction of Ordovician sea floor in the epeiric sea that flooded the Baltic Shield. Consolidated limestone, bored by organisms, becomes a discontinuity surface or hardground. Corroded crests of partly buried folds rise above the bottom: the origin of such folds is probably related to processes of submarine lithification. Total thickness of bed shown is approximately 30 cm (after Lindström, 1963).

are usually irregular on a small scale, carrying pits and borings and are further distinguished by a mineralized patina of glauconite, phosphorite, pyrite, goethite or haematite. Glauconitic skins occur atop some folds and, where the fold is broken, may coat the inside of the structure.

The skeletal constituents, commonly fragmentary, of these limestones include trilobites and nautiloid cephalopods, both of which can locally dominate the fauna, plus echinoderm fragments, ostracods, conodonts, gastropods, brachiopods and rare Bryozoa. Bryozoa and echinoderm roots have been found attached to cephalopod conchs and trilobite pygidia (Jaanusson, 1972). The limestones are heavily bioturbated biomicrites. Carbonate commonly comprises some 80% of the rock; clay minerals, occasional tiny quartz crystals and chamosite, phosphorite and glauconite grains constitute most of the remainder. Chamosite and glauconite also occur as casts of small fossils (Jaanusson, 1955, 1960).

The widespread extent of these Cambro–Ordovician facies (some 500,000 km^2, Fig. 11.40), coupled with the scarcity of sessile benthos amongst the fauna, militates strongly for a pelagic environment of deposition. Sedimentary rates, about 1 mm/10^3 years, are consistent with this interpretation. The richness in iron minerals in the limestones, taken by some as suggestive of the proximity of land, may be explained by the contribution of volcanically derived aeolian dust (Lindström, 1974). The source of the carbonate must presumably be sought in finely comminuted macro-fossil debris; Lindström (1971), for example, refers to the Lower Ordovician limestones as lithified arthropod mud, and sedimentary rates may be low enough to concede the possibility of skeletal carbonate as a sole source of sediment (see Sect. 11.4.4).

Clearly the slow accumulation of the Ordovician cephalopod limestones ensured that submarine lithification was an important process (see Sect. 11.3.3, 11.3.6); its effectiveness is readily documented by the frequency of hardgrounds. The small folds are probably also related to submarine lithification, being best interpreted as polygonal expansion structures or 'tepees' caused by the displacive growth of aragonite or high-magnesian calcite cements (Fig. 11.4) (cf. Smith, D.B., 1974) comparable structures have been illustrated from the subtidal environment in the Persian Gulf, where the sedimentary rate is low and submarine lithification takes place (Shinn, 1969). The indications of sea-floor solution in these Ordovician limestones – i.e. 'corroded' hardgrounds and crests of folds – should not be taken as evidence of depths near the calcite compensation level; solution and lithification may take place side by side in depths of a few hundred metres (Fischer and Garrison, 1967). Furthermore, as with Mesozoic pelagic limestones (see Sect. 11.4.4), much corrosion may be biologically inspired (Schlager, 1974).

CRETACEOUS CHALKS OF EUROPE, NORTH AMERICA AND MIDDLE EAST

The Glauconitic Marl: Cenomanian phosphatic–glauconitic basement beds from the Chalk of South-west England

The *Glauconitic Marl* is a thin (0.3–4.8 m) basement bed of the *Chalk* which, in its more westerly outcrops, is stratigraphically condensed (Kennedy and Garrison, 1975a). This particular facies carries everywhere a basal erosion surface transfixed by burrows (*Thalassinoides*) that pipe down overlying sediment. In basal layers locally derived glauconitic calcareous concretions can constitute a pebble bed up to 30 cm thick; the exteriors of these nodules are bored by sponges, bivalves and algae, covered by oysters and other epizoans, and superficially mineralized by brown phosphate and green glauconite. Careful inspection reveals multiple generations of boring and mineralization. Black glauconitized and phosphatized shell fragments also occur gleaned, like the concretions, from the beds below.

Above this rubbly zone, phosphate-bearing glauconitic chalks usually follow; they are intensely burrowed and contain unphosphatized sponges, inoceramids, brachiopods, gastropods, echinoids and ahermatypic corals. Originally aragonitic fossils are rendered into moulds. The phosphates occur as nodules of millimetre to centimetre scale and are coloured various shades of brown to black; mineralogically they are carbonate apatites. Many of these nodules are recognizable whole or fragmentary fossils, particularly internal moulds, and are bored or encrusted similarly to the underlying concretions; a patina of glauconite is not uncommon. The apatite shows clear evidence of replacement of carbonate and also occurs, in a minor way, as a rim cement. Multiple phases of boring, encrustation and mineralization may be deciphered from petrographic study. The glauconitic grains themselves also show several and varied growth stages.

The condensed nature of the *Glauconitic Marl*, its distinctive mineralogy, and occurrence at the base of the Chalk, allows its interpretation as a hemipelagic deposit roughly comparable with the sediments presently found on the banks off southern California, and more particularly with the relict phosphates and glauconites of the South African Agulhas Bank (see Sect. 11.3.6). The *Glauconitic Marl* clearly marks a phase of deepening and transgression as pelagic conditions spread over this part of northern Europe. Vigorous bottom currents are suggested by the reworked concretions as well as by successive generations of boring, encrustation, and mineralization. Submarine lithification was promoted in this setting.

The depth of water was probably a few tens of metres; upwelling of nutrients from deeper zones or the inherent high fertility of a transgressive sea (see Sect. 11.4.5) presumably triggered the phytoplankton growth that reached its acme during the deposition of the *Chalk* itself: there was thus a rain of organic material to the sea floor which could ultimately supply phosphate ions to replace the lithified carbonates (see Sect. 11.3.6).

Chalk of north-west Europe

'There is only one Chalk' say the textbooks and indeed this bright white sediment has many unique qualities. Developed throughout north-west Europe (Fig. 11.42), thicknesses of a few hundred metres are commonly exposed in spectacular coastal sections. The *Chalk* may be described as a friable biomicrite containing planktonic and benthonic Foraminifera, calcispheres, bivalve fragments (chiefly *Inoceramus*), echinoid plates, and locally yielding Bryozoa, sponges, brachiopods, belemnites and ammonites. Large fossils may be encrusted by epizoans. Recognizable trace fossils include *Chondrites*, *Thalassinoides* and *Zoophycos*; at most horizons the *Chalk* is strongly bioturbated. The fine fraction is essentially constituted by coccoliths and rhabdoliths; this nannoflora is different from that found in coeval oceanic and near-shore facies (Håkansson, Bromley and Perch-Nielsen, 1974; Hancock, 1975a). Coccoliths from some North Sea chalks show traces of corrosion (Hancock and Scholle, 1975). The non-carbonate components include tiny quartz crystals, small authigenic feldspars, glauconite and phosphatic flecks, plus illite and montmorillonite.

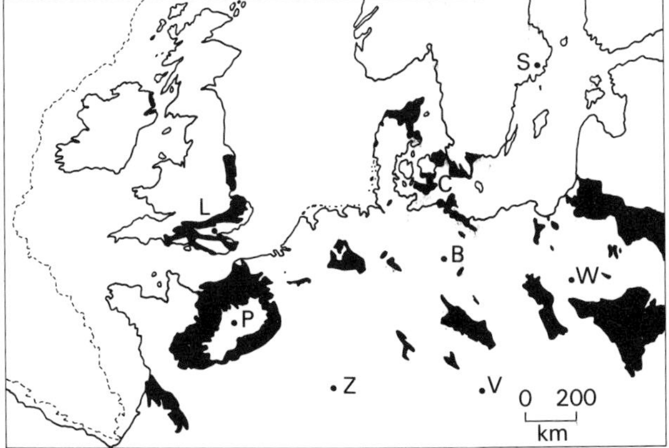

Fig. 11.42. Map of central Europe and the British Isles showing major present-day extra-Alpine outcrops of chalks and related facies. Outcrops in the Aquitaine, Paris, Mons, Danish–Swedish and other areas include calcarenitic marginal facies as well as chalks. L – London, P – Paris, S – Stockholm, C – Copenhagen, B – Berlin, W – Warsaw, Z – Zürich, V – Vienna (after Scholle, 1974).

The lower levels of the *Chalk*, if pre-Coniacian in age, are particularly rich in clay; and certain higher levels of the Formation, as they are traced north and east into the North Sea grabens, also become more argillaceous, being ultimately replaced by clays and silty clays (Hancock and Scholle, 1975). Other marginal facies variants include glauconitic and quartz sandstones and bioclastic limestones. Chert nodules (flints) are ubiquitous; the more slender varieties mimic the form of burrows, the most spectacular examples of which are 5–9-metre scale vertically orientated paramoudras (Bromley, Schulz and Peake, 1975). Some flints yield Radiolaria in their calcareous centres.

Bedding is not always obvious in normal chalk although the rows of flints generally provide an accurate trace. Local marly horizons may serve as an indicator, as do the discontinuity surfaces and hardgrounds described below. Faint parallel laminations may also be discerned in less friable chalks. Although the English *Chalk*, and indeed most chalks, are soft and powdery, when traced north from southern England the facies become harder, being finally replaced by the completely lithified *White Limestone* in Northern Ireland (Scholle, 1974). This limestone contains crushed fossils and is clearly considerably more compacted than the soft white English Chalk: the porosity of the Irish facies averages about 10% as compared with the 30–40% typical of its southerly English counterpart.

Locally interrupting the typical succession are nodular chalks, so-called omission or discontinuity surfaces, and lithified hardgrounds (Kennedy and Garrison, 1975b). At one end of the spectrum are the omission surfaces only rendered conspicuous by varying trace-fossil assemblages and a difference in sediment type above and below; at the other end are the stained and mineralized, burrowed and bored hardgrounds that may correspond with a distinct palaeontological gap (Bromley, 1975; Hancock, 1975a). Nodular chalks, which are generally associated with hardgrounds, are manifested by randomly dispersed centimetre-scale nodules floating in softer bioturbated chalk; some of the kernels possess sharp margins, others merge into the matrix. Locally, where the nodules are interleaved with marl, the structure bears an uncanny similarity to the Palaeozoic *Griotte* and *Cephalopodenkalk* of Europe and the *Adneter Kalk* and *Ammonitico Rosso* of the Tethyan Mesozoic (cf. Fig. 11.32). There are also conglomerates whose nodules are bored and encrusted by epizoans.

Nodular chalk may grade laterally into well-defined hardgrounds characterized by flat or hummocky relief: these surfaces are commonly associated with *Thalassinoides* burrows, encrusted with bivalves, serpulids, Bryozoa, ahermatypic corals, and bored by algae, fungi, bryozoans, cirrepedes, sponges, bivalves and worms (e.g. Bromley, 1970, 1975; Kennedy, 1970; Håkansson, Bromley and Perch-Nielsen, 1974). Gastropods, scaphopods and ammonites also tend to be relatively common in hardgrounds (Hancock, 1975a). Voids in these shell cavities and borings are usually filled with micrite, locally pelletal; sparry calcite is rare. Glauconitic and phosphatic mineralization typically characterize hardgrounds, the former mineral occurring as thin green replacement rims fading away downwards from the upper surface of the lithified zone. Micrite-filled microfossil chambers are also a favoured locus of glauconitization, and the mineral also occurs as sand- to pebble-sized grains. Similarly, phosphate chiefly manifests itself as replacement rims on the hardground surface itself, on burrow walls, on fossil fragments and as discrete grains including fish teeth. Where both glauconite and phosphate are present the glauconite can be shown, on petrographic grounds, to pre-date the phosphate. Silicification does not affect the hardgrounds themselves although the chalk below may contain flints. Hardgrounds can be very extensive and may be traced over distances of tens or even hundreds of kilometres.

In some places the *Chalk* is not flat-bedded but cast up into banks. Spectacular examples occur in the Turonian–Santonian of Normandy where the mounds range up to 50 m high and 1.5 km long (Kennedy and Juignet, 1974). The banks, built of typical chalk, comprise a series of overlapping lenticles delineated by hardgrounds and burrow flints. Around these banks there are slump-folded beds, breccias and traces of syn-sedimentary faulting and injection; locally the slumps carry eroded tops. The inter-bank deposits are bryozoan and echinodermal gravels. Bryozoan-rich mounds, also with peripheral slump-folds, occur in the Maastrichtian and Danian of Denmark (Håkansson, Bromley and Perch-Nielsen, 1974; Thomsen, 1976). Small banks are also known from the English Chalk (Hancock, 1975a).

One other feature worthy of note is the local occurrence, in Yorkshire and Lincolnshire, England, of a thin seam of black clay (the so-called *Black Band*) within the *Chalk* at the Cenomanian–Turonian boundary (Fig. 11.43). This clay band

Fig. 11.43. The 'Black Band': a bituminous smectitic clay, sandwiched between normal chalk at the Cenomanian Turonian boundary. The bituminous horizon contains fish remains, pyrite nodules and lacks benthos; it testifies to the deoxygenated state of much of the world ocean at this time (cf. Fig. 11.7). Key is 5.7 cm long. Yorkshire coast, England.

is bituminous (circa 1% organic carbon), contains pyrite nodules and complete fish whilst lacking benthos (Jefferies, 1963). The clay is a smectite (Schlanger and Jenkyns, 1976). This organic-rich horizon recalls similar levels in the Alps (see Sect. 11.4.4).

The composition of the *Chalk*, a biogenic sediment with at least 75% planktonic components, shows clearly that it is a pelagic deposit (e.g. Håkansson, Bromley and Perch-Nielsen, 1974; Hancock, 1975a); and the Chalk Sea covered much of northern Europe (Fig. 11.42). The fact that coccoliths in the Maastrichtian chalk are dissimilar from those that occur in coeval oceanic and nearshore sediments clearly suggests an open-marine but not fully oceanic environment, as might be expected in an epeiric sea. Sedimentation of the coccoliths (and Radiolaria) probably took place as faecal pellets, ejected by copepods and/or pelagic tunicates (cf. Sect. 11.3.1); sedimentary rates were locally as high as 15 cm/10^3 years, considerably greater than for most Recent and ancient pelagic sediments. Lack of clastics, except in marginal regions, may be related to an arid non-seasonal climate (Hancock, 1975a, b). Small quartz fragments and illite could well be aeolian; the montmorillonite may be ultimately of volcanic origin. Paucity of sedimentary deformation would suggest that the bottom slopes were very subdued.

The sediment on the sea floor was generally soft and

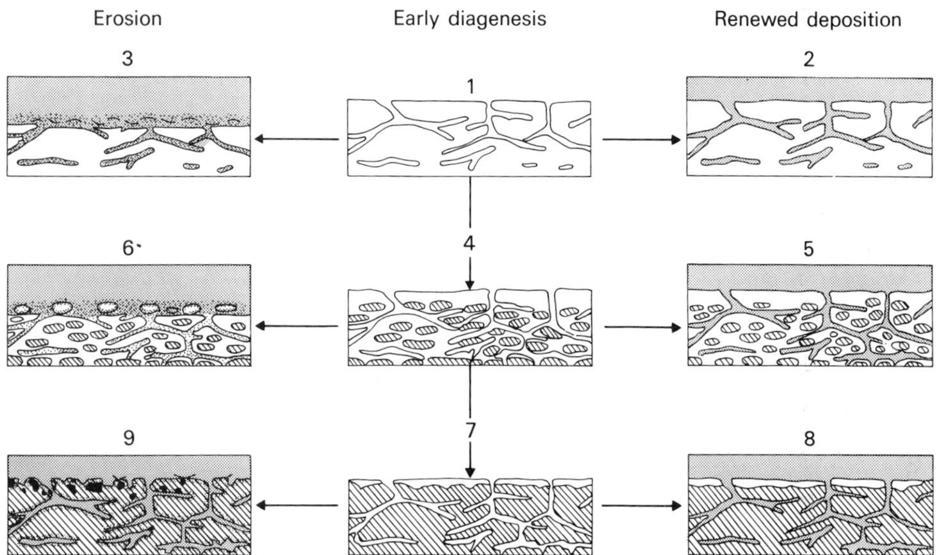

Fig. 11.44. Flow diagram showing proposed relationship between diagenesis, erosion, burial and the formation of nodular chalks and hardgrounds. (1) A pause in sedimentation leads to the development of an omission suite of trace fossils (*Thalassinoides*). (2) If buried this is preserved as an omission surface. (3) With scour an erosion surface is formed; burrows are infilled with calcarenitic chalk and the surface is overlain by a shelly lag. (4) Early diagenesis associated with a longer pause in deposition leads to the growth of calcareous nodules in soft sediment. Burrow systems are extended, the animals avoiding the nodules. (5) Burial at this stage leads to a nodular chalk. (6) If eroded, nodules are reworked and burrows truncated; the nodular chalk is overlain by a terminal intraformational conglomerate, the pebbles of which may be mineralized and bored. Burrows are filled by, and pebbles embedded in a winnowed calcarenitic chalk. (7) Prolonged diagenesis leads to a coalescence of nodules to form a continuous lithified subsurface layer; later burrows are entirely restricted to sites of pre-lithification burrows. (8) If buried, this rock band, with no signs of superficial mineralization, becomes an incipient hardground. (9) With erosion, the rock band is exposed on the sea floor, and a true hardground develops. It may become bored (borings shown in black), encrusted by epizoans, and mineralized at or below the sediment-water interface. All these processes also affect the walls of burrows (after Kennedy and Garrison, 1975b).

incoherent, for much of the fauna shows adaptive strategies, on the 'snow-shoe' principle, to prevent premature burial in the soupy ooze. Clearly, however, the bottom of the Chalk Sea was not always so treacherous; certainly below a depth of 50 cm or so the sediment was apparently firm enough to preserve the burrows of *Chondrites, Thalassinoides* and *Zoophycos* (Håkansson, Bromley and Perch-Nielsen, 1974). And, both locally and regionally, syn-sedimentary lithification resulted in the formation of nodular chalks and hardgrounds (Fig. 11.44). The prime mover behind these phenomena must have been reduced rates of sedimentation which characterized local non-subsident areas or more widespread regions during shallowing; in both cases accelerated bottom currents probably played significant roles (Kennedy and Garrison, 1975b). Low sedimentary rates favoured precipitation of a submarine (?high-magnesian calcite) cement such that crusty surfaces could be produced in as short a time as a few hundred years. The first stages of this process saw the formation of nodules which, if allowed to coalesce, produced a continuous hardground. Local current erosion excavated nodules which could then be colonized by organisms. The source of the early carbonate cements may have come from dissolution of molluscan aragonite within the top few centimetres of the sedimentary pile or from sea water itself.

Glauconitization and phosphatization apparently took place preferentially in areas which were lithified by this submarine carbonate cement. Similarly to the *Glauconitic Marl,* exposure of a bedding surface for a considerable period of time to overlying fertile waters allowed input of planktonic organic matter that encouraged sub-surface replacement by glauconite and phosphate. The initial substitution of a carbonate crust by glauconite may have been favoured by the elevated magnesium content of the calcite precipitate; after this had been consumed, or in areas where the high-magnesian calcite had earlier inverted to low-magnesian calcite, phosphate may have been the favoured replacement product.

If, as suggested above, hardgrounds correlate with periods of regression, then they are presumably the shallowest facies of the *Chalk*. Faunal and floral evidence would seem to support this, although the issue is sometimes debatable. For example, presumed algal-grazing trochid gastropods were taken by Kennedy (1970) to suggest levels less than 50 m, although this depth dependence has been disputed by Hancock (1975a). Tiny borings, attributed to algae, could suggest photic depths; although some or perhaps all of these perforations may be fungal (Bromley, 1970). Some chalk basement beds contain laminated structures that resemble algal stromatolites; and indeed the Polish Chalk contains convincing examples in one hardground sequence (Marcinowski and Szulczewski, 1972). This evidence, albeit equivocal, may be used to suggest that some hardgrounds were illuminated, and a maximum depth of 200 m is reasonable (e.g. Scholle, 1974). A compatible figure was obtained by Reid (1962) on the basis of hexactinellid sponge faunas.

Relevant to a consideration of depth are the mounds that occur in the Chalk of Normandy; these may be interpreted as accretionary bodies of lime mud stabilized by a plant covering, either algal or of marine angiosperms (Kennedy and Juignet, 1974). Photic depths are thus implied; although reference to a modern analogue would not support this (see Sect. 11.3.6). The Danish examples probably have a bryozoan frame and are hence of different origin. In both cases their primary relief on the sea floor is indicated by the soft-sediment deformation around their flanks.

If the hardgrounds were formed in depths of a few hundred metres and represent the shallowest-water facies, clearly the white Chalk and its equivalents in the North Sea – which typically lack non-depositional surfaces – were laid down at greater depths. Just how much greater is difficult to assess; those parts of the North Sea Chalk that contain corroded coccolith plates might have been laid down at depths of 1 to 2 km where bottom waters were relatively aggressive. In the more tectonically quiet parts of Europe maximum depths of about 600 m seem plausible (Hancock, 1975a); bathymetric data from sponges and Foraminifera are consistent with this figure. So, also, is the predicted sea-level rise obtained from an examination of Late Cretaceous sea-floor spreading rates (Hays and Pitman, 1973).

It must be assumed that bottom-water conditions in the Chalk Sea were generally oxidizing since bioturbation is so common throughout the sedimentary section. Yet the organic-rich, benthos-poor *Black Band* of Yorkshire and Lincolnshire clearly indicates that conditions were at times anoxic. This particular horizon, probably an altered volcanic ash-fall, must have accumulated so rapidly that abundant organic material could be entombed within it. This bed, and the more widespread glauconite and phosphate mineralization, must be considered as varied sedimentary responses to oxygen-depleted waters prevalent in the Late Cretaceous oceans (see Sect. 11.3.1).

Chalk of the Western Interior, North America

The Upper Cretaceous rocks of the Western Interior outcrop, in very similar facies, from Mexico to Canada and are developed in a variety of lithologies including coals, sandstones, shales, locally containing diatoms and Radiolaria (Wall, 1975) and, in some areas, chalks and limestones: total thicknesses attained are in the 2–5 km range. A lateral variation from carbonate to clastic facies (Fig. 11.45) is also seen, in a vertical sense, as sedimentary cycles (Kauffman, 1969). The obviously pelagic

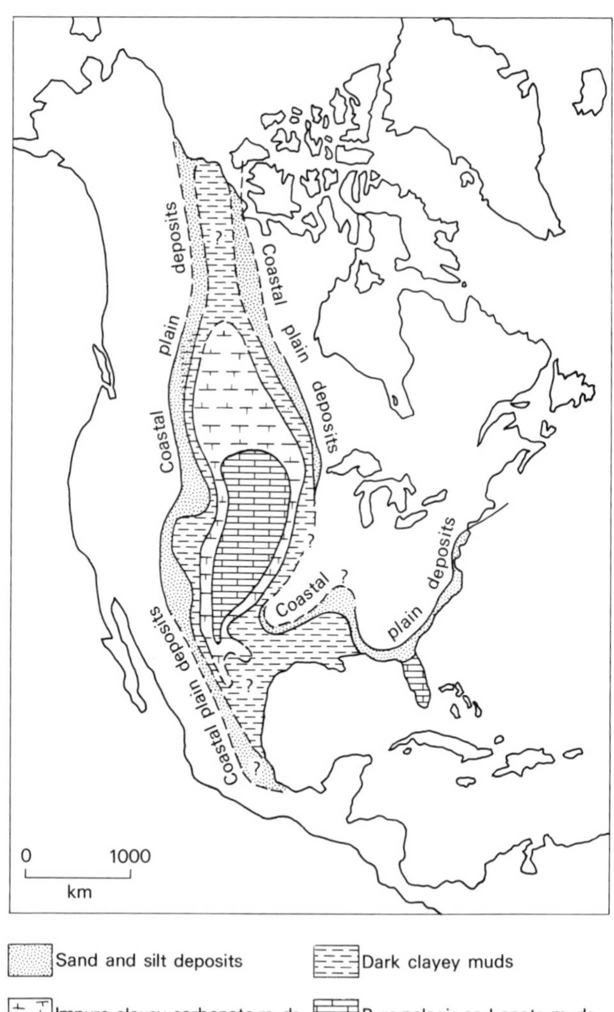

Fig. 11.45. Distribution of sediments in the Western Interior epeiric seaway during peak transgression of the Greenhorn marine cycle (Early Turonian) (after Kauffman, 1969).

facies, which occur at several stratigraphical horizons, include chalks, locally shell-rich, chalky limestones and shaly chalks; some of these facies have exact correlatives in Europe (Hancock, 1975b).

The Cenomanian to Turonian *Greenhorn Limestone,* a few tens of metres thick and punctuated by bentonites, comprises an olive-grey to olive-black series of burrow-mottled chalky limestones, locally concretionary, and laminated shaly chalks, these latter containing elevated quantities of quartz, clay minerals, organic carbon and pyrite (Hattin, 1975a). Small patchily silicified zones are present in the limestones. The macrofauna includes ammonites as moulds, which are preferentially concentrated in the limestones, and bivalves, chiefly inoceramids; large shells are locally encrusted with epizoans. Recognizable trace fossils include *Planolites, Chondrites* and *Thalassinoides*. Planktonic and benthonic Foraminifera are abundant and locally concentrated in laminae; fish remains and calcispheres are tolerably common. Pyrite, haematite and silica may occlude microfossil tests as well as the more common micrite or sparite. Inoceramid and foraminiferal biosparites occur at some levels, particularly in basal parts of the section where they may be cross-bedded. The microstructure of the chalky deposits is typically pelletal; the pellets themselves, generally a fraction of a millimetre or so in diameter, are composed of coccoliths and their debris (Hattin, 1975b). In the limestone beds the pellets are elliptical to circular in cross-section, whilst those in the shaly facies have clearly suffered more compaction. The limestone beds also contain uncompacted moulds of aragonitic fossils; the microstructure of these beds, unlike the shale, shows the coccoliths to be largely recrystallized. Above the Upper Cenomanian–Lower Turonian, the quantity of benthonic organisms increases, and the amount of organic matter in the sediment declines.

The lower, Coniacian part of the *Niobrara Chalk,* described by Frey (1972), is a white to grey to orange series of moderately pure chalky limestones separated by thinner beds of chalky shale; the unit is about 20 m thick. The composition of these beds is similar to the *Greenhorn Limestone*. Serpulids and bryozoans have also been recorded, as have rare ostracods; there are abundant diverse burrows in the sediment and, in calcareous skeleta, borings of clionid sponges and barnacles may be found. Unlike the *Greenhorn Limestone,* ammonites are virtually absent. Locally the beds are cross- and parallel-laminated and such current-bedded facies have been found to fill small channels.

The depositional setting of both the *Greenhorn Limestone* and the *Niobrara Chalk* was clearly a large epeiric sea of low relief stretching north–south across the continent of North America (Fig. 11.45). At times of maximum transgression, when pelagic oozes were being laid down, the seaway was some 5,000 km long and some 1,400 km wide. Clastics dominated along the margins where rapid subsidence prevented their reaching the interior of the marine area; pelagic conditions were thus favoured, except when the seas withdrew during regression. As in the European Chalk, the transport of coccoliths to the sea bottom was probably effected by rapidly sinking faecal pellets (Hattin, 1975b). Interspersed with this biogenic carbonate were influxes of terrigenous clay and volcanic ash. The sea bottom was apparently soupy for much of the time, but the Greenhorn Sea, unlike the Niobrara, must have seen the formation of some lithified layers where ammonites and inoceramids could concentrate and survive uncrushed. These zones of submarine lithification – as with European hardgrounds – were probably favoured environmental niches for many organisms including those of apparent pelagic habitat. The lack of such horizons in the Niobrara may explain its extreme paucity of ammonites. The only hard substrates available were the shells of bivalves, on which oysters could attach themselves.

Oxygen-poor bottom waters, here as elsewhere (see Sect. 11.3.1), characterized the Greenhorn Sea during part of Cenomanian–Turonian time, accounting for organic-rich sediments and a paucity of benthonic fauna throughout the Western Interior Seaway (Frush and Eicher, 1975). The basal Greenhorn facies and the *Niobrara Chalk* were clearly laid down in more turbulent oxygenated water; the number of cross-laminated beds and small channels, possibly of tidal origin, confirm this (Frey, 1972; Hattin, 1975b).

The depth of water of these pelagic carbonates has been disputed. On the basis of the foraminiferal faunas (planktonic versus benthonic) and palaeoslope estimates Eicher (1969) suggested 500–1,000 m for the Greenhorn Sea. Hattin (1975a), however, arguing on the basis of stratigraphical, palaeoecological and tectonic considerations, suggested more modest figures in the 30–90 m range. For the *Niobrara Chalk* Frey (1972), on the basis of sedimentological and palaeontological features, suggested a deepening throughout the succession, and conjured with figures in the 100–200 m range. These shallower figures seem intrinsically more likely.

Clearly there was a general similarity between Upper Cretaceous pelagic rocks of the Western Interior and north-west Europe. The main differences were apparently the greater influence of clastics in the North American facies, and the fact that in this area the climate was typically continental whereas in north-west Europe it was non-seasonal and arid (Hancock, 1975b).

Chalk of the Middle East

In parts of the Middle East, rocks of Santonian through Maastrichtian age are developed as chalks (Fig. 11.46). These pelagic facies, more lithified at the base, are underlain by shoal-water carbonates and capped by shales, further chalks, and limestones rich in chert (Flexer, 1968, 1971; Schneidermann,

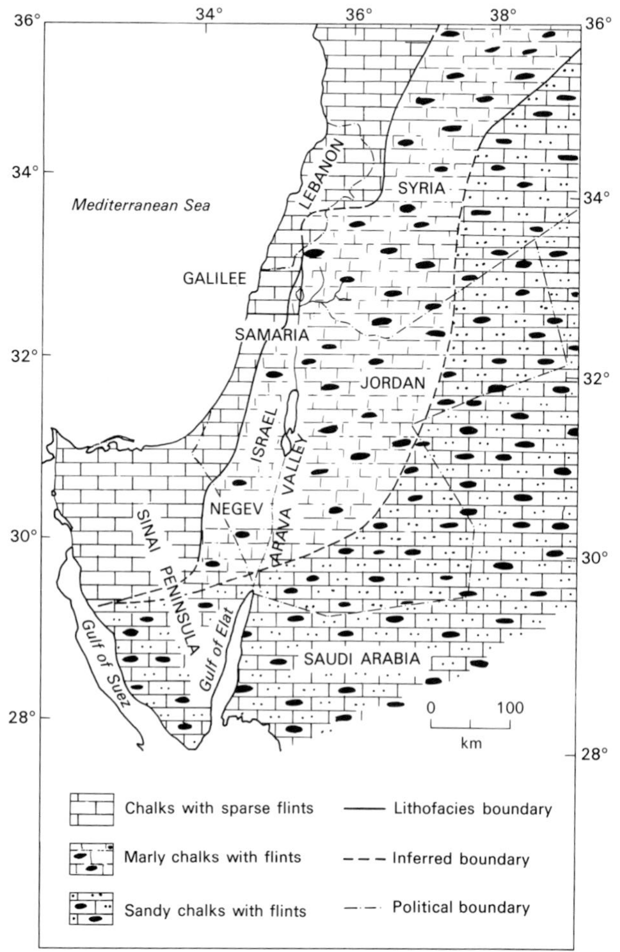

Fig. 11.46. Distribution of Upper Cretaceous chalks and associated facies in the Middle East (after Flexer, 1971).

1970). Thicknesses of the chalks are in the two-hundred metre range but vary laterally. Marginal facies include sandstones, conglomerates and a range of shallow-water carbonates. These chalks are strictly comparable with their European and North American counterparts; they were formed in a fertile epeiric sea that deepened to the north-west.

PALAEO-OCEANOGRAPHIC PERSPECTIVE

The absence of Recent pelagic epeiric sediments and their rarity in the geological column clearly shows that they are a function of atypical oceanographic conditions. Both the Silurian and Cretaceous examples described above may be simply related to eustatic sea-level rise and consequent transgression (e.g. Lindström, 1971; Håkansson, Bromley and Perch-Nielsen, 1974). Flooding of extensive continental areas was probably aided by the presence of bevelled landscapes.

An intriguing palaeo-oceanographic problem is the control behind the change-over from Cambrian black shales to Ordovician cephalopod limestones on the Baltic Shield, given that they they are both pelagic or hemipelagic facies (Lindström, 1963). An explanation of this clearly hinges directly on the sources of carbonate for the limestones, and such sources must be related to evolutionary and ecological successes of the biota; further than that it is difficult to go.

The prevalence of chalk facies is clearly related to paucity of clastics and high organic productivity of phytoplankton. Epeiric seas are likely to be fertile since continentally linked marine areas are usually adequately supplied with nutrients (Koblenz-Mischke, Volkovinsky and Kabanova, 1970; Menzel, 1974). High organic productivity and anoxic waters tend to go hand in hand and this must explain the local occurrence of bituminous facies and deposits of phosphate and glauconite whose genesis is favoured in reducing conditions (see Sect. 11.3.1, 11.3.6). The fact that the very same hardground mineralization occurs in the Silurian cephalopod limestones and the *Chalk* might suggest that the former were also deposited in waters where productivity was high. Perhaps this is even a faint pointer to the existence of a Palaeozoic calcareous nannoflora that has left no recognizable mark. What is clearly established, however, is the link between epeiric pelagic carbonates, eustatic sea-level rise, transgressions, and the formation of bituminous and glauconitic/phosphatic deposits.

11.5 CONCLUSIONS

Early concepts of pelagic sedimentation embodied the assumption that the open sea was the site of tranquil, unchanging particle-by-particle deposition where the resultant sediments were 'piled in horizontal layers and . . . spread over marvelous distances' (Walther in Twenhofel, 1926). Such assumptions are clearly not warranted; they hinge on the idea of a static globe. We are now aware that there have been many and varied changes in the physical and chemical state of the oceans during Phanerozoic time: controlling factors include the evolutionary and ecological successes and failures of particularly planktonic biota, eustatic rises and falls in sea level, fluctuating climatic conditions and continental drift. It is the juggling of these variables, which themselves may be interdependent, that governs the nature of pelagic facies. More than any other sediments, perhaps, the genesis of such deposits is held in the grip of fundamental earth processes which, moreover, but rarely allow the ultimate preservation of the pelagic record.

FURTHER READING

Hsü K.J. and Jenkyns H.C. (Eds.) (1974) *Pelagic Sediments: on Land and under the Sea*. Spec. Publs int. Ass. Sediment, **1**, 447 pp.

Goldberg E.D. (Ed.) (1974) *The Sea,* **v5,** Marine Chemistry, 895 pp. Wiley–Interscience.

Riley J.P. and Chester R.L. (Eds.) (1976) *Chemical Oceanography,* **v5,** 401 pp. 2nd edn, Academic Press.

Cook H.E. and Enos P. (Eds.) (1977) *Deep-water carbonate environments*. Spec. Publs Soc. econ. Paleont. Miner., **25,** 336 pp.

CHAPTER 12 Deep Clastic Seas

12.1 HISTORY OF THE TURBIDITY CURRENT THEORY

The modern study of deep-sea clastic sediments originated with the concept of turbidity currents, i.e. undercurrents flowing as a result of excess density caused by suspended sediment. Interest in these currents derives from their capacity to transport coarse sediment to the deep-sea environment. Prior to the development of the turbidity current theory in the late 40s and early 50s, it was widely held that only pelagic shales indicate a deep-sea environment and that conglomerates and sandstones characterize shallow water and continental environments. Consequently, the sandstone/shale alternations in flysch successions were attributed, in one theory, to vertical tectonic oscillations (Vassoevich, 1948).

The turbidity current theory originated from several different lines of research. The existence of density undercurrents produced in part by sediment suspensions was recognized in studies of Swiss lakes during the 1880s and of North American reservoirs during the 1930s (Forel, 1887; Grover and Howard, 1938). Where the Rhone River enters Lake Geneva, the river water with its suspended sediment forms a density undercurrent spreading across the bottom of the lake. In Lake Mead the water of the Colorado River forms a similar undercurrent which persists for many days and over a trajectory of many kilometres. Laboratory experiments with dilute sediment suspensions were carried out to study the hydraulics of their occurrence in reservoirs (Bell, 1942).

The idea of density undercurrents produced by suspended sediment was used by Daly (1936) to account for the origin of submarine canyons. He argued that wave activity on the shelf during periods of low sea-level in Pleistocene time would have stirred up sediment to produce density currents. These would have excavated canyons while flowing down the continental slope. In support of this view, Kuenen (1937) conducted a series of classic flume experiments which showed both the possibility of dilute density flow and some of its properties. The term *turbidity current* for this type of flow was introduced by Johnson (1938). The deposit of a turbidity current has been called a *turbidite* (Kuenen, 1957a).

Initially, it was believed that turbidity currents are dilute, low density flows with specific gravities of no more than 1.02, which carry principally clay and silt. The focus of interest in turbidity currents was on their competence to erode, not on their capacity to transport and deposit sediment. This emphasis changed when Kuenen (1950b), in a new series of flume experiments, showed that turbidity currents of high density, with specific gravities between 1.5–2.0, could carry not only clay and silt, but also sand. The significance of high density turbidity currents to transport and deposit sands was recognized by Migliorini when Kuenen presented his experimental results at the International Geological Congress in London in 1948.

This set the stage for the epoch making paper by Kuenen and Migliorini on 'Turbidity currents as a cause of graded bedding' (1950). In it the authors reported yet another series of flume experiments in which graded bedding was produced by high density turbidity currents. The experimental evidence was combined with field evidence to suggest that many graded sandstones in ancient clastic successions such as the Macigno in the Apennines were deposited by high density turbidity currents. This theory of turbidity currents carrying sands into the deep-sea environment eliminated the need for vertical tectonics to produce the sandstone/shale alternations in flysch successions.

The turbidity current theory solved several apparent anomalies of sand deposition. Graded sand layers, which had been encountered in cores taken from the bottom of the Atlantic Ocean, could now be understood as emplaced by turbidity currents (Ericson, Ewing and Heezen, 1951). The occurrence of sandstones containing displaced faunal elements, intercalated between shales containing deep water foraminiferal assemblages in the Plio-Pleistocene Ventura Basin of southern California, could be interpreted as turbidity current resedimentation of coarse, shallow-water clastics into a deep-water environment (Natland and Kuenen, 1951). Furthermore, in one of the most intricate geological detective cases ever, a turbidity current was traced as the cause of breaks in submarine telegraph cables off the Grand Banks after an earthquake in the area in 1929 (Heezen and Ewing, 1952) (see 12.2.4).

The turbidity current theory has revolutionized the study of

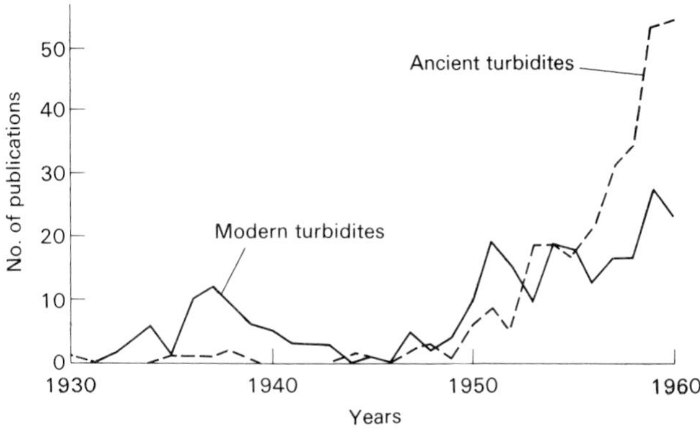

Fig. 12.1. The number of publications per year from 1930 till 1960 dealing with turbidity currents and turbidites (based on Kuenen and Humbert, 1964).

clastic sediments. It stimulated oceanographic studies of deep-sea environments. Further elaboration of sedimentary structures in turbidites by both laboratory experiments and field observations provided new tools for flysch facies analysis and for palaeogeographic reconstructions of fold belts. It profoundly affected thinking on geosynclinal development (Chapter 14). The success of the turbidity current theory has been likened to a scientific revolution (Walker, 1973) (Fig. 12.1).

12.2 PROCESSES OF CLASTIC SEDIMENT TRANSPORT IN THE DEEP SEA

12.2.1 Mass-gravity transport: general

In shallow-water environments much of the sediment is transported by *fluid flow,* i.e. the fluid medium, propelled by the force of gravity, moves the sediment. The sediment load is moved grain-by-grain by bottom traction and in suspension. However, in deep-sea environments much of the sediment moves by *mass-gravity transport,* under the direct influence of the force of gravity and, if mixed with water, it is the sediment that moves the fluid. Generally, this process takes place spasmodically, i.e. intermittently and catastrophically by the rapid downslope displacement of large masses of sediment.

Mass-gravity transport can be explained in terms of the mechanics of sediment failure and the geological conditions that trigger failure (Moore, 1961; Morgenstern, 1967). When sediment is deposited on a slope surrounding a deep-sea basin, e.g. on the upper continental slope or in the head of a submarine canyon, it will move downslope only when the shear stress exerted by the force of gravity exceeds the shear strength of the sediment. The shear strength is a function of the cohesion between the grains plus the intergranular friction. In order for slope failure to occur a certain thickness of sediment

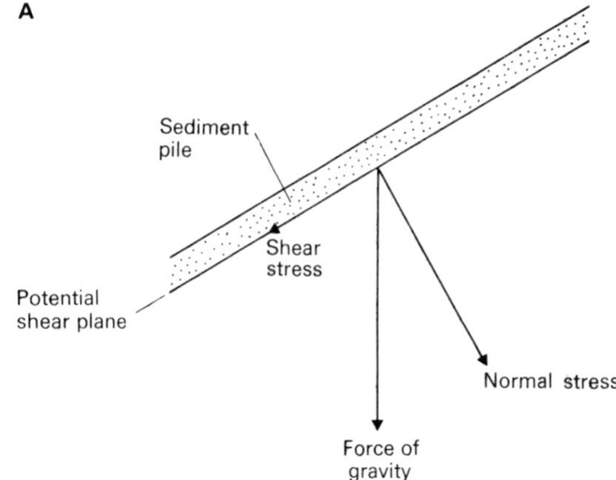

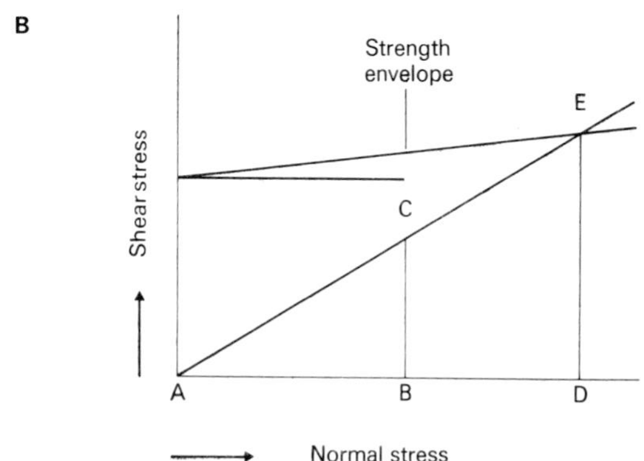

Fig. 12.2. A. Vector diagram of the weight of a sediment pile on a slope. B. Graphic presentation of the shear strength of a sediment (after Moore, 1961).

is needed. In Fig. 12.2A it is shown that the force of gravity or vertical pressure divides into two vectors, the normal stress perpendicular to the slope surface and the shear stress directed downslope. In Fig. 12.2B these are shown as AB and BC respectively. Also shown is the strength (or failure) envelope of the sediment which is defined as the line of maximum shear stress values for every value of normal stress. It can be seen that for a value of vertical stress greater than AE the shear stress will be greater than DE and failure along a slip plane will result (reviews by Hoedemaeker, 1973; Carter, 1975a).

Sediment slope failure will result either from an increase in shear stress or, conversely, from a decrease in shear strength. An increase in shear stress can result from several causes. (1) A steepening of the slope caused by undercutting by waves or currents, by slope failure further downslope etc. (2) Thickening of the sediment pile by deposition. A decrease in shear strength can result from the following causes. (1) An increase in pore fluid pressure leading to sediment fluidization. This can be caused by the fabric collapse of sediment initially deposited with a metastable packing, by the compaction of porous layers between relatively impermeable layers, or by compaction of permeable layers as such. Excess pore fluid pressure can diminish the shear strength of a layer to virtually zero and may take from a few minutes to several hours to dissipate. (2) Thixotropic behaviour, mostly in muddy sediments, in which the gel-sol transition takes place. Fluidization and thixotropy tend to be induced by strain caused by mechanical impact or shock. These are produced by such phenomena as earthquakes, tsunami waves, storm waves, or possibly also turbidity currents. The weight of rapidly deposited sediments may exert a similar strain effect.

In brief, mass-gravity transport is the predominant process of clastic resedimentation to the deep sea because on most slopes sediment strength requires that a substantial thickness and weight of sediment be accumulated before failure occurs, and the geological conditions that trigger failure tend to be periodic.

Mass-gravity transport is commonly referred to as *resedimentation,* even though in a strict sense resedimentation can refer also to transfer of sediment from one shallow water sediment trap to another. Mass-gravity transport is a term which comprises all processes of downslope movement of sediment masses under the direct action of gravity. The various processes can be classified on the basis of the internal disaggregation of the moving sediment mass. The classification presented here is synthesized from Dott (1963) and Middleton and Hampton (1976). Only *subaqueous mass-gravity transport* is considered.

In order of increasing internal disaggregation the following processes of subaqueous mass gravity transport can be recognized (Fig. 12.3). (1) *Rock fall,* in which lithified, often large rock fragments move by free fall. This process tends to be associated with debris flow. (2) *Sliding* and *slumping,* in which a commonly semi-consolidated sediment mass moves along a basal plane of failure while retaining some internal (bedding) coherence. In recent usage, sliding and slumping are nearly synonymous, though sliding emphasizes the lateral displacement of the sediment mass, whereas slumping emphasizes its internal deformation or can refer to rotational movements. (3) *Sediment gravity flow* (synonyms: mass flow, sediment flow), a general term for flow of mixtures of sediment and fluid in which bedding coherence is destroyed and the individual grains move in a fluid medium and propel it. Four types of sediment gravity flow can be distinguished based on the mechanism of grain support (Fig. 12.4), viz. *debris flow* (clasts supported by matrix), *grain flow* (by grain-to-grain interactions), *fluidized sediment flow* (by escaping pore fluids), and *turbidity flow* (by fluid turbulence).

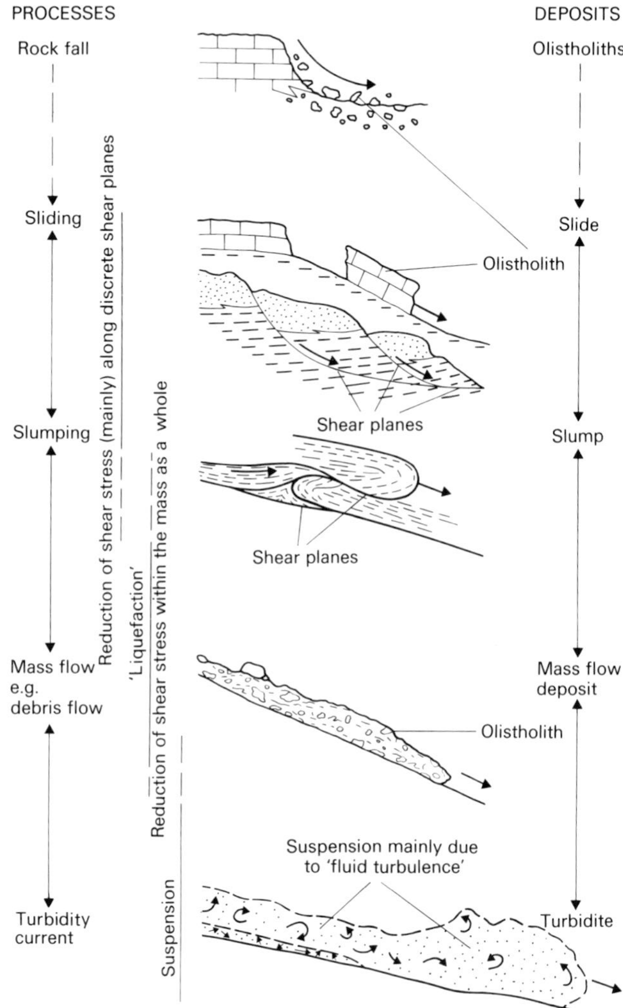

Fig. 12.3. Processes of mass-gravity transport and their deposits (after Kruit, Brouwer, Knox, Schöllnberger and van Vliet, 1975).

These conceptually distinct processes are likely to occur together during any single event of mass gravity transport (Fig. 12.5). In sediment gravity flows, grains may move not only in suspension but also as bed load by traction. For example, traction may operate during the immediately predepositional stage of turbidity flow and produce or modify some of the structures and textures in the final deposit (Sanders, 1965). Deep-sea sediments may be redistributed by *indigenous bottom currents* which move sediment largely by fluid flow as bed load.

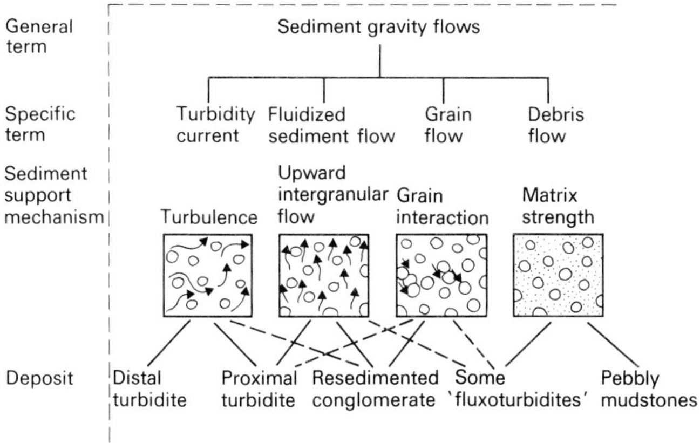

Fig. 12.4. Classification of sediment gravity flows based on the mechanism of grain support (after Middleton and Hampton, 1976).

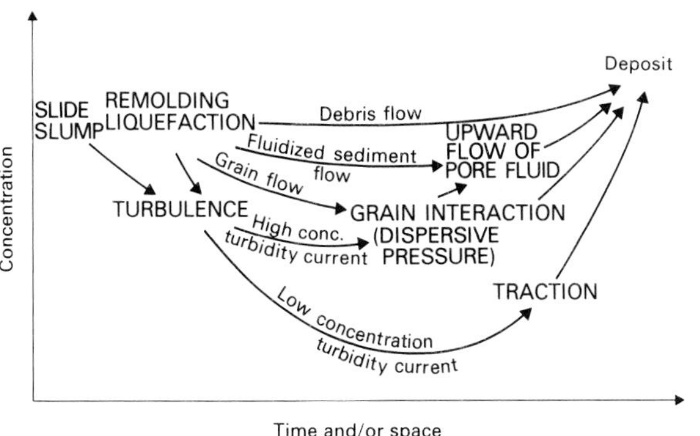

Fig. 12.5. Hypothetical interrelation of different processes in a single event of mass-gravity transport (from Middleton and Hampton, 1976).

12.2.2 **Slumping**

The extent to which slumping operates as a process of mass transport in the deep sea has been a matter of controversy. Early observations seemed to indicate that slumping can occur on slopes as gentle as about one degree (review by Morgenstern, 1967). Moore (1961), however, has argued from a quantitative study of the shear strength of marine sediments that slumping is largely restricted to areas of rapid deposition such as deltaic and canyon head environments, and that slumping seldom occurs on the open shelf and in the deep sea. He also argued that if slumping does occur on a relatively gentle, stable slope, it does so by *progressive slumping*, i.e. a local slump slides down from a steep slope, and overrides a sediment pile on a stable slope. This induces progressive downslope failure along a slip plane, having caused an increase in pore pressure and a decrease in shear strength along it.

Table 12.1 Dimensions of submarine slumps (after Roberts, 1972)

	Slope (degrees)	Volume (m^3)	Average thickness (m)	Maximum thickness (m)
Magdalena River Delta	2	3×10^{8}	20	60
Mississippi River Delta	0·5	4×10^{7}	10	20
Suva, Fiji	3	$1{\cdot}5\times10^{8}$	30	100
Sagami Wan	11	7×10^{10}	–	–
Scripps Canyon	6–8	$10^{4}-10^{5}$	4	6
Kidnapper's Slump	1–4	8×10^{9}	25	50
Grand Banks	3	$7{\cdot}6\times10^{11}$	350	–
Valdez, Alaska	6	$7{\cdot}5\times10^{7}$	–	–
Rockall, upper slump	2	$3{\cdot}0\times10^{11}$	265	332

The development of the continuous seismic profiler has made possible the mapping of sub-bottom structures of sea floor sediments. This technique has shown that slumping does occur on gentle slopes and that the slumping phenomenon is widespread in the deep-sea, especially on the upper continental slope (Lewis, 1971; Roberts, 1972). The dimensions of single slump masses can be very large, and their volumes can be of the order of hundreds of cubic kilometres (Table 12.1). Large-scale slumping is known to have been triggered by earthquakes of great magnitude. The Grand Banks slump of 1929 was caused by an earthquake of magnitude 7.2 (Richter scale). The occurrence of slumping is not only a function of the seismicity of an area, but also of such interrelated factors as sediment shear strength, lithology, rate of deposition, slope, indigenous current system, etc. (review by Morgenstern, 1967).

A large slump on a gentle deep-sea slope shows the following morphology (Fig. 12.6) (Lewis, 1971). The *head* is characterized by tensional structures, e.g. a large rotational slump scar

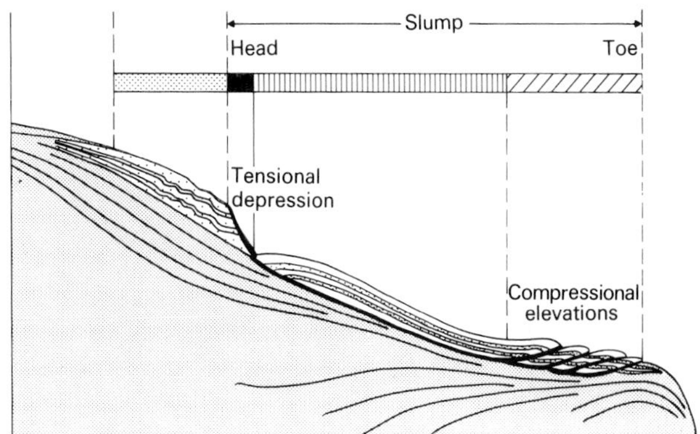

Fig. 12.6. Diagrammatic cross-section of a submarine slump on a gentle slope (from Lewis, 1971).

and bed deficiency. Above the head area *retrogressive slumping* may have occurred, i.e. the upslope propagation of unstable slump scar surfaces and their successive failure. The main *body* of the slump mass can be relatively undisturbed. The *toe* area is characterized by compressional structures, i.e. thrusting and overriding of beds.

Large slump sheets have been described from ancient deep-sea clastic successions, some with volumes of the order of 10 cubic kilometres (Rupke, 1976a). Such large slump occurrences may be used as indicators of palaeoseismicity.

Early interest in slumping phenomena in ancient sediments centred on the question whether the deformational structures are produced by sedimentary slumping (sliding at the sediment/water interface) or tectonically (within a sediment column undergoing tectonic deformation). A classic debate on this question between Jones (1954) and Boswell (1953) preceded the widespread recognition of sedimentary slumping in ancient deep-sea sediments. Jones advocated that deformational structures in Silurian clastics in Wales are sedimentary in origin, whereas Boswell argued for a tectonic origin. In another study, a spectrum of deformational structures from purely sedimentary to purely tectonic has been recognized (Fig. 12.7) (Maas, 1974).

Lists of features that characterize sedimentary slumps have been compiled (Kuenen, 1953; Helwig, 1970), among which are the following (Fig. 12.8). (1) The deformed beds occur as a zone between undisturbed beds. (2) The upper contact of the zone of deformed beds is welded, i.e. a depositional fit occurs between the irregularities of its upper surface and the base of the overlying bed. (3) Fold anticlines may be eroded at the upper surface. (4) The preferred orientation of fold axes, if present, may be unrelated to the tectonic strike. (5) Within a single slump the structural style may be irregular and a wide range of deformational structures may occur. However, structural style alone is not a reliable criterion to discriminate between sedimentary and tectonic deformational structures (Woodcock, 1976a). In general, structural style in slumps may vary according to such factors as lithology of the sediments, degree of induration, thickness of the slumping mass, or distance of transport (Morris, 1971; Rupke, 1976a).

Additional interest in slumping in ancient deep-sea sediments has come from their role in generating turbidity currents (review by Morgenstern, 1967). The intensified study of ancient deep-sea clastics since 1950 has led to many descriptions of slump occurrences and their structures. Based on the degree of internal deformation, slumps can be divided into (1) *coherent*, in which the slump has moved largely as an undeformed block along a basal plane of décollement (synonym: slide), and (2) *incoherent*, in which movement has been more pervasive and disruption of beds has occurred (Fig. 12.8). A variety of deformational structures has been recognized, viz. several types of folds, thrusts, balls, hook-

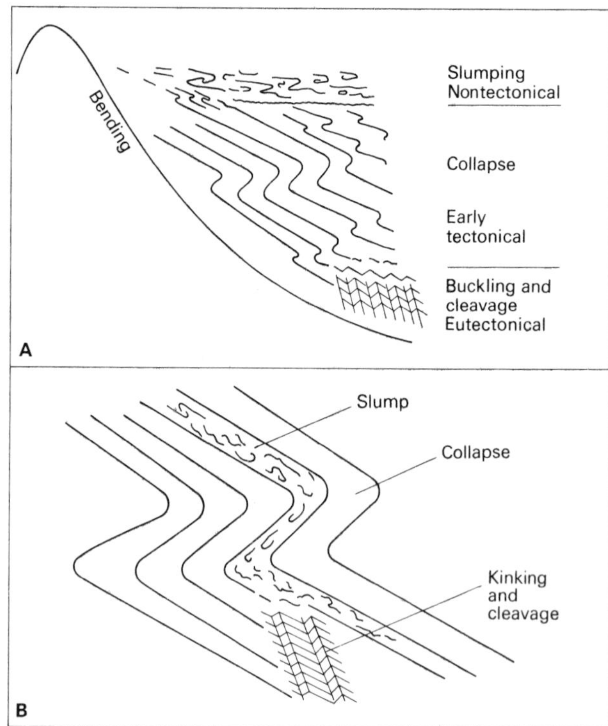

Fig. 12.7. Concurrence in (a) time and (b) space of a spectrum of deformational structures from sedimentary to tectonic (from Maas, 1974).

shaped overfolds, rotational slump scars, etc. (Dzulynski and Walton, 1965).

The axes of slump folds in many slump occurrences are preferentially oriented (Rupke, 1976a; Woodcock, 1976b). The direction of slumping is generally assumed to be perpendicular to the mean of the azimuths of the slump fold axes and is determined from the sense of overturning (or facing direction) of the folds (Fig. 12.9). However, in a slumping sediment mass, friction along the margin of the mass or internal obstructions may cause slump folds to rotate so that their long axes are turned parallel to the downslope movement. In such instances the axes of slump folds may seem randomly oriented, but the direction of the slumping can be determined using the method of the separation angle. This method, developed by Hansen (1971) from a tundra sod slide, uses the fact that axes of asymmetric slump folds in stereographic projection may show planar preferred orientation which defines the slump slip plane (Fig. 12.10). Within the slip plane, fold axes are divided into two groups with opposite rotational sense. The angle within the slip plane which separates the two groups is called the *separation angle*. Hansen showed that its bisector, the

Fig. 12.8. Slump sheet between undisturbed turbidite beds from the Eocene flysch of the southwestern Pyrenees (from Rupke, 1976b).

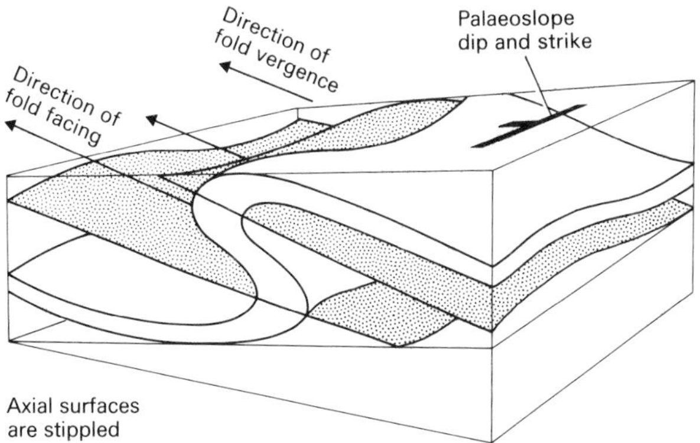

Fig. 12.9. Relationship between the geometry of slump folds and the direction of slumping (from Woodcock, 1976b).

separation line, virtually coincided with the known slip line of the tundra sod slide. The method of the separation angle has been successfully used in ancient slump occurrences (Stone, 1976).

12.2.3 **Debris flow**

Knowledge of debris flows and their deposits stems from two rather separate traditions. First, knowledge of the mechanics of flow derives from the study of subaerial debris flows, and from related theoretical work (Johnson, 1970; Hampton, 1972; review by Middleton and Hampton, 1976). Second, knowledge of the geological significance of debris flow deposits in ancient deep-sea clastics derives primarily from their study in the Upper Cretaceous to Miocene flysch in the Northern Apennines of Italy where they have been named *olisthostromes*.

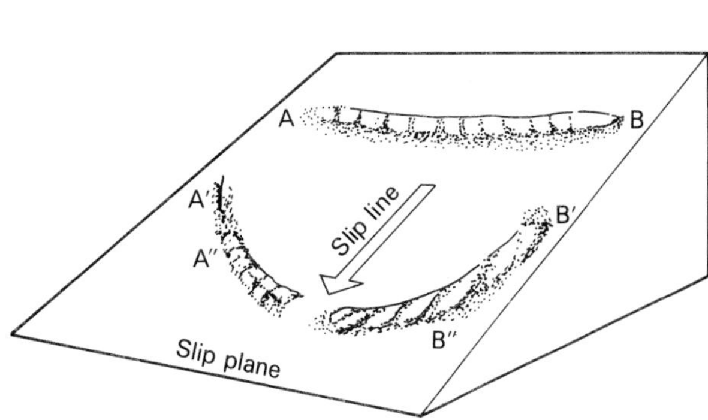

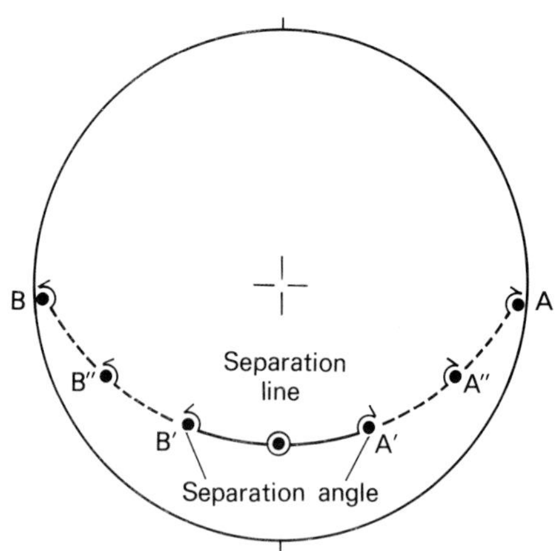

Fig. 12.10. The method of the separation angle for the determination of the slumping direction. Asymmetric fold AB slumps down slip plane to A′B′. The curved fold axis lies in the slip plane. In stereo projection, plots of axes (solid circles) define the arc of the slip plane (dashed line). Arrows indicate the sense of rotation of fold. The separation line is parallel to the true slip line (from Stone, 1976).

Interest has focused on their lithology and their significance for the tectonic history of the depositional basin (reviews by Abbate, Bortolotti and Passerini, 1970; and by Elter and Trevisan, 1973). Both traditions of study are comprehensively discussed by Hoedemaeker (1973).

A Newtonian fluid such as water does not exhibit the property of strength. However, highly concentrated sediment flows no longer behave like a Newtonian fluid, because they exhibit strength, i.e. they offer resistance to shear stress up to a certain yield value above which flow occurs. Debris flows are highly concentrated, non-Newtonian sediment dispersions with a yield strength. They also possess high viscosity and their mode of flow is laminar. Most debris flows are composed of clasts carried by a watery mud matrix. The specific density of the matrix is high and can be up to 2.5. The clast support mechanism consists of the yield strength of the matrix plus the buoyancy of the clasts exerted by the matrix. Debris flow can take place on gentle slopes of less than one or two degrees. Flow velocity is variable, but is assumed to be faster than soil creep. When the downslope pull of gravity no longer exceeds the shear strength of the debris mass, the flow 'freezes', i.e. comes in its entirety to a sudden halt.

The composition of debris flow deposits ranges from mudstone containing only a few clasts, to a bouldery mass containing little mudstone. The thickness of individual flows can be several metres, possibly much more in exceptional instances. The deposits are internally structureless; the clasts are poorly sorted and may range in size from coarse sand to boulders; they form a matrix-supported framework; the fabric is random though in places clasts may be aligned with their long axes parallel to the flow direction (Fig. 12.11). Reverse grading at the base of debris flow deposits has been reported (Fisher, 1971). Debris flows do not abrade the sedimentary floor over which they travel unless a rigid plug forms inside the

Fig. 12.11. Structures and textures of deposits from single mechanism mass-gravity flows. No vertical scale is implied (from Middleton and Hampton, 1976).

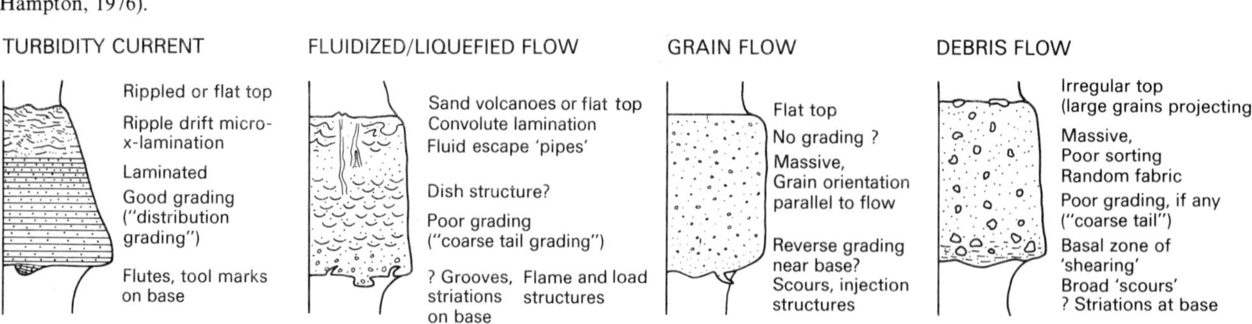

flow which then may be rafted along the floor to produce slide marks (summary by Middleton and Hampton, 1976).

Debris flow deposits are known from modern deep-sea environments where they may travel several hundreds of kilometres and cover areas of many thousands of square kilometres (Embley, 1976; Moore, Curray and Emmel, 1976).

Both intra- and extra-formational debris flow deposits have been described from ancient deep-sea clastic successions. *Pebbly mudstones* which consist of a muddy matrix with dispersed clasts have been attributed to the mixing of intraformational gravel and mud layers, and their emplacement has been ascribed to debris flows (Crowell, 1957). Similarly chaotic deposits, viz. *wildflysch*, have been attributed to debris flows. Olisthostromes have also been interpreted as emplaced by rock fall, sliding, and associated debris flows (Görler and Reutter, 1968; Conaghan, Mountjoy, Edgecombe, Talent and Owen, 1976). The term olisthostrome was originally given to a wide spectrum of redeposited clastics, but has since been redefined to apply to chaotic deposits emplaced by debris flows and related mass gravity processes which are composed of extraformational material or which contain exotic clasts which are older than the enclosing sedimentary sequence (Fig. 12.12)

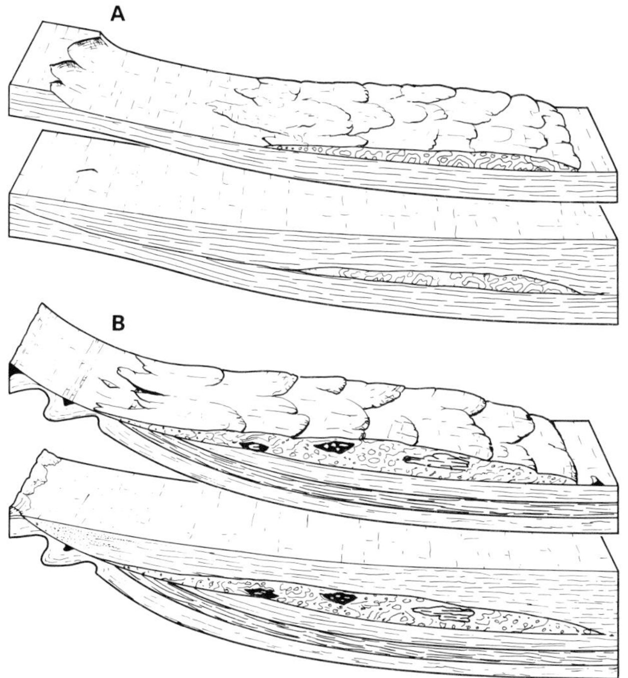

Fig. 12.12. The difference between (A) slumping (sediments derived from the same formation) and (B) olisthostrome emplacement (sediments derived from extra-formational, older sources) (after Elter and Trevisan, 1973).

(Abbate, Bortolotti and Passerini, 1970). The clasts, which may be of gigantic dimensions, are called *olistholiths*.

Olisthostromes are normally taken to indicate tectonically active phases of ancient deep-sea basins. Their lithology may be similar to that of chaotic occurrences produced by either tectonic shearing during subduction in deep-sea trenches or during gravity gliding of nappes. Such tectonically produced chaotic deposits have been called *mélanges* (Hsü, 1974).

12.2.4 **Turbidity currents: high-density**

EXPERIMENTS AND THEORY

Kuenen and Migliorini (1950) visualized turbidity currents as high-density suspensions of sand and mud with specific gravities of up to 1.5–2.0. When the flow decelerates and fluid turbulence diminishes, a turbidy current becomes overloaded causing the coarsest grains to settle out. This reduces the density of the current, further deceleration results, and the next coarsest fraction settles out. In the manner of this simplified model, vertically and horizontally graded beds are produced. Later theoretical and experimental studies, especially those by Middleton (1966), have increased our knowledge of the complex fluid dynamic behaviour of turbidity currents (see also Kuenen, 1966; review by Middleton, 1970).

In the flume experiments by Kuenen and Middleton, turbidity currents were produced from prefabricated suspensions released from a bucket or a lock. These experiments have therefore not elucidated the genesis of the suspensions. It is widely held that turbulent sediment suspensions form from slides (Morgenstern, 1967), but other types of mass-gravity transport such as debris flows may be precursors to turbidity currents (Fig. 12.5) (Hampton, 1972). Van der Knaap and Eijpe (1968) produced experimental turbidity currents starting with a clay layer draped over an inclined core of sand. Shock applied to the tank liquefied the sand and gravitational sliding developed along the plane of permeability contrast between clay and sand. The sliding clay layer mixed with water and developed into a (low-density) turbidity current. It has been argued that turbulence and sediment suspension may be increased by a *hydraulic jump* (Van Andel and Komar, 1969; Komar, 1971). This jump may take place at an initial decrease in slope, e.g. at the mouth of a submarine canyon, where the gravitational energy of a fast, dense sediment gravity flow may in part be converted leading to an increase in the thickness and the turbulence of the flow and a decrease in its density by water entrainment.

Two modes of fluid flow are recognized, viz. laminar and turbulent. *Laminar flow* is characterized by smooth, linear streamlines. In *turbulent flow* the streamlines are irregular and convoluted, and some fluid motion takes place perpendicular to the primary flow direction. Laminar flow changes into turbulent flow when the product of the flow velocity and the depth of flow surpasses a critical value for a given fluid viscosity.

In turbidity currents the sediment support mechanism which keeps the sediment particles in suspension is provided primarily by the upward component of fluid turbulence, which is mainly sustained by friction at the boundary between the flow and both the floor and the ambient fluid. It has been argued that turbidity flow can be sustained in the form of *auto-suspension* (Bagnold, 1962; review by Middleton, 1970). Auto-suspension is a state of dynamic equilibrium in which (1) the excess density of the suspended sediment propels the flow, (2) the flow generates friction and fluid turbulence, and (3) the turbulence keeps the sediment particles in suspension, etc., i.e. a complete feed-back loop. All that is needed to keep the loop intact is that the loss of energy by friction be compensated for by a gain in gravitational energy as the flow travels downslope. In this theoretical model it is possible for a turbidity current to travel over long distances without appreciable erosion or deposition as long as the slope remains constant (see also Middleton and Hampton, 1976).

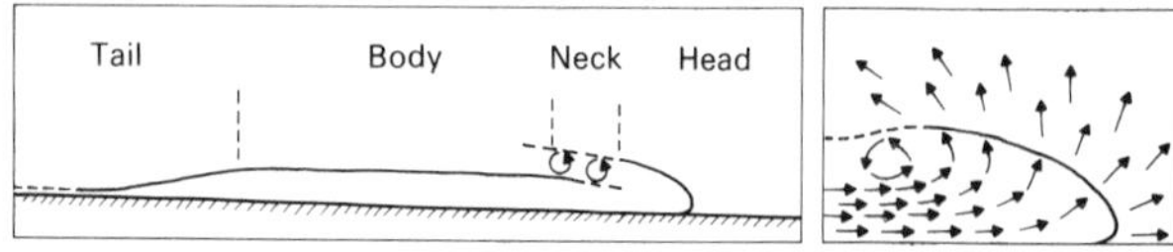

Fig. 12.13. Longitudinal anatomy of a turbidity current, with the flow pattern in and around the head (from Middleton and Hampton, 1976).

Experiments have shown that turbidity currents develop a characteristic longitudinal anatomy of head, neck, body and tail (Fig. 12.13) (Middleton, 1966; Middleton and Hampton, 1976). The head of a turbidity current has a characteristic shape and flow pattern. In plan view, the head appears lobate with local divergences of flow direction (Allen, 1971b). Inside the head a forward and upward sweeping, circulatory flow pattern exists. The coarsest grains tend to become concentrated in the head. The body is the part behind the head where the flow is almost uniform in thickness. Deposition may take place from the body while the head still acts erosively. The *tail* is the part where the flow thins rapidly and becomes very dilute. Mixing between the flow and the ambient fluid produces a dilute entrained layer. Komar (1972) has calculated that on slopes greater than 0.0022 the head is thicker than the body, whereas on lesser slopes the body is thicker than the head. This is important for the type of sediment overflow in channelized environments. Mixing of the flow with water, loss of sediment by deposition and by flow separation in the *neck* will slacken and eventually stop a turbidity current. Kuenen (1967) has calculated that for an average turbidity current most coarse sediment will be deposited in a time-span of hours, though complete settling of the fine-grained tail may take a week.

TURBIDITY CURRENTS IN NATURE

Turbidity currents in the oceans are several orders of magnitude larger than those produced in laboratory flumes. The extent to which the experimental results can be applied to turbidity currents in nature is therefore somewhat problematic. An attempt to generate a high-density turbidity current on the sea floor off San Diego, California, failed (Buffington, 1961).

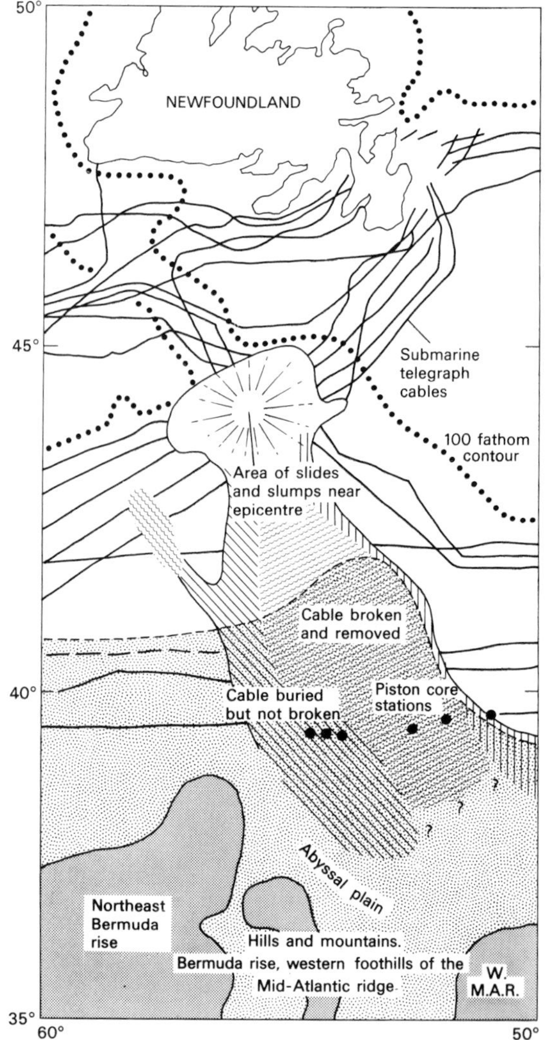

Fig. 12.14. The Grand Banks turbidite of 1929 (after Heezen and Hollister, 1971).

The closest observations of high-density turbidity currents in nature are those on breakages of submarine cables. The classic example is the breakage of telegraph cables off the Grand Banks following an earthquake in the area in 1929. Heezen and Ewing (1952), who studied the case some two decades after the event, regarded the breakage as a true-scale experiment in turbidity current activity.

In the Grand Banks area, off Newfoundland, a thick sequence of sediment was deposited by the St. Lawrence River during Pleistocene times of low sea-level. The earthquake of 1929 triggered slope failure in this sediment pile. Following the earthquake, the submarine telegraph cables, lying down-slope from the epicentre, were broken in sequence from shallow to deep. A seismic and coring study of the region revealed that in the area of the epicentre a slump had been triggered, some 100 km long and wide and some 400 m thick, which had moved downslope over a distance of several hundreds of metres (Fig. 12.14). Along the slump margin several telegraph cables had been broken. A bed of graded silt containing shallow water faunal elements was found emplaced downslope from the area of the epicentre. Heezen and Ewing (1952) interpreted the cable breaks as produced by the slump and an ensuing turbidity current which travelled downslope for hundreds of kilometres on to Sohm Abyssal Plain (Fig. 12.14). Kuenen (1952) calculated that the volume of wet sediment carried by the Grand Banks turbidity current may have been of the order of 100 km^3 spread over an area of some 100,000 km^2. This estimate was confirmed by further oceanographic research in the region (Heezen, Ericson and Ewing, 1954). The cables broken by the slump and by the ensuing turbidity current served as stop watches from which the downslope flow velocity has been calculated (Fig. 12.15). It appears that the Grand Banks turbidity current may have reached a maximum velocity of some 70 km per hour (Menard, 1964).

The volume of wet sediment of the Grand Banks turbidite is exceptionally large. A single, uncompacted turbidite bed in the Adriatic Sea, for example, has a volume of approximately 0.08 km^3 (van Straaten, 1967).

The estimated high velocity of turbidity flow in the deep sea, higher than that of river flow and more like that of strong winds, has met with scepticism. This scepticism was not only generated by geological argument, but also by psychological inertia as it seemed difficult to magine that the deep sea, until recently visualized as the realm of darkness and quiet, would be disturbed by such catastrophic events. Alternative explanations for the submarine cable breaks have been proposed, e.g. progressive spontaneous liquefaction of sediment (Terzaghi, 1956). However, further study of other instances of submarine cable breaks has given support to the turbidity current interpretation (Heezen, 1963). Well documented examples are from the coast of Algeria where the Orléansville earthquake of 1954 triggered the breakage of submarine cables and the

Fig. 12.15. The velocity of the Grand Banks turbidity current, calculated from the cable breaks (compare with Fig. 12.14) (after Heezen and Hollister, 1971).

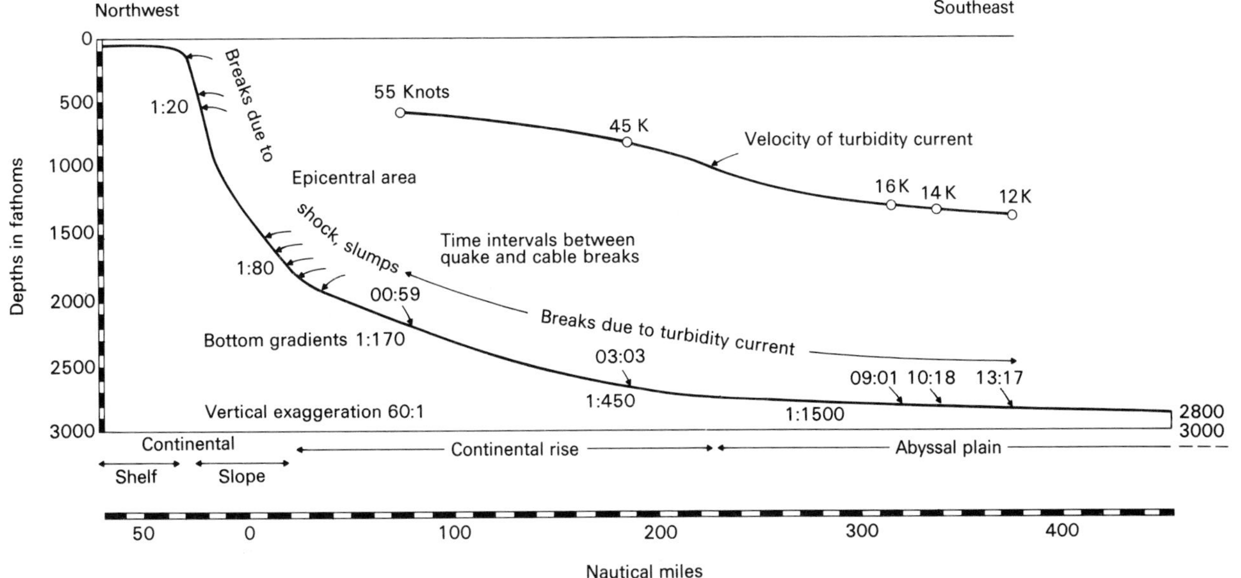

emplacement of a graded sand bed on the Balearic Abyssal Plain (Heezen and Ewing, 1955), from the canyon systems off the mouths of the Congo and Magdalena rivers, and in the Western New Britain Trench (summary by Heezen and Hollister, 1971). Moreover, in these instances velocities of turbidity flow of the order of tens of kilometres per hour have been calculated. Cable breaks and turbidity currents are not only triggered by earthquakes, but also by high river discharge.

The width and thickness of turbidity currents in the oceans and the distances they travel can be deduced from the depositional topography they produce. The natural levees of submarine channels are believed to be deposited by overflow from channelized turbidity currents (see 12.5). A transverse cross-section of a channel must therefore correspond to that of an average turbidity current. The width of deep-sea channels is generally a few kilometres and the height of levees above the channel floor may be up to some 200 m (Komar, 1969; Griggs and Kulm, 1970; Nelson and Kulm, 1973). The length of deep-sea channels, indicative of the distance turbidity currents may travel, can be as much as some 4,000 km (Chough and Hesse, 1976).

The frequency with which turbidity currents are generated and turbidites are emplaced in any particular locality in the deep sea depends on such factors as the nature of the area from where turbidity currents originate, proximity of the area of deposition to the source, seismicity of the source area, and sea-level. River-generated turbidity currents produced during periods of high river discharge (see 12.2.5) may occur as often as once every two years (Heezen and Hollister, 1971). These turbidity currents are generally low density flows. Sands accumulated in the heads of submarine canyons may be flushed out and turbidity currents develop at the same high frequency (Reimnitz, 1971). In proximal environments of deep-sea fans, turbidites may be emplaced approximately once every ten years (Gorsline and Emery, 1959; Nelson, 1976). However, a distal basin plain environment, in any particular locality, receives a turbidite approximately once every thousand years, though this frequency may vary a great deal (Fig. 12.19) (Gorsline and Emery, 1959; summary in Rupke and Stanley, 1974; Rupke, 1975). Rise of sea-level lowers the frequency of turbidity currents, especially those which are shelf- and slope-generated (Heezen and Hollister, 1964b; Nelson, 1976).

TURBIDITES

In the two decades following the 1950 paper by Kuenen and Migliorini many studies have described and classified structures associated with turbidite beds. One reason for so many of these studies is that several of the structures are indicators of palaeocurrents. Some of the classic studies are doctoral dissertation projects (e.g. ten Haaf, 1959; Bouma, 1962; Enos, 1969). In the early 1960s much of the knowledge of structures in turbidites was summarized in well-illustrated publications (e.g. Dzulynski and Walton, 1965). Many structures have been reproduced in laboratory experiments (Dzulynski and Walton, 1963; Kuenen, 1965). Some of the main points of the work on structures and textures in ancient turbidites are briefly dealt with below. (For longitudinal, downcurrent variations in turbidite beds see 12.6.3.)

The characteristic assemblage of structures and textures of a turbidite bed consists of (1) sole structures, (2) vertical grading, (3) a predictable sequence of internal structures, and (4) a relatively high percent of matrix (Fig. 12.11). The structures can be subdivided into sedimentary and biogenic.

Current-produced, or primary, sedimentary structures on the sole of turbidite beds consist mainly of scour marks (e.g. flutes) and tool marks (e.g. grooves). Flute and groove marks have proved particularly useful as indicators of palaeocurrents. The moulds of these structures are eroded by turbidity currents in what usually is a surface of hemi-pelagic mud, or mud deposited by the tail of the previous turbidity current. Shortly after being eroded, the moulds are cast usually with sand carried by the turbidity current which had earlier eroded the moulds.

Deposition of a relatively dense sand bed over a less dense mud bed produces a reverse density gradient, and post-depositional, or secondary, load structures may form (Anketell, Cegla and Dzulynski, 1970).

One of the most significant results of the study of structures in turbidites has been the *Bouma sequence* which recognizes that the various internal sedimentary structures of turbidites occur in a predictable vertical sequence (Bouma, 1962). A

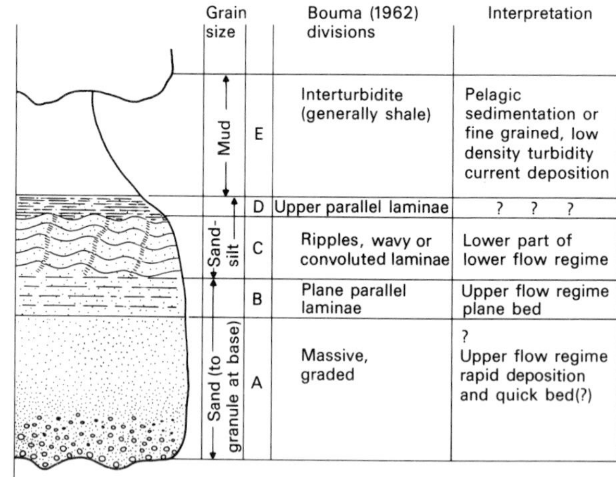

Fig. 12.16. The Bouma sequence of structural divisions in a turbidite bed and its flow regime interpretation (from Middleton and Hampton, 1976).

Thickness of beds (Cm)	Number of beds in section A	B	C	Post-depositional burrows: Granularia			Fucusopsis			Phycosiphon			Scolicia			Neonereites		
0 – 1	21	–	3	6	–	–	–	–	–	4	–	1	10	–	2	3	–	–
1 – 2	41	12	16	19	3	5	–	–	–	13	2	8	12	7	6	5	1	2
2 – 3	33	14	18	18	7	8	–	–	–	12	1	7	4	5	1	4	2	3
3 – 4	19	17	9	15	9	7	1	–	–	9	4	1	–	4	–	2	1	2
– 5	18	14	7	15	7	4	1	–	–	9	3	4	1	3	1	1	–	1
– 6	17	15	8	10	14	4	–	–	–	4	4	6	1	–	1	1	–	–
– 7	12	11	10	11	11	9	1	–	–	5	1	6	–	–	–			
– 8	10	11	8	7	11	4	–	–	–	6	–	2	1	–	1			
– 9	12	14	6	10	13	4	–	–	–	2	–	2						
–10	12	5	2	12	4	2	1	–	–	6	–	–						
–12	13	17	5	11	16	3	–	–	–	3	1	3						
–14	7	16	4	7	16	4	1	–	–	1	–	–						
–16	7	5	5	6	5	5	–	–	–									
–18	7	6	1	7	6	1	–	1	–									
–20	8	8	–	6	8	–												
–25	14	10	–	12	10	–												
–30	9	10	1	7	9	1												
–35	10	6	–	10	6	–												
–40	4	3	–	4	3	–												
–45	7	7	–	7	7	–												
–50	4	7	–	3	7	–												
–60	8	3	–	8	3	–												
–80	7	5	–	7	5	–												
–100	2	3	–	2	3	–												
–150	12	8	–	12	7	–												
–200	4	7	–	3	7	–												
SUM	655			493			6			130			60			28		

Thickness of beds (Cm)	Pre-depositional burrows: Lorenzinia						Regular Palaeodictyon			Irregular Palaeodictyon			Scolicia			Ceratophycus		
0 – 1	3	–	–	2	–	–	4	–	–	–	–	–	2	–	–	1	–	–
1 – 2	9	–	–	3	–	–	4	1	5	3	1	–	1	–	–	12	1	–
2 – 3	5	–	1	1	–	–	7	2	6	1	2	–	5	2	–	8	–	–
3 – 4	5	2	–	2	–	–	2	3	7	2	–	–	2	1	–	4	3	–
– 5	3	–	–	1	–	–	4	1	6	3	3	–	1	3	–	1	2	–
– 6	4	4	1	1	–	–	1	2	3	2	2	1	4	1	–	2	5	–
– 7	–	–	–	–	–	–	4	1	5	1	1	–	5	2	–	3	–	–
– 8	–	–	–	–	1	–	2	–	6	1	–	–	1	2	–	2	3	–
– 9	1	–	2	–	–	–	1	–	3	1	1	–	4	1	–	3	2	–
–10	2	–	–	1	–	–	–	–	–	2	1	–	2	1	–	–	1	–
–12	5	1	–	–	–	–	6	7	3	2	1	–	4	4	–	1	1	–
–14	–	–	–	1	–	–	2	7	3	–	3	–	1	3	–	–	3	–
–16	2	–	–	1	–	–	2	–	–	2	–	–	3	1	2	1	–	–
–18	2	1	–	1	–	–	–	–	–	–	2	–	3	1	–	2	–	–
–20	2	–	–	1	–	–	4	1	–	–	–	–	2	–	–	–	1	–
–25	5	–	–	1	–	–	–	–	–	2	–	–	5	1	–	3	–	–
–30	3	1	–	–	1	–	–	–	–	–	1	–	3	1	–	1	1	–
–35	–	1	–	1	–	–	–	1	–	–	–	–	2	1	–	–	1	–
–40	1	–	–	–	1	–	1	1	–	1	1	–	2	–	–	–	1	–
–45	–	–	–	–	1	–	–	–	–	–	1	–	–	–	–	–	1	–
–50	1	1	–	–	1	–	–	–	–	–	–	–	1	1	–	–	–	–
–60	1	–	–	2	1	–	1	–	–	2	–	–	3	–	–	3	1	–
–80	–	–	–				–	–	–	–	–	–	1	–	–	–	–	–
–100	–	–	–				–	–	–	–	–	–	–	–	–	–	–	–
–150	2	–	–				–	1	–	–	1	–	2	–	–	–	–	–
–200	1	–	–				1	–	–	1	1	–	–	2	–	1	–	–
SUM	72			24			121			49			89			75		

Fig. 12.17. Occurrence of trace fossils in flysch sandstones of northern Spain. Post-depositional burrows show a relationship between type of trace fossil and bed thickness, whereas no such relationship exists for pre-depositional burrows (from Seilacher, 1962).

complete sequence consists of five successive structural divisions (Fig. 12.16), viz. A (massive or graded), B (parallel-laminated), C (cross-laminated or convoluted), D (parallel-laminated), and E (pelitic). Complete Bouma sequences are relatively rare, but the divisions seldom occur in a reversed order. Often the base of the sequence is absent, BCDE, CDE, and DE. Additional significance of the Bouma sequence was recognized when bed forms were related to flow regime (Harms and Fahnestock, 1965; Walker, 1965; Walton, 1967) and the Bouma sequence was interpreted in terms of waning flow regime of the depositing turbidity current. Divisions A and B reflect upper, and division C lower flow regime conditions, whereas division D is believed to form from suspension when the current is no longer able to develop bed forms (Fig. 12.16).

In some turbidites the C division may show convolute laminations. Their origin has been a source of controversy (Dzulynski and Walton, 1965), but probably involves a combination of fluidization of freshly deposited C division sand and silt, coupled with current drag.

A sequence of structural divisions has also been identified in turbidites in cores from the ocean floor (van Straaten, 1964; review by Walker, 1970). Textures of turbidites have received less attention than their structures, mainly because their study requires fairly laborious thin section petrography. Different types of grading have however been documented, e.g. distribution grading (the entire grain-size distribution shows a vertical trend), coarse-tail grading (only the coarse fraction shows a vertical variation), and reverse grading (see 12.2.6) (Middleton, 1966; Middleton and Hampton, 1976). Grading is best shown by changes in the grain-size distribution in successive samples. It can also be effectively shown by the coarsest single or coarsest few grains per sample (Parkash and Middleton, 1970; Rupke, 1976b).

Fabric studies have shown that elongate particles (sand grains, plant fragments, graptolite remains) are often current-aligned (Colton, 1967). From the base to the top of a turbidite the alignment can increasingly diverge from the orientation of sole marks (Scott, 1967; Parkash and Middleton, 1970). Such divergences may be due to a meandering flow pattern in turbidity currents (Fig. 12.43). Grain imbrication dipping upcurrent also occurs (McBride, 1962).

Many ancient turbidites have a greywacke texture, i.e. contain from approximately 15% to over 40% matrix. The matrix content is inversely proportional to grain size. Turbidites in cores from the ocean floor contain at most approximately 20% matrix, commonly much less (Kuenen, 1966; Moore, 1974), and the question whether the matrix of ancient greywackes is primary or secondary has been referred to as 'the greywacke problem' (Cummins, 1962). It has been argued that matrix might be introduced in porous sand beds during compaction by upward migrating pore water (Emery, 1964). A more likely alternative is that part of the matrix is primary and part produced by diagenic alteration or low-grade metamorphism of unstable mineral grains (Cummins, 1962; Kuenen, 1966).

Turbidites may be both silici- and bioclastic. Carbonate turbidites are well known from modern and ancient deep-sea basins (Meischner, 1964; Rusnak and Nesteroff, 1964; Davies, 1968). The carbonate materials derive from reefs or carbonate banks bordering the turbidite basin and the bioclastic sands are composed of skeletal parts of shallow water fauna. Much less commonly, turbidites may be composed of pelagic carbonates, resedimented from submarine highs into neighbouring basins (van Andel and Komar, 1969). Diagenetic alteration may obscure structures in ancient carbonate turbidites.

Biogenic structures, mostly identified as casts on the soles of turbidites, can be either pre- or post-depositional (ten Haaf, 1959). Seilacher (1962) has shown that post-depositional sole trails are present only at the base of the relatively thin beds, up to a thickness characteristic of each type of trace fossil (Fig. 12.17). This relationship indicates that turbidites are rapidly deposited. Seilacher (1967) also showed that trace fossil communities are bathymetry controlled (see 12.6.1). In addition, turbidites may contain shallow-water faunal elements, whereas shale layers between the turbidites may contain a deep-water benthonic and pelagic fauna (Natland and Kuenen, 1951; Weidmann, 1967).

Fig. 12.18. Sources, avenues and modes of transport of sand (solid arrows) and mud (dotted arrows) in California Borderland Basins (from Moore, 1969).

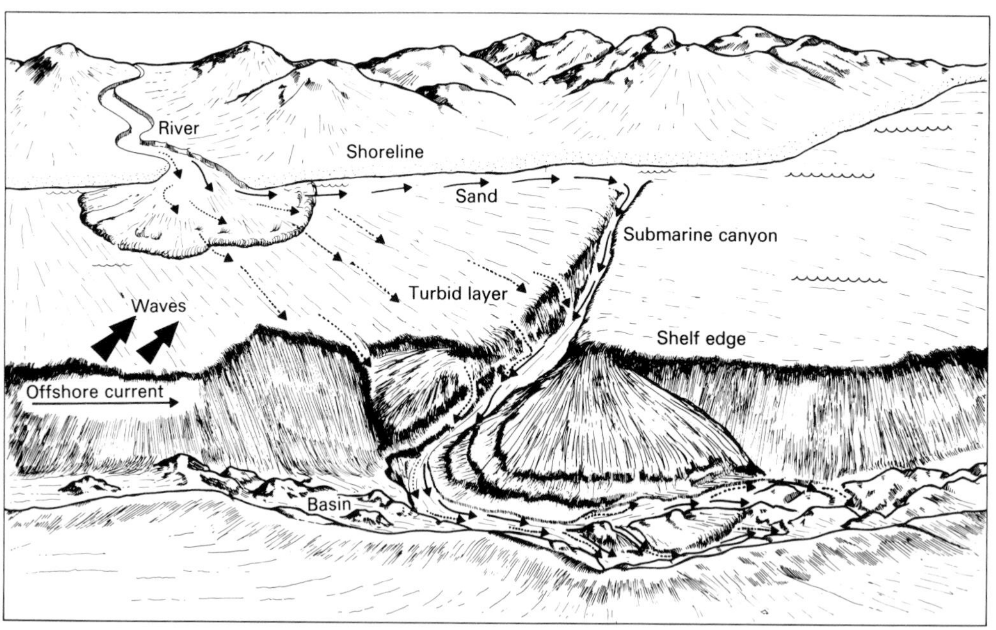

12.2.5 Turbidity currents: low-density

Before 1950 turbidity currents had been regarded as low-density flows generated by storm waves near the shelf edge (see 12.1) (Daly, 1936). For some time after 1950, high-density, high-velocity turbidity currents and the origin of graded sands and sandstones with their great variety of structures preoccupied most workers in deep-sea clastics. Low-density turbidity currents and the origin of mud and shale layers were ignored. However, as the fascination for turbidite sands and sandstones diminished in the late 1960s, renewed interest in low-density turbidity currents emerged. In a classic study of the California Borderland Basins, Moore (1969) argued that the fine-grained sediments in these basins were deposited by low-density, low-velocity flows which he called *turbid layers* (Fig. 12.18).

High-density turbidity currents are spasmodic, whereas low-density ones tend to be relatively slow and long-lived. Their velocities may measure 1.8 km per hour (Shepard, McCoughlin, Marshall and Sullivan, 1977). Low-density flows are generally thought to carry largely clay- and silt-sized particles. They may produce parallel-lamination (Bouma division D) (Rupke and Stanley, 1974). Such low-density turbidity currents can be generated by (1) storm waves on the shelf, stirring up sediment to produce a turbid layer, (2) the infusion of a muddy stream into a lake or into the sea, and (3) the development of a dilute tail to a high-density turbidity current (Fig. 12.18) (Moore, 1969; Rupke and Stanley, 1974). Stream- or river-generated, low-density turbidity currents (category 2) may be seasonally controlled and occur at times of high river discharge.

In the deep waters off the eastern continental margin of North America a semi-permanent bottom cloud of clay-sized sediment in turbulent suspension, called a *nepheloid layer*, may be of importance in deep-sea mud deposition (Ewing and Thorndike, 1965). Light scattering measurements show that at depths below 3,000 m the nepheloid layer exists over much of the north-east Atlantic Ocean floor where it has an average thickness of 1 km (Fig. 12.19) (Eittreim, Ewing and Thorndike, 1969; Eittreim and Ewing, 1972). The mostly terrigenous clay particles form modest concentrations of 0.2 mg/l which may originate from a combination of several current types, e.g. turbidity currents which inject fine sediments into the deep bottom waters, and geostrophic contour currents (see 12.2.7) which may re-suspend fine sediments by bottom erosion (Fig. 12.19). Nepheloid layers have also been identified in other ocean basins (summary by Kelling and Stanley, 1976).

The capacity of a low-density turbidity current to deposit an appreciable mud or shale layer has been questioned. Simply stated, the problem is that the current density required to transport an appreciable quantity of mud would generate a current velocity too high to allow for the deposition of clay-sized particles (Dzulynski, Ksiazkiewicz and Kuenen, 1959). However, transport of cohesive sediment like clay differs from granular transport. In cohesive sediment transport the effective particle size may increase as a result of flocculation, and deposition of clay particles may occur at relatively high velocities by adhesion of particles onto a muddy surface (Hesse, 1975). In addition, ponding and the development of low-density return flows may allow exceptionally thick mud layers to form (Van Andel and Komar, 1969; Scholle, 1971; Rupke, 1976b).

A large number of mud and shale turbidites has been reported from both modern and ancient deep-sea clastic

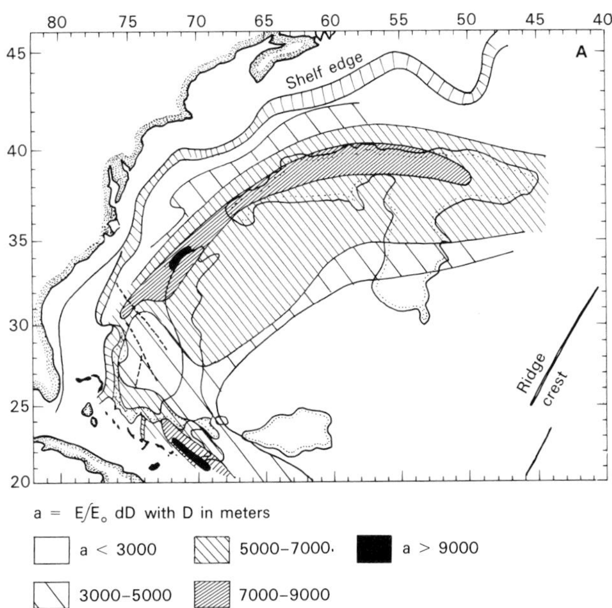

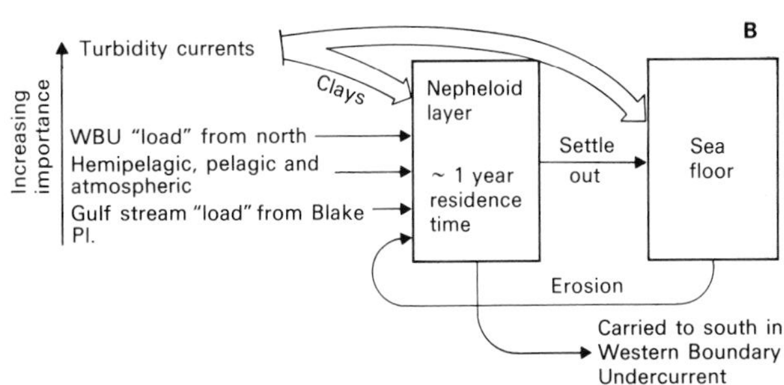

Fig. 12.19. (A) Concentration of suspended matter (a) in the nepheloid layer in the northwestern Atlantic Ocean. (B) Hypothetical input-output scheme for the suspended matter of a nepheloid layer (after Eittreim and Ewing, 1972).

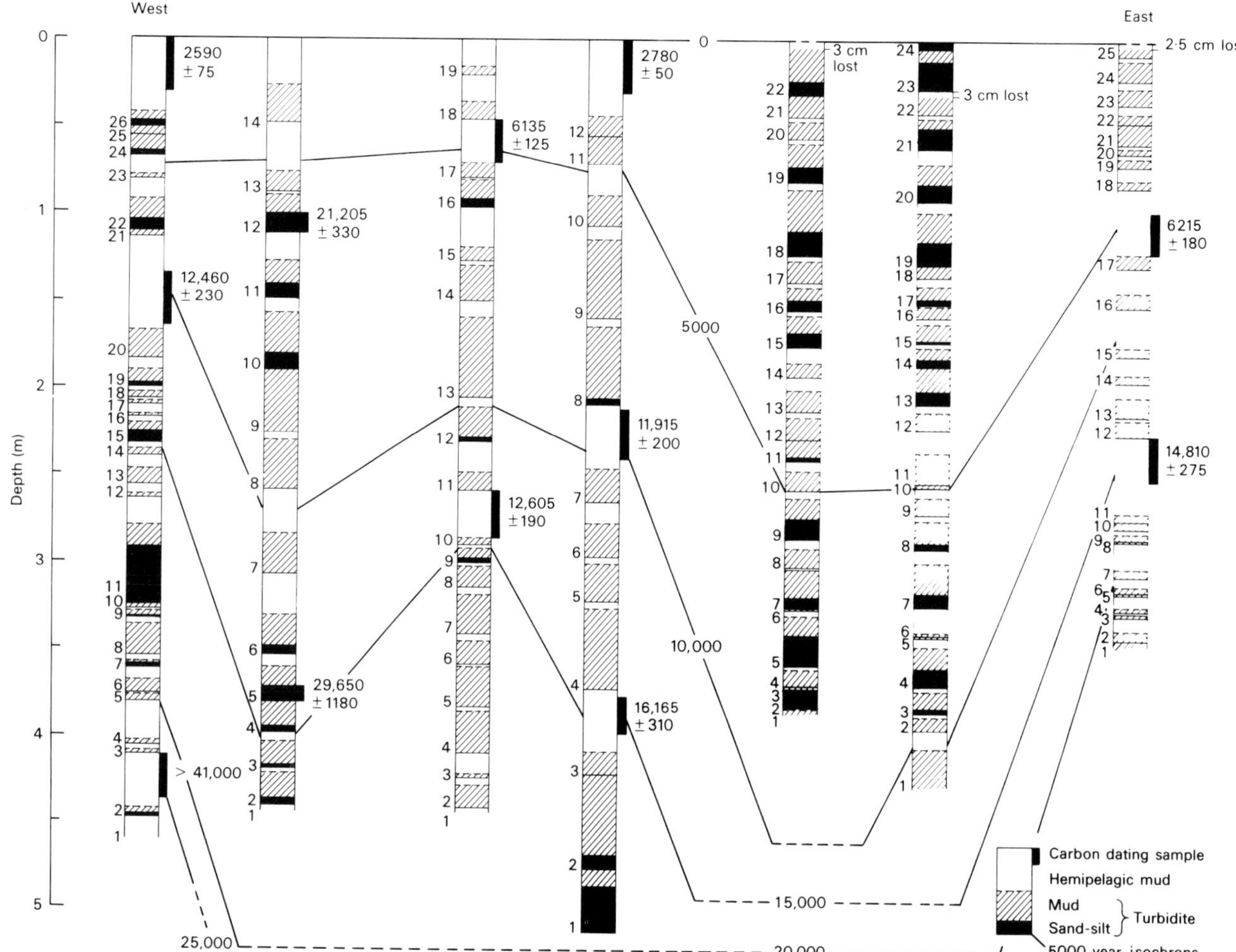

Fig. 12.20. Lithology of cores along a 1,500 km wide traverse across the southern part of the Balearic Abyssal Plain, western Mediterranean Sea. The cores are composed of (1) sand-silt turbidites, (2) silt-mud turbidites and (3) hemi-pelagic mud layers. Turbidites are numbered on left side of cores. Note anomalous C-14 ages of redeposited turbidites in one core which prove resedimentation from older, shallow water clastics (after Rupke, 1975).

successions (review by Rupke and Stanley, 1974). Initially, in the years following 1950, only sands and sandstones were believed to be turbidites. The muds and shales were interpreted as (hemi-) pelagic background sediments. Gradually, evidence showed that part of the fine-grained layers in turbidite successions has a turbidite origin as well, deposited from the dilute tail of the turbidity currents that formed the underlying coarse-grained beds. In a series of cores from the Balearic Abyssal Plain more than half of the sediment thickness consists of turbidite mud (Fig. 12.20) (Rupke and Stanley, 1974; Rupke, 1975). The characteristic properties of the turbidite mud layers are: (1) their lower part may show delicate lamination, (2) they are graded, (3) they are homogeneous and contain virtually no sand-sized particles (Fig. 12.21). In contrast, the hemi-pelagic mud layers are: (1) structureless and bioturbated, (2) they contain sand-sized particles (up to some 15%) which consist largely of skeletal remains of Foraminifera and pteropods, (3) their sorting is poor and their grain size distribution is essentially lognormal (Fig. 12.21) (Rupke, 1975). In a turbidite succession from the Cretaceous flysch of the Eastern Alps some 80% of the fine-grained beds have been attributed to turbidity currents (Hesse, 1975).

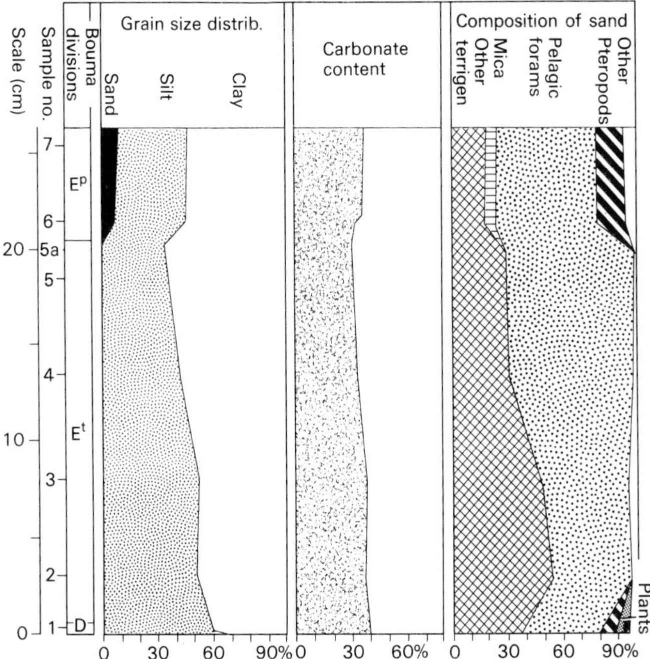

Fig. 12.21. Properties of a mud turbidite composed of a silt lamina (Bouma division D), followed by a turbidite mud layer (division E^t), overlain by a hemi-pelagic mud layer (E^p). Note the textural and compositional grading in the turbidite mud layer. The absolute % of sand in D and E^t is too small to be plotted (after Rupke, 1975).

12.2.6 **Other mass flow processes**

PROBLEMATIC BEDS

When the turbidity current theory was first applied to the study of ancient deep-sea clastic successions, it appeared that certain thick-bedded and coarse-grained beds do not fit the model of a typical turbidite. Kuenen (1958) speculated that the depositional process of these beds combined a slide and a turbidity current. He called such beds *fluxo-turbidites*. Similar, though not always identical, beds have been found by others and variously called fluxo-turbidites (Dzulynski, Ksiazkiewicz and Kuenen, 1959), proximal turbidites (Walker, R.G., 1967), grain flow deposits (Stauffer, 1967), etc. Stauffer first gave a detailed description of the sedimentary features of such beds (Fig. 12.22). Subsequent descriptions have been added (Corbett, 1972; van de Kamp, Harper, Conniff and Morris, 1974; Rupke, 1977).

Several terms have been in use for the depositional process(es) of these problematic beds, e.g. watery slide, inertia flow, grain flow, etc. (reviews by Walker, 1970; Carter, 1975a). A lucid review of these processes is given by Middleton and Hampton (1976) and their terminology is followed here. Collectively, the features of these beds suggest that the depositional process(es) involve(s) two types of grain support mechanism, viz. grain-to-grain interactions (grain flow) and escaping pore fluids (fluidized sediment flow) (Fig. 12.4). A distinction has been made between liquefied flow (when grains settle downwards displacing the fluid upwards) and fluidized flow (when the fluid moves upward) (Lowe, 1976b). The former process may be much more significant as a process of resedimentation than the latter. These conceptually distinct single mechanism flows are likely to operate together in any event of mass gravity transport and closely intertwine with both slumping and turbidity flow (Fig. 12.5) (Middleton and Hampton, 1976; Lowe, 1976).

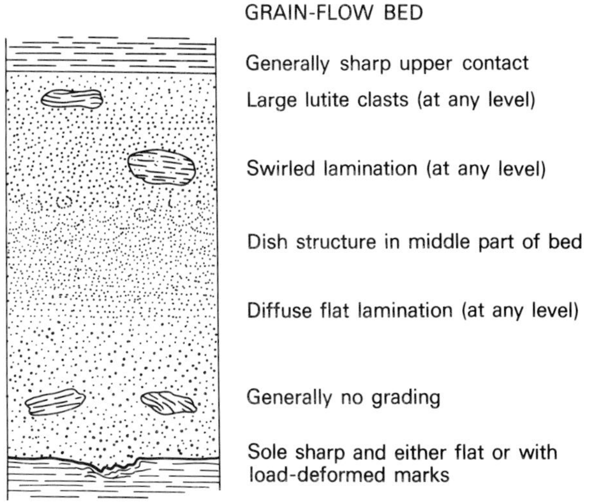

Fig. 12.22. Generalized sequence of structures in a 'grain flow' bed (after Stauffer, 1967).

GRAIN FLOW

The concept of grain flow stems to a large extent from the theoretical and experimental work of Bagnold (1954). In experiments in which he sheared granular, cohesion-less sediment, Bagnold measured a dispersive pressure produced by momentum change when grains interact and bounce off one another. This dispersive pressure can act as a sediment support mechanism allowing a cohesion-less sediment mass to flow. The angle of slope needed for sustained grain flow is high, i.e. some 18° or higher (review by Middleton, 1970). This suggests that pure grain flow is a very localized process in the deep sea. Small-scale sand avalanches in the heads of submarine canyons have been interpreted as grain flows (Shepard and Dill, 1966).

Deposition takes place by the freezing or mass emplacement of a grain layer several grains thick. Depositional features are (Fig. 12.11): massive bedding, sharp base and top, reverse grading, and the absence of tractional structures (Middleton and Hampton, 1976). Reverse grading may be produced by a kinetic sieve effect, i.e. during flow small grains fall downward in between large grains and gradually displace the large grains upward (Middleton, 1970). Other mechanisms to produce reverse grading, such as Bernoulli's principle, have been proposed (Fisher and Mattinson, 1968).

FLUIDIZED SEDIMENT FLOW

Fluidization of silt and sand involves the collapse of a metastable fabric. The grains no longer form a supportive framework, but are in part supported by the pore fluid. The grains become suspended and the sediment strength is reduced to zero. Loosely packed silt and sand are especially susceptible to fluidization. Gravel is usually highly permeable, fluid escapes readily, and the weight of the clasts offers a relatively strong resistance to lift. In clay the cohesive forces resist fluidization (Middleton and Hampton, 1976).

Fluidized sand behaves like a fluid of high viscosity. It can flow rapidly down a relatively gentle slope of 2° or 3°. Excess pore fluid pressures which keep the grains suspended may dissipate quickly (from several minutes to a few hours). The dissipation time is a function of bed thickness and grain-size (Middleton, 1969). Deposition occurs by a gradual freezing from bottom to top and little grain segregation occurs. The features commonly produced are (Fig. 12.11): poor grading, sharp bottom and top, no tractional structures (though diffuse lamination may develop), basal load casts, fluid escape structures such as dish and pillar structures, sand volcanoes and convolute laminations (Lowe, 1975b; Middleton and Hampton, 1976).

TRANSPORT OF GRAVEL

Descriptive sequences, akin to Bouma sequences in sandy turbidites, have been proposed for resedimented gravel and conglomerate beds (Davies and Walker, 1974; Walker, 1975, 1977). The features used to characterize the sequences are: the presence or absence of stratification, the type of grading (normal, reverse, or grading absent), and the presence or absence of a preferred clast fabric. A preferred clast fabric involves an upcurrent imbrication of clasts and a preferred orientation, either with long axes parallel to the current direction or, less frequently, with long axes perpendicular to the current direction (Fig. 12.23). In the latter instance, a phase of bed load traction occurred immediately prior to deposition (Rupke, 1977).

Fig. 12.23. Organized conglomerate from the Upper Carboniferous flysch of the Cantabrian Mountains, northern Spain, showing imbrication associated with reverse grading.

Walker (1975, 1977) proposes four descriptive sequences (Fig. 12.49). (1) The *inverse-to-normally graded* sequence which is characterized by both inverse and normal grading and by a preferred clast fabric. Stratification is absent. This type of bed is believed to have been deposited by a sediment gravity flow with both fluid turbulence and grain-to-grain interaction as clast support mechanisms (Davies and Walker, 1974). (2) The *graded* sequence in which inverse grading is rare or poorly developed and stratification is absent. In thinner beds of smaller clast size (pebble and granule grade) the graded-bed sequence may pass into (3) the *graded-stratified* sequence, characterized by stratification, normal grading and a preferred clast fabric. This type is believed to have been deposited from suspension, but with bed load traction becoming increasingly important in the upper part of the bed. (4) The fourth sequence is the *disorganized* bed type which differs from debris flow deposits in that the framework is clast supported (Fig. 12.49).

12.2.7 Thermo-haline ocean bottom currents

In the 1960s, after a decade of expanding support for the turbidity current theory, scattered resistance broke out against its uncritical application. Indigenous ocean bottom currents were proposed as an alternative process to transport terrigenous clastics from the continental shelf through submarine canyons onto the ocean floor (e.g. Hubert, 1964). In defence of turbidity currents, Kuenen (1967) presented much of

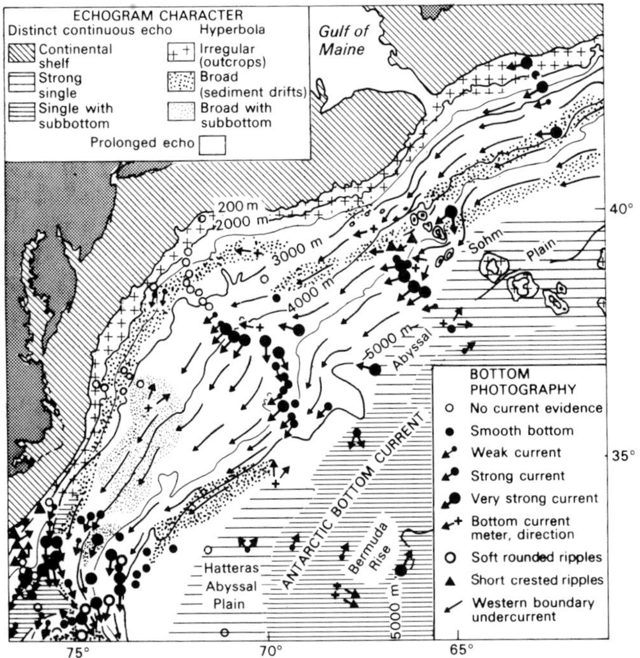

Fig. 12.24. Direction and strength of the Western Boundary Undercurrent, a geostrophic contour current which shapes the continental rise prism off western North America (after Heezen and Hollister, 1971).

the available evidence for turbidity current emplacement of deep-sea sands and sandstones. Nevertheless, evidence has accumulated through the 1960s for the existence of ocean bottom currents capable of transporting mud, silt, and sand, especially on the continental rise (Heezen and Hollister, 1964a; Heezen, Hollister and Ruddiman, 1966). The main evidence comes from bottom photographs and seismic records showing bed forms produced by prolonged shifting of bed load by traction (reviews by Heezen and Hollister, 1971; Bouma and Hollister, 1973).

Ocean bottom currents may be generated by winds, tides, or, most voluminously, by thermo-haline density differences. The importance of bottom currents in submarine canyons has been amply demonstrated (Shepard and Dill, 1966; Shepard, 1973b) and large thermo-haline current systems have been measured near the bottom of the deep-sea. In the polar regions the cold temperatures produce dense, cold and relatively saline waters which sink and move as underflows to equatorial regions. As a result of the Coriolis effect, the underflows are deflected to the west and may form *geostrophic contour following currents*. A well known example is the Western Boundary Undercurrent of Arctic water which has hugged the continental rise of eastern North America since the Early Tertiary (Fig. 12.24) (Heezen and Hollister, 1971; summary by Kelling and Stanley, 1976).

Contour currents are permanent, relatively slow moving fluid flows with variable velocities which may reach values of 2 km per hour. They produce a suite of bed forms ranging from small current ripples to large sediment waves or dunes with amplitudes of tens of metres and wavelengths of thousands of metres. Contour currents may transport and redeposit mud, silt and sand. The deposits are called *contourites*. Their features are distinct from those of normal turbidites though they may considerably overlap with those of thin-bedded turbidites (Heezen and Hollister, 1964a; Bouma, 1972; Bouma and Hollister, 1973), viz. (1) relatively thin-bedded, normally less than 5 cm thick, (2) sharp bottom and top, (3) ubiquitously laminated, the cross-laminations being placered by heavy minerals, (4) relatively fine-grained, normally silty or muddy, (5) well sorted with a low matrix percent of some 0–5%, (6) normal or reverse graded, and (7) with a dispersal pattern parallel to the basin margin. Contourites are normally not encountered in cores from abyssal plains, but occur commonly in those from continental rise environments. Contourites have been reported from ancient deep-sea clastic successions (Bouma, 1972; Bein and Weiler, 1976).

12.3 DISCOVERY OF DEEP-SEA CLASTIC ENVIRONMENTS

The deep-sea floors off the east and the west coast of North America differ in significant ways. As a result, oceanographic research in east and west coast institutions through the 50s and 60s led to two separate traditions in deep-sea clastic research, referred to here as the *Atlantic* and *Pacific tradition*. In the Atlantic tradition abyssal plains were considered to be the main deep-sea environment of clastic deposition. In the Pacific tradition emphasis was on submarine canyons, valleys, channels and deep-sea fans.

12.3.1 The Atlantic tradition: abyssal plains

The Atlantic model of deep-sea clastic sedimentation originated with the discovery in 1947 that vast regions of the North Atlantic Ocean floor are flat, nearly level plains (Tolstoy and Ewing, 1949). At about the same time, the theory of high-density turbidity currents was proposed (see 12.1). The Hudson Submarine Canyon was found to end in a flat, broad topographic feature where sediment cores were found to contain graded sand beds with shallow water faunal elements. This combination of flat topography and resedimented sands suggested that abyssal plains are constructed by turbidity currents (Ericson, Ewing and Heezen, 1951; Heezen, Ewing and Ericson, 1951). This theory was supported by the fact that many abyssal hills and seamounts protrude abruptly from these plains and appear to be partly buried (Ericson, Ewing and Heezen, 1952). Mapping of other abyssal plains in the Atlantic and other ocean basins confirmed the theory (Heezen, Tharp and Ewing, 1959).

The Atlantic tradition culminated with the first leg of the JOIDES expedition in 1968 off the Atlantic and Gulf coasts of North America. It was demonstrated that reflective fill below abyssal plains from as far back as the Eocene consists of terrigenous sands of turbidite origin (Burk, Ewing, Worzel, Beall, Berggren, Bukry, Fischer and Pessagno, Jr., 1969; Beall and Fischer, 1969).

12.3.2 The Pacific tradition: channels and fans

The Pacific model of deep-sea clastic sediments grew to a large extent out of the study of submarine canyons. Bathymetric work off the west coast of North America revealed the existence of channels with natural levees on the sea floor away from the mouths of submarine canyons. Initially, the leveed deep-sea channels were interpreted as subaerial channels formed during times of low sea-level (Shepard, 1951). As channels were found at great depths turbidity currents were invoked to produce the leveed deep-sea channels, removing the need to lower sea-level to such depths (Menard and Ludwick, 1951). Further bathymetric work off the Monterey Submarine Canyon revealed the existence of a delta-like fan with leveed channels. Based primarily on the association of a fan with channels, it was argued that the deep-sea fan was deposited by turbidity currents (Dill, Dietz and Stewart, 1954).

In an early classic on deep-sea sedimentation, Menard (1955) related the presence and absence of abyssal plains in the north-east Pacific Ocean to the respective absence or presence of sediment traps and barriers which prevent turbidity currents from reaching a particular region of the deep-sea floor. In addition, Menard described deep-sea fans at the mouths of many submarine canyons off the west coast of North America. He noticed that many of the associated deep-sea channels hook leftward (looking down-slope), a feature he attributed to the Coriolis effect (see 12.5). In a later paper, Menard (1960b) presented the first block diagram of deep-sea fans in a study of the Monterey and Delgada fans off central California.

In spite of the early, well documented deep-sea fan studies, channels and fans were effectively ignored as a model of deep-sea clastic deposition through the 50s and 60s, and the Atlantic abyssal plain model dominated thinking about deep-sea clastic sediments, even in Menard's own book on the marine geology of the Pacific (1964). The Pacific deep-sea fan model began to attract attention only after the culmination of the Atlantic tradition around 1970. At this time, a number of studies of west coast fans appeared, viz. of the La Jolla Fan (Shepard, Dill and von Rad, 1969), of the San Lucas Fan (Normark, 1970), of the Astoria Fan (Nelson, Carlson, Byrne and Alpha, 1970) and of the Redondo Fan (Haner, 1971).

The Atlantic model was not only based on the discovery of abyssal plains. It was also influenced by experiments on turbidity currents as sheet flows (Kuenen and Menard, 1952), and by the example of the Grand Banks turbidity current which was triggered on the upper continental slope (Figs. 12.14; 12.15). The Pacific tradition, influenced by the discovery of deep-sea channels, viewed turbidity currents primarily as channelized flows, triggered in the heads of submarine canyons. Since about 1970, much of deep-sea clastic research has been pre-occupied with the fan model and with channelized flow of turbidity currents. However, a balanced, comprehensive theory of deep-sea clastic sedimentation must consider both fans and abyssal plains as depositional environments.

12.4 MODERN BASIN PLAINS

12.4.1 General characteristics and classification

Abyssal plains are flat regions of basin floors with gradients of less than 1 : 1,000 (Heezen and Laughton, 1963). However, since the depth of basins with such flat floors is not always abyssal, the more general term of *basin plain* (or turbidite plain) is preferable.

Basin plains function as the ultimate trap for sediments eroded from subaerial or submarine highs, the most extensive basin plains being located seaward of the major drainage basins of the world (Fig. 12.25). Seismic records of basin plains show rugged basement topography smoothed by horizontally stratified turbidite fill which has buried pre-existing topographic irregularities of the basin floor (Fig. 12.26). A single basin plain may be fed by several sources, e.g. submarine canyons, deep-sea channels, or from the surrounding basin slopes (Heezen and Laughton, 1963; Horn, Ewing and Ewing, 1972).

The maximum size of basin plains may be several thousands of kilometres in length and hundreds of kilometres in width (Fig. 12.25). Many are elongate, although equidimensional and irregular shapes also occur. Sediment thicknesses below basin plains commonly are a few hundred metres, though in some cases thicknesses of 1,500–2,000 m have been recorded (Collette, Ewing, Lagaay and Truchan, 1969; Horn, Ewing and Ewing, 1972).

The lithofacies of basin plains generally lacks abrupt lateral or vertical changes. The three main sediment types of basin plain sequences are (Fig. 12.20): (1) turbidite sand or silt beds, (2) turbidite mud beds, and (3) hemi-pelagic muds (Rupke and Stanley, 1974; Rupke, 1975), though most authors do not distinguish between the two mud types. The turbidite sand-silt beds range in thickness from a few millimetres to several metres. Individual beds can be very extensive and have been

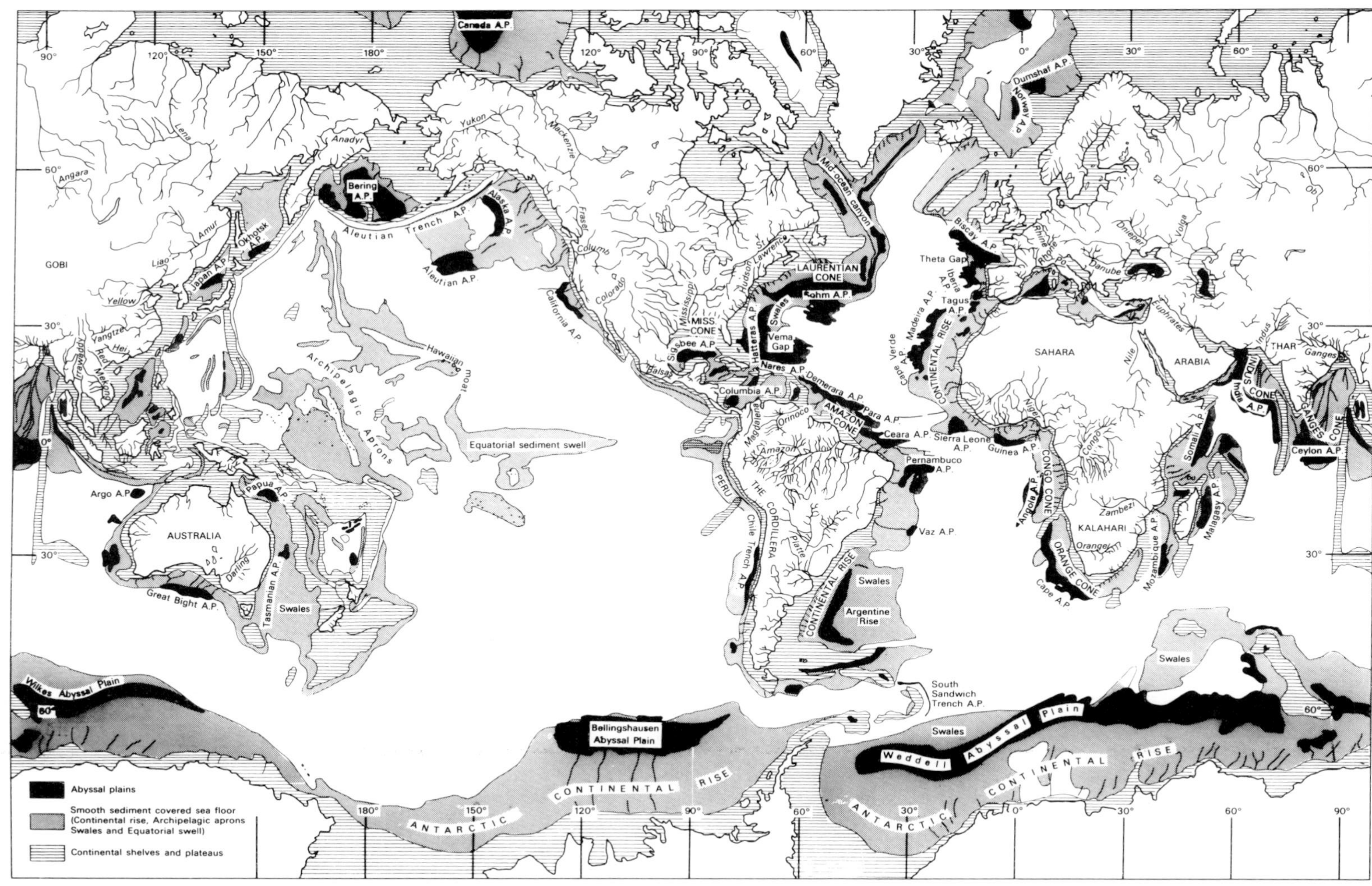

Fig. 12.25. World distribution of major deep-sea basin plains and abyssal cones (from Heezen and Hollister, 1971).

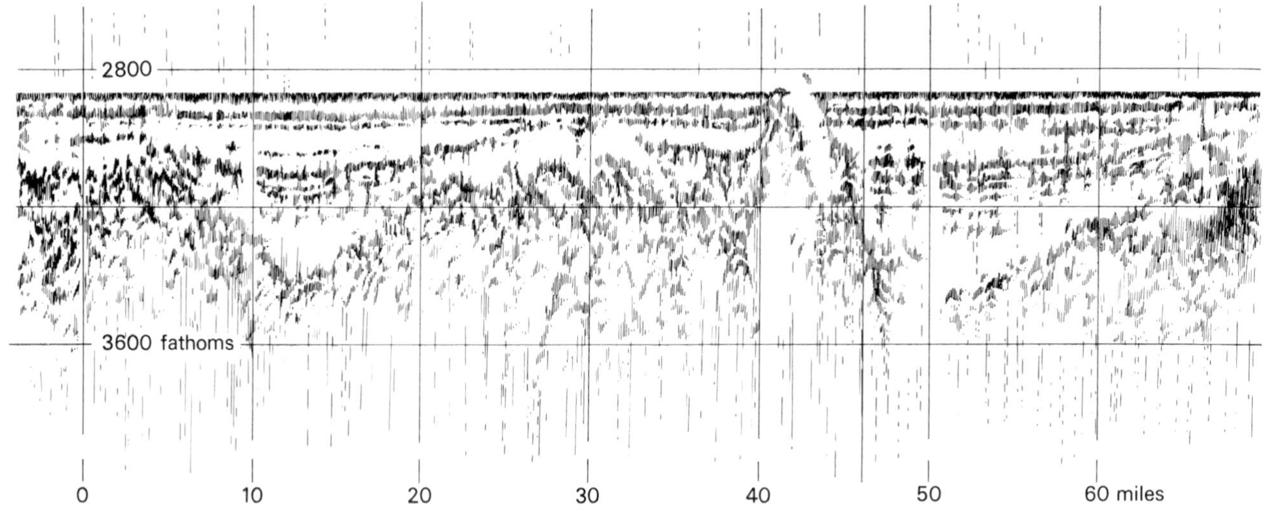

Fig. 12.26. Seismic reflection profile across part of the Madeira Abyssal Plain, showing reflective turbidite fill which has smoothed over basement topography (after Heezen and Hollister, 1971).

correlated over distances of the order of 100–200 km (Conolly and Ewing, 1967; Bennets and Pilkey, 1976). Poorly graded and graded sand beds of medium thickness are deposited near the mouths of submarine canyons and other sediment sources. Locally, gravel has been found. Well graded coarse silt beds of medium thickness are deposited along the main axes of sediment dispersal across the basin plains. At the downcurrent margin of basin plains thin, graded, medium to fine silt beds are deposited. The predominance of a particular sediment source may determine the overall gradient and dispersal pattern (Horn, Ewing and Ewing, 1972).

Basin plain sediments are little if at all affected by synsedimentary deformation, sliding and slumping being largely restricted to basin slope and rise environments (Horn, Ewing and Ewing, 1972). However, sediments are disturbed by organisms (bioturbation) (Heezen and Hollister, 1971), by differential compaction over basement irregularities (Collette, Ewing, Lagaay and Truchan, 1969), or by salt diapirism (see 12.4.4).

The geometry of basin plain successions and their internal facies organization are affected by several factors, such as (1) *in space*, the physiography of the depositional basin, specifically the basin floor topography and the sediment sources, and (2) *in time*, sea level fluctuations and the tectonic history of both the depositional and drainage basins. Features such as shape of the sediment bodies, dispersal pattern of turbidites, composition of clastics, etc. vary with basin physiography. Basin plains are therefore classified here using the criterion of basin physiography, closely following Heezen and Laughton (1963): (1) open oceans, (2) deep-sea trenches, (3) enclosed seas, (4) borderland basins, (5) miscellaneous categories, e.g. oceanic ridge basins and lakes. Other clastic basin classifications have used the criteria of composition (terrigenous versus bioclastic) (Rusnak and Nesteroff, 1964), of geometry (open versus enclosed) (Nelson and Kulm, 1973), and of depth (above or below carbonate compensation depth) (Hesse, 1975).

Eustatic sea level fluctuations may affect both the composition and the frequency of turbidity currents in the deep-sea. The glacio-eustatic rise in sea-level towards the close of the Pleistocene caused a decrease in grain-size, thickness and frequency of turbidity currents in most deep-sea basins. Large parts of basin plains have not received terrigenous clastics during the Holocene and generally turbidity current frequency has decreased by about an order of magnitude (Heezen and Hollister, 1964b; Horn, Ewing and Ewing, 1972). A greater abundance of unweathered feldspars has been documented for arkosic turbidites deposited during low sea-level phases (Damuth and Fairbridge, 1970).

The tectonic nature and history of the basin may affect the composition of the clastics (Horn, Delach and Horn, 1969), the degree of tectonic and sedimentary preservation of the sedimentary succession, and the nature of the basin fill sequence (see Chapter 14).

12.4.2 Open oceans

Basin plains in open oceans are most extensively developed in the North and South Atlantic Ocean (Fig. 12.25). In most instances, their shape is elongate with the long axis of the basin plain parallel to the continental margin. On the landward side the plains are bordered by the continental rise whence the terrigenous sediments are derived (Fig. 12.27). On the seaward side the plains encroach upon abyssal hill provinces. Average rates of sedimentation are of the order of a few centimetres per thousand years. The flat surface of the plains may be interrupted by protruding abyssal hills or seamounts (Fig. 12.26) (Heezen, Tharp and Ewing, 1959; Heezen and Laughton, 1963). Based on the criterion of source, one can distinguish: (1) primary basin plains which receive terrigenous sediment directly from a marginal source, (2) secondary basin plains which are located seaward of primary ones, are surrounded on all sides by abyssal hills, and are fed through an abyssal gap by turbidity current overflow from a primary plain, and (3) relict basin plains which are cut off from any terrigenous sediment source. Categories (2) and (3) are comparatively rare (Fig. 12.25).

A typical *primary basin plain* is the Hatteras Abyssal Plain which has served as a model for ancient deep-sea clastic successions (Fig. 12.27) (Horn, Ewing and Ewing, 1972). Hatteras Abyssal Plain is elongate, having a length of approximately 1,000 km and an average width of approximately 200 km. The maximum thickness of the sedimentary fill is some 350 m. The main sources of terrigenous sediment are at the northern end. Gradient and main sediment dispersal are from north to south, and thickness of the sedimentary fill, individual bed thickness and grain-size decrease in downcurrent and across-current direction. The Hatteras Abyssal Plain is a good example of an elongate basin, its long axis parallel to the continental margin, its main sediment supply at one end, and with a distinct longitudinal transition from proximal to distal facies (Horn, Ewing and Ewing, 1972).

An example of a *secondary basin plain* is Nares Abyssal Plain which is fed through the Vema Gap at the downcurrent end of Hatteras Abyssal Plain (Fig. 12.25) (Heezen and Laughton, 1963; Horn, Ewing and Ewing, 1972). The secondary plains slope away from the abyssal gap, through which terrigenous sediment is supplied by turbidity current overflow, after the turbidity currents have traversed a primary plain for distances up to 1,000 km or more. As a result, the turbidite fill of secondary plains consists mostly of graded silts and muds.

An example of a *relict basin plain* is the Aleutian Abyssal Plain in the north-east Pacific (Fig. 12.25) (Hamilton, 1967; Horn, Ewing and Ewing, 1972). The plain has no connection at present with a source of terrigenous sediment, and the turbidite succession is overlain by some 100 m of pelagic sediment. The construction of the Aleutian Abyssal Plain took place before the Aleutian Trench developed trapping terrigenous sediment.

The abundance of basin plains in the Atlantic, and their scarcity along the perimeter of the Pacific (Fig. 12.25) is due to two main features. (1) The Atlantic continental margins are predominantly passive margins, towards which many large drainage basins of the world empty (Inman and Nordstrom, 1971). (2) The Pacific continental margins are predominantly active margins, characterized by arcs and trenches which act as barriers and traps to terrigenous sediments.

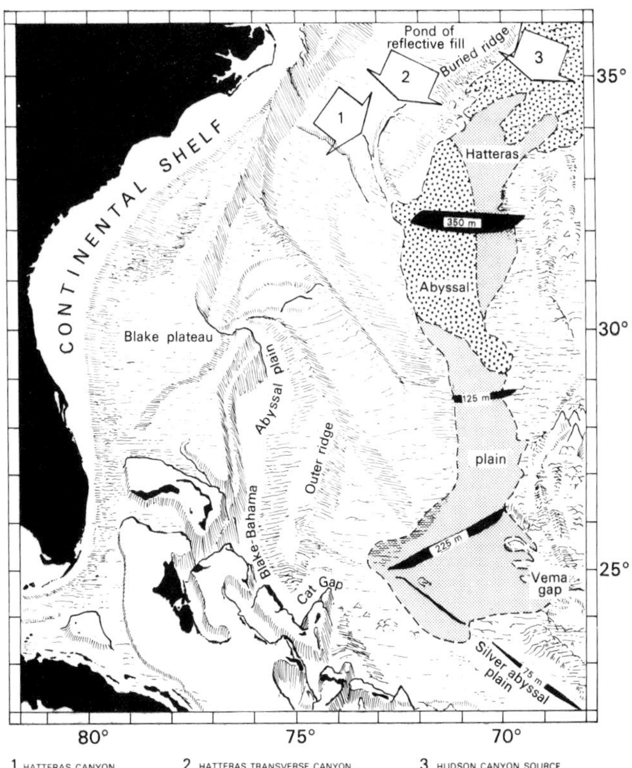

Fig. 12.27. Distribution of turbidites on Hatteras Abyssal Plain. Heavy stipples: areas of sand-silt deposition; light stipples: areas of silt deposition (after Horn, Ewing and Ewing, 1972).

12.4.3 **Deep-sea trenches**

The floors of deep-sea trenches may consist wholly or in part of basin plains. The first trench basin plain was discovered in the Puerto Rico Trench (Ewing and Heezen, 1955; Conolly and Ewing, 1967). Similar plains have been found in other deep-sea trenches, in particular those along the eastern and northern margin of the Pacific (Fig. 12.25), viz. the Peru–Chile Trench (Galli-Olivier, 1969; Scholl, Christensen, von Huene and Marlow, 1970), the Middle America Trench (Ross, 1971), and the Aleutian Trench (von Huene and Shor, 1969; Piper, von Huene and Duncan, 1973). Most recently, as part of the JOIDES programme (e.g. leg 31), trenches in the western Pacific have received attention (Moore and Karig, 1976) with

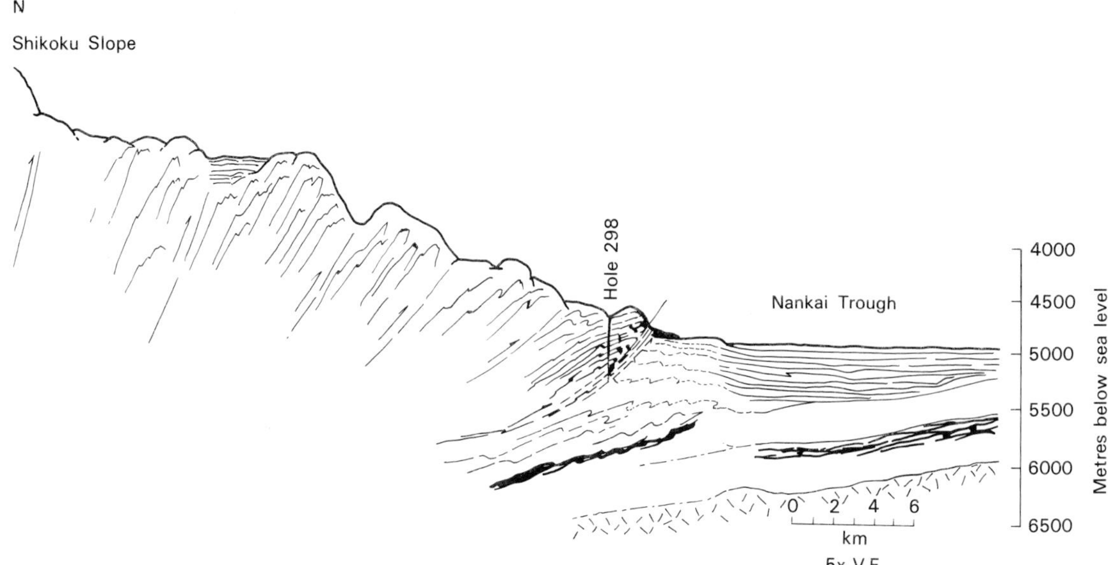

Fig. 12.28. Transverse cross-section through the lower, inner wall and floor of the Nankai Trench, north-eastern Pacific. Note perched basin (after Ingle, Karig *et al.,* 1973).

the interest particularly directed towards the effect of subduction on the sedimentary fill (see Chapter 14).

Trenches are the deepest ocean basins, elongate, with a steep (an average of 10°) slope toward the land and a more gentle slope (an average of 5°) toward the sea. Seismic and volcanic activity mark the landward side of the trench and both the land- and seaward flanks are faulted. As a result, perched basins exist on the slopes which may act as local sediment traps (Fig. 12.28) (Moore and Karig, 1976). Tectonic activity has also broken the trench floor into compartments.

As a result of the compartmentation, parts of deep-sea trenches may be sediment starved. The basin plains in most trenches are very elongate, up to hundreds of kilometres in length and at most only a few tens of kilometres wide (Figs. 12.25; 12.28; 12.29). The thickness of the sedimentary fill is generally a few hundreds of metres, but a thickness of some 2 km has been recorded. In transverse cross-section the sedimentary fill is wedge-shaped, the thickest side being located on the landward side (Fig. 12.28). The rate of sedimentation in most trenches is high, approximately 1 m per 1,000 years (Piper, von Huene and Duncan, 1973). However, the rate of sedimentation may be much lower where the continental source area has an arid climate (Galli-Olivier, 1969; see also Burk, ed., 1974).

Proximity of source area, high relief of source area and the presence of volcanic activity provide compositionally immature clastics to the trench basin plain. Their dispersal by

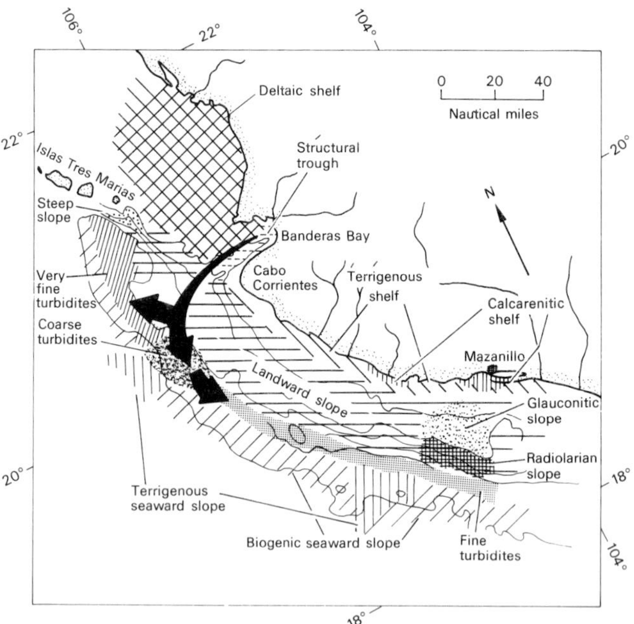

Fig. 12.29. Distribution of sedimentary facies in the northern part of the Middle America Trench. Large arrow shows sediment dispersal (after Ross, 1971).

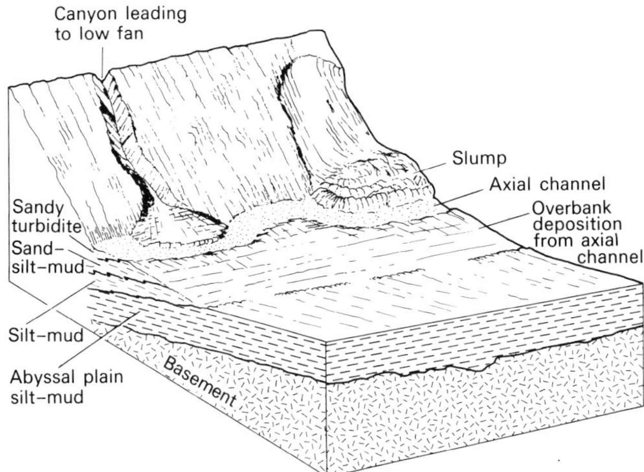

Fig. 12.30. Sedimentation in the Aleutian Trench (from Piper, von Huene and Duncan, 1973).

turbidity currents is predominantly longitudinal (Fig. 12.29). Distinct marginal facies may develop alongside the basin plain. In the Middle America Trench terrigenous silty clay predominates on the landward flank, whereas on the seaward flank biogenic sediment accumulates (Fig. 12.29) (Ross, 1971). On the landward side of the basin plain, parallel to its longitudinal axis, a deep-sea channel may exist. On the Aleutian Abyssal Plain the channel, which has a levee on its seaward side, has a width of 2.5–6 km (Fig. 12.30). Sand is present on the channel floor, whereas silts and muds are deposited on the levee and on the trench floor. No evidence exists for indigenous bottom currents. Tectonic activity and steep inner trench walls have led to the emplacement of slope sediment on to the basin plain by slumping (Fig. 12.30). The slumping direction is generally perpendicular to the predominant longitudinal dispersal of turbidites (Piper, von Huene and Duncan, 1973). Subduction in a trench may result in the stratigraphic superposition of (1) open ocean basin plain sediments, (2) trench basin plain sediments consisting of a fine-grained outer- and a coarse-grained inner-plain facies, and (3) a fine-grained, slumped slope facies, possibly containing olistholiths (Fig. 12.30).

12.4.4 **Enclosed seas: mediterraneans and marginal seas**

Well known examples of basin plains covering the central part of enclosed sea basins are the Sigsbee Abyssal Plain in the Gulf of Mexico (Davies, 1968) and the Balearic Abyssal Plain in the Western Mediterranean Sea (Horn, Ewing and Ewing, 1972; Rupke and Stanley, 1974). Similar basin plains have been recorded in marginal, behind-arc basins, e.g. the Bering Sea Abyssal Plain (Horn, Ewing and Ewing, 1972), and the Japan Sea Abyssal Plain (Ludwig, Murauchi and Houtz, 1975) (Fig. 12.25).

As a result of the enclosed nature of the basin, the basin plains are fed by turbidity currents from widely varied sources (Fig. 12.31). The deepest part of the basin is generally the central part where gradients are very low or where the floor may be level. The dispersal pattern of turbidites on the basin plain is approximately centripetal and ponding of turbidity currents is common (Hersey, 1965). Turbidites from both silici- and bio-clastic sources may interdigitate, as in the Sigsbee Abyssal Plain (Fig. 12.31) (Davies, 1968). Holocene sedimentation rates range from 10–20 cm per 1,000 years, but during the last glacial phase these figures were several times higher. Salt diapirism may disrupt the sedimentary sequence (Hersey, 1965; Burk, Ewing, Worzel, Beall, Berggren, Bukry, Fischer and Pessagno, Jr., 1969) and produce abyssal hills which obstruct turbidity sheet flow and act as barriers.

12.4.5 **Borderland basins**

The California Borderland is indented by some 14 small deep-sea basins. Their long axes are parallel to the coastline and they are arranged in both an along-shore and an off-shore succession (Emery, 1960a). The basins are believed to be fault bounded, originating within the tectonic regime of the San Andreas Fault (Crowell, 1974a). They are relatively small (of the order of tens of kilometres) and shallow (hundreds of metres). The floors of the basins are occupied by nearly flat basin plains, surrounded by both deep-sea fans and slumped slope sediments or aprons (Fig. 12.32) (Gorsline and Emery, 1959; Moore, 1969). In general, the thickness and grain-size of turbidite sand-silt beds decreases away from the source, though thickening towards the central part may occur, indicative of ponding. Slopes with fault scarps of up to 30°, neighbouring the basin plains, facilitate the emplacement of slumped aprons (Gorsline and Emery, 1959).

12.4.6 **Miscellaneous**

A variety of relatively small basins exists which do not belong to any of the foregoing categories, but which do have basin plains constructed by turbidity currents. On the flanks of the Mid-Atlantic Ridge small fault basins are present which act as sediment traps in which the pelagic sediments from adjacent highs may be concentrated by turbidity current resedimentation (van Andel and Komar, 1969) (see Chapters 11 and 14). On the oceanic side of the Great Bahama Bank a number of re-entrants exist (Tongue of the Ocean, Exuma Sound, Columbus Basin) which receive biogenous detritus by turbidity current resedimentation from their upper slopes (Fig. 11.18). Deep-sea fans are virtually absent, but clear basin plains have developed

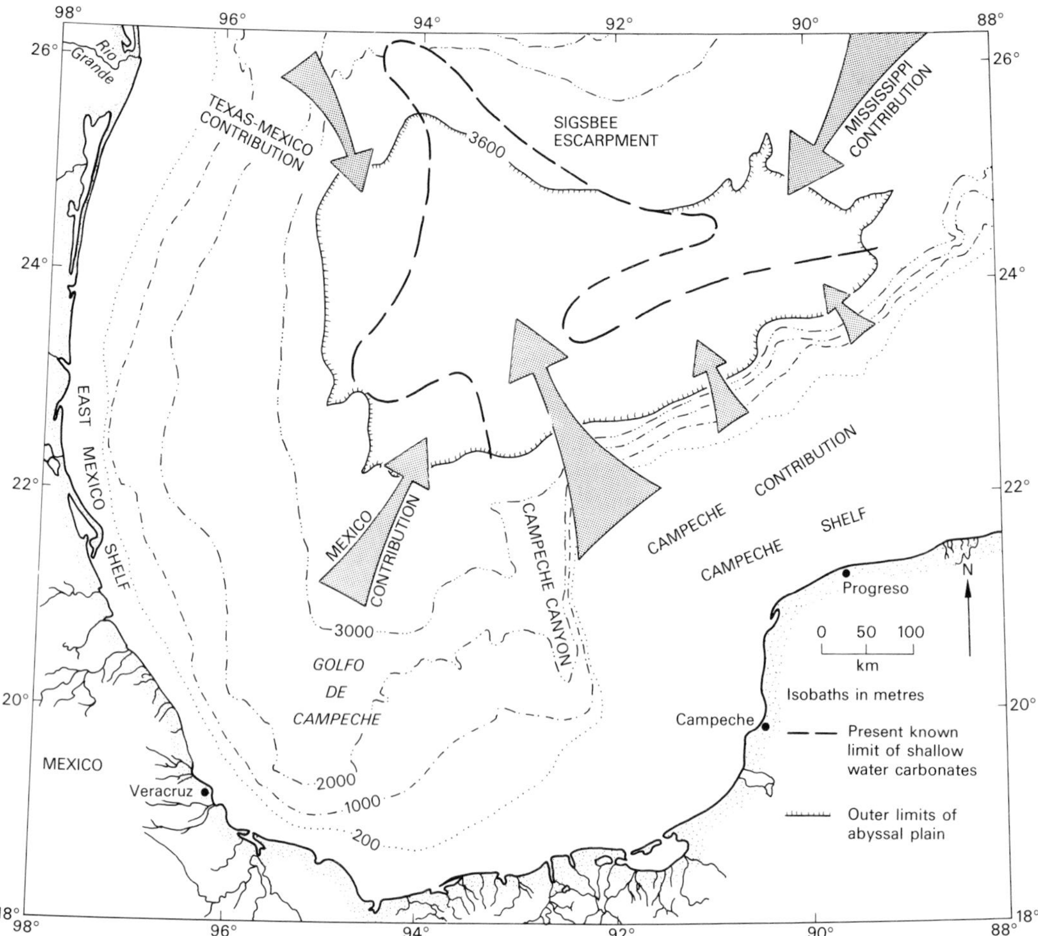

Fig. 12.31. Centripetal distribution of both silici- and bioclastic turbidites on Sigsbee Abyssal Plain, Gulf of Mexico (after Davies, 1968).

(Rusnak and Nesteroff, 1964; Bornhold and Pilkey, 1971). Basin plains are also known from epicontinental seas (van Straaten, 1970), from inland seas (Ross, 1974), from lakes (Sturm, 1975) and from fjords (Holtedahl, 1975).

12.5 MODERN TURBIDITE FANS

Seaward of most submarine canyons and major river deltas large sedimentary cones are situated on the deep-sea floor adjacent to the slope break. Initially, the term *deep-sea delta* was given to these sedimentary cones, based on the belief that they had formed as subaerial deltas during times of glacio-eustatic low sea level (Shepard, 1952). However, these cones are formed by turbidity currents and in many respects are the submarine equivalents of alluvial fans which also form at a slope break. The early studies of these sedimentary cones were of examples off the Pacific coast of North America where they have formed at the mouths of submarine canyons, and have been called *deep-sea fans* (Menard, 1955). In the Atlantic, large sedimentary cones exist off major river deltas (Fig. 12.25) and these have been called *abyssal cones* (Ewing, Ericson and Heezen, 1958). Collectively, all types of fans and cones formed in the deep sea (or in lakes) by resedimentation of clastics will be referred to here as *turbidite fans*.

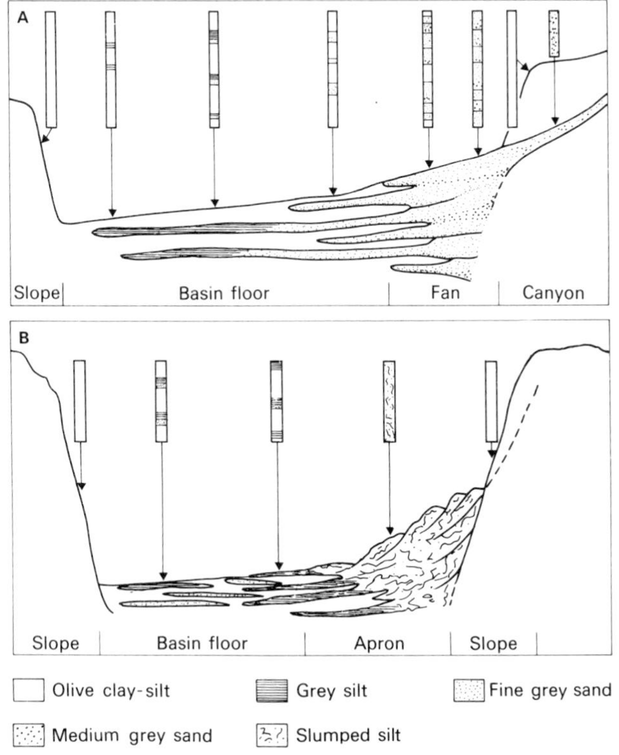

Fig. 12.32. Basin floor to basin margin traverses for (A) canyon-fan and (B) slope-apron margins from California Borderland Basins (after Gorsline and Emery, 1959).

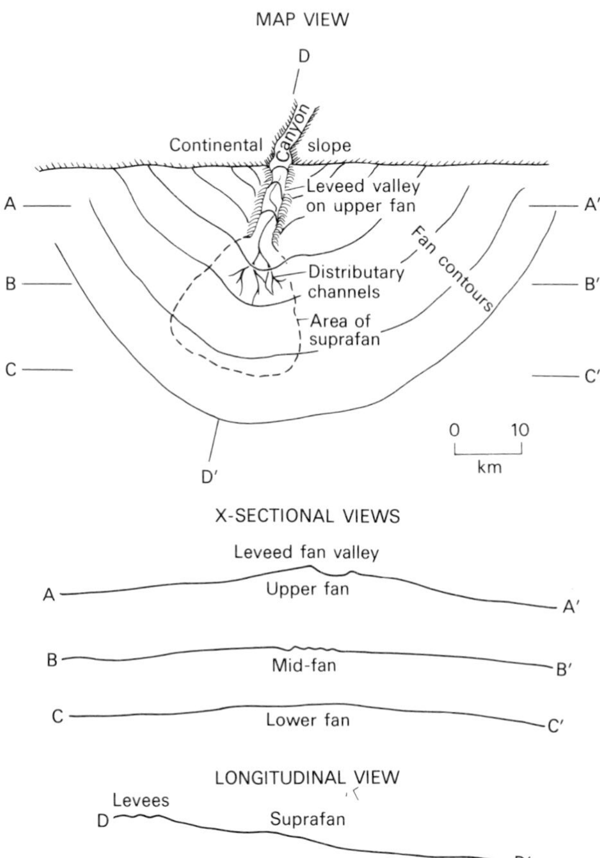

Fig. 12.33. Model of the physiography of a turbidite fan, based largely on the characteristics of deep-sea fans off the west coast of North America (from Normark, 1970).

12.5.1 General characteristics and classification

Until the late 1960s the deep-sea channels associated with fans attracted more attention than the fans themselves (Normark and Piper, 1969). Fans were visualized as massive half-cones of undifferentiated sediment (Menard, 1960b; Burke, 1967). Experiments were conducted to show that fans form because turbidity currents, leaving the mouths of submarine canyons, spread out into fan-shaped flows (Hampton and Colburn, 1967). Around 1970, however, in several studies of fans off the Pacific coast of North America it was recognized that the fan surface consists of several distinct depositional environments and that fan growth takes place by an intricate interplay between mass flow processes and the depositional topography of the fan surface (Shepard, Dill and von Rad, 1969; Nelson, Carlson, Byrne and Alpha, 1970; Normark, 1970; Haner, 1971).

In transverse profile, a fan surface includes (1) channels, (2) levees and (3) interchannel areas (Fig. 12.33). In radial profile, one may distinguish between (1) an upper fan environment (or inner fan), (2) a middle fan environment with suprafan lobes and (3) a lower fan environment (or outer fan) (Fig. 12.33). This tripartite division of depositional environments is only distinct in those instances where the middle fan is characterized by a depositional bulge (or suprafan) (Nelson, Carlson, Byrne and Alpha, 1970; Normark, 1970; Nelson and Kulm, 1973; Nelson, 1976). In some turbidite fans, in particular the abyssal cones, the division into upper, middle and lower fan is not distinct, though the Amazon Cone does show the tripartite division (Damuth and Kumar, 1975).

The *upper fan environment* is ideally characterized by a concave-up profile with rugged topography and the presence of a main *fan valley* (Fig. 12.33; 12.34). The fan valley may be straight or sinuous. It has levees which may be from a few tens of metres to over 200 m above the valley floor. The valley floor itself is depositional and may be elevated above the adjacent

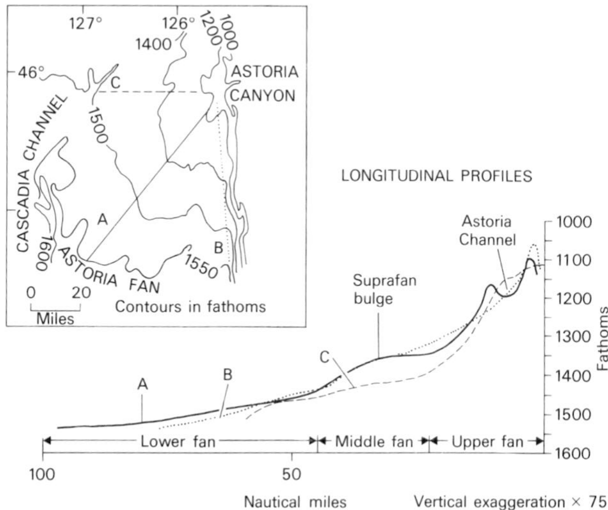

Fig. 12.34. Longitudinal profiles across the Astoria Fan, off Oregon. Note the suprafan bulge (after Nelson, 1976).

fan surface by many tens of metres. The valley width ranges from approximately 0.1–10 km. The *middle fan environment* is characterized by a convex-up profile with hummocky topography where the main fan valley splits up into many distributaries, called *fan channels* (Fig. 12.33). The channels may meander or braid, be active or abandoned. Their axial depth may be several tens of metres and their width up to about 1 km. At the termination of the channels, in the lower part of the middle fan, depositional lobes occur. The *lower fan environment* has a concave-up profile and smooth topography with numerous small channels without levees (Fig. 12.34). The boundary between the middle and lower fan environments is gradual and indistinct. In general, fan valleys and channels can be (1) depositional, (2) erosional and (3) mixed depositional-erosional (Fig. 12.35) (Nelson and Kulm, 1973).

Mass-gravity transport other than turbidity currents is largely restricted to the upper fan. There are two types of sediment dispersal by turbidity currents across the fan (Fig. 12.36). Much of the coarse-grained sediment is transported radially through channels and is deposited as thick elongate sand bodies within distributaries or as sand lobes at their terminations, building outward and sideways. Fine-grained sediment may be transported laterally by overflow on to levees and into interchannel areas.

Sand–shale ratios are high inside channels and low in interchannel areas. The overbank muds are characterized by a relatively high percent of platy constituents (e.g. plant fragments, mica flakes) which become concentrated in the dilute tail of turbidity currents (Nelson and Kulm, 1973; Nelson and Nilson, 1974; Nelson, 1976). It has been argued that some of the fan sands may be emplaced by indigenous bottom currents (Huang and Goodell, 1970).

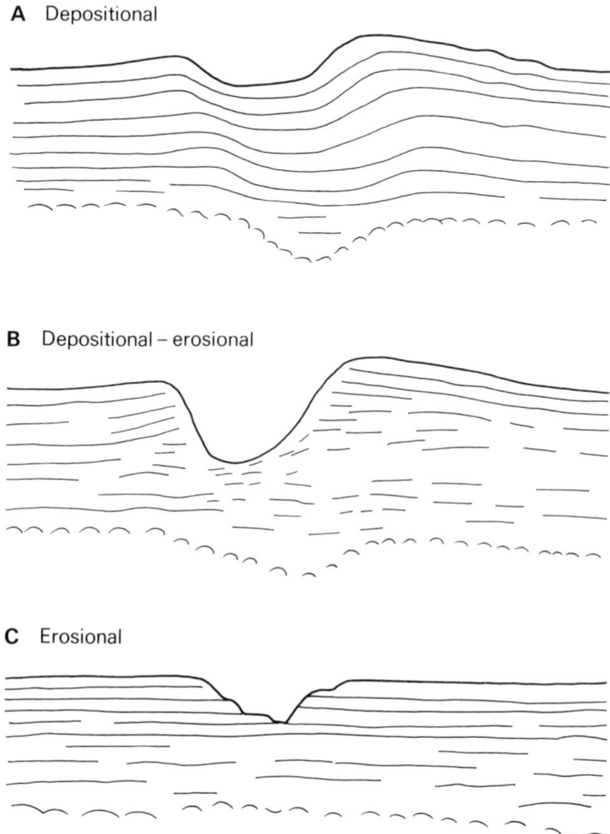

Fig. 12.35. Three types of deep-sea channels, drawn from seismic reflection profiles (from Nelson and Kulm, 1973).

In the upper fan, channel sands generally are thick immature turbidites characterized by poorly developed Bouma divisions (T(A–E)). In the middle and lower fan the channel sands are medium thick mature turbidites T(ABCDE) or T(BCDE). On the levees and in interchannel areas thin turbidite silts T(CDE) or T(DE) are deposited with the silts becoming coarser downfan as channel depth decreases. In contrast to basin plains, the fan surface is characterized by the juxtaposition of high and low flow regime turbidites, the contrast between sediment types increasing in an up-fan direction (Fig. 12.36) (Haner, 1971).

Deposition of turbidite sands is localized in channels and in prograding depositional lobes, the lateral shifting of which produces a fan-shape. Shifting may take place either gradually or catastrophically by avulsion in which an old distributary

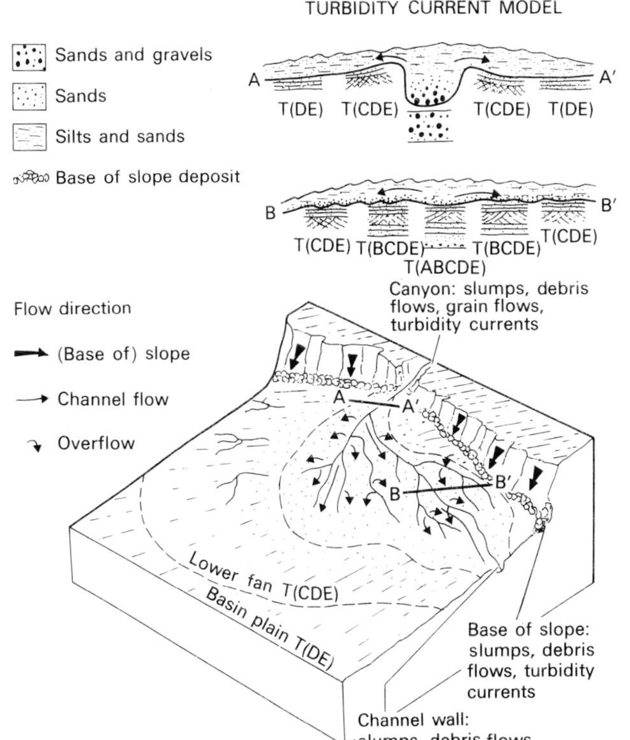

Fig. 12.36. Bi-modal channel and overbank deposition by turbidity currents across a turbidite fan. The model is largely based on the Astoria Fan (from Nelson and Kulm, 1973).

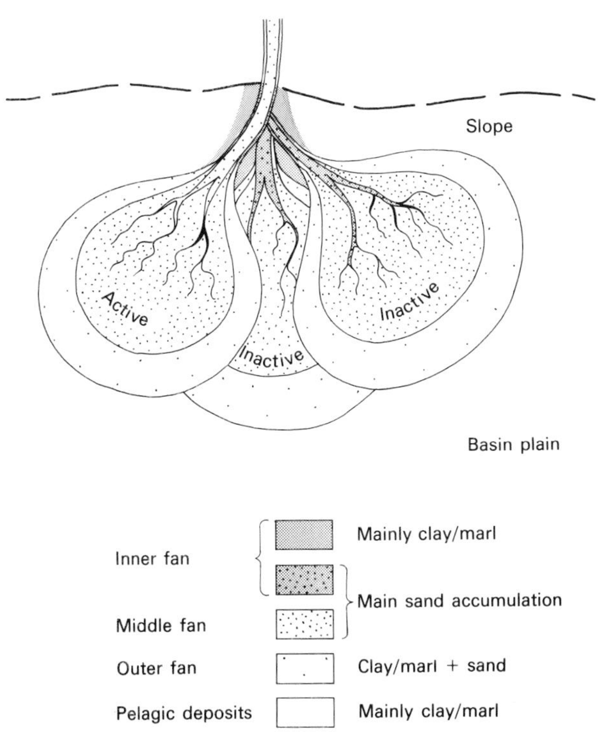

Fig. 12.37. Model of fan growth by progradation and lateral avulsion of an active, major fan lobe (after Kruit, Brouwer, Knox, Schöllnberger and van Vliet, 1975).

system is abandoned and a new one formed (Fig. 12.36; 12.37). As a result, turbidite sands become emplaced on top of interchannel muds. JOIDES drilling in the Bengal Cone (leg 22) has shown that since the Middle Miocene four major episodes of turbidite sand deposition have occurred (Fig. 12.38). The turbidite silty sand and silt sequences, some 100–200 m thick, alternate with sequences of clayey silt and nanno ooze. The alternations may be the result of either lateral shifting, possibly avulsion, of channels and lobes, or sea-level fluctuations and tectonic activity (Moore, Curray, Raitt and Emmel, 1974).

Factors that influence the growth of a turbidite fan as a whole are the physiography of the depositional basin, latitude, sea level fluctuations, the tectonic history of the depositional and drainage basins, the nature of the sediment source, and the presence or absence of indigenous bottom currents.

Topographic features of the sea floor such as abyssal hills, seamounts, mid-ocean ridges or oceanic fracture zones may affect the shape of a turbidite fan. Basin topography has been used to divide fans into open basin fans and restricted basin fans (Nelson and Kulm, 1973). Latitude may affect fan growth as a result of the Coriolis effect. In the northern hemisphere the first order influence of the Coriolis effect is that turbidity currents build higher levees on the right side of deep-sea channels (Fig. 12.35). The difference of levee height may range from a few tens of metres to some 100 m. The second order influence of the Coriolis effect is that avulsion and new channels tend to develop towards the left because the left levee is lower. In the Astoria deep-sea fan, the distributary system has shifted from right to left (Fig. 12.36). In the southern hemisphere the effect is the reverse (Menard, 1955; Komar, 1969; Griggs and Kulm, 1970; Chough and Hesse, 1976). The effect is not distinct in all channels and fans, and differences in height between left and right levees can also be produced by the centrifugal force operating on turbidity currents in curved stretches of channels when the higher levee is on the outside bend (Buffington, 1952).

The present-day surface morphology of most turbidite fans is the result of clastic resedimentation during Pleistocene intervals of glacio-eustatic low sea level, when compositionally

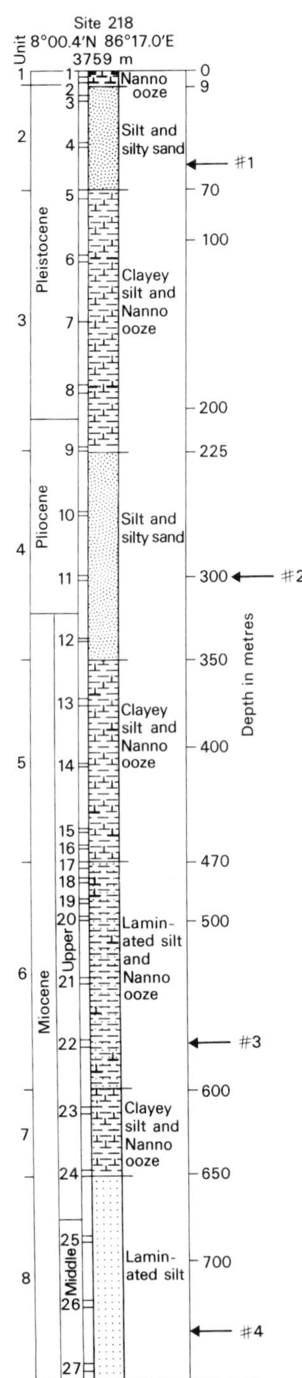

Fig. 12.38. Stratigraphic log of site 218, leg 22, of JOIDES drilling on the Bengal Cone (compare with Fig. 12.39). Note four phases of active turbidity current deposition (after Moore, Curray, Raitt and Emmel, 1974).

immature river detritus was dumped in or near the heads of submarine canyons and delta-front troughs. The Holocene transgression has moved the loci of river deposition to the inner shelf where sediment is now trapped. As a result, canyon heads are either sediment-starved or they receive compositionally mature sediments moved by longshore drift. Much of the present-day surface of fans is covered by a (hemi-)pelagic mud blanket, especially in interchannel areas. Holocene turbidity current activity is largely restricted to channelized low density transport (Fig. 12.40) which aggrades channels with fine-grained turbidites. Rates of sedimentation since the Pleistocene have decreased from up to 100 cm per 1,000 years to as low as 5 cm per 1,000 years (Damuth and Kumar, 1975; Nelson, 1976).

The influence of tectonism on the growth of fans is illustrated by the Bengal Cone where deposition began after the Indian plate collided with Asia during the Middle Eocene to Late Miocene (Moore, Curray, Raitt and Emmel, 1974).

The initial concern of turbidite fan study was to demonstrate that fans have several fundamental features in common which can be combined to construct a unified fan model (Normark, 1970; Nelson and Nilson, 1974; Damuth and Kumar, 1975). However, fans may differ in many respects, e.g. size, grain-size, nature and number of sediment sources, presence or absence of indigenous erosional currents, and these features can be used to study variability of fans and to distinguish different fan types. A classification according to sediment source would follow the historic precedent of delta-associated fan work in the Atlantic, and submarine canyon-associated fan work in the Pacific (Heezen and Menard, 1963). Turbidite fans are accordingly classified here as follows. (1) *Abyssal cones,* which are formed off the delta-front troughs of major river deltas. (2) *Deep-sea fans,* which are formed off the mouths of submarine canyons. (3) *Short headed delta-front fans,* which are the subaqueous part of short headed deltas, fed primarily by river-generated turbidity currents. (4) *Continental rise fans,* formed in the regime of redepositing geostrophic contour currents. (5) *Mixed type fans,* which combine features of any of the other types.

12.5.2 **Abyssal cones**

Examples of abyssal cones in open ocean basins are the Mississippi Cone (Ewing, Ericson and Heezen, 1958; Stuart and Caughy, 1976) the Congo Cone (Heezen, Menzies, Schneider, Ewing and Granelli, 1964), the Ganges (or Bengal) Cone (Moore, Curray, Raitt and Emmel, 1974; Curray and Moore, 1974), the Indus Cone (Jipa and Kidd, 1974), and the Amazon Cone (Damuth and Kumar, 1975) (Fig. 12.25). The Mediterranean Sea has the Rhône Cone (Menard, Smith and Pratt, 1965) and the Nile Cone (Maldonado and Stanley, 1976).

Abyssal cones develop off the deltas of major rivers with

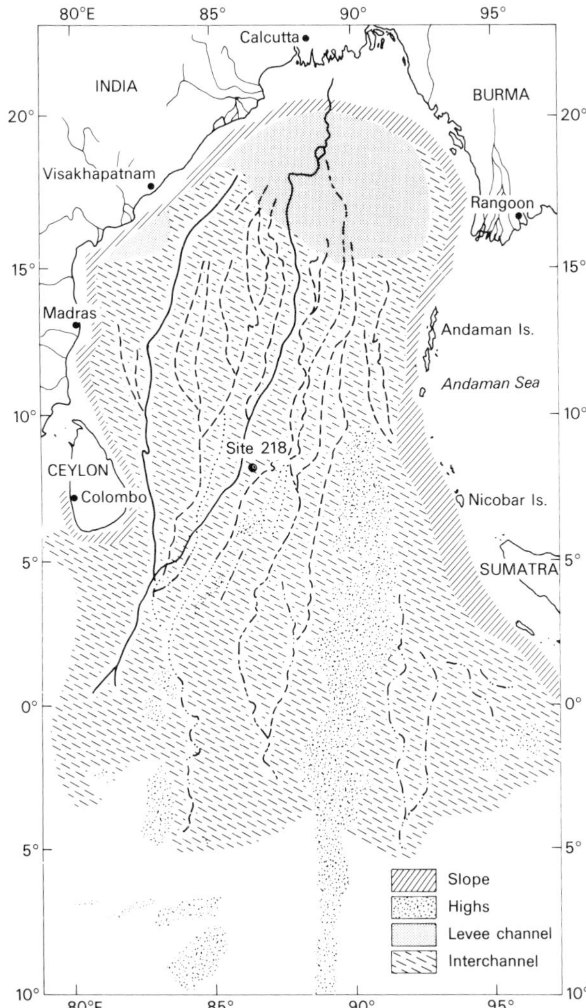

Fig. 12.39. Generalized distribution of facies across the Bengal Cone and adjacent areas (after Curray and Moore, 1974).

large drainage basins and voluminous suspended load (Fig. 12.25). The cones are fed from the delta through one major and several minor delta-front troughs. Instead of a distinct apex, a broad head grades imperceptibly into the continental slope (Fig. 12.39). A broad front of sediment supply to the abyssal cones is maintained as a result of changes in river discharge, the abandonment of feeder troughs, the erosion of new troughs or the re-activation of old ones (Ewing, Ericson and Heezen, 1958; Menard, Smith and Pratt, 1965; Reimnitz and Gutierrez-Estrada, 1970). Abyssal cones widen seaward and generally grade into large abyssal plains. The radial dimensions of cones range from some 300 km (Rhône Cone) to some 3,000 km (Bengal Cone). The upper cone can attain a thickness of over 10 km (Moore, Curray, Raitt and Emmel, 1974; Damuth and Kumar, 1975). Channels extend over most of the fan surface (Fig. 12.39). Sediments are predominantly fine-grained due to the absence of a very coarse size grade in the detritus of large rivers. Only very locally has gravel been encountered. The topography of abyssal cones may be severely disrupted as a result of very large slumps of tens of kilometres in length and width (Walker and Massingill, 1970).

12.5.3 Deep-sea fans

The classic examples of deep-sea fans are those off the west coast of North America (refs. in 12.3.2). They develop off submarine canyons which function as funnels for sediments which are derived either from a connected river or from longshore drift or both (Fig. 12.40). In most instances the

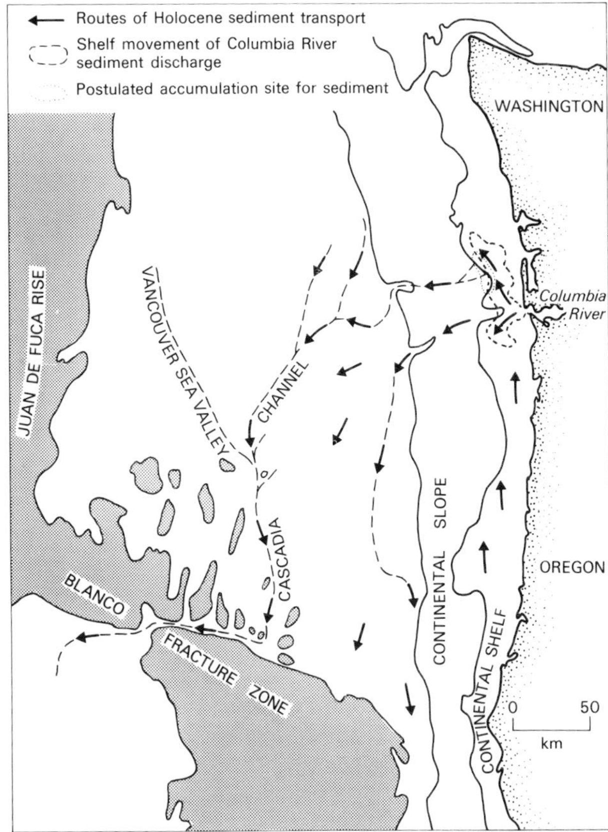

Fig. 12.40. Holocene shallow- and deep-water sediment dispersal in the Astoria Fan region, showing river supply, longshore drift and channelized resedimentation across the fan (from Griggs and Kulm, 1970).

connected rivers are small and do not form subaerial deltas. The mouth of a feeder canyon is a fixed point source and consequently deep-sea fans have a distinct apex, though contributary canyons may blur the fan shape or produce coalescing fans (Normark, 1970). On the seaward side of deep-sea fans large basin plains seldom exist, but small ponded plains may occur. The radii of deep-sea fans range from a few tens of kilometres for California Borderland fans to some 300 km for Monterey Fan (Nelson and Kulm, 1973). Sediment thickness is at most of the order of 1 km. The grain-size of deep-sea fans is generally coarser than that of abyssal cones, gravel beds being extensive in the upper fan.

12.5.4 **Short headed delta-front fans**

Examples have been described from Swiss lakes (Sturm, 1975) and from North American lakes (Nelson, 1967). Delta-front fans develop as the subaqueous turbidite part of a short-headed delta. A short-headed delta forms where a stream flows down steep gradients from nearby headwaters and debouches at the margin of a depositional basin, e.g. a lake. The delta-front and pro-delta deposits are predominantly emplaced by high- and low-density turbidity currents. The turbidity currents are in most instances river-generated during periods of high run-off or by landslides in the drainage basin (Sturm, 1975). Delta-front fans are relatively small, some 10 km in radial dimension at most, and thin, in the order of 100 m or less. Channels occur on the delta-front, up to some 200 m in width and 30 m in depth. In most instances they originate some way down the delta slope and may continue on to the basin plain. A lateral division of channel, levee and interchannel occurs. In the channels sands predominate, on the levees laminated muds with occasional sands, and in the interchannel areas muds (Sturm, 1975).

12.5.5 **Continental rise fans**

The eastern continental margin of North America is incised by several major submarine canyons (Shepard and Dill, 1966). However, turbidite fans are either absent or poorly developed (Fig. 12.25). The foot of the continental slope merges with a relatively smooth continental rise, shaped by geostrophic contour currents (Fig. 12.24). The virtual absence of turbidite fans on the continental rise has been attributed to (1) the absence of a sharp break in gradient between continental slope and ocean floor (Nelson, Carlson, Byrne and Alpha, 1970) and (2) to the low sedimentary preservation potential of turbidite fans in an environment of indigenous bottom currents capable of re-working sand-grade material (Normark, 1970). Two poorly developed continental rise fans have been described, viz. the Hatteras Fan (Cleary and Conolly, 1974) and the Laurentian Fan (Stowe, 1975). Evidence exists for the re-working of fan sediments by the Western Boundary Undercurrent. High-constructional periods of glacio-eustatic low sea-level have alternated with destructional periods during high sea-level.

12.5.6 **Mixed-type fans**

Mixed-type fans develop as hybrids of more than one fan type. An example is the Hope-Liguanea Fan in the Yallahs Basin, off the south-east of Jamaica. Here a delta-front fan (Hope) has merged with a canyon-related deep-sea fan (Liguanea) (Burke, 1967) (Fig. 14.28).

12.6 FLYSCH FACIES

12.6.1 **Early interest: definition and bathymetry**

In contrast to the work on modern deep-sea clastic environments which emanated largely from North American oceanographic institutions, most of the early work on ancient deep-sea clastic facies was done in Europe.

From the start, the notion of a deep-sea clastic facies, formed by turbidity current resedimentation, was enmeshed in a problem of definition. Several terms were used to describe deep-sea clastic facies, e.g. graded greywacke (a petrographic term), Kulm or Macigno (local time-rock terms) or Flysch (used as a petrographic term, a time-rock or also as a recurrent facies term) (Hsü, 1970). The term *flysch* gradually became accepted as a facies term to describe thick successions of redeposited deep-sea clastics. Flysch definition, however, remained a subject of much controversy, particularly the tectonic context of flysch deposition. More recently, plate tectonic theory has shown that deep-sea clastic facies may exist in a variety of tectonic settings (Chapter 14).

Much confusion arose from the fact that some authors believed that isolated sedimentary structures (e.g. graded bedding, flute marks) are diagnostic of turbidites and that isolated turbidites are diagnostic of flysch facies. This belief was at the root of early objections to the turbidity current theory of flysch deposition (Mangin, 1962). The objections were based on the fact that graded bedding and flute marks had been found in shallow water molasse facies (de Raaf, 1964). Clearly, isolated sedimentary structures such as graded bedding and flute marks may be produced by types of episodic flow other than turbidity currents. Turbidite beds may be deposited in shallow water. The only depth connotation of a turbidite bed is that a certain amount of subaqueous gradient was required to allow the development of the turbidity current and that deposition was below wave base to ensure preservation.

Deep-sea depth indicators for flysch facies are the following.

(1) The only modern environments where thick successions of redeposited clastics accumulate are in the deep-sea. (2) Trace fossil communities exist of which the habitat is controlled by bathymetry-related factors (Fig. 9.28) (Seilacher, 1962, 1967). The following communities have been identified: shallow shelf assemblage (*Cruziana*), a littoral to bathyal (*Cruziana* and *Zoophycos*), and a deep-basin assemblage (*Nereites*). The latter is commonly found in flysch successions. (3) The (hemi-)pelagic shales between the turbidite sandstones have been found in many deposits to contain remains of deep-water benthonic Foraminifera (Natland and Kuenen, 1951; Weidmann, 1967).

12.6.2 **Palaeocurrent analysis**

Initially, it was believed that flysch sandstones were derived marginally, from cordilleras which flanked the depositional troughs (Kuenen and Migliorini, 1950). This model of flysch deposition was based on the contemporary geosynclinal theory which postulated downbuckling of a geosynclinal trough and concurrent rising of an adjacent geanticline. However, the usage of sedimentary structures in flysch sandstones, especially flutes and grooves, for the study of palaeocurrents led to the discovery that flysch turbidites were dispersed longitudinally, parallel to the long axis and often the tectonic strike of the flysch troughs (Fig. 12.46) (Kuenen, 1957b). Based on this discovery, a palaeocurrent model of flysch was proposed which suggested longitudinal filling of elongate basins by a major sediment source at one end plus possibly several minor marginal sources (Kuenen and Sanders, 1956; Kuenen, 1958).

The discovery of longitudinal filling was of fundamental importance to palaeogeographic reconstructions of geosynclinal belts. It stimulated palaeocurrent mapping of flysch successions and the development of flysch basin analysis (Potter and Pettijohn, 1963). The early classics of palaeocurrent analysis of flysch were by Crowell (1955) of the pre-Alpine flysch in Switzerland, by ten Haaf (1959) of the Oligo–Miocene flysch of the Northern Apennines (a longitudinal palaeocurrent pattern), by Dzulynski, Ksiaskiewicz and Kuenen (1959) of the Cretaceous and Palaeogene flysch of the Polish Carpathians (longitudinal dispersal with marginal supply), and by Wood and Smith (1959) of the Silurian Aberystwyth Grits in Wales (longitudinal dispersal). These and other studies supported the model of axial (or longitudinal) and marginal (or transverse) flysch trough fill. Through the 1960s a large number of palaeocurrent analyses of flysch successions in fold belts on both sides of the Atlantic were carried out and these served to support the model further (Fig. 12.44) (McBride, 1962; several papers in Bouma and Brouwer (ed.), 1964; Enos, 1969; several papers in Lajoie (ed.), 1970, and in Sestini (ed.), 1970).

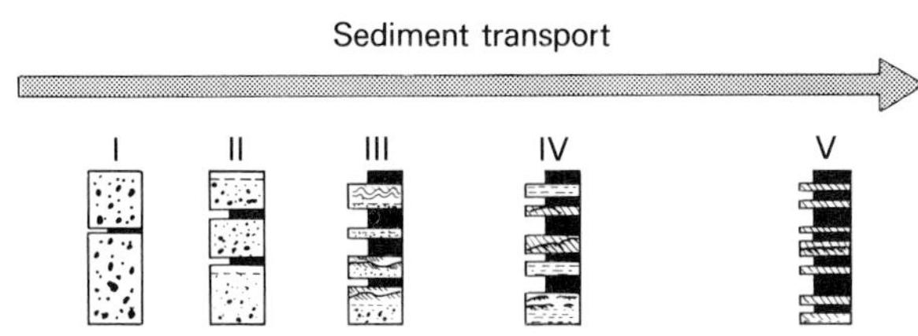

Fig. 12.41. Schematic illustration of the downcurrent decrease of bed thickness, grain size and sand-shale ratio in a turbidite sequence (from Einsele, 1963).

12.6.3 **Facies analysis: proximal versus distal**

Palaeocurrent analysis was the conceptual stepping stone to flysch facies analysis. Knowledge of the palaeocurrent pattern made it possible to trace lithological changes in a downcurrent direction and to relate these changes to depositional environments both close to the source (proximal) and further away from the source (distal). From several early studies it appeared that such features as sandstone–shale ratio, bed thickness, grain size, grading, etc. change systematically in a downcurrent direction (Fig. 12.41) (Wood and Smith, 1959; Einsele, 1963).

Based on a large number of observations and published data, Walker R.G. (1967) summarized the characteristics of proximal and distal facies. In a sequence of turbidite beds the following changes take place in downcurrent direction: sandstone shale ratio, sandstone thickness, grain size and erosive features (amalgamated sandstone, channels) all decrease; scour marks such as flutes become fewer and tool marks such as grooves more in number (Table 12.2; Fig. 12.41).

To these proximality criteria, Walker (1967) added his ABC-index. The index is based on the interpretation of the Bouma sequence in terms of flow regime. In a downcurrent direction a turbidity current deposits under increasingly lower flow regime

Table 12.2 Comparison of proximal and distal (or thin-bedded) turbidite sequences (after Walker R.G., 1967)

	PROXIMAL	DISTAL
A	Beds thick	Beds thin
B	Beds coarse grained	Beds fine grained
C	Individual sandstones often amalgamate to form thick beds	Individual sandstones rarely amalgamate
D	Beds irregular in thickness	Beds parallel-sided, regularly bedded
E	Scours, washouts and channels common	Few small scours, no channels
F	Mudstone partings between sandstones poorly developed or absent. Sand/mud ratio high	Mudstone layers between sandstones well developed. Sand/mud ratio low
G	Beds ungraded or crudely graded	Beds well graded
H	Base of sand always sharp, top often sharp, many AE sequences	Base of sand always sharp, top grades into finer sediment, AE sequences rare
I	Laminations and ripples occur infrequently	Laminations and ripples very common
J	Scour marks occur more frequently than tool marks	Tool marks occur more frequently than scour marks

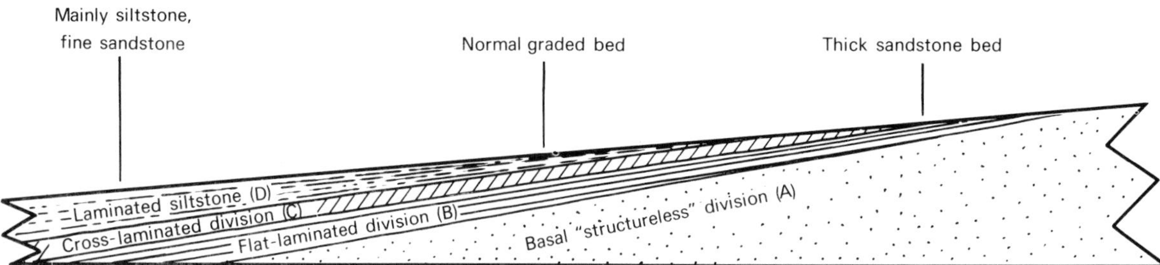

Fig. 12.42. Idealized lateral variations in a single turbidite bed (after Corbett, 1972).

conditions (Fig. 12.42). As a result, the base of the turbidite bed would be Bouma division A near the source, but change into B or C further away from the source. The percent of beds in a turbidite sequence of which the base is Bouma division A, B or C therefore expresses relative distance from the source, or proximality. The ABC-index is defined as the percent of beds beginning with division A plus half the percent of beds beginning with division B. As with other proximality indicators, the ABC-index only works for turbidite sequences deposited from a single source by longitudinal sheet flows. In turbidite fans, turbidites of both proximal and distal aspect may be directly juxtaposed (Haner, 1971) (Fig. 12.36). As a result the term 'thin-bedded turbidites' may be preferable to 'distal turbidites' (Mutti, 1977).

Further studies have confirmed and refined our knowledge of proximal-to-distal facies changes (Enos, 1969; Lovell, 1969; Thompson and Thomassen, 1969; review by Walker, 1970). In particular the study of the Ordovician Cloridorme Formation, Gaspé Peninsula, Canada, has added to our knowledge of downcurrent variations (Enos, 1969; Parkash and Middleton, 1970). Single turbidite beds can be walked out here and appear to be discontinous. For a certain bed sequence, beds disappear at a rate of 50% per 10 km. Imbrication of beds, local current by-passing and differential deposition of shale has produced an intricate mosaic of stacked beds. Most beds show no appreciable mean grain-size variations at their base in a downcurrent direction. Current-oriented features show meandering trend lines (Fig. 12.43) (Parkash and Middleton, 1970).

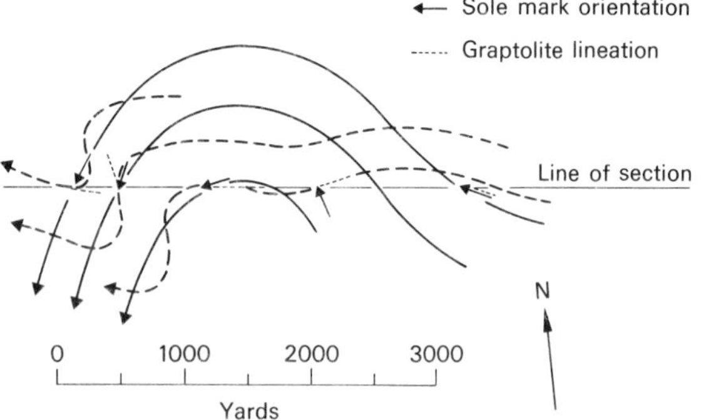

Fig. 12.43. Trend lines for sole marks and graptolite orientations in a single bed, indicating meandering of turbidity current (from Parkash and Middleton, 1970).

12.6.4 Facies classification

The early description of flysch facies emphasized its homogeneity, i.e. the absence of abrupt vertical and lateral facies changes. The following features were listed: (1) a monotonous alternation of sandstone and shale beds, (2) graded bedding of the sandstones and other features associated with turbidity current deposition, (3) lateral continuity and

Table 12.3 Facies scheme for the description of flysch (after Walker and Mutti, 1973)

BOUMA SEQUENCE NOT APPLICABLE	FACIES A – Coarse grained sandstones and conglomerates A1 Disorganized conglomerates A2 Organized conglomerates A3 Disorganized pebbly sandstones A4 Organized pebbly sandstones FACIES B – Medium-fine to coarse sandstones B1 Massive sandstones with "dish" structure B2 Massive sandstones without "dish" structure
BEDS CAN REASONABLY BE DESCRIBED USING THE BOUMA SEQUENCE	FACIES C – Medium to fine sandstones – classical proximal turbidites beginning with Bouma's division A FACIES D – Fine and very fine sandstones, siltstones – classical distal turbidites beginning with Bouma's division B or C FACIES E – Similar to D, but higher sand/shale ratios and thinner more irregular beds
BOUMA SEQUENCE NOT APPLICABLE	FACIES F – Chaotic deposits formed by downslope mass movements, e.g. slumps FACIES G – Pelagic and hemipelagic shales and marls – deposits of very dilute suspensions

regularity of bedding, (4) consistency of palaeocurrent direction over long distances and great thicknesses, and (5) absence of shallow-water features, e.g. large-scale cross-bedding or indications of subaerial exposure (Dzulynski and Walton, 1965).

Some facies inhomogeneity, expressed in differences of sandstone–shale ratio, formed the basis of a subdivision into sandy, normal, and shaly flysch. Since then, several more complex facies classifications have been proposed (review by Walker, 1970). In the past several years a facies scheme from A through G has been worked out (Table 12.3) (Mutti and Ricci-Lucchi, 1972; Walker and Mutti, 1973). The most important criteria used in facies classification are: bed thickness, grain size, sandstone–shale ratio, bedding regularity, sole marking assemblages, internal structures and textures, and palaeo-ecological indicators. The purpose of such schemes, none of which should be considered definitive, is to facilitate rapid, factual description, and comparison between flysch basins.

Fig. 12.45. Characteristic basin plain sequence from the Palaeocene flysch of the western Pyrenees, near Zumaya. Note the light coloured pelagic intervals.

12.7 ANCIENT BASIN PLAINS

12.7.1 General characteristics

In the early studies of flysch, the depositional environment was generally assumed to be a basin plain. Thus, specific criteria which characterize a basin plain facies have only been formulated since the dual model of turbidite fan and basin plain was introduced (Fig. 12.44) (Mutti and Ricci-Lucchi, 1972) and much proximal flysch was re-assigned to a turbidite fan environment.

Features which characterize a basin plain facies are the following (Fig. 12.45) (Mutti and Ricci-Lucchi, 1972; Walker and Mutti, 1973; Mutti, 1977). (1) The general aspect is one of regularly bedded, alternating sandstone and shale beds. No distinct vertical thickening- or thinning-up sequences occur (Fig. 12.52). (2) The sandstones are predominantly medium- to fine-grained turbidites beginning with Bouma divisions B or C. (3) The sandstone–shale ratio is approximately one or less, and the shale beds are well developed. (4) The upper part of the shale beds may represent a (hemi-)pelagic interval. (5) Single turbidite sandstones may be laterally very continuous. Some key beds have been traced for distances of over 200 km. (6) The palaeocurrent pattern is longitudinal or centripetal. The palaeocurrent direction in the key beds may be completely reversed compared with the beds below and above (Fig. 12.44; 12.46). This palaeocurrent reversal is in accord with the model of a flat basin floor with little or no gradient.

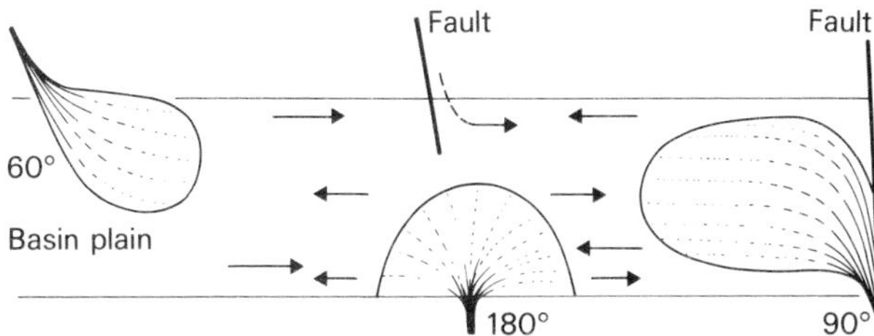

Fig. 12.44. Palaeocurrent model of a deep-sea basin with a basin plain and turbidite fans (after Ricci-Lucchi, 1975a).

12.7.2 Examples of ancient basin plains

Several flysch successions have been interpreted as trench basin plains, e.g. the Cretaceous Monte Antola flysch of the Northern Apennines (Scholle, 1971), the Cretaceous flysch belt of the Alaskan Peninsula (Moore, 1973) and the Cretaceous flysch belt of the Eastern Alps in Bavaria (Fig. 12.46) (Hesse, 1974, 1975). In the latter case, the evidence which suggests a trench is (1) lateral continuity of single turbidite beds of up to some 115 km, (2) a longitudinal palaeocurrent pattern without notable marginal supply, (3) three major palaeocurrent reversals throughout the depositional history of the succession, (4) deposition below carbonate compensation depth, a conclusion derived from the presence of carbonate-free hemi-pelagic intervals above carbonate-containing turbidite shales. In the Eocene flysch of the South-western Pyrenees a basin plain facies of a small fault basin has been identified (Rupke, 1976b). Extensive slump sheets are overlain by very thick key- or mega-beds, each of which was deposited by more than one turbidity current, simultaneously flowing on to the basin floor where they stagnated and filled small basin floor compartments (Fig. 12.47).

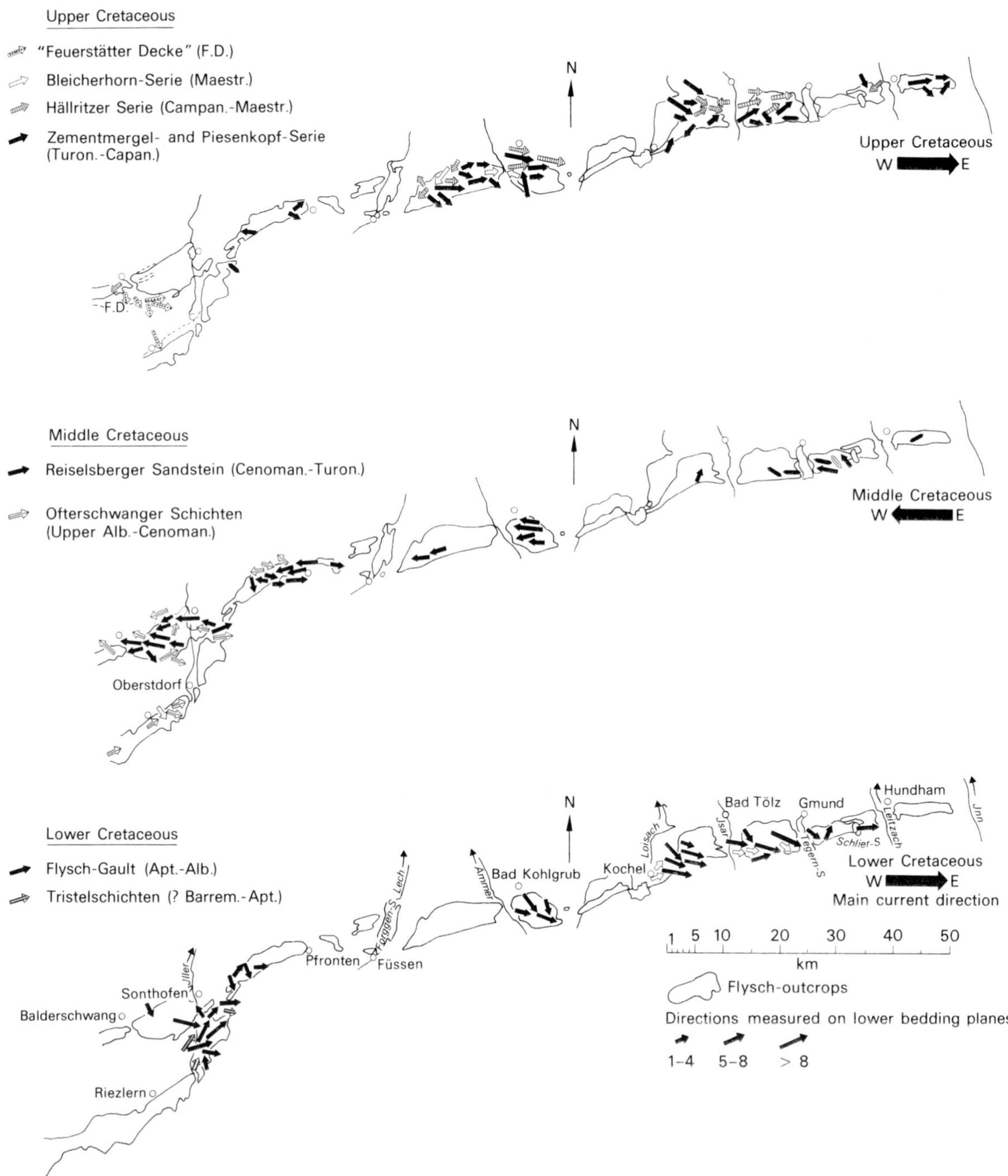

Fig. 12.46. Palaeocurrent directions in three successive formations of the Cretaceous flysch of the Eastern Alps. Note palaeocurrent reversals (from Hesse, 1974).

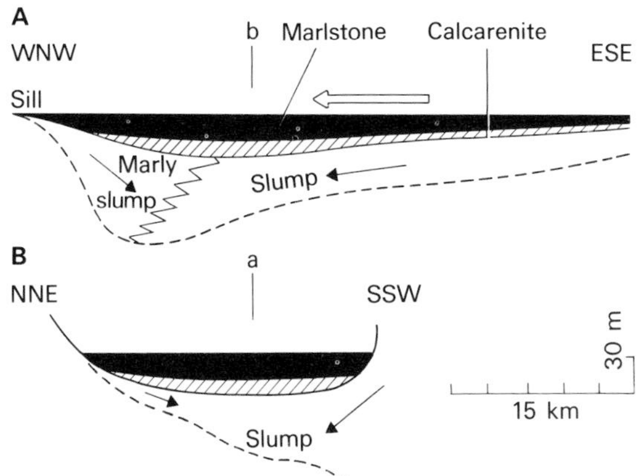

Fig. 12.47. Longitudinal (a) and transverse (b) cross-sections through a key- or mega-bed, ponded on the basin plain of a small fault basin (after Rupke, 1976b).

12.8 ANCIENT TURBIDITE FANS

12.8.1 Growth of the ancient fan model

The turbidite fan model as applied to flysch successions has grown from early allusions to the possibility of such fans in the 50s (Kuenen and Migliorini, 1950; Natland and Kuenen, 1951) to the use of the fan model in actual case studies in the 60s (Sullwold, 1960; Kelling and Woollands, 1969), to the recognition of ancient fan environments similar to those known from modern fans around 1970 (Jacka, Beck, Germain and Harrison, 1968; Mutti and Ricci-Lucchi, 1972). Use of the fan model in flysch study has grown from three roots, viz. (1) palaeocurrent analysis, (2) the study of proximal facies and (3) vertical sequence analysis.

Flysch palaeocurrent analysis produced the model of axial and marginal trough fill (Fig. 12.44). The marginally derived turbidites were found to show radial palaeocurrent patterns, indicative of the existence of fans near marginal point sources (Kruit, Brouwer, Knox, Schöllnberger and van Vliet, 1975; Ricci-Lucchi, 1975a).

The study of coarse-grained and thick-bedded sandstones and conglomerates in proximal flysch facies has shown that many of these beds are elongate bodies deposited in valleys and channels (Ricci-Lucchi, 1969; Stanley and Unrug, 1972; Davies and Walker, 1974; summary in Walker, 1975). Many of these proximal channel deposits were originally referred to as fluxo-turbidites (Dzulynski, Ksiazkiewicz and Kuenen, 1959). These coarse-grained channel deposits are believed to be among the characteristic features of fans (Nelson and Nilson, 1974; Walker, 1975, 1977).

Vertical sequence analysis of flysch successions has demonstrated that two basic types of vertical sequences may occur: (1) thickening- and coarsening-up and (2) thinning- and fining-up (Angelucci, De Rosa, Fierro, Gnaccolini, La Monica, Martinis, Parea, Pescatore, Rizzini and Wezel, 1967; Ricci-Lucchi, 1969, 1975b; Walker, 1970; Mutti and Ricci-Lucchi, 1972; Walker and Mutti, 1973; Rupke, 1977). Initially, vertical sequences were interpreted as the result of tectonic oscillations or eustatic sea level fluctuations. These interpretations have been particularly popular with the Russian flysch workers. Among the other theories to explain vertical sequences are changes in source area, in sediment supply, and the effect of basin floor topography (Sagri, 1972). More recently, sequences in flysch have been interpreted using sedimentological analogies from fluvial and deltaic environments such as lateral channel migration, channel aggradation, and delta lobe progradation. Similar processes may take place on turbidite fans and distinct vertical sequences in flysch are now believed to characterize fan facies (Mutti and Ricci-Lucchi, 1972; Walker and Mutti, 1973; Mutti, 1974; Ricci-Lucchi, 1975b; Rupke, 1977). In general, the trend in sequence analysis has been away from tectonic and towards sedimentational interpretations.

12.8.2 General characteristics

The historic roots of the ancient fan model have grown into the main identifying features of fan successions. (1) *Palaeocurrents*. A fan shape can normally not be demonstrated in ancient fans. The closest indication of such a geometry is the presence of a radial palaeocurrent pattern (Sullwold, 1960; Rupke, 1977) (Fig. 12.57). (2) *Lithology*. Near the source there may be a predominance of thick-bedded, coarse-grained mass flow deposits other than turbidites such as debris flow, grain flow or fluidized sediment flow deposits (Fig. 12.48). The size grade may be gravel or boulders. Abrupt facies changes occur in a direction perpendicular to the radial palaeocurrent directions from elongate sandstone and conglomerate bodies to fine-grained silt- and mudstones (Nelson and Nilson, 1974; Walker, 1975; Rupke, 1977).

(3) *Sequences*. Within ancient fan successions, sequences of several orders of magnitude can be demonstrated (Ricci-Lucchi, 1975; Rupke, 1977). At one extreme is the basin fill sequence, generally several kilometres thick and composed of basin plain sediments, passing up into fan, slope and finally shelf sediments. At the other extreme is the individual graded bed, generally a few centimetres or decimetres thick. The former is primarily tectonic in origin, the latter sedimentational. In between these extremes, sequences of a few hundreds or tens of metres exist which are primarily produced by the interaction of the depositional topography of the fan surface with the depositional processes (Fig. 12.49). Thickening-up sequences reflect localized deposition by

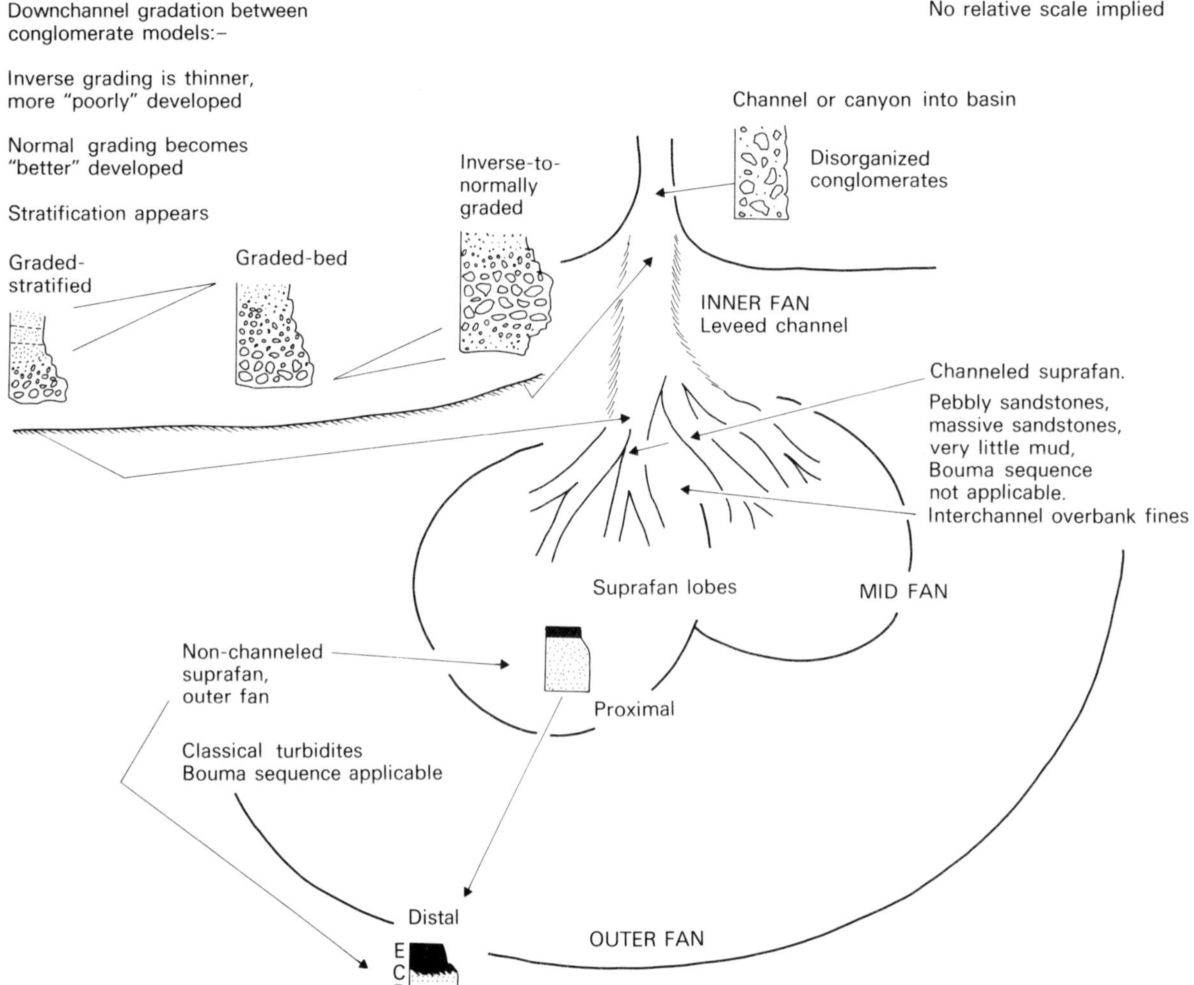

Fig. 12.48. Turbidite fan model, showing the distributions of various coarse- and fine-grained bed types (after Walker, 1975, 1977).

progradation of depositional lobes (Fig. 12.50). Thinning-up sequences characterize the lateral migration or gradual abandonment of a channel (Fig. 12.51). By analogy with models of modern turbidite fans, particularly those off the west coast of North America, ancient fans have been interpreted in terms of upper, middle and lower fan facies (Mutti and Ricci-Lucchi, 1972; Walker and Mutti, 1973). An alternative interpretation is in terms of a distributary inner fan and a prograding outer fan (Mutti, 1977).

The *upper fan facies association* is generally characterized by a thick-bedded, coarse-grained lenticular sandstone facies and a laminated or bioturbated mudstone facies (Fig. 12.52). They represent respectively channel and inter-channel deposits. Thin-bedded sand- or siltstone beds alternating with thin mudstone interbeds represent the levee facies. The palaeocurrent directions in the levee sequences diverge from those in the channel sequences (Walker and Mutti, 1973; Nelson and Nilson, 1974; Rupke, 1977).

The *middle fan facies association* is generally characterized by thinning-up sequences in its upper part (distributary channel) and thickening-up sequences in its lower part (prograding depositional lobes) (Fig. 12.52). Medium- to coarse-grained turbidite sandstones are the predominant bed type. The turbidites are laterally discontinuous and their palaeocurrents

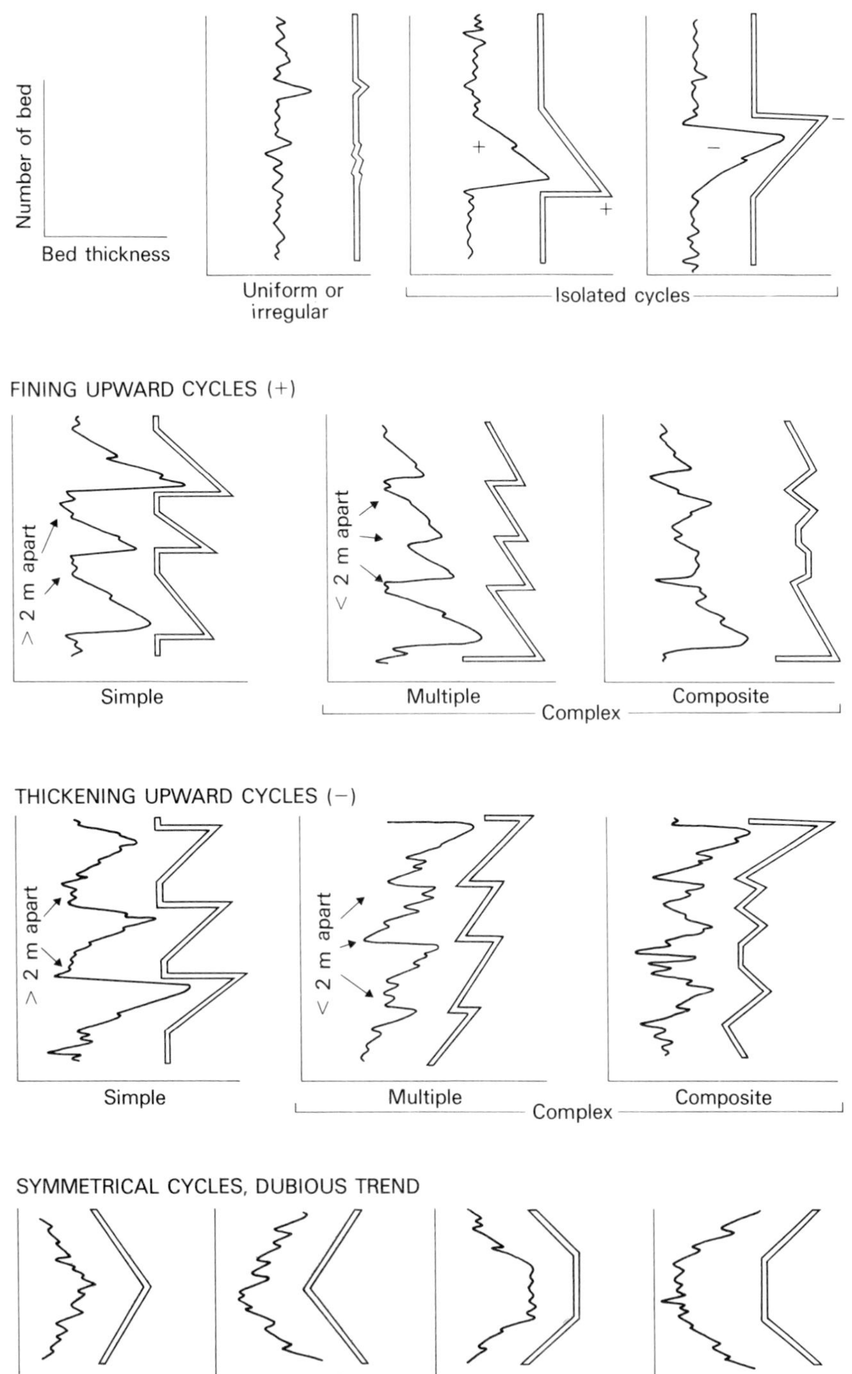

Fig. 12.49. Types of vertical sequences of the order of 10–100 m in ancient turbidite fans (from Ricci-Lucchi, 1975b).

Fig. 12.50. Thickening-up sequence from the Upper Carboniferous flysch of the Cantabrian Mountains, Spain (from Rupke, 1977).

become increasingly divergent in a downslope direction. Interchannel or -lobe areas are characterized by finer-grained beds and by occasional crevasse-splay sandstones. The *outer fan facies association* is characterized by an absence of lateral facies changes and a gradual thickening-up. The predominant bed type is that of medium-grained turbidite sandstones which may merge with basin plain turbidites (Fig. 12.52) (Walker and Mutti, 1973; Nelson and Nilson, 1974; Mutti, 1977).

Progradation of the entire active fan produces a characteristic vertical sequence which both thickens and coarsens up (Fig. 12.59). Progradation and loss of depositional slope may lead to the aggradation of channels and to subsequent avulsion and the growth of a new distributary system with depositional lobes, lateral to the old one. The new active fan will prograde over the muds of the previously inactive part of the fan. A complete progradational sequence of an entire active fan lobe consists therefore of (1) a mudstone sequence, (2) sandstone sequences from the outer and middle fan facies, and (3) a thick-bedded, coarse-grained sequence of the upper fan facies, possibly associated with a levee sequence (Fig. 12.59) (Rupke, 1977).

Fig. 12.51. Thinning-up sequence from the Eocene flysch of the southwestern Pyrenees, Spain.

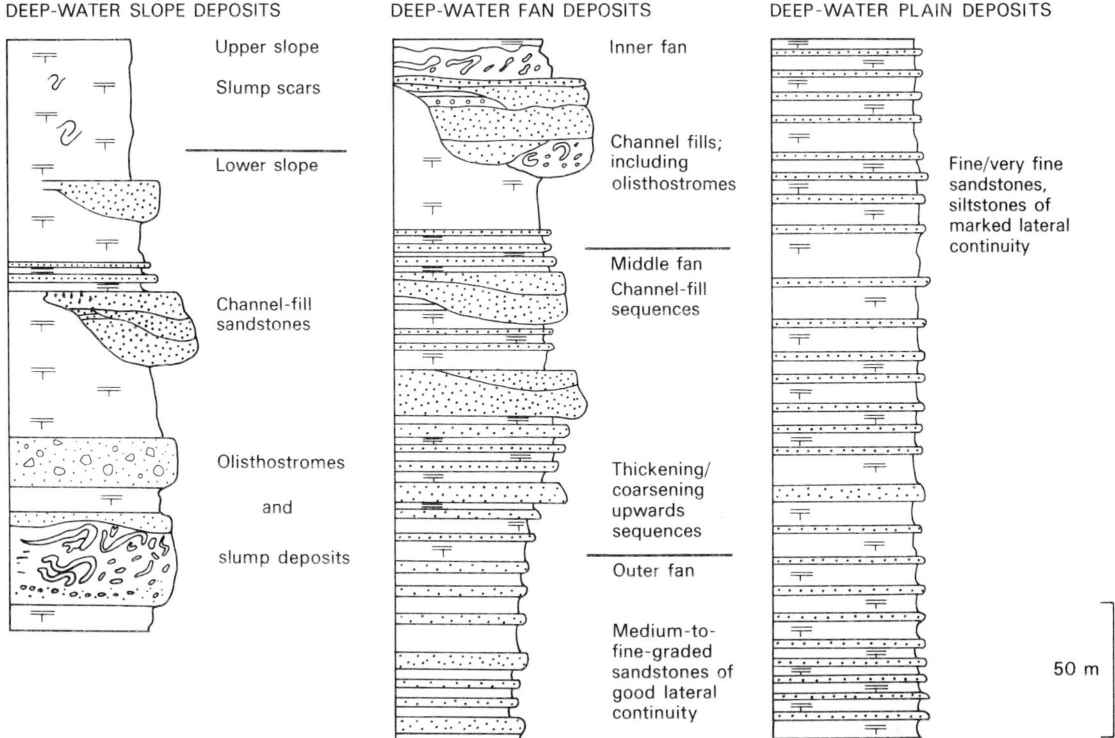

Fig. 12.52. Sequences characteristic of slope, fan and basin plain facies (after Mutti and Ricci-Lucchi, 1972).

12.8.3 **Examples of ancient turbidite fans**

Some ancient fans are associated with shallow-water facies in a progradational, regressive sequence (Walker, R.G., 1966, 1967; van de Graaff, 1971; Klein, de Melo and Della Favera, 1972; van de Kamp, Harper, Conniff and Morris, 1974). The fan model has been applied to a Namurian basin fill sequence in the Pennines, northern England where basin shales pass into turbidites of distal and subsequently of proximal aspect. Deep water channels in the proximal facies are believed to be connected with channels in the overlying fluvial delta which cut the intervening delta slope (Walker, 1966) (see Chapter 6). In the Eocene-Oligocene of the Santa Ynez Mountains, California, several basin fill sequences pass from mudstones through turbidite sandstones into shallow water facies (van de Kamp, Harper, Conniff and Morris, 1974). Variations in the lateral and vertical relationships of thick-bedded (distal) facies suggest three facies models.

Type 1 (Fig. 12.53) lacks coarse-grained turbidites and has only thin-bedded turbidites which pass directly up into shallow marine sandstones. This relationship is interpreted as progradation of a small delta-front fan into relatively shallow water under stable tectonic conditions.

Type 2 (Fig. 12.54) is similar to the Namurian fan of northern England in that coarse-grained turbidites overlie thin-bedded turbidites. This succession is related to slope instability at a delta front.

Type 3 (Fig. 12.55) has coarse-grained turbidites and conglomerate wedges separated from the thin-bedded turbidites. In turn the turbidite sandstones are separated from the shallow water sandstones by a non-depositional slope. This succession is thought to be due to slope instability produced by basin margin faulting.

In the Upper Pennsylvanian of the Midland Basin, north-central Texas, a subsurface fan has been identified based on electrical logs, cores and isolith and isopach maps (Galloway and Brown, 1973). Sandstone wedges, broadening downslope, are divided into (1) shelf-margin sandstones which grade up-slope into shelf limestones, (2) slope-trough (channel) sandstones of coarse-grained (proximal) facies, (3) slope

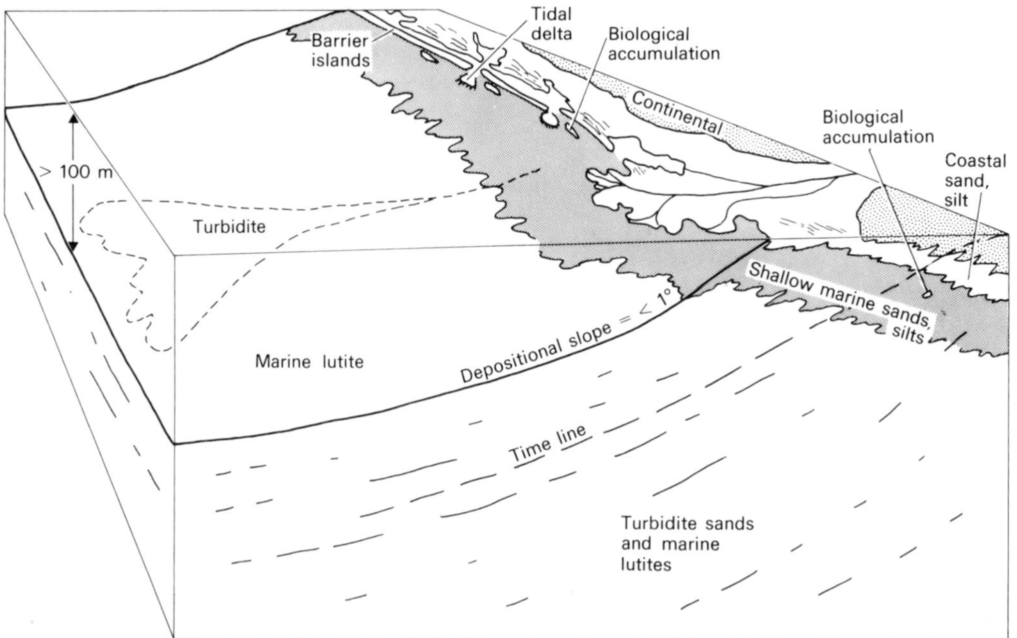

Fig. 12.53. Facies interrelationships in the Eocene–Oligocene of the Santa Ynez Mountains, California, developed under relatively stable basin conditions (from van de Kamp, Harper, Conniff and Morris, 1974).

Fig. 12.54. Facies interrelationships in the Eocene–Oligocene of the Santa Ynez Mountains, California, developed on an unstable slope produced by delta progradation into progressively deeper water (from van de Kamp, Harper, Conniff and Morris, 1974).

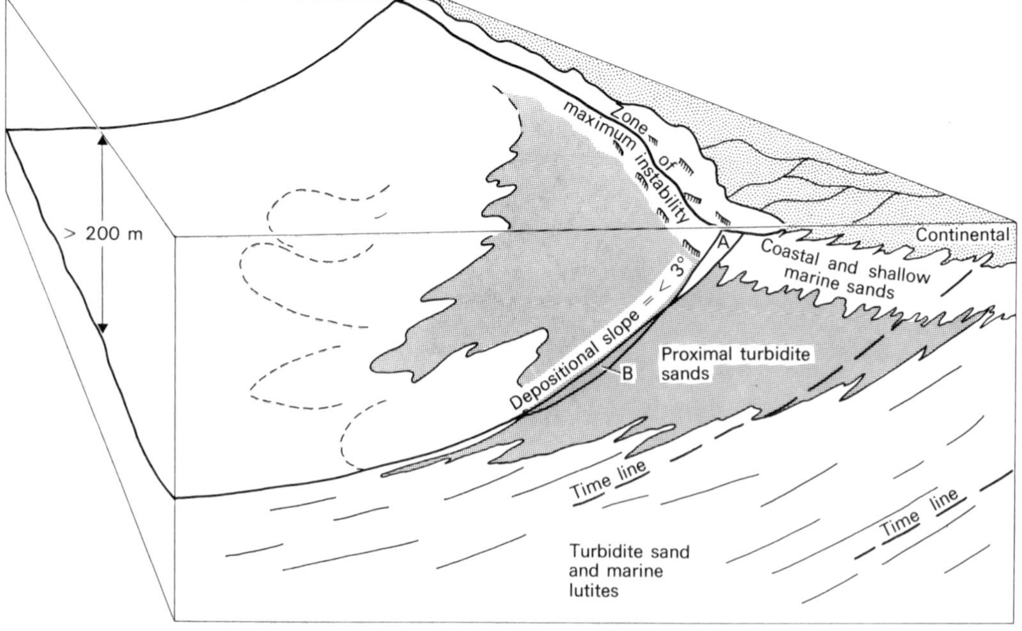

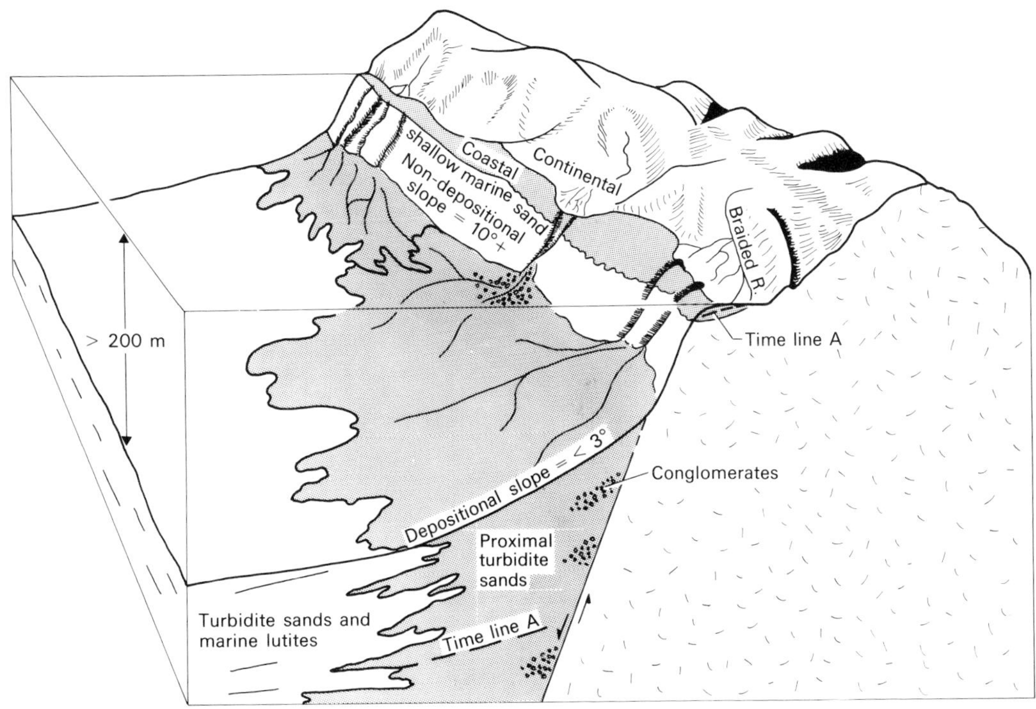

Fig. 12.55. Facies interrelationships in the Eocene–Oligocene of the Santa Ynez Mountains, California, developed under conditions of an unstable slope resulting from faulting of the basin margin (from van de Kamp, Harper, Conniff and Morris, 1974).

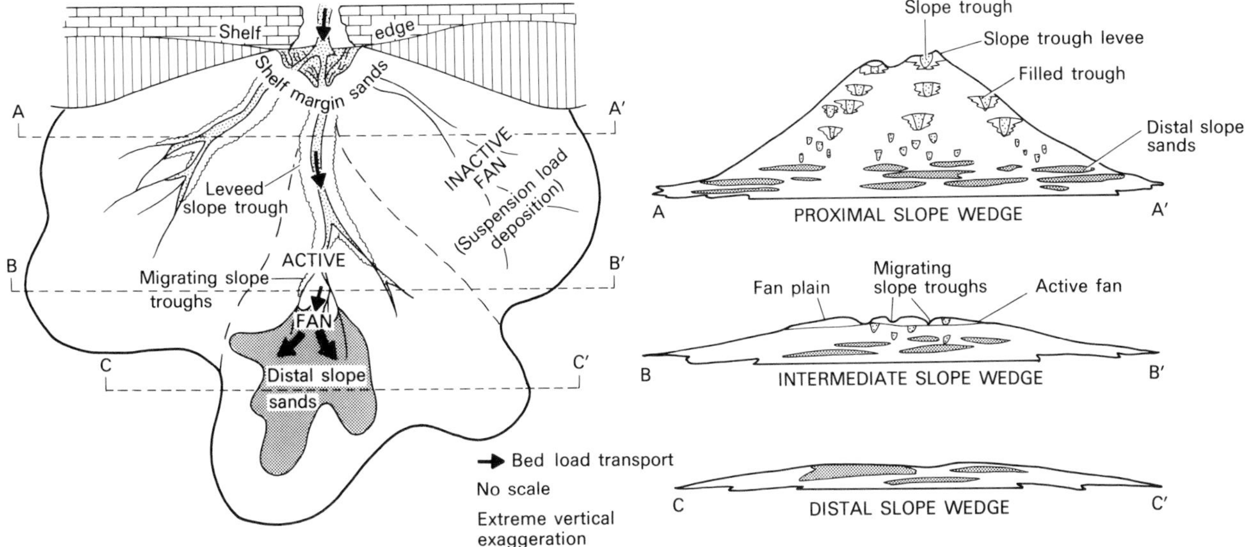

Fig. 12.56. Fan model for the Upper Pennsylvanian Cisco Group, north-central Texas (from Galloway and Brown, 1973).

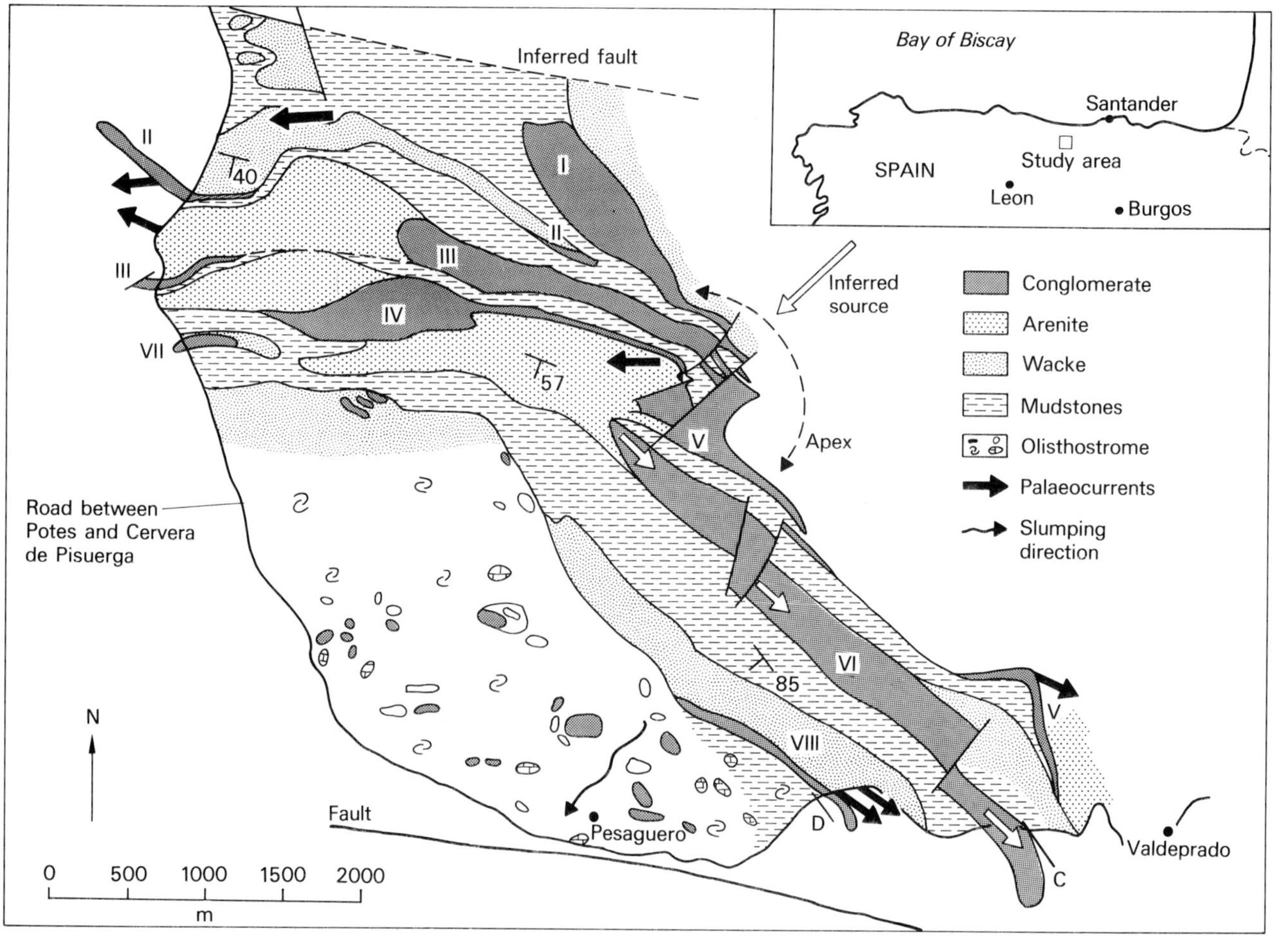

Fig. 12.57. The Upper Carboniferous Pesaguero Fan from the Cantabrian Mountains, Spain. The sequence dips more or less vertically and youngs towards the southwest. Note the radial palaeocurrent pattern (after Rupke, 1977).

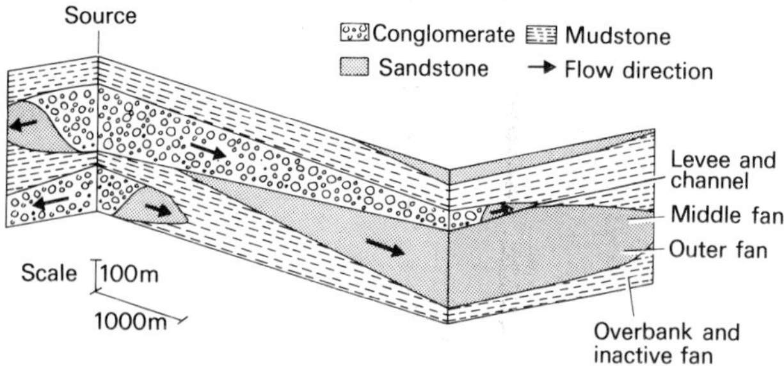

Fig. 12.58. Model of the three-dimensional shape and interrelation of the three main facies types of the Pesaguero Fan. A facies triplet (mudstone blanket, sandstone lobe, conglomerate tongue) represents a complete cycle of progradation of a major fan lobe. A new lobe forms by lateral avulsion (compare with Fig. 12.37) (after Rupke, 1977).

sandstones of thin-bedded (distal) facies, (4) slope mudstones which occur between the channels, and (5) limestone aprons which were emplaced by debris flow from the carbonate shelf edge (Fig. 12.56).

The fan was closely related to a high-constructive fluvio-deltaic system, but was separated from it by a carbonate shelf. As the delta advanced and retreated it is thought to have alternately prograded as far as the shelf edge so that sands could be directly transported to the fan head, and subsequently retreated when sands were transported to the fan head by channels crossing the carbonate shelf. Over 70% of the total volume of sands were stored in the fan system, the fluvio-deltaic sands being comparatively thin.

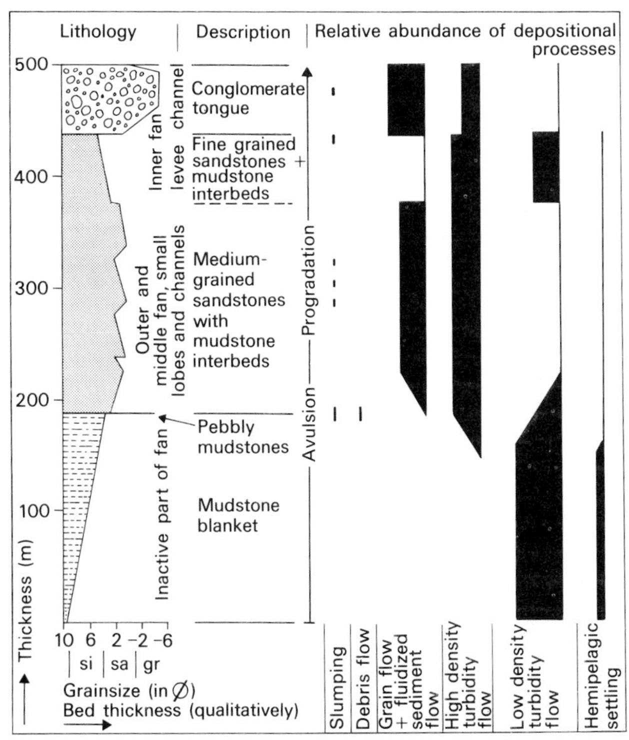

Fig. 12.59. Model of a complete sequence of progradation of a major fan lobe, combined with avulsion. Note the overall shift from low density turbidity flow to gradually denser flows upwards through the progradation sequence (from Rupke, 1977).

A similar model has been developed for the Upper Carboniferous of the Cantabrian Mountains, northern Spain (van de Graaff, 1971). A particular problem here is that the coarsening-upward quartzose sequence of the fan is overlain by thin-bedded turbidites which are of lithic composition. The former may be interpreted as derived from the shelf where current action had matured the sand, whereas the latter may have been deposited on a delta-front directly by a river (river-generated).

An example of a deep-sea fan which developed in a small fault basin is the Pesaguero Fan in the Upper Carboniferous of the Cantabrian Mountains, northern Spain (Fig. 12.57) (Rupke, 1977). No vertical passage into a shelf or a delta can be documented. The fan is composed of several sequences, each a few hundreds of metres thick and consisting of a mudstone blanket, a sandstone lobe and a conglomerate tongue (Fig. 12.57; 12.58). Each of these facies triplets represents a complete sequence of progradation of an active part of the fan (sandstone and conglomerate) over an inactive part (mudstone) (Fig. 12.59). Sedimentary processes vary systematically through the progradation sequence, from low-density turbidity currents in the inactive fan facies, through high-density turbidity currents in the outer and middle fan facies, to other mass flow processes in the upper fan facies (Fig. 12.59). New facies triplets form by avulsion and progradation. The effect of a rise in sea level on the facies organization of the Pesaguero Fan may be indicated by the fact that in an upward direction lithic wacke sandstones (immature river detritus) are replaced by quartz arenites (mature longshore drift) and subsequently by a mudstone blanket across the entire fan surface (high sea level) (Fig. 12.53). The fan succession is overlain by an olisthostrome which may reflect progradation of the slope facies and tectonic activity (Fig. 12.52; 12.59) (Rupke. 1977).

In the late Jurassic to early Cretaceous of East Greenland, fining-upward cycles, several hundred metres thick, have been interpreted by Surlyk (1978) as the deposits of coalescent deep-sea fans formed at the base of eastward-facing fault scarps which margin north–south orientated troughs. The cycles are thought to correspond to major phases of faulting and rotational tilting of fault blocks, the landward dips of which result in vertical aggradation rather than fan progradation as is commonly the case in rift-fault situations. The characteristics of this *tilted fault block* model are ubiquitous fining-upward sequences, a lack of coarsening/thickening-upward sequences, and an absence of a well-defined canyon cut into fine-grained slope deposits.

FURTHER READING

DOTT R.H. and SHAVER R.H. (Eds.) (1974) *Modern and ancient geosynclinal sedimentation.* SEPM Spec. Pub. no. **19,** pp. 56–127.

LAJOIE J. (Ed.) (1970) *Flysch sedimentology in North America.* Geol. Assoc. Can. Spec. Pub. **7,** 272 pp.

HEEZEN B.C. and HOLLISTER CH.D. (1971) *The Face of the Deep.* Oxford University Press, New York, 659 pp.

Turbidites and deep water sedimentation. SEPM Pacific Section, Short Course, Anaheim 1973. 157 pp.

CHAPTER 13 Glacial Environments

13.1 HISTORICAL BACKGROUND

The bitter controversy over the origin of the drifts, the superficial deposits containing large blocks of varied rock types found over broad areas of Europe and N. America, was resolved during the 19th century by the gradual acceptance of the glacial theory. This theory stated that extensive ice sheets and glaciers transported and deposited drift during periods of colder climate (Flint, 1971, pp. 11–15).

Until the 1950s geologists used the glacial theory to infer that drift-like rocks or mixtites in ancient sequences had been deposited by either glaciers or icebergs. More and more ancient tillites were thus identified, sometimes associated with striated pavements or pebbles, sometimes with evenly stratified shales interpreted as varves. As a result, most geological periods were considered to include an ice age (Coleman, 1926).

An alternative explanation for the origin of mixtite (lithified mixton) was introduced with the turbidity current theory which emphasized submarine slumping and mass flow (Crowell, 1957). Mixtites could now be interpreted in terms of non-glacial mechanisms and a glacial origin could not be inferred just from a rock's texture (Dott, 1961). Non-glacial mixtites formed by debris flows are also known from the alluvial fan and deep-sea fan environments (Chapters 3 and 12).

Thus, alleged glacial rocks had to be carefully re-examined and some geologists tended to favour either a glacial or a mass-flow origin. Others attempted to determine criteria for diagnosing a glacial origin (e.g. Harland, Herod and Krinsley, 1966; Flint, 1975). Many such criteria were suggested, for example, poor sorting, presence of exotic clasts, striations on clasts and pavement, great lateral extent and dropstones in laminated shale.

To some extent the problem of a glacial versus a non-glacial origin for a particular deposit has distracted attention from developing and applying conceptual models. The glacial marine models of Carey and Ahmed (1961) were applied to ancient glacial deposits by Reading and Walker (1966). However, the many subsequent applications of these models (e.g. Casshyap, 1969; Lindsey, 1971; Aalto, 1971) have not significantly increased our understanding of the processes which form glacial facies and structures. Observations on ancient deposits have seldom been aimed at adding to our understanding of the glacial environment.

13.2 PRESENT-DAY GLACIERS

About 10% of the Earth's surface is covered by glacial ice. During the Quaternary glaciation, maximum coverage was about 30% (Flint, 1971, p. 80) and the resulting sediments were distributed over a large portion of the Earth's surface. In addition to direct glacial products such as till, which form within glaciated areas, proglacial deposits such as outwash, loess and muds with dropstones also cover substantial areas marginal to glaciated regions. Moreover, large ice sheets influenced sedimentation over the whole globe, by causing changes in climate, sea level and oceanic circulation patterns.

Glacial ice extends over both land and sea, in areas of low and high relief, and low and high altitude. Thus, glacial deposits can be preserved in both marine and terrestrial settings and in a variety of tectonic situations.

Glaciers are nourished above the snow line, where snow accumulation exceeds melting. In general, temperatures are low and precipitation high. The accumulating snow is buried and compacted by subsequent snowfalls and refreezing of percolating water where a summer thaw occurs. The density increases as air is gradually pressed out. The resulting glacial ice consists of interlocking ice crystals with isolated air bubbles and is effectively impermeable to both air and water (Paterson, 1969).

A glacier or ice sheet is a mass of ice flowing under the force of its own weight. There is a constant exchange of mass and heat between the glacier and both the atmosphere above and the bed or water body below. The variation in glacier mass controls the areal extent and thickness of the ice. The thermal regime is thought to influence many geologically significant glacial processes. The continual transfer of mass and heat, which ultimately responds to variations in climate, controls the sedimentary balance between erosion and deposition and the nature of the sedimentary processes.

The relative inaccessibility of the glacial environment has impeded the development of a satisfactory general theory of sedimentation. Furthermore, glaciers move slowly compared to other agents of sedimentation, and longer periods of observations may be required to reach well-founded conclusions about the mechanisms of glacial sedimentation. Particularly hazardous is the reconstruction of vanished glaciers by interpreting their deposits.

13.2.1 **Glacier flow**

Glacial ice deforms like a crystalline solid near its melting point. The flow behaviour of ice in a glacier can be simply described by two equations. One relates the deformation of the ice (shear strain) to the deforming force (shear stress), the other describes the distribution of shear stress in the glacier. By mapping the latter, one may calculate the shear strain, and thus the velocity for any point in the glacier. The greatest shear strain is found in the basal part of the glacier, while the maximum velocity occurs at the top of the glacier.

Some glaciers also slide over the bed, in which case the total surface velocity is the sum of the basal sliding velocity and the velocity due to internal strain. A temporary, catastrophic increase in basal sliding results in glacier surging. Such glaciers have velocities of the order of 100 times the normal.

13.2.2 **Thermal regime**

A fundamental classification of glaciers is based on the thermal condition or regime of the ice. This is determined by the net heat budget, which is the difference between the amount of heat lost and gained by the glacier over a given period of time. *Cold glaciers* are those whose temperature is below the pressure melting point. Free water does not exist in the glacier. *Temperate glaciers* are at the pressure melting point and free water exists, for example, in englacial or subglacial tunnels. Some glaciers contain both temperate and cold ice.

The thermal state of ice at the base is of greater sedimentological importance than that of the ice within the glacier. *Dry-based glaciers* are thermally cold at the base, and they are generally considered to be frozen to the bed (rather than sliding over it) as the adhesive strength of the glacier-bed contact is greater than the shear strength of the ice. *Wet-based glaciers* are thermally temperate at the base and slide over the substrate, separated from it by a thin film of water.

The supply and removal of heat from the glacier is directly related to climate. A cold climate (for example, continental) contributes cold snow to the glacier and surrounds it with cold air. A warm climate (for example, maritime) leads to summer thaws during which water percolates into the glacier or through it via tunnels. Seasonal variations affect, at the most, only the upper 20 m of the glacier. The two other main sources of heat are geothermal heat, which is capable of melting about 6 mm of ice per year, and frictional heat generated by both internal

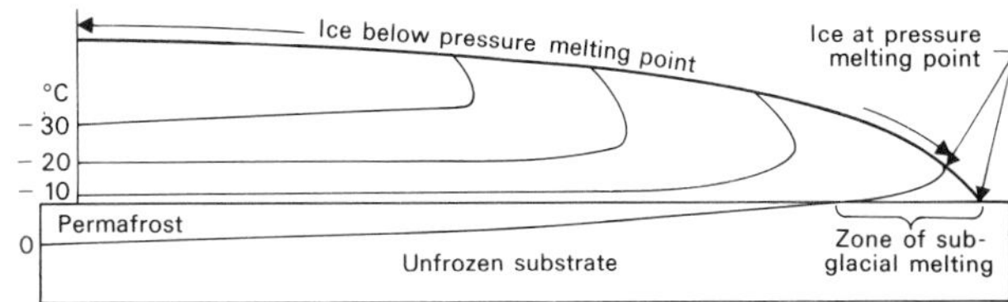

Fig. 13.1. Temperature distribution in a large ice sheet consisting of mainly cold, dry-based ice. The marginal part is temperate and wet-based (based on Hooke, 1977).

strain and basal sliding (Paterson, 1969). Heating and cooling also take place by advection, the transport of cool or warm masses of ice within the glacier.

In a temperate glacier, which is at the pressure melting point, there is no significant vertical temperature gradient. Geothermal heat and heat generated by friction are used in melting basal ice. Cold glaciers, whether wet- or dry-based, have a thermal gradient which shows a general increase in temperature downwards and in the direction of flow (Fig. 13.1, Hooke, 1977). Thus a largely cold glacier may have a temperate base at the margin.

13.2.3 **Mass balance**

Continuing isostatic adjustments of recently deglaciated areas, earlier eustatic changes, and historical records of advancing and retreating ice fronts indicate that geologically rapid fluctuations in glacier mass can take place. *Accumulation* includes all material added to the glacier, predominantly by snowfall. *Ablation* refers to material removed from the glacier, by basal ice melting, runoff, evaporation and iceberg calving.

The difference between the amounts of accumulation and ablation over a given time is the net mass balance, or net mass budget. A glacier in which accumulation is equal to ablation (net mass balance = 0) will have a constant mass, with a corresponding thickness and area. A change in mass balance changes the dimensions, determining whether the glacier margin is advancing, retreating or stationary.

13.3 GLACIAL AND RELATED ENVIRONMENTS

The *glacial* environment proper embraces all areas which are in direct contact with glacial ice. It includes three zones, two of which are important for sedimentation. The *basal* or *subglacial* zone is the lower part of the glacier which is influenced by contact with the bed. Both erosion and deposition can take place. The *supraglacial* zone is the upper surface of the glacier where seasonal influence is strong. Deposition may occur in areas of ablation, chiefly at the glacier margin. The marginal

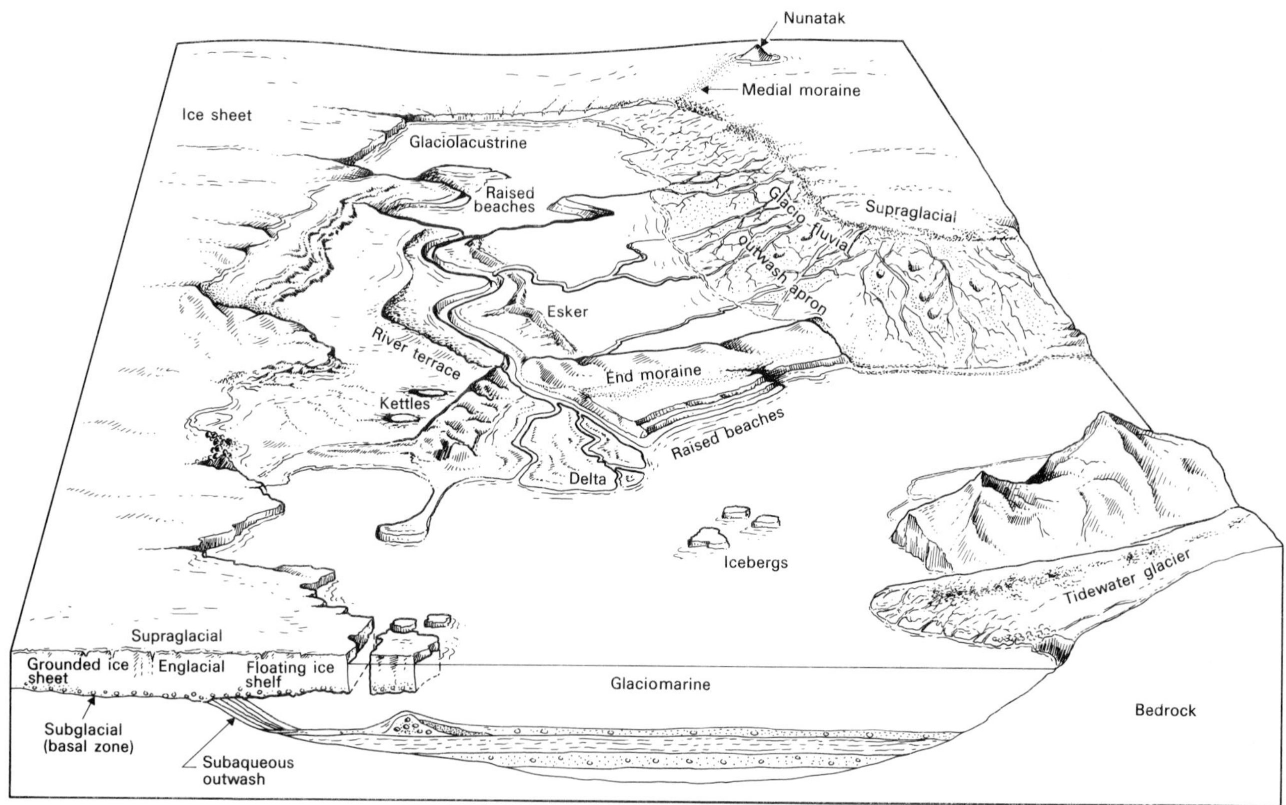

Fig. 13.2. Glacial environments and associated landforms typical of glaciated areas, including glaciofluvial outwash fan, ice-contact end moraine, glacial lake- and sea-bottom deposits, raised beaches and marine delta.

supraglacial zone grades into the *ice-contact* part of the proglacial zone. The interior or *englacial* zone does not play a major role in sedimentation.

Glaciers in contact with the bed are *grounded,* whether above or below sea level. Around its margin, and strongly influenced by the ice, is the *proglacial* environment, including the *glaciomarine* environment, where glacier ice floats over a layer of sea water, as well as the *glaciofluvial* and *glaciolacustrine* environments. Buried stagnant ice, which is no longer flowing with the glacier, may form in abundance along a terrestrial glacial/proglacial boundary. Overlapping with the proglacial environment is the *periglacial* environment, which is influenced by the distinctive climatic zones adjacent to an ice sheet (Fig. 13.2). The characteristic sediment of the glacial environment is till, a glacially deposited mixton. A mixton is a sediment composed of a mixture of gravel, sand and mud-sized particles, mixtite being the lithified equivalent (Schermerhorn, 1966). Sediment derived by the reworking of till forms in the proglacial environments. Periglacial processes further modify previously deposited sediments, whether glacial or non-glacial in origin.

13.3.1 The basal zone

Processes occurring in the basal zone are thought to depend upon the thermal state. A dry-based glacier is frozen to the substrate. A wet-based glacier has three possible conditions: net melting, net freezing, or equilibrium (Boulton, 1972).

The absence of differential movement between the glacier base and the bed of a dry-based glacier might suggest that neither deposition nor erosion is likely. However, it is possible that erosion will occur if a subglacial layer composed of frozen material, such as permafrost, is frozen on to the base of the glacier and is incorporated into the glacier base. This process is thought to occur mainly at the marginal zone of a glacier or ice sheet (Boulton, 1972, 1975). Plastic flow in the basal zone may lead to rapid disintegration of poorly consolidated materials.

The behaviour of the contact zone between the glacier bed and the glacier base may also be influenced by ice thickness, pore-water pressure, glacier bed relief and characteristics of the debris assemblage (Boulton, 1972, 1975). It has been suggested that these factors strongly control the amount of friction between the glacier bed and base. An increase in friction tends to favour deposition of basal till and deformation of the substrate, if this is composed of poorly or unconsolidated non-frozen materials. Frozen sediments have a higher strength than glacier ice, but may be moved by the glacier where they overlie unfrozen, weak materials.

An increase in ice thickness increases friction with the bed, favouring deposition of basal till. The greater ice overburden also increases the strength of the substrate, which inhibits deformation, especially if the angle of internal friction is large (as for example for coarse-grained sediments). Low permeability or a ponded water table may lead to high pore-water pressure in the substrate which counteracts the pressure due to ice overburden. The resulting decrease in basal friction and substrate strength inhibits till deposition and favours deformation of the substrate. Thus a well drained sand underlying a thick ice sheet would be least likely to deform, while a saturated mud beneath a thin ice sheet would deform comparatively easily. An increase in the dip of the bed in an upglacier direction or a general increase in bed roughness increases basal friction, favouring till deposition and substrate deformation. The basal debris which is in contact with the bed generates friction with the bed. The friction increases with the area of contact between a particle and the bed.

Observations of the basal zone of many glaciers suggest that temperate glaciers, such as those in western Norway and the Alps, have only a very thin layer of debris-rich ice at the base, while cold glaciers in polar areas, for example, Arctic Canada, Greenland and Svalbard, may have substantial quantities of debris extending for several to several tens of metres above the base of the glacier (Boulton, 1970). A persistent feature of the debris is that it occurs in discontinuous bands roughly parallel to the glacier base. Such bands vary in thickness and debris concentration. This debris is probably incorporated into the base of the glacier by freezing of water along the glacier base (Boulton, 1970). Net freezing of the subglacial water layer continues to add ice to the base of the glacier and lifts debris further into the glacier downstream. On the glacier bed, the alternation of sites of debris inclusion with sites of formation of clear regelation ice leads to banding. Fluting of the ice around debris particles indicates that deformation of the regelation ice takes place after inclusion into the glacier.

Meltwater tunnels at the base of glacier margins show that not all water in the basal zone is transmitted in a thin layer. Channelization of water is believed to become stable during high melting rates (Shreve, 1972). Away from the margin, such tunnels are filled with water, whose flow maintains the size and shape of the tunnel. Such subglacial streams are at least one process contributing to the formation of eskers. The pseudo-anticlinal bedding observed in some eskers may be due to secondary currents in the tunnels (Shreve, 1972).

The various thermal states possible at the glacier base, and the additional independent factors discussed above, give an indication of how complicated subglacial processes are. These factors probably vary both spatially and temporally. Simple models can be drawn to show possible positions of different thermal states at the glacier base at a given time, or the movement of zones with time (Boulton, 1972).

13.3.2 The supraglacial and ice-contact proglacial zone

At the glacier margin, ablation concentrates englacial debris on the upper surface. This debris may be derived from different sources (Fig. 13.3); debris bands (Goldthwait, 1951), debris along active thrusts (Boulton, 1972), fluid till squeezed up through fissures in the glacier (Gravenor and Kupsch, 1959), and medial and lateral moraines (Sharp, 1949). Thrusts are frequently observed at the glacier margin, possibly reflecting a longitudinal compressive stress due to freezing of the margin to the bed, a proglacial obstruction, or stagnation of glacial ice (Boulton, 1968; Flint, 1971).

The nature of glacier marginal processes is apparently related to whether the glacier margin is advancing, stationary or retreating. The front of an advancing glacier is a steep wall. As the glacier advances, frontal ablation releases debris, which falls or is washed down in front of the glacier. With continuing advance, the glacier overrides its own supraglacial deposit, which is presumably remoulded in the process. In contrast, a retreating glacier has a gentle profile. Melted out debris will be distributed over a large area, resulting in a thin deposit.

A glacier whose margin is stationary can transport supraglacial debris to the same place over a long period of time. This results in a substantial moraine, normally up to about ten metres high and several hundred metres wide. Melting out of debris along bands and thrusts produces a supraglacial mantle of ablation till as much as several metres thick (Fig. 13.3, Boulton, 1972). These accumulations rest on a melting surface of ice, are saturated with water, and often flow downslope as a viscous to fluid flow till (Goldthwait, 1951; Boulton, 1968). The development of a till mantle protects the underlying ice by decreasing the rate of ablation. Thus, large blocks of buried stagnant ice may persist long after the glacier has retreated from the area. Gradual melting of this ice will deform sediment deposited on it (Kaye, 1960).

Mobilized flow tills may flow for large distances down low slopes (Boulton, 1968) and thus become intercalated with proglacial stratified deposits of glaciofluvial or glaciolacustrine origin (Fig. 13.3). In addition, heavily laden supraglacial

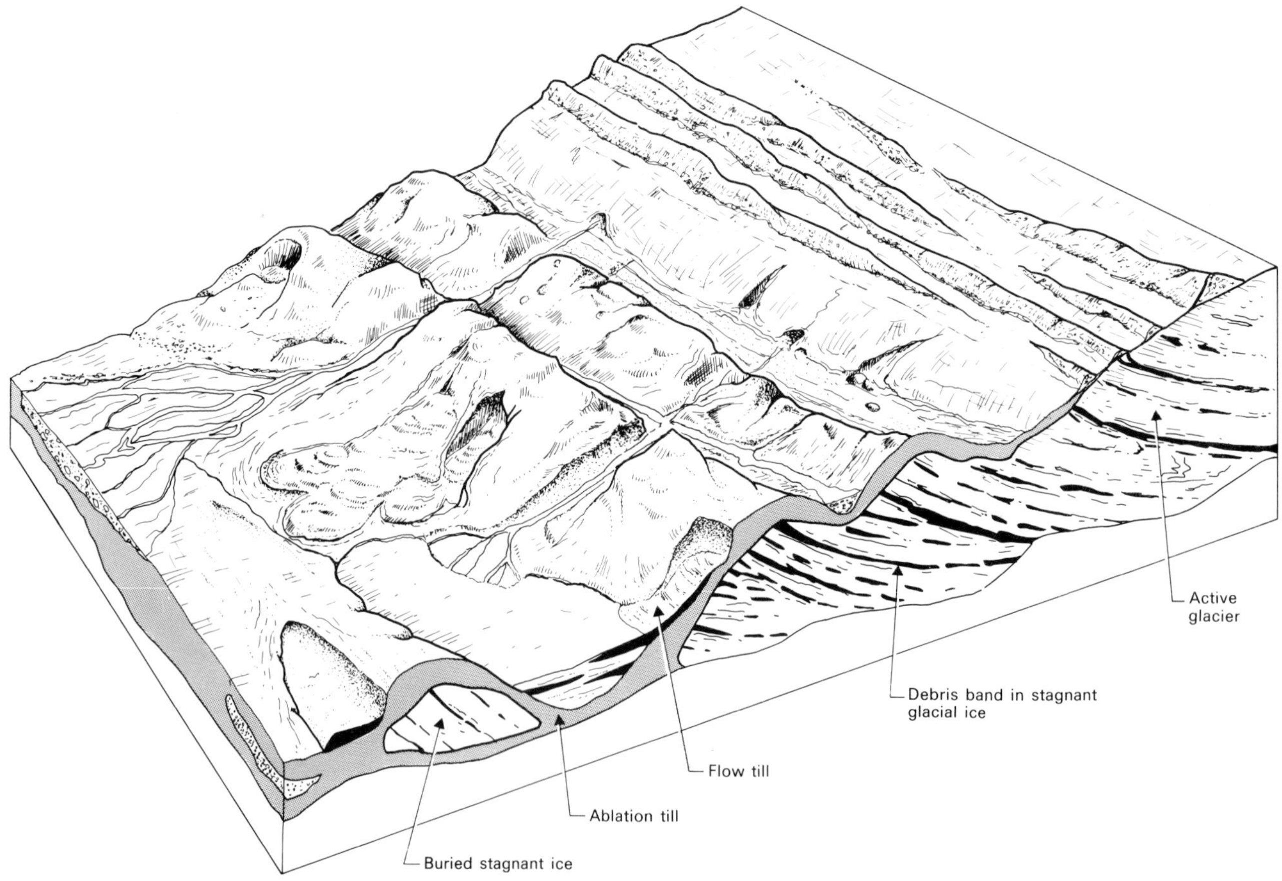

Fig. 13.3. Supraglacial to ice-contact proglacial zones of a glacier whose margin is gradually retreating, and which is thermally cold at the margin. Subglacial material is brought into the glacier by basal freezing and thrusting. This material is released at the surface as ablation till, as the enclosing glacial ice gradually melts. Ablation till is reworked by flowing meltwater, and may slump downslope to form flow tills. Beds of flow till can be intercalated with proglacial stream or lake deposits, and may be extensively reworked (modified from Boulton, 1972).

streams may rework some of the exposed till and dump this in front of the glacier, where it may be further reworked by proglacial streams. Thus a frequent feature of supraglacial/proglacial deposits is extreme variability in texture, and the lenticularity of sedimentary units.

13.3.3 **Glaciofluvial environment**

Stratified drift composed largely of gravel and sand forms adjacent to a glacier where meltwater is abundant and coarse material is available (Flint, 1971). The dominance of braided stream activity in this environment is due to the association of high slopes, variable discharges and coarse grain sizes (Fahnestock, 1963). A drift apron can form around the ice margin. Adjacent to the glacier it has certain features typical of ice-contact stratified drift (Flint, 1971): interstratified flow tills, blocks of stratified drift, often deformed and probably originally ice-cemented, and deformation due to subsequent melting of large buried ice blocks. These features decrease away from the glacier margin. Similar sediments may also form behind the glacier margin as kames, and at the glacier margin as kame terrace and marginal stream deposits (Flint, 1971). There may be a close similarity in composition between glaciofluvial outwash and till formed during the same glacial (Gillberg, 1968b). Extensive outwash fans or aprons can be expected along the margin of a stationary glacier or where extensive earlier coarse moraine has been reworked by stream activity. Elsewhere deposits are usually thin and lenticular.

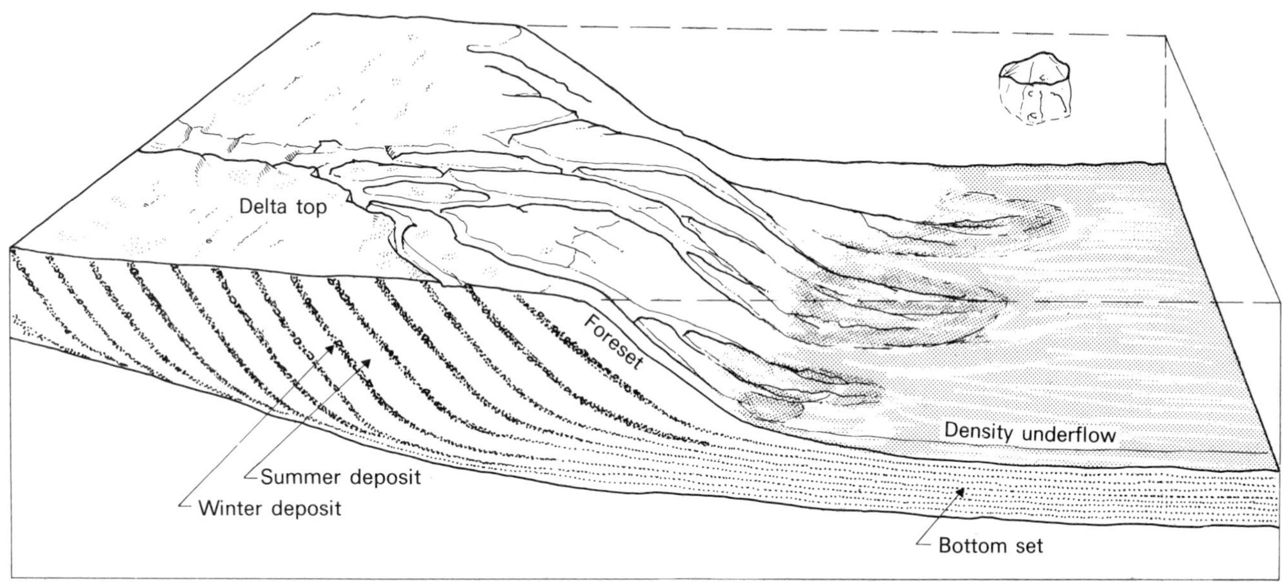

Fig. 13.4. Sedimentation in a glacial lake. During the summer, mainly climbing ripple-laminated sand is deposited on a prograding delta, while thin silt layers are deposited by a density underflow on the lake bottom. During the winter, these areas are covered with a clay drape. Icebergs, and shore and river ice may raft coarse sediment into the lake during the spring thaw and summer.

13.3.4 **Glaciolacustrine environment**

Lakes are a common feature in the proglacial landscape. Several factors may lead to lake development on a regional scale: damming of river courses by glacial ice, the reversal of regional slope by isostatic depression due to glacial buildup, and formation of irregular topography by glacially deposited or eroded landforms (Flint, 1971). Damming by glacial ice and glacial deposits is usually temporary during deglaciation, while overdeepening due to glacial scour can form more permanent lakes. Although at any one time a lake may not be very large, the lake may follow the retreating glacier margin so that, by the end of deglaciation, lake deposits or beaches may cover a large area.

Deltas formed at the margins of glacial lakes generally have steeply inclined foresets and pass downslope into fine-grained lakefloor deposits. Deltas generally occupy a small part of the entire lake area, while the remainder may be covered with lake-bottom silts and clays. Wave activity contributes to the formation of beaches around the margins of large lakes. Beach deposits are generally thin, composed of relatively well sorted sand and gravel, but the extent of their development depends on the stability of the lake level and the erodibility of the coastal materials.

Where the glacier terminates in a standing body of fresh or salt water, coarse subaqueous outwash may be deposited (Rust and Romanelli, 1975). This ranges from esker deposits formed in subglacial channels transverse to the glacier margin, to fan-like deltas (Banerjee and McDonald, 1975), to boulder moraines formed under a glacial ramp, parallel to the glacier terminus (Barnett and Holdsworth, 1974). In fresh or brackish water, these coarse deposits pass rapidly into varves.

Glacial-lake bottom deposits typically consist of varves. A varve is a type of rhythmic sedimentary deposit which forms in the course of one year (Ashley, 1975). Each varve consists of two distinct layers, a lower coarse layer composed mainly of very fine sand to silt and an upper fine layer composed mainly of clay. Occasionally a basal third unit composed of fine sand is observed. The total thickness of each varve generally decreases markedly away from the ice front. Upward thinning of varves is related to progressive glacial retreat during deglaciation. Varves are a result of two mechanisms (Fig. 13.4):

(1) Sediment-laden stream water is denser than lake water, so that the coarsest sediment is dumped in the form of a steep-fronted delta near the stream mouth, while finer sediment is transported further as a density underflow (Kuenen, 1951; Gustavson, 1975). The occasional presence of ripple cross-lamination in the lower layer indicates deposition by a bottom traction current and argues against an origin by fall-out from above.

(2) Strong seasonal variations in run-off and a winter ice cover on lakes lead to deposition of coarse sediment during the

summer and the fine sediment from suspension during the winter.

Varves formed in glacial lakes generally show a clear separation of the silt and clay fractions and are termed *diatactic* (Sauramo, 1923). In contrast, varves considered to have formed in brackish water show mixing of clay and silt in both the coarse and fine layers and are termed *symmict*.

Glaciomarine deposits lack rhythms. For example, varved lake deposits which pass gradually upward into fossiliferous laminated muds generally show a dying out of varves concomitant with the development of marine conditions (Gadd, 1971). This may be due to several factors:
(1) Cold, sediment-laden glacial meltwater, being less dense than sea water, flows over it, dispersing suspended sediment over a wide area,
(2) Clays entering saline water are flocculated and thus behave hydrodynamically more like silt particles (Sauramo, 1923) and
(3) Glacial seas may have sediment sources in addition to glacial streams and be affected by various types of currents, so that seasonal influence can be sharply reduced compared to glacial lakes.

Dropstones (clasts dropped from floating ice) are occasionally present in glaciolacustrine deposits. They do not seem to be especially abundant, possibly because calving of icebergs will not occur in relatively shallow glacial lakes (Flint, 1971). Nevertheless, seasonal lake, shore and river ice is capable of transporting some coarse material during the thawing of the lake and its tributaries.

13.3.5 **Aeolian environment**

Both climatic instability and powerful katabatic (sinking, cold air layer) winds contribute to aeolian activity around a glacier or ice sheet. Ventifacts are formed by the abrasive action of wind-transported grains. Sand reworked from glacial deposits, particularly dried out alluvial flats, is shaped into aeolian dunes. Silt is also readily picked up from fluvioglacial outwash and is deposited down-wind as loess, generally well sorted and poorly or non-stratified (Smalley, 1976).

Loess deposits can form blankets up to tens of metres thick and extending hundreds of kilometres from the ice margin. The blankets mantle topographic highs and lows and decrease in thickness and in mean grain size down-wind away from the source (Smith, 1942). The long axes of loess grains are preferentially orientated parallel to the depositing wind, and imbricated in an up-wind direction (Matalucci, Shelton and Abdel-Hady, 1969).

13.3.6 **Pedogenic environment**

If no substantial erosion or deposition takes place, soil-forming processes modify exposed sediments. These processes involve the vertical movement of carbonate, iron oxides, silicates and other chemicals in soil solutions. Compared to fresh tills, weathered tills show solution of carbonate from clasts and matrix, partial solution of silicates and disappearance of unstable minerals (Ruhe, 1965; Willman, Glass and Frye, 1966).

Physical disturbance of the upper soil layers may take place under freeze-and-thaw in a periglacial climate. Typical forms are frost wedges and involutions (Sharp, 1942). The latter can be confused with loading phenomena.

13.3.7 **Glaciomarine environment**

The glaciomarine environment includes seas bordered by tide-water glaciers or ice shelves. The ice masses influence sediment transport by rafting and by modifying normal marine processes via changes in sea-water temperature, salinity, density, and concentration of suspended sediment.

Glacial debris can be rafted by an ice shelf or by icebergs. Dropstones are substantially coarser in grain size than the matrix into which they are dropped. If the matrix is laminated, the impact of the stone may puncture or buckle the underlying laminae. During differential compaction, the laminae surrounding the stone are pressed around it. Glacial dropstones may include striated and faceted clasts, and till clots, i.e. lumps of till released by melting glacial ice (Ovenshine, 1970). The sudden overturning of an iceberg can drop masses of coarse melted-out debris from the top of the iceberg into the sea. Even during non-glacial conditions seasonal ice can transport coarse sediment.

Three distinct types of sediment are forming around tide-water glaciers in Spitsbergen (Boltunov, 1970). Beneath the terminus of the glacier, large quantities of debris are released on to the sea floor with minimal reworking. Just seaward of this is a zone with little coarse material, as this is either dropped under the glacier or carried away by icebergs (Fig. 13.5). In the outermost zone, normal sedimentation takes place, accompanied by rafting of coarse material. On the basis of observations on these and Pleistocene deposits, Boltunov contrasted basal tills and glaciomarine mixtons. Basal tills show no indigenous fauna or flora, no sorting, a preferred clast long axis orientation and no stratification. Glaciomarine deposits show molluscs buried in living position, occasional marine plant fragments, definite sorting with a higher clay content, clast long axes with random azimuthal orientation and stratification. They have a lower density, higher porosity and higher plasticity index than subglacial tills, as reported by other workers (e.g. Easterbrook, 1964).

Buoyancy exerted by the sea reduces friction at the base of the ice sheet and causes the thinning characteristic of an ice shelf. It has been suggested that patterns of sedimentation around an ice shelf are largely controlled by the thermal regime

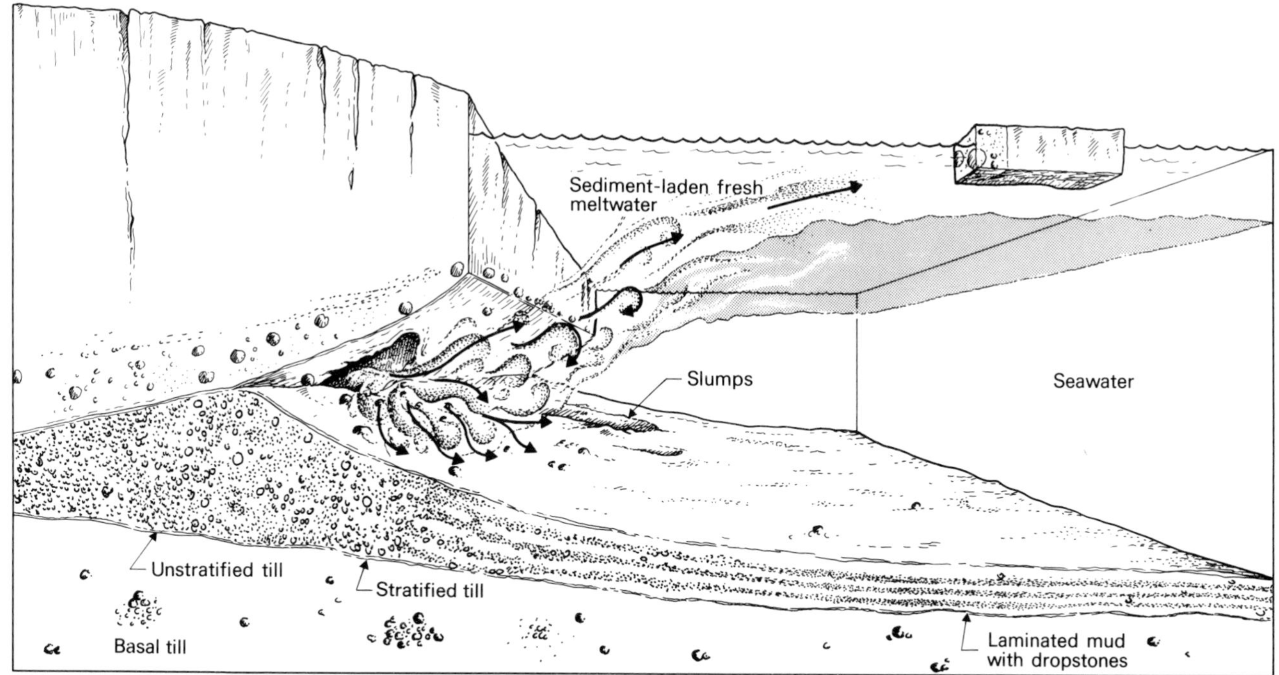

Fig. 13.5. Hypothetical model of glaciomarine sedimentation adjacent to a wet-based tidewater glacier, based on observations of Pleistocene subaqueous outwash and considerations of current dynamics of modern marine deltas. Most of the fresh glacial meltwater rises to the surface of the sea as a low density overflow layer. This layer gradually mixes with sea water while being driven by winds and currents. Silt and flocculated clays gradually fall out of suspension. Rapid mixing of fresh water and sea water adjacent to tunnel mouths may produce a high density underflow capable of transporting sand-grade sediment and occasionally coarser material.

of the ice shelf, which determines whether freezing or melting is taking place at the base (Carey and Ahmed, 1961). An important form of ablation in all ice shelves is the calving of icebergs where tidal and other stresses fracture the seaward thinning shelf ice. The stability of an ice shelf is thus greatly increased by lateral valley or coastal support and by points of grounding on the seabottom.

Additional ablation occurs by melting at the base and/or top of the ice shelf. Bottom melting of a wet-based ice shelf leads to the formation of brackish, low-density water. Bottom freezing of dry-based ice shelves may locally incorporate bottom sediments. Freezing at the base of the shelf increases the salinity and density of the remaining sea water. Masses of dense brines formed by this process may contribute to the formation of Antarctic Bottom Water (Anderson, 1972). Alternatively, these brines may result from the formation of pack ice, most of which is blown offshore (Gill, 1973).

Studies of sediments surrounding the Antarctic ice shelves have been based on relatively few, widely spaced bottom samples. Anderson (1972) related bottom sediments to inferred wet- and dry-based ice shelves fringing the Weddell Sea. Gravelly well-sorted sands adjacent to wet-based shelves pass offshore into gravelly mud (mixton), sandy mud and mud. Mud is the predominant deposit of dry-based shelves, with coarse rafted material present only in distal areas. The nature and distribution of sedimentary processes adjacent to the Ross Ice Shelf was studied by factor analysis of grain-size parameters of bottom samples (Chriss and Frakes, 1972). Adjacent to the ice shelf, faintly to non-stratified mixton is formed mainly by ice rafting. Seaward, ice rafting decreases in importance, and reworking (retransport and winnowing) has formed poorly to moderately well sorted muddy silts and fine sands. The significance of these results is brought into doubt by the widespread erosion reported by Fillon (1972).

Late Cenozoic glaciomarine deposits are occasionally fossiliferous (including various microfossils and articulated complete bivalves partly in living position) and are laminated or massive (see Sect. 13.4.3). Marine fossils in a till do not necessarily indicate a marine origin and may be found in tills containing eroded fossiliferous sediments. Models of ice-shelf deposition have predicted the formation of turbidity currents and mass flows under wet-based conditions (Carey and

Ahmed, 1961; Anderson, 1972). However, turbidites are relatively uncommon in inferred Late Cenozoic glaciomarine deposits (Barrett, 1975).

13.4 GLACIAL SEDIMENTARY FACIES

Glacial sediments in many Quaternary and older sequences can usually be grouped into more or less distinct descriptive sedimentary facies (Table 13.1). A review of the deposits associated with the early Proterozoic, late Pre-Cambrian, late Ordovician, late Palaeozoic and late Cenozoic glaciations suggests that three such facies dominate these successions: (1) massive tillite, (2) stratified conglomerate and sandstone, (3) laminated mudstone with or without dropstones. A fourth facies, beds of mixtite or tillite, occurs with facies (2) and (3). A fifth facies, banded tillite, does not seem to be widespread.

For Quaternary deposits on the continents, the interpretation of these facies is often relatively straightforward: massive till is subglacial basal till, stratified gravel and sand are glaciofluvial or ice-contact proglacial deposits, and laminated muds with varves, dropstones or diagnostic fossils are glaciomarine or glaciolacustrine. Beds of till or mixton emplaced by mass flow are interbedded either with stratified gravels in the ice-contact zone or with laminated muds in the proglacial subaqueous environment. Massive tills are also found in late Cenozoic glaciomarine deposits, associated with laminites and other marine facies.

In ancient glacial deposits, it is often difficult to determine both the precise environment and the extent of glacial influence for a particular facies. The approach followed here is to identify the various facies developed in the succession, and then to relate these to each other and to applicable sedimentary models so as to reconstruct a coherent depositional environment. The glacial facies considered below are distinguished on the basis of their descriptive characteristics. However, in order to emphasize the genetic association of the facies they are grouped into three general environments: (1) subglacial, mainly with massive tillite and occasional banded tillite, (2) supraglacial to ice marginal proglacial, with varied conglomerates and sandstones and (3) subaqueous, embracing glaciomarine laminites and massive tillite, and also glaciolacustrine varvites.

13.4.1 Subglacial deposits

Two subglacial facies are distinguished, based on their internal structure. *Massive* till, which lacks internal stratification, is the most common. *Banded* till shows a banding similar to that observed in schists and gneisses. There are also compositional distinctions between tills, often reflecting distance of transport (Francis, 1975). Material which has not been transported, but only deformed below a moving glacier is *deformation* till. A till consisting of local material only is termed *local* till. An *exotic* till consists exclusively of far-travelled materials. Most tills include both exotic and local components.

In addition to subglacial basal tills, recognition and use of subglacial deformation structures are important in determining both palaeoflow directions and subglacial conditions. Though not strictly a facies, deformation structures are treated with subglacial deposits.

MASSIVE TILLITE

Subglacially formed massive tillite is a characteristic facies of all glaciations. Most massive tillite units: (1) are texturally a

Table 13.1 Outline of the main glacial facies, their chief process of deposition, and the environments in which they form

Facies lithology	*Depositional process*	*Environments*
Massive mixtite Banded mixtite	Basal till deposited below active ice	Subglacial
Stratified sandstone and conglomerate	Deposition in flowing water	Supraglacial, englacial, subglacial, proglacial (incl. subaqueous)
Beds and lenses of mixtite in sandstone and conglomerate	Mass flow of mixton (flow till) and flowing water	Supraglacial, ice-contact proglacial (incl. subaqueous)
Rhythmically laminated siltstone and claystone	Varves, seasonal deposits	Glaciolacustrine
Randomly laminated mudstone with dropstones	Deposition from suspension and rafting by ice in agitated water	Glaciomarine
Massive mixtite	Aquatill, deposition from suspension and ice rafting in quiet water	

mixtite, (2) lack internal stratification, (3) can be traced for at least several kilometres, (4) are several metres to tens of metres thick, and (5) contain a wide range of clast types, some of which may be faceted and striated.

Regionally, tills appear to thicken from the source area to marginal areas, thinning again at the fringe. Quaternary tills of eastern Canada have been tentatively correlated for over 2,000 km (McDonald, 1971). The general abundance of locally derived material in till suggests that erodibility of substrate is a very important factor controlling till thickness. Where there is considerable local relief, glacial deposits are thicker in depressions, and thinner or absent on highs (Flint, 1971, p. 171). Massive tills occur as sheets (White G.W., 1968), tongues (Flint, 1971, pp. 152–3), or wedges (Edwards, 1975a) but locally cross-sections appear blanket-shaped (White, Totton and Gross, 1969).

Erosion at the lower contact of a massive tillite unit may be indicated by mapping on a regional scale (Banks, Edwards, Geddes, Hobday and Reading, 1971), by observations on a local scale at an outcrop (Beuf, Bijou-Duval, de Charpal, Rognon, Gabriel and Bennacef, 1971, Fig. 212), but the appearance within the till of material apparently derived from the underlying strata (Banks *et al.*, 1971), and by the presence of a striated or grooved pavement at the base of the tillite (Bigarella, Salamuni and Fuck, 1967). Where the glacial substrate is hard or compacted, the basal contact of the tillite appears sharp. Occasionally where the substrate is composed of unconsolidated or partly consolidated materials, so much of the substrate may be incorporated into the base of the ice that the erosive contact can appear gradational.

The texture of till is extremely variable. It is generally very poorly sorted with a bimodal or polymodal grain-size distribution (Karrow, 1976). An arbitrary boundary between clasts and matrix is placed at 2 mm. Over a large area, a given till unit shows systematic variations in texture and composition (Gillberg, 1965) but adjacent till sheets often have contrasting texture (usually determined for the matrix only) and composition that can be used for correlation (Fenton and Dreimanis, 1976). In field studies of consolidated tills the relative amount of clasts may be determined by counting the number of clasts in a square metre drawn on the outcrop, or by using a standard visual estimation chart.

Till composition varies according to the composition and erodibility of extrabasinal and intrabasinal sources, the distance of transport, as well as glacial factors. There are three main types of source materials: (1) extrabasinal rocks, generally crystalline or metasedimentary, (2) intrabasinal sedimentary rocks, and (3) intraformational, retransported glacial sediments, such as outwash or laminites, which are often poorly consolidated and rapidly disintegrated into component particles by glacial shear. Material thrusted up from the glacier base into the body of the glacier can be transported intact over small distances and still be recognizable. Many studies of Pleistocene tills have investigated the relationship between source materials and till composition and texture (Shepps, 1953; Gillberg, 1968a; Dreimanis and Vagners, 1971). These studies demonstrate both how clasts are abraded in the downflow direction within a given till sheet, and how matrix forms from particles of different lithologies. Abrasion of each lithologic type results in a distinct 'terminal grade' grain-size distribution.

Clasts may be faceted, striated, polished or fractured, (Holmes, 1960). The formation of striated, faceted clasts requires a hard substrate and wet-based thermal regime. The percentage of clasts which show striations varies greatly from one till unit to another, ranging from 1 or 2% to more than 30%. Glacial transport involves not only crushing, which increases angularity, but grinding, which increases roundness (Drake, 1972). Long axes of clasts in massive tills show a preferred aximuthal orientation.

Though almost entirely structureless, massive tillite contains isolated bodies of stratified sediment which were deposited in their present site. *In situ* bodies are usually composed of sandstone, parallel or cross-stratified, or conglomerate, occasionally with interbedded mudstone or mixtite. They are either tabular with a coarse lag at the base (Lindsay, 1970; Edwards, 1975a) or triangular in cross-section with conglomerate at the pointed top (Frakes and Crowell, 1967; Frakes, Figuerido and Fulfaro, 1968; Lindsay, 1970). Contacts are gradational, sharp or deformed. Most of these sediment bodies are interpreted to be subglacial or englacial stream deposits. The presence of isolated stratified and sorted sediment bodies is important evidence supporting a subglacial origin for the enclosing massive tillite.

Massive basal tillites may resemble: (1) tills resedimented by mass movement (see Sect. 13.4.4), (2) paratillite (Harland, Herod and Krinsley, 1966) or aquatillite (Schermerhorn, 1974) which forms subaqueously without any lamination (see Sect. 13.4.3) and (3) non-glacial mass-flow deposits such as olisthostromes, or fluxo-turbidites. Strong evidence for a subglacial origin is provided by a lateral extent exceeding several kilometres, the presence of isolated, *in situ* stratified bodies, an erosive base, and by distinct contacts with associated dropstone laminite. However, alternative mechanisms should be considered by placing the mixtite in other environmental or depositional contexts.

BANDED TILLITE

Banded tillite often occurs in the lower part of an otherwise massive tillite unit, associated with an erosive contact and glaciotectonic disturbance of the underlying beds. The banded appearance is due to variations in colour, composition and grain size distribution. Individual bands are several millimetres

Fig. 13.6. Banded tillite from the late Pre-Cambrian Smalfjord Formation (lower tillite), N. Norway. Light bands with clasts consist of intrabasinally derived yellow dolomitic tillite. Dark bands are purple mudstone homogenized from underlying glacial laminite.

to several tens of centimetres thick, vary in lateral continuity and are frequently folded. They are usually oriented parallel or subparallel to regional bedding, and the folds are isoclinal with axial planes subparallel to banding (Fig. 13.6).

Banded tillite has been recognized in the late Pre-Cambrian of North Norway (Edwards, 1975a) and Svalbard (Edwards, 1976b), and in the early Proterozoic Gowganda Tillite of Ontario (H.G. Reading, personal observation).

This type of banding was formed by the mixing of two different sediment populations, one exotic in origin, the other identical to the underlying subtillite sediment and of local derivation. The distinctness of the banding is therefore a function of the contrast between the two sediment types, and the degree to which mixing of these progressed. For example, in North Norway a striking banded tillite was formed when glacial debris, composed of exotic, partly consolidated light buff dolomite, was incompletely mixed with underlying soft purple muds (Fig. 13.7). In another tillite unit, banding was observed only in a coastal exposure where the bands formed by mixing of light grey-green partly consolidated sandstone, and dark grey-green partly consolidated mudstone. Following incorporation into the base of the glacier, unconsolidated sediments generally appear to have behaved like a paste, whereas partly consolidated sediments were apparently brittle and these formed both clasts and matrix in breccia-like bands. The features of this facies, especially the observed bulk intermixing of two sediment populations, suggest that locally derived sediment was incorporated *en masse* into a glacier which was carrying exotic debris. The sediments were later partly mixed and sheared into bands by glacial plastic flow. Banded tillite lacks primary sedimentary structures such as parallel lamination and ripple lamination. However, mixtite

Fig. 13.7. Context of the banded tillite shown in Fig. 13.6. At the bottom is undeformed laminated purple mudstone. This is erosively overlain by massive purple mudstone, which upward shows the appearance of lenticular bands of dolomitic tillite and large dolomite clasts. Bands dip to the right, and the inferred flow direction was to the left.

deposited by highly viscous mass-flows typically contains scattered deformed rip-up clasts with diffuse margins (Crowell, 1957) and may also show banding.

Banded till may be an early stage in the englacial disaggregation and mixing of contrasting sediment populations and it eventually forms the homogeneous glacial debris deposited as massive basal till. In Pleistocene basal tills, straight to undulating bedding planes and bands referred to as lamination (Virkkala, 1952) were considered to originate along shear planes in the base of the glacier (Dreimanis, 1976). Where such shear planes have incorporated subglacial materials, the resulting banded structure was termed 'glaciodynamic integration texture' by Lavrushin (1971).

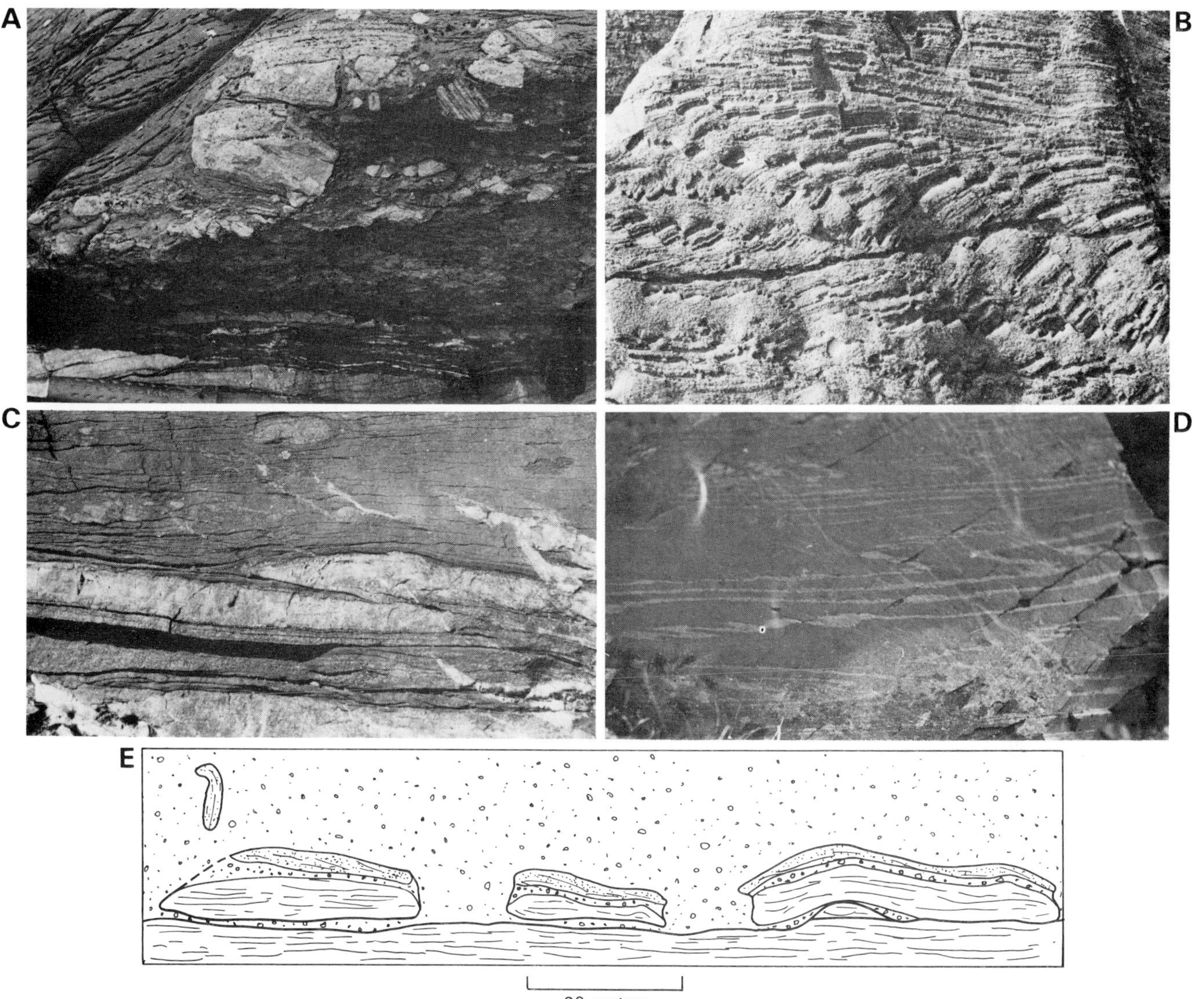

Fig. 13.8. Examples of subglacial deformation structures from the late Pre-Cambrian tillites of N. Norway. A (40 cm high): brecciated sandstone in the lower part of a basal tillite, derived from the immediately underlying sandstone beds. B (12 cm high): step faults in laminites immediately below a basal tillite. Inferred flow was to the left. C (15 cm high): small thrust fault along the lower contact of a basal tillite. D (15 cm high): tightly folded laminated mudstone. E: large rafts of preglacial and glacial sediments in the lower part of a massive basal tillite.

SUBGLACIAL DEFORMATION

Structures formed by subglacial stress may form in the glacial substrate, which in turn may be incorporated *en masse* into the glacier.

Small-scale folds and faults may be evident in stratified deposits located below or within massive tillite. Folds are often tight, with subhorizontal axial planes, and may have a sheared-out appearance. It may be difficult to infer the sense of movement from such folds. Thrust faults and normal step faults, however, are reliable indicators of movement when observed in a subglacial setting (Fig. 13.8) (Arbey, 1968, 1971; Biju-Duval, Deynoux and Rognon, 1974; Occhietti, 1973). Further deformation may lead to brecciation (e.g. Occhietti, 1973; Nystuen, 1976a) and to the gradual development of deformation till, possibly in association with banded till.

Large-scale structures include substrate deformed into folds several metres or more in height (Ochietti, 1973; Martin, 1961, 1964), large imbricated plates of substrate below or in the base of a massive till (Rognon, Bijou-Duval and de Charpal, 1972; Moran, 1971), large plate-like rafts of substrate in the base of a massive till oriented approximately parallel to regional bedding (Banham, 1975), and large blocks of substrate variably oriented and deformed within a massive till (Spencer, 1971).

13.4.2 Supraglacial and proglacial deposits, stratified conglomerate and sandstone

Conglomerate and sandstone with intercalated tillite and mudstone reflect the activity of glacial meltwater, either above, below or beyond the glacier. Facies range from slightly sorted tillites to well sorted and stratified conglomerates and sandstones. Most of these deposits form in the supraglacial to glaciofluvial environments. Here, ice-contact stratified drift and glaciofluvial outwash are laid down. The characteristic englacial or subglacial deposit formed by running water is the esker, which generally consists of a shoestring of stratified gravel and sand which may be preserved during glacial retreat (Charlesworth, 1957; Banerjee and McDonald, 1975). Where the ice front is submerged, subaqueous outwash, in the form of esker fans composed of stratified gravel and sand, may be deposited (Rust and Ronamelli, 1975; Anderson, Clark and Weaver, 1977). In this account, attention is given to the more important subaerial (i.e. supraglacial and glaciofluvial) deposits. Their main features are: (1) presence of mud, sand and gravel fractions, regardless of the presence of stratification and degree of small-scale sorting, (2) great variability in type and abundance of stratification, (3) local extent, up to several kilometres in association with glacial retreat or stagnation, but great extent, up to hundreds of kilometres, of marginal accumulations formed in front of stationary active ice, and (4) variable thickness, usually up to a few tens of metres, rarely exceeding 100 m.

The distribution of stratified drift may be topographically controlled, and depressions tend to be filled with thick deposits, for example, braided stream valley trains. In areas of low relief, aprons of stratified drift may be deposited adjacent to the glacier, beyond or as part of an end-moraine complex, forming a wedge up to several tens of kilometres wide and hundreds of kilometres long. Smaller scale ice-contact deposits, such as kames, kame terraces, and marginal channel fills tend to have a more irregular and localized distribution.

The composition of stratified drift may be similar to basal till deposited by the same glacier if it forms by partial reworking of the till, or from supraglacial deposits derived from the glacier bed (Fig. 13.3). The composition may differ if supraglacial debris has been derived from material which has undergone long englacial or supraglacial transport such as in a medial or lateral moraine, or in a supraglacial stream system.

Supraglacial and ice-contact stratified drift show great variation in texture and sorting. Sections including mixton, poorly-sorted boulder gravel, gravelly mud and well-sorted sand can form from water-logged ablation till on the changing glacier surface, downslope movement of flow tills, winnowing by melt-waters, intertonguing of flow tills of varying textures, and deformation of sediments by melting of buried ice. Away from the ice margin, stratified drift, or outwash, is progressively finer grained and increasingly stratified and sorted (Sect. 13.4.3). Glacially formed shapes and surfaces of clasts are rapidly destroyed.

A great variety of sedimentary structures can potentially form in stratified drift, due to the wide range of grain sizes, hydrodynamic conditions, and mechanisms of transport. Minimal reworking and rapid deposition are most common in the ice-contact environment. In sands, depositional structures include massive internal structure, upper flow regime parallel-lamination, cross-stratification and climbing ripple lamination. Mud drapes form in abandoned channels and pools. Channels are filled or lined with coarse material. Flow tills are interbedded with all of these deposits as lenses and beds of massive to slightly stratified mixtons.

The melting of buried stagnant ice produces structures which are unique to the ice-contact environment. Ice blocks at the surface melt to form holes which may be filled by coarse or fine sediment. Slumping or grain flow may occur along the margins of the hole. Melting of buried blocks deforms the overlying layers, forming grabens, often with a complicated fault pattern. Melting of large areas of buried ice can form lakes, cause large-scale faulting, create temporary steep slopes upon which deposition may continue, or slumping be induced, and tilt layers so that the dips exceed possible primary dips. Blocks of glacial ice would be expected to deform greater thicknesses of sediment, and lead to deeper depressions on melting than would be possible with seasonally formed ice (see Collinson, 1971a).

Due to the changing relief and the frequent presence of a frozen substrate in the ice-contact zone, frozen sediments can be eroded and deposited *en masse* into younger deposits. Such blocks commonly have deformed stratification and may contrast in grain-size to the host deposit.

Glacial outwash is similar to sediments formed in non-glacial braided streams and alluvial fans. The two main distinguishing features are: (1) deformation structures which can be related to melting of buried ice, and (2) the context.

13.4.3 Glaciomarine and glaciolacustrine deposits

Two subaqueous facies are distinguished, laminites, and massive tillites which lack internal stratification and are also termed aquatillites (Schermerhorn, 1966, 1974). Laminites include both varved deposits formed in fresh and brackish water and deposits without rhythmic structure formed in the sea. Aquatillites are marine, forming in quiet water.

LAMINITES

Laminites consist of alternating sand, silt and clay laminae, usually even and continuous in appearance. Depending on composition, texture and thickness the lamination can be either quite easy or very difficult to detect. Amongst the many types of lamination which have been described, an important distinction is between random and rhythmic lamination. A sequence showing the haphazard intercalation of laminae of different grain-size and structure is considered to be random. The regular repetition of two or three distinct types of laminae is termed rhythmic, and an individual cycle is termed a rhythm. Rhythmites are widespread in ancient glacial sequences, but it is difficult to prove that a rhythm formed in one year and is a varve (see Sect. 13.3.4). Additional important features of laminites are (1) whether grading is symmict or diatactic, (2) presence of intercalated sandstone or mixtite beds, and (3) presence of dropstones, tillite clots and conglomerate clusters (see Sect. 13.3.7). Abundant dropstones scattered throughout a laminite provide strong evidence of ice rafting, but the absence of dropstones is not evidence against glaciation as many Pleistocene rhythmites (varves) contain few or no dropstones.

In the deep oceans, massive to faintly laminated muds with rafted debris blanket huge areas: these deposits extend for several thousand kilometres from Antarctica (Lisitzin, 1972, pp. 108–113). On the continents, Pleistocene laminites formed over large areas during deglaciation and transgression: for example in the Baltic Sea, North Sea, Hudson Bay, and Lake Champlain (Flint, 1971). In these areas the rates of isostasy and sea-level rise, and ice damming led to fluctuations between fresh water and marine conditions. Such laminites may blanket huge areas and have thicknesses up to several tens of metres, but partial erosion may occur during a subsequent glacial advance.

Where laminite units overlie basal tillite, the contact in many cases is gradational. It may also be sharp, or there may be an intervening deposit of sandstone or bedded tillite.

Laminites may be confused with banded tillite, and primary current features such as ripple cross-lamination are therefore very important. The identification of dropstones is not always easy. Stones should be totally enclosed by laminae which are much thinner than the diameter of the stone. Additional evidence such as till clots and gravel clusters should be sought, as well as striated and faceted dropstones. Dropstones can also be transported into quiet water by floating vegetation and seasonal coastal and lake ice.

MASSIVE TILLITE

In some random laminites, lamination is coarse and faint, giving the deposit a till-like appearance. Such deposits may be called laminated mixtons or tills. With the complete disappearance of lamination, these grade into aquatills, massive glaciomarine tills.

Massive tills interpreted as glaciomarine deposits have been described in the Pleistocene of the Soviet Union, Spitsbergen (Boltunov, 1970) and the northern Pacific coast of N. America (Easterbrook, 1963) and in the Miocene–Pleistocene of Alaska (Miller, 1953; Plafker and Addicott, 1976). In these deposits, the massive tills are usually fossiliferous, including a varied microbiota and articulated bivalves partly in living position. They interfinger and alternate with normal marine sediments showing clear stratification and good sorting. The Alaskan mixtites are mostly blanket shaped, locally laminated internally, have large extent and are up to 200 m, though mostly several tens of metres, thick. Clasts compose 5–20% of the rock, and reach 5 m in length. Mixtite also occurs in lenticular units which grade into normal marine deposits.

Aquatills compose somewhat more than half of the glaciomarine late Cenozoic sequence in the Ross Sea shelf (Hayes, Frakes *et al.,* 1975a and b; Barrett, 1975). The sequence contains a varied biota and is rich in diatoms. Bioturbation is rare. Both laminated and massive parts of the sequence consist of poorly sorted silty clays. Clasts compose less than 1% of the sediment, those larger than 3 cm are rare. Fine lamination probably formed by bottom traction currents, while coarse stratification may in part represent turbidites. These sediments were interpreted (Barrett, 1975) to be texturally similar to the debris supplied by continental ice. However, the textures are similar to those of Pleistocene basal tills derived from lake sediments (Flint, 1971, p. 157). This suggests that in addition to rafting by icebergs, fine sediment was deposited from suspension by weak currents.

Massive glaciomarine tills, as opposed to massive basal tills, show: (1) presence of unbroken fossils, occasionally in living position, and rare bioturbation, (2) gradational contact with normal stratified sediments, (3) intercalation of turbidites or other beds of contrasting sediment types which are laterally continuous, (4) absence of isolated bodies of stratified sediment, (5) random azimuthal orientation of clasts and (6) possible finer grain size than associated basal tills.

13.4.4 **Bedded mixtite and tillite**

Beds of till may occur in proglacial deposits, intercalated with other facies. Beds are lenticular to parallel-sided and laterally continuous. They are mostly less than a metre thick, but may reach several metres or more. Internally they are mostly massive, but occasionally display faint banding or stratification. Some beds are graded, and they may contain rip-up clasts from the underlying beds.

Till or mixton beds represent either tills or mixtures of other deposits such as lacustrine clays and fluvial gravels which have been resedimented by a gravity-driven mass-flow mechanism. Such flows may vary greatly in viscosity.

In the supraglacial to ice-contact fluvioglacial zones these mixtons form as flow tills where they are deposited along with stratified sands and gravels (see Sect. 13.3.2 and 13.4.2). In the proglacial subaqueous environment they may be interbedded with laminite or aquatills, and occur with turbidites or the deposits of other gravity-driven mechanisms. These glacial mixtons share many properties with mass-flow deposits found in other environments, for example alluvial fans (Chapter 3) or submarine fans (Chapter 12).

It can be difficult to distinguish thick mixtons formed by mass-flow from massive basal tills, especially where other facies are not present between mixton beds.

13.5 ANCIENT GLACIAL FACIES

Glacial facies form the basis for documentation of the five major phases of glaciation currently known in Earth history: late Cenozoic, late Palaeozoic, late Ordovician, late Pre-Cambrian and early Proterozoic. Late Cenozoic deposits are considered here as ancient because most of the glaciers which deposited them are gone.

13.5.1 Late Cenozoic glaciation

Glaciomarine sediments from both the Ross Shelf and from southern Alaska indicate glaciation from at least the late Oligocene in Antarctica and from lower-middle Miocene in N. America (see Sect. 13.4.3). Drilling at four localities in the Ross embayment has yielded extensive cores through a mainly late Cenozoic sequence greater than 1.5 km thick (Hayes, Frakes *et al.,* 1975; Barrett, 1975). Sediment was supplied largely by wet-based tidewater glaciers, and deposited mostly below wave base. The tilted Tertiary glaciomarine deposits are truncated by an unconformity and overlain by about 20 m of structureless sediments which may be glaciomarine. These are richer in sand and silt than the underlying deposits and contain a smaller proportion of microfossils. Alternatively, they may be a basal till partly derived from the underlying sediments, as supported by overconsolidation (Anderson, Clark and Weaver, 1977). At the top of the sequence is a thin layer of marine silty clays with abundant diatoms.

Marine and glaciomarine mixtites are interbedded with conglomerates, sandstones and mudstones in the 5 km thick Yakataga Formation of southern Alaska (Plafker and Addicott, 1976). Deposition was in depths shallower than 100 m, mostly 20–60 m. Supply was by iceberg rafting, bottom currents and downslope mass-flow from tidewater glaciers in the coastal mountain range. The regional facies distribution is still poorly known.

Of Pleistocene deposits massive basal till is the most widespread facies on land areas. It extends from the interior of glaciated shields, where there are only thin remains, to middle-latitude lowlands and the edges of continental shelves where glaciers were grounded during low sea level and several thick till units may be present. The limit of glacial advance marks the boundary between regionally contrasting facies. Beyond this limit no basal till occurs, but, stratified gravels and sands, formed on outwash plains, and loess covered extensive proglacial areas (see Flint, 1945, 1959). Behind the ice terminus the facies left by the retreating ice sheets form a very complex pattern. Tunnel valleys, formed by subglacial stream erosion, are present in ice marginal areas: e.g. Minnesota (Wright, 1973), the North Sea (Jansen, 1976), and England (Woodland, 1970). Stratified gravels and sands are present over wide areas in the form of eskers, kames, terraces, meltwater channel fills, lake deltas and beaches, as well as extensive proglacial aprons associated with halts in ice retreat. Behind and along the ice limit varved silts and clays formed in ice marginal lakes (Flint, 1945, 1959; Prest, 1968; Woldstedt, 1970, 1971). In stable basins, such as Hudson Bay and the Baltic Sea, random laminites and varvites were formed in low-lying marine or brackish seas and lakes associated with isostatic depression, late-glacial transgression and ice retreat.

The facies distribution of marine influenced areas, the continental shelves and rapidly subsiding basins such as the North Sea, is less well known. Which proglacial facies forms is largely determined by whether the glacier terminus is above sea level, below sea level, or bounded by an ice shelf. This is related to the relative rates of glacial retreat, eustatic rise and isostatic uplift. Many glaciated shelves appear to have been completely covered by ice sheets which calved icebergs over the continental slopes (probable examples are the Gulf of Alaska shelf, north-eastern N. America shelf, northern North Sea, western Barents Sea shelf). Removal of debris from the ice terminus by further transport either in icebergs or in an ice shelf may be one reason for the apparent absence of end moraines at some shelf margins (Flint, 1971; Grant, 1972; King *et al.*, 1972), though this could also be due to subsequent reworking.

On the N.W. Atlantic shelf of N. America till locally extends to the shelf edge (King, Maclean and Drapeau, 1972) suggesting that large areas were blanketed by massive till, which in deeper areas is now covered by glacial and postglacial muds, and in shallow areas has been eroded away. Sparker profiling suggests that glacial deposits fill depressions (Oldale, Uchupi and Prada, 1973; Grant, 1972) and vary from 0 to 200 m averaging around 100 m thick. On Georges Bank, Schlee and Pratt (1970) mapped the limit of the last glaciation at a

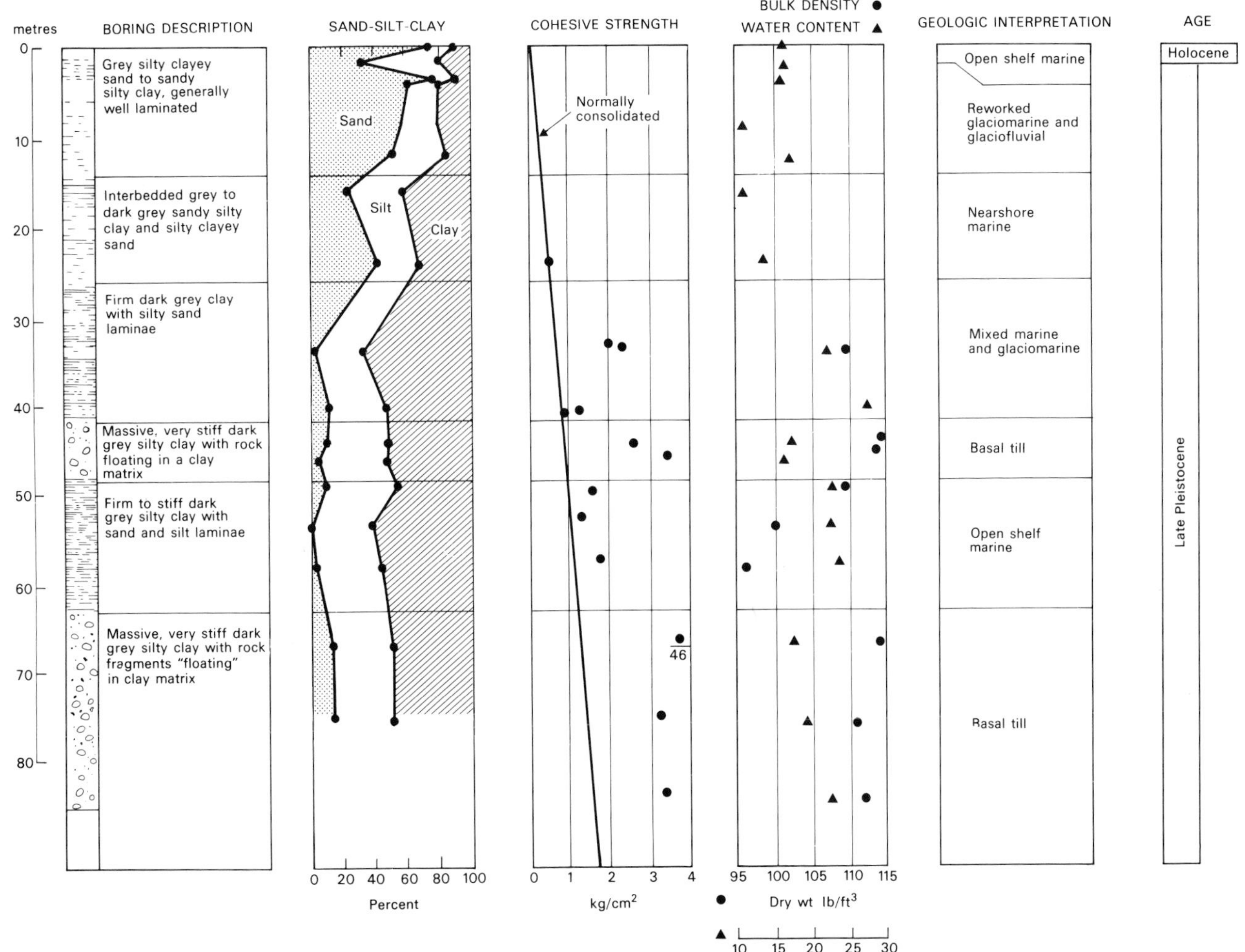

Fig. 13.9. A late Pleistocene succession in the North Sea, west of the Norwegian Channel, about 59°N, 130 m water depth (modified from Milling, 1975). The two basal till units are overconsolidated and have a higher bulk density and lower water content than the adjacent marine deposits. Milling suggested that the upper basal till was deposited beneath a grounded ice shelf as indicated by the normal consolidation of the underlying marine sediments. This contradicts the observed overconsolidation of the till unit itself. The normal consolidation of the marine sediment is probably related to high subglacial water pressure.

marked drop in quantity and grain size of gravels in bottom sediments, which took place across several kilometres.

Distinct linear ridges extending along offshore Nova Scotia for hundreds of kilometres have been interpreted as end moraines formed below sea level, possibly along the grounding zone of an ice shelf (King, Maclean and Drapeau, 1972). The moraines pass offshore into laminated silts.

The Pleistocene of the North Sea reaches a maximum thickness of about 1,000 m (Caston, 1977). The sequence is characterized by the alternation of thick massive tills with clays and silty and sandy clays (Fig. 13.9). Geotechnical tests on the deposits delineate strong, dense tills, formed subglacially, and weak, less dense clays which were marine, glaciomarine or brackish in origin (Milling, 1975). Similar lithologies are present off the Netherlands (Oele, 1969, 1971). Glaciotectonic deformation of earlier Pleistocene deposits is seen in many areas (Oele, 1971; Caston, 1974). End moraines are preserved as linear banks covered with coarse material (Veenstra, 1965).

13.5.2 Late Palaeozoic glaciation

The late Palaeozoic glaciation is widely preserved in Gondwanaland. Gondwanaland drifted across the south pole, with glacial centres developing first in South America and southern Africa during the early Carboniferous, extending to Antarctica, India and Australia during the late Carboniferous, and then, during the Permian, culminating and waning in Antarctica and Australia (Crowell and Frakes, 1975).

During the local glacial climaxes, large ice sheets developed in southern Africa, Antarctica, and southern Australia. In outlying areas, and preceding and following glacial maxima, valley glaciation was more typical, for example in western S. America, Rhodesia, and eastern Australia.

The ice sheets deposited three main facies: massive till, dropstone laminites and stratified gravels and sands (Wanless and Cannon, 1966; Hamilton and Krinsley, 1967). In southern Africa, several glacial lobes can be distinguished (Du Toit, 1921). These flowed predominantly to the south, where more and thicker massive tillite units were deposited (Du Toit, 1921; Crowell and Frakes, 1972) and marine influence increased as indicated by dropstone laminites. In Antarctica, an increase in the number and thickness of massive tillites has been interpreted as due to an ice sheet radiating out from the area of minimal sedimentation (about 100 m) into the area of maximum sedimentation (more than 1,000 m) (Lindsay, 1970). The tillite is associated with outwash sandstone and conglomerate, and shales occasionally with dropstones. In most of these areas, the number of glacial advances and retreats is uncertain but there is evidence for thirteen in Antarctica (Lindsay, 1970) and six in Victoria, Australia (Jacobsen and Scott, 1937).

Linear valleys, partly filled with glacial sediments, occur in several areas covered by late Palaeozoic ice sheets. In South Australia, valleys are as much as 600 m deep, 10 km long, and 5 km wide (Campana and Wilson, 1955). In deep valleys in the Congo, Boutakoff (1948) mapped massive basal tillite, stratified conglomerates and sandstones formed as glaciofluvial outwash, and dropstone rhythmites. Glacially modified Palaeozoic valleys in southern Africa are up to 1,500 m deep, 5 km wide and 60 km long. The bottom and sides show striated pavements and roches moutonnées, and they are filled with a variety of glacial facies (Martin, 1953, 1968; Frakes and Crowell, 1970).

Valley glaciers have been inferred for the Zambezi Basin of Rhodesia on the basis of the thin and patchy distribution of glacial sediments, especially north of the Transvaal ice centre (Bond, 1970). However, the facies, mainly massive tillite, glaciofluvial outwash and rhythmic and random laminites (Bond and Stocklmayer, 1967) are similar to areas of widespread glaciation. Facies in the Upper Carboniferous of New South Wales, Australia, indicate a different type of valley glacier development and palaeogeography. These deposits are dominated by fluvial conglomerates which lack evidence of glacial influence apart from dropstone laminites and local intercalations of mixtite with striated and faceted clasts (Whetten, 1965; White A.H., 1968). The conglomerates were interpreted as valley-train outwash deposits with small remnants of till and glaciolacustrine mud preserved from reworking.

In proglacial areas typified by high relief, glacial mixtons were mobilized and resedimented. These mixtites are associated with marine fossils and turbidities in the Andean Mountains (Frakes and Crowell, 1969), interbedded with fossiliferous marine sandstones and conglomerates in New South Wales, Australia (Lindsay, 1966; Crowell and Frakes, 1971) and interbedded with turbidities in glacial lake deposits in India (Banerjee, 1966; Frakes, Kemp and Crowell, 1975).

13.5.3 Late Ordovician glaciation

During the late Ordovician, ice sheets flowed radially outward from central Africa to South Africa (Lock, 1973), South America (Bigarella, 1973), West Africa (Tucker and Reid, 1973), northern Africa (Beuf, Bijou-Duval, de Charpal, Rognon, Gariel and Bennacef, 1971), and possibly Ethiopia (Dow, Beyth and Tsegaye Hailu, 1971).

The most extensive outcrops occur in the Sahara, where massive basal tillites rest on unconformities locally shaped into striated pavements and roches moutonnées. Partly conglomeratic stratified sandstone is the dominant facies. It forms sandy outwash fans, subglacial tunnel-valley fills, eskers, and various fluvial and aeolian sandstones, often with a periglacial overprint. Pingos and kettles, and dead-ice- freeze-and-thaw and polygonal structures associated with fossil ice-wedges have been described (Beuf *et al.*, 1971). Laminites, both poorly and distinctly laminated, occasionally have dropstones. Locally, carbonate beds and concretions and a marine biota are associated.

The vertical recurrence of up to four unconformities, each overlain by massive tillite, indicates repeated advances and retreats of the ice sheet, but rapid lateral variations prevent regional correlation of the deposits.

The abundance of sand in the matrix of the subglacial tillites and stratified deposits seems to be due partly to glaciation of a low-relief mature shield area and partly to the widespread occurrence of sandstones below the glacial rocks, which were probably only partly consolidated at the time of glaciation.

Both the facies distribution and palaeocurrent indicators in the Central Sahara suggest a northward palaeoslope. The general thickness increases from under 100 m in the south to about 100–300 m (maximum about 500 m) in the north. In the south, roches moutonnées and striated pavements are well developed, as are massive tillites and proglacial sandstones.

Periglacial features are scarce, due either to long periods of ice cover or erosion during successive glacial advances. In the north, tillites are less common, while periglacial features are better developed, intercalated with proglacial terrestrial and marine deposits. The characteristic erosive form is the palaeovalley, which developed near the margin of the ice sheet, though the precise location appears to be related to preglacial tectonic movements. In outcrop, the valleys are 2 to 10 km wide, 100 to 300 m deep, and up to 40 km long. They trend parallel to the palaeoslope, may branch upslope, and are locally overdeepened. Erosion surfaces in the glacial deposits indicate that the valleys were at least partially filled and scoured out several times in the course of the glaciation.

In the northeast, further down the palaeoslope, marine deposits are dominant, and basal tillites are absent. Dropstone laminites reach a thickness of 300 m. Ice-shelf facies have not been identified, rafting being attributed to icebergs and shore ice.

13.5.4 **Late Pre-Cambrian glaciation**

Glacial deposits of this age are known on all continents except Antarctica, and show great variations in facies and thickness. In many cases, the mixtites or tillites occur in marine-dominated successions, in contrast to the early Proterozoic, late Ordovician and late Palaeozoic glacials, which occur in terrestrial or mixed sequences. Radiometric dating suggests that late Pre-Cambrian glacials occur in two or three groups, separated by substantial gaps in time (Williams, 1975; Kröner, 1977). Within a given area, one such glacial horizon may include two or three distinct glacial formations, separated by non-glacial deposits.

Doubts as to the glacial origin of mixtite occurrences and therefore extensive glaciation at the time have been based on: (1) widespread (global) distribution (not climatically feasible), (2) low palaeomagnetic latitudes, and (3) association with dolomites. A heated controversy between those arguing for and against a major late Pre-Cambrian glaciation has resulted in several important reviews (Harland, 1964; Saito, 1969; Schermerhorn, 1974). The frequent occurrence of dropstone laminites in alleged glacial sequences (which often contain glacially striated and faceted clasts) is good reason to consider the associated facies in a glacial context.

In many sequences, mixtite is the dominant lithology (Fig. 13.10). In North Norway (Edwards, 1975b) Svalbard (Edwards, 1976) and Scotland (Spencer, 1971) erosive basal contacts, thickness, internal stratified bodies, and other features typify massive basal tillites. The tillites in north Norway are intercalated with laminites, partly containing dropstones, and with subordinate glaciofluvial and subaqueous conglomerate and sandstone (Edwards, 1975a). In vertical sections, these facies show cyclic repetition: basal tillites pass upwards into stratified proglacial deposits, which are in turn overlain erosively by a basal tillite. Each cycle is interpreted to reflect glacial advance and retreat. There is great variation in composition and texture from one basal tillite unit to the next, suggesting that successive ice advances were associated with important changes in ice flow paths. The cycles seem to be analogous to the 'drift-sheets' recognized in Pleistocene deposits, and may reflect climatic and glacial fluctuations on a glacial/interglacial scale. In other areas, mixtites occur mainly in lenticular beds, without marked erosion surfaces, occasionally showing grading, and these have been interpreted as mass-flow deposits, formed with or without glacial imprint, for example, in France (Winterer, 1964) and Central Africa (Schermerhorn and Stanton, 1963). Where lenticular mixtite beds are intercalated with dropstone laminites, and thick mixtites interpreted as basal tillites are also present in the sequence, a galcial setting can be demonstrated, for example in western N. America (Aalto, 1971) and southern Norway (Nystuen, 1976a and b).

A variety of lithologies are present as intercalations between thick massive tillites: sandstones, a few dolomites and rare dropstone laminites in East Greenland (Poulson, 1930; Poulson and Rasmussen, 1951; Spencer, 1975), mainly laminites with or without dropstones in North Norway (Edwards, 1975a), Svalbard (Wilson and Harland, 1964; Edwards, 1976b), and Virginia, eastern USA (Blondeau and Lowe, 1972), and conglomerates, sandstones and rare dropstone laminites in Scotland (Spencer, 1971, 1975). Poorly stratified siltstones in Svalbard and Finnmark have been interpreted as consolidated loesses (Edwards, 1976).

Glacial striated pavements have been reported from North Norway (Edwards, 1975b), Brazil (Isotta, Rocha-Campos and Yoshida, 1969), Western Australia (Dow, 1965) and northwest Africa (Deynoux and Trompette, 1976). A possible glaciated landscape occurs in Northern Greenland (Jepsen, 1971). A large valley approximating the present Varangerfjord, North Norway, is situated along the basement-sediment contact (Bjørlykke, 1967), and U-shaped valleys occur in the western USA (Crittenden, Stewart and Wallace, 1972).

13.5.5 **Early Proterozoic glaciation**

In North America, massive tillites and dropstone laminites extend from Wyoming to Quebec, a distance of 3,000 km, suggesting glaciation on a continental scale (Young, 1970).

In southern Ontario, pre-glacial, glacial and post-glacial sediment transport was mainly towards the south (Casshyap, 1968), and there is a general increase in thickness of the glacial Gowganda Formation in this direction (Lindsey, 1969). The pre-glacial setting was a fluvial-dominated coastal plain. The dominant glacial facies in the Gowganda Formation is massive basal tillite, which occurs as two tillite members averaging

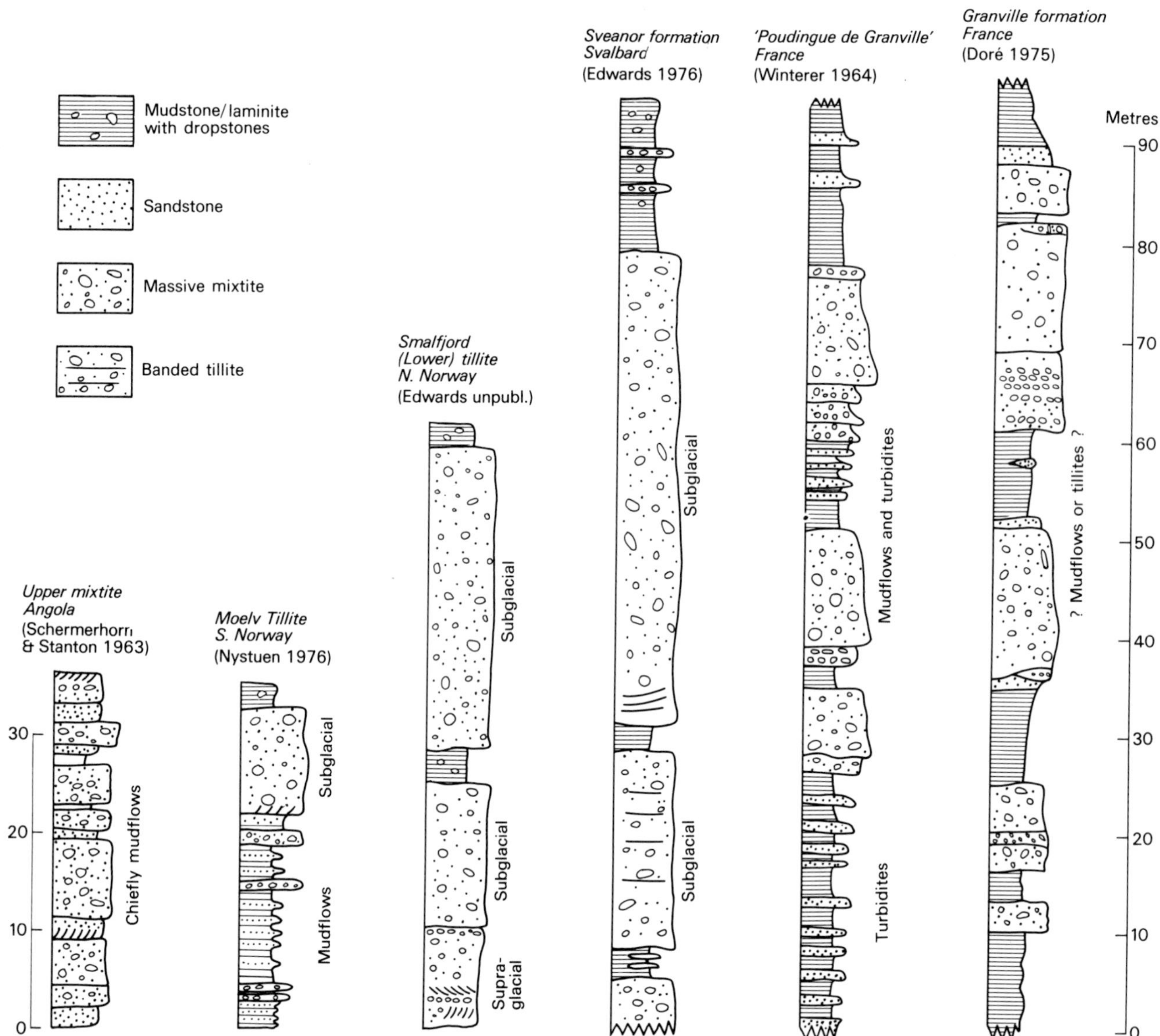

Fig.13.10. Late Pre-Cambrian sequences from different areas, claimed to be glacial by some authors, disputed by others. The interpretation provided is that of the author who measured the section (from Doré, 1975).

about 50 and 80 m thick, and which can be traced for tens of kilometres. Lindsey (1969) mapped out marine and fresh water areas by assuming that rhythmites were deposited in fresh water and random laminites in saline water. During glaciation, due to isostatic depression and rapid post-glacial transgression, marine conditions developed in the deeper southern area, with the formation of random laminites and presumed ice-shelf deposits (Casshyap, 1968; Lindsey, 1969). Ice-shelf deposition was inferred from the association of stratification, increased thickness, and great lateral continuity. To the north, rhythmites were deposited in large lakes formed in front of the retreating ice margin.

13.6 FACIES ASSOCIATIONS AND GLACIAL MODELS

The geographical distribution of glacial facies is related to climate and environment. Basal till forms by direct glacial deposition. Both varved laminites and fluvial outwash form marginal to the glacier, mostly during glacier retreat, but also during glacier advance and maximum. The development of these facies is related to the availability of sediment and meltwater and the amount of damming on the deglaciated terrain. Loess also forms marginal to the glacier, but may extend for hundreds of kilometres beyond the limit of glaciation. Random laminites and aquatills, locally including bedded mixtons, are deposited in seas adjacent to a glaciated area, and may have a very widespread distribution. Away from glacial centres, glacial influence decreases, and proglacial facies pass into normal facies.

Facies which tend to occur together are called a facies association which can be either defined on the basis of actual examples, or predicted from conceptual models (Chapter 2). For example, terrestrial associations consist entirely of facies which form in terrestrial environments, above the marine limit. Terrestrial associations can be subdivided into an inner association including basal till, a marginal association with mainly ice marginal facies, and an outer association excluding basal till, and including mainly marginal to non-glacial facies (Fig. 11A). Examples of the lateral relationship between these associations are seen in Pleistocene deposits of both central North America, along the southern boundary of the ice sheet (e.g. Frye, Willman and Black, 1965) (low relief), and the Alps (e.g. Wright, 1937) (high relief).

Marine facies associations form below the marine limit. Between the marine limit and the lower limit of subaerial conditions both marine and terrestrial facies may occur in the same area and interfinger with one another. Within the ice limit, basal till can be accompanied by both varved and random laminites, and subaerial and subaqueous outwash. Pleistocene deposits of this type are preserved in the North Sea (Milling, 1975) and may typify the continental terrace and intercratonic depressions.

Below the lower limit of terrestrial environments, varved laminites and fluvial outwash cannot form. An inner marine facies association includes basal till, random laminite and aquatill. A marginal marine association consists mainly of random laminite and/or aquatill, with varying amounts of subaqueous outwash and little basal till. This association may be forming at the present time in the Weddell Sea (Anderson, Clark and Weaver, 1977). The outer association consists of random laminite and/or aquatill, locally including or replaced by bedded mixton (Fig. 11B). It comprises the thick late Cenozoic sequence of the Ross Shelf. Additional variations in glacial marine facies develop in shallow water where normal waves and currents rework rafted materials, forming stratified muds, sands and gravels, as in the late Cenozoic Yakataga Formation of Alaska (Plafker and Addicott, 1976).

Facies associations showing alternations of basal till and proglacial facies may result from geologically rapid (tens of thousands of years in the Pleistocene) fluctuations in the extent of the ice, as exemplified by the Pleistocene glaciation of Scandinavia, North America and the Alps. Each basal till unit represents glacial expansion, while the associated proglacial (or interglacial) deposits indicate glacial withdrawal. A model glacial advance/retreat sequence includes a basal erosion surface overlain by basal till; this is locally overlain by ice marginal and proglacial deposits and possibly interglacial normal sediments (Fig. 13.11). Such facies sequences are preserved in ancient deposits (Sect. 13.5).

However, in certain areas with a favourable geographical location, and/or during times of relative climatic stability, glaciers can be maintained more or less continuously during a glaciation. The resulting deposits may show minimal facies variation, as seen for example in late Cenozoic glacial marine deposits of the Ross Sea. The presence of up to 5,000 m of glacial marine sediments in the Yakataga Formation of Alaska also probably reflects long-term glaciation in addition to high sedimentation rates (up to 5,000 m in 15 m.y. = 330 m/m.y.). In the North Sea, Quaternary sediments, in part glacial, also show a high sedimentation rate (up to 1,000 m in 2 m.y. = 500 m/m.y., Caston, 1977).

Longer, more continuous glaciation (resulting therefore in potentially thicker and more homogeneous glacial facies) is most likely to occur around glacial centres in the polar regions. Nevertheless, the absence of glaciation and glacial deposits along the northern coast of Alaska (Péwé, 1975) shows how local climatic conditions control the development of glaciers and their deposits. A better understanding of ancient glacial deposits, and the causes of glaciation, will be achieved when detailed facies analyses, coordinated with other studies, are available for all glaciations over as wide an area as possible.

A

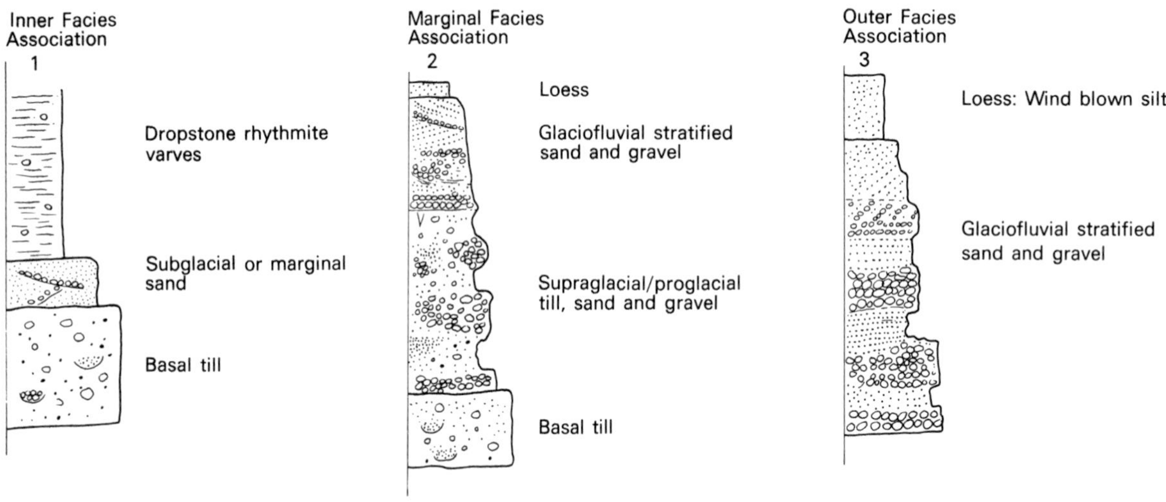

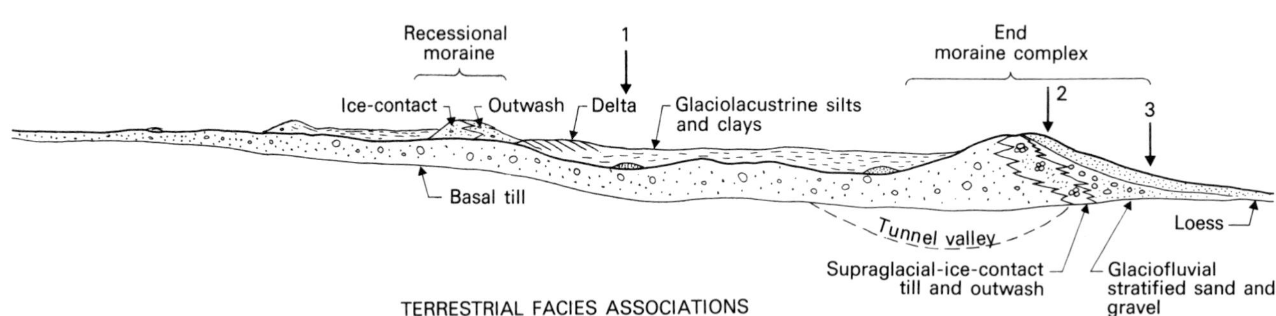

Fig. 13.11. Very simplified sketches of the facies associations and sequences which develop from the advance and retreat of an ice sheet in terrestrial (A) and marine (B) contexts. During successive advances and retreats, differential erosion of the previous deposits could occur, and

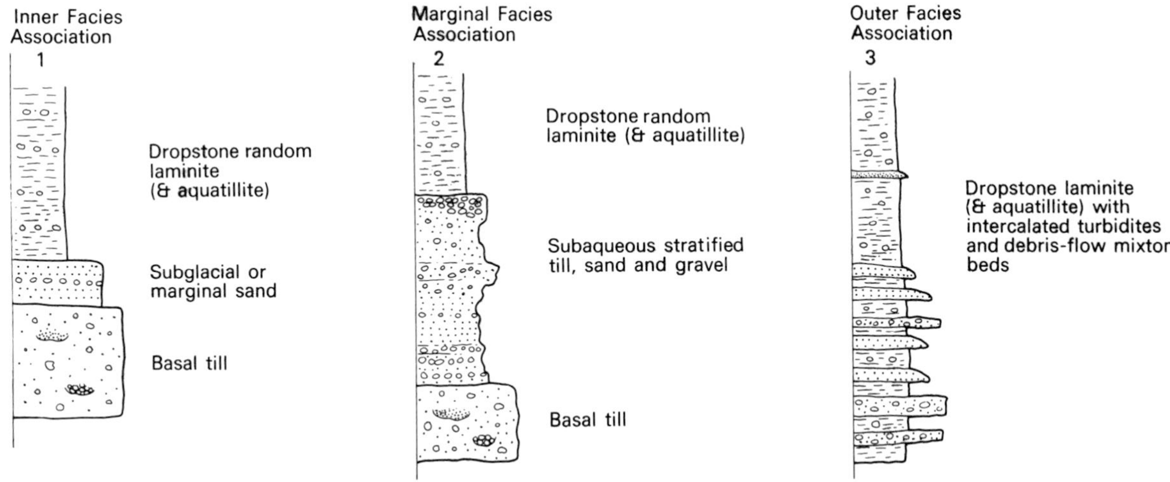

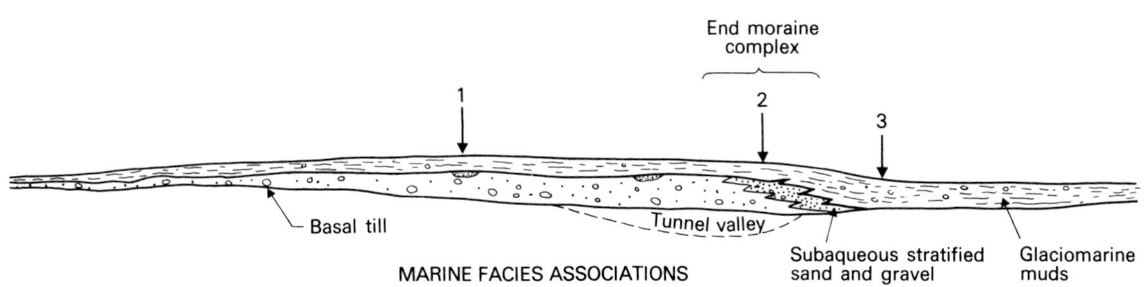

each facies would probably be deposited in somewhat different areas. The radius of the ice sheet could have been as much as 2,000 km, and the deposits are as much as 100 m thick. The characteristic facies sequences indicated may average about 5 to 50 m thick.

FURTHER READING

PATERSON W.S.B. (1969) *The Physics of Glaciers.* 250 pp. Pergamon, London.

FLINT R.F. (1971) *Glacial and Quarternary Geology,* 892 pp. Wiley, New York.

GOLDTHWAIT R.P. (Ed.) (1971) *Till: A Symposium,* 402 pp. Ohio State University Press.

JOPLING A.V. and MCDONALD B.C. (Eds.) (1975) *Glaciofluvial and Glaciolacustrine Sedimentation,* 320 pp. Spec. Publ. Soc. econ. Palaeont. Miner. Tulsa, 23.

WRIGHT A.E. and MOSELEY F. (1975) *Ice Ages: Ancient and Modern.* 320 pp. Geol. J. Spec. Issue 6, Seal House Press, Liverpool.

LEGGET R.F. (Ed.) (1976) *Glacial Till: An Inter-disciplinary Study,* 412 pp. Spec. Publ. R. Soc. Canada 12.

CHAPTER 14 Sedimentation and Tectonics

14.1 INTRODUCTION

Until the 1950s most books on sedimentation related facies to the tectonic background. For example, in Krumbein and Sloss's influential textbook *Stratigraphy and Sedimentation* (1951, 1963) the greywacke facies was related to eugeosynclines and quartz arenites to stable shelves. However, as knowledge of sedimentary processes and modern environments increased, interest diminished in the tectonic control of facies and during the 1960s few sedimentologists gave much consideration to tectonics. The advent of plate-tectonic theory has since led to a revival of interest in tectonics and sedimentation.

14.2 THE GEOSYNCLINAL THEORY

For more than 100 years the concept of the geosyncline has been used to explain the frequent association of thick sedimentary successions, folding and mountain building. Plate-tectonics has now largely superseded geosynclinal theory by suggesting that one of the most important controls on sedimentation and orogeny is the position of the sedimentary basin relative to a plate or continent–ocean boundary. However, many elements of the geosynclinal theory have been retained and it is still used to describe ancient orogenic belts. In north America and Europe it has developed along different lines, according to the emphasis given to particular geological processes.

14.2.1 Early American and European views 1859–1920

The earliest general hypothesis concerning the relationships between deformation and sedimentation was that of Hall (1859), based on geological mapping in the northern Appalachian Mountains. Hall recognized that the very thick Lower Palaeozoic succession of well-sorted sandstones, carbonates and shales, with only minor greywackes, was of shallow-water origin. He considered that subsidence during deposition was due to weight of sediment and that folding and metamorphism, but not subsequent mountain building, were related to subsidence.

Dana (1873) suggested that there was no causal relationship between subsidence and sediment accumulation, and that the sediments were derived from a postulated geanticlinal uplift. Of particular significance was Dana's suggestion that subsidence of a 'geosynclinal' belt and subsequent orogeny was the result of lateral compression, that this compression was caused by movement of ocean floor towards a continent and that 'geosynclinals' developed either beside or on a continental margin. This American view of geosynclines as asymmetric and ensialic was strongly influenced by the present position of the Appalachians and western Cordillera, on the edge of the continent.

In Europe, early ideas relating sedimentation to orogeny were 'borrowed' from North America and adjusted to explain the position of European mountain ranges and their stratigraphy. Working on ancient sediments of the Tethyan Ocean which formerly separated the Mesozoic continents of Europe and Africa, Europeans such as Suess (1875) and Neumayr (1875) considered geosynclines to be essentially symmetrical and to contain oceanic sediments.

Haug (1900) recognized that graptolitic shales in the Caledonian mountain chain and schistes à Aptychus in the Alpine mountain chain were bathyal facies and he concluded that most geosynclines were deep marine troughs. Based on the present-day intracontinental position of the Alpine chains in Europe and Asia, he considered that geosynclines developed between and on the adjacent margins of two continents, rather than on or beside a continental margin, and that during geosynclinal development sedimentation took place progressively closer to one of the continents, termed a 'foreland'. The Alpine geosyncline thus developed between and on the margin of the European foreland to the north and the African hinterland to the south. In trying to relate the North American Western Cordillera geosyncline to a similar setting Haug was forced to postulate an enormous 'Pacific continent' to the west.

Stille (1913) emphasised the role of the geanticlinal uplift of Dana, arguing that in a subsiding belt, sediment supply, facies and the subsidence itself were related to uplift of an adjacent belt. While not concerned specifically with the position of a geosyncline relative to a continent, he also recognized that folds in geosynclinal sediments were directed towards the continental 'foreland'.

14.2.2 Concepts and classification of geosynclines in Europe

After the First World War, many European geologists were occupied with the tectonics of geosynclines, particularly with the problem of the origin of the impressive nappe structures and apparent tectonic thickening of the crust in the Alps. A conflict developed between the proponents of lateral compression and the advocates of gravity tectonics. Following Termier (1902), many considered that the nappes in the Dinarides resulted from lateral compression between two continental masses, with an advancing foreland squeezing out large recumbent folds or nappes from a hypothetical root zone. Others followed Haarmann (1930) and earlier workers in Italy in favouring the idea that nappes were due to gravity gliding (van Bemmelen, 1949, 1954).

Schuchert (1923) erected the first comprehensive classification of geosynclines to include both American and European views. By analogy with present-day mountain belts he divided geosynclines into an intercontinental Mediterranean type, and an Appalachian type within but near the margin of a continent with sediment derived from a continental borderland or geanticline on the ocean side. As a subgroup of the Appalachian type he included island arc systems of East Asia.

Geosynclinal classification, based largely on deformation and magmatism, was extended by Stille (1936, 1940). He recognized *orthogeosynclines* or 'true' geosynclines, characterized by Alpine type deformation and orogeny resulting in mountain chains. He divided them into *eugeosynclines* with pre-orogenic ophiolites or greenstones interpreted as submarine flows, syn-orogenic andesites and post-orogenic granites and *miogeosynclines* without igneous rocks. He also recognized *parageosynclines,* which did not form mountain chains and were characterized by block faulting.

Following Schuchert, Stille considered that orthogeosynclines could develop either between two continents or at the boundary between continent and ocean. He noted that within an orthogeosyncline the miogeosyncline lay on the continent side of the eugeosyncline, and, following Suess, he considered that continents grew through addition of successively younger geosynclines to their margins.

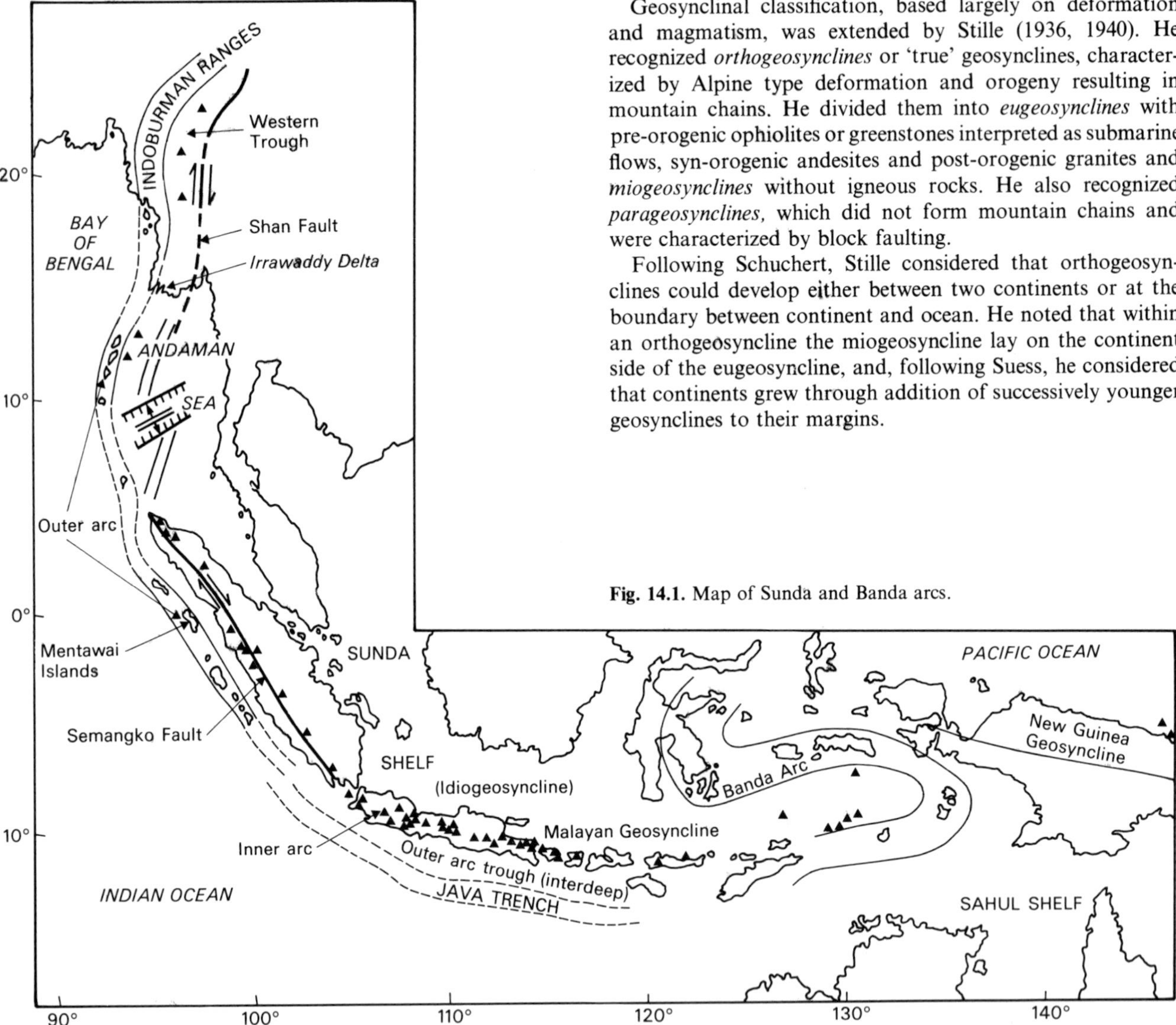

Fig. 14.1. Map of Sunda and Banda arcs.

The European view of geosynclines before and particularly after the Second World War was strongly influenced by the experiences of Dutch geologists in Indonesia. A terminology was developed based largely on comparison of the Alpine chains with the Sunda Arc (Fig. 14.1). Following Haug (1900) who had first considered the Sunda Arc as a modern geosyncline, numerous authors, in particular Rutten (1927) and Kuenen (1935), used the region as an actualistic model for hypotheses of mountain building (Umbgrove, 1949; Van Bemmelen, 1949). Within the Sunda Arc these workers recognized the volcanic island or 'inner arc' in Sumatra, on and near the southwestern edge of the Sundaland continent, and extending southeastwards as an island arc through Java to Flores. The *idiogeosyncline* of Van Bemmelen (1949) lies on the landward side of the inner arc, on the oceanic side of which a further sedimentary trough or interdeep (Van Bemmelen, 1949) is bordered by an outer arc consisting mostly of deformed sedimentary rocks with local ophiolites; this system is bordered on the Indian Ocean side by a deep submarine trench.

The tightly curved Banda Arc of Indonesia provided an alternative model to that of the Sunda Arc. Umbgrove (1938, 1949) and de Sitter (1956) noted its inter-continental position between the Sunda Shelf to the west and Sahul Shelf of Australia to the east, and considered that the arc's curvature and Miocene folding in the 'Malayan geosyncline' to the south and 'northern New Guinea geosyncline' to the north were the result of movement between the two continental areas.

The concentric distribution of arcs in Indonesia provided the basis for Van Bemmelen's 'undation theory' (1949) of gravity gliding of sedimentary rocks away from rising granitic asthenoliths, with outward growth of continents through addition of successive arcs. Although accepted by some Dutch geologists as a possible explanation of the evolution of part of the East Indies (de Sitter, 1956) the theory could not explain geosynclinal sedimentary successions elsewhere and was popular neither in Britain nor in North America.

During the period 1955 to 1965 the geosynclinal concept reached its ultimate elaboration (e.g. Trümpy, 1960). Kuendig (1959) illustrated (Fig. 14.2) a view prevalent in Europe of the sedimentary facies pattern from shelf, through slope to trough, stressing in particular the importance of ophiolites as indicators of the eugeosyncline. Aubouin (1959, 1965), whose views incorporate many ideas widely held immediately before and during the early years of the plate tectonics concept, took the Mediterranean Alpine chains, and particularly the Hellenides of Greece, as his type-geosynclines. He compared these chains with Indonesia but tried to fit the Sunda Arc into a model based on the Mediterranean chains, rather than explaining the Alps in terms of the modern Sunda Arc.

Aubouin defined an elementary geosyncline model, of which the Western Alps, Hellenides, Apennines and Carpathians each formed examples. Like Stille and many North American geologists, Aubouin divided an elementary geosyncline into a miogeosyncline and eugeosyncline, and further divided each of these into a ridge and furrow (Fig. 14.3). The miogeosyncline lay between the foreland on the outer or external side, and the eugeosyncline on the inner or internal side.

Of particular significance was Aubouin's emphasis on the

Fig. 14.2. Geosynclinal model of Kuendig (1959).

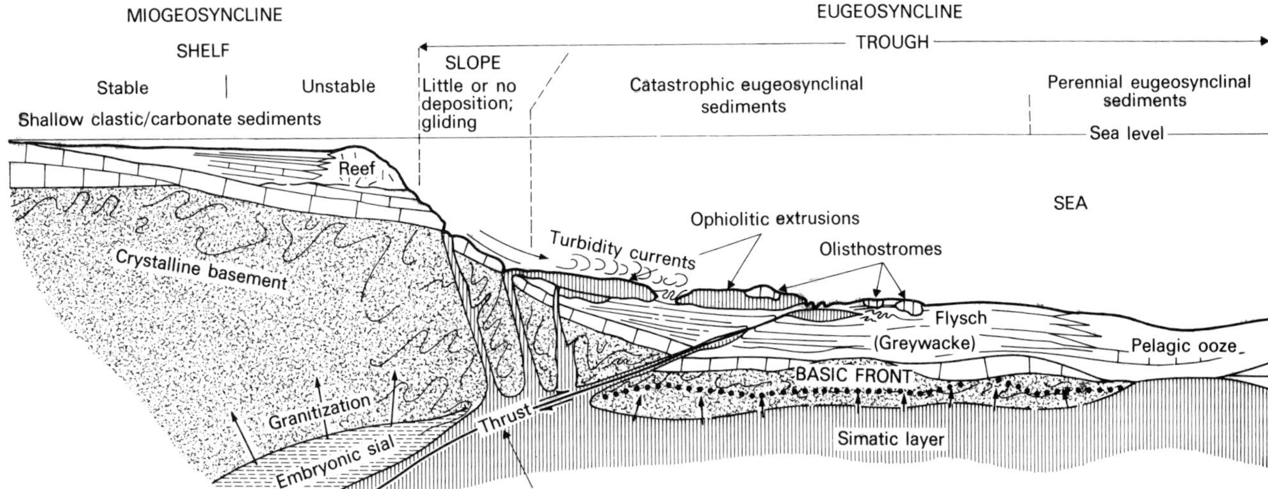

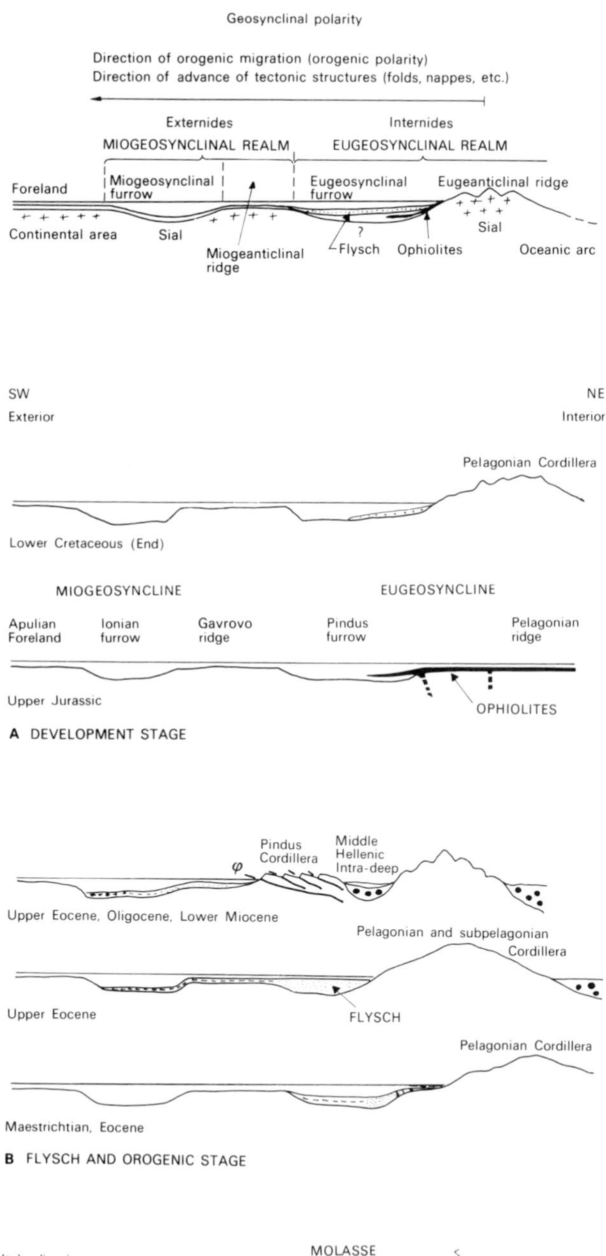

Fig. 14.3. Geosynclinal model of Aubouin (1965) showing eugeosyncline and miogeosyncline with ridges and furrows, and stages of development.

polarity of geosynclines, and his recognition both of distinct stages of development and of the migration of sedimentation and deformation towards the foreland during orogeny. However he was unable to explain satisfactorily the presence of oceanic foreland (the Indian ocean) in the Sunda Arc in the same tectonic setting as continental foreland in the Mediterranean chains, and did not fully accept his own evidence for convergence of the foreland and hinterland.

14.2.3 **Concepts and classification of geosynclines in North America**

The elaboration of geosynclinal classification culminated with the work of Kay (1947, 1951). He followed broadly the ideas of Stille on orthogeosynclines, with a miogeosyncline on the continental side of the eugeosyncline, but also stressed the significance of both greywackes and igneous rocks. In the Appalachian geosyncline (Fig. 14.4) the Champlain miogeosyncline belt was equivalent to the 'synclinal' of the folded Appalachians described by Hall, while the Magog belt to the east consisted of a much thicker succession of folded and metamorphosed sedimentary and volcanic rocks. Kay favoured an easterly or 'continental borderland' source area for the clastics of his Magog belt eugeosyncline, but considered that a volcanic cordillera within the eugeosyncline was also a possible source.

Kay divided the parageosyncline of Stille into three types: *exogeosynclines*, situated on a continental margin and receiving detritus from orogeny of ortho-geosynclines, *autogeosynclines* mostly consisting of carbonates and located within the continent, independent of the orthogeosyncline; and *zeugogeosynclines*, also within the continent and filled by erosion of intercontinental mountain chains.

Kay also erected a group of sedimentary basins considered to form late in geosynclinal development. These comprised *epieugeosynclines*, receiving detritus from orthogeosynclinal mountain ranges; *taphrogeosynclines*, related to intracontinental rift zones and block faulting, and *paraliageosynclines* located on a continental margin with the Gulf of Mexico as the type example. Taphrogeosynclines were the most widely recognized of these, and the Triassic fault troughs of New England were considered an example, with their alluvial fan conglomerates and arkosic red beds. They were sometimes incorrectly compared in tectonic setting to the Alpine molasse.

During the 1950s the turbidity current hypothesis fundamentally altered two aspects of geosynclinal models by showing that sandstones and conglomerates could be deposited in deep water. Firstly, palaeogeographic reconstructions no longer required that the lateral passage of shales into sandstones reflected the approach of a shoreline, and secondly orthogeosynclines did not require prolonged subsidence during deposition because a great thickness of mass-flow deposits

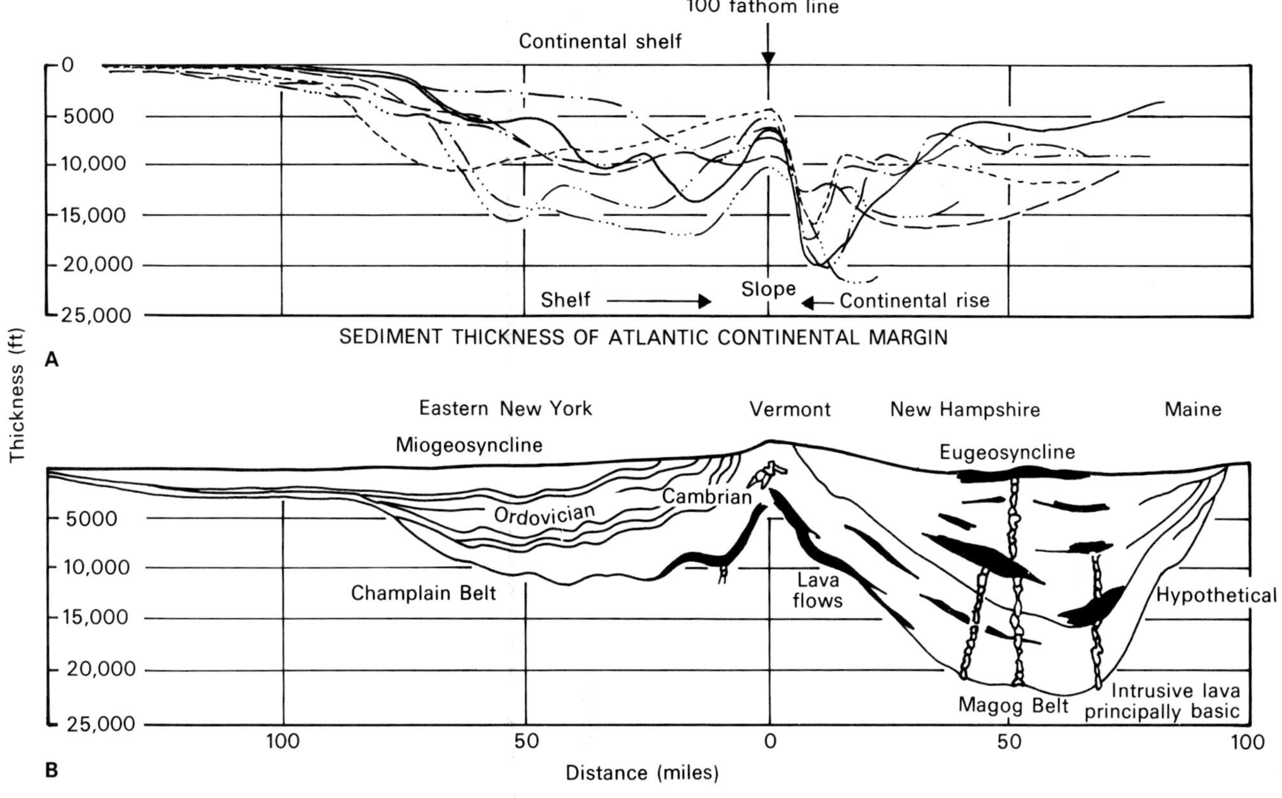

Fig. 14.4. Comparison of modern (Atlantic) continental margin geosyncline with the Ordovician of the Appalachian geosyncline of Kay (1951) (after Drake *et al.*, 1959).

could accumulate within an initially deep basin. In addition, both the evidence that turbidity currents generally flowed along rather than across geosynclinal axes and the geography of many present-day elongate basins suggested filling from one end, as well as from the sides (Kuenen, 1957b).

Drake, Ewing and Sutton (1959) showed that on the eastern margin of North America a seaward thickening continental shelf prism is bordered by a continental slope with a thin sedimentary cover. This passes oceanwards into the continental rise prism with thick successions of turbidites (Fig. 14.4).

Comparison of this Atlantic margin cross-section with schematic cross-sections of the Lower Palaeozoic of eastern North America (Kay, 1951) (Fig. 14.4) and the upper Palaeozoic of western North America (Eardley, 1947) showed convincing similarities in sedimentary thicknesses and facies, if not in magmatic associations.

In America the Atlantic margin analogy was widely accepted and modified by Dietz (1963) and Dietz and Holden (1966) who suggested that the late Mesozoic and Cenozoic miogeosynclinal succession of the eastern United States thickened seawards towards the continental shelf edge, forming a miogeocline. They compared the Atlantic miogeocline to the Lower Palaeozoic folded Appalachian geosyncline of Hall (1859), and the continental rise to the metamorphosed rocks of the crystalline Appalachians or eugeosyncline of Kay (Fig. 14.5).

14.2.4 Concepts of geosynclines and metallogenesis in U.S.S.R.

The Russian viewpoint on the position of geosynclines was summarized in 1950 by Peyve and Sinitzyn (see Aubouin, 1965 pp. 31–33), who recognized primary geosynclines, corresponding to the eugeosynclines of Stille but developed within fractures in intracontinental platforms. Secondary and residual

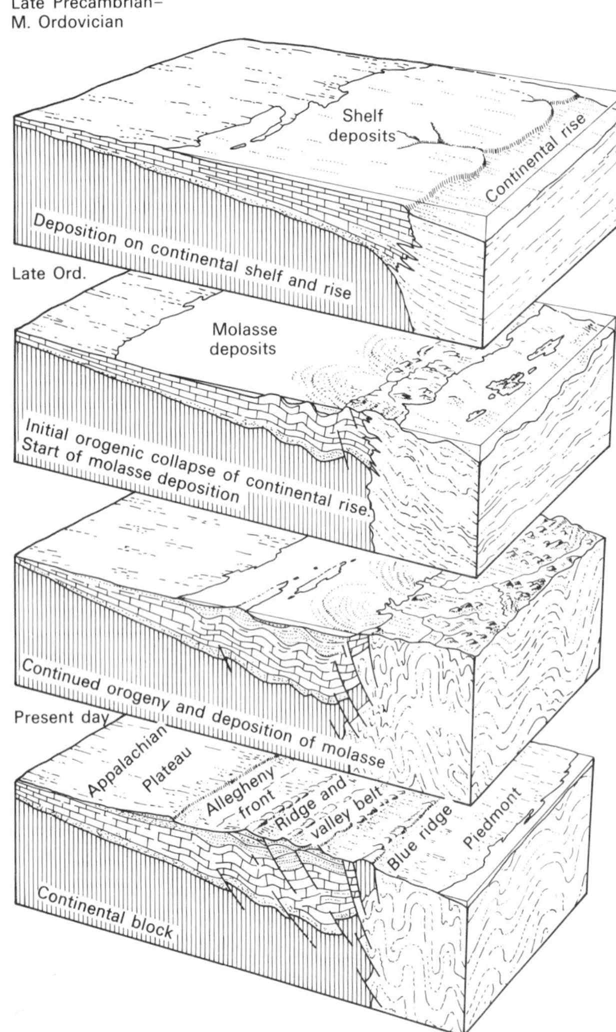

Fig. 14.5. Deposition and collapse of continental rise (from Dietz and Holden, 1966).

geosynclines were considered to develop successively following orogeny of the primary geosynclines. This intracontinental situation of geosynclines, based on the position of the Urals between the Russian and Siberian platforms, characterized Russian views on the problem throughout the 1950s and 1960s (e.g. Beloussov, 1962; Smirnov, 1968).

The Russian concept is important because a number of geologists from the USSR have attempted to interpret metallogenesis, or the origin of concentrations of metallic elements in specific areas of the earth's surface, in terms of geosynclinal evolution (Bilibin *in* McCartney and Potter, 1962; Smirnov, 1968). While the relationship of mineralization to geosynclinal evolution attracted relatively little attention outside Russia, more recent attempts to relate metal provinces and types of mineral deposit to the plate tectonic setting in which they formed have attracted widespread interest, partly as a result of the growing appreciation of the close relationship between the formation of ore bodies and that of the surrounding host rocks (Stanton, 1972). This relationship is particularly close for syngenetic sedimentary and volcanogenic mineral deposits, the origins of most of which are now commonly discussed together with deposition of the host rocks as characteristic of specific types of tectonic setting (e.g. Guild, 1974; Mitchell and Garson, 1976).

14.2.5 **Geosynclinal facies and cycles of sedimentation**

Bertrand (1897) was the first to relate sedimentation to geosynclinal growth by recognizing a geosynclinal cycle made up of four facies, termed here pre-orogenic → pre-flysch → flysch → molasse (Table 14.1). The idea was developed by several Alpine geologists and transported across the Atlantic by Van der Gracht van Waterschoot (1928) where it had considerable influence on American stratigraphers during the 1940s and 1950s (Hsü, 1973b). It also had a great effect on both geosynclinal and orogenic concepts for many decades and it is significant that individual geologists found it applicable in so many mountain belts. Yet, although there was a basic similarity of facies successions, there were also substantial variations in both the definitions and uses of the terms. By their very nature most basinal successions now exposed must pass from a basement (pre-orogenic) phase through an extending, starved (pre-flysch) phase, as the basin deepened, to a deep marine clastic (flysch) phase and finally to a continental clastic (molasse) phase. However, since the tectonic settings of basins are extremely variable, it is to be expected that there will be considerable variation in detail and it is these variations which can now partially be explained by the use of present-day tectonic analogues.

PRE-OROGENIC FACIES

This is varied and includes granitic and/or metamorphic basement, terrigenous clastics, platform carbonates, and facies attributable to a miogeosynclinal setting. It indicates that prior to initiation of the geosyncline, shallow-water sedimentation was taking place upon continental crust beside or beneath the pre-flysch.

PRE-FLYSCH

This includes a wide range of facies of which the only common factor is a stratigraphical position beneath the flysch. Primarily

Table 14.1 Facies terminology for the geosynclinal cycle (partly after Hsü, 1973b)

Bertrand (1897)	Hercynian Geosyncline of Europe	Krynine (1941)	Pettijohn (1957)	Aubouin (1965)
Grès rouges = molasse	Molasse	Post-geosynclinal Arkoses	Subgreywacke suite/Molasse	Molasse
Flysch grossier	Flysch	Geosynclinal Greywackes	Greywacke suite/Flysch	Flysch
Flysch schisteux = schistes lustrés	Bathyal lull Schwellen and Becken	?	Euxinic (evaporites)	Pre-flysch
Gneiss Cambriens	?	Early geosynclinal Carbonate/Orthoquartzite	Pre-orogenic	Carbonate Platform

it consists of fine-grained sediments, characteristically cherts, dark limestones, black shales and siltstones, occurring as two subfacies. Hercynian geologists termed the pre-flysch 'the bathyal lull' (Goldring, 1962) and divided it into a *Becken* (basin) and *Schwellen* (swell) subfacies (see Sect. 11.4.4). In the Alps the equivalents include the schistes lustrés (Bündnerschiefer or shaly flysch) and the leptogeosynclinal or starved geosynclinal facies of Trümpy (1960).

In the basins, relatively thick successions of fine-grained sediments were deposited in deep-water as hemi-pelagic sediments or low density turbidity current and other mass-flow deposits demonstrably derived from the adjacent swells as resedimented siliceous and calcareous oozes. Except for ostracods, fossils in Palaeozoic *Becken* facies are sparse and since the facies also lacks the more spectacular sandstones of the flysch it has been neglected by both palaeontologists and sedimentologists. On the swells the successions are condensed, and stratigraphically interesting faunas abound, such as Palaeozoic graptolites and goniatites, and Mesozoic ammonites.

In the pre-flysch of the Alps and in some Lower Palaeozoic successions ophiolites are abundant. Originating as a 19th century term meaning serpentinite in Greek, their association with radiolarian cherts and other deep sea sediments was first recognized by Steinmann (1905, 1927). An ophiolite comprises the upward sequence ultramafic rocks, gabbros, dykes and pillowed lavas (Penrose Conference, 1972). Bailey and McCallien (1960) gave the name Steinmann Trinity to the association radiolarite, serpentinite and 'greenstones' (basalts and gabbros). Many Alpine geologists generally considered ophiolites to be ocean floor rocks because of their association with radiolarian cherts analogous to present day radiolarian oozes. Petrologists and some stratigraphers, for example Aubouin, on the other hand thought they were emplaced in the early stage of a magmatic cycle within ensialic geosynclines.

In North America the significance of ophiolites in geosynclinal development was not recognized until recently. Kay (1951) included a wide range of volcanic rocks in his eugeosyncline without concern for their stratigraphic position.

Pettijohn's (1957) euxinic facies, though occurring in the same stage of the cycle as the pre-flysch, lacks volcanics and comprises black shales considered to have been deposited in reducing conditions as in the present Black Sea. It is significant that many pre-flysch formations lack the characteristic igneous suite.

FLYSCH

Flysch is one of the most used and abused words in geology. It has been used (Hsü, 1970) as a formation name and as a sedimentary facies, descriptivally for alternating sandstones and shales and genetically as a synonym of turbidite. It has also been used as a tecto-facies (Sect. 2.1) for the sediments deposited during orogeny (syn-orogenic), though others have argued that it is pre-paroxysmal and was formed prior to the main deformational stage. Reading (1972) suggested that the word be used for 'any thick succession of alternations of sandstone calcarenite or conglomerate with shale or mudstone, interpreted as having been deposited by turbidity currents or mass-flow in a deep water environment within a geosynclinal belt'. This definition is thus descriptive but depends on a sedimentological interpretation; it brings in the tectonic setting, but begs the question of what is a geosyncline.

De Raaf (1968) and Stanley (1970) limit the term to mass-flow deposits found in orthogeosynclines and call those found in late geosynclinal stages, and where active tectonism is not known, flysch-like or flyschoid. However, although it is important to avoid the term flysch for turbidites or turbidite-like beds found in fluvial, lacustrine or deltaic environments, we prefer to define flysch independently of tectonic setting and to qualify it when its genesis is known.

MOLASSE

The term molasse has had a history which is similar to, if slightly shorter than, that of flysch (Van Houten, 1974). It is used in a lithological sense for thick successions of sandstones and conglomerates and also for an environmental facies of continental, dominantly fluvial character. Bersier (1959), for example, described fining-upward fluvial cycles from the molasse of Switzerland before similar sequences were recognized in North America and Britain as being diagnostic of fluvial facies. Molasse has also included fluvio-deltaic and alluvial fan facies. However, since Bertrand (1897), molasse has been used as a tecto-facies for sediments which either follow or are partly contemporaneous with flysch and were deposited on the flanks of the earlier geosyncline as both a late orogenic facies now deformed and as an undeformed post-orogenic facies.

The term is often confusing because turbidites may accumulate in some post-orogenic basins such as the present Swiss lakes resulting in flysch formations within a molasse tecto-facies.

In spite of these problems molasse, like pre-flysch and flysch, is a useful term, the definition of which is less important than an understanding of the sense of which the word is used.

We apply it to any thick succession of continental deposits consisting in part of sandstones and conglomerates which were formed as a result of mountain building. The molasse facies is therefore the main diagnostic feature of orogeny in its morphogenetic sense of mountain building.

14.2.6 **Plate tectonics and geosynclines**

Up to the 1960s geosynclines were interpreted as narrow elongate troughs in which the various stages of development resulted from vertical movement, with subsidence followed by compression and uplift. Although some overall lateral movement was allowed by many geologists, the original widths of the troughs were thought to be of the same order of magnitude as the present Mediterranean rather than the Atlantic or Pacific. Other geologists however were unhappy to concede any overall lateral movement of continents and preferred to explain evidence for contraction, such as nappes, by vertical tectonics and gravity gliding.

The theory of continental drift was finally established during the late 1950s by the discovery that secular changes in palaeolatitude could only be explained by huge relative movements of continents. During the 1960s a mechanism for continental drift was provided by the ocean-floor spreading hypothesis. This postulated that oceans were created by addition of crust at oceanic rises by demonstrating that magnetic striping in oceans could be matched either side of oceanic rifts and fitted into the magnetic reversal stratigraphy known from Iceland and deep sea sediment cores. Thus a quantitative estimate could be made of the rate of spreading and hence the size of modern oceans at different stages of the Tertiary.

This resulted in an attempt by Wilson (1966) to explain geosynclines in terms of a modern spreading ocean such as the Atlantic. Earlier attempts such as Dietz's (1963) had led the way but were unsatisfactory for two reasons. There was neither seismic nor structural evidence for thrusting of the Atlantic under the American continent and secondly, it was difficult to see how magmatic activity at an active ridge could drive oceanic crust, only 5 km thick, under a neighbouring continental mass.

Wilson's cycle did not try to fit all features of geosynclines and orogenic belts into a single present-day model. By demonstrating that there could be an opening phase of extension followed by a closing stage of contraction and that the present Atlantic was simply the opening stage, while the Caledonian orogeny represented an earlier closing of a Proto-Atlantic, he explained the established geosynclinal/orogenic cycle.

However, during the 1965–68 period it was not clear where the *present* models for the contracting, compressional, stage should be found. In 1968, several authors (Morgan, 1968; Le Pichon, 1968; Isacks, Oliver and Sykes, 1968) showed firstly that it was not just thin oceanic crust that was moving *on* mantle but that the crust moved *with* the upper part of the mantle as a rigid lithosphere, roughly 100 km thick. The crust, including the continents, thus rode passively on the plates. Secondly they demonstrated how lithosphere could be consumed along a seismic Benioff zone by subduction beneath deep-sea trenches and island arcs. Thus extension at spreading centres could be accommodated by consumption at subduction zones without a global change in surface area.

The concept of plate tectonics was then applied to geosynclines by Mitchell and Reading (1969), Dewey and Bird (1970) and Dickinson (1971a). They showed that geosynclines had several modern analogues, thus explaining the variability that was known among them. By postulating that modern oceans would go through stages of opening and then closing they showed how the modern Atlantic might represent the early extensional geosynclinal phase with the ocean appearing to be symmetrical in its entirety and asymmetric if only one margin was considered. The closing stages would be represented by subduction zones of which there were three types, *island-arc* where oceanic crustal lithosphere descended beneath oceanic crustal lithosphere, *Andean* where oceanic crustal lithosphere descended beneath a continental margin and *Himalayan* where subduction had led to the collision of two continents.

14.3 MODERN PLATE TECTONIC SETTINGS

The principal difficulty in relating sedimentation to plate tectonics is that sedimentary facies are only indirectly related to those processes on which the theory is based. Plate tectonics derives from geophysical evidence of processes taking place within oceanic lithosphere, particularly seismic activity. In some cases these geophysical processes are associated with igneous activity and metamorphism, the products of which can be seen both in modern settings, such as active island-arcs, and in ancient ones, on continents. However, these products are not always preserved on continents and so the direct evidence of seismic activity is frequently absent or inconspicuous.

As has long been known to structural geologists (Anderson, 1951; Harland, 1965) the earth has three types of mobile zone (1) extensional, recognized by dykes, volcanoes and normal faults (2) contractional, recognized by folding and thrusting (3) horizontal shear zones recognized by transcurrent, strike-slip faults, frequently of great lateral extent.

Each of these zones has its own topographic expression, seismology and magmatic activity and the plate tectonic theory offers a mechanism for their formation by postulating that they form boundaries between plates, viz:

(1) divergent, where two lithospheric plates are moving apart as ocean-floor spreading takes place by accretion of new lithosphere;

(2) convergent, where one plate descends beneath another and lithosphere is consumed;

(3) strike-slip or transform, where plates move laterally past one another and where, without substantial divergence or convergence, lithosphere is conserved.

There are sound theoretical grounds, supported by much geological evidence, for thinking that each type of tectonic zone or plate junction has distinctive characteristics and for this reason these divisions can be used to classify sedimentary basins. However, it is also clear that few areas of the world fit simply into one or other of these compartments. Because movement between plates is usually oblique most plate junctions show characteristics of more than one type of zone. In addition, on one scale one type may be apparent, while on another scale another type appears to be dominant (Reading, 1972).

Strike-slip junctions, or the transform faults of oceanic areas, have in recent years not been given as much consideration as the other two types of junction. For example, possibly because the concept of ocean-floor spreading preceded that of plate tectonics, many geologists have sought an explanation of the initiation of plate patterns in the distribution of hot spots, resulting in formation of oceanic ridges. However, some patterns may be initiated at a subduction zone, perhaps by continental margin downwarping due to sedimentation. It is also possible that the mega-shearing of Carey (1958) resulted first in strike-slip faults and subsequent subduction or divergence.

As interpretation of ancient sediments is by analogy with modern counterparts, distinction between modern and ancient is essential. We apply the term 'modern' to those rocks which today lie in a tectonic position similar to that in which they were formed and 'ancient' to those rocks which have been emplaced in a different tectonic situation. For example Jurassic oceanic rocks of the present Pacific are modern and late Tertiary marine rocks now found uplifted in the Himalayas are ancient. Since there are all gradations from those which are clearly modern to those which are ancient, in many cases the distinction is arbitrary.

14.3.1 **Spreading-related settings**

Spreading-related settings are characterized by domal uplift of a central ridge, commonly with a median valley, though fast spreading ridges, such as the East Pacific Rise, lack the central rift. Uplift is due to thermal expansion indicated by high heat flow and by magmatic activity. Such settings can be divided into those where no oceanic crust has been emplaced and are therefore intracontinental such as the East African rift and Baikal rift, those where some oceanic crust has been emplaced like the Red Sea and Gulf of California and those larger oceans such as the present Atlantic and Indian oceans, where the continental margin is now distant from the oceanic ridge. Spreading-related settings also include the deep and long lasting troughs interpreted as one arm of three-armed rift systems which aborted before major lateral movement began. They are filled by major sedimentary piles and form the sites of many deltas, such as the Niger (Burke, 1976).

INTRACONTINENTAL RIFTS

The East African rift system is a major global structure nearly 3,000 km long with evidence of intermittent deformation and magmatism since the Pre-Cambrian without any apparent major lateral movement on either side (Fig. 14.6). It is characterized by a gradual topographic rise towards the central rift and a steep escarpment above the rift which is 40–50 km wide with a floor some 2 km below the level of the surrounding plateau. Within the rift are Cenozoic volcanoes and non-volcanic horsts such as Ruwenzori which rises, at a bifurcation of the rift, to nearly 4 km above the plateau level. The rift is thus discontinuous, with many transverse highs due to volcanic eruptions, differential vertical movements, and *en echelon* offsetting of the rift faults. Between these highs are a number of separate basins, the best known of which contain present-day lakes, some filled with up to 2 km of sediment.

Because of the gentle slope up to the lip of the rift the major drainage of East Africa is into rivers like the Congo and Zambesi, and away from the rift itself, where clastic sediment is limited to material which comes from the neighbouring scarps and uplifted blocks within the rift, though locally rivers such as the tributary of the Zambesi, the Luanga, use the rift as a

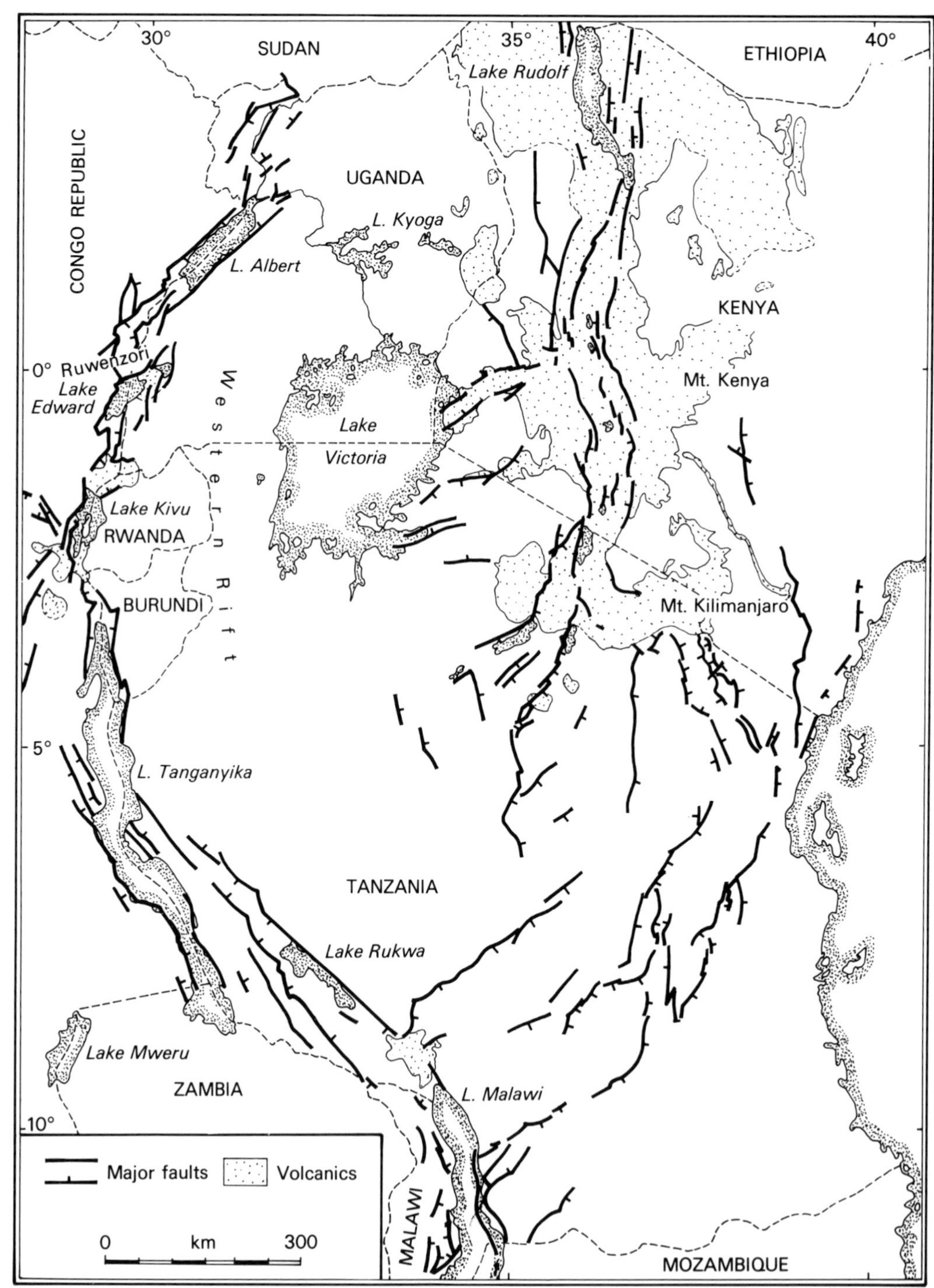

Fig. 14.6. East African rift system (from King, 1970).

channel. Hence, the dominant sediments are those of alluvial fans and lakes. Many East African lakes contain economic deposits of sodium carbonate which form as a result of volcanic activity and the hot climate (Sect. 4.2.4).

The Baikal rift system has a similar length to that of East Africa (King, 1976). It consists of linear systems of intermontane depressions, essentially graben or half-graben with one steep side, close to a major dislocation, and the other more gently sloping. The depressions occur along the crests of arched uplifts and consequently the maximum altitudes of mountain ranges are very close to the margins of the depressions. The 12 largest depressions range in length from 100 to 700 km but are only 15–18 km in width. They are mostly dry, but are partly occupied by lakes such as Lake Baikal, the world's deepest lake, which is 670 km long and over 1,700 m deep. The Baikal depression contains over 5 km of sediment, but most depressions are filled by less than 3 km of sediment. The sediments are continental, the Lower Oligocene to Lower Pliocene portion consisting of shallow lake, swamp and fluvial deposits. The later, Pliocene to Recent, portion is composed of diverse sediments resulting from active vertical movements and the mountainous relief. The volcanic rocks, which are similar in composition to those of the East African rift system but are less abundant, are largely confined to the uplifted blocks and arches.

FAILED RIFTS

Extending outwards from the interior of a continent and deepening toward the continental margin are long-lived deep linear troughs, sometimes called aulacogens (Burke, 1977) (Sect. 14.4.1). One of the earliest described examples, the Benue trough, extends northeastwards from the Gulf of Guinea and has been interpreted as a Cretaceous rift system linked to other rifts which subsequently became the South Atlantic ocean (Burke, Dessauvagie and Whiteman, 1972). It has been filled during the Tertiary by over 10 km of sediment comprising a succession which passes from submarine fan deposits, through deltaic to fluvial sediments (Fig. 14.7).

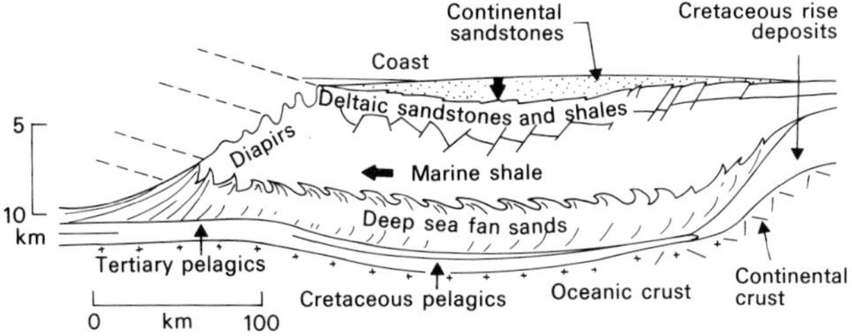

Fig. 14.7. Schematic structural cross-section of Niger delta showing sedimentary fill of Benue trough. Outward movement of marine shale leads to diapir formation (after Burke, 1972).

INTERCONTINENTAL RIFTS, PASSIVE CONTINENTAL MARGINS AND OCEAN FLOOR

Plate tectonic theory suggests that an intracontinental rift zone may develop into an intercontinental one as oceanic crust starts to be emplaced in the axial zone; this process is commonly diachronous along the length of the rift. As spreading continues, each half of the 'rift' becomes a passive continental margin, also termed trailing, inactive or 'Atlantic' type which comprises a shelf, slope and rise. Between these continental margins the ocean floor consists of abyssal plains and oceanic rises where spreading is still active.

EARLY STAGE OF INTERCONTINENTAL RIFTING

The present Red Sea is an elongate basin 2,000 km long, with shorelines 180 km apart at the northern end and 360 km at the widest point, narrowing to 28 km at its connection with the Gulf of Aden. Within the main trough is an axial trough, over 1,000 m deep, from 4 to 30 km wide within which are volcanic islands and, in the deeper parts, hot brines. Before the Pliocene, when the present episode of spreading started (Richardson and Harrison, 1976), vast outpourings of alkali olivine basalts on the marginal swells of Ethiopia and Arabia coincided with depression of the Red Sea during the late Eocene and early Oligocene. A marine incursion into this basin from the Mediterranean led to deposition of a very thick (2–5 km) evaporite-clastic sequence which continued into the Miocene (Hutchinson and Engels, 1970). Locally basaltic lavas interfinger with the sediments. This period was possibly an intracontinental stage as evidence for a pre-Pliocene spreading phase is disputed. In the Pliocene and Quaternary marine oozes were deposited and new oceanic crust, consisting of basalts, gabbros and diabases, was emplaced (Fig. 14.8) (Sect. 11.3.5).

The Gulf of California (Fig. 14.9) is at an early spreading stage, though, lying as it does at the southern end of the San Andreas fault system, the principal component of movement is lateral with the peninsula of Baja California moving northwestward relative to the mainland of North America along a series of transform faults which segment the spreading centres and form a series of individual basins. Unlike the Red Sea, which the Nile is prevented from entering by the marginal lip, the inflow of the Colorado River results in large quantities of clastic sediments accumulating at one end, though these do not reach far into the gulf, where sedimentation is dominated by pelagic sediments especially diatomites (see Chapter 11; Fig. 11.17).

Crowell (1974a) has shown how the basins between the transform faults can be looked upon as pullapart basins (Sect. 14.3.3) similar to those associated with the San Andreas fault system. The Saltern trough, at the head of the gulf, has

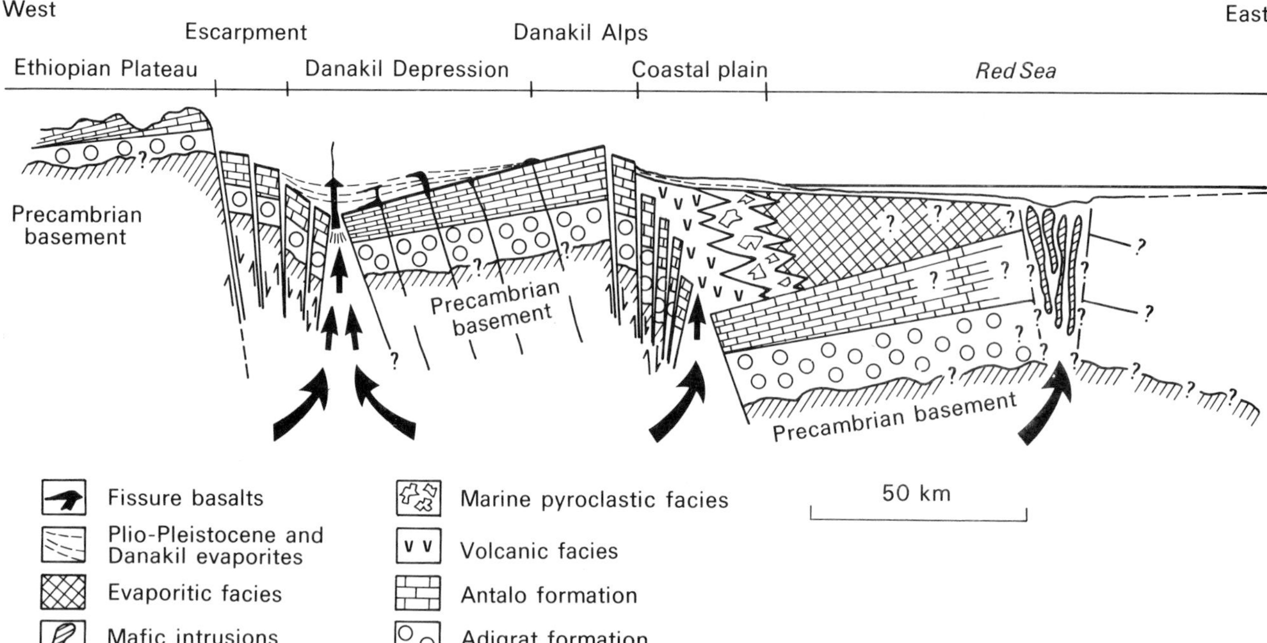

Fig. 14.8. Section across Red Sea and Danakil depression (from Hutchinson and Engels, 1970).

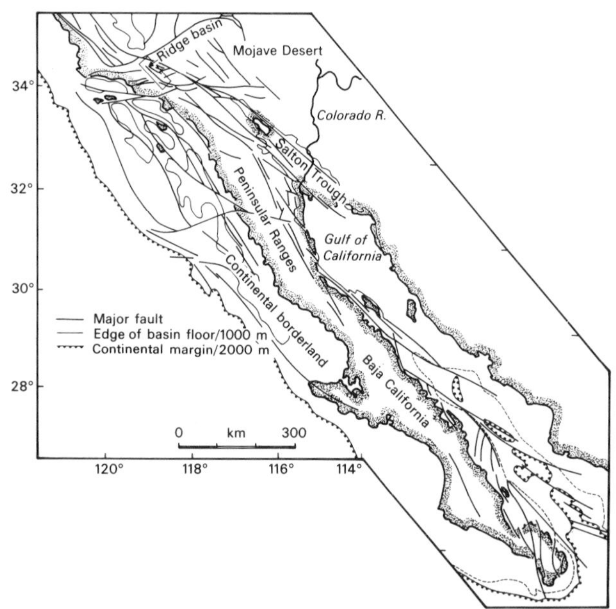

Fig. 14.9. Map of Gulf of California, the Salton trough, the Colorado river, the Ridge basin and basins of the California Borderland (after Crowell, 1974a, b).

considerable volcanic and intrusive activity, due to either a spreading centre or diapiric magmas at depth. At present, evaporites and marginal alluvial fans are forming in this trough which lies 110 m below sea level, but until the recent past huge quantities of material were brought down into the head of the gulf by an ancestral Colorado river to give a fluvio-deltaic pile perhaps as much as 6 km thick. An important aspect of the stratigraphy of these extending troughs is that the oldest units would be expected near the trough margin with the younger beds overlapping towards the centre.

LATE STAGE OF INTERCONTINENTAL RIFTING

The present Atlantic illustrates how intercontinental rifting can produce two laterally equivalent successions. Beneath the continental margin itself thick piles of continental shelf sediments overlie evaporites and red beds deposited within fault bounded troughs in the early intercontinental rift stages. Oceanward there is a lateral transition from continental rise through abyssal plain to oceanic ridge. This is matched by the vertical sequence of oceanic crust passing upwards through pelagic sediments to turbidites (Fig. 14.10).

Oceanic ridges may be highly fractured, like the mid-Atlantic ridge, or relatively smooth, like the East Pacific rise. Fractured

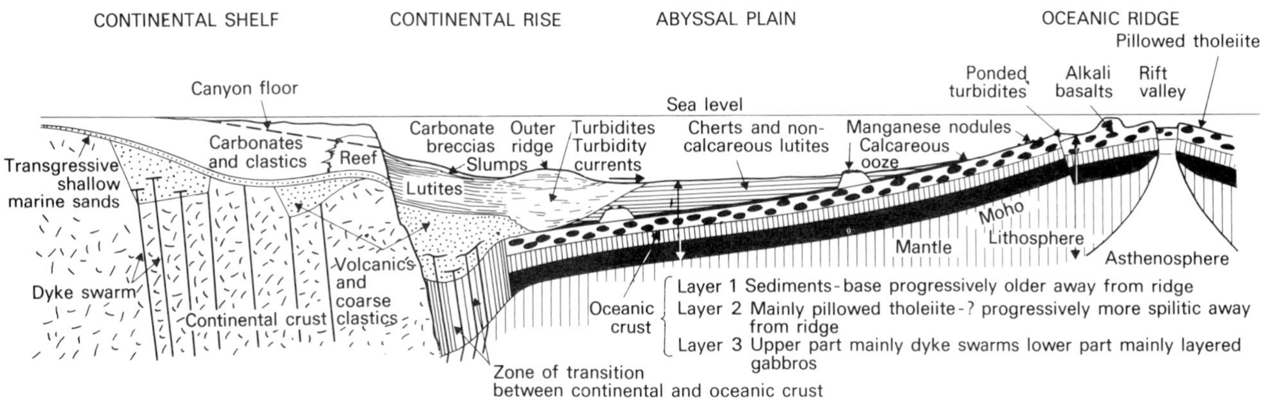

Fig. 14.10. Cross-section of the western Atlantic showing relationships of continental crust, oceanic crust and sediments (from Dewey and Bird, 1970).

ridges are block faulted due to normal faulting parallel to the ridge trend and transform faulting perpendicular to it. A complex pattern of ridges and basins develops (Fig. 14.11a). Some basins parallel the ridge such as the ponds which lie 75–100 km from the ridge axis between 22° and 23° North latitude in the Atlantic (Fig. 14.12) (Van Andel and Komar, 1969). They are approximately 10–30 km by 5–10 km, with surrounding hills rising up to 1,500 m above the valley floor, and have depths of about 4,000 m. Sediments within them are about 500 m thick and consist of fine-grained turbidites derived from the calcareous pelagic deposits of the neighbouring hills (For details see Sect. 11.3.2 and Fig. 11.11).

Other basins lie more or less transverse to the ridge axis and are related to fracture zones where the ridge axis is offset by transform faults. The Vema Fracture Zone is over 400 km long (Van Andel, von Herzen and Phillips, 1971) and consists of a central basin 3,000 to 5,000 m deep and 20 km wide margined either by the ridge crests or by high 'walls' which rise up to 3,000 m above the valley floor with gradients up to 15°. The trough itself has a rugged basement profile and contains up to 1,200 m of evenly bedded sediment with only local disturbance possibly due to slumping and including zones of basaltic cobbles at least 300 m thick (Perch-Nielsen *et al.,* 1975). The total basement relief along these fracture zones may be as much as 5,000 m. These graben-like basins and bordering ridges are characteristic of many Atlantic fracture zones (Van Andel, Rea, von Herzen and Hoskins, 1973). A detailed profile across a fracture zone (Fig. 14.11b) shows closely spaced faults and a varied pattern of lavas, intrusions, sedimentary breccias and hydrothermal deposits (Fig. 11.12); cleavage is developed along fault scarps (Arcyana, 1975; Francheteau *et al.,* 1976).

The detailed morphology of the rift valley is also very complex. In the only one surveyed in detail (Arcyana, 1975), the spacing of faults averages 50 m, with vertical scarps up to 20 m high. The sedimentary sequence is made up of breccia and talus units 50–85 m thick interbedded with pillow lava units and intruded by dykes (Fig. 14.11c).

As the new ocean floor cools and sinks to abyssal plain depths of about 4,000 m, the earlier formed ridge sediments are overlain either by pelagic sediments, whose nature depends on the local oceanic circulation and compensation depth, or by turbidites if a continental or island arc source is sufficiently close, and in many oceans a pelagic unit is overlain by turbidites. In some cases, such as the Aleutian abyssal plain, where a recent ridge has cut off the supply of turbidites, pelagic sediments again overlie the turbidites (Hamilton, 1967).

Oceanic ridges and abyssal plains contain important deposits of manganese nodules and less widespread metal-rich muds. Hydrothermal manganese oxide deposits with low trace metal contents form on topographic highs adjacent to faults near the axial rift zone of oceanic rises; hydrothermal manganese crusts are also associated with some oceanic spreading zones (see Sect. 11.3.2).

Hydrogenous nodules, some of which are of economic potential, are more or less restricted to the present ocean floors and only minor amounts are preserved either beneath the water-sediment surface or in ancient ocean floor successions. They consist largely of iron and manganese oxy-hydroxides, with higher contents of Co, Ni and Cu than hydrothermal nodules, and are widespread in parts of the Pacific, Indian and Atlantic ocean basins. The nodules form on the flanks of oceanic rises, seamounts and on abyssal plains, mostly at depths greater than 4,000 m, and are most abundant, locally forming manganese 'pavements', where strong bottom currents prevent accumulation of sediments. The only known deposits of economic interest in the immediate future are those in the

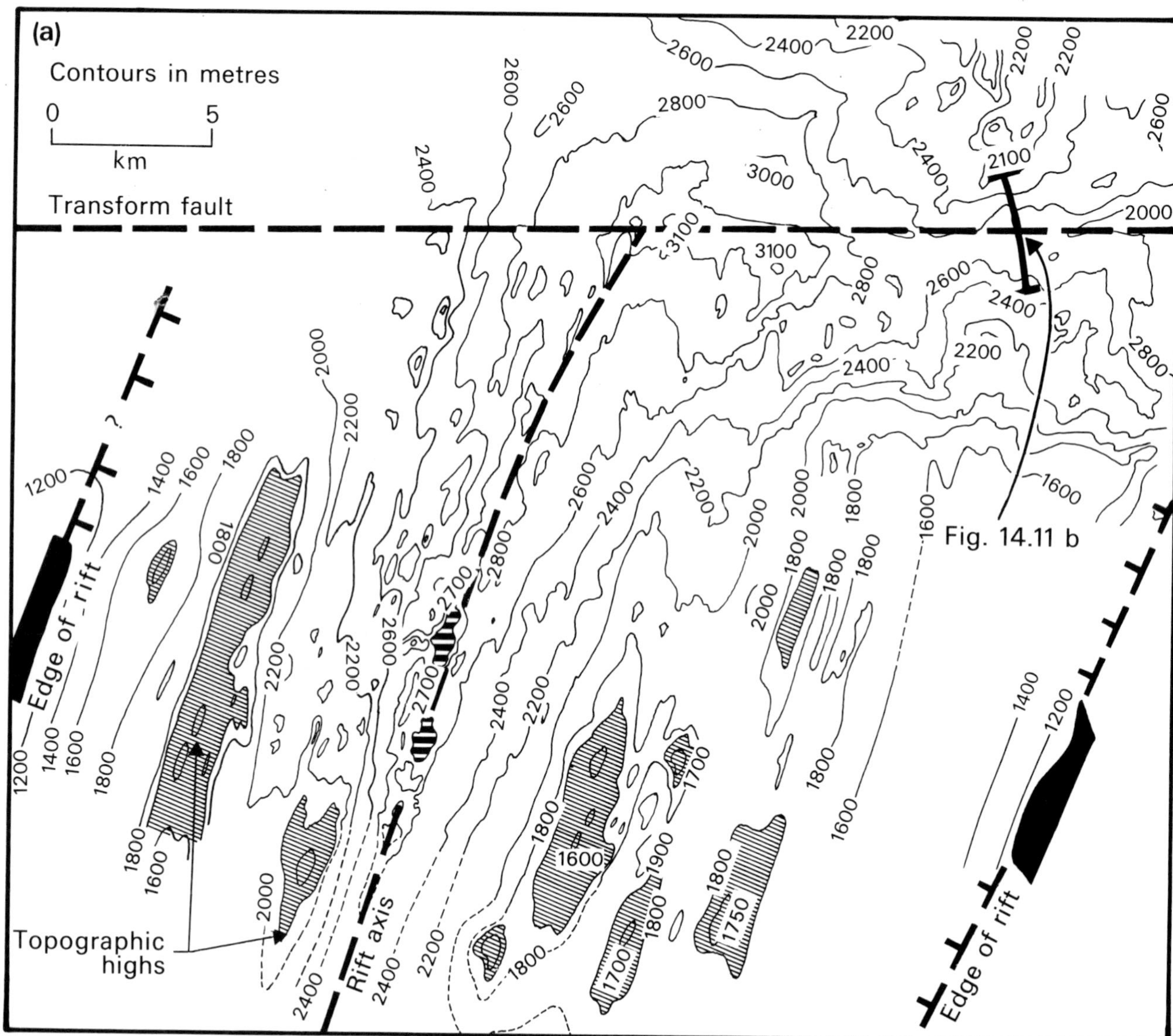

Fig. 14.11. Rift valley and transform fault zone in north Atlantic between 36° 45′ and 37°N.
(a) bathymetry
(b) section across transform fault
(c) section across part of western scarp bounding the rift floor (after Arcyana, 1975).

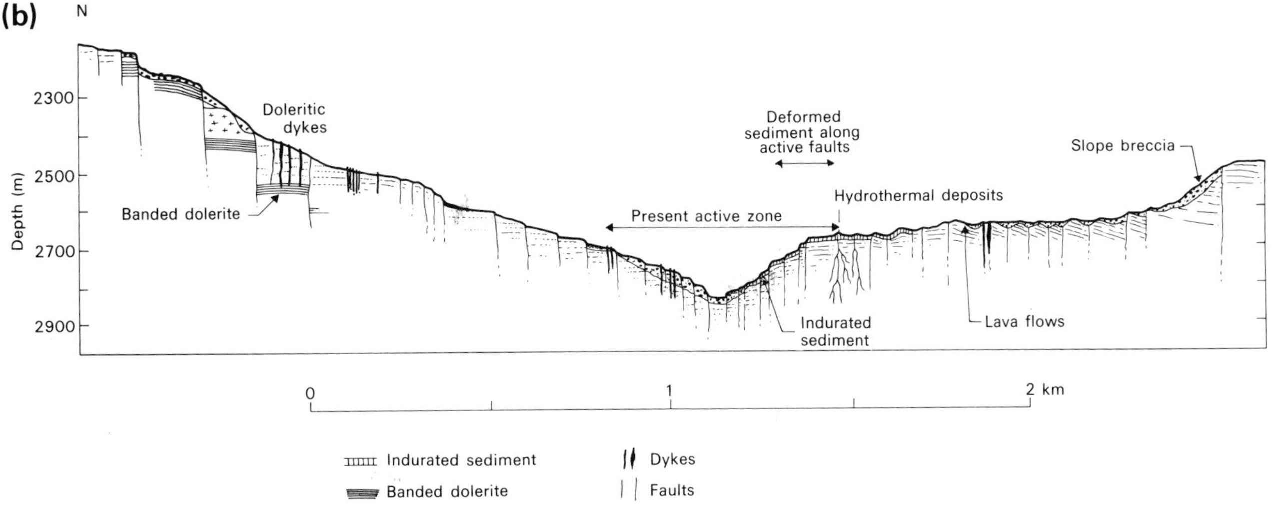

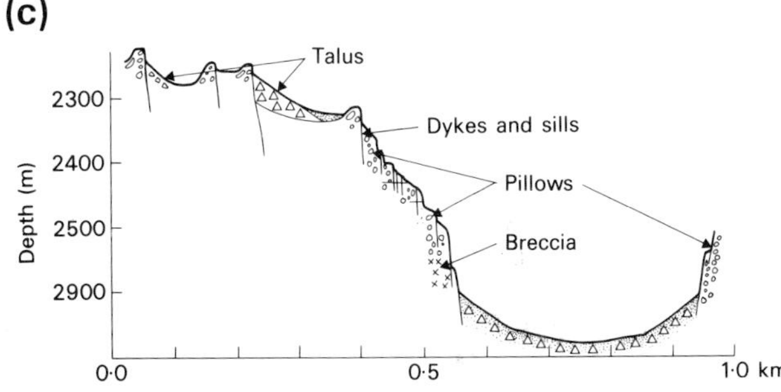

North Pacific, which could contain extractable quantities of Cu, Ni and Co, and, less probably, manganese (Granville, 1975) (see Sect. 11.3.3, 11.3.4).

Metal-rich muds have been found in the last 10 years in the Red Sea median valley deeps, on the Atlantic rise and on and adjacent to the East Pacific rise. The Red Sea zinc–copper–lead deposits, associated with hot metal-rich brines (Degens and Ross, 1969), are the only ocean floor muds of possible economic potential found so far. There is evidence that the brine pools and metalliferous sediments are related to the intersection of transform faults with the spreading ridge (Garson and Krs, 1976). Their origin is probably related to either magmatic hydrothermal or heated sea water circulation through the underlying basalts (Bignell, 1975); a possible alternative source is water circulating through both the Miocene evaporites flanking the rift and stratabound sulphides in adjacent Miocene sediments. Formation of metalliferous sediments at ocean ridge crests has been related to hydrothermal solutions from sea water percolating through underlying basaltic volcanic rocks (Corliss, 1971; Spooner and Fyfe, 1973). Other metal-rich muds possibly form at more complex ridge-related tectonic settings, for example in the Bauer Deep (Sect. 11.3.4).

Passive continental margins have a slope and rise where sedimentation at present is dominated by muds, silts, and fine sands deposited by low-density turbidity currents or by ocean bottom contour currents. It is probable that in the past, when sea level was lower and sediment was deposited by rivers close to the shelf edge, coarser clastics, deposited by turbidity currents and other mass-flow mechanisms, were more common. Large scale rotational sliding and slumping is an important feature of passive continental margins and is the cause of the absence of sediment on some parts of the continental slope (Rona, 1969) (Fig. 14.13).

Shelf deposition is controlled primarily by sedimentary processes, in particular waves, storms and tides, by the availability of detrital sediment, and by climate and the chemistry of ocean waters (see Chapters 7, 8, 9 and 10). Most shelves are bordered on one side by a coastal plain and on the other side by an ocean. However, some are more complicated, with major island source areas separating the shelf seas from the ocean as, for example, off north west Europe.

Continental shelves and coastal plains are the main environments for the accumulation of most types of surficial metallic mineral deposits. They include coastal plain and offshore deposits of tin, the latter largely restricted to drowned river valleys, and marine placer deposits of rutile, ilmenite, zircon, monazite and magnetite occurring in modern and submerged beaches, for example along the east coast of Australia.

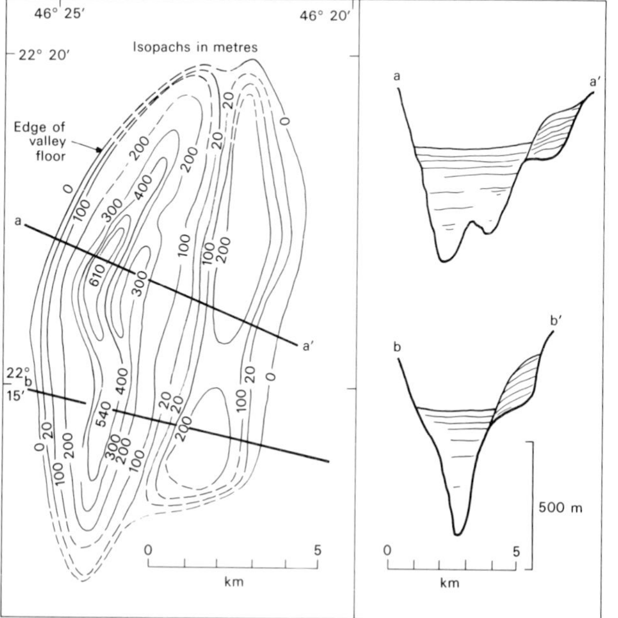

Fig. 14.12. Isopach map and cross-section of South Pond, a partly filled valley about 100 km west of the mid-Atlantic ridge axis (after Van Andel and Komar, 1969).

Beneath present continental margins, the thickness of sediment varies considerably but is greater than at one time was thought. Off North America there are several basins with 8–18 km of Mesozoic and Tertiary sediment (Sheridan, 1974; Figs. 14.14, 14.15 and 14.16). Off the British Isles, the Rockall Trough (Fig. 14.17), which is floored by oceanic basement and separates the important micro-continent of the Rockall Plateau from the true continental margin, is only one of several basins (Roberts, 1974). The origin of continental margin basins is not certain (see Bott, 1976 for discussion). They may be early, aborted, intra-continental rifts, or they may have started as pre-rift, strike-slip fault basins. They are a major target for the next stage of petroleum exploration.

14.3.2 Subduction-related settings

At convergent plate boundaries, oceanic lithosphere of the subducting plate bends downwards in a subduction zone and descends along the seismic Benioff zone, inclined beneath the volcanically active magmatic arc of the overriding plate.

FOREARC AREAS

Between the volcanic arc and the trench is a 50–400 km wide *arc-trench gap* (Dickinson, 1974a) or *accretionary prism* (Karig and Sharman, 1975) (Fig. 14.18). The arc-trench gap may have a simple slope, with, as in the Marianas, a sedimentary apron or prism not topographically apparent, or there may be a

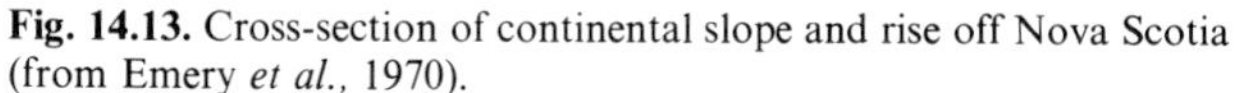

Fig. 14.13. Cross-section of continental slope and rise off Nova Scotia (from Emery *et al.*, 1970).

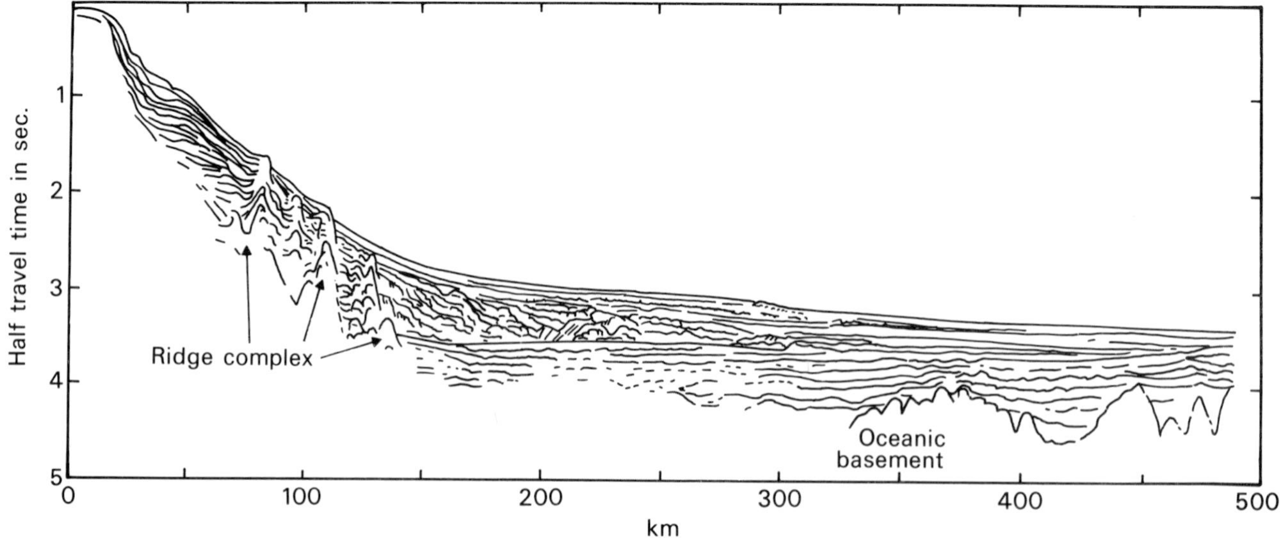

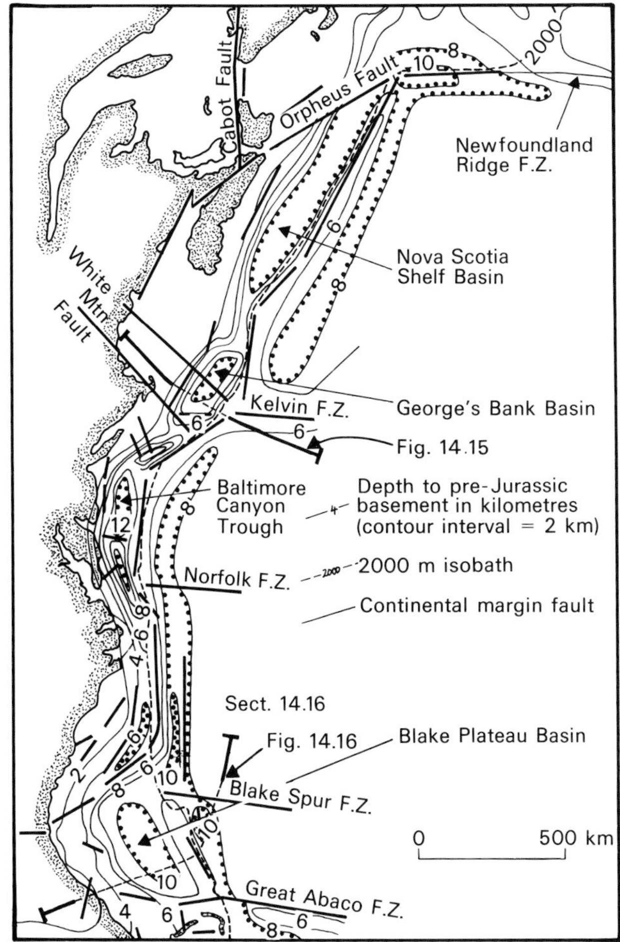

Fig. 14.14. Map of sedimentary basins in continental margin of eastern North America (after Sheridan, 1974).

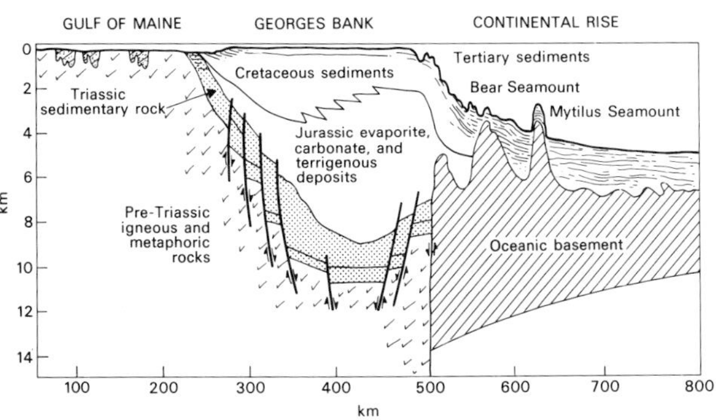

Fig. 14.15. Structural cross-section across the Gulf of Maine and Georges Bank (after Sheridan, 1974).

morphological bench or *trench-slope break* called the *outer-arc ridge* which is separated from the volcanic arc by an *outer-arc trough* or forearc basin. The outer-arc ridge is built of deformed strata and may be overlain by gently dipping sediments inclined towards the outer-arc trough. It emerges as, for example, Barbados Island in the Lesser Antilles, the Mentawai islands of the Sumatra arc (Fig. 14.1) and Middleton Island of the Aleutian arc system. In the outer-arc trough, sediments are either undeformed or gently folded.

OCEAN FLOOR, DEEP-SEA TRENCHES AND OUTER ARCS

The ocean floor seaward of many submarine trenches is made up of pelagic sediments and tuffs above tholeiitic basalt of the oceanic crust. In much of the ocean, such as those parts of the Pacific which subduct beneath the Andes and the island arcs and inter-arc troughs of the western Pacific, no sediment derived from the continent reaches the basin floor. However, continental detritus does reach the subducting oceanic crust in the Indian Ocean, off the north-east coast of South America, and in the eastern Mediterranean.

A major controversy has surrounded sedimentation, preservation and deformation of trench deposits. In their initial enthusiasm for plate-tectonics, most geologists followed the Dutch views of the 1930s that trenches were the sites of ancient geosynclines. However, seismic profiles across trenches show little deformation of the sedimentary fill and most trenches, especially the deepest which border the island arcs of the western and northern Pacific, contain only 200–500 m of sediment, mostly pelagic and hemi-pelagic (Scholl and Marlow, 1974). A few trenches, such as the southern part of the Peru–Chile trench, the Middle America trench, the eastern Mediterranean Hellenic trough and the Eastern Aleutian trench (Fig. 12.30), are partly filled by up to 2,500 m of sediment, and their bathymetry is subdued. Off the Washington–Oregon coast and east of the Lesser Antilles no trench is apparent but very thick piles of sediment occur above the subduction zone. These thicker fills are due partly to proximity to a continental source especially in areas of glacial erosion, and partly to the Pleistocene eustatic falls of sea level.

In both the deep and the shallower trenches, the maximum sediment thickness is normally about 1,000 m and, unless they are filled after subduction has ceased, they cannot be the site of thick geosynclinal successions (Von Huene, 1974). Some of these successions may have formed as thick piles of sediment, like the Bengal fan, on oceanic crust and were then carried into the trench; others may have been outer arc trough successions. However, in most geosynclinal successions the thickness is more apparent than real and is due to tectonic thickening and not original depositional thickness (see Sect. 14.4.2).

Another controversy has been the question of how much

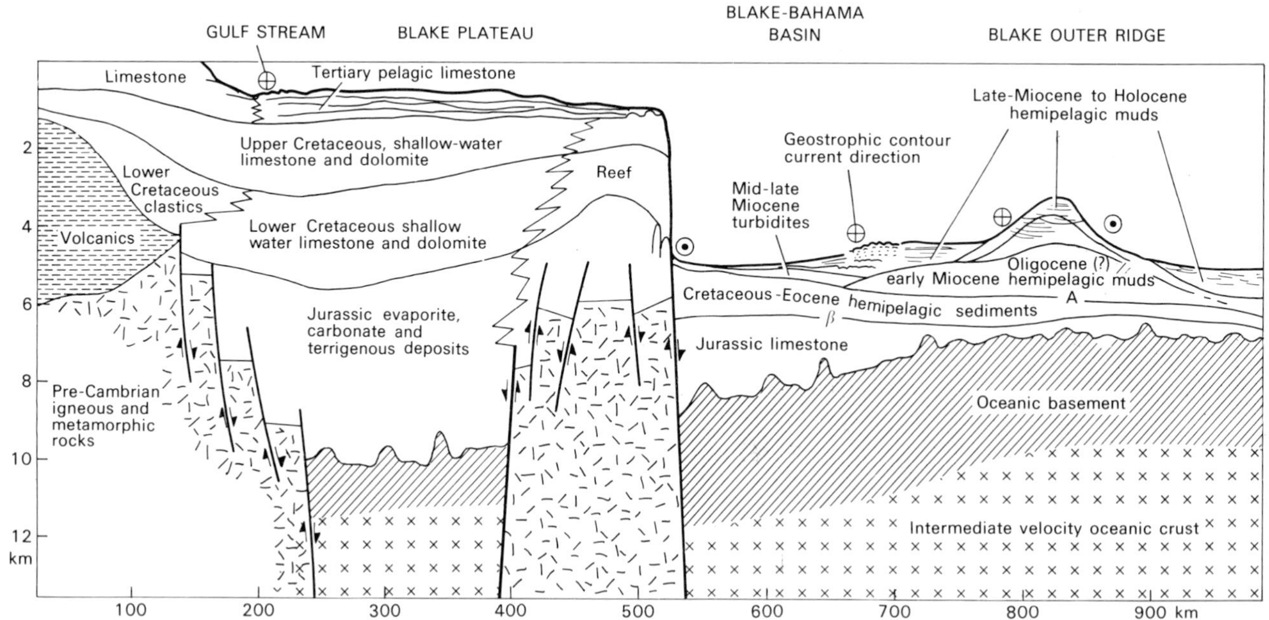

Fig. 14.16. Structural cross-section across Florida platform, Blake plateau, Blake–Bahama basin and Blake Outer ridge (after Sheridan, 1974).

sediment of the subducting plate is carried down the Benioff zone and how much is scraped off to be uplifted and incorporated in a future mountain belt. Many of those working in the Alpine and Appalachian–Caledonian mountain belts have used the evidence of ophiolites and the associated pelagic oceanic sediments in thrust sheets and nappes to construct models in which segments of oceanic crust are scraped off (Bird and Dewey, 1970). On the other hand Scholl and Marlow (1974), working in the Pacific, have pointed to the lack of ocean floor deposits in the surrounding island-arcs and in the Andes. This is surprising considering that over the last 100 million years some 2.5×10^6 km^3 of oceanic sediment should have

Fig. 14.17. Cross-section of Rockall Trough (from Roberts, 1974).

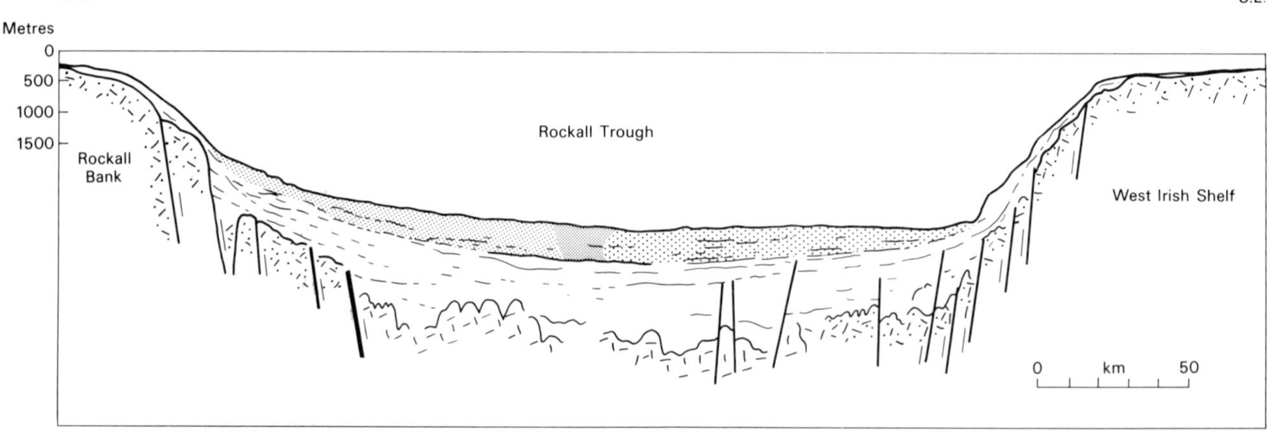

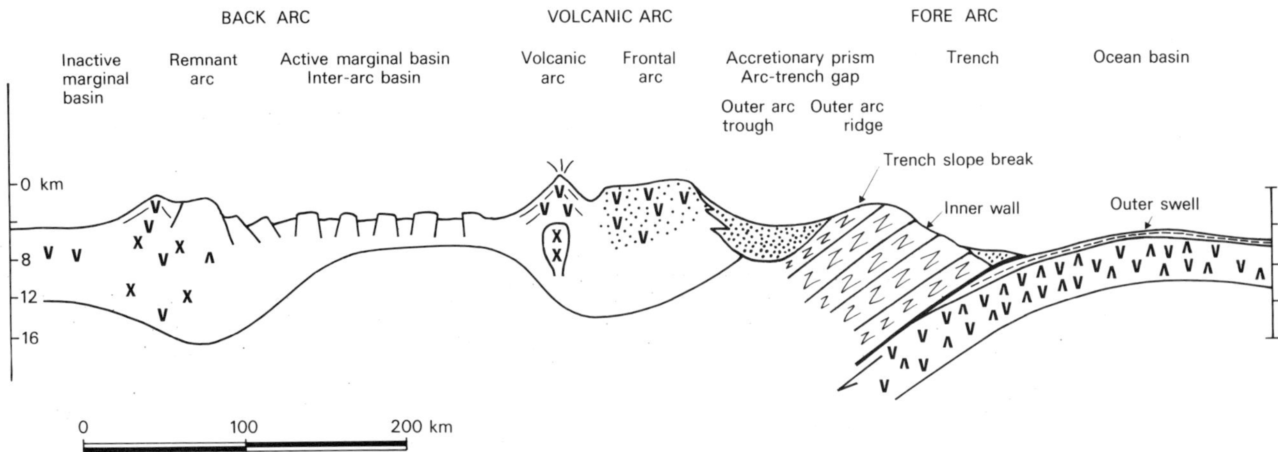

Fig. 14.18. Generalized cross-section of intra-oceanic island arc.

been carried into each 1,000 km segment of trench. This is about a third of the volume of a 1,000 km long segment of a eugeosynclinal belt 200 km wide and 20 km deep.

One explanation of this paradox is that where pelagic sediments alone overlie oceanic crust they are mostly carried down the Benioff zone (Moore, 1975). Where there is a substantial thickness of overlying turbidites these are scraped off to form an outer-arc composed of stacks of thrust sheets which individually young towards the magmatic arc but whose overall age decreases towards the trench (Fig. 14.19, 14.20). A relatively thin (< 1.0 km) turbidite wedge in a trench is sufficient to form a major outer-arc by tectonic repetition and accretion, thus progressively widening the distance between the magmatic arc and trench and so flattening the upper part of the Benioff zone (Fig. 14.19) (Kulm and Fowler, 1974; Seely, Vail and Walton, 1974; Karig and Sharman, 1975).

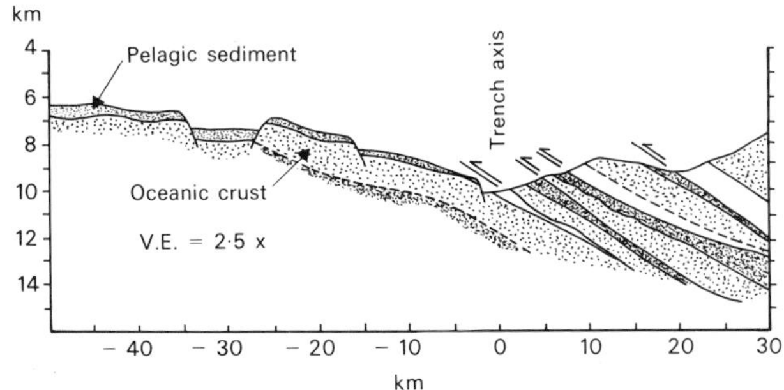

Fig. 14.19. Possible mode of tectonic accretion of thin pelagic sediment cover and oceanic crust. Slabs of the upper oceanic crust are intermittently sheared off when topographic irregularities enter the trench (from Karig and Sharman, 1975).

In some arcs, sedimentary basins develop on the inner wall of the trench (Moore and Karig, 1976) as sediments are ponded behind ridges formed by the successive slices of thrusted oceanic plate (Figs. 12.28, 14.20). These basins become larger up the slope wall and are filled by hemipelagic sediments from adjacent topographic highs, and by turbidites and other mass-flow sediments from the same highs and from neighbouring shallow-water and terrigenous sources.

OUTER-ARC TROUGHS

Outer-arc troughs are 50–100 km wide and may extend for 5,000 km. They are filled by over 5,000 m of sediments which probably overlie the tectonically emplaced outer-arc rocks stratigraphically in some arcs, but may be thrust over them in others. On the magmatic arc side the sediments often interfinger with the volcanic rocks, although in some troughs there is evidence for a tectonic discontinuity with the magmatic arc (Karig and Sharman, 1975).

There are three sources of sediments; the outer-arc, the magmatic arc and a longitudinal source from the adjacent continent. Clastic sedimentation predominates, but varies within different troughs from turbidite to deltaic and fluvial.

The best developed modern outer-arc trough is the 'interdeep' of the Sunda arc (Van Bemmelen, 1949) (Fig. 14.1). This trough lies west of the Burma–Sumatra–Java magmatic arc and east of a belt of flysch sediments and ultrabasic rocks which form the outer arc of the Arakan and the Andaman–Nicobar–Mentawai line of islands. In the north, the outer-arc trough forms the western trough of Burma where at least 8,000 m of late Cretaceous to Pliocene marine, deltaic and fluvial sediments have been deposited and which is now the site

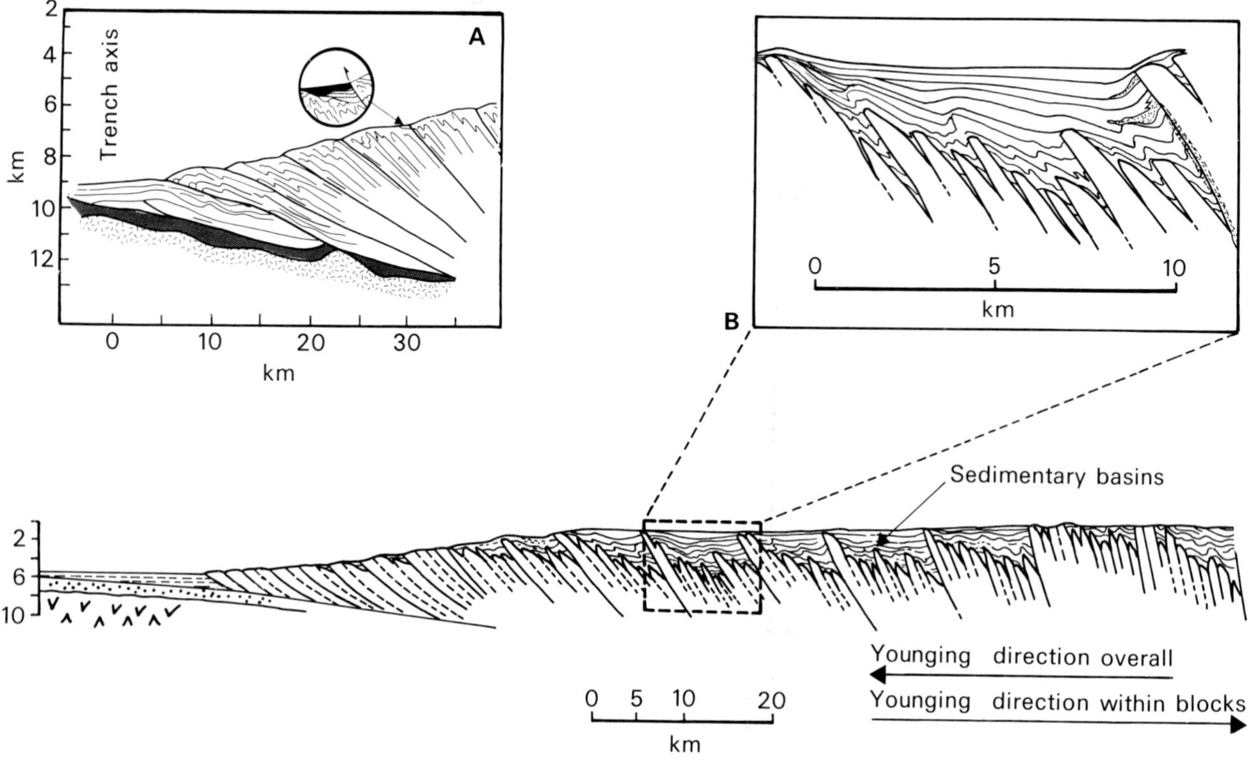

Fig. 14.20. Possible mode of tectonic accretion of thick sediment cover. The turbidite section tends to be sheared off along the weak, high porosity uppermost pelagic section and rides over the trench wedge (A). Sedimentary basins develop unconformably on the inner wall of the trench (B) (from Karig and Sharman, 1975; Moore and Karig, 1976).

of the present Chindwin and lower Irrawaddy river courses and delta (Win Swe, Thacpaw C., Nay Thaung Thaung and Kyaw Nyunt, in press).

VOLCANIC ARCS

Modern volcanic arcs can be divided into those which are now volcanically *active* and those which are now *inactive* but located in a similar tectonic position to that during volcanism.

Active arcs can themselves be subdivided into *intra-oceanic arcs* situated between areas of oceanic crust and *continental margin arcs* situated on the margin of a continent near the continent–ocean crustal boundary (Dickinson, 1974a). The nature and thickness of crust beneath the arcs varies widely both among different arcs and in some cases along an arc. Intra-oceanic arcs can be underlain by oceanic crust (e.g. southwestern Pacific arcs) or by both oceanic crust and fragments of continental crust (e.g. Japan). Continental margin arcs are underlain by continental crust of normal, greater, or less than normal thickness. They may be part of the continental mainland (e.g. the Andes) or they may, like Sumatra and Java, form a chain of islands separated from the mainland by shallow seas.

Every modern arc forms a topographic ridge, interrupted by fault-bounded basins. The active volcanic belt ranges from a single chain of volcanoes to a zone more than 150 km wide in which scattered volcanoes occur, although, at any one time, most volcanism takes place in a zone less than 50 km wide.

Inactive arcs include those which have become inactive because of cessation of subduction and *frontal arcs* (Karig, 1971) (Fig. 14.18) lying on the trench side of the active magmatic arc and commonly consisting of older volcanic rocks forming a major source of sediment, such as the western belt in the New Hebrides.

Many intraoceanic volcanic arcs, for example those of Tonga, the New Hebrides, the Lesser Antilles and the Ionian Arc, consist of subaerial and submarine volcanoes and associated sediments underlain by crust less than about 20 km thick probably consisting of oceanic crust overlain by older volcanic rocks or outer-arc successions. Lavas are predominantly low-potash basaltic or andesitic rocks although calc-alkaline and silicic lavas and tuffs are common, and ignimbrites can occur as in the Tonga arc. In tropical latitudes fringing or more commonly barrier reefs develop around the islands.

During initial stages of arc development pillow lavas erupt and locally undergo submarine gravity collapse to form thick submarine talus fans. Rapid erosion of island stratovolcanoes and bordering reefs leads to the deposition by mass-flow of extensive aprons of volcanogenic and carbonate boulder breccias, conglomerates and sandstones, commonly lacking quartz.

Pyroclastic rocks and autoclastic submarine volcanic flows are interbedded with the epiclastic sediments, but the voluminous 'agglomerates' described from many Pacific volcanic arc successions, and on which the high 'pyroclastic index' of island arc volcanoes was based, probably include abundant epiclastic mass-flow deposits; in weathered exposures indurated volcanogenic turbidites are easily mistaken for lava flows and tuffs.

In general there is probably a lateral facies change from abundant epiclastic and pyroclastic rocks and lava flows near the islands through thick wedges of conglomerate and mass-flow transported epiclastic and pyroclastic rocks to turbidites, pelagic sediments and air-transported ash deposits more than 20 to 30 kilometres from individual volcanic sources.

Intraoceanic volcanic arcs which are underlain by continental crust show sedimentary facies resembling those of continental margin arcs but the composition of the clastic material reflects the highly variable nature of the volcanic rocks, with tholeiitic lavas and calc-alkaline andesites, dacites and rhyolites both occurring in parallel belts and succeeding one another stratigraphically as a result of subduction-controlled changes in type of magmatism.

Continental margin magmatic arcs occur in more varied tectonic and geomorphic settings, with a consequently greater variation in sedimentary facies.

The volcanic rocks in the Andes and Sumatra are commonly more silicic and potassic than those of other arcs, consisting of dacites, rhyolitic ignimbrites and andesites; basaltic rocks are rare. Sediments are derived both from the volcanoes and from older uplifted metamorphic sedimentary and volcanic rocks, within which silicic plutons emplaced beneath older eroded volcanic arcs are common, as in the Peruvian Andes; they accumulate and are preserved mostly in fault troughs.

BACK-ARC AREAS

Behind all volcanic arcs a complex pattern of ridges and basins develops. Behind intra-oceanic arcs there is frequently a *remnant-arc* (also called a *dead arc* or *third arc*), (Karig, 1970) which is separated from the magmatic arc by an *interarc basin.* Behind the remnant-arc there is either another interarc basin or a *marginal basin* separating it from the continental mainland. In some arcs, particularly those underlain by fragments of continental crust, there is no remnant-arc and the marginal back-arc basin extends from the continental margin to the magmatic arc. Back-arc ensialic basins develop behind continental-margin arcs where they have been called *retroarc basins* by Dickinson (1974b).

Menard (1967) first drew attention to marginal basins as accumulators of great quantities of sediment. He showed how variable was the crust beneath them, though in general it had thicknesses approaching those of continental crust, but seismic velocities closer to that of oceanic crust. The basins were mostly interpreted either as 'oceanized' continental crust that had subsided, a view popular with Russian geologists, or, as areas of oceanic crust that were becoming more continental through sedimentation. However, more recently, Karig (1970) and Packham and Falvey (1971) produced evidence to suggest that some were formed by spreading behind the magmatic arc with consequent lateral migration of the magmatic arc away from a continental margin.

The principal evidence for back-arc spreading is the high heat flow, the presence of old (early Palaeozoic or even possibly Pre-Cambrian) continental crust in some arcs such as Japan and, behind the Scotian arc, a small spreading centre only 7–8 my. old. However, the principal features which led Karig to postulate back-arc spreading for the interarc basin behind the Tonga-Kermadec magmatic arc were the lack of sediment cover relative to that found overlying normal oceanic crust and the extensional tectonics suggested by the linear ridges and troughs with a relief of 1,000 m in the central part of the Lau-Havre interarc basin.

Karig and Moore (1975) have developed an evolutionary sedimentation model for interarc basins which lack any terrigenous input (Fig. 14.21). There are four dominant types of sediment (1) volcaniclastic debris derived from the magmatic arc, volumetrically the most important (2) montmorillonite clays derived from the volcanic chain (3) biogenic ooze (4) wind blown, continentally derived dust.

Though the basins may open symmetrically, the sedimentation pattern is asymmetric. Adjacent to the magmatic arc a volcaniclastic apron develops, possibly as a complex of submarine fans. Beyond the distal end of the apron, pelagic brown clay, distinguishable from that of deep ocean basins by its high content of montmorillonite, glass and phenocrysts, accumulates. Pelagic oozes with a high content of $CaCo_3$ are

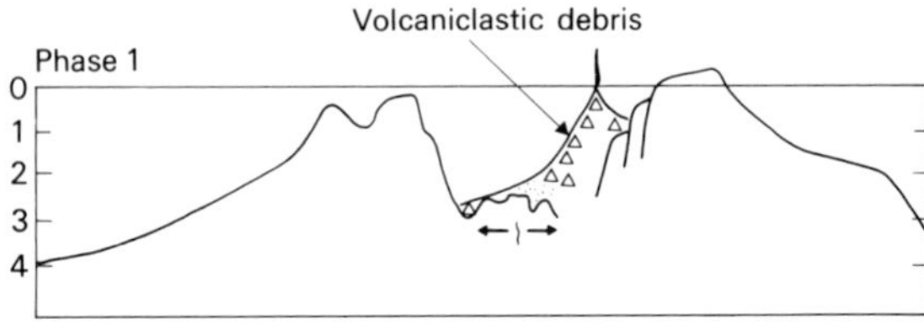

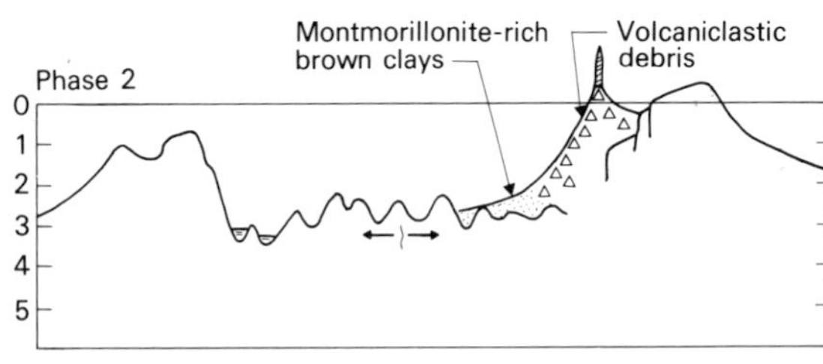

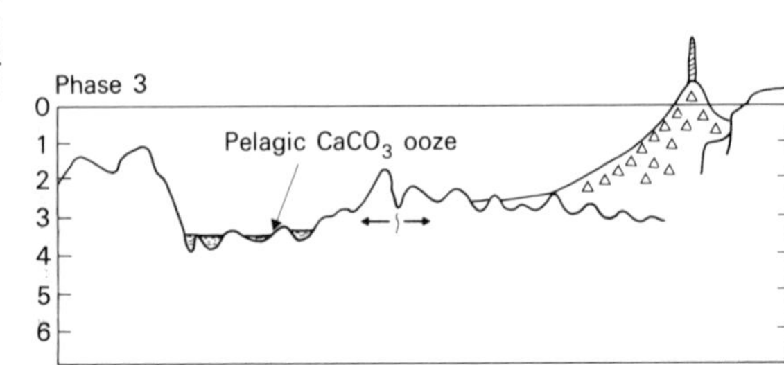

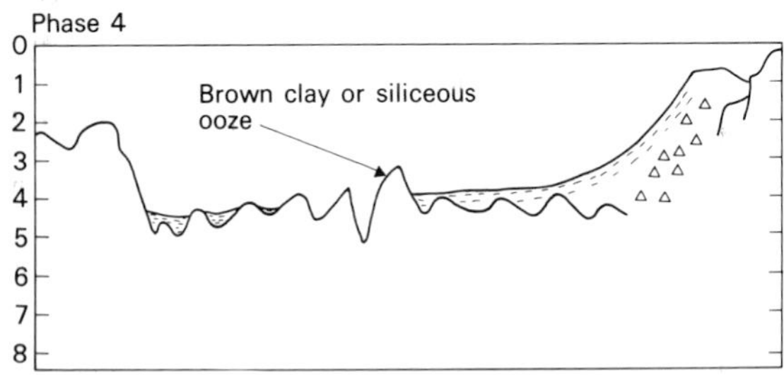

Fig. 14.21. Sedimentation model for back-arc basins (after Karig and Moore, 1975).

deposited in the distal parts of the basin until it drops below the carbonate compensation depth when brown clays or siliceous ooze accumulate.

In contrast to interarc basins, back-arc marginal basins are extremely varied and complex and no simple pattern of sedimentation exists because of the large and varied terrigenous input. The range of sedimentary facies is as large as that found in the major oceans. There are pelagic sediments overlying newly formed oceanic crust; several thousand metres of turbidites in abyssal plains; shallow and relatively deep continental shelves within which are large sedimentary basins. Some basins, such as those of the East and South China Seas, are fed by large rivers and the continental margin progrades seawards; other basins, such as the Sea of Japan, are relatively starved of terrigenous sediment and have a large biogenous component. The only possible distinctions between the facies of back-arc marginal basins and of true oceans are the probable lack in the former of significant bottom ocean current deposits and the increased abundance of volcaniclastic sediments and volcanic ash.

The Andaman Sea back-arc basin, situated between the continental margin arcs of Burma and Sumatra (Fig. 14.1) contains a deep central trough 100–200 km wide, 750 km long and 2–3,000 m deep with a central rift valley 4,000 m deep and two elongate basaltic seamount ridges (Rodolfo, 1969). These have been interpreted by Curray and Moore (1974) as the result of rifting of the Andaman Sea and spreading of oceanic-type crust in the late Cenozoic. The submarine topography is also related to transform faults which offset the spreading centre and also form the Semangko fault in Sumatra and another major transform between the Central Lowlands and Shan plateau of Burma. Whether the faults are the result of spreading or the spreading is the result of local extrusion in a pullapart basin which developed along a major strike-slip zone remains an open question. The deep central trough is margined to the east by the 250 km wide Malay Shelf which is bordered in the east by mudflats, mangrove swamps and coral reefs and in the north by the Irrawaddy delta which today brings in large quantities of silty clay.

In the Andes, the sub-Andean zone, lying east of the volcanic arc, comprises deep troughs between rapidly uplifted mountain ranges. Within these troughs alluvial fan and fluvial deposits are several thousand metres thick, and have a facies typical of molasse.

14.3.3 **Transform/strike-slip fault-related settings**

With the excitement of the discovery of the relationship between deep-focus earthquakes, present-day orogenic belts and convergent plate junctions, the possibility that orogenic belts might also be related to major strike-slip junctions became largely overlooked. Many global maps of the early 1970s, designed to portray present day orogenic belts in terms of plate-tectonics, omitted strike-slip boundaries altogether. The reasons for this neglect were probably the simplicity of the then conventional 2-dimensional model of plate-tectonics and the relative lack of igneous and metamorphic activity at these junctions where lithosphere is essentially conserved and neither created nor consumed.

Transform/strike-slip faults bound plates or blocks, between which movement is primarily lateral with only minor convergence or divergence. They may be primary or secondary. Primary faults are boundary transform faults (Gilliland and Meyer, 1976) which border major plates. They occur along fundamental breaks in the lithosphere and many have had a long history of tectonic activity. They may link a ridge with a subduction zone such as the complex fault belts which margin the Caribbean plate; they may join two subduction zones like the Alpine fault of New Zealand; they may be major oceanic fracture zones or they may connect two spreading centres, for example the San Andreas fault zone which is 1,500 km long and up to 500 km wide, although individual faults extend for no more than 100 km. The extremely complex patterns of strike-slip faults in the Eurasian continent result from continental collision (see Sect. 14.3.4).

Secondary transform/strike-slip faults are secondary features resulting from spreading (Gilliland and Meyer, 1976). In practice it is not always easy to distinguish primary boundary faults from secondary faults (e.g. the Vema fracture zone, see Sect. 14.3.1).

Although ideally blocks on either side of a strike-slip fault move laterally with no compression or extension, in practice they show some degree of either compression or extension on both a local and a global scale. Movement between plates is rarely purely lateral, divergent or convergent; it is normally oblique. Harland (1971) has distinguished two types of oblique strike-slip regimes (1) *transtension,* as at zones of ocean spreading with stepped transform faults and (2) *transpression,* as in many orogenic belts such as the Tertiary belt of Spitsbergen (Lowell, 1972).

Immediately before the development of the plate-tectonic hypothesis, Moody and Hill (1956), Carey (1958) and Raff and Mason (1961) were interpreting world tectonics in terms of strike-slip fault patterns or megashears, and it had long been known that strike-slip faults resulted in the development of some sedimentary basins. The Dead Sea formed where two parallel faults of a sinistral wrench fault system were offset (Fig. 14.22) (Quennell, 1958). In the North Island of New Zealand the curvature of the extension of the Alpine fault resulted in alternate zones of compression, resulting in uplift, and extension, leading to subsidence and the formation of sedimentary basins (Kingma, 1958a, b). The basin depths almost equalled their width and because of the proximity of contemporaneous uplift, sedimentary filling was extremely rapid.

The ideas of Quennell and Kingma have been extensively developed for the San Andreas fault system by Crowell (1974a, b) (Fig. 14.9). He has shown that the major fault system is composed of anastamosing sub-parallel faults, all with the same direction of offset. As these split and rejoin, movement along them leads either to convergence, compression and uplift or to divergence, extension and subsidence (Fig. 14.23). He has also explained the simultaneous development of the subsiding Ridge Basin and the uplift and near horizontal thrusting of the nearby Frazier Mountains by the curvature of the San Gabriel fault (Fig. 14.24). The Ridge Basin is about 50 × 20 km and has 12,000 m of stratigraphical succession in it. Yet since it forms as its margins are progressively pulled apart the actual succession at any one place in the basin centre may be no more than 4,000 m. The fill includes the Violin Breccia, a 10,000 m thick alluvial fan conglomerate at the fault margin which passes laterally over only 1,000–1,500 m into finer grained, largely lacustrine,

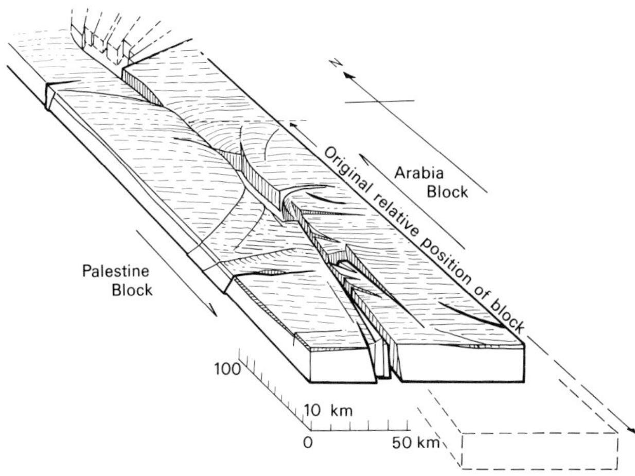

Fig. 14.22. Block diagram to show origin of the Dead Sea (from Quennell, 1958).

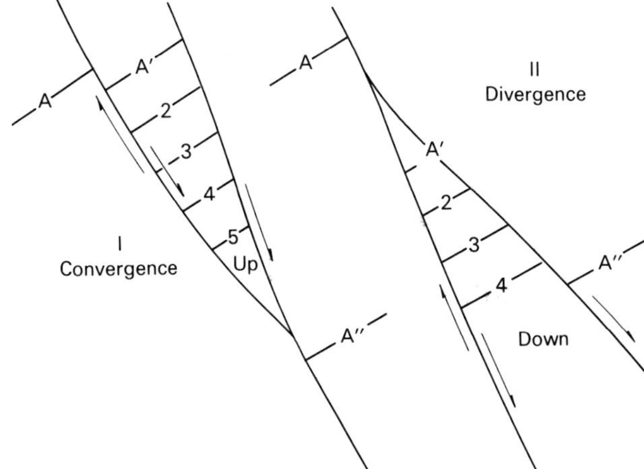

Fig. 14.23. Sketch map to show uplift at convergence and subsidence at divergence of strike-slip faults (after Crowell, 1974b).

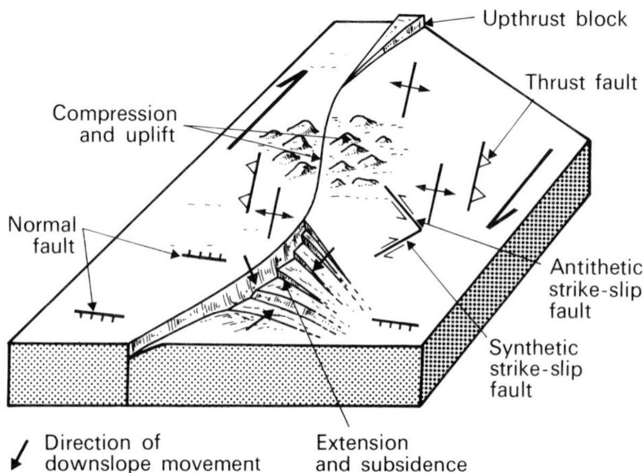

Fig. 14.24. Block diagram illustrating how the curvature of a strike-slip fault may produce closely adjacent extensional basin and uplift and erosion, with superimposed tectonic pattern (based on Kingma, 1958b; Wilcox, Harding and Seely, 1973; Crowell, 1974a).

sediments (Fig. 14.25). Further south along the San Andreas fault system and approaching the spreading centres of the Gulf of California is another pull-apart basin, the Saltern trough (Fig. 14.26) (see Sect. 14.3.1).

To the west, off southern California, lies the Continental Borderland (Fig. 14.9). It is situated on continental crust and contains some 15 basins, 1–2,000 m deep and 50 × 20 km in size. Many basins are separated by islands which rise to 1,000 m above sea level. This block and basin area is the result of movement along faults paralleling the San Andreas fault. The offshore basins are relatively starved of sediment and filled only with pelagic and hemi-pelagic fallout (see Sect. 11.3.6). Nearshore basins receive large amounts of clastic material accumulating either as submarine fans below canyons or as mass-flow deposits in aprons directly off the narrow shelf (Chapter 12.4.5, Fig. 12.32). Similar, now filled, basins of Mio–Pliocene age are found on land as the oil producing Los Angeles and Ventura basins.

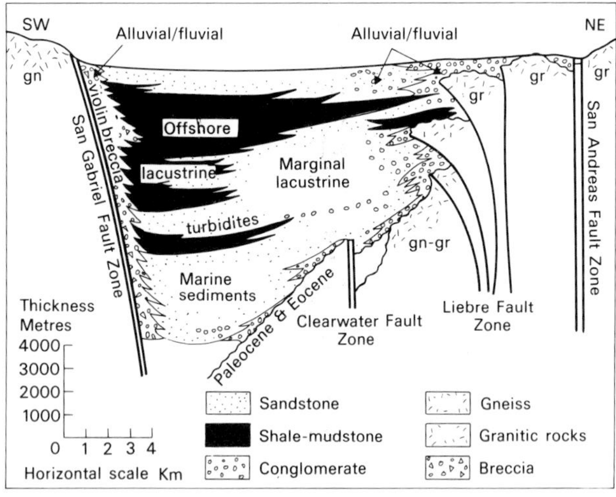

Fig. 14.25. Facies relationships in Pliocene Ridge Basin, California showing vertical transition of marine into lacustrine sediments and lateral passage of alluvial fan conglomerates through marginal lacustrine facies into offshore lacustrine facies (after Crowell, 1975; Link and Osborne, 1978).

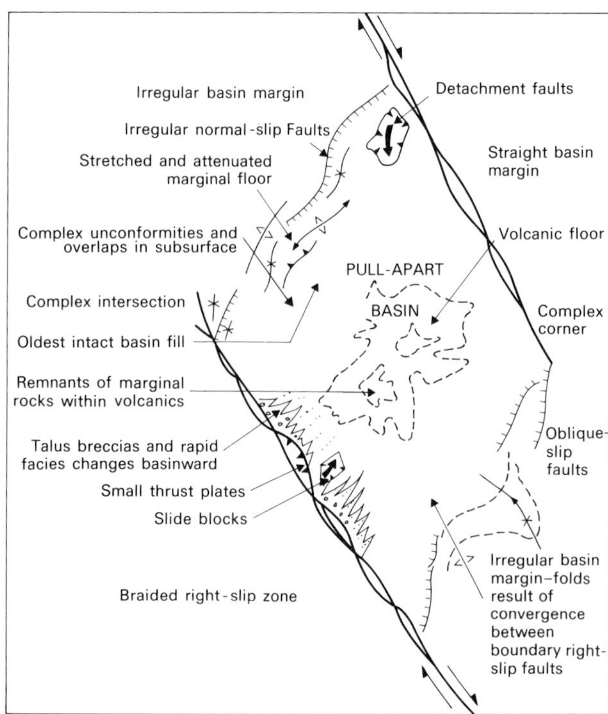

Fig. 14.26. Idealized pull-apart basin (from Crowell, 1974b).

Northwestern South America is traversed by numerous strike-slip faults which result in a complex of eroding mountains and rapidly filling basins. These range from the continental alluvial/fluvial Magdalena trough to the more varied Maracaibo basin where a large central lake (100 × 150 km) is surrounded on three sides by mountain-girt alluvial plains which are very wet and swampy in the south and semi-arid in the north. From the south-west, a large delta enters the lake which, in the north, is open to the sea. Offshore, the South Caribbean basin is separated from the Maracaibo basin by smaller basins and horsts which include the Dutch and Venezuelan Antillean islands (Fig. 14.27) (Case, 1974).

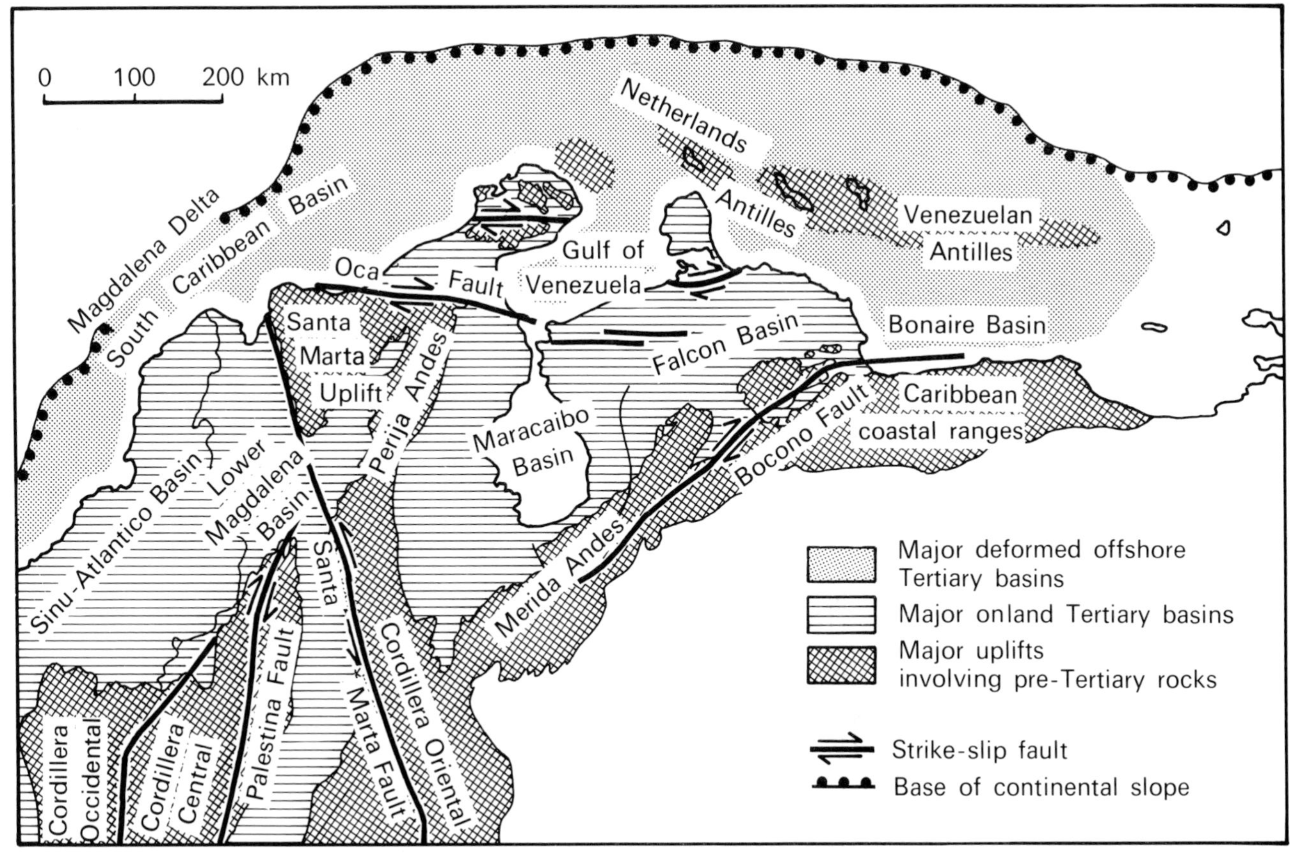

Fig. 14.27. Sketch map of basins of northern South America (after Case, 1974).

Another example is the Yallahs basin (Burke, 1967) (Fig. 14.28) situated off southern Jamaica. Environments and processes include submarine fans, basinal turbidity currents, reefs, reef debris flows, sandy quartzose shorelines and subaerial alluvial fan deltas passing directly into the sea, all within an area of about 30 × 30 km.

14.3.4 Continental collision-related settings

Continental collision results when two plates carrying either island arcs or continental crust are brought together by either subduction or strike-slip movement with a component of convergence. Since continental margins and island-arc belts are commonly highly irregular, areas of continental or arc crust first impact at one or more points. Here folding and thrusting result in uplift and the tectonic suturing of continental crust on the two plates. Between these points of collision embayments of the old ocean basin persist, termed *remnant-ocean basins* (Graham *et al.*, 1975). As collision proceeds, other basins are formed adjacent to the uplifted mountains. These lie either upon earlier partially filled remnant-ocean basins or upon continental crust. They may form upon the subducting plate, where they have been called *peripheral basins* (Dickinson, 1974b) or upon the overriding plate as later developments of Dickinson's *retroarc basins*. They may also develop as basins associated with strike-slip movements within collision belts. During collision, crust is shortened by folding, thrusting and vertical uplift accompanied by strike-slip motion to form a complex of mountain belts and sedimentary basins.

The direction of subduction and underthrusting during collision in many ancient orogenic belts can now be determined, and in these it is possible to distinguish foreland basins, situated on the subducting plate, from hinterland basins formed on the overriding plate. However, the direction of

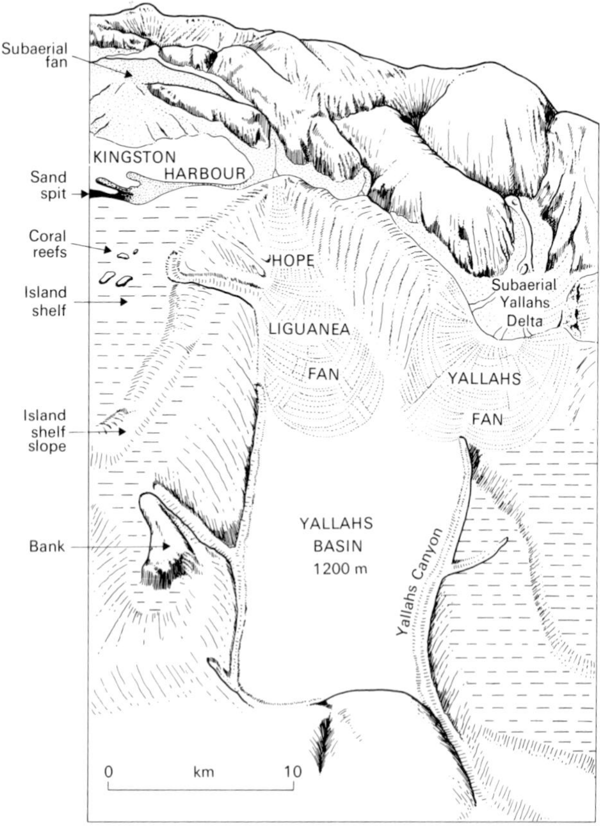

Fig. 14.28. Sketch map of Yallahs basin, Jamaica (after Burke, 1967).

subduction is not always clear and in some cases movement between plates may have been predominantly strike-slip with only minor subduction. The more general term *late-orogenic basin* is therefore used here for all basins other than remnant-basins.

REMNANT-OCEAN BASINS

The best known example of a modern remnant-ocean basin is the Bay of Bengal (Curray and Moore, 1971, 1974) (Fig. 12.39). The basin is bounded on the west by the Indian continent and is closing as a result of eastward subduction beneath the Indo-Burman Ranges and the Andaman–Nicobar and Sunda outer arc. With continued subduction the Bengal basin will eventually be closed.

During collision remnant-ocean basin deposits reach the subduction zone and mostly become scraped up to form the imbricate thrust slices of the seaward part of the outer arc. Consequently, in the early stages of collision, a major outer arc develops into a mountain range.

Drainage from this range supplies detritus to the adjacent remnant basin, resulting in a major delta system which migrates laterally together with the collision zone. Seaward of the delta the Bengal fan (see Sect. 12.5.2) has developed above the pelagic sediments of the ocean floor. With progressive subduction the fan turbidites are scraped off to form an outer arc, elevation of which may provide a further source of clastic sediment to the fan.

Thus diachronous collision of irregular margins is the direct cause of elevation in the collision zone, sedimentation in the closing remnant-basins, and scraping off to form outer arcs. In the later stages of collision remnant-basin and continental rise deposits of the passive margin are themselves scraped off and eventually thrust on to the continental margin of the subducting plate as major nappes. Following isostatic uplift in response to crustal thickening, these nappes may move further towards the foreland as gravity slides.

LATE-OROGENIC BASINS

The best known examples of present day late-orogenic basins are those of the Indian sub-continent immediately south of the actively rising Himalayas. Within the basins, which are up to 500 km wide and extend for over 1,000 km, the drainage is generally parallel to the structural trend though transverse streams also supply sediment from the side of the mountain front (Fig. 3.3). The alluvial facies are governed by both tectonics and climate with the Indus flowing through deserts and the Ganges/Brahmaputra traversing some of the wettest parts of the world. The older Siwalik trough of the Ganges has received sediments similar to those of the present alluvial valley since late Miocene times; on their northern margin these older deposits are now uplifted by folding and thrusting so that Pleistocene sediments are now at an elevation of 2,000 m. The depositional axis of the trough is migrating to the south to accommodate the thick Quaternary deposits of the present Ganges valley.

Stratabound deposits of uranium with minor vanadium occur in the late Miocene to early Pliocene Middle Siwaliks which consist mostly of fluviatile sandstones (Moghal, 1974). Ore is commonly concentrated in channel-fill conglomerates. However, because there are no extensive igneous source rocks nearby, major high-grade uranium deposits are unlikely to be present in the Siwaliks and in other foreland basins.

The Persian Gulf is another late-orogenic basin of similar size and asymmetry with a shield to the south and the active Zagros mountain range to the north. The sedimentary fill is primarily longitudinal with a fluvio-deltaic complex entering from one end of a shallow basin which is dominated by carbonates and evaporites (see Chapter 8).

The Bengal basin and Assam are an example of a late orogenic basin which passes southwards into a remnant ocean basin. The collision of the Indian continent with the Indoburman Ranges outer arc is migrating southwards, so that closure and filling are diachronous. This in turn results in diachronous uplift, erosion, sedimentation and eventual incorporation into the mountain range of the outer arc to the south.

These major peripheral basins form on the continental margin of the subducting plate and are equivalent to the foreland basins of classical geosynclinal terminology.

Basins in and around the Mediterranean are much less simple. The Po/Adriatic basin, surrounded on three sides by active mountain belts, is less than 200 km across, but nearly 1,000 km long. It is being filled by a fluvio-delta in the north under which there are over 7,000 m of Pliocene–Quaternary deposits; these pass into deep water troughs with more than 2,500 m of sediment of similar age.

In the western Mediterranean there are two main basins, the Balearic Basin in the west and the Tyrrhenian Basin between Italy and the islands of Corsica and Sardinia. In a sense they are ancient basins, although they are probably no older than the Late Tertiary. They have abyssal plains of more than 2,500 m depth which are underlain by 4–5,000 m of sediment and margined by submarine fans and deltas such as those of the Rhône and Ebro. Various mechanisms have been suggested to explain the formation of the western Mediterranean basins (Biju-Duval, Letouzey, Montadert, Courrier, Mugniot and Sancho, 1974). The Tyrrhenian Basin may have formed either by collapse of thinned continental crust, or by rotation of the Corsica–Sardinia block during sea floor spreading. The Balearic Basin is possibly a marginal basin formed earlier, during northwestward subduction beneath the Corsica–Sardinia volcanic arc. On the other hand these basins perhaps resulted from a combination of several mechanisms, thus illustrating the complexity of reality as compared with our simplified models.

Intramontane troughs of various dimensions may be found hundreds, even thousands, of kilometres from the main collision belt (Molnar and Tapponnier, 1975). These basins may be associated either with strike-slip faults or with large graben systems such as those of the Baikal and Shansi which may have resulted from the collision of India with Asia, and thus be the manifestation of collision rather than of an initial spreading stage as implied in Sect. 14.3.1.

14.4 ANCIENT PLATE-TECTONIC SETTINGS

The recognition of ancient plate-tectonic settings is rarely straightforward, although this is not always evident from the abundance of plate reconstructions of ancient orogenic belts which have appeared in the last few years. This spate of reconstructions is due to the paucity of factual constraints; the models are limited and necessarily naïve.

Criticism of the application of plate-tectonic theory does not imply that the exercise is worthless. A model based on present tectonic, sedimentary and volcanic patterns is preferable to one based on geosynclinal terminology derived from ancient rock patterns alone. However, the identification of ancient plate-tectonic settings has to be approached with caution, and with the realization that a model is at best a temporary working hypothesis.

14.4.1 **Spreading-related settings**

The most diagnostic sedimentary rocks of oceanic spreading-related settings are those due to the introduction into sea water of hydrothermal fluids, such as umbers. Rocks indicative of slow sedimentation such as cherts, black shales and nannofossil carbonates are suggestive, but also occur in other settings. They may be underlain either by oceanic crust or by older shallow-water facies of the continental margin. At the margins of the basin they may pass laterally into the evaporitic/alluvial fan suite of early stage intracontinental rifting.

Extensional tectonics give rise to small, elongate basins filled by fine-grained turbidites and other mass-flow deposits derived from small local horsts, together with hemi-pelagic sediments.

Although the main oceanic facies is typically pre-flysch, some pre-flysch associations, such as the euxinic facies, lack ophiolites or even volcanics and thus are indicative of other types of extensional basin such as those formed by strike-slip movement or back-arc spreading. The latter may however contain ophiolites and at present ocean floor generated at oceanic ridges cannot be distinguished from that formed by back-arc spreading.

INTRACONTINENTAL RIFTS

The Upper Triassic successions of the eastern margin of the United States were some of the first to be related to early Atlantic rifting (Dewey and Bird, 1970). Continental red beds with minor volcanics accumulated in a zone up to 300 km wide in which block faulting and differential subsidence were followed by regional subsidence with carbonate sedimentation.

In the Alpine–Mediterranean region Permo-Triassic successions, including fluvial red beds, evaporites, shallow-marine clastics and carbonate platform facies, developed in intra-continental rift zones prior to the emplacement of oceanic crust. Clastic sediments and evaporites are virtually absent after the Triassic because the region was submarine. The Jurassic facies largely consist of condensed *Schwellen* facies on the highs and thicker, mostly redeposited sediments, in the

Becken or basins (see Chapter 11).

A number of lead–zinc–barytes stratabound deposits may have formed within intracontinental rifts, including failed rifts, some of which are now separated by ocean floor. Examples are the zinc–lead deposits within calcareous sediments on the Red Sea coasts of Egypt and Saudi Arabia, formed in the middle Miocene before the most recent emplacement of oceanic crust, lead–zinc sulphide deposits of early Cretaceous age on the flanks of the Benue Trough, lead mineralization of similar age in the Amazon fracture zone, and in the Triassic of the Eastern Alps.

FAILED RIFTS

A number of thick successions deposited over a long period of time in major troughs have been interpreted as having formed within ancient rifts (e.g. Hoffman, Dewey and Burke, 1974). They are characterized by a lack of major deformation or orogeny and are orientated at a high angle to the trend of ancient, highly deformed fold belts. Following the Russians, these troughs have been called *aulacogens* (see Burke, 1977 for review).

One area where relatively modern graben have been interpreted as failed rifts is the northern North Sea petroleum province (Whiteman *et al.,* 1975). Alternatively Ziegler (1975) sees the same pattern as the result of a very complex tectonic history in which varying stress patterns, derived from evolving plate/plate interactions, have been influenced by earlier tectonic lines which have persisted more or less throughout the Phanerozoic.

Hercynian strike-slip movements may have resulted in the alignment of the gas fields found in aeolian sandstones of the southern North Sea, with their capping of Zechstein and Triassic evaporites (Blair, 1975) (see Chapter 5). However, the main oil fields occurring down the Central and northern (Viking) graben are the result of tensional structures probably related to the Mesozoic–Cenozoic opening of the Atlantic (Fig. 14.29A). Two major types of structural trap are known. One is the rotated fault block dipping towards the margin of the graben, found in Jurassic rocks (Fig. 14.29B). The other is the anticline draped over the earlier formed fault block and exaggerated by later movement and compaction. This characterizes Cretaceous and particularly Tertiary fields (Fig. 14.29C). In addition, movement of the underlying salt gave rise to salt-supported structures and major Cimmerian (late Jurassic and early Cretaceous) movements caused unconformity traps. The reservoirs themselves mainly occur in pelagic sediments such as the Chalk (Ekofisk), in deltaic sandstone wedges (Brent), or submarine fans (Forties).

A more ancient example is in the Canadian shield where the 12 km thick Proterozoic succession of the Athapuscow area between the Slave and Churchill Provinces has been interpreted

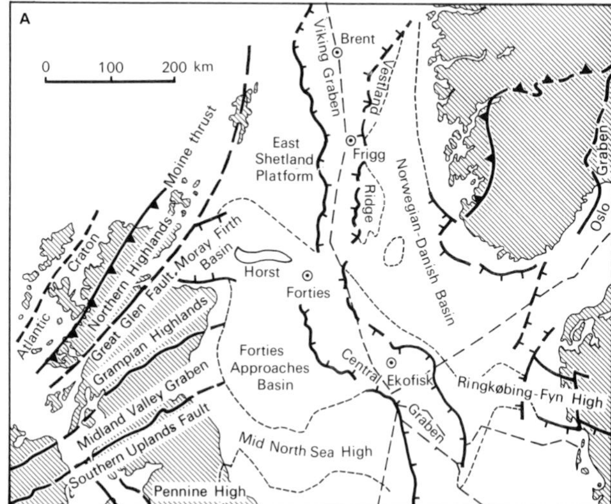

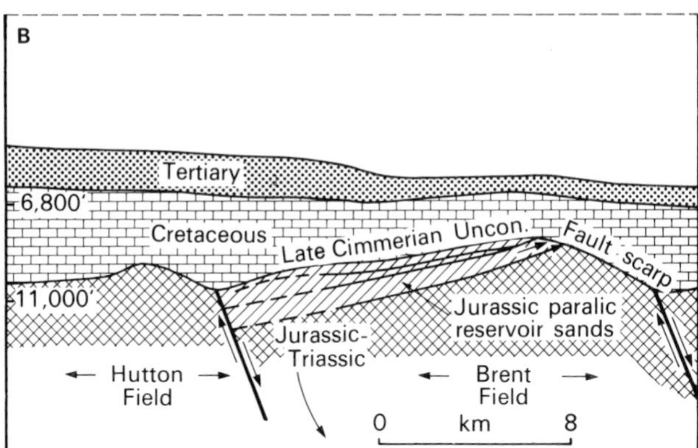

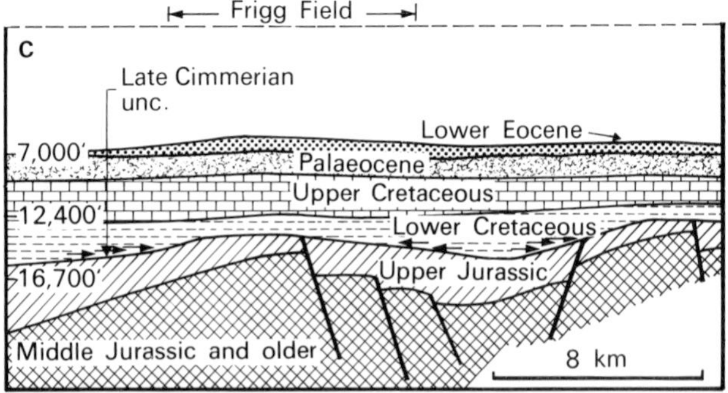

Fig. 14.29. (A) Sketch map of northern North Sea
(B) Brent oilfield. Sub-unconformity trap on a tilted fault block
(C) Frigg gasfield. Anticlinal trap enhanced by drape over deep structure (after Blair, 1975).

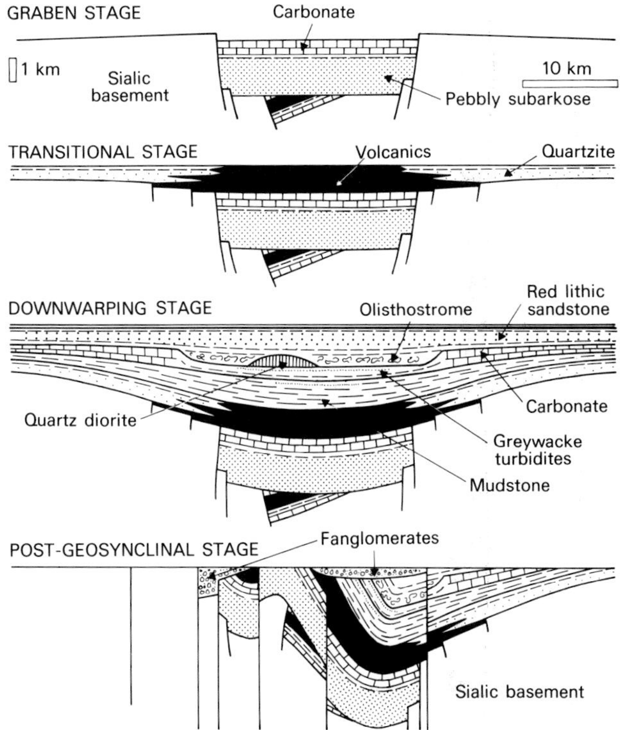

Fig. 14.30. Evolution of Pre-Cambrian Athapuscow aulacogen, Great Slave Lake, northern Canada (after Hoffman, Dewey and Burke, 1974).

as the fill of an aulacogen which developed over a period of 600 my (Hoffman *et al.,* 1974; Morton, 1974). Initially thick arkosic deposits accumulated in the subsiding graben and were mantled by dolomites; subsequently, quartzites and siltstones, overlain by volcanic rocks with a wide range in composition, accumulated both in the graben and on the platform flanks (Fig. 14.30). Renewed subsidence is indicated by the presence of mudstones and distal turbidites which thin towards the northeastern or continental end of the trough.

In southern Norway, a 1,500–3,000 m Pre-Cambrian succession of greywackes, shales, arkoses, carbonates and conglomerates has been interpreted as the fill of a rift-valley related to the opening of the 'proto-Atlantic' ocean and compared to the Jurassic North Sea rifts (Bjørlykke, Elvsborg and Høy, 1976). Coarse clastic wedges were deposited as fan-deltas along the faulted margins of a basin which is estimated to have been 60–70 km across. In the centre of the basin greywackes and shales were deposited, passing upward into shallow-water sediments, including tillites.

Many economically important Pre-Cambrian stratiform syn-sedimentary massive sulphide ores have been related to aulacogens: for example the Proterozoic silver–lead–zinc and copper ore bodies of Mount Isa in Queensland (Glikson, Derrick, Wilson and Hill, 1974), lead–zinc–silver ores of the Pre-Cambrian Sullivan Mine in British Columbia (Sawkins, 1974), and possibly the Pre-Cambrian copper ores of the Zambia–Katanga copper belt (Burke and Dewey, 1973). Many of these ore bodies are associated with volcanic or pyroclastic rocks within thick sedimentary successions.

Stratabound deposits of the uranium–vanadium–copper association account for a third of the western world's uranium resources. Although not necessarily parallel to the bedding of the sedimentary host rocks, they are confined to a particular stratigraphic unit, and hence are stratabound rather than stratiform. They are mostly considered to be epigenetic, although in some cases mineralization is thought to have taken place before significant burial of the host rocks. Deposits of this type occur at several horizons within the Athapuscow aulacogen, where the mineralization was probably related to metamorphic fluids circulating at depth in the thick prism of sediments (Morton, 1974).

Modern analogues of the early Proterozoic Witwatersrand-type gold–uranium–pyrite deposits of South Africa, Australia, and Brazil, have not been recognized. However, in South Africa, the presence of associated thick successions of fluviatile and carbonate sediments and locally volcanic rocks interbedded with the conglomeratic 'banket' ores, the unconformities within the succession, and the absence of compressional tectonics suggests some similarities with aulacogen environments, but the source of the gold was evidently a function of the earlier geological history.

The widespread stratiform Kupferschiefer copper–lead–zinc ores within black shales and argillaceous dolomites of the middle Permian Zechstein facies of Europe are distributed in a broad belt north of the Hercynian orogenic belt. Although the depositional basin coincides broadly with that of the Late Carboniferous foreland basin, it is possible that the Permian sedimentation and mineralization are related to subsequent early intracontinental rifting which resulted in the North Sea Troughs. The Kupferschiefer and Zambian copper deposits, which have some similarities, would then have formed in similar tectonic settings.

INTERCONTINENTAL RIFTS, PASSIVE CONTINENTAL MARGINS AND OCEAN FLOOR

Successions interpreted as having been formed on or close to ancient continental margins have been recognized by the presence of the mio- and eugeosynclinal couplets. The association of a laterally extensive belt of thick miogeosynclinal shallow marine clastics, shelf carbonates and coastal plain deposits with an equally thick and extensive belt of eugeosynclinal flysch and pre-flysch ophiolites and pelagic sediments now frequently juxtaposed by thrusts has been used as evidence for oceanic opening and development of an 'Atlantic type' margin (Dietz, 1963; Bird and Dewey, 1970). The Cambro–Ordovician margins of the Appalachian/Caledonian orogen have been interpreted in this way (Fig. 14.5).

The Alpine–Mediterranean region contains Mesozoic platform carbonate successions which European geologists once considered so characteristic of the early stages of their geosynclinal model. One example is the sequence of the Ionian Furrow, from neritic limestones and dolomites upward into Lower Jurassic Posidonomya shales and *Ammonitico Rosso*, overlain by pelagic cherty limestones (Aubouin, 1965). Similar successions have been described in more detail from the Triassic to Cretaceous of many parts of the Alpine–Mediterranean region (see Sect. 11.4.4). Such successions have been interpreted as the result of block faulting and differential sinking of a carbonate platform forming a basin and horst topography (Fig. 11.38). On the horsts, subsiding Upper Triassic/Lower Jurassic reefs and platform carbonates are mantled by a younger condensed sequence of red limestones and iron–manganese nodules (*Ammonitico Rosso*) while, in the basins, marls and radiolarites pass upwards into pelagic calcilutites (*Maiolica* facies) which ultimately blanketed the entire region.

The platform carbonates, some of which overlie Triassic evaporites, are considered to be continental-margin facies analogous to the Bahamian carbonates. They accumulated above intracontinental rift zone sediments and volcanics before, during and after emplacement of Tethyan Ocean crust. Differential subsidence forming basins of pelagic deposition resulted from block faulting (mostly Early Jurassic) within broad zones on the continental margins of Tethys, and perhaps also on a number of small continental fragments within the ocean (Bernoulli and Jenkyns, 1974). For example the miogeanticlinal Gavrovo ridge between the Ionian Trough to the southwest and Pindus Furrow to the northeast (Fig. 14.3) underwent slow crustal subsidence balanced by shallow-water carbonate platform sedimentation from the Late Jurassic to Early Eocene, and occupied a tectonic position analogous to the Atlantic outer shelf ridge of Dietz (1963). Block faulting and subsidence have been related to movement of the continental margins away from the spreading ocean rise.

In the Pindus Furrow, the type eugeosynclinal furrow of Aubouin, ophiolitic rocks indicating ocean floor generation are overlain by Jurassic pelagic sediments including radiolarian cherts. From the Cretaceous to Late Eocene the pelagic sediments were blanketed in flysch, deposited either on a prograding continental rise, bordering the foreland to the southwest, or more probably, because of the evidence of eastern derivation, in a closing remnant ocean basin. Therefore, as in similar tectonic settings in the Atlantic, platform sedimentation on the Gavrovo Ridge or 'outer shelf ridge' was accompanied by pelagic and flysch deposition in the adjacent ocean basin. However, a significant difference from the Atlantic model is the presence of radiolarites in the Ionian Furrow of the 'shelf'.

Stratabound deposits of lead–zinc–barytes–fluorite commonly occur within limestones, dolomite or calcareous sandstone successions of ancient continental shelves. Examples are the Lower Palaeozoic 'Mississippi Valley' deposits in the United States, the smaller deposits within Triassic carbonates in Europe, and possibly those of early Proterozoic age in South Africa (Martini, 1976). They were probably deposited from metal-rich chloride brines, circulating as a result of either magmatic heat or, more probably, tectonic movements, soon after deposition of the host rocks.

Banded iron formations, for example those of the Labrador Trough of Quebec and Newfoundland and the Hamersley province of Western Australia, are all of Early Proterozoic age. Those within the Early Proterozoic Transvaal Supergroup of South Africa possibly formed in an analogous setting to modern evaporites (Burton, 1976); alternatively they could be calcareous oolitic sediments subsequently replaced by iron and silica (Kimberley, 1974). Either explanation, together with the distribution of the deposits in narrow belts up to 1,000 km long and their association with thick carbonate sequences, suggests a tectonic setting broadly comparable to that of modern continental shelves.

Some of the modern metalliferous muds form a possible model for the late Mesozoic stratiform pyritic copper ores of Cyprus and similar on-land older deposits associated with ocean floor tholeiitic basalts elsewhere, for example the Ordovician deposits in the Notre Dame Bay area of Newfoundland (Upadhyay and Strong, 1973). Possibly pyritic sulphide bodies within pillow lavas develop beneath, and are related to the same hydrothermal solutions as overlying manganese oxides (Rona *et al.*, 1976).

Marine phosphorites, which account for 75% of the world's phosphate production, are known from continental shelf facies, although they reach their highest grade in the black shale–chert–carbonate facies of 'starved' sedimentary basins lacking clastic detritus. Most deposits formed within low latitudes in shallow-water, either in areas of upwelling cold ocean currents, or where warm currents were present and the deposits were reworked, for example the Cenozoic deposits of the eastern coast of the USA. The most favourable environments are probably the deeper parts of continental shelves, and nearshore possibly estuarine environments, of passive continental margins (Cathcart and Guildbrandsen, 1972).

14.4.2 **Subduction-related settings**

Successions interpreted as ancient equivalents of those deposited in modern subduction-related settings are included within, and in some examples comprise, the eugeosyncline of American authors. In European models they lie within but on the hinterland or internal side of the eugeosyncline.

The most diagnostic feature of an ancient subduction zone is the blueschist metamorphic facies, formation of which is widely considered to require subduction of a cold slab of oceanic lithosphere. The direction of subduction can be determined by the presence of a paired metamorphic belt, with the high-pressure/low-temperature belt on the subducting plate and the high-temperature/low-pressure belt on the overriding plate. However, the blueschist facies is infrequently found in orogenic belts.

Another diagnostic feature is the magmatic arc facies of calc-alkaline rocks, commonly with abundant andesites. Where blueschists and andesites are absent, ancient subduction may be inferred from evidence of oceanic closure, based on palaeontological or palaeomagnetic evidence for continental drift and from evidence of extensive thrusting and crustal shortening in orogenic belts.

FOREARC SETTINGS

Ocean floor, deep sea trenches and outer-arcs

Successions formed in trenches are very difficult to distinguish from those formed on oceanic crust. If oceanic crust is overlain only by pelagic sediments, these are unlikely to be preserved. However, where oceanic crust is overlain by continent-derived turbidites, or where the trench is receiving terrigenous sediments, these may be scraped off to give the characteristic fold style of an ancient outer arc as has been suggested for the orogenic belt of the Southern Uplands of Scotland (Mitchell and McKerrow, 1975; McKerrow, Leggett and Eales, 1977). The successive slicing off of sections of crust creates what until recently was an insoluble stratigraphic problem. The broad zonal stratigraphy gives a succession across the geosynclinal belt which apparently youngs oceanwards and is tremendously thick. However, detailed studies across the belt, using top and bottom criteria to indicate younging, suggest that, in spite of minor folding, the most common younging direction is towards the continent. On sedimentological evidence the succession apparently youngs towards the magmatic arc and on palaeontological evidence it apparently youngs towards the ocean. In both cases it appears to be extremely thick. Where both lines of evidence are available the paradox can be resolved by considering the belt to be an outer-arc comparable to the Arakan ranges of Burma (Fig. 14.31 cf. Fig. 14.20).

Many belts of deformed flysch with mélanges and tectonic slices of ophiolites and pelagic sediments are now tentatively interpreted as ocean floor or remnant-basin deposits deformed above a subduction zone to form an outer-arc. In these successions the stratigraphic dip is commonly towards the magmatic arc and structural vergence is away from it. The Carboniferous flysch of the Ouachita Mountains is a major belt of flysch recently re-interpreted as ocean floor and remnant-basin deposits deformed in an outer-arc within the Hercynian orogenic belt of eastern North America (Graham *et al.*, 1975).

Sediments deposited within an ancient outer-arc have so far been described only from the Mentawai Islands (Moore and Karig, 1976; Fig. 14.20) and possibly from the Franciscan complex.

The Franciscan Complex (Fig. 11.25) comprises tectonized marine sediment of Upper Jurassic to Upper Cretaceous ages,

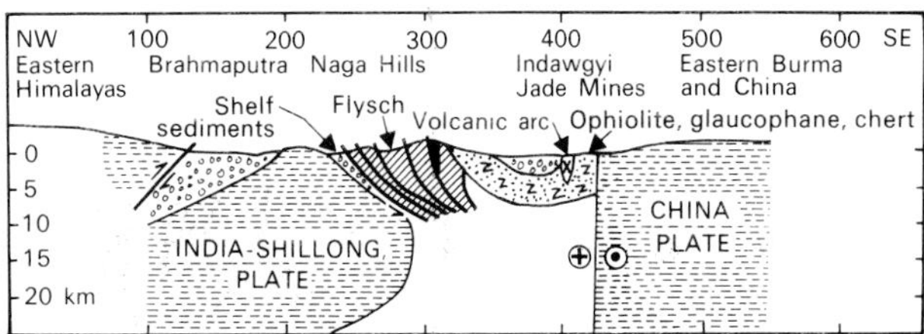

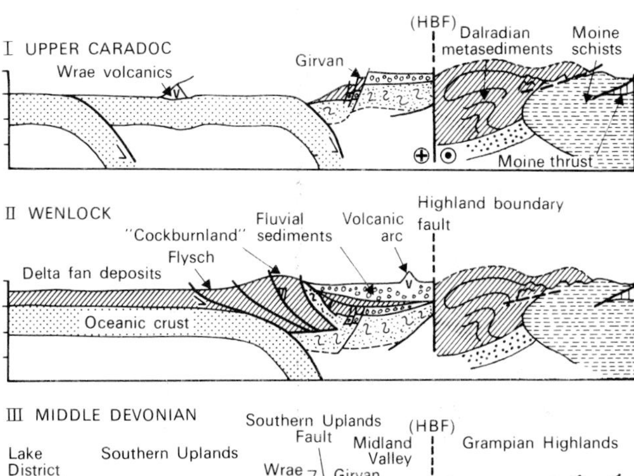

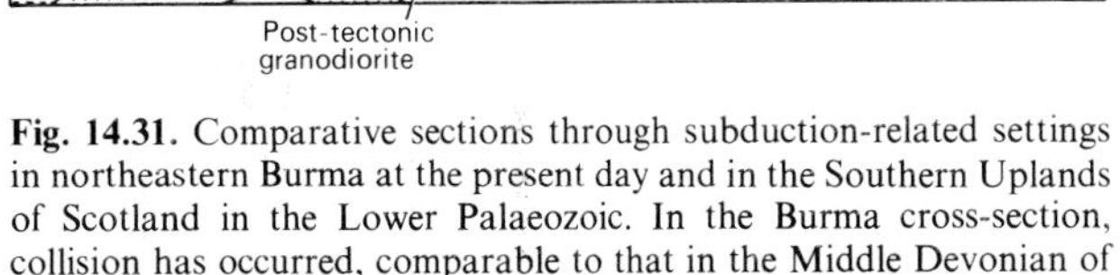

Fig. 14.31. Comparative sections through subduction-related settings in northeastern Burma at the present day and in the Southern Uplands of Scotland in the Lower Palaeozoic. In the Burma cross-section, collision has occurred, comparable to that in the Middle Devonian of Scotland (modified from Mitchell and McKerrow, 1975).

with minor igneous and metamorphic rocks. Turbidites and mudstones are predominant, with minor radiolarian chert, conglomerate and volcanic and ultramafic rocks (for details see 11.4.2). All these rock types occur both as mappable units and as blocks associated with blueschists within 'wildflysch', interpreted either as olisthostromes or as mélanges.

Most workers relate deformation of the succession to subduction beneath a trench or developing outer-arc. Subduction changed to transform motion in the middle Tertiary and fragments of older crust have been transported northwards and now lie west of the Franciscan Complex.

Cowan and Page (1975) have shown that debris-flow deposits of Late Cretaceous age accumulated in fault troughs within the highly deformed Franciscan outer-arc rocks. The debris-flows lie unconformably on outer-arc rocks and include blocks of ultrabasic and volcanic rock, glaucophane schist and chert, derived from active thrust-related fault scarps composed of older outer-arc mélanges and olisthostromes. The clasts therefore underwent two or more erosion-deposition cycles before they were eventually incorporated into a Lower Tertiary mélange.

OUTER-ARC TROUGHS

Ancient outer-arc trough successions were first recognized by Dickinson (1971b) who included as examples the Great Valley sequence of California, the Hokonui facies in New Zealand, and the Median Zone of Southwest Japan. They can be recognized by their great thickness, predominance of clastic facies, elongate basin shape, and monoclinal or synclinal folding. However, these features are not always diagnostic, and the succession may be composed of pelagic sediments like those of the Oceanic Group of Barbados (Westbrook, 1975; Fig. 11.27). The succession can most easily be recognized from the presence of adjacent parallel magmatic and outer-arc successions.

The scarcity of ancient outer-arc trough successions may reflect either the relatively limited attention which modern ones have received in the literature, or the ease with which ancient ones may be destroyed during the uplift which accompanies continental collision.

The Upper Jurassic to Cretaceous Great Valley sequence of California was deposited in a trough more than 70 km wide between the outer-arc of the Franciscan Complex to the west and magmatic arc of the Sierra Nevada to the east (Dickinson, 1971b) (Fig. 11.24). The sequence dips eastward away from the outer-arc and consists of up to 12 km of mostly volcaniclastic turbidites interpreted as submarine fan deposits derived from the Sierra Nevada. The Jurassic to late Cretaceous age of the sequence is similar to that of the Franciscan Complex. In the west the Great Valley sequence overlies ophiolites interpreted as oceanic crust (see Chapter 11) and in the east it rests unconformably on metamorphic and igneous rocks (Fig. 11.25). Many authors consider it to be thrust westwards over the Franciscan Complex, although others favour a stratigraphic contact (Maxwell, 1974).

In Scotland, the southern part of the Midland Valley, between the Southern Uplands outer-arc to the south and the Sidlaw Anticline volcanic arc to the north, has been compared in tectonic position to the modern outer-arc trough of Burma (Mitchell and McKerrow, 1975). Near the southern margin of the Midland Valley, Upper Ordovician turbidites pass northwards into thick conglomerates and limestones; the turbidites are overlain by Lower Silurian turbidite and shelf deposits succeeded by Upper Silurian non-marine and Lower Devonian fluvial 'Old Red Sandstone' deposits. There is evidence for longitudinal advance of non-marine sediments from northeast to southwest, which can be related to diachronous continental collision.

Part of the Hokonui assemblage of Permian to Jurassic age in the south of South Island and Triassic to Jurassic age in the west of North Island, New Zealand, probably comprises an ancient outer-arc trough sequence (Dickinson, 1971b; Landis and Bishop, 1972). Triassic and Jurassic rocks of the Murihiku terrain are up to 10 km thick and occupy a regional syncline, consisting of both flysch-type and shallow-water tuffaceous sediments. The situation of the Murihiku rocks between magmatic arc rocks to the west and outer-arc mélanges and schists to the east, suggests that they are outer-arc trough rocks deposited during westward subduction. Closure of the ocean basin was accompanied by collision, resulting in juxtaposition of the outer-arc trough and outer-arc rocks with continental margin rocks of the Torlesse terrain of eastern New Zealand (Blake, Jones and Landis, 1974).

MAGMATIC ARCS

Ancient magmatic arcs comprise volcanic arcs which have become either welded to or included within continents, mostly as a result either of cessation of subduction, such as some of the Palaeozoic and Mesozoic arcs in western North and South America, or of continent–continent or continent–arc collision, for example the Banda arc north of Timor (Fig. 14.1). Other ancient magmatic arcs are present in the Lake District of northern England (Fitton and Hughes, 1970), in the Cambrian succession of Tasmania, and the late Pre-Cambrian of Newfoundland (Hughes, 1970).

In many ancient arcs, the volcanic and sedimentary successions are strikingly similar to those exposed in active volcanic arcs within the oceans and comprise andesitic volcanics and volcaniclastic sediments. However, in other arcs the volcanic rocks have mostly been stripped off to expose plutons of diorite to tonalite and granodiorite composition.

It is frequently difficult to determine whether an arc developed on a continental-margin or as an intra-oceanic arc, particularly if the continental-margin arc did not form a major mountain range, as, for example, the Cretaceous arcs of the coastal belt in Peru and the central belt in Burma. Probably the most diagnostic features of ancient intra-oceanic magmatic arcs are thick submarine andesites and andesitic volcaniclastic rocks with or without silicic lavas and tuffs. Ancient continental-margin arcs on the other hand commonly, but not invariably, consist largely of subaerially erupted dacites and rhyolites interbedded with terrigenous clastics and forming successions within which are numerous unconformities. However, distinction is often only possible by using criteria external to the arc itself and thus requires an understanding of the polarity of subduction and collision.

Although magmatic arcs are known chiefly for their very large low-grade epigenetic magmatic hydrothermal porphyry copper deposits, they are also characterized by an important class of stratiform massive sulphide deposits, the Kuroko ores (Sato, 1977). These are pyritic zinc–lead–copper stratiform massive sulphides named from the type locality in Japan. The Japanese deposits lie within a belt less than 80 km wide and more than 400 km long, and probably all accumulated in the late Miocene (Ueno, 1975). They are closely associated with shallow-marine pyroclastic dacite domes, were emplaced in and on the flanks of local sedimentary basins, and locally pass laterally into mass-flow sediments bearing sulphide clasts (Horikoshi, 1969). Kuroko-type ores of Tertiary age are known in Fiji (Colley and Rice, 1975) and the Solomon Islands, and numerous deposits of this type are found in ancient magmatic arcs. Examples include those of Lower Carboniferous age in the Iberian pyrite belt, the Buchans Mine in Newfoundland, and a number of Proterozoic pyritic zinc–copper deposits.

BACK-ARC AREAS

In ancient successions it is rarely possible to distinguish between sediments of a major ocean basin and those of a back-arc marginal basin or inter-arc basin on the basis of sedimentary facies. However, where the regional palaeogeography can be reconstructed on other evidence, notably by recognition of the adjacent magmatic and remnant-arcs or the continental margin, it is possible to infer the presence of ancient marginal or inter-arc basins, for example in the late Silurian–early Devonian Tasman geosynclines of eastern Australia (Packham and Leitch, 1974). The presence of abundant volcanic material is diagnostic in some cases.

Most back-arc oceanic basins eventually close as a result of subduction of their floor and arc–arc or arc–continent collision. Their sedimentary fill is probably preserved in either remnant back-arc basins or as highly deformed successions within outer-arcs.

Continental back-arc basins are also difficult to recognize, mostly because sedimentation within them may be similar in facies to, and take place only slightly before the deposition of, overlying late orogenic basin deposits. A possible example of an ancient back-arc continental basin succession is the Silurian–Lower Devonian sequence in the Strathmore Syncline of the Midland Valley of Scotland, deposited north of the Sidlaw Anticline magmatic arc during northward subduction, prior to the Caledonian orogeny (Fig. 14.31).

14.4.3 Transform/strike-slip fault-related settings

These highly seismic settings of overall lateral motion may have, at any one place, very rapid differential vertical movement sometimes exceeding the strike-slip motion. Although crust in total is being conserved, locally there may be either great extension or compression. Thus the characteristics of transform/strike-slip-related settings are (1) rapid deposition giving huge local thicknesses of sediments (2) uplift and erosion leading to development of unconformities in close proximity to subsiding sedimentary basins (3) lateral variation in facies (4) simultaneous development of both extensional and compressional tectonics in nearby areas (5) little or no metamorphism (6) sparse igneous activity.

Few strike-slip orogenic belts lack some overall component of compression or of extension and most are therefore either transpressive (oblique compression) or transtensile (oblique extension) (Harland, 1971). They may be associated with igneous activity appropriate to either subduction or spreading.

The lack of ophiolites and metamorphism in the Spitsbergen Tertiary orogenic belt led Lowell (1972) to follow Harland (1969) in thinking that this fold belt differed fundamentally from subduction-related fold belts and was a strike-slip transpressive belt. Lowell recognized in particular the associated pattern of *en echelon* folds, which elsewhere may form petroleum traps (Harding, 1974; Wilcox, Harding and Seely, 1973). En echelon folds are associated with the San Andreas fault in California (Fig. 14.32), the Dead Sea fault and the Alpine fault of New Zealand. They are valuable indicators in cover rocks of the presence of strike-slip faults at depth, of the sense of the movement and of the existence of sedimentary, pullapart basins below the cover rocks. In older orogenic belts, however, one of the difficulties is that a pure shear with a N–S compression gives an identical structural pattern to simple shear produced by a NW–SE dextral shear pattern (Fig. 14.33).

The eastern and central Cantabrian Mountains in northern Spain have also been interpreted as a strike-slip orogenic belt which developed on continental crust in Upper Carboniferous times (Reading, 1975). The most striking feature is the evidence for vertical movements which gave rise to tectonic deformation including gravity nappe formation at the same time as newly created sedimentary basins being filled largely by conglomeratic and mass-flow deposits.

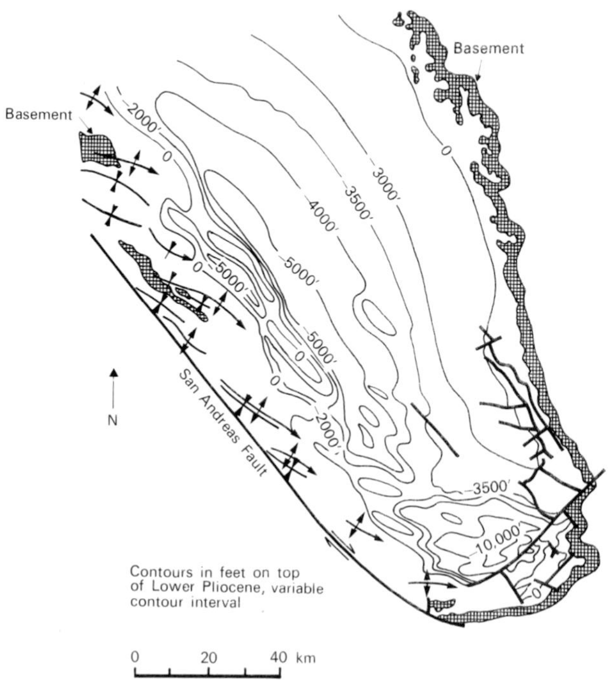

Fig. 14.32. En echelon folds associated with dextral movement on the San Andreas fault. Contours on top of Lower Pliocene indicate very rapid and variable subsidence (after Harding, 1974).

During a period of tectonic quiescence, which lasted from Pre-Cambrian to Namurian times, shallow-water sandstones, shales and limestones accumulated over a wide area. Following a cessation of clastic deposition in the early Carboniferous 'bathyal lull' when condensed 'griotte' (pelagic nodular limestone) was deposited, rapid vertical movements during the late Carboniferous allowed the accumulation of very thick successions of very varied sedimentary facies including turbidites and mass-flow conglomerates, shelf limestones, deltaic and fluvial sandstones, and alluvial fan conglomerates. Facies variations are rapid both vertically and laterally and strong angular unconformities separate successions 2–3,000 m thick. Deformation was partly contemporaneous with sedimentation, tectonic deformation often being difficult to separate from sedimentary sliding (Fig. 14.34).

In this case, as in Spitsbergen, the strike-slip orogenic belt developed in continental crust without the opening of an ocean. In other cases flysch and molasse basins may have developed in strike-slip settings following ocean closure and continental collision, for example at the late stages of the Caledonian orogeny in Great Britain and Norway (Phillips, Stillman and Murphy, 1976; Steel, 1976) and in the Alps (Gigot, Gubler and Haccard, 1975).

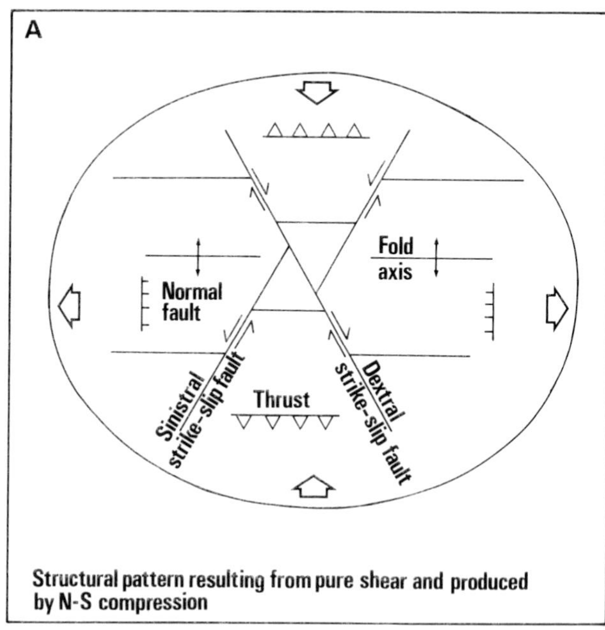

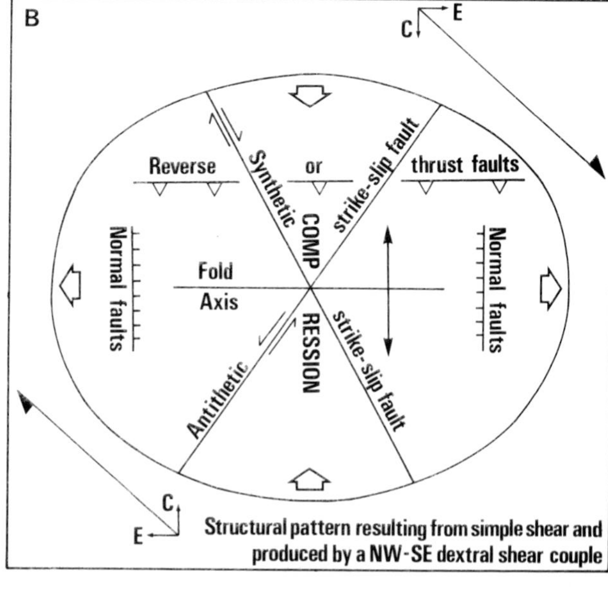

Fig. 14.33. Apparently similar structural patterns resulting from (A) pure shear produced by N–S compression (B) simple shear produced by a NW–SE dextral shear couple.

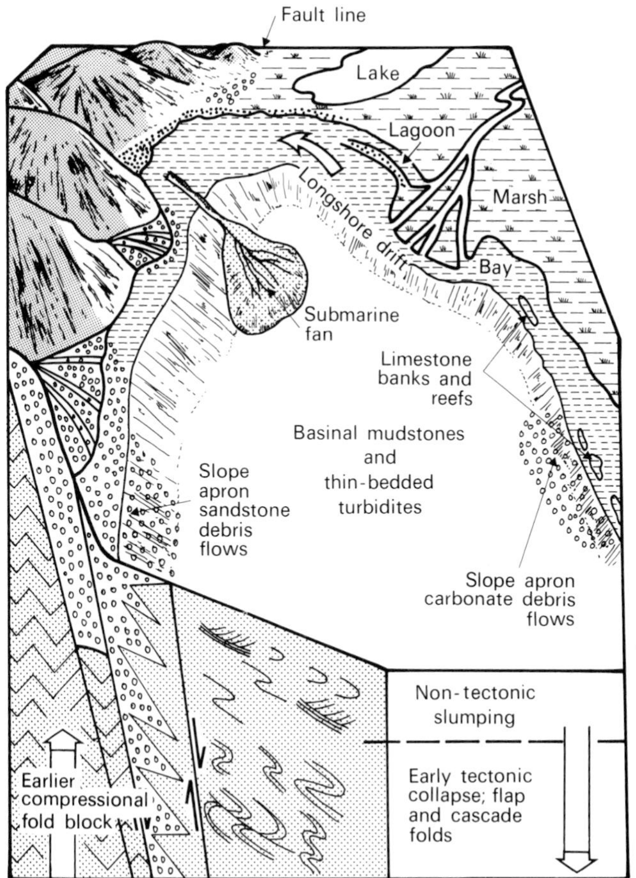

Fig. 14.34. Sedimentary and tectonic model for a strike-slip orogenic basin, based on the Hercynian Cantabrian belt of northern Spain (from Reading, 1975).

14.4.4 Continental collision-related settings

Late orogenic basin successions, in contrast to those of remnant-ocean basins, have been recognized in orogenic belts for more than 80 years, and are equivalent to the exogeosynclines of North American geologists. They consist of thick, mostly continental, clastic molasse sequences situated beside but largely above the flysch, metamorphic rocks and plutons of the collision belt or orogen. In many orogens foreland-basin successions cannot be distinguished from those formed on the hinterland or overriding plate. However, where nappes of flysch and ophiolite are thrust over a succession of non-marine clastic sediments younger than the flysch, the succession probably accumulated in a foreland basin.

Since a large number of ancient examples are known, foreland basin successions probably have a high preservation potential. They occupy a cratonic basin which subsequently became stable and they are commonly overridden by and preserved beneath nappes of older rocks on the interior side of the basin.

The Molasse of the Northern Alpine foredeep is the most useful single stratigraphic unit in determining the late orogenic history of the Central and Eastern Alps. Autochthonous early Oligocene Helvetic flysch passes up transitionally into a slope facies overlain by mostly non-marine molasse of late Oligocene to early Pliocene age up to 6,000 m thick. The Molasse comprises the stratigraphic sequence Lower Marine (including slope facies) → Lower Freshwater → Upper Marine → Upper Freshwater Molasse. Facies are variable both among and within these units but in the Freshwater units thick fanglomerates and fining-upward fluvial cycles are most characteristic, with abundant conglomerates and sandstones (Van Houten, 1974).

Füchtbauer (1967) recognized 7 coarsening-upward mega-cycles related to phases of Alpine deformation. Sedimentary facies and clast composition provide evidence for intermittent uplift of the Alpine source areas, unroofing of plutons, and advance or elevation of several nappes each of specific lithology. Stratigraphy and deformation of the Molasse also indicate migration of the depositional basin towards the foreland, and the folding of the more proximal molasse beneath advancing nappes.

Depression of a foreland basin in the early stages of collision may result in depths sufficient for turbidite accumulation (Dickinson, 1974b), and in the Alpine Molasse basin it is possible that part of the Helvetic flysch was deposited in the incipient foreland basin, during the earliest stages of continental collision. However, Molasse sedimentation post-dated the main Alpine collision, and hence the Molasse, from the Lower Freshwater unit upwards, is younger than the flysch.

In many orogens and geosynclines the polarity or direction of subduction prior to collision is uncertain and so also is the relative importance of strike-slip and subduction. An example is the late orogenic Mesozoic basin east of the Columbian orogen of western North America, which contains two coarsening-upward mega-cycles of marine to continental 'molasse' facies, one of late Jurassic to early Cretaceous, and one of late Cretaceous to Oligocene age (Eisbacher, Carrigy and Campbell, 1974). This basin was termed retro-arc by Dickinson (1974b) and considered to have formed on the hinterland as a result of late Mesozoic–early Cenozoic eastward subduction during the Pacific orogen. Alternatively, it could be argued that the basin developed during the closing stages of an early Mesozoic westward subduction and continental collision which formed the Columbian orogen. It would then have been a foreland basin, situated on the westward underthrusting foreland of the North American continent (Eisbacher, Carrigy and Campbell, 1974).

The late orogenic Ebro Basin south of the Pyrenees contains a late Eocene to late Miocene succession up to 3 km thick (Van Houten, 1974).

Conglomerates and fining-upward fluvial cycles predominate, derived partly from the rising Pyrenean chain to the north and partly from Hercynian blocks to the south. Early Miocene evaporites are interbedded with conglomerates in the interior of the basin. Although the basin has been termed a foreland basin (Van Houten, 1974), if evidence of southward subduction beneath the Pyrenees followed by collision of Northern Europe with Iberia is accepted (Dewey *et al.*, 1973), the basin developed on the overriding continental plate rather than on the foreland. A third possibility is that development of the Ebro basin and the complementary Aquitaine basin to the north was associated with major strike-slip movement along the North Pyrenean wrench fault. During the Cenozoic increasing compressive (transpressive) movements led to uplift of the Axial Zone and gravity gliding of nappes and olisthostromes into the northern margin of the Ebro basin, accompanied by sedimentation due to erosion of the rising mountain belt (Choukroune and Seguret, 1973). The Ebro basin contains stratiform copper mineralization within Oligocene sandstones (Caia, 1976) which show some similarities to the Kupferschiefer of Northwest Europe although the two basins formed in different types of tectonic setting.

The intramontane basin successions show some similarities to those of intracontinental rifts, and some, for example the Triassic succession of New England, both post-date one orogeny and pre-date the subsequent opening of an ocean. This suggests that there might be a causal relationship between collision and subsequent rifting. Possible causes include both collision-generated strike-slip faults themselves becoming oceanic fracture zones, and fundamental megashears within the lithosphere manifesting themselves both as strike-slip faults during collision and as fracture zones during ocean-floor spreading.

14.5 GEOSYNCLINAL EVOLUTION

So far we have been concerned with relating stratigraphic sequences to particular tectonic settings and plate boundaries. We shall now attempt to reconcile geosynclinal hypotheses with interpretations based on plate-tectonics, by considering only two alternative and undoubtedly simplistic models which neglect many difficulties arising from the complexity of global tectonics.

The first model is one based on the now widely accepted Wilson (1966, 1968) cycle of intracontinental rifting, generation of ocean-floor, subduction and collision (Mitchell and Reading, 1969; Dewey and Bird, 1970; Hoffman, Dewey and Burke, 1974). The second model has evolved from consideration of both modern strike-slip orogenic belts and problems that arise from the difficulties of interpretation of some ancient orogenic belts in terms of simple subduction.

14.5.1 The Wilson cycle

Deposition and deformation of many successions can be related at least in part to the cycle of oceanic opening and closing by subduction leading to continental collision. The rifting stage has already been considered in Sect. 14.3.1 and Sect. 14.4.1. Here we shall discuss first the subduction stage, then the incipient collision stage with development of remnant-basins, and finally the collision stage, involving the tectonic juxtaposition of successions following ocean closure, collision and strike-slip faulting.

A schematic cross-section, drawn to include both a passive margin and ocean basin subducting beneath a continental margin arc (Fig. 14.35A) (modified from Dewey and Bird, 1970), shows ocean-floor turbidites being scraped off above the subduction zone to form what is, at this stage, a poorly developed submarine outer-arc. At the same time subduction-related volcanism takes place in the magmatic arc on the overriding plate. Between the outer and magmatic arcs flysch and later continental facies accumulate in the inter-arc trough.

In terms of the European geosynclinal model (Aubouin, 1965), at the subduction stage the passive margin corresponds to the foreland and miogeosyncline of Dietz (1963); the ocean-basin, the incipient outer-arc and the magmatic arc are equivalent to the eugeosyncline; and the hinterland is the interior of the overriding plate. This early stage of orogenic development can be applied to the Indian Ocean–Sunda Arc area where the tectonic elements were related to those of the Mediterranean chains by Aubouin. The problem here has been that the 'foreland' west of the Sunda Arc is the Indian Ocean whereas in the Alps the 'foreland' is the north European continent. It is resolved if the Sunda Arc is considered to be at an early stage of development compared to the Alps, and thus the southern part of the Indian continent or Africa is the continental foreland.

In the second or remnant basin stage (Fig. 14.35B) some of the major sedimentary and tectonic features characteristic of eugeosynclines develop.

As the continental 'foreland' approaches the subduction zone, salients of the continental margin impinge on and may underthrust the overriding plate, resulting in elevation in isostatic response to crustal thickening. Predominantly non-volcanogenic sediments derived from the mountain ranges are trapped as submarine fan and deltaic deposits in remnant-basins; as the basins progressively close, the sediments are scraped off, in some cases, together with arc-derived volcanogenic trench deposits, forming a major outer-arc. Continental-rise deposits of the foreland margin may also be

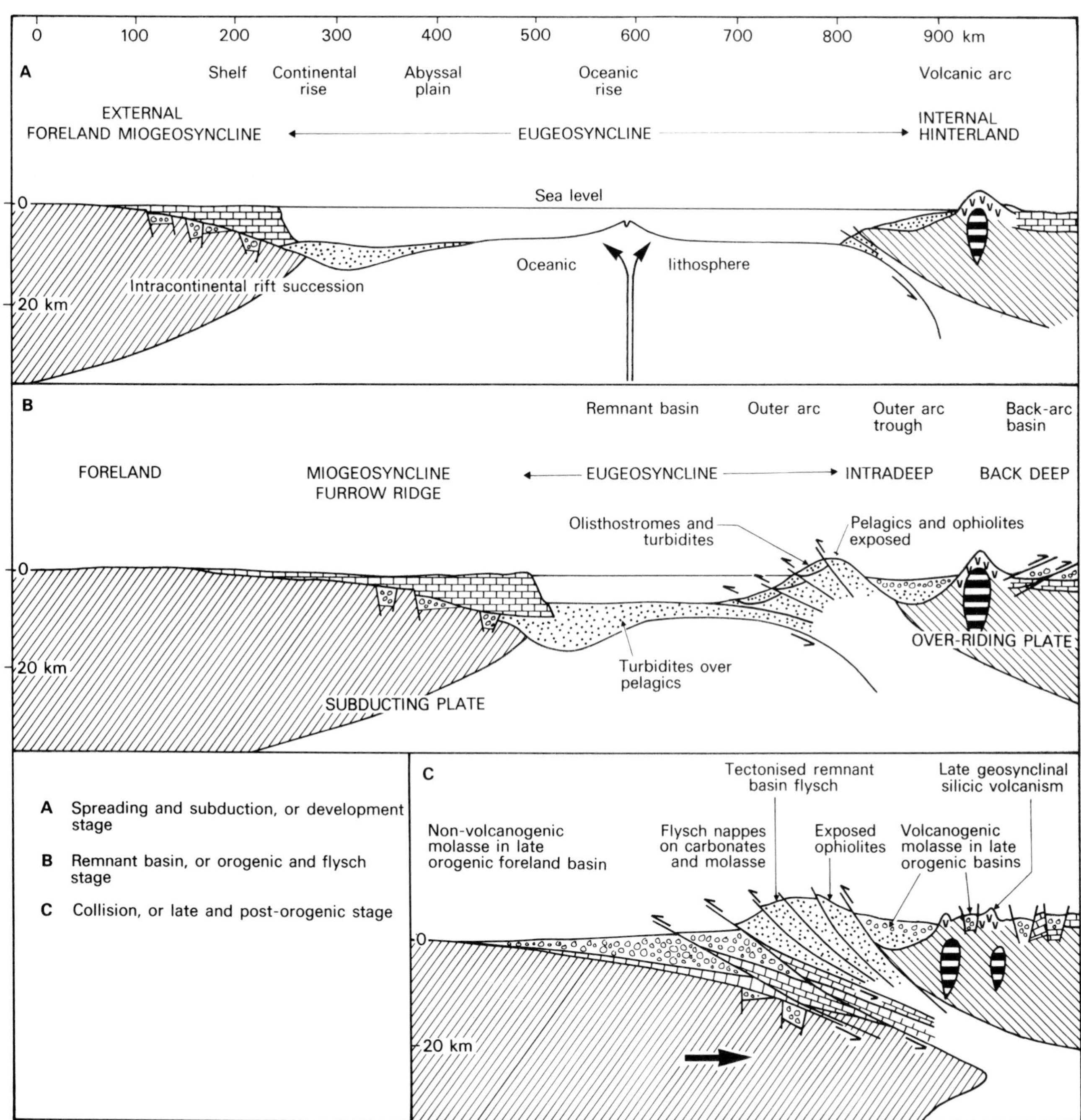

Fig. 14.35. Stages of development of European geosynclinal model of Aubouin (1965) interpreted in terms of the Wilson cycle of oceanic opening, subduction and continental collision.

tectonized and contribute to the outer-arc. Within and adjacent to the colliding continental salients major thrusts develop, dipping in the same direction as the subduction zone, and resulting in overthrusting towards the foreland. These involve crustal rocks of either the overriding plate and under-thrusting foreland, as in the Alps, or only the foreland, as in the Himalayas.

In terms of the geosynclinal theory, the remnant-basin stage corresponds to the orogenic or flysch stage (Fig. 14.35B), the onset of major flysch sedimentation on pelagic shales which overlie oceanic crust being related directly to development of remnant-basins diachronous with collision. The geosynclinal polarity of Aubouin, or migration of flysch deposition and orogeny towards the foreland, results partly from tectonic growth of outer-arcs as the foreland is subducted and underthrust beneath the outer-arc (cf. Figs. 14.3, 14.20).

In the third or collision stage (Fig. 14.35C), equivalent to the late-geosynclinal stage of the geosynclinal model, closure of remnant-basins is accompanied by depression of the under-thrusting continental foreland to form, initially, a foreland flysch basin. With rapid sedimentation the foreland basin becomes filled with thick clastic molasse, which may be overridden by flysch nappes as under-thrusting by the foreland continues; these may rest directly on the miogeosynclinal successions, resulting in tectonic juxtaposition of flysch and shallow-marine sediments. Development of new thrusts within the foreland results in crustal thickening, uplift and erosion. This exposes the deeper structural levels of the outer-arc and subduction zone, with ophiolites and mélanges, considered typical of suture zones, and which, in the geosynclinal model, are situated on the internal side of the eugeosyncline. In the later stages of collision, following uplift, tectonically emplaced nappes may undergo further transport towards the foreland as a result of gravity gliding.

Before plate-convergence finally ceases, collision-related volcanism with emplacement and exposure of late-orogenic granitic plutons may be accompanied by formation of fault troughs forming intramontane basins filled with arkosic and volcanogenic molasse.

14.5.2 Strike-slip orogenic model

In the strike-slip orogenic model curvature and offsetting of faults gives rise to pull-apart basins which extend sufficiently to allow magmatic material to rise and form spreading centres as is happening today in the Gulf of California. As these pull-apart basins extend, pre-flysch sediments are deposited, associated with magmatic activity. Laterally the pre-flysch passes into continental-margin sediments and the stratigraphical and lateral facies patterns are identical to those formed in the early stages of intra- and inter-continental rifting.

However, where the relative plate movement is essentially lateral shear rather than extension at right angles to the rift, subsequent tectonic and sedimentary events are significantly different. Not far from the pull-apart basin, compression leads to tectonic thickening of crust, uplift and erosion of terrestrial clastic sediments which will be transported into the basin, thus giving rise to deposition of flysch. As the basins fill, the flysch passes up into continental clastics similar to molasse. At the same time, due to lateral motion of blocks, the tectonic regime changes from one of extension to one of compression. Thus folding, uplift and extensive deformation, including the formation of thrust sheets and gliding nappes completes the geosynclinal/orogenic cycle.

The two models are not completely distinct and most orogenic belts contain elements of both. On the whole, the importance of strike-slip movements relative to subduction increases as continental collision proceeds. It is possible too that while subduction was more important in the Caledonian/Appalachian fold-belt the Hercynian fold-belt was dominated by strike-slip movements, though the latter model has so far received less consideration than the former.

FURTHER READING

Burk C.A. and Drake C.L. (Eds.) (1974) *The Geology of Continental Margins*. pp. 1009. Springer-Verlag, New York.

Dickinson W.R. (Ed.) (1974) *Tectonics and sedimentation*. pp. 204. Spec. Publ. Soc. econ. Paleont. Miner., **22,** Tulsa.

Dott R.H., Jr. and Shaver R.H. (Eds.) (1974) *Modern and ancient geosynclinal sedimentation*. pp. 380. Spec. Publ. Soc. econ. Paleont. Miner., **19,** Tulsa.

Bott M.H.P. (Ed.) (1976) *Sedimentary basins of continental margins and cratons*. pp. 314. Tectonophysics, 36.

CHAPTER 15 Problems and Perspectives

An intriguing problem in the history of geology is why, after the foundations of sedimentology had been laid in the 19th century, so little progress was made in the first half of this century.

Walther, by establishing his 'Law of Succession of Sedimentary Facies', had shown how the vertical sequence may be used in environmental analysis, one of the corner stones of this book. Sorby had initiated studies of sedimentary petrography, qualitative petrography, sedimentary structures, palaeocurrent analysis, diagenesis and shell structure. Van't Hoff, a leading physical chemist, had used experimental chemistry and an understanding of the phase rule to explain the origin of evaporites. Sorby's work on the petrography of chertified limestones pre-dated the first paper on igneous petrology. Van't Hoff experimented on marine evaporites before the first experimental laboratories for the study of igneous and metamorphic rocks were established.

In the early part of this century, the study of sedimentary rocks was largely undertaken by stratigraphers who were mainly concerned with correlation and who used fossils as zonal indices rather than as environmental indicators. Because the 'useful' fossil was independent of environment, facies-bound fossils were ignored despite their environmental significance. Sedimentary petrography, apart from diagenetic studies such as Cullis on Funafuti and Sander on the Triassic, also became a tool for correlation with concentration on the analysis of heavy minerals and microfacies. The need by the oil industry to characterize and interpret buried sandstones from cuttings led to the development of grain-size analysis, based on the earlier work of Udden and Wentworth. Unfortunately an increasing emphasis on methodology and statistics turned some studies of grain-size analysis into a tool for statistical manipulation of little geological relevance.

Other stratigraphers and sedimentary petrographers in the 1930s and 1940s were mainly concerned with tectonic aspects of sedimentation. In particular they used sedimentary facies such as 'flysch' and 'molasse' as indicators of contemporary tectonics and of stages in the development of geosynclinal basins and orogenic belts; or they explained sedimentary cycles in deltaic or fluvial environments as due to intermittent uplift or downwarp, rather than to sedimentary factors.

The turbidity-current concept contributed to the initiation of modern sedimentology in 1950 not only because it solved the particular problem of deep-sea sands and graded beds, but because it integrated discoveries in modern oceans, in ancient rocks and through experiments. It focused attention on process and forced geologists to look again at rocks, to examine every grain and feature and to wonder what produced them.

A major stimulus in sedimentological research has been the necessity to discover and exploit resources of coal and oil. In the 1930s and 1940s much fruitful discussion centred on the cyclic arrangement of coal-bearing strata in the Carboniferous. The initial stimulus and much of the best work on deltas has been and still is being done by oil-company geologists as an aid to predicting facies patterns in petroleum reservoir rocks. The great carbonate reservoirs of the world and particularly the problems of porosity distribution were a stimulus for studies on carbonate facies and modern carbonates in the Bahamas and Persian Gulf. The appreciation of the importance of diagenesis in both carbonates and evaporites has largely been due to the importance of these rocks as reservoirs.

Most of those who work on ancient sedimentary rocks today realize the necessity of being familiar with modern environments, although this has not always been so. In recent years the gap between physical geographers and geologists has narrowed. Geologists are looking at changes in climate, meteorological and oceanographical circulation and erosional processes as they see that these offer clues to facies changes: physical geographers are becoming concerned with depositional facies studies as they realize their importance in geomorphology. In addition the concern for the environment has brought many geologists and physical geographers together in schools of environmental studies.

Sedimentology has now come of age. It is appreciated that a geology department needs at least one sedimentologist as much as an igenous petrologist. In industry the same need is becoming apparent, since sedimentology can give answers to many of the problems posed by the mining or reservoir engineer. Given sufficient data, we can now establish the broad environment of deposition of almost all ancient sedimentary rocks, at least in the Phanerozoic. In many cases we can deduce the detailed environment and particular process or group of processes that formed the rocks. The nature of the source area, the climate, the local tectonics, the marine circulatory patterns, tidal or meteorological can be established. However, we need to identify environments, the processes which gave rise to them and the geological background with fewer data since, only rarely, are the necessary data available and, in applied geology, early predictions are essential. In the Pre-Cambrian, facies interpretation is much more difficult, even in clastic sediments. We are not always sure of distinguishing a shallow-marine environment from a fluvial one, let alone lacustrine from marine, although, curiously, if they are distinguished, some of the best detailed environmental models have been obtained from Pre-Cambrian rocks. In chemical sediments the problems are compounded but the future should see a resolution of the problems of Pre-Cambrian stromatolites, ironstones and cherts, the solution of which should add to our knowledge of the chemistry of Pre-Cambrian oceans and crust.

In ancient alluvial rocks we can often tell the approximate channel pattern, bedform types, the nature of the interfluvial area and its soils. We are thus moving slowly towards an understanding of palaeofluvial controls by an estimation of the tectonic background, climate and the river regime.

The morphology of modern deserts is well established, but in ancient desert deposits it is difficult to recognize and assess the effects of short term climatic changes such as are known to have happened in the Pleistocene. It is not easy, for example, to decide whether an alternation of sheet flood deposits and aeolian dune beds is due to occasional cloudbursts or to longer-term climatic changes.

The importance of ancient lakes has been underestimated by geologists. Because of their association with other sediments such as evaporites or turbidites or with other environments such as deltaic or fluvial, consideration of lacustrine environments as such is usually subordinated to the particular facies or environment. Geologists seem reluctant to use the word 'lake' preferring the terms 'inland sabkha', 'coal-measure swamp', 'non-marine basin', 'interfluvial area'. However, interest in modern lakes is growing because many are close to centres of population; they constitute easily accessible natural laboratories; they often highlight problems of pollution; they have unique chemical systems and facies. Interest in ancient lakes will also grow as their value as indicators of past climates and tectonics is appreciated.

In deltas we have probably advanced further than in other environments towards identifying particular types from which we can infer the kind of source area, the climate, local tectonics and the basinal current patterns. In particular we can predict sand-body trend and shape though further studies are needed in facies patterns within tidal and wave-deltas.

Shorelines in temperate climates have a long history of study by physical geographers and engineers, and consequently clastic coasts have been better known than the less hospitable semi-arid ones. Nevertheless very little of this knowledge has been applied to ancient rocks. This reflects the divergent approaches of geographer and geologist in the past. Until recently many stratigraphers required a conglomerate before they acknowledged a transgression and the myth that a major transgression must have a conglomeratic base has taken long to die. Geographers on the other hand seldom added time to their models, showed a vertical sequence, or considered what could or could not be preserved during transgression, regression or lateral migration of sub-environments. Thus few ancient examples are known. This dearth of rigorously described ancient beach/barrier shorelines is probably due to their lack of recognition so far. In the semi-arid environment a vigorous stimulus has been given by the work in the Persian Gulf, but there has been a danger that this one area might become the only model.

In the shallow marine environment the level of understanding differs considerably between carbonates and clastics probably reflecting the more equable climate in which modern carbonates can be studied. With carbonates there has been a tradition of matching the modern with the ancient, using petrographically based microfacies, since the 1930s. In clastics the idea that modern studies aid understanding of the ancient has been seriously applied only in the last 15 years. The result is that our sedimentary facies models are incomplete. We need many more individual studies in modern seas and case studies of the ancient before we can claim, as for deltas, that we have a comprehensive framework of process-based models into which we can fit the ancient facies patterns. Even when this has been achieved, non-actualistic models will have to be generated since modern shallow seas are so unlike many of the past.

Although research on pelagic sediments goes back 100 years, some of the most exciting work in the oceans in recent years has been the discovery of turbidity currents and bottom ocean currents. We still need detailed descriptions of the morphology of modern submarine fans, of the currents which transport and deposit sediment and of the sequences to be found. Many classical flysch areas need to be re-interpreted in terms of fan/basin-plain models. Very large abyssal cones are particularly difficult to identify in the ancient record. We need more detailed accounts of individual examples, less generalized than they have been in the past, with facies tied closely to process and palaeo-flow. Ancient submarine fans should be linked with their source area to determine the type of shoreline or shallow marine environment into which they passed. In this way the sedimentological, tectonic, climatic or eustatic factors which governed the type of fan or basin-plain can be differentiated.

Knowledge of pelagic sediments has probably grown faster in the last 10 years than that of any other sediment group because of the Deep Sea Drilling Project. Synthesis at the moment is very limited since the amount of data is enormous.

Although progress is being made in the comparison of modern with ancient pelagic sediments, the difficulty is that the particular type of sedimentary facies is so influenced by biological evolution and the nature of oceanic circulation and chemistry at various times in the past. Thus the gathering of ancient facies models has started only in the last year or two.

The study of glacial facies and facies models is at an early phase. Terrestrial Pleistocene glacial deposits have been well described for the purposes of correlation and unravelling the geological history of an area but comparatively little attention has been paid to sedimentary structures and facies relationships. There has also been general ignorance of glacial marine deposits until, in the past five years, studies in Antarctica and the Gulf of Alaska have shown how varied were the nature and duration of glaciation. Studies of ancient glacial deposits have concentrated on the question of a glacial versus non-glacial origin instead of developing detailed and coherent sedimentary models.

There is need for investigation of late Cenozoic glacial facies both in marine and terrestrial environments and in various tectonic settings, in combination with detailed correlation and chronology. The resulting models, when compared with studies of older glaciations, will allow the geologist to distinguish depositional, tectonic, temporal, local geographical and worldwide climatic controls of glacial sedimentation. An understanding of many ancient glaciations will be needed to determine satisfactorily the causes of glaciation, and possibly predict, or even influence, glacial trends in the future.

In the field of tectonics, understanding of present global plate movements has helped enormously in relating present sedimentary facies to particular tectonic regimes. Nevertheless it has also shown that there is no unique match of facies to tectonic environment. For example 'flysch' or 'flysch-like' facies can be formed in a variety of tectonic situations, ranging from Atlantic-type margins, to trenches, back-arc basins and small ocean basins situated on continental crust and even in lakes. In addition the global tectonic models help more to understand the tectonics of oceans than to understand the tectonics of continents. Yet most sediments are deposited on continents or at their margins. Also, sedimentologists are normally concerned with rather smaller basins than those on which the concept of plate tectonics is based. Consequently we have far to go in the matching of sedimentary facies to tectonics, though now within a firmer framework than we had in the past.

The study of several facies has been relatively neglected. Since sandstones and carbonates are well exposed, easy to study and are obvious reservoir rocks, they have received most attention. Mudstones on the other hand, have never received attention in proportion to their volume. As techniques are developed for their study and as their importance not only as petroleum source rocks but as indicators of where petroleum-bearing sedimentary basins lie, so future progress will be made in their study. Chemical rocks generally, especially ironstones, phosporites and evaporites have seldom been studied sedimentologically. The start that has been made in the past few years indicates what can be done in the future. Diagenetic studies have generally been made either as an end in themselves or as a necessary task before unravelling the original composition of the sediment. In future, diagenesis will be used more as an indicator of the depositional environment and as a tool for understanding overburden pressures and temperatures, ground water movement and the stratigraphical and structural history of an area after deposition.

In the industrial application of sedimentology the need is for integrated studies by engineers, seismologists and sedimentologists. The problem of the stacking of facies bodies in time and space has scarcely been tackled. However, the increasing use of cores, especially orientated cores, the use of seismic and petrophysical data through acoustic impedance contrast studies enables 3-dimensional models to be developed to a degree of sophistication unthinkable a few years ago and seldom possible by outcrop geology.

One final word of caution. The present is not a master key to all past environments although it may open the door to a few. The majority of past environments differ in some respect from modern environments. We must therefore be prepared, and have the courage, to develop non-actualistic models unlike any that exist today.

FURTHER READING

Ginsburg R.N. (Ed.) (1973) *Evolving Concepts in Sedimentology*. pp. 191. Johns Hopkins University Press, Baltimore.

References

The section(s) in which each reference appears is given on a separate line below the reference.

AALTO K.R. (1971) Glacial marine sedimentation and stratigraphy of the Toby Conglomerate (Upper Proterozoic), Southeastern British Columbia, Northwestern Idaho and Northeastern Washington. *Can. J. Earth Sci.*, **8**, 753–787.
13.1, 13.5.4

ABBATE E., BORTOLOTTI V. & PASSERINI P. (1970) Olistostromes and olistoliths. *Sedim. Geol.*, **4**, 521–557.
12.2.3.

ABBATE E., BORTOLOTTI V. & PASSERINI P. (1972) Studies on mafic and ultramafic rocks: 2 – Paleogeographic and tectonic considerations on the ultramafic belts in the Mediterranean area. *Boll. Soc. geol. ital.*, **91**, 239–282.
11.4.1, 11.4.2.

ACKERS P. & CHARLTON F.G. (1971) The slope and resistance of small meandering channels. *Inst. Civ. Engrs. Proc.*, Suppl. **15**, Paper 7362S, 349–370.
3.10.1, Fig. 3.53.

AGASSIZ L. (1869) Report on the deep-sea dredgings in the Gulf Stream during the third cruise of the U.S. Steamer *Bibb. Bull. Mus. Comp. Zool. Harvard Coll.*, **1**, 363–386.
11.1.2.

AGASSIZ A. (1894) A reconnaisance of the Bahamas and of the elevated reefs of Cuba in the steam yacht *Wild Duck*, January to April, 1893. *Bull. Mus. Comp. Zool. Harvard Coll.*, **26**, 1–203.
10.1.1.

AGASSIZ A. (1896). The elevated reefs of Florida. *Bull. Mus. Comp. Zool. Harvard Coll.*, **28**, 29–62.
10.1.1.

AGER D.V. (1963) *Principles of Palaeoecology*, pp. 371. McGraw-Hill.
9.1.1, 9.8.

ALLEN J.R.L. (1963) The classification of cross-stratified units, with notes on their origin. *Sedimentology*, **2**, 93–114.
3.9.3.

ALLEN J.R.L. (1964) Studies in fluviatile sedimentation: Six cyclothems from the Lower Old Red Sandstone, Anglo-Welsh Basin. *Sedimentology*, **3**, 163–198.
2.2.2, 3.1, 3.9.2, 3.9.4, Fig. 3.24, 3.49.

ALLEN J.R.L. (1965a) Fining upwards cycles in alluvial successions. *Geol. J.*, **4**, 229–246.
2.2.2, 3.9.1, 3.9.4, Fig. 3.47.

ALLEN J.R.L. (1965b) The sedimentation and palaeogeography of the Old Red Sandstone of Anglesey, North Wales. *Proc. Yorks. geol. Soc.*, **35**, 139–185.
3.9.3, 3.9.4, 3.10.3, Fig. 3.43, 3.54.

ALLEN J.R.L. (1965c) A review of the origin and characteristics of Recent Alluvial Sediments. *Sedimentology*, **5**, 89–191.
3.5, 3.9.4, Fig. 3.30, 3.46.

ALLEN J.R.L. (1965d) Late Quaternary Niger delta, and adjacent areas: sedimentary environments and lithofacies. *Bull. Am. Ass. Petrol. Geol.*, **49**, 547–600.
6.2, 6.3.1, 6.5.1, 6.5.2, Fig. 6.15, 6.29.

ALLEN J.R.L. (1965e) Coastal geomorphology of eastern Nigeria: beach ridge barrier islands and vegetated tidal flats. *Geol. Mijnb.*, **44**, 1–21.
7.2, 7.2.2, 7.2.5.

ALLEN J.R.L. (1966) On bedforms and palaeocurrents. *Sedimentology*, **6**, 153–190.
3.3.2.

ALLEN J.R.L. (1968) *Current Ripples*, pp. 433. North-Holland, Amsterdam.
1.1, 3.4.1.

ALLEN J.R.L. (1970a) Studies in fluviatile sedimentation: A comparison of fining-upwards cyclothems with special reference to coarse-member composition and interpretation. *J. sedim. Petrol.*, **40**, 298–323.
3.5.2, 3.9.4.

ALLEN J.R.L. (1970b) A quantitative model of grain size and sedimentary structures in lateral deposits. *Geol. J.*, **7**, 129–146.
3.5.2, 3.9.4, Fig. 3.24.

ALLEN J.R.L. (1970c) *Physical Processes of Sedimentation*, pp. 248. Allen & Unwin, London.
1.1.

ALLEN J.R.L. (1971a) Instantaneous sediment deposition rates deduced from climbing-ripple cross-lamination. *J. geol. Soc. Lond.*, **127**, 553–561.
3.9.3.

ALLEN J.R.L. (1971b) Mixing at turbidity current heads, and its geological implications. *J. sedim. Petrol.*, **41**, 97–113.
12.2.4.

ALLEN J.R.L. (1973) Features of cross-stratified units due to random and other changes in bed forms. *Sedimentology*, **20**, 189–202.
9.10.1.

ALLEN J.R.L. (1974a) Studies in fluviatile sedimentation: implications of pedogenic carbonate units, Lower Old Red Sandstone, Anglo-Welsh outcrop. *Geol. J.*, **9**, 181–208.
3.9.2, Fig. 3.42.

ALLEN J.R.L. (1974b) Studies in fluviatile sedimentation: lateral variation in some fining upwards cyclothems from the Red Marls, Pembrokeshire. *Geol. J.*, **9**, 1–16.
3.9.4.

ALLEN J.R.L. & COLLINSON J.D. (1974) The superimposition and classification of dunes by unidirectional aqueous flows. *Sedim.*

Geol., **12,** 169–178.
3.4.2.

Allen J.R.L. & Friend P.F. (1968) Deposition of the Catskill Facies, Appalachian region: with notes on some other Old Red Sandstone Basins. In: *Late Paleozoic and Mesozoic Continental Sedimentation, northeastern North America* (Ed. by G. de V. Klein), pp. 21–74, *Spec. Paper Geol. Soc. Am.*, **106.**
3.9.3.

Amiel A.J. & Friedman G.M. (1971) Continental Sabkha in Arava Valley between Dead Sea and Red Sea. *Bull. Am. Ass. Petrol. Geol.*, **55,** 581–592.
5.2.6.

Andel Tj. H. Van (1964) Recent marine sediments of Gulf of California. In: *Marine Geology of the Gulf of California* (Ed. by Tj. H. van Andel and G.G. Shor, Jr.), pp. 216–310. *Mem. Am. Ass. Petrol. Geol.*, **3.**
11.3.5.

Andel Tj. H. Van (1967) The Orinoco delta. *J. sedim. Petrol.*, **37,** 297–310.
6.2.

Andel Tj. H. Van (1975) Mesozoic/Cenozoic compensation depth and the global distribution of calcareous sediments. *Earth Planet. Sci. Letts.*, **26,** 187–194.
11.3.1, Fig. 11.6.

Andel Tj. H. Van & Bowin C.O. (1968) Mid-Atlantic Ridge between 22° and 23° north latitude and the tectonics of mid-ocean rises. *J. geophys. Res.*, **73,** 1279–1298.
10.3.2.

Andel Tj. H. Van & Curray J.R. (1960) Regional aspects of modern sedimentation in northern Gulf of Mexico and similar basins, and palaeogeographic significance. In: *Recent Sediments, northwest Gulf of Mexico* (Ed. by F.P. Shepard, F.B. Phleger and Tj. H. van Andel), pp. 345–364. Am. Ass. Petrol. Geol., Tulsa.
6.2, Fig. 6.24.

Andel Tj. H. Van & Komar P.D. (1969) Ponded sediments of the Mid-Atlantic Ridge between 22° and 23° north latitude. *Bull. geol. Soc. Am.*, **80,** 1163–1190.
11.3.2, 12.2.4, 12.2.5, 12.4.6, 14.3.1, Fig. 14.12.

Andel Tj. H. Van, Heath G.R. & Moore T.C. (1976) Cenozoic history and palaeo-oceanography of the central equatorial Pacific Ocean. *Mem. geol. Soc. Am.*, **143,** pp. 134.
11.3.2, 14.3.1.

Andel Tj. H. Van, Von Herzen R.P. & Phillips J.D. (1971) The Vema Fracture zone and the tectonics of transverse shear zones in oceanic crustal plates. *Mar. geophys. Res.*, **1,** 261–283.
14.3.1.

Andel Tj. H. Van, Rea D.K., Von Herzen R.P. & Hoskins H. (1973) Ascension Fracture Zone, Ascension Island, and the Mid-Atlantic Ridge. *Bull. geol. Soc. Am.*, **84,** 1527–1546.
11.3.2, 14.3.1.

Andel Tj. H. Van, Heath G.R. *et al.* (1973) *Initial Reports of the Deep Sea Drilling Project*, **16,** pp. 949. U.S. Government Printing Office, Washington.
11.3.3.

Anderson E.M. (1951). *The Dynamics of Faulting and Dyke Formation with Applications to Britain*, pp. 206. 2nd ed. Oliver & Boyd, Edinburgh.
14.3.

Anderson F.W. & Bazley R.A.B. (1971) The Purbeck beds of the Weald (England). *Bull. geol. Surv. Gt. Br.*, **34,** 1–25.
8.7.2.

Anderson J.B. (1972) Nearshore glacial-marine deposition from modern sediments of the Weddell Sea. *Nature*, **240,** 189–192.
13.3.7.

Anderson J.B., Clark H.C. & Weaver F.M. (1977) Sediments and sediment processes on high latitude continental shelves. *Proc. Offshore Technology Conference*, pp. 91–94. Houston, 1977.
13.4.2, 13.5.1, 13.6.

Anderson R.V.V. (1933) The diatomaceous and fish-bearing Beida Stage of Algeria. *J. Geol.*, **41,** 673–698.
11.4.3.

Anderton R. (1976) Tidal shelf sedimentation: an example from the Scottish Dalradian. *Sedimentology*, **23,** 429–458.
9.1.1, 9.8, 9.10.1, 9.10.2, 9.10.3, 9.10.4, 9.11.2, Fig. 9.31, 9.33.

Andrews P.B. (1970) Facies and genesis of a hurricane washover fan, St. Joseph Island, Central Texas Coast. *Rep. Invest. Bur. econ. Geol.*, **67,** pp. 147. Austin, Texas.
7.2.5, Fig. 7.27.

Andrews J.E., Packham G. *et al.* (1975) *Initial Reports of the Deep Sea Drilling Project*, **30,** pp. 753. U.S. Government Printing Office, Washington.
11.3.3.

Andri E. & Fanucci F. (1975) La resedimentazione dei Calcari Calpionelle liguri. *Boll. Soc. geol. ital.*, **94,** 915–925.
11.4.2.

Angelucci A., De Rosa E., Fierro G. *et al.* (1967) Sedimentological characteristics of some Italian turbidites. *Geol. Romana*, **6,** 345–420.
12.8.1.

Anketell J.M., Cegla J. & Dzulynski S. (1970) On the deformational structures in systems with reversed density gradients. *Ann. Soc. Geol. Pologne*, **40,** 3–30.
12.2.11.

Arbey F. (1968) Structures et dépôts glaciaires dans l'Ordovicien terminal des chaînes d'Ougarta (Sahara algérien). *C.r. hebd. Séanc. Acad. Sci., Paris*, **266,** 76–78.
13.4.1.

Arbey F. (1971) Glacio-tectonique et phénomènes glaciaires dans les dépôts siluro-ordoviciens des Monts d'Ougarta (Sahara algérien), *C.r. hebd. Séanc. Acad. Sci., Paris*, **273,** 854–857.
13.4.1.

Arcyana (1975) Transform fault and rift valley from bathyscaphe and diving saucer. *Science*, **190,** 108–116.
14.3.1, Fig. 14.11.

Armstrong-Price (1955) Correlation of shoreline types with offshore bottom conditions. *A. & M. College of Texas, Dept. of Oceanog.*, Proj. **63.**
7.1.

Arndorfer D.J. (1973) Discharge patterns in two crevasses in the Mississippi River delta. *Mar. Geol.*, **15,** 269–287.
6.5.1.

Arrhenius G. (1952) Sediment cores from the East Pacific. *Repts. Swedish Deep Sea Exped.* (1947–1948), **5,** 228 pp.
11.1.1, 11.3.4.

Arrhenius G. (1963) Pelagic sediments. In: *The Sea* (Ed. by M.N. Hill), Vol. 3, pp. 655–727. Interscience, New York.
11.2, 11.3.4.

Arthurton R.S. (1973) Experimentally produced halite compared with Triassic layered halite-rock from Cheshire, England. *Sedimentology*, **20,** 145–160.
5.3.4, 8.3.2.

Arthurton R.S. & Hemingway J.E. (1972) The St. Bees Evaporites – a

carbonate evaporite formation of upper Permian age in west Cumberland, England. *Proc. Yorks. geol. Soc.*, **38,** 565–592.
8.4.1.

ARX W.S. VON (1962) *An Introduction to Physical Oceanography*, pp. 422. Addison-Wesley Pub. Co. Inc.
9.4.1.

ASHLEY G.M. (1975). Rhythmic sedimentation in glacial lake Hitchcock, Massachusetts–Connecticut. In: *Glaciofluvial and Glaciolacustrine Sedimentation* (Ed. by A.V. Jopling & B.C. McDonald), pp. 304–320. *Spec. Publ. Soc. econ. Paleont. Miner.*, **23,** Tulsa.
13.3.4.

ASQUITH D.O. (1970) Depositional topography and major marine environments, late Cretaceous, Wyoming. *Bull. Am. Ass. Petrol Geol.*, **54,** 1184–1224.
6.7.2.

ASQUITH D.O. (1974) Sedimentary models, cycles and deltas, Upper Cretaceous, Wyoming. *Bull. Am. Ass. Petrol. Geol.*, **58,** 2274–2283.
6.7.2.

AUBOUIN J. (1959) Place des Hellénides parmi les édifices structuraux de la Méditerranée orientale (2e thèse, Paris, 1958). *Ann. Géol. Pays Helléniques*, **10,** 487–525.
14.2.2.

AUBOUIN J. (1965) Geosynclines: *Developments in Geotectonics*, pp. 335. Elsevier, Amsterdam.
14.2.2, 14.2.4, 14.4.1, 14.5.1, Fig. 14.3, 14.35.

AUDLEY-CHARLES M.G. (1965) A geochemical study of Cretaceous ferromanganiferous sedimentary rocks from Timor. *Geochim. Cosmochim. Acta*, **29,** 1153–1173.
11.4.2.

AUDLEY-CHARLES M.G. (1972) Cretaceous deep-sea manganese nodules on Timor: implications for tectonics and olistostrome development. *Nature, Phys. Sci.*, **240,** 137–139.
11.4.2.

AUMENTO F. (1969) The Mid-Atlantic Ridge near 45°N; V: Fission-track and ferromanganese chronology. *Can. J. Earth Sci.* **6,** 1431–1440.
11.3.2.

AUMENTO F., LAWRENCE D.E. & PLANT A.G. (1968) The ferromanganese pavement on San Pablo Seamount. *Pap. Geol. Surv. Canada*, 68–32, pp. 30.
11.3.3.

AXELSSON V. (1967) The Laitaure Delta, a study of deltaic morphology and processes. *Geogr. Annalr.*, **49A,** 1–127.
4.2.3.

BAARS D.L. (1961) Permian blanket sandstones of Colorado Plateau. In: *Geometry of Sandstone Bodies* (Ed. by J.A. Peterson and J.C. Osmond), pp. 179–207. Am. Ass. Petrol. Geol., Tulsa.
5.3.2, 5.3.6.

BADER R.G. *et al.* (1970) *Initial Reports of the Deep Sea Drilling Project*, **4,** pp. 753. U.S. Government Printing Office, Washington.
11.4.2.

BAGNOLD R.A. (1941) *The Physics of Blown Sand and Desert Dunes*, pp. 265. Methuen, London.
5.2.4, Fig. 5.8.

BAGNOLD R.A. (1954). Experiments on a gravity-free disperson of large solid spheres in a Newtonian fluid under shear. *Proc. Roy. Soc. London, ser. A*, **225,** 49–63.
12.2.6.

BAGNOLD R.A. (1960) Some aspects of river meanders. *Prof. Pap. U.S. geol. Surv.*, **282-E,** pp. 10.
3.5.2.

BAGNOLD R.A. (1962) Auto-suspension of transported sediment; turbidity currents. *Proc. Roy. Soc. London, ser. A.*, **265,** 315–319.
12.2.4.

BAILEY E.B. & MCCALLIEN W.J. (1960) Some aspects of the Steinmann Trinity: mainly chemical. *Quart. J. geol. Soc. Lond.*, **116,** 365–395.
14.2.5.

BAILEY E.H., BLAKE M.C. & JONES D.L. (1970) On-land Mesozoic oceanic crust in California Coast Ranges. *Prof. Pap. U.S. geol. Surv.*, **700-C,** 70–81.
11.4.1, 11.4.2, Fig. 11.24.

BAILEY E.H., IRWIN W.P. & JONES D.L. (1964) Franciscan and related rocks, and their significance in the geology of Western California. *Bull. Calif. Div. Mines Geol.*, **183,** pp. 177.
11.4.2.

BALDWIN C.T. (1977) The stratigraphy and facies associations of trace fossils in some Cambro-Ordovician rocks of north-west Spain. In: *Trace Fossils* (Ed. by T.P. Crimes & J.C. Harper), pp. 9–40. *Geol. J. Spec. Issue.* **9.**
9.8.

BALDWIN C.T. & JOHNSON H.D. (1977) Sandstone mounds and associated facies sequences in some late Precambrian and Cambro-Ordovician inshore tidal flat lagoonal deposits. *Sedimentology*, **24,** 801–818.
9.11.1.

BALL M.M. (1967) Carbonate sand bodies of Florida and the Bahamas. *J. sedim. Petrol.*, **37,** 556–591.
10.2.2, Fig. 10.10.

BAŁUK W. (1971) Lower Tortonian chitons from the Korytnica Clays, southern slopes of the Holy Cross Mountains. *Acta geol. Pol.*, **21,** 449–471 (in Polish with English abstract and résumé).
10.3.2.

BANDEL K. (1974) Deep-water limestones from the Devonian-Carboniferous of the Carnic Alps, Austria. In: *Pelagic Sediments: on Land and under the Sea* (Ed. by K.J. Hsü and H.C. Jenkyns). *Spec. Publ. int. Ass. Sediment.*, **1,** 93–115.
11.4.4.

BANERJEE I. (1966) Turbidites in a glacial sequence: a study from the Talchir Formation, Raniganj Coalfield, India. *J. Geol.*, **74,** 593–606.
13.5.2.

BANERJEE I. & MCDONALD B.C. (1975) Nature of esker sedimentation. In: *Glaciofluvial and Glaciolacustrine Sedimentation* (Ed. by A.V. Joplin & B.C. McDonald), pp. 132–154. *Spec. Publ. Soc. econ. Paleont. Miner.*, **23,** Tulsa.
13.3.4, 13.4.2.

BANHAM P.H. (1975) Glacitectonic structures: a general discussion with particular reference to the contorted drift of Norfolk. In: *Ice Ages: Ancient and Modern* (Ed. by A.E. Wright and F. Moseley). pp. 69–94. *Geol. J. Spec. Issue*, **6.**
13.4.1.

BANKS N.L. (1973a) Tide-dominated offshore sedimentation, Lower Cambrian, North Norway. *Sedimentology*, **20,** 213–228.
9.8, 9.10.2, 9.10.3.

BANKS N.L. (1973b) Innerelv Member: Late Precambrian marine shelf deposit, East Finnmark. *Norg. geol. Unders.*, **288,** 7–25.
9.1.1, 9.11.5, Fig. 9.53.

BANKS N.L. (1973c) Falling-stage features of a Precambrian braided stream: criteria for subaerial exposure. *Sedim. Geol.*, **10,**

147–154.
3.9.3.

Banks N.L. & Collinson J.D. (1974) Discussion of 'Some sedimentological aspects of planar cross-stratification in a sandy braided river'. *J. sedim. Petrol.*, **44**, 265–267.
3.4.3, 3.9.4.

Banks N.L., Edwards M.B., Geddes W.P., Hobday D.K. & Reading H.G. (1971) Late Precambrian and Cambro-Ordovician sedimentation in East Finnmark. In: *The Caledonian Geology of Northern Norway* (Ed. by D. Roberts and M. Gustavson), pp. 197–236. *Norg. geol. Unders.*, **269**.
3.9.3, 13.4.1.

Barnett D.M. & Holdsworth G. (1974) Origin, morphology, and chronology of sublacustrine moraines, Generator Lake, Baffin Island, Northwest Territories, Canada. *Can. J. Earth Sci.*, **11**, 380–408.
13.3.4.

Barrell J. (1912) Criteria for the recognition of ancient delta deposits. *Bull. geol. Soc. Am.*, **23**, 377–446.
6.2, Fig. 6.1.

Barrell J. (1914) The Upper Devonian delta of the Appalachian geosyncline. *Am. J. Sci.*, **37**, 225–253.
6.2.

Barrett P.J. (1975) Textural characteristics of Cenozoic preglacial and glacial sediments at Site 270, Ross Sea, Antarctica. In *Initial Reports of the Deep Sea Drilling Project*, **28** (D.E. Hayes, L.S. Frakes *et al.*), pp. 757–766. U.S. Government Printing Office, Washington.
13.3.7, 13.4.3, 13.5.1.

Barrett T.J. & Spooner E.T.C. (1977) Ophiolitic breccias associated with allochthonous oceanic crustal rocks in the East Ligurian Apennines, Italy – a comparison with observations from rifted oceanic ridges. *Earth Planet. Sci. Letts.*, **35**, 79–91.
11.4.2.

Barthel K.W. (1970) On the deposition of the Solnhofen lithographic limestone (Lower Tithonian, Bavaria, Germany). *Neues Jb. Geol. Paläontol. Abh.*, **135**, 1–18.
10.4.2.

Bartlett G.A. & Greggs R.G. (1970) The Mid-Atlantic Ridge near 45°00′ North; VIII: Carbonate lithification on oceanic ridges and seamounts. *Can. J. Earth Sci.*, **7**, 257–267.
11.3.2, 11.3.3.

Bates C.C. (1953) Rational theory of delta formation. *Bull. Am. Ass. Petrol. Geol.*, **37**, 2119–2162.
6.3.1, 6.5.2, Fig. 6.17.

Bathurst R.G.C. (1958) Diagenetic fabrics in some British Dinantian limestones. *Geol. J.*, **2**, 11–36.
10.1.1.

Bathurst R.G.C. (1959) Diagenesis in Mississippian calcilutites and pseudo-breccias. *J. sedim. Petrol.*, **29**, 365–376.
10.1.1.

Bathurst R.G.C. (1964) Diagenesis and palaeoecology: a survey. In: *Approaches to Paleoecology* (Ed. by J. Imbrie and N.D. Newell), pp. 319–344. Wiley, New York.
10.1.1.

Bathurst R.G.C. (1967) Sub-tidal gelatinous mat, sand stabilizer and food, Great Bahama Bank. *J. Geol.*, **75**, 736–738.
10.2.2.

Bathurst R.G.C. (1971) *Carbonate sediments and their diagenesis. Developments in Sedimentology*, **12**, pp. 620. Elsevier, Amsterdam.
1.1, 10.2.4, Fig. 10.20a, 10.22.

Bathurst R.G.C. (1975) *Carbonate sediments and their diagenesis:* second enlarged edition, pp. 658. Elsevier, Amsterdam.
10.1.1, 10.1.2, 10.2.2.

Baudrimont R. & Degiovanni C. (1974) Interprétation paléo-écologique des diatomites du Miocène supérieur de l'Algérie occidentale. *C.r. hebd. séanc. Acad. Sci. Paris*, **D279**, 1337–1340.
11.4.3.

Beadle L.C. (1974) *The Inland Waters of Tropical Africa*, pp. 365. Longman, London.
4.2.1, 4.2.2, 4.3.3, Fig. 4.1, 4.2, 4.3.

Beall A.O. Jr. (1968) Sedimentary processes operative along the western Louisiana shoreline. *J. sedim. Petrol.*, **38**, 869–877.
7.4.

Beall A.O. & Fischer A.G. (1969) Sedimentology. In: *Initial Reports of the Deep Sea Drilling Project*, **I** (M. Ewing, J.L. Worzel *et al.*), pp. 521–593. U.S. Government Printing Office, Washington.
12.3.1.

Beardsley R.C. & Butman B. (1974) Circulation on the New England Continental Shelf: Response to strong winter storms. *Geophys. Res. Lett.*, **1**, 181–184.
9.6.3.

Beatty C.B. (1974) Debris flows, alluvial fans and a revitalised catastrophism. *Zeit. Geomorph. Suppl. Bd.* **21**, 39–51.
3.2.4.

Beaumont E. De (1845) *Leçons de Géologie Protique*, pp. 223–252. Paris
7.2.7.

Beaumont P. (1972) Alluvial fans along the foothills of the Elburz Mountains, Iran. *Palaeogeogr. Palaeoclim. Palaeoecol.*, **12**, 251–274.
3.2.2.

Beckinsale R.P. (1969) River Regimes. In: *Water, Earth and Man* (Ed. by R.J. Chorley), pp. 455–471. Methuen, London.
3.10.1.

Beckmann J.P. (1953). Die Foraminiferen der Oceanic Formation (Eocaen-Oligocaen) von Barbados, Kl. Antillen. *Eclog. Geol. Helv.*, **46**, 301–412.
11.4.2.

Bein A. & Weiler Y. (1976) The Cretaceous Talme Yafe Formation: a contour current shaped sedimentary prism of calcareous detritus at the continental margin of the Arabian Craton. *Sedimentology*, **23**, 511–532.
12.2.7.

Belderson R.H. & Stride A.H. (1966) Tidal current fashioning of a basal bed. *Mar. Geol.*, **4**, 237–257.
9.1.1, 9.5.1, Fig. 9.34.

Bell H.S. (1942) Density currents as agents for transporting sediment. *J. Geol.*, **50**, 512–547.
12.1.

Beloussov V.V. (1962) *Basic Problems in Geotectonics*, pp. 809. McGraw-Hill, London.
14.2.4.

Belt E.S. (1965) Stratigraphy and paleography of Mabou Group and related Middle Carboniferous facies, Nova Scotia, Canada. *Bull. geol. Soc. America*, **76**, 777–802.
4.3.4.

Belt E.S. (1975) Scottish Carboniferous cyclothem patterns and their palaeoenvironmental significance. In: *Deltas, Models for Exploration* (Ed. by M.L. Broussard), pp. 427–450. Houston Geological Society, Houston.
6.3.

Belt E.S., Freshney E.C. & Read W.A. (1967) Sedimentology of Carboniferous Cementstone facies, British Isles and eastern

Canada. *J. Geol.*, **75**, 711–721.
4.3.4.

Bemmelen R.W. Van (1949) *The Geology of Indonesia*, pp. 997. Nijhoff, The Hague.
14.2.2, 14.3.2.

Bemmelen R.W. Van (1954) *Mountain Building*, pp. 177. Nijhoff, The Hague.
14.2.2.

Bennetts K.R.W. & Pilkey O.H. (1976) Characteristics of three turbidites, Hispaniola-Caicos Basin. *Bull. geol. Soc. Am.*, **87**, 1291–1300.
12.4.1.

Benson W.E. & Sheridan R.E. *et al.* (1976). In the North Atlantic: Deep-sea drilling. *Geotimes*, **21**, 3, 23–26.
11.3.6.

Bentley A. (1970) *Sedimentation studies in the Lower Greensand. A Report on the Sedimentology of the Aptian and Lower Albian strata near Leighton Buzzard, Bedfordshire.* Unpubl. B.Sc. Thesis, University of Oxford.
Fig. 9.30.

Bentor Y.K. (1961) Some geochemical aspects of the Dead Sea and the question of its age. *Geochim. Cosmochim Acta*, **25**, 239–260.
4.2.1, 4.2.4.

Berg R.R. & Davies D.K. (1968) Origin of Lower Cretaceous Muddy Sandstone at Bell Creek Field, Montana. *Bull. Am. Ass. Petrol. Geol.*, **52**, 1888–1898.
7.3, 7.3.1.

Berg R.R. (1975) Depositional environment of Upper Cretaceous Sussex Sandstone, House Creek Field, Wyoming. *Bull. Am. Ass. Petrol. Geol.*, **59**, 2099–2110.
9.10.5, Fig. 9.43, 9.44.

Berger W.H. (1970) Planktonic Foraminifera: selective solution and the lysocline. *Mar. Geol.*, **8**, 111–138.
11.3.1, Fig. 11.1.

Berger W.H. (1971) Sedimentation of planktonic Foraminifera. *Mar. Geol.*, **11**, 325–358.
11.3.1, Fig. 11.1.

Berger W.H. (1974) Deep-sea sedimentation. In: *The Geology of Continental Margins* (Ed. by C.A. Burk and C.L. Drake), pp. 213–241. Springer-Verlag, New York.
11.2, 11.3.4, Tab. 11.2, 11.3.

Berger W.H. (1976) Biogenous deep-sea sediments: production, preservation and interpretation. In: *Chemical Oceanography* (Ed. by J.P. Riley and R. Chester), 2nd edn, Vol. 5, pp. 265–388. Academic Press, London.
11.3.1.

Berger W.H. & Johnson T.C. (1976) Deep-sea Carbonates: dissolution and mass movement on Ontong Java Plateau. *Science*, **192**, 785–787.
11.3.3.

Berger W.H. & Soutar A. (1970) Preservation of plankton shells in an anaerobic basin off California. *Bull. geol. Soc. Am.*, **81**, 275–282.
11.3.6.

Berger W.H. & Von Rad U. (1972) Cretaceous and Cenozoic sediments from the Atlantic Ocean. In: *Initial Reports of the Deep Sea Drilling Project*, **14** (D.E. Hayes, A.C. Pimm *et al.*), pp. 787–954. U.S. Government Printing Office, Washington.
11.3.5, Fig. 11.6.

Berger W.H. & Winterer E.L. (1974). Plate stratigraphy and the fluctuating carbonate line. In: *Pelagic Sediments: on Land and under the Sea* (Ed. by K.J. Hsü and H.C. Jenkyns), pp. 11–48. *Spec. Publs. int. Ass. Sediment.*, **1**.
11.3.1, 11.4.2, Fig. 11.1, 11.4, 11.5, 11.6.

Berggren W.A. & Hollister C.D. (1977) Plate tectonics and paleocirculation – commotion in the ocean. *Tectonophys.*, **38**, 11–48.
11.3.1.

Bernard H.A. (1965) A resumé of river delta types. *Bull. Am. Ass. Petrol. Geol.* (Abs.), **47**, 334–335.
6.3.

Bernard H.A. & LeBlanc R.J. (1965) Resumé of the Quaternary geology of the north-eastern Gulf of Mexico Province. In: *The Quaternary of the United States* (Ed. by H.E. Wright Jr. and D.G. Frey), pp. 137–185. Princeton Univ. Press. Princeton.
3.6.2.

Bernard H.A. & Major C.F. Jr. (1963) Recent meander belt deposits of the Brazos River: an alluvial 'sand' model. *Bull. Am. Ass. Petrol. Geol.*, **47**, 350.
3.1, 3.5.2, 3.9.4.

Bernard H.A., LeBlanc R.J. & Major C.F. Jr. (1962) Recent and Pleistocene geology of southeast Texas. *Geol. Gulf Coast and Central Texas and guidebook of excursion*, pp. 175–225. Houston Geol. Soc.
7.2, Fig. 7.30.

Bernard H.A., Major C.F. Jr. & Parrott B.S. (1959) The Galveston Barrier Island and environs – a model for predicting reservoir occurrence and trend. *Trans. Gulf-Cst. Ass. geol. Socs.*, **9**, 221–224. 221–224.
7.2.6.

Bernard H.A., Major C.F. Jr., Parrott B.S. & LeBlanc R.J. (1970) Recent sediments of southeast Texas. A field guide to the Brazos alluvial and deltaic plains and the Galveston Barrier Island Complex. *Guidebook* No. **11**, pp. 16. *Bur. econ. Geol.* Austin, Texas.
3.5.2.

Berner R.A. (1971) *Principles of Chemical Sedimentology*, pp. 240. McGraw-Hill, New York.
1.1, 10.1.1, 10.1.2.

Berner R.A. & Morse J.W. (1974) Dissolution kinetics of calcium carbonate in sea water. IV. Theory of calcite dissolution. *Am. J. Sci.*, **274**, 108–134.
10.1.2.

Berner R.A. & Wilde P. (1972) Dissolution kinetics of calcium carbonate in sea water. I. Saturation state parameters for kinetic calculations. *Am. J. Sci.*, **272**, 826–839.
10.1.2.

Bernoulli D. (1964) Sur Geologie des Monte Generoso (Lombardische Alpen). *Beitr. geol. Karte Schweiz*, **118**, 134.
11.4.4.

Bernoulli D. (1972) North Atlantic and Mediterranean Mesozoic facies; a comparison. In: *Initial Reports of the Deep Sea Drilling Project*, **9** (C.D. Hollister, J.I. Ewing *et al.*), pp. 801–871. U.S. Government Printing Office, Washington.
11.4.4, Fig. 11.20.

Bernoulli D. & Jenkyns H.C. (1970) A Jurassic basin: the Glasenbach Gorge, Salzburg, Austria. *Verh. geol. Bundesanst. Wien*, **1970**, 504–531.
11.4.4, Fig. 11.33, 11.37, 11.38.

Bernoulli D. & Jenkyns H.C. (1974) Alpine, Mediterranean, and Central Atlantic Mesozoic facies in relation to the early evolution of the Tethys. In: *Modern and Ancient Geosynclinal Sedimentation* (Ed. by R.H. Dott and R.H. Shaver), pp. 129–160. *Spec. Publ.*

Soc. econ. Paleont. Miner., **19,** Tulsa.
10.2.5, 11.4.2, 11.4.4, 14.4.1.

BERNOULLI D. & WAGNER C. (1971) Subaerial diagenesis and fossil caliche deposits in the Calcare Massiccio Formation (Lower Jurassic, Central Apennines, Italy). *Neues Jahrb. Geologie und Paläontologie, Abh.*, **138,** 135–149.
10.2.5.

BERRISFORD C.D. (1969) Biology and zoogeography of the Vema Seamount: a report on the first biological collection made on the summit. *Trans. R. Soc. S. Afr.*, **38,** 387–398.
11.3.3.

BERSIER A. (1959) Séquence détritiques et divagations fluviales. *Ecolog. geol. Helv.*, **51,** 854–893.
3.1, 14.2.5.

BERRYHILL H.L. JR., DICKINSON K.A. & HOLMES C.W. (1969) Criteria for recognising ancient barrier coastlines (abstr.). *Bull. Am. Ass. Petrol. Geol.*, **53,** 706–707.
7.2.

BERTRAND M. (1897) Structure des Alpes françaises et récurrence de certains faciès sédimentaires. *VIe Int. geol. Congr.* (*Zürich*), 161–177.
14.2.5.

BEUF S., BIJU-DUVAL B., DE CHARPAL O., ROGNON P., GARIEL O. & BENNACEF A. (1971) *Les Grès du Paléozoique Inferieur au Sahara*, pp. 464. Editions Technip, Paris.
3.9.3, 13.4.1, 13.5.3.

BIGARELLA J.J. (1972) Eolian environments: their characteristics, recognition and importance. In: *Recognition of Ancient Sedimentary Environments* (Ed. by J.K. Rigby and W.K. Hamblin), pp. 12–62. *Spec. Publ. Soc. econ. Paleont. Miner.*, **16,** Tulsa.
5.1, 5.2.3, 5.2.4.

BIGARELLA J.J. (1973) Paleocurrents and the problem of continental drift. *Geol. Rdsch.*, **62,** 447–477.
13.5.3.

BIGARELLA J.J., BECKER R.D. & DUARTE G.M. (1969) Coastal dune structures from Paraná (Brazil). *Mar. Geol.*, **7,** 5–55.
5.2.4, 7.2.2, Fig. 5.7.

BIGARELLA J.J., SALAMUNI R. & FUCK R. (1967) Striated surfaces and related features developed by the Gondwana Ice Sheets (State of Paraná, Brazil). *Palaeogeogr. Palaeoclim. Palaeoecol.*, **3,** 265–276.
13.4.1.

BIGNELL R.D. (1975) Timing, distribution and origin of submarine mineralisation in the Red Sea. *Trans. Inst. Mining Metall.*, **84,** B1–B6.
14.3.1.

BIJU-DUVAL B., DEYNOUX M. & ROGNON P. (1974) Essai d'interprétation des 'fractures en gradins' observées dans les formations glaciaires Précambriennes et Ordoviciennes du Sahara. *Rev. Géogr. phys. géol. dyn.*, **16,** 503–512.
13.4.1.

BIJU-DUVAL B., LETOUZEY J., MONTADERT L., COURRIER P., MUGNIOT J.F. & SANCHO J. (1974) Geology of the Mediterranean Sea Basins. In: *The Geology of Continental Margins* (Ed. by C.A. Burk and C.L. Drake), pp. 695–731. Springer-Verlag, New York.
14.3.4.

BIRD E.C.F. (1973) Australian coastal barriers. In: *Barrier Islands* (Ed. by M.L. Schwartz), pp. 410–426. Benchmark Papers in Geology Series, **9.** Dowden, Hutchinson and Ross, Stroudsburg.
7.2.

BIRD J.M. & DEWEY J.F. (1970) Lithosphere plate-continental margin tectonics and the evolution of the Appalachian Orogen. *Bull. geol. Soc. Am.*, **81,** 1031–1060.
14.3.2, 14.4.1.

BIRKELAND P.W. (1974) *Pedology, Weathering and Geomorphological Research*, pp. 285. Oxford University Press, New York.
3.6.2.

BIRKENMAYER K., GASIOROWSKI S.M. & WIESER T. (1960) Fragments of exotic rocks in the pelagic deposits of the Bathonian of the Niedzica Series (Pieniny Klippen-Belt, Carpathians). *Ann. Soc. géol. Pol.*, **30,** 29–57.
11.4.4.

BISCHOFF J.L. & ROSENBAUER R.J. (1977) Recent metalliferous sediment in North Pacific manganese nodule area. *Earth Planet. Sci. Letts.*, **33,** 379–388.
11.3.4.

BISHOP W.F. (1968) Petrology of Upper Smackover limestone in north Haynesville field, Claiborne parish, Louisiana. *Bull. Am. Ass. Petrol. Geol.*, **52,** 92–128.
10.2.5, Fig. 10.29a.

BITTERLI P. (1965). Bituminous intercalations in the Cretaceous of the Breggia River, S. Switzerland. *Bull. Ver. Schweiz. Petrol.-Geol. Ing.*, **31,** 179–185.
11.4.4.

BJØRLYKKE K. (1967) The Eocambrian 'Reusch moraine' at Bigganjargga and the geology around Varangerfjord, Northern Norway. *Norg. geol. Unders.*, **251,** 18–44.
13.5.4.

BJØRLYKKE K. (1974) Depositional history and geochemical composition of Lower Palaeozoic epicontinental sediments from the Oslo Region. *Norg. geol. Unders.*, **305,** 81 pp.
11.4.4.

BJØRLYKKE K., ELVSBORG A. & HØY T. (1976) Late Precambrian sedimentation in the central sparagmite basin of south Norway. *Norsk. geol. Tidsskr.*, **56,** 233–290.
14.4.1.

BLACK M. (1933a) The algal sediments of Andros Island, Bahamas. *Phil. Trans. R. Soc.*, London, ser. B., **222,** 165–192.
10.1.1.

BLACK M. (1933b) The precipitation of calcium carbonate on the Great Bahama Bank. *Geol. Mag.*, **70,** 455–466.
10.1.1.

BLACK M., HILL M.N., LAUGHTON A.S. & MATTHEWS D.H. (1964) Three non-magnetic seamounts off the Iberian coast. *Quart. J. geol. Soc. London.*, **120,** 477–517.
11.3.6.

BLACKWELDER E. (1928) Mudflow as a geological agent in semi-arid mountains. *Bull. geol. Soc. Am.*, **39,** 465–480.
3.2.4.

BLAIR D.G. (1975) Structural styles in North Sea oil and gas fields. In: *Petroleum and the Continental Shelf of North-west Europe*, Vol. I *Geology* (Ed. by A.W. Woodland), pp. 327–338. Applied Science Publishers, Barking.
14.4.1, Fig. 14.29.

BLAKE M.C. JR. & JONES D.L. (1974) Origin of Franciscan melanges in northern California. In: *Modern and Ancient Geosynclinal Sedimentation* (Ed. by R.H. Dott Jr. and R.H. Shaver), *Spec. Publ. Soc. econ. Paleont. Miner.*, **19,** Tulsa.
11.4.2, Fig. 11.25.

BLAKE M.C., JONES D.L. & LANDIS C.A. (1974) Active continental margins: comparisons between California and New Zealand. In: *The Geology of Continental Margins* (Ed. by C.A. Burk and C.L. Drake), pp. 853–872. Springer-Verlag, Berlin.
14.4.2.

BLATT H., MIDDLETON G.V. & MURRAY R.C. (1972) *Origin of Sedimentary Rocks,* pp. 634. Prentice-Hall, New Jersey.
1.1, 2.1.2, 9.8.

BLISSENBACH E. (1954) Geology of alluvial fans in semi-arid regions. *Bull. geol. Soc. Am.,* **65,** 175–190.
3.2.4.

BLONDEAU K.M. & LOWE D.R. (1972) Upper Precambrian glacial deposits of the Mount Rogers Formation, Central Appalachians, U.S.A. *24th Int. geol. Congr.,* **1,** 325–332.
13.5.4.

BLUCK B.J. (1964) Sedimentation of an alluvial fan in Southern Nevada. *J. sedim. Petrol.,* **34,** 395–400.
3.2.4, 3.2.6, 3.8.1.

BLUCK B.J. (1965) The sedimentary history of some Triassic conglomerates in the Vale of Glamorgan, South Wales. *Sedimentology,* **4,** 225–245.
3.8.1, Fig. 3.37.

BLUCK B.J. (1967) Deposition of some Upper Old Red Sandstone conglomerates in the Clyde area: A study in the significance of bedding. *Scott. J. Geol.,* **3,** 139–167.
3.8.1, Fig. 3.35, 3.39.

BLUCK B.J. (1971) Sedimentation in the meandering River Endrick. *Scott. J. Geol.,* **7,** 93–138.
3.5.2.

BLUCK B.J. (1974) Structure and directional properties of some valley sandur deposits in Southern Iceland. *Sedimentology,* **21,** 533–554.
3.3.1, 3.3.2, Fig. 3.11, 3.13.

BLUCK B.J. (1976) Sedimentation in some Scottish rivers of low sinuosity. *Trans. R. Soc. Edinburgh,* **69,** 425–456.
3.3.1, 3.3.3.

BOCCALETTI M. & MANETTI P. (1972) Traces of lower-middle Liassic volcanism in the crinoidal limestones of the Tuscan sequence in the Montemerano area (Grosseto, northern Apennines). *Eclog. Geol. Helv.,* **65,** 119–129.
10.2.5.

BOERSMA J.R. (1970) *Distinguishing Features of Wave-ripple Cross-stratification and Morphology.* Doctoral thesis, University of Utrecht, pp. 65.
9.11.1, Fig. 7.46, 9.47, 9.48.

BOILLOT G., BOYSSE P. & LAMBOY M. (1971) Morphology, sediments and Quaternary history of the continental shelf between the Straits of Dover and Cape Finisterre. In: *The Geology of the East Atlantic Continental Margin, 3. Europe* (Ed. by F.M. Delany), pp. 79–90. Rept. no. 70/15, Inst. Geol. Sci., London.
10.3.

BOLLI H.M., RYAN W.B.F. *et al.* (1975) Basins and margins of the eastern South Atlantic. *Geotimes,* **20,** 6, 22–24.
11.3.3.

BOLTUNOV V.A. (1970) Certain earmarks distinguishing glacial and moraine-like glacialmarine sediments, as in Spitsbergen. *Internat. Geology Rev.,* **12,** 204–211.
13.3.7, 13.4.3.

BONATTI E. (1975) Metallogenesis at oceanic spreading centres. In: *Annual Revs Earth Planet. Sci.,* **3** (Ed. by F. Donath, F.G. Stehli and G.W. Wetherill), pp. 401–431. Annual Reviews Inc., Palo Alto, California.
11.3.2, Fig. 11.12.

BONATTI E. & JOENSUU O. (1966) Deep-sea iron deposits from the South Pacific. *Science,* **154,** 643–645.
11.3.2.

BONATTI E., HONNOREZ J. & GARTNER S. (1973) Sedimentary serpentinites from the Mid-Atlantic Ridge. *J. sedim. Petrol.,* **43,** 728–735.
11.3.2.

BONATTI E., KRAEMER T. & RYDELL H. (1972) Classification and genesis of submarine iron-manganese deposits. In: *Ferromanganese Deposits on the Ocean Floor* (Ed. by D. Horn), pp. 146–166. National Science Foundation, Washington, D.C.
11.3.2.

BONATTI E., HONNOREZ J., JOENSUU O. & RYDELL H. (1972). Submarine iron deposits from the Mediterranean Sea. In: *The Mediterranean Sea: a Natural Sedimentation Laboratory* (Ed. by D.J. Stanley), pp. 701–710. Dowden, Hutchinson and Ross, Stroudsburg.
11.3.5.

BONATTI E., ZERBI M., KAY R. & RYDELL H. (1975) Metalliferous deposits from the Apennines ophiolites: Mesozoic equivalents of modern deposits from oceanic spreading centers. *Bull. geol. Soc. Am.,* **87,** 83–94.
11.4.2.

BONATTI E., EMILIANI C., FERRARA G., HONNOREZ J. & RYDELL H. (1974) Ultramafic carbonate breccias from the equatorial Mid-Atlantic Ridge. *Mar. Geol.,* **16,** 83–102.
11.3.2, 11.4.2.

BOND G. (1970) The Dwyka Series in Rhodesia. *Proc. Geol. Ass.,* **81,** 463–472.
13.5.2.

BOND G. & STOCKLMAYER V.R.C. (1967) Possible ice-margin fluctuations in the Dwyka Series in Rhodesia. *Palaeogeogr. Palaeoclim. Palaeoecol.* **3,** 433–446.
13.5.2.

BONNEY T.G. (1904) Editor. *The Atoll of Funafuti.* R. Soc. London.
10.1.1.

BOOTHROYD J.C. (1972) Coarse-grained sedimentation on a braided outwash fan, Northeast Gulf of Alaska, pp. 127. *Tech. Rept. No. 6, C.R.D.* Coastal Research Division, U. of South Carolina, Columbia.
3.3.1, 3.3.3, 3.4.2, 3.8.1, Fig. 3.14.

BOOTHROYD J.C. & ASHLEY G.M. (1975) Process, bar morphology and sedimentary structures on braided outwash fans, North-eastern Gulf of Alaska. In: *Glaciofluvial and Glaciolacustrine Sedimentation* (Ed. by A.V. Jopling and B.C. McDonald), pp. 193–222. *Spec. Publ. Soc. econ. Paleont. Miner.,* **23,** Tulsa.
3.4.2, 3.8.1.

BOOTHROYD J.C. & HUBBARD D.K. (1976) Genesis of bedforms in mesotidal estuaries. In: *Estuarine Research,* Vol. II, Geology and Engineering (Ed. by L.E. Cronin), pp. 217–234. Academic Press, New York.
7.5.

BORCH C.C. VON DER & TRUEMAN N.A. (1974) Dolomitic basal sediments from northern end of Ninety-east Ridge. In: *Initial Reports of the Deep Sea Drilling Project,* **22** (C.C. Von Der Borch and J.G. Sclater *et al.*), pp. 477–483. U.S. Government Printing Office, Washington.
11.3.3.

BORCHERT H. & MUIR R.O. (1964) *Salt Deposits. The Origin Metamorphism and Deformation of Evaporites,* pp. 338. Van Norstrand, London.
8.5.1, Fig. 8.36.

BORNHOLD B.D. & PILKEY O.H. (1971) Bioclastic turbidite sedimentation in Columbus Basin, Bahamas. *Bull. geol. Soc. Am.,* **82,** 1341–1354.
11.3.6, 12.4.6.

BOSELLINI A. & GINSBURG R.N. (1971) Form and internal structure of recent algal nodules (rhodolites) from Bermuda. *J. Geol.*, **79,** 669–682.
11.3.3.

BOSELLINI A. & WINTERER E.L. (1975) Pelagic limestone and radiolarite of the Tethyan Mesozoic: a genetic model. *Geology*, **3,** 279–282.
11.4.2.

BOSENCE D.W.J. (1973) Facies relationships in a tidally influenced environment. *Geol. Mijnb.*, **52,** 63–67.
7.7.

BOSENCE D.W.J. (1976) Ecological studies on two unattached coralline algae from Western Ireland. *Palaeontology*, **19,** 365–395.
10.3.1, 10.3.2, Fig. 10.35a, 10.35b.

BOSTRÖM K. (1973) The origin and fate of ferromanganoan active ridge sediments. *Stockh. Contr. Geol.*, **27,** 149–243.
11.3.2.

BOSWELL P.H.G. (1953) The alleged subaqueous sliding of large sheets of sediments in the Silurian rocks of North Wales. *Liverpool Manchester geol. J.*, **1,** 148–152.
12.2.2.

BOTT M.H.P. (1976) Mechanisms of basin subsidence – an introductory review. In: *Sedimentary Basins of Continental Margins and Cratons* (Ed. by M.H.P. Bott), pp. 1–4. *Tectonophysics*, **36.**
14.3.1.

BOULTON G.S. (1968) Flow tills and related deposits on some Vestspitsbergen Glaciers. *J. Glaciol.*, **7,** 391–412.
13.3.2.

BOULTON G.S. (1970) On the origin and transport of englacial debris in Svalbard glaciers. *J. Glaciol.*, **9,** 213–229.
13.3.1.

BOULTON G.S. (1972) The role of thermal regime in glacial sedimentation. *Spec. Publ. Inst. Brit. Geogr.*, **4,** 1–19.
13.3.1, 13.3.2, Fig. 13.3.

BOULTON G.S. (1975) Processes and patterns of subglacial sedimentation: a theoretical approach. In: *Ice Ages: Ancient and Modern* (Ed. by A.E. Wright and F. Moseley). *Geol. J. Spec. Issue*, **6,** 7–42.
13.3.1.

BOUMA A.H. (1962) *Sedimentology of some flysch deposits: A graphic approach to facies interpretation*, pp. 168. Elsevier, Amsterdam.
2.2.1, 3.9.2, 12.2.4.

BOUMA A.H. (1972) Fossil contourites in Lower Niesenflysch, Switzerland. *Jour. sedim. Pet.*, **42,** 917–921.
12.2.7.

BOUMA A.H. & BROUWER A. (Eds.) (1964) *Turbidites*, pp. 264. Elsevier, Amsterdam.
12.6.2.

BOUMA A.H. & HOLLISTER C.D. (1973) Deep ocean basin sedimentation. *Turbidites and Deep Water Sedimentation.* Soc. econ. Paleont. Miner., Pacific Section, Short Course, Anaheim, 79–118.
12.2.7.

BOUTAKOFF N. (1948) Les formations glaciaires et postglaciaires fossilifères, d'âge permocarbonifère (Karroo inférieur) de la Région de Walikale (Kivu, Congo Belge). *Mem. Inst. Geol. Univ. Louvain*, **9,** 1–122.
13.5.2.

BOUYSSE P., HORN R., LAPIERRE F. & LE LANN F. (1976) Etude des grands bancs de sable du sud-est de la Mer Celtique. *Mar. Geol.*, **20,** 251–275.
9.5.3.

BOYD D.R. & DYER B.F. (1964) Frio barrier bar system of south Texas. *Trans. Gulf-Cst. Ass. geol. Socs.*, **14,** 309–322.
6.7.4.

BRADLEY W.H. (1964) Geology of Green River Formation and associated Eocene rocks in Southwestern Wyoming and adjacent parts of Colorado and Utah. *Prof. Pap. U.S. geol. Surv.*, **496-A,** pp. 86.
4.3.2.

BRADLEY W.H. (1973) Oil shale formed in desert environment: Green River Formation, Wyoming. *Bull. geol. Soc. Am.*, **84,** 1121–1124.
4.3.2.

BRADLEY W.H. & EUGSTER H.P. (1969) Geochemistry and paleolimnology of the trona deposits and associated authigenic minerals of the Green River Formation of Wyoming. *Prof. Pap. U.S. geol. Surv.*, **496-B,** pp. 71.
4.3.2.

BRAITSCH O. (1971) *Salt Deposits–Their Origin and Composition*, pp. 297. Springer-Verlag, Berlin.
8.5.1.

BRAMLETTE M.N. (1946) The Monterey Formation of California and the origin of its siliceous rocks. *Prof. Pap. U.S. geol. Surv.*, **212,** pp. 57.
11.4.3.

BRAMLETTE M.N. (1958) Significance of coccolithophorids in calcium carbonate deposition. *Bull. geol. Soc. Am.*, **69,** 121–126.
11.4.4.

BRAMLETTE M.N. (1961) Pelagic sediments. In: *Oceanography* (Ed. by M. Sears), pp. 345–366. *Publs. Am. Ass. Advnt. Sci.*, **67,** Washington, D.C.
11.2.

BRASIER M.D. (1976) Early Cambrian intergrowths of archaeocyathids, *Renalcis*, and pseudostromatolites from S. Australia. *Palaeontology*, **19,** 223–248.
10.4.2.

BRENNER R.L. & DAVIES D.K. (1973) Storm-generated coquinoid sandstone: genesis of high energy marine sediments from the Upper Jurassic of Wyoming and Montana. *Bull. geol. Soc. Am.*, **84,** 1685–1698.
9.1.1, 9.9, 9.10.5, 9.11.2, Fig. 9.41, 9.42.

BRENNER R.L. & DAVIES D.K. (1974) Oxfordian sedimentation in Western Interior United States. *Bull. Am. Ass. Petrol. Geol.*, **58,** 407–428.
9.10.5, Fig. 9.42.

BRENNINKMEYER B. (1976) Sand fountains in the surf zone. In: *Beach and Nearshore Sedimentation* (Ed. by R.A. Davis Jr. and R.L. Ethington), pp. 69–91. *Spec. Publ. Soc. econ. Paleont. Miner.*, **24,** Tulsa.
7.2.1.

BRICE J.C. (1964) Channel patterns and terraces of the Loup Rivers in Nebraska. *Prof. Pap. U.S. geol. Surv.*, **422-D,** pp. 41.
3.4.

BRIDGE J.S. (1975) Computer simulation of sedimentation in meandering streams. *Sedimentology*, **22,** 3–44.
3.5.2.

BRIDGES E.M. (1970) *World Soils*, pp. 89. Cambridge University Press, Cambridge.
3.6.2.

BRIDGES P.H. (1972) The significance of toolmarks on a Silurian erosional furrow. *Geol. Mag.*, **109,** 405–410.
9.10.3.

BRIDGES P.H. (1975) The transgression of a hard substrate shelf: the Llandovery (Lower Silurian) of the Welsh Borderland. *J. sedim. Petrol.*, **45,** 79–94.
9.9., 9.11.2, 9.11.3.

BRIDGES P.H. (1976) Lower Silurian transgressive barrier islands, southwest Wales. *Sedimentology*, **23**, 347–362.
7.3.2, Fig. 7.40.

BRIDGES P.H. & LEEDER M.R. (1976) Sedimentary model for intertidal mudflat channels with examples from the Solway Firth, Scotland. *Sedimentology*, **23**, 533–552.
7.6.

BRINKMANN R. (1933) Über Kreuzschichtung in deutschen Buntsandsteinbecken. *Göttinger Nachr., Math.-physik.*, Kl. IV, Fachgr. IV, Nr. **32**, 1–12.
5.1.

BROECKER W.S. & BROECKER S. (1974) Carbonate dissolution on the western flank of the East Pacific Rise. In: *Studies in Paleo-oceanography* (Ed. by W.W. Hay), pp. 44–57. *Spec. Publ. Soc. econ. Paleont. Miner.*, **20**, Tulsa.
11.3.2.

BROMLEY R.G. (1967) Marine phosphorites as depth indicators. *Mar. Geol.*, **5**, 503–509.
9.3.6.

BROMLEY R.G. (1970) Borings as trace fossils and *Entobia cretacea* Portlock as an example. In: *Trace Fossils* (Ed. by T.P. Crimes and J.C. Harper). *Geol. J. Spec. Issue*, **3**, 49–90.
11.4.5.

BROMLEY R.G. (1975) Trace fossils at omission surfaces. In: *The Study of Trace Fossils* (Ed. by R.W. Frey), pp. 399–428. Springer-Verlag, New York.
11.4.5.

BROMLEY R.G., SCHULTZ M-G & PEAKE N.B. (1975) Paramoudras: giant flints, long burrows and the early diagenesis of chalks. *Biol. Skr. Dansk Vidensk. Selsk.*, **20**, 10, pp. 31.
11.4.5.

BROOKFIELD M.E. (1977) The origin of bounding surfaces in ancient aeolian sandstones. *Sedimentology*, **24**, 303–332.
5.3.2.

BROUSSARD M.L. (Ed.) (1975) *Deltas; Models for Exploration*, pp. 555. Houston Geol. Soc., Houston.
1.1.

BROUWER H.A. (1953) Rhythmic depositional features of the east Surinam coastal plain. *Geol. Mijnb.*, **15**, 226–236.
7.4.

BROWN L.F. JR., CLEAVES A.W. II & ERXLEBEN A.W. (1973) Pennsylvanian depositional systems in north-central Texas: a guide for interpreting terrigenous clastic facies in a cratonic basin. *Guidebook* No. **14**, pp. 122. *Bur. econ. Geol.* Austin, Texas.
3.9.4, 6.7.2, 6.7.4, 6.8.3.

BROWN P.R. (1963) Algal limestones and associated sediments in the basal Purbeck of Dorset. *Geol. Mag.*, **100**, 565–573.
8.4.2.

BRUCE C.H. (1973) Pressured shale and related sediment deformation: mechanism for development of regional contemporaneous faults. *Bull. Am. Ass. Petrol. Geol.*, **57**, 878–886.
6.8.2, Fig. 6.55.

BRUNSTROM R.G.W. & WALMSLEY P.J. (1969) Permian evaporites in the North Sea Basin. *Bull. Am. Ass. Petrol. Geol.*, **53**, 870–883.
Fig. 8.38.

BRUUN P. (1962) Sea level rise as a cause of shore erosion. *Proc. ASCE J. Waterw. Harbors Div.*, **88**, 117–130.
7.2.6, Fig. 7.31.

BRYANT W.R., MEYERHOFF A.A., BROWN N.K., FURRER M.A., PYLE T.E. & ANTOINE J.W. (1969) Escarpments, reef trends, and diapiric structures, eastern Gulf of Mexico. *Bull. Am. Ass. Petrol. Geol.*, **53**, 2506–2542.
Fig. 10.41a.

BUCHHEIM H.P. & SURDAM R.C. (1977) Fossil catfish and the depositional environment of the Green River Formation, Wyoming. *Geology*, **5**, 196–198.
4.3.2.

BUDINGER T.G. (1967) Cobb Seamount. *Deep-Sea Res.*, **14**, 191–201.
11.3.3.

BUFFINGTON E.C. (1952) Submarine 'natural levees'. *J. Geol.*, **60**, 473–479.
12.5.1.

BUFFINGTON E.C. (1961) Experimental turbidity currents on the sea floor. *Bull. Am. Ass. Petrol. Geol.*, **45**, 1392–1400.
12.2.4.

BULL W.B. (1964) Geomorphology of segmented alluvial fans in Western Fresno County, California. *Prof. Pap. U.S. geol. Surv.*, **352-E**, 89–129.
3.2.2, 3.2.4.

BULL W.B. (1968) Alluvial fan. In: *Encyclopedia of Geomorphology* (Ed. by R.W. Fairbridge), pp. 7–10. Reinhold, New York.
3.2.2.

BULL W.B. (1972) Recognition of alluvial-fan deposits in the stratigraphic record, In: *Recognition of Ancient Sedimentary Environments* (Ed. by K.J. Rigby & W.K. Hamblin), pp. 68–83. *Spec. Publ. Soc. econ. Paleont. Miner.*, **16**, Tulsa.
3.2.4, 5.2.6.

BULLER A.T. (1969) *Source and distribution of calcareous skeletal components in recent marine carbonate sediments, Mannin and Clifden Bays, Connemara, Ireland.* Unpubl. Ph.D. Thesis, Reading University.
10.3.2.

BUMPUS D.F. (1973) A description of circulation on the continental shelf of the east coast of the United States. *Prog. Oceanogr.*, **6**, 117–157.
9.4.4.

BUNTING B.T. (1967) *The Geography of Soils* (2nd Edn), pp. 213. Hutchinson, London.
3.6.2.

BURGESS I.C. (1961) Fossil soils of the Upper Old Red Sandstone of South Ayrshire. *Trans. geol. Soc. Glasgow*, **24**, 138–163.
3.9.2.

BURK C.A. & DRAKE C.L. (Eds.) (1974). *The Geology of Continental Margins*, pp. 1009. Springer, New York.
12.4.3.

BURK C.A., EWING M., WORZEL J.L. *et al.* (1969) Deep-sea drilling into the Challenger Knoll, Central Gulf of Mexico. *Bull. Am. Ass. Petrol. Geol.*, **53**, 1338–1347.
12.3.1, 12.4.4.

BURKE K. (1967) The Yallahs Basin: a sedimentary basin southeast of Kingston, Jamaica. *Mar. Geol.*, **5**, 45–60.
12.5.1, 12.5.6, 14.3.3, Fig. 14.28.

BURKE K. (1972) Longshore drift, submarine canyons, and submarine fans in development of Niger delta. *Bull. Am. Ass. Petrol. Geol.*, **56**, 1975–1983.
Fig. 14.7.

BURKE K. (1976) Development of graben associated with the initial ruptures of the Atlantic Ocean. In: *Sedimentary Basins of Continental Margins and Cratons* (Ed. by M.H.P. Bott), pp. 93–112. *Tectonophysics*, **36**.
14.3.1.

BURKE K. (1977) Aulacogens and continental breakup. *Ann. Rev. Earth*

Planet. Sci., **5,** 371–396.
14.3.1, 14.4.1.

BURKE K. & DEWEY J.F. (1973) Plume-generated triple junctions: key indicators in applying plate tectonics to old rocks. *J. Geol.*, **81,** 406–433.
14.4.1.

BURKE K., DESSAUVAGIE T.E.J. & WHITEMAN A.J. (1971) Origin of the Gulf of Guinea and geological history of the Benue Depression and Niger delta. *Nature*, **233,** 51–55.
6.3.1.

BURKE K., DESSAUVAGIE T.F.J. & WHITEMAN A.J. (1972) Geological history of the Benue Valley and adjacent areas. In: *African Geology* (Ed. by A.J. Whiteman and T.F.J. Dessauvagie), pp. 187–206. Ibadan, 1970.
14.3.1.

BURNETT W.C. (1977) Geochemistry and origin of phosphorite deposits from off Peru and Chile. *Bull. geol. Soc. Am.*, **88,** 813–823.
11.3.6.

BURST J.F. (1968a) 'Glauconite' pellets; their mineral nature and application to stratigraphic interpretation. *Bull. Am. Ass. Petrol. Geol.*, **42,** 310–327.
9.8.

BURST J.F. (1968b) Mineral heterogeneity in 'glauconite' pellets. *Am. Miner.*, **43,** 481–497.
9.8

BURTON A. (1976) Iron-formation as an end member in carbonate sedimentary cycles in the Transvaal Supergroup, South Africa. *Econ. Geol.*, **71,** 193–201.
14.4.1.

BUSCH D.A. (1953) The significance of deltas in subsurface exploration. *Tulsa Geol. Soc. Digest*, **21,** 71–80.
6.2.

BUSCH D.A. (1975) Influence of growth faulting on sedimentation and prospect evaluation. *Bull. Am. Ass. Petrol. Geol.*, **59,** 217–230.
6.8.2.

BUSH P.R. (1970) Chloride-rich brines from sabkha sediments and their possible role in ore formation. *Trans. Inst. Min. Metall. B*, **79,** 137–144.
7.4.4, 8.2.4, 8.2.6, Tab. 8.2.

BUTLER G.P. (1969) Modern evaporite deposition and geochemistry of coexisting brines, the sabkha, Trucial Coast, Arabian Gulf. *J. sedim. Petrol.*, **39,** 70–89.
8.2.6.

BUTLER G.P. (1970) Holocene gypsum and anhydrite of the Abu Dhabi sabkha, Trucial Coast: an alternative explanation of origin. In: *Third Symposium on Salt* (Ed. by J.L. Race and L.F. Dellwig), pp. 120–152. Northern Ohio Geol. Soc., Cleveland, Ohio.
8.2.6, 8.4.2.

BUTLER J.B. (1975) The West Sole Gas-field. In: *Petroleum and the Continental Shelf of North West Europe*. Vol. I *Geology* (Ed. by A.W. Woodland), pp. 213–219. Applied Science Publishers, Barking.
5.1.

BUURMAN P. (1975) Possibilities of palaeopedology. *Sedimentology*, **22,** 289–298.
3.9.2.

BUURMAN P. & JONGMANS A.G. (1975) The Neerrepen Soil, an early Oligocene podzol with a fragipan and gypsum concretions from Belgian and Dutch Limburg. *Pedologie*, **25,** 105–117.
3.9.2.

BYRNE J.V., LEROY D.O. & RILEY C.M. (1959) The chenier plain and its stratigraphy, southwestern Louisiana. *Trans. Gulf-Cst. Ass. geol. Socs.*, **9,** 1–23.
7.4.

BYRNE R.J., BULLOCK P. & TYLER D.G. (1976) Response characteristics of a tidal inlet: a case study. In: *Estuarine Research*. Vol. II, *Geology and Engineering* (Ed. by L.E. Cronin), pp. 201–216. Academic Press Inc., London.
7.2.4.

CADY W.M. (1975) Tectonic setting of the Tertiary volcanic rocks of the Olympic Peninsula, Washington. *J. Res. U.S. geol. Surv.*, **3,** 573–582.
11.4.2, Fig. 11.26.

CAIA J. (1976) Paleogeographical and sedimentological controls of copper, lead and zinc mineralisations in the Lower Cretaceous sandstones of Africa. *Econ. Geol.*, **71,** 409–422.
14.4.4.

CALVERT S.E. (1966a) Accumulation of diatomaceous silica in the sediments of the Gulf of California. *Bull. geol. Soc. Am.*, **77,** 569–596.
11.3.5, 11.4.3, Fig. 11.17.

CALVERT S.E. (1966b) Origin of diatom-rich, varved sediments from the Gulf of California. *J. Geol.*, **74,** 546–565.
11.3.5, 11.4.3.

CALVERT S.E. & PRICE N.B. (1977) Shallow-water continental margin and lacustrine nodules: distribution and geochemistry. In: *Marine Manganese Deposits* (Ed. by G.P. Glasby), pp. 45–86. Elsevier, Amsterdam.
4.2.4.

CAMERON W.M. & PRITCHARD D.W. (1963) Estuaries. In: *The Sea* (Ed. by M.N. Hill), **2,** pp. 306–324. John Wiley, New York.
7.5.

CAMPANA B. & WILSON R.B. (1955) Tillites and related glacial topography of South Australia. *Eclog. geol. Helv.*, **48,** 1–30.
13.5.2.

CAMPBELL C.V. (1971) Depositional model – Upper Cretaceous Gallup beach shoreline, Ship Rock area, northwestern New Mexico. *J. sedim. Petrol.* **41,** 395–409.
7.3.1, Fig. 7.37.

CAMPBELL C.V. (1976) Reservoir geometry of a fluvial sheet sandstone. *Bull. Am. Ass. Petrol. Geol.*, **60,** 1009–1020.
3.9.4, Fig. 3.45, 3.50.

CAMPISI B. (1962) Una formazione diatomitica nell'Altipiano di Gangi (Sicilia). *Geol. Rom.*, **1,** 283–288.
11.4.3.

CANT D.J. & WALKER R.G. (1976) Development of a braided-fluvial facies model for the Devonian Battery Point Sandstone, Quebec. *Can. J. Earth Sci.*, **13,** 102–119.
2.1.2, 3.9.4, Fig. 2.4, 3.51.

CAREY S.W. (1958) A tectonic approach to continental drift. In: *Continental Drift, a Symposium* (Ed. by S.W. Carey), pp. 177–355. University of Tasmania, Hobart.
14.3, 14.3.3.

CAREY S.W. & AHMED N. (1961) Glacial marine sedimentation. In: *Geology of the Arctic*, **2** (Ed. by G.O. Raasch), pp. 865–894. Univ. Toronto Press, Toronto.
13.1, 13.3.7.

CAREY W.C. & KELLER M.B. (1957) Systematic changes in the beds of alluvial rivers. *J. Hydraul. Proc. Am. Soc. Civ. Engrs.*, **83,** Paper 1331, pp. 24.
3.4.2.

CARLSTON C.W. (1965) The relation of free meander geometry to stream discharge and its geomorphic implications. *Am. J. Sci.*, **263**, 864–885.
3.5.1, 3.5.2.

CARTER R.M. (1975a) A discussion and classification of subaqueous mass-transport with particular application to grain-flow, slurry-flow and fluxoturbidites. *Earth-Sci. Rev.*, **11**, 145–177.
12.2.1, 12.2.6.

CARTER R.M. (1975b) Mass emplaced sand-fingers at Mararoa construction site, southern New Zealand. *Sedimentology*, **22**, 275–288.
3.2.4.

CARVER R.E. (1968) Differential compaction as a cause of regional contemporaneous faults. *Bull. Am. Ass. Petrol. Geol.*, **52**, 414–419.
6.8.2.

CASE J.E. (1974) Major basins along the continental margin of northern South America. In: *The Geology of Continental Margins* (Ed. by C.A. Burk and C.L. Drake), pp. 733–741. Springer-Verlag, New York.
14.3.3, Fig. 14.27.

CASEY R. & GALLOIS R.W. (1973) The Sandringham Sands of Norfolk. *Proc. Yorks. geol. Soc.*, **40**, 1–22.
9.8, 9.10.1.

CASSHYAP S.M. (1968) Huronian stratigraphy and paleocurrent analysis in the Espanola–Willisville area, Sudbury District, Ontario, Canada. *J. sedim. Petrol.*, **38**, 920–942.
13.5.5.

CASSHYAP S.M. (1969) Petrology of the Bruce and Gowganda Formations and its bearing on the evolution of Huronian sedimentation in the Espanola–Willisville area, Ontario (Canada). *Palaeogeogr. Palaeoclim. Palaeoecol.*, **6**, 5–36.
13.1.

CASTELLARIN A. (1970) Evoluzione paleotettonica sinsedimentaria del limite tra 'piattaforma veneta' e 'bacino lombardo' a nord di Riva del Garda. *Giorn. Geol.*, ser. 2, **38**, 11–212 (1972).
11.4.4.

CASTELLARIN A., DEL MONTE M. & FRASCARI F.R.S. (1971) Cosmic fallout in the 'hard grounds' of the Venetian region. *Giorn. Geol.*, ser. 2, **39**, 333–346 (1974).
11.4.4.

CASTON V.N.D. (1972) Linear sand banks in the southern North Sea. *Sedimentology*, **18**, 63–78.
9.5.2, 9.5.3, Fig. 9.15.

CASTON V.N.D. (1974) Bathymetry of the northern North Sea: knowledge is vital for offshore oil. *Offshore*, **34**, 76–84.
13.5.1.

CASTON V.N.D. (1977) Quaternary deposits of the central North Sea: 1. A new isopachyte map of the Quaternary of the North Sea. *Rep. Inst. geol. Sci.*, **77/11**, 1–8.
13.5.1, 13.6.

CATHCART J.B. & GUILBRANDSEN R.A. (1972) Phosphate deposits. *Prof. Pap. U.S. geol. Surv.*, **830**, 515–525.
14.4.1.

CAYEUX L. (1935) *Les Roches Sédimentaires de France; Roches carbonatées*, pp. 463. Masson, Paris.
10.1.1.

CEBULSKI D.E. (1969) Foraminiferal populations and faunas in barrier-reef tract and lagoon, British Honduras. In: *Other papers on Florida and British Honduras. Mem. Am. Ass. Petrol. Geol.*, **11**, 311–328.
10.2.4.

CHAPMAN R.E. (1973) *Petroleum Geology: a Concise Study*, pp. 304. Elsevier, Amsterdam.
6.8.2.

CHARLESWORTH J.K. (1957) *The Quaternary Era with Special Reference to its Glaciation*, pp. 1700. Edward Arnold, London.
13.4.2.

CHASE T.E., NEWHOUSE D.A., LONG B.J., CROCKER W.L., HYDOCK L., HALLMAN C.M., WOOD T.C., PALUSO P.R., GLIPTIS M. & PINE J.S. (undated). Topography of the oceans with Deep-Sea Drilling Project sites through Leg 44. Geologic Data Center, Scripps Inst. Oceanography.
Fig. 11.8, 11.9, 11.10.

CHAVE K.E. (1967) Recent carbonate sediments – an unconventional view. *A.G.I. Counc. Educ. Geol. Sci. Short Rev.*, **7**, 200–204.
10.1.2.

CHEN C. (1964) Pteropod ooze from Bermuda Pedestal. *Science*, **144**, 60–62.
11.3.3.

CHIEN N. (1961) The braided stream of the Lower Yellow River. *Scientia Sinica*, **10**, 734–754.
3.4.

CHIPPING D.H. (1971) Paleoenvironmental significance of chert in the Franciscan Formation of Western California. *Bull. geol. Soc. Am.*, **82**, 1707–1712.
11.4.2.

CHISHOLM I.C. (1977) Growth faulting and sandstone deposition in the Namurian of the Stanton syncline, Derbyshire. *Proc. Yorks. geol. Soc.*, **41**, 305–323.
6.8.3.

CHOUGH S. & HESSE R. (1976) Submarine meandering talweg and turbidity currents flowing for 4,000 km in the Northwest Atlantic Mid-Ocean Channel, Labrador Sea. *Geology*, **4**, 529–533.
12.2.4, 12.5.1.

CHOUKROUNE P. & SEGURET M. (1973) Tectonics of the Pyrenees: role of compression and gravity. In: *Gravity and Tectonics* (Ed. by K.A. de Jong & R. Scholten), pp. 141–156. John Wiley, New York.
14.4.4.

CHRISS T. & FRAKES L.A. (1972) Glacial marine sedimentation in the Ross Sea. In: *Antarctica Geology and Geophysics* (Ed. by R. Aidie), pp. 747–762. Comm. Antarctic Res., Oslo.
13.3.7.

CHURCH M. (1972) Baffin Island Sandurs: A study of Arctic fluvial processes. *Bull. geol. Surv. Can.*, **216**, pp. 208.
3.3.1.

CHURKIN M. JR. (1974) Paleozoic marginal ocean–basin–volcanic arc systems in the Cordilleran Foldbelt. In: *Modern and Ancient Geosynclinal Sedimentation* (Ed. by R.H. Dott, Jr. and R.H. Shaver), pp. 174–192. *Spec. Publ. Soc. econ. Paleont. Miner.*, **19**, Tulsa.
11.4.2.

CITA M.B., BIGIOGGERO B. & FERRARIO A. (1975) Micrometeorites in the 'glacial' Pleistocene of the Mediterranean Ridge. *Boll. Soc. geol. ital.*, **94**, 877–887.
11.3.5.

CLEARY W.J. & CONOLLY J.R. (1974) Hatteras deep-sea fan. *J. sedim. Petrol.*, **44**, 1140–1154.
12.6.5.

CLIFTON H.E. (1969) Beach lamination – nature and origin. *Mar. Geol.*, **7**, 553–559.
7.2.2.

CLIFTON H.E. (1976) Wave-formed sedimentary structures – a

conceptual model. In: *Beach and Nearshore Sedimentation* (Ed. by R.A. Davis Jr. and R.L. Ethington), pp. 126–148. *Spec. Publ. Soc. econ. Paleont. Miner.*, **24**, Tulsa.
7.2.3.

CLIFTON H.E., HUNTER R.E. & PHILLIPS R.L. (1971) Depositional structures and processes in the non-barred, high energy nearshore. *J. sedim. Petrol.*, **41**, 651–670.
7.2.2, 7.2.3, 7.3.1, 9.6.2, Fig. 7.12, 7.13, 7.39.

CLIFTON H.E., PHILLIPS R.L. & HUNTER R.E. (1973) Depositional structures and processes in the mouths of small coastal streams, southwestern Oregon. In: *Coastal Geomorphology* (Ed. by D.R. Coates), pp. 115–140. Publications in Geomorphology, State University of New York, Binghamton.
7.2.2.

CLEMMEY H. (1974) Sedimentary geology of a late Pre-Cambrian copper deposit at Kitwe, Zambia. In: *Gisements Stratiformes et Provinces Cuprifères* (Ed. by P. Bartholomé), pp. 255–265.
8.4.1.

CLOUD P.E. (1955) Bahama Banks west of Andros Island. *Bull. geol. Soc. Am.*, **66**, 1542 (abs.).
10.1.1.

COCKS L.R.M. & MCKERROW W.S. (1978) Cambrian, Ordovician, Silurian. In: *The Ecology of Fossils* (Ed. by W.S. McKerrow), pp. 52–124. Duckworth, London.
9.12.

COLACICCHI R., PASSERI L. & PIALLI G. (1975) Evidences of tidal environment deposition in the Calcare Massiccio Formation (Central Apennines – Lower Lias). In: *Tidal Deposits: a Casebook of Recent Examples and Fossil Counterparts* (Ed. by R.N. Ginsburg), pp. 345–353. Springer-Verlag, Berlin.
10.2.5, Fig. 10.26a, 10.26b.

COLEMAN A.P. (1926) *Ice Ages Recent and Ancient*, pp. 296. Macmillan, London.
13.1.

COLEMAN J.M. (1969) Brahmaputra River: Channel processes and sedimentation. *Sedim. Geol.*, **3**, 129–239.
3.4, 3.4.1, 3.4.2, 3.4.3, 3.6.1, 3.9.4, 6.3.1, 6.5.1, 6.5.2, Fig. 3.16, 3.32.

COLEMAN J.M. & GAGLIANO S.M. (1964) Cyclic sedimentation in the Mississippi river delta plain. *Trans. Gulf-Cst. Ass. geol. Socs.*, **14**, 67–80.
6.2, 6.5.1, Fig. 6.10, 6.36.

COLEMAN J.M. & WRIGHT L.D. (1975) Modern river deltas: variability of processes and sand bodies. In: *Deltas, Models for Exploration* (Ed. by M.L. Broussard), pp. 99–149. Houston Geol. Soc. Houston.
6.2, 6.3, 6.4, 6.5.1, 6.5.2, 9.3.1, Fig. 6.4, 6.21.

COLEMAN J.M., GAGLIANO S.M. & SMITH W.G. (1970) Sedimentation in a Malaysian high tide tropical delta. In: *Deltaic Sedimentation Modern and Ancient* (Ed. by J.P. Morgan), pp. 185–197. *Spec. Publ. Soc. econ. Paleont. Miner.*, **15**, Tulsa.
9.3.1.

COLEMAN J.M., GAGLIANO S.M. & WEBB J.E. (1964) Minor sedimentary structures in a prograding distributary. *Mar. Geol.*, **1**, 240–258.
6.2, 6.5.1, Fig. 6.9.

COLEMAN J.M., SUHAYDA J.N., WHELAN T. & WRIGHT L.D. (1974) Mass movement of Mississippi river delta sediments. *Trans. Gulf-Cst. Ass. geol. Socs.*, **24**, 49–68.
6.3.1, 6.5.2, 6.8.1., 6.8.2, 6.8.2, Fig. 6.52.

COLEMAN P.J. (1968) Tsunamis as geological agents. *J. geol. Soc. Aust.*, **15**, 267–273.
9.4.

COLLEY H. & RICE C.M. (1975) A Kuroko-type ore deposit in Fiji. *Econ. Geol.*, **70**, 1373–1386.
14.4.2.

COLLETTE B.J., EWING J.I., LAGAAY R.A. & TRUCHAN M. (1969) Sediment distribution in the oceans: the Atlantic between 10° and 19°N. *Mar. Geol.*, **7**, 279–345.
12.4.1.

COLLINSON J.D. (1968) Deltaic sedimentation units in the Upper Carboniferous of northern England. *Sedimentology*, **10**, 233–254.
3.9.4.

COLLINSON J.D. (1969) The sedimentology of the Grindslow Shales and the Kinderscout Grit: a deltaic complex in the Namurian of northern England. *J. sedim. Petrol.*, **39**, 194–221.
6.7.1, 6.7.2, 6.7.4, Fig. 6.51.

COLLINSON J.D. (1970a) Deep channels, massive beds and turbidity current genesis in the Central Pennine Basin. *Proc. Yorks. geol. Soc.*, **37**, 495–519.
6.7.2.

COLLINSON J.D. (1970b) Bedforms of the Tana River, Norway. *Geogr. Annalr.*, **52-A**, 31–56.
3.3.1, 3.4, 3.4.1, 3.4.2, 3.8.1, 9.10.1, Fig. 3.18, 3.19.

COLLINSON J.D. (1971a) Some effects of ice on a river bed. *J. sedim. Petrol.*, **41**, 557–564.
13.4.2.

COLLINSON J.D. (1971b) Current vector dispersion in a river of fluctuating discharge. *Geol. Mijnb.*, **50**, 671–678.
3.3.2.

COLLINSON J.D. (1972) The Røde Ø Conglomerate of Inner Scoresby Sund and the Carboniferous(?) and Permian rocks west of the Schuchert Flod. *Meddr. om Grønland*, Bd. **192**, Nr. 6, 1–48.
3.2.1, 3.8, 3.8.1, 5.3.6, Fig. 3.34.

COLLINSON J.D. & BANKS N.L. (1975) The Haslingden Flags (Namurian G_1) of south-east Lancashire: bar finger sands in the Pennine basin. *Proc. Yorks. geol. Soc.*, **40**, 431–458.
6.7.4, Fig. 6.47, 6.50.

COLOM G. (1952) Aquitanian–Burdigalian diatom deposits of the North Betic Strait. *J. Paleont.*, **26**, 867–885.
11.4.3, Fig. 11.29.

COLTER V.S. & BARR K.W. (1975) Recent developments in the geology of the Irish Sea and Cheshire Basins. In: *Petroleum and the Continental Shelf of North West Europe*: Vol. **1**, *Geology* (Ed. by A.W. Woodland), pp. 61–73. Applied Science Publishers, Barking.
5.3.6.

COLTON G.W. (1967) Late Devonian current directions in western New York, with special reference to *Fucoides graphica*. *J. Geol.*, **75**, 11–22.
12.2.4.

CONAGHAN P.J., MOUNTJOY E.W., EDGECOMBE D.R. *et al.* (1976) Nubrigyn algal reefs (Devonian), eastern Australia: allochthonous blocks and megabreccias. *Bull. geol. Soc. Am.*, **87**, 515–530.
12.2.3.

CONNOLLY J.R. & EWING M. (1965) Ice-rafted detritus as a climatic indicator in Antarctic deep-sea cores. *Science*, **150**, 1822–1824.
11.3.1.

CONOLLY J.R. & EWING M. (1967) Sedimentation in the Puerto Rico Trench. *J. sedim. Petrol.*, **37**, 44–59.
12.4.1, 12.4.3.

CONOLLY J.R. & VAN DER BORCH C.C. (1967) Sedimentation and physiography of the sea floor south of Australia. *Sedim. Geol.*, **1**, 181–220.
10.1.2.

COOK D.O. (1969) Occurrence and geologic work of rip currents in

southern California. *J. sedim. Petrol.*, **39,** 781–786.
7.2.2.

Cook A.C. (1974) Report on the petrography of a Paleocene brown coal sample from the Ninety-east Ridge, Indian Ocean. In: *Initial Reports of the Deep Sea Drilling Project,* **22** (C.C. Von Der Borch and J.G. Sclater *et al.*), pp. 477–483. U.S. Government Printing Office, Washington.
11.3.3.

Cook D.O. & Gorsline D.S. (1972) Field observations of sand transport by shoaling waves. *Mar. Geol.*, **13,** 31–56.
7.2.2.

Cook H.E. & Taylor M.E. (1977) Comparison of continental slope and shelf environments in the Upper Cambrian and lowermost Ordovician of Nevada. In: *Deep-water Carbonate Environments* (Ed. by H.E. Cook and P. Enos), pp. 51–81. *Spec. Publ. Soc. econ. Paleont. Miner.*, **25,** Tulsa.
11.4.4, Fig. 11.30.

Cooke R.U. & Warren A. (1973) *Geomorphology in Deserts*, pp. 374. Batsford, London.
3.6.2, 5.2.1, 5.2.4, 5.2.5, Fig. 5.4.

Coombs D.S., Landis C.A., Norris R.J., Sinton J.M., Borns D.J. & Craw D. (1976) The Dun Mountain Ophiolite Belt, New Zealand, its tectonic setting, constitution and origin with special reference to the southern portion. *Am. J. Sci.*, **276,** 561–603.
11.4.1, 11.4.2.

Cooper W.S. (1958) Coastal sand dunes of Oregon and Washington. *Mem. geol. Soc. Am.*, **72,** 1–169.
7.2.2.

Cooper W.S. (1967) Coastal sand dunes of California. *Mem. geol. Soc. Am.*, **104,** 1–131.
7.2.2.

Copper P. (1974) Structure and development of Early Palaeozoic reefs. *Proc. of 2nd Internat. Coral Reef Symposium,* **1,** 365–386.
10.4.1.

Corbett K.D. (1972) Features of thick-bedded sandstones in a proximal flysch sequence, Upper Cambrian, southwest Tasmania. *Sedimentology,* **19,** 99–114.
12.2.6, Fig. 12.42.

Corliss J.B. (1971). The origin of metal-bearing submarine hydrothermal solutions. *J. geophys. Res.*, **76,** 8128–8138.
10.3.2, 11.3.2, 14.3.1.

Cornish V. (1914) *Waves of Sand and Snow*, pp. 383. T. Fisher Unwin, London.
5.2.4.

Correns C.W. (1939) Pelagic sediments of the North Atlantic Ocean. In: *Recent Marine Sediments* (Ed. by P.D. Trask). *Am. Ass. Petrol. Geol.*, 373–395.
11.1.1.

Costello W.R. & Walker R.G. (1972) Pleistocene Sedimentology; Credit River, Southern Ontario: A new component of the braided river model. *J. sedim. Petrol.*, **42,** 389–400.
3.8.1, Fig. 3.41.

Cotton C.A. (1952) Criteria for the classification of coasts. *17th Int. Geog. Cong. Abs.*, **15.**
7.1.

Cowan D.S. & Page B.M. (1975) Recycled Franciscan material in Franciscan melange west of Paso Robles, California. *Bull. geol. Soc. Am.*, **86,** 1089–1095.
14.4.2.

Cramer H.R. (1969) Evaporites – a selected bibliography. *Bull. Am. Ass. Petrol. Geol.*, **53,** 982–1011.
8.5.1.

Crampton C.B. & Carruthers R.G. (1914) *The geology of Caithness. Mem. Geol. Surv. Scotland,* pp. 194. H.M. Stationery Office, London.
4.3.5.

Creager J.S. & Sternberg R.W. (1972) Some specific problems in understanding bottom sediment distribution and dispersal on the continental shelf. In: *Shelf Sediment Transport: Process and Pattern* (Ed. by D.J.P. Swift, D.B. Duane and O.H. Pilkey), pp. 347–362. Dowden, Hutchinson & Ross, Stroudsburg, Pennsylvania.
9.2.

Crittenden M.D. Jr., Stewart J.H. & Wallace C.A. (1972) Regional correlation of Upper Precambrian strata in western North America. *24th Int. geol. Congr.*, **1,** 334–341.
13.5.4.

Croft A.R. (1962) Some sedimentation phenomena along the Wasatch Mountain Front. *J. geophys. Res.*, **67,** 1511–1524.
3.2.4, 3.2.6, 3.8.1.

Cromwell J.E. (1971) Barrier coast distribution; a world-wide survey (abstr.). *Natl. Coastal Shallow Water Res. Conf.*, Abst. No. **2,** 50.
7.2.

Cronan D.S. (1976) Basal metalliferous sediments from the eastern Pacific. *Bull. geol. Soc. Am.*, **87,** 928–934.
11.3.2.

Cronan D.S. & Moorby S.A. (1976) Preliminary results of renewed investigations of manganese nodules and encrustations in the Indian Ocean. In: *Marine Geological Investigations in the Southwest Pacific and Adjacent Areas* (Ed. by G.P. Glasby and H.R. Katz), pp. 118–123. *Tech. Bull. Comm. co-ord. joint Prospect., Econ. Soc. Comm. Asia and Pacific (U.N.)* No. 2.
11.3.4, Fig. 11.15.

Cronan D.S. & Tooms J.S. (1969) The geochemistry of manganese nodules and associated pelagic deposits from the Pacific and Indian Oceans. *Deep-Sea Res.*, **16,** 335–359.
11.3.3, 11.3.4.

Cronan D.S., Damiani V., Kinsman D.J.J. & Thiede J. (1974) Sediments from the Gulf of Aden and western Indian Ocean. In: *Initial Reports of the Deep Sea Drilling Project,* **24** (R.L. Fisher *et al.*), pp. 1047–1110. U.S. Government Printing Office, Washington.
10.3.2.

Crowell J.C. (1955) Directional-current structures from the prealpine flysch, Switzerland. *Bull. geol. Soc. Am.*, **66,** 1351–1384.
12.6.2.

Crowell J.C. (1957) Origin of pebbly mudstones. *Bull. geol. Soc. Am.*, **68,** 993–1010.
12.2.3, 13.1, 13.4.1.

Crowell J.C. (1974a) Sedimentation along the San Andreas Fault, California. In: *Modern and Ancient Geosynclinal Sedimentation* (Ed. by R.H. Dott Jr. and R.H. Shaver), pp. 292–303. *Spec. Publ. Soc. econ. Paleont. Miner.*, **19,** Tulsa.
12.4.5, 14.3.1, 14.3.3, Fig. 14.9, 14.24.

Crowell J.C. (1974b) Origin of late Cenozoic basins in southern California. In: *Tectonics and Sedimentation* (Ed. by W.R. Dickinson), pp. 190–204. *Spec. Publ. Soc. econ. Paleont. Miner.*, **22,** Tulsa.
11.4.3, 14.3.3, Fig. 14.9, 14.23, 14.26.

Crowell J.C. (1975) The San Gabriel fault and Ridge Basin, southern California. In: *San Andreas Fault in Southern California* (Ed. by J.C. Crowell), pp. 208–233. *Spec. Rept. California Div. Mines Geol.*, **118**.

Fig. 14.25.

CROWELL J.C. & FRAKES L.A. (1971) Late Paleozoic Glaciation: Part IV, Australia. *Bull. geol. Soc. Am.*, **82**, 2515–2540.
13.5.2.

CROWELL J.C. & FRAKES L.A. (1972) Late Paleozoic Glaciation: Part V, Karroo Basin, South Africa. *Bull. geol. Soc. Am.*, **83**, 2887–2912.
13.5.2.

CROWELL J.C. & FRAKES L.A. (1975) The Late Paleozoic Glaciation. In: *Gondwana Geology* (Ed. by K.S.W. Campbell), pp. 313–331. Austral. Nat. Univ. Press, Canberra.
13.5.2.

CULBERTSON W.C. (1971) Stratigraphy of the trona deposits in the Green River Formation, southwest Wyoming. *Wyoming Univ. Contr. Geol.*, **10**, 15–23.
4.3.2.

CULLIS C.G. (1904) The mineralogical changes in the cores of the Funafuti borings. In: *The Atoll of Funafuti* (Ed. by T.G. Bonney), pp. 392–420. R. Soc., London.
10.1.1.

CUMMINS W.A. (1962) The greywacke problem. *Liverpool Manchester geol. J.*, **3**, 51–72.
12.2.4.

CURRAY J.R. (1960) Sediments and history of the Holocene transgression, continental shelf, Gulf of Mexico. In: *Recent Sediments, Northwest Gulf of Mexico* (Ed. by F.P. Shepard, F.B. Phleger and Tj.H. Van Andel), pp. 221–266. Am. Ass. Petrol. Geol.
9.6.1.

CURRAY J.R. (1964) Transgressions and regressions. In: *Papers in Marine Geology* (Ed. by R.L. Miller), pp. 175–203. Macmillan, New York.
9.2, 9.3.3.

CURRAY J.R. (1965) Late Quaternary history, continental shelves of the United States. In: *The Quaternary of the United States* (Ed. by H.E. Wright and D.G. Frey), pp. 723–735. Princeton Univ. Press, New Jersey.
9.2.

CURRAY J.R. (1969) History of continental shelves. In: *The New Concepts of Continental Margin Sedimentation: application to the geological record* (Ed. by D.J. Stanley), pp. JC-6-1–JC-6-7. American Geological Institute, Washington.
9.3.3.

CURRAY J.R. (1975) Marine sediments, geosynclines and orogeny. In: *Petroleum and Global Tectonics* (Ed. by S. Judson and A.J. Fischer), pp. 157–222. Princeton Univ. Press, New Jersey.
9.13.

CURRAY J.R. & MOORE D.G. (1964) Holocene regressive littoral sand, Costa de Nayarit, Mexico. In: *Deltaic and Shallow Marine Deposits* (Ed. by L.M.J.U. Van Straaten), pp. 76–82. Developments in Sedimentology, I. Elsevier, Amsterdam.
7.2.6.

CURRAY J.R. & MOORE D.G. (1971) Growth of the Bengal deep-sea fan and denudation in the Himalayas. *Bull. geol. Soc. Am.*, **82**, 563–572.
14.3.4.

CURRAY J.R. & MOORE D.G. (1974) Sedimentary and tectonic processes in the Bengal deep-sea fan and geosyncline. In: *The Geology of Continental Margins* (Ed. by C.A. Burk and C.L. Drake), pp. 617–627. Springer-Verlag, Berlin.
12.5.2, 14.3.2, 14.3.4, Fig. 12.39.

CURRAY J.R., EMMEL F.J. & CRAMPTON P.J.S. (1969) Holocene history of a strand plain, lagoonal coast, Nayarit, Mexico. In *Coastal Lagoons – a Symposium* (Ed. by A.A. Castanares and F.B. Phleger), pp. 63–100. Universidad Nacional Autónoma, Mexico.
7.2.6, Fig. 7.29.

CURTIS R., EVANS G., KINSMAN D.J.J. & SHEARMAN D.J. (1963) Association of dolomite and anhydrite in the recent sediments of the Persian Gulf. *Nature*, **197**, 679–680.
8.1.1.

DALEY B. (1973a) The palaeoenvironment of the Bembridge Marls (Oligocene) of the Isle of Wight, Hampshire. *Proc. Geol. Assoc.*, **84**, 83–94.
4.3.6.

DALEY B. (1973b) Fluvio-lacustrine cyclothems from the Oligocene of Hampshire. *Geol. Mag.*, **110**, 235–242.
4.3.6.

DALY R.A. (1936) Origin of submarine 'canyons'. *Am. J. Sci.*, **31**, 401–420.
1.1, 12.1, 12.2.5.

DAMUTH J.E. & FAIRBRIDGE E.W. (1970) Equatorial Atlantic deep-sea arkosic sands and ice-age aridity in tropical South America. *Bull. geol. Soc. Am.*, **81**, 189–206.
12.4.1.

DAMUTH J.E. & KUMAR N. (1975) Amazon Cone: morphology, sediments, age, and growth pattern. *Bull. geol. Soc. Am.*, **86**, 863–878.
12.5.1, 12.5.2.

DANA J.D. (1873) On some results of the earth's contraction from cooling, including a discussion of the origin of mountains and the nature of the earth's interior. *Am. J. Sci.*, **3**, **5**, 423–443; **6**, 6–14, 104–115, 161–171.
14.2.1.

D'ANGLJEAN B.F. (1967) Origin of marine phosphorites off Baja California. *Mar. Geol.*, **5**, 15–44.
11.3.6.

D'ARGENIO B., DE CASTRO P., EMILIANI C. & SIMONE C. (1974) Bahamian and Appenninic limestones of identical lithofacies and age. *Bull. Am. Ass. Petrol. Geol.*, **59**, 524–530.
10.2.5.

DAVIDSON-ARNOTT R.G.D. & GREENWOOD B. (1974) Bedforms and structures associated with bar topography in the shallow water wave environment, Kouchibouguac Bay, New Brunswick, Canada. *J. sedim. Petrol.*, **44**, 698–704.
7.2.3.

DAVIDSON-ARNOTT R.G.D. & GREENWOOD B. (1976) Facies relationships on a barred coast, Kouchibouguac Bay, New Brunswick, Canada. In: *Beach and Nearshore Sedimentation* (Ed. by R.A. Davis Jr. and R.L. Ethington), pp. 149–168. *Spec. Publ. Soc. econ. Paleont. Miner.*, **24**, Tulsa.
7.2.1, 7.2.3, Fig. 7.14.

DAVIES D.K. (1968) Carbonate turbidites, Gulf of Mexico. *J. sedim. Petrol.*, **38**, 1100–1109.
12.2.4, 12.4.4, Fig. 12.31.

DAVIES D.K., ETHRIDGE F.G. & BERG R.R. (1971) Recognition of barrier environments. *Bull. Am. Ass. Petrol. Geol.*, **55**, 550–565.
7.3, 7.3.1, Fig. 7.35, 7.36.

DAVIES G.R. (1970) Algal-laminated sediments, Gladstone embayment, Shark Bay, Western Australia. *Mem. Am. Assoc. Petrol. Geol.*, **13**, 169–205.
10.4.2.

DAVIES I.C. & WALKER R.G. (1974) Transport and deposition of resedimented conglomerates: the Cap Enragé Formation, Cambro-

Ordovician, Gaspé, Quebec. *J. sedim. Petrol.*, **44**, 1200–1216.
12.2.6, 12.8.1.

Davies J.L. (1964) A morphogenetic approach to world shorelines. *Zeits. Geomorph.*, **8** (Sp. No.), 127–142.
7.1, Fig. 7.1.

Davies T.A. & Gorsline D.S. (1976) Oceanic sediments and sedimentary processes. In: *Chemical Oceanography* (Ed. by J.P. Riley and R. Chester), 2nd Edn, **5**, pp. 1–80. Academic Press, London.
11.3.1, Fig. 11.3, 11.11.

Davies T.A., Weser O.E., Luyendyk B.P. & Kidd R.B. (1975) Unconformities in the sediments of the Indian Ocean. *Nature*, **253**, 15–19.
11.3.1.

Davies T.A., Hay W.W., Southam J.R. & Worsley T.R. (1977) Estimates of Cenozoic oceanic sedimentation rates. *Science*, **197**, 53–55.
11.3.1.

Davies T.A., Luyendyk B.P. *et al.* (1974) *Initial Reports of the Deep Sea Drilling Project*, **26**, pp. 1129. U.S. Government Printing Office, Washington.
11.3.3.

Davis C.A. (1900) A contribution to the natural history of marl. *J. Geol.*, **8**, 485–497.
4.2.4.

Davis E.F. (1918) The radiolarian rocks of the Franciscan Group. *Univ. Calif. Publs. Bull. Dep. Geol.*, **11**, 235–432.
11.1.2, 11.4.2.

Davis R.A. & Ethington R.L. (Eds) (1976) *Beach and Nearshore Sedimentation*, pp. 187. *Spec. Publ. Soc. econ. Paleont. Miner.*, **24**, Tulsa.
1.1.

Davis R.A. Jr. & Fox W.T. (1972) Coastal processes and nearshore sand bars. *J. sedim. Petrol.*, **42**, 401–412.
7.2.2.

Davis R.A., Fox W.T., Hayes M.O. & Boothroyd J.C. (1972) Comparison of ridge and runnel systems in tidal and non-tidal environments. *J. sedim. Petrol.*, **32**, 413–421.
7.2.2, Fig. 7.9.

Decandia F.A. & Elter P. (1972) La 'zona' ofiolitifera del Bracco nel settore compreso fra Levanto e la Val Graveglia (Appennino ligure). *Memorie Soc. geol. ital.*, **11**, 503–530.
11.4.2.

Deegan C.E. (1973) Tectonic control of sedimentation at the margin of a Carboniferous depositional basin in Kirkcudbrightshire. *Scott. J. Geol.*, **9**, 1–28.
3.8.1.

Defant A. (1958) *Ebb and Flow. The Tides of Earth, Air and Water*, pp. 121. Univ. Michigan Press.
9.4.2.

Degens E.T. (1965) *Geochemistry of Sediments*, pp. 342. Prentice-Hall, New Jersey.
1.1.

Degens E.T. & Ross D.A. (Eds) (1969) *Hot Brines and Recent Heavy Metals in the Red Sea*, pp. 600. Springer-Verlag, New York.
11.3.5, 14.3.1.

Degens E.T., Williams E.G. & Keith M.L. (1957) Environmental studies of carboniferous sediments, I, Geochemical criteria for differentiating marine and fresh-water shales. *Bull. Am. Ass. Petrol. Geol.*, **41**, 2427–2455.
9.8.

Denny C.S. (1965) Alluvial fans in the Death Valley Region, California and Nevada. *Prof. Pap. U.S. geol. Surv.*, **466**, pp. 62.
3.2.2.

Denny C.S. (1967) Fans and pediments. *Am. J. Sci.*, **265**, 81–105.
3.2.2, 3.2.5.

Deventer J. Van & Postuma J.A. (1973) Early Cenomanian to Pliocene deep-marine sediments from North Malaita, Solomon Islands. *J. geol. Soc. Aust.*, **20**, 145–150.
11.4.2.

Dewey J.F. (1974) Continental margins and ophiolite obduction: Appalachian–Caledonian system. In: *The Geology of Continental Margins* (Ed. by C.A. Burk and C.L. Drake), pp. 933–950. Springer-Verlag, Berlin.
11.4.2.

Dewey J.F. & Bird J.M. (1970) Mountain belts and the new global tectonics. *J. geophys. Res.*, **75**, 2625–2647.
14.2.6, 14.4.1, 14.5, 14.5.1, Fig. 14.10.

Dewey J.F. & Bird J.M. (1971) Origin and emplacement of the ophiolite suite: Appalachian ophiolites in Newfoundland. *J. geophys. Res.*, **76**, 3179–3206.
11.4.1, 11.4.2.

Dewey J.F., Pitman III W.C., Ryan W.B.F. & Bonnin J. (1973). Plate tectonics and the evolution of the Alpine System. *Bull. geol. Soc. Am.*, **84**, 3137–3180.
14.4.4.

Deynoux M. & Trompette R. (1976) Late Pre-Cambrian mixtites: glacial and/or nonglacial? Dealing especially with the mixtites of West Africa. *Am. J. Sci.*, **276**, 1302–1315.
13.5.4.

Dickinson K.A. (1971) Grain size distribution and the depositional history of northern Padre Island, Texas. *Prof. Pap. U.S. geol. Surv.*, **750C**, C1–C6.
7.2.3, 7.2.6, Fig. 7.17, 7.23.

Dickinson K.A., Berryhill H.L. Jr. & Holmes C.W. (1972) Criteria for recognising ancient barrier coastlines. In: *Recognition of Ancient Sedimentary Environments* (Ed. by J.K. Rigby and W.K. Hamblin), pp. 192–214. *Spec. Publ. Soc. econ. Paleont. Miner.*, **16**, Tulsa.
7.2.6.

Dickinson W.R. (1971a) Plate tectonic models of geosynclines. *Earth Planet. Sci. Letts.*, **10**, 165–174.
14.2.6.

Dickinson W.R. (1971b) Clastic sedimentary sequences deposited in shelf, slope and trough settings between magmatic arcs and associated trenches. *Pacific Geology*, **3**, 15–30.
14.4.2.

Dickinson W.R. (1974a) Sedimentation within and beside ancient and modern magmatic arcs. In: *Modern and Ancient Geosynclinal Sedimentation* (Ed. by R.H. Dott Jr. and R.H. Shaver), pp. 230–239. *Spec. Publ. Soc. econ. Paleont. Miner.*, **19**, Tulsa.
14.3.2.

Dickinson W.R. (1974b) Plate tectonics and sedimentation. In: *Tectonics and Sedimentation* (Ed. by W.R. Dickinson), pp. 1–27. *Spec. Publ. Soc. econ. Paleont. Miner.*, **22**, Tulsa.
14.3.2, 14.3.4, 14.4.4.

Dieleman P.J. & Ridder N.A. de (1964) Studies of salt and water movement in the Bol Guini Polder, Chad Republic. *Int. Inst. for Land Reclamation and Improvement, Wageningen. Bull.*, **5**, 1–40.
4.2.1.

Dietz R.S. (1963) Collapsing continental rises: an actualistic concept of geosynclines and mountain building. *J. Geol.*, **71**, 314–333.

14.2.3, 14.2.6, 14.4.1, 14.5.1.

DIETZ R.S. & HOLDEN J.C. (1966) Miogeoclines (miogeosynclines) in space and time. *J. Geol.*, **74**, 566–583.
14.2.3, Fig. 14.5.

DIETZ R.S., HOLDEN J.C. & SPROLL W.P. (1973) Geotectonic evolution and subsidence of Bahama Platform. *Bull. geol. Soc. Am.*, **81**, 1915–1928.
10.2.4.

DILL R.F., DIETZ R.S. & STEWART H. (1954) Deep-sea channels and delta of the Monterey submarine canyon. *Bull. geol. Soc. Am.*, **65**, 191–194.
12.3.2.

DILLON W.P. (1970) Submergence effects on a Rhode Island barrier and lagoon and inferences on migration of barriers. *J. Geol.*, **78**, 94–106.
7.2.6.

DINELEY D.L. & WILLIAMS B.P.J. (1968) Sedimentation and paleoecology of the Devonian Escuminac Formation and related strata, Escuminac Bay, Quebec. In: *Late Paleozoic and Mesozoic Continental Sedimentation, Northeastern North America* (Ed. by G. de V. Klein), pp. 241–264. *Spec. Pap. geol. Soc. Am.*, **106.**
4.3.4, Fig. 4.13.

DINGLE R.V. (1965) Sand waves in the North Sea mapped by continuous reflection profiling. *Mar. Geol.*, **3**, 391–400.
9.5.2.

DINGLE R.V. (1974) Agulhas Bank phosphorites: a review of 100 years of investigation. *Trans. geol. Soc. S. Afr.*, **77**, 261–264.
11.3.6.

DOEGLAS D.J. (1962) The structure of sedimentary deposits of braided rivers. *Sedimentology*, **1**, 167–190.
3.3.1.

DOLAN R. & FERM J.C. (1968) Concentric landforms along the Atlantic coast of the United States. *Science*, **159**, 627–629.
7.2.2.

DOLAN R. (1971) Coastal landforms: crescentic and rhythmic. *Bull. geol. Soc. Am.*, **82**, 177–180.
7.2.2.

DONALDSON A.C. (1974) Pennsylvanian sedimentation of central Appalachians. *Spec. Pap. geol. Soc. Am.*, **148**, 47–78.
6.7.4.

DONALDSON A.C., MARTIN R.H. & KANES W.H. (1970) Holocene Guadalupe delta of Texas Gulf Coast. In: *Deltaic Sedimentation Modern and Ancient* (Ed. by J.P. Morgan and R.H. Shaver), pp. 107–137. *Spec. Publ. Soc. econ. Paleont. Miner.*, **15**, Tulsa.
6.3.1, 7.2.5.

DONOVAN R.N. (1973) Basin margin deposits of the Middle Old Red Sandstone at Dirlot, Caithness. *Scott. J. Geol.*, **9**, 203–212.
4.3.5.

DONOVAN R.N. (1975) Devonian lacustrine limestones at the margin of the Orcadian Basin, Scotland. *J. geol. Soc. Lond.*, **131**, 489–510.
4.3.5, Fig. 4.14, 4.15, 4.16.

DONOVAN R.N. & ARCHER R. (1975) Some sedimentological consequences of a fall in the level of Haweswater, Cumbria. *Proc. Yorks. geol. Soc.*, **40**, 547–562.
4.2.3.

DONOVAN R.N. & FOSTER R.J. (1972) Subaqueous shrinkage cracks from the Caithness Flagstone Series (Middle Devonian) of northeast Scotland. *J. sedim. Petrol.*, **42**, 309–317.
4.3.5.

DONOVAN R.N., FOSTER R.J. & WESTOLL T.S. (1974) A stratigraphical revision of the Old Red Sandstone of North-eastern Caithness. *Trans. Roy. Soc. Edin.*, **69**, 167–201.
4.3.5.

DORÉ F. (1975) Excursion XIII, Origine glaciaire de 'Poudingue de Granville'? In: *Voyage D'Études no.* **1**, *Normandie: Baie du Mont-Saint Michel et Massif Armoricain* (Ed. by C. Larsonneur and F. Doré), pp. 118–124. 9th Int. Congr. Sedimentol., Nice, 1975.
Fig. 13.10.

DÖRJES J. (1971) Der Golf von Gaeta (Tyrrhenisches Meer). IV. Das makrobenthos und seine Küstenparallele zonierung. *Senckenberg. Marit.*, **3**, 203–246.
9.3.5.

DÖRJES J. (1972) Georgia Coastal Region, Sapelo Island, U.S.A.: Sedimentology and Biology. VII. Distribution and zonation of macrobenthic animals. *Senckenberg. Marit.*, **4**, 183–216.
9.3.5.

DOTT R.H. JR. (1961) Squantum 'Tillite', Massachusetts – Evidence of glaciation or subaqueous mass movements? *Bull. geol. Soc. Am.*, **72**, 1289–1306.
13.1.

DOTT R.H. JR. (1963) Dynamics of subaqueous gravity depositional processes. *Bull. Am. Ass. Petrol. Geol.*, **47**, 104–129.
12.2.1.

DOTT R.H. JR. (1969) Circum-Pacific Late Cenozoic structural rejuvenation: implications for sea-floor spreading. *Science*, **166**, 874–876.
11.3.5, 11.4.3.

DOTT R.H. JR. (1974) Cambrian tropical storm waves in Wisconsin. *Geology*, **2**, 243–246.
9.10.2, 9.13.

DOTT R.H. JR. & BATTEN R.L. (1971) *Evolution of the Earth*, pp. 649. McGraw-Hill, New York.
9.10.2, Fig. 9.3.2.

DOTT R.H. JR. & SHAVER R.H. (Eds) (1974) *Modern and Ancient Geosynclinal Sedimentation*, pp. 380. *Spec. Publ. Soc. econ. Paleont. Miner.*, **19**, Tulsa.
1.1.

DOVETON J.H. (1971) An application of Markov Chain analysis to the Ayrshire Coal Measures succession. *Scott. J. Geol.*, **7**, 11–27.
2.1.2.

DOW D.B. (1965) Evidence of a Late Pre-Cambrian glaciation in the Kimberley region of Western Australia. *Geol. Mag.*, **102**, 407–414.
13.5.4.

DOW D.B., BEYTH M. & TSEGAYE HAILU (1971) Paleozoic glacial rocks recently discovered in northern Ethiopia. *Geol. Mag.*, **108**, 53–60.
13.5.3.

DRAKE C.L., EWING M. & SUTTON G.H. (1959) Continental margin and geosynclines: the east coast of North America, north of Cape Hatteras. In: *Physics and Chemistry of the Earth*, **3** (Ed. by L.H. Ahrens, F. Press, S.K. Runcorn and H.C. Urey), pp. 110–198. Pergamon Press, Oxford.
14.2.3, Fig. 14.4.

DRAKE D.E. (1976) Suspended sediment transport and mud deposition on continental shelves. In: *Marine Sediment Transport and Environmental Management* (Ed. by D.J. Stanley and D.J.P. Swift), pp. 127–158. John Wiley, New York.
9.3.1.

DRAKE D.E., KOLPACK R.L. & FISCHER P.J. (1972) Sediment transport on the Santa Barbara–Oxnard shelf, Santa Barbara Channel, California. In: *Shelf Sediment Transport: Process and Pattern* (Ed. by D.J.P. Swift, D.B. Duane and O.H. Pilkey), pp. 307–331. Dowden, Hutchinson & Ross, Stroudsburg.

9.3.1.

DRAKE L.D. (1972) Mechanisms of clast attrition of basal till. *Bull. geol. Soc. Am.*, **83**, 2159–2166.
13.4.1.

DRAPER L. (1967) Wave activity at the sea-bed around northwestern Europe. *Mar. Geol.*, **5**, 133–140.
9.1.1.

DREIMANIS A. (1976) Tills: their origin and properties. In: *Glacial Till* (Ed. by R.F. Legget), pp. 11–49. *Royal Soc. Can. Spec. Publ.*, **12.**
13.4.1.

DREIMANIS A. & VAGNERS U.J. (1971) Bimodal distribution of rock and mineral fragments in basal tills. In: *Till: A Symposium* (Ed. by R.P. Goldthwait), pp. 237–250. Ohio State Univ. Press, Columbus.
13.4.1.

DROOGER C.W. (Ed.) (1973) *Messinian Events in the Mediterranean*, pp. 272. North Holland, Amsterdam.
8.5.5.

DUANE D.B., FIELD M.E., MEISBURGER E.P., SWIFT D.J.P. & WILLIAMS S.J. (1972) Linear shoals on the Atlantic inner continental shelf, Florida to Long Island. In: *Shelf Sediment Transport: Process and Pattern* (Ed. by D.J.P. Swift, D.B. Duane and O.H. Pilkey), pp. 447–498. Dowden, Hutchinson & Ross, Stroudsburg.
9.5.3, 9.6.3, Fig. 7.11, 9.22.

DUFF K. (1975) Palaeoecology of a bituminous shale – the Lower Oxford Clay of central England. *Palaeontology*, **18**, 443–482.
9.12.

DUFF P. MCL. D., HALLAM A. & WALTON E.K. (1967) *Cyclic Sedimentation*, pp. 280. Elsevier, Amsterdam.
2.3.

DUFF P. MCL. D. & WALTON E.K. (1962) Statistical basis for cyclothems: a quantitative study of the sedimentary succession in the East Pennine Coalfield. *Sedimentology*, **1**, 235–255.
2.1.2.

DUNHAM R.J. (1962) Classification of carbonate rocks according to depositional texture. In: *Classification of Carbonate Rocks* (Ed. by W.E. Ham), pp. 108–121. *Mem. Am. Ass. Petrol. Geol.*, **1**, Tulsa.
10.1.1.

DUNHAM R.J. (1969) Vadose pisolite in the Capitan reef (Permian), New Mexico and Texas. In: *Depositional Environments in Carbonate Rocks* (Ed. by G.M. Friedman), pp. 182–191. *Spec. Publ. Soc. econ. Paleont. Miner.*, **14**, Tulsa.
8.2.7.

DUNHAM R.J. (1970) Stratigraphic reefs versus ecologic reefs. *Bull. Am. Ass. Petrol. Geol.*, **54**, 1931–1932.
10.2.5.

DUNHAM R.J. (1972) Guide for study and discussion for individual reinterpretation of the sedimentation and diagenesis of the Permian Capitan geologic reef and associated rocks, New Mexico and Texas. In: *Permian Basin Section, Soc. econ. Paleont. Miner. Publ.*, **72–14**, pp. 235.
10.4.2, Fig. 10.44.

DUPEUBLE P.A., REHAULT X., AUZIETRE J.L., DUNAND C.P. & PASTOURET L. (1976) Résultats de dragages et essai de stratigraphie des bancs de Galice, et des montagnes de Porto et de Vigo (Marge occidentale Ibérique). *Mar. Geol.*, **22**, M37–M49.
11.3.6.

DYMOND J., CORLISS J., HEATH G.R., FIELD C., DASCH J. & VEEH H. (1973) Origin of metalliferous sediments from the Pacific Ocean. *Bull. geol. Soc. Am.*, **84**, 3355–3372.
11.3.2.

DZULYNSKI S. & WALTON E.K. (1963) Experimental production of sole markings. *Trans. Edinburgh geol. Soc.*, **19**, 279–305.
12.2.4.

DZULYNSKI S. & WALTON E.K. (1965) *Sedimentary Features of Flysch and Greywackes*, pp. 274. Elsevier, Amsterdam.
12.2.2, 12.2.4, 12.6.4.

DZULYNSKI S., KSIAZKIEWICZ M. & KUENEN PH.H. (1959) Turbidites in flysch of the Polish Carpathian Mountains. *Bull. geol. Soc. Am.*, **70**, 1089–1118.
12.2.5, 12.2.6, 12.6.2, 12.8.1.

EAGAR R.M.C. & SPEARS D.A. (1966) Boron contact in relation to organic carbon and to palaeosalinity in certain British Upper Carboniferous sediments. *Nature*, **209**, 177–181.
9.8.

EARDLEY A.J. (1947) Paleozoic Cordilleran Geosyncline and related orogeny. *J. Geol.*, **55**, 309–342.
14.2.3.

EASTERBROOK D.J. (1963) Late Pleistocene glacial events and relative sea level changes in the northern Puget Lowland. *Bull. geol. Soc. Am.*, **7**, 1465–1484.
13.4.3.

EASTERBROOK D.J. (1964) Void ratios and bulk densities as means of identifying Pleistocene tills. *Bull. geol. Soc. Am.*, **75**, 745–750.
13.3.7.

EDWARDS M.B. (1975a) Late Precambrian subglacial tillites, North Norway. *9th Int. Congr. Sedimentol., Nice 1975. Thème 1*, 61–66.
13.4.1, 13.5.4.

EDWARDS M.B. (1975b) Glacial retreat sedimentation in the Smalfjord Formation, Late Pre-Cambrian, North Norway. *Sedimentology*, **22**, 75–94.
13.5.4.

EDWARDS M.B. (1976a) Growth faults in Upper Triassic deltaic sediments, Svalbard. *Bull. Am. Ass. Petrol. Geol.*, **60**, 341–355.
6.8.3.

EDWARDS M.B. (1976b) Sedimentology of Late Precambrian Sveanor and Kapp Sparre Formations at Aldousbreen, Wahlenbergfjorden, Nordaustlandet. *Norsk Polarinstitut Årbok. 1974*, 51–61.
13.4.1, 13.5.4.

EICHER D.L. (1969). Paleobathymetry of the Cretaceous Greenhorn Sea in eastern Colorado. *Bull. Am. Ass. Petrol. Geol.*, **53**, 1075–1090.
11.4.5.

EISBACHER G.H., CARRIGY M.A. & CAMPBELL R.B. (1974) Paleodrainage pattern and late-orogenic basins of the Canadian Cordillera. In: *Tectonics and Sedimentation* (Ed. by W.R. Dickinson), pp. 143–166. *Spec. Publ. Soc. econ. Paleont. Miner.*, **22**, Tulsa.
14.4.4.

EINSELE G. (1963) Über Art und Richtung der Sedimentation im klastischen rheinischen Oberdevon (Famenne). *Abhandl. Hess. Landesamtes Bodenforsch.*, **43**, 60.
12.6.3, Fig. 12.41.

EITTREIM S. & EWING M. (1972) Suspended particulate matter in the deep waters of the North American Basin. In: *Studies in Physical Oceanography* (Ed. by A.L. Gordon), **2**, 123–168. Gordon & Breach, New York.
12.2.5.

EITTREIM S., EWING M. & THORNDIKE E.M. (1969) Suspended matter

along the continental margin of the North American Basin. *Deep-Sea Research*, **16**, 613–624.
12.2.5, Fig. 12.19.

El-Ashry M.T. & Wanless H.R. (1965) Birth and early growth of a tidal delta. *J. Geol.*, **73**, 404–406.
7.2.4.

Elliott R.E. (1968) Facies, sedimentation successions and cyclothems in productive coal measures in the East Midlands, Great Britain. *Mercian Geologist*, **2**, 351–372.
3.9.2.

Elliott T. (1974a) Abandonment facies of high-constructive lobate deltas, with an example from the Yoredale Series. *Proc. Geol. Ass.*, **85**, 359–365.
6.7.3.

Elliott T. (1974b) Interdistributary bay sequences and their genesis. *Sedimentology*, **21**, 611–622.
6.5.1, 6.7.4, Fig. 6.8.

Elliott T. (1975) The sedimentary history of a delta lobe from a Yoredale (Carboniferous) cyclothem. *Proc. Yorks. geol. Soc.*, **40**, 505–536.
6.7.1, 7.3.1, Fig. 6.44, 7.38.

Elliott T. (1976a) Upper Carboniferous sedimentary cycles produced by river-dominated, elongate deltas. *J. Geol. Soc.*, **132**, 199–208.
2.1.2, 6.7.2, 6.7.4, Fig. 6.48.

Elliott T. (1976b) The morphology, magnitude and regime of a Carboniferous fluvial-distributary channel. *J. sedim. Petrol.*, **46**, 70–76.
6.7.1.

Elter P. (1972) La zona ofiolitifera del Bracco nel quadro dell' Appennino settentrionale. *Guida alle escursioni, 66 Contr. Soc. geol. ital.*, Pisa, 63 pp.
11.4.2, Fig. 11.21.

Elter P. & Trevisan L. (1973) Olistostromes in the tectonic evolution of the Northern Apennines. In: *Gravity and Tectonics* (K.A. de Jong and R. Scholten), pp. 175–188. John Wiley, New York.
12.2.3, Fig. 12.12.

Embley R.W. (1976) New evidence for occurrence of debris flow deposits in the deep sea. *Geology*, **4**, 371–374.
12.2.3.

Emelyanov E.M. (1972) Principal types of recent bottom sediments in the Mediterranean Sea: their mineralogy and chemistry. In: *The Mediterranean Sea: a Natural Sedimentation Laboratory* (Ed. by D.J. Stanley), pp. 355–386. Dowden, Hutchinson & Ross, Stroudsburg.
11.3.5, Fig. 11.16.

Emery K.O. (1952) Continental shelf sediments off southern California. *Bull. geol. Soc. Am.*, **63**, 1105–1108.
9.1.1, 9.2.

Emery K.O. (1960a) Basin plains and aprons off southern California. *J. Geol.*, **68**, 464–479.
12.4.5.

Emery K.O. (1960b) *The Sea off Southern California*, pp. 366. John Wiley, New York.
11.3.6.

Emery K.O. (1964) Turbidites – Pre-Cambrian to Present. *Studies Oceanogr., Tokyo*, 486–495.
12.2.4.

Emery K.O. (1968) Relict sediments on continental shelves of the world. *Bull. Am. Ass. Petrol. Geol.*, **52**, 445–464.
9.1.1, 9.2, 9.6.2, 10.3.2.

Emery K.O. (1976) Perspectives of shelf sedimentology. In: *Marine Sediment Transport and Environmental Management* (Ed. by D.J. Stanley and D.J.P. Swift), pp. 581–592. John Wiley, New York.
9.1.1.

Emery K.O. & Uchupi E. (1972) Western North Atlantic Ocean: topography, rocks, structure, water, life and sediments. *Mem. Am. Ass. Petrol. Geol.*, **17**, pp. 532.
11.3.2, 11.3.3, 11.3.4, 11.3.6, Fig. 11.18.

Emery K.O., Tracey J.I. & Ladd H.S. (1954) Geology of Bikini and nearby atolls. *Prof. Pap. U.S. geol. Surv.*, **260-A**, pp. 265.
10.1.1.

Emery K.O., Uchupi E., Phillips J.D., Bowin C.O., Bunce E.T. & Knott S.T. (1970) Continental rise off eastern North America. *Bull. Am. Ass. Petrol. Geol.*, **54**, 44–108.
Fig. 14.13.

Enos P. (1969) Anatomy of a flysch. *J. sedim. Petrol.*, **39**, 680–723.
12.2.4, 12.6.2, 12.6.3.

Enos P. (1974) Reefs, platforms and basins of Middle Cretaceous in northeast Mexico. *Bull. Am. Ass. Petrol. Geol.* **58**, 800–809.
11.4.4, Fig. 11.3.9.

Enos P. (1977) Tamabra Limestone of the Poza Rica Trend, Cretaceous, Mexico. In: *Deep water Carbonate Environments* (Ed. by H.E. Cook and P. Enos), pp. 273–314. *Spec. Publ. Soc. econ. Paleont. Miner.*, **25**, Tulsa.
11.4.4.

Enos P. & Freeman T. (in press) Shallow-water limestones from the Blake Nose, Sites 390 and 392. In: *Initial Reports of the Deep Sea Drilling Project*, **44** (W.E. Benson, R.E. Sheridan *et al.*). U.S. Government Printing Office, Washington.
11.3.6.

Ericson D.B., Ewing M. & Heezen B.C. (1951) Deep-sea sands and submarine canyons. *Bull. geol. Soc. Am.*, **62**, 961–965.
12.1, 12.3.1.

Ericson D.B., Ewing M. & Heezen B.C. (1952) Turbidity currents and sediments in North Atlantic. *Bull. Am. Ass. Petrol. Geol.* **36**, 489–511.
12.3.1.

Ernst W. (1970) *Geochemical Facies Analysis*. pp. 152. Elsevier, Amsterdam.
9.8.

Erxleben A.W. (1975) Depositional systems in Canyon Group (Pennsylvanian System) North-Central Texas. *Rep. Invest. Bur. econ. Geol.*, **82**, 75. Austin, Texas.
6.7.2, 6.7.3, 6.7.4, 6.8.3.

Eugster H.P. (1969) Inorganic bedded cherts from the Magadi area, Kenya. *Contr. Mineral. and Petrol.*, **22**, 1–31.
4.2.4, 4.3.2.

Eugster H.P. (1970) Chemistry and origin of brines of Lake Magadi, Kenya. In: *Mineralogy and Geochemistry of Non-marine Evaporites* (Ed. by B.A. Morgan), pp. 215–235. *Spec. Pap. Mineralog. Soc. Am.*, **3.**
4.2.1, 4.3.2.

Eugster H.P. (1973) Experimental geochemistry and the sedimentary environment: Van't Hoff's study of marine evaporites. In: *Evolving Concepts in Sedimentology* (Ed. by R.N. Ginsburg), pp. 38–65. John Hopkins University Press, Baltimore.
8.5.1.

Eugster H.P. & Hardie L.A. (1975) Sedimentation in an ancient playa-lake complex: The Wilkins Peak Member of the Green River Formation of Wyoming. *Bull. geol. Soc. Am.*, **86**, 319–334.
4.3.2, Fig. 4.6, 4.7.

Eugster H.P. & Smith G.I. (1965) Mineral equilibria in the Searles

Lake evaporites, California. *J. Petrol.*, **6,** 473–522.
4.2.4.

EUGSTER H.P. & SURDAM R.C. (1973) Depositional environment of the Green River Formation of Wyoming: A preliminary report. *Bull. geol. Soc. Am.*, **84,** 1115–1120.
4.3.2.

EVAMY B.D. (1973) The precipitation of aragonite and its alteration to calcite on the Trucial Coast of the Persian Gulf. In: *The Persian Gulf* (Ed. by B.H. Purser), pp. 329–342. Springer-Verlag, Berlin.
8.2.4.

EVAMY B.D. & SHEARMAN D.J. (1965) The development of overgrowths from echinoderm fragments. *Sedimentology*, **5,** 211–233.
10.1.1.

EVAMY B.D. & SHEARMAN D.D. (1969) Early stages in development of overgrowths on echinoderm fragments in limestones. *Sedimentology*, **12,** 317–322.
10.1.1.

EVANS G. (1965) Intertidal flat sediments and their environments of deposition in the Wash. *Quart. J. geol. Soc. Lond.*, **121,** 209–245.
7.6, Fig. 7.46.

EVANS G. (1975) Intertidal flat deposits of the Wash, western margin of the North Sea. In: *Tidal Deposits: A Casebook of Recent Examples and Fossil Counterparts* (Ed. by R.N. Ginsburg), pp. 13–20. Springer-Verlag, Berlin.
7.6.

EVANS G., MURRAY J.W., BIGGS H.E.J., BATE R. & BUSH P. (1973) The oceanography, ecology, sedimentology and geomorphology of parts of the Trucial Coast barrier island complex, Persian Gulf. In: *The Persian Gulf* (Ed. by B.H. Purser), pp. 233–277. Springer-Verlag, Berlin.
8.2.1, 8.2.3, 8.4.2.

EVANS W.B. (1970) The Triassic Salt deposits of northwestern England. *Quart. J. geol. Soc. Lond.*, **126,** 103–122.
5.3.4.

EVANS W.B., WILSON A.A., TAYLOR B.J. & PRICE D. (1968) *Geology of the Country around Macclesfield, Congleton, Crewe and Middlewich, Mem. Geol. Surv. Gt. Br.*, pp. 328. H.M. Stationery Office, London.
5.3.4.

EWING M. & HEEZEN B.C. (1955) Puerto Rico Trench, topographic and geophysical data. In: *Crust of the Earth* (Ed. by A. Poldervaart), pp. 255–268. *Spec. Pap. geol. Soc. Am.*, **62.**
12.4.3.

EWING M., ERICSON D.B. & HEEZEN B.C. (1958) Sediments and topography of the Gulf of Mexico. In: *Habitat of Oil* (Ed. by L.G. Weeks), pp. 995–1053. *Am. Assoc. Petrol. Geologists,* Tulsa.
12.5, 12.5.2.

EWING M. & THORNDIKE E.M. (1965) Suspended matter in deep ocean water. *Science,* **147,** 1291–1294.
12.2.5.

EYNON G. & WALKER R.G. (1974) Facies relationships in Pleistocene outwash gravels, Southern Ontario: a model for bar growth in braided rivers. *Sedimentology,* **21,** 43–70.
3.8.1, Fig. 3.40.

FAHNESTOCK R.K. (1963) Morphology and hydrology of a glacial stream – White River, Mount Rainier, Washington. *Prof. Pap. U.S. geol. Surv.*, **422-A,** pp. 70.
3.3.1, 13.3.3.

FAIRBRIDGE R.W. & HILLAIRE-MARCEL C. (1977) An 8,000-yr palaeoclimatic record of the 'Double-Hale' 45-yr solar cycle. *Nature,* **268,** 413–416.
7.2.6.

FANNIN N.G.T. (1969) Stromatolites from the Middle Old Red Sandstone of Western Orkney. *Geol. Mag.*, **106,** 77–88.
4.3.5.

FENTON M.M. & DREIMANIS A. (1976) Methods of stratigraphic correlation of till in central and western Canada. In: *Glacial Till* (Ed. by R.F. Legget), pp. 67–82. *Royal Soc. Canada Spec. Publ.*, **12.**
13.4.1.

FERGUSON J. & LAMBERT I.B. (1972) Volcanic exhalations and metal enrichments at Matupi Harbour, New Britain, T.P.N.G. *Econ. Geol.*, **67,** 25–37.
11.3.5.

FERM J.C. (1962) Petrology of some Pennsylvanian sedimentary rocks. *J. sedim. Petrol.*, **32,** 104–123.
7.3.

FERM J.C. (1970) Allegheny deltaic deposits. In: *Deltaic Sedimentation Modern and Ancient* (Ed. by J.P. Morgan and R.H. Shaver), pp. 246–255. *Spec. Publ. Soc. econ. Paleont. Miner.*, **15,** Tulsa.
6.7.3, 6.7.4, Fig. 6.43.

FERM J.C. & CAVAROC V.V. JR. (1968) A non-marine model for the Allegheny of West Virginia. In: *Late Paleozoic and Mesozoic Continental Sedimentation, Northeastern North America* (Ed. by G. de V. Klein), pp. 1–19. *Spec. Paper geol. Soc. Am.*, **106.**
6.7.1.

FETH J.H. (1964) Review and annotated bibliography of ancient lake deposits (Pre-Cambrian to Pleistocene) in the Western States. *Bull. U.S. geol. Surv.*, **1080,** pp. 119.
4.3.1, 4.3.2.

FIELD M.E. & DUANE D.B. (1976) Post-Pleistocene history of the United States inner continental shelf: significance to the origin of barrier islands. *Bull. geol. Soc. Am.*, **87,** 691–702.
7.2.7, 9.6.3.

FIELD R.M. & HESS H.H. (1933) A bore hole in the Bahamas. *Trans. Am. Geophys. Union,* **14,** 234–245.
10.1.1.

FILIPESCO M.-G. (1931) Sur les roches siliceuses d'origine organique et chimique de l'Oligocène des Carpathes roumaines. *C.r. hebd. séanc. Acad. Sci., Paris,* **192,** 1040–1042.
11.4.3.

FILLON R.H. (1972) Evidence from the Ross Sea for widespread submarine erosion. *Nature,* **238,** 40–42.
13.3.7.

FISCHER A.G. (1961) Stratigraphic record of transgressing seas in the light of sedimentation on the Atlantic coast of New Jersey. *Bull. Am. Ass. Petrol. Geol.*, **45,** 1656–1666.
7.2.6, Fig. 7.31.

FISCHER A.G. (1964) The Lofer cyclothems of the Alpine Triassic. In: *Symposium on Cyclic Sedimentation* (Ed. by D.F. Merriam), pp. 107–149. *Bull. geol. Surv. Kansas,* **169.**
10.2.5, Fig. 10.25a, 10.25b.

FISCHER A.G. (1975) Tidal deposits, Dachstein Limestone of the North-Alpine Triassic. In: *Tidal Deposits: a casebook of recent examples and fossil counterparts* (Ed. by R.N. Ginsburg), pp. 235–242. Springer-Verlag, Berlin.
10.2.5, Fig. 10.25a.

FISCHER A.G. & ARTHUR M. (1977) Secular variations in the pelagic realm. In: *Deep-water Carbonate Environments* (Ed. by H.E. Cook and P. Enos), pp. 19–50. *Spec. Publ. Soc. econ. Paleont. Miner.*, **25,** Tulsa.
11.3.1, 11.4.3, 11.4.4.

FISCHER A.G. & GARRISON R.E. (1967) Carbonate lithification on

the sea floor. *J. Geol.*, **75**, 488–496.
11.3.3, 11.4.5.

FISCHER A.G., HONJO S. & GARRISON R.E. (1967) Electron micrographs of limestones and their nannofossils. *Monogr. Geol. Paleont.* Vol. 1 (Ed. by A.G. Fischer), pp. 141. Princeton University Press, Princeton.
11.4.4.

FISHER R.V. (1971) Features of coarse-grained, high-concentration fluids and their deposits. *J. sedim. Petrol.*, **41**, 916–927.
12.2.3.

FISHER R.V. & MATTINSON J.M. (1968) Wheeler Gorge turbidite-conglomerate series, California; inverse grading. *J. sedim. Petrol.*, **38**, 1013–1023.
12.2.6.

FISHER W.L. (1969) Facies characterisation of Gulf Coast Basin delta systems with some Holocene analogues. *Trans. Gulf-Cst. Ass. geol. Socs.*, **19**, 239–261.
6.7.4, Fig. 6.46.

FISHER W.L. & MCGOWEN J.H. (1967) Depositional systems in the Wilcox Group of Texas and their relationship to occurrence of oil and gas. *Trans. Gulf-Cst. Ass. geol. Soc.*, **17**, 105–125.
6.7.3, 6.7.4, Fig. 6.34, 6.45.

FISHER W.L., BROWN L.F., SCOTT A.J. & MCGOWEN J.H. (1969) Delta systems in the exploration for oil and gas. *Bur. econ. Geol.*, Univ. Texas, Austin, pp. 78.
6.2, 6.3, 6.4, 6.5.1, 6.7.3, 6.7.4, Fig. 6.3, 6.13, 6.17, 6.32, 6.35.

FISHER W.L., PROCTOR C.V. JR., GALLAWAY W.E. & NAGLE J.S. (1970) Depositional systems in the Jackson Group of Texas – their relationship to oil, gas and uranium. *Trans. Gulf-Cst. Ass. geol. Socs.*, **20**, 234–261.
7.4.

FISK H.N. (1944) *Geological investigation of the alluvial valley of the lower Mississippi River.* Mississippi River Commission, Vicksburg, pp. 78.
6.2.

FISK H.N. (1947) *Fine Grained Alluvial Deposits and their Effects on Mississippi River Activity*, pp. 82. Mississippi River Commission, Vicksburg, Miss.
3.5.2, 3.5.3, 6.2, Fig. 3.31.

FISK H.N. (1952a) Geological investigations of the Atchafalaya Basin and the problem of Mississippi River diversion. *U.S. Army Corps. Engin. Waterways Expt. St., Vicksburg, Miss.*, pp. 145.
6.6.

FISK H.N. (1952b) Mississippi River Valley geology: relation to river regime. *Trans. Am. Soc. Civ. Engrs.*, **117**, 667–682.
3.5, 3.5.2.

FISK H.N. (1955) Sand facies of recent Mississippi delta deposits. *Wld. Petrol. Cong.*, Rome, 377–398.
6.2, 6.5.2, 6.6, 7.4.

FISK H.N. (1959) Padre Island and the Laguna Madre flats, coastal south Texas. *National Academy of Science–National Research Council, Second Coastal Geography Conference*, pp. 103–151.
7.2.5.

FISK H.N. (1960) Recent Mississippi River sedimentation and peat accumulation. *Congr. Avan. Études Stratigraph. Géol. Carbonifère, Compte Rend.*, **4**, Heerlen, 1958, 187–199.
6.2.

FISK H.N. (1961) Bar finger sands of the Mississippi delta. In *Geometry of Sandstone Bodies – a Symposium* (Ed. by J.A. Peterson and J.C. Osmond), pp. 29–52. Am. Ass. Petrol. Geol., Tulsa.
6.2, 6.5.2, 6.8.2, Fig. 6.22.

FISK H.N., MCFARLAN E. JR., KOLB C.R. & WILBERT L.J. JR. (1954) Sedimentary framework of the modern Mississippi delta. *J. sedim. Petrol.*, **24**, 76–99.
6.2, 6.5.1, 6.5.2, Fig. 6.19.

FITTON J.G. & HUGHES D.J. (1970) Volcanism and plate tectonics in the British Ordovician. *Earth-Planet Sci. Letts.*, **8**, 223–228.
14.4.2.

FLACH K.W., NETTLETON W.D., GILE L.H. & CADY J.G. (1969) Pedocementation: Induration by silica, carbonates, and sesquioxides in the Quaternary. *Soil Sci.*, **107**, 422–453.
3.6.2.

FLEXER A. (1968) Stratigraphy and facies development of Mount Scopus Group (Senonian–Paleocene) in Israel and adjacent countries. *Israel J. Earth-Sci.*, **17**, 85–114.
11.4.5.

FLEXER A. (1971) Late Cretaceous palaeogeography of northern Israel and its significance for the Levant Geology. *Palaeogeogr., Palaeoclimat., Palaeoecol.*, **10**, 293–316.
11.4.5, Fig. 11.46.

FLINT R.F. (1945) Glacial map of North America. *Spec. Pap. geol. Soc. Am.*, **60**, Pt. 1, Glacial Map; Pt. 2, Explanatory Notes, pp. 37.
13.5.1.

FLINT R.F. (1959) Glacial map of the United States east of the Rocky Mountains: scale 1:750,000. *Geol. Soc. America*, 2 sheets.
13.5.1.

FLINT R.F. (1971) *Glacial and Quaternary Geology*, pp. 892. John Wiley, New York.
13.1, 13.2, 13.3.2, 13.3.3, 13.3.4, 13.4.1, 13.4.3, 13.5.1.

FLINT R.F. (1975) Features other than diamict as evidence of ancient glaciations. In: *Ice Ages: Ancient and Modern* (Ed. by A.E. Wright and F. Moseley), pp. 121–136. *Geol. J. Spec. Issue*, **6.**
13.1.

FOLGER D.W. (1970) Wind transport of land-derived mineral, biogenic, and industrial matter over the North Atlantic. *Deep-Sea Res.*, **17**, 337–352.
11.3.2.

FOLK R.L. (1959) Practical petrographic classification of limestones. *Bull. Am. Ass. Petrol. Geol.*, **43**, 1–38.
10.1.1.

FOLK R.L. (1971) Longitudinal dunes of the northwestern edge of the Simpson Desert, Northern Territory, Australia, 1, Geomorphology and grain size relationships. *Sedimentology*, **16**, 5–54.
5.2.4.

FOLK R.L. (1973) Carbonate petrography in the post-Sorbian age. In: *Evolving Concepts in Sedimentology* (Ed. by R.N. Ginsburg), pp. 118–158. John Hopkins University Press, Baltimore.
10.1.1.

FOLK R.L. & MCBRIDE E. (1976a) Possible pedogenic origin of Ligurian ophicalcite: a Mesozoic calichified serpentinite. *Geology*, **4**, 327–332.
11.4.2.

FOLK R.L. & MCBRIDE E.F. (1976b) The Caballos Novaculite revisited, Part 1: origin of novaculite members. *J. sedim. Petrol.*, **46**, 659–669.
11.4.4.

FOLK R.L. & PITTMAN J.S. (1971) Length-slow chalcedony: a new testament for vanished evaporites. *J. sedim. Petrol.*, **41**, 1045–1058.
8.4.2.

FOREL F.A. (1887) Le ravin sous-lacustre du Rhône dans le lac Léman. *Bull. Soc. Vaudoise Sci. Nat.*, **23**, 85–107.
12.1.

FORRISTALL G.Z. (1974) Three-dimensional structure of storm-generated currents. *J. geophys. Res.*, **79**, 2721–2729.
9.4.3.

FRANCHETEAU J., CHOUKROUNE P., HEKINIAN R., LE PICHON X. & NEEDHAM H.D. (1976) Oceanic fracture zones do not provide deep sections in the crust. *Can. J. Earth Sci.*, **13**, 1223–1235.
14.3.1.

FRAKES L.A. & CROWELL J.C. (1967) Facies and paleogeography of Late Paleozoic diamictite, Falkland Islands. *Bull. geol. Soc. Am.*, **78**, 37–58.
13.4.1.

FRAKES L.A. & CROWELL J.C. (1969) Late Paleozoic Glaciation: Part I, South America. *Bull. geol. Soc. Am.*, **80**, 1007–1042.
13.5.2.

FRAKES L.A. & CROWELL J.C. (1970) Late Paleozoic Glaciation Part II, Africa exclusive of Karroo Basin. *Bull. geol. Soc. Am.*, **81**, 2261–2286.
13.5.2.

FRAKES L.A., FIGUEIREDO F.P.M. DE & FULFARO V. (1968) Possible fossil eskers and associated features from the Paraná Basin, Brazil. *J. sedim. Petrol.*, **38**, 5–12.
13.4.1.

FRAKES L.A., KEMP E.M. & CROWELL J.C. (1975) Late Paleozoic glaciation: Part IV, Asia. *Bull. geol. Soc. Am.*, **86**, 454–464.
13.5.2.

FRANCIS E.A. (1975) Glacial sediments: a selective review. In: *Ice Ages: Ancient and Modern* (Ed. by A.E. Wright and F. Moseley), pp. 43–68. *Geol. J. Spec. Issue*, **6.**
13.4.1.

FRANKE W., EDER W. & ENGEL W. (1975) Sedimentology of a Lower Carboniferous shelf margin. (Velbert Anticline, Rheinisches Schiefergebirge, W. Germany). *Neues Jb. Geol. Pälaont., Abh.*, **150**, 314–353.
11.4.4.

FRANZINI M., GRATZIU C. & SCHIAFFINO L. (1968). Sedimenti silicei non detritici dell' Appennino centro settentrionale, 1. La Formazione dei diaspri di Reppia (Genova). *Memorie Soc. tosc. Sci. nat.*, a, **75**, 154–203.
11.7.2.

FRAZIER D.E. (1967) Recent deltaic deposits of the Mississippi delta: their development and chronology. *Trans. Gulf-Cst. Ass. geol. Socs.*, **17**, 287–315.
6.5.2, 6.6, Fig. 6.34.

FRAZIER D.E. & OSANIK A. (1961) Point-bar deposits; Old River Locksite, Louisiana. *Trans. Gulf-Cst. Ass. geol. Socs.*, **11**, 121–137.
3.5.2, 3.9.4.

FRAZIER D.E. & OSANIK A. (1969) Recent peat deposits – Louisiana coastal plain. In: *Environments of Coal Deposition* (Ed. by E.C. Dapples and M.E. Hopkins), pp. 63–85. *Spec. Paper geol. Soc. Am.*, **114.**
6.6.

FRESHNEY E.C. (1970) Cyclical sedimentation in the Petrockstow Basin. *Proc. Ussher Soc.*, **2**, 179–189.
4.3.6.

FRESHNEY E.C. & FENNING P.J. (1967) The Petrockstow Basin. *Proc. Ussher Soc.*, **1**, 278–280.
4.3.6.

FREY R.W. (1971) Ichnology – the study of fossil and recent lebensspuren. In: *Trace Fossils* (Ed. by B.F. Perkins), pp. 91–125. Louisiana State Univ., Misc. Publ., 71–1.
9.8.

FREY R.W. (1972). Paleoecology and depositional environment of Fort Hays Limestone Member, Niobrara Chalk (Upper Cretaceous), West-central Texas. *Paleont. Contrib. Univ. Kansas*, **58**, 77.
11.4.5.

FREY R.W. (1975) *The Study of Trace Fossils. A synthesis of principles, problems and procedures in Ichnology*, pp. 562. Springer-Verlag.
9.8.

FREY R.W. & HOWARD J.D. (1969) A profile of biogenic sedimentary structures in a Holocene barrier island – salt marsh complex, Georgia. *Trans. Gulf-Cst. Ass. geol. Socs.*, **19**, 427–444.
7.2.3, 7.2.5.

FREY R.W. & MAYOU T.V. (1971) Decapod burrows in Holocene barrier island beaches and washover fans. *Senckenberg. Marit.*, **3**, 53–77.
7.2.3, 7.2.5.

FRIEDMAN G.M. (1959) Identification of carbonate minerals by staining methods. *J. sedim. Petrol.*, **29**, 87–97.
10.1.1.

FRIEDMAN G.M. (1961) Distinction between dune, beach and river sands from their textural characteristics. *J. sedim. Petrol.*, **31**, 514–529.
5.2.4.

FRIEDMAN G.M. (1965) Occurrence and stability relationships of aragonite, high-magnesian calcite and low-magnesian calcite under deep-sea conditions. *Bull. geol. Soc. Am.*, **76**, 1191–1196.
11.3.5.

FRIEND P.F. & MOODY-STUART M. (1970) Carbonate deposition on the river floodplains of the Wood Bay Formation (Devonian) of Spitsbergen. *Geol. Mag.*, **107**, 181–195.
3.9.2.

FRUSH M.P. & EICHER D.L. (1975) Cenomanian and Turonian Foraminifera and palaeoenvironments in the Big Bend region of Texas and Mexico. In: *The Cretaceous System in the Western Interior of North America* (Ed. by W.G.E. Caldwell), pp. 277–301. *Spec. Pap. geol. Ass. Canada*, **13.**
11.4.5.

FRYE J.C., WILLMAN H.B. & BLACK F.B. (1965) Outline of glacial geology of Illinois and Wisconsin. In: *The Quaternary of the United States*, pp. 43–61 (Ed. by H.E. Wright Jr. and D.G. Frey).
13.6.

FUGANTI A. (1966) Il tettonismo cretacico e la deposizione degli "scisti neri uraniferi" nel Trentino. In: *Atti del Symposium Internazionale sui giacimenti minerari delle Alpi*, Vol. II, pp. 341–346. Trento-Mendola.
11.4.4.

FULLER J.G.C.M. & PORTER J.W. (1969) Evaporite formations with petroleum reservoirs in Devonian and Mississippian of Alberta, Saskatchewan and North Dakota. *Bull. Am. Ass. Petrol. Geol.*, **53**, 909–926.
8.4.1, 8.5.2, Fig. 8.34.

FÜCHTBAUER H. (1967) Die Sandsteine in der Molasse nördlich der Alpen. *Geol. Rdsch.*, **56**, 266–300.
14.4.4.

FÜRSICH F.T. (1976) Fauna substrate relationships in the Corallian of England and Normandy. *Lethaia*, **8**, 151–172.
9.8.

FÜRSICH F.T. (1977) Corallian (Upper Jurassic) marine benthic associations from England and Normandy. *Palaeontology*, **20**, 337–385.
9.8, 9.12, Fig. 9.27.

GADD N.R. (1971) Pleistocene geology of the central St. Lawrence Lowland, pp. 153. *Mem. geol. Surv. Brch. Can.*, **359.**
13.3.4.

GADOW S. (1971) Der Golf von Gaeta (Tyrrhenisches Meer). I. Die sedimente. *Senckenberg. Marit.*, **3**, 103–133.
9.3.5.

GADOW S. & REINECH H.E. (1969) Ablandiger sand transport bei sturmfluten. *Senckenberg. Marit.*, **1**, 63–78.
9.4.3, 9.11.5.

GAGLIANO S.M. & VAN BEEK J.L. (1970) Geologic and geomorphic aspects of deltaic processes, Mississippi delta system. In: *Hydrologic and Geologic Studies of Coastal Louisiana*, Report **No. 1**, pp. 140. Centre for Wetland Resources, Louisiana State Univ.
Fig. 6.11.

GAGLIANO S.M. & VAN BEEK J.L. (1973) Environmental management in the Mississippi delta system. *Trans. Gulf-Cst. Ass. geol. Socs.*, **23**, 203–209.
6.1.

GAGLIANO S.M. & VAN BEEK J.L. (1975) An approach to multiuse management in the Mississippi delta system. In: *Deltas, Models for Exploration* (Ed. by M.L. Broussard), pp. 223–238. Houston Geological Society, Houston.
6.1, 6.5.1.

GAGLIANO S.M., LIGHT P. & BEEKER R.E. (1971) Controlled diversions in the Mississippi River delta system: an approach to environmental management. In: *Hydrologic and Geologic Studies of Coastal Louisiana*, Report **No. 8**, pp. 134. Centre for Wetland Resources, Louisiana State Univ.
6.5.1.

GALÁCZ A. & VÖRÖS A. (1972) Jurassic history of the Bakony Mountains and interpretation of principal lithological phenomena. *Földtani Közlöny*, **102**, 122–135.
11.4.4.

GALLI-OLIVIER C. (1969) Climate: a primary control of sedimentation in the Peru-Chile Trench. *Bull. geol. Soc. Am.*, **80**, 1849–1852.
12.4.3.

GALLOWAY W.E. (1968) Depositional systems of the Lower Wilcox Group, north-central Gulf Coast Basin. *Trans. Gulf-Cst. Ass. geol. Socs.*, **28**, 275–289.
6.7.4, Fig. 6.44.

GALLOWAY W.E. (1975) Process framework for describing the morphologic and stratigraphic evolution of the deltaic depositional systems. In: *Deltas, Models for Exploration* (Ed. by M.L. Broussard), pp. 87–98. Houston Geological Society, Houston.
6.2, 6.3, 6.4, Fig. 6.5.

GALLOWAY W.E. & BROWN L.F. JR. (1973) Depositional systems and shelf-slope relations on cratonic basin margin, Uppermost Pennsylvanian of north-central Texas. *Bull. Am. Ass. Petrol. Geol.*, **57**, 1185–1218.
6.7.2, 6.7.4, 12.8.3, Fig. 6.41, 12.56.

GALVIN C.J. JR. (1968) Breaker type classification on three laboratory beaches. *J. geophys. Res.*, **73**, 3651–3659.
7.2.1.

GARDNER J.V. (1975) Late Pleistocene carbonate dissolution cycles in the Eastern equatorial Atlantic. In: *Dissolution of Deep-sea Carbonates* (Ed. by W.V. Sliter, A.H.H. Bé and W.H. Berger). *Spec. Publ. Cusham Fndn. Foraminifer. Res.*, **13**, 129–141.
11.3.4.

GARDNER L.R. (1972) Origin of the Mormon Mesa Caliche, Clark County, Nevada. *Bull. geol. Soc. Am.*, **83**, 143–156.
3.6.2.

GARRELS R.M. & MACKENZIE F.T. (1971) *Evolution of Sedimentary Rocks*, pp. 397. Norton, New York.
1.1.

GARRETT P. (1970) Phanerozoic stromatolites: Non-competitive ecologic restriction by grazing and burrowing animals. *Science*, **169**, 171–173.
8.2.5.

GARRISON R.E. (1972) Inter- and intra-pillow limestones of the Olympic Peninsula, Washington. *J. Geol.*, **80**, 310–322.
11.4.2.

GARRISON R.E. (1973) Space-time relations of pelagic limestones and volcanic rocks, Olympic Peninsula, Washington. *Bull. geol. Soc. Am.*, **84**, 583–594.
11.4.2.

GARRISON R.E. (1974) Radiolarian cherts, pelagic limestones and igneous rocks in eugeosynclinal assemblages. In: *Pelagic Sediments: on Land and under the Sea* (Ed. by K.J. Hsü and H.C. Jenkyns), pp. 367–399. *Spec. Publ. int. Ass. Sediment.*, **1**.
11.4.2, Fig. 11.11.

GARRISON R.E. (1975) Neogene diatomaceous sedimentation in East Asia: a review with recommendations for further study. *Tech. Bull. Comm. co-ord. joint Prospect., Econ. Soc. Comm. Asia and Pacific, (U.N.)*, **9**, 57–69.
11.4.3, Fig. 11.26.

GARRISON R.E. & BAILEY E.H. (1967) Electron microscopy of limestones in the Franciscan Formation of California. *Prof. Pap. U.S. geol. Surv.*, **575-B**, B94–B100.
11.4.2.

GARRISON R.E. & FISCHER A.G. (1969) Deep-water limestones and radiolarites of the Alpine Jurassic. In: *Depositional Environments in Carbonate Rocks* (Ed. by G.M. Friedman), pp. 20–56. *Spec. Publ. Soc. econ. Paleont. Miner.*, **14**, Tulsa.
11.4.4.

GARRISON R.E., HEIN J.R. & ANDERSON F.T. (1973) Lithified carbonate sediment and zeolitic tuff in basalts, Mid-Atlantic Ridge. *Sedimentology*, **20**, 399–410.
11.3.2.

GARRISON R.E., SCHLANGER S.O. & WACHS D. (1975) Petrology and palaeogeographic significance of Tertiary nannoplankton-foraminiferal limestones, Guam. *Palaeogeogr. Palaeoclim. Palaeoecol.*, **17**, 49–64.
11.4.2.

GARRISON R.E., ROWLAND S.M., HORAN L.J. & MOORE J.C. (1975) Petrology of siliceous rocks recovered from marginal seas of western Pacific, Leg 31, Deep Sea Drilling Project. In: *Initial Reports of the Deep Sea Drilling Project*, **21** (D.E. Karig, J.C. Ingle Jr. *et al.*), pp. 519–530. U.S. Government Printing Office, Washington.
11.3.5.

GARSON M.S. & KRS M. (1976) Geophysical and geological evidence of the relationship of Red Sea transverse tectonics to ancient fractures. *Bull. geol. Soc. Am.*, **87**, 169–181.
14.3.1.

GARTNER S. & GENTILE R. (1972) Problematic Pennsylvanian coccoliths from Missouri. *Micropaleontology*, **18**, 401–404.
11.4.4.

GASS I.G., SMITH A.G. & VINE F.J. (1975) Origin and emplacement of ophiolites. In: *Geodynamics Today*, pp. 54–64. The Royal Society, London.
11.4.2.

Gebelein C.D. (1969) Distribution, morphology, and accretion rate of recent subtidal algal stomatolites, Bermuda. *J. sedim. Petrol.*, **39**, 49–69.
8.2.5.

Germann K. (1971) Mangan-Eisen-führende Knollen und Krusten in jurassichen Rotkalken der Nördlichen Kalkalpen. *Neues Jb. Geol. Paläont.*, Mh., **1971**, 133–156.
11.4.4.

Gevirtz J.L. & Friedman G.M. (1966) Deep-sea carbonate sediments of the Red Sea and their implications on marine lithification. *J. sedim. Petrol.*, **36**, 143–151.
11.3.5.

Gignoux M. (1936) *Géologie Stratigraphique*, 2nd Edn, pp. 709. Masson et Cie, Paris.
11.1.2.

Gigot P., Gubler Y. & Haccard D. (1975) Relations entre depôts et tectonique synsedimentaire (en extension et en compression). Examples pris dans des bassins tertiaires des Alpes du Sud et de Haute-Provence, *9th Int. Congr. Sedimentol., Nice 1975. Thème 4(1)*, 157–162.
14.4.3.

Gilbert G.K. (1885) The topographic features of lake shores. *Ann. Rept. U.S. geol. Surv.*, **5**, 75–123.
4.1, 4.2.3, 6.2, 7.2.7, Fig. 6.1.

Gilbert G.K. (1890) Lake Bonneville *Mon. U.S. geol. Surv.*, **1**, pp. 438.
6.2.

Gile L.M., Peterson F.F. & Grossman R.B. (1966) Morphological and genetic sequences of carbonate accumulation in desert soils. *Soil Sci.*, **101**, 347–360.
3.6.2, Tab. 3.1.

Gill A.E. (1973) Circulation and bottom water production in the Weddell Sea. *Deep-Sea Res.*, **20**, 111–140.
13.3.7.

Gillberg G. (1965) Till distribution and ice movements on the northern slopes of the South Swedish Highlands. *Geol. Føren. Stockh. Førh.*, **86**, 433–484.
13.4.1.

Gillberg G. (1968a) Distribution of different limestone material in till. *Geol. Føren. Stockh. Førh.*, **89**, 401–409.
13.4.1.

Gillberg G. (1968b) Lithological distribution and homogeneity of glaciofluvial material. *Geol. Føren. Stockh. Førh.*, **90**, 189–204.
13.3.3.

Gilliland W.N. & Meyer G.P. (1976) Two classes of transform faults. *Bull. geol. Soc. Am.*, **87**, 1127–1130.
13.3.3.

Gingerich P.D. (1969) Markov analysis of cyclic alluvial sediments. *J. sedim. Petrol.*, **39**, 330–332.
2.1.2.

Ginsburg R.N. (1956) Environmental relationships of grain size and constituent particles in some south Florida carbonate sediments. *Bull. Am. Ass. Petrol. Geol.*, **40**, 2384–2427.
Fig. 10.20b, 10.21b.

Ginsburg R.N. (1957) Early diagenesis and lithification of shallow water carbonate sediments in south Florida. In: *Regional Aspects of Carbonate Deposition* (Ed. by R.J. Le Blanc and J.G. Breeding), pp. 80–99. *Spec. Publ. Soc. econ. Paleont. Miner.*, **5**, Tulsa.
10.1.1.

Ginsburg R.N. (1964) South Florida carbonate sediments. *Guidebook for Field Trip no.* **1**, *Geol. Soc. Am. Convention 1964*, pp. 72. Geol. Soc. Am., New York.
10.1.1.

Ginsburg R.N. (1972) Editor. *South Florida Carbonate Sediments*, Sedimenta II, pp. 72. University of Miami.
10.4.1.

Ginsburg R.N. (1975) *Tidal Deposits: A Casebook of Recent Examples and Fossil Counterparts*. pp. 428. Springer-Verlag, Berlin.
10.2.5.

Ginsburg R.N. & Hardie L.A. (1975) Tidal and storm deposits, northeastern Andros Island, Bahamas. In: *Tidal Deposits: A Casebook of Recent Examples and Fossil Counterparts* (Ed. by R.N. Ginsburg), pp. 201–208, Springer-Verlag, Berlin.
10.2.4, Fig. 10.7A, 10.7B.

Ginsburg R.N. & James N.P. (1974) Holocene carbonate sediments of continental shelves. In: *The Geology of Continental Margins* (Ed. by C.A. Burk and C.L. Drake), pp. 137–155. Springer-Verlag, Berlin.
10.2.1, 10.2.2, 10.2.3, 10.2.4, Fig. 10.17, 10.21A, 10.27A, 10.27B.

Ginsburg R.N. & Lowenstam H.A. (1958) The influence of marine bottom communities on the depositional environment of sediments. *J. Geol.*, **66**, 310–318.
10.4.2.

Ginsburg R.N., Shinn E.A. & Schroeder J.H. (1968) Submarine cementation and internal sedimentation within Bermuda reefs. *Spec. Pap. geol. Soc. Am.*, **115**, 78–79.
10.1.1.

Giresse P. & Odin G.S. (1973). Nature minéralogique et origine des glauconies du plateau continental du Gabon et du Congo. *Sedimentology*, **20**, 457–488.
11.3.6.

Glasby G.P. (Editor) (1977) *Marine Manganese Deposits*, pp. 523. Elsevier, Amsterdam.
11.3.1.

Glasby G.P. (1978) Deep-sea manganese nodules in the stratigraphic record: evidence from DSDP cores. *Mar. Geol.*, **28**, 51–64.
11.4.2.

Glassley W. (1974). Geochemistry and tectonics of the Crescent Volcanics, Olympic Peninsula, Washington. *Bull. geol. Soc. Am.*, **85**, 785–794.
11.4.2.

Glennie K.W. (1970) Desert sedimentary environments. *Developments in Sedimentology, No.* **14**, pp. 222. Elsevier, Amsterdam.
3.4.4, 3.6.2, 5.2.4, 5.2.6, 5.3.2, 8.2.7, Fig. 5.1, 5.5, 5.6, 5.8.

Glennie K.W. (1972) Permian Rotliegendes of Northwest Europe interpreted in the light of modern desert sedimentation studies. *Bull. Am. Ass. Petrol. Geol.*, **56**, 1048–1071.
5.3.2, 5.3.4, 5.3.6, Fig. 5.15, 5.16.

Glennie K.W. & Evans G. (1976) A reconnaissance of the recent sediments of the Ranns of Kutch, India. *Sedimentology*, **23**, 625–647.
8.1.1.

Glikson A.Y., Derrick G.M., Wilson I.H. & Hill R.M. (1974) Structural evolution and geotectonic nature of the Middle Proterozoic Mount Isa Fault trough, northwestern Queensland. *Bur. Miner. Res., Geol. Geophys.*, pp. 18.
14.4.1.

Goldberg E.D. (1954) Marine geochemistry, I. Chemical scavengers of the sea. *J. Geol.*, **62**, 249–265.
11.3.2.

Goldring R. (1962) The Bathyal Lull: Upper Devonian and Lower Carboniferous sedimentation in the Variscan geosyncline. In: *Some Aspects of the Variscan Fold Belt* (Ed. by K. Coe), pp. 75–91.

Manchester University Press, Manchester.
14.2.5.

GOLDRING R. (1965) Sediments into rock. *New Scientist*, **26**, 863–865.
2.2.3.

GOLDRING R. & BRIDGES P. (1973) Sublittoral sheet sandstones. *J. sedim. Petrol.*, **43**, 736–747.
9.1.1, 9.9, 9.10.3, 9.11.2.

GOLDRING W. (1938) Algal barrier reefs in the lower Ozarkian of New York with a chapter on the importance of coralline algae as reef builders through the ages. *New York State Mus. Bull.*, **315**, 5–75.
10.4.2.

GOLDSMITH V., BYRNE R.J., SALLENGER A.H. & DRUCKER D.M. (1976) The influence of waves on the origin and development of the offset coastal inlets of the southern Delmarva Peninsula, Virginia. In: *Estuarine Research*, Vol. II Geology and Engineering (Ed. by L.E. Gronin), pp. 183–200. Academic Press, London.
7.2.4.

GOLDTHWAIT R.P. (1951) Development of end moraines in East-Central Baffin Island. *J. Geol.*, **59**, 567–577.
13.3.2.

GOLE C.V. & CHITALE S.V. (1966) Inland delta building activity of Kosi River. *J. Hydraul. Proc Am. Soc. civ. Engrs.*, **92**, 111–126.
3.2.1, 3.2.3, 3.4, 3.5, 3.9.4, Fig. 3.3.

GOLUBIC S. (1976) Organisms that build stromatolites. In: *Stromatolites* (Ed. by M.R. Walter), pp. 113–148. Elsevier, Amsterdam.
8.2.6.

GOMBERG D.N. (1973) Drowning of the Floridan platform margin and formation of a condensed sedimentary sequence (Abstract). *Geol. Soc. Am.* (*Abstract with Programs*), **5**, 640.
11.3.6.

GORSLINE D.S. & EMERY K.O. (1959) Turbidity-current deposits in San Pedro and Santa Monica basins off southern California. *Bull. geol. Soc. Am.*, **70**, 279–290.
12.2.4, 12.4.5, Fig. 12.32.

GORSLINE D.S. & MILLIGAN D.B. (1963) Phosphatic lag deposits along the margin of the Pourtales Terrace. *Deep-Sea Res.*, **10**, 259–262.
11.3.6.

GOUDARZI G.H. (1970) Non-metallic mineral resources: Saline deposits, silica sand, sulfur and trona. *Prof. Pap. U.S. geol. Surv.*, **660**, 83–93.
5.2.6.

GOUDIE A. (1973) *Duricrusts in Tropical and Subtropical Landscapes*, pp. 174. Clarendon Press, Oxford.
3.6.1, 3.6.2.

GOULD H.R. & MCFARLAN E. (1959) Geological history of the chenier plain, southwestern Louisiana. *Trans. Gulf-Cst. Ass. geol. Socs.*, **9**, 261–270.
7.4, Fig. 7.41.

GRAAF F.R. VAN DE (1972) Fluvial deltaic facies of the Castlegate Sandstone (Cretaceous), East Central Utah. *J. sedim. Petrol.*, **42**, 558–571.
6.7.1.

GRAAFF W.J.E. VAN DE (1971) Three Upper Carboniferous, limestone-rich, high-destructive, delta systems with submarine fan deposits, Cantabrian Mountains, Spain. *Leidse geol. Meded.*, **46**, 157–235.
12.8.3.

GRABAU A.W. (1904) On the classification of sedimentary rocks. *Am. Geologist*, **33**, 228–247.
10.1.1.

GRABAU A.W. (1913) *Principles of Stratigraphy*, pp. 1185. A.G. Seiler and Co., New York.
10.1.1, 11.1.2.

GRACHT W.A.J.M. VAN DER VAN WATERSCHOOT (1928) The problem of continental drift. In: *Theory of Continental Drift*, pp. 1–75, 197–226. *Am. Ass. Petrol. Geol.*, Tulsa.
14.2.5.

GRAHAM S.A., DICKINSON W.R. & INGERSOLL R.V. (1975) Himalayan-Bengal model for flysch dispersal in the Appalachian-Ouachita system. *Bull. geol. Soc. Am.*, **86**, 273–286.
14.3.4, 14.4.2.

GRANT A.C. (1972) The continental margin off Labrador and eastern Newfoundland – morphology and geology. *Can. J. Earth Sci.*, **9**, 1394–1430.
13.5.1.

GRANVILLE A. (1975) The recovery of deep-sea minerals: problems and prospects. *Minerals Sci. Engng.*, **3**, 170–188.
14.3.1.

GRAVENOR C.P. & KUPSCH W.O. (1959) Ice-disintegration features in Western Canada. *J. Geol.*, **67**, 48–64.
13.3.2.

GREENSMITH J.T. (1965) *Petrology of the Sedimentary Rocks*, pp. 408 (4th Edn). Allen & Unwin, Hemel Hempstead.
1.1.

GREENSMITH J.T. & TUCKER E.V. (1968) The origin of Holocene shell deposits in the chenier plain facies of Essex (Great Britain). *Mar. Geol.*, **7**, 403–425.
7.4.

GREENSMITH J.T. & TUCKER E.V. (1975) Dynamic structures in the Holocene chenier plain setting of Essex, England. In: *Nearshore Sediment Dynamics and Sedimentation* (Ed. by J.R. Hails and A. Carr), pp. 251–272. John Wiley, New York.
7.4.

GREER S.A. (1975) Sandbody geometry and sedimentary facies at the estuary-marine transition zone, Ossabaw Sound, Georgia: a stratigraphic model. *Senckenberg. Mar.*, **7**, 105–135.
7.5.1, Fig. 7.44.

GREGORY J.L. (1966) A Lower Oligocene delta in the sub-surface of south-eastern Texas. *Trans. Gulf-Cst. Ass. geol. Socs.*, **16**, 227–241.
6.7.4.

GREINER H.R. (1962) Facies and sedimentary environments of Albert Shale, New Brunswick. *Bull. Am. Ass. Petrol. Geol.*, **46**, 219–234.
4.3.4.

GRESSLY A. (1838) Observations géologiques sur le Jura Soleurois. *Neue Denkschr. allg. schweiz, Ges. ges. Naturw.*, **2**, 1–112.
2.1.1.

GRIFFIN J.J., WINDOM H. & GOLDBERG E.D. (1968) The distribution of clay minerals in the world ocean. *Deep-Sea Res.*, **15**, 433–459.
11.3.4.

GRIGGS G.B. & KULM L.D. (1970) Sedimentation in Cascadia Deep-Sea Channel. *Bull. geol. Soc. Am.*, **81**, 1361–1384.
12.2.4, 12.5.1, Fig. 12.40.

GROEN P. (1967) *The Waters of the Sea*, pp. 328. Van Nostrand Co. Ltd., London.
9.4.1.

GROSS M.G. (1965) Carbonate deposits on Plantagenet Bank near Bermuda. *Bull. geol. Soc. Am.*, **76**, 1283–1290.
11.3.3.

GROVE A.T. (1970) The rise and fall of Lake Chad. *Geogr. Mag.*, **42**, 432–439.
4.2.1.

GROVER N.C. & HOWARD C.S. (1938) The passage of turbid water through Lake Mead. *Trans. Am. Soc. Civil Engrs.*, **103,** 720–790.
12.1.

GUBER A.L. (1969) Sedimentary phosphate method for estimating palaeosalinities: a palaeontological assumption. *Science,* **166,** 744–746.
9.8.

GUILD P.W. (1974) Distribution of metallogenic provinces in relation to major earth features. *Schriftenr. Erdwiss. Komm. Oesterr. Akad. Wiss.,* **1,** 10–24.
14.2.4.

GULBRANDSEN R.A. (1969). Physical and chemical factors in the formation of marine apatite. *Econ. Geol.,* **69,** 365–382.
11.3.6, 11.4.4.

GUNATILAKA A. (1975) Some aspects of the biology and sedimentology of laminated algal mats from Mannar Lagoon, north-west Ceylon. *Sedim. Geol.,* **14,** 275–300.
8.2.5, Tab. 8.1.

GUSTAVSON T.C. (1975) Sedimentation and physical limnology in proglacial Malaspina Lake, South-eastern Alaska. In: *Glaciofluvial and Glaciolacustrine Sedimentation* (Ed. by A.V. Jopling & B.C. McDonald), pp. 249–263. *Spec. Publ. Soc. econ. Paleont. Miner.,* **23,** Tulsa.
13.3.4.

GWINNER M.P. (1976) Origin of the Upper Jurassic limestones of the Swabian Alb (Southwest Germany). *Contr. Sedimentology,* **5,** 1–75.
10.4.2.

GÖRLER K. & REUTTER K-J. (1968) Entstehung und Merkmale der Olisthostrome. *Geol. Rdsch.,* **57,** 484–514.
12.2.3.

HAAF E. TEN (1959) *Graded Beds of the Northern Apennines,* pp. 102. Thesis, University of Groningen.
12.2.4, 12.6.2.

HAARMANN E. (1930) *Die Oszillationstheorie, eine Erklärung der Krustenbewegungen von Erde und Mond,* pp. 260. Ferdinand Enke Verlag, Stuttgart.
14.2.2.

HADLEY M.L. (1964) Wave induced bottom currents in the Celtic Sea. *Mar. Geol.,* **2,** 164–167.
9.1.1.

HAGEN G.M. & LOGAN B.W. (1975) Prograding tidal-flat sequences: Hutchinson Embayment, Shark Bay, Western Australia. In: *Tidal Deposits: A Casebook of Recent Examples and Fossil Counterparts* (Ed. by R.N. Ginsburg), pp. 215–222. Springer-Verlag, Berlin.
10.2.5.

HAILS J.R. (1968) The late Quaternary history of part of the mid-north coast, New South Wales, Australia. *Trans. Inst. Br. Geogr.,* **44,** 133–149.
7.2.

HAILS J.R. & HOYT J.H. (1969) An appraisal of the evolution of the lower coastal plain of Georgia, U.S.A. *Trans. Inst. Br. Geogr.,* **46,** 53–68.
7.2.2.

HAJASH A. (1975) Hydrothermal processes along mid-ocean ridges: an experimental investigation. *Contr. Miner. Petrol.,* **53,** 205–226.
11.3.2.

HALL J. (1859) Description and figures of the organic remains of the lower Helderberg Group and the Oriskany Sandstone. *Natural History of New York, Palaeontology,* pp. 544. Geol. Surv., Albany, N.Y., 3.
14.2.1, 14.2.3.

HALLAM A. (1975) *Jurassic Environments,* pp. 269. Cambridge Univ. Press, Cambridge.
9.12.

HALLAM A. & SELLWOOD B.W. (1976) Middle Mesozoic sedimentation in relation to tectonics in the British area. *J. Geol.,* **84,** 301–321.
9.12, Fig. 9.54.

HALPERN D. (1976) Structure of a coastal upwelling event observed off Oregon during July 1973. *Deep-Sea Res.,* **23,** 495–508.
9.6.2.

HAM W.E. (Editor) (1962) *Classification of Carbonate Rocks – a Symposium,* 279 pp. *Mem. Am. Ass. Petrol. Geol.,* **1,** Tulsa.
1.1, 10.1.1.

HAMILTON E.L. (1967) Marine geology of abyssal plains in the Gulf of Alaska. *J. geophys. Res.,* **72,** 4189–4213.
12.4.2, 14.3.1.

HAMILTON W. (1970) The Uralides and the motion of the Russian and Siberian platforms. *Bull. geol. Soc. Am.,* **81,** 2553–2576.
11.4.2.

HAMILTON W. & KRINSLEY D. (1967) Upper Paleozoic glacial deposits of South Africa and Southern Australia. *Bull. geol. Soc. Am.,* **78,** 783–800.
13.5.2.

HAMPTON M.A. (1972) The role of subaqueous debris flow in generating turbidity currents. *J. sedim. Petrol.,* **42,** 775–793.
12.2.3, 12.2.4.

HAMPTON M.A. & COLBURN I.P. (1967) Directional features of an experimental turbidite fan. A report of progress. *J. sedim. Petrol.,* **37,** 509–513.
12.5.1.

HANCOCK J.M. (1975a) The petrology of the Chalk. *Proc. Geol. Ass.,* **86,** 499–535.
11.4.5.

HANCOCK J.M. (1975b) The sequence of facies in the Upper Cretaceous of Northern Europe compared with that in the Western Interior. In: *The Cretaceous System in the Western Interior of North America* (Ed. by W.G.E. Caldwell), pp. 83–118. *Spec. Pap. geol. Ass. Canada,* **13.**
11.4.5.

HANCOCK J.M. & SCHOLLE P.A. (1975) Chalk of the North Sea. In: *Petroleum and the Continental Shelf of North-west Europe,* Vol. 1, Geology (Ed. by A.W. Woodland), pp. 413–425. Applied Science Publishers, Barking.
11.4.5.

HANER B.E. (1971) Morphology and sediments of Redondo Submarine Fan, southern California. *Bull. geol. Soc. Am.,* **82,** 2413–2432.
12.3.2, 12.5.1, 12.6.3.

HANLEY J.H. & STEIDTMANN J.R. (1973) Petrology of limestone lenses in the Casper Formation, Southernmost Laramie Basin, Wyoming and Colorado. *J. sedim. Petrol.,* **43,** 428–434.
4.3.6.

HANSEN E. (1971) *Strain Facies.* Springer-Verlag, New York.
12.2.2.

HÄNTZSCHEL W. (1975) Trace fossils and problematica. In: *Treatise on Invertebrate Palaeontology* (Ed. by C. Teichert), Part W, Miscellanea, Supplement **1,** pp. 269. Geol. Soc. Am., and Univ. Kansas Press.
9.8.

HARBAUGH J.W. & BONHAM-CARTER G. (1970) *Computer Simulation in*

Geology, pp. 98. Wiley-Interscience, New York.
2.1.2.

HARDER E.C. (1919) Iron-depositing bacteria and their geologic relations. *Prof. Pap. U.S. geol. Surv.*, **113,** pp. 89.
4.2.4.

HARDER H. (1970) Boron content of sediments as a tool in facies analysis. *Sedim. Geol.*, **4,** 153–175.
9.8.

HARDIE L.A., SMOOT J.P. & EUGSTER H.P. (1978) Saline lakes and their deposits: a sedimentological approach. In: *Modern and Ancient Lake Sediments.* (Ed. by A. Matter and M. Tucker), pp. 7–41. *Spec. Publ. int. Ass. Sediment.*, **2,** Blackwell Scientific Publs.
4.2.1, 4.2.4, 5.2.7.

HARDING T.P. (1974) Petroleum traps associated with wrench faults. *Bull. Am. Assoc. Petrol. Geol.*, **58,** 1290–1304.
14.4.3, Fig. 14.32.

HARLAND W.B. (1964) Critical evidence for a great Infra-Cambrian glaciation. *Geol. Rdsch.*, **54,** 45–61.
13.5.4.

HARLAND W.B. (1965) The tectonic evolution of the Arctic–North Atlantic region. *Phil. Trans. R. Soc., Ser. A,* **258,** 59–75.
14.3.

HARLAND W.B. (1969) Contribution of Spitsbergen to understanding of tectonic evolution of North Atlantic region. In: *North Atlantic – Geology and Continental Drift: a Symposium* (Ed. by M. Kay), pp. 817–851. *Mem. Am. Assoc. Petrol. Geol.*, **12,** Tulsa.
14.4.3.

HARLAND W.B. (1971) Tectonic transpression in Caledonian Spitzbergen. *Geol. Mag.*, **108,** 27–42.
14.3.3, 14.4.3.

HARLAND W.B., HEROD K. & KRINSLEY D.H. (1966) The definition and identification of tills and tillites. *Earth Sci. Rev.*, **2,** 225–256.
13.1, 13.4.1.

HARLETT J.C. & KULM L.D. (1973) Suspended sediment transport on the northern Oregon continental shelf. *Bull. geol. Soc. Am.*, **83,** 3815–3826.
9.6.2.

HARMS J.C. (1974) Brush Canyon Formation, Texas: a deep-water density current deposit. *Bull. geol. Soc. Am.*, **85,** 1763–1784.
10.7.2.

HARMS J.C. (1975) Stratification produced by migrating bed forms. In: *Depositional Environments as Interpreted from Primary Sedimentary Structures and Stratification Sequences,* pp. 45–61. *Soc. econ. Paleont. Miner., Short Course* **2,** Dallas, Texas.
9.9, 9.11.1, 9.11.2.

HARMS J.C. & FAHNESTOCK R.K. (1965) Stratification, bed forms, and flow phenomena (with an example from the Rio Grande). In: *Primary Sedimentary Structures and their Hydrodynamic Interpretation* (Ed. by G.V. Middleton), pp. 84–115. *Spec. Publ. Soc. econ. Paleont. Miner.*, **12,** Tulsa.
2.2.1, 3.4.1, 9.9., 12.2.4.

HARMS J.C., MACKENZIE D.B. & MCCUBBIN D.G. (1963) Stratification in modern sands of the Red River, Louisiana. *J. Geol.*, **71,** 566–580.
3.5.2, Fig. 3.25.

HARMS J.C., SOUTHARD J., SPEARING D.R. & WALKER R.G. (1975) *Depositional environments as interpreted from primary sedimentary structures and stratification sequences. Lecture Notes: Soc. econ. Paleont. Miner., Short Course* **2.** Dallas, Texas, pp. 161.
1.1, 3.4.1.

HARRISON W., NORCROSS J.J., PORE N.A. & STANLEY E.M. (1967) Shelf waters off the Chesapeake Bight. *Environ. Sci. Services Admin. Prof. Paper* **3,** 1–82.
9.6.3.

HARTMANN M. (1964) Zur Geochemie von Mangan und Eisen in der Ostsee. *Meyniana,* **14,** 3–20.
11.4.4.

HARVEY J.G. (1976) *Atmosphere and Ocean: Our Fluid Environments,* pp. 143. Artemis Press, Sussex.
Fig. 9.8, 9.10.

HATCH F.H., RASTALL R.H. & BLACK M. (1938) *The Petrology of the Sedimentary Rocks* (3rd Edn), pp. 383. Allen and Unwin, London.
11.1.2.

HATCHER P.G. & SEGAR D.A. (1976) Chemistry and continental sedimentation. In: *Marine Sediment Transport and Environmental Management* (Ed. by D.J. Stanley and D.J.P. Swift), pp. 461–477. John Wiley & Sons, Inc., New York.
9.3.6, Fig. 9.6.

HATTIN D.E. (1975a) Stratigraphy and depositional environment of Greenhorn Limestone (Upper Cretaceous) of Kansas. *Bull. Kansas geol. Surv.*, **209,** 1–128.
11.4.5.

HATTIN D.E. (1975b) Petrology and origin of faecal pellets in Upper Cretaceous strata of Kansas and Saskatchewan. *J. sedim. Petrol.*, **45,** 686–696.
11.4.5.

HAUG E. (1900) Les géosynclinaux et les aires continentales. Contribution a l'étude des regressions et des transgressions marines. *Bull. Soc. géol. France,* **28**(3), 617–711.
14.2.1, 14.2.2.

HAWKINS L.K. (1969) Visual observations of manganese deposits on the Blake Plateau. *J. geophys. Res.*, **74,** 7009–7017.
11.3.6.

HAY R.L. (1968) Chert and its sodium silicate precursors in sodium carbonate lakes of East Africa. *Contr. Miner. and Petrol.*, **17,** 255–274.
4.2.4, 4.3.2.

HAYES D.E., FRAKES L.A. *et al.* (1975a) Sites 270, 271, 272. In: *Initial Reports of the Deep Sea Drilling Project,* **28** (D.E. Hayes, L.A. Frakes *et al.*), pp. 211–334. U.S. Printing Office, Washington.
13.4.3, 13.5.1.

HAYES D.E., FRAKES L.A. *et al.* (1975b) Site 273. In: *Initial Reports of the Deep Sea Drilling Project,* **28** (D.E. Hayes, L.A. Frakes *et al.*), pp. 335–368. U.S. Government Printing Office, Washington.
13.4.3.

HAYES M.O. (1967a) Hurricanes as geological agents: case studies of Hurricanes Carla, 1961, and Cindy, 1963. *Rep. Invest. Bur. econ. Geol.*, Austin, Texas, **61,** 54.
7.2.1, 7.2.5, 9.3.1, 9.4.3, 9.6.1, 9.11.5.

HAYES M.O. (1967b) Relationship between coastal climate and bottom sediment type on the inner continental shelf. *Mar. Geol.*, **5,** 111–132.
9.3.4, Fig. 9.2.

HAYES M.O. (1976) Morphology of sand accumulation in estuaries: an introduction to the symposium. In: *Estuarine Research,* Vol. **II** Geology and Engineering (Ed. by L.E. Gronin), pp. 3–22. Academic Press, London.
7.1, 7.2.3, 7.2.5, Fig. 7.2, 7.3, 7.28, 7.43.

HAYES M.O. & KANA T.W. (1976) *Terrigenous Clastic Depositional Environments* – Some Modern Examples, pp. I-131, II-184, *Tech. Rept.* **11**-CRD, Coastal Res. Div., Univ. South Carolina.
7.1, 7.2.2, 7.2.4, 7.2.5, Fig. 7.10.

HAYNES C.V. JR. (1968) Geochronology of late-Quaternary alluvium.

In: *Means of Correlation of Quaternary Successions* (Ed. by R.B. Morrison and H.E. Wright Jr.), pp. 591–615. Univ. Utah Press, Salt Lake City.
3.6.2.

Hays J.D. & Pitman W.C. III (1973) Lithospheric plate motions, sea-level changes and climatic and ecological consequences. *Nature,* **246,** 18–22.
11.4.5.

Hays J.D., Cook H.E. III *et al.* (1972) *Initial Reports of the Deep Sea Drilling Project,* **9,** pp. 1205. U.S. Government Printing Office, Washington.
11.3.4.

Heath G.R. (1969) Mineralogy of Cenozoic deep-sea sediments from the equatorial Pacific Ocean. *Bull. geol. Soc. Am.,* **80,** 1997–2018.
11.3.4.

Heath G.R. (1974) Dissolved silica and deep-sea sediments. In: *Studies in Paleo-oceanography* (Ed. by W.W. Hay), pp. 77–93. *Spec. Publ. Soc. econ. Paleont. Miner.,* **20,** Tulsa.
11.3.4.

Heckel P.H. (1972) Recognition of ancient shallow marine environments. In: *Recognition of Ancient Sedimentary Environments* (Ed. by J.K. Rigby and W.K. Hamblin), pp. 226–296. *Spec. Publ. Soc. econ. Paleont. Miner.,* **16,** Tulsa.
9.1, 9.10.2, Fig. 9.26.

Heckel P.H. (1974) Carbonate buildups in the geological record: a review. In: *Reefs in Time and Space* (Ed. by L.F. Laporte), pp. 90–154. *Spec. Publ. Soc. econ. Paleont. Miner.,* **18,** Tulsa.
10.4.1, 10.4.2, Fig. 10.38.

Hedberg H.D. (1937) Stratigraphy of the Rio Querecual section of north-eastern Venezuela. *Bull. geol. Soc. Am.,* **48,** 1971–2204.
11.4.4.

Hedberg H.D. (1974) Relation of methane generation to undercompacted shales, shale diapirs and mud volcanoes. *Bull. Am. Ass. Petrol. Geol.,* **58,** 661–673.
6.8.1.

Hedgepeth J.W. (1957) Treatise on marine ecology and palaeoecology. Vol. 1: Ecology. *Mem. Geol. Soc. Am.,* **67,** pp. 1296.
9.1.1.

Heezen B.C. (1963) Turbidity currents. In: *The Sea* (Ed. by M.N. Hill), **3,** 742–775. Wiley, New York.
12.2.4.

Heezen B.C. & Ewing M. (1952) Turbidity currents and submarine slumps, and the 1929 Grand Banks earthquake. *Am. J. Sci.,* **250,** 849–873.
12.1, 12.2.4.

Heezen B.C. & Ewing M. (1955) Orléansville earthquake and turbidity currents. *Bull. Am. Ass. Petrol. Geol.,* **39,** 2505–2514.
12.2.4.

Heezen B.C. & Hollister C.D. (1964a) Deep-sea current evidence from abyssal sediments. *Mar. Geol.,* **1,** 141–174.
12.2.7.

Heezen B.C. & Hollister C.D. (1964b) Turbidity currents and glaciation. In: *Problems in Palaeoclimatology* (Ed. by A.E.M. Nairn), pp. 99–112. Interscience, London.
12.2.4, 12.2.7, 12.4.1.

Heezen B.C. & Hollister C.D. (1971) *The Face of the Deep,* pp. 659. Oxford Univ. Press, New York.
11.3.3, 12.2.4, 12.2.7, 12.4.1, Fig. 12.14, 12.15, 12.24, 12.25, 12.26.

Heezen B.C. & Laughton A.S. (1963) Abyssal plains. In: *The Sea* (Ed. by M.N. Hill), **3,** 312–364. Wiley, New York.
12.4.1, 12.4.2.

Heezen B.C. & Menard H.W. (1963) Topography of the deep-sea floor. In: *The Sea* (Ed. by M.N. Hill), **3,** 233–280. John Wiley, New York.
12.5.1.

Heezen B.C. & Rawson M. (1977) Visual observations of contemporary current erosion and tectonic deformation on the Cocos Ridge Crest. *Mar. Geol.,* **23,** 173–196.
11.3.3.

Heezen B.C., Ericson D.B. & Ewing M. (1954) Further evidence for a turbidity current following the 1929 Grand Banks earthquake. *Deep-sea Res.,* **1,** 193–202.
12.2.4.

Heezen B.C., Ewing M. & Ericson D.B. (1951) Submarine topography in the North Atlantic. *Bull. geol. Soc. Am.,* **62,** 1407–1410.
12.3.1.

Heezen B.C., Hollister C.D. & Ruddiman W.F. (1966) Shaping of the continental rise by deep geostrophic contour currents. *Science,* **152,** 502–508.
12.2.7.

Heezen B.C., Menzies R.J., Schneider E.D. *et al.* (1964) Congo Submarine Canyon. *Bull. Am. Ass. Petrol. Geol.,* **48,** 1126–1149.
12.5.2.

Heezen B.C., Tharp M. & Ewing M. (1959) The floors of the oceans. *Spec. Pap. geol. Soc. Am.,* **65,** 122.
12.3.1, 12.4.2.

Heezen B.C., Matthews J.L., Catalano R., Natland J., Coogan A., Tharp M. & Rawson M. (1973) Western Pacific guyots. In: *Initial Reports of the Deep Sea Drilling Project,* **20,** (B.C. Heezen, I.D. MacGregor *et al.*), pp. 653–723. U.S. Government Printing Office, Washington.
11.3.3.

Heim A. & Gansser A. (1939) Central Himalaya, geological observations of the Swiss Expedition, 1936. *Mem. Soc. Helv. Sci. nat.,* **73,** pp. 245.
11.4.4.

Hekinian R. & Hoffert M. (1975) Rate of palagonitization and manganese coating on basaltic rocks from the rift valley in the Atlantic Ocean near 36° 50′ N. *Mar. Geol.,* **19,** 91–109.
11.3.2.

Helwig J. (1970) Slump folds and early structures, northeastern Newfoundland Appalachians. *J. Geol.,* **78,** 172–187.
12.2.2.

Helwig J. (1972) Stratigraphy. sedimentation, paleogeography and paleoclimates of Carboniferous ('Gondwana') and Permian of Bolivia. *Bull. Am. Ass. Petrol. Geol.,* **56,** 1008–1033.
5.1.

Hersey J.B. (1965) Sedimentary basins of the Mediterranean Sea. In: *Submarine Geology and Geophysics* (Ed. by W.F. Whittard and R. Bradshaw) pp. 75–89. Butterworth, London.
12.4.4.

Hertweck G. (1972) Distribution and environmental significance of Lebens-spuren and *in situ* skeletal remains. *Senckenberg. Mar.,* **4,** 125–167.
7.2.3, 9.3.5, Fig. 7.15.

Hesse R. (1974) Long-distance continuity of turbidites: possible evidence for an Early Cretaceous trench-abyssal plain in the East Alps. *Bull. geol. Soc. Am.,* **85,** 859–870.
12.7.2, Fig. 12.46.

Hesse R. (1975) Turbiditic and non-turbiditic mudstone of Cretaceous

flysch sections of the East Alps and other basins. *Sedimentology*, **22,** 387–416.
12.2.5, 12.4.1, 12.7.2.

HICKIN E.J. (1974) The development of meanders in natural river channels. *Am. J. Sci.*, **274,** 414–442.
3.5.2.

HIGGINS G.M., AHMAD M. & BRINKMAN R. (1973) The Thal Interfluve, Pakistan, Geomorphology and depositional history. *Geol. Mijnb.*, **52,** 147–155.
3.6.1, 3.6.2.

HILL D. (1972) Archaeocyatha. In: *Treatise on Invertebrate Paleontology, Part E* (Ed. by C. Teichert), pp. 158. Geol. Soc. Am. and Univ. Kansas Press.
10.4.2.

HILL D. (1974) An introduction to the Great Barrier Reef. *Proc. Second Int. Coral Reef Symp.* (Ed. by A.M. Cameron, B.M. Campbell, A.B. Cribb, R. Endean, J.S. Jell, O.A. Jones, P. Mather and F.H. Talbot), **2,** 723–731.
10.2.2, 10.2.4.

HILL E.W. & HUNTER R.E. (1976) Interaction of biological and geological processes in the beach and nearshore, northern Padre Island, Texas. In: *Beach and Nearshore Sedimentation* (Ed. by R.A. Davis Jr. and R.L. Ethington), pp. 169–187. *Spec. Publ. Soc. econ. Paleont. Miner.*, **24,** Tulsa.
7.2.3, Fig. 7.17.

HINDE G.J. (1890) Notes on Radiolaria from the Lower Palaeozoic rocks (Llandeilo-Caradoc) of the south of Scotland. *Ann. Mag. nat. Hist.*, Ser. **5,** 6, 40–59.
11.4.2.

HINE A.C. (1976) Bedform distribution and migration patterns on tidal deltas in the Chatham Harbor estuary, Cape Cod, Massachusetts. In: *Estuarine Research,* Vol. II Geology and Engineering (Ed. by L.E. Gronin), pp. 235–252. Academic Press, London.
7.2.4, 7.2.5, Fig. 7.22.

HJULSTRÖM F. (1952) The geomorphology of the alluvial outwash plains (sandurs) of Iceland and the mechanism of braided rivers. *8th Gen. Assembly & Proc. 17th Internat. Congress, Internat. Geograph. Union,* 337–342.
3.3.1.

HOBDAY D.K. & HORNE J.C. (1977) Tidally influenced barrier island and estuarine sedimentation in the Upper Carboniferous of southern West Virginia. *Sedim. Geol.*, **18,** 97–122.
7.3.1.

HOBDAY D.K. & ORME A.R. (1974) The Port Durnford Formation: A major Pleistocene barrier-lagoon complex along the Zululand coast.*Trans. Geol. Soc. S. Afr.*, **77,** 141–149.
7.3.2.

HOBDAY D.K. & READING H.G. (1972) Fair weather versus storm processes in shallow marine sand bar sequences in the late Pre-Cambrian of Finnmark, North Norway. *J. sedim. Petrol.*, **42,** 318–324.
9.1.1, 9.8, 9.10.2, 9.10.5, Fig. 9.39.

HOEDEMAEKER PH.J. (1973) Olisthostromes and other delapsional deposits, and their occurrence in the region of Moratalla (Prov. Murcia, Spain). *Scripta Geologica,* **19,** 207.
12.2.1, 12.2.3.

HOFMANN H.J. (1969) Attributes of stromatolites. *Pap. geol. Surv. Canada,* **69–39,** 58.
11.3.3.

HOFFMAN P. (1974) Shallow and deepwater stromatolites in Lower Proterozoic platform-basin facies change, Great Slave Lake, Canada. *Bull. Am. Ass. Petrol. Geol.*, **58,** 856–867.
10.4.2, Fig. 10.49.

HOFFMAN P., DEWEY J.F. & BURKE K. (1974) Aulacogens and their genetic relation to geosynclines, with a Proterozoic example from Great Slave Lake, Canada. In: *Modern and Ancient Geosynclinal Sedimentation* (Ed. by R.H. Dott Jr. and R.H. Shaver), pp. 38–55. *Spec. Publ. Soc. econ. Paleont. Miner.*, **19,** Tulsa.
14.4.1, 14.5, Fig. 14.30.

HOLEMAN J.N. (1968) The sediment yield of major rivers of the world. *Water Resources Res.*, **4,** 737–747.
9.3.1.

HOLLIDAY D.W. (1971) Origin of Lower Eocene gypsum–anhydrite rock, southeast St. Andrew, Jamaica. *Trans. Inst. Min. Metall. B,* **81,** B305–B315.
8.4.1.

HOLLIDAY D.W. & SHEPHARD-THORNE E.R. (1974) Basal Purbeck evaporites of the Fairlight Borehole, Sussex. *Rep. Inst. Geol. Sci.*, No. **74/4,** pp. 14.
8.4.2.

HOLLINGWORTH S.E. (1965) In discussion of G.P. Butler and others: Recent evaporite deposits along the Trucial Coast of the Arabian Gulf. *Proc. geol. Soc. Lond.*, **1623,** 92–93.
5.3.4.

HOLLISTER C.D., EWING J.I. *et al.* (1972) *Initial Reports of the Deep Sea Drilling Project,* **9,** pp. 1077. U.S. Government Printing Office, Washington.
11.3.1, 11.3.5, 11.4.4.

HOLMES C.D. (1960) Evolution of till-stone shapes, Central New York. *Bull. geol. Soc. Am.*, **71,** 645–660.
13.4.1.

HOLSER W.T. (1966) Diagenetic polyhalite in Recent salt from Baja California. *Am. Mineral.*, **51,** 99–109.
8.1.1.

HOLTEDAHL H. (1975) The geology of the Hardangerfjord, west Norway. *Norg. geol. Unders.*, **323,** pp. 87.
12.4.6.

HOOKE R. LE B. (1965) *Alluvial fans,* pp. 192. Ph.D. Thesis, California Institute of Technology, Pasadena.
3.2.2.

HOOKE R. LE B. (1967) Processes on arid-region alluvial fans. *J. Geol.*, **75,** 438–460.
3.2.3, 3.2.4, Fig. 3.2, 3.5, 3.6.

HOOKE R. LE B. (1972) Geomorphic evidence for Late-Wisconsin and Holocene tectonic deformation, Death Valley, California. *Bull. geol. Soc. Am.*, **83,** 2073–2098.
5.2.6.

HOOKE R. LE B. (1977) Basal temperatures in polar ice sheets: a qualitative review. *Quat. Res.*, **7,** 1–13.
13.2.2, Fig. 13.1.

HOPSON C.A., FRANO C.J., PESSAGNO E.A. JR. & MATTINSON J.M. (1975) Preliminary report and geologic guide to the Jurassic ophiolite near Point Sal, southern California Coast. *71st Ann. Meeting, Cordilleran Section, Geol. Soc. Am.*, Field Trip 5, pp. 36.
11.4.2.

HORIKOSHI E. (1969) Volcanic activity related to the formation of the Kuroko-type deposits in the Kosaka District, Japan. *Mineralium Deposits,* **4,** 321–345.
14.4.2.

HORN D.R., DELACH M.N. & HORN B.M. (1969) Distribution of volcanic ash layers and turbidites in the North Pacific. *Bull. geol. Soc. Am.*, **80,** 1715–1724.

12.4.1.

HORN D.R., EWING J.I. & EWING M. (1972) Graded-bed sequences emplaced by turbidity currents north of 20°N in the Pacific, Atlantic and Mediterranean. *Sedimentology*, **12**, 247–275.
12.4.1, 12.4.2, 12.4.4, Fig. 12.27.

HORNE J.C. & FERM J.C. (1976) *Carboniferous depositional environments in the Pocahontas Basin, Eastern Kentucky and Southern West Virginia: A Field Guide*, pp. 129. Department of Geology, University of South Carolina.
6.7.1, 6.7.2, 6.7.4, 7.3.1.

HOROWITZ A. & CRONAN D.S. (1976) The geochemistry of basal sediments from the North Atlantic Ocean. *Mar. Geol.*, **20**, 205–228.
11.3.2.

HOUBOLT J.J.H.C. (1968) Recent sediments in the southern bight of the North Sea. *Geol. Mijnb*, **47**, 245–273.
9.1.1, 9.5.1, 9.5.2, 9.5.3, 9.10.1, Fig. 9.8, 9.13, 9.14.

HOUBOLT J.J.H.C. & JONKER J.B.M. (1968) Recent sediments in the eastern part of the Lake of Geneva (Lac Léman). *Geol. Mijnb.*, **47**, 131–148.
4.2.3.

HOUTEN F.B. VAN (1964) Cyclic lacustrine sedimentation, Upper Triassic Lockatong Formation, Central New Jersey and adjacent Pennsylvania. In: *Symposium on Cyclic Sedimentation* (Ed. by D.F. Merriam), pp. 495–531. *Bull. geol. Surv. Kansas*, **169.**
2.2.1, 4.3.3, Fig. 4.10, 4.11.

HOUTEN F.B. VAN (1965) Composition of Triassic Lockatong and associated formations of Newark Group, Central New Jersey and adjacent Pennsylvania. *Am. J. Sci.*, **263**, 825–863.
4.3.3.

HOUTEN F.B. VAN (1972) Iron and clay in tropical Savanna alluvium, northern Columbia: A contribution to the origin of red beds. *Bull. geol. Soc. Am.*, **83**, 2761–2772.
5.3.1.

HOUTEN F.B. VAN (1974) Northern Alpine molasse and similar Cenozoic sequences of Southern Europe. In: *Modern and Ancient Geosynclinal Sedimentation* (Ed. by R.H. Dott Jr. and R.H. Shaver), pp. 260–273. *Spec. Publ. Soc. econ. Paleont. Miner.*, **19**, Tulsa.
14.2.5, 14.4.4.

HOWARD J.D. (1971) Comparison of the beach to offshore sequence in modern and ancient sediments. In: *Recent Advances in Palaeoecology and Ichnology*, pp. 148–183. Amer. Geol. Inst.
7.2.2, 7.3.1.

HOWARD J.D. (1972) Trace fossils as criteria for recognizing shorelines in the stratigraphic record. In: *Recognition of Ancient Sedimentary Environments* (Ed. by J.K. Rigby and W.K. Hamblin), pp. 215–225. *Spec. Publ. Soc. econ. Paleont. Miner*, **16**, Tulsa.
7.3.1, 9.9.

HOWARD J.D. & FREY R.W. (1975a) Introduction. *Senckenberg. Mar.*, **7**, 1–32.
7.5.1.

HOWARD J.D. & FREY R.W. (1975b) Regional animal-sediment characteristics of Georgia estuaries. *Senckenberg. Mar.*, **7**, 33–104.
7.5.1.

HOWARD J.D. & REINECK H.E. (1972a) Conclusions. *Senckenberg. Mar.*, **4**, 217–223.
7.2.3.

HOWARD J.D. & REINECK H.E. (1972b) Physical and biogenic sedimentary structures of the nearshore shelf. *Senckenberg. Mar.*, **4**, 81–123.
7.2.2, 7.2.3, Fig. 7.16.

HOWARD J.D., FREY R.W. & REINECK H.E. (1972) Introduction. *Seckenberg. Mar.*, **4**, 3–14.
7.2.3, Fig. 7.15.

HOWARD J.D., ELDERS C.A. & HEINBOKEL J.F. (1975) Animal–sediment relationships in estuarine point bar deposits, Ogeechoe River–Ossabaw Sound, Georgia. *Senckenberg. Mar.*, **7**, 181–203.
Fig. 7.45.

HOWARD J.D., REMMER G.H. & JEWITT J.L. (1975) Hydrography and sediments of the Duplin River, Sapelo Island, Georgia. *Senckenberg. Mar.*, **7**, 237–256.
7.5.1.

HOWITT F. (1964) Stratigraphy and structure of the Purbeck inliers of Sussex (England). *Quart. J. geol. Soc. Lond.*, **120**, 77–113.
8.4.2.

HOYT J.H. (1962) High angle beach stratification, Sapelo Island, Georgia. *J. sedim. Petrol.*, **32**, 309–311.
7.2.2.

HOYT J.H. (1967) Barrier island formation. *Bull. geol. Soc. Am.*, **78**, 1125–1136.
7.2.7.

HOYT J.H. (1969) Chenier versus barrier: genetic and stratigraphic distinction. *Bull. Am. Ass. Petrol. Geol.*, **53**, 299–306.
7.4, Fig. 7.42.

HOYT J.H. & HENRY V.J. (1967) Influence of island migration on barrier island sedimentation. *Bull. geol. Soc. Am.*, **78**, 77–86.
7.2.4, Fig. 7.20.

HOYT J.H. & WEINER R.J. (1963) Comparison of modern and ancient beaches, central Georgia coast. *Bull. Am. Ass. Petrol. Geol.*, **47**, 529–531.
7.2, 7.2.2.

HSÜ K.J. (1970) The meaning of the word flysch – a short historical search. In: *Flysch Sedimentology in North America* (Ed. by J. Lajoie), pp. 1–11. *Spec. Pap. Geol. Soc. Canada*, **7.**
12.6.1, 14.2.5.

HSÜ K.J. (1971) Franciscan melanges as a model for eugeosynclinal sedimentation and under-thrusting tectonics. *J. geophys. Res.*, **76**, 1162–1169.
11.4.2.

HSÜ K.J. (1972) Origin of saline giants: a critical review after the discovery of the Mediterranean Evaporite. *Earth Sci. Rev.*, **8**, 371–396.
8.5.1, 8.5.2, 8.5.5, Fig. 8.42, 8.43.

HSÜ K.J. (1973a) The desiccated deep-basin model for the Messinian events. In: *Messinian Events in the Mediterranean* (Ed. by C.W. Drooger), pp. 60–67. North Holland, Amsterdam.
8.5.5.

HSÜ K.J. (1973b) The Odyssey of Geosyncline. In: *Evolving Concepts in Sedimentology* (Ed. by R.N. Ginsburg), pp. 66–92. The Johns Hopkins University Press, Baltimore.
14.2.5, Tab. 14.1.

HSÜ K.J. (1974) Melanges and their distinction from olistostromes. In: *Modern and Ancient Geosynclinal Sedimentation* (Ed. by R.H. Dott and R.H. Shaver), pp. 321–333. *Spec. Publ. Soc. econ. Paleont. Miner.*, **19**, Tulsa.
12.2.3.

HSÜ K.J. (1976) Paleoceanography of the Mesozoic Alpine Tethys. *Spec. Pap. geol. Soc. Am.*, **170**, pp. 44.
11.4.2, 11.4.4.

HSÜ K.J. & JENKYNS H.C. (Eds) (1974) Pelagic sediments: on Land and under the Sea. *Spec. Publ. int. Ass. Sediment.*, **1**, pp. 447.

1.1.

Hsü K.J. & Montadert L. *et. al.* (1975). *Glomar Challenger* returns to the Mediterranean Sea. *Geotimes,* **20,** 8, 16–19.
11.3.5.

Hsü K.J., Montadert L., Bernoulli D., Cita M.B., Erickson A., Garrison R.E., Kidd R.B., Mèlierés X., Müller C. & Wright R. (1977) History of the Mediterranean salinity crisis. *Nature,* **267,** 399–403.
8.5.5, 11.3.5, 11.4.3.

Huang T.C. & Goodell H.G. (1970 Sediments and sedimentary processes of eastern Mississippi Cone, Gulf of Mexico. *Bull. Am. Ass. Petrol. Geol.,* **54,** 2070–2100.
12.5.1.

Hubert J.F. (1964) Textural evidence for deposition of many western North Atlantic deep-sands by ocean bottom currents rather than turbidity currents. *J. Geol.,* **72,** 757–785.
12.2.7.

Hubert J.F., Butera J.G. & Rice R.F. (1972) Sedimentology of Upper Cretaceous Cody-Parkman delta, southwestern Powder River Basin, Wyoming. *Bull. geol. Soc. Am.,* **83,** 1649–1670.
6.7.1, 6.7.2, 6.7.3.

Hubert J.F., Reed A.A. & Carey P.J. (1976) Paleogeography of the East Berlin Formation, Newark Group, Connecticut Valley. *Am. J. Sci.,* **276,** 1183–1207.
4.3.3, Fig. 4.12.

Huddle J.W. & Patterson S.H. (1961) Origin of Pennsylvanian underclay and related seat rocks. *Bull. geol. Soc. Am.,* **72,** 1643–1660.
3.9.2.

Huene R. Von & Shor G.G. Jr. (1969) The structure and tectonic history of the Eastern Aleutian Trench. *Bull. geol. Soc. Am.,* **80,** 1889–1902.
12.4.3.

Huene R. Von (1974) Modern trench sediments. In: *The Geology of Continental Margins* (Ed. by C.A. Burk and C.L. Drake), pp. 207–211. Springer-Verlag, New York.
14.3.2.

Hughes C.J. (1970) The late Pre-Cambrian Avalonian orogeny in Avalon, southeast Newfoundland. *Am. J. Sci.,* **269,** 183–190.
14.4.2.

Hülsemann T. & Emery K.O. (1961) Stratification in recent sediments of Santa Barbara Basin as controlled by organisms and water characteristics. *J. Geol.,* **69,** 279–290.
11.3.6.

Hunt C.B. (1972) *Geology of Soils,* pp. 344. Freeman, San Francisco.
3.6.2, Fig. 3.33.

Hunt C.B., Robinson T.W., Bowles W.A. & Washburn A.L. (1966) Hydrological Basin, Death Valley, California. *Prof. Pap. U.S. geol. Surv.,* **494B,** pp. 138.
5.2.6, 5.3.4, Fig. 5.9, 5.10, 5.11.

Hunt R.E., Swift D.J.P. & Palmer H. (1977) Constructional shelf topography, Diamond Shoals, North Carolina. *Bull. geol. Soc. Am.,* **88,** 299–311.
9.6.3, 9.10.2.

Hunter R.E. (1976) Comparison of eolian and subaqueous sand-flow cross-strata (Abstract). *Bull. Am. Ass. Petrol. Geol.,* **60,** 683–684.
5.2.4.

Hunter R.E. (1977) Basic types of stratification in small eolian dunes. *Sedimentology,* **24,** 361–388.
5.2.4.

Hunter W. & Parkin D.W. (1961) Cosmic dust in Tertiary rock and the lunar surface. *Geochim. Cosmochim. Acta,* **24,** 32–39.
11.4.2.

Hutchinson G.E. (1957) *A Treatise on Limnology.* Volume 1: Geography, Physics and Chemistry, pp. 1015. Wiley, New York.
4.2.1.

Hutchinson R.W. & Engels G.G. (1970) Tectonic significance of regional geology and evaporite lithofacies in northeastern Ethopia. *Phil. Trans. R. Soc. A,* **267,** 313–329.
14.3.1, Fig. 14.8.

Håkansson E., Bromley R. & Perch-Nielsen K. (1974) Maastrichtian chalk of north-west Europe – a pelagic shelf sediment. In: *Pelagic Sediments: on Land and under the Sea* (Ed. by K.J. Hsü and H.C. Jenkyns), pp. 211–233. *Spec. Publ. int. Ass. Sediment.,* **1.**
11.4.5.

Illing L.V. (1954) Bahamian calcareous sands. *Bull. Am. Ass. Petrol. Geol.,* **38,** 1–95.
10.1.1, 10.2.2.

Imbrie J. & Buchanan H. (1965) Sedimentary structures in modern carbonate sands of the Bahamas. In: *Primary Sedimentary Structures and their Hydrodynamic Interpretation* (Ed. by G.V. Middleton), pp. 34–52. *Spec. Publ. Soc. econ. Paleont. Miner.,* **12,** Tulsa.
3.4.1.

Imbrie J. & Newell N. (1964) *Approaches to Palaeoecology,* pp. 432. John Wiley, New York.
9.1.1, 9.8.

Ingle J.C. (1966) *The Movement of Beach Sand,* pp. 221. New York: American Elsevier.
7.2.1, Fig. 7.4, 7.5.

Ingle J.C. Jr. (1973) Summary comments on Neogene biostratigraphy, physical stratigraphy, and paleo-oceanography in the marginal northeastern Pacific Ocean. In: *Initial Reports of the Deep Sea Drilling Project,* **18** (L.D. Kulm, R. Von Huene *et al.*), pp. 949–960. U.S. Government Printing Office, Washington.
11.4.3.

Ingle J.C. (1975) Summary of Late Paleogene-Neogene insular stratigraphy, paleobathymetry, and correlations, Philippine Sea and Sea of Japan Region. In: *Initial Reports of the Deep Sea Drilling Project,* **31** (D.E. Karig, J.C. Ingle *et al.*), pp. 837–855. U.S. Government Printing Office, Washington.
11.4.3.

Ingle J.C., Karig D.E., Bouma A.H. *et al.* (1973) Leg 31, Western Pacific floor. *Geotimes,* **18(10),** 22–25.
Fig. 12.28.

Inman D.L. & Nordstrom C.E. (1971) On the tectonic and morphologic classification of coasts. *J. Geol.,* **79,** 1–21.
12.4.2.

Isacks B., Oliver J. & Sykes L.R. (1968) Seismology and the new global tectonics. *J. geophys. Res.,* **73,** 5855–5899.
14.2.6.

Isotta C.A.L., Rocha-Campos A.C. & Yoshida R. (1969) Striated pavement of the Upper Pre-Cambrian glaciation in Brazil. *Nature,* **222,** 466–468.
13.5.4.

Jaanusson V. (1955) Description of the microlithology of the Lower Ordovician limestones between the *Ceratopyge* shale and the *Platyrus* limestone of Böda Hamm. *Bull. geol. Inst. Univ. Uppsala,* **35,** 153–173.
11.4.5.

Jaanusson V. (1960) The Viruan (Middle Ordovician) of Öland. *Bull.*

geol. Inst. Univ. Uppsala, **38,** 207–288.
11.4.5.

Jaanusson V. (1961) Discontinuity surfaces in limestones. *Bull. geol. Inst. Univ. Uppsala*, **40,** 221–241.
11.4.5.

Jaanusson V. (1971) Constituent analysis of an Ordovician limestone from Sweden. *Lethaia*, **5,** 217–237.
11.4.5.

Jacka A.D., Beck R.H., St. Germain L.C. *et al.* (1968) Permian deep-sea fans of the Delaware Mountain Group (Guadalupian), Delaware Basin. In: *Guadalupian Facies, Apache Mountain Area, West Texas* (Ed. by B.A. Silver), pp. 49–67. Symposium and Guidebook, 1968 Field Trip, Permian Basin Section. *Soc. econ. Paleont. Miner. Pub.*, 68–11.
12.8.1.

Jackson E.D., Silver E.A. & Dalrymple G.B. (1972) Hawaiian Emperor chain and its relation to Cenozoic circum-Pacific tectonics. *Bull. geol. Soc. Am.*, **83,** 601–618.
11.3.3.

Jackson R.G. II (1975a) *A Depositional Model of Point Bars in the Lower Wabash River*, pp. 263. Ph.D. Thesis, University of Illinois at Urbana-Champaign.
3.5.2, Fig. 3.29.

Jackson R.G. II (1975b) Velocity–bed-form–texture patterns of meander bends in the Lower Wabash River of Illinois and Indiana. *Bull. geol. Soc. Am.*, **86,** 1511–1522.
3.5.2, 3.5.3, Fig. 3.28.

Jackson R.G. II (1976) Depositional model of point bars in the Lower Wabash River. *J. sedim. Petrol.*, **46,** 579–594.
3.9.4, Fig. 3.28.

Jackson S.A. & Beales F.W. (1967) An aspect of sedimentary basin evolution: the concentration of Mississippi Valley-type ores during late stages of diagenesis. *Bull. Can. Petrol. Geol.*, **15,** 383–433.
8.5.2.

Jacobson R. & Scott T.R. (1937) The geology of Korpukerrimul Creek area, Victoria. *Roy. Soc. Victoria*, **50,** 110–156.
13.5.2.

James N.P., Ginsburg R.N., Marszalek D.S. & Choquette P.W. (1976) Facies and fabric specificity of early subsea cements in shallow Belize (British Honduras) reefs. *J. sedim. Petrol.*, **46,** 523–544.
10.1.1.

James N.P. & Kobluk D.R. (1978) Lower Cambrian patch reefs and associated sediments: southern Labrador, Canada. *Sedimentology*, **25,** 1–35.
10.4.2.

Jamieson E.R. (1971) Paleoecology of Devonian reefs in western Canada. In: *Reef Organisms through Time*, Proc. J.: 1300–1340. North Am. Paleont. Convention, Chicago, 1969.
10.4.2, Fig. 10.46.

Jansen J.H.F. (1976) Late Pleistocene and Holocene history of the northern North Sea, based on acoustic reflection records. *Neth. J. Sea Res.*, **10,** 1–43.
9.5.3, 13.5.1.

Jefferies R.P.S. (1963) The stratigraphy of the *Actinocamax plenus* Subzone (Turonian) in the Anglo-Paris Basin. *Proc. Geol. Ass.*, **74,** 1–34.
11.7.5.

Jenkyns H.C. (1970) Fossil manganese nodules from the west Sicilian Jurassic. *Eclog. Geol. Helv.*, **63,** 741–774.
11.4.4.

Jenkyns H.C. (1971a) The genesis of condensed sequences in the Tethyan Jurassic. *Lethaia*, **4,** 327–352.
11.4.4.

Jenkyns H.C. (1971b) Speculations on the genesis of crinoidal limestones in the Tethyan Jurassic. *Geol. Rdsch.*, **60,** 471–488.
11.4.4.

Jenkyns H.C. (1972) Pelagic 'oolites' from the Tethyan Jurassic. *J. Geol.*, **80,** 21–33.
11.4.4.

Jenkyns H.C. (1974) Origin of red nodular limestones (Ammonitico Rosso, Knollenkalke) in the Mediterranean Jurassic: a diagenetic model. In: *Pelagic Sediments: on Land and under the Sea* (Ed. by K.J. Hsü and H.C. Jenkyns), pp. 249–271. *Spec. Publ. int. Ass. Sediment.*, **1.**
11.4.4.

Jenkyns H.C. (1976) Sediments and sedimentary history of the Manihiki Plateau, South Pacific Ocean. In: *Initial Reports of the Deep Sea Drilling Project*, **33** (S.O. Schlanger, E.D. Jackson *et al.*), pp. 873–890. U.S. Government Printing Office, Washington.
11.3.3.

Jenkyns H.C. (1977) Fossil nodules. In: *Marine Manganese Deposits* (Ed. by G.P. Glasby), pp. 85–108. Elsevier, Amsterdam.
11.4.2, 11.4.4.

Jenkyns H.C. & Hardy R.G. (1976) Basal iron-titanium-rich sediments from Hole 315A (Line Islands, Central Pacific). In: *Initial Reports of the Deep Sea Drilling Project*, **33** (S.O. Schlanger, E.D. Jackson *et al.*), pp. 833–836. U.S. Government Printing Office, Washington.
11.3.3.

Jenkyns H.C. & Hsü K.J. (1974) Pelagic sediments: on Land and under the Sea – an introduction. In: *Pelagic Sediments: on Land and under the Sea* (Ed. by K.J. Hsü and H.C. Jenkyns), pp. 1–10. *Spec. Publ. int. Ass. Sediment.*, **1.**
11.1.2, 11.4.2.

Jennings J.N. & Sweeting M.M. (1961) Caliche pseudo-anticlines in the Fitzroy Basin, Western Australia. *Am. J. Sci.*, **259,** 635–639.
3.6.2.

Jepson H.F. (1971) The Precambrian, Eocambrian and early Palaeozoic stratigraphy of the Jørgen Brønlund Fjord area, Peary Land, North Greenland, pp. 42. *Meddr. om Grønland*, **192.**
13.5.4.

Jerzyiewicz T. (1968) Sedimentation of the Younger Sandstones of the Intrasudetic Cretaceous Basin. *Geologica Sudetica*, **IV,** 409–462.
9.10.4.

Jipa D. & Kidd R.B. (1974) Sedimentation of coarser grained interbeds in the Arabian Sea and sedimentation processes of the Indus Cone. In: *Initial Reports of the Deep Sea Drilling Project* (R.B. Whitmarsh, O.E. Weser, D.A. Ross *et al.*), **23,** 471–495, U.S. Government Printing Office, Washington.
12.5.2.

Johnson A.M. (1970) *Physical Processes in Geology*, pp. 577. Freeman, Cooper & Co., San Francisco.
3.2.4, 12.2.3.

Johnson D. (1938) The origin of submarine canyons. *J. Geomorphology*, **1,** 230–243.
12.1.

Johnson D.A. (1974) Deep Pacific circulation: intensification during the Early Cenozoic. *Mar. Geol.*, **17,** 71–78.

11.3.1.
Johnson D.A. & Johnson T.C. (1970) Sediment redistribution by bottom currents in the central Pacific. *Deep-Sea Res.*, **17**, 157–169.
11.3.4.
Johnson D.A. & Lonsdale P.F. (1976) Erosion and sedimentation around Mytilus Seamount, New England continental rise. *Deep-Sea Res.*, **23**, 429–440.
11.3.3.
Johnson D.W. (1919) *Shore Processes and Shoreline Development*, pp. 584. John Wiley, New York.
7.1.
Johnson H.D. (1975) Tide- and wave-dominated inshore and shoreline sequences from the late Precambrian, Finnmark, north Norway. *Sedimentology*, **22**, 45–73.
7.3.2, 7.7.
Johnson H.D. (1977a) Shallow marine sand bar sequences: an example from the late Precambrian of North Norway. *Sedimentology*, **24**, 245–270.
9.7, 9.10.5, Fig. 9.39.
Johnson H.D. (1977b) Sedimentation and water escape structures in some late Precambrian shallow marine sandstones from Finnmark, North Norway. *Sedimentology*, **24**, 389–411.
9.10.3.
Johnson J.W. & Eagleson P.S. (1966) Coastal process. In: *Estuarine and Coastline Hydrodynamics* (Ed. by A.J. Ippen), pp. 404–492. McGraw-Hill, New York.
7.2.2.
Johnson M.A. & Belderson R.H. (1969) The tidal origin of some vertical sedimentary changes in epicontinental seas. *J. Geol.*, **77**, 353–357.
9.10.1.
Johnson M.A. & Stride A.H. (1969) Geological significance of North Sea sand transport rates. *Nature*, **224**, 1016–1017.
9.1.1, 9.10.1.
Johnson R.G. (1974) Particulate matter at the sediment–water interface in coastal environments. *J. Mar. Res.*, **32**, 313–330.
9.3.5.
Johnston W.A. (1921) Sedimentation of the Fraser River delta. *Mem. Can. geol. Surv.*, **125**, 46.
6.2.
Johnston W.A. (1922) The character of the stratification of the sediments in the Recent delta of the Fraser River, British Columbia, Canada. *J. Geol.*, **30**, 115–129.
6.2.
Jones B.F. (1965) The hydrology and mineralogy of Deep Springs Lake, Inyo County, California. *Prof. Pap. U.S. geol. Surv.*, **502-A**, pp. 56.
4.3.2.
Jones C.M. (1977) The effects of varying discharge regimes on bed form sedimentary structures in modern rivers. *Geology*, **5**, 567–570.
3.4.2, 3.4.3.
Jones D.L., Bailey E.H. & Imlay R.W. (1969) Structural and stratigraphic significance of the *Buchia* Zones in the Colyear Springs–Paskenta area, California. *Prof. Pap. U.S. geol. Surv.*, **647-A**, 1–24.
11.4.2.
Jones J.B. & Segnit E.R. (1971) The nature of opal. I. Nomenclature and constituent phases. *J. geol. Soc. Aust.*, **18**, 57–68.
11.3.4.
Jones O.T. (1954) The characteristics of some Lower Paleozoic marine sediments. *Proc. Roy. Soc. London, Ser. A*, **222**, 327–332.
12.2.2.
Jong J.D. De (1971) The scenery of the Netherlands against the background of Holocene geology: a review of the recent literature. *Revue Géogr. phys. Géol. Dyn.*, **13**, 143–162.
7.2.
Jopling A.V. & McDonald B.C. (Eds) (1975) Glaciofluvial and glaciolacustrine sedimentation, pp. 320. *Spec. Publ. Soc. econ. Paleont. Miner.*, **23**, Tulsa.
1.1.
Jukes-Browne A.J. & Harrison J.B. (1892) The geology of Barbados. Part 2. The Oceanic deposits. *Quart. J. geol. Soc. Lond.*, **48**, 170–226.
11.1.2, 11.4.2.

Kamp P.C. Van De, Harper J.D., Conniff J.J. & Morris D.A. (1974) Facies relations in the Eocene-Oligocene in the Santa Ynez Mountains, California. *J. geol. Soc. London*, **130**, 545–565.
12.2.6, 12.8.3, Figs. 12.53, 12.54, 12.55.
Kanes W.H. (1970) Facies and development of the Colorado river delta in Texas. In: *Deltaic Sedimentation, Modern and Ancient* (Ed. by J.P. Morgan and R.E. Shaver), pp. 78–106. *Spec. Publ. Soc. econ. Paleont. Miner.*, **15**, Tulsa.
6.3.1, 6.5.2, 7.2.5.
Karcz I. (1972). Sedimentary structures formed by flash floods in Southern Israel. *Sedim. Geol.*, **7**, 161–182.
3.4.4.
Karig D.E. (1970) Ridges and trenches of the Tonga-Kermadec island arc system. *J. geophys. Res.*, **77**, 239–254.
14.3.2.
Karig D.E. (1971) Origin and development of marginal basins in the Western Pacific. *J. geophys. Res.*, **76**, 2542–2561.
14.3.2.
Karig D.E. & Moore G.F. (1975) Tectonically controlled sedimentation in marginal basins. *Earth Planet. Sci. Letts.*, **26**, 233–238.
14.3.2, Fig. 14.21.
Karig D.E. & Sharman III G.F. (1975) Subduction and accretion in trenches. *Bull. geol. Soc. Am.*, **86**, 377–389.
14.3.2, Fig. 14.19, 14.20.
Karig D.E., Peterson M.N.A. & Shor G.G. Jr. (1970) Sediment-capped guyots in the Mid-Pacific Mountains. *Deep-Sea Res.*, **17**, 373–378.
11.3.3.
Karig D.E. & Ingle J.C. Jr. *et al.* (1975) *Initial Reports of the Deep Sea Drilling Project*, **31**, pp. 927. U.S. Government Printing Office, Washington.
11.3.5.
Karrow P.F. (1976) The texture, mineralogy, and petrography of North American tills. In: *Glacial Till* (Ed. by R.F. Legget), pp. 83–98. *Spec. Publ. Royal Soc. Can.*, **12.**
13.4.1.
Kastner M., Keene J.B. & Gieskes J. (1977) Diagenesis of siliceous oozes. I. Chemical controls on the rate of opal-A to opal-CT transformation – an experimental study. *Geochim. Cosmochim. Acta*, **41**, 1041–1059.
11.3.4.
Kaszap A. (1963). Investigations on the microfacies of the Malm Beds of the Villány Mountains. *Ann. Univ. Scient. Budapestinensis R. Eötvös Nom. Sect. Geol.*, **6**, 47–57.
11.4.4.
Kauffman E.G. (1967) Coloradoan macroinvertebrate assemblages, Central Western Interior United States. *Paleoenvironments of the*

Cretaceous Seaway – a Symposium, pp. 67–143. Colorado School of Mines, Golden.
9.12.

KAUFFMAN E.G. (1969) Cretaceous marine cycles of the Western Interior. *Mountain Geologist*, **6**, 227–245.
11.4.5, Fig. 11.45.

KAUFFMAN E.G. (1974) Cretaceous assemblages, communities, and associations: Western Interior United States and Caribbean Islands. In: *Principles of Benthic Community Analysis* (Ed. by A.M. Zeigler, K.R. Walker, E.J. Anderson, E.G. Kauffman, R.N. Ginsburg and N.P. James), 12.1–12.25. Sedimenta IV. The University of Miami.
9.12.

KAUFFMAN E.G. (1976) Deep-sea Cretaceous macrofossils: Hole 317A, Manihiki Plateau. In: *Initial Reports of the Deep Sea Drilling Project*, **33** (S.O. Schlanger, E.D., Jackson *et al.*), pp. 503–535. U.S. Government Printing Office, Washington.
11.3.3.

KAUFFMAN E.G. & RUNNEGAR B. (1975) *Atomodesma* (Bivalvia), and Permian species of the United States. *J. Paleont.*, **49**, 23–41.
11.4.2.

KAUFFMAN E.G. & SOHL N.F. (1974) Structure and evolution of Caribbean Cretaceous rudist frameworks. *Festschrift fur Hans Kugler* (Ed. by P. Jung), pp. 1–80. Mus. Nat. Hist. Basel, Switzerland.
10.4.2.

KAY M. (1947) Geosynclinal nomenclature and the craton. *Bull. Am. Ass. Petrol. Geol.*, **31**, 1289–1293.
14.2.3.

KAY M. (1951) North American geosynclines. *Mem. geol. Soc. Am.*, **48**, pp. 143.
14.2.3, 14.2.5, Fig. 14.4.

KAY M. (1975) Campbellton sequence: manganiferous beds adjoining Dunnage Melange, northeastern Newfoundland. *Bull. geol. Soc. Am.*, **86**, 105–108.
11.4.1.

KAYE C.A. (1960) Surficial geology of the Kingston Quadrangle, Rhode Island. *Bull. U.S. geol. Surv.*, **107-I**, 341–396.
13.3.2.

KEANY J., LEDBETTER M., WATKINS N. & HUANG T.-C. (1976) Diachronous deposition of ice-rafted debris in sub-Antarctic deep-sea sediments. *Bull. geol. Soc. Am.*, **87**, 873–882.
11.3.1.

KEITH M.L. & DEGENS E.T. (1959) Geochemical indicators of marine and fresh-water sediments. In: *Researches in Geochemistry* (Ed. by P.H. Abelson), pp. 38–61. John Wiley, New York.
9.8.

KELLING G. (1968) Patterns of sedimentation in Rhondda Beds of South Wales. *Bull. Am. Ass. Petrol. Geol.*, **52**, 2369–2386.
3.9.4.

KELLING G. & GEORGE G.T. (1971) Upper Carboniferous sedimentation in the Pembrokeshire coalfield. In: *Geological Excursions in South Wales and the Forest of Dean* (Ed. by D.A. Bassett and M.G. Bassett), pp. 240–259. Geol. Ass. South Wales Group, Cardiff.
6.7.2, Figs. 6.39, 6.40, 6.42.

KELLING G. & MULLIN P.R. (1975) Graded limestones and limestone-quartzite couplets: possible storm-deposits from the Moroccan Carboniferous. *Sedim. Geol.*, **13**, 161–190.
9.11.3.

KELLING G. & STANLEY D.J. (1976) Sedimentation in canyon, slope, and base-of-slope environments. In: *Marine Sediment Transport and Environmental Management* (Ed. by D.J. Stanley and D.J.P. Swift), pp. 379–435. John Wiley, New York.
12.2.4, 12.2.7.

KELLING G. & WOOLLANDS M.H. (1969) The stratigraphy and sedimentation of the Llandoverian rocks of the Rhayader district. In: *The Pre-Cambrian and Lower Palaeozoic Rocks of Wales* (Ed. by A. Wood), pp. 255–281. Univ. of Wales Press, Cardiff.
12.8.1.

KENDALL C.G. ST. C. & SKIPWITH P.A.d'E. (1968) Recent algal mats of a Persian Gulf lagoon. *J. sedim. Petrol.*, **38**, 1040–1058.
8.2.5, Tab. 8.1.

KENDALL C.G. ST. C. & SKIPWITH P.A.d'E. (1969a) Holocene shallow water carbonate and evaporite sediments of Khor al Bazam, Abu Dhabi, Southwest Persian Gulf. *Bull. Am. Ass. Petrol. Geol.*, **53**, 841–849.
10.2.5.

KENDALL C.G. ST. C. & SKIPWITH P.A.d'E. (1969b) Geomorphology of a recent shallow water carbonate province: Khor al Bazam, Trucial Coast, Southwest Persian Gulf. *Bull. geol. Soc. Am.*, **80**, 865–891.
10.2.5.

KENNEDY W.J. (1970) Trace fossils in the chalk environment of south-east England. In: *Trace Fossils* (Ed. by T.P. Crimes and J.C. Harper), pp. 263–282. *Geol. J. Spec. Issue*, **3**, Seel House Press, Liverpool.
11.4.5.

KENNEDY W.J. & GARRISON R.E. (1975a) Morphology and genesis of nodular phosphates in the Cenomanian Glauconitic Marl of south-east England. *Lethaia*, **8**, 339–360.
11.4.5.

KENNEDY W.J. & GARRISON R.E. (1975b) Morphology and genesis of nodular chalks and hardgrounds in the Upper Cretaceous of southern England. *Sedimentology*, **22**, 311–386.
11.4.5. Fig. 11.44.

KENNEDY W.J. & JUIGNET P. (1974) Carbonate banks and slump beds in the Upper Cretaceous (Upper Turonian-Santonian) of Haute Normandie, France. *Sedimentology*, **21**, 1–42.
11.4.5.

KENNETT J.P. & WATKINS N.D. (1975) Deep-sea erosion and manganese nodule development in the south-east Indian Ocean. *Science*, **188**, 1011–1013.
11.3.4, Fig. 11.15.

KENNETT J.P. & WATKINS N.D. (1976) Regional deep-sea dynamic processes recorded by Late Cenozoic sediments of the south-east Indian Ocean. *Bull. geol. Soc. Am.*, **87**, 321–339.
11.3.1.

KENNETT J.P., BURNS R.E., ANDREWS J.E., CHURKIN M., DAVIES T.A., DUMITRICA P., EDWARDS A.R., GALEHOUSE J.S., PACKHAM G.H. & VAN DER LINGEN G.J. (1972) Australian-Antarctic continental drift, paleocirculation changes and Oligocene deep-sea erosion. *Nature, Phys. Sci.*, **239**, 51–55.
11.3.1, 11.4.2, Fig. 11.23.

KENNETT J.P., HOUTZ R.E., ANDREWS P.B., EDWARDS A.R., GOSTIN V.A., HAJOS M., HAMPTON M., JENKINS D.G., MARGOLIS S.V., OVENSHINE A.T. & PERCH-NIELSEN K. (1975) Cenozoic paleoceanography in the southwest Pacific Ocean, Antarctic glaciation, and the development of the circum-Antarctic current. In: *Initial Reports of the Deep Sea Drilling Project*, **29** (J.P. Kennett, R.E. Houtz *et al.*), pp. 1155–1169. U.S. Government Printing Office, Washington.
11.3.4.

KENT P.E. & WALMSLEY P.J. (1970) North Sea progress. *Bull. Am. Ass. Petrol. Geol.*, **54,** 168–181.
5.3.4.

KENYON N.H. (1970) Sand ribbons of European tidal seas. *Mar. Geol.*, **9,** 25–39.
9.1.1, 9.5.1, Fig. 9.11.

KENYON N.H. & STRIDE A.H. (1970) The tide-swept continental shelf sediments between the Shetland Isles and France. *Sedimentology*, **14,** 159–173.
9.1.1, 9.5, 9.5.2, Fig. 9.12.

KIMBERLEY M.M. (1974) Origin of iron ore by diagenetic replacement of calcareous oolite. *Nature*, **250,** 319–320.
14.4.1.

KING B.C. (1970) Vulcanicity and rift tectonics in East Africa. In: *African Magmatism and Tectonics* (Ed. by T.N. Clifford and I.G. Gass), pp. 263–283. Oliver & Boyd, Edinburgh.
Fig. 14.6.

KING B.C. (1976) The Baikal Rift. *J. geol. Soc. Lond.*, **132,** 348–349.
14.3.1.

KING L.H., MACLEAN B. & DRAPEAU G. (1972) The Scotian Shelf submarine end-moraine complex. *24th Int. Geol. Congr.*, **8,** 237–249.
13.5.1.

KINGMA J.T. (1958a) The Tongaporutuan sedimentation in Central Hawke's Bay. *New Zealand J. Geol. Geophys.*, **1,** 1–30.
14.3.3.

KINGMA J.T. (1958b) Possible origin of piercement structures, local unconformities and secondary basins in the Eastern Geosyncline, N.Z. *New Zealand J. Geol. Geophys.*, **1,** 269–274.
14.3.3, Fig. 14.24.

KINSMAN D.J.J. (1964) Reef coral tolerance of high temperatures and salinities. *Nature*, **202,** 1280–1282.
8.2.2.

KINSMAN D.J.J. (1966) Gypsum and anhydrite of recent age, Trucial Coast, Persian Gulf. In: *Second Symposium on Salt*. Vol. I (Ed. by J.L. Rau), pp. 302–326. Northern Ohio Geol. Soc., Cleveland, Ohio.
8.2.6, Figs. 8.2, 8.15, 8.16.

KINSMAN D.J.J. (1967) Huntite from a carbonate-evaporite environment. *Am. Mineral.*, **52,** 1332–1340.
8.2.6.

KINSMAN D.J.J. (1969a) Interpretation of Sr^{2+} concentrations in carbonate minerals and rocks. *J. sedim. Petrol.*, **39,** 486–508.
10.1.2.

KINSMAN D.J.J. (1969b) Modes of formation, sedimentary associations, and diagnostic features of shallow-water supratidal evaporites. *Bull. Am. Ass. Petrol. Geol.*, **53,** 830–840.
8.2.6.

KINSMAN D.J.J. (1974) Calcium sulphate minerals of evaporite deposits: their primary mineralogy. In: *Fourth Symposium on Salt*, Vol. I (Ed. by A. Coogan), pp. 343–348. Northern Ohio Geol. Soc., Cleveland, Ohio.
8.2.6, 8.4.2.

KINSMAN D.J.J. (1975) Salt floors to geosynclines. *Nature*, **255,** 375–378.
11.3.5.

KINSMAN D.J.J. (1976) Evaporites: relative humidity control of primary mineral facies. *J. sedim. Petrol.*, **46,** 273–279.
8.2.6.

KINSMAN D.J.J. & PARK R.K. (1976) Algal belt and coastal sabkha evolution, Trucial Coast, Persian Gulf. In: *Interpreting Stromatolites* (Ed. by M.R. Walther), pp. 421–433. Elsevier, Amsterdam.
8.2.4, 8.2.5, 8.4.2, Tab. 8.1.

KLEIN G. DE V. (1962) Triassic sedimentation, Maritime Provinces, Canada. *Bull. geol. Soc. Am.*, **73,** 1127–1146.
4.3.3.

KLEIN G. DE V. (1963) Bay of Fundy intertidal zone sediments. *J. sedim. Petrol.*, **33,** 844–854.
7.6.

KLEIN G. DE V. (1967a) Paleocurrent analysis in relation to modern marine sediment dispersal patterns. *Bull. Am. Ass. Petrol. Geol.*, **51,** 366–382.
9.8.

KLEIN G. DE V. (1967b) Comparison of recent and ancient tidal flat and estuarine sediments. In: *Estuaries* (Ed. by G.H. Lauff), pp. 207–218. Am. Ass. Adv. Sci., Washington.
7.6.

KLEIN G. DE V. (1969) Deposition of Triassic sedimentary rocks in separate basins, eastern North America. *Bull. geol. Soc. Am.*, **80,** 1825–1832.
4.3.3.

KLEIN G. DE V. (1970a) Depositional and dispersal dynamics of intertidal sand bars. *J. sedim. Petrol.*, **40,** 1095–1127.
7.7, 9.10.1.

KLEIN G. DE V. (1970b) Tidal origin of a PreCambrian Quartzite – the Lower Fine Grained Quartzite (Middle Dalradian) of Islay, Scotland. *J. sedim. Petrol.*, **40,** 973–985.
9.10.1.

KLEIN G. DE V. (1971) A sedimentary model for determining paleotidal range. *Bull. geol. Soc. Am.*, **82,** 2585–2592.
7.7, 9.10.2.

KLEIN G. DE V. (1975) Depositional facies of leg 30, Deep Sea Drilling Project Sediment Cores. In: *Initial Reports of the Deep Sea Drilling Project*, **30** (J.E. Andrews, G. Packham *et al.*), pp. 423–442. U.S. Government Printing Office, Washington.
11.3.3.

KLEIN G. DE V., DE MELO U. & DELLA FAVERA J.C. (1972) Subaqueous gravity processes on the front of Cretaceous deltas, Recôncavo Basin, Brazil. *Bull. geol. Soc. Am.*, **83,** 1469–1492.
6.8.3, 12.8.3.

KLIMEK K. (1972) Present day fluvial processes and relief of the Skeidarársandur Plain (Iceland). *Pol. Acad. Nauk. Geogr. Studies*, **94,** pp. 139 (Polish with English summary).
3.3.1.

KLIMEK K. (1974a) The retreat of alluvial river banks in the Wisloka Valley (South Poland). *Geogr. Polonica.*, **28,** 59–75.
3.5.2.

KLIMEK K. (1974b) The structure and mode of sedimentation of the flood-plain deposits in the Wistoka Valley (South Poland). *Studia Geomorphologica Carpatho-Balcanica*, **8,** 135–151.
3.6.1.

KLINGSPOR A.M. (1969) Middle Devonian Muskeg evaporites of Western Canada. *Bull. Am. Ass. Petrol. Geol.*, **53,** 927–948.
8.5.2.

KNAAP W. VAN DER & EIJPE R. (1968) Some experiments on the genesis of turbidity currents. *Sedimentology*, **11,** 115–124.
12.2.4.

KOBLENTZ-MISCHKE O.J., VOLKOVINSKY V.V. & KABANOVA J.G. (1960). Plankton primary production of the World Ocean. In: *Scientific Explorations of the South Pacific* (Ed. by W.S. Wooster), pp. 183–193. *Nat. Acad. Sci.*, Washington.

11.4.5.

Kolb C.R. & Dornbusch W.K. (1975) The Mississippi and Mekong deltas – a comparison. In: *Deltas, Models for Exploration* (Ed. by M.L. Broussard), pp. 193–207. Houston Geological Society, Houston.
6.5.2.

Komar P.D. (1969) The channelized flow of turbidity currents with application to Monterey deep-sea fan channel. *J. geophys. Res.*, **74,** 4544–4558.
12.2.4, 12.5.1.

Komar P.D. (1971) Hydraulic jumps in turbidity currents. *Bull. geol. Soc. Am.*, **82,** 1477–1488.
12.2.4.

Komar P.D. (1972) Relative significance of head and body spill from a channelized turbidity current. *Bull. geol. Soc. Am.*, **83,** 1151–1156.
12.2.4.

Komar P.D. (1976a) Nearshore currents and sediment transport, and the resulting beach configuration. In: *Marine Sediment Transport and Environmental Management* (Ed. by D.J. Stanley and D.J.P. Swift), pp. 241–254. John Wiley, New York.
7.2.1.

Komar P.D. (1976b) Evaluation of longshore current velocities and sand transport rates produced by oblique wave approach. In: *Beach and Nearshore Sedimentation* (Ed. by R.A. Davis Jr. and R.L. Ethington), pp. 48–53. *Spec. Publ. Soc. econ. Paleont. Miner.*, **24,** Tulsa.
7.2.1.

Komar P.D. (1976c) *Beach Processes and Sedimentation*, pp. 429. Prentice-Hall, New Jersey.
7.2.1.

Komar P.D. (1976d) The transport of cohesionless sediments on continental shelves. In: *Marine Sediment Transport and Environmental Management* (Ed. by D.J. Stanley and D.J.P. Swift), pp. 107–125. John Wiley, New York.
9.4.3.

Komar P.D. & Inman D.L. (1970) Longshore sand transport on beaches. *J. geophys. Res.*, **75,** 5914–5927.
7.2.1.

Komar P.D., Kulm L.D. & Harlett J.C. (1974) Observations and analysis of bottom nepheloid layers on the Oregon continental shelf. *J. Geol.*, **82,** 104–111.
9.6.2.

Komar P.D., Neudeck R.H. & Kulm L.D. (1972) Observations and significance of deep water oscillatory ripple marks on the Oregon continental shelf. In: *Shelf Sediment Transport: Process and Pattern* (Ed. by D.J.P. Swift, D.B. Duane and O.H. Pilkey), pp. 601–619. Dowden, Hutchinson & Ross, Stroudsburg.
9.1.1, 9.6.1, 9.6.2.

Köppen J.W. & Wegener A. (1924) *Die Klimate der Geologischen Vorzeit*, pp. 225. Gebrüder Borntraeger, Berlin.
5.1.

Kornicker L.S. & Bryant W.R. (1969) Sedimentation on the continental shelf of Guatemala and Honduras. *Mem. Am. Ass. Petrol. Geol.*, **11,** 244–257.
10.2.4.

Kraft J.C. (1971) Sedimentary facies patterns and geologic history of a Holocene marine transgression. *Bull. geol. Soc. Am.*, **82,** 2131–2158.
7.2, 7.2.6, Fig. 7.33.

Kraft J.C., Biggs R.B. & Halsey S.D. (1973) Morphology and vertical sedimentary sequence models in Holocene transgressive barrier systems. In: *Coastal Geomorphology*, Proc. 3rd Am. Geomorph. Symp. Series (Ed. by D.R. Coates), pp. 321–354. State University, New York.
7.2.6, Fig. 7.33.

Krauskopf K. (1956) Factors controlling the concentrations of thirteen rare metals in sea water. *Geochim. cosmochim. Acta*, **9,** 1–32.
11.3.2.

Krebs W. (1972) Facies and development of the Meggen Reef (Devonian, West Germany). *Geol. Rdsch.*, **61,** 647–671.
11.4.4.

Krebs W. (1974) Devonian carbonate complexes of Central Europe. In: *Reefs in Time and Space* (Ed. by L.F. Laporte), pp. 155–208. *Spec. Publ. Soc. econ. Paleont. Miner.*, **18,** Tulsa.
10.4.2.

Krigström A. (1962) Geomorphological studies of Sandur Plains and their braided rivers in Iceland. *Geogr. Annlr.*, **44,** 328–346.
3.3.1, Fig. 3.10.

Krinsley D.H. & Doornkamp J.C. (1973) Atlas of sand surface textures, pp. 91. *Cambridge Earth Sci. Ser.*
5.2.4.

Kröner A. (1977) Non-synchroneity of late Precambrian glaciations in Africa. *J. Geol.*, **85,** 289–300.
13.5.4.

Kruit C. (1955) Sediments of the Rhône delta. Grain size and microfauna. *Ned. Geol. Mijnb. Genoot. Verh. Geol. Ser.*, **15,** 357–514.
6.5.1, 6.5.2.

Kruit C., Brouwer J., Knox G., Schöllnberger W. & Vliet van A. (1975) Une excursion aux cônes d'alluvions en eau profonde d'âge Tertiaire près de San Sebastian (province de Guipúzcoa, Espagne). *9th Int. Congr. Sedimentol., Nice 1975, excursion* **23,** pp. 75.
12.8.1, Figs. 12.3, 12.37.

Krumbein W.C. & Sloss L.L. (1963) *Stratigraphy and Sedimentation*, pp. 660. W.H. Freeman, San Francisco.
2.1.1, 14.1.

Krystyn L. (1971) Stratigraphie, Fauna und Fazies der Klaus-Schichten (Dogger/Oxford) in den östlichen Nordalpen. *Verh. geol. Bundesanst., Wien*, **1971,** 486–509.
11.4.4.

Kuendig E. (1959) Eu-geosynclines as potential oil habitats. *Proc. 5th World Petrol. Congr. Sect.*, **1,** 1–13.
14.2.2, Fig. 14.2.

Kuenen Ph.H. (1935) Geological interpretation of the bathymetrical results, *Snellius Expedition. Sci. Results Snellius Expedition Eastern Pt. East-Indian Archipelago, 1929–1930*, **1,** pp. 124.
14.2.2.

Kuenen Ph.H. (1937) Experiments in connection with Daly's hypothesis on the formation of submarine canyons. *Leidse geol. Meded.*, **8,** 327–335.
1.1, 12.1.

Kuenen Ph.H. (1950a) *Marine Geology*, pp. 568. John Wiley, New York.
11.2.

Kuenen Ph.H. (1950b) Turbidity currents of high density. *18th Intl. geol. Congr., London, 1948; Rept., pt.* **8,** 44–52.
1.1, 12.1.

Kuenen Ph.H. (1951) Mechanics of varve formation and the action of turbidity currents. *Geol. Føren. Stockh. Førh.*, **73,** 69–84.
13.3.4.

Kuenen Ph.H. (1952) Estimated size of the Grand Banks turbidity

current. *Am. J. Sci.*, **250**, 874–884.
12.2.4.

KUENEN PH.H. (1953) Graded bedding, with observations on Lower Paleozoic rocks of Britain. *Verhandel. koninkl. Ned. Akad. Wetenschap., Afdel. Natuurk., Sect. I*, 20(3), 1–47.
12.2.2.

KUENEN PH.H. (1957a) Sole markings of graded graywacke beds. *J. Geol.*, **65**, 231–258.
12.1.

KUENEN PH.H. (1957b) Longitudinal filling of oblong sedimentary basins. *Verhandel. koninkl. Ned. geol. mijnbouwk. Genoot., geol Ser.*, **18**, 189–195.
12.6.2, 14.2.3.

KUENEN PH.H. (1958) Problems concerning source and transportation of flysch sediments. *Geol. Mijnb.*, **20**, 329–339.
12.2.6, 12.6.2.

KUENEN PH.H. (1965) Experiments in connection with turbidity currents and clay suspensions. In: *Geology and Geophysics, Colston Research Society 17th Symposium, Colston Papers*, **17**, pp. 47–74. Butterworth, London.
12.2.4.

KUENEN PH.H. (1966) Matrix of turbidites: experimental approach. *Sedimentology*, **7**, 267–297.
12.2.4.

KUENEN PH.H. (1967) Emplacement of flysch-type sand beds. *Sedimentology*, **9**, 203–243.
12.2.4, 12.2.7.

KUENEN PH.H. & HUMBERT F.L. (1964) Bibliography of turbidity currents and turbidites. In: *Turbidites* (Ed. by A.H. Bouma and A. Brouwer), pp. 222–246. Elsevier, Amsterdam.
Fig. 12.1.

KUENEN PH.H. & MENARD H.W. (1952) Turbidity currents graded and non-graded deposits. *J. sedim. Petrol.*, **22**, 83–96.
12.3.2.

KUENEN PH.H. & MIGLIORINI C.I. (1950) Turbidity currents as a cause of graded bedding. *J. Geol.*, **58**, 91–127.
1.1, 12.1, 12.2.4, 12.6.2, 12.8.1.

KUENEN PH.H. & SANDERS J.E. (1956) Sedimentation phenomena in Kulm and Flözleeres graywackes, Sauerland and Oberharz, Germany. *Am. J. Sci.*, **254**, 649–671.
12.6.2.

KUGLER H. (1953) Jurassic to Recent sedimentary environments in Trinidad. *Bull. Ver. Schweiz. Petrol. Geol. Ing.*, **20**, 27–60.
11.4.4.

KÜHN R. (1952) Reaktionen zwischen festen, insbesondere ozeanischen Salzen. *Heidelberger Beitr. z. Miner.*, **3**, 147–170.
8.5.3.

KUIPERS E.P. (1971) Transition from fluviatile to tidal marine sediments in the Upper Devonian of the Seven Heads Peninsula. *Geol. Mijnb.*, **50**, 443–450.
7.7.

KULM L.D. & FOWLER G.A. (1974) Oregon continental margin structure and stratigraphy: a test of the imbricate thrust model. In: *The Geology of Continental Margins* (Ed. by C.A. Burk and C.L. Drake), pp. 261–283. Springer-Verlag, New York.
14.3.2.

KULM L.D., ROUSCH R.C., HARLETT J.C., NEUDECK R.H., CHAMBERS D.M. & RUNGE E.J. (1975) Oregon continental shelf sedimentation: interrelationships of facies distribution and sedimentary processes. *J. Geol.*, **83**, 145–176.
9.1.1, 9.6.2, Fig. 9.18, 9.19.

KUMAR N. & SANDERS J.E. (1974) Inlet sequence: a vertical succession of sedimentary structures and textures created by the lateral migration of tidal inlets. *Sedimentology*, **21**, 491–532.
7.2.4, 7.2.6, Fig. 7.21.

KUMAR N. & SANDERS J.E. (1976) Characteristics of shoreface deposits: modern and ancient. *J. sedim. Petrol.*, **46**, 145–162.
7.2.2, 9.6.1.

LADD H.S. (Ed.) (1957) Treatise on marine ecology and palaeoecology; Palaeoecology. *Mem. geol. Soc. Am.*, **67, vol. 2,** pp. 1077.
9.1.1, 9.8.

LADD H.S. (1971) Existing reefs – geological aspects. In: *Reef Organisms through Time. North Am. Paleont. Convention, Chicago, 1969, Proc. J.*, 1273–1300.
10.2.2, 10.4.1, 10.4.2.

LADD H.S., TRACEY J.I., WELLS J.W. & EMERY K.O. (1950) Organic growth and sedimentation on an atoll. *J. Geol.*, **58**, 410–425.
10.1.1.

LAFON G.M. & MACKENZIE F.T. (1974) Early evolution of the oceans; a weathering model. In: *Studies in Paleo-oceanography* (Ed. by W.W. Hay), pp. 205–218. *Spec. Publ. Soc. econ. Paleont. Miner.*, **20**, Tulsa.
10.2.1.

LAGAAIJ R. & KOPSTEIN F.P.H.W. (1964) Typical features of a fluviomarine offlap sequence. In: *Deltaic and Shallow Marine Deposits* (Ed. by L.M.J.U. van Straaten), pp. 216–226. Elsevier.
6.5.2.

LAJOIE J. (Ed.) (1970) *Flysch Sedimentology in North America*, pp. 242. *Geol. Assoc. Canada, Spec. Pap.*, **7.**
12.6.2.

LAMBRICK H.T. (1967) The Indus flood-plain and the 'Indus' civilisation. *Geogr. J.*, **133**, 483–495.
3.6.1, 3.6.2.

LAMING D.J.C. (1966) Imbrication, paleocurrents and other sedimentary features in the lower New Red Sandstone, Devonshire, England. *J. sedim. Petrol.*, **36**, 940–959.
3.8, 3.8.1, 3.9.2, 5.3.6.

LANCELOT Y. (1973) Chert and silica diagenesis in sediments from the Central Pacific. In: *Initial Reports of the Deep Sea Drilling Project*, **17** (E.L. Winterer, J.I. Ewing *et al.*), pp. 377–405. U.S. Government Printing Office, Washington.
11.3.3, 11.3.4.

LANCELOT Y. & SEIBOLD E. *et al.* (1975). The Eastern North Atlantic. *Geotimes*, **20**, 7, 18–21.
11.3.5.

LANDIS C.A. (1974) Stratigraphy, lithology, structure, and metamorphism of Permian, Triassic, and Tertiary rocks between the Mararoa River and Mount Snowdon, western Southland, New Zealand. *J. Roy. Soc. N.Z.*, **4**, 229–251.
11.4.2.

LANDIS C.A. & BISHOP D.G. (1972) Plate tectonics and regional stratigraphic-metamorphic relations in the southern part of the New Zealand geosyncline. *Bull. geol. Soc. Am.*, **83**, 2267–2284.
14.4.2.

LANE E.W. (1957) A study of the shape of channels formed by natural streams flowing in erodible material. *U.S. Army Corps of Engineers, Missouri River Division, Omaha, Nebraska. Sediment Series*, **9**, pp. 106.
Fig. 3.53.

LANE E.W. & EDEN E.W. (1940) Sand waves in the lower Mississippi

River. *J. western Soc. of Engrs.*, **45**, 281–291.
3.5.2.

LANGBEIN W.B. & LEOPOLD L.B. (1966) River meanders – theory of minimum variance. *Prof. Pap. U.S. geol. Surv.*, **422-H**, pp. 15.
3.5.1.

LANGFELDER J., STAFFORD D. & AMEIN M. (1968) *A Reconnaissance of Coastal Erosion in North Carolina*, pp. 127. Dept. Civil Eng., North Carolina State University, Raleigh, N.C.
Fig. 9.25.

LANGHORNE D.N. (1973) A sandwave field in the Outer Thames Estuary, Great Britain. *Mar. Geol.*, **14**, 129–143.
9.5.1.

LAPORTE L.F. (1971) Paleozoic carbonate facies of the Central Appalachian shelf. *J. sedim. Petrol.*, **41**, 724–740.
10.2.5, Figs. 10.27a, 10.27b.

LAPORTE L.F. (1975) Carbonate tidal flat deposits of the Early Devonian Manlius Formation of New York State. In: *Tidal Deposits: A Casebook of Recent Examples and Fossil Counterparts* (Ed. by R.N. Ginsburg), pp. 243–250. Springer-Verlag, Berlin.
10.2.5, Figs. 10.27a, 10.27b, 10.28.

LATTMAN L.H. & LAUFFENBURGER S.K. (1974) Proposed role of gypsum in the formation of caliche. *Z. Geomorph.*, **Suppl. Bd. 20**, 140–149.
3.6.2.

LAUGHTON A.S. (1967) Underwater photography of the Carlsberg Ridge. In: *Deep-Sea Photography* (Ed. by J.B. Hersey), *Johns Hopkins Oceanographic Studies*, **3**, pp. 191–205. Johns Hopkins Press, Baltimore.
11.3.2.

LAURY R.L. (1971) Stream bank failure and rotational slumping: preservation and significance in the geologic record. *Bull. geol. Soc. Am.*, **82**, 1251–1266.
3.5.2, 3.9.3, 6.5.1, Fig. 6.7.

LAVRUSHIN YU. A. (1971) Dynamische Fazies und Subfazies der Grundmoräne. *Zeitschr. Angew. Geol.*, **17**, 337–343.
13.4.1.

LE BLANC R.F. & HODGSON W.D. (1959) Origin and development of the Texas shoreline. *Trans. Gulf-Cst. Ass. geol. Soc.*, **9**, 197–220.
7.2.

LEEDER M.R. (1973) Fluviatile fining-upwards cycles and the magnitude of palaeochannels. *Geol. Mag.*, **110**, 265–276.
3.5.1.

LEEDER M.R. (1974) Lower Border Group (Tournaisian) fluvio-deltaic sedimentation and palaeogeography of the Northumberland Basin. *Proc. Yorks. geol. Soc.*, **40**, 129–180.
3.9.2.

LEEDER M.R. (1975) Pedogenic carbonate and flood sediment accretion rates: a quantitative model for alluvial, arid-zone lithofacies. *Geol. Mag.*, **112**, 257–270.
3.9.2.

LEEDER M.R. & BRIDGES P.H. (1975) Flow separation in meander bends. *Nature*, **253**, 338–339.
3.5.2.

LEES A. (1961) The Waulsortian 'Reefs' of Eire: a carbonate mudbank complex of Lower Carboniferous age. *J. Geol.*, **69**, 101–109.
10.4.2.

LEES A. (1964) The structure and origin of the Waulsortian (Lower Carboniferous) 'Reefs' of west-central Eire. *Phil. Trans. Roy. Soc. London, Ser. B.* **247, 740**, 483–531.
10.4.2.

LEES A. (1975) Possible influences of salinity and temperature on modern shelf carbonate sedimentation. *Mar. Geol.*, **19**, 159–198.
10.1.2, 10.3, Figs. 10.1a, 10.1b, 10.2, 10.3, 10.4, 10.5.

LEES A. & BULLER A.T. (1972) Modern temperate water and warm water shelf carbonate sediments contrasted. *Mar. Geol.*, **13**, 1767–1773.
10.1.2, 10.3.2.

LEES A., BULLER A.T. & SCOTT J. (1969) *Marine Carbonate Sedimentation Processes, Connemara, Ireland*, pp. 64. Reading Univ. Geol. Rep., 2, Reading.
10.3, 10.3.1, Fig. 10.34.

LEGGETT R.F., BROWN R.J.E. & JOHNSTON G.H. (1966) Alluvial fan formation near Aklavik, Northwest Territories, Canada. *Bull. geol. Soc. Am.*, **77**, 15–30.
3.2.1.

LEIGHTON M.W. & PENDEXTER C. (1962) Carbonate rock types. In: *Classification of Carbonate Rocks: A Symposium* (Ed. by W.E. Ham), pp. 33–61. *Mem. Am. Ass. Petrol. Geol.*, **1**, Tulsa.
10.1.1.

LEMOINE M. (1972) Eugeosynclinal domains of the Alps and the problem of past oceanic areas. In: *24th Int. geol. Congr.*, Montreal, Sect. 3, 476–485.
11.4.2.

LEOPOLD L.B. & WOLMAN M.G. (1957) River channel patterns: braided meandering and straight. *Prof. Pap. U.S. geol. Surv.*, **282-B**, pp. 85.
3.1, 3.3.1, 3.5.1, 3.10.1, Fig. 3.53.

LEOPOLD L.B. & WOLMAN M.G. (1960) River meanders. *Bull. geol. Soc. Am.*, **71**, 769–794.
3.5.1, 3.5.2.

LEWIS K.B. (1971) Slumping on a continental slope inclined at 1°–4°. *Sedimentology*, **16**, 97–110.
12.2.2, Fig. 12.6.

LINDSAY D.A. (1969) Glacial sedimentology of Precambrian Gowganda Formation, Ontario, Canada. *Bull. geol. Soc. Am.*, **80**, 1685–1702.
13.5.5.

LINDSAY D.A. (1971) Glacial marine sediments in the Precambrian Gowganda Formation at Whitefish Falls, Ontario, Canada. *Palaeogeogr. Palaeoclim. Palaeoecol.*, **9**, 7–25.
13.1.

LINDSAY J.F. (1966) Carboniferous subaqueous mass-movements in the Manning-Macleay Basin, Kempsey, New South Wales. *J. sedim. Petrol.*, **36**, 719–732.
13.5.2.

LINDSAY J.F. (1968) The development of clast fabric in mudflows. *J. sedim. Petrol.*, **38**, 1232–1253.
3.8.1.

LINDSAY J.F. (1970) Depositional environment of Paleozoic glacial rocks in the central Transantarctic Mountains. *Bull. geol. Soc. Am.*, **81**, 1149–1172.
13.4.1, 13.5.2.

LINDSTRÖM M. (1963) Sedimentary folds and the development of limestone in an Early Ordovician Sea. *Sedimentology*, **2**, 243–292.
11.4.5, Fig. 11.41.

LINDSTRÖM M. (1971) Vom Anfang, Hochstand und Ende eines Epikontinentalemeeres. *Geol. Rdsch.*, **60**, 419–438.
11.4.5.

LINDSTRÖM M. (1974) Volcanic contribution to Ordovician pelagic sediments. *J. sedim. Petrol.*, **44**, 287–291.
11.4.5, Fig. 11.40.

LINK M.H. & OSBORNE R.H. (1978) Lacustrine facies in the Pliocene

Ridge Basin Group: Ridge Basin, California. In: *Modern and Ancient Lake Sediments* (Ed. by A. Matter and M. Tucker), pp. 167–185. *Spec. Publ. int. Ass. Sediment.*, **2**, Blackwell Scientific.
Fig. 14.25.

LIPMANN F. (1973) *Sedimentary Carbonate Minerals*, pp. 228. Springer-Verlag, Berlin.
10.1.1.

LISITZIN A.P. (1971) Distribution of siliceous microfossils in suspension and in bottom sediments. In: *The Micropalaeontology of Oceans* (Ed. by B.M. Funnell and W.R. Riedel), pp. 173–195. Cambridge University Press, Cambridge.
11.3.4, 11.3.5.

LISITZIN A.P. (1972) *Sedimentation in the World Ocean*. pp. 218. *Spec. Publ. Soc. econ. Paleont. Miner.*, **17**, Tulsa.
13.4.3.

LOCHMAN-BALK C. (1971) The Cambrian of the craton of the United States. In: *Cambrian of the New World* (Ed. by C.H. Holland), pp. 79–167. John Wiley, New York.
9.10.2.

LOCK B.E. (1973) The Ordovician ice age in South Africa. *Geol. Mag.*, **110**, 372–376.
13.5.3.

LOGAN B.W., HOFFMAN P. & GEBELEIN C.F. (1974) Algal mats, cryptalgal fabrics and structures, Hamelin Pool, Western Australia. *Mem. Am. Ass. Petrol. Geol.*, **22**, 140–194.
4.3.2, 8.2.5, Tab. 8.1, Fig. 8.11.

LOGAN B.W., REZAK R. & GINSBURG R.N. (1964) Classification and environmental significance of algal stromatolites. *J. Geol.*, **69**, 517–533.
10.2.5.

LOGAN B.W., HARDING J.L., AHR W.M., WILLIAMS J.D. & SNEAD R.G. (1969) Carbonate sediments and reefs, Yucatan shelf, Mexico. *Mem. Am. Ass. Petrol. Geol.*, **11**, 1–198.
10.2.3, Fig. 10.18, 10.19.

LOHMANN G.P. (1973) Stratigraphy and sedimentation of deep-sea Oceanic Formation on Barbados, West Indies. *Bull. Am. Ass. Petrol. Geol.*, **57**, 791 (Abstract).
11.4.2.

LONG J.T. & SHARP R.P. (1964) Barchan-dune movement in the Imperial Valley, California. *Bull. geol. Soc. Am.*, **75**, 149–156.
5.2.4.

LONSDALE P. (1975). Sedimentation and tectonic modification of Samoan archipelagic apron. *Bull. Am. Ass. Petrol. Geol.*, **59**, 780–798.
11.3.4.

LONSDALE P. (1976) Abyssal circulation of the south-east Pacific and some geological implications. *J. geophys. Res.*, **81**, 1163–1176.
11.3.1, 11.3.4.

LONSDALE P. & MALFAIT B. (1974) Abyssal dunes of foraminiferal sand on the Carnegie Ridge. *Bull. geol. Soc. Am.*, **85**, 1697–1712.
11.3.3.

LONSDALE P.F., NORMARK W.R. & NEWMAN W.A. (1972) Sedimentation and erosion on Horizon Guyot. *Bull. geol. Soc. Am.*, **83**, 289–316.
11.3.3.

LOREAU J.P. & PURSER B.H. (1973) Distribution and ultrastructure of Holocene ooids in the Persian Gulf. In: *The Persian Gulf* (Ed. by B.H. Purser), pp. 279–328. Springer-Verlag, Berlin.
8.2.3, 10.2.2, Fig. 8.4.

LOTZE F. (1938) *Steinsalz und Kalisalze*, pp. 936. Borntraeger, Berlin.
8.5.3.

LOVELL J.P.B. (1969) Tyee Formation: a study of proximality in turbidites. *J. sedim. Petrol.*, **39**, 935–953.
12.6.3.

LOWE D.R. (1975a) Regional controls on silica sedimentation in the Ouachita System. *Bull. geol. Soc. Am.*, **86**, 1123–1127.
11.4.4, Fig. 11.31.

LOWE D.R. (1975b) Water escape structures in coarse-grained sediments. *Sedimentology*, **22**, 157–204.
12.2.6.

LOWE D.R. (1976a) Nonglacial varves in lower member of Arkansas Novaculite (Devonian), Arkansas and Oklahoma. *Bull. Am. Ass. Petrol. Geol.*, **30**, 2103–2116.
11.4.4.

LOWE D.R. (1976b) Subaqueous liquefied and fluidized sediment flows and their deposits. *Sedimentology*, **23**, 285–308.
12.2.6.

LOWELL J.D. (1972) Spitsbergen Tertiary orogenic belt and the Spitsbergen fracture zone. *Bull. geol. Soc. Am.*, **83**, 3091–3102.
14.3.3, 14.4.3.

LOWENSTAM H.A. (1950) Niagaran reefs in the Great Lakes area. *J. Geol.*, **58**, 430–487.
Fig. 10.47a.

LOWENSTAM H.A. & EPSTEIN S. (1954) Palaeotemperatures of the post-Aptian Cretaceous as determined by the oxygen isotope method. *J. Geol.*, **62**, 207–248.
10.3.2.

LUDWICK J.C. (1970) Sand waves and tidal channels in the entrance to Chesapeake Bay. *Virginia J. Sci.*, **21**, 178–184.
9.6.3.

LUDWIG W.J., MURAUCHI S. & HOUTZ R.E. (1975) Sediments and structure of the Japan Sea. *Bull. geol. Soc. Am.*, **86**, 651–664.
12.4.4.

LUNDELL L.L. & SURDAM R.C. (1975) Playa-lake deposition: Green River Formation, Piceance Creek Basin, Colorado. *Geology*, **3**, 493–497.
4.3.2.

LUSTIG L.K. (1965) Clastic sedimentation in Deep Springs Valley, California. *Prof. Pap. U.S. geol. Surv.*, **352F**, 131.
3.8.1.

MAAS K. (1974) The geology of Liébana, Cantabrian Mountains, Spain: deposition and deformation in a flysch area. *Leidse geol. Meded.*, **49**, 379–465.
12.2.2, Fig. 12.7.

MACKAY J.R. (1963) The Mackenzie delta area, N.W.T. *Geol. Surv. Canada Geog. Branch. Mem.*, **8**, pp. 202.
6.5.1.

MACKENZIE D.B. (1972) Tidal sand flat deposits in Lower Cretaceous Dakota Group near Denver, Colorado. *The Mountain Geologist*, **8**, 141–150.
7.6, 7.7.

MACKENZIE D.B. (1975) Tidal sand flat deposits in Lower Cretaceous Dakota Group near Denver, Colorado. In: *Tidal Deposits: A Casebook of Recent Examples and Fossil Counterparts* (Ed. by R.N. Ginsburg), pp. 117–126. Springer-Verlag, Berlin.
7.7.

MADDOCK T. JR. (1969) The behaviour of straight open channels with movable beds. *Prof. Pap. U.S. geol. Surv.*, **622A**, pp. 70.
3.4.1.

MADSEN O.S. (1976) Wave climate of the continental margin: elements

of its mathematical description. In: *Marine Sediment Transport and Environmental Management* (Ed. by D.J. Stanley and D.J.P. Swift), pp. 65–87. John Wiley, New York.
7.2.1, 9.4.3.

MAIKLEM W.R. (1971) Evaporative drawdown – a mechanism for water-level lowering and diagenesis in the Elk Point Basin. *Bull. Can. Petrol. Geol.*, **19,** 487–503.
8.5.2, Fig. 8.35.

MALARODA R. (1962) Gli hard grounds al limite tra Cretaceo ed Eocene nei Lessini occidentali. *Memorie Soc. geol. ital.*, **3,** 111–135.
11.4.4.

MALDONADO A. (1975) Sedimentation, stratigraphy and development of the Ebro delta, Spain. In: *Deltas, Models for Exploration* (Ed. by M.L. Broussard), pp. 312–338. Houston Geological Society, Houston.
6.5.1, 6.5.2.

MALDONADO A. & STANLEY D.J. (1976) The Nile Cone: submarine fan development by cyclic sedimentation. *Mar. Geol.*, **20,** 27–40.
12.5.2.

MANGIN J.PH. (1962) Traces de pattes d'oiseaux et flute-casts associés dans un 'facies flysch' du Tertiaire pyrénéen. *Sedimentology,* **1,** 163–166.
3.9.2, 12.6.1.

MANTEN A.A. (1971) *Silurian Reefs of Gotland,* pp. 539. Elsevier, Amsterdam.
10.4.2.

MARCINOWSKI R. & SZULCZEWSKI M. (1972) Condensed Cretaceous sequence with stromatolites in the Polish Jura Chain. *Acta Geol. Pol.*, **22,** 515–539.
11.4.5.

MARIE J.P.P. (1975) Rotliegendes stratigraphy and diagenesis. In: *Petroleum and the Continental Shelf of North West Europe,* Volume **1,** *Geology* (Ed. by A.W. Woodland), pp. 205–210. Applied Sci. Publishers, Barking.
5.3.6.

MARLOWE J.I. (1971) Dolomite, phosphorite and carbonate diagenesis on a Caribbean seamount. *J. sedim. Petrol.*, **41,** 809–827.
11.3.3.

MARR J.E. (1929) *Deposition of the Sedimentary Rocks,* pp. 245. Cambridge University Press, Cambridge.
11.1.2.

MARTIN H. (1953) Notes on the Dwyka succession and on some Pre-Dwyka valleys in South West Africa. *Trans. geol. Soc. S. Afr.*, **56,** 37–42.
13.5.2.

MARTIN H. (1961) The hypothesis of continental drift in the light of recent advances of geological knowledge in Brazil and South West Africa. *Trans. geol. Soc. S. Afr.*, **64,** pp. 47.
13.4.1.

MARTIN H. (1964) The directions of flow of the Itararé ice sheets in the Paraná Basin, Brazil. *Bol. Paranaense de Geografia 1964,* 25–76.
13.4.1.

MARTIN H. (1968) Paläomorphologische Formelemente in den Landschaften Südwest-Afrikas. *Geol. Rdsch.*, **58,** 121–128.
13.5.2.

MARTINI J.E.J. (1976) The fluorite deposits in the Dolomite Series of the Marico District, Transvaal, South Africa. *Econ. Geol.*, **71,** 625–635.
14.4.1.

MASSARI F. & MEDIZZA F. (1973) Stratigrafia e paleogeografia del Campaniano-Maastrichtiano nelle Alpi Meridionali. *Memorie Ist. Geol. Miner. Padova,* **28,** pp. 62
11.4.4.

MATALUCCI R.V., SHELTON J.W. & ABDEL-HADY M. (1969) Grain orientation in Vicksburg Loess. *J. sedim. Petrol.*, **39,** 969–979.
13.3.5.

MATHEWS W.H. & SHEPARD F.P. (1962) Sedimentation of the Fraser River delta: British Columbia. *Bull. Am. Ass. Petrol. Geol.*, **46,** 1416–1443.
6.8.2.

MATTER A. & GARDNER J.V. (1975) Carbonate diagenesis at Site 308, Kōko Guyot. In: *Initial Reports of the Deep Sea Drilling Project,* **32** (R.L. Larson, R. Moberly *et al.*), pp. 521–528. U.S. Government Printing Office, Washington.
11.3.3.

MATTHEWS J.L., HEEZEN B.C., CATALANO R., COOGAN A., THARP M., NATLAND J. & RAWSON M. (1974) Cretaceous drowning on mid-Pacific and Japanese guyots. *Science,* **184,** 462–464.
11.3.3.

MATTHEWS R.K. (1966) Genesis of Recent lime mud in British Honduras. *J. sedim. Petrol.*, **36,** 428–454.
10.1.2, 10.2.4.

MATTHEWS V. & WACHS D. (1973) Mixed depositional environments in the Franciscan geosynclinal assemblage. *J. sedim. Petrol.*, **43,** 516–517.
11.4.2.

MATTI J.C., MURPHY M.A. & FINNEY S.C. (1975). Silurian and Lower Devonian basin and basin-slope limestones, Copenhagen Canyon, Nevada. *Spec. Pap. geol. Soc. Am.*, **159,** pp. 48.
11.4.4.

MATTINSON J.M. (1975) Early Palaeozoic ophiolite complexes of Newfoundland: isotopic ages of zircons. *Geology,* **3,** 181–183.
11.4.2.

MAXWELL A.E. *et al.* (1970) *Initial Reports of the Deep Sea Drilling Project,* **3,** pp. 806. U.S. Government Printing Office, Washington.
11.3.3.

MAXWELL J.C. (1974) Anatomy of an orogen. *Bull. geol. Soc. Am.*, **85,** 1195–1204.
14.4.2.

MAXWELL W.G.H. (1968) *Atlas of the Great Barrier Reef,* pp. 258. Elsevier, Amsterdam.
10.2.2, 10.2.4, Fig. 10.11, 10.23b, 10.23c.

MAXWELL W.G.H. & SWINCHATT J.P. (1970) Great Barrier Reef: variation in a terrigenous carbonate province. *Bull. geol. Soc. Am.*, **81,** 691–724.
Fig. 10.23b, 10.23c.

MCBRIDE E.F. (1962) Flysch and associated beds of the Martinsburg Formation (Ordovician), central Appalachians. *J. sedim. Petrol.*, **32,** 39–91.
12.2.4, 12.6.2.

MCBRIDE E.F. (1970) Stratigraphy and origin of the Maravillas Formation, Upper Ordovician, West Texas. *Bull. Am. Ass. Petrol. Geol.*, **54,** 1719–1745.
11.4.4.

MCBRIDE E.F. & HAYES M.O. (1962) Dune cross-bedding on Mustang Island, Texas. *Bull. Am. Ass. Petrol. Geol.*, **46,** 546–551.
7.2.2.

MCBRIDE E.F. & THOMSON A. (1970) The Caballos Novaculite, Marathon region, Texas. *Spec. Pap. geol. Soc. Am.*, **122,** pp. 129.
11.4.4.

MCBRIDE E.F., WEIDIE A.E. & WOLLEBEN J.A. (1975) Deltaic and

associated deposits of Difunta Group (Late Cretaceous to Palaeocene), Parras and La Popa Basins, northwestern Mexico. In: *Deltas, Models for Exploration* (Ed. by M.L. Broussard), pp. 485–522. Houston Geological Society, Houston.
6.7.1, 6.7.2.

McCabe P.J. (1975) *The Sedimentology and Stratigraphy of the Kinderscout Grit Group (Namurian, R_1) between Wharfedale and Longdendale*. Ph.D. Thesis, University of Keele, pp. 172.
3.9.4, Fig. 3.52.

McCabe P.J. (1977) Deep distributary channels and giant bedforms in the Upper Carboniferous of the Central Pennines, northern England. *Sedimentology*, **24**, 271–290.
3.9.4, 6.7.1.

McCabe P.J. & Jones C.M. (1977) The formation of reactivation surfaces within superimposed deltas and bed forms. *J. sedim. Petrol.*, **47**, 707–715.
3.4.2.

McCartney W.D. & Potter R.F. (1962) Mineralisation as related to structural deformation, igneous activity and sedimentation in folded geosynclines. *J. Can. Mining*, **83**, 83–87.
14.2.4.

McCave I.N. (1968) Shallow and marginal marine sediments associated with the Catskill Complex in the Middle Devonian of New York. *Spec. Pap. geol. Soc. Am.*, **106**, 75–107.
9.1.1.

McCave I.N. (1970) Deposition of fine-grained suspended sediment from tidal currents. *J. geophys. Res.*, **75**, 4151–4159.
9.3.1, 9.5.1, 9.5.2, 9.10.1.

McCave I.N. (1971a) Wave-effectiveness at the sea-bed and its relationship to bed-forms and deposition of mud. *J. sedim. Petrol.*, **41**, 89–96.
9.1.1, 9.3.1, 9.5.1, 9.5.2.

McCave I.N. (1971b) Sand waves in the North Sea off the coast of Holland. *Mar. Geol.*, **10**, 199–225.
9.1.1, 9.5.1, Fig. 9.9.

McCave I.N. (1971c) Mud in the North Sea. In: *North Sea Science. NATO North Sea Science Conference* (Ed. by E.D. Goldberg), pp. 75–100.
9.1.1, 9.5.1, 9.5.2.

McCave I.N. (1972) Transport and escape of fine-grained sediment from shelf areas. In: *Shelf Sediment Transport: Process and Pattern* (Ed. by D.J.P. Swift, D.B. Duane and O.H. Pilkey), pp. 225–248. Dowden, Hutchinson & Ross, Stroudsburg.
9.3.1, 9.5.1, Fig. 9.1.

McCave I.N. (1973) The sedimentology of a transgression: Portland Point and Cooksburg Members (Middle Devonian), New York State. *J. sedim. Petrol.*, **43**, 484–504.
9.10.1, 9.10.2.

McDonald B.C. (1971) Late Quaternary stratigraphy and deglaciation in eastern Canada. In: *The Late Cenozoic Glacial Ages* (Ed. by K.K. Turekian), pp. 331–353.
13.4.1.

McDonald B.C. & Banerjee I. (1971) Sediments and bedforms on a braided outwash plain. *Can. J. Earth Sci.*, **8**, 1282–1301.
3.3.1, 3.3.2.

McDowell J.P. (1960) Cross-bedding formed by sand waves in Mississippi River point bar deposits. *Bull. geol. Soc. Am.*, **71**, 1925.
3.5.2, 3.5.3.

McGowen J.H. & Garner L.E. (1970) Physiographic features and stratification types of coarse-grained point bars: modern and ancient examples. *Sedimentology*, **14**, 77–111.
3.5.2, 3.9.4, Fig. 3.26, 3.27.

McGowen J.H. & Groat C.G. (1971) Van Horn Sandstone, West Texas: an alluvial fan model for mineral exploration. *Report of Investigations*, **72**, pp. 57. Bureau of Economic Geology, Univ. of Texas, Austin.
3.8, 3.8.1, Figs. 3.36, 3.38.

McGowen J.H. & Scott A.J. (1976) Hurricanes as geologic agents on the Texas Coast. In: *Estuarine Research*, Vol. II Geology and Engineering (Ed. by L.E. Cronin), pp. 23–46. Academic Press, London.
7.2.1, 7.2.3, Fig. 7.25.

McKee E.D. (1945) Small-scale structures in the Coconino Sandstone of northern Arizona. *J. Geol.*, **53**, 316–320.
5.3.2.

McKee E.D. (1957) Primary structures in some recent sediments. *Bull. Am. Ass. Petrol. Geol.*, **41**, 1704–1747.
Fig. 7.8.

McKee E.D. (1966) Structures of dunes at White Sands National Monument, New Mexico (and comparison with structures of dunes from other selected areas). *Sedimentology*, **7**, 1–69.
5.2.4, Fig. 5.6.

McKee E.D. & Bigarella J.J. (1972) Deformational structures in Brazilian coastal dunes. *J. sedim. Petrol.*, **42**, 670–681.
5.2.4.

McKee E.D., Chronic J. & Leopold E.B. (1959) Sedimentary belts in Lagoon of Kapingamarangi Atoll. *Bull. Am. Ass. Petrol. Geol.*, **43**, 501–562.
Fig. 10.13.

McKee E.D., Crosby E.J. & Berryhill H.L. Jr. (1967) Flood deposits, Bijou Creek, Colorado, June 1965. *J. sedim. Petrol.*, **37**, 829–851.
3.4.4, 3.6.1.

McKee E.D., Douglass J.R. & Rittenhouse S. (1971) Deformation of leeside laminae in eolian dunes. *Bull. geol. Soc. Am.*, **82**, 359–378.
5.2.4.

McKee E.D. & Tibbitts G.C. Jr. (1964) Primary structure of a seif dune and associated deposits in Libya. *J. sedim. Petrol.*, **34**, 5–17.
5.2.4, Fig. 5.8.

McKelvey V.E. (1967) Phosphate deposits. *U.S. geol. Surv. Bull.*, **1252-D**, pp. 21.
9.8.

McKerrow W.S. (1978) (Ed.) *The Ecology of Fossils*, pp. 384. Duckworth, London.
9.12.

McKerrow W.S., Leggett J.K. & Eales M.H. (1977) Imbricate thrust model of the Southern Uplands of Scotland. *Nature*, **267**, 237–239.
14.4.2.

McKinney T.F., Stubblefield W.L. & Swift D.J.P. (1974) Large scale current lineations on the Great Egg Shoal retreat massif, New Jersey shelf: investigations by sidescan sonar. *Mar. Geol.*, **17**, 79–102.
9.6.3.

McManus D.A. (1970) Criteria of climatic change in the inorganic components of marine sediments. *Quatern. Res.*, **1**, 72–102.
9.3.4.

McManus D.A. (1975) Modern versus relict sediment on the continental shelf. *Bull. geol. Soc. Am.*, **86**, 1154–1160.
9.2.

McMurty G.M. & Burnett W.C. (1975) Hydrothermal metallogenesis in the Bauer Deep of the south-eastern Pacific. *Nature*, **254**,

42–43.
11.3.4.

Meckel L.D. (1975) Holocene sand bodies in the Colorado delta area, northern Gulf of California. In: *Deltas, Models for Exploration* (Ed. by M.L. Broussard), pp. 239–265. Houston Geological Society, Houston.
6.5.1, 9.5.2.

Meischner K.D. (1964) Allodapische Kalke, Turbidite in riffnahen Sedimentations-Becken. In: *Turbidites* (Ed. by A.H. Bouma and A. Brouwer), pp. 156–191. Elsevier, Amsterdam.
12.2.4.

Meischner D. (1971) Clastic sedimentation in the Variscan Geosyncline east of the River Rhine. In: *Sedimentology of Parts of Central Europe* (Ed. by G. Müller), pp. 9–43. *Guidebook, VIII Int. Sedim. Congr.*, Kramer, Frankfurt.
11.4.4.

Melson W.G. & Thomson G. (1973) Glassy abyssal basalts, Atlantic sea floor near St. Paul's Rocks: petrography and composition of secondary clay minerals. *Bull. geol. Soc. Am.*, **84**, 703–716.
10.3.2.

Menard H.W. (1955) Deep-sea channels, topography and sedimentation. *Bull. Am. Ass. Petrol. Geol.*, **39**, 236–255.
12.3.2, 12.5, 12.5.1.

Menard H.W. (1960a) Consolidated slabs on the floor of the eastern Pacific. *Deep-Sea Res.*, **7**, 35–41.
11.3.2.

Menard H.W. (1960b) Possible pre-Pleistocene deep sea fans off central California. *Bull. geol. Soc. Am.*, **71**, 1271–1278.
12.3.2, 12.5.1.

Menard H.W. (1964) *Marine Geology of the Pacific*, pp. 271. McGraw Hill, New York.
11.3.4, 12.2.4.

Menard H.W. (1967) Transitional types of crust under small ocean basins. *J. geophys. Res.*, **72**, 3061–3073.
14.3.2.

Menard H.W. & Ludwick J.C. (1951) Applications of hydraulics to the study of marine turbidity currents. *Spec. Publ. Soc. econ. Paleont. Miner.*, **2**, 2–13.
12.3.2.

Menard H.W., Smith S.M. & Pratt R.M. (1965) The Rhône deep-sea fan. In: *Submarine Geology and Geophysics* (Ed. by W.F. Whittard and R. Bradshaw), pp. 271–284. Butterworth, London.
12.5.2.

Menzel D.W. (1974) Primary productivity, dissolved and particulate organic matter and the sites of oxidation of organic matter. In: *The Sea*, Vol. 5 (Ed. by E.D. Goldberg), pp. 659–679. Wiley-Interscience, New York.
11.4.5.

Menzies R.J., George R.Y. & Rowe G.T. (1973) *Abyssal Environment and Ecology of the World Oceans*, pp. 488. John Wiley, New York.
9.3.5.

Mergner H. (1971) Structure, ecology and zonation of Red Sea reefs (in comparison with South Indian and Jamaican reefs). In: *Regional Variation in Indian Ocean Coral Reefs* (Ed. by D.R. Stoddart and M. Yonge), pp. 141–161. Symp. Zoo. Soc. Lond., **28.**
Fig. 10.40b.

Mero J.L. (1965) *The Mineral Resources of the Sea*, pp. 312. Elsevier, Amsterdam.
11.3.3, 11.3.4.

Mesolella K.J., Robinson J.D. & Ormiston A.R. (1974) Cyclic deposition of Silurian carbonates and evaporites in Michigan Basin. *Bull. Am. Ass. Petrol. Geol.*, **58**, 34–62.
Fig. 10.47a.

Meyerhoff A.A. (1970) Continental drift: implications of paleomagnetic studies, meteorology, physical oceanography and climatology. *J. Geol.*, **78**, 1–51.
5.1.

Meylan M.A. (1976) A comparison of the morphology and mineralogy of manganese nodules from the southwestern Pacific Basin and northeastern equatorial Pacific. In: *Marine Geological Investigations in the Southwest Pacific and Adjacent Areas* (Ed. by G.P. Glasby and H.R. Katz), pp. 92–98. *Tech. Bull. Comm. co-ord. joint Prospect., Econ. Soc. Comm. Asia and Pacific*, (U.N.), **2.**
11.3.4.

Miall A.D. (1970) Devonian alluvial fans, Prince of Wales Island, Arctic Canada. *J. sedim. Petrol.*, **40**, 556–571.
3.8.1.

Miall A.D. (1973) Markov chain analysis applied to an ancient alluvial plain succession. *Sedimentology*, **20**, 347–364.
2.1.2.

Miall A.D. (1976) Palaeocurrent and palaeohydraulic analysis of some vertical profiles through a Cretaceous braided stream deposit, Banks Island, Arctic Canada. *Sedimentology*, **23**, 459–483.
3.10.3.

Miall A.D. (1977) A review of the braided-river depositional environment. *Earth-Sci. Rev.*, **13**, 1–62.
3.1.

Michelson P.C. & Dott R.H. Jr. (1973) Orientation analysis of trough cross stratification in Upper Cambrian sandstones of western Wisconsin. *J. sedim. Petrol.*, **43**, 784–794.
9.10.2.

Middleton G.V. (1965) (Ed.) *Primary sedimentary structures and their hydrodynamic interpretation*, pp. 265. *Spec. Publ. Soc. econ. Paleont. Miner.*, **12**, Tulsa.
1.1, 3.1.

Middleton G.V. (1966) Experiments on density and turbidity currents. I. Motion of the head. *Can. J. Earth Sci.*, **3**, 523–546.
12.2.4.

Middleton G.V. (1969) Turbidity currents and grain flows and other mass movements down slopes. In: *The New Concepts of Continental Margin Sedimentation* (Ed. by D.J. Stanley). GM-A-1 to GM-B-14. Am. geol. Inst. Short Course Notes.
12.2.6.

Middleton G.V. (1970) Experimental studies related to flysch sedimentation. In: *Flysch Sedimentology in North America* (Ed. by J. Lajoie), pp. 253–272. *Spec. Pap. geol. Assoc. Can.*, **7.**
12.2.4, 12.2.6.

Middleton G.V. (1973) Johannes Walther's law of correlation of facies. *Bull. geol. Soc. Am.*, **84**, 979–988.
2.1.2.

Middleton G.V. & Hampton M.A. (1976) Subaqueous sediment transport and deposition by sediment gravity flows. In: *Marine Sediment Transport and Environmental Management* (Ed. by D.J. Stanley and D.J.P. Swift), pp. 197–218. John Wiley, New York.
3.2.4, 12.2.1, 12.2.3, 12.2.4, 12.2.6, Figs. 12.4, 12.5, 12.11, 12.13, 12.16.

Miller D.J. (1953) Late Cenozoic marine glacial sediments and marine terraces of Middleton Island, Alaska. *J. Geol.*, **61**, 17–40.
13.4.3.

Milliman J.D. (1969) Four southwestern Caribbean atolls. *Atoll Res. Bull.*, **129**, 26.

10.2.2.

MILLIMAN J.D. (1971) Carbonate lithification in the deep sea. In: *Carbonate Cements* (Ed. by O.P. Bricker), *Studies in Geology*, **19**, 95–102. Johns Hopkins University, Baltimore.
11.3.3, 11.3.5, 11.3.6.

MILLIMAN J.D. (1972) Atlantic continental shelf and slope of the United States – petrology of the sand fraction of sediments, northern New Jersey to southern Florida. *Prof. Pap. U.S. geol. Surv.*, **529-J**, 40.
Figs. 10.16a, 10.16b.

MILLIMAN J.D. (1974) *Marine Carbonates*, pp. 375. Springer-Verlag, Berlin.
10.1.1.

MILLIMAN J.D. (1975) Dissolution of aragonite, Mg-calcite, and calcite in the North Atlantic Ocean. *Geology*, **3**, 461–462.
11.3.1.

MILLIMAN J.D. & MÜLLER J. (1973) Precipitation and lithification of magnesian calcite in the deep-sea sediments of the eastern Mediterranean Sea. *Sedimentology*, **20**, 25–49.
11.3.5.

MILLIMAN J.D., PILKEY O.H. & ROSS D.A. (1972) Sediments of the continental margin of the eastern United States. *Bull. geol. Soc. Am.*, **83**, 1315–1334.
10.2.3, Figs. 10.16a, 10.16b.

MILLING M.E. (1975) Geological appraisal of foundation conditions, Northern North Sea. *Oceanology International 1975* (conference papers), 310–319.
13.5.1, 13.6, Fig. 13.9.

MITCHELL A.H.G. (1970) Facies of an Early Miocene volcanic arc, Malekula Island, New Hebrides. *Sedimentology*, **14**, 201–243.
11.4.2.

MITCHELL A.H.G. & GARSON M.S. (1976) Mineralisation at plate boundaries. *Minerals Sci. Engng.*, **8**, 129–169.
14.2.4, Fig. 14.31.

MITCHELL A.H.G. & MCKERROW W.S. (1975) Analogous evolution of the Burma orogen and the Scottish Caledonides. *Bull. geol. Soc. Am.*, **86**, 305–315.
14.4.2, Fig. 14.31.

MITCHELL A.H.G. & READING H.G. (1969) Continental margins, geosynclines and ocean floor spreading. *J. Geol.*, **77**, 629–646.
14.2.6, 14.5.

MOBERLY R.M. & LARSON R.L. (1975) Mesozoic magnetic anomalies, oceanic plateaus, and seamount chains in the north-western Pacific Ocean. In: *Initial Reports of the Deep Sea Drilling Project*, **32** (R.L. Larson and R. Moberly *et al.*), 945–957. U.S. Government Printing Office, Washington.
11.3.3, Fig. 11.13.

MOGHAL M.Y. (1974) Uranium in Siwalik Sandstones, Sulaiman Range, Pakistan. In: *Formation of Uranium Ore Deposits*, pp. 383–403. Proc. Int. Atomic Energy Agency, Vienna.
14.3.4.

MOLENGRAAF G.A.F. (1915) On the occurrence of nodules of manganese in Mesozoic deep-sea deposits from Borneo, Timor, and Rotti, their significance and mode of formation. *Proc. Sect. Sci. K. ned. Akad. Wet.*, **18**, 415–430.
11.1.2, 11.4.2.

MOLENGRAAF G.A.F. (1922) On manganese nodules in Mesozoic deep-sea deposits of Dutch Timor. *Proc. Sect. Sci. K. ned. Akad. Wet.*, **23**, 997–1012.
11.1.2, 11.4.2.

MOLNAR P. & TAPPONNIER P. (1975) Cenozoic tectonics of Asia: effects of a continental collision. *Science*, **189**, 419–426.
14.3.4.

MONTE M. DEL, GIOVANELLI G., NANNI T. & TAGLIAZUCCA M. (1976) Black magnetic spherules in condensed sediments from topographic highs. *Arch. Met. Geophy. Biokl.*, ser. A, **25**, 151–157.
11.3.5.

MONTY C. (1967) Distribution and structure of recent stromatolitic algal mats, eastern Andros Island, Bahamas. *Annls. Soc. geol. Belg.*, **90**, 57–93.
8.2.5.

MOODY J.D. & HILL M.J. (1956) Wrench fault tectonics. *Bull. geol. Soc. Am.*, **67**, 1207–1246.
14.3.3.

MOODY-STUART M. (1966) High- and low-sinuosity stream deposits, with examples from the Devonian of Spitsbergen. *J. sedim. Petrol.*, **36**, 1102–1117.
3.9.3, 3.9.4, Fig. 3.44.

MOOERS C.N.K. (1976) Introduction to the physical oceanography and fluid dynamics of continental margins. In: *Marine Sediment Transport and Environmental Management* (Ed. by D.J. Stanley and D.J.P. Swift), pp. 7–21. John Wiley, New York.
9.4.3.

MOORE D. (1959) Role of deltas in the formation of some British Lower Carboniferous cyclothems. *J. Geol.*, **67**, 522–539.
2.2.2.

MOORE D.G. (1961) Submarine slumps. *J. sedim. Petrol.*, **31**, 343–357.
12.2.1, 12.2.2, Fig. 12.2.

MOORE D.G. (1969) Reflection profiling studies of the California continental borderland: structure and Quaternary turbidite basins. *Spec. Pap. geol. Soc. Am.*, **107**, pp. 142.
12.2.5, 12.4.5, Fig. 12.18.

MOORE D.G., CURRAY J.R. & EMMEL F.J. (1976) Large submarine slide (olistostrome) associated with Sunda Arc subduction zone, northeast Indian Ocean. *Mar. Geol.*, **21**, 211–226.
12.2.3.

MOORE D.G., CURRAY J.R., RAITT R.W. & EMMEL F.J. (1974) Stratigraphic-seismic section correlations and implications to Bengal Fan history. In: *Initial Reports of the Deep Sea Drilling Project*, **22** (C.C. von der Borch, J.G. Sclater *et al.*), 403–412. U.S. Government Printing Office, Washington D.C.
12.5.1, 12.5.2, Fig. 12.38.

MOORE D.G. & SCRUTON P.C. (1957) Minor internal structures of some recent unconsolidated sediments. *Bull. Am. Ass. Petrol. Geol.*, **41**, 2723–2751.
9.6.1.

MOORE D.R. & BULLIS R.W. (1960) A deep-water coral reef in the Gulf of Mexico. *Marine Sci. & Caribbean Bull.*, **10**, 125–128.
10.4.1.

MOORE G.F. & KARIG D.E. (1976) Development of sedimentary basins on the lower trench slope. *Geology*, **4**, 693–697.
14.3.2, 14.4.2.

MOORE J.C. (1973) Cretaceous continental margin sedimentation, southwestern Alaska. *Bull. geol. Soc. Am.*, **84**, 595–614.
12.7.2.

MOORE J.C. (1974) Turbidites and terrigenous muds, DSDP leg 25. In: *Initial Reports of the Deep Sea Drilling Project* **25**, (E.S.W. Simpson, R. Schlich *et al.*), 441–479. U.S. Government Printing Office, Washington, D.C.
12.2.4.

MOORE J.C. (1975) Selective subduction. *Geology*, **3**, 530–532.
14.3.2.

MOORE J.C. & KARIG D.E. (1976) Sedimentology, structural geology, and tectonics of the Shikiku subduction zone, southwestern Japan. *Bull. geol. Soc. Am.*, **87**, 1259–1268.
12.4.3.

MOORE T.C. JR. (1970) Abyssal hills in the equatorial Pacific: sedimentation and stratigraphy. *Deep-Sea Res.*, **17**, 573–593.
11.3.4.

MOORE W.S. & VOGT P.R. (1976) Hydrothermal manganese crusts from two sites near the Galapagos spreading axis. *Earth Planet. Sci. Letts.*, **29**, 349–356.
11.3.2.

MOORES E.M. & VINE F.J. (1971) The Troodos Massif, Cyprus and other ophiolites as oceanic crust: evaluation and implications. *Phil. Trans. R. Soc. A*, **268**, 443–466.
11.4.1, 11.4.2.

MORAN S.R. (1971) Glaciotectonic structures in drift. In: *Till: A Symposium* (Ed. by R.P. Goldthwait), pp. 127–148. Ohio State Univ. Press, Columbus.
13.4.1.

MORGAN J.P. (1961) Mudlumps at the mouths of the Mississippi River. In: *Genesis and Paleontology of the Mississippi River Mudlumps.* Louisiana Dept. Conservation Geol. Bull., **35**, pt. 1, pp. 116.
6.8.2.

MORGAN J.P. (Ed.) (1970) Deltaic sedimentation; modern and ancient, pp. 312. *Spec. Publ. Soc. econ. Paleont. Miner.*, **15**, Tulsa.
1.1, 6.3.1.

MORGAN J.P., COLEMAN J.M. & GAGLIANO S.M. (1963) Mudlumps at the mouth of South Pass, Mississippi River; sedimentology, paleontology, structure, origin and relation to deltaic processes. *Louisiana State Univ. Coastal Studies Ser.*, **10**, pp. 190.
6.8.2.

MORGAN J.P., COLEMAN J.M. & GAGLIANO S.M. (1968) Mudlumps: diapiric structures in Mississippi delta sediments. In: *Diapirism and Diapirs* (Ed. by J. Braunstein and G.D. O'Brien), pp. 145–161. *Mem. Am. Ass. Petrol. Geol.*, **8**, Tulsa.
6.8.2, Fig. 6.54.

MORGAN J.P. & MCINTIRE W.G. (1959) Quaternary geology of the Bengal Basin, East Pakistan and India. *Bull. geol. Soc. Am.*, **70**, 319–342.
6.3.1.

MORGAN W.J. (1968) Rises, trenches, great faults and crustal blocks. *J. geophys. Res.*, **73**, 1959–1982.
14.2.6.

MORGENSTERN N. (1967) Submarine slumping and the initiation of turbidity currents. In: *Marine Geotechnique* (Ed. by A.F. Richards), pp. 189–220. Univ. Illinois Press, Urbana.
12.2.1, 12.2.2, 12.2.4.

MORRIS K. (1977) A model for deposition of bituminous shales in the Lower Toarcian. *Symp. sur la séd. du Jurassique W. européen.* Ass. des Sed. Français (Abstract), pp. 39.
9.12, Fig. 9.55.

MORRIS R.C. (1971) Classification and interpretation of disturbed bedding types in Jackfork flysch rocks (Upper Mississippian) Ouachita Mountains, Arkansas. *J. sedim. Petrol.*, **41**, 410–424.
12.2.2.

MORTON R.A. (1978) Large-scale rhomboid bed forms and sedimentary structures associated with hurricane washover. *Sedimentology*, **25**, 183–204.
7.2.5.

MORTON R.A. & DONALDSON A.C. (1974) Sediment distribution and evolution of tidal deltas along a tide-dominated shoreline, Ivachapreague, Virginia. *Sedim. Geol.*, **10**, 285–300.
7.2.5, Fig. 7.24.

MORTON R.D. (1974) Sandstone-type uranium deposits in the Proterozoic strata of Northwestern Canada. In: *Formation of Uranium Ore Deposits*, pp. 255–273. Proc. Int. Atomic Energy Agency, Vienna.
14.4.1.

MOSS A.J. (1962) The physical nature of common sandy and pebbly deposits, Part I. *Am. J. Sci.*, **260**, 337–373.
5.2.4.

MOSS A.J. (1963) The physical nature of common sandy and pebbly deposits, Part II. *Am. J. Sci.*, **261**, 297–343.
5.2.4.

MRAKOVICH J.V. & COOGAN A.H. (1974) Depositional environment of the Sharon Conglomerate Member of the Pottsville Formation in northeastern Ohio. *J. sedim. Petrol.*, **44**, 1186–1199.
3.9.3.

MÜLLER G. (1969) Sedimentary phosphate method for estimating palaeosalinities: limited applicability. *Science*, **163**, 812–813.
9.8.

MÜLLER J. & FABRICIUS F. (1974) Magnesian-calcite nodules in the Ionian deep sea: an actualistic model for the formation of some nodular limestones. In: *Pelagic Sediments: On Land and Under the Sea* (Ed. by K.J. Hsü and H.C. Jenkyns), pp. 249–271. *Spec. Publ. int. Ass. Sediment.*, **1.**
11.3.5.

MULTER H.G. (1971) *Field Guide to Some Carbonate Rock Environments, Florida Keys and Western Bahamas*, pp. 158. Fairleigh Dickinson University, Madison, New Jersey.
10.2.4, Figs. 10.20a, 10.21d.

MURATA K.J. & LARSON R.R. (1975) Diagenesis of Miocene siliceous shales, Temblor Range, California. *J. Res. U.S. geol. Surv.*, **3**, 553–556.
11.4.3.

MURRAY J. (1890) The Maltese Islands, with special reference to their geological structure. *Scott. geogr. Mag.*, **6**, 449–488.
11.1.2, 11.4.3.

MURRAY J. & HJÖRT J. (1912) *The Depths of the Ocean*, pp. 821. Macmillan, London.
11.1.1, Fig. 11.2.

MURRAY J. & RENARD A.F. (1884) On the microscopic characters of volcanic ashes and cosmic dust, and their distribution in deep-sea deposits. *Proc. R. Soc. Edinb.*, **12**, 474–495.
11.3.4.

MURRAY J. & RENARD A.F. (1891) Report on deep-sea deposits based on specimens collected during the voyage of H.M.S. *Challenger* in the years 1873–1876. In: *'Challenger' Reports*, pp. 525. H.M.S.O., Edinburgh.
11.1.2, 11.2, 11.3.4, Tab. 11.1.

MURRAY J.W. & WRIGHT C.A. (1971) The Carboniferous Limestone of Chipping Sodbury and Wick, Gloucestershire. *Geol. J.*, **7**, 255–270.
10.2.5, Fig. 10.33.

MUTTI E. (1974) Examples of ancient deep-sea fan deposits from circum-Mediterranean geosynclines. In: *Modern and Ancient Geosynclinal Sedimentation* (Ed. by R.H. Dott and R.H. Shaver), pp. 92–105. *Spec. Publ. Soc. econ. Paleont. Miner.*, **19**, Tulsa
12.8.1.

MUTTI E. (1977) Distinctive thin-bedded turbidite facies and related depositional environments in the Eocene Hecho Group (South-central Pyrenees, Spain). *Sedimentology*, **24**, 107–132.
12.6.3, 12.7.1, 12.8.2.

MUTTI E. & RICCI-LUCCHI F. (1972) Le torbiditi dell'Appennino settentrionale: introduzione all'analisi di facies. *Mem. Soc. geol. Ital.*, **11**, 161–199.
12.6.4, 12.7.1, 12.8.1, 12.8.2, Fig. 12.52.

NAGTEGAAL P.J.C. (1969) Sedimentology, palaeoclimatology and diagenesis of post-Hercynian continental deposits in the south-central Pyrenees, Spain. *Leidse geol. Meded.*, **42**, 143–238.
3.9.3.

NAIDU A.S. & MOWATT T.C. (1975) Depositional environments and sediment characteristics of the Colville and adjacent deltas, north Arctic Alaska. In: *Deltas, Models for Exploration* (Ed. by M.L. Broussard), pp. 283–309. Houston Geological Society, Houston.
6.5.1.

NAMI M. (1976) An exhumed Jurassic meander belt from Yorkshire, England. *Geol. Mag.*, **113**, 47–52.
3.9.4.

NANZ R.H. JR. (1954) Genesis of Oligocene sandstone reservoir, Seeligson field, Jim Wells and Kleberg Counties, Texas. *Bull. Am. Ass. Petrol. Geol.*, **38**, 96–117.
6.2.

NARAYAN J. (1971) Sedimentary structures in the Lower Greensand of the Weald, England, and Bas-Boulonnais, France. *Sedim. Geol.*, **6**, 73–109.
9.1.1, 9.10.1, 9.10.2, 9.10.3.

NATLAND J.H. (1973) Basal ferromanganoan sediments at DSDP sites 183, Aleutian abyssal plain, and site 192, Meiji Guyot, northwest Pacific, leg 19. In: *Initial Reports of the Deep Sea Drilling Project*, **19** (J.S. Creager, D.W. Scholl *et al.*), pp. 629–640. U.S. Government Printing Office, Washington.
11.3.3.

NATLAND M.L. & KUENEN PH.H. (1951) Sedimentary history of the Ventura Basin, Calif., and the action of turbidity currents, pp. 76–107. *Spec. Publ. Soc. econ. Paleont. Miner.*, **2**, Tulsa.
12.1, 12.2.7, 12.6.1, 12.8.1.

NEAL J.T. (1965) Giant dessication polygons of Great Basin playas. *Air Force Cambridge Research Laboratories, Environmental Research Papers*, **123**, pp. 30.
Fig. 5.9.

NEDECO (1959) *River Studies and Recommendations on Improvement of Niger and Benue*, pp. 1000. North Holland: Amsterdam.
3.4, 3.4.1.

NEDECO (1961) *The waters of the Niger delta*, The Hague, pp. 317.
6.5.1, 6.5.2, Fig. 6.12.

NEEV D. & EMERY K.O. (1967) Dead Sea: depositional processes and environments of evaporites, pp. 147. *Bull. Israel geol. Surv.*, **41.**
4.2.4.

NEILL C.R. (1969) Bed forms in the Lower Red Deer River, Alberta. *J. Hydrol.*, **7**, 58–85.
3.4, 3.4.2.

NELSON B.W. (1967) Sedimentary phosphate method for estimating palaeosalinities. *Science*, **158**, 917–920.
9.8.

NELSON B.W. (1970) Hydrography, sediment dispersal and recent historical development of the Po River delta, Italy. In: *Deltaic Sedimentation Modern and Ancient* (Ed. by J.P. Morgan and R.H. Shaver), pp. 152–184. *Spec. Publ. Soc. econ. Palaeont. Miner.*, **15**, Tulsa.
6.5.2, Fig. 6.23.

NELSON C.H. (1967) Sediments of Crater Lake, Oregon. *Bull. geol. Soc. Am.*, **79**, 833–848.
12.5.4.

NELSON C.H. (1976) Late Pleistocene and Holocene depositional trends, processes, and history of Astoria deep-sea fan, northeast Pacific. *Mar. Geol.*, **20**, 129–173.
12.2.4, 12.5.1, Fig. 12.34.

NELSON C.H., CARLSON P.R., BYRNE J.V. & ALPHA T.R. (1970) Development of the Astoria canyon-fan physiography and comparison with similar systems. *Mar. Geol.*, **8**, 259–291.
12.3.2, 12.5.1, 12.5.5.

NELSON C.H. & KULM L.D. (1973) Submarine fans and channels. In: *Turbidites and Deep Water Sedimentation*, pp. 39–78. Soc. econ. Paleont. Miner., Pacific Section, Short Course, Anaheim.
12.2.4, 12.3.1, 12.5.1, 12.5.3, Figs. 12.35, 12.36.

NELSON C.H. & NILSON T. (1974) Depositional trends of modern and ancient deep-sea fans. In: *Modern and Ancient Geosynclinal Sedimentation* (Ed. by R.H. Dott and R.H. Shaver), pp. 54–76. *Spec. Publ. Soc. econ. Paleontol. Miner.*, **19.**
12.5.1, 12.8.1, 12.8.2.

NESTEROFF W.D., WEZEL F.C. & PAUTOT G. (1973) Summary of lithostratigraphic findings and problems. In: *Initial Reports of the Deep Sea Drilling Project*, Vol. XIII (W.B.F. Ryan, K.J. Hsü *et al.*), pp. 1021–1040. U.S. Government Printing Office, Washington.
11.3.5.

NEUMANN A.C., KOFOED J.W. & KELLER G.H. (1977) Lithoherms in the Straits of Florida. *Geology*, **5**, 4–10.
11.3.6.

NEUMANN A.C. & LAND L.S. (1975) Lime mud deposition and calcareous algae in the Bight of Abaco, Bahamas: a budget. *J. sedim. Petrol.*, **45**, 763–786.
11.4.4.

NEUMAYR M. (1875) *Erdgeschichte*, **1**, pp. 364. Bibliographisches Inst., Leipzig.
14.2.1.

NEUMAYR M. (1887) *Erdgeschichte. Erster Band, Allgemeine Geologie*, pp. 653. Bibliographisches Inst., Leipzig.
11.1.2.

NEWELL N.D., IMBRIE J., PURDY E.G. & THURBER D.L. (1959) Organism communities and bottom facies, Great Bahama Bank. *Bull. Am. Museum Nat. Hist.*, **117**, 177–228.
10.2.4.

NEWELL N.D., RIGBY J.K., FISCHER A.G., WHITEMAN A.J., HICKOX J.E. & BRADLEY J.S. (1953) *The Permian Reef Complex of the Guadalupe Mountains Region, Texas and New Mexico*, pp. 236. W.H. Freeman, San Francisco.
10.4.2.

NEWELL N.D., RIGBY J.K., WHITEMAN A.J. & BRADLEY J.S. (1951) Shoal-water geology and environments, eastern Andros Island, Bahamas. *Bull. Am. Mus. Nat. Hist.*, **97**, 1–29.
10.1.1.

NEWTON R.S. (1968) Internal structure of wave-formed ripple marks in the nearshore zone. *Sedimentology*, **11**, 275–292.
9.11.1.

NEWTON R.S., SEIBOLD E. & WERNER F. (1973) Facies distribution patterns on the Spanish Sahara continental shelf mapped with side-scan sonar. *Meteor. Forsch. Eng.*, **C15**, 55–77.
9.4.1, 9.10.3.

NIEDORADA A.W. (1972) Waves, currents, sediments and sand bars associated with low energy coastal environments. *Trans. Gulf-Cst. Ass. geol. Socs.*, **22**, 229–239.
7.2.2.

NIEDORADA A.W. & TANNER W.F. (1970) Preliminary study of transverse bars. *Mar. Geol.*, **9**, 41–62.
7.2.2.

NIILER P.P. (1975) A report on the continental shelf circulation and coastal upwelling. *Rev. Geophys. Space Phys.*, **13**, 609–614.
9.4.1.

NILSEN T.H. (1968) The relationship of sedimentation to tectonics in the Solund Devonian district of south-western Norway. *Norg. geol. Unders.*, **259**, pp. 108.
3.8.1.

NILSEN T.H. (1969) Old Red sedimentation in the Beulandet-Vaerlandet Devonian district, Western Norway. *Sedim. Geol.*, **3**, 35–57.
3.8.1.

NILSSON H.D. (1973) Sand bars along low energy coasts. Part I. Multiple parallel sand bars of southeastern Cape Cod Bay. In: *Coastal Geomorphology* (Ed. by D.R. Coates), pp. 99–102. Publications in Geomorphology, State University of New York, Binghamton.
7.2.2.

NIO S-D (1976) Marine transgressions as a factor in the formation of sand wave complexes. *Geol. Mijnb.*, **55**, 18–40.
9.10.1, 9.10.4.

NIPKOW H.F. (1928) Über das Verhalten des Skelette planktischer Kieselalgen im geschichteten Tiefenschlamm des Zürich und Baldegersees. *Rev. Hydrobiol.*, **4**, 71–120.
4.2.4, 4.3.5.

NOBEL J.P.A. (1970) Biofacies analysis, Cairn formation of Miette reef complex (Upper Devonian), Jasper National Park, Alberta. *Bull. Can. Petrol. Geol.*, **18**, 493–543.
10.4.2.

NOCERA DI S. & SCANDONE P. (1977) Triassic nannoplankton limestones of deep basin origin in the central Mediterranean region. *Palaeogeogr. Palaeclimat. Palaeoecol.*, **21**, 101–111.
11.4.4.

NORMARK W.R. (1970) Growth patterns of deep-sea fans. *Bull. Am. Ass. Petrol. Geol.*, **54**, 2170–2195.
12.3.2, 12.5.1, 12.5.3, 12.5.5, Fig. 12.33.

NORMARK W.R. & PIPER D.J.W. (1969) Deep-sea fan-valleys, past and present. *Bull. geol. Soc. Am.*, **80**, 1859–1866.
12.5.1.

NORRIS R.M. (1964) Sediments of Chatham Rise. *Bull. N.Z. Dept. sci. ind. Res.*, **159**, 1—39.
11.3.6.

NORRIS R.M. & NORRIS K.S. (1961) Algodones Dunes of Southeastern California. *Bull. geol. Soc. Am.*, **72**, 605–620.
5.2.4.

NORRMAN J.O. (1964) Lake Vättern: investigations on shore and bottom morphology. *Geogr. Annalr.*, **46A**, 1–238.
4.2.1, 4.2.2.

NORTHOLT A.J.G. & HIGHLEY D.E. (1975) Gypsum and anhydrite. *Mineral Dossier*, **13**, pp. 38. H.M. Stationery Office, London.
8.4.2.

NOTA D.J.E. (1958) *Reports of the Orinoco Shelf Expedition*, **2**, pp. 98. H. Veenman en Zönen, Wageningen.
6.8.2.

NYSTUEN J.P. (1976a) Facies and sedimentation of the Late Precambrian Moelv Tillite in the eastern part of the Sparagmite region, Southern Norway. *Norg. geol. Unders.*, **329**, 1–70.
13.4.1, 13.5.4.

NYSTUEN J.P. (1976b) Late Precambrian Moelv Tillite deposited on a discontinuity surface associated with a fossil ice wedge, Rendalen, Southern Norway. *Norsk geol. Tidsskr.*, **56**, 29–56.
13.5.4.

OCAMB R.D. (1961) Growth faults of south Louisiana. *Trans. Gulf-Cst. Ass. geol. Socs.*, **11**, 139–175.
6.8.2, Fig. 6.55.

OCCHIETTI S. (1973) Les structures et déformations engendrées par les glaciers. Essai de mise au point. I. Déformations et structures glaciotectoniques. *Rev. Géogr. Montreal.*, **27**, 365–380.
13.4.1.

ODER C.R.L. & BRUMGARNER J.G. (1961) Stromatolitic bioherms in the Maynardville (Upper Cambrian) limestone, Tennessee. *Bull. geol. Soc. Am.*, **72**, 1021–1028.
10.4.2.

OEHLER J.H. (1975) Origin and distribution of silica lepispheres in porcelanite from the Monterey Formation of California. *J. sedim. Petrol.*, **45**, 252–257.
11.4.3.

OELE E. (1969) The Quaternary geology of the Dutch part of the North Sea, North of the Frisian Isles. *Geol. Mijnb.*, **48**, 467–480.
13.5.1.

OELE E. (1971) The Quaternary geology of the southern area of the Dutch part of the North Sea. *Geol. Mijnb.*, **50**, 461–474.
13.5.1.

OERTAL G.F. (1972) Sediment transport on estuary entrance shoals and the formation of swash platforms. *J. sedim. Petrol.*, **42**, 858–863.
7.2.4, 9.6.3.

OERTAL G.F. (1976) Ebb tidal deltas of Georgia estuaries. In: *Estuarine Research*, Vol. II Geology and Engineering (Ed. by L.E. Cronin), pp. 267–276. Academic Press, London.
7.2.4.

OERTEL G.F. & HOWARD J.D. (1972) Water circulation and sedimentation at estuary entrances on the Georgia coast. In: *Shelf Sediment Transport: Process and Pattern* (Ed. by D.J.P. Swift, D.B. Duane and O.H. Pilkey), pp. 411–427. Dowden, Hutchinson & Ross, Stroudsburg.
9.6.3.

OFF T. (1963) Rhythmic linear sand bodies caused by tidal currents. *Bull. Am. Ass. Petrol. Geol.*, **47**, 324–341.
9.5.

OGNIBEN L. (1957) Petrografia della Serie Solfifera siciliana e considerazioni geologiche relative. *Mem. Descr. Carta. geol. Italia*, **33**, pp. 275.
11.4.3.

OLDALE R.N., UCHUPI E. & PRADA K.E. (1973) Sedimentary framework of the western Gulf of Maine and the southeastern Massachusetts offshore area. *Prof. Pap. U.S. geol. Surv.*, **757.**
13.5.1.

OOMKENS E. (1966) Environmental significance of sand dikes. *Sedimentology*, **7**, 145–148.
5.2.6.

OOMKENS E. (1967) Depositional sequences and sand distribution in a deltaic complex. *Geol. Mijnb.*, **46**, 265–278.
6.2, 6.5.1, 6.5.2, 7.2.6, Fig. 6.25.

OOMKENS E. (1970) Depositional sequences and sand distribution in the post-glacial Rhône delta complex. In: *Deltaic Sedimentation Modern and Ancient* (Ed. by J. P. Morgan and R. H. Shaver), pp. 198–212. *Spec. Publ. Soc. econ. Palaeont. Miner.*, **15**, Tulsa.
6.5.2, 7.2.6.

OOMKENS E. (1974) Lithofacies relations in the Late Quaternary Niger delta complex. *Sedimentology*, **21**, 195–222.
6.2, 6.5.1, 6.5.2, Figs. 6.6, 6.14, 6.28, 6.31.

OOMKENS E. & TERWINDT J.H.J. (1960) Inshore estuarine sediments in the Haringvliet, Netherlands. *Geol. Mijnb.*, **39**, 701–710.
6.5.1, 7.5.1.

OPDYKE N.D. & RUNCORN S.K. (1960) Wind direction in the western United States in the Late Palaeozoic. *Bull. geol. Soc. Am.*, **71**, 959–972.
5.1.

OPEN UNIVERSITY (1976) *Sedimentary Basin Case Study: The Western Canadian Sedimentary Basin*, pp. 68. The OU Press, Milton Keynes.
8.5.2.

ORE H.T. (1963) Some criteria for recognition of braided stream deposits. *Wyoming Univ. Dept. Geol. Contrib. to Geol.*, **3**, 1–14.
3.1, 3.3.1.

ORME A.R. (1973) Barrier and lagoon systems along the Zululand coast, South Africa. In: *Coastal Geomorphology* (Ed. by D.R. Coates), pp. 181–217. Publications in Geomorphology, State University of New York, Binghamton.
7.2.2.

ORR W.N. (1972) Pacific northwest siliceous phytoplankton. *Palaeogeogr. Palaeoclimat. Palaeoecol.*, **12**, 95–114.
11.4.3.

OSGOOD R.G. (1970) Trace fossils of the Cincinnati area. *Palaeont. Americana*, **6**, 281–444.
9.8.

OTVOS E.G. (1970) Development and migration of barrier islands, northern Gulf of Mexico. *Bull. geol. Soc. Am.*, **81**, 241–246.
7.2.7.

OVENSHINE A.T. (1970) Observations of iceberg rafting in Glacier Bay, Alaska, and the identification of ancient ice-rafted deposits. *Bull. geol. Soc. Am.*, **81**, 891–894.
13.3.7.

OVERSBY B.S. (1971) Palaeozoic plate tectonics in the southern Tasman geosyncline. *Nature, Phys. Sci.*, **234**, 45–47.
11.4.2.

PACKHAM G.H. & FALVEY D.A. (1971) An hypothesis for the formation of marginal seas in the western Pacific. *Tectonophysics*, **11**, 79–109.
14.3.2.

PACKHAM G.H. & LEITCH E.C. (1974) The role of plate tectonic theory in the interpretation of the Tasman Orogenic Zone. In: *The Tasman Geosyncline – a Symposium* (Ed. by A.K. Denmead, G.W. Tweedale and A.F. Wilson), pp. 129–154. Geol. Soc. Aust. Qd. Div.
14.4.2.

PALMER H.D. (1964) Marine geology of Rodriguez Seamount. *Deep-Sea Res.*, **11**, 737–756.
11.3.3.

PARK C.F. & MACDIARMID R.A. (1975) *Ore Deposits* (*Third Edition*), pp. 529. Freeman, San Francisco.
8.2.6.

PARK R.K. (1973) *Algal Belt Sedimentation of the Trucial Coast, Arabian Gulf.* Ph.D. Thesis, University of Reading, England.
8.2.5, Figs. 8.9, 8.10.

PARK R.K. (1976) A note on the significance of lamination in stromatolites. *Sedimentology*, **23**, 379–393.
8.2.5, Fig. 8.12.

PARK R.K. (1977) The preservation potential of some recent stromatolites. *Sedimentology*, **24**, 485–506.
8.2.5, 8.5.2.

PARKASH B. & MIDDLETON G.V. (1970) Downcurrent textural changes in Ordovician turbidite greywackes. *Sedimentology*, **14**, 259–293.
12.2.4, 12.6.3, Fig. 12.43.

PARKER R.H. (1960) Ecology and distributional patterns of marine macro-invertebrates, northern Gulf of Mexico. In: *Recent Sediments, Northwest Gulf of Mexico* (Ed. by F.P. Shepard, F.B. Phleger and T.H. van Andel), pp. 302–337, *Am. Ass. Petrol. Geol.*, Tulsa.
7.2.5.

PARKER R.J. (1975) The petrology and origin of some glauconitic and glauco-conglomeratic phosphorites from the South African continental margin. *J. sedim. Petrol.*, **45**, 230–242.
11.3.6.

PARKER R.J. & SIESSER W.G. (1972) Petrology and origin of some phosphorites from the South African continental margin. *J. sedim. Petrol.*, **42**, 434–440.
11.3.6.

PARROT J.F. & DELAUNE-MAYÈRE M. (1974) Les terres d'ombre du Bassit (nordouest Syrien). Comparaison avec les termes similaires du Troodos (Chypre). *Cah. ORSTOM, sér. Géol.*, **6**, 147–160.
11.4.2.

PASSERI L. & PIALLI G. (1972) Facies lagunari nel Calcare Massiccio dell'Umbria occidentale. *Boll. Soc. Geol. Ital.*, **90**, 481–507.
10.2.5.

PATERSON W.S.B. (1969) *The Physics of Glaciers*, pp. 250. Pergamon Press, London.
13.2, 13.2.2.

PATTERSON R.J. & KINSMAN D.J.J. (1977) Marine and continental ground water sources in a Persian Gulf coastal sabkha. In: *Studies in Geology*, pp. 381–397. Am. Ass. Petrol. Geol., **4**, Tulsa.
8.2.6, Figs. 8.17, 8.18, 8.19, Tab. 8.3.

PAULUS F.J. (1972) The geology of site 98 and the Bahama platform. In: *Initial Reports of the Deep Sea Drilling Project* (C.D. Hollister, J.I. Ewing *et al.*), pp. 877–897. U.S. Government Printing Office, Washington.
11.3.6.

PAUTOT G. & MELGUEN M. (1975) Deep bottom currents, sedimentary hiatuses and polymetallic nodules. In: *Marine Geological Investigations in the Southwest Pacific and Adjacent Areas* (Ed. by G.P. Glasby and H.R. Katz), pp. 54–61. *Tech. Bull. Comm. Co-ord. joint Prospect., Econ. Soc. Comm. Asia and Pacific* (*U.N.*), **2.**
11.3.4.

PEACH B.N. & HORNE J. (1899) The Silurian rocks of Great Britain. Vol. 1, Scotland. *Mem. geol. Surv. Great Britain*, pp. 749.
11.4.1.

PEARCE J.A. & CANN J.R. (1971) Ophiolite origin investigated by discriminant analysis using Ti, Zr and Y. *Earth Planet. Sci. Letts.*, **12**, 339–349.
11.4.1.

PEARCE J.A. & CANN J.R. (1973) Tectonic setting of basic volcanic rocks determined using trace element analyses. *Earth Planet. Sci. Letts.*, **19**, 290–300.
11.4.1.

PEARSON E.F. & HANLEY J.H. (1974) Significance of thin carbonates in interpreting the depositional environments of thick clastic sequences. *Wyoming Univ. Dept. Geol., Contrib. to Geol.*, **13**, 63–66.
4.3.6.

PEDLEY H.M. (1976) A palaeoecological study of the Upper Coralline limestone, Terebratula-Aphelesia bed (Miocene, Malta) based on

bryozoan growth-form studies and brachiopod distributions. *Palaeogeogr. Palaeoclim. Paleaoecol.*, **20,** 209–234.
10.3.2.

PEDLEY H.M., HOUSE M.R. & WAUGH B. (1976) The geology of Malta and Gozo. *Proc. Geol. Ass.*, **87,** 325–341.
10.3.2, 11.4.3.

PEDLEY H.M. & WAUGH B. (1976) Easter field meeting to the Maltese Islands, report by the Directors. *Proc. Geol. Ass.*, **87,** 343–358.
10.3.2.

PEEL R.F. (1960) Some aspects of desert geomorphology. *Geography*, **45,** 241–262.
5.2.3.

PENROSE CONFERENCE (1972) Ophiolites. *Geotimes*, **17(12),** 24–25.
14.2.5.

PEPPER J.F., DEWITT W. JR. & DEMAREST D.F. (1954) Geology of the Bedford shale and Berea sandstone in the Appalachian Basin. *Prof. Pap. U.S. geol. Surv.*, **259,** pp. 111.
6.2.

PERCH-NIELSEN K., SUPKO P.R., BOERSMA A., BONATTI E., CARLSON R.L., DINKELMAN M.G., FODOR R.V., KUMAR N., MCCOY F., NEPROCHNOV Y.P., THIEDE J. & ZIMMERMAN H.B. (1975) Leg 39 examines facies change in South Atlantic. *Geotimes*, **20**(**3**), 26–28.
14.3.1.

PESSAGNO E.A. JR. (1973) Age and geologic significance of radiolarian cherts in the California coast ranges. *Geology*, **1,** 153–156.
11.4.2.

PETERSEN G.G.J. (1913) Valvation of the sea, II. The animal communities of the sea-bottom and their importance for marine zoogeography. *Rep. Dan. Biol. Stn.*, **21,** 1–44.
9.3.5.

PETTIJOHN F.J. (1949) *Sedimentary Rocks*, pp. 526. Harper and Row, New York.
10.1.1.

PETTIJOHN F.J. (1957) *Sedimentary Rocks*, pp. 718 (2nd Edn). Harper, New York.
14.2.5.

PETTIJOHN F.J. (1975) *Sedimentary Rocks*, pp. 628(3rd Edn). Harper and Row, New York.
1.1.

PETTIJOHN F.J., POTTER P.E. & SIEVER R. (1973) *Sand and Sandstone*, pp. 617. Springer-Verlag, Berlin.
9.8.

PÉWÉ T.L. (1975) Quaternary geology of Alaska. *Prof. Pap. U.S. geol. Surv.*, **835,** pp. 139.
13.6.

PEYVE A.V. & SINITZYN V.M. (1950) Certains problèmes fondamentaux de la doctrine des géosynclinaux. *Izv. Akad. Nauk. S.S.S.R., Ser. Geol.*, **4,** 28–52.
14.2.4.

PHILLIPS W.E.A., STILLMAN C.J. & MURPHY T. (1976) A Caledonian plate tectonic model. *J. geol. Soc. Lond.*, **132,** 579–609.
14.4.3.

PHLEGER F.B. (1960) Sedimentary patterns of microfaunas in northern Gulf of Mexico. In: *Recent Sediments, Northwest Gulf of Mexico* (Ed. by F.P. Shepard, F.B. Phleger and Tj. H. van Andel), pp. 267–301. Am. Ass. Petrol. Geol., Tulsa.
7.2.5.

PHLEGER F.B. (1965) Sedimentology of Guerrero Negro Lagoon, Baja California, Mexico. In: *Geology and Geophysics, Colston Research Society 17th Symposium, Colston Papers*, **17,** pp. 205–237. Butterworth, London.
7.2.5.

PHLEGER F.B. (1969) A modern evaporite deposit in Mexico. *Bull. Am. Ass. Petrol. Geol.*, **53,** 824–829.
8.1.1.

PICARD M.D. & HIGH L.R. JR. (1968) Sedimentary cycles in the Green River Formation (Eocene), Uinta Basin, Utah. *J. sedim. Petrol.*, **38,** 378–383.
4.3.2.

PICARD M.D. & HIGH L.R. JR. (1972) Criteria for recognizing lacustrine rocks. In: *Recognition of Ancient Sedimentary Environments* (Ed. by J.K. Rigby and W.K. Hamblin), pp. 108–145. *Spec. Publ. Soc. econ. Paleont. Miner.*, **16,** Tulsa.
4.3.1.

PICARD M.D. & HIGH L.R. JR. (1973) Sedimentary structures of ephemeral streams. *Developments in Sedimentology*, No. **17,** pp. 223. Elsevier, Amsterdam.
3.4.4.

PICHON J-F. & LYS M. (1976) Sur l'existence d'une série du Jurassique supérieur à Crétacé inférieur, surmontant les ophiolites dans les collines de Krapa (Massif du Vourinos, Grèce). *C.r. hebd. séanc. Acad. Sci., Paris, D*, **282,** 523–526.
11.7.2.

PICHON X. LE (1968) Sea-floor spreading and continental drift. *J. geophys. Res.*, **73,** 3661–3697.
14.2.6.

PIERCE J.W. & COLQUHOUN D.J. (1970) Holocene evolution of a portion of the North Carolina coast. *Bull. geol. Soc. Am.*, **81,** 3697–3714.
7.2.7.

PILKEY O.H. (1964) The size distribution and mineralogy of the carbonate fraction of the United States south Atlantic shelf and upper slope sediments. *Mar. Geol.*, **2,** 121–136.
10.2.3.

PILKEY O.H. & RUCKER J.B. (1966) Mineralogy of Tongue of the Ocean sediments. *J. Mar. Res.*, **24,** 276–285.
11.4.4.

PIPER D.J.W. (1975) A reconnaissance of the sedimentology of Lower Silurian mudstones, English Lake District. *Sedimentology*, **22,** 623–630.
11.4.4.

PIPER D.J.W., VON HUENE R. & DUNCAN J.R. (1973) Late Quaternary sedimentation in the active eastern Aleutian Trench. *Geology*, **1,** 19–22.
12.4.3, Fig. 12.30.

PIPER D.Z. (1974) Rare earth elements in ferromanganese nodules and other marine phases. *Geochim. Cosmochim. Acta*, **38,** 1007–1022.
11.3.3.

PIPER D.Z., VEEH H.H., BERTRAND W.G. & CHASE R.L. (1975) An iron-rich deposit from the northeast Pacific. *Earth Planet. Sci. Letts.*, **26,** 114–120.
11.3.3.

PIRINI RADRIZANNI C. (1971) Coccoliths from Permian deposits of eastern Turkey. In: *Proc. 2nd Planktonic Conference* (Ed. by A. Farinacci), pp. 993–1001. Edizioni Technoscienza, Rome.
11.4.4.

PITTY A.F. (1971) *Introduction to Geomorphology*, pp. 526. Methuen, London.
2.3.5.

PLAFKER G. & ADDICOTT W.O. (1976) Glaciomarine deposits of Miocene through Holocene age in the Yakataga Formation along

the Gulf of Alaska Margin. In: *Recent and Ancient Sedimentary Environments in Alaska* (Ed. by T.P. Miller), pp. Q1–Q23. Alaska geol. Soc., Anchorage, Alaska.
13.4.3, 13.5.1, 13.6.

PLAYFORD P.E. & COCKBAIN A.E. (1969) Algal stromatolites: deepwater forms in the Devonian of Western Australia. *Science,* **165,** 1008–1010.
10.4.2.

PORRENGA D.H. (1967) Glauconite and chamosite as depth indicators in the marine environment. *Mar. Geol.,* **5,** 495–501.
9.3.6.

POSTMA H. (1967) Sediment transport and sedimentation in the marine environment. In: *Estuaries* (Ed. by G.D. Lauff), pp. 158–180. Am. Assoc. Adv. Sci., Washington D.C.
7.5.

POTTER P.E. & PETTIJOHN F.J. (1963) *Paleocurrents and Basin Analysis,* pp. 296. Springer-Verlag, Berlin.
1.1, 12.6.2.

POULSEN C. (1930) Contributions to the stratigraphy of the Cambro-Ordovician of East Greenland. *Medd. om. Grønland,* **74,** 297–316.
13.5.4.

POULSEN C. & RASMUSSEN H.W. (1951) Geological map (scale 1:50,000), and description of Ella Ø. *Medd. om. Grønland,* **151,** Nr. 5, pp. 36.
13.5.4.

PRATT R.M. (1968) Atlantic continental shelf and slope of the United States – physiography and sediments of the deep-sea basin. *Prof. Pap. U.S. geol. Surv.,* **529-B,** 1–44.
11.3.3.

PRATT R.M. (1971) Lithology of rocks dredged from the Blake Plateau. *S.-East. Geol.,* **13,** 19–38.
11.3.6.

PRATT R.M. & HEEZEN B.C. (1964) Topography of the Blake Plateau. *Deep-Sea Res.,* **11,** 721–728.
11.3.6.

PRATT W.L. (1963) Glauconite from the sea floor off southern California. In: *Essays in Marine Geology in Honor of K.O. Emery* (Ed. by T. Clements, R.E. Stevenson and D.M. Halmos), pp. 97–119. Univ. S. California Press.
11.3.6.

PREST V.K. *et al.* (1968) Glacial map of Canada. Scale 1:5,000,000. *Geol. Surv. Canada,* Map 1253A, pp. 30.
13.5.1.

PRETIOUS E.S. & BLENCH T. (1951) Final report on special observations of bed movement in the Lower Fraser River at Ladner Reach during 1950 freshet. *Nat. Res. Council Canada, Vancouver,* pp. 12.
3.4.2, Fig. 3.17.

PRICE W.A. (1955) Environment and formation of the chenier plain. *Quaternaria,* **2,** 75–86.
7.4.

PRITCHARD D.W. (1955) Estuarine circulation patterns. *Proc. Amer. Soc. Civil Eng.,* **81** (separate 717).
7.5.

PRITCHARD D.W. (1967) What is an estuary: physical viewpoint. In: *Estuaries* (Ed. by G.D. Lauff), pp. 3–5. Am. Assoc. Adv. Sci., Washington D.C.
7.5.

PRYOR W.A. & AMARAL E.J. (1971) Large-scale cross-stratification in the St. Peter Sandstone. *Bull. geol. Soc. Am.,* **82,** 239–244.
9.10.3, 9.10.4.

PSUTY N.P. (1966) The geomorphology of beach ridges in Tabasco, Mexico. *Louisiana State Univ. Coast. Stud. Ser.,* **18,** 51.
6.5.2, Fig. 6.27.

PUGH M.E. (1968) Algae from the lower Purbeck limestones of Dorset. *Proc. Geol. Ass.,* **79,** 513–523.
8.4.2.

PUIGDEFABREGAS C. (1973) Miocene point bar deposits in the Ebro Basin, Northern Spain. *Sedimentology,* **20,** 133–144.
3.9.4.

PURDY E.G. (1961) Bahamian oolite shoals. In: *Geometry of Sandstone Bodies* (Ed. by J.A. Peterson and J.C. Osmond), pp. 53–62. Am. Ass. Petrol. Geol., Tulsa.
10.1.1.

PURDY E.G. (1963a) Recent calcium carbonate facies of the Great Bahama Bank. I. Petrography and reaction groups. *J. Geol.,* **71,** 334–355.
10.1.1, 10.2.1, 10.2.4, Fig. 10.21c.

PURDY E.G. (1963b) Recent calcium carbonate facies of the Great Bahama Bank. II. Sedimentary facies. *J. Geol.,* **71,** 472–497.
10.1.1, 10.2.2, 10.2.4, Fig. 10.21c.

PURDY E.G. (1974a) Reef configurations: cause and effect. In: *Reefs in Time and Space* (Ed. by L.F. Laporte), pp. 9–76. *Spec. Publ. Soc. econ. Paleont. Miner.,* **18,** Tulsa.
10.2.1, 10.2.2, 10.2.3, 10.2.4, 10.4.1, Fig. 10.39.

PURDY E.G. (1974b) Karst-determined facies patterns in British Honduras: Holocene carbonate sedimentation model. *Bull. Am. Ass. Petrol. Geol.,* **58,** 825–855.
Figs. 10.24a, 10.24b.

PURSER B.H. (1973a) Sedimentation around bathymetric highs in the southern Persian Gulf. In: *The Persian Gulf* (Ed. by B.H. Purser), pp. 157–178. Springer-Verlag, Berlin.
10.1.1, 10.4.1.

PURSER B.H. (1973b) (Ed.) *The Persian Gulf: Holocene Carbonate Sedimentation and Diagenesis in a Shallow Epicontinental Sea,* pp. 471. Springer-Verlag, Berlin.
8.1.1.

PURSER B.H. & EVANS G. (1973) Regional sedimentation along the Trucial Coast, S.E. Persian Gulf. In: *The Persian Gulf* (Ed. by B.H. Purser), pp. 211–232. Springer-Verlag, Berlin.
Figs. 8.1, 8.2.

PURSER B.H. & LOREAU J.P. (1973) Aragonitic, supratidal encrustations on the Trucial Coast, Persian Gulf. In: *The Persian Gulf* (Ed. by B.H. Purser), pp. 343–376. Springer-Verlag, Berlin.
8.2.7.

QUENNELL A.M. (1958) The structural and geomorphic evolution of the Dead Sea Rift. *Quart. J. geol. Soc. Lond.,* **114,** 1–24.
14.3.3, Fig. 14.22.

RAAF J.F.M. DE (1964) The occurrence of flute-casts and pseudomorphs after salt crystals in the Oligocene 'grès à ripple-marks' of the southern Pyrenees. In: *Turbidites* (Ed. by A.H. Bouma and A. Brouwer), pp. 192–198. Elsevier, Amsterdam.
3.9.2, 12.6.1.

RAAF J.F.M. DE (1968) Turbidites et associations sédimentaires apparentées. *Proc. Koninkl. Nederlandse Akad. Wetensch.,* **71,** 1–23.
14.2.5.

RAAF J.F.M. DE & BOERSMA J.R. (1971) Tidal deposits and their sedimentary structures. *Geol. Mijnb.,* **50,** 479–504.
6.5.1, 7.5.1, 7.7, 9.1.1, 9.10.1, 9.10.2.

RAAF J.F.M. DE, BOERSMA J.R. & GELDER A. VAN (1977) Wave-generated structures and sequences from a shallow marine succession, Lower Carboniferous, County Cork, Ireland. *Sedimentology,* **24,** 451–483.
9.11.1, 9.11.3, 9.11.4. Figs. 9.47, 9.51, 9.52.

RAAF J.F.M. DE, READING H.G. & WALKER R.G. (1965) Cyclic sedimentation in the Lower Westphalian of north Devon, England. *Sedimentology,* **4,** 1–52.
2.1.2, 6.7.2, 6.7.4. Figs. 2.1, 2.2.

RAD U. VON (1974) Great Meteor and Josephine Seamounts (eastern North Atlantic): composition and origin of bioclastic sands, carbonate and pyroclastic rocks. *'Meteor' Forschungsergebnisse,* **C-19,** 1–61.
11.3.3.

RADWANSKI A. (1968) Lower Tortonian transgression onto the Miechow and Cracow Uplands. *Acta Geol. Pol.,* **18,** 387–446 (in Polish with English abstract and résumé).
10.3.2, Fig. 10.37a, 10.37b.

RADWANSKI A. (1969) Lower Tortonian transgression onto the southern slopes of the Holy Cross Mountains. *Acta Geol. Pol.,* **19,** 137–164 (in Polish with English abstract and résumé).
10.3.2, Figs. 10.37a, 10.37b.

RADWANSKI A. (1973) Lower Tortonian transgression onto the south-eastern and eastern slopes of the Holy Cross Mountains. *Acta Geol. Pol.,* **23,** 375–434 (in Polish with English abstract and résumé).
10.3.2.

RADWANSKI A. & SZULCZEWSKI M. (1965) Jurassic stromatolites of the Villány Mountains (southern Hungary). *Ann. Univ. Scient. Budapestinensis R. Eötvos Nom., Sect. Geol.,* **9,** 87–107.
11.4.4.

RAFF A.D. & MASON R.G. (1961) Magnetic survey off the west coast of North America, 40°N latitude to 52°N latitude. *Bull. geol. Soc. Am.,* **72,** 1267–1270.
14.3.3.

RAHN P.H. (1967) Sheetfloods, streamfloods and the formation of pediments. *Ann. Assoc. Am. Geogr.,* **57,** 593–604.
3.2.4.

RAMSAY A.T.S. (1973) A history of organic siliceous sediments in oceans. In: *Organisms and Continents through Time* (Ed. by N.F. Hughes), pp. 199–234. *Spec. Pap. Palaeont.,* **12.**
11.3.1, 11.3.4, Fig. 11.14.

RAMSBOTTOM W.H.C. (1970) Carboniferous faunas and palaeogeography in the south-west England region. *Proc. Ussher Soc.,* **2,** 144–157.
Fig. 10.32.

RAMSBOTTOM W.H.C. (1973) Transgressions and regressions in the Dinantian: a new synthesis of the British Dinantian stratigraphy. *Proc. Yorks. geol. Soc.,* **39,** 567–607.
10.2.5, Fig. 10.32.

RAUP D.M. & STANLEY S.M. (1971) *Principles of Paleontology,* pp. 388. W.H. Freeman, San Francisco.
9.8.

RAYNER D.H. (1963) The Achanarras Limestone of the Middle Old Red Sandstone, Caithness, Scotland. *Proc. Yorks. geol. Soc.,* **34,** 117–138.
4.3.5.

REA D.J. (1976) Analysis of a fast-spreading rise crest: the East Pacific Rise, 9° to 12° south. *Mar. geophys. Res.,* **2,** 291–313.
11.3.2.

READ W.A. (1969) Analysis and simulation of Namurian sediments in central Scotland using a Markov-process model. *J. int. Ass. mathl. Geol.,* **1,** 199–219.
2.1.2.

READING H.G. (1964) A review of the factors affecting the sedimentation of the Millstone Grit (Namurian) in the Central Pennines. In: *Deltaic and Shallow Marine Deposits* (Ed. by L.M.J.U. Van Straaten), pp. 26–34. Elsevier.
6.7.4, Fig. 6.50.

READING H.G. (1971) Sedimentation sequences in the Upper Carboniferous of northwest Europe. C.r. 6e Congr. Int. Strat. Géol. Carbonif., Sheffield 1967, IV, 1401–1412.
6.2.

READING H.G. (1972) Global tectonics and the genesis of flysch successions. *24th Int. geol. Cong. Proc. Sect.,* **6,** 59–66.
14.2.5, 14.3.

READING H.G. (1975) Strike-slip fault systems; an ancient example from the Cantabrians. *9th Int. Cong. Sedimentol, Nice 1975. Thème 4(2),* 289–292.
14.4.3, Fig. 14.34.

READING H.G. & WALKER R.G. (1966) Sedimentation of Eocambrian Tillites and associated sediments in Finnmark, Northern Norway. *Palaeogeogr. Palaeoclim. Palaeoecol.,* **2,** 177–212.
13.1.

REEVES C.C. JR. (1970) Origin, classification and geologic history of caliche on the southern High Plains, Texas, and eastern New Mexico. *J. Geol.,* **78,** 352–362.
3.6.2.

REICHE P. (1938) An analysis of cross-lamination, the Coconino Sandstone. *J. Geol.,* **46,** 905–932.
5.1, 5.3.2.

REID R.E.H. (1962) Sponges and the chalk rock. *Geol. Mag.,* **99,** 273–278.
11.4.5.

REIMNITZ E. (1971) Surf-beat origin for pulsating bottom currents in the Rio Balsas submarine canyon, Mexico. *Bull. geol. Soc. Am.,* **82,** 81–89.
12.2.4.

REIMNITZ E. & GUTIERREZ-ESTRADA M. (1970) Rapid changes in the head of the Rio Balsas submarine canyon system, Mexico. *Mar. Geol.,* **8,** 245–258.
12.5.2.

REINECK H.E. (1955) Haftrippeln und Haftwarzen, Ablagerungsformen von Flugsand. *Senckenberg. leth.,* **36,** 347–357.
5.2.4.

REINECK H.E. (1958) Longitudinale schrägschicht im Watt. *Geol. Rdsch.,* **47,** 73–82.
7.6.

REINECK H.E. (1963) Sedimentgefüge in Bereich der südlichen Nordsee. *Abh. senckenbergische naturforsch. Ges.,* **505,** 138.
7.6, 9.1.1, 9.5.1, 9.10.1.

REINECK H.E. (1967) Layered sediments of tidal flats, beaches and shelf bottoms of the North Sea. In: *Estuaries* (Ed. by G.D. Lauff), pp. 191–206. Am. Assoc. Adv. Sci., Washington D.C.
7.6, Fig. 7.48.

REINECK H.E. (1972) Tidal flats. In: *Recognition of Ancient Sedimentary Environments* (Ed. by J.K. Rigby and W.K. Hamblin), pp. 146–159. *Spec. Publ. Soc. econ. Palaeont. Miner.,* **16,** Tulsa.
7.6.

REINECK H.E. & SINGH I.B. (1971) Der Gulf von Gaeta (Tyrrhenisches Meer) III. Die gefuge von vorstrandund schelfsedimenten. *Senckenberg. Mar.,* **3,** 135–183.

7.2.3, Fig. 7.18.

REINECK H.E. & SINGH I.B. (1972) Genesis of laminated sand and graded rhythmites in storm-sand layers of shelf mud. *Sedimentology*, **18,** 123–128.
9.3.1, 9.4.3, 9.6.2.

REINECK H.E. & SINGH I.B. (1973) *Depositional Sedimentary Environments – With Reference to Terrigenous Clastics*, pp. 439. Springer-Verlag, Berlin.
1.1, 7.6, 9.9, 9.11, Figs. 7.18, 7,47, 7.49, 9.5, 9.46, 9.48.

REINECK H.E. & WUNDERLICH F. (1968) Classification and origin of flaser and lenticular bedding. *Sedimentology*, **11,** 99–104.
7.6, 9.9, 9.10.1.

REINECK H.E., GUTMANN W.F. & HERTWECK G. (1967) Das schlickgebiet südlich Helgoland als Beispiel rezenter schelfablagerungen. *Senckenberg. leth.*, **48,** 219–275.
9.5.1, Fig. 9.5.

RENZ O. (1973) Two lamellaptychi (Ammonoidea) from the Magellan Rise in the Central Pacific. In: *Initial Reports of the Deep Sea Drilling Project*, **17** (E.L. Winterer, J.I. Ewing *et al.*), pp. 377–405. U.S. Government Printing Office, Washington.
11.3.3.

REVELLE R.R. (1944) Marine bottom samples collected in the Pacific Ocean by the *Carnegie* on its seventh cruise. *Publ. Carnegie Instn.*, **556,** 1–180.
11.1.1, 11.2.

REVELLE R.R., BRAMLETTE M.N., ARRHENIUS G. & GOLDBERG E.D. (1955) Pelagic sediments of the Pacific. In: *The Crust of the Earth* (Ed. by A. Poldervaart), pp. 221–236. *Spec. Paps. geol. Soc. Am.*, **62.**
11.2.

REYNOLDS R.L. JR. (1965) The concentration of boron in Precambrian seas. *Geochim. Cosmochim. Acta*, **29,** 1–16.
9.8.

RHOADS D.C. (1972) Mass properties, stability and ecology of marine muds related to burrowing activity. *Mar. Geol.*, **13,** 391–406.
9.3.5.

RICCI-LUCCHI F. (1969) Channelized deposits in the Middle Miocene flysch of Romagna (Italy). *Giorn. Geol., S.*, **2,** 36, 203–282.
12.8.1.

RICCI-LUCCHI F. (1975a) Sediment dispersal in turbidite basins: examples from the Miocene of northern Apennines. *9th Int. Congr. Sedimentol., Nice 1975, Thème 5(2)*, 347–352.
12.8.1, 12.8.2, Fig. 12.44.

RICCI-LUCCHI F. (1975b) Depositional cycles in two turbidite formations of the northern Apennines (Italy). *J. sedim. Petrol.*, **45,** 3–43.
12.8.1, Fig. 12.49.

RICH J.L. (1942) The face of South America: an aerial traverse. *Spec. Publ. Am. Geogr. Soc.*, **26,** pp. 229.
3.4.

RICHARDS A.F. & PARKS J.M. (1976) Marine geotechnology. In: *The Benthic Boundary Layer* (Ed. by I.N. McCave), pp. 157–182. Plenum Press, New York.
9.3.5.

RICHARDSON E.S. & HARRISON C.G.A. (1976) Opening of the Red Sea with two poles of rotation. *Earth Planet. Sci. Letts.*, **30,** 135–142.
14.3.1.

RICHTER R. (1931) Tierwelt und Umwelt im Hunsrückschiefer; zur Entstelung eines schwarzen Schlammsteins. *Senckenberg.*, **13,** 299–342.
9.8.

RICHTER-BERNBERG G. (1955) Stratigraphische Gliederung des deutschen Zechsteins. *Z. Dtsch. Geol. Ges.*, **105,** 843–854.
8.5.3, Fig. 8.37.

RICKARDS R.B. (1964) The graptolitic mudstone and associated facies in the Silurian strata of the Howgill Fells. *Geol. Mag.*, **101,** 435–451.
11.4.4.

RIGBY L.K. & HAMBLIN W.K. (Eds) (1972) *Recognition of Ancient Sedimentary Environments*, pp. 340. *Spec. Publ. Soc. econ. Paleont. Miner.*, **16,** Tulsa.
1.1.

RILEY J.P. & CHESTER R. (1971) *Introduction to Marine Chemistry*, pp. 465. Academic Press, New York.
Fig. 9.6.

ROBERTS D.G. (1972) Slumping on the eastern margin of the Rockall Bank, North Atlantic Ocean. *Mar. Geol.*, **13,** 225–237.
12.2.2, Tab. 12.1.

ROBERTS D.G. (1974) Structural development of the British Isles, continental margin, and the Rockall plateau. In: *The Geology of Continental Margins* (Ed. by C.A. Burke and C.L. Drake), pp. 343–359. Springer-Verlag, New York.
14.3.1, Fig. 14.17.

ROBERTS H.H., CRATSLEY D.W. & WHELAN T. (1976) Stability of Mississippi delta sediments as evaluated by analysis of structural features in sediment borings. *Offshore Tech. Conf. Paper*, No. **OTC 2425,** pp. 14.
Fig. 6.51.

ROBERTSON A.H.F. (1975) Cyprus umbers: basalt-sediment relationships on a Mesozoic ocean ridge. *J. geol. Soc. Lond.*, **131,** 511–531.
11.4.2.

ROBERTSON A.H.F. (1976) Origin of ochres and umbers: evidence from Skouriotissa, Troodos Massif, Cyprus. *Trans. Inst. Min. Metall.*, B, **85,** 245–251.
11.4.2.

ROBERTSON A.H.F. & FLEET A.J. (1976) The origins of rare earths in metalliferous sediments of the Troodos Massif, Cyprus. *Earth Planet. Sci. Letts.*, **28,** 385–394.
11.4.2.

ROBERTSON A.H.F. & HUDSON J.D. (1973) Cyprus umbers: chemical precipitates on a Tethyan ocean ridge. *Earth Planet. Sci. Letts.*, **18,** 93–101.
11.4.2.

ROBERTSON A.H.F. & HUDSON J.D. (1974). Pelagic sediments in the Cretaceous and Tertiary history of the Troodos Massif, Cyprus. In: *Pelagic Sediments: on Land and under the Sea* (Ed. by K.J. Hsü and H.C. Jenkyns), pp. 403–436. *Spec. Publ. int. Ass. Sediment.*, **1.**
11.4.2, Fig. 11.22.

ROBINSON A.H.W. (1966) Residual currents in relation to sandy shoreline evolution of the East Anglian coast. *Mar. Geol.*, **4,** 57–84.
9.5.3, 9.10.5.

ROBINSON P. (1973) Palaeoclimatology and Continental Drift. In: *Implications of Continental Drift to the Earth Sciences*, Vol. 1 (Ed. by D.H. Tarling and S.K. Runcorn), pp. 451–476. Academic Press, London.
5.1.

ROCHE M.A. (1970) Evaluation des pertes du Lac Tchad par abandon superficiel et infiltrations marginales. *Cah. ORSTOM Sér. Géol.*, **11,** 67–80.
4.2.1.

RODGERS J. (1968) The eastern edge of the North American continent

during the Cambrian and Early Ordovician. In: *Studies of Appalachian Geology, Northern and Maritime* (Ed. by E. Zen), pp. 141–149. Wiley-Interscience, New York.
11.4.4.

RODOLFO K.S. (1969) Sediments of the Andaman Basin, northeastern Indian Ocean. *Mar. Geol.*, **7,** 371–402.
14.3.2.

ROESCHMANN G. (1971) Problems concerning investigations of paleosols in older sedimentary rocks, demonstrated by the example of Wurzelböden of the Carboniferous system. In: *Paleopedology, Origin, Nature and Dating of Paleosols* (Ed. by D.H. Yaalon), pp. 311–320. Internat. Soc. of Soil Sci. and Israel Univ. Press, Jerusalem.
3.9.2.

ROGNON P., BIJU-DUVAL B. & DE CHARPAL O. (1972) Modèles glaciaires dans l'Ordovicien supérieur saharien: phases d'érosion et glaciotectonique sur la bordure nord des Eglab. *Revue Géogr. phys. geol. dyn.*, **14,** 507–527.
13.4.1.

RONA P.A. (1969) Middle Atlantic continental slope of United States: deposition and erosion. *Bull. Am. Ass. Petrol. Geol.*, **53,** 1453–1465.
14.3.1.

RONA P.A. (1973) Worldwide unconformities in marine sediments related to eustatic changes of sea level. *Nature, Phys. Sci.*, **244,** 25–26.
11.3.1, 11.3.4.

RONA P.A., HARBISON R.N., BASSINGER B.G., SCOTT R.B. & NALWALK A.J. (1976) Tectonic fabric and hydrothermal activity of Mid-Atlantic ridge crest (lat. 26°N). *Bull. geol. Soc. Am.*, **87,** 661–674.
14.4.1.

ROSS D.A. (1971) Sediments of the northern Middle America Trench. *Bull. geol. Soc. Am.*, **82,** 303–322.
12.4.3, Fig. 12.29.

ROSS D.A. (1974) The Black Sea. In: *The Geology of Continental Margins* (Ed. by C.A. Burk and C.L. Drake), pp. 669–682. Springer-Verlag, New York.
12.4.6.

RUHE R.V. (1965) Quaternary Palaeopedology. In: *The Quaternary of the United States* (Ed. by H.E. Wright and D.G. Frey), pp. 755–764. Princeton University Press, Princeton.
13.3.6.

RUPKE N.A. (1975) Deposition of fine-grained sediments in the abyssal environment of the Algéro-Balearic Basin, Western Mediterranean Sea. *Sedimentology*, **22,** 95–109.
12.2.5, 12.4.1, Figs. 12.20, 12.21.

RUPKE N.A. (1976a) Large-scale slumping in a flysch basin, southwestern Pyrenees. *J. geol. Soc. London*, **132,** 121–130.
12.2.2.

RUPKE N.A. (1976b) Sedimentology of very thick calcarenite-marlstone beds in a flysch succession, southwestern Pyrenees. *Sedimentology*, **23,** 43–65.
12.2.4, 12.2.5, 12.7.2, Figs. 12.8, 12.47.

RUPKE N.A. (1977) Growth of an ancient deep-sea fan. *J. Geol.*, **85,** 725–744.
12.2.6, 12.8.1, 12.8.2, 12.8.3, Fig. 12.50, 12.57, 12.58, 12.59.

RUPKE N.A. & STANLEY D.J. (1974) Distinctive properties of turbiditic and hemipelagic mud layers in the Algéro-Balearic Basin, Western Mediterranean Sea. *Smithsonian Contributions to the Earth Sciences*, **13,** 40 pp.
12.2.4, 12.2.5, 12.4.1, 12.4.4.

RUSNAK G.A. (1960) Sediments of Laguna Madre, Texas. In: *Recent Sediments, Northwest Gulf of Mexico* (Ed. by F.P. Shepard, F.B. Phleger and T.H. van Andel), pp. 153–196. Am. Ass. Petrol. Geol., Tulsa.
7.2.5.

RUSNAK G.A. & NESTEROFF W.D. (1964) Modern turbidites: terrigenous abyssal plain versus bioclastic basin. In: *Papers in Marine Geology, Shepard Commemorative Volume* (Ed. by R.L. Miller), pp. 448–507. Macmillan, New York.
12.2.4, 12.4.1, 12.4.6.

RUSSELL J.S. & MOORE A.W. (1972) Some parameters of gilgai microrelief. *Soil Science*, **114,** 82–87.
3.6.2.

RUSSELL R.J. (1936) Physiography of the lower Mississippi River delta. In: *Reports on the Geology of Plaquemines and St. Bernard Parishes*. Louisiana Dept. Conservation. *Geol. Bull.*, **8,** 3–199.
6.2.

RUSSELL R.J. & HOWE H.V. (1935) Cheniers of Southwestern Louisiana. *Geogr. Rev.*, **25,** 449–461.
7.4.

RUSSELL R.J. & RUSSELL R.D. (1939) Mississippi River delta sedimentation. In: *Recent Marine Sediments* (Ed. by P.D. Trask), pp. 153–177. Am. Ass. Petrol. Geol., Tulsa.
6.2.

RUST B.R. (1972a) Structure and process in a braided river. *Sedimentology*, **18,** 221–245.
3.3.1, 3.3.2, 3.8.1, Figs. 3.8, 3.12.

RUST B.R. (1972b) Pebble orientation in fluviatile sediments. *J. sedim. Petrol.*, **42,** 384–388.
3.3.2.

RUST B.R. & ROMANELLI R. (1975) Late Quaternary subaqueous outwash deposits near Ottawa, Canada. In: *Glaciofluval and Glaciolacustrine Sedimentation* (Ed. by A.V. Jopling and B.C. McDonald), pp. 177–192. *Spec. Publ. Soc. econ. Paleont. Miner.*, **23,** Tulsa.
13.3.4, 13.4.2.

RUST I.C. (1977) Evidence of shallow marine and tidal sedimentation in the Ordovician Graafwater Formation, Cape Province, South Africa. *Sedim. Geol.*, **18,** 123–133.
7.7.

RUTTEN L.M.R. (1927) *Voordrachten over de Geologie van Nederlandsch Oost-Indie*, pp. 839. Wolters, Groningen.
14.2.2.

RYAN W.B.F., HSÜ K.J. *et al.* (1973) *Initial Reports of the Deep Sea Drilling Project*, **13,** 1447 pp. U.S. Government Printing Office, Washington.
8.5.5.

SAGRI M. (1972) Rhythmic sedimentation in the turbidite sequences of the Northern Apennines (Italy). *24th Int. geol. Congr. Proc., Sect.* **6,** 82–88.
12.8.1.

SAILA S.B. (1976) Sedimentation and food resources: animal–sediment relationships. In: *Marine Sediment Transport and Environmental Management* (Ed. by D.J. Stanley and D.J.P. Swift), pp. 479–492. John Wiley, New York.
9.3.5.

SAITO R. (1969) Glacial problems of late Pre-Cambrian eon. *Kumamoto J. Sci. Ser. B, Sect. 1, Geology*, **8,** 7–44.
13.5.4.

SANDBERG P.A. (1975) New interpretations of Great Salt Lake ooids and of ancient non-skeletal carbonate mineralogy. *Sedimentology*, **22**, 497–538.
10.2.1.

SANDERS J.E. (1965) Primary sedimentary structures formed by turbidity currents and related sedimentation mechanisms. In: *Primary Sedimentary Structures and their Hydrodynamic Interpretation* (Ed. by G.V. Middleton), pp. 192–219. *Spec. Publ. Soc. econ. Paleontal. Miner.*, **12**, Tulsa.
12.2.1.

SANDERS J.E. (1968) Stratigraphy and primary sedimentary structures of fine-grained, well-bedded strata, inferred lake deposits, Upper Triassic, Central and Southern Connecticut. In: *Late Paleozoic and Mesozoic Continental Sedimentation, Northeastern North America* (Ed. by G. de V. Klein), pp. 265–305. *Spec. Pap. geol. Soc. Am.*, **106.**
4.3.3.

SANDERS J.E. & KUMAR N. (1975a) Evidence of shoreface retreat and in-place 'drowning' during Holocene submergence of barriers, shelf off Fire Island, New York. *Bull. geol. Soc. Am.*, **86**, 65–76.
7.2.2, 7.2.6, Fig. 7.34.

SANDERS J.E. & KUMAR N. (1975b) Holocene shoestring sand on inner continental shelf off Long Island, New York. *Bull. Am. Ass. Petrol. Geol.*, **59**, 997–1009.
7.2.6.

SARTORI R. (1974) Modern deep-sea magnesian calcite in the central Tyrrhenian Sea. *J. sedim. Petrol.*, **44**, 1313–1322.
11.3.5.

SATO T. (1977) Kuroko deposits: their geology, geochemistry and origin. In: *Volcanic Processes in Ore Genesis* (Ed. by M.J. Jones), pp. 153–161. *Spec. Publ. geol. Soc. Lond.*, **7.** London.
14.4.2.

SAUNDERS J.B. (1965) Field trip guide, Barbados. In: *Trans. 4th Caribbean Geological Conference, Trinidad* (Ed. by J.B. Saunders), pp. 433–449.
11.4.2.

SAURAMO M. (1923) Studies on the Quaternary varve sediments in southern Finland. *Bull. Comm. Geol. de Finlande*, **60**, pp. 164.
13.3.4.

SAYLES F.L., KU T.-L. & BOWLES P.C. (1975) Chemistry of ferromanganoan sediments of the Bauer Deep. *Bull. geol. Soc. Am.*, **86**, 1422–1431.
11.3.4.

SAWKINS F.J. (1974) Massive sulphide deposits in relation to geotectonics. *Geol. Assoc. Canada/Mineral. Assoc. Canada* (abs), St. John's, Newfoundland, pp. 81.
14.4.1.

SCHÄFER W. (1962) *Actuo-Paläontologie nach Studien in der Nordsee*, pp. 666. Kramer, Frankfurt a/Main.
10.2.2.

SCHÄFER W. (1972) *Ecology and Palaeoecology of Marine Environments* (Trans. by I. Oertel; Ed. by G.Y. Craig), pp. 568. Oliver & Boyd, Edinburgh.
1.1, 9.1.1, 9.3.5.

SCHENK P.E. (1969) Carbonate-sulphate–redbed facies and cyclic sedimentation of the Windsorian stage (middle Carboniferous), Maritime Provinces. *Can. J. Earth Sci.*, **6**, 1037–1066.
8.4.1, 8.4.3, Figs. 8.31, 8.32.

SCHERMERHORN L.J.G. (1966) Terminology of mixed coarse-fine sediments. *J. sedim. Petrol.*, **36**, 831–835.
13.3, 13.4.3.

SCHERMERHORN L.J.G. (1974) Late Precambrian mixtites: glacial and/or nonglacial? *Am. J. Sci.*, **274**, 673–824.
13.4.1, 13.4.3, 13.5.4.

SCHERMERHORN L.J.G. & STANTON W.I. (1963) Tilloids in the West Congo Geosyncline. *Quart. J. geol. Soc. Lond.*, **119**, 291–241.
13.5.4.

SCHLAGER W. (1969) Das Zusammenwirken von Sedimentation und Bruchtektonik in den triadischen Hallstätterkalken der Ostalpen. *Geol. Rdsch.*, **59**, 289–308.
11.4.4.

SCHLAGER W. (1974) Preservation of cephalopod skeletons and carbonate dissolution on ancient Tethyan sea floors. In: *Pelagic Sediments: on Land and under the Sea* (Ed. by K.J. Hsü and H.C. Jenkyns), pp. 49–70. *Spec. Publ. int. Ass. Sediment.*, **1.**
11.4.4, 11.4.5, Tab. 11.4.

SCHLAGER W. & BOLZ H. (1977) Clastic accumulation of sulphate evaporites in deep water. *J. sedim. Petrol.*, **47**, 600–609.
8.5.3.

SCHLAGER W. & SCHLAGER M. (1973) Clastic sediments associated with radiolarites (Tauglboden–Schichten, Upper Jurassic, Eastern Alps). *Sedimentology*, **20**, 65–89.
11.4.4.

SCHLANGER S.O. & DOUGLAS R.G. (1974) The pelagic ooze-chalk-limestone transition and its implications for marine stratigraphy. In: *Pelagic Sediments: on Land and under the Sea* (Ed. by K.J. Hsü and H.C. Jenkyns), pp. 117–148. *Spec. Publ. int. Ass. Sediment.*, **1.**
11.3.3.

SCHLANGER S.O., JACKSON E.D. *et al.* (1976) *Initial Reports of the Deep Sea Drilling Project*, **33**, pp. 973. U.S. Government Printing Office, Washington.
11.3.3.

SCHLANGER S.O. & JENKYNS H.C. (1976) Cretaceous oceanic anoxic events: causes and consequences. *Geol. Mijnb.*, **55**, 179–184.
11.3.1, 11.4.5, Fig. 11.7.

SCHLEE J.S. & PRATT R.M. (1970) Atlantic continental shelf and slope of the United States – gravels of the northeastern part. *Prof. Pap. U.S. geol. Surv.*, **529-H**, pp. 39.
13.5.1.

SCHMALZ R.F. (1969) Deep-water evaporite deposition: a genetic model. *Bull. Am. Ass. Petrol. Geol.*, **53**, 798–823.
8.5.2, 8.5.4, Fig. 8.41.

SCHNEIDERMANN N. (1970) Genesis of some Cretaceous carbonates in Israel. *Israel. J. Earth Sci.*, **19**, 97–115.
11.4.5.

SCHOLL D.W. & CREAGER J.S. (1973) Geologic synthesis of Leg 19 (DSDP) results: far north Pacific and Aleutian Ridge, and Bering Sea. In: *Initial Reports of the Deep sea Drilling Project*, **19**, (J.S. Creager, D.W. Scholl *et al.*), pp. 897–913. U.S. Government Printing Office, Washington.
11.3.5.

SCHOLL D.W. & MARLOW M.S. (1974) Sedimentary sequence in modern Pacific trenches and the deformed circum-Pacific eugeosyncline. In: *Modern and Ancient Geosynclinal Sedimentation* (Ed. by R.H. Dott Jr. and R.H. Shaver), pp. 193–211. *Spec. Publ. Soc. econ. Paleont. Miner.*, **19**, Tulsa.
14.3.2.

SCHOLL D.W., CHRISTENSEN M.N., VON HUENE R. & MARLOW M.S. (1970) Peru-Chile trench sediments and sea floor spreading. *Bull. geol. Soc. Am.*, **81**, 1339–1360.
12.4.3.

SCHOLLE P.A. (1971) Sedimentology of fine-grained deep water

carbonate turbidites, Monte Antola flysch (Upper Cretaceous), northern Apennines, Italy. *Bull. geol. Soc. Am.*, **82,** 629–658.
12.2.5, 12.7.1.

SCHOLLE P.A. (1974) Diagenesis of Upper Cretaceous chalks from England, Northern Ireland, and the North Sea. In: *Pelagic Sediments: On Land and Under the Sea* (Ed. by K.J. Hsü and H.C. Jenkyns), pp. 117–148. *Spec. Publ. int. Ass. Sediment.*, **1.**
11.4.5, Fig. 11.42.

SCHOLLE P.A. & KINSMAN D.J.J. (1974) Aragonitic and high-Mg calcite caliche from the Persian Gulf – a modern analogy for the Permian of Texas and New Mexico. *J. sedim. Petrol.*, **44,** 904–916.
8.2.7.

SCHOLLE P.A. & KLING S.A. (1972) Southern British Honduras: lagoonal coccolith ooze. *J. sedim. Petrol.*, **42,** 195–204.
10.1.2, 10.2.4.

SCHÖTTLE M. & MÜLLER G. (1968) Recent carbonate sedimentation in the Gnadensee (Lake Constance) Germany. In: *Recent Developments in Carbonate Sedimentology in Central Europe* (Ed. by G. Müller and G.M. Friedman), pp. 148–156. Springer-Verlag Berlin.
4.2.4.

SCHREIBER B.C., FRIEDMAN G.M., DECIMA A. & SCHREIBER E. (1976) Depositional environments of Upper Miocene (Messinian) evaporite deposits of the Sicilian Basin. *Sedimentology*, **23,** 729–760.
8.5.5, Fig. 8.44.

SCHUCHERT C. (1923) Sites and natures of the North-American geosynclines. *Bull. geol. Soc. Am.*, **34,** 151–260.
14.2.2.

SCHUMM S.A. (1963) Sinuosity of alluvial rivers on the Great Plains. *Bull. geol. Soc. Am.*, **74,** 1089–1100.
3.10.1.

SCHUMM S.A. (1968) River adjustment to altered hydrologic regimen–Murrumbidgee River and paleochannels, Australia. *Prof. Pap. U.S. geol. Surv.*, **598,** pp. 65.
3.10.1.

SCHUMM S.A. (1969) River Metamorphosis. *J. Hydraul. Proc. Am. Soc. civ. Engrs.*, **95,** 255–273.
3.10.1, 3.10.2.

SCHUMM S.A. (1971a) Fluvial geomorphology: the historical perspective. In: *River Mechanics* (H.W. Shen, Editor and Publisher), chapter 4, pp. 30. Fort Collins, Colorado.
3.3.

SCHUMM S.A. (1971b) Fluvial geomorphology: channel adjustment and river metamorphosis. In: *River Mechanics* (H.W. Shen, Editor and Publisher), chapter 5, pp. 22. Fort Collins, Colorado.
Fig. 3.20.

SCHUMM S.A. & KAHN H.R. (1972) Experimental study of channel patterns. *Bull. geol. Soc. Am.*, **83,** 1755–1770.
3.5.1, 3.9.4, 3.10.1.

SCHUMM S.A. & LICHTY R.W. (1963) Channel widening and floodplain construction along Cimarron River in southwestern Kansas. *Prof. Pap. U.S. geol. Surv.*, **352-D,** 71–88.
3.10.2.

SCHWARTZ M.L. (1973) *Barrier Islands*, pp. 451. Dowden, Hutchinson and Ross, Stroudsburg.
7.2.7.

SCHWARTZ R.K. (1975) Nature and genesis of some washover deposits. *U.S. Army Corps. Engin. Coastal Eng. Res. Centre Tech. Mem.*, **61,** 98.
7.2.5, Fig. 7.26.

SCHWARZ H.-U., EINSELE G. & HERM D. (1975) Quartz-sandy, grazing-contoured stromatolites from coastal embayments of Mauritania, West Africa. *Sedimentology*, **22,** 539–561.
8.2.5, Tab. 8.1.

SCLATER J.G. & FISHER R.L. (1974) Evolution of the East Central Indian Ocean, with emphasis on the tectonic setting of the Ninetyeast Ridge. *Bull. geol. Soc. Am.*, **85,** 683–702.
11.3.3.

SCLATER J.G., ANDERSON R.N. & BELL M.L. (1971) Elevation of ridges and evolution of the central eastern Pacific. *J. geophys. Res.*, **76,** 7888–7915.
11.3.3, 11.4.2.

SCOFFIN T.P. (1971) The conditions of growth of the Wenlock reefs of Shropshire (England). *Sedimentology*, **17,** 173–219.
10.4.2.

SCOTT K.M. (1967) Intra-bed palaeocurrent variations in a Silurian flysch sequence, Kirkcudbrightshire, Southern Uplands of Scotland. *Scott. J. Geol.*, **3,** 268–281.
12.2.4.

SCOTT M.R., SCOTT R.B., RONA P.A., BUTLER L.W. & NALWALK A.J. (1974) Rapidly accumulating manganese deposit from the median valley of the Mid-Atlantic Ridge. *Geophys. Res. Letts.*, **1,** 355–358.
11.3.2.

SCOTT R.B., MALPAS J., RONA P.A. & UDINTSEV G. (1976) Duration of hydrothermal activity at an oceanic spreading center, Mid-Atlantic Ridge (lat. 26°N). *Geology*, **4,** 233–236.
11.3.2.

SCOTT R.W. & WEST R.R. (1976) *Structure and Classification of Palaeocommunities*, pp. 291. Dowden, Hutchinson & Ross, Stroudsburg.
9.1.1, 9.8.

SCRUTON P.C. (1956) Oceanography of Mississippi delta sedimentary environments. *Bull. Am. Ass. Petrol. Geol.*, **40,** 2864–2952.
6.5.2.

SCRUTON P.C. (1960) Delta building and the delta sequence. In: *Recent Sediments, Northwest Gulf of Mexico* (Ed. by F.P. Shepard and Tj.H. van Andel), pp. 82–102. Am. Ass. Petrol. Geol., Tulsa.
6.4, 6.6, Fig. 6.37.

SEELY D.R., VAIL P.R. & WALTON G.G. (1974) Trench slope model. In: *The Geology of Continental Margins* (Ed. by C.A. Burk and C.L. Drake), pp. 249–260. Springer-Verlag, New York.
14.3.2.

SEILACHER A. (1962) Paleontological studies on turbidite sedimentation and erosion. *J. Geol.*, **70,** 227–234.
12.2.4, 12.6.1, Fig. 12.17.

SEILACHER A. (1964) Biogenic sedimentary structures. In: *Approaches to Palaeoecology* (Ed. by J. Imbrie and N. Newell), pp. 296–316. John Wiley, New York.
9.8, 9.9.

SEILACHER A. (1967) Bathymetry of trace fossils. *Mar. Geol.*, **5,** 413–428.
9.8, 12.2.4, 12.6.1, Fig. 9.28.

SEILACHER A. (1973) Biostratinomy: the sedimentology of biologically standardized particles. In: *Evolving Concepts in Sedimentology* (Ed. by R.N. Ginsburg), pp. 159–177. Johns Hopkins University Press, Baltimore.
9.8, 9.9.

SELLEY R.C. (1968) A classification of palaeocurrent models. *J. Geol.*, **76,** 99–110.
9.8.

SELLEY R.C. (1969) Studies of sequence in sediments using a simple

mathematical device. *J. geol. Soc. Lond.*, **125,** 557–581.
2.1.2, Fig. 2.3.

SELLEY R.C. (1970) *Ancient Sedimentary Environments,* pp. 237. Chapman & Hall, London.
1.1.

SELLEY R.C. (1972) Diagnosis of marine and non-marine environments from the Cambro-Ordovician sandstones of Jordan. *J. geol. Soc. Lond.*, **128,** 135–160.
3.9.4.

SELLEY R.C. (1976a) *An Introduction to Sedimentology,* pp. 408 Academic Press, London.
2.2.1, 8.4.1.

SELLEY R.C. (1976b) Subsurface environmental analysis of North Sea sediments. *Bull. Am. Ass. Petrol. Geol.*, **60,** 184–195.
9.8.

SELLI R. (1970) Ricerche geologiche preliminari nel Mar Tirreno. *Giorn. Geol.*, Ser. 2, **37/1,** 1–249.
11.3.5.

SELLWOOD B.W. (1972a) Regional environmental changes across a Lower Jurassic stage-boundary in Britain. *Palaeontology,* **15,** 127–157.
9.12.

SELLWOOD B.W. (1972b) Tidal flat sedimentation in the Lower Jurassic of Bornholm, Denmark. *Palaeogeogr. Palaeoclimat. Palaeoecol.*, **11,** 93–106.
7.7.

SELLWOOD B.W. (1975) Lower Jurassic tidal flat deposits, Bornholm, Denmark. In: *Tidal Deposits: A Casebook of Recent Examples and Fossil Counterparts* (Ed. by R.N. Ginsburg), pp. 93–101. Springer-Verlag, Berlin.
7.7.

SELLWOOD B.W. (1978) Jurassic. In: *The Ecology of Fossils* (Ed. by W.S. McKerrow), pp. 204–279. Duckworth, London.
9.12.

SENIN Y.M. (1975) The climatic zonality of the recent sedimentation on the West African shelf. *Oceanology,* **14,** 102–110.
9.3.4.

SENN A. (1940) Paleogene of Barbados and its bearing on history and structure of Antillean-Caribbean region. *Bull. Am. Ass. Petrol. Geol.*, **24,** 1548–1610.
11.4.2.

SERVANT M. & SERVANT S. (1970) Les formations lacustres et les diatomées du quaternaire recent du fond de la cuvette tchadienne. *Rev. Géogr. phys. Géol. dynam.*, **13,** 63–76.
4.2.1.

SESTINI G. (1970) *Development of the Northern Apennines Geosyncline,* pp. 203–647. *Sedim. Geol.*, **4.**
12.6.2.

SESTINI G. (1973) Sedimentology of a paleoplacer: The gold-bearing Tarkwaian of Ghana. In: *Ores in Sediments* (Ed. by G.C. Amstutz and A.J. Bernard), pp. 275–305. Springer-Verlag, Berlin.
3.8.1.

SEYFRIED W. & BISCHOFF J.L. (1977) Hydrothermal transport of heavy metals by seawater: the role of seawater/basalt ratio. *Earth Planet. Sci. Letts.*, **34,** 71–77.
11.3.2.

SHANTZER E.V. (1951) Alluvium of river plains in a temperate zone and its significance for understanding the laws governing the structure and formation of alluvial suites. *Tr. Inst. Geol. Nauk. Akad. Nauk., S.S.S.R.* Geol. Ser. **135,** 1–271 (in Russian).
3.4.1.

SHARMA G.D. (1975) Contemporary epicontinental sedimentation and shelf grading in the southeast Bering Sea. *Spec. Pap. geol. Soc. Am.*, **151,** 33–48.
9.6.1.

SHARMA G.D., NAIDU A.S. & HOOD D.W. (1972) Bristol Bay: a model contemporary graded shelf. *Bull. Am. Ass. Petrol. Geol.*, **56,** 2000–2012.
9.6.1, Fig. 9.16.

SHARP R.P. (1942) Periglacial involutions in northeastern Illinois. *J. Geol.*, **50,** 113–133.
13.3.6.

SHARP R.P. (1949) Studies of superglacial debris on valley glaciers. *Am. J. Sci.*, **247,** 289–315.
13.3.2.

SHARP R.P. (1963) Wind Ripples. *J. Geol.*, **71,** 617–636.
5.2.4.

SHARP R.P. (1966) Kelso dunes, Mojave Desert, California. *Bull. geol. Soc. Am.*, **77,** 1045–1074.
5.2.4.

SHARP R.P. & NOBLES L.H. (1953) Mudflow of 1941 at Wrightwood, Southern California. *Bull. geol. Soc. Am.*, **64,** 547–560.
3.2.4, 3.8.1.

SHAVER R.H. (1974) Silurian reefs of northern Indiana: reef and inter-reef macrofaunas. *Bull. Am. Ass. Petrol. Geol.*, **58,** 934–956.
10.4.2.

SHAW A.B. (1964) *Time in Stratigraphy,* pp. 363. McGraw-Hill, New York.
9.1.

SHEARMAN D.J. (1966) Origin of marine evaporites by diagenesis. *Trans. Inst. Min. Metall. B.*, **75,** 208–215.
2.2.1, 8.2.8, 8.4.1, 8.4.2, Fig. 8.23.

SHEARMAN D.J. (1970) Recent halite rock, Baja California, Mexico. *Trans. Inst. Min. Metall. B.*, **79,** 155–162.
8.1.1, 8.3.2, 8.5.1, 8.5.3, Figs. 8.21, 8.22.

SHEARMAN D.J. & FULLER J.G.C.M. (1969) Anhydrite diagenesis, calcitization, and organic laminites, Winnipegosis formation, Middle Devonian, Saskatchewan. *Bull. Can. Petrol. Geol.*, **17,** 496–525.
8.5.2, Tab. 8.3.

SHELL OIL CO. (1975) *Stratigraphic Atlas of North and Central America.* Shell Oil Co. Exploration Dept., Houston.
Figs. 10.41B, 10.47A.

SHELTON J.W. & NOBLE R.L. (1974) Depositional features of braided-meandering stream. *Bull. Am. Ass. Petrol. Geol.*, **58,** 742–749.
3.5, 3.10.2.

SHEPARD F.P. (1932) Sediments on continental shelves. *Bull. geol. Soc. Am.*, **43,** 1017–1034.
9.1.1.

SHEPARD F.P. (1948) *Submarine Geology,* pp. 348. Harper and Bros., New York.
11.1.1, 11.2.

SHEPARD F.P. (1951) Transportation of sand into deep water. *Spec. Publ. Soc. econ. Paleont. Miner.*, **2,** 53–65.
12.3.2.

SHEPARD F.P. (1952) Composite origin of submarine canyons. *J. Geol.*, **60,** 84–96.
12.5.

SHEPARD F.P. (1955) Delta front valleys bordering Mississippi distributaries. *Bull. geol. Soc. Am.*, **66,** 1489–1498.
6.8.2.

SHEPARD F.P. (1960) Gulf Coast barriers. In: *Recent Sediments,*

Northwest Gulf of Mexico (Ed. by F.P. Shepard, F.B. Phleger and Tj.H. van Andel), pp. 197–220. Am. Ass. Petrol. Geol., Tulsa.
7.2, 7.2.4.

SHEPARD F.P. (1963) *Submarine Geology*, 2nd edn, pp. 557. Harper and Row, New York.
7.1, 11.2.

SHEPARD F.P. (1973a) Sea floor off Magdalena delta and Santa Marta area, Columbia. *Bull. geol. Soc. Am.*, **84**, 1955–1972.
6.8.2.

SHEPARD F.P. (1973b) *Submarine Geology*, 3rd edn., pp. 551. Harper and Row, New York.
11.2, 12.2.7.

SHEPARD F.P. & DILL R.F. (1966) *Submarine Canyons and Other Sea Valleys*, pp. 381. Rand McNally, Chicago.
12.2.6, 12.2.7, 12.5.5.

SHEPARD F.P. & INMAN D.L. (1950) Nearshore circulation related to bottom topography and wave refraction. *Trans. Am. geophys. Union*, **31**, 555–565.
7.2.1, Fig. 7.6.

SHEPARD F.P. & MOORE D.G. (1960) Bays of Central Texas Coast. In: *Recent Sediments, Northwest Gulf of Mexico* (Ed. by F.P. Shepard, F.B. Phleger and Tj.H. van Andel) pp. 117–152. Am. Ass. Petrol. Geol., Tulsa.
7.2.5.

SHEPARD F.P., DILL R.F. & HEEZEN B.C. (1968) Diapiric intrusions in foreset slope sediments off Magdalena delta, Colombia. *Bull. Am. Ass. Petrol. Geol.*, **52**, 2197–2207.
6.8.2.

SHEPARD F.P., DILL R.F. & VON RAD U. (1969) Physiography and sedimentary processes of La Jolla submarine fan and fan-valley, California. *Bull. Am. Ass. Petrol. Geol.*, **53**, 390–420.
12.3.2, 12.5.1.

SHEPARD F.P.,MCLOUGHLIN P.A., MARSHALL N.F. & SULLIVAN G.G. (1977) Current-meter recordings of low-speed turbidity currents. *Geology*, **5**, 297–301.
12.2.5.

SHEPPS V.C. (1953) Correlation of tills of northeastern Ohio by size analysis. *J. sedim. Petrol.*, **23**, 34–48.
13.4.1.

SHERIDAN R.E. (1974) Atlantic continental margin of North America. In: *The Geology of Continental Margins* (Ed. C.A. Burke and C.L. Drake), pp. 391–407. Springer-Verlag, New York.
10.2.4, 11.3.6, 14.3.1, Figs. 14.14, 14.15, 14.16.

SHINN E.A. (1963) Spur and groove on the Florida reef tract. *J. sedim. Petrol.*, **33**, 291–303.
10.2.4.

SHINN E.A. (1968a) Practical significance of birdseye structures in carbonate rocks. *J. sedim. Petrol.*, **38**, 215–223.
8.4.2.

SHINN E.A. (1968b) Selective dolomitization of recent sedimentary structures. *J. sedim. Petrol.*, **38**, 612–616.
10.2.2, 10.2.4.

SHINN E.A. (1969) Submarine lithification of Holocene carbonate sediments in the Persian Gulf. *Sedimentology*, **12**, 109–144.
8.2.4, 10.1.1, 11.4.5.

SHINN E.A., GINSBURG R.N. & LLOYD R.M. (1965) Recent supratidal dolomite from Andros Island, Bahamas. In: *Dolomitization and Limestone Diagenesis*: a symposium (Ed. by R.C. Pray and R.C. Murray), pp. 112–123. *Spec. Publ. Soc. econ. Paleont. Miner.*, **13**, Tulsa.
8.2.6.

SHINN E.A., LLOYD R.M. & GINSBURG R.N. (1969) Anatomy of a modern carbonate tidal flat, Andros Island, Bahamas. *J. sedim. Petrol.*, **39**, 1202–1228.
10.2.4, Figs. 10.8A, 10.8B, 10.9A, 10.9B, 10.9C.

SHORT K.C. & STAUBLE A.J. (1967) Outline of the geology of the Niger delta. *Bull. Am. Ass. Petrol. Geol.*, **51**, 761–779.
6.8.2.

SHOTTON F.W. (1937) The lower Bunter Sandstones of north Worcestershire and east Shropshire. *Geol. Mag.*, **74**, 534–553.
5.1, 5.3.2.

SHOTTON F.W. (1956) Some aspects of the New Red desert in England. *Liverpool Manchester geol. J.*, **1**, 450–465.
5.2.4, 5.3.2.

SHREVE R.L. (1972) Movement of water in glaciers. *J. Glaciol.*, **11**, 205–214.
13.3.1.

SIESSER W.G. (1972) Limestone lithofacies from the South African continental margin. *Sedim. Geol.*, **8**, 83–112.
11.3.6.

SIMONS D.B., RICHARDSON E.V. & NORDIN C.F. (1965) Sedimentary structures generated by flow in alluvial channels. In: *Primary Sedimentary Structures and their Hydrodynamic Interpretation* (Ed. by G.V. Middleton), pp. 34–52. *Spec. Publ. Soc. econ. Paleont. Miner.*, **12**, Tulsa.
9.9, 9.11.2.

SIMPSON F. (1975) Marine lithofacies and biofacies of the Colorado Group (Middle Albian to Sautonian) in Saskatchewan. *Spec. Pap. geol. Ass. Canada*, **13**, 553–587.
9.12, 9.13.

SIMSON E.S.W. & HEYDORN A.E.F. (1965) Vema Seamount. *Nature*, Lond., **207**, 249–251.
11.3.3.

SINGH I.B. (1972) On the bedding in the natural-levee and point-bar deposits of the Gomti River, Uttar Pradesh, India. *Sedim. Geol.*, **7**, 309–317.
3.6.1.

SITTER L.U. DE (1956) *Structural Geology*, pp. 552. McGraw-Hill, New York.
14.2.2.

SKELTON P.W. (1976) Functional morphology of the Hippuritidae. *Lethaia*, **9**, 83–100.
10.4.2.

SMALE D. (1973) Silcretes and associated silica diagenesis in southern Africa and Australia. *J. sedim. Petrol.*, **43**, 1077–1089.
3.6.2.

SMALLEY I.J. (1966) The properties of glacial loess and the formation of loess deposits. *J. sedim. Petrol.*, **36**, 669–676.
5.2.5.

SMALLEY I.J. (1976) *Loess Lithology and Genesis*, pp. 429. Dowden, Hutchinson & Ross, Stroudsburg.
13.3.5.

SMALLEY I.J. & VITA-FINZI C. (1968) The formation of fine particles in sandy deserts and the nature of 'desert' loess. *J. sedim. Petrol.*, **38**, 766–774.
5.2.5.

SMIRNOV V.I. (1968) The sources of ore-forming fluids. *Econ. Geol.*, **63**, 380–389.
14.2.4.

SMITH A.G., BRIDEN J.C. & DREWRY G.E. (1973) Phanerozoic world maps. In: *Organisms and Continents through Time* (Ed. by N.F. Hughes), pp. 1–39. *Spec. Paps. Palaeont.*, **12.**

10.3.2.
SMITH D.B. (1971) Possible displacive halite in the Permian Upper Evaporite Group of northeast Yorkshire. *Sedimentology*, **17,** 221–232.
5.3.4.
SMITH D.B. (1972) The Lower Permian in the British Isles. In: *Rotliegend: Essays on European Lower Permian* (Ed. by M. Falke), pp. 1–33. E.J. Brill, Leiden.
5.3.6.
SMITH D.B. (1973) The origin of the Permian Middle and Upper Potash deposits of Yorkshire, England: an alternative hypothesis. *Proc. Yorks. geol. Soc.*, **39,** 327–346.
8.5.3.
SMITH D.B. (1974). Sedimentation of Upper Artesia (Guadalupian) cyclic shelf deposits of northern Guadalupe Mountains, New Mexico. *Bull. Am. Ass. Petrol. Geol.*, **58,** 1699–1730.
11.4.5.
SMITH D.B., BRUNSTROM R.G.W., MANNING P.I., SIMPSON S. & SHOTTON F.W. (1974) A correlation of Permian rocks in the British Isles. *J. geol. Soc. Lond.*, **130,** 1–45.
8.5.3.
SMITH D.J. & HOPKINS T.S. (1972) Sediment transport on the continental shelf of Washington and Oregon in light of recent current measurements. In: *Shelf Sediment Transport: Process and Pattern* (Ed. by D.J.P. Swift, D.B. Duane and O.H. Pilkey), pp. 143–180. Dowden, Hutchinson & Ross, Stroudsburg.
9.4.3, 9.6.2, Fig. 9.17.
SMITH G.D. (1942) Illinois loess – Variations in its properties and distribution: a pedologic interpretation. *Univ. of Illinois. Agric. Expt. Sta. Bull.*, **490,** 139–184.
13.3.5.
SMITH J.D. (1969) Geomorphology of a sand ridge. *J. Geol.*, **77,** 39–55.
9.2.
SMITH N.D. (1970) The braided stream depositional environment: Comparison of the Platte River with some Silurian clastic rocks, North-Central Appalachians. *Bull. geol. Soc. Am.*, **81,** 2993–3014.
3.3.1, 3.4, 3.4.1.
SMITH N.D. (1971) Transverse bars and braiding in the Lower Platte River, Nebraska. *Bull. geol. Soc. Am.*, **82,** 3407–3420.
3.4, 3.4.1, 3.4.2.
SMITH N.D. (1972) Some sedimentological aspects of planar cross-stratification in a sandy braided river. *J. sedim. Petrol.*, **42,** 624–634.
3.4.3, 3.9.4.
SMITH N.D. (1974) Sedimentology and bar formation in the Upper Kicking Horse River, a braided outwash stream. *J. Geol.*, **82,** 205–224.
3.3.1, 3.8.1, Fig. 3.9.
SMITH P.B. (1968) Paleoenvironment of phosphate-bearing Monterey Shale in Salinas Valley, California. *Bull. Am. Ass. Petrol. Geol.*, **52,** 1785–1791.
11.4.3.
SMITH R.L. (1974) A description of currents, winds, and sea level variations during coastal upwelling off the Oregon coast, July–August 1972. *J. geophys. Res.*, **79,** 435–443.
9.6.2.
SONU C.J. & VAN BEEK J.L. (1971) Systematic beach changes in the outer banks, North Carolina. *J. Geol.*, **74,** 416–425.
7.2.1.
SORBY H.C. (1851) On the microscopical structure of the calcareous grit of the Yorkshire Coast. *Quart. J. geol. Soc. Lond.*, **7,** 1–6.
10.1.1.
SORBY H.C. (1859) On the structures produced by the currents present during the deposition of stratified rocks. *Geologist*, **2,** 137–147.
1.1.
SORBY H.C. (1879) Anniversary address of the President: structure and origin of limestones. *Proc. geol. Soc. Lond.*, **35,** 56–95.
1.1.
SOREM R.K. & GUNN D.W. (1967) Mineralogy of manganese deposits, Olympic Peninsula, Washington. *Econ. Geol.*, **62,** 22–56.
11.4.2.
SOUTAR A. & ISAACS J.D. (1974) Abundance of pelagic fish during the 19th and 20th centuries as recorded in anaerobic sediment off the Californias. *Fishery Bull.*, **72,** 257–273.
11.3.6.
SPEARING D.R. (1975) Shallow marine sands. In: *Depositional Environments as Interpreted from Primary Sedimentary Structures and Stratification Sequences*, pp. 103–132. *Soc. econ. Paleont. Miner. Short Course*, **2.**
Fig. 9.45.
SPEARING D.R. (1976) Upper Cretaceous Shannon Sandstone: an offshore, shallow-marine sand body. *Wyoming Geol. Ass. Guidebook*, 28th Field Conf., pp. 65–72.
9.8, 9.10.5, Fig. 9.45.
SPENCER A.M. (1971) Late Pre-Cambrian glaciation in Scotland. pp. 100. *Mem. geol. Soc. Lond.*, **6.**
13.4.1, 13.5.4.
SPENCER A.M. (1975) Late Pre-Cambrian glaciation in North Atlantic region. In: *Ice Ages: Ancient and Modern* (Ed. by A.E. Wright and F. Moseley), pp. 217–240. *Geol. J. Spec. Issue*, **6.**
13.5.4.
SPENCER-DAVIES P., STODDART D.R. & SIGEE D.C. (1971) Reef forms of Addu Atoll, Maldive Islands. In: *Regional Variation in Indian Ocean Coral Reefs* (Ed. by D.R. Stoddart and M. Yonge), pp. 217–259. Symp. Zool. Soc. Lond., **28.**
10.2.2, Figs. 10.12A, 10.12B.
SPOONER E.T.C. & FYFE W.S. (1973) Sub-sea-floor metamorphism, heat, and mass transfer. *Contr. Miner. Petrol.*, **42,** 287–304.
11.4.2, 14.3.1.
STAHL L., KOCZAN J. & SWIFT D.J.P. (1974) Anatomy of a shoreface-connected ridge system on the New Jersey shelf: implications for the genesis of the shelf surficial sand sheet. *Geology*, **2,** 117–120.
7.2.6.
STANLEY D.J. (Ed.) (1969) *The New Concepts of Continental Margin Sedimentation*. pp. 400. Am. geol. Inst., Washington D.C.
1.1.
STANLEY D.J. (1970) Flyschoid sedimentation on the outer Atlantic margin off northeast North America. In: *Flysch Sedimentology in North America* (Ed. by J. Lajoie), pp. 179–210. *Spec. Pap. geol. Ass. Can.*, **7.**
14.2.5.
STANLEY D.J. & SWIFT D.J.P. (Eds.) (1976) *Marine Sediment Transport and Environmental Management*. Wiley, New York.
1.1.
STANLEY D.J. & UNRUG R. (1972) Submarine channel deposits, fluxoturbidites and other indicators of slope and base-of-slope environments in modern and ancient marine basins. In: *Recognition of Ancient Sedimentary Environments* (Ed. by J.K. Rigby and W.K. Hamblin), pp. 287–340. *Spec. Publ. Soc. econ. Paleont. Miner.*, **16,** Tulsa.
12.8.1.
STANLEY D.J., FENNER P. & KELLING G. (1972) Currents and sediment

transport at Wilmington Canyon shelfbreak, as observed by underwater television. In: *Shelf Sediment Transport: Process and Pattern* (Ed. by D.J.P. Swift, D.B. Duane and D.H. Pilkey), pp. 621–644. Dowden, Hutchinson & Ross, Stroudsburg.
9.3.5.

Stanley D.J., Krinitzsky E.L. & Campton J.R. (1966) Mechanics of Mississippi River bank failure in deltaic plain sediments at Fort Jackson, Louisiana. *Bull. geol. Soc. Am.*, **77**, 859–866.
6.5.1.

Stanley G.M. (1955) Origin of playa stonetracks, Racetrack Playa, Inyo County, California. *Bull. geol. Soc. Am.*, **66**, 1329–1350.
5.2.6.

Stanton R.L. (1972) *Ore Petrology*, pp. 713. McGraw-Hill, New York.
14.2.4.

Stapor F.W. Jr. (1975) Holocene beach ridge plain development, northwest Florida. *Zeit. Geomorph.*, **22**, 116–144.
7.2.6.

Stauffer P.H. (1967) Grain flow deposits and their implications, Santa Ynez Mountains, California. *J. sedim. Petrol.*, **37**, 487–508.
12.2.6, Fig. 12.22.

Steel R.J. (1974) New Red Sandstone floodplain and piedmont sedimentation in the Hebridean Province. *J. sedim. Petrol.*, **44**, 336–357.
3.8, 3.8.1, 5.3.6, Fig. 3.39.

Steel R.J. (1976) Devonian basins of Western Norway – Sedimentary response to tectonism and to varying tectonic context. *Tectonophysics*, **36**, 207–224.
3.8, 3.8.1, 14.4.3.

Steel R.J., Nicholson R. & Kalander L. (1975) Triassic sedimentation and palaeogeography in Central Skye. *Scott. J. Geol.*, **11**, 1–13.
3.8.1.

Steel R.J. & Wilson A.C. (1975) Sedimentation and tectonism (?Permo–Triassic) on the margin of the North Minch Basin, Lewis. *J. geol. Soc. London.*, **131**, 183–202.
3.8.1.

Steinmann G. (1905) Geologische Beobachtungen in den Alpen. II. Die Schardtsche Überfaltungs-theorie und die geologische Bedeutung der Tiefseeabsätze und der ophiolithischen Massengesteine. *Ber. naturf. Ges. Freiburg*, **16**, 18–67.
11.1.2, 14.2.5.

Steinmann G. (1925) Gibt es fossile Tiefseeablagerungen von erdgeschichtliche Bedeutung? *Geol. Rdsch.*, **16**, 435–468.
11.1.2.

Steinmann G. (1927) Die Ophiolithischen Zonen in den Mediterranen Kettengebirgen. pp. 637–667. *C.R. Intern. geol. Congr.*, **14**, Madrid.
14.2.5.

Sternberg R.W. & Larsen L.H. (1976) Frequency of sediment movement on the Washington continental shelf: A note. *Mar. Geol.*, **21**, M37–M47.
9.6.2.

Stevens R.K. (1970) Cambro–Ordovician flysch, sedimentation and tectonics in west Newfoundland and their possible bearing on a proto-Atlantic Ocean. In: *Flysch Sedimentology in North America* (Ed. by J. Lajoie), pp. 165–179. *Spec. Pap. geol. Ass. Canada*, **7.**
11.4.2.

Stewart A.D. (1969) Torridonian rocks of Scotland reviewed. In: *North Atlantic – Geology and Continental Drift* (Ed. by Marshall Kay), pp. 595–698. Memoir, **12.** Am. Ass. Petrol. Geol.
3.10.3.

Stewart F.H. (1963) The Permian Lower Evaporites of Fordon in Yorkshire. *Proc. Yorks. geol. Soc.*, **34**, 1–44.
8.5.3

Stewart H.B. & Jordan G.F. (1964) Underwater sand ridges on George's Shoal. In: *Papers in Marine Geology* (Ed. by R.L. Miller), pp. 102–116. Macmillan, New York.
9.2.

Stewart J.H. & Poole F.G. (1974) Lower Paleozoic and uppermost Pre-Cambrian Cordilleran miogeocline, Great Basin, western United States. In: *Tectonics and Sedimentation* (Ed. by W.R. Dickinson), pp. 28–57. *Spec. Publ. Soc. econ. Paleont. Miner.*, **22**, Tulsa.
11.4.2.

Stille H. (1913) *Evolution und Revolutionen in der Erdgeschichte*. pp. 32. Borntraeger, Berlin.
14.2.2.

Stille H. (1936) *Wege und Ergebnisse der geologisch-tektonischen Forschung*. pp. 77–97. 25 Jahr, Kaiser Wilhelm Ges., 2.
14.2.2.

Stille H. (1940) *Einführung in den Bau Nordamerikas*. pp. 717. Borntraeger, Berlin.
14.2.2.

Stockman K.W., Ginsburg R.N. & Shinn E.A. (1967) The production of lime mud by algae in south Florida. *J. sedim. Petrol.*, **37**, 633–648.
10.1.2, 10.2.2.

Stoddart D.R. (1969) Ecology and morphology of Recent coral reefs. *Biol. Rev.*, **44**, 433–498.
10.4.1.

Stoertz G. & Ericksen G.E. (1974) Geology of salars in Northern Chile. *Prof. Pap. U.S. geol. Surv.*, **811**, pp. 65.
5.2.6.

Stoffers P. & Ross D.A. (1974) Sedimentary history of the Red Sea. In: *Initial Reports of the Deep Sea Drilling Project*, **23** (R.B. Whitmarsh, O.E. Weser, D.A. Ross *et al.*), pp. 849–865, U.S. Government Printing Office, Washington.
11.3.5.

Stokes W.L. (1961) Fluvial and eolian sandstone bodies in Colorado Plateau. In: *Geometry of Sandstone Bodies* (Ed. by J.A. Peterson and J.C. Osmond), pp. 151–178. Am. Assoc. Petrol. Geol. Tulsa.
3.9.4, 5.3.5.

Stokes W.L. (1968) Multiple parallel-truncation bedding planes – a feature of wind deposited sandstone. *J. sedim. Petrol.*, **38**, 510–515.
5.3.2.

Stone B.D. (1976) Analysis of slump slip lines and deformation fabric in slumped Pleistocene lake beds. *J. sedim. Petrol.*, **46**, 313–325.
12.2.2, Fig. 12.10.

Stowe D.A.V. (1975) *The Laurentian Fan: Late Quaternary Stratigraphy*. Unpubl. Report, Univ. of Dalhousie, pp. 82.
12.5.5.

Straaten L.M.J.U. Van (1951) Texture and genesis of Dutch Wadden Sea sediments. *Proc. 3rd Internat. Congress Sedimentology, Netherlands*, 225–255.
3.9.3.

Straaten L.M.J.U. Van (1954) Composition and structure of Recent marine sediments in the Netherlands. *Leidse. geol. Meded.*, **19**, 1–110.
7.6.

Straaten L.M.J.U. Van (1959) Littoral and submarine morphology of the Rhône delta. In: *Proc. 2nd Coastal Geog. Conf.*: Baton Rouge, Louisiana State Univ., Natl. Acad. Sci. Nat. Research Council (Ed. by R.J. Russell), pp. 223–264.

6.5.2.
STRAATEN L.M.J.U. VAN (1960) Some recent advances in the study of deltaic sedimentation. *Liverpool Manchester geol., J.*, **2,** 411–445.
6.5.2.
STRAATEN L.M.J.U. VAN (1961) Sedimentation in tidal flat areas. *J. Alberta Soc. Petrol. Geologists*, **9,** 203–226.
7.2.2, 7.6.
STRAATEN L.M.J.U. VAN (1964) Turbidites in the southeastern Adriatic Sea. In: *Turbidites* (Eds. A.H. Bouma and A. Brouwer), pp. 142–147. Elsevier, Amsterdam.
12.2.4.
STRAATEN L.M.J.U. VAN (1965) Coastal barrier deposits in south and north Holland – in particular in the area around Scheveningen and IJmuiden. *Meded. Geol. Stricht.* NS**17,** 41–75.
7.2, 7.2.6.
STRAATEN L.M.J.U. VAN (1967) Turbidites, ash layers and shell beds in the bathyal zone of the southeastern Adriatic Sea. *Rev. Géograph. Phys. Géol. Dyn.*, **9,** 219–240.
12.2.4.
STRAATEN L.M.J.U. VAN (1970) Holocene and late-Pleistocene sedimentation in the Adriatic Sea. *Geol. Rdsch.*, **60,** 106–131.
12.4.6.
STRAATEN P. VAN & TUCKER M.E. (1972) The Upper Devonian Saltern Cove goniatite bed is an intraformational slump. *Palaeontology*, **15,** 430–438.
11.4.4.
STRAKHOV N.M. (1970) *Principles of Lithogenesis*. Volume 3, pp. 577. Oliver and Boyd, Edinburgh.
8.3.2.
STRIDE A.H. (1963) Current swept floors near the southern half of Great Britain. *Quart. J. geol. Soc. Lond.*, **119,** 175–199.
9.1.1, 9.5.1.
STRIDE A.H. (1965) Periodic and occasional sand transport in the North Sea. *La Revue Pétrolière*, Int. Cong., *'Le Pétrole et la Mer,'* Sect. 1, **no. 3,** pp. 4.
Fig. 9.12.
STRIDE A.H. (1970) Shape and size trends for sand waves in a depositional zone of the North Sea. *Geol. Mag.*, **107,** 469–477.
9.5.1.
STRIDE A.H. (1974) Indications of long term tidal control of net sand loss or gain by European coasts. *Estuarine & Coastal Mar. Sci.*, **2,** 27–36.
9.5.2, 9.5.3.
STUART C. & CAUGHY C.A. (1976) Form and composition of the Mississippi Fan. *Trans. Gulf-Cst. Ass. geol. Socs.*, **26,** 333–343.
12.5.2.
STUBBLEFIELD W.L., LAVELLE J.W., SWIFT D.J.P. & MCKINNEY T.F. (1975) Sediment response to the present hydraulic regime on the central New Jersey shelf. *J. sedim. Petrol.*, **45,** 337–358.
9.5.3, 9.6.3.
STUBBLEFIELD W.L. & SWIFT D.J.P. (1976) Ridge development as revealed by sub-bottom profiles on the central New Jersey shelf. *Mar. Geol.*, **20,** 315–334.
Fig. 9.21.
STURANI C. (1971) Ammonites and stratigraphy of the 'Posidonia alpina' beds of the Venetian Alps (Middle Jurassic, mainly Bajocian). *Memorie Ist. geol. miner. Univ. Padova*, **28,** 1–190.
11.4.4.
STURANI C. & SAMPÒ M. (1973) Il Messiniano inferiore in facies diatomitica nel bacino terziaro piemontese. *Memorie Soc. geol. ital.*, **12,** 335–357.
11.4.3.
STURM M. (1975) Depositional and erosional sedimentary features in a turbidity current controlled basin (Lake Brienz). *9th Int. Congr. Sedimentol., Nice 1975, Thème, 5(2)*, 385–390.
12.4.6, 12.5.4.
SUESS E. (1875) Die Entstehung der Alpen, pp. 168. W. Braumüller, Vienna.
14.2.1.
SULLWOLD H.H. (1960) Tarzana Fan, deep submarine fan of Late Miocene age, Los Angeles county, California. *Bull. Am. Ass. Petrol. Geol.*, **44,** 433–457.
12.8.1, 12.8.2.
SUNDBORG Å. (1956) The River Klarälven: A study of fluvial processes. *Geogr. Annalr.*, **38,** 127–316.
3.1, 3.5.2, Fig. 3.23.
SURDAM R.C. & EUGSTER H.P. (1976) Mineral reactions in the sedimentary deposits of the Lake Magadi region, Kenya. *Bull. geol. Soc. Am.*, **87,** 1739–1752.
4.2.1, 4.2.4.
SURDAM R.C. & WOLFBAUER C.A. (1975) Green River Formation, Wyoming: A playa-lake complex. *Bull. geol. Soc. Am.*, **86,** 335–345.
4.3.2, Figs. 4.5, 4.8, 4.9.
SURLYK F. (1978) Submarine fan sedimentation along fault scarps on tilted fault blocks (Jurassic–Cretaceous boundary, East Greenland). *Bull. Grønlands geol. Unders.*, **128,** pp. 108.
12.8.3.
SURLYK F. & CHRISTIANSON W.K. (1974) Epifaunal zonation on an Upper Cretaceous rocky coast. *Geology*, **2,** 529–534.
10.3.1, Fig. 10.36A.
SUTTON R.H., BOWEN Z.P. & MCALESTER A.L. (1970) Marine shelf environments of the Upper Devonian Sonyea Group of New York. *Bull. geol. Soc. Am.*, **81,** 2975–2992.
9.9.
SWARBRICK E.E. (1967). Turbidite cherts from northeast Devon. *Sedim. Geol.*, **1,** 145–158.
11.4.4.
SWETT K., KLEIN G. DE VRIES & SMIT D.E. (1971) A Cambrian tidal sand body – the Eriboll Sandstone of Northwest Scotland: an ancient-recent analog. *J. Geol.*, **79,** 400–415.
9.1.1, 9.10.2.
SWETT K. & SMITT D.E. (1972) Palaeogeography and depositional environments of the Cambro–Ordovician shallow marine facies of the North Atlantic. *Bull. geol. Soc. Am.*, **83,** 3223–3248.
9.8, 9.10.2, 9.10.3.
SWIFT D.J.P. (1968) Coastal erosion and transgressive stratigraphy. *J. Geol.*, **76,** 444–456.
7.2.6.
SWIFT D.J.P. (1969a) Inner shelf sedimentation: processes and products. In: *The New Concepts of Continental Margin Sedimentation: Application to the Geological Record* (Ed. by D.J. Stanley), pp. DS-4-1–DS-4-46. American Geological Institute, Washington.
9.1.1, 9.2.
SWIFT D.J.P. (1969b) Outer shelf sedimentation: processes and products. In: *The New Concepts of Continental Margin Sedimentation: Application to the Geological Record* (Ed. by D.J. Stanley), pp. DS-5-1–DS-5-26. American Geological Institute, Washington.
9.1.1, 9.4.1.
SWIFT D.J.P. (1970) Quaternary shelves and the return to grade. *Mar.*

Geol., **8,** 5–30.
9.1.1, 9.2.

Swift D.J.P. (1972) Implications of sediment dispersal from bottom current measurements; some specific problems in understanding bottom sediment distribution and dispersal on the continental shelf: A discussion of two papers. In: *Shelf Sediment Transport: Process and Pattern* (Ed. by D.J.P. Swift, D.B. Duane and O.H. Pilkey), pp. 363–371. Dowden, Hutchinson & Ross, Stroudsburg.
9.6.3.

Swift D.J.P. (1974) Continental shelf sedimentation. In: *The Geology of Continental Margins* (Ed. by C.A. Burke and C.L. Drake), pp. 117–135. Springer-Verlag, Berlin.
9.2, 9.5.2, 10.2.3, Fig. 9.20.

Swift D.J.P. (1975a) Barrier island genesis: evidence from the Middle Atlantic Shelf of North America. *Sedim. Geol.*, **14,** 1–43.
7.2.1, 7.2.2, 7.2.6, 7.2.7, Fig. 7.31.

Swift D.J.P. (1975b) Tidal sand ridges and shoal-retreat massifs. *Mar. Geol.*, **18,** 105–134.
9.5.3, 9.7.

Swift D.J.P. (1976a) Coastal sedimentation. In: *Marine Sediment Transport and Environmental Management* (Ed. by D.J. Stanley and D.J.P. Swift), pp. 255–310. John Wiley, New York.
7.2.1, 7.2.2, Figs. 7.19, 9.25.

Swift D.J.P. (1976b) Continental shelf sedimentation. In: *Marine Sediment Transport and Environmental Management* (Ed. by D.J. Stanley and D.J.P. Swift), pp. 311–350. John Wiley, New York.
9.2.

Swift D.J.P., Duane D.B. & McKinney T.F. (1973) Ridge and swale topography of the Middle Atlantic Bight, North America: secular response to the Holocene hydraulic regime. *Mar. Geol.*, **15,** 227–247.
9.1.1, 9.5.3, 9.6.3, Figs. 9.20, 9.23.

Swift D.J.P., Heron S.D. Jr. & Dill C.E. (1969) The Carolina Cretaceous: petrographic reconnaissance of a graded shelf. *J. sedim. Petrol.*, **39,** 18–33.
9.8.

Swift D.J.P., Holliday B., Avignone N. & Shideler G. (1972a) Anatomy of a shoreface ridge system, False Cape, Virginia. *Mar. Geol.*, **12,** 59–84.
9.1.1, 9.6.1, 9.6.3.

Swift D.J.P., Kofoed J.W., Saulsbury F.P. & Sears P. (1972b) Holocene evolution of the shelf surface, central and southern Atlantic shelf of North America. In: *Shelf Sediment Transport: Process and Pattern* (Ed. by D.J.P. Swift, D.B. Duane and O.H. Pilkey), pp. 499–574. Dowden, Hutchison & Ross, Stroudsburg.
9.5.3, 9.6.3, Fig. 9.24.

Swift D.J.P., Stanley D.J. & Curray J.R. (1971) Relict sediments on continental shelves: a reconsideration. *J. Geol.*, **79,** 322–346.
9.1.1, 9.2, 9.6.1, Fig. 9.7.

Swinchatt J.P. (1965) Significance and constituent composition, texture, and skeletal breakdown in some Recent carbonate sediments. *J. sedim. Petrol.*, **35,** 71–90.
Fig. 10.21A.

Szulczewski M. (1968) Slump structures and turbidites in the Upper Devonian limestones of the Holy Cross Mts. *Acta Geol. Pol.*, **18,** 303–324.
11.4.4.

Szulczewski M. (1971) Upper Devonian conodonts, stratigraphy and facial development in the Holy Cross Mts. *Acta Geol. Pol.*, **21,** 1–129.
11.4.4.

Szulczewski M. (1973) Famennian–Tournaisian neptunian dykes and their conodont fauna from Dalnia Hill. *Acta Geol. Pol.*, **23,** 15–59.
11.4.4.

Sykes R.M. (1974) Sedimentological Studies in Southern Jameson Land, East Greenland. 1. Fluviatile sequences in the Kap Stewart Formation (Rhaetic–Hettangian). *Bull. geol. Soc. Denmark*, **23,** 203–212.
3.9.4.

Taliaferro N.L. (1933) The relation of volcanism to diatomaceous and associated siliceous sediments. *Univ. Calif. Publs. Bull. Dep. Geol.*, **23,** 1–56.
11.4.3.

Tankard A.J. & Hobday D.K. (1977) Tide-dominated back-barrier sedimentation, early Ordovician Cape Basin, Cape Peninsula, South Africa. *Sedim. Geol.*, **18,** 135–159.
7.7.

Tanner W.F. (1965) Upper Jurassic paleogeography of the Four Corners Region. *J. sedim. Petrol.*, **35,** 564–574.
5.3.2, Fig. 5.12.

Tanner W.F. (1976) Tectonically significant pebble types: sheared, pocked and second-cycle examples. *Sedim. Geol.*, **16,** 69–83.
3.2.6.

Taylor J.C.M. & Colter V.S. (1975) Zechstein of the English Sector of the southern North Sea Basin. In: *Petroleum and the Continental Shelf of North-west Europe*. Vol. 1. Geology (Ed. by A.W. Woodland), pp. 325. Applied Science Publishers, London.
8.5.3, Figs. 8.39, 8.40.

Taylor J.C.M. & Illing L.V. (1969) Holocene intertidal calcium carbonate cementation. Quatar, Persian Gulf. *Sedimentology*, **12,** 69–107.
10.1.1.

Teichert C. (1958a) Cold and deep-water coral banks. *Bull. Am. Ass. Petrol. Geol.*, **43,** 1064–1082.
10.4.1.

Teichert C. (1958b) Concept of Facies. *Bull. Am. Ass. Petrol. Geol.*, **42,** 2718–2744.
2.1.1.

Teisseyre A.K. (1975) Pebble fabric in braided stream deposits with examples from Recent and 'frozen' Carboniferous channels (Intrasudetic Basin, Central Sudetes). *Geologia Sudetica*, **10,** 7–56.
3.8.1.

Termier P. (1902)Quatar, coupes à travers les Alpes franco–italiennes. *Bull. Soc. géol. France*, **2,** 411–432.
14.2.2.

Teruggi M.E. & Andreis R.R. (1971) Micromorphological recognition of paleosolic features in sediments and sedimentary rocks. In: *Paleopedology: Origin, Nature and Dating of Paleosols* (Ed. by D.H. Yaalon), pp. 161–172. Internat. Soc. of Soil. Sci. and Israel Univ. Press, Jerusalem.
3.9.2.

Terwindt J.H.J. (1971a) Sand waves in the Southern Bight of the North Sea. *Mar. Geol.*, **10,** 51–67.
9.1.1, 9.5.1.

Terwindt J.H.J. (1971b) Lithofacies of inshore estuarine and tidal inlet deposits. *Geol. Mijnb.*, **50,** 515–526.
6.5.1, 7.5.1.

Terzaghi K. (1956) Varieties of submarine slope failures. *Proc. 8th Texas Conf. Soil Mech. Found. Eng.*, **52,** 41 pp.
12.2.4.

TEXTORIS D.A. & CAROZZI A.V. (1964) Petrography and evolution of Niagaran (Silurian) reefs, Indiana. *Bull. Am. Ass. Petrol. Geol.*, **48,** 397–426.
10.4.2.

THIEL G.A. (1933) A correlation of marl beds with types of glacial deposits. *J. Geol.*, **38,** 717–728.
4.2.4.

THOMSEN E. (1976) Depositional environment and development of Danian bryozoan biomicrite mounds (Karlby Klint, Denmark). *Sedimentology*, **23,** 485–509.
11.4.5.

THOMPSON A.F. & THOMASSEN M.R. (1969) Shallow to deep water facies development in the Dimple Limestone (Lower Pennsylvanian), Marathon region, Texas. In: *Depositional Environments in Carbonate Rocks* (Ed. by G.M. Friedman), pp. 58–78. *Spec. Publ. Soc. econ. Paleont. Miner.*, **14,** Tulsa.
12.6.3.

THOMPSON D.B. (1969) Dome-shaped aeolian dunes in the Frodsham Member of the so-called 'Keuper' Sandstone Formation (Scythian–?Anisian: Triassic) at Frodsham, Cheshire (England). *Sedim. Geol.*, **3,** 263–289.
5.3.2, Fig. 5.13.

THOMPSON D.B. (1970) Sedimentation of the Triassic (Scythian) Red Pebbly Sandstone in the Cheshire Basin and its margins. *Geol. J.*, **7,** 183–216.
3.9.2, 3.9.4, 5.3.6.

THOMPSON R.W. (1968) Tidal flat sedimentation on the Colorado River delta. *Mem. geol. Soc. Am.*, **107,** 1–133.
7.6, 8.1.1, 8.3.1.

THOMPSON R.W. (1975) Tidal flat sediments of the Colorado River delta, northwestern Gulf of California. In: *Tidal Deposits: A Casebook of Recent Examples and Fossil Counterparts* (Ed. by R.N. Ginsburg), pp. 57–65. Springer-Verlag, Berlin.
7.6.

THOMPSON W.O. (1937) Original structures of beaches, bars and dunes. *Bull. geol. Soc. Am.*, **48,** 723–752.
7.2.2.

THORSON G. (1957) Bottom communities (sublittoral or shallow shelf). *Mem. geol. Soc. Am.*, **67,** 461–534.
9.3.5, 7.8.

THURSTON D.R. (1972) Studies on bedded cherts. *Contr. Miner. Petrol.*, **36,** 329–334.
11.4.2.

TODD T.W. (1968) Dynamic diversion: influence of longshore current–tidal flow interaction on chenier and barrier island plains. *J. sedim. Petrol.*, **38,** 734–746.
7.4.

TOIT A.L. DU (1921) The Carboniferous glaciation of South Africa. *Trans. Proc. geol. Soc. S. Afr.*, **24,** 188–227.
13.5.2.

TOLSTOY I. & EWING M. (1949) North Atlantic hydrography and the Mid-Atlantic Ridge. *Bull. geol. Soc. Am.*, **60,** 1527–1540.
12.3.1.

TOOMS J.S., SUMMERHAYES C.P. & CRONAN D.S. (1969) Geochemistry of marine phosphate and manganese deposits. In: *Ann. Rev. Oceanogr. Mar. Biol.*, **7** (Ed. by H. Barnes), pp. 49–100. Allen and Unwin, London.
11.3.6.

TOWNSON W.G. (1971) *Facies Analysis of the Portland Beds*. Unpubl. D.Phil. thesis, Univ. Oxford.
Fig. 10.30B.

TOWNSON W.G. (1975) Lithostratigraphy and deposition of the type Portlandian. *J. geol. Soc. Lond.*, **131,** 619–638.
8.4.2, 10.2.5, Figs. 10.31A, 10.31B.

TROWBRIDGE A.C. (1930) Building of Mississippi delta. *Bull. Am. Ass. Petrol Geol.*, **14,** 867–901.
6.2.

TRUEMAN A.E. (1946) Stratigraphical problems in the Coal Measures of Europe and North America. *Quart. J. geol. Soc. Lond.*, **102,** xlix–xciii.
2.2.2.

TRÜMPY R. (1960) Paleotectonic evolution of the Central and Western Alps. *Bull. geol. Soc. Am.*, **71,** 843–908.
11.1.2, 14.2.2, 14.2.5.

TUCHOLKE B. & VOGT P. *et al.* (1975) Glomar Challenger drills in the North Atlantic. *Geotimes*, **20,** 12, 18–21.
11.3.3.

TUCKER M.E. (1969) Crinoidal turbidites from the Devonian of Cornwall and their palaeogeographical significance. *Sedimentology*, **13,** 281–290.
11.4.4.

TUCKER M.E. (1973a). Sedimentology and diagenesis of Devonian pelagic limestones (Cephalopodenkalk) and associated sediments of the Rhenohercynian Geosyncline, West Germany. *Neues Jb. Geol. Paläont., Abh.*, **142,** 320–350.
11.4.4.

TUCKER M.E. (1973b) Ferromanganese nodules from the Devonian of the Montagne Noire (S. France) and West Germany. *Geol. Rdsch.*, **62,** 137–153.
11.4.4.

TUCKER M.E. (1974) Sedimentology of Palaeozoic pelagic limestones: the Devonian Griotte (Southern France) and Cephalopodenkalk (Germany). In: *Pelagic Sediments: on Land and Under the Sea* (Ed. by K.J. Hsü and H.C. Jenkyns). pp. 71–92. *Spec. Publ. int. Ass. Sediment*, **1.**
11.4.4.

TUCKER M.E. & KENDALL A.C. (1973). The diagenesis and low-grade metamorphism of Devonian styliolinid-rich pelagic carbonates from West Germany: possible analogues of Recent pteropod oozes. *J. sedim. Petrol.*, **43,** 672–687.
11.4.4.

TUCKER M.E. & REID P.C. (1973) The sedimentology and context of Late Ordovician glacial marine sediments from Sierra Leone, West Africa. *Palaeogeogr. Palaeoclim. Palaeoecol.*, **13,** 289–307.
13.5.3.

TURMEL R.J. & SWANSON R.G. (1969) Evolution of Rodriguez Bank, a modern carbonate mound. Cited in: *Field Guide to Some Carbonate Rock Environments* (Comp. by G. Multer (1971)), pp. 82–86. Farleigh Dickinson University, Madison, New Jersey.
10.2.2.

TURMEL R.J. & SWANSON R.G. (1976) The development of Rodriguez Bank, a Holocene mudbank in the Florida reef tract. *J. sedim. Petrol.*, **46,** 497–518.
10.2.2.

TURNBULL W.J., KRINITZSKY E.L. & WEAVER F.S. (1966) Bank erosion in soils of the Lower Mississippi valley. *Soil. Mech. and Foundat. Proc. Amer. Soc. Civil Eng.*, **92,** 121–136.
3.52, 6.5.1, Figs. 3.21, 6.7.

TWENHOFEL W.H. (1926) *Treatise on Sedimentation*, pp. 661. Williams and Wilkins Co., Baltimore.
11.1.2, 11.2, 11.5.

TWIDALE C.R. (1972) Landform development in the Lake Eyre region,

Australia. *Geogr. Rev.*, **62,** 40–70.
5.2.6.

Ueno H. (1975) Duration of the Kuroko mineralisation episode. *Nature,* **253,** 428–429.
14.4.2.

Uffenorde H. (1976) Zur Entwicklung des Warsteiner Karbonat-Komplexes im Oberdevon und Unterkarbon (Nördliches Rheinisches Schiefergebirge). *Neues Jb. Geol. Paläont. Abh.*, **152,** 75–111.
11.4.4.

Umbgrove J.H.F. (1938) Geological history of East Indies. *Bull. Am. Ass. Petrol. Geol.*, **22,** 1–70.
14.2.2.

Umbgrove J.H.F. (1949) *Structural History of the East Indies*. pp. 76. Cambridge University Press, London.
14.2.2.

UNDP–UNESCO (1976) Proceedings of Seminar on Nile Delta sedimentology, Alexandria.
6.1.

Upadhyay H.D. & Strong D.F. (1973) Geological setting of the Betts Cove Copper Deposits, Newfoundland: an example of ophiolite sulfide mineralisation. *Econ. Geol.*, **68,** 161–167.
14.4.1.

Vai G.B. & Ricci Lucchi F. (1977) Algal crusts, autochthonous and clastic gypsum in a cannibalistic evaporite basin: a case history from the Messinian of Northern Apennines. *Sedimentology,* **24,** 211–244.
8.5.5, Figs. 8.45, 8.46.

Vann J.H. (1959) The geomorphology of the Guiana coast. *Proc. 2nd Coast. Geomorph. Conf.*, 153–188.
7.4.

Vassoevich N.S. (1948) *Le Flysch et les Méthodes de son Étude*. pp. 216. Gostoptekhizdat, Leningrad (translated into French by Bureau de recherches géologiques et minières).
12.1.

Vassoevich N.S. (1951) *Les conditions de la formation du Flysch*, pp. 240. Gostoptekhizdat, Leningrad (translated into French by Bureau de recherches géologiques et minières).
2.2.2.

Vaughan T.W. (1910) A contribution to the geologic history of the Floridian plateau. *Papers Tortugas Lab. Carnegie Inst. Wash. Publ.*, **133,** 99–185.
10.1.1.

Veen J. van (1935) Sand waves in the southern North Sea. *Int. Hydrograph. Rev.*, **12,** 21–29.
9.1.1.

Veen J. van (1936) *Onderzoekingen in de Hoofden*. Algemene Landsdrukkerij, The Hague, pp. 252.
9.1.1.

Veen F.R. van (1975) Geology of the Leman Gas-field. In: *Petroleum and the Continental Shelf of North West Europe. Volume 1. Geology* (Ed. by A.W. Woodland), pp. 322–331. Applied Science Publishers, Barking.
5.1, 5.3.6.

Veenstra H.J. (1965) Geology of the Dogger Bank area, North Sea. *Mar. Geol.*, **3,** 245–262.
13.5.1.

Verwey J. (1952) On the ecology of distribution of cockles and mussels in the Dutch Wadden Sea, their role in sedimentation and the source of their food supply, with a short review of the feeding behaviour of bivalve molluscs. *Arch. Néeland. Zool.*, **10,** 171–239.
9.3.5.

Virkkala K. (1952) On the bed structure of till in Eastern Finland. *Comm. Geol. de Finlande Bull.*, **157,** 97–109.
13.4.1.

Visher G.S. (1965) Fluvial processes as interpreted from ancient and recent fluvial deposits. In: *Primary Sedimentary Structures and their Hydrodynamic Interpretation* (Ed. by G.V. Middleton), pp. 116–132. *Spec. Publ. Soc. econ. Paleont. Miner.*, **12,** Tulsa.
2.2.1, 3.9.4.

Voo R. Van Der & French R.B. (1974) Apparent polar wandering for the Atlantic-bordering continents: late Carboniferous to Eocene. *Earth Sci. Rev.*, **10,** 99–119.
10.3.2.

Wachs D. & Hein J.R.(1974) Petrography and diagenesis of Franciscan Limestones. *J. sedim. Petrol.*, **44,** 1217–1231.
11.4.2.

Wachs D. & Hein J.R. (1975) Franciscan limestones and their environments of deposition. *Geology,* **3,** 29–33.
11.4.2.

Waddell E. (1976) Swash-groundwater-beach profile interactions. In: *Beach and Nearshore Sedimentation* (Ed. by R.A. Davis Jr and R.L. Ethington), pp. 115–125. *Spec. Publ. Soc. econ. Palaeont. Miner.*, **24,** Tulsa.
7.2.1.

Wagner C.W. & Van Der Togt C. (1973) Holocene sediment types and their distribution in the Southern Persian Gulf. In: *The Persian Gulf* (Ed. by B.H. Purser), pp. 123–155. Springer-Verlag, Berlin.
8.2.1.

Wakeel S.K. El (1964) Chemical and mineralogical studies of siliceous earth from Barbados. *J. sedim. Petrol.*, **34,** 687–690.
11.1.2, 11.4.2.

Wakeel S.K. El & Riley J.P. (1961) Chemical and mineralogical studies of fossil red clays from Timor. *Geochim. Cosmochim. Acta* **24,** 260–265.
11.1.2, 11.4.2.

Walker C.T. & Price N.B. (1963) Departure curves for computing palaeosalinity from boron in illites and shales. *Bull. Am. Ass. Petrol. Geol.*, **47,** 833–841.
9.8.

Walker J.R. & Massingill J.V. (1970) Slump features on the Mississippi fan, northeastern Gulf of Mexico. *Bull. geol. Soc. Am.*, **81,** 3101–3108.
12.5.2.

Walker K. & Laporte L.F. (1970) Congruent fossil communities from Ordovician and Devonian carbonates of New York. *J. Paleont.*, **44,** 928–944.
10.2.5.

Walker R.G. (1965) The origin and significance of the internal sedimentary structures of turbidites. *Proc. Yorks. geol. Soc.*, **35,** 1–32.
2.2.1, 12.2.4.

Walker R.G. (1966) Shale Grit and Grindslow Shales: transition from turbidite to shallow water sediments in the Upper Carboniferous of northern England. *J. sedim. Petrol.*, **36,** 90–114.
6.7.2, 6.7.4, 12.8.3, Fig. 6.50.

WALKER R.G. (1967) Turbidite sedimentary structures and their relationship to proximal and distal depositional environments. *J. sedim. Petrol.*, **37**, 25–43.
2.2.1, 12.2.6, 12.6.3, 12.8.3, Tab. 12.2.

WALKER R.G. (1970) Review of the geometry and facies organization of turbidites and turbidite-bearing basins. In: *Flysch Sedimentology in North America* (Ed. by J. Lajoie), pp. 219–251. *Spec. Pap. geol. Ass. Canada*, **7.**
12.2.4, 12.2.6, 12.6.3, 12.6.4, 12.8.1.

WALKER R.G. (1973) Mopping-up the turbidite mess. In: *Evolving Concepts in Sedimentology* (Ed. by R.N. Ginsburg), pp. 1–37. Johns Hopkins Univ. Press, Baltimore.
12.1.

WALKER R.G. (1975) Generalized facies models for resedimented conglomerates of turbidite association. *Bull. geol. Soc. Am.*, **86,** 737–748.
12.2.6, 12.8.1, 12.8.2, Fig. 12.48.

WALKER R.G. (1976) Facies models: 1. General introduction. *Geosci. Can.*, **3,** 21–24.
2.2.1.

WALKER R.G. (1977) Deposition of upper Mesozoic resedimented conglomerates and associated turbidites in southwestern Oregon. *Bull. geol. Soc. Am.*, **88,** 273–285.
12.2.6, 12.8.1.

WALKER R.G. & MUTTI E. (1973) Turbidite facies and facies associations. In: *Turbidites and Deep Water Sedimentation*, pp. 119–157. Soc. econ. Paleont. Mineral. Pacific Section, Short Course, Anaheim.
12.6.4, 12.7.1, 12.8.1, 12.8.2, Fig. 12.48, Tab. 12.3.

WALKER T.R. (1967) Formation of red beds in ancient and modern deserts. *Bull. geol. Soc. Am.*, **78,** 353–368.
3.2.5, 3.6.2, 3.9.2.

WALKER T.R. (1974) Formation of red beds in moist tropical climates: a hypothesis. *Bull. geol. Soc. Am.*, **85,** 633–638.
5.3.1.

WALKER T.R. & HARMS J.C. (1972) Eolian origin of Flagstone beds, Lyons Sandstone (Permian), Type area, Boulder County, Colorado. *Mountain Geologist*, **9,** 279–288.
5.3.2, Fig. 5.14.

WALL J.H. (1975) Diatoms and radiolarians from the Cretaceous system of Alberta – a preliminary report. In: *The Cretaceous System in the Western Interior of North America*, pp. 391–410 (Ed. by W.G.E. Caldwell). *Spec. Pap. geol. Ass. Canada*, **13.**
11.4.5.

WALTHER J. (1894) *Einleitung in die Geologie als Historische Wissenschaft*, **Bd. 3.** Lithogenesis der Gegenwart, pp. 535–1055. Fischer Verlag, Jena.
2.1.2.

WALTHER J. (1897) Ueber Lebensweise fossiler Meeresthiere. *Z. dt. geol. Ges.*, **49,** 209–273.
11.1.2.

WALTHER J. (1911) The origin and peopling of the deep sea. *Am. J. Sci.*, **31,** 55–64.
11.1.2.

WALTHER J. (1924) *Das Gesetz der Wüstenbildung in gegenwart und vorzeit*, pp. 421. Von Quelle und Meyer, Leipzig.
5.1.

WANLESS H.R. & CANNON J.R. (1966) Late Paleozoic glaciation. *Earth Sci. Rev.*, **1,** 247–286.
13.5.2.

WANNER J. (1931) De Stratigraphie van Nederlandsch Oost-Indie: Mesozoicum. *Leidse. Geol. Meded.*, **5,** 567–610.
11.4.4.

WALTER M.R. (1976) *Stromatolites*, pp. 790. Elsevier, Amsterdam.
10.1.1.

WALTERS J.E. (1959) Effect of structural movement on sedimentation in the Pheasant–Francitas area, Matogorda and Jackson counties, Texas. *Trans. Gulf.-Cst. Ass. geol. Socs.*, **9,** 51–58.
6.8.2.

WALTON E.K. (1967) The sequence of internal structures in turbidites. *Scott. J. Geol.*, **3,** 306–317.
12.2.4.

WARDLAW N.C. & SCHWERDTNER W.M. (1966) Halite-anhydrite seasonal layers in Middle Devonian Prairie evaporite formation, Saskatchewan, Canada. *Bull. geol. Soc. Am.*, **77,** 331–342.
8.3.2, 8.5.2.

WARRINGTON G. (1974) Les évaporites du Trias Britannique. *Bull. Soc. géol. France*, **16,** 708–723.
5.3.4.

WASS R.E., CONOLLY J.R. & MACINTYRE R.J. (1970) Bryozoan carbonate sand continuous along southern Australia. *Mar. Geol.*, **9,** 63–73.
10.1.2.

WASSON R.J. (1974) Intersection point deposition on alluvial fans: an Australian example. *Geogr. Annalr.*, **56A,** 83–92.
3.2.4.

WATERHOUSE J.B. (1964) Permian stratigraphy and faunas of New Zealand. *Bull. geol. Surv. N.Z.*, **72,** 101.
11.4.2.

WATERS R.A. (1970) The Variscan structure of eastern Dartmoor. *Proc. Ussher Soc.*, **2,** 191–197.
11.4.4.

WATSON W.N.B. (1967/68) Sir John Murray – a chronic student. *Univ. Edinburgh Jl.*, **23,** 123–137.
11.1.1.

WATTS N.L. (1976) Paleopedogenic palygorskite from the basal Permo-Triassic of northwest Scotland. *Am. Mineralogist*, **61,** 299–302.
3.9.2.

WAUGH B. (1970) Petrology, provenance and silica diagenesis of the Penrith Sandstone (Lower Permian) of northwest England. *J. sedim. Petrol.*, **40,** 1226–1240.
3.9.2.

WEBB J.E., DÖRJES D.J., GRAY J.S., HESSLER R.R., VAN ANDEL T.H., RHOADS D.D., WERNER F., WOLFF T. & ZIJLSTRA J.J. (1976) Organism sediment relationships. In: *The Benthic Boundary Layer* (Ed. by I.N. McCave), pp. 273–295. Plenum Press, New York.
9.3.5, Figs. 9.3, 9.4.

WEBER K.J. (1971) Sedimentological aspects of oilfields of the Niger delta. *Geol. Mijnb.*, **50,** 569–576.
6.2, 6.3.1, 6.5.1, 6.5.2, 6.8.2.

WEBER K.J. & DAUKORU E. (1975) Petroleum geology of the Niger delta. *Proc. 9th World Petrol. Conf.*, pp. 14.
6.3.1, 6.8.1, 6.8.2, Fig. 6.54.

WEIDMANN M. (1967) Petite contribution à la connaissance du flysch. *Bull. Lab. Géol. Minéral. Géophys. Mus. Géol. Univ. Lausanne*, **166,** pp. 6.
12.2.4, 12.6.1.

WEIGEL R.L. (1964) *Oceanographical Engineering*, pp. 532. Prentice-Hall, New Jersey.
9.6.3.

WEILER Y., SASS E. & ZAK I. (1974) Halite oolites and ripples in the Dead Sea, Israel. *Sedimentology*, **21,** 623–632.
4.2.4, Fig. 4.4.

WEIMER R.J. (1973) A guide to Uppermost Cretaceous stratigraphy, Central Front Range, Colorado; deltaic sedimentation, growth faulting and early Laramide crustal movement. *Mountain Geologist*, **10**, 53–97.
6.8.3.

WENDT J. (1969) Foraminiferen 'Riffe' im Karnischen Hallstätter Kalk des Feuerkogels (Steiermark, Österreich). *Paläont. Z.*, **43**, 177–193.
11.4.4, Fig. 11.35.

WENDT J. (1970) Stratigraphische Kondensation in triadischen und jurassichen Cephalopodenkalken der Tethys. *Neues Jb. Geol. Paläont., Mh.*, **1970**, pp. 433–448.
11.4.4, Fig. 11.34.

WENDT J. (1971) Genese und Fauna submariner sedimentärer Spaltenfüllungen im mediterranen Jura. *Palaeontographica*, A, **136**, 122–192.
11.4.4, Fig. 11.36.

WENDT J. (1973) Cephalopod accumulations in the Middle Triassic Hallstatt-Limestone of Jugoslavia and Greece. *Neues Jb. Geol. Paläont., Mh.*, **1973**, pp. 624–640.
11.4.4.

WENDT J. (1974) Encrusting organisms in deep-sea manganese nodules. In: *Pelagic Sediments: on Land and under the Sea* (Ed. by K.J. Hsü and H.C. Jenkyns), pp. 437–447. *Spec. Publ. int. Ass. Sediment.*, **1**.
11.3.3, 11.3.6.

WEST I.M. (1964) Evaporite diagenesis in the lower Purbeck beds of Dorset. *Proc. Yorks. geol. Soc.*, **34**, 315–330.
8.4.2.

WEST I.M. (1965) Macrocell structure and enterolithic veins in British Purbeck gypsum and anhydrite. *Proc. Yorks. geol. Soc.*, **35**, 47–58.
8.4.2.

WEST I.M. (1975) Evaporites and associated sediments of the basal Purbeck formation (Upper Jurassic) of Dorset. *Proc. Geol. Ass.*, **86**, 205–225.
8.4.2, 10.2.5, Figs. 8.26, 8.30.

WEST I.M., BRANDON A. & SMITH M. (1968) A tidal flat evaporitic facies in the Viséan of Ireland. *J. sedim. Petrol.*, **38**, 1079–1093.
8.4.1.

WESTBROOK G.K. (1975) The structure of the crust and upper mantle in the region of Barbados and the Lesser Antilles. *Geophys. J.R. astr. Soc.*, **43**, 201–242.
11.4.2, 14.4.2, Fig. 11.27.

WETHEY D.S. & PORTER J.W. (1976) Sun and shade differences in productivity of reef corals. *Nature*, **262**, 281–282.
10.4.1.

WEYL P.K. (1967) The solution behaviour of carbonate materials in sea water. *Studies Tropical Oceanog., Univ. Miami*, **5**, 178–228.
10.2.2.

WHETTEN J.T. (1965) Carboniferous glacial rocks from the Werrie Basin, New South Wales, Australia. *Bull. geol. Soc. Am.*, **76**, 43–56.
13.5.2.

WHITE A.H. (1968) The glacial origin of Carboniferous conglomerates west of Barraba, New South Wales. *Bull. geol. Soc. Am.*, **79**, 675–686.
13.5.2.

WHITE G.W. (1968) Pleistocene deposits in the north-western Allegheny Plateau, U.S.A. *Quart. J. geol. Soc. Lond.*, **124**, 131–151.
13.4.1.

WHITE G.W., TOTTON S.M. & GROSS D.L. (1969) Pleistocene stratigraphy of north-western Pennsylvania. *Bull. Pennsylvania geol. Surv.*, **G.55**, 88 pp.
13.4.1.

WHITEMAN A., NAYLOR D., PEGRUM R. & REES G. (1975) North Sea troughs and plate tectonics. *Tectonophysics*, **26**, 39–54.
14.4.1.

WILCOX R.E., HARDING T.P. & SEELY D.R. (1973) Basic wrench tectonics. *Bull. Am. Ass. Petrol. Geol.*, **57**, 74–96.
14.4.3, Fig. 14.24.

WILKINSON B.H. (1975) Matagorda Island, Texas: the evolution of a Gulf Coast barrier complex. *Bull. geol. Soc. Am.*, **86**, 959–967.
7.2.6.

WILLIAMS G.D. & STELCK C.R. (1975) Speculations on the Cretaceous palaeogeography of North America. *Spec. Pap. geol. Ass. Canada*, **13**, 1–20.
Fig. 9.45.

WILLIAMS G.E. (1966) Palaeogeography of the Torridonian Applecross Group. *Nature*, **209**, 1303–1306.
3.10.3.

WILLIAMS G.E. (1969) Characteristics and origin of a Pre-Cambrian pediment. *J. Geol.*, **77**, 183–207.
3.8, 3.8.1.

WILLIAMS G.E. (1971) Flood deposits of the sand-bed ephemeral streams of Central Australia. *Sedimentology*, **17**, 1–40.
3.4.4.

WILLIAMS G.E. (1975) Late Precambrian glacial climate and the earth's obliquity. *Geol. Mag.*, **112**, 441–465.
13.5.4.

WILLIAMS P.F. & RUST B.R. (1969) The sedimentology of a braided river. *J. sedim. Petrol.*, **39**, 649–679.
3.3.1.

WILLMAN H.B., GLASS H.D. & FRYE J.C. (1966) Mineralogy of glacial tills and their weathering profiles in Illinois: Part II. Weathering profiles. *Ill. State geol. Surv. Circ.*, **400**, pp. 76.
13.3.6.

WILLS L.J. (1929) *Physiographical Evolution of Britain*, pp. 376. Arnold, London.
5.1.

WILLS L.J. (1970) The Triassic succession in the central Midlands in its regional setting. *Quart. J. geol. Soc. Lond.*, **126**, 225–283.
3.9.2, 5.3.3, 5.3.4.

WILSON C.B. & HARLAND W.B. (1964) The Polarisbreen series and other evidences of late Pre-Cambrian ice ages in Spitsbergen. *Geol. Mag.*, **101**, 198–219.
13.5.4.

WILSON I.G. (1972) Aeolian bedforms – their development and origins. *Sedimentology*, **19**, 173–210.
5.2.4, Fig. 5.3.

WILSON I.G. (1973) Ergs. *Sedim. Geol.*, **10**, 77–106.
5.2.3, 5.2.4, 5.2.7, Fig. 5.2.

WILSON J.L. (1969) Microfacies and sedimentary structures in 'deeper-water' lime mudstones. In: *Depositional Environments in Carbonate Rocks* (Ed. by G.M. Friedman), pp. 4–19. *Spec. Publ. Soc. econ. Paleont. Miner.*, **14**, Tulsa.
11.4.4.

WILSON J.L. (1970) Depositional facies across carbonate shelf margins. *Trans. Gulf-Cst. Ass. geol. Socs.*, **20**, 229–233.
Fig. 10.14.

WILSON J.L. (1974) Characteristics of carbonate platform margins. *Bull. Am. Ass. Petrol. Geol.*, **58**, 810–824.
Fig. 10.14.

WILSON J.L. (1975) *Carbonate Facies in Geologic History*, pp. 471. Springer-Verlag, Berlin, Heidelberg, New York.

1.1, 10.1.1, 10.2.2, 10.2.5, 10.4.1, 10.4.2, 11.4.4, Figs. 10.14, 10.29b, 10.42, 10.45, 10.47a, 10.48, 10.50.

WILSON J.T. (1966) Did the Atlantic close and then re-open? *Nature, Lond.*, **211,** 676–681.
14.2.6, 14.5.

WILSON J.T. (1968) Static or mobile earth: the current scientific revolution. *Proc. Am. philos. Soc.*, **112,** 309–320.
14.5.

WILSON M.J. (1965) The origin and geological significance of the South Wales underclays. *J. sedim. Petrol.*, **35,** 91–99.
3.9.2.

WIN SWE, THACPAW C., NAY THAUNG THAUNG & KYAW NYUNT (in press) *Geology of Part of the Chundwin Basin of the Central Burma Belt.* 7th Burma Res. Congr., Rangoon.
14.3.2.

WINDOM H. (1975) Eolian contributions to marine sediments. *J. sedim. Petrol.*, **45,** 520–529.
11.3.4.

WINDOM H.L. (1976) Lithogenous material in marine sediments. In: *Chemical Oceanography*, 2nd Edn, Vol. 5, pp. 103–135. Academic Press, London.
11.3.1.

WINTERER E.L. (1964) Late Precambrian pebbly mudstones in Normandy, France: tillite or tilloid? In: *Problems in Palaeoclimatology* (Ed. by A.E.M. Nairn), pp. 159–178. Interscience, London.
13.5.4.

WINTERER E.L., LONSDALE P.F., MATTHEWS J.L. & ROSENDAHL B.R. (1974) Structure and acoustic stratigraphy of the Manihiki Plateau. *Deep-Sea Res.*, **21,** 793–814.
11.3.3.

WINTERER E.L., EWING J.I. *et al.* (1973) *Initial Reports of the Deep Sea Drilling Project*, **17,** pp. 930. U.S. Government Printing Office, Washington.
11.3.3.

WISE S.W. JR. & WEAVER F.M. (1973) Origin of cristobalite-rich Tertiary sediments in the Atlantic and Gulf coastal plain. *Trans. Gulf-Cst. Ass. geol. Socs.*, **23,** 305–323.
11.4.3.

WISE S.W. JR. & WEAVER F.M. (1974) Chertification of oceanic sediments. In: *Pelagic Sediments: on Land and under the Sea* (Ed. by K.J. Hsü and H.C. Jenkyns), pp. 301–326. *Spec. Publ. int. Ass. Sediment.*, **1.**
11.3.3, 11.3.4.

WOLDSTEDT P. (1970) *International Quaternary Map of Europe, Sheet* **6,** København. Bundesanstalt für Bodenforschung, Hannover.
13.5.1.

WOLDSTEDT P. (1971) *International Quaternary Map of Europe, Sheet* **7,** Moskva. Bundesanstalt für Bodenforschung, Hannover.
13.5.1.

WOLMAN M.G. & LEOPOLD L.B. (1957) River flood plains; some observations on their formation. *Prof. Pap. U.S. geol. Surv.*, **282-C,** 87–107.
3.6.1.

WOOD A. & SMITH A.J. (1959) The sedimentation and sedimentary history of the Aberystwyth grits (Upper Llandoverian). *Quart. J. geol. Soc. Lond.*, **114,** 163–195.
12.6.2, 12.6.3.

WOOD G.V. & WOLFE M.J. (1969) Sabkha cycles in the Arab/Darb formation off the Trucial Coast of Arabia. *Sedimentology*, **12,** 165–191.
8.4.1.

WOODCOCK N.H. (1976a) Structural style in slump sheets: Ludlow series, Powys, Wales. *J. geol. Soc. Lond.*, **132,** 399–415.
12.2.2.

WOODCOCK N.H. (1976b) Ludlow series slumps and turbidites and the form of the Montgomery Trough, Powys, Wales. *Proc. Geol. Ass.*, **87,** 169–182.
12.2.2, Fig. 12.9.

WOODLAND A.W. (1970) The buried tunnel-valleys of East Anglia. *Proc. Yorks. geol. Soc.*, **37,** 521–578.
13.5.1.

WOOLNOUGH W.C. (1930) The influence of climate and topography in the formation and distribution of products of weathering. *Geol. Mag.*, **67,** 123–132.
3.9.2.

WRAY J.L. (1971) Algae in reefs through time. In: *Reef organisms through time*, pp. 1358–1373. *Proc. J. North Am. Paleont. Convention, Chicago*, **1969.**
10.4.2.

WRAY J.L. (1977) *Calcareous Algae.* pp. 186. Elsevier, Amsterdam.
10.4.2.

WRIGHT H.E. (1956) Origin of the Chuska Sandstone, Arizona–New Mexico: a structural and petrographic study of a Tertiary aeolian sediment. *Bull. geol. Soc. Am.*, **67,** 413–434.
5.3.2.

WRIGHT H.E. (1973) Tunnel valleys, glacial surges, and subglacial hydrology of the Superior Lobe, Minnesota. *Mem. geol. Soc. Am.*, **136,** 251–276.
13.5.1.

WRIGHT L.D. & COLEMAN J.M. (1973) Variations in morphology of major river deltas as functions of ocean wave and river discharge regimes. *Bull. Am. Ass. Petrol. Geol.*, **57,** 370–398.
6.2, 6.3, 6.3.1, 6.4, 6.5.1.

WRIGHT L.D. & COLEMAN J.M. (1974) Mississippi River mouth processes: effluent dynamics and morphologic development. *J. Geol.*, **82,** 751–778.
6.5.1, 6.5.2, Figs. 6.16, 6.18.

WRIGHT L.D., COLEMAN J.M. & THOM B.G. (1973) Processes of channel development in a high tide range environment: Cambridge Gulf–Ord River delta, western Australia. *J. Geol.*, **81,** 15–41.
6.5.2, 7.5.

WRIGHT L.D., COLEMAN J.M. & THOM B.G. (1976) Sediment transport and deposition in a macrotidal river channel: Ord River, western Australia. In: *Estuarine Research, Vol. II Geology and Engineering* (Ed. by L.E. Cronin), pp. 309–322. Academic Press, New York.
7.5.

WRIGHT P. (1976) A cine-camera technique for process measurement on a ridge and runnel beach. *Sedimentology*, **23,** 705–712.
7.2.1.

WRIGHT W.B. (1937) *The Quaternary Ice Age*, pp. 478. Macmillan, London.
13.6.

WUNDERLICH F. (1972) Beach dynamics and beach development. *Senckenberg. Mar.*, **4,** 47–79.
7.2.2, Fig. 7.47.

YAALON D.H. (Ed.) (1971) Paleopedology: origin, nature and dating of paleosols. *Internat. Soc. of Soil Sci.*, pp. 350. Israel Univ. Press.
3.9.2.

YAALON D.H. & DAN J. (1974) Accumulation and distribution of loess-derived deposits in the semi-desert and desert fringe areas of Israel. *Z. Geomorph., N.F. Suppl. Bd.* **20,** 91–105.
3.6.2, 5.2.5.

YAALON D.H. & GINSBOURG D. (1966) Sedimentary characteristics and climatic analysis of easterly dust storms in the Negev (Israel). *Sedimentology,* **6,** 315–332.
5.2.5.

YEATS R.S. & HART S.R. *et al.* (1976) *Initial Reports of the Deep Sea Drilling Project,* **34,** pp. 814. U.S. Government Printing Office, Washington.
11.3.4.

YOUNG G.M. (1970) An extensive early Proterozoic glaciation in North America? *Palaeogeogr. Palaeoclim. Palaeoecol.,* **7,** 85–101.
13.5.5.

YOUNG R. (1976) *Sedimentological studies in the Upper Carboniferous of north-west Spain and Pembrokeshire.* Unpub. D.Phil. Thesis., Univ. of Oxford.
3.9.3.

ZAKOWA H. (1970) The present state of the stratigraphy and paleogeography of the Carboniferous of the Holy Cross Mts. *Acta Geol. Pol.,* **20,** 4–31.
11.4.4.

ZAMARREÑO I. (1972) Las litofacies carbonatadas del Cambrico de la zona cantabrica (N.W. Espana) y su distribucion paleogeografica. *Trab. Geol.,* **5,** 1–118.
11.4.4.

ZANKL H. (1971) Upper Triassic carbonate facies in the northern Limestone Alps. In: *Sedimentology of Parts of Central Europe* (Ed. by G. Müller), pp. 147–185. *Guidebook VIII Int. sedim. Congr.,* Kramer, Frankfurt.
10.4.2, 11.4.4, Fig. 10.43.

ZELENOV K.K. (1964) Iron and manganese in exhalations of the submarine Banu Wahu volcano (Indonesia). *Dokl. Acad. Sci. U.S.S.R., Earth Sciences Section,* **155,** 1317–1320 (pp. 94–96 in translations published by American Geological Institute).
11.3.5.

ZIEGLER A.M., COCKS L.R.M. & BAMBACH R.K. (1968) The composition and structure of Lower Silurian marine communities. *Lethaia,* **1,** 1–27.
9.8.

ZIEGLER A.M., COCKS L.R.M. & MCKERROW W.S. (1968) The Llandovery transgression of the Welsh Borderland. *Palaeontology,* **11,** 736–782.
9.12.

ZIEGLER A.M. & MCKERROW W.S. (1975) Silurian marine red beds. *Am. J. Sci.,* **275,** 31–56.
11.4.4.

ZIEGLER A.M., NEWALL G., HALLECK M.S. & BAMBACH R.K. (1971) Repeated community sediment patterns in the Silurian of the northern Appalachian Basin. *Geol. Soc. Am., Abstracts with programs,* **3,** 760–761.
9.12.

ZIEGLER A.M., WALKER K.R., ANDERSON E.J., KAUFFMAN E.G., GINSBURG R.N. & JAMES N.P. (1974) *Principles of Benthic Community Analysis* (*Notes for a Short Course*). Sedimenta IV, Comparative Sedimentology Laboratory. Univ. of Miami.
10.2.5.

ZIEGLER W.H. (1975) Outline of the geological history of the North Sea. In: *Petroleum and the Continental Shelf of North-West Europe. Volume 1. Geology* (Ed. by A.W. Woodland), pp. 165–187. Applied Science Publishers, Barking.
14.4.1.

Index

Numbers in bold type indicate illustrations and tables which relate to the index entry